Smart Innovation, Systems and Technologies

Volume 233

The Smart Innovation, Systems and Technologies book series encompasses the topics of knowledge, intelligence, innovation and sustainability. The aim of the series is to make available a platform for the publication of books on all aspects of single and multi-disciplinary research on these themes in order to make the latest results available in a readily-accessible form. Volumes on interdisciplinary research combining two or more of these areas is particularly sought.

The series covers systems and paradigms that employ knowledge and intelligence in a broad sense. Its scope is systems having embedded knowledge and intelligence, which may be applied to the solution of world problems in industry, the environment and the community. It also focusses on the knowledge-transfer methodologies and innovation strategies employed to make this happen effectively. The combination of intelligent systems tools and a broad range of applications introduces a need for a synergy of disciplines from science, technology, business and the humanities. The series will include conference proceedings, edited collections, monographs, handbooks, reference books, and other relevant types of book in areas of science and technology where smart systems and technologies can offer innovative solutions.

High quality content is an essential feature for all book proposals accepted for the series. It is expected that editors of all accepted volumes will ensure that contributions are subjected to an appropriate level of reviewing process and adhere to KES quality principles.

Indexed by SCOPUS, EI Compendex, INSPEC, WTI Frankfurt eG, zbMATH, Japanese Science and Technology Agency (JST), SCImago, DBLP.

All books published in the series are submitted for consideration in Web of Science.

More information about this series at http://www.springer.com/series/8767

Yuzo Iano · Osamu Saotome · Guillermo Kemper ·
Ana Claudia Mendes de Seixas ·
Gabriel Gomes de Oliveira
Editors

Proceedings of the 6th Brazilian Technology Symposium (BTSym'20)

Emerging Trends and Challenges in Technology

Springer

Editors
Yuzo Iano
Faculty of Electrical and Computer Engineering
UNICAMP
Campinas, São Paulo, Brazil

Guillermo Kemper
Univers. Peruana de Ciencias Aplicadas
Lima, Peru

Gabriel Gomes de Oliveira
Universidade Estadual de Campinas
Campinas/SP, Brazil

Osamu Saotome
Divisão de Engenharia Eletrônica
Instituto Tecnológico de Aeronáutica
São José dos Campos, São Paulo, Brazil

Ana Claudia Mendes de Seixas
PUC Campinas
Pontifícia Universidade Católica de Ca
Campinas, Brazil

Associate-Editors
Alex Midwar Rodriguez
School of Electrical and Computer Engineering
UNICAMP
Campinas, São Paulo, Brazil

Maria CeciliaLuna Alvarado
School of Electrical and Computer Engineering
UNICAMP
Campinas, São Paulo, Brazil

Diego Pajuelo Castro
School of Electrical and Computer Engineering
UNICAMP
Campinas, São Paulo, Brazil

Pablo David Minango Negrete
School of Electrical and Computer Engineering
UNICAMP
Campinas, São Paulo, Brazil

Gabriel Caumo Vaz
School of Electrical and Computer Engineering
UNICAMP
Campinas, São Paulo, Brazil

ISSN 2190-3018 ISSN 2190-3026 (electronic)
Smart Innovation, Systems and Technologies
ISBN 978-3-030-75679-6 ISBN 978-3-030-75680-2 (eBook)
https://doi.org/10.1007/978-3-030-75680-2

This Springer imprint is published by the registered company Springer Nature Switzerland AG
The registered company address is: Gewerbestrasse 11, 6330 Cham, Switzerland

Foreword

With great satisfaction, I write this Foreword for the Proceedings of the 6th Brazilian Technology Symposium – Emerging Trends and Challenges in Technology (BTSym'20), held virtually, for the first time, at the PUC Campinas University, Brazil, in October 2020 and the UPC University, Peru, in December 2020. This event is in its sixth edition and has consolidated to become an excellent opportunity for researchers, professors, and students to present and discuss the results of their research works.

In the 2020 edition, the BTSym activities could not be accomplished in person due to the Covid-19 pandemic. However, the event has been characterized since its first edition by the broad scope of the areas exposed and, within a virtual environment, it was possible to expand our network of researchers and encourage them to expose their papers, which deal with current and priority topics for Brazilian and world technological development, including subjects related to the various branches of innovation in industrial processes, robotics, telecommunications, buildings, urban infrastructure, product development, and biomedicines.

Events such as BTSym are an essential part of the research and innovation process. Firstly, these events contribute to the promotion of research activities, which are key to a country's technological development. The dissemination of research results, as promoted by BTSym, contributes to the transformation of research findings into technological innovation. In addition, these events facilitate the sharing of findings, leading eventually to the formulation of research networks, which accelerate the achievement of new results. Therefore, I would like to congratulate the BTSym General Chair, Prof. Dr. Yuzo Iano, and his group of collaborators for the important initiative of organizing the BTSym'20 and for providing the opportunity for authors to present their work to a wide audience through this publication. Finally, I congratulate the authors for the high-quality work presented in these proceedings.

Alex Midwar Rodriguez Ruelas
Proceedings Chair of Brazilian
Technology Symposium

Preface

This book contains the Proceedings of the 6th Brazilian Technology Symposium – Emerging Trends and Challenges in Technology, held in Brazil in October 2020 and Peru in December 2020.

The Brazilian Technology Symposium is an excellent forum for presentations and discussions of the latest results of projects and development research in several areas of knowledge, in scientific and technological scope, including smart designs, sustainability, inclusion, future technologies, architecture and urbanism, computer science, information science, industrial design, aerospace engineering, agricultural engineering, biomedical engineering, civil engineering, control and automation engineering, production engineering, electrical engineering, chemical engineering, and probability and statistics.

This event seeks to bring together researchers, students, and professionals from the industrial and academic sectors, seeking to create and/or strengthen the linkages between issues of joint interest. Participants were invited to submit research papers with methodologies and results achieved in scientific level research projects, completion of course work for graduation, dissertations, and theses.

The 56 full chapters accepted for this book were selected from 267 submissions, and, in each case, the authors were guided by an experienced researcher with a rigorous peer-view process. Among the main topics covered in this book, we can highlight manufacturing processes, lean manufacturing, industrial costing models, sustainability and productivity, circular economy, workplace safety, control systems, Internet of Things, cyber-physical systems, Transportation Management System (TMS), logistic services analysis, digital supply chain, socio-economic impacts of technologies 4.0, robotics applications, artificial neural networks, Big Data, deep learning, computational vision, cybersecurity, soft-computing methodologies, technologies applied to cities development, smart cities, energy sustainability, Building Information Modeling (BIM), environment analysis, technologies applied to health, biomedical innovations, socio-economic impacts of COVID-19, technologies applied to education, academic development, civil aviation studies, and much more.

We hope you enjoy and take advantage of this book and feel motivated to submit your papers, in the future, to Brazilian Technology Symposium.

Best wishes,
Gabriel Gomes de Oliveira
Technical Program and Finance Chair
of Brazilian Technology Symposium

Acknowledgements

Our appreciation goes to a lot of colleagues and friends who assisted in the development of this book, Proceedings of the 6th Brazilian Technology Symposium – Emerging Trends and Challenges in Technology (BTSym'20).

First of all, I would like to thank all the members of the Organizing and Executive Committee for the commitment throughout the year. Several meetings were held, and many challenges were overcome for the accomplishment of the BTSym 2020. Also, and with great merit, I would like to thank all the Scientific and Academic Committee and Technical Reviewers Committee members for their excellent work, which was essential to ensure the quality of our peer-review process, collaborating with the visibility and technical quality of the BTSym 2020.

The Brazilian Technology Symposium is an event created by the Laboratory of Visual Communications of the Faculty of Electrical and Computer Engineering of the University of Campinas (UNICAMP). In this way, I would like to thank the PUC Campinas and UPC Universities, especially for supporting and hosting the BTSym'20 and BTSym'20 Satellite, respectively, which was fundamental for the successful accomplishment of the events.

Finally, on behalf of Prof. Yuzo Iano, the General Chair of the Brazilian Technology Symposium, I thank all the authors for their participation in the BTSym'20; We sincerely hope to have provided a very useful and enriching experience in the personal and professional lives of everyone. Our special thanks go to Professors Ana Cláudia Seixas, Cláudia Cotrim Pezzuto, Guillermo Leopoldo Kemper Vásquez, and Carlos Raymundo Ibañes; Without their efficiency and hardworking, our events could not have been accomplished, so the BTSym'20 and BTSym'20 Satellite certainly have much to thank them.

Best wishes,
Gabriel Caumo Vaz
Institutional Relationship Chair
of Brazilian Technology Symposium

Contributors

Organizing Committee

Alex Rodriguez Ruelas	LCV/DECOM/FEEC/UNICAMP - Proceedings Chair
Alysson Gomes de Oliveira	LCV/DECOM/FEEC/UNICAMP - Marketing Chair
Ana Cláudia Seixas	LCV/DECOM/FEEC/UNICAMP - Vice-Associate-General Chair BTSym
Claudia Cotrim Pezzuto	PUC/UNICAMP - Vice-Associate-General Chair BTSym
David Minango	LCV/DECOM/FEEC/UNICAMP - Institutional Relationship Chair
Diego Arturo Pajuelo Castro	LCV/DECOM/FEEC/UNICAMP - Proceedings Chair
Gabriel Gomes de Oliveira	LCV/DECOM/FEEC/UNICAMP - Technical Program and Finance Chair
Lisber Arana	LCV/DECOM/FEEC/UNICAMP - Institutional Relationship Chair
Maria Cecilia Luna	LCV/DECOM/FEEC/UNICAMP - Proceedings Chair
Osamu Saotome	ITA - Associate-General Chair BTSym
Rangel Arthur	FT/UNICAMP - Vice-General Chair BTSym
Yuzo Iano	LCV/DECOM/FEEC/UNICAMP - General Chair BTSym & WSGE

Executive Committee

Abel Dueñas Rodríguez	LCV/DECOM/FEEC/UNICAMP - Midia Chair
Airton Vegette	LCV/DECOM/FEEC/UNICAMP - Institutional Relationship Chair

Angélica F. G.	LCV/DECOM/FEEC/UNICAMP - Institutional Relationship Chair
Daniel B. Katze	LCV/DECOM/FEEC/UNICAMP - Institutional Relationship Chair
Daniellle Thiago Ferreira	LCV/DECOM/FEEC/UNICAMP - Editorial Committe Chair
Elizangela Santos Souza	LCV/DECOM/FEEC/UNICAMP - Editorial Committe Chair
Gabriel Caumo Vaz	LCV/DECOM/FEEC/UNICAMP - Institutional Relationship Chair
Jennifer Chuin Lee	LCV/DECOM/FEEC/UNICAMP - Designer Chair
João Carlos Gabriel	Universidade Presbiteriana Mackenzie - Campinas - Vice-Associate-General Chair BTSym
Leticia Cursi	LCV/DECOM/FEEC/UNICAMP - Institutional Relationship Chair
Lucas Alves	LCV/DECOM/FEEC/UNICAMP - Institutional Relationship Chair
Luiz Vicente F. de Mello Filho	Universidade Presbiteriana Mackenzie - Campinas - Vice-Associate-General Chair BTSym
Mariana Melo	LCV/DECOM/FEEC/UNICAMP - Institutional Relationship Chair
Paulo Roberto dos Santos	UniMetrocamp - Vice-Associate-General Chair BTSym
Raquel J. Lobosco	UFRJ - Vice-Associate-General Chair BTSym
Thais Paiao	LCV/DECOM/FEEC/UNICAMP - Institutional Relationship Chair
Telmo Cardoso Lustosa	LCV/DECOM/FEEC/UNICAMP - Local Arrangements Chair
Ubiratan Matos	LCV/DECOM/FEEC/UNICAMP - Institutional Relationship Chair

Scientific and Academic Committee

Alessandra Cristina Santos Akkari	Universidade Presbiteriana Mackenzie
Ana Cláudia Seixas	LCV/DECOM/FEEC/UNICAMP
Angela del Pilar Flores Granados	FEA/UNICAMP

Antonio Carlos Demanboro	PUC CAMPINAS
Celso Iwata Frison	PUC/Minas-Poços de Caldas
Cláudia Cotrim Pezzuto	PUC CAMPINAS
David Bianchini	LCV/DECOM/FEEC/UNICAMP
Edgard Luciano Oliveira da Silva	Universidade Estadual do Amazonas (UEA)
Edwin Valencia Castillo	Universidad Nacional de Cajamarca
Ernesto Karlo Celi Arevalo	UNPRG, Lambayeque, Perú
Erwin Junger Dianderas Caut	Instituto de Investigaciones de la Amazonía Peruana IIAP
Fábio Menegatti de Melo	PUC CAMPINAS
Grimaldo Wilfredo Quispe Santivañez	UERJ
Hugo Enrique Hernandez Figueroa	DECOM/FEEC/UNICAMP
Janito Vaqueiro Ferreira	DMC/FEM/UNICAMP
Jessie Leila Bravo Jaico	UNPRG, Lambayeque, Perú
João Carlos Gabriel	Universidade Presbiteriana Mackenzie
José Hiroki Saito	UFSCAR
Lia Toledo Moreira Mota	PUC CAMPINAS
Lucielen Santos	PURG
Luiz Vicente F. de Mello Filho	Universidade Presbiteriana Mackenzie
Marcos Fernando Espindola	IFSP São Paulo
Maria Thereza de Moraes Gomes Rosa	Universidade Presbiteriana Mackenzie
Marina Lavorato de Oliveira	PUC CAMPINAS
Néstor Adolfo Mamani Macedo	Universidad Nacional Mayor de San Marcos
Paulo Roberto dos Santos	UniMetrocamp
Osamu Saotome	ITA
Rangel Arthur	FT/UNICAMP
Raquel J. Lobosco	UFRJ
Silva Neto	UERJ
Suelene Silva Mammana	Universidade Presbiteriana Mackenzie
Talía Simões dos Santos	FT/UNICAMP
Telmo Cardoso Lustosa	LCV/DECOM/FEEC/UNICAMP
Victor A. M. Montalli	Faculdade São Leopoldo Mandic
Victor Murray	Universidad de Ingenieria y Tecnologia – UTEC

Technical Reviewers Committee

Abel Alejandro Dueñas Rodriguez	LCV/DECOM/FEEC/UNICAMP
Adao Boava	Universidade Federal de Santa CatarinaUFSC
Agord de Matos Pinto Júnior	DESIF/FEEC/UNICAMP
Airton José Vegette	LCV/DECOM/FEEC/UNICAMP
Alessandra Cristina Santos Akkari	Universidade Presbiteriana Mackenzie
Alex R. Ruelas	LCV/DECOM/FEEC/UNICAMP
Alex Restani Siegle	LCV/DECOM/FEEC/UNICAMP
Alysson Gomes De Oliveira	LCV/DECOM/FEEC/UNICAMP
Amilton da Costa Lamas	PUC CAMPINAS
Ana Cláudia Seixas	LCV/DECOM/FEEC/UNICAMP
Angela del Pilar Flores Granados	FEA/UNICAMP
Antônio José da Silva Neto	IPRJ/UERJ
Celso Fabrício Correia de Souza	LCV/DECOM/FEEC/UNICAMP
Cláudia Cotrim Pezzuto	PUC CAMPINAS
Daniel Katz Bonello	LCV/DECOM/FEEC/UNICAMP
Daniel Rodrigues Ferraz Izario	LCV/DECOM/FEEC/UNICAMP
Daniela Helena Pelegrine Guimarães	EEL/USP
David Allan Ibarra	Universidad de las Fuerzas Armadas ESPE
David Bianchini	LCV/DECOM/FEEC/UNICAMP
David Minango	LCV/DECOM/FEEC/UNICAMP
Diego Arturo Pajuelo	LCV/DECOM/FEEC/UNICAMP
Douglas do Nascimento	Marie Skłodowska-Curie Actions (MSCA)
Edgard Luciano Oliveira da Silva	EST/UEA
Edson Camilo	Eldorado Institute
Euclides Lourenço Chuma	LCV/DECOM/FEEC/UNICAMP
Everton Dias de Oliveira	UNIMEP
Fabiana da Silva Podeleski	UNISAL
Fábio Menegatti de Melo	PUC CAMPINAS
Francisco Fambrini	UFSCAR
Gabriel Caumo Vaz	LCV/DECOM/FEEC/UNICAMP
Gabriel Gomes de Oliveira	LCV/DECOM/FEEC/UNICAMP
Guilherme Barbosa Lopes Júnio	UFPE
João Carlos Gabriel	Universidade Presbiteriana Mackenzie
Josué Marcos de Moura Cardoso	LCV/DECOM/FEEC/UNICAMP

Juan Minango Negrete	LCV/DECOM/FEEC/UNICAMP
Jullyane Figueiredo	UFSC
Leonardo Bruscagini de Lima	LCV/DECOM/FEEC/UNICAMP
Leticia Dias Gomes	UDESC
Lisber Arana Hinostrosa	LCV/DECOM/FEEC/UNICAMP
Lucas Heitzmann Gabrielli	FEEC/UNICAMP
Luigi Ciambarella Filho	Universidade Veiga de Almeida/Develop Biotechnology
Luis Fernando Gonzalez	KonkerLabs
Luiz Antonio Sarti Junior	UFSCAR
Luiz Vicente Figueira de Mello Filho	Universidade Presbiteriana Mackenzie
Marcelo Jara	Eldorado Institute
Marcos Fernando Espindola	IFSP São Paulo
Maria Cecilia Luna	LCV/DECOM/FEEC/UNICAMP
Maria Thereza de Moraes Gomes Rosa	Universidade Presbiteriana Mackenzie
Miriam Tvrzska de Gouvea	Universidade Presbiteriana Mackenzie
Murilo Cesar Perin Briganti	LCV/DECOM/FEEC/UNICAMP
Osamu Saotome	ITA
Polyane Alves Santos	Instituto Federal Da Bahia
Rangel Arthur	INOVA/FT/UNICAMP
Raquel Jahara Lobosco	Federal University of Rio de Janeiro
Ricardo Barroso Leite	LCV/DECOM/FEEC/UNICAMP
Roger Prior Gregio	LCV/DECOM/FEEC/UNICAMP
Rosivaldo Ferrarezi	UNIP
Suelene Silva Piva	Universidade Presbiteriana Mackenzie
Telmo Cardoso Lustosa	LCV/DECOM/FEEC/UNICAMP
Victor Angelo Martins Montalli	Faculdade São Leopoldo Mandic - SLMANDIC

Contents

Emerging Trends in Human Smart and Sustainable Future of Cities

Technology and Inclusive Education, Hybrid Web Application - Entertainment and Learning

Karine Izario[1], Daniel Izario[2](✉), Yuzo Iano[2], João Brancalhone[3], Gabriel Gomes[2], and Diego Pajuelo[2]

[1] Educational Anhanguera, Indaiatuba, SP, Brazil
[2] University of Campinas, Campinas, SP, Brazil
yuzo@decom.fee.unicamp.br, oliveiragomesgabriel@ieee.org
[3] National Telecommunications Institute, Santa Rita do Sapucai, MG, Brazil

Abstract. Considering the present moment in the Brazilian society in which schools are legally required to accept all kinds of students, it has appeared opportunities for development and uplift of cases regarding the study of the web application's implementation focused on inclusive education in schools with the use of hybrid technology. They can be used in all kinds of disabilities in which computers and gadgets can be required or not, bringing satisfactory results to children's learning and entertainment.

Keywords: Accessibility · Education · Entertainment · Inclusion · Learning · Technology · Web application

1 Introduction

For many years during Brazilian and global history, several models and structures were created to integrate the education of disabled children in the society, but governments used to ignore inclusion questions. Until few years, it used to exist separated schools for children regarded as normal and, separately, special schools for disabled children with hearing impairment as well as physical, visual, mental, and even multiple disabilities [1].

However, the people's consciousness changed, new laws appeared, and questions rose, letting society change. Hence, inclusive education has appeared, breaking the existing segregated system and enclosing inclusive education into the regular school model. The main objective is the interaction and social integration of students with disabilities [2], as in Fig. 1.

The whole present scenario key point is the obligation in which schools must accept every student, regarded as normal or with disabilities, adapting the school environment, providing capacitation and proper training to its staff to be ready to all day to day activities [1], even existing several failures in the learning and entrainment of children with disabilities, people considering the implementation of a web application.

Y. Iano et al. (Eds.): BTSym 2020, SIST 233, pp. 3–9, 2021.
https://doi.org/10.1007/978-3-030-75680-2_1

Fig. 1. Illustration showing an example of integration and coexistence of students with disabilities [2].

2 Methodology

The hybrid applications are developed in web technology, using, in most cases, programming languages such as HTML5 [3], CSS3 [4] and JavaScript [5], which are based in web view systems, part of Google Chrome technology that allows apps to exhibit web content besides of the aggregation of several advantages such as the facility and agility in the development, use of the same code to several platforms, simple and common maintenance to all the creation platforms and wide variety of available plug-ins in the market.

Scientifically established, the use of technology with disabled children is an effective way of achieving progress in treatments and providing entertainment as doing new activities, the brain may present a better performance in its abilities, improving difficulties. That said, hybrid applications considered several disabilities to entertain and teach towards digital games, answers, questions, children's stories, music, and several other activities, the most appropriate individually considered to each disability [2, 6].

As previously mentioned, the following details of the different kind of disabilities will be explained below:

2.1 Visual

The visual impairment refers to an irreversible situation in which the visual response is reduced by congenital or hereditary causes, even after clinical or surgical treatment or the use of ordinary glasses [7]. According to the WHO - World Health Organization, the impairment degrees are classified as one of the following possibilities:

1. Low Sight (mild, moderate or severe): Stage in which can be used auxiliary instruments, such as glasses, contact lens, magnifiers, canes, and location training;
2. Near Blindness: Stage in which distinguishing shade and light is still possible, but where braille, audio description, and canes may be used as well as location and mobility training;
3. Blindness: Stage in which there is no light perception, where braille, canes as well as location and mobility training are mandatory;

2.2 Hearing

The hearing impairment is defined as the difference between an individual's performance and the normal hearing ability established by ANSI - American National Standards Institute [8]. Generally, a normal audition refers to the normal capacity of detecting sounds up to 20dB [7]. The hearing impairments can be categorized in the following cases:

1. Above 0 to 24 dB - Normal hearing;
2. Above 25 to 40 dB - Mild hearing impairment (mild deafness);
3. Above 41 to 55 dB - Moderate hearing impairment (moderate deafness);
4. Above 56 to 70 dB - Pronounced hearing impairment (pronounced deafness);
5. Above 71 to 90 dB - Severe hearing impairment (severe deafness);
6. Above 91 dB - Profound hearing impairment (profound deafness);
7. Anacusis (complete deafness);

2.3 Mental

Mental impairment is a term used to define brain disorders that compromise cognitive functions. The children with the mentioned impairment present intellectual function patterns below the normality, featuring IQ scores below 70, which led to difficulties in their adaptive behavior [7]. Those scores can be divided into four other sub-categories:

1. F 70 - Mild Mental impairment (IQ 50 till 69);
2. F 71 - Moderate Mental Impairment (IQ 36 till 49);
3. F 72 - Severe Mental Impairment (IQ 20 till 35);
4. F 73 - Profound Mental Impairment (IQ below 20).

2.4 Physical

Physical impairment is a term used to define the limitation in mobility or general motor coordination. There can be several origins for it, e.g., neurological or neuromuscular disorders, congenital malformation, or acquired characteristics such as hydrocephaly (fluid accumulation in the cranial box) or cerebral palsy [7].

The children with this kind of impairment usually have difficulties in writing, as motor coordination may be affected. The learning in these cases may take longer than usual, but, except for severe brain injuries, language is acquired without major drawbacks [7].

The physical impairments may be classified into different categories, such as paraplegia (the motion loss in the lower limbs), tetraplegia (the motion loss of the four limbs), and hemiplegia (the motion loss in just one body hemisphere). Other classifications may apply in cases such as amputations, cerebral palsy, and ostomy (abdominal opening for feeding inlets use) [7].

2.5 Multiple

Multiple disabilities are the ones in which two or more parts of the body may be affected, associating different disabilities with a wide range of combinations. It may lead to a serious situation [7], which can be classified as:

1. Deafblindness: Single impairment in which there is visual and hearing loss leading into special necessities regarding communication and necessary adaptations to reach educational, recreational, or social vocational goals to achieve a better world comprehension [7];
2. Multiple sensorial impairments: It is the presence of hearing or visual impairment associated with other physical and mental impairments as well as other disorders (neurological, emotional, of language as well as of global development) which can lead to delays in the educational, vocational, social and emotional development bringing difficulties to the individual self-sufficiency [7];

It is important to remind that many children with disabilities may use gadgets attached to their computers, totems, and tablets to achieve better resourcefulness. At present, there are a lot of independent tools which can be used to facilitate disabled people's lives, allowing more active participation in multiple different activities, presenting in this way, a better quality of life [6, 9].

In the web application developed for the study, all functionalities are attached to the same place, none of them being returned independently as the ones currently in use. There are some tools put in context to assist children with visual impairment, such as an advanced text description using a speaker and printer in braille.

In the case of hearing impairment, there is attached a voice synthesizer with representative images and the translation of texts from Portuguese to LIBRAS (*Língua Brasileira de Sinais*, in English, Brazilian Sign Language). The motor impairment, the one considered with the most complex adaptation into hybrid models, can use the device's camera to capture the eyes' movement, transferring them into a movement tracer developed in JavaScript, acting intelligently, allowing the children to play, write, reading, communicate.

3 Results

It is important to highlight that the results were positive and exhilarating during the project, and it was based on the PMBOK - Project Management Body of Knowledge [10]. That said, it was studied the day to day of 5 children to implement a solution better and closely understand each one of their necessities.

Many tests were performed to grant that leisure and learning were complete and significant to inclusive education. The first test considered the internal use as how hybrid applications would react to the processing available in the several devices currently in the market, as it is critical to keep several functionalities working with an internal processor.

In the second test, five separated children were separately taken to have the functionalities tested, resulting in smiles from them, indicating the first version's success.

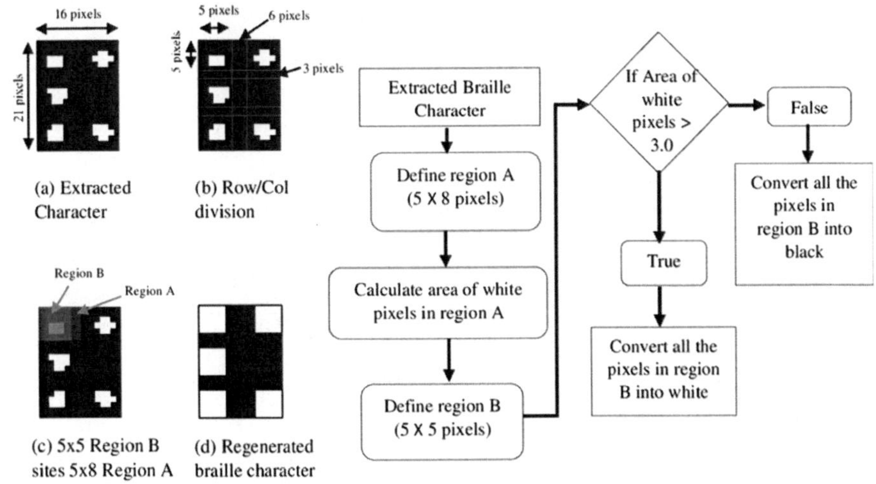

Fig. 2. Diagram for new Braille algorithm operation.

However, two failures were found in the application, which led to corrections in both the Braille and Libras algorithm, in Fig. 2.

Psychologists and teachers were consulted to find the best interaction between the children and the new digital devices in the scholar day-to-day and normally keep other children's activities. The level of concentration in classes improved as in Fig. 3.

Another considered aspect is the analysis of the household environment's application, improving familiar relationships and day to day entertainment, helping in the homework execution either, bringing equality to other students regarding it. With this test, leisure time at home improved the time dedicated to homework execution, as in Fig. 4.

The value to its implementation in schools, considering high-quality equipment and supposed adaptors to plug them and the gadget itself, is around R$ 74.000,00 in the case of 5 disabled children in the school. It is a low value compared to similar governmental projects developed inside schools.

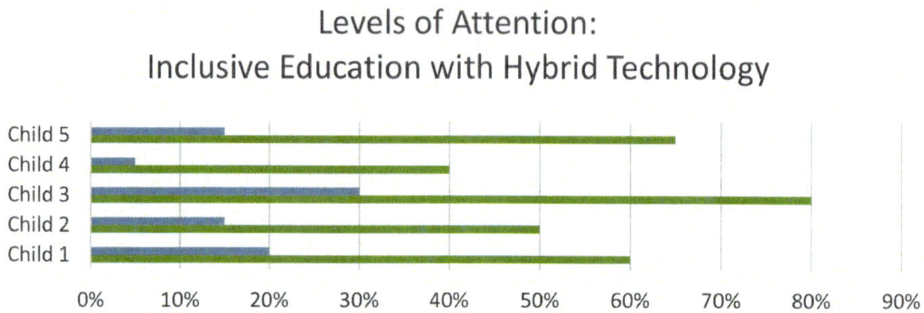

	Child 1	Child 2	Child 3	Child 4	Child 5
Before	20%	15%	30%	5%	15%
After	60%	50%	80%	40%	65%

Before After

Fig. 3. Graphic presentation of the change in the concentration levels after the deployment of the hybrid application.

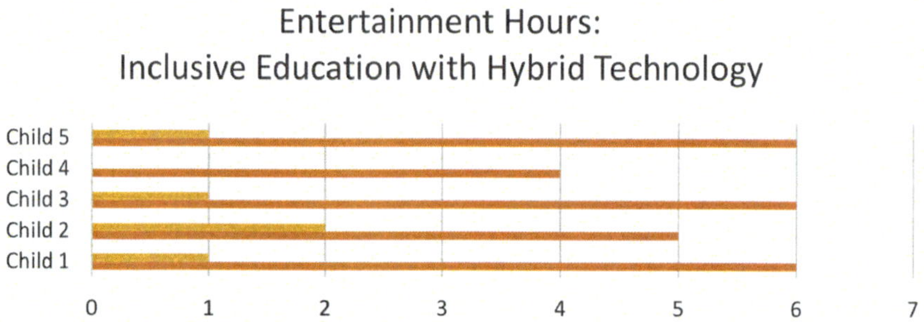

	Child 1	Child 2	Child 3	Child 4	Child 5
Before (h)	1	2	1	0	1
After (h)	6	5	6	4	6

Before (h) After (h)

Fig. 4. Graphic presentation of the gains in leisure time after the hybrid application deployment.

4 Conclusion

The project development to inclusive education, with the implementation of a hybrid application in web language, presented great results, mainly on the emotional side, allowing the relationship with disabled children, which is compelling, but it requires much attention.

Notable changes were noticed when it was delivered to schools. Besides behavior changes, in which other children started to help others with disabilities explore the new things, it brought new friendships and a unit spirit. The most important presented result was the increase in the learning and entertainment levels at home and the school.

Concerning the deployment value, even considered high, it is much inferior to the one quoted for development by the government and private schools.

References

1. Mazzotta MJS, Antino MEF (2011) Inclusão social de pessoas com deficiências e necessidades especiais: cultura, educação e lazer. Saúde e Sociedade, vol 20, no 2, São Paulo. https://doi.org/10.1590/S0104-12902011000200010
2. Mantoan TEM (2003) Inclusão escolar - O que é? Por quê? Como fazer? Moderna - 1ª edn, Unicamp. ISBN 85-16-03903-X
3. Meyer J (2011) Guia essencial do Html 5. Ciência Moderna. ISBN 8539901153
4. Castro E, Hyslop B (2013) HTML5 e CSS3: guia prático e visual. Alta Books - 7ª edn. ISBN 8576088037
5. Duckett J (2016) Javascript e Jquery: desenvolvimento de interfaces web interativas. Alta Books - 1ª edn. ISBN 8576089459
6. Parente R (2018) O que as novas tecnologias podem fazer pela educação inclusiva. PORVIR - Inovações em Educação. https://porvir.org/o-que-as-novas-tecnologias-podem-fazer-pela-educacao-inclusiva
7. Cotomacci G (2007) LAZER: direitos e acessibilidade das pessoas com deficiência ou necessidades especiais. Unicamp. https://www.bibliotecadigital.unicamp.br/document/?down=000437308
8. American National Standards Institute. https://www.ansi.org/
9. Ortega G (2019) Educação inclusiva: como a tecnologia ajuda nesse processo. Escolas Disruptivas. https://escolasdisruptivas.com.br/metodologias-inovadoras/educacao-inclusiva-como-a-tecnologia-ajuda-nesse-processo
10. Project Management Institute (2018) Um guia do conhecimento em gerenciamento de projetos (Guia PMBOK). Project Management Institute - 6ª edn. ISBN 1628251018

UGVs - Applications in the Smart Cities (Angular 2+ and .Net Core 3+)

Daniel Izario[1](✉), Yuzo Iano[1], João Brancalhone[2], Karine Izario[3], Gabriel Gomes[1,2,3], and Diego Pajuelo[1,2,3]

[1] University of Campinas, Campinas, SP, Brazil
yuzo@decom.fee.unicamp.br, oliveiragomesgabriel@ieee.org
[2] National Telecommunications Institute, Santa Rita do Sapucai, MG, Brazil
[3] Educational Anhanguera, Indaiatuba, SP, Brazil

Abstract. Many recent activities require robotic systems such as the UGVs - Unmanned Ground Vehicles, terrestrial robots developed as an extension of human resources. In this way, they can have their basic motion mechanisms such as acceleration, braking, and speed controlled remotely. With a higher goal, the implementation can come with sensing and real-time monitoring systems, providing tools for its use in smart cities, using a web application to its front and back end management. This ideation model is developed in a diagram, presenting how the system can be used in smart cities.

Keywords: .Net Core 3+ · Angular 2+ · Arduino · Back end · Front end · Smart cities · UGVs

1 Introduction

In order to achieve the development of UGVs - Unmanned Ground Vehicles, many systems need to be considered. First and foremost, the automatic control, in which the commands are sent to the mechanism, to perform its designed actions and attend the solicitations of data measurements. That is performed along with other systems such as the mechanic and the communication system, which involve the transmission and reception of data and the electric system, supplying other previously mentioned systems [1].

Based on the implementation needs of the smart cities in activities such as monitoring remote and, in some cases, also with high risks areas or assisting the surveillance in corporations, condos, technological pars among others, it was applied computerized techniques in the systems automation in order to reduce risks to humans and let the project as a whole more viable to application. In the first implementation and routine tests step, two prototypes, a remote control, and a simple mechanical arm able to catch small objects were developed, as in Fig. 1.

There are multiple applications to the projects, but it will be presented the applications of use in smart cities, though, where the management must be present to integrate the society to the technology better. Hence, all the vehicles contain sensors able to monitor

Y. Iano et al. (Eds.): BTSym 2020, SIST 233, pp. 10–16, 2021.
https://doi.org/10.1007/978-3-030-75680-2_2

and alert if a specific region is at risk of flooding or firing. That is just possible thanks to the real-time data surveillance in parameters such as luminosity, temperature, level of carbon dioxide, humidity, among others [2]. They are later used in a web application, developed in Angular 2+ (front end) and .Net Core 3+ (back end).

Fig. 1. Prototypes developed for the first phase of implementation and testing

All the developed vehicle functionalities commands are performed using a remote control or a web application throw the radiofrequency (RF) use. A 360° camera support was installed in front of the vehicle to follow the vehicle path in real-time [2].

2 Methodology

The main challenge in creating UGVs is the ability to provide a viable relation between weight and autonomy. Hence, the best choice was an aluminum-made chassis, both very resistant and light, offering an efficient cost-benefit ratio. The wheel/motor set was attached with bolt and nut into the vehicle's structure in a plane, using aluminum strips to protect the motor's cowl. The attachment also allows the range spin in 360°. As commands are received, this mentioned displacement allows the vehicle rotation in distinct directions [1].

All the engines were connected throw an L293d Shield - H bridge, which controls the vehicles' motors and the camera support, this circuit/shield being the one which has presented the best drive features. Either adequate to the speed control and directions of the used motors, easing its mobility during the movements in different angles and directions: forward, reverse, left, and right [3].

Since all the vehicles contain cameras able to send real-time images to the web application, it is an increasing at the humans' safety as, for example, mitigating the death risk when monitoring a risk area, using the vehicle, instead of a human to perform the same activity, but giving the person a more distant image. It is necessary to remind either that the human visual system has a remarkable capacity in recognizing patterns, although barely processing the massive volume of information present in an image as well as its multiple distortions and degradations, inherent of the acquisition, transmission, and visualizing processes, which limit, even more, this human eye capacity. Considering it, besides watching live images, it is possible to save, execute, digitally process, and filter them to further analysis [4, 5].

2.1 RF Communication

One key point is communication, and there are several ways of creating an Arduino based system, the most common being the use of shields, such as Bluetooth and RF (radiofrequency). Bluetooth communication is widely used in the Arduino, figuring it as a cheap and straightforward mode of sending and receiving remote information. As a disadvantage, it just allows the data exchange at a distance of up to 10 m [6].

Considering the short-range limitation, the UGVs use their own RF transmitter and receiver module attached to its antennas. There are also multiple kinds of antennas, all of them executing an essential task in the signal receiving/transmitting networks: It is performed towards them the energy transfer from the transmitter to the communication medium and the receiver. The system's efficiency is up to the radiant factors or the receptions connected to it, which led to the development of a wide variety of antennas, the most commonly used the monopole (Fig. 2) and the dipole [6].

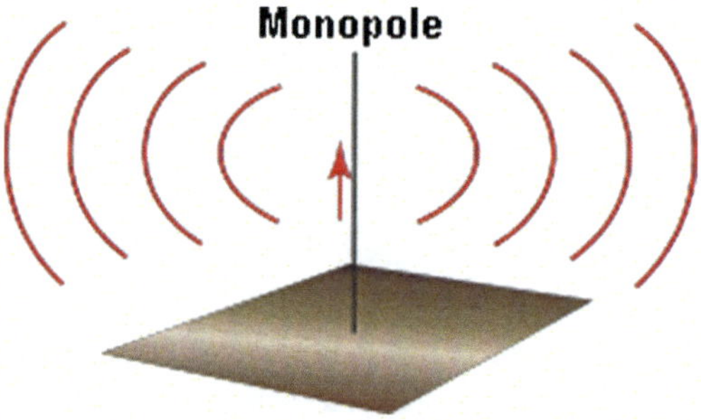

Fig. 2. Example of a monopole antenna

The monopole antenna is widely used in the present time, more specifically the monopole quarter of wave. As a feature, it presents the omnidirectional pattern, continually keeping the reception of the signal in any position. The quarter of wave monopole antenna must stay in an earth plane, whence it derives its polarization. The monopoly antenna must have its polarization vertically about its earth plane, either [6].

As it requires a code to execute multiple movements, it will be sent a binary represented number to each direction. Each execution is generated by a millisecond given time, which the demodulator will interpret as a gap of time to the execution, all obtained in the oscilloscope AGILENT 1000MHz DSO3102A as presented in Fig. 3, visualizing two different outbound signal with specific direction from an Arduino port in (a) and without movements in (b).

a)

b)

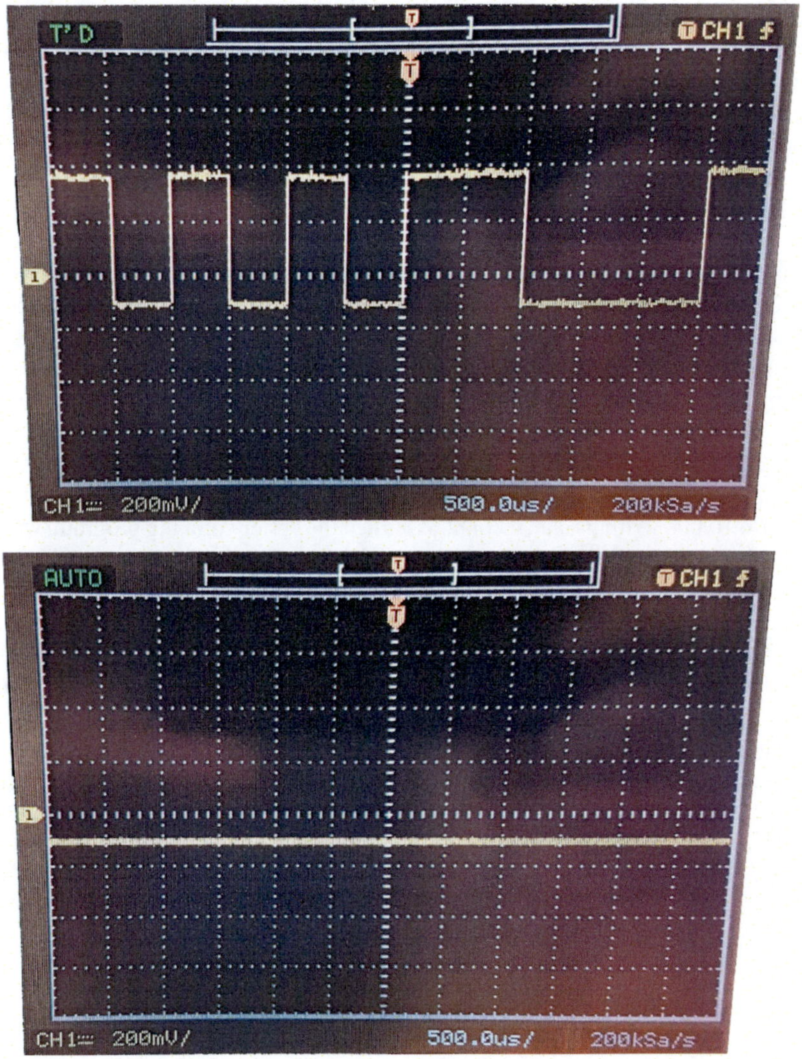

Fig. 3. Oscilloscope: (a) Motion detected (bits) (b) Absence of movement (bits)

2.2 Remote Data

The Web application (Angular 2+ and .Net Core 3+) construe the selected information and look for a specific file in the cloud's directory, which is a pre-installed software in the machine in order to run along with the Arduino and with the obtained information from the sensors installed in the UGVs. The Arduino is charged with a code able to take measures and sending them through a USB - Universal Serial Bus output to a computer connected to the board, which can capture this data and exhibit them as numeric data in Arduino software interface (Serial Monitor).

It is either possible, besides the simple numeric data, the use of the PLX-DAQ software, which allows sending data to an Excel sheet. With the data in the sheet, it is possible to use the software facilities in order to build, among others, a graphic with the received information in parameters such as temperature, luminosity, level of carbon dioxide, humidity, among others [7].

There are also other possibilities to store this data, later exhibiting them in a sheet, such as the use of a USB output or remotely throw the RF itself. The first option is through the use of EEPROM Arduino memory, which requires the individual recording of the bytes. It requires either, also, to read it, an EEPROM library pre-installed in the IDE - Integrated Development Environment, and, when using the read command, it is informed the EEPROM address. The problem regarding this option is the slow data writing and reading (approximately 3ms), despite it, existing the guaranty of reliability up to 100.000 writings though [7].

Another possibility is the use of FLASH memory, mostly available in the internal Arduino's structure. In this case, it also has the presence of some problems. It accepts just up to 10.000 writings, and it contains, either, the system running program, which can receive interferences in the case of the input of invalid data, letting it to work improperly.

It is either possible to use an SD card. Even the Arduino board not presenting an SD connector itself, there are multiple shields able to work as interface, such as the Ethernet Shield. To use the SD, it is necessary to add the SD library, pre-installed in the IDE, in the Arduino code. This recording memory shown method, using sensors data, is very useful in case of later use in an Excel sheet [7].

Lately, there is the possibility of directly sending data throw RF to a web server connected to the application, avoiding any conflict or loss of information, but that is the most expensive method.

2.3 Management and Integration in the Society

To public agencies transform an ordinary city into a smart one, better manage its resources, and better assist its population, a diagram in 4 sequential steps was implemented and developed as in Fig. 4.

The first step, referred to as (1), is the base where the available and necessary resources will be used in step (2). In the next step, the society undergoes a research, in which the UGV with the better application in the local society is chosen. Step (3), called integration step, assures, if the integration steps were followed, the mitigation of errors, allowing the establishment of an action plan. Lastly, there is step (4), called the project implementation phase, in which the project is built and integrated into the

TIC ("*Tecnologia da Informação e Comunicação*", in English, Information and Communication Technologies - ICT). With it, society is integrated into the public services' network through the internet and web applications, following all the data in real-time, mainly those captured and monitored by the vehicles.

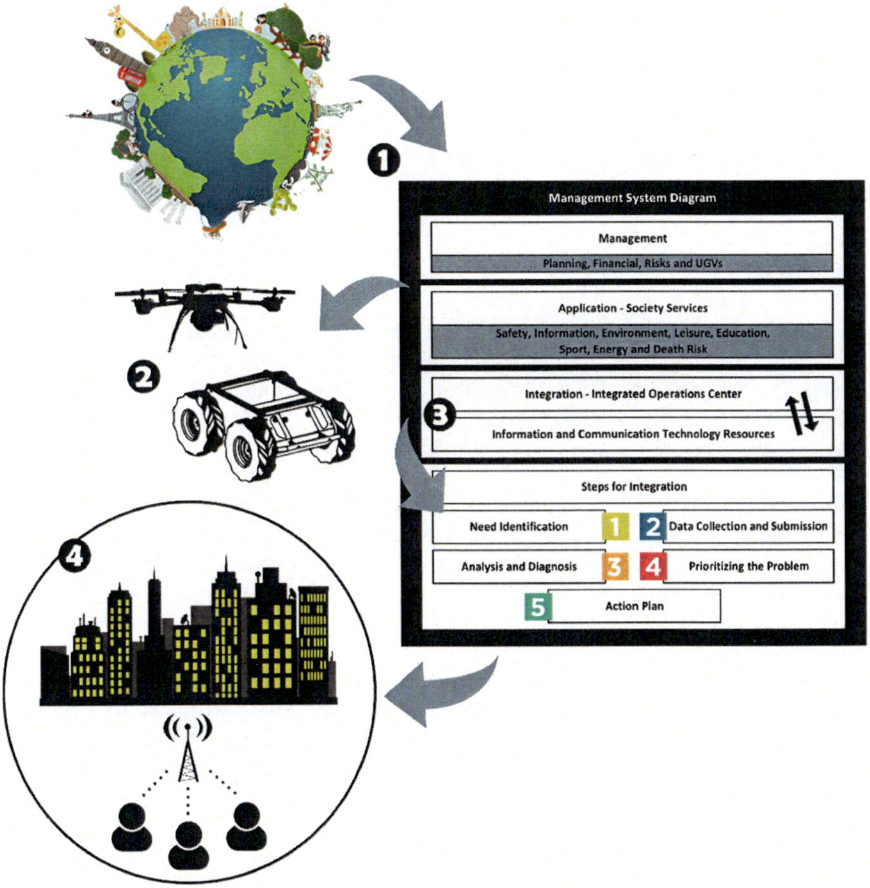

Fig. 4. Management and integration diagram

3 Conclusion

The use of the application in the smart cities was a breakthrough, bringing a more sustainable environment, able to better manage the cities and assist the whole society ethically. The UGVs can be used in many strategic points, which can be monitored or have its uncertainness better assured, bringing greater safety to society without major risks to their lives.

The ability of remote management of the application, as well as the exhibition of real-time results, bring much greater satisfaction in living in a city. All of it, thanks to a robust and reliable application developed in Angular 2+ and .Net Core 3+.

Besides all the software configurations, the UGVs hardware can keep themselves electrified during the actions performed, owning solar panels able to supply itself with some efforts. The architecture of it was carefully studied, considering all the shields' power consumption and the proper balance of the VCC and GND connections to keep the circuit balanced on both sides. There are also the antennas, implemented in order to provide safe and stable communication.

There is also the possibility of creating other UGVs, enclosing the prototypes' development developed to any kind of landscape as well as military, domestic, hospital, and business applications.

References

1. Pauli E, Uliana FS (1996) Mecânica: noções básicas de elementos de máquinas. Senai/CST - Companhia Siderúrgica de Tubarão. https://www.abraman.org.br/arquivos/72/72.pdf
2. Santos TA (2011) ROVIM - Robô de vigilância de instalações militares - comunicações e posto de controle. https://fenix.tecnico.ulisboa.pt/downloadFile/395143514515/dissertação.pdf
3. Monk S (2014) Programação com arduino II: passos avançados com Sketches. Bookman. 247 p
4. Lopes ES, Lima IMS, Gonçalves TC (2012) A Importância de detecção de gases para prevenção de danos à segurança, meio ambiente e saúde. Rev. Divulgação Proj. Univ. Petrobras IF Fluminense 2(1):301–304. https://essentiaeditora.iff.edu.br/index.php/BolsistaDeValor/article/viewFile/2431/1319
5. Solomon C (2013) Fundamentos de Processamento digital de imagens: uma abordagem prática com exemplos em MATLAB. LTC. 281 p
6. Ribeiro JAJ (2012) Engenharia de antenas: fundamentos. Projetos e Aplicações. Érica. 584 p
7. Dworakowski LA, Hartmann AM, Kakuno EM, Dornelas PFT (2016) Uso da plataforma arduino e do software PLX-DAQ para construção de gráficos de movimento em tempo real. Rev Brasileira - Ensino Física 38(3). https://www.scielo.br/scielo.php?script=sci_arttext&pid=S1806-11172016000300603

Health 4.0: A Conceptual Approach to Evaluate the Application of Digital Technologies in the Healthcare Field

Luana Spósito Valamede and Alessandra Cristina Santos Akkari(✉)

Mackenzie Presbyterian University, Campinas, Brazil
alessandra.akkari@mackenzie.br

Abstract. In the face of facts that challenge public health globally, such as population aging and pandemic, the health sector is of particular importance. Integrating processes, devices, workers, and services through large automated networks, Fourth Industrial Revolution strongly encourages intelligent manufacturing associated with disruptive changes, impacting sectors of the economy such as the healthcare segment. Therefore, the term Health 4.0 is emerging connected with advances in applying digital technologies in management, productivity, research, and development in the medical industry. This paper aimed to characterize Heath 4.0 through the design of a conceptual approach to evaluate the application of digital technologies in the health segment. A literature study process was used to identify the Industry 4.0 principles emphasizing the healthcare sector based on a three-step methodology, including observation, categorization, and association phases. Also, a technological mapping was developed to qualify the concept of Health 4.0 according to the leading technologies 4.0. It was identified that the Internet of Things, Big Data Analytics, Automated Guided Vehicles, Additive Manufacturing, and Augmented Reality are supporting health digitization. A multilevel circular diagram was developed to highlight the gains of each technology 4.0 in the health domain, suggesting the increase of interconnectivity between agents and devices, analysis of medical data in real-time, interventions based on parameters, heath customization, professional training, among others. The standardization of information and technological mapping developed in this research can contribute from an industrial and academic perspective to the area.

Keywords: Health 4.0 · Industry 4.0 · Digital technologies

1 Introduction

Circumstances that challenge global public health, such as population aging and pandemic scenarios, encourage in-depth discussions about the need to reinvent and adapt health systems based on the pillars of flexibility, autonomy, and focus on the patient. In this perspective, studies pointed out that a technology-driven approach is promising in order to optimize social indicators associated with the healthcare sector [1, 2].

In fact, personalized procedures with high quality, shorter delivery times, and customized technical installations are essential points that are not solved by applying only

Y. Iano et al. (Eds.): BTSym 2020, SIST 233, pp. 17–24, 2021.
https://doi.org/10.1007/978-3-030-75680-2_3

traditional manufacturing technologies. To achieve patient-centered care, health professionals need to use digital technologies, which can be adapted according to different demands of each individual, providing flexible access and involvement with systems and services. Thus, technologies do not replace human interactions but facilitate connections among people who provide and those who receive care and services, focusing on time and health processes organization [3].

In this context, Industry 4.0 can provide interesting resources and new opportunities for patient care focuses on innovative design and intelligent devices manufacturing to attend to different patient needs with precision and high quality. Being an industrial revolution suitable for automation, Industry 4.0 can perform multiple functions when adopted in the medical sector, such as maximizing productivity and material management, applying accurate control processes for complex surgeries, detecting patient data to determine relative information to support health professionals, and basing medical researches and products development. All innovations applied in the healthcare segment, originated from Industry 4.0 pillars, attributes, and technologies, are related to the so-called Health 4.0 term that is emerging in the literature [3–5].

This paper aimed to characterize the domain of Heath 4.0 through the design of a conceptual approach to the application of emerging technologies in the healthcare field. To this end, an exploratory and descriptive study, with a qualitative approach and bibliographic strategy was developed based on a three-phase methodology. This chapter's main contribution refers to the standardization of information and technological mapping, enabling a multilevel view of the new concept of Health 4.0.

2 Theoretical Background

From a historical perspective, similar to industrial revolutions, the evolution of the healthcare sector is also characterized by generations. Occurred from 1970 until 1990, Health 1.0 presented the emergence of modular Information Technology (IT) systems, but with preliminary tests and limited resources. Therefore, this first phase's health processes were based on highly trained doctors' knowledge and medicines based on natural substances. The next 15 years comprised Health 2.0, with IT systems operating in networks and with technologies development for medical tracking as well as advanced imaging systems (such as x-ray diagnostics). From 2005 until 2015, computational methods and data processing systems were applied, which popularized the use of Electronic Health Records (EHR). The EHR technology promotes a digital version of a patient's medical record, including medical history, tests performed, and other medical information that can be instantly and safely available to other users. EHR systems network, combined with other technologies, contribute to health professionals' performance promoting real-time data share, classifying Health 3.0 [6].

These technological advances, which marked the previous revolutions in the health area, can be integrated with the Fourth Industrial Revolution's new digital tools. This association, focusing on collaboration, coherence, and data analysis, characterizes the current Health 4.0. With the terms Digital Health, M-Health, E-Health, and Smart Health as synonyms, Health 4.0 can provide the virtualization to support medical care personalization in real-time for patients, health professionals, and other stakeholders, making health care more predictive and individualized. Thus, the digital transformation from Industry 4.0 affects the health care space far beyond the product, encompassing medical services and health data collection, processing, and sharing [5, 6].

In the global context, Industry 4.0 refers to a new production philosophy that is in its embryonic phase based on the Internet of Things (IoT), Internet of Services (IoS), Cyber-Physical Systems (CPS), and smart factories [5]. Nowadays, the growing pressures for medical services are demanding an increasing amount of deployment of digital health resources especially associated with technologies 4.0. A new trend for health market development is guided by the consumer, professional, and macro levels, with the consumer level being considered the innovation engine to enhance digital development for the entire segment. This level includes easy access to information and communication technologies, data network availability with flexible access to the Internet for healthcare solutions, and gamification through new devices, such as smartphones and tablets [4].

3 Methodology

A three-phase methodology [7] was applied, including observation, categorization, and association phases, as summarized in Fig. 1. In the first phase, a systematic literature review was carried out to identify the principles, attributes, characteristics, and guiding concepts of Industry 4.0, emphasizing the healthcare sector. The databases Web of Science, ScienceDirect, Scopus, and PubMed were used with time restriction from 2013 to 2020.

The search fields analyzed to select the papers were limited to the abstract, title, and keywords, resulting in the following search character string *Title-Abstract-Keyword*, which included the terms "Health 4.0", "Industry 4.0 applied in health", "Cyberphysical System", "Emerging Technologies" and "Attributes 4.0". Only articles in English were considered to compose this study. Criteria for inclusion and exclusion of articles, definition of information to be extracted from studies, analysis, discussion and presentation of results were considered as literature review stages.

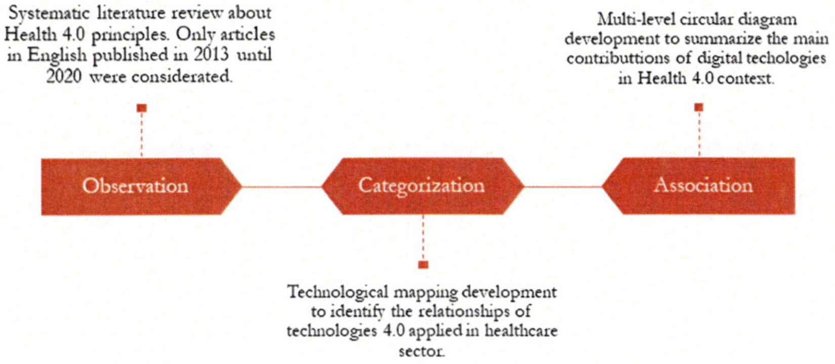

Fig. 1. Research methodology stages applied in this study

In the categorization stage, in order to qualify the concept of Health 4.0 according to the leading technologies 4.0, a technological mapping was developed and the relationships of the digital technologies in the health sector were identified. The mapping prioritized the most cited technologies in the literature according to the healthcare application.

Finally, in the association phase, a multilevel circular diagram was created to illustrate and synthesize the Industry 4.0 performance in the health sector, pointing out the three main applications of each emergent technology analyzed in the Health 4.0 context.

4 Results and Discussion

The literature review pointed out that Health 4.0 concept can be defined as a healthcare domain strategy derived from the Industry 4.0 context. It is a phenomenon that improves services provision and increases connectivity among all stakeholders in the value chain, applying the newest digital technologies. Health 4.0 aims to promote advanced virtualization to achieve integrated, flexible, and real-time processes for patients, doctors, and other professionals, as well as formal and informal healthcare providers. Health services personalization will be achieved through the massive use of technologies 4.0 [5, 8–10].

Focusing on a digitally connected healthcare system, Health 4.0 works on its development to improve the services provided to patients in order to solute their specific patient needs with high speed and quality. In this sense, Health 4.0 can select and promote data value more consistently and effectively. With data portability, it is possible to identify areas for improvement and allow assertive decision making. Data portability also enables both patients and healthcare professionals to have access to medical information anytime and anyplace. This essential factor leads to optimized analyzes, differential diagnoses, and predictive, timely, and innovative medical responses [5, 6, 9].

Health 4.0 also aims at interconnectivity among the main pillars in healthcare in addition to personalized patient care. In order to avoid doctor-patient communication imbalance, the medical sector must improve the relationship between people, products, devices, and services. In this scope, technological trends implementation, such as IoT, will radically change healthcare worldwide. The environment 4.0 facilitates remote surgery, hybrid operations, centralized access to health data, service, and training upgrades. Another advantage is the e-commerce portals, which provide consumer digital interfaces [11].

4.1 Technological Mapping: Technologies 4.0 Applied in the Healthcare Field

The technological mapping pointed out the leading technologies 4.0 that support the digital transformation in the health domain, encompassing the Internet of Things (IoT), Big Data and Analytics, Autonomous Robots and Vehicles (AGVs), Additive Manufacturing (AM), and Augmented Reality (AR). It was observed that Industry 4.0 works to promote health digitalization applying emerging technologies that interpret clinical, medical, and laboratory data, as well as automatize manual processes in general health environments.

Therefore, as shown in Fig. 2, a multilevel circular diagram was proposed to demonstrate the applications of technologies 4.0 most associated with gains in health management. Process qualification improvement; training and agility of doctors and other health professionals; patient data share and analysis; efficient use of resources; and patient health deterioration minimization stood out.

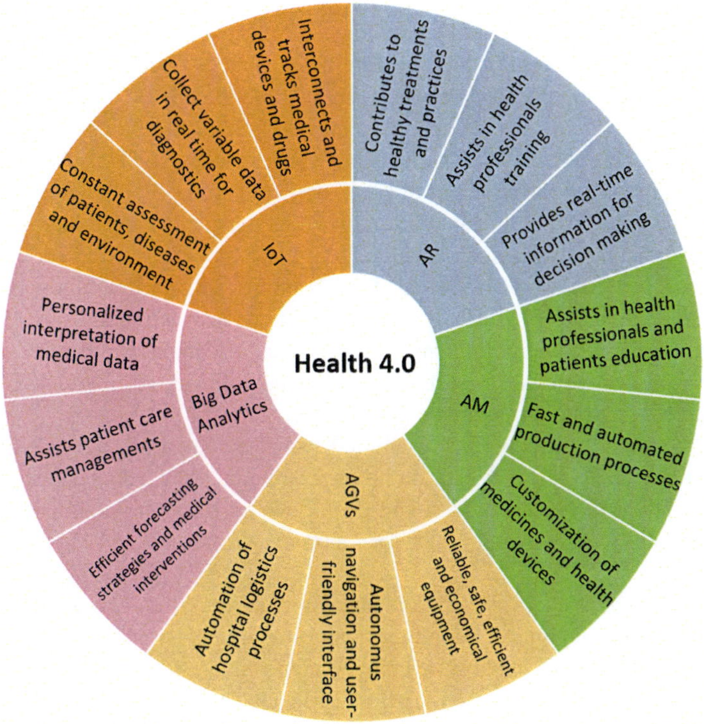

Fig. 2. Proposed multilevel circular diagram that represents the main contributions of emergent technologies to Health 4.0

Considering key gains for the health digitalization, the analysis showed that the IoT is linked to the connectivity between the different actors and devices of the process, real-time data collection, and continuous evaluation for feedback. Big Data and Analytics relates to the assertive interpretation of a large volume of patient data, medical forecasts, and interventions based on parameters. AGV is linked to hospital logistics and desirable attributes to be incorporated into the health system, such as reliability and economy. AM collaborates to automate the process and personalize products and medical devices, also showing interest in health professionals' training. Finally, the AR can collaborate for the assertiveness of the patient treatment, medical decision process, and the training of professionals.

IoT represents the integration of physical and digital objects that communicate and co-operate with each other and with humans in real-time over a wireless network. Featuring a range of applications in the health sector, this technology 4.0 can connect the Internet with medical devices and collect personal and treatment patient data, which are essential in monitoring, diagnosis, and testing processes. Acting together with Radio Frequency Identification (RFID) technology, IoT assists in producing, distributing, and tracking medical devices and medicines. With this integration, it will be possible to transmit information about medicines to a public database, enabling a patient or a health professional to access and check a specific medicine label data to easily identify possible counterfeits [12, 13].

Considering IoT as an essential technology to work with diseases management and health education, the concept of Medical Internet of Things (MIoT) emerges. Through MIoT, disease education data can be easily sent to patients' mobile terminals. This data sharing makes individuals receive updated status about their health conditions, as well as promotes associations between patients and medications. MIoT also facilitates the monitoring and real-time assessment of physiological parameters and diseases, which support early detections of clinical deterioration. Patients can control their health tests using their cell phones, while doctors can regularly monitor their results [5, 9].

Healthcare services can also be considerably improved with the joint performance of Cloud Computing (CC) with IoT and contribute to its continuous and systematic innovation in a Big Data environment. Since CC has a scalable capacity to store and process data, this technology transmits application and infrastructure services to various stakeholders with particular requirements.

In fact, the integration between CC and IoT provides new storage, processing, scalability, and networking capabilities. Combining communication technologies, interconnected applications, sensors, devices, and people would form an intelligent system to monitor, track, and store patient registers for continuous health care and Big Data analysis [14].

The Big Data technologies are designed innovatively, once they can process information with high speed and greater visibility, supporting organizations in decision-making processes. Through this technology 4.0, the efficiency and effectiveness of forecasting medical strategies, clinical treatments, health services, and health policies can be optimized, since this emergent tool will allow access to well-treated and high-quality data [3, 5].

Receiving data from mobile health apps or connected devices, Big Data Analytics can relate with each patient, resulting in a customized interaction. This singularized data interpretation will generate accurate individual results, as well as minimize costs for the health system through forecasts customization, early diagnosis, and improved decision support during the cyclic processes of continuous assessment. The health data treatment guarantees a fast and assertive identification of medicines safety signs and their possible adverse effects. Using stratifying populations methodologies, detailed responses can be obtained to treat specific patient subgroups [15].

Due to the need for more efficient and productive health systems capable of treating more patients without increasing costs, another technology 4.0 that extrapolates the factory floor and is highlighted in the medical environment comprehends in AGVs. The automatic navigation innovations, associated with user-friendly interfaces as well as with new materials and technologies, made AGVs, initiated developed to act in manufacturing production flows, been used also in medical field.

Since being adopted in real-time according to demands, transports, and movements of a hospital environment, AGVs prioritize the most important tasks to be completed first, which saves health professionals' time and intensifies a direct patient care. Associated with high reliability, safety, efficiency, and economy, these robots have sensors that detect obstacles early, guaranteeing a safe stop before reaching them. Besides, AGVs can carry heavy and hazardous wastes, operating all day long [14, 16].

AM, especially three-dimensional printing techniques, is also expanding rapidly in healthcare. This emergent technology consists of automated computer processes that manufacture objects layer by layer through the addition of raw material. This production method has revolutionized how medical operations are performed, covering the healthcare areas such as cardiology, neurosurgery, ophthalmology, dentistry, and prosthetic and implant markets, for example [17].

3D printing benefits in the health area are not limited to the customization of medical products, medicines, and devices, but also favor the cost-benefit ratio, productivity as well as design and manufacturing democratization. Printed models of internal organs and anatomical parts can reproduce the size, weight, densities, colors, and texture in detail. This allows surgeons to test complicated procedures on these 3D models first, as well as significantly improves the knowledge, management, and confidence of health students [18].

Also used as a learning tool, AR has been applied to prepare doctors, nurses, allied health professionals, and health students in different procedural skills. Students learn clinical practices in safe and controlled work environments while being supervised, as this technology 4.0 works with virtual objects overlapping or composed of the real environment. Since portable devices, such as a tablet or smartphone, are the most used hardware for AR, this technology can be easily accessed and does not require specialized equipment for its use [19].

5 Conclusion

The technological mapping pointed out the leading technologies 4.0 associated with health management gains, demonstrating that IoT, Big Data and Analytics, AGVs, AM, and AR support health digitization. It is noteworthy that approaches, such as user-centered design, focused on human factors, are aligned with the scope of Health 4.0. Thus, it was identified that the new Health 4.0 concept is associated with stakeholders' realignment to achieve the patient centralization in health system, enabling the production management of medical devices and attending particular requests according to mass customization. Also, applying technologies 4.0, different challenges regarding the price, speed and complexity can be optimized, as well as interconnectivity can be increased and the relationship between patients and health professionals can be strengthened, helping patients to understand their own personal health. Due to the need for a patient-centered approach, health stakeholders will need to promote a significant cultural change in the current medical system, adjusting themselves according to new environments, technologies, and regulations. In addition, health professionals must adopt new technologies and learn about them in an already highly pressured environment. Ownership of patients' personal data is also another challenge, as well as the divergences generated between the patient's symptoms description versus the technological description. As digital technologies are being improved and implemented, solutions to these challenges must also be addressed, aiming to become Health 4.0 a reality.

References

1. Wang CJ, Ng CY, Brook RH (2020) Response to COVID-19 in Taiwan: big data analytics, new technology, and proactive testing. JAMA 323(14):1341–1342
2. Pham Q, Nguyen DC, Huynh-The T, Hwang W, Pathirana PN (2020) Artificial intelligence (AI) and big data for coronavirus (COVID-19) pandemic: a survey on the state-of-the-arts. Preprints 2020040383
3. Javaid M, Haleem A (2019) Industry 4.0 applications in medical field: a brief review. Curr Med Res Pract 8
4. Gentner A, Elbel G (2014) Perspektive E-Health. Consumer-Lösungen als Schlüssel zum Erfolg? Deloitte Studienrei 1–22
5. Monteiro ABC, França RP, Estrela VV, Iano Y, Khelassi A, Razmjooy N (2019) Health 4.0: applications, management, technologies and review. Med Technol J 2(2):262–276
6. Kumari A, Tanwar S, Tyagi S, Kumar N (2018) Fog computing for healthcare 4.0 environment: opportunities and challenges. Comput Electr Eng 72(September):1–13
7. Prasetyo YA, Arman AA (2017) Group management system design for supporting society 5.0 in smart society platform. In: 2017 International conference on information technology systems and innovation (ICITSI). IEEE, Bandung, pp 398–404
8. Chute C, French T (2019) Introducing care 4.0: an integrated care paradigm built on industry 4.0 capabilities. Int J Environ Res Public Health 16(12):2247
9. Bause M, Khayamian Esfahani B, Forbes H, Schaefer D (2019) Design for health 4.0: exploration of a new area. In: Proceedings of the design society: international conference on engineering design, vol 1, no 1, pp 887–896
10. Thuemmler C, Bai C (eds) (2017) Health 4.0: how virtualization big data are revolutionizing healthcare. Springer, Cham
11. Hosseini M (2015) What will the future look like under industry 4.0 and digital transformation in the healthcare space? Rol Berger Strateg Consult Stuttgart
12. Hermann M, Pentek T, Otto B (2016) Design principles for industrie 4.0 scenarios: a literature review. In: 49th Hawaii international conference on system science. IEEE, Koloa
13. Silveira F, Neto IR, Machado FM (2019) Analysis of industry 4.0 technologies applied to the health sector: systematic literature review. Occup Environ Saf Heal 202(March):701–709
14. Punetha D, Kumar N, Mehta V (2013) Development and applications of line following robot based health care management system. Int J Adv Res Comput Eng Technol 2(8):2446–2450
15. Koutkias VG, Jaulent MC (2015) Computational approaches for pharmacovigilance signal detection: toward integrated and semantically-enriched frameworks. Drug Saf 38:219–232
16. Pedan M, Gregor M, Plinta D (2017) Implementation of automated guided vehicle system in healthcare facility. Proc Eng 192:665–670
17. Aimar A, Palermo A, Innocenti B (2019) The role of 3D printing in medical applications: a state of the art. J. Healthc Eng 2019:1–10
18. Dodziuk H (2016) Applications of 3D printing in healthcare. Pol J Cardio-Thorac Surg 13(3):283–293
19. Moro C, Stromberga Z, Raikos A, Stirling A (2017) The effectiveness of virtual and augmented reality in health sciences and medical anatomy. Anat Sci Educ 10:549–559

Work, Occupational Stress and Job Satisfaction: A Focus on Brazilian Millennials

Gabriela Manzatto Roveri[1], Igor Polezi Munhoz[2], and Alessandra Cristina Santos Akkari[1](✉)

[1] Mackenzie Presbyterian University, Campinas, Brazil
alessandra.akkari@mackenzie.br
[2] Federal Institute of Education, Science and Technology of Sao Paulo (IFSP), Sao Paulo, Brazil
igor.munhoz@ifsp.edu.br

Abstract. Generational cohort literature points out that multigenerational workforce has different values, beliefs, expectations, and sociodemographic characteristics, leading to diversity in the workplace. To guarantee sustainable employee well-being and improve business performance, the management of stress and job satisfaction at work becomes essential, considering the generational profiles. Therefore, this paper aimed to evaluate occupational stress and job satisfaction in Brazilian Millennials (Generation Y), considering individuals from the State of Sao Paulo. For this purpose, a validated Job Satisfaction Scale (JSS) and a validated Work Stress Scale (WSS) were applied and descriptive and inferential statistical methods were used for data treatment. A low level of occupational stress was identified in the sample ($1{,}64 \pm 0{,}82$) and the satisfaction of respondents with leadership (4.96 ± 1.83) was highlighted, contrasting with the dissatisfaction with the salary (3.83 ± 1.72) and with the promotions (3.72 ± 1.73). Cramer's V coefficient pointed out an association of greater magnitude between Generation Y and job stress ($V = 0.622$) and a moderate association between this group and the five dimensions of job satisfaction ($0.378 \leq V \leq 0.493$). Spearman's correlation test demonstrated a negative statistical correlation between stress and satisfaction at work, suggesting the promotion of policies, practices, and organizational culture that develop a healthy work environment of mutual respect and joint effort in order to maximize satisfaction and, therefore, reduce stress in Millennials.

Keywords: Generation Y · Millennials · Job stress · Job satisfaction

1 Introduction

The challenges that contemporary world history has been experiencing require a new view by the organization to manage diversity in the workplace [1]. Considering generational groups, sociodemographics, and cultural variables, values, backgrounds, and expectations need to be considered in the organizational environment to guarantee a healthy workplace and enhance business performance. In fact, changes in generations lead leaders to continuous reviews regarding their philosophy, mission, values and practices to optimize their teams' results [2].

Y. Iano et al. (Eds.): BTSym 2020, SIST 233, pp. 25–32, 2021.
https://doi.org/10.1007/978-3-030-75680-2_4

Regarding the work environment and multigenerational workforce, stress management and job satisfaction require in-depth analysis, as these parameters are directly linked to the employee's sustainable well-being and productivity, impacting the company's results in an increasingly competitive scenario [3].

Millennials, also known as Generation Y, refer to the most recent generation to enter the workforce, stimulating empirical studies to characterize this generational profile [2, 4]. In the literature, there are studies referring to the mapping of job satisfaction and occupational stress in parallel, as well as research that conceptually explores and characterizes the Millennials. However, there are few works that assess stress and satisfaction in the Millennial Generation from an association approach, nor studies that analyze whether there is a correlation between these two variables in the workplace.

Therefore, considering the definitions of Zemke et al. [5] for the generational cohort, this paper aimed to map the level of work stress and job satisfaction in Millennials (1981–2000), making a cut of Brazilian workers in the state of Sao Paulo. This chapter's main contribution refers to the identification of stressors and satisfaction factors that can guide organizational practices to promote a healthy work environment of mutual respect, commitment, and joint effort.

This work passed and was approved by the faculty's Ethics Committee.

2 Theoretical Background

According to Howe and Strauss [6], generations are characterized by individuals who share the history they lived, considering a specific period of time. In a complementary approach, generations are socially marked by the stage of life, conditions, and technology, involving experiences that have passed and influenced their beliefs, values, and behaviors [2]. Therefore, a generational cohort can be defined as a group of people who share historical and social experiences from the same perspective [4].

Considering the main generation cohorts at work (i.e., Baby Boomers, Generation X, and Millennials), Generation Y, the focus of this paper, cover the population born between the years 1980 to 2000 [5] and is commonly composed of individuals who value work-life balance and freedom, as well as who accepts cultural, religious and other types of diversity. Also, these individuals have multi-tasking abilities and they are used to technology, communication and the digital environment [2, 7].

The current scenario forces Millennials to think differently, influencing the change in the understanding of career, stability, and organizational life in relation to their predecessors. However, even with differentiated priorities, Generation Y still plans their career, aspires to a promising professional future, and how to conquer it [4].

From an organizational perspective, companies realize that factors such as effective leadership and job satisfaction can influence the organization's success [1]. Therefore, in order to understand the well-being of a multigenerational workforce, the mapping of stress and satisfaction at work is increasingly taking on special significance [3].

Occupational stress can be defined based on three aspects, including (i) stressful stimuli, referring to the impulses of the organizational environment that require coping skills from the employee; (ii) responses to stressful events, which include both physiological responses and responses to professional performance; and stress-response stimuli, which consider the entire process of workload and impact on employees [8].

Also, occupational stress is contemplated by several organizational stressors, which can be physical or psychosocial in nature. The first type involves factors related to the workplace, such as lighting and noise. The second type involves indirect factors of the workplace, such as relationship, autonomy, and control of functions [8, 9].

Job satisfaction is associated with a positive feeling of the individual reflecting the degree of alignment between personal values and interests and what is offered by the organization. In this context, there are variables linked to the work environment, encompassing the salary, opportunities for growth, recognition, competition; as well as variables related to labor agents, including colleagues and leadership [10].

When these factors are in balance, that is, with good management, incentives for employees, good relationships, among others, there is a demonstration of employee satisfaction, representing greater protection for mental health, and consequently, low occupational stress. In contrast, when these factors are not in balance, job dissatisfaction can have consequences on individual life, as well as on the individual's mental and physical health, leading to stress [11]. Dissatisfaction can also point to absenteeism, turnover, prolonged delays or unauthorized pauses, drop in productivity, protests or strikes, and dissatisfaction with the importance of work in the individual's life [9].

3 Methodology

It is an exploratory and descriptive research with applied nature and a quantitative approach. As for the procedures, a longitudinal survey was conducted, based on obtaining data from a certain group of individuals, without the supervision of responses and for a specific time (June/2019). The study started from a bibliographic review and the terms "Millennial Generation", "Millennials", "Job satisfaction", "Occupational Stress", "Generations", and "Generation Y" were used to search in Web of Knowledge, Scopus, Science Direct, and Scielo databases. Criteria for inclusion and exclusion of articles, definition of the information to be extracted from the articles, analysis, discussion, and presentation of the results will be considered as stages of the literature review.

3.1 Data Collection and Sampling

Data collection was based on applying two research instruments previously validated in Portuguese, encompassing the Job Satisfaction Scale (JSS) and the Work Stress Scale (WSS), both in their reduced version. The JSS instrument was validated by Siqueira [12], representing a multidimensional view of occupational satisfaction, according to five theoretical dimensions (salary, colleagues, leadership, promotion, and task nature) and the use of a 7-point Likert scale. The WSS instrument was developed and validated by Paschoal and Tamayo [13], based on theories about organizational stressors and reactions to occupational stress. This instrument considers the relationship between a stressor and the employee's reaction to it, using a 5-point Likert scale. For saving time and resources, non-probability convenience sampling was used, restricting the results of this study to the selected sample.

The self-administered questionnaire was designed and sociodemographic questions (gender, age, marital status, and education background) were included to characterize the sample. The instrument was digitally distributed, associated with an invitation letter that contained the research's objective and the Informed Consent Form. It was explained about the guarantee of the participant's anonymity, giving the individual the option to respond or not to the instrument according to the general guidelines of the Research Ethics Committee of Mackenzie Presbyterian University.

3.2 Statistical Treatment of Data

Initially, the data were submitted for a descriptive statistical analysis through measures of centrality and dispersion, using Microsoft Excel. With the cutoff points defined, the stress levels were categorized into three stages, comprising low score (1.00 to 2.00), medium score (2.01 to 2.99), and high score (3.00 to 5.00) [13]. For the interpretation of job satisfaction data, the dimensions scores were categorized in dissatisfaction (values between 1.0 and 3.9), indifference (values between 4.0 and 4.9), and satisfaction (values between 5.0 and 7.0) [12].

Then, the data were submitted to the treatment of statistical inference, using IBM SPSS software and considering a significance level of 5% (p-value). Spearman's correlation test was applied, adopting a null hypothesis (H0) and an alternative hypothesis (H1), with H0 corresponding to the existence of a correlation between stress and each dimension of satisfaction, while H1 corresponded to the lack of correlation between stress and each dimension of satisfaction. For this purpose, the mean scores of the instruments were used and the rejection of the null hypothesis automatically led to the alternative hypothesis analysis.

Focusing on the Millennium generation, Cramer's V coefficient test was also applied to identify the association between Generation Y and the stress level, as well as the association between Generation Y and the five dimensions of job satisfaction.

4 Results and Discussion

Regarding the sample characterization, the study had the participation of 89 respondents who were employed at the time of the survey, all of whom were individuals from companies located in the State of Sao Paulo (Brazil). Survey participants were classified into the generational cohorts, according to Zemke et al. [5]. Approximately 85% of respondents were from the Millennial Generation, followed by 9% from Generation X, and 6% from the Baby Boomer Generation. In the sample, the female gender (60.67%), singles (77.53%), and individuals with incomplete higher education (42.70%); complete higher education (28.09%) or post-graduation (24%) prevailed. Participants mostly occupied positions as interns (44.90%) or analysts (29.20%), with only 15.70% in coordination or supervision positions. This finding was reflected in the financial gain, so that the salary range up to 3,000.00 was the most reported (57.30%), followed by the range from 3,001.00 to 5,000.00 (21.30%) Brazilian Real. Only 21.35% of the sample has more than one job, of which 63.16% represented the Millennials.

A low to medium degree of job stress was identified (1.64 ± 0.82). According to the analysis of the job stressors (Fig. 1) dominant in the sample, it was observed that the respondents feel uncomfortable with the lack of training for professional improvement

(34.83%); irritated by the difficulty of disclosing organizational decisions (33.70%); there are not many prospects for career growth (33.70%); and there is not enough time to carry out the requested workload (31.46%).

Far from becoming a stressor, only a minority agree that there is a lack of autonomy in the execution of tasks (11.22%), competition in the workplace (14.60%), and a lack of confidence in the leadership in relation to employees (15.73%). In fact, there is a greater chance of having a low level of stress in individuals who work in an environment with autonomy and trust [2, 9].

Furthermore, it is noteworthy that studies in the literature point to higher levels of occupational stress in professionals with greater responsibility and directly involved in decision-making, such as managers from different areas [14, 15].

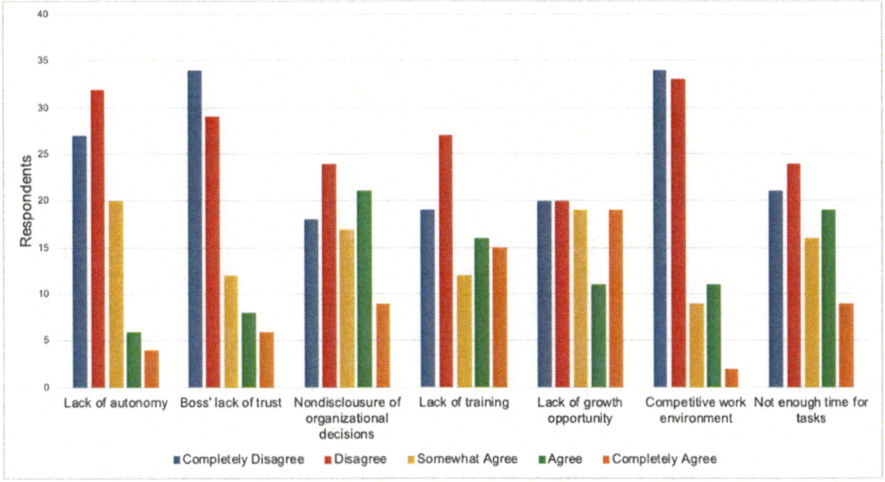

Fig. 1. Occupational stress factors pointed out by the respondents.

In reference to work satisfaction (Fig. 2), the sample indicated a state of indifference with the leadership (4.96 $\pm$ 1.83), with colleagues (4.85 $\pm$ 1.63), and with task nature (4.47 $\pm$ 1.61). Although these indexes do not express full satisfaction, the results are expressive mainly related to the leadership, suggesting that employees are satisfied when the organization has an effective people management process and understands them as partners of the company for collective construction. Also, healthy relationships in the workplace can be associated with job satisfaction.

Dissatisfaction was obtained for the dimensions of salary (3.83 $\pm$ 1.72) and promotion (3.72 $\pm$ 1.73), although the centrality measure was shown to be close to what was expected for the indifference status. In fact, according to Fantini and Souza [16], financial security proves to be an important factor for the satisfaction of Brazillian Millennials, suggesting that the company should be concerned with this issue in order to ensure the contentment of this generation.

The high standard deviation obtained in the dimensions of job satisfaction indicates the lack of homogeneity in the employees' perception. Therefore, the leadership needs to

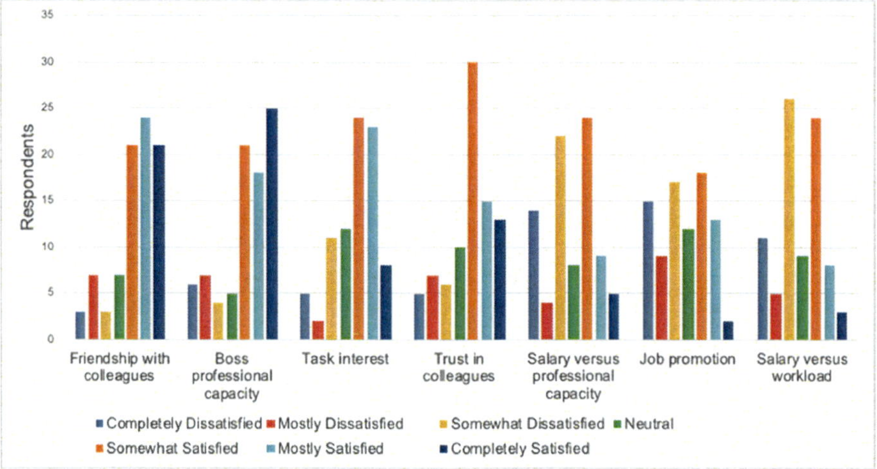

Fig. 2. Occupational satisfaction factors pointed out by the respondents.

maintain a close relationship with their team to assess each one's perception and ensure a healthy workplace.

The Spearman correlation test (Table 1) was applied considering the entire sample. It was found that occupational stress has a negative correlation with the five dimensions of job satisfaction, encompassing stress and satisfaction with colleagues ($p = 0.000$); stress and satisfaction with leadership ($p = 0.000$); stress and satisfaction with the task nature ($p = 0.000$); stress and satisfaction with promotions ($p = 0.000$), and stress and satisfaction with salary ($p = 0.026$). The coefficient values point to a degree of association between weak to moderate.

Table 1. Spearman correlation test.

		Dimensions				
		Satisfaction with				
		Colleagues	Leadership	Task nature	Promotion	Salary
Job Stress	Coefficients*	−0.382	−0.455	−0.411	−0.391	−0.241
	p-value**	0.000	0.000	0.000	0.000	0.026

*Spearman correlation
**Statistical significance

The negative correlation suggests that job stress is inverse to the degree of occupational satisfaction, that is, the lower the level of stress, the greater the level of job satisfaction. This finding is corroborated by Lelis et al. [14] that applied JSS and WSS instruments to 114 Brazilian commercial managers, concluding that, statistically, both variables are associated and there is a negative correlation.

The data obtained lead to organizational reflections about practices and policies that aim to investigate and promote factors that maximize job satisfaction. Thus, there is a reduction in occupational stress, valuing the worker's well-being and a positive environment.

Finally, the association between generations and the levels of job stress and work satisfaction were assessed. An association with a stronger magnitude was identified between Generation Y and the stress level ($V = 0.622$). Regarding to job satisfaction, moderate associations were observed among Millennials and the degree of satisfaction with colleagues ($V = 0.493$), followed by the level of satisfaction with the task nature ($V = 0.480$), with promotions ($V = 0.470$), with the leadership ($V = 0.441$), and with the salary ($V = 0.378$). The data obtained can be discussed based on the peculiar profile of Millennials. Generation Y gives importance to human relations, including co-workers; they like freedom to express themselves; seek motivating and challenging projects in their job; they value personal satisfaction, often above professional achievement, and want meaningful work.

5 Conclusion

This study pointed to a low level of occupational stress, possibly due to the fact that respondents occupy positions that are not directly associated with decision making since more than half of the sample was comprised of interns or analysts. Regarding job satisfaction, it was identified points to be improved by the organization since none of the dimensions reached the average score corresponding to worker satisfaction. A negative statistical correlation between occupational stress and each of the dimensions of job satisfaction was obtained, suggesting the promotion of culture and organizational policies that maximize occupational satisfaction in order to also reduce stress at work. Considering Millennials, a stronger association of this generation with the level of job stress was found, as well as a moderate association of this generation with the levels of job satisfaction, and these points can be explained according to the generational profile. The mapping of stress and satisfaction in the workplace, mainly focusing on the Millennial Generation, contributes to an in-depth discussion about the importance of respecting the attributes, culture, values, and expectations of generations in the organization, entering into the scope of diversity management.

References

1. Shaban A (2016) Managing and leading a diverse workforce: one of the main challenges in management. Proc – Soc Behav Sci 230:76–84
2. Cvenkel N (2020) Well-being in the workplace: governance and sustainability insights to promote workplace health. In: Approaches to global sustainability, markets, and governance. Springer, Singapore
3. Kowalski THP, Loretto W (2017) Well-being and HRM in the changing workplace. Int J Hum Resour Manage 28(16):2229–2255
4. Lissitsa S, Kol O (2019) Four generational cohorts and hedonic m-shopping: association between personality traits and purchase intention. Electron Commer Res

5. Zemke R, Raines C, Filiczak B (1999) Generations at work: managing the clash of veterans, baby boomers, xers, and nexters in your workplace. American Management Association, New York
6. Howe N, Strauss W (1991) Generations: the history of America's future, 1584 to 2069. William Morrow & Company, New York
7. Çelikdemir DZ, Tukel I (2015) Incorporating ethics into strategic management with regards to generation Y's view of ethics. Proc – Soc Behav Sci 207:528–535
8. Spector PE, Jex S (1998) Development of four self-report measures of job stressors and strain: Interpersonal conflict at work scale, organizational constraints scale, quantitative workload inventory, and physical symptoms inventory. J Occup Health Psychol 3:356–367
9. Luz CMDR, de Paula SL, Oliveira LMB (2018) Organizational commitment, job satisfaction and their possible influences on intent to turnover. Rev Gestão 25(1):84–101
10. Baotham S, Hongkhuntod W, Rattanajun S (2010) The effects of job satisfaction and organizational commitment on voluntary turnover intentions of Thai employees in the new university. Rev Bus Res 10(1):73–82
11. Henne D, Locke E (1985) Job dissatisfaction: what are the consequences? Int J Psychol 20:221–240
12. Siqueira MMM (2008) Medidas do comportamento organizacional: ferramentas de diagnóstico e de gestão. Artmed, Porto Alegre
13. Paschoal T, Tamayo A (2004) Validação da escala de estresse no trabalho. Estudos Psicol 9(1):45–52
14. Lelis JWF, dos Santos NMBF, Akkari ACS, Munhoz IP (2019) Occupational stress and job satisfaction among Brazilian managers: the case of commercial area. In: Reis J, Pinelas S, Melão N (eds) Industrial engineering and operations management II. IJCIEOM 2018. Springer proceedings in mathematics & statistics, vol 281. Springer, Cham
15. Zille LP, Braga CD, Marques AL (2014) Estresse no trabalho: um desafio para os gestores das organizações brasileiras. REGE Rev Gestão 21(3):401–413
16. Fantini CA, Souza NCDSD (2015) Análise dos fatores motivacionais das gerações baby boomers, X, Y e Z e as suas expectativas sobre carreira profissional. Rev IPecege 1(3/4):126–145

Indicators for Assessing Sustainability and Productivity in Companies with Implementation of Industry 4.0 in MERCOSUR

Barbara Socorro Barbosa de Freitas, Igor Polezi Munhoz(✉), and Fernando Cesar Mendonça

Federal Institute of Education, Science and Technology of Sao Paulo (IFSP), Sao Paulo, Brazil
barbara.freitas@aluno.ifsp.edu.br, {igor.munhoz, fernando.mendonca}@ifsp.edu.br

Abstract. The concept of sustainable development is the incorporation of economic, social, and environmental dimensions in the search for development. Considering that Industry 4.0 is defining a new era of production, this article's main objective is to present the sustainability indicators that best describe the impact of the adoption process of Industry 4.0 in MERCOSUR. The indicators were selected according to the integrative review method, a process based on selecting and systematizing the analysis of pre-existing works. The search for articles and academic dissertations was performed using the keywords "sustainability", "sustainability indicators", and "Sustainability in industry 4.0", in English and Portuguese, in order to find national and international references on the topic. The indicators selected were ETHOS, GM Metrics for Sustainable Manufacturing, INMETRO, and the ISO 14031 Environment Performance Evaluation standard, considering their potential scope and applicability in the MERCOSUR member countries in terms of sustainability. The main contribution of this article refers to the mapping of performance indicators that can guide the practices adopted by companies when implementing Industry 4.0.

Keywords: Sustainability · Industry 4.0 · MERCOSUR

1 Introduction

Starting in 1987, the Brundtland Commission conceptualized what sustainability would be: a big umbrella that shelters the needs of economic development, social development, and environmental preservation [1]. With the public formalization of what would be the foundations of sustainable development, Agenda 21 emerges and, with it, the series of ISO 14000 standards. The first aims to guide States in decision-making processes to mitigate global warming impacts, and the second aims for the same use, but at the corporate level [2]. However, within the corporate microcosm, the demands are diverse and specific; therefore, several sustainability indicators emerge.

Y. Iano et al. (Eds.): BTSym 2020, SIST 233, pp. 33–40, 2021.
https://doi.org/10.1007/978-3-030-75680-2_5

Thus, this work's first task was to study sustainability indicators that best adapt to the context of Industry 4.0, since the fourth industrial revolution is part of this scenario in which it is necessary to deal with the resource scarcity aspect, under either environmental, legal or consumers pressure. In a second section, the aim was to seek indicators that could potentially describe MERCOSUR countries' reality, which, now, are not part of the group of Industry 4.0 pioneers.

The concept of Industry 4.0 may vary between authors, but the cutting-edge technology required to achieve it is a constant among them. The countries identified as pioneers of this revolution are, not by chance, holders of such technology. However, this work aims to analyze its possible impacts in terms of sustainability and production performance.

The main approach present in this new industry concept is to make communication and cooperation among people, machines, products, and logistics systems, using production systems with embedded technology, composed of integrated sensors and intelligent actuators, composing the cyber-physical system [3].

In addition to the high demand for technology, Industry 4.0 may also be a great ally of sustainability indicators, since sustainability can offer benefits for both companies and society, in a way that we can consider a company sustainable when it takes care of the environment, of the stakeholder and seeks to improve its reputation, which enables decision making for better performance, such as lower emission of gases that cause the greenhouse effect, better use of the raw material and assist with the disposal of highly polluting elements to places less harmful to the environment [4].

The Organization for Economic Cooperation and Development (OECD) defines that an indicator must be understood as a parameter, and it can be a figure that allows obtaining information about the situation of an event. For example, there are social indicators that are complex to measure and are linked to historical, cultural, and political factors.

The main objective of this article is to present the sustainability indicators that best describe the impact of the adoption process of Industry 4.0 in MERCOSUR, using a systemic procedure for collecting and analyzing data based on the integrative review method.

2 Theoretical Framework

The concept of sustainability is formed as an economic, social, and environmental triangle and expresses the persistence of an uncertain future with needs and characteristics that involve the ecosystem and the human subsystem.

The origin of the concept of sustainability occurred in the "Brundtland Report" in 1980 in the event "The World Conservation Energy", in which sustainability was seen as the development that meets the needs of the moment, but without compromising the possibility of the next generations to also satisfy their needs [5].

Therefore, linked to the concept of sustainability is the concept of sustainable development: the incorporation of economic, social, and environmental dimensions in the search for development. With the development of the concept of sustainability, indicators and indices have emerged to assess the performance of people, companies, governments, and the world, as a whole.

There are indicators that can be measured quantitatively and others qualitatively. Following, there is a presentation of three possible sustainability indicators that can be used in the conduction of this work [6]:

- The economic dimension of sustainable development: this indicator is influenced by the size of the economy, the economy's productive structure, and the specialization extent of the economy. For example, we have the GDP (Gross Domestic Product), which measures a given region's economic growth through private consumption plus the total investments made in the region, government spending, and exports, discounting imports. Given the limitations that some authors point to the use of GDP concerning sustainable development, other indicators can be used, such as the Genuine progress indicator, which is based on consumption or savings expenses.
- The social dimension of sustainable development: this indicator is influenced by the size of the population, its structure, and social cohesion. One of the indicators we have as a reference is the Human Development Index (HDI), which measures a given society's human and economic development, although other relevant indexes can also be applied.
- The environmental dimension of sustainable development: this indicator includes the presence of environmental items, such as the use of water resources, the quality of soil and air, and biodiversity. The Ecological Footprint is suggested as an indicator, which presents the biosphere's regeneration capacity based on human consumption. The measurement unit is the "global hectare", which is a unique measure of global productivity in the search for resources in different countries.

The concept of Industry 4.0 emerged in the Industry-Science Research Alliance report, presented in 2013 at Hannover Messe, an industrial business fair and considered the greatest exponent in Industry 4.0 today. Among the various definitions that the term presents, it is worth highlighting the worldwide trend of implementation of Industry 4.0, mainly in countries with advanced industrialization, defining a new era of production [7].

Therefore, factory automation, using technologies such as the Internet of Things (IoT - Internet of Things), is associated with concepts such as Big Data, Intelligent Robotics (Intelligent Robotics), Industrial Automation, and others. With the large volume and data flow, significant changes in the production process are also inherent, connecting technologies, work data, inputs, and energy use.

The use of IoT combined with the Internet of Services (IoS) in the production process will bring to the fourth industrial revolution since the use of IoT allows sensors, actuators, telephones, and other systems and objects to interact through "intelligent" components and create the cyber-physical network in which the systems cooperate using the internet infrastructure [8].

3 Methodology

The article was developed according to the integrative review method, a process based on selecting and systematizing the analysis of pre-existing works on a given subject [5]. Its procedure is fully described in Table 1.

Table 1. Integrative review.

Steps	Description
1st step	- Definition of the problem - Formulation of a question for the research - Definition of a search strategy - Definition of the descriptors - Definition of the database
2nd step	- Use of the database - Search of the studies with the database with the inclusion and exclusion criteria
3rd step	- Reading of the abstracts, keywords, and title of the publications - Organization of the shortlisted studies - Identification of the selected studies
4th step	- Elaboration and use of the summary matrix - Categorization and analysis of the information - Elaboration of an individual library - Critical analysis of the selected studies
5th step	- Discussion of the results
6th step	- Setting-up of a document that fully describes the review - Proposals for future studies

Source: Authors (2020)

The first step was to define the approach to the theme of "sustainability" as one of the possible factors impacted by the adoption of Industry 4.0 while still outlining the research project. Subsequently, articles about sustainability indicators in the context of Industry 4.0 were selected as a consolidation of the second step. The third step consisted of searching for characteristics that would best describe the reality addressed by the work, such as the choice of national indicators and major international referenced indicators [9]. The fourth and fifth steps were carried out simultaneously, using each indicator's available manuals and their reports to better understand its functionality. The sixth step is the result of this article presented here.

The search for academic papers was carried out in the Google Scholar databases, given their open consultation, both for articles and academic dissertations, using the keywords "sustainability", "sustainability indicators" and "Sustainability in industry 4.0", in English and Portuguese, in order to find national and international references on the topic.

Based on the results obtained, a table was constructed covering the articles' perspectives and dissertations on each indicator. Subsequently, a refinement of the indicators was carried out based on the availability of information on the MERCOSUR countries in their databases.

4 Results and Discussion

From the integrative review proposed in the methodology, the list made of an analysis of 24 sustainability indicators, 11 are detailed in Table 2. Table 2 contains the indicators that simultaneously comply with the three pillars of sustainability - social, economic, and environmental - and do not present restrictions on data access or high complexity of analysis [9]. The criteria used to evaluate the advantages and disadvantages were based on the author's assessment considering each index's available information.

From the survey presented in Table 2, it was possible to analyze each indicator from their manuals, but now seeking the selection of the ones that may be associated with the MERCOSUR countries, which drove to the ETHOS, GM Metrics for Sustainable Manufacturing, INMETRO, and ISO 14031 Environment Performance Evaluation indicators.

The Ethos indicator integrates social, business, and sustainable behaviors, proposing measures for their optimization, but does not aim to measure performance or classify companies as sustainable. Through an online platform, the organization invites companies to answer an annual questionnaire that, in the end, proposes planning based on the quantitative analysis generated and allows creating comparisons between associated companies. Participants companies may choose one out of four formats of the questionnaire according to their reality, whose range from a minimal sense of the sustainable context to the optimization of sustainable perspectives plus an analysis of pioneering issues. Beyond those facts, the indicator's selection was also determined by its presence in Brazil and its association with Argentina, Paraguay, and even Bolivia, Ecuador, and Peru.

The GM Metrics for Sustainable Manufacturing indicator has seven working subgroups - energy consumption and environmental impact, energy reliability, changes in the consumer's view, the image transmitted by the brand, and the development of better production systems and sustainable technologies, which guarantee its complexity. This selected indicator due to the auto industry's high presence in South American countries and it is one of the industrial segment pioneers in Industry 4.0.

The INMETRO (National Institute of Metrology, Quality, and Technology) indicators are inside some criteria, and these are inside seven major principles, which are: compliance with legislation; sustainable management of natural resources; preservation, conservation, and recovery of biodiversity, improvement, and maintenance of environmental quality; appreciation and well-being of workers; environmental, economic and social development of the regions where the activities of the organization are located and promotion of technological innovation. The criteria specify aspects of a given point of the principle and then point out its indicators. Although it is a Brazilian state initiative, it is somewhat generic in its definitions, precisely to better deal with possible particularities, even allowing a possible application to other countries' scenarios.

Finally, the ISO 14031 (Environment Performance Evaluation) standard not only provides indicators but also instructs the measurement, analysis, evaluation, and performance processes, in addition to proposing greater complexity to NBR ISO 14001, with the insertion of the entire cycle product life, from the production chain to the end of its consumption. The intrinsic objective of global standardization can make it highly complex, but, for comparative purposes, it is possible to shape it to each reality.

Table 2. Sustainability indexes, advantages, and disadvantages.

Index	Description	Advantages	Disadvantages
Composite Sustainable Development Index	Created by Krajnc and Glavic in 2005 to assess the three pillars of sustainability	It can be adapted according to the scenario of each company	It is just a method for handling information from other indexes
Coral	Evaluates the level of corporate sustainability with a focus on strategic planning	It can be adapted according to the scenario of each company	It is just a method for handling information from other indexes
Environmental Pressure Indicators for the European Union (EPI-EU)	It assesses negative human actions that impact the environment	It explains the European vision of sustainability, which is a reference and an Industry 4.0 pioneer	Depicts only European Union countries
Ethos	Indexes are suggested in the three dimensions of sustainability, with greater emphasis on the social aspect	Depicts a national scenario of companies interested in sustainable development	Limited to Brazil and companies associated with ETHOS
ETHOS SEBRAE	A tool to assess social responsibility and sustainable development in micro and small companies	Depicts a national scenario of micro and small companies interested in sustainability	Limited to Brazil and companies associated with ETHOS - Sebrae
Ford Product Sustainability Index	Model focused on the automotive industry, mainly on product life cycle assessment	Presence of Industry 4.0 in automobile companies, which have agreements with South American countries	Limited to the automotive niche and low complexity with only 8 indicators
GM Metrics for Sustainable Manufacturing	Created by General Motors, indexes are divided into 6 categories: environmental impact, energy consumption, personal well-being, occupational safety, manufacturing costs, and waste management	It interconnects the pillars, considering what the impact on one pillar can affect the other	Limited to the automotive niche

(*continued*)

Table 2. (*continued*)

Index	Description	Advantages	Disadvantages
INMETRO	A platform of seven principles, each consisting of a set of indicators, which serve as a reference for assessing sustainability	It also has a performance index that can easily be linked to sustainability's	Not found
ISO 14031 Environment Performance Evaluation	International standard with specifications for organizations to develop their own sustainability index	Easy access to information	Not found
ITT Flygt Sustainability Index	Applied to evaluate the economic, social, and environmental pillars within companies	Assemble your own index, following the assembly instructions	Limited to the local reality of Sweden
OECD Sustainable Manufacturing Toolkit	Created by the OECD to monitor environmental conditions in the organization's member countries	Consisted of model countries in the scope of sustainability	OECD does not have MERCOSUR members

Source: Authors (2020)

The ETHOS indicator is already applied in MERCOSUR member countries, such as Brazil. GM's indicator also considers the reality of MERCOSUR since GM has subsidiaries in member countries. The INMETRO indicator was developed by the Brazilian government, which is a member of MERCOSUR and ISO 14031, which originated NBR ISO 14001, is already applied in the Brazilian context. Thus, because these indicators are already used in MERCOSUR member countries, they indicate their possibility of adaptation and use in the local context.

Based on the integrative review, 11 indicators were identified and discussed that showed potential (adaptation to the organization, ease of access to information, and interconnection between the pillars of sustainability) and some limitations (regional application and business segment application). However, it was observed that all the indicators mentioned have the common characteristic of promoting a culture of sustainability in the adhering organizations.

Therefore, studies aimed at grouping and analyzing data on productivity and sustainability indicators become a facilitator in order to drive companies or countries, optimizing results, especially in view of the new Industry 4.0 paradigm.

5 Conclusion

Considering that the main objective of this article is to present the sustainability indexes that can impact the process of adopting Industry 4.0 in MERCOSUR, it is concluded that the proposed objective was achieved, so that the ETHOS, GM Metrics for Sustainable Manufacturing, INMETRO, and the ISO 14031 Environment Performance Evaluation standard were selected according to their due complexity for analysis, considering their potential scope and applicability in the MERCOSUR member countries in terms of sustainability.

Therefore, the previous indexes' selection took into consideration the accessibility of the data from the integrative review of its information, as well as the main advantages and disadvantages of each index. It is a highly complex topic, so new studies are recommended, such as developing a case study on the application of indicators in a company that uses the concept of Industry 4.0, seeking to analyze the relationship between sustainability and Industry 4.0.

References

1. ONU. https://nacoesunidas.org/acao/meio-ambiente
2. Seiffert M (2011) ISO 14001 Sistemas de gestão ambiental. 4th edn. Atlas S.A., São Paulo
3. Astarloa A, Bidarte U, Jiménez J, Zuloaga A, Lázaro J (2016) Intelligent gateway for Industry 4.0 – compliant production. In: IECON 2016, 42nd annual conference of the IEEE industrial electronics society, Florence, pp 4902–4907
4. Almeida F (2002) O bom negócio da sustentabilidade. Nova Fronteira, São Paulo
5. Botelho LLR, Cunha CCA, Macedo M (2011) The integrative review method in organizational studies. Gestão e Soc 5(11):121–136
6. D'Avila GLR (2010) Análise comparativa entre indicadores para a sustentabilidade na dimensão macro. Trabalho de Conclusão de Curso em Engenharia de Produção, Universidade de São Paulo, São Carlos
7. Rocha MFM, Oliveira KF, Munhoz IP, Akkari ACS (2019) Industry 4.0: technology mapping and the importance of cognitive ergonomics. Int J Adv Eng Manage Sci 5(5):296–303
8. Brettel M, Friederichsen N, Keller M, Rosenberg M (2014) How virtualization, decentralization and network building change the manufacturing landscape: an industry 4.0 perspective. Int J Mech Ind Sci Eng 8(1):37–44
9. Schules MV (2018) Proposta de diagnóstico para adoção das tecnologias da indústria 4.0 em um processo produtivo com base em indicadores de sustentabilidade: um estudo de caso. Dissertação de Mestrado em Engenharia de Produção, Universidade Federal do Paraná, Curitiba

Characterization of Digital Supply Chain

A Focus on Configuration and Resources

Gabriela Rodrigues Souza, Luana Spósito Valamede, and Alessandra Cristina Santos Akkari(✉)

Mackenzie Presbyterian University, Campinas, Brazil
alessandra.akkari@mackenzie.br

Abstract. Currently, the development of digital technologies associated with the Fourth Industrial Revolution has been driving the formation of interconnected and intelligent systems, bringing reflections on the configuration of traditional supply chains (SC) in the face of an increasingly competitive market. Therefore, this paper aimed to characterize the digital supply chain (DSC), considering frameworks in the literature and the application of technologies 4.0. An exploratory and descriptive study, with a qualitative approach and using the bibliographic strategy, was developed to this end. Among the results, it was identified three relationships for the development of DSC, consisting of (i) Industry 4.0 and digitalization as a facilitator of DSC; (ii) SC innovation and transparency through Industry 4.0 tools; (iii) and a mutual correlation between Industry 4.0 and DSC. Based on the interpretation of these perspectives, it was concluded that organizations need to view their SC under new perspectives by applying emerging technologies to guarantee the flexibility, integration, and efficiency of their chains in the digital age.

Keywords: Supply chain · Industry 4.0 · Digital supply chain

1 Introduction

Regarding the management of the processes and interactions that configure a value chain, Supply Chain (SC) aims to preserve and establish connections between stakeholders, ranging from logistics activities to data sharing [1]. However, classic SCs have infrastructures that are no longer self-sufficient in the digital age, requiring technological innovations that guarantee greater flexibility, integration, and efficiency [2].

In this scope, digital technologies developed from the Fourth Industrial Revolution have excelled [3, 4]. In fact, the introduction of emerging technologies in current manufacturing systems can influence the entire existing SC, innovating the means by which products are designed, produced, and transported to the customer, causing profound changes in classic SC [5].

Therefore, in order to remain sustainable in the competitive market, as well as to guarantee new growth opportunities, companies are becoming aware and developing action plans to optimize their current business models, highlighting the Digital Supply Chain (DSC) as an opportunity to add value to organizational processes. In the emerging

Y. Iano et al. (Eds.): BTSym 2020, SIST 233, pp. 41–47, 2021.
https://doi.org/10.1007/978-3-030-75680-2_6

digital model, organizations have countless technologies coming from Industry 4.0 that assist in the transformation process of traditional SC. However, companies need to improve the level of technical adaptability to be able to implement them in their systems so that they can thus move towards DSC [6].

Considering that organizations' dominant focus is to preserve and invigorate their vital competencies, it is important to standardize information about attributes and characteristics of technologies from Industry 4.0 and point out trends and interactions in SC [7, 8].

This paper aims to characterize the DSC, addressing its pillars and technological integration to assess how the productive chain is behaving in the face of the Fourth Industrial Revolution. For this purpose, an exploratory and descriptive research with a qualitative approach and bibliographic strategy was conducted. This article contributes to the discussion of SC redesigns in order to promote its alignment with the digital age based on structure and resources.

2 Theoretical Background

The definition of SC is related to the management of all the interactions that make up an organization's production chain. Regarding configurations, the SC covers the main business processes; the agents, from the supplier to the customer; the links between these agents; and the flow of information and material [9]. Figure 1 exemplifies a generic SC flow with the main steps.

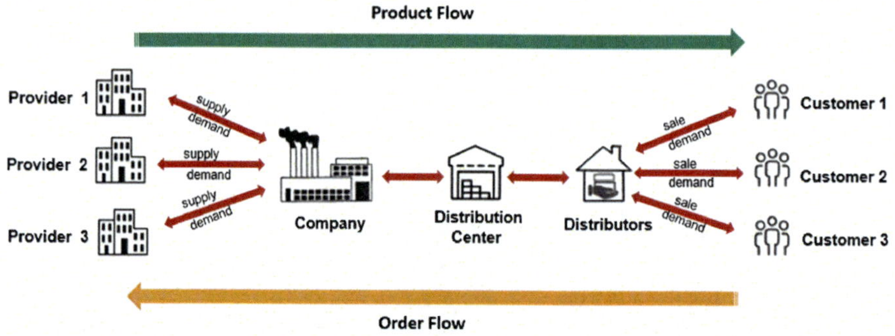

Fig. 1. Composition of the main flows of a SC. Source: Authors (2020)

Nowadays, organizations face situations never seen before in SC scope, such as the possibility of dispatching fleets of trucks and vehicles and locating them through a portable device [2]. The factory, which previously existed only in the physical environment, currently can also exist in the virtual environment, allowing customer demand monitoring in real-time [10]. Innovations like these, which directly impact SC, are being feasible due to the technological impulse brought by Industry 4.0.

In fact, with the introduction of the Fourth Industrial Revolution, production systems are moving towards an advanced stage of organizational development and value chain management, overcoming productivity paradigms and changing the entire logic of the manufacturing field [10].

Therefore, smart factories contribute to a holistic approach to organizational processes, which are no longer analyzed individually and become integrated from an end-to-end perspective. This fact leads companies to rethink the way they manage their SC. Given these trends, the digitalization of SC is shown as a way for the improvement of SC, allowing chains to deal with new requirements and to be more integrated and flexible based on the application of a wide range of new technologies [1].

In the DSC concept, companies must restructure their SC as a digital supply network that unites not only products, services, and flows but also people, data, and finances, allowing interactions that are increasingly interconnected and accessible [2].

3 Methodology

This article's research methodology was composed of the phases of observation, categorization, and association [11]. As an exploratory and descriptive research, the first phase consisted of a literature review for structuring information to identify the main characteristics, guiding concepts, and research trends of SC and Industry 4.0.

In this step, the academic databases Web of Science, ScienceDirect, and Scopus were used, and a time constraint (2011–2020) was applied. The search fields used to choose the articles were limited to the Abstract, Title, and Keywords, resulting in the following search string: Title-Abstract-Keyword ("Supply Chain", "Digital Supply Chain", "Digitization", "Digital Technologies", "Circular Economy", "Industry 4.0", and "Supply Chain Management"). The choice to use these keywords was due to the fact that they were selected to make part of the twenty words most frequently associated with SC digitalization. Also, only articles in English were considered to compose this study.

In the categorization phase, the attributes of SC and Industry 4.0 were analyzed, classified, and grouped. Thus, it was possible to structure traditional SC as well as identify configurations and resources most used for the digital transformation of classic SC, focusing on the principles and technologies provided by Industry 4.0.

Finally, the association phase aimed to combine both concepts already structured to analyze the existing relations between SC and Industry 4.0. Through the points of synergy evaluated, it was possible to discuss the DSC approach, as well as the restrictions and difficulties of integration.

4 Results and Discussions

The literature review pointed out that classic SCs operate with a clear objective, through structures and guidelines that define the best way forward. However, traditional SC consists of a series of poorly integrated steps, which do not act with a real-time perspective and depend on various electronic processes. On the other hand, DSC offers disseminated data, cooperation, and communication on digital platforms, resulting in greater efficiency and security.

The conceptual approach proposed by Büyüközkan and Göçer [2] was very cited in the literature for the development of DSC based on three stages and ten resources associated with digital solutions for traditional SC. According to the authors, the framework to develop a DSC includes digitization, technologies, and SC management.

Complementarily, it was identified, in the framework of Queiroz et al. [12], the composition of DSC capacities aligned with the application of digital technologies. Thus, the study proposes seven capabilities associated with policy development, digital integration, and smart production, which are supported by six technologies such as Big Data, Cloud Computing, and Internet of Things (IoT), among others.

There is an interaction between smart solutions, smart factories, and smart innovations, which lead to process optimization, digitization of the structure, and improvement of the chain management. Thus, there is an interconnection between processes, services, and customers supported by the application of digital technologies.

Digitization was driven by the significant increase in the automation of processes and systems, growth in the number of data, and continued exchange of information between devices [13]. As a result of this scenario, companies are being encouraged to seek tools and resources to support the digitization process within organizations in a way that it happens gradually and constantly [12].

Associated with digitization, the technological infrastructure was identified as fundamental for the organization to configure and operate the DSC [14]. These technologies can act as a DSC simplifier due to their relevance in digitization [12]. In general, the insertion of new technologies in the production system offers a set of risks and vulnerabilities.

Therefore, the process of implementing technologies must take into account and determine the essential phases to establish the technological infrastructure. Once the organization is able to stipulate the nature of the infrastructure during the implementation stages, it is also able to define technological priorities. The technology's implementation methodology should also contain details of the activities that need to be developed and which tools and devices should be acquired during the implementation phase. Another point that must be taken into account during this period is the human and technological relationship. Thus, it is essential to prepare employees to take control of new resources through training and interaction [15].

In fact, according to Queiroz et al. [12] framework, to support the progress of all the transition phases of a DSC, the capacity of employees is indispensable and can be pointed out as a critical resource.

In addition to the pillars mentioned above, redesigning a SC to achieve a strategic purpose involves dealing with several decisions at a complex level. It is essential to understand the possible problems between the various components that interact in a SC, since one step depends on the other. Therefore, the management of all stages of the production process must be improved by digitizing and implementing emerging technologies to optimize processes and reduce existing gaps [2, 16].

It was also identified that the redesign of the SC requires an in-depth discussion on resources and integration. When integrated and managed in the right way, the resources are transformers in SC, as it can combine lots of information from own sources and market, investigate the root cause of the problems, demand forecasts through advanced analyzes, thus enabling the forecasts to be more accurate and problems anticipated.

Organizations that have DSC are able to guarantee a competitive advantage in the market. Thus, it can be seen how the resources associated with digital solutions contribute to the transformation of traditional SC.

Innovation is an essential resource for DSC and allows the combination of advanced manufacturing techniques, aiming to create an interconnected organization, with the ability to drive other intelligent actions in the physical world [1]. Companies are dealing with a plethora of data that is generated all the time. In this way, intelligence is a resource used in DSC that allows information to be actionable quickly, increasing the interaction between man-machine and improving innovations [17].

Through the use of various digital resources, organizations that have a global connection enjoy highly evolved operations models, which can change functions and services according to customer demand and visualize processes in real-time, which makes SC dynamic, reliable, and interactive [17].

In addition to SC's global connectivity, speed is another challenging resource for companies, as the time required in managing information and transporting products and materials is hugely relevant for DSC [2, 17].

It is noted that the digitization of SC entails the need for functional agility and adaptability to new situations. With that, DSCs can react to devastating problems. Flexibility is a feature present in the DSC, as it can predict and react effectively and efficiently, minimizing impacts through the use of software that efficiently connects information.

Therefore, all information connected to the DSC must be continuously updated. Thus, the seventh resource is characterized by real-time inventory, which makes warehouse management feasible, monitors inventory, and monitors customer demand by using tools that have advanced technologies.

Although the initial investment for introducing advanced technologies is high and the payback is not so fast, it will present a significant decline for the company over time. Therefore, the cost-benefit feature must be considered in a DSC, given that all areas of the production chain can be reduced with the introduction of innovative technologies [2].

Also, DSC is characterized by transparency, which contributes to the coordination of all its links. Thus, this resource allows an orderly flow within the SC because, if there is not, the fluidity will be interrupted in some step of the chain, blocking other steps. With that, the DSC will be prepared to adjust to different situations.

Finally, DSC must be ecological, making it relevant for organizations to emphasize the importance of not only following environmental laws and making their workers aware, as well as promoting processes that are capable of innovating in the destination of components that are unusable during production [2].

5 Conclusion

It was identified that the introduction of new technologies is transforming the world, contributing to it being dynamic, interconnected, and competitive. For this reason, new terms for SC are emerging, such as DSC. From a structural point of view, it was analyzed that the cooperation between manufacturers, co-workers, and customers is extremely important to increase the transparency of all the phases that structure the SC. In addition, due to the introduction of automation and digitalization of processes and systems, traditional SC management is being changed. As a result, organizations should take a new look at digital technologies because new opportunities for innovation in SC arise through them. For DSC to be present, organizations must restructure their SC as a digital supply network that unites not only products, services, and flows but also people, data, and finance. As for resources, it was noted that DSC provides a large amount of data and interactions in real-time, in addition to providing collaboration and assistance in unlimited superior communication, which guarantees greater productivity, reliability, and speed in all stages of the system. Therefore, DSC aims to innovate business processes through technologies 4.0, and different frameworks in the literature can contribute to DSC's configuration and resources.

References

1. Nasiri M, Ukko J, Saunila M, Rantala T (2020) Managing the digital supply chain: the role of smart technologies. Technovation 96–97:102121
2. Büyüközkan G, Göçer F (2017) Digital supply chain: literature review and a proposed framework for future research. Comput Ind 97:157–177
3. Dalenogare LS, Benitez GB, Ayala NF, Frank AG (2018) The expected contribution of Industry 4.0 technologies for industrial performance. Int J Prod Econ 204:383–394
4. Wang L, Törngren M, Onori M (2015) Current status and advancement of cyberphysical systems in manufacturing. J Manuf Syst 37:517–527
5. Kagermann H (2015) Change through digitization—value creation in the age of industry 4.0. In: Albach H, Meffert H, Pinkwart A, Reichwald R (eds) Management of Permanent Change. Springer Gabler, Wiesbaden, pp 23–45
6. Frank AG, Dalenogare LS, Ayala NF (2019) Industry 4.0 technologies: implementation patterns in manufacturing companies. Int J Prod Econ 210:15–26
7. Thames L, Schaefer D (2016) Software-defined cloud manufacturing for industry 4.0. Proc CIRP 52:12–17
8. Okorie O, Salonitis K, Charnley F, Moreno M, Turner F, Tiwari A (2018) Digitisation and the circular economy: a review of current research and future trends. Energies 11:3009
9. Bienhaus F, Haddud A (2018) Procurement 4.0: factors influencing the digitisation of procurement and supply chains. Bus Process Manag J 24(4):965–984
10. Alicke K, Rachor J, Seyfert A (2017) Supply chain 4.0 – the next-generation digital supply chain. McKinsey. https://www.mckinsey.com/businessfunctions/operations/our-insights/supply-chain-40--the-next-generationdigital-supply-chain
11. Prasetyo YA, Arman AA (2017) Group management system design for supporting society 5.0 in smart society platform. In: International conference on information technology systems and innovation (ICITSI). IEEE, Bandung, pp 398–404
12. Queiroz MM, Pereira SCF, Telles R, Machado MC (2019) Industry 4.0 and digital supply chain capabilities: a framework for understanding digitalisation challenges and opportunities. Benchmark Int J
13. Prakash S, Singh R, Gunasekaran A, Ghadimi P (2018) Supply chain management, industry 4.0, and the circular economy. Resour Conserv Rcycl J
14. Pramanik HS, Kirtania M, Pani AK (2019) Essence of digital transformation—manifestations at large financial institutions from North America. Futur Gener Comput Syst 95:323–343

15. Najmi A, Shakouri HG, Nazari S (2016) An integrated supply chain: a large scale complementarity model for the biofuel markets. Biomass Bioenergy 86:88–104
16. Schmidt B, Wallenburg CM, Rutkowsky S, Einmahl S, Petersen I, Klötzke F (2015) Digital supply chains: increasingly critical for competitive edge. A.T. Kearney's European Excellence in Supply Chain Management. https://docplayer.net/3304945-Digital-supply-chains-increasingly-critical-for-competitive-edge.html
17. Schrauf S, Berttram P (2016) Industry 4.0: how digitization makes the supply chain more efficient, agile, and customer-focused. PWC Report. https://www.strategyand.pwc.com/gx/en/insights/2016/digitization-moreefficient.html

A Conceptual Design of the Competences Circle for the Project Manager 4.0

Isabela Coppi and Alessandra Cristina Santos Akkari(✉)

Mackenzie Presbyterian University, Campinas, Brazil
isabela.coppi@siemens.com, alessandra.akkari@mackenzie.br

Abstract. The advent of digital technologies has led to the emergence of a new industrial scenario, where machines and processes are connected and communicate through the exchange of information in real-time. Project management is inserted in this scenario of changes linked to the Fourth Industrial Revolution and is considered a strategic area in organizations, directly linked to the company's competitiveness. In this context, project management will be impacted by significant changes in the digital age, bringing reflections on traditional management and the project manager's role. This paper aimed to propose a conceptual design of the project manager's competencies in the digital era. To this end, an exploratory and descriptive study was developed, with a qualitative approach and a bibliographic strategy. The study pointed out that the project leader inserted in the context 4.0 will need to develop skills related to flexibility and adaptation to changes, knowledge linked to the application of technologies 4.0, and attitudes that promote problem-solving through teamwork and cooperation. This study's results can mainly benefit organizations and managers in order to direct the formation of new skills for professionals in the project area.

Keywords: Project manager · Project management · Industry 4.0

1 Introduction

The spread of digital technologies, driven by the Fourth Industrial Revolution, has impacted different fields of the economy as well as several departments of the company. In this perspective, project management, which is a strategic and transversal area in organizations, is becoming increasingly challenging due to the various elements that structure the new technologies of Industry 4.0. Therefore, the traditional view of project management requires discussions and reflections in order to adapt to the various variables inherent in the digital age.

According to Simion et al. [1], project management in the era of Industry 4.0, or project management 4.0, can be represented as a practical activity that encompasses the entire process of planning, organizing, coordinating, and controlling projects using essentially digital technological tools.

The restructuring of project management processes leads to the need to develop new competencies in the project manager. In the role of coordinating the entire progress of

Y. Iano et al. (Eds.): BTSym 2020, SIST 233, pp. 48–54, 2021.
https://doi.org/10.1007/978-3-030-75680-2_7

the project, the project manager must also adapt to this new management, especially in relation to information technologies. This improvement in their intellectual capacity is due to the fact that Industry 4.0 projects are more complex and autonomous due to the use of autonomous systems that make decisions for themselves, such as autonomous robots, which make them complex to manage [2].

This paper aimed to propose a conceptual design of the project manager's competencies in the digital era, contemplating the pillars of knowledge, skills, and attitudes. To this end, an exploratory and descriptive study was developed, with a qualitative approach and bibliographic strategy based on a three-phase methodology. This paper's main contribution is to direct the training and improvement of skills of project manager 4.0 to optimize their performance in the face of a new industry paradigm that will impact the project area.

2 Theoretical Background

A project is a temporary effort developed in order to produce a single product, service, or result [3]. The objective of a project is strategically aligned with the business, as its results can contribute to the growth and recognition of companies. The union of the projects' results with the business objectives allows the organization to be more competitive in its markets, considering that the project was concluded within the defined quality standards [4].

Thus, project management can be defined as a process of continuous control of the achievement of project objectives or its deliverables. Project management is the effective application of tools, techniques, knowledge, and skills in order to achieve the desired objective. The practice of management means that there is a targeting of the available resources in order to achieve the project objective according to time, quality, and cost. According to Watt [5], there are categories of knowledge in project management that can be called integration management, the scope, the schedule, costs, quality, resources, and risks. Also, the PMBoK [6] indicates ten areas of project management knowledge that, in addition to integration, scope, schedule, costs, quality, resources, and risks, also include the management of communications, acquisitions, and project stakeholders, these being the focus of this work.

Therefore, the project manager is the person responsible for applying knowledge and skills to achieve the project's objectives. Thus, it has some attributions such as the ability to identify project needs, the analysis to establish clear and attainable objectives, the management of stakeholders' expectations, and the balance between quality, scope, time, and cost [7]. So, the project manager is responsible for coordinating a team and capturing the best way for each resource to deliver results.

Viana [8] points out the desirable attributes for the project manager. Among the knowledge, there is a subdivision between organizational knowledge, referring to the knowledge of the organization's practices, human resources (HR) systems, administrative implications before decisions, and technical knowledge, referring to the domain of technical and specialization areas and research methods.

Regarding skills, there are those related to command, linked to leadership, ability to plan and control, ability to allocate resources, and decision-making ability, among other skills, such as teamwork, creativity, personal relationship, and ability to act with consistency and clarity.

Finally, attitudes are divided into positioning in relation to external and internal factors - such as interaction with staff, ambition of work, and discipline - and action strategies - such as the methodology for reading technical texts or attacking the problem according to the literature.

Also, according to Liikamaa [9], it is possible to characterize the project manager by means of four figures, highlighting his role as leader, manager, facilitator, and mentor, in addition to being a good communicator.

The project manager as a leader has the function of having the flexibility to respond to uncertain situations with the least possible stress. The manager's perspective can refer to being continuously monitoring project performance and intervening when necessary. The facilitator's role comprises the effective actions taking, in addition to making room for team decisions. Finally, concerning the role of mentor, the project manager must promote the absorption of knowledge and experiences by the entire project team and guide them, integrating with the attribution of a good communicator, in order to keep everyone updated with the information from the project [9].

Project Managers must continuously develop and improve skills due to the increasing demands on their work, mainly due to their communicative character.

In relation to soft skills, one of the skills most mentioned refers to the management of changes. Due to the high speed employed in the Industry 4.0 universe, where project stakeholders will have real-time information through tools that promote a high data flow, project managers must develop skills such as fast and effective decision making, maintaining the balance in the same way [9].

As for hard skills, the project manager must develop skills aimed at understanding physical cyber systems, an environment in which industry 4.0 develops primarily. In addition, knowledge related to Big Data and predictive algorithms can assist the manager in managing projects mainly in relation to achieving objectives, as this individual will be able to analyze data and future scenarios and thus be informed when making decisions at the present moment that may impact the future of the project [9].

3 Methodology

This research's methodological approach included three phases, encompassing observation, categorization, and association [10].

The observation phase referred to the study's exploratory stage, developing a literature review on topics related to the role of the project manager and its various functions in project management in the context of the Fourth Industrial Revolution. Information was collected from databases such as Scientific Electronic Library Online (SciELO), Elsevier Journals, Scopus, among others. Priority was given to articles published between the years 2015 to 2019. In this study, the words "project management", "project manager", "skills project manager", "industry 4.0", "industrie 4.0" and "project management 4.0" were used primarily. The use of such keywords aimed to capture as many studies as possible and, thus, collaborate so that this work included a complete study on the topic and synthesized all the necessary information. Criteria of inclusion or exclusion of articles and definition of the information to be extracted were considered as part of the literature review process.

Regarding to project management, the literature review focused on understanding the project definition and its respective importance, to, then, understand the project management process under a traditional view, as well as its life cycle, its interfaces, and restrictions. In addition, analyze the role of the Project Manager in the management process.

Regarding Industry 4.0, it was developed about its evolution in parallel to the Industrial Revolutions, as well as its main characteristics. Thus, the pillars, attributes, and technologies involved in this scenario were identified and characterized.

In the categorization phase, based on the in-depth review developed in the observation phase, the interface between project management and the Fourth Industrial Revolution was mapped to then proceed to the stage of elaboration of relationships between concepts. Based on the pillars of knowledge, skills, and attitudes, and based on the observation phase results, it was possible to relate and categorize the project manager's primary skills in the scenario of Industry 4.0 according to each of the pillars.

Finally, in the association phase, based on the categorization carried out in the previous stage, project manager 4.0's main competencies were proposed, covering this professional's main knowledge, skills, and attitudes.

4 Results and Discussion

In the context of the transformation of project management concerning Industry 4.0, different skills are demanded from the project manager, focused not only on traditional management but also on the flexibility 4.0 industry and its aspects.

The competencies' design pointed out, in total, 7 knowledge, 5 skills, and 5 attitudes related to project manager 4.0. The main competencies are related to the affinity with information technologies, robotics, and cloud architectures. Regarding skills, flexibility, adaptation to changes, and decision making stand out. Finally, attitudes are based on leadership, cooperation, and problem-solving.

The transformation of the industry to the environment 4.0 requires that professionals who interact in this environment be able to be more efficient, more flexible, agile, and, therefore, more competitive since people and machines will be connected [11]. For the project manager, this situation is no different. In addition to its competencies applied to traditional project management, the conversion to the dynamic of Industry 4.0 improves its profile of knowledge, skills, and attitudes.

Initially, in relation to the set of knowledge, it was identified the necessity of project manager to be informed about cloud architectures, given that the vast majority of the information of the project will be available on cloud platforms. In this way, a certain affinity with information technologies is created and it becomes possible to apply expertise and use viable technologies [12].

Combined with knowledge of cloud architectures and affinity with information technologies, it is essential that the project manager also knows about data protection and security relating to cybersecurity technology. The manager may be able to understand how data is protected and formulate protection mechanisms when necessary.

The data, stored and digitally protected, can be subject to analysis in order to direct its processing. The project manager must be able to extract useful values from this essentially voluminous and varied data set (in reference to Big Data), thereby processing the data that will actually add value to the project management processes and minimizing wastes of time and resources.

Consequently, the industry's new vision will provide to the employees a transformation in work processes and business models that will involve new technologies, making processes more interconnected and complex. This change will not only affect production but also the way of working in all dimensions [12]. Therefore, the project manager will need to develop a vision of process optimization, focusing mainly on adapting the existing project management processes in this new work dynamic. To meet future demand, processes must be connected and more agile, eliminating gaps in existing processes.

Even in the face of all technological change, people management is still part of the project manager's knowledge profile. Changing the work dynamics and the flexibility provided by the Industry 4.0 requires that the manager correctly direct its human resources in order to distribute the necessary skills to the entire team, making people absorb all knowledge and be able to perform diverse tasks. The work profile will undergo changes, so that the project manager in his/her role as leader must ensure that the team is qualified to manage the projects' dynamic.

Finally, the project manager must obtain knowledge related to robotics and artificial intelligence (AI), as the projects in the digital age will interact with mechanisms that involve AI, such as robots and autonomous vehicles, in addition to virtual simulation software [13].

Regarding the abilities of Project Manager 4.0, decision-making is an important factor in the management process, as the decision is made from a global view of all processes, which are composed of data that will be processed quickly and efficiently. In view of this situation, the manager must analyze all the available information and make the appropriate decisions in a timely manner and with greater speed.

Due to the massive production of project data linked to Industry 4.0, the manager must have the ability to adapt to changes. Projects will be more dynamic and flexible on their own, which requires the project leader to keep up with changes without significant losses in the process and, mainly, in the project results. The flexibility ability is directly related to adaptation to changes because, through a flexible posture, it is possible to adapt. Also, knowledge of the technologies involved helps interpret data and directs possible action plans for the project.

When coordinating the project team, the manager must establish communication channels so that the information flows easily and that all members of the team have access to the data they need. In the context of Industry 4.0, communication channels may be more diverse, for example, via chats, audio, and video connections in different media and locations. Communication management will be improved from emails to real-time communications, which can be analyzed simultaneously to inform project decisions.

Kiesel and Wolpers [14] cite a case in which a machine operator can comment directly on the process that was planned for him, and this information can be sent directly to the project manager, who has the power to replan his tasks, in this way reducing wastes of time.

Critical thinking helps the project manager to assess what are the impacts generated by the different situations of a project in the context of Industry 4.0. For example, after analyzing the data available at a given point in the project, the manager must think critically in order to diagnose any flaws that may be impacting, that is, critical thinking promotes reflection on the direction of the project and its consequences.

As in traditional project management, having a leader's attitude guarantees the project manager a position that allows the team to work according to a direction and a greater focus on the project's objectives. Especially in projects involving Industry 4.0, due to its dynamic nature, the Project Manager must ensure that the team is aligned with the moment's purposes, even if they are subject to change. In this way, once the leadership practices are applied, the team will be result-oriented, thus focusing on their tasks and project objectives.

Teamwork in environments where Industry 4.0 projects will be involved is essential for the progress of the project. Due to flexibility and constant changes, the team must work as a team, helping each other when necessary, which also relates to cooperation. In this situation, the Project Manager must act as a leader and promote a harmonious environment that is susceptible to team relations.

Thereby, the project manager 4.0 must improve its circle of competencies in relation to traditional project management. The dynamism of the projects demands that managers have the capacity to be flexible to changes, as well as understand how data is generated and analyzed in the project. Having an affinity with information technologies, project manager 4.0 is able to communicate better with the team and have data on his project in real-time, thus improving the controls used.

Despite the view that the Industry 4.0 scenario is highly automated, the project manager must still have support from his team in interactions with new technologies. Thus, communication is still considered a crucial skill in this scenario.

Essentially, the profile of Project Manager 4.0 is formed by a combination of technical skills, focused on information technologies (such as data processing), and behavioral skills, involving the leadership of the project team in a volatile environment. Indeed, Akkari et al. [15] affirm the importance of improving behavioral skills in the face of Industry 4.0, which will demand the development of even more accurate social and emotional skills, such as communication, motivation, and collaboration.

5 Final Considerations

The present work's objective can be considered as achieved, since it presented a proposal for the competence circle of the project manager 4.0, which indicates which skills, abilities, and attitudes should be improved or developed in the project leaders.

Among the competencies proposed, information technology knowledge stands out, such as cloud architectures, robotics, AI, and data security, due to the highly virtuous projects. Regarding attitudes, in addition to those that were already part of traditional project management, emphasis is placed on problem solving and cooperation, since the project team must be able to solve problems quickly, in addition to members cooperating with each other. The skills of adapting to changes and flexibility consolidate the main objective of project management 4.0, developing the ability to react to changes quickly and without losses.

References

1. Simion C, Popa S, Albu C (2018) Project management 4.0 – project management in the digital era. In: Popa I, Dobrin C, Ciocoiu CN (eds) 12th International management conference. Editura ASE, Bucharest, pp 93–100
2. Win T, Moon Kham NS (2018) Transformation of project management in industry 4.0. In: Popa I, Dobrin C, Ciocoiu CN (eds) 12th International management conference. Editura ASE, Bucharest, pp 37–44
3. Project Management Instituite. https://brasil.pmi.org/brazil/AboutUS/WhatIsProjectManagement.aspx
4. Dandaro F, Tonani FR, De Carvalho DO (2016) Gestão de projetos como estratégia organizacional. Rev Eletrôn eF@tec 6(1):15
5. Watt A (2014) Project management, 2nd edn. BCcampus, British Columbia
6. Project Management Instituite (2017) A guide to the project management body of knowledge (PMBOK Guide). 6th edn. Project Management Institute, Newton Square
7. Do Valle A, Cierco A, Soares C, Finocchio Jr. J (2010) Fundamentos do gerenciamento de projetos. Editora FGV, Rio de Janeiro
8. Viana C (2012) A gestão de projetos no âmbito da administração pública federal: uma visão estratégica. In: V congresso de gestão pública. CONSAD, Brasília, pp 2–21
9. Liikamaa K (2015) Developing a project manager's competencies: a collective view of the most important competencies. Proc Manuf 3:681–687
10. Prasetyo Y, Arman A (2017) Group management system design for supporting society 5.0 in smart society platform. In: 2017 International conference on information technology systems and innovation (ICITSI). IEEE, Bandung, pp 398–404
11. Cerezo-Narvaez A, Otero-Mateo M, Pastor-Fernandez A (2017) Development of professional competences for industry 4.0 project management. In: Bousonville T, Melo T, Rezg N, Vernadat F (eds) 7th International conference on industrial engineering and systems management. IESM, Saarbrucken, pp 487–492
12. Prifti L, Knigge M, Kienegger H, Krcmar H (2017) A competency model for "industrie 4.0" employees. In: Leimeister JM, Brenner W (eds) Proceedings der 13. Internationalen Tagung Wirtschaftsinformatik. WI, St. Gallen, pp 46–60
13. Radziwon A, Bilberg A, Bogers M, Madsen E (2014) The smart factory: exploring adaptive and flexible manufacturing solutions. Proc Eng 69:1184–1190
14. Kiesel M., Wolpers M (2015) Educational challenges for employees in project-based Industry 4.0 scenarios. In: Proceedings of the 15th international conference on knowledge technologies and data-driven business. ACM, New York, p 41
15. Akkari A, da Rocha M, De Farias Novaes R (2019) Cognitive ergonomics and the industry 4.0. In: Iano Y, Arthur R, Saotome O, Estrela VV, Loschi H (eds) Brazilian Technology Symposium. Springer, Cham, pp 275–280

Evaluating of Incapacitation of the Civil Aviation Flight Crew

Nikolai I. Plotnikov(✉)

Scientific Research Project Civil Aviation Institute «AviaManager», 630078 Novosibirsk, Russian Federation
am@aviam.org

Abstract. In this paper, the analysis of research on the subject of loss of capacitation of the flight crew of civil aviation (CA) is performed. There is a complete loss of capacitation (Incapacitation) and a temporary or partial loss of capacitation (Impairment). The loss of performance of the pilot (Pilot Incapacitation) (PI) and especially all the crew members (All Pilot Incapacitation) (API) is a critical, extremely dangerous event. The development of the API subject method in flight can be carried out in technical and organizational directions. Technical measures to prevent aviation accidents (AA) from dangerous API events may include: a) further saturation of the automation of modern aircraft and their ability to complete the flight automatically without human intervention; b) development of technical means of controlling the aircraft from the ground and remote flight control. Organizational measures can be aimed at reserving piloting functions: training in automated approach and landing of cabin crew; providing preferential or free flights to persons of flight specialty.

Keywords: Incapacitation · Impairment · Civil aviation · Flight crew · Simulator experiment · Emergency delegation

1 Introduction

In this paper, the analysis of research on the subject of loss of capacitation of the flight crew of civil aviation (CA) is performed. A method of emergency delegation (ED) in case of loss of flight crew capacitation has been developed. The purpose of the study is to assess the risks of successful completion of the flight when delegating control of the aircraft to a third party in the event of loss of crew capacitation. An event simulation was performed in training experiments. As a result of the experiment, conclusions were made about the possibility of safely completing the flight with the support of an experienced pilot from the ground to piloting the aircraft by a third-party person: a passenger or flight attendant. The concept of ED is introduced. The aim of the experiment is to inform air transport consumers about the reliability of automation of modern airliners, to reduce the aerophobia of flying and potential passengers. The development of the method requires careful structuring of the subject. A brief comment concerning ED in the theory of delegation is made. The content of the problem and the most plausible and possible procedures are compiled.

Y. Iano et al. (Eds.): BTSym 2020, SIST 233, pp. 55–65, 2021.
https://doi.org/10.1007/978-3-030-75680-2_8

2 Overview of the Subject

There is a total loss of performance (Incapacitation) and temporary or partial loss of performance (Impairment). The loss of performance of the pilot (Pilot Incapacitation) (PI) and especially all the crew members (All Pilot Incapacitation) (API) is a critical, extremely dangerous event. In case of a partial loss, the pilot can perform some functions, such as reading the checklist, maintaining communication. In case of complete loss of capacitation, the individual is not able to perform the duties. The lack of adequate response in standard communications with air traffic control (ATC) services and operating procedures of crew members is a sign of loss of capacitation, for example, Boeing-737 Helios Airways on August 14, 2005.

If the pilot and crew lose their ability to work, the imperative of survival sets the task of ED of control of the aircraft. The ED cases and forced control of the aircraft by a third party are rare but real events. Such events served as material for the military pilot the Canadian writer Arthur Hailey, who put on a TV show in 1956, later reworked into the story "Flight into danger", which describes how both pilots lost consciousness due to poisoning by onboard food in flight. One of the passengers, the main character of the story, John Spencer, a retired military pilot who flew many years ago, is forced to take control of the plane. He landed the plane with support from the ground at Vancouver airport. The book is reissued [1].

The first plane crash, the causes of which were tried to establish as a result of PI in flight, occurred in 1911. Later, the cause of the accident was attributed to pilot error [2]. For the period 1993–1998, 50 events were recorded: 11 partial and 39 full PI, which is the frequency of 0.013 and 0.045 per 100 thousand flight hours, respectively. Two events ended not fatal aviation accidents (AA), which makes the probability $p < 0.001$. The frequency of PI in the age of pilots 45–59 years is more than twice as high as in the age group 25–44 years [3].

The largest share of common causes of PI is a loss of consciousness and heart attacks. Full PI occurs in cases of hypoxia, cardiovascular disease, loss of consciousness, diabetes, reactions to medications, and injuries. Partial PI occurs in respiratory and gastrointestinal diseases, food poisoning, acute fatigue, kidney attacks, and vision deficiencies. It is significant that there is a larger number of complete PI, although logically, it should be the opposite [3]. This can be explained by a) the statistical sample of this analysis may have been insufficient; b) full PI is a more serious event and is documented more strictly; c) partial PI events can be hidden by pilots due to the threat of decommissioning [4].

The loss of performance of one of the crew members is an emergency event with a frequency of approximately one case per 2 million flight hours. Combined analyses of full and partial PI indicate frequencies of 0.019–0.059 per 100,000 flight hours [5, 6]. For military aviation, this figure is significantly higher. The frequency of AA due to PI in [7] is determined by the value of 0.14 in 20 cases out of 146; in [8], it is determined by the value of 0.28 in 20 cases out of 59. The differences in values can be explained by the fact that studies were for military (possibly fighter) aviation, and the reasons for PI were indicated more often. In the study [9], 1300 simulator experiments were performed with various input of PI cases in flight. The critical level of risk for flight safety was set in 15 out of 500 exercises, of which eight cases (1.6%) were recorded as AA. In [10] and

[11], a series of simulator experiments with full and partial PI in flight was performed. It was found that the average detection time of the pilot's PI in the crew is 1.5 min. The fourth part (25%) of training sessions with the introduction of PI ended in AA. Surveys of 5,000 respondents revealed confessions about 2,000 PI incidents in flight in which a safety threat was observed in four percent (4%) of incidents.

In the course of professional training and training of professional pilots on simulators, exercises are practiced to simulate the event of loss of capacitation of one of the crew members. The conclusions in [12] state that AAs due to PI are less than one percent. Determining the causes of AA by PI is a difficult task since these facts are hidden in a chain of other causes. The method for determining PI can be formed from an aviation medical research, simulator experiments, survey data, and the study of data on the end of the flight career of pilots. As a rule, data analysis does not reveal the direct facts of the causes of AA and attribute them to PI in flight. When establishing PI, the terms "possible", "probable", and "determined" are used.

3 All Pilot Incapacitation Emergency Delegation Experiment

API ED was initiated by TV "Galileo" and fulfilled in Aeroflot airline in 2012 [https://yandex.ru/video/preview/text/galileo the flight attendant controls the plane]. The experiment contains the following scenario. In the flight of a modern multi-seat commercial aircraft, both pilots of the two-member crew suddenly lose their ability to work. The plane is attended by cabin crew members who do not have flight training - flight attendants. Flight attendants discover and establish the failure of the cockpit crew. Direct appeal "are there any airline employees among you?", "does anyone know how to fly a plane?" it turns out that there are people of the flight profession among the passengers. If available, they are invited to the cockpit to act as an emergency understudy for the crew. In the absence, this role is performed by flight attendants. After communicating with the ground and informing about an emergency, the flight is controlled by an instructor pilot, who informs and gives instructions about piloting actions. As a result, the plane with an inoperable crew under the guidance of a flight attendant or passenger will land from the ground.

There are three options for flight control in the cockpit. Piloting: a) flight attendant, b) passenger - a fan of virtual computer aerobatics; c) passenger - a military pilot. For the greatest reality (surprise, emergency), there were no preliminary agreement or setup conversations with the participants of the experiment for the role of piloting the aircraft. The instructor sets the conditions of the visual flight rules (VFR), and the complexity of the situation is different for each of the participants. The instructor is separated by a screen from the actions of the pilot in the cockpit and could only see the position of the aircraft on display like an air traffic controller. The main result of the experiment is proof of a plausible and possible successful completion of API ED, Table 1.

The pilot-instructor chose different tactics of communication, behavior, and flight conditions for piloting in three flight variants. For the flight attendant, the time was unlimited, and communication was built on every connection to complete mutual understanding through many clarifications and repeated questions. The instructor chose the language, using as many special terms as possible. He started the aircraft along the route, allowing to find the controls and, acting on them, manage to configure the aircraft in

Table 1. Simulator experiment for API ED

Piloting	Task	Conditions	Time, minutes	Outcome
Stewardess	Horizontal flight, automated approach, landing	Simple, VFR	~25	Safe landing, without any consequences
Passenger with an assistant	Visual manual approach and landing	No communication, autopilot doesn't work, strong crosswind	~7	Safe landing, without any consequences
Passenger military pilot with an assistant	Visual manual landing	No communication, the autopilot, flight and navigation instruments do not work, and the maximum permissible crosswind	~5	Safe landing, side roll-out, possible damage to the landing gear

terms of altitude, speed, approach trajectory, and corresponding mechanization, landing gear release, reverse activation, and automatic braking. For fans of virtual computer piloting, a time-compressed scenario of the urgency of the event was chosen based on the assumption that they have skills in piloting simulators. An even more stressful scenario was chosen for the military pilot. The calculation was that he is an active fighter pilot with the actual skills and reaction of piloting modern supersonic aircraft.

The validity of an API event can be much more complex. The direct appeal to passengers and informing them about the situation of the crew's inability to work is suitable for simulating reality and filming the simulator experiment. But such treatment is doubtful in reality due to the unpredictable behavior of passengers. Rational construction of the experiment would require establishing the same conditions and the same sequential complication, entering failures for all participants. Perhaps the differences in flight scenarios, the time limit for a military pilot affected the differences in the results of flight outcomes for each event. Landing with the maximum allowable crosswind are not frequent events, and damage from chassis damage (possibly tires) may be calculated risks. The tactics chosen by the instructor concentrated the experiment in time and complexity.

4 Calculation of PI and API Events vs AA Probabilities

The calculation of AA probabilities requires stable statistical frequencies of events, which are considered insufficient and unrepresentative in research on the subject of PI and API. In these circumstances, probability measures and likelihood measures are used to describe the subject area under study. In relation to probabilistic measures, these measures are less clear. To calculate PI events, it is used interval probability estimates, probability estimates, and plausible reasoning.

Due to the vagueness of identification of partial and full PI events, their plausible joint frequency was established in the confidence intervals of 0.013–0.059 for civil aviation and 0.14–0.28 for military aviation per 100,000 flight hours. In the global AA statistics, it is used the measure of risk per million flights. For developed countries, it is equal to about one accident per 1.5–2 million flights, or the AA probability value of about $P = 10^{-6}$ for all the combined causes of AA, attributed to the person, machine, and environment. To take the average flight duration of four hours for the CA and one hour for the military, these frequencies are approximately the same and equal to the average value of about three percent (3%) of all causes of AA with a probability less than $P = 10^{-7}$. The calculation of the probabilities of AA values for PI and API based on probability measures and plausible reasoning is presented in Table 2.

Table 2. Calculation of probabilities of AA values for PI and API

Loss of capacitation	Pilot PI	Crew API
Impairment	$P = 10^{-5}$	$P = 10^{-6}$
Incapacitation	$P = 10^{-7}$	$P = 10^{-8}$

It is assumed that impairment has frequencies two orders of magnitude higher than incapacitation. The basis for this statement is the data from surveys and questionnaires, in which the facts of incomplete performance according to the confessions of pilots are indicated in two out of five cases. Then the probability of impairment is about $P = 10^{-5}$. The API events are unique, and there are no statistical studies available. We assume that the simultaneous loss of capacitation of all (two) crew members is an event much rarer than the PI of a single crew member. The basis for this reasoning is the fact that the main causes of PI are medical factors: heart attack or stroke in two pilots at once is an order of magnitude less likely. In the case of fatigue, depressurization, and hypoxia, of course, the PI and API events are close or identical in frequency.

5 Developing an Emergency Delegation Method

5.1 The Main ED Method Concept

The concept of ED is introduced for critical situations to make decisions to change the delegation strategy completely. ED is identical to a dynamic reservation of technical objects. This problem makes it necessary to develop a management method and standard actions for such events. The problem is that the subject of the malfunction is fuzzy. Partial and complete failure is not easy to distinguish. The measure of the threshold or level of partial inactivity of a crew member when there is a threat to flight safety is also difficult to establish. If an event is associated with a cabin depressurization, it can be classified as extremely rare and extremely critical. But if partial inactivity is due to fatigue, then this is the usual state of the pilot. Non-trivial is the method of establishing inoperable inside the flight crew or cabin crew members, flight attendants.

An API event is extremely complex and depends on many unequal conditions, circumstances, and factors. It is not possible to check all the components of the API event, scenarios for how the event will flow, and flight outcome options, even in a series of simulator experiments, due to an infinitely large combination of factors. Therefore, it is necessary to prepare a preliminary theoretical content of the API, in which the most plausible, possible, and probable conditions and factors are structured.

To make up the content of the research method, it is necessary to make various selections of plausible events. To do this, it is guided by an analysis of the known facts of the events of the loss of crew performance, common sense, and plausible reasoning. The result of this work should be the preparation of the most possible scenario of the event and the calculation of the probability of variants of the flight outcome. Thus, it is necessary to move from the initial fuzzy content of empirical data: from likelihood measures to probability measures.

The development of the method can be carried out in technical and organizational directions. Technical measures to prevent AA from dangerous API events may include: a) the creation of "smart" modern aircraft, the development of automation, their ability to complete the flight automatically without human participation; b) similar development of technical means of controlling the aircraft from the ground and remote flight control. Organizational measures can be aimed at reserving piloting functions: training in automated approach and landing of cabin crew; providing preferential or free flights to persons with a flight specialty.

The key competencies of API event participants for a successful flight outcome are a) the presence or absence (hazardous) of a person with flight training among passengers and flight attendants onboard the aircraft. The key factor is the presence or absence (hazardous) of an instructor pilot at the ATC points with actual professional experience in this aircraft to assist the pilot.

The effectiveness of communication between the ground and the flight depends on how successfully the pilot-instructor can transfer the necessary competencies to the person who is behind the control of the aircraft. A rationally constructed communication procedure can have the following structure.

A. TRAINING: (1) informing the pilot-instructor of the piloting about the location of instruments and controls, (2) receiving messages by the piloting and confirming the understanding, (3) establishing the correct understanding by the pilot-instructor;
B. INDICATION: (4) transmission of the instruction by the pilot-instructor about the action, (5) action of the piloting, confirmation of the action and about the changes of flight parameters, (6) execution control by the pilot-instructor.

In communication procedures, messages are repeated until complete understanding is reached.

5.2 Composition of the Method Content and Method Development

The complexity of an API event requires a heuristic approach to creating the most truthful scenarios. The components of a complex API event are installed heuristically, based on experience, and expert-based on knowledge. The composition content can be designed for scenarios based on the greatest likelihood of occurrence of the elements of the event and that they lead to a favorable outcome; the scenario of probable events of this type require the highest data completeness, suggesting relative independence and equivalence of the component of the complex event.

For example, whether there is an experienced pilot-instructor with current experience of this type of aircraft and active knowledge of the cabin and equipment. If not, how long does it take to get it to the ATC communication console? A small part of the most intensive and important components can be taken into account in an event without taking into account the rest. If any objects are not taken into account, they can be added later when detected. Each object is structured into components that are reviewed and commented on. A detailed assessment is made, and a plausible component is named, which may be part of the most likely event. After careful preparation of plausible content, one can begin to describe the most likely ED event. The details of the method were previously drafted in the author's work [13].

5.3 Interpretation of the Most Likely ED Event

The method of plausible reasoning allows creating the following possible PI scenario of the most likely ED event. On a commercial multi-seat passenger plane, in a horizontal flight 40–60 min after takeoff, both pilots lose their capacitation due to food poisoning. Flight attendants discover the fact for five minutes in the course cockpit crew service. As a result of the discussion of the emergency by the cabin crew, the senior flight attendant establishes a connection with the ATC. After 10 min from the beginning of detection of API, the organization of emergency management from the ground begins. According to the recommendations of the earth, the flight attendant, introducing as the captain of the aircraft on the internal radio, briefly informs passengers about the flight and invites his colleagues, active or retired pilots, to the cockpit for friendly communication. A retired pilot with an in-flight break work for three years is discovered among the passengers. He notifies the flight attendants of his desire to enter the flight crew cabin. He is taken to the cabin and initiated into the emergency situation. The pilot does not have experience flying this type of aircraft. After 30 min from the beginning of API detection, the pilot controls the aircraft by commands from the ground of an instructor pilot with relevant experience for this aircraft. The pilot-instructor takes tactics: training, indicating, controlling. After one hour, the flight ends well with a rough landing at the reserve airfield, having received damage, without breakdowns. Passengers, crew members entirely healthy.

An event can contain only three key and critical components: (1) a modern aircraft with high automation; (2) a flight attendant; and (3) an instructor pilot with up-to-date experience in direct ATC communication. The probability of the presence and composition of these components is the highest. Since the simulator experiment proved, the possibility and likelihood of successful completion of the flight, standards, and regulations for emergency actions can be formed, Table 3.

Table 3. Structure and content of the ED method

Objects	Components, options	Content comments	Probable event
Aircraft	Multiseat passenger aircraft of commercial civil aviation (CA) Aircraft of general civil aviation (GA)	CA. A high level of automation decisively and multiply increases the chances of a successful flight outcome. The largest volumes of world air transportation. The risk is the highest in terms of damage GA. The largest number of aircraft and flight frequencies. The risk in terms of damage is less. Aircraft automation is significantly lower than on a commercial aircraft. This minimizes the chances of a safe control of a crew backup from the ground	Commercial passenger aircraft
Flight cruise phase	Takeoff, climb, horizontal flight, descent, approach, landing	The establishment of the fact of the flight crew API by the cabin crew or the ATC service at all stages, except for the stage of cruising horizontal flight, significantly reduces the chances of a successful flight outcome	Flight cruise phase
Flight rules	VFR IFR	IFRs reduce the chances of a successful flight	VFR
Meteo conditions	Simple, complex	Complex weather conditions reduce the chances of a successful flight	Simple
Cockpit crew	The number of crew members	Two, three; reserve crew for trans-continental flights; multi-member crews in charter, air cargo transportation	Two-member crew
	The composition of the crew	Pilots, specialists of other flying specialties: navigator, flight engineer, radio operator, flight operator	Pilots
Extraordinary stunt double	Flight attendant	One or two as piloting. Possibly with elementary flight training: glider, hang glider, GA aircraft	Two flight attendants
	Passenger	CA pilot with experience in this type of aircraft CA pilot with no experience in flying this type of aircraft CA pilot with a break in flight operations for 3 months, a year, three years for 10 years Test pilot of civil aircraft The military pilot of transport aviation Military fighter pilot Military test pilot	CA pilot with a break in-flight work
Incapacitation	Depressurization, acute illness, fatigue, food poisoning, hijacking	Disability due to pilot disease has the highest frequency Food poisoning is in second place Disability due to illness of two pilots is less likely In case of illegal seizure, it is assumed first to eliminate terrorists from the control of the Armed Forces. There were such events	Food poisoning

(continued)

Table 3. (*continued*)

Objects	Components, options	Content comments	Probable event
	Establishment of the fact of inoperability Identification of Incapacitation	Inside the flight crew: if the pilot does not respond to the flight crew's requests more than twice Regulations, standards of communication frequencies for flight and cabin crew How do flight attendants establish the fact of inoperability, how long it will take. According to the results [9, 10], the average time of a PI event is set at 1.5 min. The identification time of the API event is not known; it must be determined experimentally	Five minutes
Message	Establishing a connection with the ground	Depends on how familiar the flight attendants are with communications, or how quickly they can figure out communications. Time	Five minutes
ATC reaction	Adoption of an emergency broadcast Emergency response	Standards and regulations for action Search for experts of emergency management and notification of emergency response services Instructions to the flight attendants of the aircraft	Ten minutes
Pilot instructor	The presence of pilot on ATC tower. Actual experience. Standards of actions	Availability: standards for the presence of workover ATC; a pilot with up-to-date experience in flying an aircraft type corresponding to an emergency; action standards for events	Pilot instructor with up-to-date experience, 10 min
Cabin crew actions	Decision-making of flight attendants	How flight attendants need to identify the flight specialty among passengers. Direct contact or covert polling tactics to prevent passengers from panicking. Time	Covert tactics, 10 min
The outcome of the flight	Options	(1) happy, standard; (2) deviations from the standard approach and landing: rough touchdown, lateral, longitudinal roll-out, without damage or breakdown of the aircraft; (3) the same as in (2) with damage and breakage; (4) an accident with the destruction of the aircraft, minor injuries without human casualties; (5) disaster, fatal	Option 2. 20 min
		Total	60 min

6 Conclusion

The existing description of the PI subject is blurry and needs to be more clearly defined. The basis for the determination can be the outcome of the flight: if the flight ended safely, then the pilot and crew were able to work properly. PI studies confirm the "U-profile" of the pilot's age-related dependability. In older age groups of pilots, the probability of PI is more than twice as high as the probability of PI at a young age. According to the conclusions of [3], statistical samples of most PI studies cannot be considered representative and sufficient. It is argued that the risks of AA due to PI may be lower due to errors in retrospective assessments and analyses, erroneous perceptions, and judgments of pilots about the facts of their actual capacitation [14].

However, there are grounds for claiming that incomplete (partial, temporary) performance is a much more frequent state of the professional activity of the pilot than it is officially recorded and documented. Serious PI facts in flight and out of flight can be hidden by an individual and are multiple times higher than statistical studies and analyses.

Partial PI may be more dangerous for a favorable end to the flight. Heart attacks are life-threatening, but they may still be partially functional. In the study [3], two AA occurred in cases of partial rather than complete PI: (1) due to acute fatigue, and (2) during a rough landing by a pilot who was wearing mono visual contact lenses.

The current operational workload and flight fatigue play a crucial role in the probability of PI and API. PI should be considered in the context of crew capacitation. If the PI of a crew member leads to an increase in the unacceptable workload on the piloting pilot, the crew dependability resources are discriminated against. The PI event goes to the API event.

The results of the performed API simulator experiment in flight and numerous experiments of training sessions show the likelihood of possible successful completion of ED situation in case of loss of crew capacitation. The cabin depressurization has a decisive impact on the appearance of PI and API. The development of the method can be carried out in solving the problems of automated event detection and aircraft control from the ground.

This paper describes the heuristic content of an extremely complex event, tasks, and method of API ED and forced control of the aircraft by a third party. The heuristic approach is the only possible way to start forming this subject of research and development. To promote the method, an extended model of experimental simulations is needed. The result of such research is an analysis and method for developing standards and guidelines for joint actions of people on board in a state of extreme danger and ground-based flight operations and emergency services.

References

1. Hailey A, Castle D (2010) Na grani katastrofy [On the verge of disaster]. «ACT, ACT Moscow» edition. 224 p. (in Russian)
2. Parmet A, Underwood-Ground K (1987) Reported in-flight incapacitation: the early birds of 1911. Aviat Space Environ Med 58:276–278
3. DeJohn CA, Wolbrink AM, Larcher JG (2004) In-Flight Medical Incapacitation and Impairment of US Airline Pilots: 1993 to 1998. DOT/FAA/AM-04/16 Final Report, 25 p
4. Lane JC (1971) Risk of in-flight incapacitation of airline pilots. Aero Med 42(12):1319–1321
5. Kleeman J (1986) ATC training: realism and training effectiveness. Controller 24(30):13–18
6. McCormick TJ, Lyons TJ (1991) Medical causes of in-flight incapacitation: USAF experience 1978–1987. Aviat Space Environ Med 62:884–887
7. Rayman R, McNaughton GB (1983) Sudden incapacitation: USAF experience, 1970–80. Aviat Space Environ Med 54(2):161–164
8. Rayman R (1973) Sudden incapacitation in flight: 1 Jan 1966–30 Nov 1971. Aero Med 44(8):953–955

9. Chapman P (1984) The consequences of in-flight incapacitation in civil aviation. Aviat Space Environ Med 55:497–500
10. Harper C, Kidera G, Cullen J (1970) Study of simulated airline pilot incapacitation; Phase 1 – obvious and maximal loss of function. Aero Med 41(10):1139–1142
11. Harper C, Kidera G, Cullen J (1971) Study of simulated airline pilot incapacitation: phase II. Subtle or partial loss of function. Aero Med 42(9):946–948
12. Froom P, Benbassat J, Gross M et al (1988) Air accident, pilot experience, and disease-related in-flight sudden incapacitation. Aviat Space Environ Med 59:278–281
13. Plotnikov NI (2013) Resurcy pilota. Nadezhnost. Monographia. [Pilot resources. Dependability. Monograph], Novosibirsk, Russia, AviaManager Publ. 264 p. (in Russian)
14. Stone A, Shiffman S (2002) Capturing momentary, selfreport data: a proposal for reporting guidelines. Ann Behav Med 24(3):236–243

Identification of Bacteria in Hospital Environments by Fluorescence Spectroscopy

Henri Alves de Godoy[1(✉)], Rodrigo Bueno de Oliveira[2], Rafael Yuri Sano[2], Talita Mazon[3], Aline Macedo Faria[3], Adriane Elisabete Costa Antunes[4], Fernando Moreira Simabuco[4], and Rangel Arthur[1]

[1] School of Technology, University of Campinas (UNICAMP), Limeira, SP, Brazil
henri@unicamp.br, rangel@ft.unicamp.br

[2] School of Medical Sciences, University of Campinas (UNICAMP), Campinas, SP, Brazil
rbo@unicamp.br, r228298@dac.unicamp.br

[3] Centro de Tecnologia da Informação Renato Archer (CTI), Campinas, SP, Brazil
{talita.mazon,aline.faria}@cti.gov.br

[4] School of Applied Sciences, University of Campinas (UNICAMP), Limeira, SP, Brazil
{adriane,ferms}@unicamp.br

Abstract. Hospital-Acquired Infection (HAI) is a public health issue that is well known and discussed by health and scientific communities. Hand washing by healthcare workers following guidelines provided by the World Health Organization (WHO) is an effective practice in combating nosocomial infections and, consequently, helps prevent cross-contamination of patients in a hospital environment. This article discusses the use of fluorescence spectroscopy as a non-invasive technique that can be used to find bacteria in samples collected when exposed to ultraviolet light. When analyzed using computer programs, results of fluorescence present to healthcare workers a rapid diagnosis in case of nosocomial infections. As a result of the research, we establish a need for enhancement the bacteria detection by spectroscopy using biomarkers with the help of nanotechnology. It can transform smartphones into portable spectrometers to reduce operating costs and bring the technique closer to the patient in a hospital environment.

Keywords: Bacteria · Nosocomial infection · Spectroscopy · Biosensor

1 Introduction

The act of hand hygiene during care for the patient is considered the best way to avoid cross-contamination, which is the transfer of microorganisms between patients and health workers [1]. There is evidence that health workers' education and training in correct and systematic hand hygiene are effective. However, the monitoring and retraining plan often becomes flawed, promoting variable results depending on professionals' adherence [2–4].

The need for an analysis of the pathogen identification processes such as fungi, bacteria, and viruses in an automated way and with rapid diagnosis has led this research area to grow over the last few years.

Y. Iano et al. (Eds.): BTSym 2020, SIST 233, pp. 66–71, 2021.
https://doi.org/10.1007/978-3-030-75680-2_9

Several studies of fluorescence spectroscopy show the viability for diagnosing the presence of bacteria, fungi, and viruses. Research data [5] indicates that this technique is a non-invasive tool where bacteria, such as *Escherichia coli*, *Enterococcus faecalis*, and *Staphylococcus aureus*, could be quickly identified with a sensitivity of more than 90%. The method saves time between 18 and 24 h by eliminating the need for cultures in laboratories.

This work aims to carry out a review study on fluorescence spectrography techniques exploring the intrinsic characteristics existing in bacteria (label-free) without the need to use high-cost commercial equipment. This article also seeks to contribute to the development of portable and flexible devices integrated into the patient's place of care, also called point-of-care (POC), to reduce nosocomial infections and achieve greater compliance with the hygiene of health workers [6–8].

2 Related Work

In our skin, we have two types of bacteria categories, resident and transient. Colonies of transient bacteria are the most susceptible to hand washing removal, as they live on the skin surface, more specifically in the epidermis. They are acquired through direct contact with contaminated people, patients, and objects, and depending on the number of microorganisms and the skin moisture, they can become persistent, such as *Staphylococcus aureus*.

According to a study published in the WHO Guide [9], *E. faecalis* and *E. faecium* bacteria survived for at least 60 min on the fingertips with and without gloves. The likelihood of contamination by healthcare workers when lifting a patient, measuring their temperature or blood pressure, and touching parts of the body must be considered when performing any medical procedure.

The existence of the autofluorescence property in bacteria makes it possible to classify them in genus, species, and group, using spectral analysis. A substance with fluorescent properties is tryptophan, an amino acid that, when excited by the emission of ultraviolet light with a maximum radiation peak of 280 nm, emits fluorescence at a wavelength of 350 nm [10].

Spectroscopic analysis has become an excellent tool in detecting pathogens, as pointed out [11] in concluding that it is a promising low-cost technique and rapid data processing. The study of spectral fingerprints for the detection, differentiation, and *characterization* of *Escherichia coli*, *Salmonella enterica*, *Pseudomonas aeruginosa*, *Bacillus subtilis*, and *Staphylococcus epidermidis* bacteria was explored in the research by [12], using the fluorescence spectroscopy method.

The autofluorescence property of nicotinamide adenine dinucleotide phosphate (NAPDH) that acts as an intrinsic fluorophore marker, without the need for labeling, reinforces the thesis of a valuable and non-invasive image analysis technique to study the metabolism and bacterial behavior [13, 14].

Pseudomonas aeruginosa is a bacteria present in humans that can infect blood, skin, bones, ears, eyes, urinary tract, heart valves, lungs, and excretes an extracellular compound called Pyoverdine that has fluorescent characteristics four times more intense than the tryptophan. When excited by UV light at 395 nm, it emits fluorescence with a peak at 455 nm, according to [15], corresponding to the spectrum of light visible by humans, which varies between 400–750 nm and can be easily captured by devices image analysis.

Conventional methods for detecting bacteria's presence are time-consuming processes in the laboratory and take days to obtain a result. As an example, the PCR (Polymerase Chain Reaction) and ELISA (Enzyme-Linked Immunosorbent Assay), which present the need to use high-cost instruments, well-trained teams, and bacteria concentration between 10^6–10^7 CFU/mL (colony-forming units/mL), the result delivered in a few hours or days, as they require a waiting period for the completion of the enzymatic reactions [16].

Staphylococcus aureus is one of the most common bacteria in hospitals with high mortality rates. The development of a biosensor for the detection of *S. aureus* is necessary to fast assist patients' diagnosis during their hospital stay. Studies on the development of paper biosensors by researchers in [17] served to detect a small number of colony-forming units of bacteria of 7, 40, and 100 CFU/mL in foods such as ground meat, milk, and lettuce. The results presented in this case, visible to the naked eye, were quantified through the use of image analysis software, presenting the results in minutes.

3 Methodology

The methodology used in this work will be a more in-depth and conceptual analysis. The results of this analysis will be grouped in a table comparing the most common methods and the main features and the detection times of the pathogens with the most common incidence in hospital environments.

To obtain the results of the existing studies from the year 2000 on the techniques of bacteria identification by fluorescence spectroscopy, a survey of data from the world scientific production was carried out in different scientific bases (SciELO, PubMed, Scopus, Google Scholar, Springerlink and IEEE) involving keywords and search strings: (Health AND Diagnostic AND Spectroscopy AND Fluorescence AND Bacteria).

The articles selected in the searches that contributed with information about the use of devices such as smartphones and that involve the subject of biomarkers in bacteria and studies with spectroscopic analysis in the same scientific databases already mentioned used the keywords and search strings: (Smartphones AND Biosensor * AND Detection AND Bacteria AND Nano*).

4 Results and Discussion

The urgency in combating nosocomial infections caused by *S. aureus* and the development of rapid, simple, and economical tests is highlighted by [16] in his research. The tests performed proved to be adequate for the detection of *S. aureus* in food samples for diagnosis at the site with the capture of the images by a smartphone and then transferred to a computer using a Bluetooth wireless network for analysis and quantitative processing, reaching a detection limit of this method at 10^3 CFU/mL.

Therefore, there is a need to improve the fluorescence spectroscopy technique for detecting microorganisms, reduce the high cost of equipment, and speed up patients' diagnoses during their hospital stay.

The introduction of the concept of biomarkers for the detection of *S. aureus* bacteria and viruses has advanced as an instrument for reading data and real-time detection of clinical samples with the help of nanotechnology in the development of sensitive components and matrices for the detection of pathogens.

The research [18] comments on traditional bacteria detection techniques that are based on colony-count culture and time-consuming and costly. A fast and growing development in the use of nanoparticles in biosensors production is seen as a promising tool in the search for methods for the diagnosis of bacterial infections to be applied in the hospital. The studies also demonstrated the detection of bacteria *E. coli* at different times of 5, 15, and 30 min according to the variation in the number of colony-forming units CFU/mL.

Another study [19] demonstrates the need for a rapid diagnosis of bacterial infections in the bloodstream to increase the survival rate and apply the correct antibiotic treatment. The use of nanoparticles is again combined with a portable system to increase the bacteria's signal so that it can be detected quickly and ultra-sensitive. In the study, the biosensor detected bacteria in the blood plasma in about 30 min, with sensitivity in the range between 10 and 10^5 CFU/mL for *Pseudomonas aeruginosa* and *Staphylococcus aureus*.

The existing methods have limitations and there is a need for them to be tested on a large scale and in hospital environments during medical practice and not just in

Table 1. Comparison of the most common pathogen detection methods and their main features and detection time. Developed by the authors.

Method	Pathogens	Time	Main features	Ref
Traditional PCR-ELISA	*E. coli* *S. aureus* *Salmonella spp.* Influenza virus	Hours or days	High-cost Low sensitivity Requires amplification Requires many reagents High sample concentration Time-consuming Requires training	[20] [21] [22] [23] [24]
Fluorescence spectrography	*E. coli* *E. faecalis* *S. aureus* *S. typhimurium* *S. epidermidis* *S. carnosus* *C. difficile* *K.pneumoniae* *P. aeruginosa*	10 min to 6 h	High-cost equipment Without reagents Non-invasive No sample culture required Characterizes several species of pathogens Without using markers Method not applied on a large scale Requires laboratories	[5] [10] [11] [12] [13] [14] [15] [25]
Biosensors	*E. coli* *S. aureus* *P. aeruginosa* H1N1 HPV Zika virus	5 to 30 min	Low cost Rapid detection Without amplification Simple operation High sensitivity Use of real samples No training required	[16] [17] [16] [18] [19] [26] [27]

laboratories. From the researched results, we created Table 1 to better understand the main features according to the currently existing techniques and the detection times.

5 Conclusions

The fluorescence spectroscopy technique has proved to be excellent for the detection of microorganisms; however, the high cost of equipment and the need for operation in laboratories prevents its use in inpatient care facilities. The integration with portable devices is seen as a viable and promising solution of low cost and rapid detection.

The versatility of today's smartphones, transforming them into a spectrometer for the detection of fluorescent markers present in bacteria or viruses, is seen as a viable and promising solution of low cost and rapid detection. However, many clinical tests are still needed to validate and make the spectroscopy technique reliable mainly for its use in hospitals.

The detection technique using biosensors proves to be a practice with the shortest detection time and that does not require technical training and the operation is simple and friendly. The possibility of coupling with smartphones brings the technique closer to patients by extending testing beyond the hospital environment. The incorporation of nanotechnology to biosensors allows a fast and highly sensitive analysis of reduced dimensions, which can facilitate the development of increasingly portable devices.

References

1. Pettit D (2001) Improving adherence to hand hygiene practice; a multidisciplinary approach. Emerg Infect Dis 7(2):234–244
2. Kampf G, Löffler H, Gastmeier P (2009) Hand hygiene for the prevention of nosocomial infections. Deutsches Aerzteblatt Online 106(40):649–656
3. Petersson LP et al (2014) Portable UV light as an alternative for decontamination. Am J Infect Control 42(12):1334–1336
4. Jelden KC, Gibbs SG, Smith PW, Hewlett A, Iwen PC, Schmid KK, Lowe JJ (2017) Ultraviolet (UV)-reflective paint with ultraviolet germicidal irradiation (UVGI) improves decontamination of nosocomial bacteria on hospital room. J Occup Environ Hygiene 14(6):456–460
5. Giana HE et al (2003) Rapid identification of bacterial species by fluorescence spectroscopy and classification through principal components analysis. J Fluorescence 13(6):489–493
6. Baslyman M, Rezaee R, Amyot D, Mouttham A, Chreyh R, Geiger G, Stewart A, Sader S (2015) Real-time and location-based hand hygiene monitoring and notification: proof-of-concept system and experimentation. Pers Ubiquit Comput 19(3):667–688
7. Mondol MAS, Stankovic JA (2015) Harmony: a hand wash monitoring and reminder system using smart watches. In: Proceedings of the 12th international conference on mobile and ubiquitous systems: computing, networking and services
8. Majeed Q, Hbail H, Chalechale A (2015) A comprehensive mobile e-healthcare system. In: 2015 7th conference on information and knowledge technology, IKT 2015, pp 1–4
9. Pittet D, Allegranzi B, Boyce J (2009) The world health organization guidelines on hand hygiene in health care and their consensus recommendations. Infect Control Hospit Epidemiol 30(7):611–622

10. Ammor MS (2007) Recent advances in the use of intrinsic fluorescence for bacterial identification and characterization. J Fluorescence
11. Shahzad A et al (2009) Emerging applications of fluorescence spectroscopy in medical microbiology field. J Transl Med 7:1–6
12. Bhattacharjee A, Datta R, Gratton E, Hochbaum AI (2017) Metabolic fingerprinting of bacteria by fluorescence lifetime imaging microscopy. Sci Rep 7(1):1–10
13. Belal T, Romdhane K, Jean-Louis B, Tahar B, Eric D, Franoise L (2011) Optical fiber-based synchronous fluorescence spectroscopy for bacterial discrimination directly from colonies on agar plates. Anal Methods 3(1):133–143
14. Vishwanath K, Ramanujam N (2011) Fluorescence spectroscopy in vivo. In: Encyclopedia of analytical chemistry
15. Dartnell LR et al (2013) Fluorescence characterization of clinically important bacteria. PLoS One 8:e75270
16. Zheng X, Wang Y, Bu S, Chen Z, Wan J (2019) Point-of-care detection of 16S rRNA of Staphylococcus aureus based on multiple biotin-labeled DNA probes. Mol Cell Probes 47(July):101427
17. Suaifan G, Alhogail S, Zourob M (2016) Rapid and low-cost biosensor for the detection of Staphylococcus aureus. Biosens Bioelectron 90:230–237
18. Mathelié-Guinlet M, Cohen-Bouhacina T, Gammoudi I, Martin A, Béven L, Delville MH, Grauby-Heywang C (2019) Silica nanoparticles-assisted electrochemical biosensor for the rapid, sensitive and specific detection of Escherichia coli. Sens Actuat B: Chem 292(April):314–320
19. Lee CW, Chang HY, Wu JK, Tseng FG (2019) Ultra-sensitive electrochemical detection of bacteremia enabled by redox-active gold nanoparticles (raGNPs) in a nano-sieving microfluidic system (NS-MFS). Biosens Bioelectron 133(January):215–222
20. Munch M, Nielsen LP, Handberg KJ, Jørgensen PH (2001) Detection and subtyping (H5 and H7) of avian type A influenza virus by reverse transcription-PCR and PCR-ELISA. Arch Virol 146(1):87–97
21. Daly P, Collier T, Doyle S (2002) PCR-ELISA detection of Escherichia coli in milk. Lett Appl Microbiol 34(3):222–226
22. Li Y, Cao L, Zhang C, Chen Q, Lu F, Bie X, Lu Z (2013) Development and evaluation of a PCR-ELISA assay for the detection and quantification of Cronobacter spp. Int Dairy J 33(1):27–33
23. Kahya S, Guran HS, Yilmaz O (2016) PCR and ELISA for staphylococcal enterotoxins and detection of some exotoxins from Staphylococcus spp. strains by PCR. Medycyna Weterynaryjna 72(1):28–33
24. Hu J, Huang R, Wang Y, Wei X, Wang Z, Geng Y, Jing J, Gao H, Sun X, Dong C, Jiang C (2018) Development of duplex PCR-ELISA for simultaneous detection of Salmonella spp. and Escherichia coli O157: H7 in food. J Microbiol Methods 154(136):127–133
25. Gosnell ME, Anwer AG, Mahbub SB, Menon Perinchery S, Inglis DW, Adhikary PP, Jazayeri JA, Cahill MA, Saad S, Pollock CA, Sutton-McDowall ML, Thompson JG, Goldys EM (2016) Quantitative non-invasive cell characterization and discrimination based on multispectral autofluorescence features. Sci Rep 6(1):23453
26. Faria AM, Mazon T (2019) Early diagnosis of Zika infection using a ZnO nanostructures-based rapid electrochemical biosensor. Talanta 203(January):153–160
27. Kizek R, Krejcova L, Michalek P, Merlos Rodrigo M, Heger Z, Krizkova S, Vaculovicova M, Hynek D, Adam V (2015) Nanoscale virus biosensors: state of the art. Nanobiosens Dis Diagnosis 4:47

Transportation Management System (TMS) Use in the Automotive Parts Industry

Jonatas Ribas Santos(✉) and Suelene Silva Piva

Mackenzie Presbyterian University, Campinas, Brazil
suelene.piva@mackenzie.br

Abstract. Efficient transportation management in a business' logistics processes is crucial to ensure competitive advantages. Therefore, the use of tools that ensure data visibility and freight costs reduction has become increasingly common. In this sense, this study intends to analyze the effects of implementing the Transportation Management System (TMS) tool in the logistics management of a large enterprise that operates in the automotive parts industry. With the use of TMS, it was found that it was possible to reduce the execution time of the month-end process by 3 days. Moreover, the hiring of a logistics operator, supplier of the tool, has resulted in a gross revenue increase from 129 to 134 million Brazilian reais and a decrease of 3.8% in freight costs. In conclusion, by monitoring the implementation of TMS, it was possible to obtain insights into the functionality of logistics in a large company, providing evidence on the importance of using technological resources to automate logistics processes seeking competitiveness in the industry.

Keywords: Logistics 4.0 · Transportation Management System · TMS · Information Technology

1 Introduction

Currently, the demand for customizable products with ever shorter life cycles is a factor that has had remarkable growth in the manufacturing industry, posing significant challenges for businesses. In this regard, the role of logistics is fundamental to keep organizations active in the industry, integrating essential activities to ensure the delivery of goods with speed and quality. Initially, logistics developed during wartime, with a need for strategic planning because of scarce resources. Currently, logistics is not only associated with military motions but also embraces the manufacturing industry [1].

In parallel, the fast-growing advances in Information Technology (IT) with the arrival of Industry 4.0 brought a wide range of opportunities for organizations. In this scenario, logistics is addressed as "logistics 4.0" and seeks faster processes in which information is shared in real-time. Thus, efficient logistics must depend on the use of technological applications, such as the Transportation Management System (TMS) [2].

In this sense, this study is related to monitoring and analyzing the impacts of a major TMS implementation project on a large automotive parts manufacturer. To carry out the project, the studied company hired a logistic operator to provide the tool and sought a

Y. Iano et al. (Eds.): BTSym 2020, SIST 233, pp. 72–79, 2021.
https://doi.org/10.1007/978-3-030-75680-2_10

5% reduction in logistics costs of domestic road freight. The data presented in this work were provided exclusively for academic purposes on the anonymity condition.

Based on the literature that states that the use of TMS provides control over the logistics operation, providing increased agility in obtaining and viewing data, the first hypothesis (H1) that this study proposes to verify is defined.

- H1: TMS allows greater practicality and visibility in the month-end closing process of the company.

Also, based on the proposal of the project to implement the logistic operator that foresees a 5% reduction in the costs of domestic road transport, the second hypothesis (H2) is defined.

- H2: hiring the logistics operator results in a 5% reduction in domestic road transport costs.

2 Logistics and Its Evolution

2.1 Concepts

Nowadays, logistics have a commercial function, which aims to provide goods or services on-site, in time and quantity needed to meet consumer demand. Thus, logistics has gained a new dimension, moving from a military strategy developed during wartime to a business strategy, uniting the acquisition, handling, storage, and delivery of goods [3].

Transportation is seen as part of logistics, attributed to the movement of goods from one location to another, and characterized as the most important element of the supply chain, accounting for about two-thirds of total logistics costs in businesses. In logistics, the basic modes of transportation are classified as road, rail, air, water, and pipeline [1].

2.2 Logistics Operators

Logistics operators are defined as suppliers that meet most –if not all– the logistics needs of their contractors. The outsourcing of operations eliminates the need for companies to invest in their assets, increasing operational practices efficiency and reducing freight costs.

Thus, logistics operators are classified into three groups, (1) operators based on operational assets that use their own assets to carry out the logistics operation; (2) administrative operators who do not possess their own assets but offer administrative technological systems; (3) integrated operators that are a combination of the previous groups, offering both their assets as well as technological systems [4].

2.3 Industry 4.0 and Its Impact on Logistics

The 4th Industrial Revolution, also referred to as "Industry 4.0" and presented in 2011 in Germany, has a goal to develop smarter, more flexible, and dynamic factories that are connected, allowing devices at different locations to communicate with real-time data transmission [5].

Overall, the concept of Industry 4.0 is to integrate innovative information and communication technologies to promote an intelligent network of products and processes in the industry can be divided into three main pillars: Cyber-physical Systems (CPS), Big Data, and the Internet of Things (IoT) [6].

With a rising demand for more interconnected companies that offer customizable products while maintaining product quality and speed of delivery, the concept of Logistics 4.0 was created, linking the technologies developed with Industry 4.0 and logistics, supported mainly by the pillars of CPS and IoT [7].

3 Transportation Management System (TMS)

The TMS is a software that integrates supply chain management and is centered in logistic transportation, which allows users to view and have total control over the logistic operation, designed to integrate the different stages that make up the freight transportation process contributing to the execution and control of activities. In this sense, TMS functionalities are divided into three groups [8].

- Planning and execution: determine the ideal transport mode, optimized route, and cargo consolidation for better use of vehicle capacity.
- Monitoring and control: provide real-time status information of vehicles and cargo transported by GPS tracking.
- Auditing and support to commercial negotiations: stores a history of approved freight rates that seek to simplify the process of freight auditing.

For companies seeking to integrate TMS into their logistics management, the selection of the software supplier is delicate since it involves high set-up costs [9]. For this, the authors emphasize the importance of benchmarks with the supplier's current customers to evaluate positive and negative points of TMS usability, considering the costs of personnel training, system maintenance, and licenses involved with implementing the tool.

4 Case Study

The target object of this study is one of five sites of a multinational mechanical transmissions manufacturer company, which began its activities in Brazil in the 1960s in the State of São Paulo. On this site, the domestic road freight was conducted by multiple carriers before the hiring of the logistics operator.

Seeking to provide transparency in data analysis regarding road freight costs, the company implemented TMS in its management. For this purpose, through bidding, which comprises selecting service providers or manufacturing suppliers, based on multiple criteria of choice [10], an integrated logistics operator was hired as the supplier of the tool.

In parallel to providing TMS, the contracted logistics operator was also made responsible for assuming the domestic road freight service that was previously conducted by the carriers. With this strategy, the company estimated a 5% decrease in freight costs.

The financial analysis of the monthly freight costs is carried out through the calculation of the P indicator, which refers to the proportion between the sum of freight fees paid and the company's sales throughout the month, as per (1).

$$P = \sum freight\,fees / \sum sales \tag{1}$$

This indicator is an evaluator of the carriers' performance regarding freight charges, considered as "predicted" a value of up to 1.5%.

5 Methodology

The main objective of this study is to monitor and analyze the impacts of implementing TMS in an automotive parts manufacturer, to present the benefits encountered and verify whether TMS has led to the expected improvements described in the literature, and to verify whether the replacement of carriers by the logistics operator resulted in an effective reduction in freight costs.

Therefore, regarding methodological processes, this study is characterized as exploratory, with an empirical basis, using a case study for data collection and analysis. To achieve its objective, the methodology was divided into five sections:

1. At first, theoretical studies were conducted on topics related to logistics.
2. Then, the scenario pre-TMS implementation was presented, analyzing the month-end procedures of data collection and freight costs calculations.
3. Posteriorly, the bidding process for the selection of the TMS supplier was presented.
4. Successively, the scenario post-TMS implementation was presented, highlighting its functionalities and benefits encountered.
5. Finally, a financial analysis of the impact on freight charges due to the contracting of the logistic operator was carried out.

The financial analysis was carried out, dividing the chronology of the TMS implementation project into three phases: Pre-TMS, phase-in, and Post-TMS.

6 Results and Discussion

6.1 Pre-TMS Implementation Scenario

In this scenario, for data collection and freight costs calculation, at the end of each month, the logistics team contacted the carriers individually by e-mail, requesting each to send a report containing details of all shipments made within 30 days.

All reports were received in Excel format; however, each carrier followed its standard report, containing unnecessary information for analysis. Therefore, at first, it was necessary to filter the received worksheets so that only relevant data were selected for analysis. Once filtered, the data regarding freight cost and value of the transported goods were compiled from the carriers' reports and inserted into a different template form.

With the template report filled-in, it was forwarded to the site controller for validation and comparison between the total freight cost calculated and the forecasted cost planned to verify if there were any significant divergences between what was planned and what was being spent.

In Latin America, the company uses the road, water, and air transportation modes for cargo freight. Among the accounts that make up the company's total logistical cost, domestic road freight is what makes up the largest portion of this sum, as shown in Table 1.

Table 1. Breakdown of company logistics costs in 2018.

Accounts	Total costs (in BRL)	Percentage of total (%)
Domestic road freight	15,393,164	47
Customs broker	8,706,987	26
Freight agent	4,759,794	14
Other	4,183,560	13
Overall total	33,044,505	100

Given the above, the comparison between the Brazilian sites reveals that, in 2018, Valinhos represented 60% of the total value spent on domestic road freight. Thus, it is possible to identify the scope set by the company for the TMS implementation project, which has opted to focus initially on domestic road freight management at the Valinhos site since it has the highest costs.

6.2 Logistic Operator Selection and Contracting

The process of selecting a supplier for a technological solution such as TMS is delicate and requires close attention; therefore, the company launched a market bidding process for the TMS supplier selection. Thus, as a criterion of selection, participants of the process must: (1) be an integrated logistics operator; (2) have had a prior relationship with the company; (3) have the infrastructure and know-how for absorbing the company's domestic road freight scope.

Thus, the company invited six logistics operators to participate in the bidding process, introducing participants to the project's scope, and conducting benchmarks with clients. As a result of the bidding process, company management selected one of the participants, a Brazilian corporation in operation for over 20 years, mainly by considering the commercial proposal presented and for seemingly having greater engagement and commitment to the initiative.

With this, the project implementation timeline was set, initiating the pilot tests in June 2019. During the months of July to August, started the phase-in, when the shipments began to be gradually carried out by the logistic operator until, finally, in September 2019, 100% of domestic road freight was under the responsibility of the contracted logistic operator.

6.3 Post-TMS Implementation Scenario

Using TMS to manage company logistics has ensured increased transparency of domestic road freight data, thus contributing to the practicality of logistics activities.

After initial registration, users can access the tool and extract reports regarding freight transport operations from the online system. The TMS control panel allows for reports to be issued considering shipping, as shown in Fig. 1 or by invoice identification number.

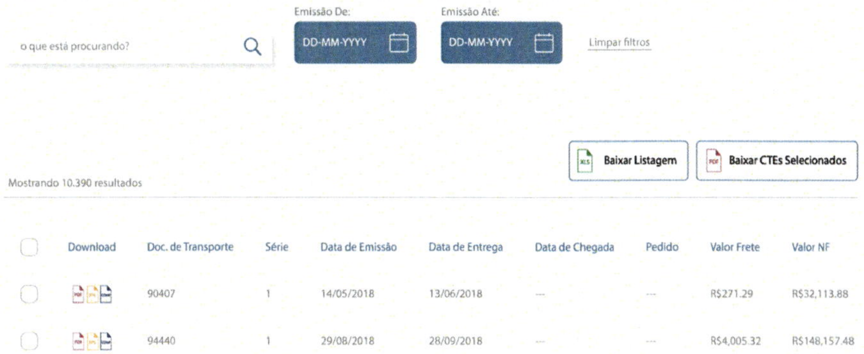

Fig. 1. TMS dashboard interface.

It is worth mentioning that both panels provide a range of filters that assist the search for a specific shipment or a list of shipments made within a period. In addition, TMS also allows users to upload documents into the system and download standardized reports with detailed shipping information.

Another feature available in the implemented TMS is the financial dashboard, which allows users to view the proportion between the company's revenues and freight costs in a particular period displayed in a graph.

With the use of TMS, the month-end process of data collection and freight costs calculation started being carried out in an automated manner, making use of reports and dashboards provided by the tool. Therefore, the key aspects raised by TMS users that show improvement in the process are:

- Decrease in the duration of the month-end process of data collection and calculation of freight costs, from a 4-day to a 1-day cycle for completion, eliminating the need of contacting the supplier to obtain data.
- Practicality in using the tool for its intuitive and user-friendly interface.
- Increased reliability since all data compilation is handled by the system.
- Cloud storage of important documents and receipts.

These benefits attest to those indicated by Festa and Assumpção [8] that mention the availability of accurate freight data and support for performance control as advantages of TMS. Therefore, it is possible to affirm that the presented hypothesis H1 is sustained since TMS has allowed a previously lengthy and highly susceptible to errors process to be carried out in a practical and fast manner, eliminating unnecessary steps and presenting a more trustworthy result.

6.4 Financial Analysis

Besides providing reports in an automated way through the TMS tool, the implementation project of the logistic operator in the company also foresaw a 5% reduction in road freight costs through the replacement of the carriers by the contracted logistic operator.

With this, Table 2 displays a breakdown of average values of invoicing and freight rates paid over the months of implementation of the project in 2019 to calculate the P indicators that measure the performance of the logistics operator regarding freight charges.

Table 2. Average P value throughout TMS implementation phases in 2019.

Phase	Months	Avg. revenue (in BRL)	Avg. freight cost (in BRL)	Avg. P value (%)
Pre-TMS	Jan.-May	21,008,442	266,107	1.3
Phase-in	Jun.- Aug.	19,834,145	306,308	1.6
Post-TMS	Sep.- Dec.	33,422,340	371,707	1.1

From September to December, with the project already consolidated, a tendency to decrease the P indicator can be noted, averaging 1.1%. Even with an increase in shipments during this period, the index has remained on a downward trend.

In conclusion, when comparing the months of September to December 2019 to the same period of the preceding year, there is a saving of approximately BRL 57,000 in freight costs for the company. This slight reduction occurs even with a 12% increase in the number of shipments made during the same periods. It is observed yet, a 3.7% increase in its gross income, totaling BRL 4.7 million above the previous year.

It is possible to state that there was a reduction in freight costs for the company; however, different from the hypothesis H2, which supports a 5% reduction in freight costs, the introduction of the operator during this period resulted in a reduction of approximately 3.8%.

7 Final Considerations

The development of this study allowed an in-depth analysis of how TMS can enhance the logistics management of an automotive company. The idea of implementing the tool is to allow increased visibility and availability of accurate freight data so that the month-end process of calculating domestic road transport costs could be carried out in a fast, easy, and efficient way.

Thus, users' positive perception of the tool's functionality supported the statements in the literature. It was also shown that the contracting of the logistic operator led to a slight reduction in freight costs; therefore, it is considered a positive factor. The centralization of the domestic road transport operation in a single supplier has provided simplified logistic management.

Given the results, it was found that the use of technological resources seeking the automation of logistics processes allowed operations to be carried out more efficiently, avoiding time and resource losses, grating the company's greater performance, and ensuring a more prominent position in the industry.

References

1. Novaes AG (2007) Logística e gerenciamento da cadeia de distribuição, 11th edn. Elsevier, Rio de Janeiro
2. Barreto L, Amaral A, Pereira T (2017) Industry 4.0 implications in logistics: an overview. In: Manufacturing engineering society international conference. Elsevier, Pontevedra, pp 1245–1252
3. Nunes KM, Júnior JRLP, Costa LFA, de Souza MCA, de Alencar DB, Sanches AE (2019) Proposal for the implementation of a transport management system in a Manaus conveyor. J Eng Technol Ind Appl (5), 68–74
4. Ribeiro PCC, Ferreira KA (2002) Logística e transporte: uma discussão sobre os modais de transporte e o panorama brasileiro. In: XXII ENEGEP. Anais do ENEGEP, Curitiba
5. Becker A, Schneider A, Ercico J, Werlang R (2018) Os conceitos da indústria 4.0 associados a abordagem da capacidade dinâmica. Anais Engenharia Prod (2), 123–136
6. Sakurai R, Zuchi JD (2018) The industrial revolutions up to industry 4.0. Rev Interface Tecnol 15:480–491
7. Motta MJ, Lusvarghi GC (2018) Logistics 4.0: challenges and opportunities in the management of modern supply chain. In: 18° CONIC. Anais do CONIC-SEMESP, São Paulo
8. Festa E, Assumpção MRP (2010) A contribuição do TMS (transport management system) no desempenho do fluxo logístico na rota São Paulo-Manaus. In: XXX ENEGEP. Anais do ENEGEP, São Carlos
9. Morettin AA, Lotierso A, Vasconcelos FW (2012) Identificação do processo de implantação de um sistema de gerenciamento de transporte. In: IX SEGET. Anais do SEGET, Resende
10. Logweb portal. https://www.logweb.com.br/

Modified Control Charts Monitoring Long-Term Semiconductor Manufacturing Processes

Jorge M. de Souza[1](✉), Giovanni M. de Holanda[1], Hingmar A. Henriques Jr.[2], and Rafael H. Furukawa[2]

[1] FITec - Technological Innovations, Campinas, SP, Brazil
{jmdsouza,gholanda}@fitec.org.br
[2] Brasil Components - Multilaser, Extrema, MG, Brazil
{hingmar.henriques,rafael.furukawa}@multilaser.com.br

Abstract. Statistical methods have been systematically adopted in efforts to improve process efficiency and semiconductor manufacturing quality. Some modeling advances have been provided to meet business peculiarities and specific characteristics of this industry. This paper presents an approach to address long-term *Pp* (*Ppk*) indices using a modified control chart. The idea is to apply this model to dynamically obtain the ppm (parts per million) and the acceptance limits given some parameters: long-term target, the probability of rejecting a good item, and the upper and lower specification limits. In a way that the cost of identifying and correcting special causes should be compatible with the cost of off-target products and with the corporate quality policy. Such an approach was developed to integrate a statistical module for monitoring and controlling of equipment set-up of a Brazilian semiconductor manufacturing company.

Keywords: Statistical process control · Modified control chart · Semiconductor industry

1 Introduction

As all industry, the semiconductor manufacturing sector aims to minimize cost and maximize quality and reliability. Quality can be achieved with stable and well-controlled processes in such a way that the increased quality and productivity are related to reduced process and product variability. In order to achieve this objective, the use of statistical methods is essential in efforts to improve process efficiency and increase manufacturing quality [1].

Several statistical and simulations methods have been applied to monitor and control manufacturing processes, considering peculiarities of the industries [2, 3], and more recently in the context of Industry 4.0 [4]. In terms of statistical process control (SPC), and more specifically concerning control charts, many studies and techniques are found in literature, for example, [5–8].

In the Integrated Circuit (IC) industry, it is not today that the SPC has been adopted in the manufacturing process, and in turn, the singularities of this process have been transforming the SPC itself [9]. Moreover, in the last decades, with the technological advance

Y. Iano et al. (Eds.): BTSym 2020, SIST 233, pp. 80–87, 2021.
https://doi.org/10.1007/978-3-030-75680-2_11

of semiconductor manufacturing, wafers for instance, new techniques and approaches have been shown to be necessary for monitoring production data [10–12].

An important activity in IC manufacturing is to measure and calibrate each equipment of the process chain before a lot production is carried out. Each equipment has a set of parameters whose calibration is based on a set of test measurements or a subgroup of observations, aiming at reducing variability based on the equipment specification and control acceptance limits. The subgroups' size is generally small to optimize cost and test duration, and the variability within subgroups is measured by the short-term standard deviation [13].

The variation of equipment parameters affects the product quality, and many factors can influence the variability during the production run-time, such as raw material quality, production equipment fatigue, variations in equipment set-up, etc. In this context, control charts are very often used to evaluate the control acceptance limits to monitor the variability across the production, and despite some controversies in SPC [14], there are different control charts adapted for long and short production runs.

Process capability indices may be used to quantify the short-term (*Cp, Cpk*) and long-term (*Pp, Ppk*) variations. However, such indices do not directly express the percentage of nonconforming items. Generally, these figures are expressed by parts per million (ppm) of the produced items without meeting the specifications. For symmetrical limits around the mean, a target $Cp = 2$ implies a severe control of identifying and correcting special causes (out-of-control points) that will lead to a 0.002 ppm and, consequently, more costly control activities. However, if a limit of 1 ppm is accepted, the control activities are less costly (less out-of-control points) and lower specified, $Cp = 1.63$. In this sense, the cost of identifying and correcting special causes should be compatible with the cost of off-target products.

This paper presents an approach to address long-term *Pp* (*Ppk*) indices using a modified control chart. The idea is to apply this model to dynamically obtain the ppm and the acceptance limits given some parameters: long-term target, the probability of rejecting a good item, and the upper and lower specification limits. Such an approach was developed to integrate an SPC module to monitor and control equipment set-up of a Brazilian semiconductor manufacturing company – Multilaser, located in Extrema-MG. The SPC module is an R&D project being carried out under the grant of the PADIS (Program of Support for the Technological Development of the Semiconductor and Displays Industry) of the Brazilian Ministry of Science, Technology, and Innovations.

The paper is structured as follows. Section 2 brings some discussion arising from the literature review. Section 3 presents the modified chart control, and Section 4 illustrates its application on set-up data of a die-attach equipment. Finally, some concluding remarks are shown in the last section.

2 Some Considerations on Literature Review

Process capability index [15, 16] (or just capability) is an indication of how well the equipment calibration meets the production quality requirements. It is a tolerance index to verify if the standard deviation process spread is well in-side the upper/lower specification limits. Kotz and Jonhson [17] presented a literature survey covering the period

1992–2000, including the most used process capability indices; Wu et al. [18] added the developments in the period 2002–2006; and de-Felipe & Benedito [19] provide a more recent review.

There are many examples in the literature on applying the statistical control process techniques in industrial manufacturing, for example, [3–6, 9–12]. Some authors introduce new kinds of analysis in order to assess the effects of peculiarities in production behavior. Godina et al. [20], for instance, compare two normality tests to observe if the accuracy increases when Anderson-Darling test is used instead of the control chart.

However, a less addressed issue is how to detect that the process is in statistical control, i.e., "the probability distribution representing the quality characteristic is constant over Time" [14]. Real examples show processes exhibiting out-control points and high capability using XBar-R technique, for example. These contradictory control indications suggest that a modified control chart should be preferred. In [15], it is analyzed the impact of the inaccuracy of the measurement process causing tighter control limits compared to process variation.

Amongst the studies addressing the modified control chart applied, Mottonen et al. [15] assume that the measurement process is accurate. The capability index is used to derive a capability-based standard deviation for the XBar chart, as far as the process is considered controlled using the R chart results. In [5], the authors consider that the test measurements are controlled and the XBar out-of-control points should be verified for an assignable cause, and in [7], measurement data is normal distributed. For Hsieh et al. [6], the control chart deals with "defect clustering" in IC wafers that, of course, do not exhibit Poisson-based statistics. It is a search for control and not for defects.

At Multilaser, the IC manufacturing process control is undergone for each equipment, which is part of the manufacturing chain. The equipment set-up is based on a set of measured parameters and is granted/accepted when they exhibit small variation. The variation is monitored using the performance measure knowing the process upper and lower specifications for each equipment and the respective standard deviation. If the set-up is successful, the mean value of the measure set is used to control the equipment capability in the manufacturing chain.

In this sense, a specific model able to consider the particularities of this manufacturing process becomes an alternative to be investigated in order to derive the ppm and the acceptance limits for long-term analysis given the general target *Ppk*, the probability of rejecting a good item, the fraction of nonconforming item and the upper and lower specification limits.

3 Formulation of the Proposed Approach

3.1 Normal Probability Plot

As already stated, one important activity in IC manufacturing is to measure and calibrate the parameters of each equipment before a lot production is carried out. The real example that will illustrate the analysis across the paper is the calibration data of a long production of a Die-Attach equipment parameter. The discussion is general and can be applied to the overall process production.

The record covers one year and a half of consistent calibration data. For each sample m the number of observations n is 5. In this case, the XBar-R (process range) is well adapted to control the quality characteristics if its observation average for each sample, $\overline{X}(j), j = 1 \ldots m$, is normally distributed.

An easy way to test against normality, it to construct a normal probability plot. This can be done by plotting the observation average, $\overline{X}(j)$, against the standardized normal, z_j, obtained from the cumulative standard normal distribution [21], as expressed in Eq. (1).

$$\frac{j - 0,5}{m} = P(Z \leq z_j) = \int_{-\infty}^{Z_j} Norm(0, 1) \quad (1)$$

Where m is the number of samples, in this case, $m = 121$.

The observation average $\overline{X}(j)$ should first be arranged in ascending order. The normal plot is shown in Fig. 1, making it evident that the normal distribution is an appropriate model. This is based on the recorded data of the parameter that will be used in the case study section. For the other parameters, the normal distribution still holds.

3.2 Formulation of the Modified Control Chart Limits

The long-term standard deviation, s, is expressed by Eq. (2)

$$S = \sqrt{\frac{\sum_{i=1}^{m} \left(\overline{X}(i) - \overline{X}\right)^2}{(m - 1)}} \quad (2)$$

The parameter m is the sample size, $\overline{X}(i)$ is the ith observation average and $\overline{X} = \frac{\sum_{i=1}^{m} \overline{X}(i)}{m}$, the sample mean. The sample size m should be ≥ 20.

If the specification limits are symmetrical around the mean, *Cp* and *Pp* are used otherwise *Cpk* and *Ppk*. The capability *Cp* (*Cpk*) is based on the short-term standard deviation while the performance *Pp* (*Ppk*) on the long-term standard deviation. The formulation addresses the long-term run using the more general index *Ppk*.

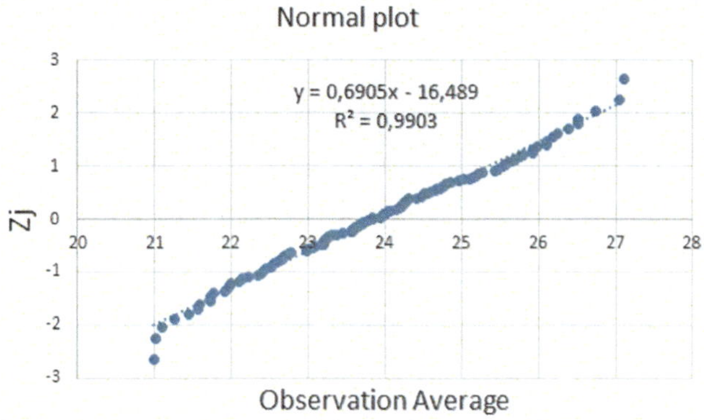

Fig. 1. Normal probability plot.

The modified control charts assume that the actual process mean is in the interval $\overline{X}l \leq \overline{X} \leq \overline{X}u$, which depends on the process nonconforming fraction: $\overline{X}l = LSL + Z_\delta\, s$ and $\overline{X}u = USL - Z_\delta\, s$, where USL is the Upper Specification Limit and LSL, the Lower Specification Limit. Z_δ is expressed by Eq. (3)

$$Z_\delta = \frac{\min(USL - \overline{X}u, \overline{X}l - LSL)}{s} \tag{3}$$

Given the *Ppk*, the relation *Ppk*, Z_δ can be derived as in Eq. (4):

$$Ppk = \frac{\min(USL - \overline{X}u, \overline{X}l - LSL)}{3s} \tag{4}$$

And $Z_\delta = 3Ppk$. Formulation of the modified control chart upper (UCL) and lower (LCL) limits is expressed by Eq. (5):

$$UCL = USL - \left(Z_\delta - \frac{Z_\alpha}{\sqrt{n}}\right), LCL = LSL + \left(Z_\delta - \frac{Z_\alpha}{\sqrt{n}}\right) \tag{5}$$

Where n is the number of observations. The fraction of nonconforming (*FNC*) is expressed by Eq. (6) and ppm is obtained by multiplying *FNC* by one million.

$$FNC = 2\left(1 - \int_{-\infty}^{Z_\delta} Norm(0, 1)\right) \tag{6}$$

Where $Z_\delta = 3Ppk$, and the error is expressed by Eq. (7):

$$Error\ Type\ I = 2(1 - \int_{-\infty}^{Z_\alpha} Norm(0, 1)) \tag{7}$$

If the accepted ppm is given, the above equations are used to derive *FNC*, Z_δ, *Ppk*, and the control limits UCL and LCL. The method is illustrated by the flowchart shown in Fig. 2, which represents the process control based on ppm specification.

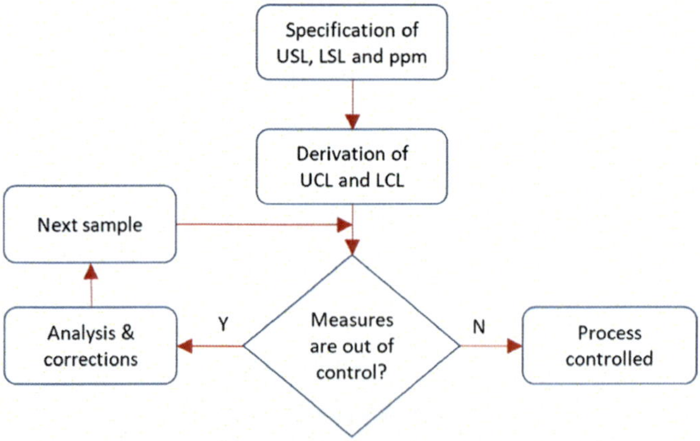

Fig. 2. Process control based on ppm specification.

4 A Production Case Study

The practical study is for parameter with asymmetrical specification limits. The process variability is controlled since *Ppk* > 1.65 calculated over the recorded period for a sliding window of size 20, as can be noted in Fig. 3.

Fig. 3. Long-term run performance.

The objective is to make a trade-off between the number of out-of-control points (correcting special causes) and the number of nonconforming items (discarded item). All the examples consider Error Type I = 0.1%.

In Fig. 4(a), the Normal density functions are plotted, and Fig. 4(b), depicts the corresponding XBar control chart. For the determined control limits plotted in Fig. 4(b): i) there are 8 out-of-control points that should be analyzed, ii) $Ppk = 2$, and iii) the fraction of nonconforming item is 0,0000002% and ppm is 0.002.

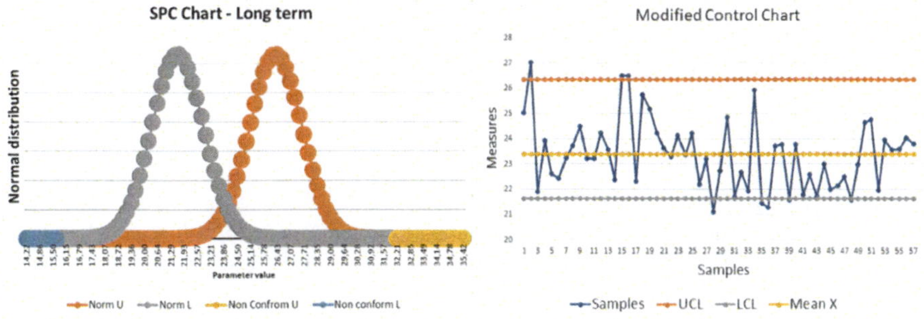

(a) **(b)**

Fig. 4. **(a)** Normal density function, for $Ppk = 2$. **(b)** XBar control chart, for $Ppk = 2$.

The plot for a less performing $Ppk = 1.7$ is shown in Fig. 5(a) and Fig. 5(b). For this case, considering the control limits plotted, the results are: i) the variability is controlled, there is no out-of-control points, ii) $Ppk = 1.7$, and iii) the fraction of nonconforming item = 0.000034% and ppm = 0,34. This result confirms those presented in Fig. 3, i.e., *Ppk* is always above 1.65.

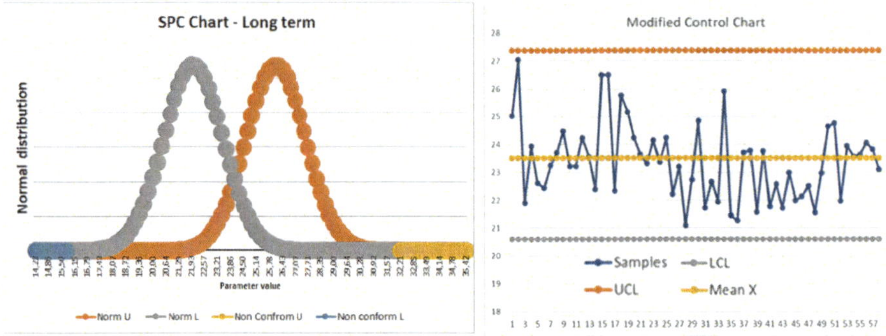

(a) (b)

Fig. 5. (**a**) Normal density function, for $Ppk = 1.7$. (**b**) XBar control chart, for $Ppk = 1.7$.

The proposed approach can support a quantitative decision based on the trade-off costs between the number of special cases to be analyzed before restarting the production process and the number of item nonconforming that must be discarded. It can also be used to guide a process quality aimed at reducing the variability of a production by tightening the control limits, based on a planned *Ppk* increase or a ppm decrease by quality steps. To attain required quality by analyzing the special causes in just one step can lead to long production interruption until all the causes are explained and corrected.

5 Final Remarks and Conclusion

This paper addresses the situation where the process exhibits many out-control points but has high performance. To deal with these contradictory control indications, a modified control chart is proposed where the control limits are based on the long-term performance *Ppk* and the corresponding number of off-target products based on the ppm specification. The control chart method should take the *Ppk* specification into account.

In this sense, the cost of identifying and correcting special causes should be compatible with the cost of off-target products.

Acknowledgments. The authors would like to thank the Brazilian Ministry of Science, Technology, and Innovations for this project's financial support through the PADIS (Program of Support for the Technological Development of the Semiconductor and Displays Industry).

References

1. May GS, Spanos CJ (2006) Fundamentals of semiconductor manufacturing and process control. Wiley-Interscience, New Jersey
2. Peres FAP, Fogliatto FS (2018) Variable selection methods in multivariate statistical process control: a systematic literature review. Comput Ind Eng 115:603–619. https://doi.org/10.1016/j.cie.2017.12.006

3. Madanhirea I, Mbohwab C (2016) Application of statistical process control (SPC) in manufacturing industry in a developing country. Proc CIRP 40:580–583. https://doi.org/10.1016/j.procir.2016.01.137
4. Zhong RY, Xu X, Klotz E, Newman ST (2017) Intelligent manufacturing in the context of industry 4.0: a review. Engineering 3:616–630. https://doi.org/10.1016/J.ENG.2017.05.015
5. Sindhumol MR, Gallo M, Srinivasan MR (2018) Monitoring industrial process using a robust modified mean chart. Aust J Stat 48(1):1–13. https://doi.org/10.17713/ajs.v48i1-1.765
6. Hsieh KL, Tong LI, Wang MC (2007) The application of control chart for defects and defect clustering in IC manufacturing based on fuzzy theory. Expert Syst Appl 32:765–776. https://doi.org/10.1016/j.eswa.2006.01.050
7. Adekeye KS (2012) Modified simple robust control chart based on median absolute deviation. Int J Stat Probab 1(2):91–95
8. Ma X, Zhang L, Hu J, Palazoglu A (2018) A model-free Shewhart individuals control chart for autocorrelated data. In: Proceedings of the 10th IFAC international symposium on advanced control of chemical processes
9. Spanos CJ (1992) Statistical process control in semiconductor manufacturing. Proc IEEE 80(6):819–830. https://doi.org/10.1109/5.149445
10. Wang R, Zhang L, Chen N (2019) Spatial correlated data monitoring in semiconductor manufacturing using gaussian process model. IEEE Trans Semicond Manuf 32(1):104–111. https://doi.org/10.1109/TSM.2018.2883763
11. Liu ZJ, Qian HHT, Liu ME (2018) Advanced process equipment matching methodology in semiconductor manufacturing. In: Proceedings of the 2018 China semiconductor technology international conference (CSTIC), pp 1–4. https://doi.org/10.1109/CSTIC.2018.8369285
12. Moyne J, Iskandar J (2017) Big data analytics for smart manufacturing: case studies in semiconductor manufacturing. Processes 5(39). https://doi.org/10.3390/pr5030039
13. Holanda GM, Souza JM (2020) Capability based control charts: an experimental approach to manufacturing process. In: Proceedings of the 8th workshop on probabilistic and statistical methods. UFSCar, São Carlos, Brazil
14. Woodall WH (2000) Controversies and contradictions in statistical process control. J Qual Technol 32(4):341–350. https://doi.org/10.1080/00224065.2000.11980013
15. Mottonen M, Belt P, Harkonen J, Haapasalo H, Kess P (2008) Manufacturing process capability and specification limit. Open Ind Manuf Eng J 1:29–36
16. Montgomery DC (1997) Introduction to statistical quality control. Wiley, Hoboken
17. Kotz S, Johnson NL (2002) Process capability indices: a review, 1992–2000. J Qual Technol 34(1):2–19. https://doi.org/10.1080/00224065.2002.11980119
18. Wu CW, Pearn WL, Kotz S (2009) An overview of theory and practice on process capability indices for quality assurance. Int J Prod Econ 117:338–359
19. de-Felipe D, Benedito E (2017) A review of univariate and multivariate process capability indices. Int J Adv Manuf Technol 92:1687–1705. https://doi.org/10.1007/s00170-017-0273-6
20. Godina R, Pimentel C, Silva FJG, Matias JCO (2018) Improvement of the statistical process control certainty in an automotive manufacturing unit. Proc Manuf 17:729–736
21. Montgomery DC, Runger GC (1999) Applied statistics and probability for engineers. Wiley, Hoboken

Periurban Settlements

A Discussion on Water Sustainability Indicators

Jakeline Pertile Mendes(✉), Denise Helena Lombardo Ferreira, and Cibele Roberta Sugahara

Pontifical Catholic University of Campinas, Campinas, Brazil
jakeline.pm@puccampinas.edu.br, {lombardo, cibelesu}@puc-campinas.edu.br

Abstract. Water is an essential natural resource for food production. Since the need for agricultural demand has increased in recent times due to population growth, water demand has also grown, making appropriate water management relevant. In this sense, the sustainability indicators were developed, aiming to expand and improve the assessment and monitoring of environmental actions through metrics, aiming to improve decision-making in the management of this resource. To this end, the present research aims to discuss the indicators of water sustainability and the importance of its development and applicability in water management in periurban settlements. As a method, a bibliographic study was carried out on material already prepared, consisting of books and scientific articles on water management and sustainability indicators, based on agroecological activities carried out in the Milton Santos periurban settlement. As a result, it is noted the importance of listing the works developed in periurban settlements for the regions in which they are inserted, giving a new meaning to the use of these spaces, especially in agricultural matters, as well as regarding the management and use of the water and the land.

Keyword: Sustainability indicators · Water sustainability · Periurban settlements · Water resources · DPSIR model

1 Introduction

Currently, sustainability has been widely discussed in social, environmental, and economic aspects. Many researchers consider it essential to include these three pillars in order to leverage, disseminate, and enrich the interdisciplinary contributions to sustainability, which requires aligning human practices and the way of being and living, considering the possible limits of each biome, to guarantee the needs of present and future generations [1]. Coupled with sustainability, one must not fail to list the development, which has other unexplored studies, considered central themes of development: cultural creativity and social morphogenesis [2]. In this sense, sustainable development should be explored as a possibility to guarantee socio-political changes without compromising the ecological and social systems that sustain communities [3].

Y. Iano et al. (Eds.): BTSym 2020, SIST 233, pp. 88–99, 2021.
https://doi.org/10.1007/978-3-030-75680-2_12

With technological advances and population growth in the world in recent years, interest in discussing the use of water resources has increased, especially in view of the demand for water supply. The work [4] considers it a great challenge, due to the estimate of the population of urban agglomerations throughout historical periods, precisely because of the complexity of the urban growth process, which, according to National Water Agency [5], are factors that threaten the balance, especially in urban areas, whose economic growth increases the water demand, in addition to causing climatic changes and extreme hydrological events.

Given the issue of water and large urban agglomerations, from the 1990s onwards, social researchers began to notice the beginning of a process of recomposing rural spaces, with a demographic increase and a variety of occupations, which went beyond agricultural activities [6]. In this sense, from the perspective of world urbanization, consistent population estimates are prepared, over historical periods, in human settlements [4].

Considering that Brazil has its social base composed mainly of "peasants, family farmers, agroecological producers, collectors of recyclable materials, artisanal producers, and cultural collectives" [7, p. 01]. Since a large part of the settlers is included in this social base, with agricultural activity as a source of survival, there is a need to address the issue of public policies as a way to promote institutional transformation and affirmative actions that favor the weakest and quietest segments of the nation, most of them hard-working, "deprived of decent work opportunities and livelihoods, and condemned to waste their lives in the daily struggle for survival" [8, p. 27].

Since it is not only industrial or chemical agriculture practices that affect species extinction and ecological processes but also climate change [9], there is a need to discuss sustainability, aiming at the compatibility of economic growth with environmental preservation, encompassing social justice [10].

Guided by the fact that agriculture corresponds globally to almost 70% of all water withdrawals, and that up to 95% in developing countries, being the activity with the greatest use of water [11], and one of the main activities of the less favored classes, it is noted that this type of agriculture, "family-based and diversified, lacks emergency policies to continue fulfilling its role of guaranteeing real food" [12, p. 03].

To this end, it is essential to create policies for the consolidation and modernization of family farming, being a strategy for stimulating rural development, based on the pluriactivity of the rural population [8].

In this sense, the biggest challenge is to develop and apply sustainability indicators in settlements, especially periurban settlements, which are located in highly populated and industrialized areas, where the water demand is greater. Sustainability indicators are widely advocated by scientists, as they facilitate the monitoring and management of sustainability, especially in decision making. One of the most used methods for environmental issues is the DPSIR (Driving Force-Pressure-State-Impact-Response) as a system of sustainability indicators [13].

2 Objective

This research aims to discuss water sustainability indicators and the importance of their development and applicability in water management in periurban settlements.

3 Methodology

This study aimed to identify the importance of the applicability of water sustainability indicators in periurban settlements. For this, a bibliographic study was carried out, which, according to [14], is based on material already prepared, mainly composed of books and scientific articles. In this study, water management and sustainability indicators were explored, based on agroecological activities carried out by the Milton Santos periurban settlement, characterizing the research as exploratory, which according to [15] has the purpose of gathering information about an object, delimiting the research field, with the objective of verifying its conditions.

4 Theoretical Reference

4.1 Water Sustainability

Water sustainability management is essential, as it ensures the integration of social, economic, and environmental issues in all stages of water resource management [16]. The sustainability of the use of water resources becomes the least of the issues if the pressure of human activities is considered to be increasingly intense, which also harms the quality of the water, to the point that, for the water to become drinking water, increasingly advanced and expensive treatment is necessary, in addition to other factors that must be guided, such as the reduction of species of aquatic plants and animals in rivers and lakes [17]. Freshwater is the most important resource for humanity, crossing all social, economic, and environmental activities, and it is a condition for all life on our planet, a factor that facilitates or limits any social and technological development, a possible source of well-being or misery, cooperation or conflict [18].

In this sense, guaranteeing the attendance to the growing demand for a limited resource in quantity in the territory, in which several times, the impacted by the quality comes from anthropic activities [19], it is considered fundamental the good management of water and other natural resources as an essential part of environmental sustainability, social equity, and quality of life for future generations [20]. Taking into account that the world has approximately 1.400 million km^3 of water, where only 0.003% of this large amount, totaling 45,000 km^3, are freshwater resources, which theoretically is water suitable for consumption, hygiene, agricultural use, and industrial [11], it can be considered that, worldwide, there is a scenario of water complexity, yet considering that many times part of this water availability is not accessible and suitable for human consumption.

For millennia, humanity has been striving to explain the issue of water resources for domestic water supply, irrigation, and limit flood losses [20]. Based on the consensus that the approach to water security should be global, organizations from all over the world started to follow paths and raise the hypotheses, following existing research on crises arising from the water resource, with a view to guaranteeing "meeting demand increasing of a limited resource in quantity in the territory and, many times, impacted in quality by human activities" [19, p. 72].

In this context, [21] highlights the importance of thinking about actions to ensure water security understood as the availability of an adequate quantity and quality of water for sustainable socioeconomic development, livelihoods, health, and ecosystems. Water security consists primarily of covering the essential needs of everyday life, food, and health: water for food production and improving agricultural yields; safe, clean water to help control waterborne diseases, which remain a major cause of death [22].

According to [23], the first intergovernmental declaration on water security was made in 2000, with the Ministerial Declaration of the Second World Water Forum. In Brazil, the concept of water security began to be formulated after the creation of the "Water Law" in 1997, which culminated in the National Water Agency in 2000. However, the concept of water security was consolidated through the National Water Security Plan, formulated based on the severe water crisis that hit Brazil's southeast. The water crisis affected Brazil, mainly between the years 2012 and 2017, caused mainly by factors such as urban agglomerations, incorrect use and occupation of the soil, increased water demands to supply various segments, and poor water infrastructure, associated with the scarcity of rainfall, among others. These factors brought to light the urgency in water treatment and the search for water management [5].

Brazil is one of the wealthiest countries in water resources, accounting for 12 to 14% of the world's freshwater. However, this resource is not evenly distributed since more than half of this resource is located in the Amazon watershed, a region with low population density. On the other hand, only 1.6% of the water is found in the state of São Paulo, where a quarter of the country's population lives [24].

There is no doubt that water security must be an absolute priority in Brazil, as it is closely linked to economic development and people's well-being [25], being essential for all sectors of society, economic and social, it can be concluded that it is the base of natural resources that the world depends on [23]. It is essential to highlight that although structural interventions are important, in the National Water Security Plan, there is a bias centered on the human being that goes in the opposite direction to the ecosystemic approach [26].

According to [23], achieving water security depends on maintaining the hydrological function cycle, implementing reliable engineering systems, formulating risk mitigation and awareness plans, and formulating legal interventions, reasonable policies, and effective management systems. For [22], it requires, above all, political will and courage to face issues and recognize the importance of healthy systems, as if they were the backbone of society. [22] complements by stating that investment in healthy systems is vital to obtain the strength and resilience to overcome water challenges.

Factors of disequilibrium in the water balance, together with the lack of planning, coordinated institutional actions, investments in water infrastructure and sanitation, lead to scenarios of Water Insecurity, instigating crises, such as those already faced by Brazil in the last decade [5].

Addressing water issues, [27] defends the importance of water security, defined as an acceptable level of risks related to water for humans and ecosystems, together with water availability in quality and quantity for the means subsistence. For [5], water security is considered an indispensable condition for social and economic development, especially when it is verified the impacts caused by the extreme hydrological events that occurred in the current decade in Brazil.

To explain the issue of water sustainability in the world, it is possible to see in Fig. 1 how freshwater is being used, checking its different uses in developed and developing countries.

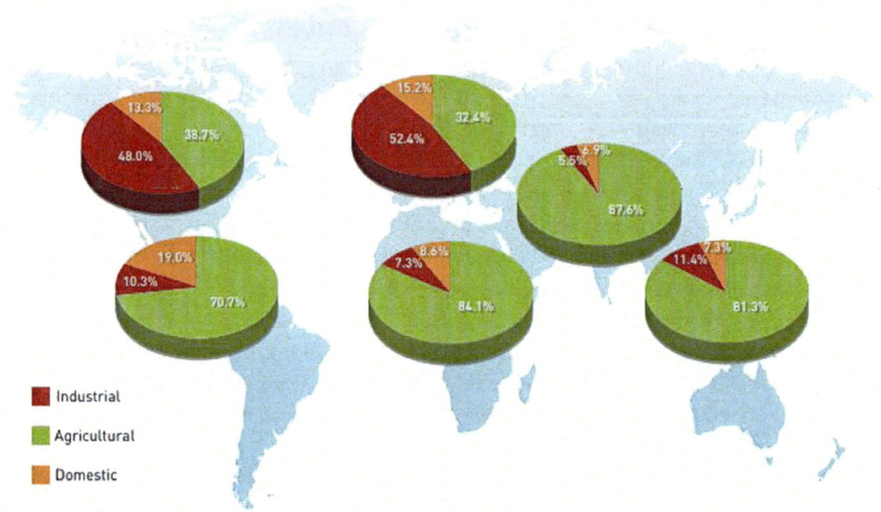

Fig. 1. Illustration of freshwater use subdivided by developed and developing countries. Source: [11].

It is noted that developing countries use freshwater, mostly for agricultural practices. The developed countries use water predominantly for the industrial sector.

In this sense, the need for developing countries to leverage advances for the improvement and implementation of agroecology is emphasized, as one of the ecological benefits of this practice is listed in the reduction of water and soil pollution and preservation of biodiversity [9]. The same authors also reinforce that agroecological practices contribute to the recovery of watersheds, as they reduce the dependence on external inputs and are energy saving. Furthermore, the objective of improving the material wealth of societies must be negotiated within the limits imposed by the availability and sustainability of the water resource and balanced with the cultural and spiritual values of water [20].

4.2 Sustainability Indicators

Taking into account the issue of municipalities, which in particular have estimates of population growth, and as consequence restrictions on water availability, means must be raised to reverse the growing trend of per capita use [28]. In order to improve water management, sustainability indicators were established with "communication as the main function, which should allow or promote the exchange of information on the subject they address" [29, p. 05].

According to [30], the term indicator comes from the Latin verb "indicate", which means to point or disclose, announce or make public, or even estimate or put a price. For

[31], the main objective of the indicator is to aggregate and quantify information so that its importance becomes more evident. According to the author, they simplify information about complex phenomena, thus trying to improve the communication process. The indicators point, indicate, approximate, translate the social dimensions of interest into operational terms, according to the previously defined theoretical or political choices [32].

The concept of water safety indicators, according to [26], is related to management measures, which aim to ensure sustainable access to water resources in qualitative and quantitative areas, guaranteeing and human well-being, the conservation of ecosystems, and socioeconomic development.

In order to improve water management, sustainability indicators were established having "communication as the main function, which should allow or promote the exchange of information on the subject they address" [29, p. 05].

The composition of environmental indicators, according to [29], follows three main objectives, which are: 1. provide information on environmental problems in order to allow policymakers to value seriousness; 2. support in the development of policies, establishing priorities, and identifying the main factors that pressure the environment; and finally, 3. monitor the effects of policy responses.

Among the environmental indicators, we can mention the intensity of use of forest resources and water resources [33]. Another example is the river water quality, presented by two parameters (oxygen and nitrate content) for selected rivers [17]. In addition, the OECD proposes physical input and output tables, tracking production, and transforming the use of each resource across the economy, which provides an analytical tool to assess the impact of sectoral economic activity on the resource [33]. The data are shown for representative locations river mouth or at the downstream border, providing a summary view of the pollution load and cleaning efforts in the upstream watershed [17].

As basic relations for the calculation of indicators of the sustainable use of natural resource quantities, the flow of these resources is taken into account [33]. It is also interesting to relate some indicators to an optimal national connection rate, considering national specificities, such as the population in remote areas [17]. As examples, one can ascertain supplementary information, sewage connection rates, and public expenditures on wastewater treatment [17, p. 41].

In short, to monitor water resources, it is necessary to ascertain: 1. Pressures/Intensity of use of water resources, considering the captures/available resources; 2. Conditions/Frequency, taking into account the duration and extent of water scarcity; 3. Answers/Prices of water, giving importance to the tariffs of users for sewage treatment [33].

5 Results

5.1 Characterization of the Milton Santos Settlement

The Milton Santos Settlement started formally in 2006, where INCRA classified it as PDS (Sustainable Development Projects) [34]. In Brazil, PDS's started in the 2000s and are configured in settlement projects established to develop environmentally differentiated activities, targeting traditional populations such as riverine people, extractive communities, among others [35]. "Enrolling in this PDS modality, the Milton Santos Settlement houses 68 families divided into one-hectare lots" [6, p. 498] such that "on average, each unit is occupied by 3.2 people, which makes it possible to estimate that approximately 220 people live in the settlement" [36, p. 03]. However, other families arrived in the surroundings of the settlement, constituting the local dynamism. Consequently, many other problems were generated, such as the lack of water, services such as health and garbage collection etc., once the public power does not adequately assist these families [37]. As [38] clarify, in the PDS, the individual land is added to the collective land, subject to numerous restrictions, however with promising opportunities.

The Milton Santos Settlement belongs to the Piracicaba, Capivari and Jundiaí Watersheds complex, more specifically the aquifer recharge area of the Salto Grande dam, which supplies the cities of Americana and Sumaré. However, this dam is characterized by polluted waters, which justifies the great interest in the ecological protection of this region [39].

Figure 2 highlights the Milton Santos Settlement location, close to the cities of Americana and Cosmópolis. The Jaguari River belonging to the Piracicaba, Capivari, and Jundiaí Watersheds, supplies the settlement. Figure 2 highlights Usina Ester, an organization that stood out due to the occurrence of some conflicts with the settlement. In the surroundings of the settlement, sugar cane monoculture is predominant.

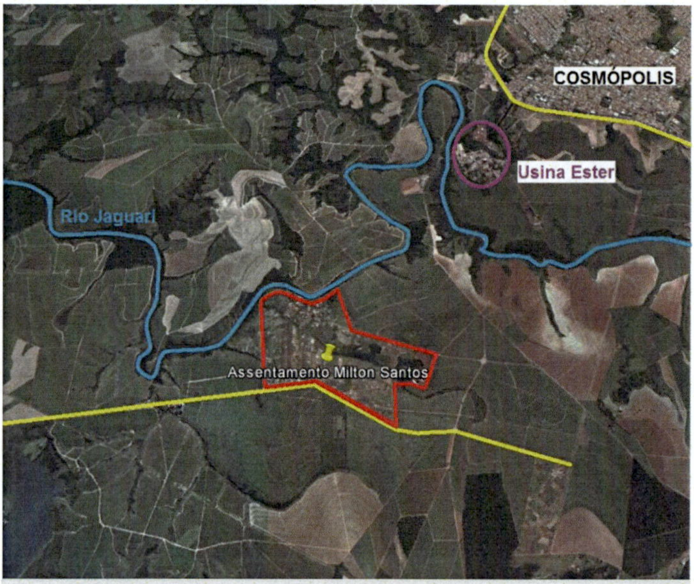

Fig. 2. Location of Milton Santos Settlement. Source: [40].

[36] found that in 2010, at the Milton Santos Settlement, there were 70 settled families, with family farming as the main source of income and practicing on average seven different crops in each lot, with manioc being the predominant culture.

As an economic practice, the Milton Santos Settlement is a local pole in food production, mainly vegetable gardens, eggs, and free-range chickens. However, not all settlers develop significant agricultural production, and some of them do not use this practice [6]. Mostly the agricultural activity is practiced by the members who have fewer chances of insertion in the traditional labor market, and it allows a job occupation that values these individuals. In addition, there is the complementation of sufficient income for these families to be located above the line of poverty [41].

A survey carried out in 2012 showed that agricultural production is the primary source of income for 51% of households, compared to 45% as a second source of income, in addition to contributing significantly to the food security of the place [41].

Despite being the main agricultural activity, the production of vegetables is limited by the lack of water in the plots, where the limiting factors for the practice of agriculture, according to 56% of the settlers is access to water, and soil fertility is considered low by 21% of settlers [41]. Water is the most significant limiting factor in agricultural activity, according to settlers. However, factors considered non-limiting in terms of land use include terrain topography, soil depth, and water infiltration into the soil [41]. The biggest impasses regarding the development and expansion of agricultural activities are in the access to water for irrigation and difficulty in accessing rural credit lines, aggravated by low soil fertility, technical assistance, and poor infrastructure [41].

Despite the adversities, the Milton Santos Settlement has good conditions in developing pluriactivity due to the urban labor market being close, which also facilitates the flow of agricultural products. Besides, families declare that 30% of their food is produced on the farm itself, leveraging the issue of food security for families [41].

In view of the above, as an alternative for some settled families, there was the implementation of Agroecological Systems, which is considered one of the most sustainable practices in all aspects and has the purpose of rescuing the human dignity of farmers, who, over time have domesticated plants and animals, maintaining much of the genetic diversity used by humans [9]. Thus, the settlers believe that these agroecological practices are much less aggressive to the environment than monocultures, such as sugar cane crops around the settlement, as there is the use of many chemical, agricultural inputs, which causes contamination of the soil and water sources [39].

In this sense, it aims to emphasize the importance of encouraging agroecological practices, based on the development of incentive policies, which are not only social policies but going into the question of land reform and other measures such as "support for family farming appear as important leverages to the development strategy" [8, p. 126]. To ascertain the satisfaction of a community, "one can raise the question of what is the minimum number of characteristics that would be sufficient to describe the dynamics of natural systems" [42, p. 40], understanding that in terms of sustainability, there are multiple dimensions of well-being and prosperity, in addition to material wealth [20].

Local measures overcome the difficulty with the availability of water in the Milton Santos settlement in order to capture and store water for the elementary activities of the settlers. Wells are drilled in the settlement, but water analysis shows contamination with heavy metals, fluorine, and fecal coliforms in one of these wells. There are cisterns in some of the settlers' houses. In addition, there is a water stream that cuts through the settlement, but it is worth mentioning that although this stream has organic pollution, it is used for the irrigation of a community garden [36].

The research carried out by [36] identified that the water supply in the settlers' homes is done through two water tanks and, due to their insufficiency, often compromises household consumption and/or irrigation.

5.2 Selection of Indicators (DPSIR Model)

Considering what was covered in the topic on sustainability indicators, it will discuss one of the most used models for systematic environmental impasses, which is the Sustainability Indicators System Driving Force-Pressure-State-Impact-Response (DPSIR, Driver-Pressure-State-Impact-Response). To better understand this nomenclature, created by Anthony Friend and David Rapport in 1979, by Statistics Canada, the PSR model was based on the SR model (Stress-Response) [13]. But, according to [29], to understand the DPSIR dynamics, it is equally useful to understand the relationship between some of the acronyms: compatibility for economic activities is in the relationship between 'D' and 'P', which is linked to an eco-efficiency function of the technology and related systems in use, that is, if eco-efficiency is improving there will be less 'P', and more 'D' [29]. Similarly, the relationship between Impacts on human beings or ecosystems occurs, and the 'S' depends on the load capacities and thresholds for these systems [29, p. 05].

Figure 3 highlights how to assess the state of the environment, together with the factors that describe the pressures on it and shape its current state and the possible responses to face the various problems.

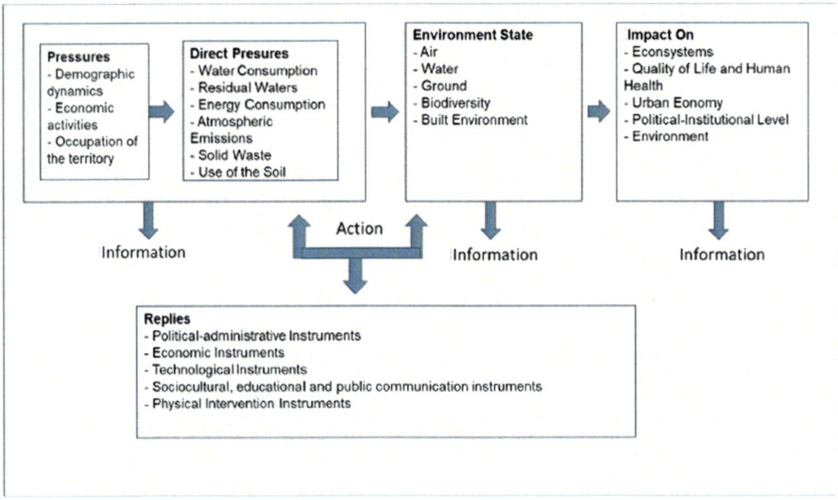

Fig. 3. PEIR matrix and the interactions of urban-environmental components. Source: [43, p. 16].

The DPSIR structure is useful for describing the relationships between the origins and the consequences of environmental problems [29] and may apply to water resources management. In addition, the importance and need to ascertain and explain the applicability of the DPSIR Sustainability Indicators System to improve the management of water resources, such as risk prevention and improvement in decision making, are noted.

6 Final Considerations

Periurban settlements can be seen as a restructuring of rural spaces close to cities as they are growing. In this way, there is a need to leverage studies from surveying and systematized observation of these communities under various aspects, such as social, ecological, and economic.

The present study showed the possibility of the excluded from society permeating a more dignified life in the face of their social and economic ideals. But for this, there is a need to expand research, especially concerning water management, as there is instability of these resources in periurban settlements. In addition, as they have agroecology as their main source of income, which demands the need for water for irrigation, these settlements need support through public and scientifically based policies so that water insecurity does not occur, and, as a consequence, food and economic insecurity for the settlers.

Intensifying studies on sustainability indicators for a specific application in periurban settlements can benefit management and decision making regarding the use of water resources, as well as the use of land in these communities. These studies may also enable the development of communities similar to the settlements due to their capacity to expand, adapt, and replicate the application of water sustainability indicators and other activities.

Acknowledgments. This study was partially funded by the Coordination for the Improvement of Higher Education Personnel - Brazil (CAPES) - Finance Code 001.

References

1. Boff L (2012) Sustentabilidade: O que é – O que não é. Ed. Vozes, Petrópolis, RJ
2. Veiga JE (2010) Desenvolvimento Sustentável, o desafio do século XXI. Editora Garamond, Rio de Janeiro
3. Jacobi P (2003) Educação ambiental, cidadania e sustentabilidade. Cadernos Pesquisa 118:189–205
4. UN United Nations (2018) World urbanization prospects. https://population.un.org/wup/General/DefinitionIssues.aspx/
5. ANA Agência Nacional de Águas (2019) Plano Nacional de Segurança Hídrica. ANA, Brasília. https://arquivos.ana.gov.br/pnsh/pnsh.pdf
6. Retière M, Marques PEM (2019) A justiça ecológica em processos de reconfiguração do rural: estudo de casos de neorrurais no estado de São Paulo. Rev Econ Sociol Rural 57(3):490–503. https://doi.org/10.1590/1806-9479.2019.184109
7. Schiochet VA (2020) Construção de uma Economia Solidária para superar a crise. Análise: Friedrich Ebert Stiftung Brasil. São Paulo, pp 1–3. https://library.fes.de/pdf-files/bueros/brasilien/16369.pdf

8. Sachs I (2008) Desenvolvimento includente, sustentável, sustentado. Garamond, Rio de Janeiro
9. Nodari RO, Guerra MP (2015) A agroecologia: estratégias de pesquisa e valores. Estudos Avançados 29(83):183–207. https://doi.org/10.1590/s0103-40142015000100010
10. Campos MVCV, Ribeiro MMR, Vieira ZCL (2013) A gestão de recursos hídricos subsidiada pelo uso de indicadores de sustentabilidade. Rev Brasil Recurs 19(2):209–222
11. FAO Food and Agriculture Organization of the United Nations (2020) Water at a Glance. The relationship between water, agriculture, food security and poverty. https://www.fao.org/3/ap505e/ap505e.pdf
12. Maluf RS (2020) Alimentação e Fome: agir na emergência e construir outro país. Friedrich Ebert Stiftung Brasil. São Paulo, pp 1–3. https://library.fes.de/pdf-files/bueros/brasilien/16162.pdf
13. Felinto CMR, Ribeiro MMR, Braga CFC (2019) Aplicação do modelo força motriz-pressão-estado-impacto-resposta (FPEIR) para gestão dos recursos hídricos em joão pessoa-PB. Rev DAE 67(218):118–136. https://doi.org/10.4322/dae.2019.038
14. Gil AC (2002) Como elaborar projeto de pesquisa, 4th edn. Atlas, São Paulo
15. Severino AJ (2007) Metodologia do trabalho científico, 23rd edn. Cortez, São Paulo
16. Sun S, Wang Y, Liu J, Cai H, Wu P, Geng Q, Xu L (2016) Sustainability assessment of regional water resources under the DPSIR framework. J Hydrol 532:140–148
17. OCDE Organization for Economic Co-operation and Development (2001) Environmental indicators: towards sustainable development. OCDE Publishing. https://www.OCDE.org/site/worldforum/33703867.pdf
18. UNESCO United Nations Educational Scientific and Cultural Organization (2020) Water Security. https://en.unesco.org/themes/water-security
19. Melo MC, Johnsson RMF (2017) O conceito emergente de segurança hídrica. Sustentare 1(1):72–92. https://periodicos.unincor.br/index.php/sustentare/article/view/4325/pdf_5
20. GWP/OCDE Global Water Partnership/Organization for Economic Co-operation and Development (2015) Securing water, sustaining growth. https://www.gwp.org/globalassets/global/about-gwp/publications/the-global-dialogue/securing-water-sustaining-growth.pdf
21. WWC World Water Council (2020) Water and Climate Change. https://www.worldwatercouncil.org/en/water-and-climate-change
22. WWC World Water Council (2013) A pact for a water security world – 2013 - 2015 Strategy. https://www.worldwatercouncil.org/sites/default/files/Offcial_docs/WWC_Strategy_2013-2015_-A_Pact_for_Water_Security.pdf
23. INPE Instituto Nacional de Pesquisas Espaciais (2012) Segurança hídrica para um planeta sob pressão: transição para a sustentabilidade: desafios interligados e soluções. Transição para a sustentabilidade: desafios interligados e soluções. The International Project Office of the Global Water System Project. https://www3.inpe.br/igbp/arquivos/Water_FINAL_LR-portugues.pdf
24. The World Bank (2018) Recovering water: a results-based approach to water supply and sanitation in Brazil's São Paulo State. https://www.worldbank.org/en/results/2018/05/07/approach-to-water-supply-and-sanitation-brazil-sao-paulo-state
25. CEBDS Conselho Empresarial Brasileiro pelo Desenvolvimento Sustentável (2015) Gerenciamento de Riscos Hídricos no Brasil e o setor empresarial: desafios e oportunidades
26. Machado FH (2018) Proposição de Indicadores de Segurança Hídrica: Seleção, Validação e Aplicação na Bacia Hidrográfica do Rio Jundiaí-Mirim, Jundiaí – SP, Brasil. 255 fls. Tese (Doutorado) - Curso de Ciências Ambientais, Universidade Estadual Paulista "Júlio de Mesquita Filho", Sorocaba. https://hdl.handle.net/11449/153669
27. Bakker K (2012) Water security: research challenges and opportunities. Science 337(6097):914–915

28. SNIS Sistema Nacional de Informações sobre Saneamento (2018) Diagnóstico dos serviços de água e esgoto. https://www.snis.gov.br/diagnostico-anual-agua-e-esgotos/diagnostico-dos-servicos-de-agua-e-esgotos-2018
29. EEA European Environment Agency (1999) Environmental indicators: typology and overview. Technical report no. 25. Luxembourg, Office for Official Publications of the European Communities
30. Hammond A, Adriaanse A, Rodenburg E, Bryant D, Woodward R (1995) Environmental indicators: a systematic approach to measuring and reporting on environmental policy performance in the context of sustainable development. World Resources Institute, Washington, D.C. https://pdf.wri.org/environmentalindicators_bw.pdf
31. Van Bellen HM (2004) Indicadores de sustentabilidade: um levantamento dos principais sistemas de avaliação. Cadernos EBAPE.BR II(1):01–14
32. Jannuzzi PM (2014) Indicadores para diagnóstico, monitoramento e avaliação de programas sociais no Brasil. Rev Serv Público – Brasília 56(2):37–160
33. OCDE Organization for Economic Co-operation and Development (2003) OECD environmental indicators: development, measurement and use. Reference Paper, Paris. https://www.oecd.org/environment/indicators-modelling-outlooks/24993546.pdf
34. INCRA Instituto Nacional de Colonização e Reforma Agrária (2017) Painel de Assentamentos: informações gerais. Incra, Brasília
35. INCRA Instituto Nacional de Colonização e Reforma Agrária (2020) Modalidades de Projetos criados pelo Incra. Incra, Brasília. https://www.incra.gov.br/pt/?option=com_content&view=article&id=142
36. Marques PEM, Oliveira, LJ, Nemoto E, Antier C (2010) Condições de trabalho e planos de reconstrução de vida: o caso do assentamento Milton Santos em Americana/SP. Anais do IV Simpósio sobre Reforma Agrária e Assentamentos Rurais: controvérsias e alternativas de desenvolvimento. UNIARA, Araraquara
37. Silva RA (2015) Jovens do Assentamento Milton Santos: Sentidos do trabalho e da educação, Brasil. 159 fls. Dissertação (Mestrado) - Curso de Educação, Universidade Federal de São Carlos – UFSCAR Sorocaba
38. Mendes JF, Porro NSM (2015) Social conflicts in times of environmentalism: living law rights to land in settlements with a conservationist focus. Ambiente Soc XVIII(2):93–110
39. Marques PEM, Gaspari L, Almeida B (2017) Organização de Controle Social (OCS) e engajamento agroecológico das famílias do assentamento Milton Santos no estado de São Paulo. Estud Soc Agric 25(3):545–560
40. Colégio Santa Clara (2015) Assentamento Milton Santos: slides para disciplina de geografia. https://www.colegio-santaclara.com.br/disciplinas/Geo/Assentamento_Milton_Santos_8oano_2015.pdf
41. Gaspari LC, Khatounian CA, Marques PEM (2018) O papel da agricultura entre as famílias pluriativas assentadas em região metropolitana: o caso do Assentamento Milton Santos em Americana e Cosmópolis/SP. Rev NERA, article no 21, (41):85–101
42. Rapport D, Friends A (1979) Towards a comprehensive framework for environmental statistics: a stress-response approach. [S. l.]: Statistics Canada. https://projects.eionet.europa.eu/leac/library/background-papers-and-publications/rapport-friend_stress-response_statcanada_11-5101979.pdf/download/en/1/Rapport-Friend_Stress-Response_StatCanada_11-5101979.pdf?action=view
43. PNUMA Programa das Nações Unidas para o Meio Ambiente (2004) Metodologia para a elaboração de Relatórios GEO Cidades. Manual de Aplicação. Versão 2. Programa das Nações Unidas para o Meio Ambiente Escritório Regional para a América Latina e o Caribe. PNUMA, Rio de Janeiro

Phytoextraction of Heavy Metals from the Soil of Aurora-Patricia Mining Environmental Liability by Herbaceous Species *Carex mandoniana, Equisetum bogotense,* and *Muehlenbeckia tamnifolia*, La Encañada-Peru 2020

Cristina Carmela López Rodríguez[1(✉)], Carlita Roxana Izquierdo Ramírez[1], Leoncio Jaime Lanfranco Colina[1], Jackelin Estefani Ciriaco Mosqueira[1], Marieta Eliana Cervantes Peralta[1], and Marco Alfredo Sánchez Peña[1,2]

[1] Ingeniería Ambiental, Universidad Privada del Norte, Av. Vía de Evitamiento Norte s/n cuadra 15, Cajamarca, Peru
{marieta.cervantes,marco.sanchez}@upn.pe
[2] Facultad de Ciencias de la Salud, Departamento Académico de Ciencia Biológicas, Universidad Nacional de Cajamarca, Av. Atahualpa Km. 3, Cajamarca, Peru

Abstract. The Cajamarca region has a large number of mining environmental liabilities; one of them is the former mining company Aurora - Patricia, which operated in the small town of Polloc. Since 1985 its operations have been stopped. Based on this, we determined the phytoextraction capacity of 13 heavy metals in three herbaceous species that grow in this place with the highest rate of importance value. To do this, we applied the stratified random sampling and square meter methods, where it was possible to identify 42 herbaceous species in 10 plots, of which *Equisetum bogotense*, *Muehlenbeckia tamnifolia* and *Carex mandoniana* were chosen. The concentration of heavy metals was analyzed from both contaminated (environmental liability) and non-contaminated (around the liability) soils of La Encañada district, as well as in herbaceous species to compare them. The highest correlation between variables was obtained from *Carex mandoniana* species (0.995), followed by *Equisetum bogotense* (0.975) and, on the other hand, *Muehlenbeckia tamnifolia* presented a negative inverse relationship. The Bioconcentration Factor was calculated where it turned out that the herbaceous species are exclusive; however, the Translocation Factor showed that *Equisetum bogotense* and *Carex mandoniana* are phyto-stabilizers while *Muehlenbeckia tamnifolia* is hyper-accumulative of heavy metals.

Keywords: Phytoextraction · Mining environmental liability · Heavy metals · Bioconcentration factor · Translocation factor · Phyto-stabilization · Hyperaccumulation

Y. Iano et al. (Eds.): BTSym 2020, SIST 233, pp. 100–109, 2021.
https://doi.org/10.1007/978-3-030-75680-2_13

1 Introduction

Mining causes a series of conflicts, the most repeated of them are non-compliance with environmental obligations derived from legislation and instruments, violation of population's fundamental rights, and health damage by direct and indirect contact with the impacted areas [1]. The costs of not facing the environmental impact from the beginning of a mining project can be highly damaging if appropriate technology is not applied to prevent pollution or a regulatory framework that works correctly [2]. Environmental liabilities are defined as: "those facilities, effluents, emissions, remains, or deposits of waste produced by mining operations, which are currently abandoned or inactive and are a permanent and potential risk to population's health, surrounding ecosystem and property" [3].

In Peru, legislation before 1993 did not consider environmental regulations. This produced great pollution in areas close to mining operations, especially in soils; as a result, there are currently 8,448 mining environmental liabilities with 4,353 categorized as high risk [4]. The surviving native species in contaminated soils with metals have developed special physiological mechanisms to tolerate their presence because they live in very limited areas and can be directly affected by mining activities. Conservation of these species is very important due to their use in environmental phytotechnologies of revegetation, phytostabilization, and phytoextraction of contaminated environments [5].

The danger of heavy metals lies in the fact that they can be degraded neither chemically nor biologically. Also, they tend to be bio-accumulated and bio-magnified, causing toxic effects of different kinds [6]. Phytoextraction or phytoaccumulation consists of the absorption of contaminating metals by roots of plants and their accumulation in stems and leaves [7]. This work determines the importance value index of the herbaceous species that grow in Aurora-Patricia mining environmental liability and evaluates the capacity of phytoextraction heavy metals in three species *Carex mandoniana*, *Equisetum bogotense,* and *Muehlenbeckia tamnifolia,* in stem, root, and leaf; obtaining the correlation between variables, Bioconcentration Factor and Translocation Factor.

2 Materials and Methods

2.1 Study Area

Aurora mining concession, previously in charge of Minera Monterrico S.A, which was a Cajamarca's city company, operated from approximately 1960 to 1980 under the charge of Lanfranco family, in the small town of Polloc, La Encañada district, at an altitude of 3,150 and 3,350 m above sea level, with a warm and humid climate, it was a disseminated polymetallic artisanal mining (gold, silver, and copper), in which approximately 20 tons of mineral were daily extracted. It had a sinkhole of 480 m and 700 m, which was distributed in 8 galleries and chimneys, and the mining concession covered a total area of 1000.37 Ha. The inhabitants know that exploitation has occurred since 1905 or perhaps before in this area, and it was called "El Gallo" sinkhole [8]. This area is currently uninhabited; however, productive activities from inhabitants such as ranching and agriculture are very close to their remains. A water stream passes through a completely collapsed sink, but it is exposed to the environment, dragging all the material. This situation is aggravated during the rainy season because there is a landslide of the sinkhole and the deposit of the leveling. It directly affects the entire surrounding ecosystem and, indirectly, people around the place.

2.2 Data Collection, Sampling, and Analysis

The current research is non-experimental in which phenomena are observed as they occur in their natural context to be posteriorly analyzed. Basic data is quantitative, and the design is cross-sectional or transactional in the correlational - causal classification. This was made to describe the existing relationship of concentration of heavy metals between the soil and the herbaceous species with the highest value of importance that thrives in the mining environmental liability [9].

The topographic survey method used in the study area was the planimetric method. Pearson's correlation coefficient was calculated to determine the degree of dependence and association between the variables. The method used for collecting herbaceous species was the square meter, which consists of small plots of fixed dimensions, where the sample units are formed of a 1m × 1m square, separated every 10 cm from side to side, forming inside 100 squares in total. The data was recorded on field cards. A botanical press prototype was created to identify species by a specialist. To determine the importance value index, the Aleph Population System program (a program developed by the science department of the National University of Cajamarca) was used.

For herbaceous species, we used the "Flora and Vegetation Inventory Guide" [10], and stratified random sampling was used to collect qualitative and quantitative information in small representative areas. For the soil, we used the "Guide for Soil Sampling" [11], which indicates that you must require up to a depth of approximately one meter for taking surface samples. We performed composite samples using the quartering method. A total of 16 samples were collected, in which 14 of them correspond to herbaceous species with the highest value of importance, and the minimum quantity of 200 g was taken for root, stem, and leaves matter. About 500 g to 1 kg were extracted from the soil in 2 areas, both from the impacted and non-impacted areas. Finally, 13 metals - Al, Cr, As, Ba, Cd, Hg, Ni, Pb, Cu, Fe, Mg, Mn, and Zn - were analyzed.

Chemical analyses were carried out in the "Servicios Analíticos Generales SAC" laboratory, which has ISO/IEC - 1702 accreditation before INACAL - DA (National Accreditation Organization) and IAS (International Accreditation Service), under the NTP - ISO/IEC 17025: 2006, with registration N° LE - 047. The methods used to analyze collected samples were: Soil, Method 200.7 Rev.4.4 EMMC. Version (1994) [12]; for the plant tissue EPA Method 200.3 and 200.7 [13].

3 Results and Discussion

In our study area of 3679.98 m^2, 42 herbaceous species have been identified, in total 1215 individuals in the 10 plots with a density of 121.5, where three species with the highest rate of importance value, whose value indicates that they have adapted better and dominated the community, can be seen in Fig. 1. These are *Equisetum bogotense* or by its common name "Cola de Caballo" with 55.27, *Muehlenbeckia tamnifolia* or "Bejuco colorado", "Anguvena" with 34.91, and *Carex mandoniana* or "Canela" with 32.93.

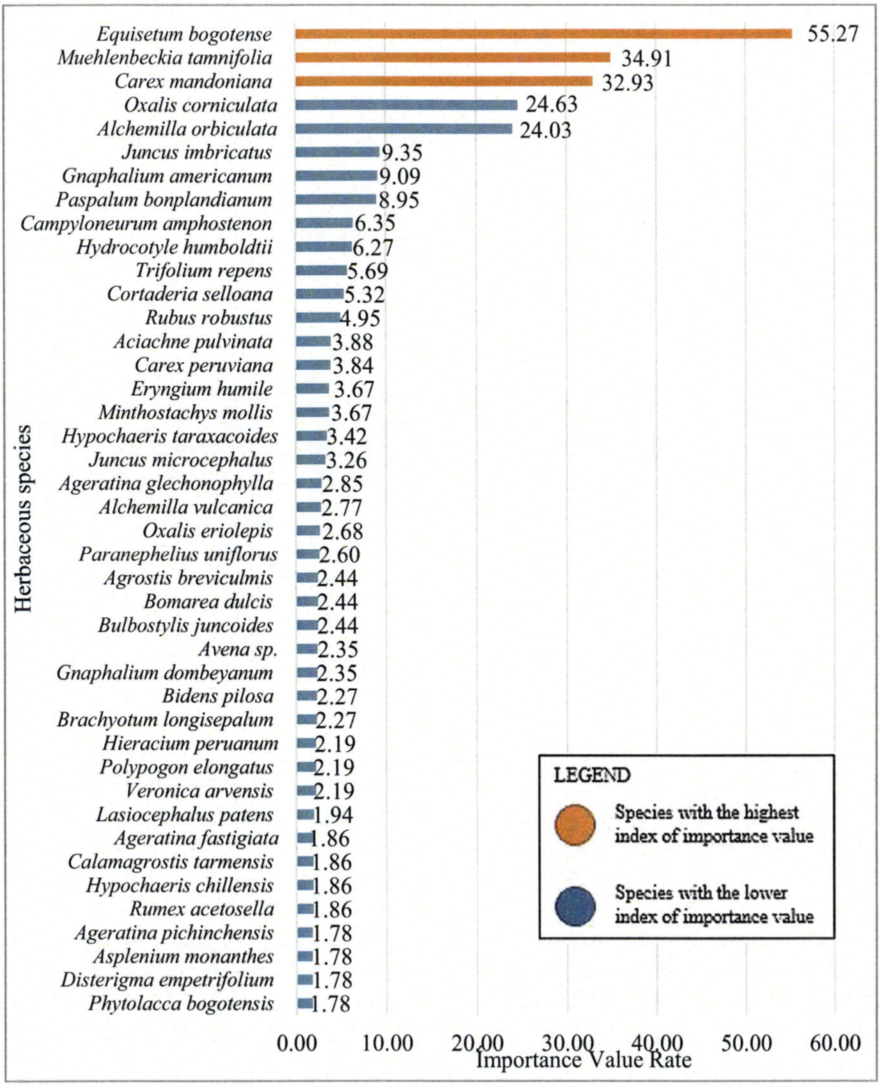

Fig. 1. Importance value of herbaceous species in the mining environmental liability.

In Table 1, we can see from the soil samples that 11 out of the 13 heavy metals (As, Ba, Cd, Hg, Ni, Pb, Cu, Fe, Mg, Mn, and Zn) in the liability soil exceed in concentration those of the uncontaminated soil and it only happens differently with aluminum and chrome. According to the Environmental Quality Standards for soil, 2017 [14], the arsenic, cadmium, and lead concentrations of the soil sample of the liability surpass the values of regulation, and, in the case of Barium and Mercury, the two samples of both the environmental liability and the uncontaminated soil are well under what regulation determines.

Table 1. Concentration of heavy metals in the soil of the environmental liability and the uncontaminated soil.

Metal (mg.kg^{-1})	Environmental liability soil	Uncontaminated soil
Aluminum (Al)	5617.1	7808.2
Arsenic (As)	367.5	4.8
Barium (Ba)	109.3	16.1
Cadmium (Cd)	4.31	1.63
Chromium (Cr)	2.07	7.76
Copper (Cu)	94.5	20.6
Iron (Fe)	>20000	10732.7
Mercury (Hg)	0.8	<0.1
Magnesium (Mg)	544	382.9
Manganese (Mn)	414.17	134.86
Nickel (Ni)	2.59	2.34
Lead (Pb)	118.73	7.52
Zinc (Zn)	147.1	22.7

Each metal and each contaminating chemical element have a mechanism of action and a preferred accumulation site [15]. The means to metabolize these compounds for plants are sulfate conjugation and glutathione conjugation [16]. Some plants act as bio-indicators of heavy metal presence by selectively retaining it, while others behave as bio-accumulators by tolerating its presence and accumulating it in their structure, giving manner to bioconcentration through the trophic chain [17].

In Table 2, we identified that the highest accumulation of heavy metals in mg.kg^{-1}of the *Equisetum bogotense* species is in the root. These concentrations are Al (381.1), As (11.3), Cd (0.34), Cr (0.48), Cu (31.6), Fe (1002.4), Mg (121.2), Ni (1.37), and Pb (7.14). In the case of species *Carex mandoniana*, most of the metals are accumulated in its root, whose concentrations are Al (480.1), As (38.6), Ba (7.2), Cd (0.41), Cr (0.58), Cu (22.5), Fe (2071.6), Ni (0.58), Pb (10.23). In *Muehlenbeckia tamnifolia,* it is observed that it accumulates metals in both the stem and the leaf in low quantities, but a little more in the stems which concentrations are Ba (0.9), Hg (0.2), Mn (37.34), Pb (0.16), Zn (4.9).

The physiological alterations found by mercury and cadmium in plants (Medicago sativa) were: growth inhibition, the formation of hydrogen peroxide, lipid peroxidation, changes in the concentrations of GSH and homoglutathione (hGHS), the activity of antioxidant enzymes, and cell death [18]. In other words, free radicals (ROS) have been generated when the species are exposed to this type of metal. In the case of the three herbaceous species extracted from environmental liabilities, a difference in size was observed with those extracted from the uncontaminated soil as they were smaller than the others. Electron leakage from respiratory chain complexes and subsequent ROS formation can cause damage to specific complex subunits and contribute to the long-term accumulation of mitochondrial dysfunction [19].

A balance between production and removal of ROS allows a normal cell function, while an imbalance causes oxidative stress with pathological consequences [20]. Therefore, species that generate free radicals are going to have problems at different scales of their tissues. However, species develop antioxidant substances as defense [21]. This means some species adapt when they are exposed to heavy metals, such as those analyzed in the environmental liability.

There is a very strong relationship between plants with antioxidant capacity and their anti-inflammatory activity [22]. For example, the species *Equisetum bogotense* or "Cola de Caballo" by its common name is known for its medicinal properties. It is used as a hemostatic, antiseptic, infused against bladder and kidney stones. It is also a diuretic and has antitumor activity in leukemia, astringent in dysentery, and was formerly used to clean silver [23, 24].

The results of Bioconcentration Factor (BCF) in the case of the quotient of herbaceous species and the soil of the environmental liability are less than 1 for both root BCF and aerial BCF in all species; this means that the plants are exclusive. However, for the quotients of the herbaceous species with the uncontaminated soil, *Carex mandoniana* has a bio-concentration factor greater than 1 in both aerial part and the root; therefore, this species accumulates Barium but under natural conditions.

In the case of Translocation Factor, ratios resulting from the aerial part between the root of the herbaceous species extracted from the Environmental Liability give us the following: *Equisetum bogotense* effectively transfers metals such as Ba, Hg, Mn, and Zn, being hyper-accumulative of these in the aerial part; however, the other nine metals give less than one, which means that it is phytostabilizer for Al, As, Cd, Cr, Cu, Fe, Mg, Ni, and Pb. The species *Carex mandoniana* is a hyperaccumulator of Mg, Mn, Zn, and is a phytostabilizer for Al, As, Ba, Cd, Cr, Cu, Fe, Hg, Ni, and Pb. For the species *Muehlenbeckia tamnifolia*, there is a hyperaccumulation of Al, Ba, Cr, Cu, Fe, Hg, Mn, Mg, Pb, and Zn, and the other metals such as As, Cd, and Ni are not being accumulated. This means that there are more metals in its aerial part than at its root. In both cases, the environmental liability and uncontaminated soils, *Equisetum bogotense, and Carex mandoniana* are more phytostabilizers, and *Muehlenbeckia tamnifolia* is hyper-accumulative.

Table 3 shows the correlation between the herbaceous species and the soil of environmental liability. It is found that the one with the greatest relationship between these variables is the *Carex mandoniana* species (0.995), and it is followed by *Equisetum bogotense* species (0.975). On the other hand, *Muehlenbeckia tamnifolia* species has a negative inverse relationship with −0.005; the same happens in the relationship with the uncontaminated soil but with coefficients with lower values. In other words, the values of metals of herbaceous species of the environmental liability are more related to the concentration in the soil of the environmental liability.

Table 2. Concentration of heavy metals in herbaceous species extracted from Aurora-Patricia Environmental Liability and those extracted from uncontaminated soil

Species	Metal ($mg.kg^{-1}$)	Environmental liability			Uncontaminated		
		Stem	Root	Leaf	Stem	Root	Leaf
Equisetum bogotense	Al	182	381.1	–	16.2	53.1	–
	As	4	11.3	–	<0.1	<0.1	–
	Ba	3.6	2.9	–	11.4	8	–
	Cd	0.26	0.34	–	<0.05	<0.05	–
	Cr	0.21	0.48	–	0.19	0.48	–
	Cu	13.9	31.6	–	4.8	3.2	–
	Fe	301.4	1002	–	25.8	50.2	–
	Hg	0.1	0.1	–	<0.1	<0.1	–
	Mg	114.9	121.2	–	132.9	166.5	–
	Mn	120	67.58	–	9.83	11.15	–
	Ni	0.76	1.37	–	<0.07	0.11	–
	Pb	3.64	7.14	–	0.09	0.17	–
	Zn	82.8	47.7	–	5.6	3.6	–
Carex mandoniana	Al	422.4	480.1	–	45.5	178.8	–
	As	16.6	38.6	–	<0.1	<0.1	–
	Ba	4	7.2	–	20.9	22.1	–
	Cd	0.37	0.41	–	0.05	0.1	–
	Cr	0.47	0.58	–	0.39	0.31	–
	Cu	13.8	22.5	–	2.1	5.7	–
	Fe	1019	2072	–	49.3	405.9	–
	Hg	<0.1	0.1	–	0.2	<0.1	–
	Mg	258.9	114	–	211.7	139.4	–
	Mn	157	71.23	–	126	59.66	–
	Ni	0.57	0.58	–	<0.07	0.16	–
	Pb	6.69	10.23	–	<0.06	0.26	–
	Zn	34.8	22.2	–	4.3	6.4	–
Muehlenbeckia tamnifolia	Al	7.8	6.8	14.1	6.9	6.8	55.7
	As	<0.1	<0.1	0.2	0.3	0.4	1
	Ba	0.9	0.4	0.5	2.1	1.1	2.7
	Cd	<0.05	<0.05	<0.05	0.13	0.29	0.35

(*continued*)

Table 2. *(continued)*

Species	Metal (mg.kg^{-1})	Environmental liability			Uncontaminated		
		Stem	Root	Leaf	Stem	Root	Leaf
	Cr	0.23	0.31	0.12	0.26	0.53	0.08
	Cu	4.7	1.7	5.6	7.1	6.1	8.4
	Fe	21.9	12.3	32.9	34.5	34.1	89.1
	Hg	0.2	0.1	<0.1	<0.1	<0.1	<0.1
	Mg	183	337.1	281.3	126.8	249.6	451.2
	Mn	37.34	1.47	3.91	110.1	25.58	750.9
	Ni	<0.07	<0.07	<0.07	0.27	0.13	0.38
	Pb	0.16	<0.06	0.08	0.37	<0.06	1.2
	Zn	4.9	1	2.2	12.9	8.3	38

In Table 3, the sign " +" means that it is direct positive, indicating there is a relationship between variables, and the sign "−" means that it is inverse negative, indicating that there is no relationship between the variables; for the correlations between metals in plants with metals in the soil.

Table 3. Coefficients of correlation between the concentrations of metals in herbaceous species and from the soil, in the environmental liability (EL) and in the uncontaminated (UC)

Herbaceous species	EL coefficient	UC coefficient
Equisetum bogotense	+0.975	+0.211
Carex mandoniana	+0.995	+0.748
Muehlenbeckia tamnifolia	−0.005	−0.024

4 Conclusions and Recommendations

We determined the phytoextraction capacity of heavy metals on the soil of Aurora-Patricia environmental liability by herbaceous species with the highest value of importance. We noticed that most metals are found in greater quantity in species extracted from the environmental liability, and this dependence has also been seen in the correlation coefficients. However, the Bioconcentration Factor of these species does not show that they are bioaccumulators, but the Translocation Factor shows that heavy metals do have a relationship between their plant tissues. This means that they have adapted to the environment and have generated antioxidant compounds.

It is recommended to use *Carex mandoniana* species for phytoremediation processes since this is the one that extracts metals the most. Conversely, *Equisetum bogotense* is not recommended to be used in this process as it is a medicinal plant and can affect the trophic chain, but it is necessary to expand studies on its antioxidant capacity, and *Muehlenbeckia tamnifolia* can be used for mine closure processes as it does not accumulate toxic metals in its tissues and grows on slopes.

This study has made it possible to identify native species with potential use in environmental remediation. Likewise, it is recommended to carry out subsequent studies of remain identified species.

References

1. Cuentas M (2001) La Minería Peruana y sus Recursos. Facultad de Ingeniería de Minas – FIM, Puno
2. Environmental Law Institute (2000) Prevención de la contaminación minera: propuesta de un Marco Común para las Américas, Washington, DC. https://www.eli.org/sites/default/files/eli-pubs/d10_02s.pdf
3. Ministerio del Ambiente (2004) Ley N° 28271: Ley que regula los pasivos ambientales de la actividad minera. Artículo 2, Perú
4. Medina K, Montano Y (2014) Determinación del factor de bioconcentración y traslocación de metales pesados en el *Juncus arcticus Willd.* y *Cortaderia rudiuscula Stapf*, de áreas contaminadas con el pasivo ambiental minero alianza - Áncash 2013 (tesis de pregrado). Universidad Nacional "Santiago Antúnez de Mayolo"
5. Becerril J, Barrutia O, García J, Hernández A, Olano J, et al (2003) Especies nativas de suelos contaminados por metales: aspectos ecofisiológicos y su uso en fitorremediación, vol 16. Ecosistemas, pp 1–16
6. Ministerio de Agricultura, Pesca y Alimentación (2017) Metales pesados. Ministerio para la Transición Ecológica (s.f), España. https://www.miteco.gob.es/es/calidad-y-evaluacion-ambiental/temas/atmosfera-y-calidad-del-aire/emisiones/prob-amb/metales_pesados.aspx
7. Delgadillo A, González C, Prieto F, Villagómez J, Acevedo O (2011) Fitorremediación: una alternativa para eliminar la contaminación. Trop Subtrop Agroecosyst 14(2). https://www.scielo.org.mx/scielo.php?script=sci_arttext&pid=S1870-04622011000200002
8. Consultores Mineros Luis Salazar Suero y Asociados (1985) Examen del Prospecto Aurora-Patricia Polloc, Encañada, Cajamarca. Minera Monterrico S.A, Lima, pp 1–30
9. Universidad Privada del Norte (2019) Guía de investigación cuantitativa. Edic 2:5–8
10. Ministerio del Ambiente (2015) Guía de inventario de la flora y vegetación. Lima. Dirección General de Evaluación, Valoración y Financiamiento del Patrimonio Natural
11. Ministerio del Ambiente (2014) Guía para el muestreo de suelos. MAVET IMPRESIONES E.I.R.L, Lima
12. United States Environmental Protection Agency (1994) Method 200.7 Rev.4.4 EMMC. Determination of metals and trace elements in water and waste by inductively coupled plasma atomic emission spectrometry
13. United States Environmental Protection Agency (1994) Method 200.3, Rev.1, April 1991. Metals, total recoverable in biological tissues/Method 200.7, Rev.4.4. EMMC
14. Ministerio del Ambiente (2017) Aprueban Estándares de Calidad Ambiental (ECA) para Suelo. El Peruano. https://sinia.minam.gob.pe/normas/aprueban-estandares-calidad-ambiental-eca-suelo-0
15. Eróstegui C (2009) Contaminación por metales pesados. Rev Cient Ciencia Méd 12(1). https://www.scielo.org.bo/scielo.php?script=sci_arttext&pid=S1817-74332009000100013
16. Moreno D (2003) Toxicología ambiental: Evaluación de riesgo para la salud humana. In: Antonio García Bage (ed) vol 1. España, pp 53–65

17. Posada M, Arroyave M (2006) Efectos del mercurio sobre algunas plantas acuáticas tropicales. Revista EIA. 6. https://www.scielo.org.co/scielo.php?script=sci_arttext&pid=S1794-12372006000200006
18. Ortega C (2006) Caracterización de respuestas de estrés oxidativo inducidas por Cadmio y Mercurio en plantas. Una metodología de estudio a tiempos cortos (tesis de doctorado). Universidad Autónoma de Madrid. https://repositorio.uam.es/handle/10486/2552?show=full
19. Chosky K, Boylston W, Rabek J, Widger W, Papaconstantinou J (2004) Oxidatively damaged proteins of heart mitochondrial electron transport complexes. In: Science direct, vol 1688, pp 95–101. https://www.sciencedirect.com/science/article/pii/S0925443903001881
20. Carvajal C (2019) Especies reactivas del oxígeno: formación, función y estrés oxidativo. Med Legal Costa Rica 36(1). https://www.scielo.sa.cr/scielo.php?script=sci_arttext&pid=S1409-00152019000100091
21. Nuñez R, Garateix A, Laguna A, Fernández M, Ortiz E, Llanio M et al (2006) Caribbean marine biodiversity as a source of new compounds of biomedical interest and others industrial applications. Pharmacol Online 3:111–119
22. García L, Rojo D, García L, Hernández M (2002) Plantas con propiedades antiinflamatorias. Rev Cubana Invest Bioméd 21(3):214–216
23. Navas L (1976) Flora de la Cuenca de Santiago. Andres Bello. 2:48
24. Restrepo M, Quintero P, Fraume N (2005) El milagro de las plantas: aplicaciones medicinales y orofaríngeas: manual. Editor San Pablo 4:76

Identification of the Trophic State of the San Nicolas Lake, Namora-Peru Through the Quantification of Chlorophyll *a*, Nitrates, Phosphates and Transparency

Jackelin Estefani Ciriaco Mosqueira[1(✉)], Leoncio Jaime Lanfranco Colina[1], Cristina Carmela López Rodríguez[1], Carlita Roxana Izquierdo Ramírez[1], Magda Rosa Velásquez Marín[2], and Marco Alfredo Sánchez Peña[1,3]

[1] Ingeniería Ambiental, Universidad Privada del Norte, Av. Vía de Evitamiento Norte s/n cuadra 15, Cajamarca, Peru
marco.sanchez@upn.pe
[2] Colegio de Ingenieros del Perú - Cajamarca, Jr. Alfonso Ugarte 973, Cajamarca, Peru
mvelasquezm@cip.org.pe
[3] Facultad de Ciencias de la Salud, Departamento Académico de Ciencias Biológicas, Universidad Nacional de Cajamarca, Av. Atahualpa Km. 3, Cajamarca 06003, Peru

Abstract. Eutrophication is a global problem that has been progressively destroying aquatic ecosystems; due to this, it was decided to evaluate and analyze one of the most representative lakes in the region: The San Nicolas lake, which is located in Namora district, province and department of Cajamarca in the north of Peru. The purpose of this study was the determination of the trophic state of this lake through the quantification of chlorophyll *a*, nitrates, phosphates, and transparency. Six sampling stations were determined; from them, samples were taken at 0 m, 1.5 m, and 3 m depth with a Van Dorn hydrographic bottle, and the parameters were analyzed in-situ, also including transparency with a Secchi Disc. After analyzing the samples, were determined: The Trophic State Index (TSI) of Carlson, modified by Aizaki et al., resulted in a TSI = 64.72 (eutrophic) for chlorophyll *a* and a TSI = 66.55 (eutrophic) for water clarity; and The Nutrient Eutrophication Index (EI) according to Karydis et al., which resulted in an EI = 2.53 (oligotrophic) for nitrates and an EI = 3.35 (mesotrophic) for phosphates.

Keywords: Trophic state · Eutrophication · San Nicolas lake · Chlorophyll *a* · Nutrients · Transparency

1 Introduction

Perú has more than 12000 lakes and lagoons, of which more than 283 belong to Cajamarca. Currently, these lakes occupy an area of 1010.9 ha and store approximately 121310400 m^3 of water [1]. Many of them are being affected due to eutrophication, and this problem is described as the excessive enrichment of nutrients in the waters,

Y. Iano et al. (Eds.): BTSym 2020, SIST 233, pp. 110–120, 2021.
https://doi.org/10.1007/978-3-030-75680-2_14

which implies a great increase in the aquatic flora, which when dying are going to be at the bottom of the water bodies generating organic waste, this waste decomposes and consumes a large percentage of dissolved oxygen, so that the aquatic flora and fauna are affected and, in some cases, die from suffocation [2].

The San Nicolas lake is one of the most representative lakes in the Cajamarca region. It has an approximate area of 579900 m^2, a volume of 6 958800 m^3, and a depth of 12 m, making it the largest lentic water body in the region. It is in a closed geographical space, where there are no water sources that flow into it; however, since there are no studies on its hydrographic function, the fact that it is connected underground and derives its waters to the Cajamarquino River is not ruled out [3].

In the areas of influence of this lake, population growth is observed, which is reflected by an increase in the development of typical activities of the region, such as agriculture and livestock, which are part of the most polluting anthropogenic activities. Therefore, they provide a high degree of fertilizers and organic remains from animals and plants. Furthermore, each of these activities is considered difficult to control and debug [4]. There are also tourist activities and some harmful customaries such as the hand laundry washing directly in the water body, which is concerning since the domestic detergents present in the water environment are found among the organic pollutants of great magnitude at a world level [5] because they are too dangerous due to their toxicity and the large contribution of phosphorus that they generate. These substances are soluble in water, which, when already used, is discharged into rivers and lakes or, in some cases, infiltrates the land and causes an accelerated eutrophication process.

All this causes changes in the appearance of the lake, where an increase in aquatic plants is observed, which could confirm that there is eutrophication.

This research was carried out due to the concern generated by the above-mentioned contaminating aspects, having as objective the identification of the trophic state of the San Nicolas lake through the quantification of chlorophyll *a*, nitrates, phosphates, and transparency. The trophic state of lakes is described as a fundamental concept in the management of them and means the relationship that exists between the state of the nutrients and the increase of the organic matter in the water body [6]; this seeks to make known how the San Nicolas lake is being affected according to the current degree of eutrophication; so that in the future measures are implemented to reduce and prevent this problem, since the consequences of high-level eutrophication are serious. For example, fish mortality; loss of the quality of the aquatic environment, which prevents its recreational, tourist, and domestic use; and may even end up converting a large part of the water body into firm land.

2 Materials and Methods

This research is a non-experimental-quantitative type with a descriptive-correlational design because none of the variables were deliberately manipulated since only the trophic state was observed, analyzed, and described in a way already established through numerical information. The linear relationship between the main parameters was also analyzed.

The study was carried out in the development of a single monitoring campaign, where six sampling stations were established based on the extension of the lake and in areas where there are constant activities; three stations (D, E, and F) near the shores where activities such as laundry washing, agriculture, and livestock are carried out; and three stations in the center (A, B, and C) where tourism takes place, such as boating and fishing.

The coordinates of the sampling stations are shown in Table 1.

Table 1. UTM coordinates of the sampling stations.

Sampling station	Coordinate x	Coordinate y	Zone
A	793605	9199151	17 M
B	793298	9199213	17 M
C	792962	9199371	17 M
D	792573	9199505	17 M
E	793874	9199116	17 M
F	793713	9199407	17 M

From each station, three water samples were extracted at different depths (0 m, 1.5 m, and 3 m) using a Van Dorn bottle of the Wildco brand of 3.2 L and following the procedure of taking samples at different depths using the hydrographic bottle, described by the National Water Authority in the National Protocol for Monitoring the Quality of Surface Water Resources [7]. In total, 18 samples were collected. Each one of them was analyzed in the laboratory to determine the concentrations of chlorophyll *a*, nitrates, and phosphates. At the same time, transparency was measured in-situ with the Secchi Disk method, described by Global Learning and Observation to Benefit the Environment in the Water Transparency Protocol [8], where it is indicated that the use of this methodology is the best way to measure transparency in deep and calm water bodies. The pH, temperature, dissolved oxygen, and electrical conductivity were also measured in-situ using a HACH - HQ40d-Multi portable multiparameter.

The quantification of chlorophyll-*a* was determined according to the trichromatic method described by the Polytechnic University of Valencia in "Spectrophotometric determination: Trichromatic method" [9], using a HACH - DR 3900 spectrophotometer. For nitrates and phosphates, the procedure described by the Hach company in Water Analysis Handbook [10] was carried out, and a HACH - DR 900 colorimeter was used.

Two methodologies were used to determine the trophic state of the lake, which are described below.

2.1 Trophic State Index (TSI) of Carlson, Modified by Aizaki et al. [11]

Index proposed by Carlson in 1977 and modified by Aizaki et al. in 1981. It is one of the most widely used today, and it is obtained through transparency, chlorophyll-*a*, and total phosphorus values. Their values indicate four levels: Oligotrophic (TSI < 30),

Mesotrophic (30 < TSI < 60), Eutrophic (60 < TSI < 90), and Hypereutrophic (90 < TSI < 100) [6].

The equation to calculate the TSI of the clarity of the water (SD) (m):

$$TSI_{SD} = 10 \times (2.46 + \frac{3.76 - 1.57Ln(SD)}{Ln2.5}) \quad (1)$$

The equation to calculate the TSI of chlorophyll *a* (Chl *a*) (mg.m^{-3}):

$$TSI_{Chla} = 10 \times (2.46 + \frac{Ln(Chl\,a)}{Ln2.5}) \quad (2)$$

2.2 Nutrient Eutrophication Index (EI) According to Karydis et al. [12]

Index proposed by Karydis et al. in 1983. It is based on the specificity for each nutrient, considers the contribution of the nutrient in different areas, is dimensionless, and is applied in various types of water. It is highly sensitive to the effects of eutrophication, and it is easy to manipulate data and perform its calculation. It categorizes the eutrophication level of water according to three levels: Oligotrophic (EI < 3), Mesotrophic (3 < EI < 5), and Eutrophic (EI > 5) [6].

The equation to calculate EI:

$$EI = \frac{C}{C - \log X} + \log A \quad (3)$$

Where:

- A = Number of sampling stations during the study period.
- C = Logarithm of total nutrient concentration during the study period. That is the sum of the Xij concentrations of the nutrient obtained at each of the Ai stations during the Mj sampling.
- X = Total nutrient concentration at the Ai station during the study period. That is the sum of the nutrient concentrations obtained at the Ai station during the Mj sampling.

3 Results and Discussion

3.1 Parameters In-Situ

The parameters measured in-situ were chosen because of their important relationship with eutrophication indicators since they are altered when there is an abnormal or strange presence of the indicator values. The results obtained are shown in Table 2.

The San Nicolas lake has an average pH value of 8.67, which indicates the presence of alkaline waters. This alkaline level promotes the growth of phytoplankton and results in the proliferation of aquatic plants [13]. It is mainly determined by the vital activity of aquatic organisms and the carbon balance. The surface waters with considerable amounts of phytoplankton that feed on CO_2 manage to increase the pH of the water during the

Table 2. Values of the parameters analyzed in-situ of the San Nicolas lake.

Parameter	Sampling station	Depth			Average
		0 m	1.5 m	3 m	
pH	A	8.68	8.68	8.66	8.67
	B	8.68	8.67	8.70	8.68
	C	8.71	8.70	8.74	8.72
	D	8.68	8.67	8.64	8.66
	E	8.66	8.64	8.65	8.65
	F	8.75	8.65	8.42	8.61
		Total average			8.67
Temperature (°C)	A	19.50	19.90	20.00	19.80
	B	19.80	20.30	20.30	20.13
	C	19.80	20.00	20.10	19.97
	D	20.60	21.20	21.20	21.00
	E	19.90	20.00	21.60	20.50
	F	19.90	20.00	21.00	20.30
		Total average			20.28
Dissolved oxygen ($mg.L^{-1}$)	A	5.97	5.92	5.76	5.88
	B	6.10	6.05	6.04	6.06
	C	6.07	6.00	5.96	6.01
	D	6.17	5.91	5.51	5.86
	E	6.63	5.89	5.84	6.12
	F	6.56	6.09	5.13	5.93
		Total average			5.98
Electrical conductivity ($\mu s.cm^{-1}$)	A	303.00	302.00	298.00	301.00
	B	300.00	298.00	297.00	298.33
	C	297.00	296.00	294.00	295.67
	D	295.00	294.00	287.00	292.00
	E	303.00	298.00	293.00	298.00
	F	297.00	295.00	293.00	295.00
		Total average			296.67
Transparency (m)	A				0.85
	B				0.85
	C				0.90
	D				1.07
	E				1.10
	F				0.95
		Total average			0.95

thermal stratification [14]. It is the result of differential heating of the water of the lake [15]. Besides, CO_2 has a direct relationship with pH. It is evidenced in photosynthesis processes because if the CO_2 reduces in water, the pH levels rise, and in the breathing process, where the reverse happens. On the other hand, its slight variability is also related to the nitrification process because it lowers the pH, as it is shown in stations D, E, and F, where there is a greater number of nitrates.

The temperature of the lake has an average value of 20.28 °C, this parameter always influences fertility [16], and it is very important because it is linked to the ability to accelerate and reduce chemical reactions, including the cycles of the degradation of matter present in water [17]. On the other hand, it can be said that the values have a direct relationship with the irradiation that the lake receives due to being surface waters [14]; for this reason, it can be seen that the values increase at a greater depth of the water. As the temperature rises, the number of microorganisms increases, and the dissolved gases decrease [4]. Despite this, in the San Nicolas lake, the levels of chlorophyll-*a* are not related to temperature but depend on the availability of nutrients.

The value that represents the dissolved oxygen of the San Nicolas lake is, on average, 5.98 mg.L^{-1}. The values of each station decrease according to the sampling depth; this is because it is obtained from the photosynthesis that plants do with chlorophyll; since this process is greater on the surface, the concentration of this parameter will be higher at this level, while it will be lower at levels close to the bottom due to the oxidation processes of organic matter [18] since the death and decomposition of fast-growing aquatic plants increases the turbidity of the water and reduces the dissolved oxygen [19].

The average value of the electrical conductivity of the lake is 296.67 μs.cm^{-1}; it is due to the electrolytes dissolved in the water, where the drained soil, the mineralogical composition, the contact time, the dissolved gases, the pH, and everything that manages to affect the solubility of salts [14]. The values of each station decrease with increasing depth, which indicates the decrease in the amount of total dissolved solids since they maintain a direct relationship by the capacity attributed to the water to transmit electric current as a function of the concentration of ions [17].

The transparency of the San Nicolas lake is on average 0.95 m, and it represents the amount of light that passes into the water body. The depth that sunlight reaches will indicate how deep aquatic plants can develop [8]. Therefore, it is an important parameter for the development of aquatic life. It also allows us to analyze the level of eutrophication concerning phytoplankton.

3.2 Lab Results

The parameters that were analyzed in the laboratory are considered the main indicators of eutrophication problems since the activities that are carried out on the shores of San Nicolas lake increase the levels of these parameters and directly influence the acceleration of the eutrophication process; for this reason, they are relevant in the study of the trophic state of this lake. The results are shown in Table 3.

The chlorophyll-*a* from the San Nicolas lake, on average, has a concentration of 58.50 mg.m^{-3}, which indicates the primary production referred to the biological activity of the lake and depends mainly on the availability of light and nutrients. The values with respect to depths (0 m = 67.81 mg.m^{-3}, 1.5 m = 30.65 mg.m^{-3}, and 3 m = 77.03 mg.m^{-3}) clearly indicate that chlorophyll-*a* is clearly related to nitrate values that, like it, are less than 1.5 m and greater than 3 m. Generally, the constancy of the proportions of carbon, nitrogen, and phosphorus is assumed as part of the cell, and therefore its stoichiometric relationships are also constant. However, these proportions depend on the species, cell size, physiological state, presence of nutrients, light and temperature conditions, etc. [20]. It means that the phytoplankton of San Nicolas lake,

Table 3. Values of the parameters analyzed in the laboratory.

Parameter	Sampling station	Depth			Average
		0 m	1.5 m	3 m	
Chlorophyll *a* ($mg.m^{-3}$)	A	17.62	37.19	34.47	29.76
	B	22.54	27.81	27.78	26.04
	C	22.81	29.95	25.60	26.12
	D	24.59	24.47	30.92	26.66
	E	41.55	30.86	28.78	33.73
	F	277.73	33.62	314.62	208.66
	Total average	67.81	30.65	77.03	58.50
Nitrates ($mg.L^{-1}$)	A	1.00	0.85	0.95	0.93
	B	0.95	0.80	0.95	0.90
	C	1.85	1.00	1.25	1.37
	D	1.10	1.10	1.20	1.13
	E	1.55	1.55	1.20	1.43
	F	1.65	1.25	2.75	1.88
	Total average	1.35	1.09	1.38	1.28
Phosphates ($mg.L^{-1}$)	A	4.10	5.70	4.75	4.85
	B	4.10	4.70	4.35	4.38
	C	3.35	6.25	4.50	4.70
	D	5.75	5.00	5.65	5.47
	E	6.15	7.70	5.35	6.40
	F	6.05	4.30	12.00	7.45
	Total average	4.92	5.61	6.10	5.54

concerning phosphates and nitrates, absorbs a greater amount of nitrogen in the form of nitrates, being the nitrogen an important element of its cell with 7.2%, while phosphorus only occurs in 1%.

Chlorophyll-*a* presents a strong positive correlation with phosphates ($r = 0.814$) and nitrates ($r = 0.815$); it is explained through the necessity for phytoplankton for these nutrients since they act as fertilizers and serve as their food. On the other hand, it keeps a weak positive correlation with transparency ($r = 0.009$) because when the levels of chlorophyll-*a* increase, transparency does not do the same concerning it but tends to decrease.

The average value of nitrates in the San Nicolas lake is 1.28 $mg.L^{-1}$; these nitrates can enter natural waters from several sources [16], such as the dissolution of rocks and minerals, decomposition of plant and animal materials, and contamination by both agricultural and industrial effluents. In surface waters that are not contaminated, the values are generally not greater than 10 $mg.L^{-1}$ [14]. The highest value belongs to station F, which is close to the shore, where the introduction of this compound is faster and easier, since it may be due to the filtration of nitrogen compounds from the soil or the direct disposal of detergents and/or fertilizers, which makes them become a source of diffuse pollution.

The nitrate concentrations related to depths (0 m = 1.35 $mg.L^{-1}$, 1.5 m = 1.09 $mg.L^{-1}$, and 3 m = 1.38 $mg.L^{-1}$) indicate that the depth of 1.5 m presents the lowest value and the depth of 3 m the largest, which is related to chlorophyll-*a* concentrations as mentioned before. Besides, these values could be linked to the production of certain types of species that cause nitrate levels to behave in this way according to depth.

The average phosphate value of the San Nicolas lake is 5.54 $mg.L^{-1}$. They are the most important compounds that phosphorus forms because they are present in energy transfer processes, such as metabolism and photosynthesis. The phosphate is a relatively immobile element and may be carried to streams through soil erosion and storm runoffs from overfertilized or excessively fertilized agricultural fields, nurseries, lawns, and orchards [16].

The phosphate concentrations concerning the depths (0 m = 4.92 $mg.L^{-1}$, 1.5 m = 5.61 $mg.L^{-1}$, and 3 m = 6.10 $mg.L^{-1}$) increase at a greater depth of sampling; this is explained starting from the phosphorus cycle since when the dead phytoplankton decomposes, phosphorus is introduced into the aquatic system [20], this phosphorus is in the form of phosphates. Besides, the decomposition of the organic matter of the phytoplankton tends to be in lower depths since, when dying, it tends to descend.

The nitrates and phosphates have a strong positive correlation ($r = 0.872$) because the two can be found in the same substances. For example, fertilizers, detergents, and pesticides, which can be introduced in the same way and together into the lake. While the transparency presents a moderate positive correlation with nitrates ($r = 0.371$) and phosphates ($r = 0.548$), it means that, when the concentrations of nitrates and phosphates rise, the transparency tends to be kept to a certain degree because, if the concentration of nutrients is higher, the proliferation of phytoplankton in the lake will be greater. If it agglomerates on the surface, it prevents the passage of light, and therefore the transparency values of the water decrease.

3.3 Identification of the Trophic State

Two different indices were used to identify the trophic state in San Nicolas lake: The Trophic State Index (TSI) of Carlson, modified by Aizaki et al. [11], based on the chlorophyll-*a* and the clarity of the water (transparency) and The Nutrient Eutrophication Index according to Karydis et al. [12], based on the nitrates and phosphates; this is because both are related according to the parameters or indicators they use since the eutrophication process is caused by the big amount of nutrients (nitrates and phosphates mainly), which stimulate the growth of phytoplankton (chlorophyll-*a* indicator), which causes the loss of transparency of the water [17]. The results of the two methodologies mentioned are shown in Table 4 and Table 5, respectively.

For the development of this methodology, Eqs. 1 and 2 described above were used. The TSI, according to clarity, resulted in 66.55 and according to chlorophyll-*a* in 64.72. Both TSIs have a eutrophic level, which indicates a high content of nutrients that facilitate

Table 4. TSI values, according to Carlson, modified by Aizaki et al. [11] based on the clarity and chlorophyll *a* from the San Nicolas lake.

Sampling station	TSI_{SD}	TSIChl *a*
A	68.42	61.63
B	68.42	60.17
C	67.44	60.21
D	64.48	60.43
E	64.00	63.00
F	66.51	82.89
Average	66.55	64.72

the proliferation of algae, that, when they die, decompose and consume the oxygen in the water. It makes it difficult for light to penetrate the lakes and these lakes have living being's characteristic of waters with low oxygen levels (barbels, tench, worms, etc.). These living beings are typical in lakes of low and warm areas [18]. The low oxygen levels are due to the large amount of vegetative matter and make aquatic life difficult.

Table 5. EI values, according to Karydis et al. [12] based on nitrates and phosphates from the San Nicolas lake.

Sampling station	$EI_{Nitrates}$	$EI_{Phosphates}$
A	2.27	3.17
B	2.24	3.05
C	2.60	3.13
D	2.42	3.33
E	2.65	3.57
F	3.01	3.86
Average	2.53	3.35

To find the EI values, Eq. 3 was used for each of the nutrients.

According to nitrates, the EI resulted in 2.53, indicating an oligotrophic state; this is the first level of the trophic state. Oligotrophic lakes are poor in nutrients. These lakes characteristically support relatively few aquatic plants (algal populations are low) and often do not contain productive fisheries. The deepest layers are likely to contain oxygen throughout the year [15]. They usually have a high level of transparency, which allows solar radiation to penetrate.

The EI, according to phosphates, resulted in 3.35, indicating a mesotrophic state. Mesotrophic lakes are moderately fertile lakes that may support abundant aquatic plant growths and productive fisheries. Occasional oxygen depletion may occur [15].

In both cases, the minimum values correspond to station B and the maximum to station F; these results refer to the location of the stations according to the proximity of the nutrient insertion since the lake's shores are closer to the activities that leave these compounds as waste, therefore, their accumulation in these areas is greater.

4 Conclusions

The trophic state of the San Nicolas lake based on the quantification of chlorophyll-*a* and transparency is eutrophic through the Trophic State Index (TSI) of Carlson, modified by Aizaki et al., while based on the quantification of nitrates and phosphates, it is oligotrophic-mesotrophic respectively, through the Nutrient Eutrophication Index (EI) according to Karydis et al. The difference in the results of the indices represents the degree of impact that each one of the indicators has on the process of eutrophication of the lake. Chlorophyll-*a* and transparency represent a major problem, followed by phosphates, and finally nitrates. Although the four are closely related and nitrates and phosphates are the main causes of the proliferation of aquatic plants, it may be that their growth is also linked to other factors.

Even though the nutrients analyzed represent a lower trophic index, a representative analysis of them should be carried out from time to time, and measures should be taken to not increase the current concentration of them; this can be done through an environmental management system and good environmental practices; since they are influenced by the activities carried out in the lake, mostly the activities that take place on the shores, such as agriculture, livestock, clothes washed by hand, etc.

The consequences that the process of eutrophication can present not only affect the locality where the problem is generated but directly affects the global destruction of the environment. Therefore, it is the commitment of all to care for and preserve our natural environment.

References

1. Estudio Hidrológico de la Región Cajamarca. https://geoservidorperu.minam.gob.pe/geoservidor/Archivos/Mapa/Cajamarca/Memoria_Descriptiva_Hidrologia.pdf
2. Bermúdez M (2010) Contaminación y turismo sostenible, CETD S.A., México
3. Roncal M, Barrantes M (2013) Laguna San Nicolás, Namora, Cajamarca, Perú. 1a edn. Universidad Nacional de Cajamarca, Perú
4. Mateu A (2016) Eutrofización en aguas superficiales (Proyecto de grado), Escuela Politécnica Superior de Gandia. Universidad Politécnica de Valencia (UPV), Gandia, Valencia, España
5. Insua D, Pérez C, Morales A, Valera Z, Meneses M (2010) Evaluación ecotoxicológica de detergentes comerciales y naturales, como criterio de contaminación ambiental. REDVET 11(03B):1–9
6. Moreno D, Quintero J, López A (2010) Métodos para identificar, diagnosticar y evaluar el grado de eutrofia. Contactos 78(1):25–33
7. del Agua AN (2016) Protocolo nacional para el monitoreo de la calidad de los recursos hídricos superficiales. Ministerio de Agricultura y Riego, Perú
8. Global Learning and Observation to Benefit the Environment (2005) Protocolo de Transparencia del Agua. Área de hidrología, Argentina

9. Determinación espectrofotométrica de clorofila: Método tricromático. Universidad Politécnica de Valencia, 23 July 2015. https://www.upv.es/visor/media/a5ff708b-56e5-4809-9d17-90755b8b9a61/c
10. Hach company (2019) Water analysis handbook. 10th edn. Hach, USA
11. Aizaki M, Otsuki A, Fukushima T, Hosomi M, Muraoka K (1981) Application of Carlson's trophic state index to Japanese lakes and relationships between the index and other parameters. SIL Proc 21(1):675–681. https://doi.org/10.1080/03680770.1980.11897067
12. Karydis M, Ignatiades L, Moschopoulou N (1983) An index associated with nutrient eutrophication in the marine environment. Estuarine Coast Shelf Sci 16(3):339–344. https://doi.org/10.1016/0272-7714(83)90151-8
13. George D, Heaney S (1978) Factors influencing the spatial distribution of phytoplankton in a small productive lake. J Ecol 66(1):133–155. https://doi.org/10.2307/2259185
14. Marín R (2010) Características físicas, químicas y biológicas de las aguas, España
15. Hughes P (1993) Hydrology, water quality, trophic status, and aquatic plants of Fowler Lake, Wisconsin. US Department of the interior, Madison, Wisconsin, USA
16. Khan F, Ansari A (2005) Eutrophication: an ecological vision. Botanical Rev 71(4):449–482. https://doi.org/10.1663/0006-8101(2005)071[0449:EAEV]2.0.CO;2
17. Maldonado D (2018) Determinación del estado trófico de la laguna de San Miguel Almaya. Capulhuac, Estado de México (Tesis de licenciatura), Facultad de Planeación Urbana y Regional. Universidad Autónoma del Estado de México, México
18. Moreta J (2008) La eutrofización de los lagos y sus consecuencias. Ibarra 2008 (Trabajo de grado), Escuela de Nutrición y Salud Comunitaria. Universidad Técnica del Norte (UTN), Ibarra, Ecuador
19. Rao C (1998) Environmental pollution control engineering. New Age International, New Delhi
20. Monerris M, Dómenech P (2010) Modelación de la calidad del agua. 1a edn. Universidad Politécnica de Valencia, España

Use of Building Information Modeling (BIM) as a Tool Applied to Construction Site Design

Nathan Santos Teixeira, Luiz Antonio Sarti Junior(✉), and Sheyla Mara Baptista Serra

Federal University of São Carlos, São Carlos, SP, Brazil
{nathansantosteixeira,luiz.sarti}@estudanteufscar.br, sheylabs@ufscar.br

Abstract. In the construction industry, physical flow or transportation activities still represent a considerable portion of the total activities necessary for developing the projects. This paper aims to improve the physical arrangement of the construction site and are increasingly needed for logistic and safety of the work. In this study, we used Building Information Modeling (BIM) technology associated with the simplified Systematic Layout Planning (SLP) method to design the construction site. The research method was the case study, making a comparative approach to reality with a model proposed. Thus, through BIM-4D modeling, an evaluation of the results was performed. As a result, we obtained a measurement of the necessary storage space and reduced 48% in the distance traveled to transport blocks and 27% to transport mortars and grouts. The conclusions are that qualitative gains related to the ease of visualization of the model by the different agents involved were obtained. Also, was verified the interoperability of the software as planning and design of the layout satisfactory.

Keywords: Building information modeling · Construction site · Construction site layout · Civil construction

1 Introduction

Due to the growing demand for more sustainable projects, it is increasingly necessary to manage the available resources and tools that help in the organization of the work environment. In this sense, the search for efficient models is increasingly present in the academic and scientific circles [1].

One strategy has been the concept of adding value to works in Brazilian industry construction, even if it traditionally presents itself with lower levels of productivity than other industries, such as automotive. Industry construction is recognized as a sector with waste levels due to the low industrialization of the components and the absence of complementary projects, mainly production projects [2].

For Ciampa [3], activities that generate value to the product within a construction site are often related to operational processes and can be measured to identify improvement opportunities. Thus, it is necessary studies that reduce the percentage of flow activities,

Y. Iano et al. (Eds.): BTSym 2020, SIST 233, pp. 121–129, 2021.
https://doi.org/10.1007/978-3-030-75680-2_15

which do not add value, because these, although indispensable, can represent time and resources beyond what is necessary. For better analysis, one can use the principles of lean construction.

Another contribution is the inclusion of building information modeling (BIM), which can be an alternative for the construction site's initial design due to interoperability between existing designs, more accessible visualization, and integration with the work's planning. Also, the use of BIM-3D and BIM-4D models can become necessary for the decision-making process [4, 5].

Some authors like Eastman et al. [6] and Koo and Fischer [7] define 4D modeling as a process of association between a 3D model of the construction product with a plan of time-distributed activities.

During the enterprise's project process, it is possible to make decisions that significantly impact the time and cost of the work more effectively. Fabricio and Melhado [8] state that, as in other industries, the design process (layout) in building construction is an essential step towards obtaining a higher quality product. Complementary designs can be created aimed at the development and organization of the production process, in the case of civil construction, construction sites.

According to Bargstädt [9], the activities inherent in the construction site require a set of information, which must have a high level of reliability and must be provided at the appropriate time. For management, only geometric information is not sufficient [9]. It is necessary to have the knowledge, such as specifications on the material type to be used and parameters such as strength and performance.

Tommelein [10] states that 4D modeling can be used for various purposes, such as visualization during design, construction, and marketing, the study of alternatives, and the definition of assembly sequencing.

Currently, research has been carried out in search of methods of management of construction sites, such as the management of the construction site through repetitive projects [11], construction site management of industrialized systems [12], and dynamic layout planning [13]. Other researchers work in research focused on developing 4D models for the management of construction sites [14, 15].

In Brazil, most construction sites are not standardized and have variable structures throughout construction. Each phase of the design of the construction site has specific needs and elements, and integrated logistic studies are required. Thus, it is necessary to use techniques and tools to design and manage physical space efficiently and safely.

In this context, the present study presents a proposal to improve the layout of the construction site design of the construction observed, using the systematic concepts of organization of physical arrangements and the modeling of construction information (BIM) as a tool.

2 Materials and Methods

The research method adopted for the present study was the Case Study complemented by an intervention or proposal for improvement for the studied project. For this, a building was selected with a predefined construction site design, with construction in progress and access to the information needed to carry out the study. Thus, the research adopted was constructing an educational building, with 221.13 m^2, comprising four research rooms, an entrance hall, a pantry, a service area, and two toilets. The building has only one floor, built-in structural masonry, characterized as a rationalized construction system. The cover was made with metallic tiles. Subsequently, the construction underwent an addition of about 70 m^2 through a contract amendment. Such expansion was part of the present analysis.

Survey information about the documentation and conditions surrounding was carried out. The following sources were identified: design of the provisional facilities for the construction site, physical-financial schedule of the building, operational planning of activities, and photographic records.

2.1 Modeling the Improvement Proposal

After evaluating the proposed provisional facilities' design, a 3D model was created, comparing it with the actual construction site. Visits were made at the construction site to identify storage locations and material delivery strategies. In possession of the data, the modeling of the proposal to improve the existing construction site design was initiated, focusing on the application of lean construction concepts, with the aid of BIM, Autodesk Revit 2019[1] (for 3D modeling), and Autodesk Navisworks 2020[2] (for integration with the 4D work schedule).

The choice of such software started from the following premises: availability of free licenses for use (student), the similarity between commands with Autodesk AutoCAD, and ease of interoperability between the software. Therefore, the 4D model's creation was based on the conception of the construction site based on the Systematic Layout Planning (SLP) method, the physical schedule, and the physical flows of material and people.

2.2 Analysis of the Results Obtained

After modeling the proposed solution, analyses were performed to verify the gains and difficulties due to the introduction of BIM technology in the construction site's design. For this, it was determined the space needed for the storage of materials and components and the routes traveled, and the ease of visualization and understanding of the design proposal.

[1] https://www.autodesk.com.br/products/revit/overview.

[2] https://www.autodesk.com.br/products/navisworks/overview.

Next, a comparative analysis of the proposed model and the study site was performed, focusing on resolving the attention found in the actual construction site. Also, the use of BIM tools was evaluated in their independent use and together, verifying the interoperability between them. In possession of the analyses, a comparison of the results was made to determine how the proposed model improved design.

3 Results

3.1 Current Site Layout

During the visit, the services were in the execution phase of the structure and seal. The main facilities were the living area made of wood with plywood, warehouse in a steel container, the place for disposal of aggregates, concrete mixer, stock of blocks, stock of anchors, and bucket.

With the positioning of the concrete mixer plant and the block stock, it was possible to calculate the distance to the point of application, allowing the comparison between the distances and consequent verification of reducing the transport flow activity for transport mortar. In this sense, for the masonry service, Fig. 1(a) presents the actual layout and in Fig. 1(b) the layout with the proposed improvements. Table 1 shows a comparison of the calculated distances.

Table 1. Comparison of distances for transporting blocks, mortar, and grout

Input	Current distance (m)	Proposal distance (m)	Reduction
Ceramic brick	33.6 (BC)	17.5 (B′C′)	48 %
Mortar/grout	27.3 (AC)	20.0 (A′C′)	27 %

In Table 1, there was a 48% reduction in the distance for the blocks comparing the actual (BC stretch) and proposed (stretch B′C′) and a 27 % reduction for the transport of laying and grout mortar.

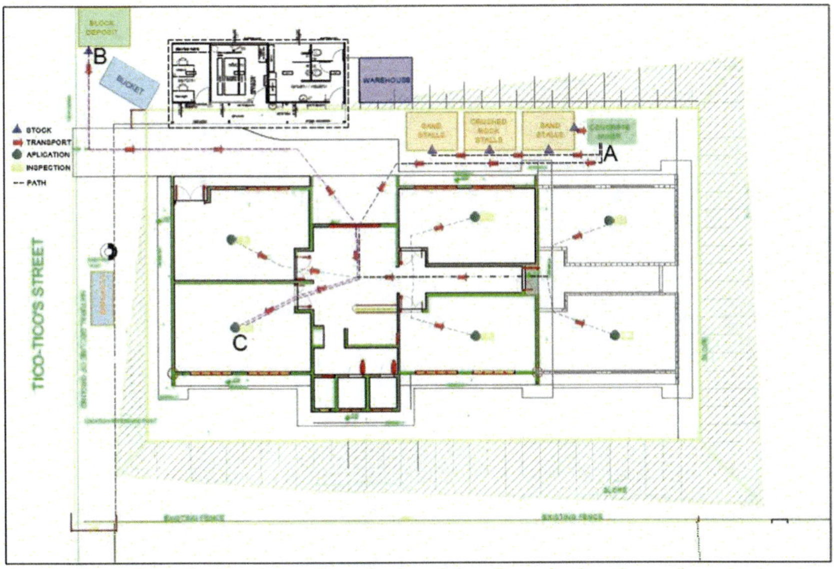

(a) real situation

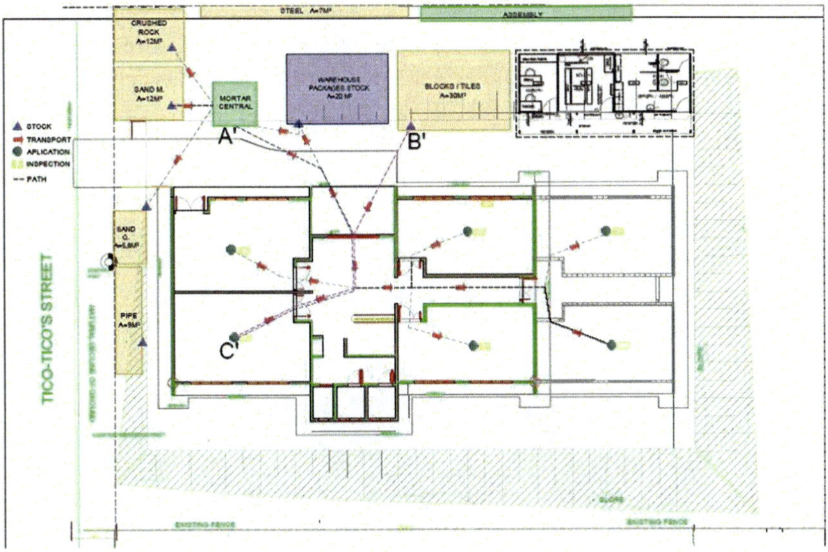

(b) proposed design

Fig. 1. Proposal for improvement in the layout of the construction site

3.2 Modeling Proposed Improvement

It was started through the 3D modeling of the construction that will be executed to improve the proposal based on the contractor's designs. Figure 2(a) presents the initial base plant (without the additive), used as the basis for the 3D model, while in Fig. 2(b) the foundation and Fig. 2(c) the 3D model.

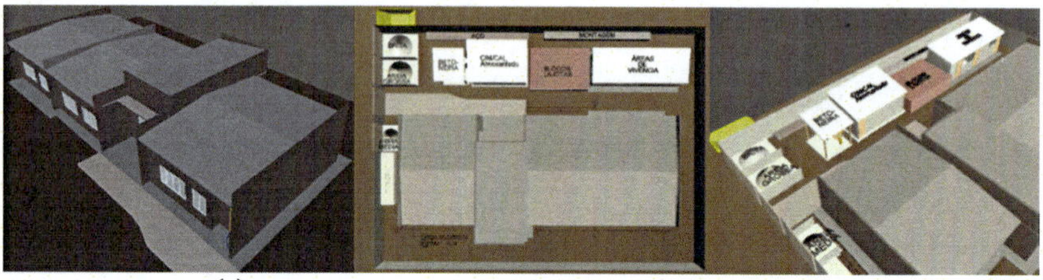

(a) (b) (c)

Fig. 2. 3D modeling

After modeling the construction elements, the spatial distribution of the necessary spaces was performed within the construction site's space. For this, the simplified SLP methodology [16] was first used to position the processing areas and related inventories, trying to reserve a privileged position to these elements due to constant use. Thus, we sought to reduce transportation activities, considered unproductive, according to lean construction principles.

It was found that it would be necessary to rewind the siding about 2 meters in the north direction of the construction site, releasing more space for the steel processing plants in construction sites bays of aggregates. This change made it possible, for example, to install two of the three bays of aggregates side by side, on the front, facilitating the discharge of this material that is characterized by the weight and specific storage conditions. The installation of the steel assembly and storage area was placed nearby and made possible by this change.

The living areas of the real construction site that occupied the land entrance were moved to a region more from the construction site's entrance and the direct access to the building. Figure 3(a) presents a top view of the modeled construction site, and Fig. 3(b) shows a picture of the prospective site.

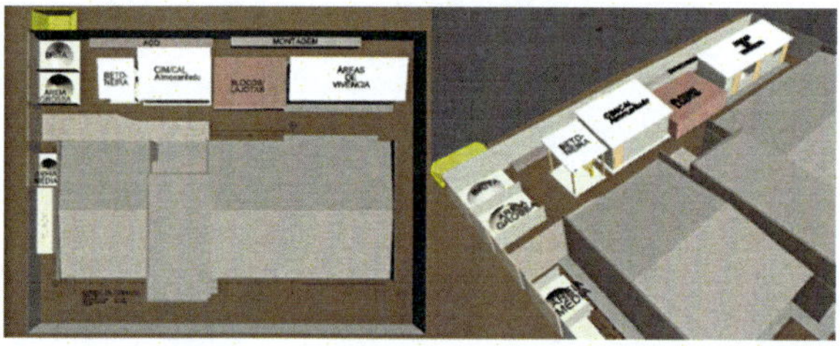

(a) (b)

Fig. 3. 3D modeled construction site

To the 3D model's integration, it was necessary to adapt the deliveries of materials based on what was observed in the construction site. For this, the physical-financial plan provided was used as the basis, critically analyzing the work's executive sequence. For the preparation of the schedule, the MS Project[3] software was used due to compatibility with Navisworks. In this program, planning information and the 3D model were linked, generating a 4D model.

Following the guidelines of the simplified SLP method, the preferred interconnection chart was drawn up. With this chart, you can determine what type of relationship exists between the construction site elements. For elaborating and validating the letter, we sought a Master of Works with more than 20 years of experience and currently works with large residential developments with extensive experience.

To the preferred interconnection chart, only four divisions were used, which was: very important (A), important (I), despicable (U), and undesirable (X). Also, the reasons or justifications for such classifications were determined. The motifs were represented by numbers in the diagram, according to the experience of the Master of Works. Table 2 shows a cutout of the chart, where it is possible to identify constraints in the positioning of the elements.

Table 2. Adaptation clipping of the preferential interconnection chart

Regions	Central		Stock		
	Steel	Mortar	Molds	Steel	Aggregates
Steel central		–	–	–	–
Mortar central	U		–	–	–
Molds stock	U	U		–	–
Steel stock	A, 1, 2	U	U		–
Aggregates stock	U	A, 1, 2	U	U	
Bagged stock	U	A1, 2	U	U	U
Tub stock	U	U	U	U	U
Ceramic brick stock	U	U	U	U	U
Warehouse	I, 1	U	U	U	U

In this case, according to Table 2, it is possible to notice that the proximity relationship between steel and mortar plants is classified as indifferent. It is also important that the steel stock is close to the steel plant for two reasons: (1) provide optimization of material flow without interference and (2) using joint personnel in both locations.

The 4D simulation was carried out in possession of the 3D models, construction site, and construction, and the project schedule was carried out. For this, the model created in Revit was imported into the Navisworks environment. After the import, the set of elements was designed and the parts of the model according to the selection's purpose. In this case, the parts correspond to the activities of the schedule.

3 https://www.microsoft.com/pt-br/project.

In this step, we sought to name these sets according to the name of the corresponding task in MS Project to make possible the use of the Navisworks Auto-Attach Using Rules tool. It allows the association of sets to tasks automatically, reducing effort and preventing any linking errors. After adjustments to the configuration, it was possible to start the simulation, stop it, and change the settings at any time. Figure 4 shows some screenshots of the building and construction site simulation 4D. After the adjustments in the simulation and verification of the failures, it was possible to conceive a construction site proposal more consistent with the idealized scenario.

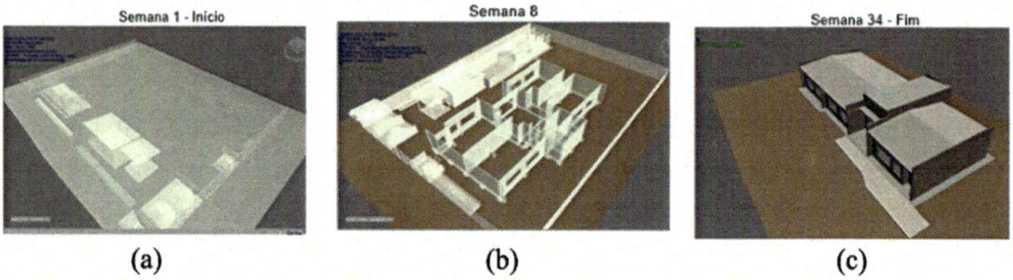

Fig. 4. 4D Modeling 4D: (a) week 1-begin, (b) Week 8, (c) Week 34-end

4 Conclusions

An analysis proposing improvements in the construction site's layout was performed. For this, we used the BIM-4D, map flowchart, and SLP method. It was found that the planning of the provisional facilities, with the appropriate dimensioning, provides increasingly efficient and safe physical arrangements.

Corroborating the study by Wang et al. [17] claims that logistics planning of space over time are critical pieces for an efficient construction project.

It is believed that the objective of the present study was achieved since it was possible to analyze the real construction site and verify its inefficiencies and propose improvements through the construction site design generated with the help of BIM and other tools. In the results obtained, reductions of 48% in the distance traveled to transport blocks and 27% for mortars and grout were observed.

As in Zouein and Tommelein [18], this work's results are the reductions in transport distances possible with elaborating the construction site layout and logistics flow studies. Consequently, improvements in performance, safety, and productivity will be obtained. It was also possible to observe decreases in interferences, with the more linear flow in the tasks, greater ease of visualization of the construction site, confirming the hypothesis that it is an excellent value in the layout design. The results obtained to corroborate the concepts of lean construction listed by Koskela [19], showing the actuality and relevance of Lean thinking in product design. However, developing the construction site design can be considered trying. Still, it helps in the early identification of misconceptions for the building phase, where there is less effort, more significant possibility of alteration, and fewer physical resources applied.

Acknowledgment. The authors are grateful to Coordination for the Improvement of Higher Education Personnel - Brazil (CAPES) - Financing Code 001.

References

1. Fecomercio. Growing demand for sustainable architecture in Brazil. https://www.fecomercio.com.br/noticia/cresce-demanda-pela-arquitetura-sustentavel-no-brasil
2. Construct (2016). Waste in civil construction: impacts on the environment. Brazil. https://www.fecomercio.com.br/noticia/cresce-demanda-pela-arquitetura-sustentavel-no-brasil
3. Ciampa D (1991) The CEO's role in the time-based competition. Blackburn, JD
4. Hashtagbim (2015) BIM from 3D to 7D. Ceará. https://hashtagbim.wordpress.com/2015/10/12/bim-do-3d-ao-7d/
5. Sienge (2019) BIM dimensions. Brazil. https://www.sienge.com.br/blog/dimensoes-do-bim/
6. Eastman CM, Teicholz P, Sacks R, Liston K (2011) BIM handbook: a guide to building information modeling for owners, managers, designers, engineers and contractors. Wiley, Hoboken
7. Koo B, Fischer M (2000) Feasibility study of 4D CAD in commercial construction, vol 998. CIFE, Stanford, p 135
8. Fabricio MM, Melhado SB (2007) The project in architecture and civil engineering and the performance in multidisciplinary teams. Tópos J 1(2):11–28
9. Bargstädt HJ (2015) Challenges of BIM for construction site operations. Proc Eng 117:52–59
10. Tommelein ID (2003) Acknowledging variability and uncertainty in product and process development. Swets & Zeitlinger, Lisse, Netherlands
11. Hajdasz M (2014) Flexible management of repetitive construction process by an intelligent support system. Expert Syst Appl 41(4):962–973
12. Ismail F, Baharuddin HEA, Marhani MA (2013) Factors towards site management improvement for industrialized building system (IBS) construction. Proc-Soc Behav Sci 85:43–50
13. Andayesh M, Sadeghpour F (2013) Dynamic site layout planning through minimization of total potential energy. Autom Constr 31:92–102
14. Russel A, Staub-French S, Tran N, Wong W (2009) Visualizing high-rise building construction strategies using linear scheduling and 4D CAD. Autom Constr 18:219–236
15. Moon H, Kim H, Kim C, Kang L (2014) Development of a schedule workspace interference management system simultaneously considering the overlap level of parallel schedules and workspaces. Autom Constr 39:93–105
16. Muther R, Wheeler JD (2000) Systematic and simplified layout planning. IMAM, São Paulo, 50p
17. Wang HJ, Zhang JP, Chau KW, Anson M (2004) 4D Dynamic management for construction planning and resource utilization. Autom Constr 13(5):575–589
18. Zouein PP, Tommelein ID (2001) Improvement algorithm for limited space scheduling. J Constr Eng Manage 127(2):116–124
19. Koskela L (1992) Application of the new production philosophy to construction. Stanford University, Stanford

Current Situation of COVID-19 and Its Impact on Phubbing and Student Performance

Patricia Uceda, Laura Bazán(✉), and Rosa López

Perú Environmental Engineering, Universidad Privada del Norte, Av. Vía de Evitamiento Norte s/n cuadra 15, Cajamarca, Peru
{patricia.uceda,rosa.lopez}@upn.edu.pe, laura.bazan@upn.pe

Abstract. The use of smartphones and the Internet increasingly affects people's lives, thereby establishing new forms of communication. Digital tools are necessary for performance and social interaction. This research study sought to identify the impact of the COVID-19 pandemic on the performance of the first year of computational systems engineering at the Universidad Privada del Norte – Cajamarca, specifically with regard to phubbing. The study was nonexperimental in nature, followed a correlate descriptive approach, and was of longitudinal design with two observations (cycles 2019-2 and 2020-1). The Chotpitayasunondh and Douglas measurement scale was used for the assessment of phubbing, which was validated in pilot testing ($\alpha = 0.8512$). Academic performance was obtained from the final averages of the Programming subject, where the phubbing survey was applied with the voluntary participation of students. Correlation results determined no positive relationship between phubbing level and performance level; hence, the variable situation of the COVID-19 pandemic, which modified the academic space, was analyzed in accordance with the needs associated with this environment and the life of students. The level of phubbing in students decreased from 40.33% to 35.51%, while their performance level increased from 12.98 and 13 to 14.48 and 15 in terms of average and median, respectively. The aspects of phubbing were evaluated with a coherent perception, reaffirming the validity of the instrument. Educational policies that promote positive smartphone use are required to improve academic performance.

Keywords: Smartphone · Phubbing · COVID-19 · Academic performance

1 Introduction

With the invention of the Internet and smartphones, new forms of communication have been established, necessitating the use of digital tools for performance, and have expanded the limits of social interaction. With use comes misuse, and despite their advantages, some people have adopted particular styles that negatively influence their everyday activities [1].

Internet addiction and excessive use of smartphones and social networks are problems on the rise [2]. Phubbing is a term used to describe the behavior of ignoring a person in favor of using a smartphone, paying more attention to an electronic device, and

Y. Iano et al. (Eds.): BTSym 2020, SIST 233, pp. 130–136, 2021.
https://doi.org/10.1007/978-3-030-75680-2_16

ignoring the people around oneself [3]. The phenomenon of phubbing is related to smartphone dependence; showing continual despair regarding draining of one's phone battery; the constant fear of missing the experiences that others are living out and sharing on networks; and even putting one's life at risk when using a smartphone while walking or driving [4].

According to Velthoven and Powell [5], phubbing is identified as the problematic use of a smartphone, where technology allows easy access, the option to escape daily life in the blink of an eye, be able to exist freely in anonymity, and frequently receive alerts and messages, all connected to the Internet. Capilla and Cubo [6] connect phubbing to the problematic use of smartphones in young people, particularly university students. There is nothing wrong with using a smartphone for a reason; however, it is important to understand its limits and exercise caution. As the saying goes, too much of anything is bad [7].

In their research study, Abeele and Postman [8] reveal that looking at one's phone during a conversation and interaction damages a relationship more than looking at a newspaper. In addition, glancing at a phone hinders integration while listening when speaking. Smartphone addiction among medical students was estimated to be around 36.8%. Phubbing was reported by 37.6% of participants, with over 60% reporting excessive use. Education policymakers and educators must develop some strategies that promote smartphones' positive use to improve academic performance [9].

Espada, Orgilés, Piqueras, and Morales [10] mention that home quarantine is an unprecedented measure that can have a negative impact on physical well-being. Medina and Jaramillo [11] point out that in order to avoid falling into depressive states or symptoms, education from family members is as important as that from state institutions. In addition, Piña [12] indicates that paying attention to social networks regarding updated information on the pandemic offers solutions through the networks with the empowerment to contact family and friends, which includes enabling people in quarantine to update loved ones about their situation.

COVID-19 has immobilized more than 95% of schools worldwide. However, the coronavirus is going to lead to some positive outcomes. The most important aspect is that it has revived family bonds and fortunately led to better relationships between family members, allowing them to spend quality time at home. This also brings new challenges and proposals, which is the reason for the necessity to see how digital education has aligned with educational systems [13].

Thus, this research study sought to identify the impact of the COVID-19 pandemic on performance and phubbing among first-year students in the Computer Systems Engineering program between 2019 and 2020, given the nature of remote education and the increased need for smartphones and the Internet.

2 Method

The study began by evaluating phubbing and performance in first-year students, with average ages between 17 and 25 years. Because of the COVID-19 pandemic, the same students who now participated at the university were reevaluated remotely, given the restrictions of social isolation.

The nature of the research work was nonexperimental, with a descriptive approach correlated with a longitudinal design using two observations—one in the regular cycle (CR) 2019–2 and the other in the regular cycle (CR) 2020–1—adapted to the remote environment. To evaluate phubbing, the Chotpitayasunondh and Douglas [14] measurement scale was used. This is an English-language instrument that was translated and adapted into Spanish and validated in a pilot test, obtaining $\alpha = 0.8512$. This instrument consists of 15 construct items, as well as a scale of 7 scores: 1-never, 2-rarely, 3-occasionally, 4-sometimes, 5-frequently, 6-usually, and 7-always.

The phubbing evaluation was regrouped into four classification groups: null level (never), low level (rarely, occasionally), medium level (sometimes, frequently), and high level (usually, always).

The performance was observed using the final averages in the Programming course. We worked with 61 volunteer students in their first year of the Computer Systems Engineering program at Universidad Privada del Norte, Cajamarca campus. This work passed and was approved by the faculty's Ethcs Committee.

3 Results

The first phubbing observation (Fig. 1a) resulted in 64% low level, followed by 34% medium level and 2% high level. An average of 40.33% phubbing was obtained. The construct with the highest phubbing score was "I put my smartphone where I can see it", followed by "I'm worried about missing something important if I don't check my smartphone". The construct with the lowest value was "It irritates me if others ask me to put down my smartphone and talk to them", followed by "I feel good when I stop concentrating on others and instead pay attention to my smartphone".

The first observation on student performance (Fig. 2a) was obtained with an average of 12.98 and a median of 13.

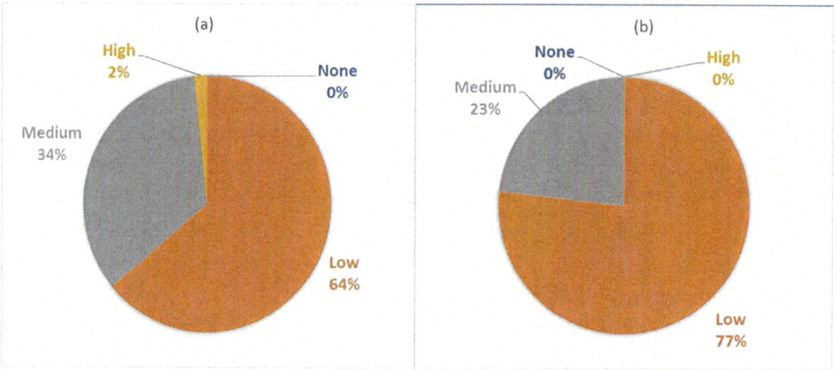

Fig. 1. Phubbing level results among first-year Computer Systems Engineering students at UPN in a) Observation 1 (CR 2019-2) and b) Observation 2 (CR 2020-1).

The second observation of phubbing (Fig. 1b) resulted in 77% low level, followed by 23% medium level. An average of 35.51% phubbing was obtained.

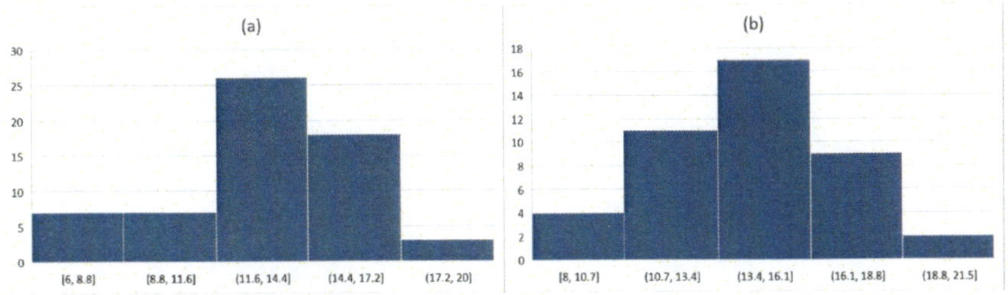

Fig. 2. Results for the performance level, average, and median in the Programming course among first-year students in the Computer Systems Engineering program at UPN in a) Observation 1 (CR 2019-2) and b) Observation 2 (CR 2020-1).

The construct with the highest phubbing rating was again "I put my smartphone where I can see it", followed by "I'm worried about missing something important if I don't check my smartphone". The construct with the lowest value was also the same "It irritates me if others ask me to put down my smartphone and talk to them", followed by "I feel good when I stop concentrating on others and instead pay attention to my smartphone".

The first observation on student performance (Fig. 2b) was obtained with an average of 14.48 and a median of 15.

The correlation results (Figs. 3a and 3b) revealed no positive relationship between the level of phubbing and performance level. Thus, the uncontrolled variable was analyzed, namely, the COVID-19 pandemic situation, which changed the academic space, the needs associated with this environment, and students' lives.

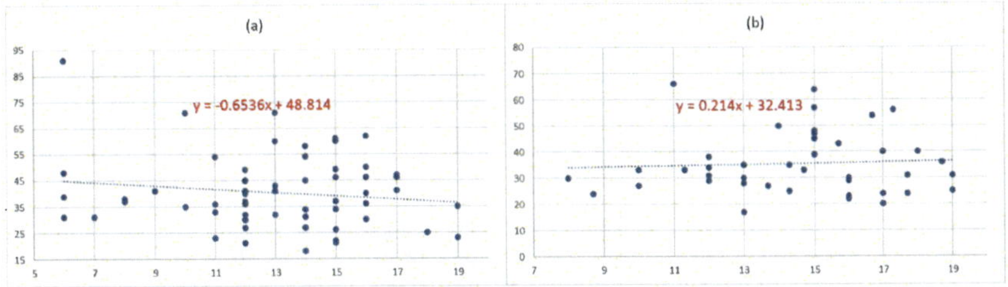

Fig. 3. Correlation results between performance level and phubbing level among first-year Computer Systems Engineering students at UPN in a) Observation 1 (CR 2019-2) and b) Observation 2 (CR 2020-1).

In general, the level of phubbing decreased from 40.33% to 35.51%, while the level of student performance increased from 12.98 and 13 to 14.48 and 15 as average and median values, respectively (Figs. 1a, 1b, 2a, and 2b).

The evaluation graph of the constructs that determine the level of phubbing, from the items with the lowest value to the ones with the highest value, shows a decrease between observations, with similar behavior and correspondence (Fig. 4).

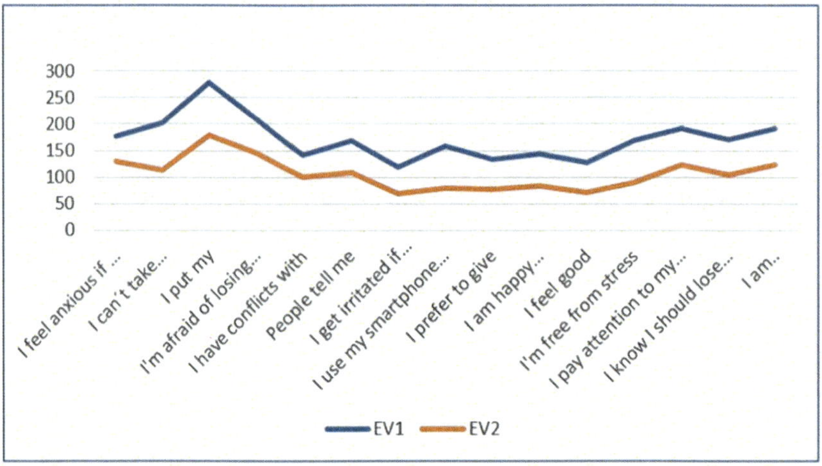

Fig. 4. Construct assessment - phubbing level among first-year Computer Systems Engineering students at UPN in Observation 1 (CR 2019-2) and Observation 2 (CR 2020-1).

4 Discussion

Based on the results obtained, phubbing aspects in both observations are the same, with a consistent perception. The statements are confirmed to have been understood, reaffirming the instrument's validity.

A phubbing level of 35.51% was obtained, a result that is very similar to the 37.6% obtained by Karki, Singh, Paudel, Khatiwada, and Timilsina [9], coinciding with the need to establish education policies that promote the positive use of smartphones for improved academic performance.

While Vega and Arencibia [2] indicated that Internet addiction, as well as excessive smartphone and social network use, are growing problems, the decreased level of phubbing and increased level of performance show that before the pandemic, smartphones were more of a distracting element, only to later become an indispensable tool for learning and work. This moves the trend toward a problem in decline.

With regard to what was mentioned by Barrios, Béjar, and Cauchos [3], the pandemic has strengthened the need and value of the family as a social institution when compared with those of smartphones or the Internet, establishing a space for communication and mutual support, demonstrated in the decreased phubbing measurement criteria.

In agreement with Piña [12], using smartphones and other tools to support remote education due to the pandemic improves affiliation and interaction, unlike the results offered by Abeele and Postma [8].

Like Diez, Rivas, Neira, and Ostoja [13] and Sanz [15], COVID-19 has positive aspects, such as improved academic performance and family relationships, which are demonstrated in the phubbing results in both research studies and the decrease in negative activities indicated [1, 5].

Smartphone dependence, as Elhai, Levine, and Hall [4] would have it, is moving toward conscious use, as pointed out by Espada, Orgilés, Piqueras, and Morales [10], where home quarantine could change and have a positive impact by allowing people to take advantage of smartphone use for academic purposes. While the level of phubbing has decreased as an expression of Internet, smartphone, and social network addiction, it is still not possible to determine what Capilla and Cubo [6] indicate regarding its impact on psychological well-being. In agreement with Piña [12], using smartphones and other tools to support remote education due to the pandemic improves affiliation and interaction, unlike the results offered by Abeele and Postma [8].

Given the assertion made by Jundante [7] on using smartphones for a purpose, even though it is not directly related to phubbing, the pandemic has positively influenced its management, demonstrating academic improvement among students.

5 Conclusions

The situation that emerged because of the COVID-19 pandemic influenced the decreased phubbing level (from 40.33% to 35.51%) in addition to increased performance in the Programming course (from 12.98 and 13 to 14.48 and 15 as average and median values, respectively).

Given the pandemic's remote nature, demand for smartphones and the Internet, the proper use of technological resources at home, and the value of technology for families have gained strength. The level of phubbing and smartphone dependence, and what it shows about other people's lives, is getting better in this situation of the pandemic.

Finally, establishing educational strategies that promote the positive use of smartphones to improve academic performance is necessary.

References

1. Orellano C (2016) Probable pathological behaviors associated with the use of cell phones and the internet. Neuro-Psychiatry J 79(3):192–193
2. Vega-Almeida R, Arencibia R (2019) The dark side of the internet. Span Mag Sci Doc 42(4):251
3. Barrios-Borjas D, Bejar-Ramos V, Cauchos-Mora V (2017) Excessive use of smartphones/cellphones: phubbing and nomophobia. Neuro-Psychiatry Chilean Mag 55(3):205–206
4. Elhai JD, Levine J, Hall B (2019) Problematic smartphone use and mental health problems: current state of research and future directions. Dusunen Adam: J Psychiatry Neurol Sci 32(1):1
5. Van M, Powell J (2018) Problematic smartphone use: digital approaches to an emerging public health problem
6. Capillaía E (2015) Phubbing, connected to the network and disconnected from reality
7. Jundante J (2016) Phubbing: its impact on human relationship. Roy Thai Navy Med J 43(2):75–79
8. Vanden P, Postma-Nilsenova M (2018) More than just gaze: an experimental vignette study examining how phone-gazing and newspaper-gazing and phubbing- while-speaking and phubbing-while-listening compare in their effect on affiliation. Commun Res Rep 35(4):303–313
9. Karki S, Singh J, Paudel G, Khatiwada S, Timilsina S (2020) How addicted are newly admitted undergraduate medical students to smartphones: a cross-sectional study from Chitwan medical college, Nepal. BMC Psychiatry 20(1):1–7

10. Espada J, Orgilés M, Piqueras J, Morales A (2020) Good practices in child and adolescent psychological care before COVID-19. Clin Health 31(2):109–113
11. Medina R, Jaramillo-Valverde L (2020) El COVID-19: Quarantine and it's psychological impact on the population. SciELO Preprints
12. Piña-Ferrer L (2020) El COVID 19: Psychological impact on human beings. Interdisc Health Sci Health Life 10 4(7):188–199
13. Diez T (2020) Digital transformation in education in times of COVID-19
14. Chotpitayasunondh V, Douglas K (2018) Measuring phone snubbing behavior: development and validation of the Generic Scale of Phubbing (GSP) and the Generic Scale of Being Phubbed (GSBP). Comput Hum Behav 88:5–17
15. Sanz J (2020) Use de ICT in educational guidance in times of COVID-19. AOSMA Mag 28:88–91

Identification of Herbaceous Flora with a Greater Value of Importance and Evaluation of Its Phytoremediator Capacity in Contaminated Soils, Tumbacucho – Peru

Jhony Miguel Lezama Oribe[1], Dante Orlando Saldaña Vega[1], Magda Rosa Velásquez Marin[1,2](✉), and Marco Alfredo Sánchez Peña[1,3]

[1] Universidad Privada del Norte, Av. Vía de Evitamiento Norte s/n block 15, Cajamarca, Peru
mvelasquezm@cip.org.pe, marco.sanchez@upn.pe

[2] Colegio de Ingenieros del Peru, Jr. Alfonso Ugarte 973, Cajamarca, Peru

[3] Facultad de Ciencias de la Salud, Departamento Academico de Ciencias Biologicas, Universidad Nacional de Cajamarca, Av. Atahualpa Km.3, Cajamarca, Peru

Abstract. A study was carried out to identify herbaceous flora with greater importance value and evaluate its potential capacity of phytoremediation in mining-contaminated soils. The study analyzed three soil samples, two root samples, two stem samples of each species of higher importance value in an area of eighty square meters. Therefore, the bioconcentration factor (BCF) was calculated for the root of the species *Paspalum tuberosum,* which is silver accumulator and is excluding arsenic, copper, lead, antimony, zinc, aluminum, and iron. The *Pennisetum clandestinium* is a silver accumulator and is exclusive of arsenic, copper, lead, antimony, aluminum, and iron. The *Trifolium repens* showed a BCF in the root that excludes all metals. The concentration of heavy metals in the aerial part of the *Paspalum tuberosum, Pennisetum clandestinium,* and *Trifolium repens,* presented an area (BCF) excluding all metals. The heavy metal translocation factor (TF) in *Paspalum tuberosum* and *Pennisetum clandestinium* does not effectively transfer heavy metals from the root to the aerial part of the plant. *Trifolium repens* showed a hyperaccumulation (TF) of silver, arsenic, copper, and zinc.

Keywords: Translocation · Bioconcentration · Phytostabilization · Herbaceous flora

1 Introduction

The large-scale consumption of natural resources as a source of raw material began with the second industrial revolution. These activities were gradually generating different types of waste, which, at first, did not seem to present any danger because they trusted in the assimilative capacity of the environment; however, from the 1960s on, many people became aware of the environmental pollutants [1].

Nowadays, we are trying to prevent environmental pollution and the depletion of natural resources through environmental management tools. We are aware of the dangers

Y. Iano et al. (Eds.): BTSym 2020, SIST 233, pp. 137–147, 2021.
https://doi.org/10.1007/978-3-030-75680-2_17

that can be generated in the ecosystem, and the survival of the human species is in danger [2].

In Peru, mining is one of the most important activities, with mining concessions occupying 20.42% of its territory [3]. Before the 1990s, this activity was carried out without considering environmental management instruments, which has caused the generation of environmental liabilities, a situation that changed with the issue of environmental decree DL. No. 613, and later in 2001 with the issue of decree law No. 27446, of the National System for Environmental Impact Assessment, which was modified by Legislative Decree No. 1078 in 2008.

Mining activities generate a series of waste substances that impact the air, soil, water, flora, and fauna due to the concentration of heavy and non-heavy metals. Heavy metals are characterized by having an atomic density greater than 4 g/cm^3 [4]; for this, it is required that they receive a treatment to prevent negative environmental impacts.

By 2020, the Ministry of Energy and Mining registered the amount of 7,956 mining environmental liabilities throughout the national territory [5]. In the province of Hualgayoc, Cajamarca region, there are 1,237 environmental liabilities [6], most of them due to mining developed more than 20 years ago. One of the main problems that arise is water pollution of the water sources around the area. This water is meant for the consumption of around 16,000 inhabitants and the areas surrounding these communities, located two and a half hours from Cajamarca city [7].

Currently, the government is implementing remediation projects for mining environmental liabilities. The wild plants used in this research are an alternative solution; since they can absorb the metals present in the soil, although this will depend on the plant species [8], organic matter, pH, and redox potential of the soil [9]. Besides, phytoremediation is a low-cost method that does not negatively impact the environment [10].

It is recommended to use phytoremediation in large areas due to its low ecological impact and because it does not require mechanical equipment [11]; this method is replacing traditional technologies [12]. However, phytoremediation is a slow process and requires more time than other processes to finish [11] and to identify the plant species capable of phytostabilization or photo-extracting heavy metals, depending on the objective.

Due to the aforementioned, the objective of this study is to identify the herbaceous flora of Tumbacucho with the highest value of importance and evaluate its potential capacity for phytochemical remediation in soils contaminated by mining activities.

2 Methodology

The identification of herbaceous species with the highest important value was carried out in the Tombacucho area. These selected species were assessed for their phytoremediation capacity in contaminated soils. Analyzing the accumulation of metals either at the root or aerial part of the plant, as well as identifying whether the species is phytoextractor and phytostabilization type through the Translocation Factor and Bioconcentration Factor, respectively.

The linear correlation statistical method was used, which made it possible to determine the relationship between the concentration of heavy metals in the wild flora and soil.

2.1 Calculation of Concentration Ratios

Bioconcentration Factor (BCF). To find the BCF, the concentration of the metal in the plant tissue (mg/kg) is divided by the concentration of the metal in the soil (mg/kg); using the following formulas [13]:

$$BCF\ Root = [metal]root/[metal]floor \tag{1}$$

$$BCF\ Air = [metal]aerial/[metal]floor \tag{2}$$

Where:

- BCF Root = Bioconcentration factor at the root of the plant;
- BCF Air = Bioconcentration factor in the aerial part of the plant;
- [Metal] root = Concentration of the metal only at the root of the plant in mg/kg;
- Metal] aerial = concentration of the metal only in the aerial part of the plant in mg/kg;
- [Metal] soil = Concentration of metal in the soil in mg/kg.

BCF on the aerial part of the plant:

- If BCF Air < 1, the plant is exclusive;
- If $1 <$ BCF Air < 10, the plant is accumulator;
- If BCF Air > 10, the plant is hyperaccumulator.

BCF at the root part of the plant:

- If BCF Root < 1, the plant is exclusive;
- If $1 <$ BCF Root < 10, the plant is accumulator;
- If BCF Root > 10, the plant is hyperaccumulator.

Translocation Factor (TF). For calculating the translocation factor, the concentration of the metal in the analyzed tissue in the aerial part is divided by the concentration obtained in the root [13]

$$TF = [metal]aerial/[metal]root \tag{3}$$

Where:

- [Metal] root = metal concentration in the root part of the plant tissue in mg/kg;
- [Metal] aerial = concentration in the aerial part of the plant tissue in mg/kg.

If the value obtained in the division is greater than 1, this indicates that the plant transfers the metal to the aerial part effectively. However, if the value obtained in the division is less than 1, this tells us that the plant does not transfer the metal to the aerial part effectively.

2.2 Test Method

The samples were analyzed in an accredited laboratory in the city of Cajamarca under the EPA 200.7 method. Rev 5.0 2001 [14].

3 Results and Discussion

The most important species in Tumbacucho were: the *Paspalum tuberosum* with an importance value of 87.88, the *Pennisetum* clandestinum with an importance value of 68.28, and *Trifolium repens* with an importance value of 16.54.

Figure 1 shows that the species *Paspalum tuberosum* and *Pennisetum clandestinum* have a higher concentration of Ag at the root than the aerial part of the plant, indicating low metal mobility.

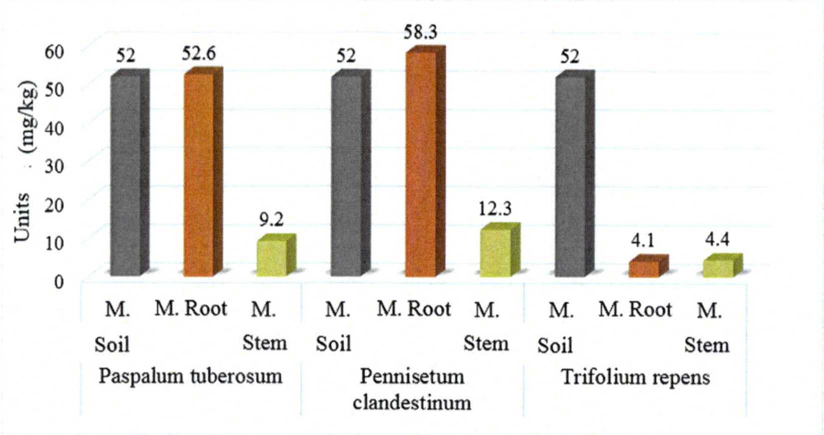

Fig. 1. Average Ag concentrations in samples of herbaceous flora and contaminated soil.

Figure 2 shows the concentrations of As in the Trifolium repens are higher in the aerial part of the plant than at the root of the plant, indicating the mobility of the metal towards the aerial part of the plant.

In Fig. 3, the species *Paspalum tuberosum* and *Pennisetum clandestinum* have a higher concentration of Cu at the root than in the aerial part, indicating low metal mobility. In contrast, the species Trifolium repens obtained a higher concentration of copper in the plant's aerial tissues, which indicates that the plant moves the metal towards the aerial parts.

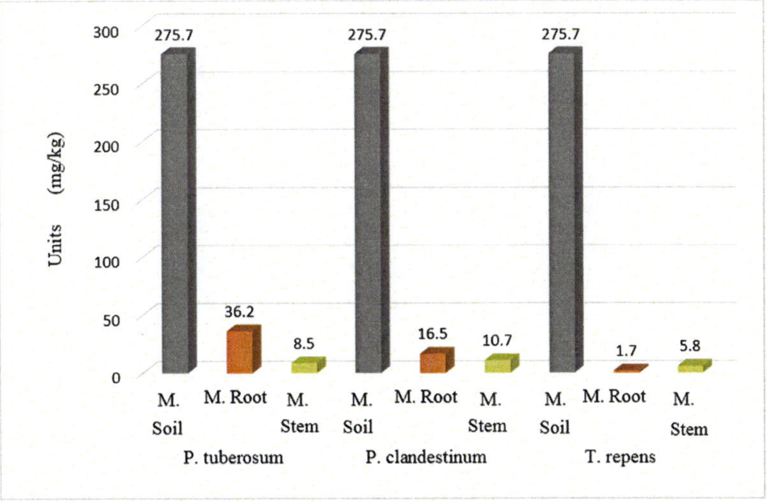

Fig. 2. Average As concentrations in samples of herbaceous flora and contaminated soil.

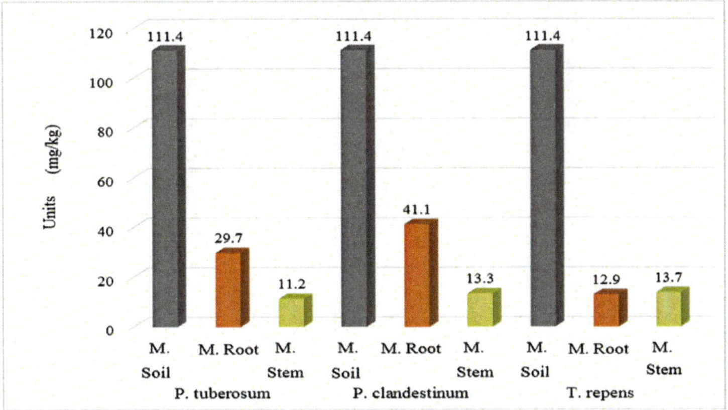

Fig. 3. Average Cu concentrations in samples of herbaceous flora and contaminated soil.

In Fig. 4, the species *Paspalum tuberosum, Pennisetum clandestinum,* and *Trifolium repens* have a higher concentration of Pb at the root than in the aerial part, indicating low metal mobility.

In Fig. 5, a species *Paspalum tuberosum* presents a higher concentration of Sb at the root (35.7 mg/kg) than in the aerial part of the plant (8.6 mg/kg), indicating that the species has potential for the stabilization of metals at its root since its translocation factor is less than 1 [13]. Similarly, the concentration of Sb in the species *Pennisetum clandestinum* and *Trifolium repens* have a higher concentration at the root than in the aerial part of the plant, indicating low mobility.

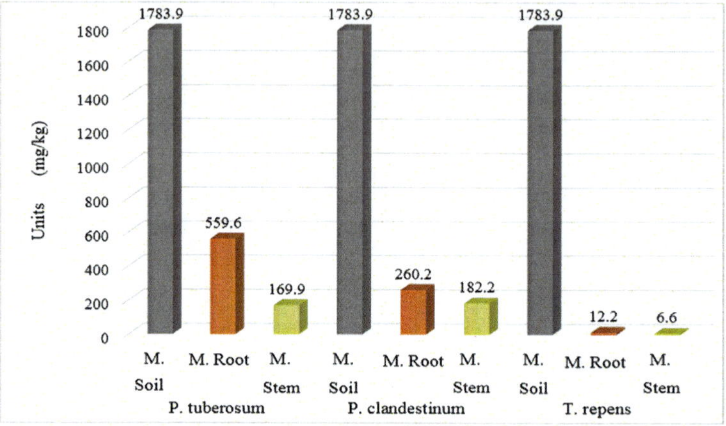

Fig. 4. Average Pb concentrations in herbaceous flora and contaminated soil samples.

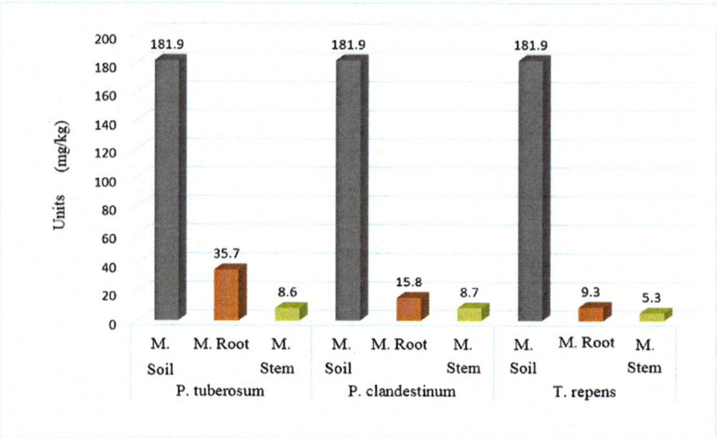

Fig. 5. Average Sb concentrations in samples of herbaceous flora and contaminated soil.

In Fig. 6, the species *Trifolium repens* and *Pennisetum clandestinum* show higher Zn concentrations in the aerial part compared to the root of the plant, meaning that the species have the potential to hyperaccumulate metals in the aerial part.

Besides, *Paspalum tuberosum* has a 384 mg/kg concentration in the aerial part and a concentration of 418.7 mg/kg, which corresponds to the part of the root. The concentration of metals in the aerial part between the concentration of metals at the root of the plant is less than 1, indicating that the species *Paspalum tuberosum* only has the metals phytostabilization potential at its roots.

Figure 7 shows that the concentration of Al in *Paspalum tuberosum* is higher at the root (222.7 mg/kg) than in the aerial part of the plant (95.1 mg/kg), meaning that its metals phytostabilization potential is found in its roots.

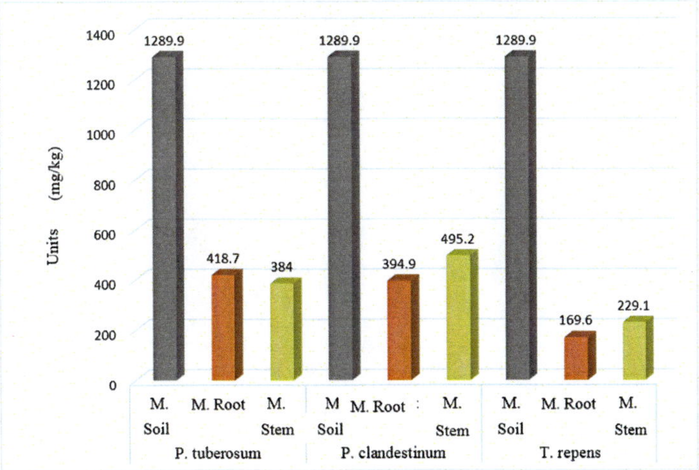

Fig. 6. Average Zn concentrations in samples of herbaceous flora and contaminated soil.

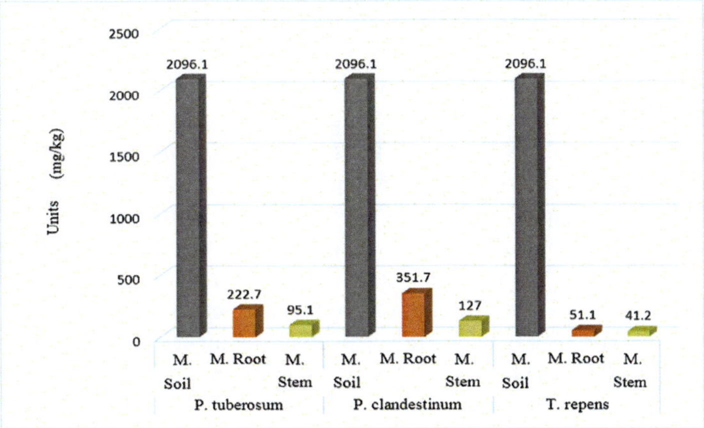

Fig. 7. Al average concentrations in samples of herbaceous flora and contaminated soil.

Figure 8 shows that the species *Paspalum tuberosum, Pennisetum clandestinum,* and *Trifolium repens* concentrate the greatest amount of Fe at the root than in the aerial part, indicating low metal mobility.

The following correlation coefficients were obtained according to the average concentration of metals in soil and plant.

- Correlation coefficient between soil metal concentration ratios and plant metal concentration ratios of the *Paspalum tuberosum*:
  $r = 0.994$ (very high positive direct correlation).
- Correlation coefficient between soil metal concentration ratios and plant metal concentration ratios of the *Pennisetum clandestinum*:

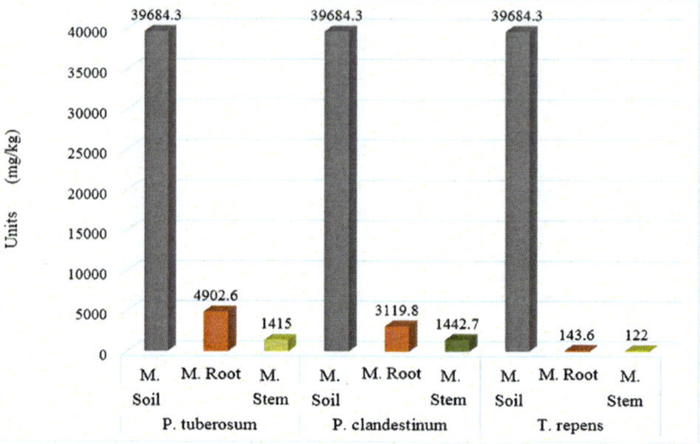

Fig. 8. Average Fe concentrations in grassland and contaminated soil samples.

$r = 0.987$ (very high positive direct correlation).

- Correlation coefficient between soil metal concentration ratios and plant metal concentration ratios of the *Trifolium repens*:
  $r = 0.460$ (moderate positive direct correlation)

The Pearson's correlation coefficient was considered to determine linear correlations in order to identify the relationship between the concentration of heavy metals in soil and the concentration of heavy metals in the species *Paspalum tuberosum, Pennisetum clandestinum,* and *Trifolium repens.* According to Ndako [15], the Pearson's correlation coefficient expresses or measures the degree of association or affinity between related variables. This indicates that the correlation coefficients were directly and very high to moderate positive, indicating that the relationship between these variables is given in a directly proportional way; therefore, as one of them increases, the other will increase. Consequently, the species *Paspalum tuberosum* has the highest correlation value with an r coefficient of 0.994, showing a better relationship between the concentrations of metals in the soil and metals in the plant.

The species studied in this research are part of the natural flora of high Andean ecosystems. These species have adapted to the extreme conditions present in the area due to the high load of metallic minerals and acidic pH of the soil [16]. The processes of adaptation of the studied species have taken place during the last 60 years of a stoppage in mining activities in the study area. That is, these species have repopulated the affected areas over time and have not been used in human remediation processes. Therefore, in natural processes, the absorption of minerals is a frequent phenomenon in several species studied in high Andean areas as observed in the research of Cahuana & Aduviri [17] with the species *Anomobryum prostratum, Marchantia Polymorpha.* Jara et al. [18] analyzed the species *Calamagrostis rigida* (*Kunth*) *Trin. ex Steud.* (*Poaceae*), and *Myriophyllum quitense Kunth,* and so we could mention much more research.

The plants in natural conditions, when they die, the nutrients that constitute them are returned to the soil, and other plants in their environment reuse these. The same happens with the heavy metals in their composition. It is important to mention that the

Trifolium repens is a cosmopolitan species and forms part of the diet of animals such as cattle, horses, and sheep, so we should avoid its use in the process of remediation of areas contaminated with heavy metals [19].

The authors consider that the species *Paspalum tuberosum* and *Pennisetum clandestinium*, being Andean species of wide distribution in Andean Pajonal ecosystems, can be applied in the remediation process because they are phytostabilizers.

It is necessary to develop future investigations that allow us to better understand the different stages of the plants in order to make an effective use of their properties. The absorption capacity of the plant depends on factors such as the ph of the soil, the rhizosphere, and the concentration of the metal in the soil [20].

4 Conclusions

The translocation factor (TF) and bioconcentration factor (BCF) were determined, of three species of herbaceous flora with the highest value of importance, the concentration of heavy metals in the roots of *Paspalum tuberosum*, which showed a Root BCF equal to 1.01 of Ag, meaning that the plant is Ag accumulator. Concerting other metals such as Cu, As, Sb, Pb, Zn, Al, and Fe, the Root BCF is less than 1, which indicates that the plant is exclusive to the mentioned metals.

The *Pennisetum clandestinium* showed a Root BCF equal to 1.12 Ag, which means that the plant is an accumulator of Ag. Concerning other metals such as Cu, As, Sb, Pb, Zn, Al, and Fe, the Root BCF is less than 1, indicating that the plant is exclusive to such metals. The *Trifolium repens* showed a BCF Root less than 1, indicating that the mentioned plant is exclusive to all metals.

The concentration of heavy metals in the aerial part of the *Paspalum tuberosum, Pennisetum clandestinium,* and *Trifolium repens* showed an aerial BCF less than 1, indicating that these plants are exclusive to all metals.

The heavy metal translocation factor (TF) in *Paspalum tuberosum* and *Pennisetum clandestinium* showed a TF less than 1, meaning that these plants do not effectively move heavy metals from the root tissue to the tissue of the aerial part of the plant, so their potential is to phytostabilize metals at their roots. Finally, the *Trifolium repens* showed a TF of Ag 1.07, As 3.41, Cu 1.06, and Zn 1.35, which means that the plant effectively moves heavy metals from the root tissue to the tissue of the aerial part of the plant [21], so its potential is to hyperaccumulate metals in the aerial part of the plant. About other metals such as Pb, Sb, Al, and Fe, the BCF Root, BCF aerial, and TF are less than 1, which means its potential to phytostabilize metals lies in its roots.

References

1. Martell N (2014) Accumulation of heavy metals in Beta vulgaris L. and Lolium perennial L. of Cuemanco soils. Fes Zaragoza, Mexico
2. Santillan T (2019) Los urgentes problemas ambientales de Hualgayoc, Asociación SER, 24 Enero 2019
3. Lapa J (2017) Lo que los ojos no ven. Grupo Editorial Arteidea, Lima

4. Onakpa MM, Njan AA, Kalu OC (2018) Una revisión de la contaminación por metales pesados de cultivos alimentarios en Nigeria. An Glob Health 84(3):488–494. https://doi.org/10.29024/aogh.2314
5. Ministerio del Ambiente de Perú (2020) Resolución Ministerial 238-2020-minem/dm. Aprueban actualización del Inventario Inicial de Pasivos Ambientales Mineros. Diario Oficial El Peruano. https://www.minem.gob.pe/_legislacionM.php?idSector=1&idLegislacion=13355
6. Martel G (2014) Analysis of the factors that influence the socioeconomic conflict of the Tantahuatay metallurgical mining project in Cajamarca department. Lima
7. Cornelis R (2007) General chemistry, sampling, analytical methods, and speciation. In: Handbook on the toxicology of metals. Elsevier, p 18. https://hdl.handle.net/1854/LU-369861
8. Panxeu G, Zhang Y, Xie H, Wei J, Zhang X, Wang J, Lou X (2020) Effect of cornstalk biochar on phytoremediation of Cd-contaminated soil by Beta vulgaris var. cicla L. Ecotoxicol Environ Saf 205(2020):111144. https://doi.org/10.1016/j.ecoenv.2020.111144
9. Zhang Z, Wu X, Tu C, Huang X, Zhang J, Fang H, Huo H, Lin C (2020) Relationships between soil properties and the accumulation of heavy metals in different *Brassica campestris L.* growth stages in a Karst mountainous area. Ecotoxicol Environ Saf 206(2020):111150. https://doi.org/10.1016/j.ecoenv.2020.111150
10. Sarwar N, Imran M, Shaheen MR, Ishaque W, Kamran MA, Matloob A, Rehim A, Hussain S (2017) Phytoremediation strategies for soils contaminated with heavy metals: modifications and future perspectives. Chemosphere 171:710–721. https://doi.org/10.1016/j.chemosphere.2016.12.116
11. Antoniadi V, Levizou F, Shaheen S, Sik Y, Sebastian A, Baum C, Prasad M, Wenzel W, Rinklebe J (2017) Trace elements in the soil-plant interface: phytoavailability, translocation, and phytoremediation–a review. Earth-Sci Rev 171(2017):621–645. https://doi.org/10.1016/j.earscirev.2017.06.005
12. Lacalle M, Aparcio J, Artetxe U, Urionabarrenetxea E, Polti M, Soto M, Garbisu C (2020) Gentle remediation options for soil with mixed chromium (VI) and lindane pollution: biostimulation, bioaugmentation, phytoremediation and vermiremediation. Heliyon 6:e04550. https://doi.org/10.1016/j.heliyon.2020.e04550
13. Olivares E, Peña E (2009) Bioconcentration of mineral elements in Amaranthus dubius (bledo, pyre), growing wild in crops in Miranda state, Venezuela, and used in food. https://www.cabdirect.org/cabdirect/abstract/20093351236
14. US EPA (2001) Method 200.7: Determination of Metals and Trace Elements in Water and Wastes by Inductively Coupled Plasma-Atomic Emission Spectrometry. Revision 5.0. Cincinnati, OH. https://www.epa.gov/sites/production/files/2015-08/documents/method_200-7_rev_5_2001.pdf
15. Ndako J, Olisa J, Ozoadibe O, Dojumo V, Fajobi V, Akinwami J (2020) Evaluation of the association between malaria infection and electrolyte variation in patients: use of Pearson correlation analytical technique. Inf Med Unlock 21(2020):100437. https://doi.org/10.1016/j.imu.2020.100437
16. Fondo Nacional del Ambiente (2005) Inventario, Diagnostico y priorización de los pasivos ambientales del Río Llaucano-Hualgayoc, Fondo Nacional del Ambiente Perú, Lima
17. Cahuana L, Aduvire O (2019) Bioacumulación de metales pesados en tejidos de vegetación acuática y terrestre evaluados en áreas donde existen pasivos ambientales mineros en el Perú. Rev Medio Ambiente Mineria 4:19–36
18. Jara-Peña E et al (2017) Acumulación de metales pesados en Calamagrostis rigida (Kunth) Trin. ex Steud. (Poaceae) y Myriophyllum quitense Kunth (Haloragaceae) evaluadas en cuatro humedales altoandinos del Perú. Arnaldoa 24(2):583–598
19. Hazrat A, Ezzat K, Ikram I (2019) Environmental Chemistry and Ecotoxicology of HazardousHeavy Metals: Environmental Persistence. Toxicityand Bioaccumulation. J Chem 2019(6730305):14

20. Rengel Z (2015) Availability of Mn. J soil sci Plant Nutr 15(2):397–409
21. Baker AJM, Brooks RR (1989) Terrestrial higher plants which hyperaccumulate metallic elements—a review of their distribution, ecology and phytochemistry. https://www.researchgate.net/publication/247713966_Terrestrial_Higher_Plants_Which_Hyperaccumulate_Metallic_Elements_A_Review_of_Their_Distribution_E

IoT- Based Indicator for Industrial Accident Risks

Igor T. T. Teixeira and Frank Herman Behrens(✉)

Pontifical Catholic University of Campinas, Campinas, SP 13086-900, Brazil
igor.ttt1@puccampinas.edu.br, frank@puc-campinas.edu.br

Abstract. The manufacturing industry is one of the most dangerous in terms of work safety. Among different kinds of accidents in industrial environments, the biggest causes involving humans are related to machines and equipment used for manufacturing. Although there are standards and regulations for machines' safe operation, some specific criteria could only be identified by specialists and managers in Environment, Health, and Safety (EHS). The internet of things (IoT) is a fundamental technology for Industry 4.0, bringing many benefits for automation and process control. Despite the increasing automation leading to a decrease in manual work, there is still a considerable presence of employees subject to accident risks. This work proposes the use of physical variables collected on machines in a production line to create a safety risk indicator. Considering that these variables are available in IoT-based monitoring systems, a method of analyzing accident risks based on multi-variable graphs obtained from the normalization of the monitored variables is proposed. This risk display method is believed to assist in safety analyzes by operators and specialists in a Safety Management System (SMS).

Keywords: Accident risks · Industrial automation · Internet of Things · Safety

1 Introduction

The industry has rapidly advanced production and world consumption, becoming one of the biggest players in the economic market. Many technologies were created and improved because of industrial advances, so that the industry's rapid adaptation allowed its evolution from the first industrial revolution to the recent boom in disruptive technologies. Despite this, some factors remained indispensable; the main one is the human presence in the manufacturing processes. The following sections introduce the current status on occupational safety as well as the concepts of IoT, IIoT, Cyber-Physical Systems, and Industry 4.0.

1.1 Industrial Safety

The industry environment is one of the most hazardous places for occupational accidents, and, according to International Labor Organization (ILOSTAT), 27% of the occupational injury cases happen in manufacturing industries [1]. In Brazil, this number rises to 28%, according to the National Classification of Economic Activities (CNAE) report [2].

Y. Iano et al. (Eds.): BTSym 2020, SIST 233, pp. 148–155, 2021.
https://doi.org/10.1007/978-3-030-75680-2_18

In fact, the manufacturing industries are among the most dangerous, considering that their work environments contain several types of risk, such as heavy cargo handling, high-temperature exposure, handling chemical agents, among others. A survey by the Occupational Safety and Health Observatory (Smartlab) found that 15% of accidents that lead to absence from work are caused by the use of machinery and equipment within the industry [3].

The ISO12100 and NR12 standards, for example, aim to structure the analysis and management of risks in machinery, establishing safety guidelines for workers and improving working conditions [4, 5]. Despite this, the regulations and guidelines presented in these standards do not meet specific criteria; therefore, there is a need to adapt them in the context of each industry sector.

In large corporations, this issue is addressed by the Environment, Health and Safety (EHS) area, which among other activities, is responsible for analyzing accident risk and preparing a technical document for diagnosing the machine, cross-checking the items described in NR12. Although supported by technical criteria, risk analysis depends on a specialist's view and observational reports from the employees. In other words, data entry is majorly manual, which can lead to poor data quality and inconsistencies in investigations; therefore the risks are not properly mitigated.

1.2 IoT, IIoT, CPS and Industry 4.0

Although well-established among researchers and enthusiasts of technology area, the concept of the Internet of Things (IoT) remains modern, once it can be understood as an objects' network or interconnected physical devices, capable not just to collect variables from an environment but also interact with it and with the real world through Cyber-Physical Systems (CPS). In addition, IoT is based on existent telecommunication technologies and infrastructures to enable access to data and information through internet applications [6].

Connecting unconventional objects to the internet offers the opportunity to improve industrial monitoring and management and, with this, a new subgroup, known as Industrial Internet of Things (IIoT), appears, opening the way for the improvement of manufacturing processes toward a more efficient and sustainable production. This way, the Industry 4.0 concept comes from the union between IIoT and CPS [7].

The biggest gain in IIoT is the process automation. Though this reduces the number of manual tasks, it is important to say that people will remain in industrial environments in the future [8]. Therefore, it becomes necessary to incorporate specific concerns related to work safety into IIoT, given that this topic is still poorly studied in the literature [9].

In this context, this work proposes to evaluate how IoT can be applied specifically in the analysis and management of industrial risks. To achieve this goal, we start from the premise of an industry with a functioning IoT infrastructure and based on the knowledge acquired in previous projects and research. With that, it is possible to structure a proposal for using data from a machine and the workstation to obtain a work safety indicator. In addition, the initial results are promising for the construction of a risk reporting method based on monitoring variables in the context of the IoT.

2 Literature Review

Before starting the methodology development, a general review of the available literature on risk management systems and IoT applications within the scope of work safety was carried out. Such studies help us understand the essential topics used in these themes, providing the basis for developing new projects and solutions.

2.1 Safety and Risk Management Systems

In general, safety is a concept related to the occurrence of unexpected events, such as risks of accidents and losses of various kinds (lives, assets, investment). Words like system and management are widely used together with organization, control, and leadership. Therefore, a Security Management System (SMS), as a combination of these three terms, can be understood as a means of managing risks in an environment.

According to [10], an SMS can be explained systematically from the point of view of a PDCA cycle (PLAN, DO, CHECK, and ACT). The PLAN stage is subdivided into Policy & Plan and Information & Reporting systems. In the DO stage, we have the Operation & Process and Risk Management systems. In the CHECK step, there is the Monitoring system. Finally, in the ACT stage, we have the Maintenance and Training & Learning systems.

The Information & Reporting system is supported by a comprehensive analysis using mathematical models and statistical tools to obtain a relationship between the data and the possible risks to the environment. The Monitoring system aims to observe the state of an organization's safety. Real-time information on safety performance can be obtained during the monitoring process, which acquires data from sensors on machines or operators [10].

As part of an SMS, Kudryavtsev et al. [11] present an example of a Risk Management System (RMS) based on statistical methods, which assesses risk as the composition of two indices: the probability of occurrence and the severity of an accident. After adding the multiplication of the indices for each of the criteria parameters by type of accident, a scale of 0 to 10 is obtained that classifies the risk among five intervals: negligible, low, medium, significant, and high. The RMS proposed by [11] not only defines guidelines for risk management, but also allows to analyze the risk status of an industry quickly and objectively, enabling assertive decision-making to prevent emergencies.

2.2 IoT Safety Applications

The use of IoT technologies, such as Wireless Sensor Network (WSN), Radio Frequency Identification (RFID), Bluetooth Low Energy (BLE), and Near Field Communication (NFC), proves to be advantageous in the detection, identification, and communication of risks in industrial environments, so that together with an SMS they can fill the gaps in security management.

In [12], several examples are presented of how sensor-based safety management systems can improve risk management in a construction site, continuously capturing dangerous situations and providing more accurate data, which can also be used as security indicators. According to the research in [12], most studies focus on the development of monitoring systems by obtaining and processing certain types of information from a work environment. Thus, future projects should aim to create a structure to combine different types of data, as well as investigating methods for quantifying safety risks, to automate and document near accident indicators.

The study presented by Gnoni et al. [13] provides an example of how IoT technologies can be integrated for security management in an industrial plant. A digital platform for risk management was developed, showing the results obtained from three different solutions, based on the Seveso Directive for the prevention of serious accidents. The prototype comprises a smart tag application to obtain maintenance data, a virtual aging sensor for equipment, and the monitoring of safe work in inspection activities. In the end, a tool was developed capable of monitoring, controlling, and inspecting dangerous equipment as well as the working environment, in addition to warning and making workers aware of the risks [13].

3 Development

Based on the premise of an industry with an already installed IoT infrastructure, containing sensors and actuators, networked equipment, and monitoring systems connected to the internet, this work proposes an SMS indicator based on the data captured and transmitted by this network in order to find situations that represent an imminent risk for operators of a workstation. The objective is to verify if it is possible to build an SMS based on the quantifiable information obtained by the IoT monitoring application.

3.1 A Case Study

At a workstation in industry X, a metal trimming process is carried out on a production line made up of cutting machines of various types. Although most machines are automated, some require manual control, and all depend on some kind of human intervention, both to trigger commands and to handle the product in the various stages of the process.

Considering the situation above, there are several types of accident risks inherent to a particular Y machine. Some examples are cutting, burns, injuries caused by projected parts or components, exposure to excessive noise, electric shock, among others. Although the machine complies with current regulations and is properly protected by safety devices, there are still chances of accident occurrence. For this reason, sensors are installed in key points of this machine, as well as the workstation, to monitor the physical variables directly or indirectly related to the risks already mentioned. Figure 1 shows examples of the variables monitored as a function of time. The vertical scale is dimensionless as explained in Sect. 3.2.

Some variables, such as temperature, are clearly associated with some type of risk, in this case, the risk of burning the operator when touching a certain unprotected part. Other variables such as vibration may not be directly related to risk, but they represent some inappropriate behavior of the machine, such as the poor fitting of a component or a function configured above normal limits, which may be indirectly associated with risk.

On the other hand, there are other types of variables categorized as behavioral variables. An operator's distance from his machine and the number of operators within a restricted space are examples of behavioral variables. These variables give rise to the concept of acceptable risk when the operator, through his behavior, positions himself at an adequate distance using the Personal Protective Equipment (PPE) necessary for the task. The operator's behavior can also minimize or even eliminate a risk when he is away from the machine's danger zone.

It is up to a safety specialist to raise probable risk-generating situations on the machines and define the set of variables that contribute to detecting a risk, based on existing norms or rules developed by the specialist himself, inferred by his knowledge of the occurrences of machine accidents.

3.2 Data Normalization

The physical variables captured from the monitored machine are expressed by numbers in varying ranges of value, with different scales, covering quite different amplitudes. For example, the electric current can be measured from 0 to 10 A, the temperature from −20 to 100 °C, and the rotation speed from 0 to 3600 RPM. These differences in value hinder the global observation of variables and the creation of more precise criteria for detecting the associated risk. Thus, data normalization is performed so that all monitored variables are expressed in the same scale and in the same range of numerical values.

The first step in this process is to identify the normal operating conditions of the analyzed machine. These conditions can be interpreted as nominal values of the data, and upon these values, the parameters and restrictions for the risk assessment method will be obtained. Table 1 presents examples of nominal values.

Table 1. Examples of nominal values for machine variables

Variable	Unit	Nominal value	Sensor range	Normalized value
Temperature	°C	40	0–100	4.0
Current	A	8	0–30	2,7
Speed	Rpm	400	0–1000	4.0
Vibration	g	0,5	0–1	5.0
Pressure	psi	50,5	0–100	5.1
Noise	dB	30	0–100	3.0

The normalization of the monitored variables is achieved considering a scale from 0 to 10 through the following calculation: the measured value is multiplied by 10 and divided by the maximum value of the respective sensor range. In Table 1, the nominal values were also normalized using the same calculation.

At the end of the normalization process, the variables will not be expressed in their respective units but on a dimensionless scale that will be used to classify the risk. Normalization allows the observation of all data in the same graph, as shown in Fig. 1.

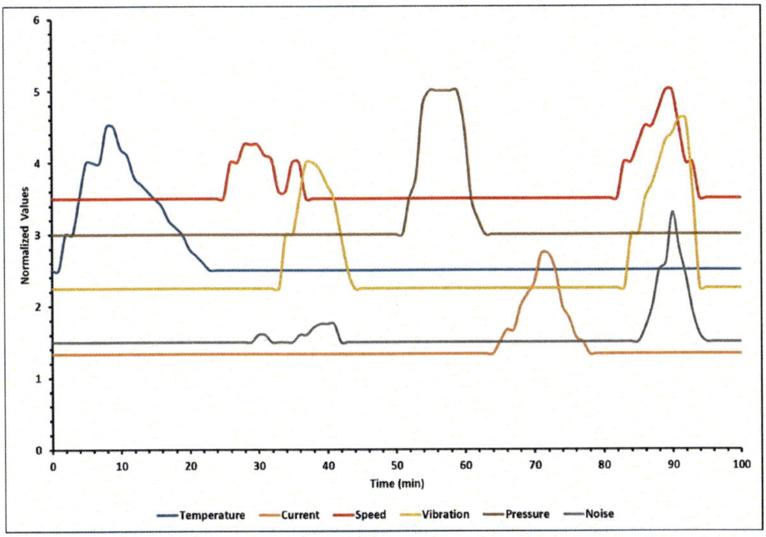

Fig. 1. Physical variables obtained from a machine

4 Results and Discussions

The visual management of variables in industries is one of the most used ways to display indexes and results for the management, as it seeks a simple understanding of all areas and departments. When it comes to the EHS area, displaying data using self-explanatory graphics makes it easier to understand the information and helps experts and other industry employees to know and monitor risks in different situations.

Figure 2 shows a radar graph plotted from some samples of the normalized variables shown in Fig. 1. Note that there are two ways of presenting the risk: one is by connecting the points of each normalized variable, forming characteristic polygonal figures; another is through a simple color code, representing a rating scale.

The presentation of risk by geometric shapes can lead to two interpretations: one considers the composition of certain variables to indicate a specific type of risk, and the other focuses on the arrangement of all variables for viewing the general safety picture, as shown in Fig. 2. In both interpretations, the variables are the same, as well as their nominal values, so it can be said that there is a polygon in the graph that represents the controlled level of risk or the safety limit.

In Fig. 2, the green polygon corresponds to the safe or acceptable risk level, as well as all points and shapes inside it are interpreted as normal, that is, with acceptable or low risk. The yellow and orange polygons, on the other hand, have points above the limit that indicate unusual situations, ranking the medium and high-risk scales, respectively.

So far, risk and its representation consider only physical variables, but in reality, the polygons shown in Fig. 2 will also be influenced by behavioral variables. Although hidden in visual management, they are of great relevance in the classification of risk. An example of this can be seen in the red polygon, in which all points are at expected levels except for the point corresponding to the electric current, which is much higher

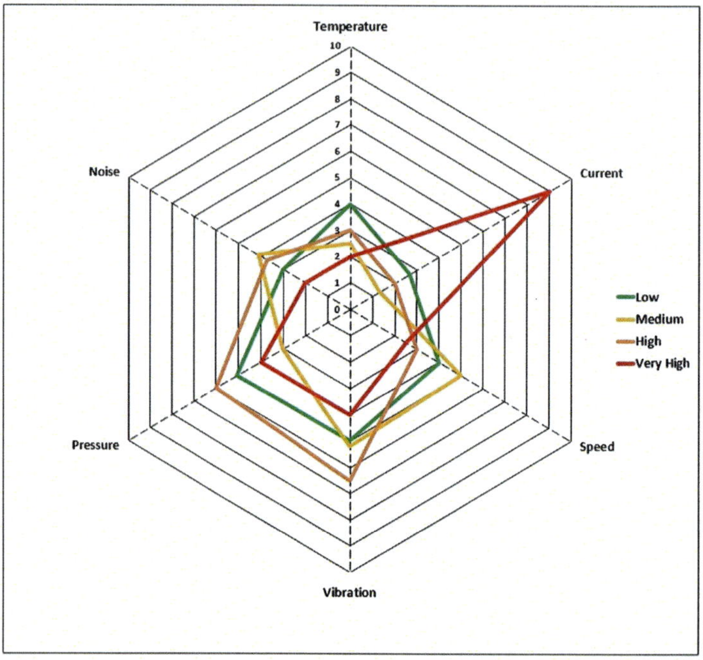

Fig. 2. Radar graph for visual management of risk

than the normalized limit. In this case, although it seems an exaggeration to classify the situation as a very high risk, it makes sense when the operator is supposed to approach a machine operating at a high level of current consumption.

5 Conclusions

This work carried out an initial assessment of the application of IoT as part of an SMS for security management in industrial environments. It was considered, as a case study, a machine belonging to a metal trimming process, to exemplify a real situation in a manufacturing industry, with the aim of discovering ways to display the risk of accidents in a simple and effective way.

To this end, risk identification was carried out based on the interpretation of physical variables obtained by sensors added to the machine and, subsequently, normalized in relation to their nominal operating values. EHS specialists are needed to define the degree of risk associated with the levels of normalized variables.

The visual management of the data plotted on a radar chart, as shown in Fig. 2, demonstrated its potential to assist operators, specialists, and managers in identifying dangerous situations, enabling better decision-making to mitigate risks. In addition to the physical variables, this method allows considering behavioral variables, which affect the shape of the polygons plotted on a radar chart.

It is important to note that the results obtained so far are preliminary, and future activities are planned to continue the development of a method for the detection, analysis, and management of accident risks based on IoT.

References

1. Nations U International Labour Organization (ILOSTAT). https://ilostat.ilo.org/
2. IBGE National Classification of Economic Activities (CNAE). https://concla.ibge.gov.br/busca-online-cnae.html?view=estrutura
3. MPT & OIT Occupational Health and Safety Observatory (SmartLab). https://smartlabbr.org/sst
4. International Organization for Standardization (2015) Safety of machinery—General principles for design—Risk assessment and risk reduction. https://www.iso.org/standard/51528.html
5. Tribunal Superior do Trabalho (TST) NR 12 - (Safety at Work in Machinery and Equipment). https://www.tst.jus.br/institucional
6. Gubbi J, Buyya R, Marusic S, Palaniswami M (2013) Internet of Things (IoT): a vision, architectural elements, and future directions. Futur Gener Comput Syst 29:1645–1660. https://doi.org/10.1016/j.future.2013.01.010
7. Sisinni E, Saifullah A, Han S et al (2018) Industrial internet of things: challenges, opportunities, and directions. IEEE Trans Ind Informatics 14:4724–4734. https://doi.org/10.1109/TII.2018.2852491
8. Gabriel M, Pessl E (2016) Industry 4.0 and sustainability impacts: critical discussion of sustainability aspects with a special focus on future of work and ecological consequences. Ann Fac Eng Hunedoara 14:131
9. Badri A, Boudreau-Trudel B, Souissi AS (2018) Occupational health and safety in the industry 4.0 era: a cause for major concern? Saf Sci 109:403–411. https://doi.org/10.1016/j.ssci.2018.06.012
10. Li Y, Guldenmund FW (2018) Safety management systems: a broad overview of the literature. Saf Sci 103:94–123. https://doi.org/10.1016/j.ssci.2017.11.016
11. Kudryavtsev SS, Yemelin PV, Yemelina NK (2018) The development of a risk management system in the field of industrial safety in the Republic of Kazakhstan. Saf Health Work 9:30–41. https://doi.org/10.1016/j.shaw.2017.06.003
12. Asadzadeh A, Arashpour M, Li H et al (2020) Sensor-based safety management. Autom Constr 113:103128. https://doi.org/10.1016/j.autcon.2020.103128
13. Gnoni MG, Bragatto PA, Milazzo MF, Setola R (2020) Integrating IoT technologies for an "intelligent" safety management in the process industry. Proc Manuf 42:511–515. https://doi.org/10.1016/j.promfg.2020.02.040

Failure Management in a Gold Mill Tailings Treatment Process

Mylena Vílchez-Torres[1](✉), Carlos Gastón Guevara Alejabo[1], Carlos Alberto Silvera Peña[1], and Roxana Elizabeth Mestanza Cacho[2]

[1] Universidad Privada del Norte, Jr. Irene Silva 231, Cajamarca, Peru
mylena.vilchez@upn.edu.pe
[2] Universidad Nacional de Cajamarca, Av. Atahualpa Km. 3, Cajamarca, Peru
remestanza@unc.edu.pe

Abstract. Despite various existing asset management methodologies, there are still many areas where they are yet to be implemented. This is mainly because they are not customized methodologies and their implementation, in many cases, generates several theoretical-practical gaps and inconsistencies, which hinder their application. Therefore, we must continue to deepen our understanding of these methodologies' application within several contexts, especially in the circumstances wherein machinery plays an important role in achieving machinery-intensive process deliverables. Accordingly, this study discusses a failure management case focusing on the machinery employed in the gold mill tailing treatment process at a gold mining process plant. This study was conducted using three of the most renowned asset management methodologies, i.e., reliability-centered maintenance (RCM), total productive maintenance (TPM), and prognostics and health management for maintenance (PHM), to produce a methodological failure management process specifically adapted to the plant situation. Through this process, the overall equipment effectiveness (OEE) increased from 31% to 87%, thus contributing to broadening the practical understanding of these methodologies.

Keywords: Failure management · TPM · RCM · PHM · Process plant · JKD

1 Introduction

Over the years, several methodologies for improving production process equipment performance have been developed [1] to achieve production objectives and safety and quality requirements at the lowest possible cost [1–5]. However, implementing a maintenance plan is difficult as several obstacles must be overcome if the methodologies are not entirely relevant to the reality in which they are intended to be applied. These methodologies steer the development of maintenance plans for equipment as well as decision-making on the strategies to be implemented [2]. Although some methodologies have been successfully applied in different companies, there is still a wide area where they are yet to be implemented. This may be because their application steps do not always constitute custom designs, often resulting in complex or incomplete implementations or in implementations that are only viable within the theoretical field, thus

Y. Iano et al. (Eds.): BTSym 2020, SIST 233, pp. 156–167, 2021.
https://doi.org/10.1007/978-3-030-75680-2_19

generating steep gaps between science and business [1, 2], feelings of rejection and frustration, and poor business competitiveness. This situation warrants further research on the use of these methodologies for solving real problems [6], widening our understanding of their machinery management applications within multiple contexts. Hence, the main contribution of this research study lies in addressing the constant failures reported for the gold mill tailing treatment process at a mining company. Here a failure management process was modeled after three of the most widely renowned maintenance management tools available: reliability-centered maintenance (RCM), prognostics and health management for maintenance (PHM), and total productive maintenance (TPM), thus crafting a more comprehensive methodological solution directed from the theoretical-practical standpoint [1, 7].

2 Methodology and Related Theories

RCM, TPM, and PHM are all renowned methodologies seeking to achieve the best possible equipment performance rates within the production processes. However, each of these uses its own distinctive approach. The following paragraphs provide an overview of each methodology and discuss their individual contributions to the methodological process used in this case study.

2.1 RCM

This methodology focuses on reliability [2], although technically, it is actually more focused on availability [8]. It has been mostly used in extremely critical sectors, but its general dissemination has been affected by a lack of accuracy in its steps and tools. The study used a 10-step RCM process developed for process plants in India. The first three stages collect detailed information on the system and its functions to construct adequate criteria for addressing the most significant failures. In stages 4–7, all failure modes are assessed to select those that would be critical to the system and determine the best courses of action. In stage 8, detailed procedures are developed for the tasks arising from the logical process sequence. Moreover, the last steps provide an assessment of the actions undertaken [2].

2.2 TPM

Japanese companies developed this tool in an effort to achieve zero failures, losses, breakdowns, defects, accidents, and pollution by merging preventive maintenance, production system maintenance, maintenance prevention, reliability engineering, and autonomous maintenance [9] into a shared responsibility system between the production and maintenance areas of the company [6]. The methodology permeates the entire organization, starting with a strong commitment from senior management, and seeks to consolidate itself as part of the corporate culture. Because there is no consensus on its implementation steps [10], this study used a 12-stage model divided into four phases [6, 10], even when some authors propose seven phases [9]. The first two phases work on pre-project aspects, the third focuses on the implementation itself, wherein preventive actions are taken, mainly through the incorporation of autonomous maintenance and training sessions, execution of planned maintenance actions, and early management. Finally, the last phase deals with the establishment of the continuous improvement cycle, Plan-Do-Check-Act (PDCA) [6, 10]. Despite its benefits, this methodology cannot yet address several issues, such as the fact that its preventive approach is not always the best option, and it sometimes even generates higher costs rather than addressing the failure itself. As TPM seeks to maximize operational productivity as measured by overall equipment effectiveness (OEE) [8], the Japanese Institute of Plant Maintenance (JIPM) developed technical and support pillars for its implementation [5, 9]. In addition, TPM relies on several prioritization-action and cause-effect (CE) analysis tools and programs such as zero breakdowns aimed exclusively at removing failures [6].

2.3 PHM

PHM is a methodology that comes from the medical sciences [7] and is mostly used in aeronautics. This methodology is recognized for its emphasis on data-driven diagnostics and prognostics of equipment degradation and decision-making processes. Although it has been recently favored by modern data management trends, such as big data and machine learning [4–7, 11], it still faces many challenges such as forecasting limitations in their practical applications, thus leading to situations wherein researchers still only manage at a theoretical level. For example, situations requiring large amounts of historical data, wherein information collection may be difficult and expensive, or models requiring real-time data, which are complicated by requiring systems capable of generating this type of information [4–7, 12]. Finally, regarding its status, some authors argue that PHM can be better described as a policy rather than as a methodology [13].

2.4 Case Study Failure Management

Table 1 lists the step sequence and methodologies that influence each step the most:

Table 1. Failure management steps for the gold mill tailings treatment process.

Steps	TPM	RCM	PHM
Commitment	*	*	
Process analysis		*	
Classification by failure modes		*	*
Prioritization	*	*	
Cause-Effect (CE) analysis	*		*
Implementation of corrective measures	*		
Post analysis	*		

Commitment. For the project to be successful, full commitment from all stakeholders is essential, starting with senior management. TPM and RCM promote this aspect more consistently [5, 6, 8, 9, 14].

Process Analysis. Understanding its configuration and the machinery intervention mechanism is critical. In this sense, RCM emphasizes system knowledge [2, 8].

Classification by Failure Modes. RCM better addresses this step, but PHM handles it in a more advanced manner through clustering algorithms [7, 13]. Regardless of the option used, failure models must be grouped, especially in scenarios where their identification may be difficult, such as in multiple failure scenarios associated with multiple causes [1, 6].

Prioritization. TPM addresses the issue in terms of selecting equipment exhibiting chronic, severe [5], or critical [9] losses. Conversely, RCM establishes criticality through a weighting method based on the impacts exerted by the failures on processes, safety, availability, and costs [2]. For this study, we decided to use a jack-knife diagram (JKD) as this diagram had already been used successfully by researchers on previous occasions and sorts failures into acute, chronic, and acute-chronic. The JKD version used in this study classifies failure modes according to the process downtime generated. Here, the most critical failures are those located in the upper right quadrant (acute-chronic), as they cause the longest downtimes more frequently. The next critical failures are in the upper left quadrant (acute), which are failures that cause downtimes but not frequently. Finally, we have failures in the lower right quadrant (chronic). These failures do not cause many downtimes but are frequent. Equation 1 expresses service downtimes due to maintenance actions for repairing failures [8–16]:

$$Downtime\,(i) = n(i) \times MTTR\,(i), \tag{1}$$

Where Downtime (i) is the idle time associated with the (i)th failure code, n (i) is the number of times the failure occurs, and MTTR is the mean time to repair, given, in turn, by Eq. 2 [8, 14].

$$MTTR = \frac{TF}{Number\ of\ Faults} \tag{2}$$

Where TF is the time required to repair the faults.

The limit value for acute and chronic failures is given by Eqs. 3 and 4 [15]:

$$Limit\,(MTTR) = \frac{D}{N}, \tag{3}$$

$$Limit\,(n) = \frac{N}{C}, \tag{4}$$

Where D is the sum of downtimes, N is the sum of ni, and C is the number of distinct failure codes.

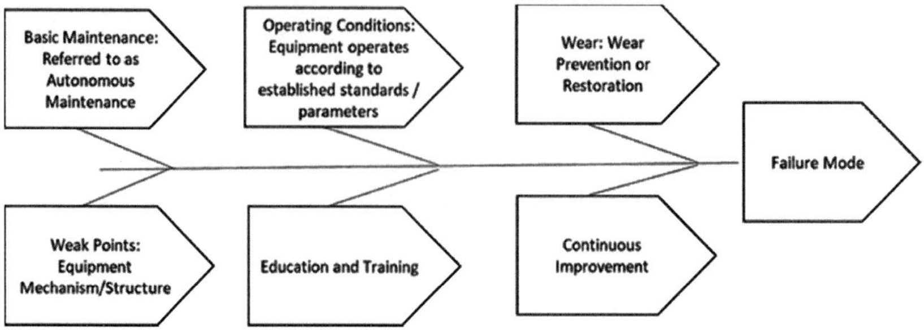

Fig. 1. Ishikawa diagram used for the CE analysis of failure modes.

CE Analysis. The Ishikawa diagram is a good tool for this purpose. However, the 5M are designed for manufacturing processes, which do not adapt well to the circumstances of this study [3]. Hence, the Ishikawa analysis areas were replaced with the six key focus areas from the TPM zero-breakdown program (Fig. 1) [6]:

Furthermore, instead of resorting to brainstorming, the PHM approach was selected, as it collects condition-based management (CBM) data, contextual data (operations, conditions, and regimes), and the maintenance data proposed by CRM [13].

Implementation of Corrective Measures. These measures were also taken from the TPM zero-breakdown program, which lists more precise actions for eliminating failures [6].

Post Analysis. TPM offers a more specific indicator for measuring equipment performance and the effect of interventions, known as OEE, as depicted in Eq. 5 [6–9, 14]:

$$OEE = OA \times OP \times OQ \tag{5}$$

Where OA is operational availability, OP is operational performance, and OQ operational quality.

Maintenance availability is given by Eq. 6 [12]:

$$A = \frac{MTBF}{MTBF + MTTR} \tag{6}$$

Where A is the availability and MTBF is the mean time between failures, as obtained from Eq. 7 [14].

$$MTBF = \frac{Operating\ Time}{Number\ of\ Faults} \tag{7}$$

Then OP and OQ are given by Eqs. 8 and 9 [14]:

$$OP = \frac{Time\ of\ Correct\ Operation}{Operating\ Time} \tag{8}$$

$$OQ = \frac{Planned\ Pumped\ Volume}{Pumped\ Volume}. \tag{9}$$

As part of OQ, TPM usually considers the number of conforming parts against the total number of parts, but the nature of this process is different, and the quality depends on whether the process is able to pump an optimal flow, framed within a parameter that keeps the process operational. Finally, Table 2 gives OEE classification according to the percentage reached and its possible consequences [17]:

Table 2. OEE Classification

OEE	Qualifier	Consequences
<65%	Unacceptable	Important economic losses, low competitiveness
≥65%–<75%	Regular	Economic losses. Acceptable only if it is part of an improvement process
≥75%–<85%	Acceptable	Slight economic losses. Slightly low competitiveness
≥85%–<95%	Good	Good competitiveness. "World Class" values
≥95%	Excellent	Excellent competitiveness

3 Case Study Results

The study was developed within the process area of an open-pit gold mine at 3,600 m above sea level, and the results below represent the first six months of application in 2019. The enterprise manager authorized this study, and the data were provided exclusively for academic purposes on the anonymity condition.

3.1 Commitment

Despite previous experiences, we were able to secure commitment from the process manager, process maintenance superintendent, supervisor, maintenance and engineering planner, hygienist, and other stakeholders in the process maintenance and operations division, who enabled the rest of the steps through the creation of a committee.

3.2 Process Analysis

The gold mill tailing treatment process is the process that fulfills the function of pumping the waste from the leached pulp treatment process towards the plant's residues deposit [18]. It pushes gold mill tailings thickener waste, which is the last stage of the counter-current decantation circuit used to recover mineral ore, toward a tailings pond, where it is diluted using low-cyanide process water before being pumped into a waste reservoir through a system of pumps and pipes. For the pilot plan, four Warman 10/8 ST-AHP pumps were selected. These pumps are critical because they set the pace of the process. Although the company does not use OEE, it was calculated to establish a baseline. The resulting average was 31%, which is well below the 90% threshold established by the committee, thus deeming it unacceptable.

3.3 Classification by Failure Modes

In 2018, 42 failures were reported, which, after being aggregated, provided C = 19 failure modes, N = 42 h, and D = 205.04 h.

3.4 Prioritization

The JKD denoted two failure modes of interest: solution leaks through oakums, deemed a chronic failure, but which, according to the plans, are within their use cycle, and solution leaks through pump housing seals, deemed an acute-chronic failure that is causing the largest number of downtimes (Fig. 2):

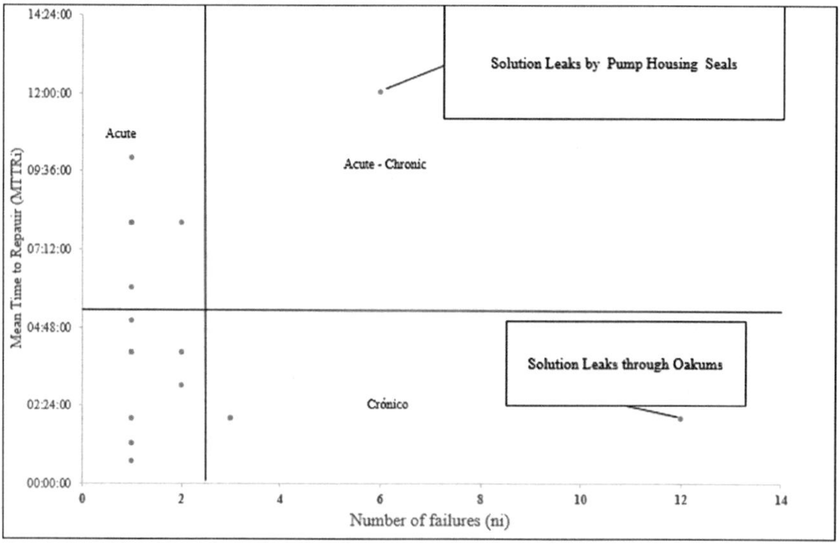

Fig. 2. Jack knife diagram for the prioritization of failure modes.

3.5 CE Analysis

A CE analysis was conducted for the solution leaks through pump housing seals, which was selected as our main failure of interest. Focusing on the six aspects of the zero-breakdown program, the data analysis evidenced premature wear in internal pump components, which needed to be replaced after an average of 1,400 h when the initial projection was 3,000. Regarding basic maintenance, there was no practice of delegating maintenance functions to operations personnel. The operating conditions reported daily vibration levels exceeding 0.5 mm/s (gE), which is the maximum acceptable operation level according to the manual. This vibration is caused by cavitation, which causes accelerated wear of the wetted parts. In addition, the pumped material is slurry, fluids with high contents of solids, which wear out internal components. The equipment manual indicates that the percent of solids should not exceed 69%; however, the process design allows up to 71% of solids by weight. The system must work at a minimum solution tank level of 55% to avoid cavitation, but we found several process downtimes records due to lower levels. The pumps are designed to pump between 738 and 986 m3 per hour; however, the reports evidenced both lower and higher flow rates, which also contributes to premature internal component wear. Regarding weak points, we detected that the internal components were metallic. Nevertheless, other options are currently available in the market that could improve the internal structure of the pump. The company has a complete training program but does not include any training related to autonomous maintenance. As observed in Fig. 3, the premature wear of internal components is the main cause associated with solution leakage from pump housing seals (Fig. 3).

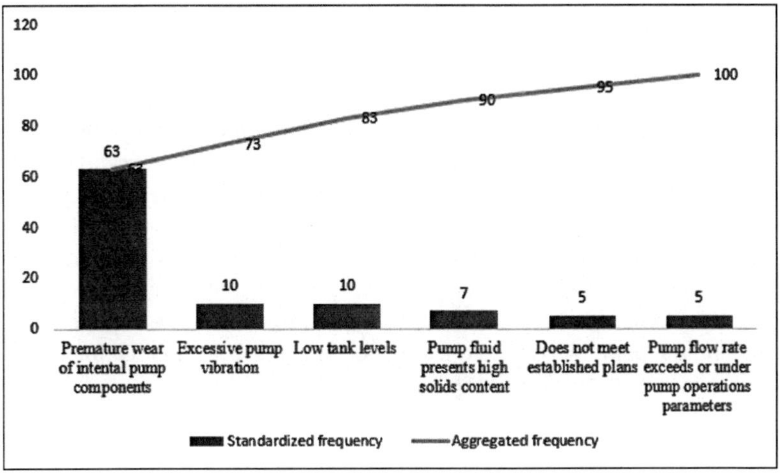

Fig. 3. Prioritization of failure causes.

3.6 Implementation of Corrective Measures

Working together with the logistics and purchase departments, the committee managed to replace all internal metallic components with rubber components. In the tests performed with the vendor, a metallic suction disk exhibited a weekly wear ratio of 1.6 mm from its original thickness of 50 mm, whereas a rubber suction disk, under the same operating conditions, presented a weekly wear ratio of 1.44 mm from its original 45 mm thickness. Similarly, a metallic coating presented a weekly wear ratio of 1.3 mm, whereas a rubber coating presented a weekly wear of 0.70 mm. Through this improvement, the company reached the 3,000 expected hours of operation and saved USD 20,425 in component replacement parts (Table 3).

Table 3. Savings generated by purchasing Rubber Components.

Complete material description	Weir pump metal component price (USD)	Weir pump rubber component price (USD)
Suction disc; Warman 10/8F Pump	7 831	2 599
Press disc	4 025	3 621
Discharge joint	525	1 035
5-blade impeller	11 016	7 463
Volute	14 257	2 511
Total	37 654	17 229
Savings = 20 425 USD		

Autonomous maintenance was implemented by assigning responsibilities through a daily checklist, including leak inspections, pitting inspections, operation parameter checking, lubrication and greasing verification, equipment cleaning, and minor repairs such as adjusting seal guards or tightening joints. In addition, a pump checklist was implemented in each shift to monitor operating parameters. This checklist included solids percentage, flow rate in m^3/h, motor noise, and tank levels. Finally, the following training modules were added to the usual training program for process workers: principles of pump operation, pumping system operation parameters, basic lubrication and failure analysis principles, and issue reporting.

3.7 Post Analysis

Although the 90% goal was not achieved, the average OEE reached 87%, improving the rating and also yielding world-class values at a good level (Fig. 4).

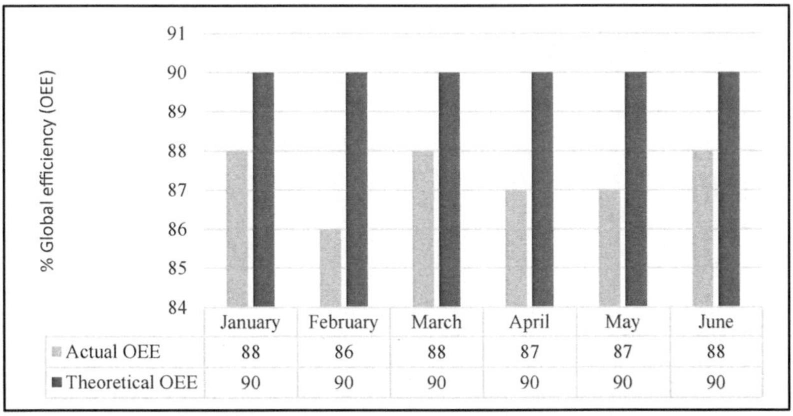

Fig. 4. Post-intervention OEE - January–June, 2019.

4 Discussion

Managing failures can be challenging, especially in contexts such as the one described in this study. Here, the process uses complex and expensive machinery from start to finish, and this machinery is constantly exposed to demanding jobs such as pumping sludge and other critical mining operation processes [1] such as the treatment of gold mill tailings. Consequently, multiple failure scenarios, whose management is not always addressed by existing theoretical and practical methodological proposals, constantly arise [2].

In fact, as evidenced in this study, the methodologies usually respond to the what (question) through their implementation steps but do not necessarily provide a response for the how because the contexts in which they were originally developed or implemented differ from those in which they are reproduced. As an example, we can mention the CE analysis promoted by TPM through the Ishikawa diagram [6], which is undoubtedly a good failure detection tool, but it does not quite fit our case study.

For instance, in the "materials" pillar, the diagram describes defects in parts or finished products [12], which is not compatible with the treatment of gold mill tailings, where the concept of finished pieces does not exist. This same aspect was revealed when determining the quality for OEE calculation, which was again referred to in terms of finished parts [6, 14]. However, in our case, quality can be explained by the extent to which the process pumps the adequate volume required by the system to remove waste without generating interruptions to the main process it serves. These minor details generate failures in practice because process managers usually tend to dismiss methodologies exhibiting these inconsistencies, considering them inappropriate for their reality. In addition, each methodology, despite sharing the same purpose, contains trends such as RCM, which is associated with reliability; TPM, which is associated with autonomous and preventive maintenance [2–9]; and PHM, which is associated with big data management and equipment degradation forecasting [4, 7, 11, 13].

Hence, each methodology provides more information based on the concepts they individually promote and generates gaps when implemented. Therefore, in the study, we maintained a flexible position, not strictly adhering to a single methodology but rather toggling among three methodologies, grounding each step against the individual methodology that presented a better vision of what had to be done, and this was extremely helpful. Finally, using prioritization tools such as DJK and grouping by failure modes [2, 15, 16] proved essential for organizing information and steering improvement actions in a scenario where failures may occur at multiple points and in multiple ways, often generating competing impacts [1].

5 Conclusions

We were able to manage the failures reported in the gold mill tailings treatment process at a mining company under study through a relevant selection of contributions from different methodologies. For example, the contributions from RCM were mainly in the initial analysis phase because this methodology helped us better understand the process and machinery used, as well as in the classification of failure modes and prioritization according to criticality. TPM provided a better framework for the implementation and post-evaluation phases, while PHM provided better insights into the two moments where the process required data analysis, classifying failures, and collecting data for the CE analysis. Overall, a new custom methodological process was developed based on the particular characteristics of the plant. Finally, this approach was clearly reflected in the OEE increasing to world-class levels, which also contributed to extending our practical understanding of these methodologies.

References

1. Lundgren C, Skoogh A, Bokrantz J (2018) Quantifying the effects of maintenancea literature review of maintenance models. In: Lihui W (eds) 51st CIRP conference on manufacturing systems, vol 72. Procedia CIRP, Stockholm, pp 1–1554. https://doi.org/10.1016/j.procir.2018.03.175
2. Vishnu CR, Regikumar V (2016) Reliability based maintenance strategy selection in process plants: a case study. In: Rajesh B, Lijo P (eds) 1st Global colloquium on recent advancements and effectual researches in engineering, science and technology, RAEREST, vol 25. Procedia Technology, Kottayam, pp 1–1334. https://doi.org/10.1016/j.protcy.2016.08.211
3. Simanová Ľ, Gejdoš P (2015) The use of statistical quality control tools to quality improving in the furniture business. In: Sujová A, Krajčírová L (eds) International scientific conference: business economics and management, BEM2015, vol 34. Procedia Economics and Finance, Efes, pp 1–704. https://doi.org/10.1016/S2212-5671(15)01630-5
4. Fink O, Wang Q, Svensén M, Dersin P, Lee W-J, Ducoffe M (2020) Potential, challenges and future directions for deep learning in prognostics and health management applications. Eng Appl Artif Intell 92:103678. https://doi.org/10.1016/j.engappai.2020.103678
5. dos Reis MDOR, Godina R, Pimentel C, Silva FJG, Matias JCO (2019) A TPM strategy implementation in an automotive production line through loss reduction. In: Ryan A, Gordon S, Tiernan P (eds) 29th International conference on flexible automation and intelligent manufacturing, FAIM2019, vol 38. Procedia Manufacturing, Limerick, pp 1–1848. https://doi.org/10.1016/j.promfg.2020.01.173
6. Cuatrecasas L (2012) Organización de la producción y dirección de operaciones. Ediciones Diaz de Santos, Madrid

7. Xia T, Dong Y, Xiao L, Du S, Pan E, Xi L (2018) Recent advances in prognostics and health management for advanced manufacturing paradigms. Reliabil Eng Syst Saf 178:255–268. https://doi.org/10.1016/j.ress.2018.06.021
8. Pascual R (2009) El arte de mantener. 3ra edn. Centro de Minería Pontificia Universidad de Chile, Santiago de Chile
9. Ribeiro H (2016) TPM guia de implantação: a metodologia para o sucesso do TPM. 1ra edn. PDCA Editora, São Caetano do Sul-SP
10. Sarria Yépez MP, Fonseca Villamarín GA, Bocanegra CC (2017) Modelo metodológico de implementación de lean manufacturing. Rev EAN 83:51–71. https://doi.org/10.21158/01208160.n83.2017.1825
11. Baker W, Nixon S, Banks J, Reichard K, Castelle K (2020) Degrader analysis for diagnostic and predictive capabilities: a demonstration of progress in DoD CBM+ initiatives. In: Kilicay-Ergin N, Dagli CH (eds) Complex adaptive systems, Malvern, vol 168. Procedia Computer Science, Pennsylvania, pp 1–296. https://doi.org/10.1016/j.procs.2020.02.253
12. dos Santos Silva FJ, Viana HRG, Queiroz ANA (2016) Availability forecast of mining equipment. J Qual Maintenan Eng 22(4):418–432. https://doi.org/10.1108/JQME-12-2015-0067
13. Calabrese F, Regattieri A, Botti L, Galizia FG (2019) Prognostic health management of production systems. New proposed approach and experimental evidences. In: Dagli C, Süer G (eds) 25th International conference on production research manufacturing innovation: cyber physical manufacturing, vol 39. Procedia Manufacturing, Chicago, pp 1–1976. https://doi.org/10.1016/j.promfg.2020.01.333
14. Ribeiro IM, Godina R, Pimentel C, Silva FJG, Matias JCO (2019) Implementing TPM supported by 5S to improve the availability of an automotive production line. In: Ryan A, Gordon S, Tiernan P (eds) 29th International conference on flexible automation and intelligent manufacturing, FAIM2019, vol 38. Procedia Manufacturing, Limerick, pp 1–1848. https://doi.org/10.1016/j.promfg.2020.01.128
15. Knights P (2004) Downtime priorities, jack-knife diagrams, and the business cycle. MJ Maintenan J 17(2):15–22
16. Vílchez M, Oblitas J, Castro W (2020) Optimization of the replacement time for critical repairable components. DYNA 87(214):93–99. https://doi.org/10.15446/dyna.v87n214.84509
17. Sayuti J, Syarifuddin F (2019) Analysis of the overall equipment effectiveness (OEE) to minimize six big losses of pulp machine: a case study in pulp and paper industries. In: International conference on science and innovated engineering, I-COSINE, vol 536. IOP Publishing, Malaca, p 011001. https://doi.org/10.1088/1757-899X/536/1/011001
18. Ministerio de Energía y Minas. https://www.minem.gob.pe/descripcion.php?idSector=4&idTitular=1194

Video Transmissions in Networks with Packet Loss

A Practical Study for Teaching and Learning on Computer Networks

Rafael Fernando Diorio(✉)

Instituto Federal de Educação, Ciência e Tecnologia de São Paulo (IFSP), Av. Dr. Ênio Pires de Camargo 2971, Capivari, SP 13360-000, Brazil
rafael.diorio@ifsp.edu.br

Abstract. Packet loss is one of the main factors that can affect and/or impair the video's quality, with a direct influence on the Quality of Experience (QoE) perceived by the end-user. For this reason, investigating the effects of packet loss on the quality of the video is crucial and could lead to a better understanding of the QoE perceived by the end-user. In this context, for purposes of teaching and learning on computer networks, this paper discusses a practical study on video transmissions in networks with packet loss. For this, based on a reference scenario, a computational environment is used for purposes of experimentation and discussion, with video transmissions under normal network conditions and with packet loss emulation of 5%, 10%, 15%, and 20%. The experiments and discussions are based in a computer network laboratory carried out with second-year students of an undergraduate computer course to help them understand the effects of packet loss on the quality of the video, as well as on the QoE perceived by the end-user. This work aims to complement other approaches to teaching and learning on computer networks.

Keywords: Computer networks · Teaching and learning · Experimentation

1 Introduction

Video services occupy a substantial percentage of today's Internet traffic. On the other hand, delivering high-quality video content to users is a challenge that comprises several factors. Among these factors, we can highlight issues related to the type of video application, video traffic characteristics, network characteristics and conditions, and expectations of the end-user, among others.

In this context, as an example, concerning to type of video application and video traffic characteristics, many applications are acutely sensitive to packet delay, jitter, and loss, requiring sufficient bandwidth capacity, minimal delay and jitter, and low or no packet loss [1–3]. If one or more application requirements are not met, the video's quality can be significantly affected. Concerning network characteristics and conditions, the

Y. Iano et al. (Eds.): BTSym 2020, SIST 233, pp. 168–177, 2021.
https://doi.org/10.1007/978-3-030-75680-2_20

Internet's best-effort model does not provide, for example, guarantees of timing preservation between packets, guarantees of orderly packets delivery, guarantees of packets delivery, and guarantees of Quality of Service (QoS) for multimedia applications, not being the best packet delivery service for all purposes [4–6]. In turn, concerning user expectations, different factors influence their perception of Quality of Experience (QoE), such as system factors (e.g., bandwidth, delay, jitter, and packet loss) and human factors (e.g., user preference, motivation, and expectation), among others [7, 8]. Due to these and other factors, delivering high-quality video content to users is a challenge.

In this scenario, packet loss is one of the main factors that can affect and/or impair the video's quality, with a direct influence on the QoE perceived by the end-user. For this reason, investigating the effects of packet loss on the quality of the video is crucial and could lead to a better understanding of the QoE perceived by the end-user. Thus, several recent works discuss the effects/impacts of packet loss on the quality of the video and/or on the end user's perceived quality [9–14].

In this context, for purposes of teaching and learning on computer networks, this paper discusses a practical study on video transmissions in networks with packet loss. For this, based on a reference scenario, a computational environment is used for purposes of experimentation and discussion, with video transmissions under normal network conditions and with packet loss emulation of 5%, 10%, 15%, and 20%. The experiments and discussions are based in a computer network laboratory carried out with second-year students of an undergraduate computer course to help them understand the effects of packet loss on the quality of the video, as well as on the QoE perceived by the end-user. This work aims to contribute to other methods of teaching computer networks [15].

The rest of this paper is organized as follows: Sect. 2 describes the reference scenario explored in this work. Section 3 describes the materials and methods. Section 4 discusses the experimental results and, finally, Sect. 5 presents the conclusion and future work.

2 Reference Scenario

The reference scenario for the practical study on video transmissions in networks with packet loss discussed in this work is shown in Fig. 1.

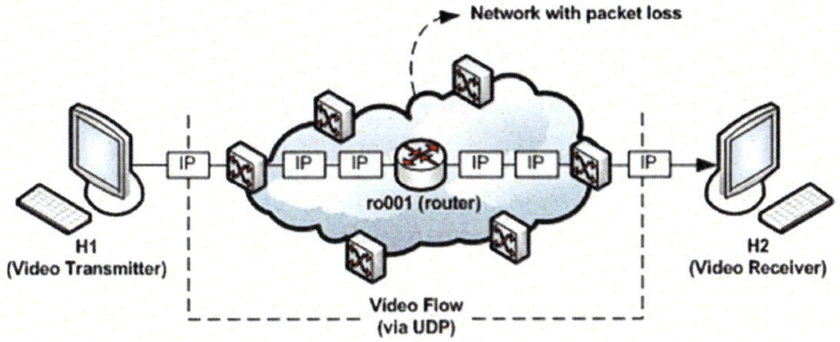

Fig. 1. Reference scenario explored in this work: The video transmitter host (H1), the video receiver host (H2), and the router host (ro001).

As shown in Fig. 1, the reference scenario explored in this work consists of three hosts: a video transmitter host (H1), a video receiver host (H2), and a router host (ro001). Both hosts are interconnected through a virtual wired network simulating a simplified Internet scenario (without configurations of Autonomous Systems, inter-AS and intra-AS routing, etc.).

In this scenario, the hosts H1 and H2 are logically configured on different networks, acting as end systems of the Internet. The network of the host H1 is 220.220.220.0/24, and the network of the host H2 is 221.221.221.0/24. The host ro001 is the router that interconnects both networks. Regarding the IP addressing settings, the host H1 was configured to use the IP address 220.220.220.220/24, and the host H2 was configured to use the IP address 221.221.221.221/24. In turn, the host ro001 was configured to use IP addresses 220.220.220.254/24 and 221.221.221.254/24.

3 Materials and Methods

3.1 Practical Study Materials

To implement the reference scenario shown in Fig. 1, the hosts H1 and H2 were configured using the Microsoft Windows 7 operating system, and the host ro001 was configured using the Linux CentOS 7 operating system. Both hosts (H1, H2, and ro001) were configured using virtual machines based on Oracle VM VirtualBox [16].

In this scenario, configurations to emulate packet loss were carried out on host ro001 using the NetEm solution [17] (via "tc" command-line tool). Initially, no packet loss emulation was configured, with video transmission under normal network conditions. The other transmissions were then carried out with packet loss emulation starting at 5%, with increments of 5% for each new experiment until the limit of 20%. Such definitions resulted in 5 different experiments, as shown in Table 1.

Table 1. Overview of the experiments.

Experiment	Characteristic
1	Video transmission under normal network conditions
2	Video transmission with 5% packet loss
3	Video transmission with 10% packet loss
4	Video transmission with 15% packet loss
5	Video transmission with 20% packet loss

In this context, no configuration to emulate packet loss was carried out on the host ro001 for experiment 1, with video transmission under normal network conditions. On the other hand, for the other experiments, considering each experiment's respective characteristics, configurations to emulate packet loss were carried out on host ro001. Thus, as an example, Table 2 shows the commands used on host ro001 to emulate packet loss, referring to experiments 2, 3, 4, and 5 shown in Table 1.

Table 2. Commands used on host ro001 to emulate packet loss referring to experiments 2, 3, 4, and 5 shown in Table 1.

```
tc qdisc del dev enp0s9 root
tc qdisc add dev enp0s9 root netem loss
PLOSS
```

The "PLOSS" directive shown in Table 2 refers to the percentage of packet loss emulation for experiments 2, 3, 4, and 5 shown in Table 1, defined as 5% for experiment 2, 10% for experiment 3, 15% for experiment 4, and 20% for experiment 5.

In this scenario, the video transmissions were carried out from H1 to H2 using the EvalVid framework [18], via UDP, and based on the same reference video (video with 500 frames, a frame rate of 50 fps, and 1280 × 720 resolution in 4:2:2 format), encoded via FFmpeg [19] with H.264/MPEG-4 AVC codec and bitrate of 9.5 Mbps.

In this context, as an example, Table 3 shows the commands used to encode the reference video, Table 4 shows the commands used to generate the reference files to verify the video quality, Table 5 shows the commands used by H1 to transmit the video content, Table 6 shows the commands used by H2 to receive the video content, and Table 7 shows the commands used to reconstruct the video received by H2 and to generate the reference file to verify its quality.

Table 3. Commands used to encode the reference video.

```
ffmpeg -i reference_video.y4m -c:v libx264 -r 50 -b:v 9.5M
reference_video.m4v
MP4Box -hint -mtu 1024 -fps 50 -add reference_video.m4v
reference_video.mp4
```

Table 4. Commands used to generate the reference files to verify the video quality.

```
ffmpeg -i reference_video.y4m reference_video.yuv
ffmpeg -i reference_video.mp4 reference_video_ref.yuv
```

Table 5. Commands used by H1 to transmit the video content.

```
WinDump.exe -n -tt -v udp port 12345 > sd
mp4trace.exe -f -t UDP -s 221.221.221.221 12345
reference_video.mp4 > st
```

Table 6. Commands used by H2 to receive the video content.

```
nc.exe -lup 12345 > null
WinDump.exe -n -tt -v udp port 12345 >
rd
```

Table 7. Commands used to reconstruct the video received by H2 and to generate the reference file to verify its quality.

```
etmp4.exe -f -x sd rd st reference_video.mp4 video_received
ffmpeg -i video_received.mp4 video_received.yuv
```

Concerning QoE, the evaluations were carried out comparing the reference file of the original video with the reference file of the received video, using the PSNR (Peak Signal-to-Noise Ratio) [7, 8] as a reference metric. As an example, Table 8 shows the command used to calculate the PSNR of the video received by H2, comparing the reference file of the original video with the reference file of the received video.

Table 8. Command used to calculate the PSNR of the video received by H2, comparing the reference file of the original video with the reference file of the received video.

```
psnr.exe 1280 720 422 reference_video.yuv video_received.yuv >
psnr_video_received.txt
```

Then, the obtained PSNR is mapped to MOS (Mean Opinion Score) [7, 8], as described in Ref. [8] and shown in Table 9.

Table 9. Mapping of PSNR to MOS [8].

MOS	Quality	Perception	PSNR (dB)
5	Excellent	Imperceptible	>37
4	Good	Perceptible	31–37
3	Fair	Slightly annoying	25–31
2	Poor	Annoying	20–25
1	Bad	Very annoying	<20

For reference purposes, the PSNR of the original video after its encoding was calculated as 38.17, with quality considered "excellent" when mapped to MOS (Table 9).

3.2 Practical Study Methodology

The practical study described in this work was carried out to complement the theoretical concepts about one of the contents of the computer networks course, in this case, related to multimedia applications. The experiments and discussions were based in a computer network laboratory carried out with second-year students of an undergraduate computer course to help them understand the effects of packet loss on the quality of the video, as well as on the QoE perceived by the end-user.

In this context, initially, the fundamental theoretical concepts related to the subject were discussed. These concepts were based on one of the books described in the bibliography of the course. For theoretical concepts, two 50-min classes were used. The concepts were presented through slides in a purely theoretical class.

Then, a practical laboratory was carried out with the students. The objective of the laboratory was to reconcile theory and practice on the theoretical concepts previously presented. For the practical laboratory, two 50-min classes were used. The practical laboratory was carried out based on procedures presented by the teacher. The procedures detailed the network environment's configuration and the commands used to carry out the practical experiment (as presented in the previous section). The experiments were carried out on computers in the school's computer labs, with the participation of 26 students, on individual computers. During the experiments, the teacher associated the practical results obtained with the theoretical concepts previously presented. The teacher also observed students' motivation and engagement, comparing a purely theoretical class with a class that combines theory and practice.

4 Results and Discussion

4.1 Experimental Results

Based on the reference scenario shown in Fig. 1, Table 10 shows the experimental results obtained during the practical study described in this work. The experimental results are based on video transmissions under normal network conditions and with packet loss emulation of 5%, 10%, 15%, and 20%.

Table 10. Experimental results obtained during video transmissions under normal network conditions and with packet loss emulation of 5%, 10%, 15%, and 20%.

Experiment	Emulated packets loss (%)	Packets sent	Packets received	Observed packet loss (%)	PSNR (dB)
1	0%	11938	11938	0%	38.17
2	5%	11938	11335	5,05%	25.46
3	10%	11938	10701	10,36%	21.46
4	15%	11938	10112	15,30%	20.52
5	20%	11938	9543	20,06%	18.89

Based on the obtained results, it is possible to observe that, for each experiment, the observed packet loss was compatible with the emulated packet loss in the network. As an example, with 0% packet loss emulation (i.e., video transmissions under normal network conditions – experiment 1), 11938 packets of video content were sent from the transmitter host to the receiver host, with no packet loss observed during the video transmission. In this experiment, no degradation was observed on the video's quality and the QoE perceived by the end-user, with a PSNR of 38.17.

In turn, in the other experiments, 11938 packets of video content were also transmitted from the transmitter host to the receiver host. However, packet losses occurred in the network, and not all packages were delivered to the final destination. In these experiments, packet losses affected the video's quality and the QoE perceived by the end-user. In this context, with 5% packet loss emulation (experiment 2), 11335 packets were delivered to the final destination, with an observed packet loss of 5.05% and PSNR of 25.46. With 10% packet loss emulation (experiment 3), 10701 packets were delivered to the final destination, with an observed packet loss of 10.36% and PSNR of 21.46. With 15% packet loss emulation (experiment 4), 10112 packets were delivered to the final destination, with an observed packet loss of 15.30% and PSNR of 20.52. Furthermore, with 20% packet loss emulation (experiment 5), 9543 packets were delivered to the final destination, with an observed packet loss of 20.06% and PSNR of 18.89.

Regarding the QoE perceived by the end-user, Table 11 shows the mapping of the PSNR values obtained in the experiments to MOS, as shown in Table 9.

Table 11. Mapping of the PSNR values obtained in the experiments to MOS.

Experiment	Emulated packet loss (%)	PSNR (dB)	MOS	Quality	Perception
1	0%	38.17	5	Excellent	Imperceptible
2	5%	25.46	3	Fair	Slightly annoying
3	10%	21.46	2	Poor	Annoying
4	15%	20.52	2	Poor	Annoying
5	20%	18.89	1	Bad	Very annoying

In this context, it is possible to observe that only in the experiment with 0% packet loss (experiment 1), no degradation in the video's quality was perceived by the end-user, with quality considered as "excellent" mapping the PSNR to MOS. On the other hand, in the other experiments, the video quality was considerably degraded, with quality considered as "Fair" in the experiment with 5% packet loss (experiment 2), "poor" in the experiments with packets losses of 10% and 15% (experiments 3 and 4), and "bad" in the experiment with 20% packet loss (experiment 5).

In this scenario, as an example, Fig. 2 shows a video frame (frame number 317) received by the end-user in the experiments with 0% packet loss emulation (left) and with 20% packet loss emulation (right).

Fig. 2. Example of video frame (frame number 317) received by the end user in the experiments with 0% packet loss emulation (left) and with 20% packet loss emulation (right).

Thus, through a practical study, it is possible to observe the effects of packet loss on the quality of the video, as well as on the QoE perceived by the end-user.

4.2 Impact of the Experiments in Education

This practical study was proposed for a computer network course with lots of theoretical concepts and little practical experiments. As a result, traditionally, there is more attention to theory teaching and less to practice teaching. For this reason, considering only the teaching activities carried out in the course, although students can obtain excellent grades in the final exams, their practice ability is weak.

In this context, the practical study was carried out to complement the theoretical concepts about one of the course contents, enabling students to better understand the subject and achieve better teaching results by means of a practical experiment.

In this scenario, based on the teacher's point of view, the following impacts were preliminarily observed through this practical study:

- The students were visibly more motivated and engaged during the practical experiment than in a purely theoretical class related to the subject;
- The students were able to apply and prove the theoretical concepts presented by the teacher through practical experiments;
- The students were able to improve their practice abilities in computer networks;
- The students were able to use other means of learning computer networks, previously limited by books described in the bibliography of the course.

It is important to highlight that future works will be carried out based on the students' point of view, complementing this work.

5 Conclusion and Future Works

In this paper, for purposes of teaching and learning on computer networks, a practical study on video transmissions in networks with packet loss was discussed. For this, based on a reference scenario, a computational environment was used for experimentation and discussion purposes.

The experiments and discussions were based in a computer network laboratory carried out with second-year students of an undergraduate computer course to help them understand the effects of packet loss on the quality of the video, as well as on the QoE perceived by the end-user. The approach was used to reconcile theory and practice on the subject.

The experience observed with students was very positive: the students were visibly more motivated and engaged during the practical experiment than in a purely theoretical class related to the subject; the students were able to apply and prove the theoretical concepts presented by the teacher through practical experiments; the students were able to improve their practice abilities in computer networks; and, the students were able to use other means of learning computer networks, previously limited by books described in the bibliography of the course.

Future works will be carried out based on the students' point of view, complementing this work. In addition, other practical studies will be proposed, explored, and evaluated.

References

1. Pande A, Ahuja V, Sivaraj R, Baik E, Mohapatra P (2013) Video delivery challenges and opportunities in 4G networks. IEEE MultiMed 20(3):88–94
2. Fan Q, Yin H, Min G, Yang P, Luo Y, Lyu Y, Huang H, Jiao L (2018) Video delivery networks: challenges, solutions and future directions. Comput Electr Eng 66:332–341
3. Zink M, Sitaraman R, Nahrstedt K (2019) Scalable 360 video stream delivery: challenges, solutions, and opportunities. Proc IEEE 107(4):639–650
4. Kurose JF, Ross KW (2013) Computer networking: a top-down approach, 6th edn. Pearson Education, New Jersey
5. Wu D, Hou YT, Zhu W, Zhang YQ, Peha JM (2001) Streaming video over the Internet: approaches and directions. IEEE Trans Circ syst Video Technol 11(3):282–300
6. Turner JS, Taylor DE (2005) Diversifying the internet. In: IEEE Global Telecommunications Conference 2005 (GLOBECOM'05). St. Louis, USA, pp 755–760
7. Yang M, Wang S, Calheiros RN, Yang F (2018) Survey on QoE assessment approach for network service. IEEE Access 6:48374–48390
8. Bouraqia K, Sabir E, Sadik M, Ladid L (2020) Quality of experience for streaming services: measurements, challenges and insights. IEEE Access 8:13341–13361
9. Frnda J, Voznak M, Sevcik L (2016) Impact of packet loss and delay variation on the quality of real-time video streaming. Telecommun Syst 62(2):265–275
10. Adeyemi-Ejeye AO, Alreshoodi M, Al-Jobouri L, Fleury M (2019) Impact of packet loss on 4K UHD video for portable devices. Multimed Tools Appl 78(22):31733–31755
11. Mrvelj Š, Matulin M (2018) Impact of packet loss on the perceived quality of UDP-based multimedia streaming: a study of user quality of experience in real-life environments. Multimed Syst 24(1):33–53
12. Korhonen J (2018) Study of the subjective visibility of packet loss artifacts in decoded video sequences. IEEE Trans Broadcast 64(2):354–366
13. Wang J, bin Hou Y (2018) Packet loss rate mapped to the quality of experience. Multimed Tools Appl 77:387–422
14. Suárez FJ, García A, Granda JC, García DF, Nuño P (2016) Assessing the QoE in video services over lossy networks. J Netw Syst Manage 24:116–139

15. Prvan M, Julije O (2020) Methods in teaching computer networks: a literature review. ACM Trans Comput Educ (TOCE) 20(3):1–35
16. Oracle VM VirtualBox. https://www.virtualbox.org
17. Netem. https://wiki.linuxfoundation.org/networking/netem
18. Klaue J, Rathke B, Wolisz A (2003) EvalVid - a framework for video transmission and quality evaluation. In: Proceedings of international conference on modelling techniques and tools for computer performance evaluation, Urbana, USA, pp 255–272
19. FFmpeg. https://ffmpeg.org

Ethical Impacts of Artificial Intelligence

Bibliometric Analysis

Vanessa Marques Paes[1](✉), Franciane Freitas Silveira[1], and Alessandra Cristina Santos Akkari[2]

[1] Federal University of ABC (UFABC), São Bernardo do Campo, Brazil
{vanessa.paes,franciane.silveira}@ufabc.edu.br
[2] Mackenzie Presbyterian University, Campinas, Brazil
alessandra.akkari@mackenzie.br

Abstract. The evolution of the industry has always brought innovations to the processes and changes in the way people deal with their activities. In the fourth industrial revolution, known as industry 4.0, intelligent data processing, the internet of things, and artificial intelligence technologies were integrated into the process of making industries more and more innovative, but, in this revolution, there are new risks that need to be evaluated before their impacts become irreversible. From this perspective, it is interesting to discuss the ethical cost associated with the application of 4.0 technologies, especially the artificial intelligence that has been used across the board in different areas of the economy and society. Therefore, this article aimed to develop a bibliometric analysis based on the ethical impacts associated with artificial intelligence. To this end, an exploratory and descriptive study was developed, based on the collection and statistical treatment of secondary data. This study's main contribution refers to pointing out trends, based on academic communication statistics through publications available for consultation, to guide the definition and implementation of principles that aim to mitigate or eliminate impacts ethical issues associated with the use of artificial intelligence.

Keywords: Industry 4.0 · Digital technologies · Ethics · Artificial intelligence

1 Introduction

The term Industry 4.0 was first cited by the German Economic Development Agency in 2011 [1], and, since then, interest in this topic has grown in several areas of the industry alongside artificial intelligence, internet initiatives things, cloud computing, and digitalization of production processes [2]. Artificial intelligence, the focus of this study, refers to the ability of a system to interpret external data correctly, using pattern recognition to develop a learning format and use that learning to achieve goals in a flexible, adaptable, and autonomous way [3].

With all the changes and adaptations in the industry to migrate to a more technological and connected environment, society also needs to adapt to a new work and lifestyle format. For this reason, the challenge of ethical and social cost becomes evident since

Y. Iano et al. (Eds.): BTSym 2020, SIST 233, pp. 178–186, 2021.
https://doi.org/10.1007/978-3-030-75680-2_21

the speed of this transformation leads to rethinking the human-machine interaction and the impacts related to all the new forms of work that appear each day [4].

Therefore, due to intense changes in the industry's work processes and methodologies, the human and ethical aspect also needs to be developed. Just as knowledge develops for technology and innovation, discussions about these new technologies' social impacts also need to be evaluated before the implementation of these new technologies. This assessment needs to be developed more intensely, as experts indicate that we need to start this revolution in the right way, with the risk of the social situation deteriorating further [5].

The concern with the ethical impacts of the use of artificial intelligence goes beyond themes related to deaths resulting from the use of autonomous vehicles. In fact, there are also risks in the most common use of this technology, as in the algorithms that make decisions in selection processes, lawsuits, and even creditworthiness, because if there is any error or prejudice inserted in the development, real people will have their lives impacted [6]. Since the technologies are being implemented at a much faster rate than the critical analysis of these risks, there are possibly several bad solutions being implemented, and, consequently, several risks arising. Moreover, the whole process of reworking and mitigating these impacts could be avoided if the whole process of evaluation and critical awareness were implemented from the beginning of each development that involves the use of artificial intelligence [7].

This article aimed to develop a bibliometric analysis based on the ethical impacts associated with the application of artificial intelligence. This study's main contribution refers to pointing out trends, based on statistics of academic communication through the publications available for consultation, to guide the definition and implementation of principles that aim to mitigate or eliminate the ethical impacts associated with the use of artificial intelligence.

2 Methodology

For the development of bibliometric analysis, the visualization structure was used, which uses five questions: Who (authors), What (keywords), Where (countries), When (years of publications), and With whom (authors' affiliation). To develop research related to the dilemma of the ethical use of artificial intelligence, this study used the Web of Science and Scopus platform, which are two databases of academic publications. To refine the search, the advanced filter of both bases was used, and the following configuration was applied:

Scopus

```
TITLE-ABS-KEY = ("ARTIFICIAL INTELLIGENCE" OR ai) AND
(ethic * OR moral *) AND (soci *)
```

Where:

- TITLE-ABS-KEY = default search fields (title, abstract and keywords)
- AND = Boolean operator "AND"
- OR = Boolean operator "OR"

Web of Science

```
TS = (("artificial intelligence" OR ai) AND (ETHIC * OR
moral *) AND soci *)
```

Where:

- TS = default search field (title, abstract and keywords)
- AND = Boolean operator "AND"
- OR = Boolean operator "OR"

The keywords used were "artificial intelligence", so the quotes were kept in the search fields to find the words in that exact order. In the word "ethic *" the asterisk was inserted to allow the inclusion of similar words that have the same meaning, such as, for example, "ethical", "ethic" and "ethics". The same strategy was used for the word "soci *", which made it possible to simultaneously search for the words "Society", "social" and "societal".

In addition to advanced research, the study period was defined, which included publications from the 2000s to 2020. Regarding the type of publication, the research was refined to include only articles. As the theme of ethics and artificial intelligence expands to several areas of knowledge, it was defined, for this study, only the areas of computer sciences, social sciences, engineering, business arts, and humanities, 168 articles were identified on the Web of Science platform and 607 articles on the Scopus platform. The distribution of publications shows that this is a subject in great ascension, with an increase in the number of publications in the last four years.

To carry out bibliometric analyses, the Bibliometrix software was used, which is a tool in R programming environment and allows the conversion, analysis, and elaboration of graphs, through the data that were exported from the Scopus and Web of Science databases [8]. Aiming to clean the data, this process was also done in the same software, and this is an advantage of this software, because all processes were done on a single platform [9]. After merging the data from both platforms, 85 duplicate records were excluded, totaling 690 articles. Different analyses were conducted, as well as Alfred Lotka's law of productivity and Bradford's law were evaluated.

3 Results and Discussion

Assessing the initial data of the research, it was evidenced that, in total, there are 1463 authors, with an average of 0.47 articles per author, with 19% of the articles being published by only one author and 81% are publications by more than one author. It is also possible to estimate that the number of authors per document is 2.12. In total, articles have 5,191 citations, which, on average, are 7.5 citations per article. The volume of citations also allows the calculation of the h-index, which in this study is 34. The h-index is an indicator of the quantity and visibility of publications. The higher the index, the greater the evaluated publications' impact, and data with values above 20 are considered good, and above 40 are exceptional [10].

In order to assess the distribution of researchers, the productivity law created by Alfred Lotka was analyzed. The study carried out for this research brings as an author with the largest number of publications Luciano Floridi, who made contributions in eleven articles, followed by Taddeo Mariarosaria, who made 6 publications. The authors' productivity can also be assessed in the timeline of Fig. 1, where Luciano Floridi also stands out.

Publications with only one author represent 91% of the total, according to data in Table 1. This assessment is also clear by comparing the values found in this study and the data from a standard curve, according to the law established by Lotka, where the coefficient n ≅ 2. The data from this research point to a coefficient n equal to 3.2, as well as a constant C equal to 0.57, suggesting that this study does not come close to Lotka's law.

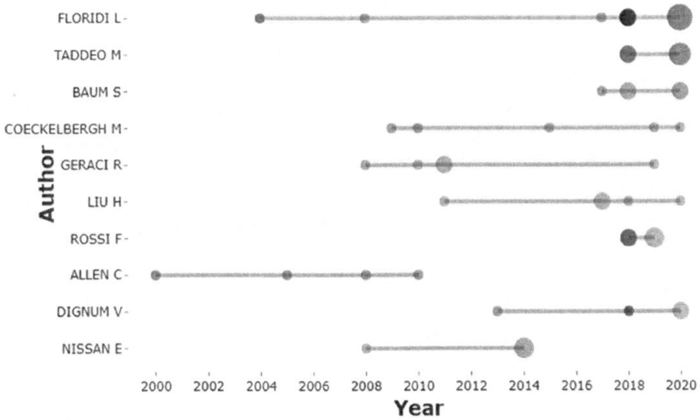

Fig. 1. Top 10 authors production over time

Table 1. Analysis of the number of authors per publication

Number of articles by the same author	Number of Authors	Frequency
1	1336	91,32%
2	90	6,15%
3	25	1,71%
4	5	0,34%
5	5	0,34%
6	1	0,07%
11	1	0,07%
Total	1463	100%

Among the most cited articles (Fig. 2), the article "On the Morality of Artificial Agents" stands out, whose author is Luciano Floridi and which has the highest number of citations, followed by "Ambient intelligence: A survey" that was written by Fariba Sadri. Among the articles evaluated, a survey was also carried out of the main journals responsible for the publications (Fig. 3), in which the AI and Society journal, with 60 publications (8.7%), stands out, followed by the journal Ethics and Information Technology with 30 published articles (4.35%). The top ten journals have 28.12% of the total publications evaluated by this study.

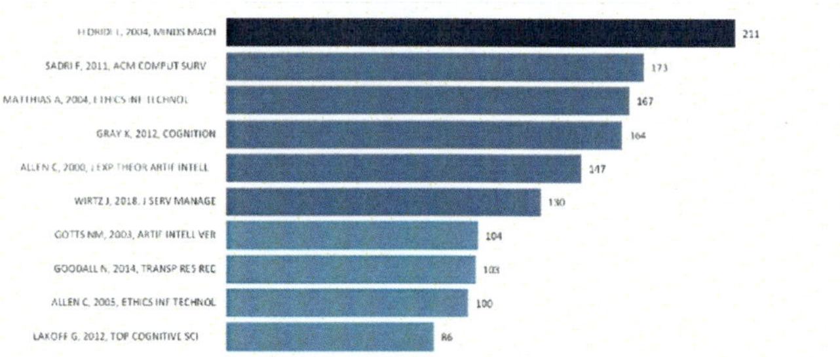

Fig. 2. Most cited documents

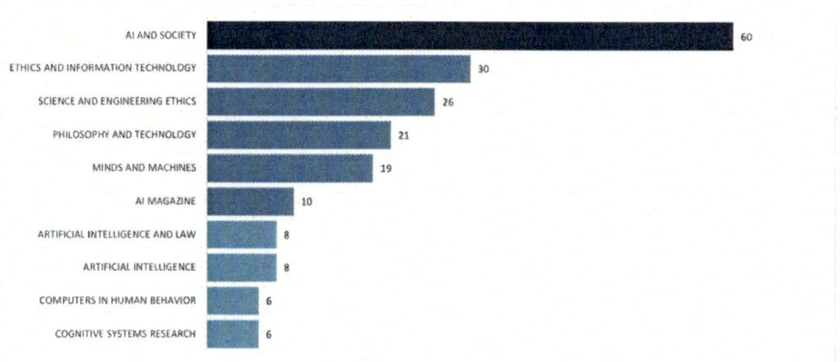

Fig. 3. Most relevant journals

According to Bradford's law, publications on a given topic follow a distribution that can be divided into three regions, with the central region having few journals, but with greater relevance to the researched topic, a second region would have a greater number of journals and about the same amount of publications and relevance. Finally, a third region where the other journals would be [11]. In this study, an approximation to Bradford's law was identified (Table 2).

Table 2. Evaluation of major journals through Bradford's Law

	Zone 1	Zone 2	Zone 3	Total
Jounals	17	128	227	372
Articles	232	231	227	690

The keyword "artificial intelligence" was used by the authors 274 times, followed by the word "ethics" which was presented 96 times. In many cases (33 articles), the acronym AI is used directly to substitute the term "artificial intelligence". When evaluating the keywords used by the authors, and refining the 20 most used words, 6 clusters were identified that are related to each other (Fig. 4). The first brown cluster brings together words related to governance and artificial intelligence. This cluster is related to the second group, orange, which has words focused on robots and machines' ethics. These two clusters come together and have links with the other four clusters, which address the topics of big data, algorithms, machine learning, privacy, human rights, and ethics.

There is a trend related to the need to create standards, rules, or laws to regulate artificial intelligence's ethical use to avoid social and ethical impacts [12]. Moreover, this need is evident in the combination of the words most used by the authors.

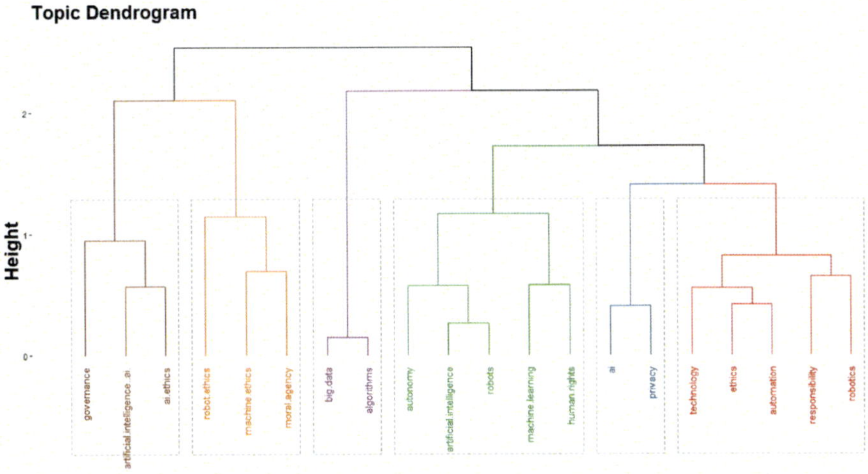

Fig. 4. Analysis of keywords by cluster formation

The countries with the highest volume of publications are the United States (191), the United Kingdom (101), and the Netherlands (34), which together account for approximately 52% of the total volume of articles surveyed. Also, a collaboration network between countries was observed, which provides the exchange of knowledge with higher collaboration levels. Among the 15 main countries, there is a representation of the continents of North America, Europe, Asia, and Oceania (Fig. 5).

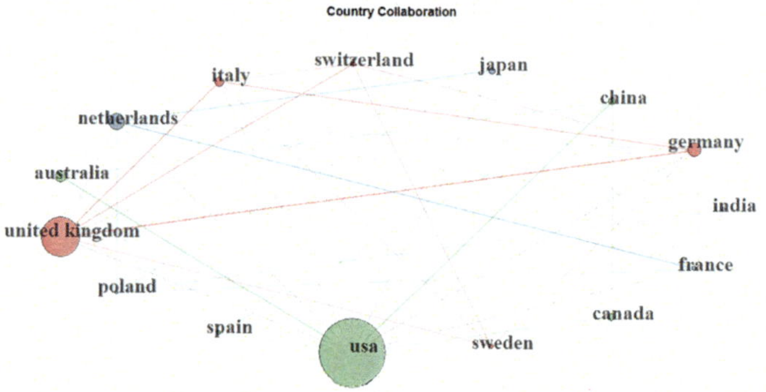

Fig. 5. Country collaboration

Co-Citation is the analysis that measures the similarity of publications and aims to assess when two articles are cited in the same publication [13]. When looking at the articles evaluated in this study, Fig. 6 shows the complexity of the links among the 30 main publications. The analysis was deepened only for articles with eight or more citation references to identify which are the strongest links. Three clusters were observed, the first of which has a strong link between the articles "The ethics of artificial intelligence" (Bos-trom, N.) and "Moral Machines" (W. Wallach) and also with the article "Human 2.0" (Kurzweil R). The second cluster included the publications "On the Morality of Artificial Agents" (Floridi L.) and "The nature, importance, and difficulty of machine ethics" (Moor J. H. E). The third cluster comprises the articles "Minds, Brains and Programs" (Searle J.) and "Computing machinery and intelligence" (Turing A. M.).

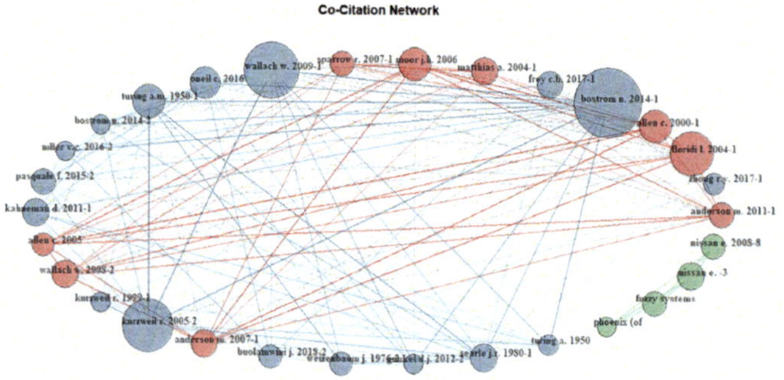

Fig. 6. Co-citation network

Finally, Bibliographic Coupling is defined as the analysis of a reference that was used by two articles, that is, the more identical citations are used, the stronger the connection between the publications [14]. Among the 30 main articles, the connections are complex, as assessed in Fig. 7.

Evaluating the connections that present eight or more coupling references, we have only two clusters. The first shows the similarity in the references of the articles "Decolonial AI: Decolonial Theory as Sociotechnical Foresight in Artificial Intelligence"

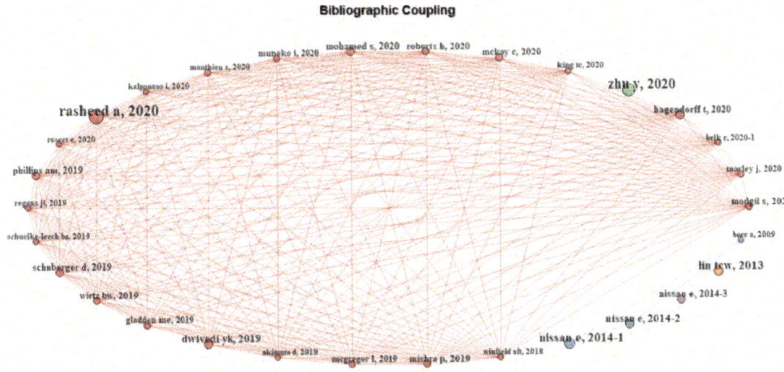

Fig. 7. Bibliographic coupling

(Mohamed S.) and "The Ethics of AI Ethics in the Evaluation of Guideline" (Hagendorff T). Moreover, the second has the articles "International Regulation of Lethal Autonomous Weapons Systems Laws Paradigms of Policy Debate in Japan" from Akimoto D and "Predicting Risk in Criminal Procedure Actuarial Tools Algorithms AI and Judicial Decision Making" from Mckay C. Most articles have different references, so the bibliographic coupling with eight or more connections has fewer connections than the co-citation network. Some studies claim that the analysis through the bibliographic coupling has greater accuracy to validate academic publications' statistical mapping. However, each method has its advantages according to each research [15].

4 Conclusion

It is evident that there is a search to understand and describe what are the ethical impacts of the use of artificial intelligence. As noted, the number of publications in recent years is significant and maintains a trend of evolution. However, this search is not yet widespread among countries, as analyzed in Fig. 5, and, in this respect, there is a great potential for improvement because ethical issues are not unique and need to be evaluated and developed in the largest number of possible regions because this is the path to a robust mapping, which enables the development of actions that mitigate negative impacts. Moreover, these actions need to be defined through critical analysis and definitions of rules and laws to ensure the correct execution, and these processes need to be defined and applied to ensure the proper use of technology.

References

1. Lennon Olsen T, Tomlin B (2019) Industry 4.0: opportunities and challenges for operations management. SSRN Electron J
2. Dastbaz M (2019) Industry 4.0 (i4.0): the hype, the reality, and the challenges ahead. In: Industry 4.0 and engineering for a sustainable future, pp 1–11
3. O'Leary DE (2013) Artificial intelligence and big data. IEEE Intell Syst 28(2):96–99
4. Rabeh M, Husam A, Mousa S (2017) The fourth industrial revolution (Industry 4.0): a social innovation perspective. Technol Innov Manag Rev 7(11):12–20

5. Wiegel V, Wallach W, Allen C (2010) Moral machines: teaching robots right from wrong. Ethics Inf Technol 12:359–361
6. Floridi L, Cowls J, Beltrametti M et al (2018) AI4People—an ethical framework for a good AI society: opportunities, risks, principles, and recommendations. Minds Mach 28:689–707
7. Floridi L (2018) Soft ethics and the governance of the digital. Philos Technol 31(1):1–8. House of Lords Artifcial Intelligence Committee. AI in the UK: ready, willing and able?
8. Aria M, Cuccurullo C (2017) Bibliometrix: an R-tool for comprehensive science mapping analysis. J Inform 11(4):959–975
9. Esfahani HJ, Tavasoli K, Jabbarzadeh A (2019) Big data and social media: a scientometrics analysis. Int J Data Netw Sci 3(3):145–164
10. Hirsch JE (2005) An index to quantify an individual's scientific research output. Proc Natl Acad Sci 102(46):16569–16572
11. Bradford SC (1985) Sources of information on specifc subjects. J Inf Sci 10(4):176–180
12. Sadri F (2011) Ambient intelligence: a survey. ACM Comput Surv 43(4):36:1–36:66
13. Small H (1973) Co-citation in the scientific literature: a new measure of the relationship between two documents. J Am Soc Inf Sci 24(4):265–269
14. Kessler MM (1963) Bibliographic coupling between scientific papers. Am Doc 14(1):10–25
15. Boyack KW, Klavans R (2010) Co-citation analysis, bibliographic coupling, and direct citation. Which citation approach represents the research front most accurately? J Am Soc Inf Sci Technol 61(12):2389–2404

Digital Transformation Model for the Reactivation of the Tourism Sector in the COVID-19 Environment of the Lambayeque Region

Roger Alarcón, Janet Aquino, Jessie Bravo(✉), Nilton Germán, and Carlos Valdivia

School of Computer and Informatics Engineering, Pedro Ruiz Gallo National University, Lambayeque, Peru
{ralarcong,jaquino,jbravo,ngerman,cvaldivias}@unprg.edu.pe

Abstract. Globally, the tourism sector has been affected by COVID-19 pandemic, which has triggered an economic and social crisis. Many countries are currently generating alternatives to promote tourism and allow tourists to return and improve the current situation. In Peru, borders were closed, and therefore the different tourist sites and activities were closed, generating unemployment in the actors involved in the tourism sector, as well as the local communities that underpinned their economy in the activities of this area. Faced with this problem, it is evident that technology and the digitization of tourism activities will allow supporting this sector, so this article proposes a digital transformation model for the reactivation of tourism, taking the Lambayeque region as a case study. We have carried out a systematic investigation of tourism information and applied interviews to specialists in the sector and surveys of the actors involved. The model is based on five dimensions: technology, management, processes, tourism product, and digital tourist. For each dimension, we describe its purpose and the proposals applying solutions and technological tools that allow digital transformation. Our model was validated through an expert judgment, the result of which was acceptable, hoping that its implementation will contribute to the reactivation of tourism.

Keywords: COVID-19 · Pandemic · Disruptive technologies · Digital transformation · Tourism · Digital tourist

1 Introduction

Tourism is one of the main sources of income in Peru and generates innumerable sources of work, and according to the Ministry of Foreign Trade and Tourism (Mincetur), Peru received 4.4 million international tourists in 2019 [1]. Tourists mainly come from Chile and the US, which together accounted for 50% of the total. Mincetur has identified 117 tourist sites, museums, and protected areas nationwide, with Lima, Cusco, Lambayeque, and Puno being the departments with the greatest presence of them [1–3].

Y. Iano et al. (Eds.): BTSym 2020, SIST 233, pp. 187–195, 2021.
https://doi.org/10.1007/978-3-030-75680-2_22

Regarding the Lambayeque region, the main places visited by foreigners were the city of Chiclayo (83.8%), Tumbas Reales de Sipan (29.8%), Lambayeque (23.8%) and Sipan (19.8%) [4].

Unfortunately, since the quarantine began, restaurants do not receive income, hotels do not have customers, archaeological centers and museums have no visitors, although the government indicated various measures, such as Reactiva Peru, Turismo Emprende, health protocols, and various awareness-raising campaigns, this is insufficient given the complexity of the sector, and tourist sites remain closed and no longer receive income [5]. Faced with this situation, the main problems encountered for the reactivation of tourism are the inadequate management of the organizations that direct the tourism sector, as there are two institutions that direct it (MINCETUR and the Ministry of Culture), the lack of execution of the strategic tourism plan (PERTUR), the lack of articulation of the unions related to this sector that allow promoting tourism, the low government budget, the formalities and cost to comply with the biosafety protocols, the low level of agreement that promotes initiatives between the private and public sectors, the fear and lack of training of the communities that intervene directly with the tourist offer, and the distrust and fear of the public to attend to those places.

The purpose of this study is to propose a digital transformation model that allows the reactivation of tourism in this environment of the pandemic in order to contribute to improving the automation and digitization of tourism processes, the country's economy, generating greater employment, especially in the communities involved, a greater visible tourist offer at the national and international level, the generation of new tourism experiences based on technology and that allows better decision-making by the entities that direct the sector. This work passed and was approved by the faculty's Ethics Committee.

2 State of the Art

Tourism is characterized by its complex interrelations system between the different factors that intervene in it [6]. According to Tafur, G. *et al.* [7], the tourism industry is currently facing a new environment derived from the transformations to which it has been subjected. As mentioned by Voronkova L. [8], in 2019, Latin American tourism assimilates the use of Internet technologies in the development of its processes, thus emerging different digital channels, mobile applications that contribute to the emergence of new forms of communication with customers. Similarly, Duarte, G. *et al.* [9] suggests that Digital Transformation goes beyond the digitization of a product, of a department; it is a change of culture throughout the company and its employees to adapt to new technologies, taking advantage of them to work on the product on customer needs. Hence, according to what Vega, E. [10] states, the use of Big Data as part of the technological trend allows focusing on the knowledge of tourist activity in key parameters, for example, length of stay, trips, number of visits, consumption and spending levels, loyalty by knowing better the customer, a better-adapted product, and a more satisfied tourist; this allows tourism companies to anticipate consumer needs. In addition, Herrero *et al.* [11] indicate that the continuous development of ICT (Information and Communication Technologies) involves more tourism industries and affects consumer behavior. This has generated the emergence of the experienced consumer; currently, he has gone from being a simple reader to interacting as a reader/writer, who has new skills and customs such as the use of forums, recommendation systems, social networks, now known as Tourist 2.0.

Finalizing, according to the diagnosis made by Mendoza, F. and García, N. [12] on the consequences derived from COVID-19 in the tourism business sector, there is an important socioeconomic crisis, for which it is necessary to guarantee to small businesses their permanence in the tourism market by promoting cooperative associationism and integration strategies member of the Chamber of Tourism with associations of restaurants, accommodation, transport associations, etc.

3 Proposal

3.1 Situational Analysis of Tourism

For the situational analysis of tourism, we have conducted interviews with four specialists in the Lambayeque region sector, who conclude as follows: The tourism sector both in the world and in Peru is one of the sectors hardest hit by the pandemic since its movement was reduced to zero, starting with airlines, land transport companies, tourist companies, hotels, tourist places, restaurants, etc. They agree with the lack of support from the central government since the resources are going to the coffers of this and they do not show the investment in their locality, as is the case of the maintenance of tourist routes; they also agree that citizen security is crucial to keep tourists protected. In addition, the pandemic environment has led organizations to reorganize and reorder their processes, mainly due to government regulations and personal health care.

The sector depends directly on Mincetur, and the municipalities require support to be able to have a digital platform that allows them to implement new tourism models. Currently, local governments do not have help from the central government. Specifically, Lambayeque must adapt to a digital tourism framework for which technology is needed to make our tourist places available to the world.

We have applied surveys to the actors involved in the tourism sector in the Lambayeque region, for which we have used as an instrument the one applied in the study of digital transformation in IEBS Business School companies [13]. The survey is divided into four aspects: In the Digital, 83.3% of organizations have begun to adapt to the digitization process, and although it is true, 66.7% consider that they are not yet at the desired level of knowledge of the use of technological tools that can support a post-Covid-19 stage. Regarding Communication and Marketing, 83.3% consider that although they do not have a defined digital marketing area, they have used social networks to communicate with their customers and promote their products and services; in addition, 66.67% informs customers about the biosecurity measures applied to their organization; finally, 50% do not measure the positioning of their companies, perhaps due to ignorance. In the Social Media aspect, 66.7% do not know or do not have a profile on the different social networks, but they are beginning to adapt to the changes required by the current environment, taking into account digital media. Finally, with regard to Human Resources, 66.67% consider that human resources are the strong pillar of companies linked to the sector with respect to innovation, disposition, creativity, etc.; however, 66.7% do not have the necessary measures for training and recruitment through virtual means.

3.2 Proposed Model

Based on the systematic review of the specialized literature, applied interviews and surveys, and the experience in the management of ICT of the research team, a digital transformation model is proposed to reactivate the tourism sector, which presents us with an overview of the main dimensions involved, favoring a global vision and considering the tourist as the main actor, therefore, it is necessary to introduce mentality changes in the participating actors, based on the use of current and emerging technologies that promote constant innovation.

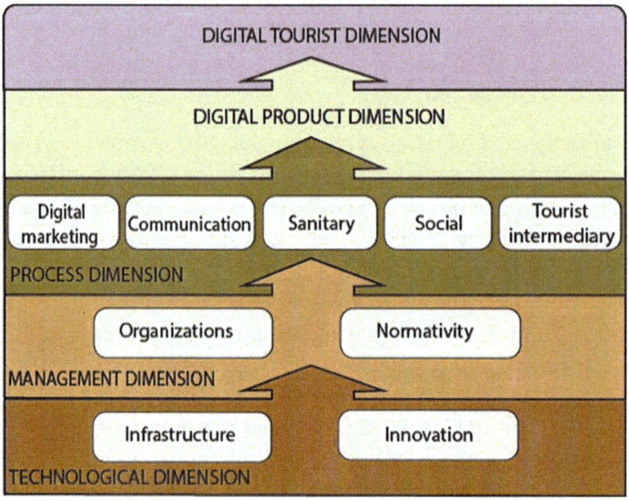

Fig. 1. Digital transformation model

The present model is conceived as a tourist experience to the experience that begins from the search for the tourist product or service, the effective planning of a tourist route, the guided or virtual visit, and the set of activities that make it up; besides, this experience is complemented with the after-sales service, based on the use of technological tools and the digitization of processes.

Our model includes five dimensions: technological dimension, management, processes, tourism product, and the digital tourist, as shown in Fig. 1.

Technological Dimension. The technological dimension includes two aspects: technological infrastructure and innovation. The technological infrastructure aims to provide the necessary technological support for the digital transformation of the tourism sector, such as servers, networks and telecommunications, storage, and information security. (See Fig. 2).

Similarly, the support for large amounts of data (Big Data) from different sources allows obtaining tacit, digitized, and transferable knowledge to all instances and services of the technological structure. Knowledge systems based on artificial intelligence (AI) will allow rapid decision-making and continuous improvement. The digital tourism

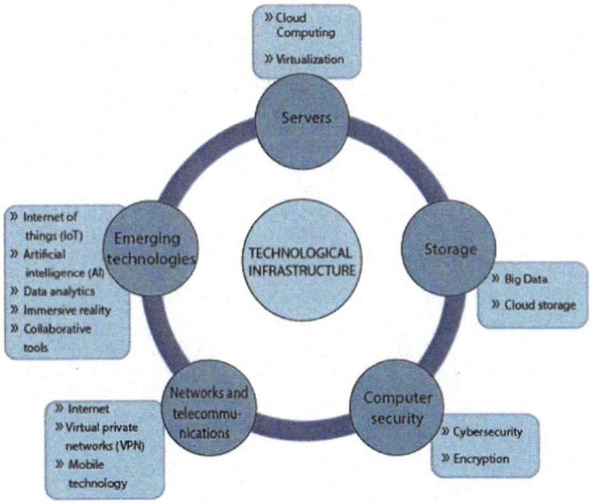

Fig. 2. Subdimension technological infrastructure

sector must include collaborative tools, management, documentation, cloud platforms (Cloud Computing) that allow universal and permanent access to new, improved services, and digitized tourism products that will be proposed.

Regarding innovation, through the generation of new proposals that improve the tourist offer. Disruptive innovation is proposed, with the use of Machine Learning applied to Big Data with predictive and prescriptive analysis, which allows proposing new and better experiences to tourists, among which the generation of smart tourist destinations is considered [14], an innovative tourist space, accessible to all, improving tourist satisfaction and the community's quality of life; also proposals for digital tourism products such as virtual visits to the different tourist places that provide an experience very close to reality.

This dimension will allow us to have the necessary technological support for the new digital tourism, in permanent innovation in accordance with the changes that the advancement of technologies involves, the new ways of doing tourism, the needs of tourists, and the changes in the ecosystem involved with the tourism sector.

Management Dimension. This dimension includes the set of norms, protocols, and standards, planned and developed in a cross-sectoral manner, with the participation of different social actors, both public and private, that seeks to favor and promote the quality of life and health of tourists and staff. They participate in the sector and materialize the right to a healthy environment through the positive transformation of social determinants. Management must be integrated and digitized in such a way that it allows timely decision-making, and all entities can contribute and improve the processes of this sector. Likewise, it contemplates the planning of the investment and its channeling to the competent entities, coordinating and collaborating as a single high-performance unit. The management must also work in a coordinated manner with the tourist intermediaries to achieve the objectives set.

The quality management of the digitized tourism process and the quality of the products and services offered are prioritized, which allows them to be characterized, analyzed, improved, and predicted in search of continuous improvement, meaning quality is the satisfaction of the needs and expectations of tourists based on the value you have perceived and received. It must also manage digital culture through standards and regulations among the digital transformation process members, sharing ideas and knowledge in an open management that empowers people, achieving differentiation and a change in the transformed processes that obtain advantages from new technologies. The technological solutions they provide are Big Data and analytics that will serve as support in making appropriate decisions for the organizations related to this sector (See Fig. 3).

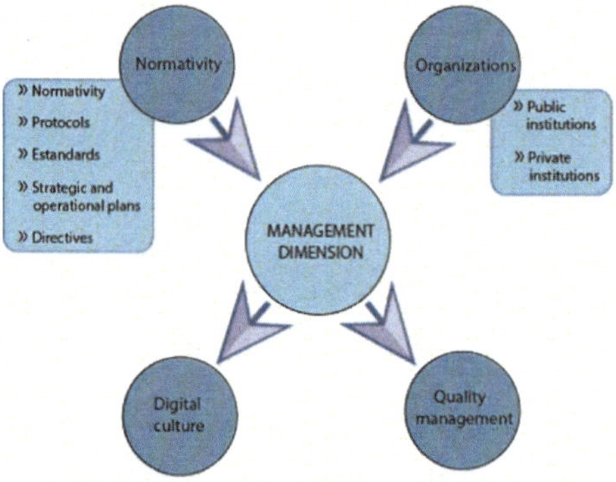

Fig. 3. Management dimension

Process Dimension. The Process dimension involves five sub-dimensions: digital marketing, communication, sanitary, social, and tourism intermediaries.

Digital Marketing. Its function is to promote tourism with a renewed and dynamic image, obtaining a global positioning as a brand, using various digital media and disruptive advertising campaigns, which reaches all the corresponding instances promptly, improving the relationship and the establishment of new efficient communication channels with tourists; such as the use of social networks, mobile applications, and post-sale channels.

Communication. Both internal and external are considered. The internal communication effectively shares all the knowledge generated among the members involved in the tourism proposal. In external communication, the information reaches all the instances involved in the digital context (stakeholders). The technologies they support are email, video-calling platforms, and corporate social networks. It is also considered a new type of tourism and tourist product, which uses digital technologies and tools, for which advertising strategies and segmented communication channels are implemented, seeking loyalty with a digital tourism experience in line with the advancement of technologies. Sales channels are also considered through a Digital Tourist Information Center that includes relevant information about the products and services of the proposed new type of digital tourism.

Sanitary. A journey is proposed from the knowledge of the potentialities and health needs to the activities that suppose a true and positive transformation. Through the training of the personnel involved, awareness campaigns, and compliance with established protocols. The technologies that support are the Internet of Things (IoT) through the development of solutions that integrate vital signs control sensors, use of geolocation and monitoring, artificial intelligence applications, guaranteeing that the tourist who arrives in the country enters healthy and come out healthy.

Social. The digital community is considered, which integrates the inhabitants who reside in the surroundings of the tourist resources or tourist offers, as well as the human resource that provides the tourist service, who must have skills and new forms of job performance with the use of technologies, forming the main axis of digital transformation. Technologies that it could support are collaborative tools, training platforms and virtual meetings, and the use of the Internet.

Tourist Intermediary. Those actors whose main participation is to establish the relationship between the tourist and the tourist product (good or service, tourist destination) are considered, not only limited to this but also can offer complementary services such as food and accommodation, among others. It comprises tour operators, tourism agencies, travel agencies, and others related to the touristic service offered. The technologies that it could support are collaborative tools, computer applications, and the use of the Internet (See Fig. 4).

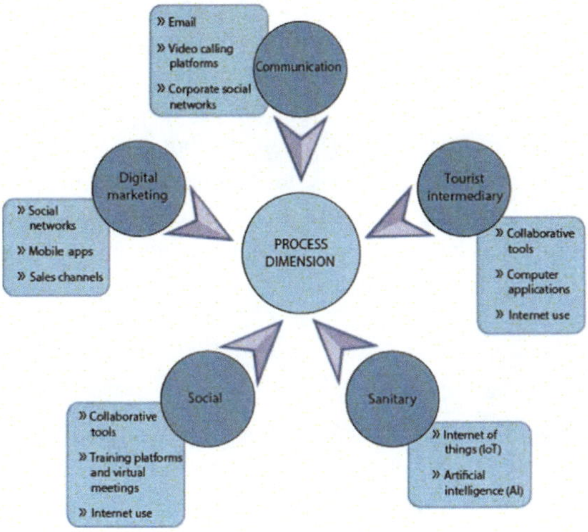

Fig. 4. Process dimension

Tourist Product Dimension. It refers to the set of characteristics linked to the tourist service that is offered. The aim is to create new digital tourism products, enhance existing ones, and promote the enhancement of tourist places. The digital tourism products to offer will be obtained and proposed from the data analytics that allows knowing the information of consumers (tourists) and the market (global digital environment), allowing them to have added values in the use of technologies, improving the tourist experience. Due to the application of data analytics, we can generate predictive models that allow the development of innovative, personalized products according to the trends of the tourism ecosystem. Innovative tourism products consider the application of virtual reality, augmented reality, artificial intelligence through assistants, and the use of QR (Quick Response) codes to access tourist information.

Digital Tourist Dimension. A digital tourist is understood to be any person who uses any of the tourist services offered through the platforms and technological solutions proposed. Find out about the tourist offer and communicate through digital media. The digital tourist has the following characteristics: hyper-connected to social networks, mastery of digital platforms and online applications, management of digital competencies, ability to search and analyze information on the Internet, values natural and cultural resources, and likes to share their experiences in digital media. The real change in digital transformation will be focused on the tourist and on the value that is delivered with disruptive technological tourism proposals.

4 Validation

To validate our digital transformation model for the reactivation of tourism, we have used the assessment of Expert Judgment, specialists related to the tourism sector, as well as in digital transformation. The rating criteria used are: sufficiency, clarity, consistency, and relevance applied to each of the dimensions of our proposed model, with a rating ranging from 1 to 4, where 1 indicates that the criteria are not met, and 4 indicates they fully comply. According to the results of the experts' judgment, we can conclude that the proposed model is ACCEPTABLE, given that it considers the tourist as the center of digital transformation, disruptive technologies, and innovation as key axes, through efficient digital marketing and communication channels that allow the promotion of tourism products and services offered and management as an important element in achieving the objectives of the tourism sector.

5 Conclusions

Taking into account that the digital transformation involves people, processes, and technology, it is concluded that the digital transformation in the tourism sector in the Lambayeque region is in its initial phase, the technology exists, but the entities involved do not yet have a culture that allows its application integrally; likewise, the tourist processes are not digitized for most of them, generating a volatile, uncertain, complex, and ambiguous scenario that does not allow to attract tourists and position tourism as one of the main axes of the country's economy.

The proposed model will allow igniting the innovation engine driven by data analysis and the use of emerging technologies aimed at re-planning business models focused on the digital tourist.

References

1. Ministerio de Comercio Exterior y Turismo del Perú (MINCETUR). https://datosturismo.mincetur.gob.pe/appdatosTurismo/Content1.html
2. Confederación Nacional de Instituiciones Empresariales Privadas (CONFIEP). https://www.confiep.org.pe/noticias/economia/turismo-en-peru/#:~:text=De%20acuerdo%20con%20el%20Ministerio,UU
3. Dirección general de investigación y estudios sobre turismo y artesania. https://www.mincetur.gob.pe/wp-content/uploads/documentos/turismo/publicaciones/Peru_Compendio_Turismo_Setiembre2019.pdf
4. Ministerio de Comercio Exterior y Turismo del Perú (MINCETUR). https://www.mincetur.gob.pe/wp-content/uploads/documentos/turismo/estadisticas/ReporteTurismoRegional/RTR_Lambayeque.pdf
5. Gestión. https://gestion.pe/blog/gestiondeservicios/2020/04/restaurantes-problemas-retos-y-soluciones-en-medio-de-la-pandemia.html/?ref=gesr
6. Cerezo Medina A (2020) Introducción al turismo y la gastronomía, Universidad de Málaga
7. Tafur Avilés G, Vélez Barros C, Alejo Machado O, Zumba Córdova M, Jacome Tapia J (2018) Desarrollo tecnológico del sector turístico en la ciudad de Guayaquil (Ecuador). Revista Espacios 39(44):3
8. Voronkova LP (2020) Transformación digital del turismo en América Latina Iberoamérica, Moscú
9. Duarte G, Albornoz E, Rangel E, Scioscia J, León F, Pinto J (2016) Importancia de la Transformación Digital en Pequeñas y Medianas Empresas de Turismo, Gestión integral administrativa, financiera y comercial de las PyME, pp 277–294
10. Vega Briceño E (2018) Turismo y Transformación Digital, Costa Rica
11. Caro Herrero J, Luque A, Zayas B (2015) Nuevas tecnologías para la interpretación y promoción de los recursos turísticos culturales. Revista de Turismo y Patrimonio Cultural 13(4):931–945
12. Félix Mendoza AG, García Reinoso N (2020) Estudio de pérdidas y estrategias de reactivación para el sector turístico por crisis sanitaria COVID-19 en el destino Manta-Ecuador. Revista Internacional de Turismo, Empresa y Territorio No 7, 4(1):79–103
13. IEBS Bussines School. https://www.iebschool.com/blog/que-es-transformacion-digital-business/
14. Castro Álvarez U, González Rodríguez JÁ (2017) Destinos turísticos inteligentes: Estrategia para el desarrollo local en países pobres? Revista Turydes: Turismo y Desarrollo 22

Review of Ultraviolet-C Light Against Coronavirus

Michel Benite Rossi[1(✉)], Sérgio Barcelos[2], Euclides Lourenço Chuma[1], Laez Barbosa da Fonseca Filho[1], and Luis Ernesto Ynoquio Herrera[1]

[1] Photonics Innovation Institute - iTech, Campinas, Brazil
euclides.chuma@ieee.org, {laez.barbosa, luis.ernesto.herrera}@ftth.com.br
[2] Fiberwork Comunicaçoes óptica Ltda, Campinas, Brazil
info@fiberwork.net

Abstract. Severe acute respiratory syndrome coronavirus (SARS-CoV2) is a new highly pathogenic virus that has induced the COVID-19 disease in over 27.6 million patients and caused over 900 thousand deaths around the world by September 8th, 2020, nine months after the first reported case in China, according to the World Health Organization. Through the realization of the research, it is implied that Ultraviolet-C (UVC) light is an efficient method for tackling the spread of coronavirus. This work comprises a systematic study of state-of-art scientific articles concerned with the application of UVC for tackling different viruses, including those from the coronavirus family and its impact on human health. While the most common germicidal wavelength, 254 nm, is harmful to humans, studies demonstrated the germicidal effect of the wavelength of 222 nm and indicated it could be safe to humans under certain exposure conditions. Also, different conditions of virus exposure to UVC light were considered and the UVC dose required for n-log reduction compared. Data from analyzed studies demonstrated that some factors such as temperature, incident wavelength, and virus family have been shown to be important to determine the inactivation dose.

Keywords: UVC · Coronavirus · Covid-19 · Photonics

1 Introduction

COVID-19, the disease caused by the SARS-CoV-2 virus, was first reported in late 2019. According to the World Health Organization (WHO) [1], by September 8th, 2020, this disease had over 27.6 million cases and caused about 900 thousand deaths worldwide. The fast-spreading of the SARS-COV-2 virus is believed to be connected to airborne and direct contact transmission [2]. Due to its high transmission rates, it is of great importance to develop methods for mitigating the virus spread and inactivating airborne as well as surface-adsorbed virus.

In recent years, methods such as pH and temperature control, chemical agents, and UVC light have been studied to fight pathogens in different conditions [3]. More specifically, UVC light, which has a well-known germicidal effect, has been used in many

Y. Iano et al. (Eds.): BTSym 2020, SIST 233, pp. 196–203, 2021.
https://doi.org/10.1007/978-3-030-75680-2_23

contexts, such as in hospitals, food disinfection, and water treatment [4]. Our attention has been driven specifically towards UVC germicidal methods relying on wavelengths ranging from 200 nm to 280 nm and using light sources such as excimer lamps, LEDs, low pressure and medium pressure vapor mercury lamps (LP-ML and MP-ML). These sources have different output powers, spectra, and efficiencies. In this review study, we show state-of-the-art scientific works using UVC methods aiming to inactivate bacteria and viruses, especially those from the Coronavirus family, including the SARS-COV-2.

2 Characterization Parameters

2.1 UVC Dose

The dose is the light energy per unit of area applied to the material sample under test. A UVC dose is given by the expression:

$$D = I \cdot t = \frac{P \cdot t}{A} \tag{1}$$

Where "I" is the UVC intensity, t is the light source irradiating time, and A is the irradiated sample area. We assume a spatially uniform irradiation distribution. The dose measurement is commonly expressed in units of mJ/cm^2. The Dose parameter does not take into account the UVC irradiation effect on the biological sample itself since this effect depends on the material interaction with the UVC electromagnetic waves.

The UVC effectiveness as a germicidal agent is due to the fact that UVC light induces a cross-link between nucleic acids, i.e., a thymine (or uracil) dimer [5], and the higher the absorbed light by the pathogen, the higher the germicidal effectiveness per irradiated energy. The germicidal effectiveness curve for E. Coli bacterium and its comparison with the ML-ML and LP-ML emission spectra can be found in [5].

The net germicidal efficiency of a UVC source is obtained by calculating the total irradiated power weighted by the effectiveness of each emitted wavelength, which can be calculated as:

$$GE = \frac{1}{\int P(\lambda) d\lambda} \cdot \int_{\lambda 0}^{\lambda} P(\lambda) \cdot EC(\lambda) \cdot d\lambda \tag{2}$$

Where GE is the net Germicidal Efficiency, EC is the effectiveness function curve, and λ is the UVC light wavelength. The EC is a function ranging from 0 to 1, where $EC(\lambda) = 1$ represents the maximum germicidal effectiveness achieved, while $EC(\lambda) = 0$ represents no germicidal effect.

To typify the UVC germicidal effect, it is important to understand the absorption curves of nucleotide bases as a function of wavelength. It is known that uridine, adenine, guanine, and cytosine absorbance peaks range from 254 nm to 270 nm. Also, these bases have increasing absorbance for wavelengths smaller than 220 nm, indicating a potentially germicidal effect in this wavelength range. The absorption curve of those nitrogenated bases can be found in [6].

Finally, it is important to highlight that there are RNA viruses and DNA viruses, which have the same nitrogenate bases, except for thymine and uracil, which are found in DNA and RNA, respectively. Due to this difference, it might be important to observe the absorption curves of thymine and uracil as it does relate to UVC efficacy against RNA and DNA viruses. According to [7], thymine and uracil absorptions peaks are 264 nm and 259 nm, respectively.

2.2 Inactivation Fraction

The Inactivation Fraction of a pathogen immersed in a sample is understood as the measurement of infectivity reduction after a given time of exposure to the disinfection method. To quantify infectivity, it is common to use the parameter "Plaque Forming Units" (PFU) to titrate the material under analysis. The PFU indicates the concentration of virus in a solution that can form plaques, or colonies, in a host cell culture per volume. For example, a viral solution with a PFU of 100 pfu/ml means that 1 ml of this solution forms 100 plaques in the host cell culture. Once the material titer is known at the starting point and after the procedure, the reduction fraction can be calculated as [8]:

$$R = \log\frac{V_1 \cdot T_1}{V_2 \cdot T_2} \tag{3}$$

Where V1 and T1 are the initial volume and titer, while V2 and T2 the final (after the procedure) volume and titer, respectively. It is usual to present these data as a function of time and express this ratio as n-log reduction. For example, if some disinfection method implies a 2-log reduction after a certain period of time, it means $V_2 T_2 = 0.01 \cdot V_1 T_1$, which means the virus concentration was reduced to 1% from its starting point, meaning a 99% virus inactivation. Similarly, 3-log reduction means 99.9% inactivation and so on.

If the solution volume in (3) is constant, then the reduction fraction can be simplified to $R = \log PFU_1/PFU_2$, since PFU_i is a common titer unit.

3 Recent Studies and Main Results

We performed a systematic review regarding UVC light as a germicidal agent. Different studies and their main results are presented as well as parameters such: necessary radiation dose (mJ/cm^2) to achieve specific reduction fractions, employed wavelength and pathogen exposure conditions are shown.

3.1 Airborne Virus Inactivation

Preventing airborne viral transmission might be an efficient method to control a disease spreading, and UVC light, commonly 254 nm lamps, can be a very useful tool for this purpose. However, 254 nm light can be a hazard to human eyes and skin [9], therefore, this sort of approach must be carefully designed in order to be safe. Three possibilities to keep this method safe are:

1. Irradiated rooms must be empty of humans and animals.
2. The UVC radiation must be contained in closed and opaque containers.

3. Radiation wavelength must be safe for both humans and animals while allowing germicidal efficiency.

In [9], Buonanno et al. demonstrated an important result in airborne virus inactivation using an unusual UVC wavelength, 222 nm. This wavelength was chosen for two reasons: first due to its germicidal potential and second due to the fact that far UVC has a very short penetration depth on skin and tear layers, representing a possibly safe wavelength for humans [9]. Different airborne viruses were studied, in particular the Human Coronavirus OC43, which is a member of the beta genus, as well as SARS-COV-2, indicating both viruses might have similar responses to this wavelength. With an irradiation intensity of –100 $\mu W/cm^2$ of 222 nm wavelength, it was demonstrated a 3-log reduction with a dose of 1.2 mJ/cm^2.

In [10], Do-Kyun Kim et al. demonstrated an almost 5-log reduction of MS2 and $Q\beta$ viruses, with 280 nm wavelength LED source using a dose of 45 mJ/cm^2. At similar conditions, the ϕX 147 virus had a 4-log reduction with 4.6 mJ/cm^2. The expressive differences in virus susceptibility to UVC were attributed to the type of virus, RNA (MS2 and $Q\beta$) or DNA (ϕX 147), since absorption peaks of uracil and thymine are different. Indeed, uracil absorbance is much smaller than thymine absorbance for 280 nm [9]. Furthermore, airborne bacteria were exposed to the same UVC light and a 2.5 to 4 log reduction was achieved with 1.5 mJ/cm^2 to 4.6 mJ/cm^2. It was said that the main difference between the susceptibility between those bacteria and viruses were their sizes, –2 μm and –30 nm, respectively. Due to their sizes and the wavelength, bacteria should be totally exposed to irradiation while the viruses could have a smaller interaction.

Figure 1 illustrates the main idea of both experiments presented in this section, which aims to inactivate airborne pathogens. Basically, this setup requires a nebulizer that can introduce water vapor inside the chamber, controlling the relative humidity and virus concentration. It is also necessary to recover the exposed airborne virus from the chamber for analysis. It is also important for the air to be properly exposed to the UVC light while flowing inside the chamber, which can be achieved by an air pump. For more detailed information about each setup characteristics and equipment used in these experiments, see references [9, 10].

3.2 Surfaces and Suspensions Virus Inactivation

It is important to highlight that different pathogens have different responses to UVC light, therefore the most efficient wavelength varies depending on the pathogen species. Said that, it is clear the importance of exposing the SARS-COV-2 to UVC light in order to assess its response and validate this germicidal method against Covid-19 Virus.

In [11], Christiane Heilingloh et al. studied the susceptibility of SARS-COV-2 to UV irradiation, including 254 nm wavelength, where the irradiation was performed in virus stock well plates with high viral concentration. The intensity of irradiated light (254 nm) was measured to be 1.94 mW/cm^2 at the top of the plates. The required dose to inactivation was calculated to be 1048 mJ/cm^2, while an older study [3] showed a 1446 mJ/cm^2 to SARS-COV inactivation in similar conditions. Comparing those results clarifies that different strains of viruses respond differently to similar irradiations, although the same order of magnitude was required for inactivation.

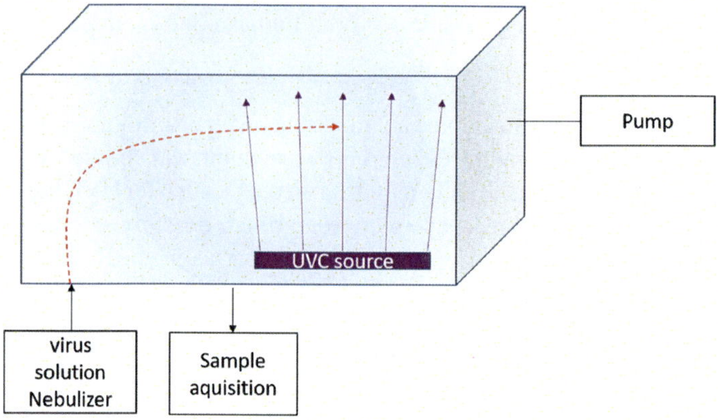

Fig. 1. Illustrative setup for airborne UVC inactivation. A virus solution nebulizer allows the introduction of water vapor with the desired virus concentration. Sample acquisition allows the recovery of air from inside the chamber for measurements and the pump allows airflow control.

David Ozog et al. demonstrated in [12] the disinfection of N95 respirators from SARS-COV2 by the use of 254 nm wavelength UVC radiation obtained by LP-ML. In this study, both sides of infected N95 respirators were irradiated separately with a dose of 1500 mJ/cm^2. After this dose of irradiation, measurements of the viral concentration of respirators were made, and was shown to be below its detection limit, therefore considered disinfected. In this study, it was not measured the minimum dose required to inactivate the virus adsorbed in the respirator; instead, it was demonstrated that 1.5 mJ/cm^2 was enough to achieve a safe reduction rate, considering the initial conditions imposed in the experiment.

A study published in 2007 by Chun-Chieh et al. [13] showed that n-log reduction of different viruses, including RNA and DNA strains, exposed to UVC light, 254 nm wavelength, depends on the applied dose and the air humidity. In order to study the virus's inactivation on surfaces by UVC light sources, different viruses' strains were spread on a surface containing a special gelatin inside an isolated chamber, allowing humidity and contamination control. The UVC light source was a lamp with a wavelength peak close to 254 nm. For air relative humidity of 55%, the dose required for 3-log reduction varied from –7 mJ/cm^2 to –25 mJ/cm^2, while for 85% relative humidity, the required dose for 3-log reduction varied from –12 mJ/cm^2 to –30 mJ/cm^2. This study shows that the air humidity might be an important factor to determine the UVC dose required for virus inactivation and suggests the higher the humidity, the smaller the virus susceptibility for a specific condition.

As we see, different UVC sources have been used, therefore, different spectrums and germicidal effectiveness were achieved. S. E. Beck. et al. [14] ran a study in which, not only germicidal efficacy of each source was evaluated, but also the energetic efficiency of each system was tested. The study compared MP-ML, LP-ML, and UVC LEDs (emission peak in 260 nm and 280 nm), for different viruses and bacteria (E. coli, MS2 Coliphage, Adenovirus 2, Bacillus pumilus Spores) in water suspension. It was shown that the germicidal efficiency of both LEDs and Mercury lamps was similar. The required dose for 3-log inactivation ranged from –10 mJ/cm^2 to –230 mJ/cm^2, depending on the pathogen. Concerning energy efficiency, it was shown that the energy required to achieve 1-log reduction using LEDs was approximately 50-fold larger than the using mercury lamps due to the low energetic efficiency of UVC LEDs.

Finally, Tomoaki F. et al. ran a study [15] to verify the germicidal efficacy and safeness for humans of the 222 nm wavelength, which can be obtained by KrCl excimer lamp and appropriated optical filters. In this study, volunteers had part of their back skin exposed to up to 500 mJ/cm^2 of 222 nm wavelength. Three parameters were analyzed:

- It was verified the erythema presence or absence in the exposed skin for specific doses.
- Cyclobutene pyrimidine dimers were measured in the exposed and non-exposed area as a DNA damage indicator.
- Bacteria measurements were conducted on the exposed and non-exposed skin.

Their results indicate that the 222 nm wavelength has a considerable germicidal effect on human skin surface and is at the same time to exposure, respecting some dose limits, corroborating results presented by Buananno [9].

The reader can read the original papers for more detailed information about each experimental setup, sample preparation, and data analysis procedures.

4 Discussing

All the different studies presented here have shown viruses or bacteria inactivation by the use of UVC radiation. The most common wavelength used was 254 nm, usually obtained from mercury lamps. Another UVC source that has been applied in the last years is LEDs [16–18], which can be found commercially, at least in the following wavelength range: from 255 nm to 285 nm.

Both DNA and RNA viruses were studied, and single-stranded RNA/DNA viruses were shown to be more susceptible to UVC irradiation than double-stranded RNA/DNA viruses [13]. Not only the genetic material influences the susceptibility of the pathogen, but also the UVC exposure conditions. While an irradiation higher than 1 J/cm^2 was required to inactivation of SARS-COV-2 from N95 respirator (–7-log reduction) [12] using a 254 nm wavelength source, less than 2 mJ/cm^2 was enough for 3-log reduction of airborne SARS-COV using a 222 nm wavelength source [9].

When comparing the UVC susceptibility of bacteria and viruses [10], it has been inferred that the particle size is an important parameter, where individuals with a size much smaller than the incident wavelength have been shown to be less susceptible than bigger ones. While [10] has studied viruses of –30 nm, SARS-COV-2 is –100 nm [18], indicating UVC dose to SARS-COV-2 inactivation may be smaller than the doses required to inactivate MS2, $Q\beta$, and ϕX 147.

Finally, unlike 254 nm, the 222 nm wavelength has been shown to be safe for human skin under certain controlled conditions [9, 15]. This safety is explained by the fact that 222 nm radiation penetration depth is considerably smaller than that of 254 nm, hence, it can penetrate neither the dead skin layer nor the tear layer but still can reach bacteria and viruses nuclei.

5 Conclusion

Based on the presented results, we can conclude that UVC electromagnetic wave ranging from 222 nm to 280 nm is efficient as a tool to viral inactivation, including SARS-COV2, which could be inactivated, when adsorbed on NR95 respirator surface, with a dose of 1500 mJ/cm^2 of 254 nm wavelength.

Beyond that, different studies demonstrated different doses for the same reduction fraction of the same pathogen in different conditions, allowing us to conclude that, so far, there is not a universal minimum required dose for viral inactivation. Instead, the inactivation dose must be calculated according to the specific exposure conditions, attacked pathogen, and applied wavelength. This means that solutions for the inactivation process must be designed for specific conditions, not generically.

Furthermore, 222 nm is a promising wavelength for germicidal purposes, since it has been suggested to be safe for human exposure under certain conditions. More studies should be developed to prove those results and, once confirming it is, in fact, safe, it could be used as an important tool for germicidal purposes in occupied spaces.

Acknowledgments. The authors would like to acknowledge the financial support of FINEP (0273-20).

References

1. World Health Organization (2020) Coronavirus disease (COVID-19), Situation reports
2. Doremalen V et al (2020) Aerosol and surface stability of SARS-CoV-2 as compared with SARS-CoV-1. N Engl J Med 382:1564–1567. https://doi.org/10.1056/NEJMc2004973
3. Darnell MER, Subbarau K, Feinstone SM, Taylor DR (2004) Inactivation of the coronavirus that induces severe acute respiratory syndrome. SARS-CoV J Virol Methods 121:85–91
4. Begum M, Hocking AD, Miskelly D (2009) Inactivation of food spoilage fungi by ultra violet (UVC) irradiation. Int J Food Microbiol 129:74–77
5. Kowalsky W (2009) Ultraviolet germicidal irradiation handbook. Springer, Berlin
6. Fischer JH (1995) Specific detection of nucleotides, creatine phosphate, and their derivatives from tissue samples in a simple, isocratic, recycling, low-volume system. LC GC Int 8:254
7. Thomas G, Sarkar N, Bányász Á, Dimitra M, Roberto I (2007) Solvent effects on the steady-state absorption and fluorescence spectra of uracil, thymine and 5-fluorouracil. Photochem Photobiol 83:595–599. https://doi.org/10.1111/j.1751-1097.2007.00052.x
8. EMEA (2008) EMEA/CHMP/BWP/398498/2005: Guideline on Virus Safety Evaluation of Biotechnological Investigational Medicinal Products. Committee for Medicinal Products for Human Use, Biologics Working Party, London
9. Buonnano M, David W, Igor S, David BJ (2020) Far-UVC light (222 nm) efficiently and safely inactivates airborne human coronaviruses. Sci Rep (Nat Public Group) 10. https://doi.org/10.1038/s41598-020-67211-2
10. Kim DK, Kang DH (2018) UVC LED irradiation effectively inactivates aerosolized viruses, bacteria, and fungi in a chamber-type air disinfection system. Appl Environ Microbiol 84. https://doi.org/10.1128/AEM.00944-18
11. Christiane HS et al (2020) Susceptibility of SARS-CoV-2 to UV irradiation. Am J Infect Control 48:1273–1275

12. David OM et al (2020) The effect of ultraviolet c radiation against different n95 respirators inoculated with SARS-CoV-2. Int J Infect Dis 100:224–229. https://doi.org/10.1101/2020.05.31.20118588
13. Chie CT, Shan LC (2007) Inactivation of viruses on surfaces by ultraviolet germicidal irradiation. J Occup Environ Hyg 4:400–405. https://doi.org/10.1080/15459620701329012
14. Beck ES et al (2017) Evaluating UV-C LED disinfection performance and investigating potential dual-wavelength synergy. Water Res 109:207–216. https://doi.org/10.1016/j.watres.2016.11.024
15. Tomoaki F et al (2020) Exploratory clinical trial on the safety and bactericidal effect of 222-nm ultraviolet C irradiation in healthy humans. PloS One 15. https://doi.org/10.1371/journal.pone.0235948
16. Irene C, José ASP, Malato S, Olivier A, Bruce J, Peter J (2015) Application of high intensity UVC-LED for the removal of acetamiprid with the photo-Fenton process. Chem Eng J 264:690–696. https://doi.org/10.1016/j.cej.2014.11.142
17. Paul ON, Yi Q, Chen G, Baoping Z, Lu Y, Liang S (2019) Comparison of the performance of pulsed and continuous UVC-LED irradiation in the inactivation of bacteria. Water Res 157:218–227. https://doi.org/10.1016/j.watres.2019.03.080
18. Bar-On YM, Flamholz A, Phillips R, Milo R (2020) SARS-CoV-2 (COVID-19) by the numbers. eLife 9:e57309

A Method to Calibrate Variable Speed Limit Control on High-Truck Share Roads

A Case Study in Brazil

Helena Tanoue Vizioli[1(✉)], Krešimir Kušić[2], Edouard Ivanjko[2], and André Luiz Cunha[1]

[1] Universidade de São Paulo, São Carlos, SP 13566-590, Brazil
{helena.vizioli,alcunha}@usp.br
[2] Faculty of Transport and Traffic Sciences, University of Zagreb, 10000 Zagreb, Croatia
{kkusic,eivanjko}@fpz.unizg.hr

Abstract. This paper focuses on the method used to calibrate a simulation of a Variable Speed Limit Control (VSLC) system implemented in a Brazilian highway with a high share of trucks and fleet heterogeneity. To achieve a validated model, crucial parameters, such as the car-following and the lane change rules, have to be adjusted, and there must be a metric for comparison, which is presented in this work. The VSLC is added to the validated model and has to be adjusted to the local situation considering the flow and average speed throughout the day. The implementation of VSLC aims to reduce congestion and homogenization of the vehicles' average speed throughout the day. The microscopic traffic simulator VISSIM and MATLAB were applied to implement and evaluate the calibrated VSLC proving its effectiveness. As a result, a valid VISSIM simulation model, parameters for VSLC calibration, and an improvement of the average speed up to 5.9% were obtained.

Keywords: Variable speed limit control · Calibration · Traffic simulation

1 Introduction

Brazil covers an area of 8.5 million km2 and has around 1.5 million kilometers of highways. In 2017, more than 75% of cargo transportation was performed through roadways using trucks [1]. Thus, it is common to have more than 25% of the traffic on highways composed of trucks, reaching more than 80% in some points [2]. This high share of trucks directly influences the traffic flow since trucks have a lower speed limit and slower dynamics than cars, causing a reduction in the average speed. Besides, the Brazilian truck fleet has different axle configuration types, some of them being heavy, causing even slower traffic dynamics. Finally, it is necessary to consider the high level of pollution generated by trucks, especially on congested highways, since in Brazil, the truck fleet age is over ten years. This old truck fleet is prone to breakdowns adding further to the congestion problem. In that sense, the study of the truck fleet's impact on roads' traffic is of significant importance.

Y. Iano et al. (Eds.): BTSym 2020, SIST 233, pp. 204–211, 2021.
https://doi.org/10.1007/978-3-030-75680-2_24

A proposal that has been studied to reduce traffic congestion is the implementation of a Variable Speed Limit Control (VSLC) system. This system varies the highway speed limit to avoid abrupt average speed changes and, consequently, congestion. A sensor is installed in the problematic area to measure the volume and average speed of vehicles. According to these data, the previous highway segment's speed limit is altered, decreasing when the volume and speed indicate the beginning of a possible traffic problem. Researchers and practitioners have studied this technology in diverse countries, such as Germany, the USA [3], and Croatia [4]. However, this solution is not very explored in Brazil, which encourages the research described in this paper.

2 Proposed Method

This paper presents an analysis of using a VSLC system in a Brazilian high- way with a high percentage of heterogeneous types of trucks in the traffic flow. Therefore, the proposed method considered the following objectives: (i) Validate the microscopic simulation to represent the local traffic behavior accurately; (ii) Calibrate the VSLC strategy to control the traffic speed; and (iii) Compare simulation scenarios with and without VSLC.

2.1 Segment Selection

The Agency of Transport of the State of São Paulo (ARTESP) provided traffic data from different highway segments, which are equipped with inductive loop traffic sensors per lane. The provided data include volume and average speed for cars and heavy vehicles. The segments between two sensors were analyzed regarding data completeness, flow, density, average speed, truck percentage, and road geometry, including the number of on- and off-ramps. The selected segment was the section from the KM 86 (sensor 21) to the KM 65 (sensor 20), at the Rodovia dos Bandeirantes (Bandeirantes Highway) between the two biggest cities in the State of São Paulo, São Paulo, and Campinas (Fig. 1).

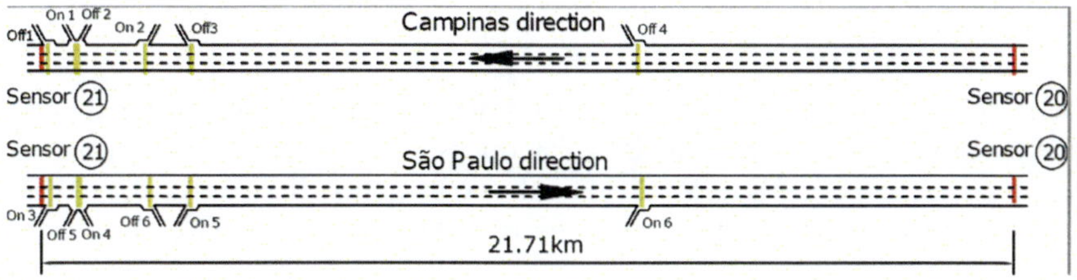

Fig. 1. Representation of the highway's segment selected

2.2 VISSIM Validation

The validation process took into account parameters already calibrated in other studies that have used the same highway in the State of São Paulo as follows.

Car-Following and Lane Change Parameters. The car-following logic in VISSIM [5] uses Wiedemann's behavior models [6, 7], which are based on an interactive process involving acceleration and deceleration of the vehicles. The logic considers two vehicles in which the following vehicle has to adapt to the leader's speed changes. The car-following logic models this behavior through parameters such as the standstill distance, the headway time, the following variation, the threshold for entering following, and the negative threshold. In turn, the lane change model used in VISSIM is based on finding an appropriate gap for a vehicle to change its current lane, as described in [8]. Regarding this, the car-following and the lane change default parameters were adjusted by those calibrated in [9], using a Genetic Algorithm (GA) to calibrate the driver behavior parameters on the same highway. In this paper, the following parameters, which were pointed out in [9] as important, were considered: the minimal headway needed, the safety distance reduction factor, and the maximum deceleration for cooperative breaking. Used values are given in Table 1.

Table 1. Car following and lane change parameters of São Paulo highways [9]

Parameter	Description	Model	Default	Best
OBSVEH	Observed Vehicles	Car following	2	4
CC0	Standstill distance	Car following	1.50	2.56
CC1	Headway time	Car following	0.90	0.75
CC2	Following variation	Car following	4.00	7.13
CC3	Threshold for entering following	Car following	−8.00	−12.92
CC4	Negative following threshold	Car following	−0.35	−0.54
MH	Min. headway - front/rear	Lane change	0.50	1.75
SD	Safety distance reduction factor	Lane change	0.60	0.39
MD	Max. Decel. For cooperative breaking	Lane change	−3.00	−1.28

Truck Performance Parameters. In VISSIM, the default values for trucks are based on German trucks. However, as the Brazilian trucks' fleet is more heterogeneous, new vehicles were created in VISSIM, dividing the default class HGV (used for a typical truck) into lightweight, midweight, heavy, and extra-heavy trucks. In [10], it is suggested that these four truck categories represent the Brazilian fleet heterogeneity in terms of the number of axles, length, weight, motor power, and pavement impact well. In [11], the truck performance model was calibrated for the same four categories, using a GA and data from the same highway. Therefore, the same set of parameters used in [11] was used in this paper, as detailed in Table 2.

Table 2. Trucks' categories calibrated in [11]

Category	Axles	Avg. length (m)	Avg. GVM (t)	Power/mass avg (kW/t)	Share (%)
Lightweight	2	6.8	6.755	14.81	36.60
Midweight	3	9.1	13.711	9.29	27.10
Heavy	4 and 5	15.5	28.583	8.79	19.86
Extra-heavy	More than 6	17.7	38.072	7.20	13.44

2.3 On- and Off-Ramps Flow Adjustment

The data provided by ARTESP did not include the on- and off-ramps volume and speed of vehicles. Thus, the on- and off-ramps' flow was estimated based on the surrounding areas' characteristics. For each on-ramp, the volume of vehicles by time interval was set, using a percentage of the volume on sensors provided by ARTESP. The on-ramp flow was 60% of the volume on sensor 20. For the off-ramps, the static vehicle routing decision of VISSIM to define the vehicles' behavior was used, keeping the majority of the vehicles on the mainline of the highway was used.

2.4 Time Periods and Taper Deceleration Lane

The simulation covered 6 h of real traffic, divided into 6 min intervals. This interval equals the input sensor data. The warm-up period was set up to 12 min, i.e., two intervals before the simulation's beginning were sufficient to reach initial values similar to the observed data. The taper deceleration lanes, in turn, are characteristically part of the off-ramps and simulate the natural driver behavior of reducing the driving speed in these places on highways.

2.5 Measure of Model Validity

To obtain a validated simulation model, it is necessary to measure its adequacy to reality. Among a list of possible measures, Hollander and Liu [12] suggested the GEH statistic for comparing pairs of observed-simulated measurements. The authors also recommended that GEH lower than 5 is an acceptable fit, whereas GEH between 5 and 10 could have corrupt data, and GEH greater than 10 is not acceptable. At least 85% of the calculated GEH values must be lower than 5 to accept the model as a valid one. This was also adopted in this paper. The GEH statistic equation is defined by:

$$GEH = \sqrt{2 * \frac{(E-V)^2}{(E+V)}} \tag{1}$$

Where E is the hourly traffic volume obtained from the microscopic simulation model and V is the observed hourly traffic data.

2.6 VSLC Calibration

The control application implemented in MATLAB aims to reduce traffic problems, decrease the speed limit when the average speed is low and the volume of vehicles is high. The first step for the implementation was defining the threshold values, i.e., the values that induce the highway speed limit change. These values are obtained based on the empirical data available, analyzing the flow-density diagram and the speed-flow dependence [13].

The flow-density diagram was plotted using all data from 2014, and the flow and critical density values were estimated from the intersection of the least- square fit for the ascending and descending curves as given in Fig. 2. One has to have in mind that critical density denotes the value where congestion starts. The real data speed-flow dependence was also plotted, and a second-degree curve was fit to these points with the restriction of containing the critical point, i.e., the critical flow previously determined and the estimated free-flow speed.

It was possible to obtain the flow values for the desired speed from the adjusted curve, which are the speed limit thresholds mentioned above. The obtained values were adapted to the specific analyzed day, and a hysteresis value was set up according to the VSLC results [14]. This value avoids abrupt and frequent speed limit change since it would be an impractical situation.

3 Result of Data Analysis

This section presents the calibration and simulation results and investigates the efficiency of VSLC through the observation of average traffic flow speed measured within the highway segment under VSLC.

3.1 Results of VISSIM Validation

The model validation results were analyzed using the GEH statistic. In Table 3, the initial values correspond to simulation using default VISSIM parameters, whereas the final GEH was using the calibrated parameters. For sensor 20, they were already good because they depend mainly on the input data, whereas sensor 21 revealed a good improvement. One has to notice that two values indicated unacceptable behavior in the initial GEH results and one a possible error. Using the calibrated parameters, only one GEH value

Table 3. Initial and final GEH values

Interval	2–3 pm	3–4 pm	4–5 pm	5–6 pm	6–7 pm	7–8 pm
Initial GEH value for sensor 20	0.9	0.5	1.2	1.7	1.3	1.1
Initial GEH value for sensor 21	5.3	0.2	4.8	11.8	10.2	2.3
Final GEH value for sensor 20	0.9	0.4	0.8	1.5	1.3	1.1
Final GEH value for sensor 21	4.7	5.0	1.6	3.9	1.2	2.7

presented the limit of acceptable and possible error. In summary, more than 85% of GEH values are under 5, meaning it is a valid simulation model of the selected highway segment.

3.2 Results of VSLC Implementation

The flow-density and the speed-flow relationship (Fig. 2) were plotted together with the least-square fit adjusted curve. With this method, the capacity was defined as 5460 veh/h, and the critical density as 53.2 veh/km, considering the three lanes. The free-flow speed was estimated as being 90 km/h, and the initial thresholds were obtained by looking at the flow values for the speeds from 50 to 100 km/h. The same procedure was then applied to the chosen day, 11th March 2014, which presented one-hour congestion. The final flow thresholds were defined as [3000; 3200; 3400; 3600; 3800; 4000] veh/h for corresponding speed values [50; 60; 70; 80; 90; 100] km/h respectively. A hysteresis was applied to prevent too large oscillations in the resulting speed limit value. Its value was obtained from an analysis of the flow value changes around the corresponding speed. The hysteresis value was determined as 250 veh/h, and the VSLC could be implemented, also considering minimal oscillations of its value.

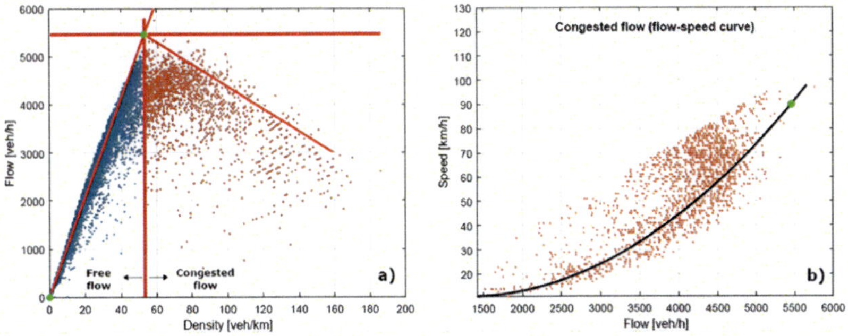

Fig. 2. Least-square fit for the flow-density point cloud (a) and adjusted curve to the data for the congested part (b)

3.3 Comparing the Improvement of Used VSLC Strategy

A ruled-based VSLC system was applied to the simulation using the parameters determined above. The software packages MATLAB and VISSIM connected in a simulation framework with a COM interface were used to collect traffic data and change the speed limit according to the VSLC rules during the simulation. The vehicles' and trucks' volume and the average vehicles' speed with and without VSLC were obtained (Fig. 3).

Analyzing the average speed's results given in Table 4, it is possible to notice the improvement, i.e., the VSLC acted to increase the average speed on the high- way, meaning that the congestion periods were reduced or dissolved. Regarding the data presented in Table 4, it is possible to see the numerical results, reaching an improvement of 1.4% on the total average speed. The maximum speed improvement happens between 7 and 8 pm, reaching 5.9%. Also, apart from the period between 2 and 3 pm, all other periods have an increased average speed.

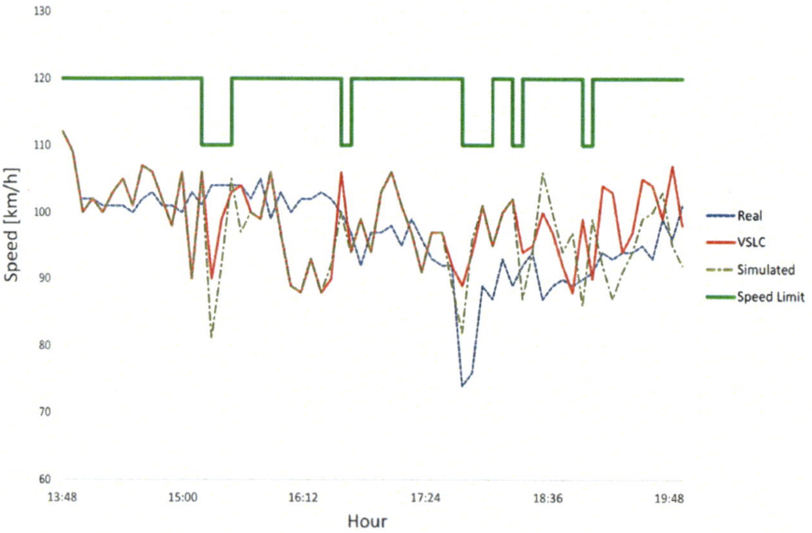

Fig. 3. Comparison of speeds on the highway

Table 4. Average speed per time interval with and without the VSLC in km/h.

Interval	2–3 pm	3–4 pm	4–5 pm	5–6 pm	6–7 pm	7–8 pm	Total
Simulated	102.4	98.2	93.4	95.9	97.6	94.6	96.9
VSLC	102.4	100.3	93.8	96.7	96.4	100.2	98.3
Improvement (%)	0	2.1	0.4	0.8	−1.2	5.9	1.4

4 Conclusion

A rule-based VSLC system was implemented using a MATLAB/VISSIM simulation framework connected with the COM interface. Real-world traffic data were used to calibrate the simulation model to represent Brazilian highways. According to the GEH statistic, calibration achieved satisfactory flow behavior, enabling the VSLC to achieve better results by using tuned parameters. A total 1.4% increase in the average speed was obtained. If the final speed analysis separated cars from trucks, this improvement could be greater since heavy vehicles have a speed limit of 80 km/h, reducing the average flow speed as a whole. It is also important to point out that the only period that the VSLC acted negatively was from 6 to 7 pm. It can be an effect of the peak-hour congestion that the implemented controller could not resolve. However, the results of the next hour showed a significant improvement, which indicates the effectiveness of VSLC.

In general, the research proved that VSLC can increase the average speed and will serve as a base for future work. Thus, the continuation of this work will consider implementing a calibration procedure of the simulation model based on the GA and a fuzzy logic based VSLC that will also be optimized using the GA.

Acknowledgment. We acknowledge the ERASMUS+ program for the International Mobility Grant, the Agência de Transporte do Estado de São Paulo (ARTESP) for providing traffic data, and the PTV Group for the VISSIM license.

References

1. Resende P (2018) Custos logísticos no Brasil 2017. https://www.fdc.org.br/conhecimento/publicacoes/relatorio-de-pesquisa-33324
2. Cunha A, Setti J (2011) Truck equivalence factors for divided, multilane highways in Brazil. In: 6th international symposium on highway capacity and quality of service Stockholm, Sweden
3. Riggins G, Bertini R, Ackaah W, Magreiter M (2016) Evaluation of driver compliance to displayed variable advisory speed limit systems: comparison between Germany and the US. Transp Res Procedia 15:640–651
4. Kušić K, Korent N, Gregurić M, Ivanjko E (2016) Comparison of two controllers for variable speed limit control. In: 2016 international symposium ELMAR. IEEE, pp 101–106
5. VISION PTV (2009) VISSIM 5.40-user manual. Product manual, PTV AG
6. Wiedemann R (1974) Simulation des Strassenverkehrsflusses. Schriftenreihe des Instituts für Verkehrswesen der Universität Karlsruhe, Band 8, Karlsruhe, Germany
7. Wiedemann R, Reiter U (1992) Microscopic traffic simulation: the simulation system MISSION, background and actual state. CEC Project ICARUS (V1052), Final Report, v. 2, Appendix A. Brussels
8. Sparmann U (1978) Lane change on two laned motorways. Forsch Strassenbau U Strasenverkehrstech 263, no 263
9. Bethonico F (2016) Calibração de simuladores microscópicos de tráfego através de medidas macroscópicas. Tese de Doutorado. Universidade de São Paulo, São Carlos, Brazil
10. Cunha A, Modotti M, Setti J (2008) Classificação de caminhões através de agrupamento por análise de cluster. XXII Congresso de Pesquisa e Ensino em Transportes, ANPET, Brazil
11. Carvalho L (2018) Calibração do modelo de desempenho de caminhões do Vissim. Tese de Doutorado. Universidade de São Paulo, São Carlos, Brazil
12. Hollander Y, Liu R (2008) The principles of calibrating traffic microssimulation models. Transportation 35:347–362
13. Dervisoglu G, Gomes G, Kwon J, Muralidharan A (2009) Automatic calibration of the fundamental diagram and empirical observations on capacity. In: Transportation research board 88th annual meeting, Washington, DC, US
14. Dadashzadeh N, Ergun M (2019) An integrated variable speed limit and ALINEA ramp metering model in the presence of high bus volume. Sustainability 11:6326. MDPI

Classification of Brain MR Images for the Diagnosis of Alzheimer's Disease Based on Features Extracted from the Three Main Brain Tissues

Vitor H. Chaves Cambui, Katia M. Poloni, Ricardo J. Ferrari(✉), and for the Alzheimer's Disease Neuroimaging Initiative

Department of Computing, Federal University of São Carlos, Rod. Washington Luis, Km 235, São Carlos, SP 13565-905, Brazil
rferarri@ufscar.br

Abstract. Alzheimer's disease (AD) is a progressive brain disorder that causes neurons to degenerate and die as the disease progress. AD is the most common cause of dementia, accounting for 60% to 80% of all cases, and has been recognized as a public health problem by the World Health Organization. In this study, we propose a method to aid in the diagnosis of AD that automatically extracts and classifies image features of the white matter (WM), gray matter (GM), and cerebrospinal fluid (CSF) tissues from the hippocampal regions. Our method uses the features as input to support vector machine (SVM) classifiers to perform the MR image classification in CN × AD and CN × MCI cases. For that, we preprocess all ADNI images and define the regions of interest for analysis. Then, we extract the GM, WM, and CSF tissues using an automated brain tissue segmentation method. Considering the intensities inside both hippocampal regions and each segmented tissue, we extract five statistical metrics from the voxel intensities inside each hippocampal region to use as features. Then, we train SVM classifiers with distinct kernels using a ten-fold nested cross-validation to perform the classification. From the classification experiments, the highest obtained AUC values for the CN × MCI and CN × AD classification cases were 0.814 and 0.922, respectively. It is important to emphasize that we obtained these results using an automated pipeline, with no human intervention, and a relatively small set of features.

Keywords: Alzheimer's diagnosis · Brain tissue segmentation · MR image classification

---

Data used in preparation of this article were obtained from the Alzheimer's Disease Neuroimaging Initiative (ADNI) database (adni.loni.usc.edu). As such, the investigators within the ADNI contributed to the design and implementation of ADNI and/or provided data but did not participate in analysis or writing of this report. A complete listing of ADNI investigators can be found at: https://adni.loni.usc.edu/wp-content/uploads/how_to_apply/ADNI_Acknowledgement_List.pdf

Y. Iano et al. (Eds.): BTSym 2020, SIST 233, pp. 212–219, 2021.
https://doi.org/10.1007/978-3-030-75680-2_25

1 Introduction

Alzheimer's disease (AD) is a progressive brain disorder that causes neurons to degenerate and die as the disease progress. AD is the most common cause of dementia, accounting for 60% to 80% of all cases, and has been recognized by the World Health Organization (WHO) as a public health problem. In fact, projections from the WHO estimate that 50 million people have dementia, and every year there are nearly 10 million new cases. The total number of people with dementia can reach 82 million in 2030 and as high as 152 million in 2050 [1].

Some AD studies presented in the literature have reported that early signs of this disease may be detected 10 to 20 years before symptoms arise [2–4], with structural brain changes that do not noticeably affect the patient. After many years of brain damage, the individuals will start to experience noticeable symptoms, such as memory loss and language problems. The observed symptoms associated with AD are caused by the damage and death of neurons (neurodegeneration), with other brain changes, including inflammation and atrophy [2].

Since there is no single test for AD, doctors use various approaches and tools to make the diagnosis. They often apply tests for assessing the individual abilities in solving problems and retrieving memories and using family history and reports from relatives or caregivers to determine changes in the individual's skills and behavior. Magnetic resonance imaging (MRI) has played an important role in the diagnosis of AD, since this imaging modality provides images with very rich anatomical details that allow the assessment of atrophies and other brain changes resulting from the disease.

Based on the above-described scenario, it is latent the need for modern solutions to take on the challenge of diagnosing AD. There are many works proposed in the literature [5–11] addressing the problem of classification of MR images in one of the three diagnostic groups: cognitively normal (CN), mild cognitive impairment (MCI), and AD. The MCI group corresponds to an intermediate stage of the disease, in which the individuals present minor cognitive issues but with a high probability of evolving to AD.

In this study, we propose a computerized method to assist in the diagnosis of AD. Our method automatically extracts and classifies image features of the white matter (WM), gray matter (GM), and cerebrospinal fluid (CSF) tissues from the hippocampal regions and use them as input to support vector machine (SVM) classifiers to perform MR brain image classification in the cases CN $\times$ AD and CN $\times$ MCI.

The rest of this paper is organized as follows. Section 2 describes all the image datasets used, Sect. 3 provides a description of all methods and processes used in this work, and Sect. 4 presents the relevant experimental results. Finally, Sect. 5 presents a conclusion and summarizes future possibilities.

2 Image Datasets

In this study, we use two image datasets; the Alzheimer's Disease Neuroimaging Initiative (ADNI) [12] and the Neuroimage Analysis Center (NAC) [13].

The ADNI dataset contains images obtained over 50 sites across the USA and Canada[1]. Due to the size and diversity of the images in this dataset, image specifications will be omitted here. However, as we will discuss in Sect. 3.1, proper standardization steps were conducted to overcome this issue. For this study, we use only images acquired using the MPRAGE sequence[2] from patients with MMSE information and age ranging between 70 and 85 years, distributed among three diagnostic groups, i.e., CN, MCI, and mild-AD. Considering these restrictions, we randomly select a total of 762 images, one image for each subject, in which the numbers of CN, MCI, and mild-AD subjects are 302, 251, and 209, respectively.

The NAC dataset consists of 149 3-D triangular meshes of distinct brain structures; they are all spatially aligned to a T1-weighted (T1-w) image of a healthy 42-year-old male subject. This dataset provides the binary masks used to delimit the hippocampal regions in this study.

3 Methods

Figure 1 illustrates the general framework of the proposed method. First, we preprocessed all ADNI images and defined the regions of interest (ROI)s - the hippocampal regions in this study. Next, we extracted the GM, WM, and CSF tissues in the intracranial brain region using the fully automated brain tissue segmentation package FMRIB's Automated Segmentation Tool (FAST). Then, considering the intensities of the voxels in the hippocampal regions of each segmented tissue, we extracted five statistical metrics, i.e., sum, mean, variance, skewness, and kurtosis to use as image features. Finally, we train SVM classifiers with distinct kernels using a ten-fold nested cross-validation to perform the brain MR image classification.

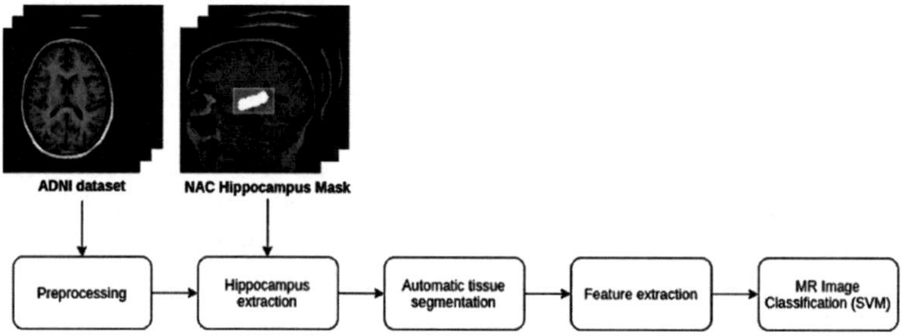

Fig. 1. Overview of the proposed method.

3.1 Preprocessing

In the preprocessing stage, all ADNI images were first processed for noise reduction using the Non-Local Means technique [14], following by bias field correction with the

[1] https://adni.loni.usc.edu/about/centers-cores/study-sites/.

[2] These MPRAGE files are considered the best in the quality ratings and have undergone preprocessing steps - https://adni.loni.usc.edu/methods/mri-tool/mrianalysis/.

N4-ITK technique [15] and image intensity standardization using the proposed method in [16], with the T1-w NAC template image [13] as a reference. After that, the spatial alignment of all study images (ADNI) was conducted using affine transformations implemented in the Nifty-Reg image registration tool [17]. Similar to the image intensity standardization stage, the T1-w NAC template image was used as the reference image. Finally, brain extraction was performed on all images using the ROBEX [18] technique.

3.2 Hippocampal Extraction

To extract each study image's hippocampal ROIs, we first spatially aligned (registered) the NAC T1-w template image with the study image using a deformable transformation [19]. In this case, the NAC T1-w image was used as a moving image and each study image was treated as a fixed image. Then, the obtained deformable transformation was applied to hippocampus mesh models to define the ROIs. Finally, we dilated the hippocampus binary masks (obtained from the mesh models) using the dilatation morphological operator [20, 21] with a sphere structuring element of radius four to encompass surroundings hippocampal structures.

3.3 Automatic Tissue Segmentation

The proper segmentation of the brain tissues is crucial since the segmented images will be used as sources of information for the feature extraction procedure. Therefore, we used the fully automated brain tissue segmentation software FAST [22] (version 4.1), which is considered the state-of-art on tissue segmentation [23] and is available on the FSL package[3]. The FAST implementation is based on a Hidden Markov Random Field Model, optimized using the Expectation-Maximization (EM) algorithm. A detailed description of the algorithm is available in [22]. The input image required by the FAST is a skull stripped MR image, which we obtained as a result of the preprocessing step using the ROBEX [18] technique. The outputs are an intensity non-uniformity corrected version of the input image and the segmented GM, WM, and CSF brain tissues. Figure 2 illustrates an example of an input MR image (first line) and the output image (second line) showing the segmented brain tissues. The segmentation data are color-coded as follows: WM is colored white, GM is light gray, and CSF is dark gray.

3.4 Feature Extraction

In this study, feature extraction was performed for each individual brain tissue within each hippocampal region. For that, both the resulting binary GM, WM, and CSF segmented images and the left (LH) and right hippocampal (RH) binary masks were used to define the regions to compute the features, i.e., GM ∩ LH, GM ∩ RH, WM ∩ LH, WM ∩ RH, CSF ∩ LH, and CSF ∩ RH. Then, we analyzed the voxel intensities inside each mask in order to describe their intensity distributions. Our analyses extracted the sum of the voxel intensities and the first four statistical moments, i.e., mean, variance, skewness, and kurtosis, resulting in 30 features.

[3] https://fsl.fmrib.ox.ac.uk/fsl/fslwiki.

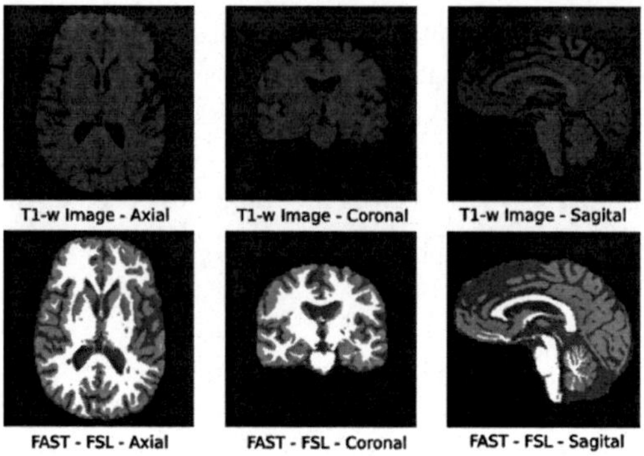

Fig. 2. Segmentation result of the three main brain tissues.

3.5 MR Image Classification

For the image classification in CN × MCI and CN × AD groups, we used the extracted features (30 in total) to train SVM classifiers with four different kernels, linear, radial basis function (RBF), polynomial, and sigmoid. Furthermore, since we have a slightly unbalanced dataset, we have automatically adjusted the model weights to be inversely proportional to class frequencies in the input data using the class_weight[4] parameter from the scikit-learn python library.

We adopted four metrics for performance evaluation, including the area of the receiver operating characteristic (ROC) curve - AUC, accuracy (ACC), balanced accuracy (BACC), and F1-score. A detailed description of these metrics can be found in [24]. Also, we considered the images from the subjects with a cognitive condition (MCI or AD) as positives samples and the images of CN subjects as negative.

We conducted the classification experiments using a ten-fold nested cross-validation. For each SVM kernel, we used a coarse to fine grid search to determine the best C parameter value that maximizes the AUC. The coarse-grid search was performed within a range of values [−5; 10] and steps of 0.5, and the finer search was conducted in the neighborhood of the best coarse-grid parameter, C, with a grid range of values $[2^{(\log2C)-2}; 2^{(\log2C)+2.1}]$ and incremental steps of 0.25.

4 Results

The highest obtained AUC values for the MR image classification experiments for the CN × MCI and CN × AD classification cases were 0.814 and 0.922, respectively. Table 1 presents the average classification results of the ten-fold nested cross-validation for the four SVM kernel functions used on the experiments mentioned above.

4 https://scikit-learn.org/stable/modules/generated/sklearn.svm.SVC.html.

Table 1. Average ± Std classification of the nested ten-fold cross-validation for each tested SVM kernel function.

	Kernel	AUC	ACC	BACC	F1-score
CN × MCI	SVM – Linear	0.814 ± 0.048	0.743 ± 0.045	0.744 ± 0.048	0.726 ± 0.057
	SVM – Polynomial	0.814 ± 0.049	0.738 ± 0.045	0.739 ± 0.048	0.72 ± 0.058
	SVM – RBF	0.804 ± 0.055	0.725 ± 0.063	0.724 ± 0.064	0.701 ± 0.074
	SVM – Sigmoid	0.814 ± 0.055	0.718 ± 0.062	0.715 ± 0.065	0.686 ± 0.077
CN × AD	SVM – Linear	0.906 ± 0.061	0.83 ± 0.083	0.829 ± 0.086	0.799 ± 0.098
	SVM – Polynomial	0.909 ± 0.058	0.834 ± 0.085	0.833 ± 0.088	0.803 ± 0.1
	SVM – RBF	0.922 ± 0.054	0.845 ± 0.06	0.839 ± 0.062	0.81 ± 0.073
	SVM – Sigmoid	0.9 ± 0.066	0.828 ± 0.074	0.826 ± 0.078	0.794 ± 0.089

5 Conclusions

As presented in the previous section, the classification results for the CN × AD case were better than the CN × MCI case, presenting higher metrics for all SVM kernels. These results can be explained due to the AD continuum phases, named the preclinical (CN), the MCI, and the AD. All these phases occur sequentially in the given order; therefore, the measurable brain changes in the AD brain are more intense than MCI [2]. Thus, the brain differences among CN and AD are more distinct than CN and MCI. Our method achieved average AUC values of 0.814 and 0.922 for the CN × MCI and CN × AD classifications, respectively, and accuracy of 0.743 and 0.845 for the same cases. These results are comparable to the state-of-art accuracy results, which vary among 75–80% for the CN × MCI case and 88–92% for the CN × AD case [25–28]. It is also important to emphasize that these results were obtained from a fully automated pipeline, with no human intervention, and with a relatively simple and small set of features. In addition, the size of the dataset used in this study is larger than most published works. Another aspect of the proposed method is that it can be used alongside other methods to build a broader and more comprehensive approach, using it as a complementary tool. Despite the results, further research must be conducted to explore new features capable of classifying more difficult cases such as MCI and AD groups.

Acknowledgments. Funding for ADNI can be found at https://adni.loni.usc.edu/about/#fund-container.

Funding Statement. The authors would like to thank the São Paulo Research Foundation (FAPESP) (grant numbers 2018/08826-9 and 2018/06049-5) and the National Council for Scientific and Technological Development (CNPq) (grant number 166082/2019-8 - PIBITI) for the financial support of this research.

References

1. World Health Organization et al (2018) Towards a dementia plan: a who guide. World Health Organization

2. Alzheimer's Association (2020) Alzheimer's disease facts and figures. Alzheimer's Dementia 16(3):391–460
3. Dubois B, Hampel H, Feldman HH, Scheltens P, Aisen P, Andrieu S, Bakardjian H, Benali H, Bertram L, Blennow K et al (2016) Preclinical Alzheimer's disease: definition, natural history, and diagnostic criteria. Alzheimer's Dementia 12(3):292–323
4. Aisen PS, Cummings J, Jack CR, Morris JC, Sperling R, Frölich L, Jones RW, Dowsett SA, Matthews BR, Raskin J et al (2017) On the path to 2025: understanding the Alzheimer's disease continuum. Alzheimer's Res Ther 9(1):60
5. Ahmed OB, Mizotin M, Benois-Pineau J, Allard M, Catheline G, Amar CB, Alzheimer's Disease Neuroimaging Initiative et al (2015) Alzheimer's disease diagnosis on structural MR images using circular harmonic functions descriptors on hippocampus and posterior cingulate cortex. Comput Med Imaging Graph 44:13–25
6. Duara R, Loewenstein DA, Potter E, Appel J, Greig MT, Urs R, Shen Q, Raj A, Small B, Barker W et al (2008) Medial temporal lobe atrophy on MRI scans and the diagnosis of Alzheimer disease. Neurology 71(24):1986–1992
7. Aderghal K, Benois-Pineau J, Afdel K, Gwenaelle C (2017) Fuseme: classification of sMRI images by fusion of deep CNNs in 2D+ ε projections. In: International workshop on content-based multimedia indexing, Florence, Italy. ACM, pp 1–7
8. Chupin M, Gérardin E, Cuingnet R, Boutet C, Lemieux L, Lehéricy S, Benali H, Garnero L, Colliot O, Alzheimer's Disease Neuroimaging Initiative et al (2009) Fully automatic hippocampus segmentation and classification in Alzheimer's disease and mild cognitive impairment applied on data from ADNI. Hippocampus 19(6):579
9. Cuingnet R, Gerardin E, Tessieras J, Auzias G, Leh'ericy S, Habert M, Chupin M, Benali H, Colliot O, Alzheimer's Disease Neuroimaging Initiative et al (2011) Automatic classification of patients with Alzheimer's disease from structural MRI: a comparison of ten methods using the ADNI database. Neuroimage 56(2):766–781
10. Poloni KM, Ferrari RJ (2018) Detection and classification of hippocampal structural changes in MR images as a biomarker for Alzheimer's disease. In: International conference on computational science and its applications, Melbourne, Australia. Springer, pp 406–422
11. Poloni KM, Villa-Pinto CH, Souza BS, Ferrari RJ (2018) Construction and application of a probabilistic atlas of 3D landmark points for initialization of BTSym2020, 129, v6:' Classification of brain MR images for the diagnosis of Alzheimer's. . . 7 8 Chaves Cambui et al. hippocampus mesh models in brain MR images. In: International conference on computational science and its applications, Melbourne, Australia. Springer, pp 310–322
12. Jack CRJ, Bernstein MA, Fox NC, Thompson G, Alexander P, Harvey et al (2017) The Alzheimer's disease neuroimaging initiative: MRI methods. J Magn Reson Imaging 27(4):685–691
13. Halle M, Talos IF, Jakab M, Makris N, Meier D, Wald L, Fischl B, Kikinis R (2017) Multi-modality MRI-based atlas of the brain. https://www.spl.harvard.edu/publications/item/view/2037
14. Buades A, Coll B, Morel J-M (2005) A review of image denoising algorithms, with a new one. Multiscale Model Simul 4(2):490–530
15. Tustison NJ, Avants BB, Cook PA, Zheng Y, Egan A, Yushkevich PA, Gee JC (2010) N4ITK: improved N3 bias correction. IEEE Trans Med Imaging 29(6):1310–1320
16. Ny'ul LG, Udupa JK, Zhang X (2000) New variants of a method of MRI scale standardization. IEEE Trans Med Imaging 19(2):143–150
17. Ourselin S, Stefanescu R, Pennec X (2002) Robust registration of multi-modal images: towards real-time clinical applications. In: Medical image computing and computer-assisted intervention. Springer, Heidelberg, pp 140–147
18. Iglesias JE, Liu CY, Thompson PM, Tu Z (2011) Robust brain extraction across datasets and comparison with publicly available methods. IEEE Trans Med Imaging 30(9):1617–1634

19. Modat M, Ridgway GR, Taylor ZA, Lehmann M, Barnes J, Hawkes DJ, Fox NC, Ourselin S (2010) Fast free-form deformation using graphics processing units. Comput Methods Programs Biomed 98(3):278–284
20. Vincent L (1991) Morphological transformations of binary images with arbitrary structuring elements. Signal Process 22(1):3–23
21. Nikopoulos N, Pitas I (1997) An efficient algorithm for 3D binary morphological transformations with 3D structuring elements of arbitrary size and shape. In: Workshop on nonlinear signal and image processing, Michigan, USA. IEEE
22. Zhang Y, Brady M, Smith S (2001) Segmentation of brain MR images through a hidden Markov random field model and the expectation-maximization algorithm. IEEE Trans Med Imaging 20(1):45–57
23. Rehman HZU, Hwang H, Lee S (2020) Conventional and deep learning methods for skull stripping in brain mri. Appl Sci 10(5):1773
24. Kelleher JD, Mac Namee B, D'arcyA (2015) Fundamentals of machine learning for predictive data analytics: algorithms, worked examples, and case studies, 1 edn. MIT Press, Cambridge
25. Liu M, Zhang J, Adeli E, Shen D (2018) Landmark-based deep multi-instance learning for brain disease diagnosis. Medical Image Anal 46
26. Lian C, Liu M, Zhang J, Shen D (2020) Hierarchical fully convolutional network for joint atrophy localization and Alzheimer's disease diagnosis using structural MRI. IEEE Trans Pattern Anal Mach Intell 24(4):880–893
27. Liu M, Li F, Yan H, Wang K, Ma Y, Shen L, Xu M, Alzheimer's Disease Neuroimaging Initiative et al (2020) A multi-model deep convolutional neural network for automatic hippocampus segmentation and classification in Alzheimer's disease. NeuroImage 208(1):116459
28. Zhang J, Liu M, An L, Gao Y, Shen D (2017) Alzheimer's disease diagnosis using landmark-based features from longitudinal structural MR images. IEEE J Biomed Health Inform 21(5):1607–1616

Mineral Nanotechnology in Circular Economy

Augusto Gonçalves Nobre(✉), José Armando Espinosa Martínez, and Odila Florêncio

Mackenzie Presbyterian University, Sao Paulo, SP, Brazil
nobre.augusto@gmail.com, odila.florencio@mackenzie.br

Abstract. The first phase of waste management was characterized by garbage disposal in dumps and its environmental remediation. The next phase was marked by recycling emergence, and, in the third phase, it was presented the specific recycling for higher added value materials. Circular Economy corresponds to the fourth phase of waste management. The Circular Economy policies have become a worldwide trend and promote the materials use maximization, aiming to reduce the unrestrained natural resources extraction, prolonging the devices' useful life, avoiding programmed obsolescence, and reducing, as much as possible, the landfills waste final deposition. These policies are expected to reduce the mineral commodities consumption, and this can seriously impact producing markets. In this scenario, mineral nanotechnology appears as an alternative for materials supply in accordance with new sustainability policies. This paper presents the evolution of the Circular Economy concept and defines mineral nanotechnology as an important raw material resource for efficient and durable device manufacture, which generates less waste in its production processes and can be reintegrated into the materials chain after losing its original function.

Keywords: Circular economy · Mineral nanotechnology · Natural resources

1 Introduction

The possibility of raw material shortages has always been a constant concern for human society, even before industrialization. Although manufacturing technologies and materials have changed throughout history, the problem of availability remains the same.

During World War II, Great Britain experienced a severe shortage of raw materials that led the country's government to establish the Ministry of Supply in August 1939, the first record of a government agency devoted to policy development to strategically deal with changes in the global material supply chain. The agency's functions included the careful management of scarce materials and the research promotion for the development of substitute materials [1].

Obviously, such scarcity scenario in the last century does not provide an exact parallel to the strategies for materials supplying in a society focused on Circular Economy that is expected for the next decades of the 21st century. However, it demonstrates the importance of policies for the strategic management of raw materials.

Y. Iano et al. (Eds.): BTSym 2020, SIST 233, pp. 220–226, 2021.
https://doi.org/10.1007/978-3-030-75680-2_26

When properly designed, government policies follow a wave pattern: efforts for reaching desired objectives are progressively increased in the early years and reduced – and subsequently stabilized – as policies become fully implemented and monitored. When new policies emerge, the pattern is repeated.

The Circular Economy is in its ascending phase, with several European Union (EU) countries creating and implementing measures for reusing, recycling, and reducing natural resource consumption, based on measures such as the 2015 Action Plan [2]. In this sense, this work aims to present mineral nanotechnology as an alternative for future raw materials supply, which is expected to perform this task with excellence. Even if the extraction of natural resources from their original sources may never be wholly stopped, mineral nanotechnology helps mineral resource consumption becoming the least harmful to the environment it can be.

Currently, there is already a success record in nano minerals obtaining, such as clay minerals [3–5], graphite [6, 7], molybdenite [8, 9], and talc [10]. These nanomaterials were obtained sustainably, allowing waste and effluent recycling. However, case studies have addressed mineral nanotechnology until now, with a lack of scientific definition. To fill this gap, this work aimed to formalize the economic and environmental principles of mineral nanotechnology based on the literature review.

2 Waste Management and Circular Economy

The Circular Economy history is linked to the evolution of waste management. Before the 1970s, waste was locally managed and, generally, this meant that it was collected and deposited in local dumps. Waste management policies were created in the early seventies when waste began to be considered a problem for the environment and the population's health. In this sense, the first phase of waste management was based on containing risk and ensuring the citizens' health. This phase was based on a policy to limit the operation of landfills and their emissions. Landfills and dumps with a clear impact on the environment were forced to close and remedy the caused impacts [11].

In the 1980s, the second phase of waste management was driven by the increasing waste volume. It became known that large volumes of waste were produced, and many of these materials were recyclable. Therefore, in this policy phase, the idea was to extract these materials from the waste stream and recycle them. The second phase led to the development of waste treatment plants and waste emission limit values policies, such as the goal of 70% recycling of the total volume of construction and demolition waste, introduced in the EU in 2008 [12].

In 2005, the European Waste Framework formalized the concept of solid waste added value and defined the 2020 recycling targets to be implemented in each EU member state. The introduction of value to waste hierarchy started the third phase of waste management. The focus of recycling activities has shifted from processing large volumes of waste to creating value through initiatives, for example, to recover precious metals from electronic waste. Thus, waste treatment activities have become more complex and intensive in investment capital [13].

The fourth phase of waste management corresponds to the Circular Economy. It is no longer driven by the waste value or recycling itself but by material restriction. This phase is based on the idea that the world economy will need certain materials that will be increasingly difficult or more expensive to obtain. Simultaneously, there will be less space for landfills and more cost and difficulty in logistics for waste management [14].

The new waste management policies must consider not only the ways in which recycled materials are used but also the period in which materials are in use. Circular Economy questions how we can increase the useful life of the objects and keep their components with high value throughout their life cycle. The political actions trend is towards the focus on waste prevention, avoiding programmed obsolescence and promoting the principle of Extended Producer Responsibility (EPR). The EPR principle was formulated and gradually introduced in several countries to encourage and hold producers accountable for considering the end of the life of their products when designing them. This is achieved by making producers responsible for managing the end of life of their products [15].

Moving from a linear economy to a circular economy requires fundamental changes in current systems of production and consumption. There is a need to change how materials are processed and used, in addition to new and more sustainable business models and policies that enable conditions that facilitate the materials' circularization.

Until recently, an 'end of line' approach was adopted, which aimed to correct problems instead of preventing their cause. If we consider the product life cycle, most of the mandatory policies worldwide aim solutions only for the end of life remediation and avoid squandering. In the other stages of the materials life cycle, most policies are voluntary, and there are fewer mandatory policy instruments [15].

Over the past 15 years, the direction of the EU's strategic resources policy has gradually shifted towards the sustainable use of natural resources, encouraging an increase in products efficiency, intensifying recycling and campaigning to prevent excessive waste generation, at the same time when it targets sustainable levels of economic growth. The EU began to promote the Circular Economy with the 2015 Action Plan. This plan aims to improve the competitiveness of EU companies, protecting industries against potential shortages and volatile raw material prices, encouraging the creation of new business opportunities and innovative forms of production and consumption [16].

The Netherlands government presented a progressive transition plan to Circular Economy, setting targets for 2030 and 2050. In 2030, the target is a 50% reduction in the use of primary raw materials, including minerals, fossil fuels, and metals. In 2050, the goal is to have a 'fully Circular Economy' in the Netherlands. The Dutch government has strategic plans to comply with this policy, which include ensuring that the raw materials in existing materials chains are used with high efficiency, there are also targets to fossil fuels and mineral raw materials consumption reduction, which, when possible, should be replaced by renewable and more widely available raw materials. Finally, there are goals of developing new production methods, promoting new forms of consumption and new ways of designing products [17].

As many countries have a significant part of their economy focused on commodities production, their companies can be strongly impacted by Circular Economy policies being implemented in their consumer markets. In this way, the incentive to mineral nanotechnology initiatives can be strategic for both the economic and environmental sustainability of raw material producing countries.

3 Nanotechnology

Nanotechnology is the branch of science that aims to develop devices, machines, products, and processes based on components made up of nanomaterials (materials that have at least one of their dimensions between 1 and 100 nm). Nanomaterials stand out for their large surface area and the possibility of, until now, unprecedented properties in material sciences. Nanotechnology can prolong the useful life of materials, making them more resistant to chemical attacks, mechanically stronger, self-cleaning, and self-healing to thermal and mechanical damage, allowing partial or total recovery of equipment components in its end of life [18].

Traditionally nanomaterials require special care for their production, such as high purity source materials and clean rooms, because most processes are sensitive to contaminants and impurities. The air molecules themselves can be incorporated into the material and adversely affect the process, yield, and product performance. Due to these technical adversities, conventional nanotechnology is currently restricted to specific applications, and nanomaterials do not behave like commodities [19].

For this set of reasons, mineral nanotechnology appears as an alternative to make nanomaterials abundantly available to industry in the context of the Circular Economy. Obtaining nanomaterials from mineral resources is based on top-down technology, in which the nanomaterial is obtained from bulk material. Thus, there is no need for chemical synthesis processes. An appropriate micronization method of the bulk material must be chosen to obtain a nanomaterial via a top-down approach.

4 Mineral Nanotechnology in Circular Economy

Minerals intrinsically have crystalline defects and compositional impurities, for this reason, nano minerals will inherit these characteristics from their bulk material. At first glance, this can be an obstacle, however, mineral nanotechnology is only concerned with the obtained material property and its potential application with purposefully least ore processing. Thus, mineral nanotechnology accepts compositional heterogeneity and structural defects if the obtained material has a consumer market and interesting industrial properties.

In this sense, the characteristics of the minerals do not need to be uniform throughout the deposit, but the way of nano mineral obtaining process must be the same to maintain economic viability.

Since nano minerals are fine-grained materials and customarily extracted in small quantities when compared to traditional mining, lesser extent mineral deposits may be commercially feasible and more sustainable in terms of space occupation and waste generation it can be.

As they are small-grained materials, aligned with the trend of technology miniaturization, with innovative properties that match the objectives of extending the products useful life, which are extracted from deposits with less waste generation and impacts on the environment, and with greater added value, the mineral nanotechnology emerges as an alternative raw material source allied with Circular Economy that will enable nano mineral producers to play an important role in the coming decades.

5 Discussion

In order to keep the material chain circular, substantially reducing the natural commodities consumption and allowing as much maintenance as possible to keep the materials in use, preventing them from becoming solid waste that requiring disposal and consequently leaving the circular context, Circular Economy aims to change the form of consumption of raw materials, the use, and the destination of materials.

In the first phase of waste management, the waste sector was considered part of the problem. Waste disposal activities were carried out in dumps and landfills that generated occupying space and health risks problems. In the second phase, the management sector became the waste owner, so they explored the possibility of recovering the residual value from materials. It was at that moment that the first recycling cooperatives emerged to profitably recycle large volumes of waste. As value became the engine of waste processing, primary producers and the largest service companies increased their interest in waste management, so, in the third phase, small local recycling companies were acquired by larger groups, and the main industrial actors entered in the waste market, aiming the recovery of greater added value waste.

The Circular Economy history (the fourth phase of waste management) had its origin in the late 1980s when many industrialized countries were struggling with waste management, landfills were becoming full, and it was not easy to open new ones. Waste incineration was becoming more common, but it was also met with much opposition. Domestic recycling had started, but the results of most initiatives were not impressive. One reason for the limited success was municipalities that did not have the money to build a convenient and effective recycling infrastructure. Add to this scenario, municipalities are usually reluctant to raise fees and taxes - and when they do - there are conflicting questions about how to use the money. Society also became aware of the fact that products were typically designed and manufactured without any planning for how they should be treated as waste. These factors conjuncture, coupled with public awareness, led to the formation of the first Circular Economy policies in the decade of 2010 and its strong incentive for implementation for the coming decades [8].

As a result of the Circular Economy policies implementation, there is a global 'risk' of natural resources consumption decline, and, consequently to this scenario, countries with mineral commodities strongly linked economy will need to adapt to new forms of consumption and mineral nanotechnology may be an important solution.

Based on successful cases of nanometric minerals (clay minerals, graphite, molybdenite, and talc) applied to nanodevices produced in an environmentally sustainable way and in accordance with the Circular Economy objectives, it is possible to observe common topics that allow the determination of mineral nanotechnology principles [3–10]. In this way, mineral nanotechnology is guided by the following principles:

1. Nanomaterials are the key to the high performance and durability product manufacture.
2. Nanomaterials must be produced from mineral resources without expensive synthesis processes.
3. It must be focused on the material property and application, regardless of its structure and composition.

4. Generate materials with greater added value than its traditional mineral commodity.
5. Having smaller spatial scale extraction operations, with less waste generation (sterile and tailings) than conventional mining.
6. Ore processing by a top-down approach must be simple and reproducible for the same mineral deposit.
7. Nano minerals property and obtaining process are more important than deposit size because they are fine-grained materials, and little ore volume can produce nanomaterials economically.

In the current mineral nanotechnology manufactured devices, the mineral resources are used as additives in metallic alloys and composites, which means their volumetric consumption is still low. However, research in the area is in rapid growth, and the manufacture of devices based on nano minerals could be a matter of time.

6 Conclusion

This work described the Circular Economy history from the waste management evolution perspective to contextualize mineral nanotechnology as a sustainable source of raw materials for the coming decades.

With Circular Economy policies implementation and, consequently, changes in the materials consumption and waste generation, the devices start to demand raw materials with new properties to prolong their useful life, facilitate their maintenance, and allow the maximum reuse of its components. In this context, nanomaterials can allow this technological development.

Conventional nanotechnology presents problems with high-cost materials and complex manufacturing processes, which slows down their widespread use. To solve this question, examples of mineral nanotechnology have appeared in the literature, with good economic and environmental results. This work compiled the common topics in these works and formulated 7 principles that guide the current mineral nanotechnology.

Acknowledgments. This study was financed in part by the Coordenação de Aperfeiçoamento de Pessoal de Nível Superior - Brasil (CAPES) - Finance Code 001.

References

1. Hurstfield J (1944) The control of british raw material supplies, 1919–1939. Econ Hist Rev 14(1):1–31
2. Milios L (2018) Advancing to a circular economy: three essential ingredients for a comprehensive policy mix. Sustain Sci 13(1):861–878
3. Batra M, Goram S, Dadarwal P, Nainwani R (2011) Nano-clay as polymer porosity reducer: a review. J Pharm Sci Technol 3(10):709–716
4. Zhou C, Zhao L, Wang A, Chen T, He H (2016) Current fundamental and applied research into clay minerals in China. Appl Clay Sci 119(1):3–7
5. Zhou C, Keeling J (2013) Fundamental and applied research on clay minerals: from climate and environment to nanotechnology. Appl Clay Sci 74(1):3–9

6. Alam S, Sharma N, Ray B, Yadav S, Biswas K (2017) Effect of graphite nanoplatelets on the mechanical properties of alumina-based composites. Ceram Int 43(14):11376–11389
7. Peng L, Xu Z, Liu Z, Wei Y, Sun H, Li Z, Zhao X, Gao C (2015) An iron-based green approach to 1-h production of single-layer graphene oxide. Nat Commun 6(1):1–9
8. Savjani N, Lewis E, Pattrick R, Haigh S, O'Brien P (2014) MoS_2 nanosheet production by the direct exfoliation of molybdenite minerals from several type-localities. RSC Adv 4(67):35609–35613
9. Jiang F, Li S, Ge P, Tang H, Khoso S, Zhang C, Yang Y, Hou H, Hu Y, Sun W, Ji X (2018) Size-tunable natural mineral-molybdenite for lithium-ion batteries toward: enhanced storage capacity and quicken ions transferring. Front Chem 6(1):389
10. Sakthivel S, Pitchumani B (2011) Production of nano talc material and its applicability as filler in polymeric nanocomposites. Part Sci Technol 29(5):441–449
11. Andrews D (2015) The circular economy, design thinking and education for sustainability. Local Econ 30(3):305–315
12. Blomsma F, Brennan G (2017) The emergence of circular economy: a new framing around prolonging resource productivity. J Ind Ecol 21(3):603–614
13. Reese M, Koch H (2005) Revising the waste framework directive basic deficiencies of European waste law and proposals for reform. J Eur Environ Plan Law 2(6):441–457
14. Potting J, Hekkert M, Worrell E, Hanemaaijer A (2017) Circular economy: measuring innovation in the product chain, 1st edn. PBL Publishers, Netherlands
15. Svensson S, Richter J, Maitre-Ekern E, Pihlajarinne T, Maigret A, Dalhammar C (2018) The Emerging' Right to Repair' legislation in the EU and the U.S. In: Going green care innovation, Austria, pp 1–18
16. European Commission (2015) Closing the loop: an EU action plan for the circular economy. Communication from the Commission to the European Parliament, the Council, the European Economic and Social Committee and the Committee of the Regions, COM(2015) 614/2, 1st edn European Commission, Brussels
17. European Environment Agency (2016) A Circular Economy in the Netherlands by 2050: Government-wide Program for a Circular Economy, 1st edn European Environment Agency, Copenhagen
18. Cao G (2004) Nanostructures and nanomaterials: synthesis, properties, and applications, 1st edn. Imperial College Press, London
19. Pokropivny V, Lohmus R, Hussainova I, Pokropivny A, Vlassov S (2007) Introduction to nanomaterials and nanotechnology, 1st edn. Tartu University Press, Tartu

Guidelines to Define a Regulatory Proposal in the Transition and Inclusion of Non-conventional Renewable Energies in Colombia and Its Role in the Development of Smart Cities

Elkin A. Caro Acevedo(✉) and Alvaro Jaramillo Duque

Universidad de Antioquia, Cl. 67 ##53-108, Medellín, Antioquia, Colombia
{ealexander.caro,alvaro.jaramillod}@udea.edu.co

Abstract. This article is based on a regulatory vision related to the transition, integration, and transformation of the Colombian electricity system for the implementation of smart grids in this country. The analysis focuses especially on the development of Smart Cities, based on social, political, environmental, and economic sustainability, showing how the regions of the United Kingdom, South Korea, and Malaga, among others, have advanced in their transition, carrying with them the infrastructure and the regulations that currently rule them.

Keywords: Smart cities · Smart grids · Renewable energy sources · Energy sustainability · Regulation

1 Introduction

Modern society faces an ever-increasing demand for energy, which involves several challenges, including the previous crisis of traditional fossil energy resources and the consequences of environmental pollution [1]. The energy transition is essential for the sustainability and evolution of cities in order to be able to label them as "smart". There is not a generic contextualized definition of "smart cities", but according to the "Vida Sostenible" Foundation [2], they are defined as: cities with urban progress aimed at sustainability, using infrastructure, innovation, and technology to respond to the basic needs of institutions, companies, and citizens, in operational, social, and environmental aspects.

Considering the former definition of a smart city, the following question is proposed: What makes a smart city? According to the CIAPE [3] (Ibero-American corporation for public and business affairs), and the article Smart and Sustainable City: Towards an inclusive innovation model [4] stands out 10 main aspects that must be taken into account for a city to be considered "smart", which are:

1. The incorporation of information and communication technology (ICT) to public services.

Y. Iano et al. (Eds.): BTSym 2020, SIST 233, pp. 227–237, 2021.
https://doi.org/10.1007/978-3-030-75680-2_27

2. Efficient public service companies aimed at implementing and using smart meters, the management and inclusion of renewable energy, efficient management and consumption of drinking water, and wastewater management, among other services.
3. Public-private partnerships are a key feature of the smart city concept.
4. Citizen security and protection.
5. Financial sustainability.
6. Citizen participation in government decisions.
7. The social capital.
8. Traffic or transportation planning: which includes the use of electric vehicles and cycle tracks, among others.
9. Ecological characteristics: Minimizing the carbon footprint and respect for the environment are essential actions.
10. The minimum population criteria for planning in the incorporation of elements in relation to the efficient energy consumption and other public services.

Given the definition mentioned formerly, it is unquestionable to talk about the importance of the electricity system, which must be efficient enough to meet the needs of the energetic demand and security.

2 The Electricity Sector in Colombia and Smart Cities

The electricity sector in Colombia plays a main role, and it is crucial for socio-economic development. This sector faces a great transformation in the generation, transmission, distribution, and efficient use of energy. Due to sustainability, demand, and economy, it has had to evolve in a short time. As shown in Fig. 1, in a traditional system, the flow of electrical energy comes from the power plants and is transmitted to substations through high-voltage transmission lines that cover long distances. Subsequently, in substations, this energy passes through voltage step-down transformers and is distributed to the final consumer, given that the energy flow always goes in the same direction.

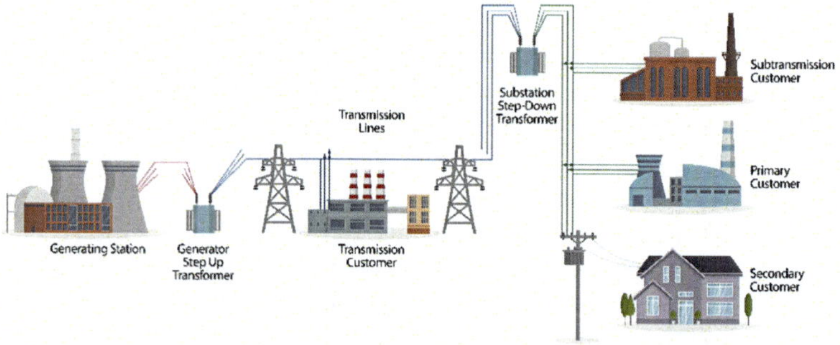

Fig. 1. Simplified diagram of the electrical power system

Along with demographic growth, the demand for electricity and the needs of both the urban and the industrial sector have changed, so the existing infrastructure has had to adapt to the rules set by the large energy generating companies, supported by the different entities regulations, leaving aside and hindering the stages for integration into the energy market of small energy producers that want to participate in the market.

The population increase has generated an impact on the sector, causing the demand for energy to increase, and the energy system must meet the needs of the population without affecting the environment and guaranteeing supply, where the demand has been projected to 2030 with three scenarios, a high one with 107,514 GWh, a medium one with 105,018 GWh and low one with 102,533 GWh (XM, demand forecast, 2019). According to the Mining-Energy Planning Unit (Unidad de Planeación Minero Energética UPME) of Colombia, it is expected to install 1.46 GW of wind plants and 0.23 GW of solar energy in the coming years to increase installed capacity, improve the complementarity of energy sources and make the system more robust [5]. These new challenges must be faced in Colombia, considering technical, political, regulatory, and environmental aspects.

Today, the electricity supply and the electricity grid need restructuring in order to adapt to the needs of a changing society on issues related to energy security, economy, and independence. In this continuous energy evolution, the type of energy sources comes to participate, playing a main role in smart cities, which have to be sustainable and energy-efficient.

Due to distributed generation, the developments and implementations in power grids have given way to more reliable and efficient grids that interact in two directions, receiving and injecting energy into the low voltage grid. This transforms the power grids structure since, instead of having the unidirectional grid, a communication and feedback relationship is created between the user and the energy supplier.

All of this has led to the future of the power grid, called the smart grid, which provides several benefits by modifying the structure and the way in which energy generation, transmission, distribution, and consumption are carried out. This new infrastructure requires changes in the rules and regulations that regulate the production and consumption of electricity that currently rules in Colombia.

In Fig. 2, the evolution of the power grid is shown in a summarized and easy-to-interpret way, showing the stages from its generation, monitoring, storage, and consumption of energy, as well as the change of a more distributed and not so centralized energy system. In the research called "Smart Grids and the role of the electric power supplier [6]", it is shown how these grids can be grouped into five blocks of functionalities, which are:

1. Advanced Measurement Infrastructure (AMI).
2. Advanced communication systems.
3. Automation of low voltage distribution.
4. Intelligent transformation centers.
5. Local and domestic area grids, highlighting the characteristics that provide the new generation, distribution, and storage of energy.

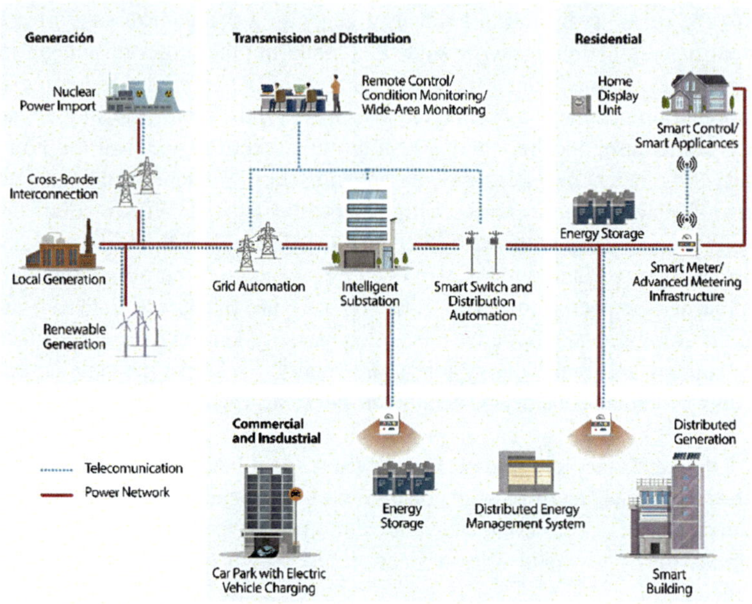

Fig. 2. Smart Grid technology overview

3 The Role of Smart Grids in the Electricity Sector

The energy transition has to go hand in hand with government entities since they must establish clear and consistent policies regarding regulations and plans for new electrical systems. These standards or regulations allow the investment to achieve innovation and transform the current electrical system, directed towards smart grids. Therefore, it is important to think about educating consumers so that they understand the regulation and benefits that this transformation will bring.

Current regulations and market systems can hinder the development of these grids; their low remuneration and financial incentives to update and improve the grid are evident, so they should offer new long-term options for investment in the construction of new systems of power generation, price guarantees, and the priority participation of the prosumer (portmanteau of the words 'provider' and 'consumer').

In accordance with the above, a series of basic principles for the development of smart grids stand out, according to the research report carried out by Orkestra Instituto Vasco De Competitividad, it summarizes the need to have mainly a regulatory, remunerative, stable, and transparent framework, followed by flexible financial incentive schemes, accompanied by balanced distribution systems on the benefits and costs of investments in smart grids between regulated and non-regulated entities of the energy system, which guarantee results in environmental terms, quality and security of energy supply and finally efficiency based on the application of regulatory regimes [6].

The improvement of smart grids is essential for social and industrial development in order to meet their energy demand; in addition to contributing to the improvement of climate change, international collaboration should be highlighted to share experiences with pilot programs in terms of regulations, implementation technology, and development of new standards in transmission networks, especially the smart ones.

4 Advanced Measurement Infrastructure

When we talk about smart grids, we cannot fail to mention the Advanced Metering Infrastructure (AMI) since both concepts are intrinsically related. The AMI is the infrastructure that allows bidirectional communication between the user and the energy service provider, using a set of software, hardware, architectures, and communication networks with the aim of improving data management.

With the most recent Colombian regulation on AMI (Resolution 131 of June 25, 2020), the national government at the head of Energy and Gas Regulation Commission (Comisión de Regulación de Energía y Gas CREG), establishes the conditions for the implementation of advanced metering infrastructure, in the provision of the home public energy service electricity in the National Interconnected System (SIN), defining and establishing those responsible for the installation, administration, operation, maintenance and change of technology. Likewise, guidelines are established regarding interoperability, usage, data protection, and remuneration in the AMI operation. In its article 33, the aforementioned resolution states that the AMI will be implemented in 4 phases of deployment. Its first phase will have a maximum duration of 6 months from the start date of the plan when all the meters are reported in the ASIC (Administrator of the Commercial Exchange System). A second phase will have a maximum duration of 12 months from the start date of the plan, when users' meters with average consumption equal to or greater than 1,000 kWh must be replaced in a month, in addition to the change of meters to AGAPE (small-scale auto generators) and AGGE (large-scale auto generators) that require it. In the third phase, international links and connections are involved, and finally, there is the fourth phase, which will have a maximum duration until December 2030, when all equipment must be installed according to the prioritization indicated in the approved plan.

Figure 3 shows a diagram with the basic components of AMI interaction between an energy producer and consumer (prosumer) user, and the distribution networks of the energy service supplier. In this scheme, the communication layer allows the prosumer to make decisions when knowing the purchase and sale price of energy can be identified.

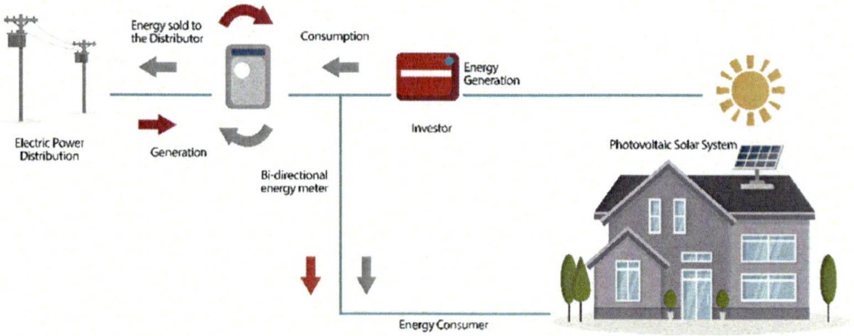

Fig. 3. Basic diagram of bidirectional communication in the AMI

5 Successful Cases

The traditional regulation favors capital investment in long-term assets. However, this must change in order to generate incentives or rewards for public service companies direct towards the development of new infrastructure that improves their performance, reduces losses, promotes consumer interaction, or encourages the objectives of sustainable development and innovations that may be carried out to promote the development of smart cities.

Several countries have already started developing the infrastructure for smart grids in order to prepare a smart and safe energy supply. A decade ago, the UK Office for Gas and Electricity Markets (Ofgem) set out to develop a new regulatory framework that would reward utility companies for their innovation by protecting consumers and ensuring that companies operate the grid efficiently and sustainably, resulting in a framework called RIIO.

The UK's RIIO (Revenue = Incentives + Innovation + Outputs) framework is the performance-based framework for setting price controls, and it is widely regarded as the most comprehensive performance-based regulatory system developed to date [7].

To better understand how RIIO works, a reference is made to four main characteristics, which were written in an article published in DEStech Transactions on Engineering and Technology research [8] called "A Comprehensive Review on the Evolution of Ofgem RIIO-ED1 Electricity Distribution Price Control Framework in United Kingdom", these particularities are:

- A multi-year rate plan
- The total expenditure approach (totex)
- Performance incentives
- Innovation Fund

RIIO is the evolution of how UK power companies are regulated; it was adopted in October 2010 and launched for electric power transmission in April 2013 and for distribution in April 2015.

Another remarkable case to name in the inclusion of the FNCER (Non-Conventional Sources of Renewable Energy) and their smart grids is the case of the Republic of Korea, in which it manifests itself as an international benchmark in the inclusion of this type of energy, using regulatory measures since 2010 and charting a national route for the large-scale implementation of smart grids, they also show millionaire savings in imports of energy resources [9, 10].

Some of the measures currently being taken by Korea to achieve these goals are:

- Investment in infrastructure for smart grids, from AMIs, to charging infrastructure for electric vehicles.
- Planning in the expansion of renewable energy supplies and electric vehicles.
- Establishment of emergency power supply systems in case the charging network for electric vehicles undergoes an increase in demand.
- To implement intelligent energy management systems.
- To carry out variable rates in real-time.
- To develop a system of standardization and certification of energy technologies to promote their commercialization, research, and development.

The implementation of a Smart City is made based on their needs and the strategic or governmental plan that they have in accordance with the growth of this. Therefore, it is not possible to make an exact comparison since all cities handle different technological developments. Malaga is an example of a city that potentiates its development in the efficient consumption of energy, with the main objective of reducing emissions of CO_2 into the atmosphere, thus the project aims to reduce the energy consumption by 20% and 6,000 tons of emissions of CO_2 into the atmosphere per year. This smart city project goes hand in hand with the consortium of 11 companies that are progressively implementing smart grids in the area of "La Misericordia" beach so that they can manage the distribution for efficient energy consumption. In addition, among other measures they want to achieve, there is the implementation of electric vehicles.

Another important example is the city of Singapore, which is considered by the World Bank as the best city to do business and the most suitable to promote business growth. Singapore is currently establishing an intelligent energy system (IES) which, through the bidirectional flow of information provided by the network, allows a reduction and efficiency of energy consumption around 3% thanks to the implementation of AMIs and the communication speed that these have through the use of optical fiber [11].

Additionally, grid stability is not an issue in Singapore as they have one of the most reliable electricity grids, with an average real outage time of less than 1 min per customer per year. These networks have an advanced supervisory control and data acquisition (SCADA) system with two-way communication channels that can automatically detect failures and interruptions in the power supply.

On the other hand, they intend to integrate a large number of power sources in their system in the plug-and-play mode (plug, connect, and use). In the case of Singapore, it consists of photovoltaic systems and small cogeneration plants. They also have created an institute called Energy Research Institute Nanyang Technological University, which will focus on areas related to solar energy, electric vehicles, and smart grids [11].

6 Political Commitment and Regulation

Given the aforementioned conditions, it is essential to raise awareness, mainly in government and business entities, about the importance of Smart Grid technologies, since they "help meet the growing needs for electricity, while minimizing the environmental impact and they contribute to efforts to limit emissions of CO_2 [12]. Furthermore, this technology comes hand in hand with non-conventional renewable energies.

The acknowledgment of the advantages that smart grids bring must be complemented with a regulatory diagnosis that considers the changes that may affect the way of regulating the FNCER. It is possible to monitor these changes from the national interconnected system's grid code established in Colombia through the Energy and Gas Regulation Commission (CREG) and the Government Resolution 25, 1995, which is articulated to the generation of regulatory decrees. The grid code of the national interconnected system is formed by the planning, connection, operation, and measurement codes [13]. All these evolutions and energy developments generate a regulatory transition, which leads to the adaptation of legal and natural persons to these new regulations.

The fundamental law in Colombia for the regulation and integration of non-conventional renewable energies to the National Energy System is Law 1715, 2014; which defines in its article 5, number 17 as Non-Conventional Sources of Renewable Energy (Fuentes No Convencionales de Energía Renovable FNCER) "those renewable energy resources available globally that are environmentally sustainable, but are not used in the country or are used marginally and are not widely traded. Biomass, small hydroelectric exploitation, wind power, geothermal power, solar power, and the seas are considered as FNCER" (Law 1715, 2014). This law aims to promote the development and use of non-conventional sources of renewable energy and their integration into the electricity market; it also includes the reduction of greenhouse gas emissions (Law 1715, 2014) [14].

The Non-Conventional Sources of Renewable Energy (FNCER), especially wind power, solar power, and small hydroelectric exploitation, have a particular property: the variability in their energy generation. This uncertainty reflects its primary source's behavior, which depends on the climatic, meteorological, and hydrological phenomena of the moment, such as solar irradiation, wind, and small water flows [15].

These potential energy sources can be divided into two groups: synchronous and non-synchronous generation sources; among the synchronous, there are "small hydroelectric exploitations, for which the rotation speed of their generators depends on the frequency of the power system." and among the non-synchronous sources, there are the "wind and solar power sources, whose generation does not respond naturally to the frequency response of the system" [16]. For wind and solar energy, the frequency must be controlled differently from hydraulic energy. It is also observed that there are technical gaps in Law 1715 of 2014 since the conditions in which the FNCER can enter operation or be connected to the electrical system are not fully regulated in a specific way.

7 Impact and Importance of the Energy Transition

Cities are the engines that aid to boost the economy of each country; they are also providers of public goods and services that are necessary for the well-being and social union of their inhabitants and visitors [16]. These cities face common challenges all around the world, and one of them is the accelerated population explosion and the displacement of more and more people to urban areas, leading to accelerated urbanization. This is a problem that makes the management of a city increasingly complex, generating impacts that are currently little considered, such as the demand for energy, mobility, urban waste, etc.

Globalization, along with the fourth industrial revolution, has led to the evolution of cities so that they are self-sustainable and generate less environmental impact. From these concepts, the essence of a Smart City was generated; this kind of cities innovate and promote education for social and industrial development. The International Telecommunications Union (ITU) [17] and a study called Smart Cities: The Main Drivers for Increasing the Intelligence of Cities [18] have identified 10 priorities for a city to be considered "smart", which closely coincide with the characteristics proposed by the Ibero-American Corporation for Public and Business Affairs (Corporación Iberoamericana de Asuntos Públicos y Empresariales CIAPE), which are:

1. Mitigation of climate change.
2. Planning and adaptation of climate change.
3. Energy efficiency.
4. Waste management.
5. Water Management.
6. Intelligent architecture and infrastructure.
7. Smart mobility.
8. Suburban areas.
9. Environmental justice and community engagement.
10. Sustainable cities.

When analyzing the 10 points, it is evident that energy is essential for the evolution towards Smart Cities; therefore, the integration of FNCER, helped by the smart grids, is a priority since they improve energy efficiency (smart metering), they serve as an engine of economic growth and offer a solution to climate change since they generate lower greenhouse gas emissions [19].

With regard to the energy transition, it is evident that our Colombian political, social, and electrical system must take into account 4 fundamental pillars for the acceleration of the energy transition, these are:

- Digitization
- Decarbonization
- Decentralization
- Demand management

When emphasizing the regulatory guidelines and their different variables and characteristics related to the implementation of smart grids for the evolution towards smart cities, it is clear that society has a fundamental role since it has to be aware and humanize with the current changes. Jointly, they must be clear about the benefits and incentives that they may have with this transition, as well as being resilient, responsible, and have a change towards efficient energy consumption. However, a deep government commitment is required, through regulatory entities, so that stable rules and regulations are instituted over time, on specific issues with the licensing of new power generation projects, especially with the FNCER, it is required to be clear about the technical requirements regarding protection and control in the safety and reliability systems when supplying energy into the grid, an evaluation of the existing electrical infrastructures, to know whether or not grids are viable for the use of smart grids, to have a clear and fair basis in the backup usage fees of the electrical grid based on the unit cost of distributed energy, to have a standardized legalization process for the grid operators regarding the participation of the prosumer, among others. Colombia must assume with all these challenges, a development path that allows accelerating the energy transition hand in hand with the implementation of Smart Cities.

8 Conclusion

For many years, the electrical system has been deployed under traditional implementations of robust generation and transmission systems up to the distribution systems, with a passive participation of the final consumer, mainly because communication or flow in the productive electrical chain moved in one direction only, from the power grid operator to the final customer. Nevertheless, the improvements in electrification, ICTs, cost reduction, massification of Distributed Energy Resources (DER), and technological trends empower the final consumer with new capacities in production and consumption (prosumer). Besides, they allow the promotion of power plants in places each time closer to the customer, which is transforming the conventional structure of the system.

Currently, the electricity sector is in an unprecedented transition, which implies not only an energy transformation but also a digital transformation that will leverage new technologies, business models, and services. Taking advantage of all this technological inertia, regulatory, political, and financial innovation is required to accelerate the adoption of these trends in the national system. Although there are policies and regulations with guidelines issued to implement and encourage their deployment in our country, it has low development rates compared to other countries in the world, even if compared with countries in the same region, in terms of massification of this type of technology.

The opportunity provided by globalization, based on digitization and technologies, founded on intelligent transportation systems, control of consumption systems in public services, and monitoring platforms which help to control citizen security, the environment, mobility, among other aspects, have been generating the base model of what we call "smart city". The communication and information protocols of IoT (Internet of Things) play an important role in improving the processes and daily work, in order that these actions are more efficient, safe, and reliable, giving the well-being of life that we all seek.

References

1. Gómez VA, Hernández C, Rivas E (2018) The influence of penetration levels of distributed generation in the energy markets. Información tecnológica 29:117–128. https://doi.org/10.4067/S0718-07642018000100117

2. Smart City | Fundación Vida Sostenible. https://www.vidasostenible.com/informes/smart-city
3. Revista digital CIAPE – alternativas estratégicas. https://revistadigital.ciape.org
4. Alvarado López RA (2018) Ciudad inteligente y sostenible: hacia un modelo de innovación inclusiva. PAAKAT: revista de tecnología y sociedad 7. https://doi.org/10.32870/pk.a7n13.299
5. Energía Eléctrica - Ministerio de Minas y Energía. https://www.minenergia.gov.co/planes-de-expansion-generacion-transmision
6. Fernández J, Menéndez J (2019) Las redes inteligentes y el papel del distribuidor de energía eléctrica. Cuadernos Orkestra 54:2019
7. Ofgem (2020) Networks explained. https://www.ofgem.gov.uk/regulating-energy-networks
8. Mujeeb A, Wang P, Hong X, Johnson A (2019) A comprehensive review on the evolution of ofgem RIIO-ED1 electricity distribution price control framework in United Kingdom. DEStech Trans Eng Technol Res. https://doi.org/10.12783/dtetr/aemce2019/29478
9. Eisenberg DA, Park J, Seager TP (2017) Sociotechnical network analysis for power grid resilience in South Korea. In: Complexity. https://www.hindawi.com/journals/complexity/2017/3597010
10. Jeon W, Mo JY (2018) The true economic value of supply-side energy storage in the smart grid environment – the case of Korea. Energy Policy 121:101–111. https://doi.org/10.1016/j.enpol.2018.05.071
11. DOMOTYS, Colado S, Gutiérrez A, Vives CJ, Valencia E (2013) Smart City: hacía la gestión inteligente. Barcelona
12. Morales Quintana B, Grandeth Salcedo NL (2012) Estado del arte de las redes inteligentes "smart grid." https://biblioteca.utb.edu.co/notas/tesis/0063151.pdf
13. Propuesta de Requerimientos Técnicos para la Integración de Fuentes de Generación no Síncrona al SIN - XM | C.N.O. https://www.cno.org.co/content/propuesta-de-requerimientos-tecnicos-para-la-integracion-de-fuentes-de-generacion-no
14. Leyes desde 1992 - Vigencia expresa y control de constitucionalidad [LEY_1715_2014]. https://www.secretariasenado.gov.co/senado/basedoc/ley_1715_2014.html
15. Torres Wilches CR, Rojas Patiño A, Higuera Rincón DA, Hernández Henríquez JD, Calle Forero JA (2017) Planeamiento estratégico del sector de las fuentes no convencionales de energía renovable en Colombia. Pontificia Universidad Católica del Perú
16. VV A (2017) Las ciudades del futuro: inteligentes, digitales y sostenibles
17. UIT-T – Medio ambiente, cambio climático y economía circular de un vistazo. https://www.itu.int:443/es/ITU-T/climatechange/Pages/ictccenv.aspx
18. Azevedo Guedes AL, Carvalho Alvarenga J, Dos Santos Sgarbi Goulart M, Rodriguez y Rodriguez MV, Pereira Soares CA (2018) Smart cities: the main drivers for increasing the intelligence of cities. Sustainability 10:3121. https://doi.org/10.3390/su10093121
19. Las redes inteligentes de energía y su implementación en ciudades sostenibles: RG-T2058. https://publications.iadb.org/es/publicacion/15319/las-redes-inteligentes-de-energia-y-su-implementacion-en-ciudades-sostenibles-rg

Morphological Change in the Mala River Basin (Lima, Peru) Applying RUSLE and Geospatial Techniques

Geraldine Roque(✉), Cyndel Santisteban, Joel Fernández, and Sissi Santos

Universidad Peruana de Ciencias Aplicadas, Av. Prolongación Primavera 2390, Santiago de Surco, Lima, Peru
{u201215069,u201324332,pccijfer,pccissan}@upc.edu.pe

Abstract. Effects of deforestation, climate change, and human actions cause significant changes in the watershed morphology due to increased sediment flow and transport. High rates of erosion have been observed as a result. The present study was carried out in the Mala River Basin of Cañete Province. Morphological change of the Mala River Basin is evident in the formation of islands and in the branched course it has adopted over the years due to erosion. The objective of this study was to analyze the morphological change resulting from soil loss during the period 1996–2016 for the Mala River Basin (Lima, Peru), applying RUSLE and geospatial techniques. RUSLE model requires five parameters (R, K, LS, C, and P), which are obtained through the geospatial and meteorological information of the main organizations of the country. Erosion rates were calculated through a map algebra with Arc-GIS software. According to the results, the average loss of soil is 248.44 t ha^{-1} $year^{-1}$ in the Mala River basin. In addition, the erosion map shows that 41.51% of the basin has extremely severe erosion levels. The results can contribute to best practices in soil conservation and land management to reduce erosion in the Mala River Basin.

Keywords: Soil erosion risk · RUSLE · GIS · Mala basin

1 Introduction

The morphological changes are caused by factors derived from the amount of flow in times of flood and low water, the geological characterization of the river, the removal of vegetation cover, erosion, and sedimentation levels, among others [1]. These problems occur in river basins around the world that are vulnerable to the effects of climate change. This is because their production systems are more sensitive to climate variations and natural phenomena, which can affect agricultural production and the safety of the population. Loss of soil and sediment transport represent imminent danger. Consequently, there is a need to increase the capacity to respond, prevent, and adapt to the effects of climate change. Besides the economic problem and low level of technology in the basins, the situation that Peru presents is vulnerability to climatic phenomena, such as the case of the extraordinary episodes caused by the Niño Phenomenon in 1998.

Y. Iano et al. (Eds.): BTSym 2020, SIST 233, pp. 238–248, 2021.
https://doi.org/10.1007/978-3-030-75680-2_28

Several studies and researches have been carried out in order to evaluate soil erosion, which is useful for planning and conservation works in a watershed [2]. Gasnari and Ramesh [3] use the Revised Universal Soil Loss Equation (RUSLE) integrated into a GIS using remote sensing data to quantify the erosion rate, which contributed to improving the farming system and soil conservation support practices. Correa et al. [4] report that the application of this method has significant potential to motivate more studies and alert about water erosion problems from the environmental and socioeconomic approach. Pérez et al. [5] indicate that to achieve a comprehensive, objective, and geospatial evaluation of the environmental problem, it is necessary to use a digital elevation model (DEM) and use tools for calculating water erosion with ArcGis software, which replaces complex mathematical systems. In Peru, basin-level scenarios have been developed due to the need for detailed information for vulnerability and adaptation studies [6]. The Mala River basin has a region that encompasses diverse changes in its geomorphology, which are evident in the formation of islands and the branching course of the river, adopted over the years.

This research evaluates the geomorphological change in the Mala River Basin from soil loss analysis applying the RUSLE model and geospatial techniques. With this methodology, the average rate of erosion is analyzed, which is validated with previous results where the USLE model is applied. The research will be a useful tool to propose prevention corrective measures and elaboration of soil conservation strategies; to guarantee the food security.

2 Study Area

The study area is located in the central part of the Peruvian coast, on the Pacific slope, in the provinces of Cañete, Huarochirí, and Yauyos, in the Lima department. The basin has a territorial perimeter of 339.05 km and an extension of 2332.08 km^2. To delimit the Mala River Basin (Fig. 1), the ArcGis Geographic Information System was used. The Mala River Basin is made up of nine sub-basins: High Mala River, Acache River, Medium high Mala, Tantara River, Medium low Mala, San Joaquin Cochas, Yauyos, Lower Mala River and Quinches River. The Mala River sub-basins belong to different provinces, and, as such, they have variable climates, from intense sunshine to heavy rain and cold weather.

3 Methodology

The applied research methodology began with the characterization of the Mala River basin. The study area was defined using a DEM. This information was processed in a geographic information system (GIS), using ArcGis software version 10.5. Then, the mathematical model RUSLE was applied to estimate soil erosion in the basin. The information required for the RUSLE model factors was obtained from daily meteorological data from the PISCO v2.0 database (Peruvian Interpolated data of the SENAMHI's Climatological and hydrological Observations), FAO's soil map, soil use, and coverage map, and land management information from the Ministry of Environment (Fig. 2). Finally, through the Raster Calculator tool, the maps are multiplied according to [7, Eq. (1)]. In

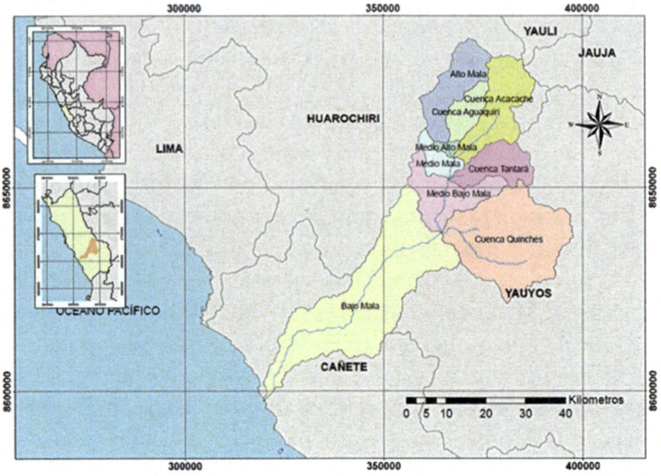

Fig. 1. Cartographic location of the study area, Mala River Basin, Lima, Peru

this way, the average erosion rate in the Mala River basin for the period 1996–2016 is determined.

The RUSLE model is the most suitable methodology for calculating soil loss in watersheds. RUSLE is based on the USLE equation, developed by Wishmeier and Smith [8], with a reformulation of the topographic factor estimates given by Renard et al. in [9, Eq. (1)]:

$$A = R \times K \times LS \times C \times P \tag{1}$$

Where A is the average soil loss rate (t ha^{-1} year^{-1}), R is the rain erosion factor (MJ mm ha^{-1} h^{-1}, year^{-1}), K is the soil erodability factor (t h MJ^{-1} mm^{-1}), LS is the dimensionless topographic factor, C and P are vegetation cover factors and conservation practices, respectively.

3.1 Description of the RUSLE Parameters

Rain Erosion Factor (R). This factor describes rain erosion as a function of the amount and intensity of rain and reflects its effect on soil erosion. To calculate the R factor (MJ mm ha^{-1} h^{-1} year^{-1}), the modified Fournier Index (MFI) developed by Arnoldus [10], which is defined in [10, Eq. (2)]:

$$R = \frac{\sum_{i=1}^{12} p_i^2}{P} \tag{2}$$

Where p_i is the monthly precipitation (mm) and P is the average annual precipitation. The spatial distribution of average annual precipitation (P) in the analyzed area was obtained by applying the kriging interpolation method [5].

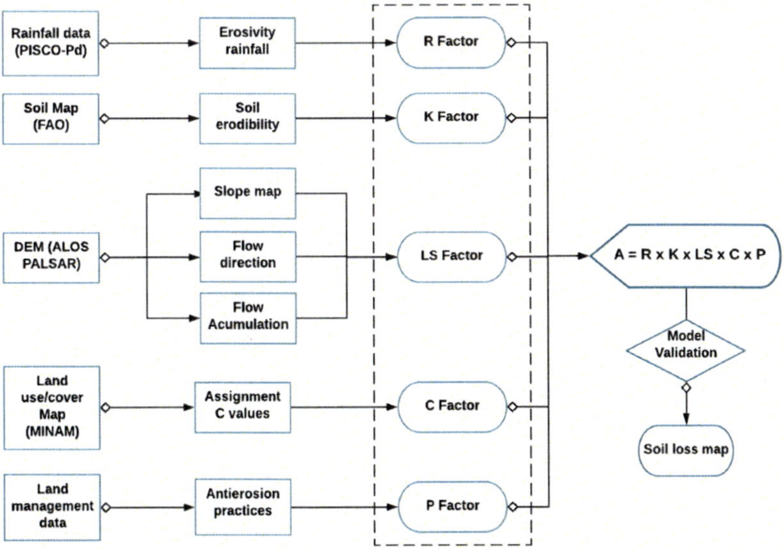

Fig. 2. Methodological flowchart of the RUSLE model

Soil Erodibility Factor (K). This factor was estimated based on soil textures, and its values reflect the rate of soil loss by rainfall-runoff erosion index. The erodability factors are shown in [7, Eq. (3)]. The K factor ($t\,hMJ^{-1}mm^{-1}$) of a soil is calculated as a function of its texture; % silt plus very fine sand, % sand, % organic matter, soil structure, and permeability [11].

$$K = \frac{(10^{-4}) \times 2.71M^{1.14} \times (12 - mo) + 4.20 \times (s - 2) + 3.23 \times (p - 3)}{100} \quad (3)$$

Where mo is the organic material, s is the code of the soil structure, p is the code of permeability, and M is the product of the size fractions of the primary particles.

Topographic Factor (LS). The topographic factor is composed of two factors: the length of the slope (L) and the inclination of the slope (S). Mitasova et al. [12] developed [12, Eq. (4)] to estimate the LS factor with GIS support, incorporating the impact of flow convergence, replacing the slope (L) length by the tributary area upstream per unit width.

$$LS = (m + 1)\times\left(\frac{Flowaccumulation \times Cellsize}{22.13}\right)^m \times (\frac{sin(Slope)}{0.0896})^n \quad (4)$$

According to De Kort [13], the RUSLE model, based on the SL factor approach, is classified as "RUSLE3D", where: Flow Accumulation x Cellsize corresponds to the tributary area upstream per unit width (m); *Slope* is the slope in degrees; m and n are empirical coefficients which were assigned 0.4 and 1.4, respectively, since they present satisfactory results for slope angles lower than 14° and length greater than 100 m [13].

Vegetable Cover Factor (C). It is an important factor since the vegetal cover intercepts the rain, increases the infiltration, and reduces the energy of the raindrops on the soil. In this study, to obtain the C factor, the shapefile provided by MINAM [14] was processed in ArcGis, which contains information on land use and vegetation cover in Peru. Then, the C values [10] were assigned according to the areas and conditions identified in the Mala River basin. The values of C in RUSLE vary from 0 to 1, where values near the unit indicate no effect of vegetation cover and therefore greater soil loss, while if the value of C decreases, the effect is positive for the soil [4].

Conservation Practices Factor (P). This factor reflects the effect of soil conservation practices that influence the reduction of the amount of soil erosion and is used to take into account the positive impacts of the application of support practices in areas influenced by agricultural activity. This factor's value is in the range of 0 to 1; the value close to 0 indicates a good conservation practice, and the value close to or equal to 1 reflects that no conservation practices are applied [3].

4 Results

4.1 Estimation of the RUSLE Factors

Determination of the Erosion Factor R. For the determination of this factor, 21 years of precipitation data from 7 rainfall stations were interpolated, which is shown in a map (Fig. 3). The average annual precipitation of the study zone is 624.45 mm. Erosivity value varies between 97.31 and 213.43 MJ mm ha^{-1} h^{-1} $year^{-1}$, and the average is 149.55 MJ mm ha^{-1} h^{-1} $year^{-1}$ classified as a high erosion rate, according to the MFI [15].

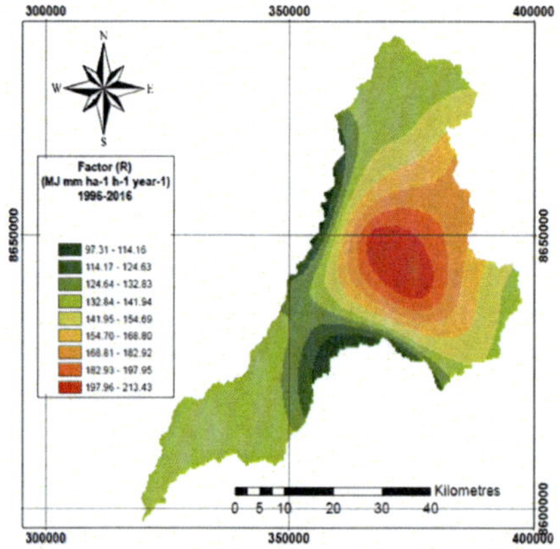

Fig. 3. Rainfall erosivity (R) factor map

Determination of the Erodibility Factor K. For the analyzed basin, the K factor was estimated based on the soil types according to the FAO methodology. The results indicate that the basin has loamy and clayey textures. The erodability value of the adapted soil by Renard et al. [16] varies from 0.26 to 0.28 t h MJ^{-1} mm^{-1}, and the average is 0.27 t h MJ^{-1} mm^{-1}. The distribution of texture in the study area was estimated by "Kriging" interpolation method [7] since it better predicts the spatial variation of soil parameters (Fig. 4).

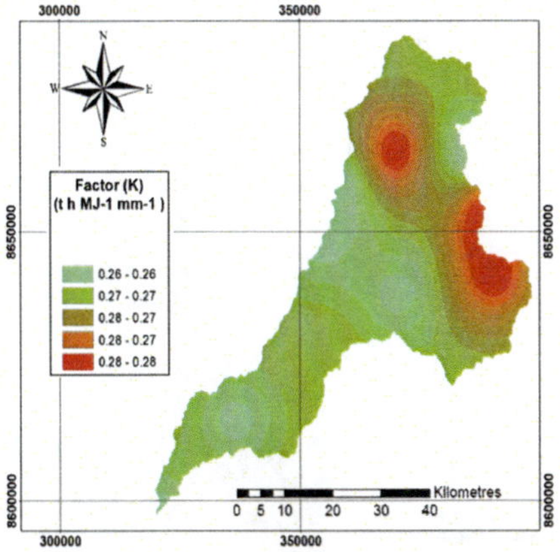

Fig. 4. Soil erodibility (K) factor map

Determination of the Length and Slope Factor LS. In order to obtain the LS factor, it was used the equation detailed in [15, Eq. (4)] with the Arc Gis raster calculator tool. The maximum value of the LS was 6411, and the minimum was 0 in the flat parts of the basin, also an average of 23.93. The corresponding map was generated for this factor (Fig. 5).

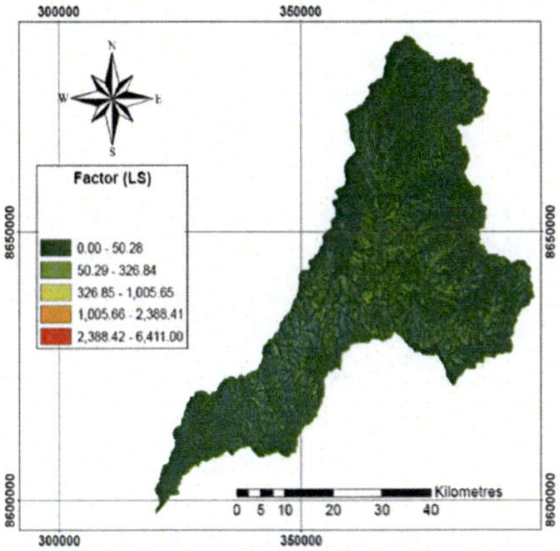

Fig. 5. Length and slope (LS) factor map

Determination of the Vegetation Cover Factor C. Determining this factor involved the use of vegetation cover database [14]. This was converted to shape format and assigned the C values established by Wishmeier and Smith [8]. Eleven types of cover were identified in the watershed, where the areas corresponding to scrub and cardinal are predominant, representing 37.53% (0.22–0.49); these covers offer greater protection to the soil. Likewise, 16.17% (0.50–0.63) corresponds to agricultural land and deserts, which are more vulnerable to erosion. The values of factor C of vegetative cover oscillate between 0 and 0.63 (Fig. 6).

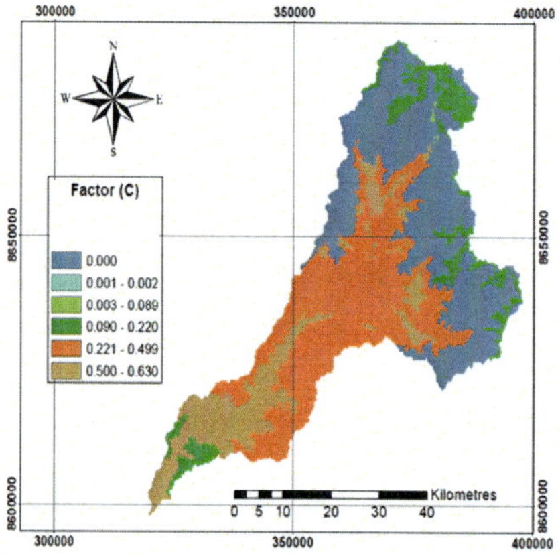

Fig. 6. Soil erodability (C) factor map

Determination of the Factor Soil Conservation Practices P. For the Mala River basin there is no data that supports conservation practices. The value was taken as 1, as referred by the National Water Authority (ANA) [1].

4.2 Estimation of Soil Erosion Rates (A)

In order to quantify and illustrate the erosion, we applied the RUSLE equation [4, Eq. (1)] in the raster calculator of the ArcGis software from the obtained maps (Fig. 7). It was found that the average erosion rate in the Mala River basin was 248.44 t ha^{-1} $year^{-1}$. The average annual soil loss distribution obtained was classified according to Li et al. [17] and Tang et al. [18] in Table 1. These results reveal that the lowest soil loss values in the area corresponding to glaciers, lakes, wetlands, and grasslands, which reaches 48.95% of the total. On the other hand, there is an extremely severe degree of erosion corresponding to 41.51% of the basin, which is distributed from areas located in the upper part of the sub-basin and, in its greatest extension, in the middle and lower sub-basin. The rates obtained for this area reflect a great susceptibility to erosion and correspond to agricultural zones and coastal desert, high Andean areas, Andean relict forest, cardonals, and shrubbery.

Table 1. Estimated soil loss 1996–2016

Loss soil (A)	Area		Classification
(t ha^{-1} $year^{-1}$)	ha	%	
0.00–5.00	114156.61	48.95	Very low
5.01–25.00	1667.43	0.71	Low
25.01–50.00	3126.09	1.34	Moderate
50.01–80.00	5245.08	2.25	Severe
80.01–150.00	12196.74	5.23	Very severe
150.01–464444.54	96816.06	41.51	Extremely severe
	233208.00	100	

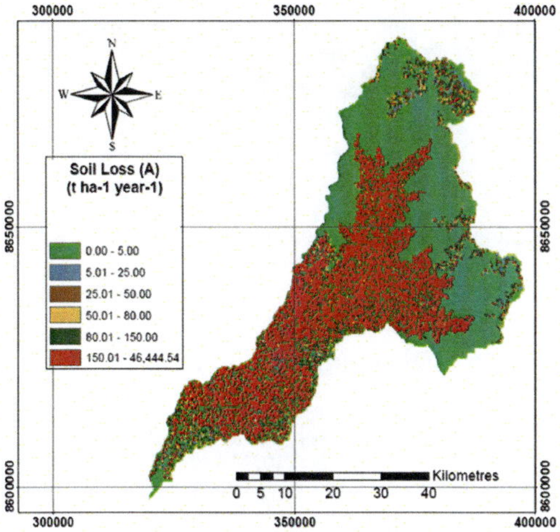

Fig. 7. Soil loss map (A)

4.3 Validation

The process of validation of the results obtained was carried out through the INGEMMET study "GEOCATMIN: Peligros geológicos en el Peru" has been consulted [19]. From this study, georeferenced information was extracted in the form of a layer concerning erosion processes in the basin. The hazards of hillside erosion and river erosion were selected, which have been identified during field and desk work according to the risk they represent in the area of analysis. The points are shown in a representative way in terms of erosion processes in the analyzed basin (Fig. 8). It can be seen that, along the middle and lower basins, there are records of river and hillside erosion in the main course of the Mala River.

In addition, it was corroborated by the official study conducted by the ANA [1], which estimated the rate of soil loss by applying the USLE method. The result of this study indicates that the basin has an average erosion rate of 326.3 t ha^{-1} $year^{-1}$, which reflects an approach with the results obtained in this research. The difference between the results may be due to the methods of analysis for each of the factors. According to this, it can be said that there is a correspondence of results, and both reflect extremely severe erosion rates, according to the classification of Li et al. [17] and Tang et al. [18].

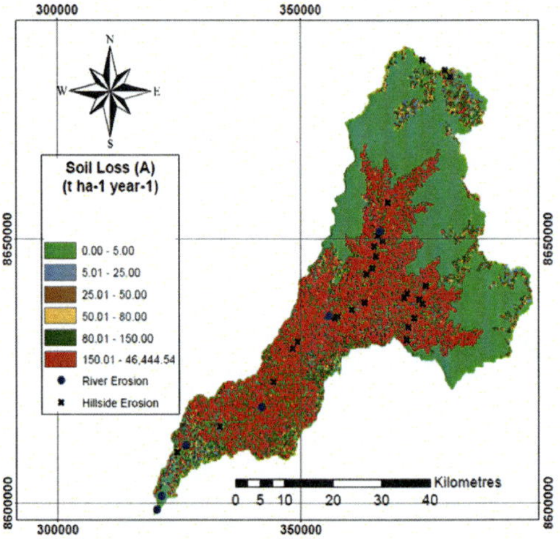

Fig. 8. Identification of points with presence of erosive processes in the map of soil loss in the Mala River Basin

Table 2. Estimated soil loss by year ranges

Years	Average Soil loss (A) (t ha^{-1} $year^{-1}$)
1996–2000	270.88
2001–2005	216.72
2006–2010	312.75
2011–2016	201.99

5 Conclusions

This article shows that the RUSLE soil erosion model can be used to analyze the erosion rate with geospatial techniques. However, the results depend on the accuracy of the input data. Based on the findings, it is concluded that the average soil loss in the Mala River Basin for the range 1996–2016 was 248.44 t ha^{-1} $year^{-1}$, a value that is classified as extremely severe. The results also indicate, as reflected in Fig. 8, that the Middle High Mala, Middle Mala, Middle Low Mala, Quinches River, and Lower Mala sub-basins have higher levels of erosion. In consequence, they are the ones at potential risk.

Additionally, the analysis was performed for the ranges of years shown in Table 2, which shows that the average soil loss is extremely severe and increases with respect to extraordinary episodes caused by the El Niño-Southern Oscillation. The assessment of geomorphological change, estimated from the Mala River Basin erosion rate, provides potential information for implementing appropriate soil conservation practices in vulnerable areas. In future investigations, it is intended to evaluate erosion rates on an annual basis in order to establish a correlation with the extreme meteorological events that take place every certain period in the analyzed country.

References

1. Autoridad Nacional del Agua (2015) Evaluación de los Recursos Hídricos en las cuencas Mala, Omas and Chilca
2. Ministerio del Ambiente (2016) El Perú y el cambio climático. Tercera comunicación nacional del Perú a la Convención Marco de las Naciones Unidas sobre el Cambio Climático. https://www.minam.gob.pe/wp-content/uploads/2016/05/Tercera-Comunicaci%C3%B3n.pdf
3. Ganasri BP, Ramesh H (2016) Soil erosión assessment using the RUSLE model using remote sensing and GIS: a case study from the Nethravathi basin. Geosci Front 7(6):953–961. https://doi.org/10.1016/j.gsf.2015.10.007
4. Correa SW, Mello CR, Chou SC, Curi N, Norton LD (2016) Soil erosion risk associated with climate change at Mantaro River basin, Peruvian Andes. CATENA 147:110–124. https://doi.org/10.1016/j.catena.2016.07.003
5. Pérez Y, Pérez J, Camejo K (2018) Procedimiento para el cálculo de la erosión hídrica a partir del MDT. X Congreso Internacional GEOMÁTICA
6. MINAM – SENAMHI (2015) Simulación de escenarios de disponibilidad hídrica en las cuencas de los ríos Urubamba, Pampas, Alto Apurímac (Cunyac) y Mala. Proyecto Tercera Comunicación Nacional. Lima: Ministerio del Ambiente, Servicio Nacional de Meteorología e Hidrología
7. Roose E (1996) Manejo de tierras - Componentes y estrategia. 70 Boletín de suelos de la FAO, Organización de las Naciones Unidas para la Agricultura y la Alimentación, Roma, Italia
8. Wischmeier WH, Smith DD (1978) Predicting los from rain erosión: A guide to conservation planning. Manual de Agricultura 282. USDA-ARS, EE. UU
9. Renard K, Foster G, Weesies G, McCool D, Yoder D (1997) Predicting Soil Erosion by Water: a guide to conservation Planning with the Revised Universal Soil Loss Equation (RUSLE). US Government Printing Office, Washington, DC
10. Arnoldus HMJ (1980) An approximation of the rainfall factor in the universal soil loss equation. In: de Boodt M, Gabriels D (eds)
11. Villatoro M, Henriquez C, Sancho F (2008) Comparación de los interpoladores IDW y Kriging en la variación espacial de PH, CA, CICE y P del suelo. Agronomía Costarricense 32(1):95–105.0
12. Mitasova H, Hofierka J, Zlocha M, Iverson LR (1996) Modelling topographic potential for erosion and deposition using GIS. Int J Geogr Inf Syst 10:629–641. https://doi.org/10.1080/02693799608902101
13. De Kort A (2013) Soil Erosion Assessment in the Dryland Areas of Bolivia Using the RUSLE 3D Model (Master thesis) Soil Physics and Land Management Group. Master of Science in International Land and Water Management at Wageningen University, The Netherlands, p 61
14. MINAM (2015) Mapa Nacional de Cobertura Vegetal: memoria descriptiva/Ministerio del Ambiente, Dirección General de Evaluación, Valoración y Financiamiento del Patrimonio Natural. Lima

15. Commission of the European Communities (1992) CORINE soil erosion risk and important land resources in the southern regions of the European Community. Commission of the European Communities, Luxembourg, vol 97, pp 127–132
16. Renard K, Freimund J (1994) Using monthly precipitation data to estimate the R-factor in the revised USLE. J Hydrol 157:287. https://doi.org/10.1016/0022-1694(94)90110-4
17. Li X, Zhang X, Zhang L, Wu B (2014) Rainfall and vegetation coupling index for soil erosion risk mapping. J Soil Water Conserv 69:213–220
18. Tang Q, Xu Y, Bennett SJ, Li Y (2015) Assessment of soil erosion using RUSLE and GIS: a case study of the Yangou watershed in the Loess Plateau China. Environ Earth Sci 73:1715–1724
19. GEOCATMIN (2012) Peligros Geológicos del Perú

Model Business Rules for Control Load through Electrical Parameters

Israel Gondres Torné(✉), Ruan Carlos Mota Teixeira, Gabrielle Stephanie Pires Mestrinho, Isaque Vilson Batista da Costa, Alison Soares de Almeida, and Evaldo Patrik dos Santos Cardoso

HUB - Technology & Innovation, Escola Superior de Tecnologia, Universidade do Estado do Amazonas, Manaus, Brazil
{itorne,rcmt.eng,gspm.eng17,ivbc.eng17,asda.ele}@uea.edu.br

Abstract. Business Rules are crucial and adaptable to different business processes; they define and limit the logic of the workflow according to the characteristics of the processes to be implemented. In this work, our attention is based on a set of rules to control loads of industrial companies that take the form of restrictions on some electrical measurements. The events foreseen in the Business Rules are in accordance with the indicators of the active power and the contracted energy demand of the customer, that is, through an algorithm that will control and manage the loads based on the events formed by the parameters monitored by the Shell. The transitions of the control are through the state machine. The program must perform the control of loads registered in the system in order to always keep consumption below the contracted demand.

Keywords: Asset administration shell · Business intelligence · Business rule · Loads

1 Introduction

Before the advent of automation and computerization of processing, the information was treated separately and unstructured; the decisions taken downstream always had a high intuition component.

Nowadays, companies in the world face different concepts and definitions of business rules, and, surprisingly, every day, they increase in a vertiginous way. Therefore, it is necessary to adapt all this information for efficient and effective management [1].

Business rules are crucial and adaptable to different business processes; they define and limit the logic of the workflow according to the characteristics of the processes to be implemented. Therefore, business rules are statements that define or restrict business conditions; they can be written in a natural language and simple enough, i.e., it can be understood between the different users, whether they are analysts, specialists, or others who should take one decision at any given time [2].

Demand-side management refers to the actions taken by the consumer to adjust the load profiles and thus obtain optimal energy costs. Therefore, demand-side management

Y. Iano et al. (Eds.): BTSym 2020, SIST 233, pp. 249–257, 2021.
https://doi.org/10.1007/978-3-030-75680-2_29

is a powerful efficiency improvement tool for both the end-user and the dealer. It can be configured to eliminate non-critical loads during situations where maximum demand threatens to increase the electric bill and reduce the company's operating benefits. All these incentives have encouraged the consumer to reach maximum levels of energy efficiency and implement demand-side management techniques [3, 4].

2 State of the Art

2.1 Asset Administration Shell

The Asset Administration Shell (AAS) is a key concept in Industry 4.0 applications. It describes a virtual asset in a standardized way, and it represents entities in the information field. To allow a meaningful interaction with 4.0 IT systems, it is assumed that assets are more than just machines, production modules, and even systems. It is also taken as products, software installations, intellectual property, or human resources. In addition to real-time parameters, production resources and asset stage must also be reported. In other words, standardizing the Asset Administration Shell reduces complexity and allows scalability.

The Reference Architecture Model for Industry 4.0, as known as RAMI 4.0, considers the stage of the asset's life cycle according to the standard IEC 62890 [5]. In practice, we must differentiate asset designs from physical assets. The phase in which the asset is being received is called "type", representing the design of an asset type. When the asset is physically produced, it becomes an identifiable instance or entity of a given "type". The relation between the type and the instance should be maintained throughout the life of the instances of the asset. In particular, type updates must be passed to the instance. As an example, there are products that might could have firmware over-the-air updating, also known as OTA [6]. The second aspect of that relation might be the feedback from the product instance to the type of product to support design and production improvements. Third, the exchange between assets from different asset classes can be used to design new products or manufacturing equipment. Finally, the relation between assets in different levels of hierarchy determines a system of subsystems, as an automation in a machine subsystem, for example.

2.2 The Business Intelligence

The Business Intelligence (BI) concept encompasses a wide range of decision support applications that enable fast, shared, and interactive access to information, as well as its analysis and manipulation; today, it is a widely used tool for discovering relationships and trends and turning large amounts of information into useful knowledge for any company or individual. Business intelligence systems have in common a set of important fundamental objectives [7]:

- *Access to reliable data* - the reliability of data, its easy integration and understanding between areas are essential for a conscious management exercise;

- *Increased business transparency and understanding* - Real-time knowledge delivery (the 'what', 'how much', 'when', 'where' and 'how') gives managers and decision-makers a perspective of the areas that they must control with full transparency and increase their understanding (the 'why');
- *Decision Support* - Only a timely understanding of reality can enable effective decision making; As such, the knowledge produced by BI systems, leveraged by current communication technologies, must support and justify the measures taken by the various actors in the management process.

2.3 Business Activity Monitoring

Applications designated as Business Activity Monitoring (BAM) are a subarea of Business Intelligence rapidly growing and technology development. Merging the analysis and reporting approach (which is the fundamental matrix of business intelligence) with the concept of business integration (which represents business process integration and automation), BAM applications deliver real-time, alert-based alerts on predefined and triggered metrics when they reach a value considered critical and require intervention by the organization's management. Metrics and their alerts should correspond to events that are relevant to the operations and objectives set [8].

Understandably, BAM technology should not work in isolation - it will only perform well when combined with multidimensional analysis or data mining features that provide the information and knowledge best suited to respond better to the alert [9].

BAM applications, therefore, have at least three essential components:

- *Data integration* - Data is collected in its source applications (databases, customer relationship management, supply chain management, internet, etc.);
- *Analysis* - Data are automatically analyzed according to established Business Rules; here, it is important to emphasize the importance of applications having user-friendly environments;
- *Information delivery* - In addition to the most common front-end application interfaces, information can be sent to mobile phones (via SMS), email, and fax, among others. The BAM functions should allow the user to signal knowledge and possibly trigger an action through the tool itself.

2.4 Business Rules

The Business Rule (BR) defines or limits some aspects of the business; that is, it is structured to control or influence the behavior of a particular business. Business rules can be seen from two perspectives: the business perspective and the information system perspective. The business perspective talks about the restrictions that apply to the behavior of people in the company in certain procedures and situations, while the information system perspective talks about the facts that are recorded as data and the restrictions on changes in value of those facts according to a specified situation [10].

The term "Business Rules" includes a wide range of rules, such as structural rules, interference rules, process rules, and others [11]. In this work, our attention is based on a set of rules to control loads of industrial companies that take the form of restrictions on some electrical measurements.

3 Materials and Methods

3.1 Asset Control Hierarchy

An asset hierarchy is a way of organizing physical assets in a set of logical groupings and subgroups, that is, the machines, equipment, and individual components of a given industry. Asset hierarchies allow an aggregated and detailed analysis of the assets' performance at various levels, from the facility level of individual equipment items (Fig. 1).

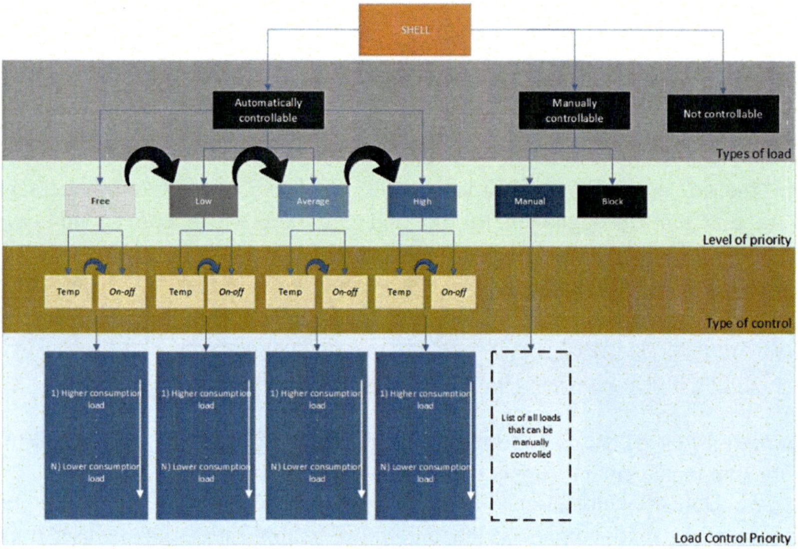

Fig. 1. Asset hierarchy.

Each asset will be grouped by four classifications:

1. **Type of load:** the loads can be classified as non-controllable (Shell that does not have a control device), manually controllable (it will be the loads that will be controlled only by manual commands as the user wishes), and automatically controllable (controlled loads autonomously through the BR, according to the events related to electrical parameters);
2. **Priority level:** In addition, there are four priority levels related to the level of impact that automatic load control can have on the production process and working conditions, these being in increasing order of priority: free for system shutdown, low priority for switched on, average priority for switched on, and high priority for switched on. Manually controlled loads also have priority criteria. There are two classifications for loads that can be operated manually: blocked (can never be activated except with the proper authorization) and manually controlled loads themselves, where, in fact, it is possible to command the charge via interface

3. **Type of control:** the project provides for two types of control: temperature and on/off, with the temperature control having less impact in relation to the on/off control;
4. **Load consumption:** the last classification criterion consists of the nominal load power. The loads with the highest consumption will be prioritized for the activation; in this specific case, only the central air and air conditioning loads will be controlled.

3.2 Asset Control Events

According to the indicators of active power and the contracted demand for energy from the client, some events are foreseen in the Business Rules. The graph in Fig. 2 illustrates the four events that the demand consumed by the industry of electronics components is susceptible to, related to the percentage of contracted demand, also listing the type of alert and the type of control of each event.

The events that involve the control of assets start from 85% of the contracted demand; thus, the algorithm has a range of 15% to control the loads without exceeding the demand. The BR provides five types of control that are related to the events mentioned above according to the demand for instantaneous active power. The controls are:

- **Normal:** loads operating according to the customer's needs without intervention by the algorithm;
- **Control 1:** free for system shutdown loads can be turned off, according to the demand-control algorithm
- **Control 2:** maintains 'control 1' and low priority for switched on priority level loads can be switched off as required by demand.
- **Control 3**: maintains the previous controls and can deactivate loads classified as average priority for switched on;
- **Control 4:** maintains the previous controls and can deactivate loads classified as high priority for switched on.

One way to manage the transitions of the control is through the state machine. Every second, the current status of the percentage of contracted demand will be checked, and, depending on the value, state transitions may occur, following the conditions established in Fig. 2.

The state machine's purpose will always be to make the system establish itself in the 'normal' state. Thus, there are two control routines for the state machine that will define the load control sequence: ramp-up and ramp-down (see Fig. 3). This routine is defined after analyzing the average of 120 samples (1 sample per second) of the active power; the counting starts from the first sample when exceeding the upper or lower state limit, if the average persists greater than the limit higher or lower than the lower limit, the trigger to change the state will be triggered. In addition, the average will define whether the trend of the demand curve is growth (ramp-up) or decrease (ramp down).

- **Ramp-up:** the control routine, when the state is in ramp-up mode, consists of deactivating or increasing the temperature gradually. The loads that their nominal power subtracts from the demand consumed will guarantee 1% below the current state's lower limit.

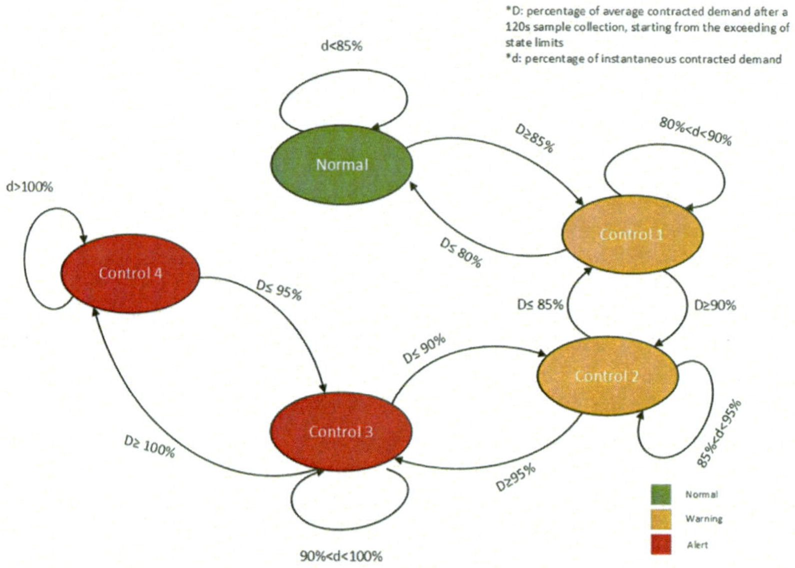

Fig. 2. State machine BR

- **Ramp-down:** when returning to the previous state, the reactivation of the source state must also be gradual so that the power of a load does not trigger the trigger for the next state, that is, the power of the charged loads added to the demand consumed must be less than 1% higher than the current state.

Some restrictions must be provided for in the algorithm, especially when the demand is stable in a state other than the 'normal' state, in which case there must be a rotation between the loads every 15 min, in order that the system does not affect the production process and the working conditions of employees.

Note: After the average of 120 samples, the algorithm predicts change for any state so that it corresponds to the percentage of the average contracted demand sampled (D). That is, the change of states does not necessarily have to be sequential. In the event of state jumps, the control priority will always be due to the loads of the priority level that occurred the jump. Example: In case of a state jump from control 1 to 3 (ramp-up), control 2 should be prioritized, and only then will the routines of control 3 occur, the same logic for ramp-down.

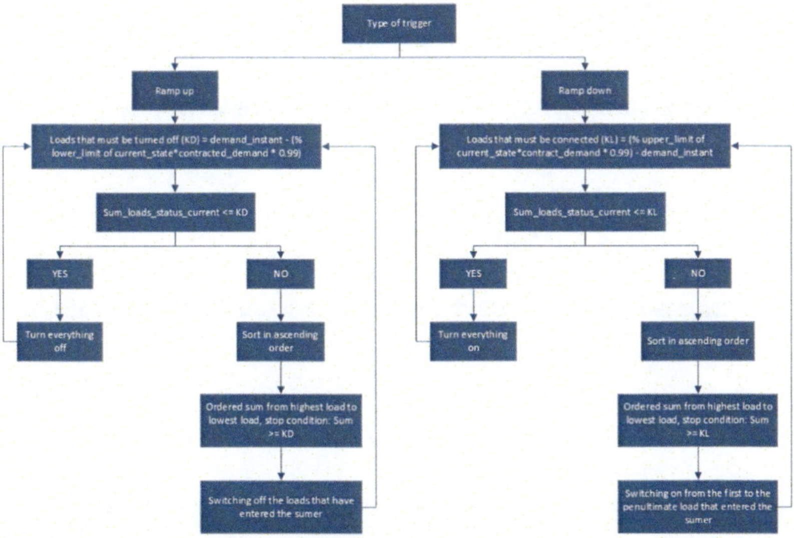

Fig. 3. Ramp up and ramp down

4 Results and Discussion

The Business Rules Algorithm is represented in the state machine diagram where it is possible to notice that two levels of control were removed, leaving only 3, against the 5 existing previously. This algorithm aims to perform load control according to a state machine. A state machine is a mathematical model used to represent computer programs in which its state varies depending on stimuli that would be the control mode of the algorithm. The stimulus of the state machine varies according to the energy consumption by the demand contracted by the consumer agent. According to the state machine's control states, the program must perform the control of loads registered in the system to always keep consumption below the contracted demand. The control of the loads must be carried out according to the state machine level and, also, respecting the load priority. For the tests and visualization of the program's actions, a log screen was developed where it is possible to view the program's control levels, which loads were turned off in addition to all the plant's consumption (Fig. 4). During the tests, the loads are being represented by a Python emulator of assets that generates voltage and current data to represent the power of each AAS.

The ramp-up routine goes from control 2 to control 3, forcing you to stimulate the Shell, thus generating a notification connected to the event screen (Fig. 5).

Finally, it is possible to verify the results of the business rules algorithm; they are represented in the graphical event screen (Fig. 6), generating alert icons in the ascending ramp and alarm in the descending ramp.

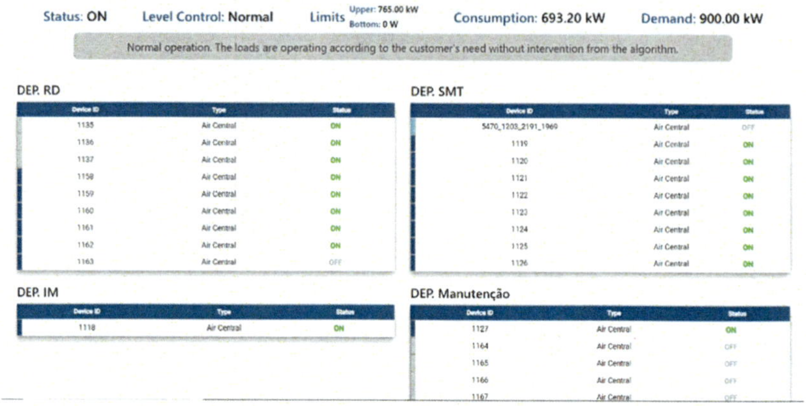

Fig. 4. Logs preview screen: Loads that have been turned on or off.

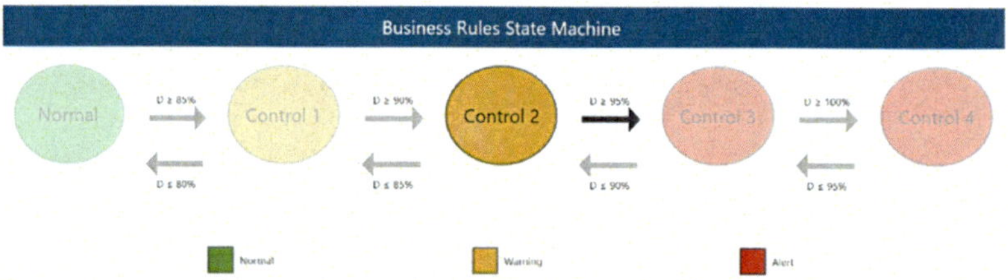

Fig. 5. Logs preview screen: State machine and its respective control levels.

Fig. 6. Graphical representation event screen.

5 Conclusion

The Algorithm business Rules is developed to supply the need to adapt all the information of the business rules for an efficient and effective management. The business rules created for automation of the charge control system can be observed in two main groups: the hierarchy of control of events, which are the cataloging of charges in four categories:

types of charge, priority level, types of control, and charge consumption and the other group is the asset control events that start from the rules and priorities that govern the system.

The integration of 5 controls named norm, which represents the stability of the charge and the control 1, 2, 3, and 4 act through the ramp-up and ramp-down processes selecting the correct charge for the operations of activating and deactivating certain devices, in addition to each state change process (alerts of 80%, 95% and 100% of the contracted demand in use) an alarm or alert message is generated for cases of exorbitant growth in demand.

References

1. Burgstaller F, Steiner D, Schrefl M (2016) Modeling context for business rule management. In: 2016 IEEE 18th conference on business informatics (CBI), on proceedings IEEE, Paris, pp 262–271
2. Deesukying J, Vatanawood W (2016) Generating of business rules for Coloured Petri Nets. In: 2016 IEEE/ACIS 15th international conference on computer and information science (ICIS), on proceedings IEEE, Okayama, pp 1–6
3. Osama M, Elshenawy AK, Elsingab M (2018) Load management optimization to reduce the demand side energy cost. In: 2018 IEEE PES/IAS PowerAfrica, on proceedings IEEE, Cape Town, pp 25–31
4. Alwan HO, Sadeghian H, Wang Z (2018) Decentralized demand side management optimization for residential and commercial load. In: 2018 IEEE international conference on electro/information technology (EIT), on proceedings IEEE, Rochester, MI, pp 0712–0717
5. IEC 62890:2020 Industrial-process measurement, control and automation - Life-cycle-management for systems and components (2020) IEC, Geneva, Switzerland
6. Kerliu K et al (2019) Secure over-the-air firmware updates for sensor networks. In: 2019 IEEE 16th international conference on mobile ad hoc and sensor systems workshops (MASSW), on proceedings IEEE, Monterey, CA, USA, pp 97–100
7. Sezões C, Oliveira J, Baptista M (2006) Business Intelligence. 2nd edn. Artes Gráficas, Lda., Porto
8. Sebu ML, Ciocârlie H (2015) Business activity monitoring solution to detect deviations in business process execution. In: 2015 IEEE 10th jubilee international symposium on applied computational intelligence and informatics, on proceedings IEEE, Timisoara, Romania, pp 437–442
9. Engel R, Krathu W, Pichler C, Zapletal M, Werthner H (2013) Towards EDI-based business activity monitoring. In: 2013 17th IEEE international enterprise distributed object computing conference workshops, on proceedings IEEE, Vancouver, Canada, pp 158–162
10. Isakh RR, Widagdo TE (2016) Business rules handling in relational databases. In: 2016 international conference on data and software engineering (ICoDSE), on proceedings IEEE, Denpasar, pp 1–6
11. Lin L, Embury SM, Warboys BC (2005) Facilitating the implementation and evolution of business rules. In: 21st IEEE international conference on software maintenance (ICSM 2005), on proceedings IEEE, Budapest, pp 063–6773

Hash Authentication VANETS Message (HAVM) Against Message Tampered (MITM Attack)

Martín Vélez Falconí(✉), Selena Jiménez Lara, and Cristhian Iza Paredes

Yachay Tech University, Hacienda San José, Urcuquí 100119, Ecuador
{martin.velez,selena.jimenez,ciza}@yachaytech.edu.ec

Abstract. Vehicle Ad-Hoc Networks (VANET) promises to enhance traffic management and road safety. However, data can be at stake by an attacker due to the dynamic network topology. The attack Man in the Middle (MITM) is probably a remarkable one in security attacks, especially tampered message attacks. This attack modifies essential information, misleading the communicated nodes, and possibly creating disasters. To address this problem, we propose a novel method for trust message validation between vehicles through the implementation of a hash algorithm, namely hash authentication VANET message (HAVM). Every node has the capacity to detect if a message has been tampered. HAVM vehicles generate a unique code per message. When this message is distributed among the VANET, the HAVM vehicles will be able to analyze whether this code belongs or not to the received message. Several simulations were carried out to evaluate the performance of HAVM. The simulation showed that a network with 70% of vehicles with HAVM, and 10% of MITM vehicles, improved the state of the Content Delivery Radio (CDR) by 14% in reliability than in a network without HAVM nodes.

Keywords: Man In The Middle (MITM) · Tampered message · Hash · Simulation · Vehicular Ad-Hoc Network (VANET) · Security

1 Introduction

Nowadays, VANET performs a remarkable aspect in improving road security, traffic efficiency, and safety. To secure a reliable transmission, vehicles should transmit secure messages. These messages share information about the status of the vehicle; thus, this shared knowledge represents a threat. Security messages must be verified in order to evade attacks. These security attacks carry out serious data corruption to the whole VANET system [1]. Sybil, black hole, denial-of-service (DoS), and MITM are the most common attacks which exploit security vulnerabilities in VANET [2].

A typical solution for these attacks is schemes based on symmetric keys, public keys, and certification signature. These algorithms are efficient in preventing external attacks [3]. Nonetheless, in the case of insider attackers, there are not standard schemes that provide security efficiency. In fact, a VANET is susceptible to attacks from insider nodes as MITM nodes. Figure 1 shows a typical MITM attack. The schema involves

Y. Iano et al. (Eds.): BTSym 2020, SIST 233, pp. 258–265, 2021.
https://doi.org/10.1007/978-3-030-75680-2_30

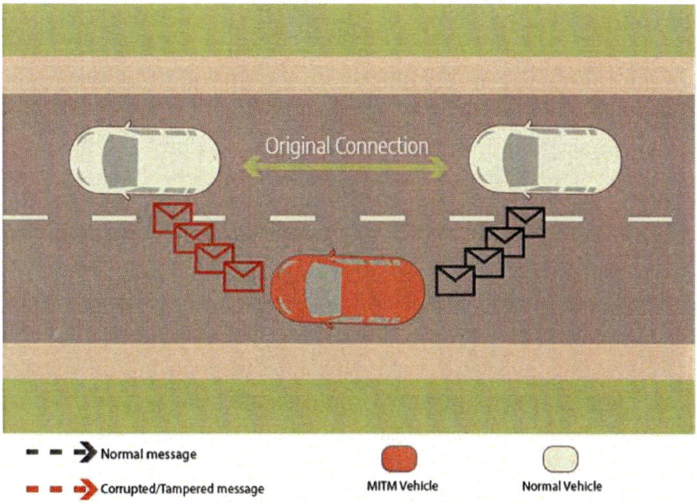

Fig. 1. Operation of the MITM attack.

two nodes, which are the victims, and a third node, which is the attacker. In the original communication channel, this attacker node has unauthorized access [4].

There are two groups within the MITM methodology; passive attacks (get the information) and active attacks (edit, delay, drop the messages). We focused on the active attacks, specifically in a message tampered in which the legitimacy of the message is compromised [5]. The goal of MITM is to corrupt the content of the received message. In the case that a MITM vehicle receives a message, these vehicles could modify relevant information. The information to be disseminated could be a blocked street, accident location, the location of the sender. Thus, this information must guarantee that it has not tampered to reply to other vehicles. However, the vehicles that do not belong to a safe network cannot detect if the information of the message was compromised, so that these vehicles will distribute the tampered message to others in the network [6, 7].

This research recommends a novel method for trust message validation between vehicles by implementing a hash algorithm, namely, hash authentication VANET message (HAVM), to avoid and contain MITM attacks. Our contributions in this paper are detailed as proceed:

1. Develop a hash authentication VANET message (HAVM) to avoid tampered messages in man in the middle attacks applied in vehicles within VANET.
2. To compare the encryption efficiency through VANET simulations to validate the security and trustworthiness of the algorithm.

The remainder of this paper is organized as follows: In Sect. 2, the related work is displayed. Section 3 provides details of our proposed method. In Sect. 4, we explain in detail the simulation environment. In Sect. 5, we include the simulation results, and finally, in Sect. 6, we present the conclusions and future work.

2 Related Work

Decentralized trust management is applicable in VANET. In these models, in order to get a distributed mutual verification, a public and private key pair is required. In the research of Wang et al. [8], this method, through the pseudonym technique, claims to preserve the privacy integrity and the toughness against tactical attacks. Nonetheless, this algorithm is deficient in analyzing new nodes integrated on the VANET because there is not a significant amount of information about the trustworthiness of the node.

Another approach for maintaining security in VANET was designed by Ali et al. [9], where the implementation of Blockchains was introduced in the network. In this research, Blockchain was proposed as an efficient Public Key Signature (CL-PKS) scheme without a certificate. This method pursues to reduce the computational cost of signature generation and verification to the minimum. A similar approach was presented by George [10], where a light-weight authentication framework was proposed, which also uses blockchain security for VANET. This process was made through updating pseudo-IDs, public keys, and private keys of the vehicles by authentication parties, such that the corrupted vehicles had their IDs revoked, and maintaining a register of credible vehicles in the network. In this model, the messages sent by the stolen vehicle continue to be distributed since they verify the authenticity of the vehicle but not of the message. In terms of trust management schemes such as node centric trust computation and data-centric trust computation, there is a Man-in-the-middle Attack Resistance trust model IN connected vehicles (MARINE) research [5]. MARINE resolves some issues generated by MITM attacks in different schemes. However, it is based on node-centric trusts with high computational costs, and it depends on statistics. Thus, the MARINE method may be unsuccessful. In essence, there are a reduced number of algorithms that can handle MITM messages. However, most of these algorithms depend on the centralized system and based on trust vehicles. In the next section, the proposed trust method HAVM is explained.

3 Proposed Method Hash Authentication VANET Message (HAVM)

The proposed method of HAVM is to implement a verification technique, the hash, inside the nodes, to identify and compare the generated hash in the messages sent within nodes. The hash function used on HAVM is natively supported in C++ through Standard Template Library (STL). In particular, the implemented hash in HAVM defines: 1) A single parameter of type Key, 2) Returns a value of type $std :: size_t$ that represents the hash value of the parameter. All explicit and partial specializations of hash provided by the standard library in C++ are Default Constructible, Copy Assignable, Swappable, and Destructible [11].

In our proposal model, MITM vehicles will tamper the original messages. Figure 2 highlights the simulation process, illustrating how the different vehicles will behave. We consider that there are three types of vehicles, I) MITM vehicles, II) Normal vehicles that are neither malicious nor HAVM implemented, III) Vehicles HAVM implemented. These

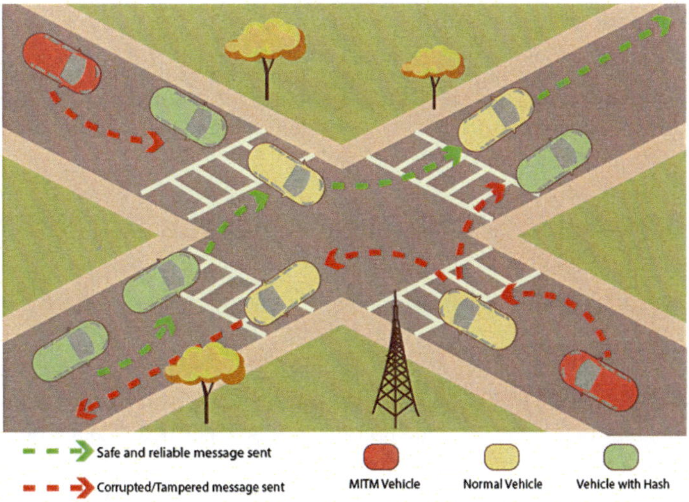

Fig. 2. Illustration of how the simulation will be developed.

different vehicles spread messages according to their behavior. Algorithm 1 summarizes the process of how the messages are identified.

```
Algorithm 1: Message identification within a vehicle

Required: Sent message(WsM ); MIT M Vehicle sends a tam-
pered message Message identification within a vehicle:
if wsm in hash verification is FALSE then
Compute MITM messages counter
end if
if hash implemented is TRUE and wsm in hash verification
is FALSE then
Compute Prevented messages counter
end if
```

HAVM operates as, the first step includes a sent message, then, if the receiver vehicle has implemented the HAVM model, the next step involves a comparison of the generated hash. If the vehicle has not the HAVM model, then it has no verification process, so it proceeds to send the message. While in the HAVM vehicle, the generated hash is compared with the hash message; if these are the same, then it proceeds to send the message since it is reliable; otherwise, the vehicle with HAVM will discard the message. Therefore, the effect that VANET with MITM vehicles generates a considerable quantity of corrupted messages that are discarded by the nodes with hash, avoiding spreading these messages deliberately on the network. Hence, corrupted content is always lost due to vehicles with hash.

3.1 A Subsection Sample

The evaluation of our proposal was based on the next three following metrics:

Content Delivery Ratio (CDR). The content delivery ratio is the number of vehicles that successfully received a legitimate message.

HAVM Prevented Ratio (HAVMPR). This metric allows evaluating the accuracy of HAVM. We compute HAVMPR according to Eq. 1.

$$HAVMPR = \frac{PreventedMessages}{TotalMessages} \quad (1)$$

Where Prevented ratio is the number of messages that were corrupted by MITM but they were detected by HAVM and stopped from being sent to neighboring nodes. Depending on the value of the prevented ratio, it can be identified whether or not HAVM is an acceptable solution to the MITM problem.

MITM Detected Ratio (MITMDR). This metric allows us to compute the percentage of tampered messages that were detected by the HAVM method. We compute MITMDR according to Eq. 2.

$$MITMDR = \frac{MITMMessages}{TotalMessages} \quad (2)$$

4 Experimental Setup

We select VEINS [12] framework which integrates SUMO [13] and OMNeT++ [14]. Furthermore, the framework includes some characteristics as network layers IEEE 1609.4 DSRC/WAVE, exhaustive models IEEE 802.11p, and support for realistic maps OpenStreetMap [15]. The performance evaluation of our scheme was provided by different simulation runs with random scenarios. Each mark in the scatter plot was calculated from the average of 10 executed simulations. The results have provided a standard error of less than 5% in a 95% confidence interval.

The efficiency of our proposal is computed based on the detection of MITM messages. Every simulation scenario is carried out with the same parameters, such as the number of vehicles, the percentage of malicious nodes, the percentage of vehicles with HAVM variating from 10% to 90%.

5 Results

The relation of the MITM messages, prevented messages, the total of messages in the network, and how these values are changing with the increase of nodes with our HAVM method is shown in Fig. 3. As the number of nodes with HAVM is increased, the corresponding number of MITM messages decreases. In this way, we are validating our proposed method. Besides, in a directly proportional relation, as the number of nodes with HAVM grows, the number of prevented messages also increases from 0% to 100%.

In a particular case, at 70% of vehicles with HAVM, 75% of MITM messages were prevented. The increasing of vehicles with HAVM results in the production of an elevated number of secure messages. In Fig. 4, the percentage of HAVMPR increases as the percentage of vehicles with HAVM. Since the objective of HAVM is to avoid the spread of corrupted data and inhibit this malicious behavior, there is an inverse proportional relation in CDR and MITMDR. This means that as the percentage of MITMDR decreases the percentage of CDR increases.

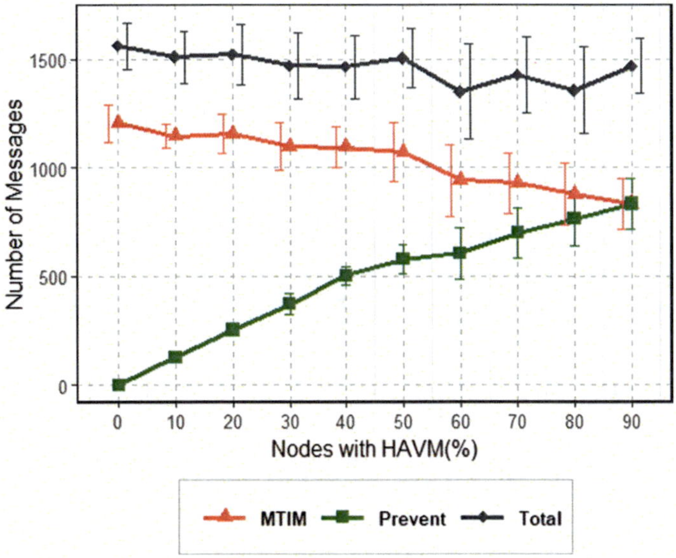

Fig. 3. Messages Prevented, MITM and Total

The application of HAVM had an important effect on VANET since the network obtained an elevated CDR in the high presence of nodes with HAVM. On the other hand, the CDR is directly affected in a network without HAVM nodes since malicious nodes adulterated content of the trustable messages. This corrupted data could not be detected by other nodes. Therefore, they resend this malicious information to other nodes. It has caused a higher ratio of garbage data that reduces the values of CDR within the network. Therefore, the results of these simulations show that through the implementation of the validation method, the messages disseminating false information among network entities are highly reduced.

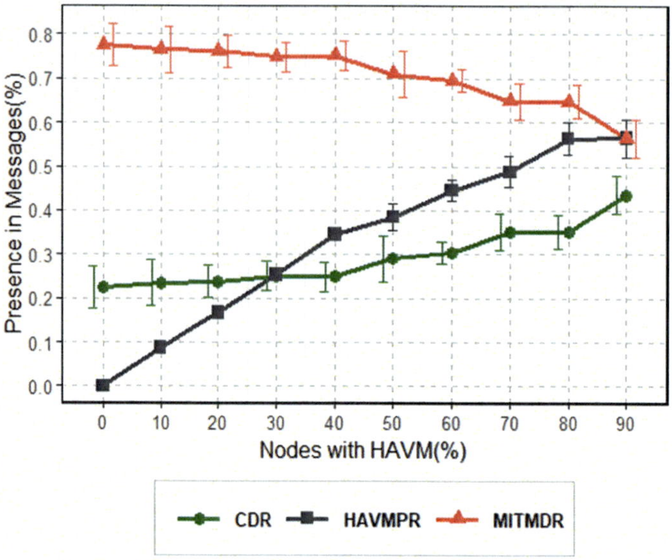

Fig. 4. Percentage of CDR, HAVMPR, MITMDR

6 Conclusion

To summarize, the paper presented an innovative verification model to improve network security in VANET. We provided a solution to the MITM problem by applying the HAVM method to prevent and reject corrupted messages. As seen in the results, HAVM is 100% efficient in detecting if a message was produced by another vehicle that has implemented HAVM. Even more, the CDR of the network tends to increase in function of the number of HAVM VANETS. Although it looks promising, the better results are achieved with a higher number of HAVM VANETS, which could be a hard task in real life. Nonetheless, the algorithm shows a good performance for a medium percentage of vehicles with HAVM.

References

1. Arif M, Wang G, Balas VE (2018) Secure VANETs: trusted communication scheme between vehicles and infrastructure based on fog computing. Stud Inform Control 27(2):235–246
2. Sakiz F, Sen S (2017) A survey of attacks and detection mechanisms on intelligent transportation systems: VANETs and IoV. Ad Hoc Netw 61:33–50
3. Lu Z, Qu G, Liu Z (2018) A survey on recent advances in vehicular network security, trust, and privacy. IEEE Trans Intell Transp Syst 20(2):760–776
4. Conti M, Dragoni N (2016) Lesyk V (2016) A survey of man in the middle attacks. IEEE Commun Surv Tutor 18(3):2027–2051
5. Ahmad F, Kurugollu F, Adnane A, Hussain R, Hussain F (2020) Marine: man-in-the-middle attack resistant trust model in connected vehicles. IEEE Internet of Things J 7(4):3310–3322
6. Stricot-Tarboton S, Chaisiri S, Ko RKL (2016) Taxonomy of man-in-the-middle attacks on https. In 2016 IEEE Trustcom/BigDataSE/ISPA. IEEE, pp 527–534

7. Hasrouny H, Samhat AE, Bassil C, Laouiti A (2017) VANET security challenges and solutions: a survey. Veh Commun 7:7–20
8. Wang J, Zhang Y, Wang Y, Gu X (2016) RPRep: a robust and privacy-preserving reputation management scheme for pseudonym-enabled VANETs. Int J Distrib Sens Netw 12(3):6138251
9. Ali I, Gervais M, Ahene E, Li F (2019) A blockchain-based certificateless public key signature scheme for vehicle-to-infrastructure communication in VANETs. J Syst Archit 99:101636
10. George SA (2020) Secure identity management framework for vehicular ad-hoc network using blockchain. Ph.D. thesis, University of Windsor (Canada)
11. Laneve C, Haveraaen M, Poletti M (2012) Concepts for C++: analysis of established and novel features
12. Sommer C, German R, Dressler F (2011) Bidirectionally coupled network and road traffic simulation for improved IVC analysis. IEEE Trans Mob Comput (TMC) 10(1):3–15
13. Lopez PA, Behrisch M, Bieker-Walz L, Erdmann J, Flötteröd Y-P, Hilbrich R, Lücken L, Rummel J, Wagner P, Wießner E (2018) Microscopic traffic simulation using sumo. In: The 21st IEEE international conference on intelligent transportation systems. IEEE
14. Varga A (2010) Omnet++. In: Modeling and tools for network simulation. Springer, pp. 35–59
15. Haklay M, Weber P (2008) Openstreetmap: user-generated street maps. IEEE Pervasive Comput 7(4):12–18

An Exploratory Analysis of COVID-19 in South America

Relationship Between Public Health Damages and Economic Impact in the Region

Santiago Pozo[1,2](✉), Génesis Carrillo[1], and Isidro R. Amaro[1]

[1] Yachay Tech University, Hacienda San José, Urcuqui 100119, Ecuador
{genesis.carrillo,iamaro}@yachaytech.edu.ec
[2] SDAS Research Group, Ibarra, Ecuador

Abstract. In the context of the COVID-19 pandemic, in the present work, we will use two main data visualization techniques, which are Cluster Analysis and Principal Component Analysis Biplot, over a dataset focused on South America; it contains variables related to both, economy and public health in South American countries; namely total cases, total tests, total deaths in public health case and GDP growth or Unemployment Rate in economic, we also considered a very important variable in this study which is the stringency index, it is the severity of the restrictions applied by each country to control the number of infected people. The mentioned analysis allowed us to lead to conclusions on which countries in South America had the best outcomes handling the pandemic in terms of public health and economic variables. By the time in which data was acquired - July 9th -, we found that Chile, Uruguay, and Paraguay could be named as the countries which better handled the pandemic in South America and will have a good economic recuperation in the upcoming years, on the other hand; Brazil, Argentina, Colombia, and Venezuela were the most affected countries due to the pandemic and will also be the ones with the slowest economic recuperation.

Keywords: SARS-CoV-2 · COVID-19 · Multivariate statistical analysis · PCA · Clustering · Biplot first section

1 Introduction

COVID-19 (Coronavirus Disease 2019) is a respiratory disease caused by the severe acute respiratory syndrome (SARS-CoV-2), which was detected in first place in December 2019 in Wuhan, China; its supposed origin are bats, and humans were infected by zoonosis; however, its real origin is still in discussion. Since then, the virus has spread to 215 countries, infected more than 13.4 million people by July 14th, 2020, killed approximately 580230 people by the same previous date, according to Worldometers [1], and put more than half of the world population in lockdown to avoid more infections and the collapse of the sanitary systems.

Even though the pandemic has come across different epicenters, being China the first, Europe after that, followed by the USA, by today, Latin America, and especially,

Y. Iano et al. (Eds.): BTSym 2020, SIST 233, pp. 266–280, 2021.
https://doi.org/10.1007/978-3-030-75680-2_31

South America, has become the epicenter of the pandemic, and the situation in the region is becoming worse every day. For example, Peru has the highest crude fatality rate in South America (371 per million), followed by Chile (370 per million) and Brazil (349 per million). However, the highest number of deaths took place in Brazil (74262), followed by Peru (12229) and Chile (7069) by July 9th, 2020. In fact, Brazil is the second most affected country in the world, having performed a lot less testing than countries like the USA, India, or the Russian Federation (first, third and fourth most affected countries in the world); this is a common factor among the countries in the region. That makes us think about the possibility of disastrous situation happening in Latin America, without even noticing it.

Therefore, some analyses are necessary to see how countries could manage the COVID-19 pandemic, mainly in the fields of economy and public health, and how they will evolve in the next months. Those analyses will make governments able to make better decisions based on previous data aroused from previous decisions. The approach made in this work tries to analyze data of public health outcomes from the countries in South America, as well as the economic outcome, trying to compare them and to visualize the effects of the pandemic in the economies of the countries in the region. Several works have been done and published related to the growing impact of the COVID-19 pandemic in different parts of the world, and also some of them have tried to analyze the impact of the virus in multiple regions of the world. Reference [2] attempts to model the COVID-19 outbreak in India, USA, China, Japan, Italy, Iran, Canada, and Germany, and to predict the impact in the upcoming months from when it was published. Also, they relate the impact of the outbreak with economic factors in India, specifically.

In addition, the study in [3] focuses on analyzing global COVID-19 data from 133 countries obtained from the Worldometer website as reported on April 17th, 2020, using machine learning techniques such as regression tree, cluster analysis, and Principal component analysis to find which covariates are related to the cumulative number of confirmed cases, if the countries are clustered with respect to the considered covariates, and if any latent factor explains the variation in the covariates. The results suggest that there are four main clusters among the countries. The first cluster consists of 8 countries in which the accumulation of cases of infection and deaths is higher. They also find that the number of tests per country variable. Therefore, they conclude that the number of tests can be vital at the individual country level, that is, at the primary level of spread, but not at the global level.

The impact of this pandemic in Latin American countries was going to be bigger than in Europe, USA, China, and others, due to the lack of preparation of most of the countries in the region in terms of their healthcare systems and their fragility [4]. In February 25th, the first case in the region was detected in Brazil [5], and just a few time later, infections were detected in all the countries in the region due to the high connectivity of São Paulo, the city where the first case was detected, and all the South American region [6].

Reference [6] shows a brief research and explanation about the impact and consequences of the arrival of the pandemic to South America. The increasing impact of the pandemic in South America and the Caribbean was studied in [7], taking data from February 26th to April 8th. Their findings about the public health outcomes are in concordance with some of our results, indicating, for example, that Chile is the country with the best numbers in tests per million/thousand, or that Brazil is the most affected country. However, they do not relate those public health variables to economic variables, which is the approximation we are presenting in the present paper.

The results of this study could guide the governments of the region in their management of the pandemic and analyze alternatives for their contingency measures by taking into consideration the results of the presented study. In other words, these results could make governments able to know if their measures are actually working or they are not.

2 Materials and Methods

2.1 Dataset

The data used for this study were retrieved from three sources which are Our World in Data [8], Worldometers [1], and the International Monetary Fund (IMF) [9]. The two first were used to acquired public health data variables, and the third source is that from which we acquired variables related to the economy of each country.

The first two sources are enormous datasets, but we have extracted the variables which are the most important for our study. Then, the public health and few economic variables extracted from the first two sources are the following: total cases per million, total deaths per million, total tests per thousand, population, population density, stringency index, extreme poverty, GDP per capita, and active cases, this last one was the only one extracted from Worldometers. Nine variables in total. Those variables are dated June 12th, when the study was performed, so we might appreciate changes if we take data from July or August. The third source was useful to extract variables related to the economy; as mentioned, this source was the IMF. The variables retrieved from this source were: GDP 2018–2019 growth, GDP 2019–2020 growth, GDP 2020–2021 growth, and unemployment rate in percentage from 2019, 2020, and 2021. Six variables in total. So, the total number of variables from both public health and economic situations from South American countries was 15 variables. Although we mentioned that the study is performed over countries from the South American region, we list them here anyway. As individuals, the countries we took for this study were the following: Argentina, Bolivia, Brazil, Chile, Colombia, Ecuador, Guyana, Paraguay, Peru, Suriname, Uruguay, and Venezuela. It is noticeable that French Guyana and the Falkland Islands were not taken into account due to the lack of data we found for those countries.

Some fields which had a lack of data were supplemented by using deductions by Principal Component Analysis from the rest of the data in the R statistical tool [10]. However, the great majority of the data used for this study was acquired from the aforementioned three sources.

2.2 Method

We divided all the data we collected into two groups in order to take advantage of them in their analysis. The first group is just public health's related variables, such as number of infected per million people, number of deaths per million people, number of tests per thousand people, and stringency index, among others. Here, we aim to analyze each country merely depending on its public health outcome and grouping them depending on that factor. On the other hand, the second group is formed by variables from both types, related to public health and related to economy; we are adding variables like GDP per capita, or the GDP growth or recession of the countries by the end of 2020.

Before the beginning of the analysis, we had to filter our data due to the large amounts of data we found on the "Our World in Data" dataset; this process was carried out in the R software. What we mean by filter the data is the process by which we just selected an exact date for all the variables in our study (July 9th), and the variables were selected manually, guided by other studies previously accomplished [3].

We used the Principal Component Analysis (PCA) and Clustering techniques to analyze and lead to conclusions from the collected data, and the same process was applied to the two data groups. First, we performed a cluster analysis to identify which countries would have a similar outcome in both scenarios, the economics and the public health, to determine the optimal number of clusters from the data, we used two different techniques, the elbow method and group dissimilarity, which in both cases provided the same result in terms of the number of clusters. After determining the optimal number of clusters, "partitioning" was performed. Then, we carried out hierarchical clustering on the same criteria to verify the results obtained with partitioning.

Later, we performed an analysis of Biplots generated from PCA in order to compare the relations between variables and individuals and their contributions to the principal factors of the model. The mainly used tool was the Principal Component Biplots. This task was performed in the statistical software R [10].

3 Results and Discussion

3.1 Public Health Situation

Clustering. First, we performed partitioning clustering, in which we used two different methods to determine the optimal number of clusters to be used. We used the elbow method and the group dissimilarity techniques.

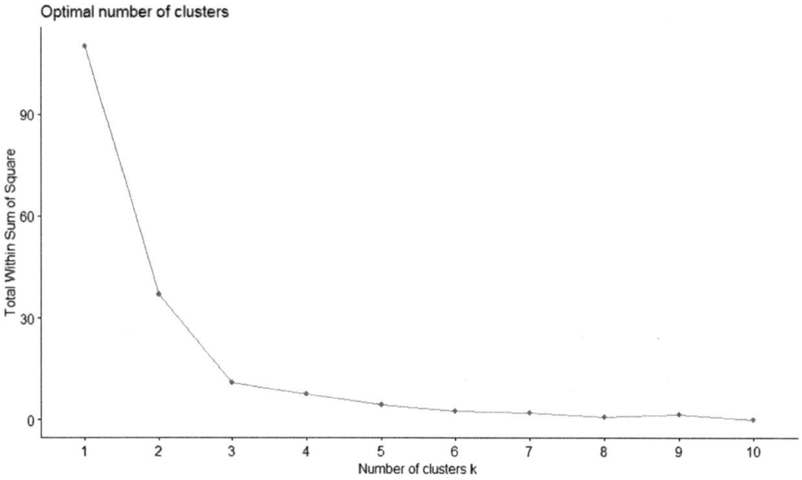

Fig. 1. Elbow method for selection of an optimal number of clusters

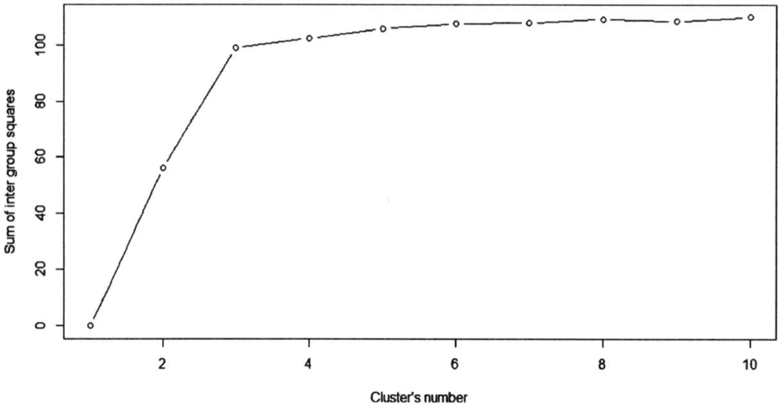

Fig. 2. Group dissimilarity for selection of an optimal number of clusters

We notice that in both plots, Fig. 1 and Fig. 2, the function stabilizes at point three; thus it is our optimal number of clusters to use in our clustering process, k-means in this case.

In Fig. 3 and Table 1, we can appreciate the clustering performed by the K-means algorithm in which we can differentiate three groups; it is noticeable that Brazil is forming a group alone, which is understandable due to the situation of Brazil being the second most affected country in the world by the pandemic. Peru and Chile are in another group, indicating that they are affected in the same proportion and the most affected countries in the region after Brazil. In the third group, we have the rest of the countries, which is reasonable knowing that Brazil, Peru, and Chile are among the world's top ten countries with the worst numbers of infected people. Probably, if we add one more group, we will have countries like Ecuador, Argentina or Colombia in it, and the rest in another group.

Table 1. Groups formed by k-means

Clusters	Countries
Cluster 1	Brazil
Cluster 2	Chile, Peru
Cluster 3	Argentina, Bolivia, Colombia, Ecuador,
	Guyana, Paraguay, Suriname, Uruguay, Venezuela

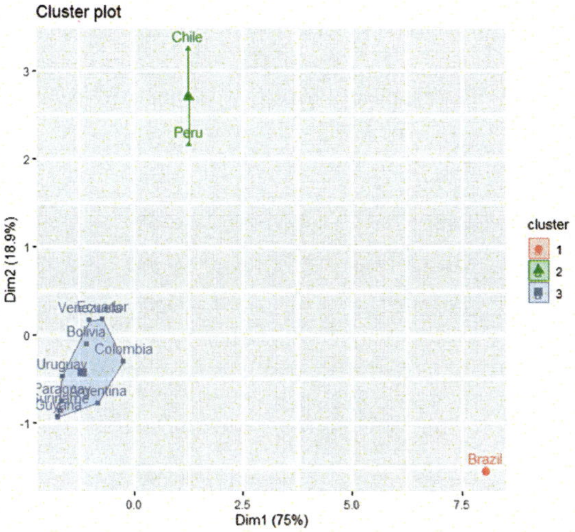

Fig. 3. K-means on public health data

Then, we did a hierarchical clustering to confirm the results obtained with partitioning. In Fig. 4, we notice that Venezuela could be separated from the rest of the countries

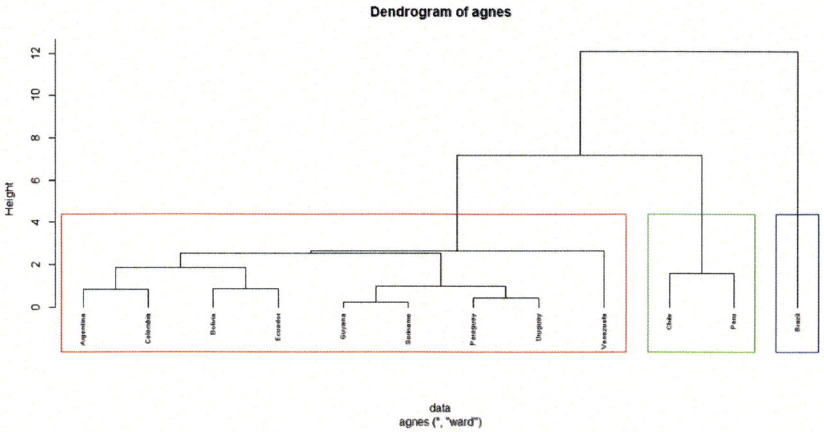

Fig. 4. Cut Hierarchical clustering

in the third group we just mentioned; however, Venezuela is there just because of the data provided by the government on tests, which is not very trustworthy. On the other hand, in Fig. 5, we can confirm the results obtained by partitioning, with three groups with the same individual in each group.

PCA and Biplot. To explain all the variability of the dataset related to public health outcomes of the countries, we just need two dimensions or components extracted from PCA; those two dimensions explain almost 94% of data variability, which is more than enough. that can be seen in Table 2.

Table 2. Dimension's cumulative percentage of variance

	Eigenvalue	Percentage of variance	Cumulative percentage of variance
comp 1	$7.50e^{00}$	$7.50e^{+01}$	75.03
comp 2	$1.89e^{00}$	$1.89e^{+01}$	93.96
comp 3	$4.48e^{-01}$	$4.48e^{00}$	98.45
comp 4	$1.048e^{-01}$	$1.048e^{00}$	99.50
comp 5	$3.80e^{-02}$	$3.80e^{-01}$	99.88
comp 6	$9.27e^{-03}$	$9.27e^{-02}$	99.97
comp 7	$2.30e^{-03}$	$2.30e^{-02}$	99.996
comp 8	$2.99e^{-04}$	$2.99e^{-03}$	99.999
comp 9	$8.55e^{-05}$	$8.55e^{-04}$	100.00
comp 10	$1.34e^{-32}$	$1.34e - 31$	100.00

Figure 5 and Fig. 6, obtained from a Principal Component Analysis, help us explain how the individual behaves and also to explain what we see in the previous clustering and

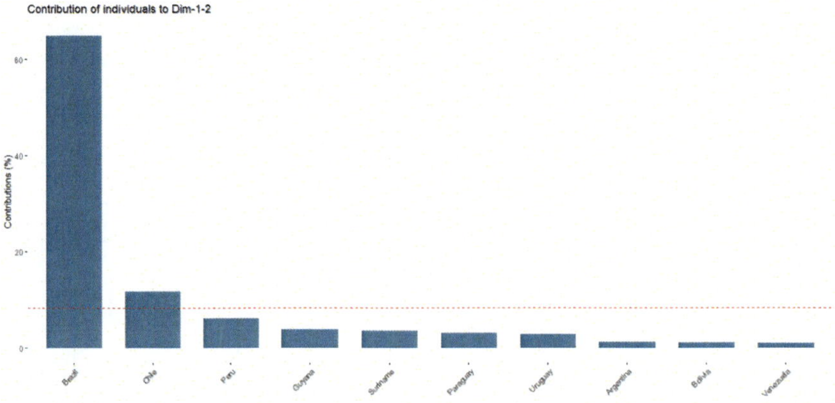

Fig. 5. Individual's contribution to 1 and 2 dimensions

in the following Biplot. In Fig. 6, we can notice that variables that contribute the most to the first two dimensions are "active cases", "total cases", and "deaths", among others. We also notice in Fig. 5 that the individual who contributes the most to the variability of data is by far Brazil, and that is why this individual forms a group alone and is that far away from the rest in the Biplots, second and third best contributors are Chile and Peru, as expected.

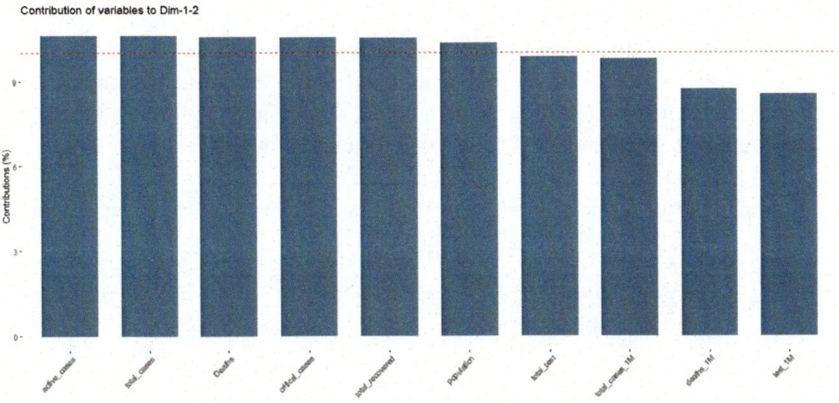

Fig. 6. Variable's contribution to 1 and 2 dimensions

It is noticeable, from Fig. 7, that countries like Uruguay, Paraguay, Suriname, or Guyana were the least impacted countries of the region in terms of public health outcomes. Although they were classified in the same group as other countries like Ecuador or Colombia, those countries have reasonable numbers in total cases or total deaths, but that is due to the lack of testing; they are inversely correlated. Although Brazil is by far the country with most cases in the region, Chile and Peru are more affected in

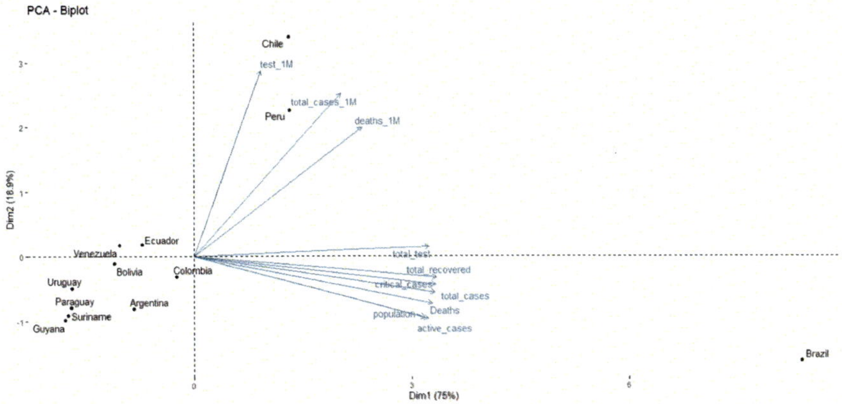

Fig. 7. Individual and variable's Biplot

terms of infections and deaths per million; in terms of test per million, both have better performances than Brazil, although they have performed a higher number of total tests.

It is important to mention, as we can see in the Biplot, that total tests and deaths have high correlations; that might be because the more test you take, the more identified deaths you have.

3.2 Economic Situation

We carried out the same steps to analyze the economic outcome through the pandemic for the countries in South America; then, for clustering, we first determine the optimal numbers of clusters to use, with the same methods, elbow and group dissimilarity. The variables added to the economic study were the GDP variance in 2019, with respect to 2018, the same with 2020 and 2021. The same for unemployment rates.

From Fig. 8 and Fig. 9, we can see that, for the same reasons mentioned before, the optimal number of clusters is five, and with that number of clusters, we performed partitioning.

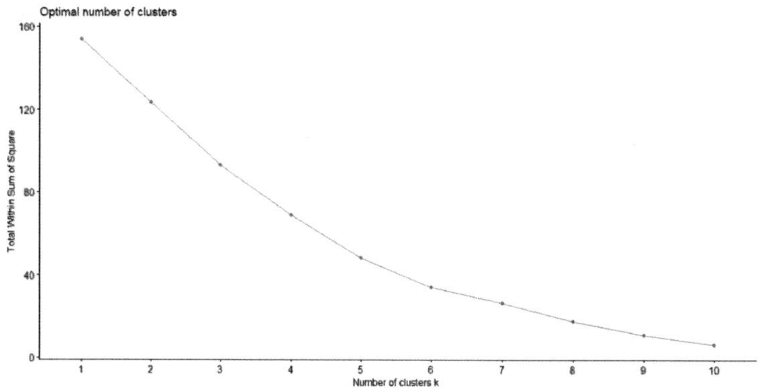

Fig. 8. Elbow method for selection of an optimal number of clusters

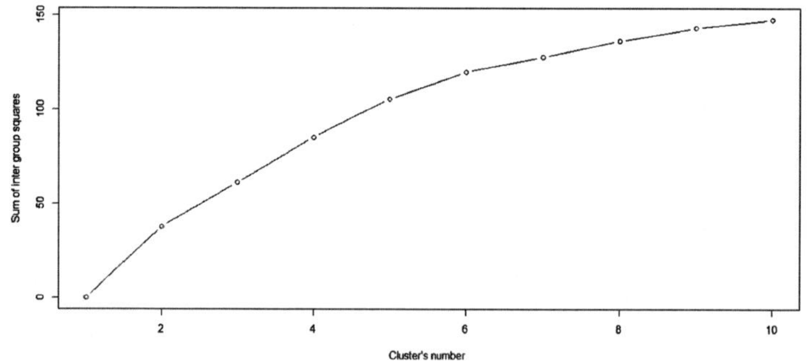

Fig. 9. Group dissimilarity for selection of an optimal number of clusters

We realize, from Fig. 10, that when economic takes part, Venezuela is alone, as expected, but we also see Brazil and Chile forming groups alone. Although Brazil could be in a group alone because of its public health situation, we cannot actually see a reason for Chile being in a group alone.

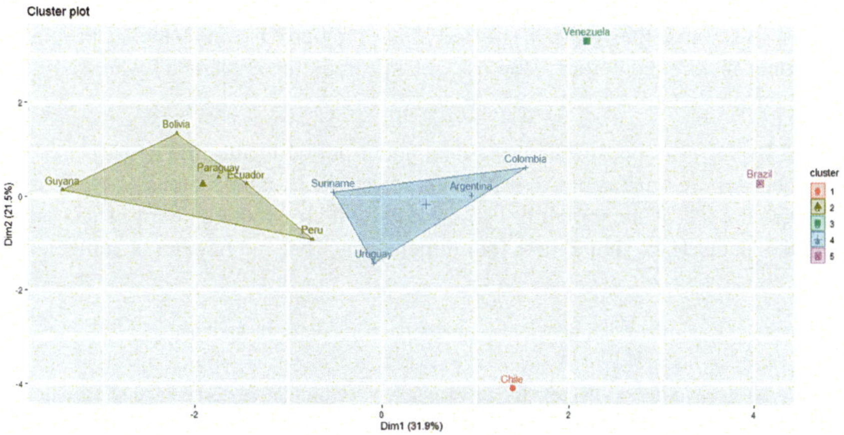

Fig. 10. K-means on public health and economic data

We can start explaining why Chile is alone in a group by looking at Fig. 11, from hierarchical clustering, which throws slightly different results from the ones obtained in partitioning. That "incongruence" is due to the goodness of fit of the data in the first two dimensions; in our partitioning plot, we see that data is plotted just over two dimensions, which explain just 53.2% of the variability of the data, that is low, and that is why hierarchical clustering and partitioning show those differences in the results. We could say that hierarchical is more accurate and take some assumptions from it.

In Fig. 11, we can differentiate five groups; we explained why Brazil and Venezuela could be in groups alone. Then, Bolivia, Ecuador, and Peru are in another group; those countries could be recognized as the ones with the worst public health outcome after

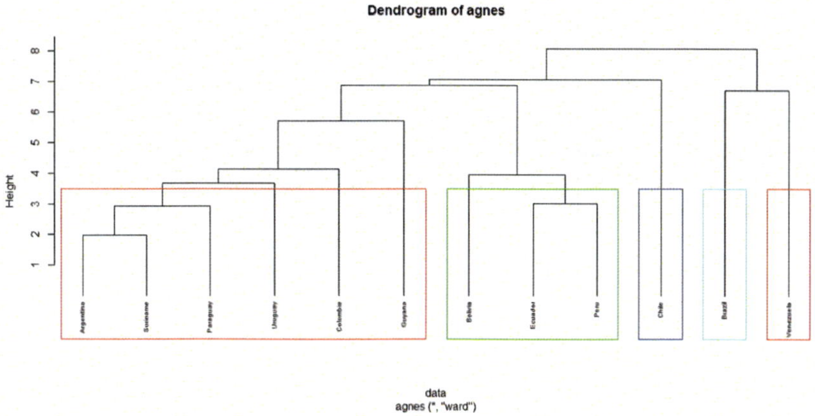

Fig. 11. Cut Hierarchical clustering on public health and economic data

Brazil, and with the worst economic outcome after Venezuela, but combining those factors and treating them as an only one. We can notice in another group countries like Argentina, Suriname, Paraguay, Uruguay, Colombia and Guyana, which can be labeled as the ones less affected in both ways, public health and economic, however, that is not equivalent to say that they are not affected at all, they are, but in a smaller proportion than the ones mentioned before. The case of Chile could be explained by analyzing the contributions in PCA.

PCA and Biplot. In opposite with the data about public health, the dataset used this time will need more dimensions to explain more variability. Dimensions one to four can explain about 85% of the variability of the data, which is enough, so to successfully analyze this dataset, we should take four dimensions. That can be seen in Table 3.

Table 3. Dimension's cumulative percentage of variance

	Eigenvalue	Percentage of variance	Cumulative percentage of variance
comp 1	4.46	31.86	31.86
comp 2	3.00	21.47	53.33
comp 3	2.39	17.13	70.47
comp 4	2.10	15.003	85.47
comp 5	0.78	5.58	91.058
comp 6	0.52	3.73	94.79
comp 7	0.34	2.49	97.28
comp 8	0.16	1.16	98.44
comp 9	0.12	0.85	99.29
comp 10	0.09	0.65	99.94
comp 11	0.007	0.05	100.00

In Fig. 12. We can see that the best contributors to the four dimensions we are analyzing are Brazil, Venezuela, and Chile, which is why they appear alone when clustering. However, it is still not very clear why Chile is alone exactly; that can be answered by looking at Table 4 and Table 5. In the second one, we can see that Chile is a great contributor to the second dimension; knowing that and looking at Table 4., we see that, among the best contributors of dimension 2, there are total cases per million, total tests per thousand, GDP 2021, and among others which are very important for Chile. So, we can name dimension 2 as the responsible for making it stay in a group alone.

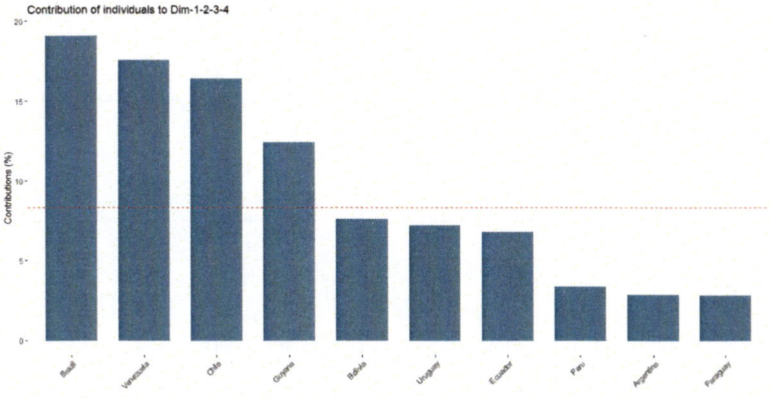

Fig. 12. Individual's contribution to 1 and 2 dimensions

Table 4. Variable's contribution to each dimension

	Dim. 1	Dim. 2	Dim. 3	Dim. 4	Dim. 5
Total cases per million	2.60	15.56	13.24	0.006	9.37
Total deaths per million	1.70	7.67	26.36	0.65	0.016
Total tests per thousand	5.13	15.86	0.18	6.30	7.78
Stringency index	2.32	11.38	5.07	14.93	0.68
Population	11.89	0.24	2.47	15.23	0.013
Population density	1.68	0.52	10.55	5.91	59.65
GDP per capita	8.97	8.13	5.20	8.17	0.036
Extreme poverty	0.008	11.99	17.98	0.14	7.89
GDP 2019	3.189	10.77	0.58	18.11	3.84
GDP 2020	7.407	0.203	2.65	10.01	2.44
GDP 2021	4.25	16.82	0.17	11.40	4.64
unemRt 2019	13.036	0.005	11.82	4.90	0.324
unemRt2020	18.66	0.80	2.28	1.30	0.74
unemRt2021	19.12	0.036	1.397	2.87	2.54

By looking at our Biplots, especially in Fig. 13, we can lead to many conclusions, such as the inverse correlation, as expected, between stringency index and the GDP for each year, that is because the stricter is the contingency measures, the more severe is the damage to the economy, and we notice it because of the angle of approximately 180° that their respective vectors form. It is also remarkable that the stringency index does not show a relation with the most important public health variables such as total deaths or total cases; that is, they are uncorrelated because the degree that their vectors form on the Biplot is approximately 90°. Also, something more or less obvious is the inverse

Table 5. Individual's contribution to each dimension

	Dim. 1	Dim. 2	Dim. 3	Dim. 4	Dim. 5
Argentina	1.85	$6.24e^{-04}$	9.60	$1.26e^{00}$	1.627
Bolivia	9.82	$5.40e^{00}$	12.84	$5.63e^{-04}$	21.052
Brazil	33.68	$2.42e^{-01}$	3.61	$3.27e^{+01}$	0.435
Chile	3.91	$5.059e^{+01}$	2.32	$1.009e^{+01}$	10.825
Colombia	4.82	$1.11e^{00}$	0.0055	$1.83e^{00}$	10.521
Ecuador	4.22	$2.121e^{-01}$	22.478	$3.88e^{00}$	40.672
Guyana	24.001	$5.072e^{-02}$	5.998	$1.283e^{+01}$	2.752
Paraguay	6.16	$4.88e^{-01}$	1.859	$2.87e^{-02}$	0.328
Peru	1.108	$2.55e^{00}$	11.31	$1.64e^{-01}$	0.011
Suriname	0.535	$2.02e^{-02}$	5.62	$9.68e^{-01}$	1.33
Uruguay	0.014	$6.12e^{00}$	23.49	$5.42e^{00}$	5.199
Venezuela	9.83	$3.31e^{+01}$	0.82	$3.07e^{+01}$	5.241

correlation between unemployment rates and the GDPs. Besides, we see that extreme poverty is directly correlated with the stringency index, which makes sense.

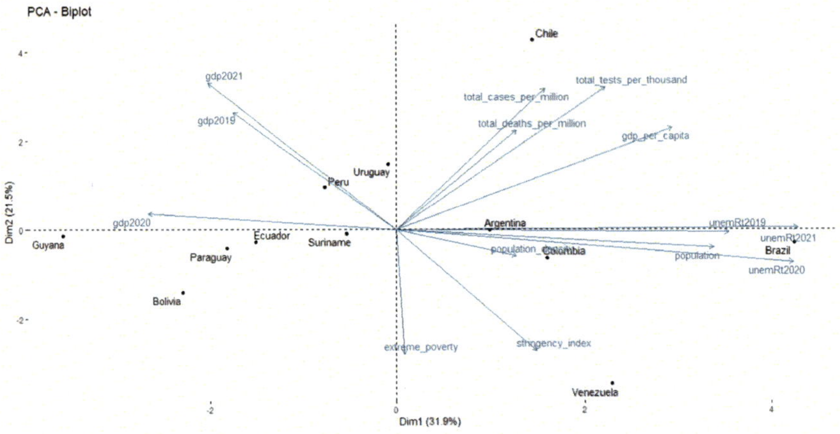

Fig. 13. Individual and variable's Biplot

A very remarkable fact is that variables like the cases per million, tests per thousand, or deaths per million are not correlated with the GDP's, and although they are correlated with unemployment rates, those correlations are almost nonexistent.

For the variables related to public health, we can replicate the results from the previous analysis from the public health only dataset. For the economic variables, we can remark Venezuela, Brazil, Argentina, and Colombia will be the countries with the slowest economic recuperation in 2021 after the pandemic; however, countries like Chile, Suriname, Peru, or Uruguay will recover faster than the others, with the last two mentioned as the fastest.

In the same way as contrasting countries with their economic recoveries, we can contrast them with their unemployment rates by 2021, being the highest those from Brazil, Venezuela, Argentina, and Colombia. And the ones with faster recovery are Chile, Suriname, Peru, and Uruguay.

4 Conclusions

Stringency index should be a variable that has an inverse correlation with the total of cases and total of deaths; however, in our region, that is not the case, and it has no correlation with that kind of variables, and also those variables did not actually have an appreciable correlation with GDPs. Knowing it, we can lead to the conclusion that the stringency index has not had a real effect on the public health outcome in our region, which might be because of the necessity of people to perceive some income through the informal economy, which is very recurrent in our region. Nonetheless, and unfortunately, the stringency index actually had a severe effect on unemployment rates and growth of GDP, leading our region to an economic recession.

Thus, it is noticeable that countries like Uruguay and Paraguay stand out for their handling of the pandemic in the region. On the other hand, Chile, Suriname, or Uruguay will be the countries with faster economic recuperation than others. Although the stringency index seemed to work in countries like China or in Europe, it does not seem to work in South America, just because our conditions are different, there are not a lot of formal employment places, and people just need to perceive incomes in order to survive. We need to look further into the contingency plans applied in South America and look for alternatives, or the situation could just become worse each day.

References

1. Coronavirus Update (Live): 69,134,368 Cases and 1,573,157 Deaths from COVID- 19 Virus Pandemic – Worldometer. https://www.worldometers.info/coronavirus/
2. Kumar A, Rani P, Kumar R, Sharma V, Purohit SR (2020) Data-driven modelling and prediction of covid-19 infection in India and correlation analysis of the virus transmission with socio-economic factors. Diabetes Metabolic Syndrome 14(5):1231–1240. https://europepmc.org/articles/PMC7347321
3. Khan M, Hossain A (2020) Countries are clustered but number of tests is not vital to predict global covid-19 confirmed cases: a machine learning approach
4. The Lancet (2019) The unfolding migrant crisis in Latin America. The Lancet 394(10213):1966. https://linkinghub.elsevier.com/retrieve/pii/S0140673619329344
5. Rodríguez-Morales AJ, MacGregor K, Kanagarajah S, Patel D, Schlagenhauf P (2020) Going global – Travel and the 2019 novel coronavirus. Travel Med Infectious Dis 33:101578. https://linkinghub.elsevier.com/retrieve/pii/S1477893920300284
6. Rodríguez-Morales AJ, Gallego V, Escalera-Antezana JP, Méndez CA, Zambrano LI, Franco-Paredes C, Suárez JA, Rodriguez-Enciso HD, Balbin-Ramon GJ, Savio-Larriera E, Risquez A, Cimerman S (2020) COVID-19 in Latin America: The implications of the first confirmed case in Brazil. Travel Med Infectious Dis 35:101613. https://linkinghub.elsevier.com/retrieve/pii/S1477893920300806

7. Simbaña-Rivera K, Gomez-Barreno L, Guerrero J, Simbaña-Guaycha F, Fernández R, López-Cortés A, Lister A, Ortiz-Prado E (2020) Interim Analy- sis of Pandemic Coronavirus Disease 2019 (COVID-19) and the SARS-CoV-2 virus in Latin America and the Caribbean: Morbidity, Mortality and Molecular Testing Trends in the Region. preprint, Public and Global Health. https://medrxiv.org/lookup/doi/10.1101/2020.04.25.20079863
8. Roser M, Ritchie H, Ortiz-Ospina E, Hasell J (2020) Coronavirus Pandemic (COVID-19). Our World in Data. https://ourworldindata.org/coronavirus
9. International Monetary Fund. https://www.imf.org/external/datamapper/
10. R: The R Project for Statistical Computing. https://www.r-project.org/

Health Impact Analysis of COVID-19 in Ecuadorian Provinces

A Statistical Approach

M. C. Sabando, Darwin Tallana-Chimarro(✉), and Isidro R. Amaro

Yachay Tech University, Hcda. San José s/n Proyecto Yachay, Urcuquí,
Imbabura 100106, Ecuador
{maria.sabando,darwin.tallana,iamaro}@yachaytech.edu.ec

Abstract. In this work, we analyze the health impact of COVID-19, focusing on the number of deaths per million inhabitants and the number of cases per million inhabitants in the 24 Ecuadorian provinces. We study this impact by considering the following aspects: the number of hospital beds per 100k inhabitants, number of intensive care units per 100k inhabitants, population density, and coloring of their cities. We use official data published by the Ministry of Public Health of Ecuador until May 28th and June 12th, 2020, and the Ecuadorian Secretariat for Risk Management (SNGRE). To carry out the analysis, we use multivariate statistical techniques: Cluster Analysis and Canonical Biplot. We found that provinces with high population density and a relatively small number of hospital beds and intensive unit care, like Pichincha and Tungurahua, had relatively good management of the pandemic until June 12th, while provinces with more advantages with respect to these variables, Morona Santiago, Galapagos, and Napo, did not show a performance better than other provinces. Finally, we encountered that the coloring system of the provinces alone does not seem to be a determinant measure in the growth rate of deaths and cases. The implemented analysis allows a periodic evaluation of the status of each province and to determine the most vulnerable provinces concerning the studied variables as well as the most affected by the virus.

Keywords: Ecuadorian provinces · COVID-19 · Canonical Biplot · Cluster analysis · Classification · Coloring

1 Introduction

In December 2019, the new coronavirus, SARS-CoV-2, triggered an outbreak in Wuhan, China, [1]. In less than three months, on March 11th, 2020, the World Health Organization (WHO) declared COVID-19 (the disease associated with SARS-CoV-2 virus) a pandemic. Nowadays, the virus continues affecting many countries around the world by changing the lifestyle of their people and affecting their economies.

Like other countries, Ecuador is also being affected by the virus. On May 4th, Ecuador's government proposed the coloring of each city [2], which consists of labeling each city with red, yellow, or green. Red means to give maximum restrictions, yellow

Y. Iano et al. (Eds.): BTSym 2020, SIST 233, pp. 281–292, 2021.
https://doi.org/10.1007/978-3-030-75680-2_32

means moderate restrictions, and green means relaxed restrictions. On May 28th, 173 cities out of 221 were in red and around 30 of them in yellow (13.57% of the total), while on June 12th, only 88 were in red and 131 in yellow (59.28% of the total) [3, 4]. This is an increment of more than 50% of the total number of cantons.

In this work, we study the health impact of COVID-19 on Ecuadorian provinces, focusing on the number of deaths per million inhabitants and the number of cases per million inhabitants through the help of multivariate statistical techniques. We perform classifications of Ecuadorian provinces by using multivariate techniques: Cluster Analysis and Canonical Biplot. Our objectives are to characterize the groups obtained in the classifications and determine the most and least vulnerable and affected provinces on both dates. Besides, due to the vast number of cantons that changed their coloring during May 28th and June 12th, we make a comparison of the groups created on both dated and analyze whether the change of coloring was a determinant measure to control the spread of the virus. This work gives a straightforward way of determining the status of each province with respect to the studied variables, which allows local and periodic control of the evolution of the pandemic. The paper's structure is as follows: in Materials and Methods, we detail the used set of data and give a brief description of Cluster Analysis and Canonical Biplot methods. Next, we introduce our results, where we present our classification plots. Finally, we show our discussion and conclusions.

2 Materials and Methods

This section describes the data, the statistical methodology, and the procedure applied for this work.

2.1 Data Description

We obtained data from each Ecuador province related to the COVID-19 virus from [5, 6]. The information is about total cases per 100k inhabitants, total deaths per 100k inhabitants, hospital beds per 100k inhabitants, intensive unit care per 100k inhabitants, population density, and coloring of the cities of each province. We divide the variables into two groups. The first group gives a notion of the preparedness or vulnerability of a province for a pandemic, while the second one represents the pandemic's effect on the provinces' population.

The first group has the variables Hospital Beds per 100k inhabitants (+), Intensive Care Units (+), and Population Density (−), while the second group has the variables Total Cases per 100k inhabitants and Total Deaths per 100k inhabitants. The values for this data were taken up to May 28th and June 12th. The symbol (+) means that while the variable takes larger values, it is better for affronting the pandemic; (−) is the opposite. The dates were chosen due to the significant increase of color change from red to yellow in the coloring system. The information about the color of the cantons of the provinces was obtained from [3, 4], whose source is [6].

2.2 Statistical Methodology

We present a brief description of the methods used in this work, emphasizing the properties of Canonical Biplot that will be useful to interpret our results.

Cluster Analysis. Cluster analysis [7] are techniques that look for patterns in data sets to group observations into clusters without any prior knowledge of the clusters. Many of these techniques are based on measures of similarity between the observations, such as correlation, cluster variability, and preliminary choice of centroids.

Canonical Biplot. The Canonical Biplot Multivariate Statistical Technique [8–13] is a complementary method to MANOVA that includes all Biplot Methods characteristics [10]. This technique allows, among other things, to search and establish differences and similarities between the groups. To describe the theoretical approach of this technique, let us consider a data matrix X of n individuals by p variables and let us assume that the n individuals are divided naturally (a priori) into K separate groups with n_k individuals, with $k = 1, 2, \ldots, K$ and $n = n_1 + \cdots + n_k$. The Canonical Biplot technique allows constructing a simultaneous representation of the rows (centroids) and columns (variables) of the matrix $\overline{X}$ of the groups' means so that the groups are separated with maximal discriminatory power, i.e., represented in the space spanned by the first canonical variables. The method considers the following matrix of the weighted groups' means.

$$Y = D_n^{1/2} \bar{X} W^{-1/2} \tag{1}$$

Its generalized singular value decomposition is given by (2).

$$Y = U D_\lambda V' \tag{2}$$

Where D_n is the diagonal matrix having the groups sample sizes.

$$W = \frac{1}{n-k}\left(X'X - \bar{X}' D_n \bar{X}\right) \tag{3}$$

(3) is the 'withing-groups' matrix. From (1) and (2), we obtain (4).

$$\bar{X} = D_n^{-1/2} Y W^{-1/2} = D_n^{-1/2} U D_\lambda V' W^{-1/2} \tag{4}$$

This decomposition allows for the construction of a biplot representation for the matrix of group means, taking the first r columns of G as row markers and the first r columns of H as column markers, given (5) and (6).

$$G = D_n^{-1/2} U D_\lambda \tag{5}$$

$$H = W^{-1/2} V \tag{6}$$

Then:

$$\bar{X} = G H' \tag{7}$$

The row markers were represented as points, and the column markers as vectors in a scatter diagram. The length of the vector representing a variable shows its variance. The larger the vector, the larger the variance. The similarity between groups can be measured as an inverse function of its distance on the graph. The angle between variables can be interpreted as an approximation of its correlation. Confidence circles are added to the group's markers in such a way that the projections of the circles onto the direction representing a variable approximate a confidence interval. Checking the overlapping of two projections allows us to establish the significance of the difference between groups over a particular variable. The ratio of the circle is calculated as $t_{n-k,\alpha}/\sqrt{n_k}$, where t_{n-k} is the critical point of a Student's t distribution with (n−K) degrees of freedom for a significance level of α. By projecting an individual on a vector, it is possible to know the values that the individual takes in a variable. If the projection falls on the origin, the observation takes a value corresponding to the average of the respective variable. Finally, the variables with vectors more parallel to a straight line connecting the centers of two groups will be the ones that influence the most in the separation of those groups. Definitions and algebraic derivations of the Canonical Biplot Analysis are discussed in [9]. Recent works on which the Canonical Biplot has been used include [14, 15].

Procedure. For each group of variables, we perform Cluster analysis and Canonical Biplot. Cluster analysis is used to classify the provinces. For Cluster analysis, we standardize the variables, then use hierarchical clustering techniques and choose the method yielding the highest cophenetic correlation coefficient, which, for the used data, is Average Linkage. Next, Canonical Biplot is implemented to determine if the obtained groups are significantly different among them and to characterize them. We make the cut for groups using the large changes in distances approach, while also considering information obtained from the Canonical Biplot analysis such as significant group differentiation, high total variance explained, and high groups and variables representation. We use the software INFOSTAT [16] to perform the hierarchical clustering and MultBiplot [17] to get the Canonical Biplot analysis.

3 Results

In this section, we present our results based on the properties of the Canonical Biplot. The percentage of the total variance explained by each factorial axis is shown in the plots obtained by using MultBiplot.

3.1 First Group of Variables

These variables' averages are 1.90 for the total number of intensive care units (ICUs), 25.47 for the total number of hospital beds, and 85.24 for the density population. The hierarchical clustering process with a cophenetic correlation coefficient of 0.845 and a cut at 2 yields the following three groups: GR1, whose elements are Napo, Morona Santiago, and Galapagos; GR2, formed by Pichincha, Tungurahua, and Guayas; and GR3, with Zamora Chinchipe, Pastaza, Carchi, Sucumbios, Esmeraldas, Orellana, Loja, El Oro, Imbabura, Cotopaxi, Chimborazo, Los Rios, Manabi, Cañar, Bolivar, Santa

Elena, Santo Domingo, and Azuay as elements. When we apply the Canonical Biplot method, we get 25.6536 for the Wilk's Lambda value (p-value: $5.9955e^{-12} < 000.1$), which means that there is a clear difference between the means of the groups.

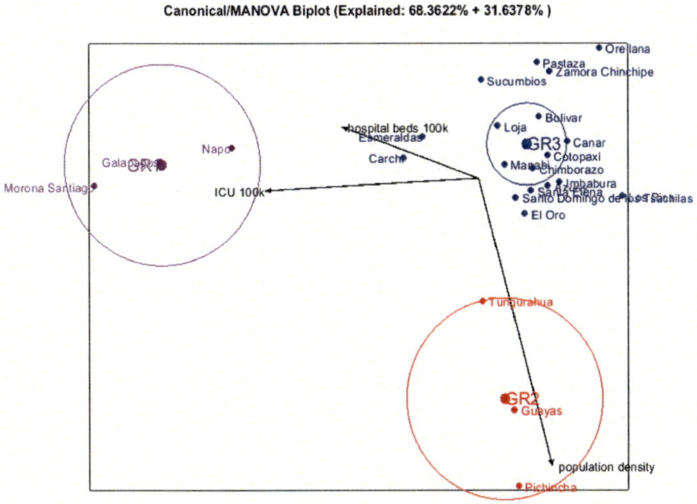

Fig. 1. Canonical Biplot for provinces on the first group of variables. GR1 in purple, GR2 in red, and GR3 in blue.

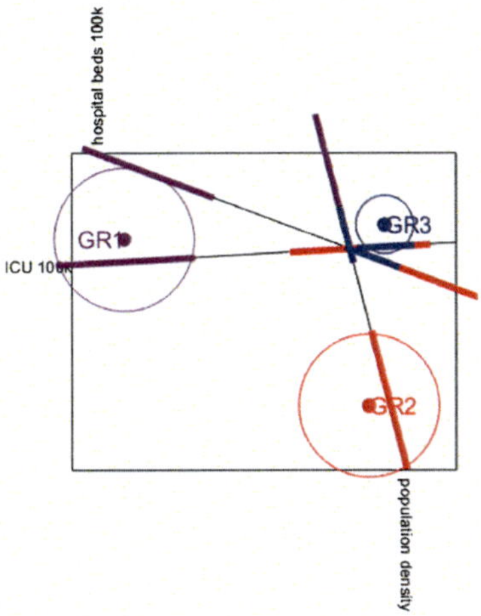

Fig. 2. Projections of the confidence circles for the first set of variables.

We see in Fig. 1 the Canonical Biplot for the first group of variables. We can see graphically that the regions of confidence of the groups are significantly different. From the angles formed by the axis, we notice that there is a low correlation between the population density and the number of ICUs/hospital beds. On the other hand, there is a high correlation between the number of ICUs and hospital beds. Thus, a province with a high number of hospital beds would also have a high number of ICUs.

To characterize the groups, we project the provinces and the means of the groups on the variables vectors in Fig. 1. We see that GR1 is the group with the highest number of ICUs and hospital beds and has provinces with a low-density population; GR2 has the highest population density and a small number of hospital beds and ICUs, and GR3 has values below but near the average for all variables. Due to this, we shall say that GR1 and GR2 are the less and the most vulnerable groups, respectively, with respect to the studied variables. Additionally, by analyzing each province, we see that Pichincha is the province with the highest population density, in contrast to Pastaza. Morona Santiago has the highest number of ICUs and hospitalization beds per 100k inhabitants, while Los Rios and Pichincha have the lowest number of ICUs and hospital beds per 100k inhabitants, respectively. Moreover, we see in Fig. 1 that the separating variable for GR1 and GR3 is the number of ICUs, while GR2 and GR3 are separated by population density.

Furthermore, we can see in Fig. 2 that GR2 and GR3 are significantly different in their population density, but not in their number of hospital beds and ICUs. In the case of GR1 and GR3, they are not so different about their population density but have marked differences in their number of hospital beds and ICUs.

3.2 Second Group of Variables

May 28th. The variables total number of confirmed cases per 100k inhabitants and total number of deaths per 100k inhabitants have an average equal to 130.15 and 16.48, respectively. For the data corresponding to these variables up to May 28th, the clustering

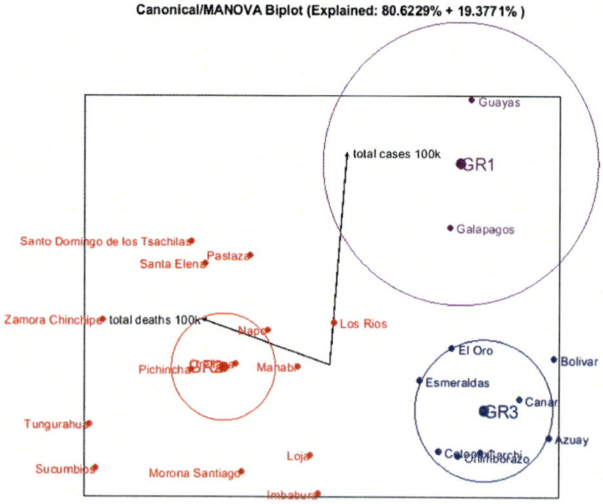

Fig. 3. Canonical Biplot for provinces on the second group of variables on May 28th. GR1 in purple, GR2 in red, and GR3 in blue.

process, with a cophenetic correlation coefficient of 0.743 and a cut at 2, yields the following groups: GR1 formed by Guayas and Galapagos; GR2 with Zamora Chinchipe, Santa Elena, Santo Domingo, Pastaza, Pichincha, Orellana, Manabi, Napo, Los Rios, Tungurahua, Sucumbios, Morona Santiago, Loja, and Imbabura; and GR3 with Esmeraldas, El Oro, Bolivar, Cotopaxi, Chimborazo, Carchi, Cañar, and Azuay.

In the Canonical Biplot, shown in Fig. 3, we observe that the chosen groups are significantly different. The Wilk's Lambda in this case is 18.9971 (p-value: $7.9837e^{-09}$ <000.1). The variable with the highest variance is the number of total confirmed cases, and the variables are not correlated. Also, we see that GR1 is the group with the highest number of cases per 100k inhabitants and a total number of deaths under the average; GR2 has the highest number of deaths and cases near the average; lastly, GR3 has the least number of deaths and cases. Due to this, we shall consider GR3 the least affected group of provinces. In addition, Zamora Chinchipe and Tungurahua stand out as the provinces with the highest number of total deaths. In contrast, Azuay has the lowest value in this variable. Also, Guayas has the highest number of reported cases. Hence, we see that Chimborazo, Cotopaxi, Azuay, and Carchi are the least affected provinces up to this date, while the more affected ones are Guayas, Zamora Chinchipe, and Tungurahua. We have that the variable total cases per 100k inhabitants is responsible for separating GR1 and GR3, while the variable number of deaths separates GR2 from GR3.

Furthermore, from Fig. 4, we have that GR2 and GR3 are not significantly different in their number of total confirmed cases but differ in the number of total deaths. Also, we see that GR1 shows similarities with GR2 and GR3 on the number of deaths and differences from them on the number of total cases.

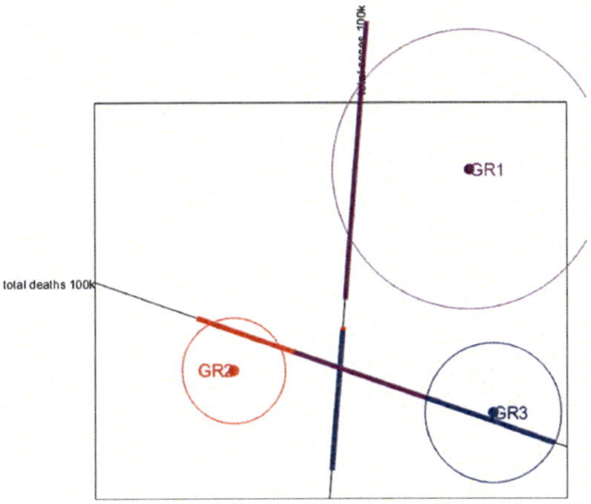

Fig. 4. Projections of the groups, May 28th.

June 12th. The variables total number of confirmed cases per 100k inhabitants and total number of deaths per 100k inhabitants have an average equal to 188.63 and 20.36, respectively. By doing cluster analysis, with a cophenetic correlation coefficient of 0.841 and a cut at 1.80, we get the following groups: GR1 with Santa Elena, GR2 has Zamora

Chinchipe, Pastaza, Orellana, Galapagos, Santo Domingo de los Tsachilas, Napo, El Oro, Los Rios, Esmeraldas, and Bolivar, and GR3 is formed by Chimborazo, Imbabura, Tungurahua, Pichincha, Cotopaxi, Cañar, Loja, Sucumbios, Carchi, Morona Santiago, and Azuay.

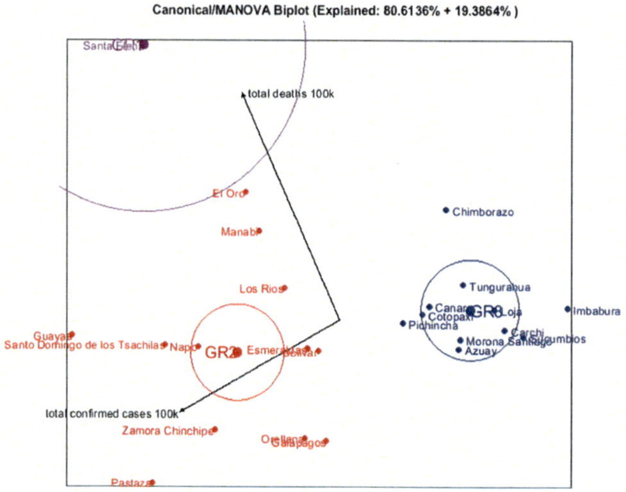

Fig. 5. Canonical Biplot for provinces on the second group of variables on June 12th. GR1 in purple, GR2 in red, and GR3 in blue.

The graph of the Canonical Biplot in Fig. 5 shows that the groups are significantly different. This is reflected in the Wilk's Lambda value, which is 20.3064 (p-value: $3.3706e^{-09} < 000.1$). The variable with the highest variance is total number of deaths per 100k inhabitants. Additionally, there is not a correlation between the variables. Regarding group characteristics, we see in Fig. 5 that GR1 is the group with the largest number of total deaths per 100k inhabitants, GR2 has values of total cases and total deaths over the mean, and GR3 has values of total cases and total deaths under the average. So, we shall say that GR3 is the least affected group of provinces. By considering the projections of the individuals on the axes, we have that the provinces with more registered deaths per 100k inhabitants are Santa Elena, El Oro, Guayas, and Manabi; and the provinces with a smaller number of confirmed deaths per 100k inhabitants are Galapagos and Orellana. Moreover, the provinces with more registered cases per 100k inhabitants are Guayas and Pastaza, while the ones with fewer cases are Sucumbios and Imbabura. Hence, the more affected provinces for the considered variables are Guayas, Santa Elena, El Oro, and Manabi, and the less affected are Imbabura, Sucumbios, and Carchi during this date. In addition, in Fig. 6, we can see that the variable that differentiates GR1 from GR2 and GR3 the most is the number of deaths, while the number of cases separates GR2 and GR3. We also observe that GR1 and GR2 are not significantly different in the number of total cases, while GR2 and GR3 are not different in the number of total deaths. On the other hand, GR1 differs from GR2 and GR3 in the total number of deaths per 100k inhabitants.

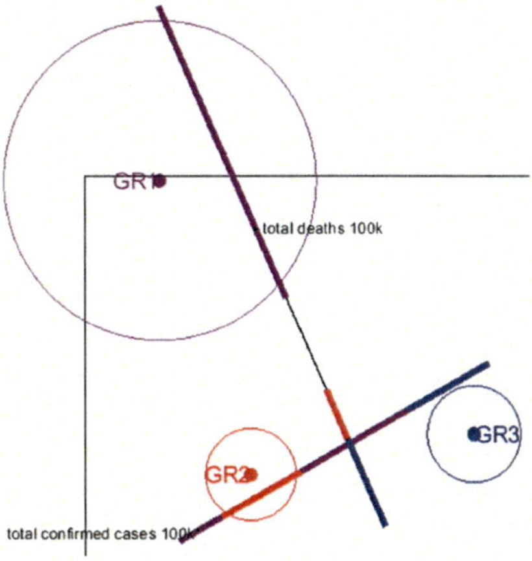

Fig. 6. Projection of the groups, June 12th.

4 Discussion

We compare the groups obtained on May 28th and on June 12th to analyze the effect that the coloring system has had in the considered period, and we establish the interaction between our two sets of variables by determining how much the vulnerability or strength of a given province influences the effects of the pandemic. From May to June, the variable with the largest variance value changes from total cases to total deaths. The lack of correlation between the variables is unchanged from one date to the other. Hence, based on the data, there is no evidence that a high number of cases are related to a high number of deaths.

4.1 Comparison from May 28th to June 12th

We analyze the behavior of some relevant provinces. We have that Guayas stays as the province with most confirmed cases per 100k inhabitants on both dates, and the province with the highest number of deaths per 100k inhabitants' changes from Zamora Chinchipe on May 28th to Santa Elena on June 12th. An interesting case is Imbabura, which goes from a high number of cases and deaths, on May 28th, to the lowest number of cases and deaths, on June 12th. Also, Carchi, Chimborazo, and Cotopaxi remain in the least affected group of provinces on both dates, although Chimborazo has increased its number of deaths per 100k hab on June 12th (see Fig. 3, Fig. 5). By considering the elements of the less affected group (GR1 on May 28th, Fig. 3), we see that El Oro, Esmeraldas, and Bolivar have moved to the group with more confirmed cases (GR2 on June 12th, Fig. 5). Additionally, we see that provinces like Morona Santiago, Sucumbios, and Loja have passed to the less affected group on June 12th, which is an improvement of their situation compared to the rest of the provinces.

4.2 Comparison Between the First and Second Group of Variables

By taking into consideration the classification presented in Fig. 1, one might expect that the provinces of the group GR2, Tungurahua, Guayas, and Pichincha, have a large number of cases and deaths per 100k inhabitants since they are elements of the most vulnerable group. On the other hand, Morona Santiago, Napo, and Galapagos (GR1 in Fig. 1) should have lower numbers of cases and deaths per 100k inhabitants because of their low population densities and the availability of beds in their health systems. In the following paragraphs, we discuss this statement.

May 28th. On this date (see Fig. 3), Tungurahua has a large number of deaths, and Guayas the largest number of cases. However, we see that Pichincha maintains cases under the average and a relatively small number of deaths with respect to Tungurahua. In the case of Galapagos, Morona Santiago, and Napo, they had a high number of cases or deaths per 100k inhabitants.

June 12th. We see in Fig. 5 that the situation of Tungurahua is better. It is classified as the least affected group (GR3), just like Pichincha. In the case of Guayas, it keeps its condition, being the province with more confirmed cases. Morona Santiago improves its status by decreasing its number of cases and deaths with respect to the mean of all the provinces. On the other hand, Napo and Galapagos maintain the same condition as on May 28th. Thus, despite the vulnerability of Pichincha and Tungurahua, these provinces have managed the pandemic relatively well. On the other hand, Napo and Galapagos did not show better results in the control of the pandemic despite their advantages.

4.3 Coloring

We discuss the influence of the coloring in the health impact of the virus on provinces mentioned in Sects. 4.1 and 4.2.

Coloring with Respect to Less and More Affected Provinces. Some of the provinces less affected by the virus, like Carchi and Cotopaxi, had more than half of their cantons in red until June 12th. However, it is also remarkable that another less affected province, Chimborazo, has more than fifty-per-cent of its cities in yellow. Another important case to discuss is Santa Elena, which has three cantons. This province changed from one canton in red (Salinas) on May 28th to zero on June 12th. Nevertheless, Salinas is the least populated city of Santa Elena. So, although the change of color in Salinas could contribute to the increase in the number of deaths, there were more factors likely involved in the sudden increase in deaths in this province.

Coloring with Respect to Less and More Advantages Provinces. Although in Pichincha, its cantons Cayambe, Mejia, Rumiñahui, and Quito changed their colors from red to yellow between May 28th and June 12th, we noticed that it did not change its status. In the case of Tungurahua, we see that it improved its conditions despite having just two of its nine cantons in red. Also, Morona Santiago improved its situation, although there were no changes in the coloring of its cantons during the dates of study. Finally, Galapagos does not change its status, even though one of its cantons changed its color from May

28th to June 12th. Therefore, we see from Section 4.2 that coloring does not have a clear role in the health impact of the virus during the dates of the study. To clarify this, it could be of interest to consider a larger period for the research or to examine more aspects of the provinces like poverty, isolation, or inequality. This study is of interest since it allows a periodic evaluation of the status of each province. In this way, it is possible to know which are the less and more affected provinces on a specific date. Additionally, it supports the process of decision making to contain the spread of the pandemic, for example, by giving a guideline to apply more preventive measures (e.g., taking more tests) in the advantaged provinces or adding more hospital beds and intensive care units in the less advantaged provinces. Moreover, it helps to clearly visualize the provinces that have been more affected by the virus and point their weaknesses. By determining these factors, it is possible to establish the provinces that need more support and in what aspects they should be supported.

5 Conclusion

In this exploratory statistical analysis of COVID-19 data, we have found that provinces with a large population density, like Pichincha, Tungurahua, and Guayas, are also locations with fewer hospital beds and ICUs than the average of all the provinces. Therefore, they were classified as the least advantaged provinces concerning our first set of variables. On the other hand, provinces with a large number of hospital beds and ICUs per 100k inhabitants, like Morona Santiago, Napo, and Galapagos, have low population densities; so, they were considered the most advantaged provinces.

We also have found that provinces in this last group do not have lower values on the second group of variables than the rest of the provinces. In the case of provinces with fewer advantages, such as Pichincha, Tungurahua, and Guayas, the situations vary. Pichincha and Tungurahua were kept in groups with cases and deaths near the averages and even under these values. Guayas shows the highest number of total confirmed cases on both dates. That is, variables showing the advantages a province had, a priori, appeared not to be a determining factor in the containment of the pandemic or the casualties from the illness with the passing of time. Additionally, the coloring system of the provinces alone does not seem to be of influence in the growth rate of deaths and cases. So, there are more factors likely to affect the growth rate of the number of cases and deaths. This analysis allows a periodic evaluation of the status of each province and to determine the most and least vulnerable provinces concerning the studied variables as well as the most and least affected by the virus.

References

1. World Health Organization: Q&A On Coronaviruses (COVID-19) (2020). https://www.who.int/emergencies/diseases/novel-coronavirus-2019/question-and-answers-hub/q-a-detail/q-a-coronaviruses
2. El Comercio Quito Tendrá Semáforo Rojo Desde El 4 De Mayo, Pero Se Amplían Horarios Para Taxis, Compras y Entregas a Domicilio. https://www.elcomercio.com/actualidad/quito-semaforo-rojo-mayo-distanciamiento.html

3. El Universo Once Cantones Más Cambiarán De Semáforo Rojo a Amarillo Entre Mañana y El 1 De Junio. https://www.eluniverso.com/noticias/2020/05/25/nota/7851258/once-cantones-cambiaran-semaforo-rojo-amarillo-manana-1-junio
4. cantones de Ecuador están en semáforo amarillo; 88 continúan en rojo. https://www.elcomercio.com/actualidad/cantones-ecuador-semaforo-amarillo-coronavirus-1.html
5. Ministerio de Salud Pública del Ecuador: Coronavirus COVID-19 (2020). https://www.salud.gob.ec/coronavirus-covid-19/
6. Servicio Nacional de Gestión de Riesgos y Emergencias. https://www.gestionderiesgos.gob.ec/
7. Kassambara A (2017) Practical Guide to Cluster Analysis in R: Unsupervised Machine Learning, 1rst edn. STHDA, New York
8. Amaro R, Vicente-Villardón J, Galindo-Villardón M (2004) MANOVA-Biplot para arreglos de tratamientos con dos factores basado en modelos lineales generales multivariantes. INTERCIENCIA 1(29)
9. Amaro R, Vicente-Villardón J, Galindo-Villardón M (2008) Contribuciones al MANOVA-Biplot: Regiones de confianza alternativas. Investigación Operacional 3(29):231–241
10. Gabriel K (1972) Analysis of meteorological data by means of canonical decomposition and Biplots. J Appl Meteorol 7(11):1071–1077
11. Gabriel K (1995) MANOVA Biplots for two-way contingency tables. In: Krzanowski, W. (ed.) Recent Advances in Descriptive Multivariate Analysis, pp 227–268. Clarendon Press, Oxford
12. Vicente-Villardón J (1992) Una alternativa a las técnicas factoriales basada en una generalización de los métodos Biplot. Universidad de Salamanca, Tesis Doctoral (1992)
13. Gower J, Hand D (1996) Biplots. Chapman and Hall, London
14. Sierra C, Ruíz-Barzola O, Menéndez M, Demey J, Vicente-Villardón J (2017) Geochemical interactions study in surface river sediments at an artisanal mining area by means of Canonical (MANOVA)-Biplot. J Geochem Explor 175:72–81
15. Iñigo A, García-Talegón J, Vicente-Palacios V, Vicente-Tavera S (2018) Canonical Biplot as a tool to detect microclimates in the inner and outer parts of El Salvador Church in Seville, Spain. Measurement 136:745–760
16. Di Rienzo J, Casanoves F, Balzarini M, Gonzalez L, Tablada M, Robledo C (2020) InfoStat versión 2020, Centro de Transferencia InfoStat, FCA, Universidad Nacional de Córdoba, Argentina. https://www.infostat.com.ar
17. Vicente-Villardón J (2010) MULTBIPLOT: A package for Multivariate Analysis using Biplots, Departamento de Estadística. Universidad de Salamanca. https://biplot.usal.es/ClassicalBiplot/index.html

ABC Costing Model for a MSE of the Metalworking Sector

Katherine Camacho[1], Jenner Saavedra[1], Yeimy Salvatierra[1], and Grimaldo Quispe[2](✉)

[1] Universidad Tecnológica del Perú, Lima, Perú
{1130221,1420414,ysalvatier}@utp.edu.pe
[2] Universidad Nacional Autónoma Alto Andina de Tarma, Junín, Perú
gquispe@unaat.edu.pe

Abstract. The metalworking sector in Peru has several production lines, which generates a high demand in the market. For this reason, an ABC Costs model has been designed as a proposal for MSEs, which allows efficient cost control through the analysis and evaluation of the relationships between the resources, activities, and products involved, thus making better decisions for the future. Therefore, the present investigation describes the cost analysis in the laminating machine line using the ABC Cost System. In this case study, the analysis was structured in two levels, First Order Directing and Second Order Directing.

Keywords: Cost driver · Resource · Activity · Cost object

1 Introduction

This work has the full consent of the "LCH GRAPHIC EQUIPMENT SAC", on behalf of the institutional representatives Donny Chávez Huamán Manager of Operations. Additionally, the author Jenner José Saavedra Rosales was responsible for managing the information collected and the experiments carried out.

The current market situation presents a relentless global competition, where technology is constantly changing, and information systems are in continuous improvement. Therefore, managers are forced to design new models to strengthen their business management [1]. These models must guarantee the efficient and effective allocation of resources and activities so that the cost information is reliable [2].

In this sense, the present research aims to present an ABC costing model, which is an information system proposed in 1980 to improve the decision-making process and provide better management strategies and improve the deficiencies of the traditional costing system [3]. Cooper and Kaplan (1988, 1991, 1992) explained the applicability of the ABC-based approach in various commercial, manufacturing, and service areas, showing that this system can be applied to most companies [4].

In this article, an ABC costing system is presented through its three principles: Resources, activities, and cost object. In this framework, the ABC system is explained, and finally, the results, conclusions, and recommendations are presented.

Y. Iano et al. (Eds.): BTSym 2020, SIST 233, pp. 293–299, 2021.
https://doi.org/10.1007/978-3-030-75680-2_33

2 Method

For the following investigation, the principles of the ABC Cost System will be used (see Fig. 1):

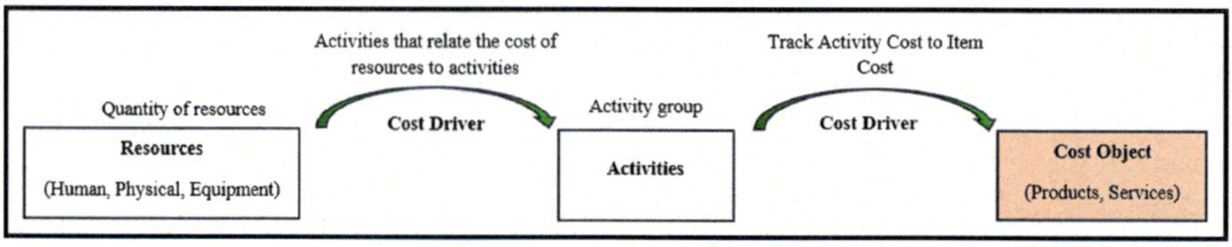

Fig. 1. Principles of the ABC cost system

2.1 Traditional Cost Accounting (TCA)

The traditional cost system was designed between 1870 and 1920, when the industry demanded a lot of labor [5]. The TCA provides a simplified view of the relationship between the resources and products of the company [6], which leads to problems of high and insufficient costs by not making correct use of the cost driver [7]. The TCA reports how and by whom the money is spent, but does not report the cost of activities or processes [8].

2.2 ABC Costing System

The ABC costing system is a quantitative and detailed cost estimation technique that allows overcoming the limitations of traditional techniques [9] since it provides more detailed information about the company's activities [10] once the activities are identified. To the activities, it is assigned a cost known as indirect costs, which allows managers to have a better view of the company's cost centers [11].

2.3 Principles of the ABC Costing System

The basic principles of the ABC cost system are three: Activities consume resources, activities have a cause, the customer or products generate different levels of activity [12], where the levels are formed into three groups: resource group, group of activities, and cost object [13], which are explained below.

Resources. They are all the necessary costs for the cost of the object or product to be determined [14].

Cost Driver. It is called the cause of cost because they transform resource expenditures into activity costs and are selected based on three criteria: easy to identify, easy to use, and easy to apply [15].

Activity. An activity is any event or transaction that is considered an important factor since it allows a cost to be generated, where all activities describe the work necessary to produce the company's cost object [16].

Cost Object. They are products manufactured, services produced, or customers served that reflect the importance of analyzing the ABC costing system [17].

3 Case Study

The metalworking company, object of this study, is dedicated to manufacturing machines for the graphic sector.

In this sense, the main problem of the company is the reduction of sales of the product that has the greatest demand, the laminating machine. Currently, sales reach S /. 566,160, while in 2014, approximately 30 machines were sold with a value of S /. 1,415,400, which means a 50% reduction in sales.

Given this, the need arises to analyze a methodology that allows evaluating the costs involved in order to precisely determine the real cost of the laminating machine and estimate a profit margin. For this, it is essential to know not only the direct materials and resources but also the related areas and the activities involved. Therefore, an analysis of the Activity-based Costing System will be applied.

In Fig. 2, the Directing of First Order and Second Order is displayed, in which the resources and activities involved for the manufacturing line of the laminating machine are shown. The arrows indicate the direct relationship that exists between each resource and the respective activity. Regarding the Second Order, the relationship that exists between all activities with cost objects (product lines) is indicated.

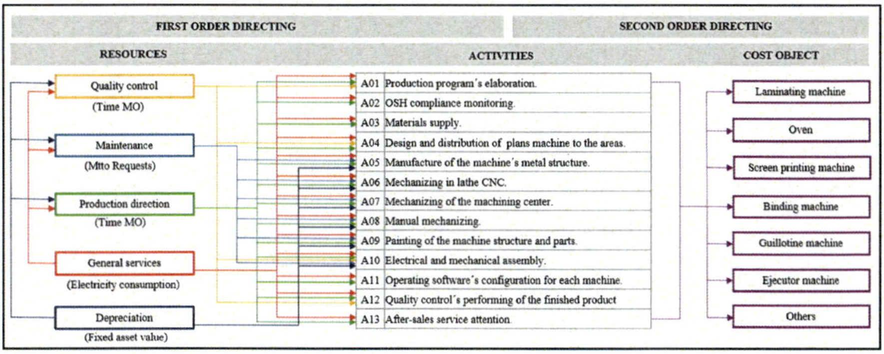

Fig. 2. First and second order directing

3.1 First Order Directing

At this first level, the relationship of indirect resources with activities is analyzed.

Resources. The resources involved in the manufacturing process of the laminating machine are:

Quality Control. It is directly related to activities A01, A04, A10, and A12, since they require verification and testing to be able to carry out the following activities.

Maintenance. This resource is related to the activities that are from the A05 to the A10 because there are involved machines, tools, and equipment, which require adequate maintenance for their correct operation.

Production Direction. It is related to the 13 activities because they are all under the constant supervision of the Production Manager.

General Services. It is related to all activities because in each one the water, electricity, or internet service intervenes.

Depreciation. This resource is related to activities that are from A05 to A10 because machines and equipment are involved there, which depreciate over time.

Activities. The main activities that have been defined for manufacturing (which are coded) are the following:

- A01. Production program's elaboration.
- A02. OSH compliance monitoring.
- A03. Materials supply
- A04. Design and distribution of plans machine to the areas
- A05. Manufacture of the machine's metal structure.
- A06. Mechanizing in lathe CNC.
- A07. Mechanizing of the machining center.
- A08. Manual mechanizing.
- A09. Painting of the machine structure and parts.
- A10. Electrical and mechanical assembly.
- A11. Operating software's configuration for each machine.
- A12. Quality control's performing of the finished product.
- A13. After-sales service attention.

3.2 Second Order Directing

In this second level, the relationship of the 13 activities with the cost objects is analyzed.

Cost Object. The cost objects (product lines) that have been selected are the six products with the highest demand in the company, and these are: laminating machine, oven, screen printing machine, binding machine, guillotine, executor, and others.

This means that, for the manufacture of any of the six machines, the 13 activities are fulfilled.

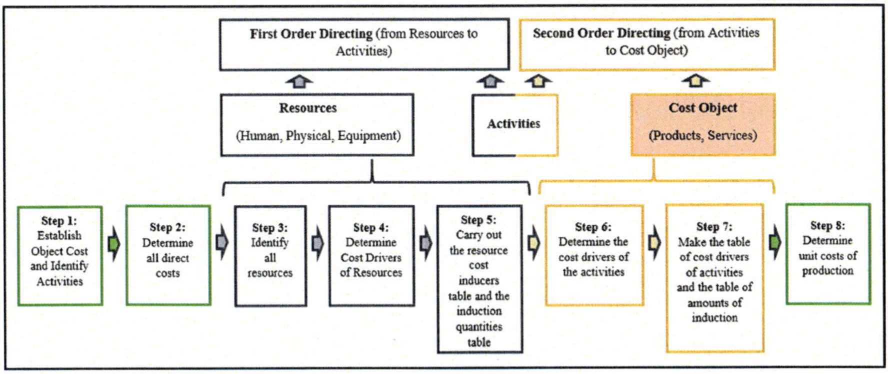

Fig. 3. Implementation of the ABC cost system

3.3 Implementation of the ABC Cost System

This system aims to provide a cost system that allows the analysis of resources, activities, and cost objects. Figure 3 presents the steps for implementation.

Step 1: Establish Object Cost and Identify Activities. It is very important to establish the Object Cost (company products), then all activities must be identified and enumerated. It is very important to consider the homogeneous activities.

Step 2: Determine All Direct Costs. ABC costing is responsible for studying indirect costs. Therefore, direct costs that are made up of direct material costs and direct labor costs must be analyzed.

Step 3: Identify All Resources. The resources need the activities to generate cost; the following resources were analyzed: Quality control, maintenance, Production management, General services, and Depreciation.

Step 4: Determine Cost Drivers of Resources. The resource cost driver allows analyzing the distribution of the resource cost to the group of activities.

Step 5: Carry Out the Resource Cost Inducers Table and the Induction Quantities Table. After making both tables, it must be verified that the costs of the resources are the same result as the cost of the activities.

Step 6: Determine the Cost Drivers of the Activities. The activity cost driver allows analyzing the distribution of the activity cost to the Object cost.

Step 7: Make the Table of Cost Drivers of Activities and the Table of Amounts of Induction. After making both tables, it must be verified that the costs of the activities have the same result as the cost of objects.

Step 8: Determine Unit Costs of Production. After analyzing the ABC costing system, the indirect costs are obtained, and with the direct costs analyzed in step 2, the unit production costs can be determined.

4 Result and Discussion

Table 1 shows the results obtained from the ABC cost system analysis, detailing the unit and total production cost (laminating machine line) expressed in nuevos soles (S /.).

Table 1. Overall results

Concept	Unit cost	Total	%
Direct materials	13,047.00	26,094.00	38.7%
Direct Labor	18,519.30	37,038.61	55.0%
Indirect costs	2,114.24	4,228.48	6.3%
TOTAL	33,680.55	67,361.09	100.0%

It is observed that the resource, direct labor, is the one that generates the highest cost (S / 185189.30), representing 55% of total costs; likewise, the cost of direct materials (S / 13047.00) also represents a high percentage of 38.7%, being direct resources those that generate the highest cost in the line of laminating machines.

The indirect costs that have been found from the ABC system represent 6.3%, having a unit cost value of S / 2114.24.

This result indicates that the indirect costs are correct; however, strategies could be applied to reduce this percentage.

5 Conclusions and Recommendations

The company's cost system analyzes direct costs (materials and direct labor) and indirect costs (materials and indirect labor) but has many shortcomings in the study of indirect costs.

The ABC system designed allows studying the deficiencies of indirect costs relying on first-order and second-order addressing, where first-order addressing allows knowing how the costs of resources are related to the costs of activities, and the addressing of Second order shows us how the costs of the activities correspond to the object costs, which allows us to know how the indirect costs are distributed throughout the ABC costing system.

The resources, activities, and cost of objects must be well defined for the ABC costing system to work properly.

It is recommended to use the ABC system as it allows a more exhaustive analysis of indirect costs related to the costs of objects of the company, and this system serves as a basis for the study of the other products of the company.

References

1. Dwivedi R, Chakraborty S (2016) Adoption of an activity based costing model in an Indian steel plant. Department of Production Engineering, Jadavpur University, Kolkata (4), 17

2. Priyatmo T, Akbar R (2019) Analysis of the prospect of implementing activity-based costing (ABC) in governmental organizations: a study at the state treasury office Jakarta IV. J Accounting Invest 20(1):1–22
3. Khan M, Batool I, Khadim Z (2018) Iqbal A (2020) Activity-based costing (ABC): implementation and success in Pakistani companies. Pakistan J Econ Stud 1(2):131–151
4. Dwivedi R, Chakraborty S (2015) An activity-based costing model for an engineering department of an Indian university. Jadavpur University, Department of Production Engineering, p 5
5. Chouhan V, Soral G, Chandra B (2017) Activity based costing model for inventory valuation. Manag Sci Lett 7:135–144
6. Emmanuel K (2014) Activity based costing (ABC) in the public sector: benefits and challenges. Probl Perspect Manag 12:4
7. Kazemi Z, Amirabadi H (2015) Activity based costing: a practical model for cost price calculation in hospitals. Indian J Sci Technol (27)
8. Mahal I, Hossain A (2015) Activity-based costing (ABC) – an effective tool for better management. Res J Finance Accounting 6:4
9. Jiran NS, Gholami H, Mahmood S, Saman MZM, Yusof NM, Draskovic V, Javovic R (2019) Application of activity-based costing in estimating the costs of manufacturing process. Transform Bus Econ (18), 2B(47B):839–860
10. Adetokunbo JO (2018) Activity based costing model for effective process analysis in nigerian healthcare management system operations. Int J Adv Trends Comput Appl (IJATCA) 3(8):23–27
11. Carg R (2016) Activity based costing-a tool to control the cost. Int J Eng Sci Res Technol (5)
12. Kannaiah D (2015) Activity based costing (ABC): is it a tool for company to achieve competitive advantage. Int J Econ Finance 7:12
13. Guo J, Shi W, Drzymalski J (2015) Activity-based costing (abc) for analyzing logistics costs in Chinese coal manufacturing industry. Western New England University, Department of Industrial Engineering and Engineering Management
14. Tong C, Wei C (2016) Application of activity-based costing on reinforcing steel bar manufacturer. Int J Organ Innov 9:1
15. Kabinlapat P, Sutthachai S (2016) An application of activity-based costing in the chicken processing industry: a case of joint products. Int Food Agribusiness Manag Rev 20:1
16. Nawar A, Shwetti M (2019) Activity Based Costing. Ministry of Higher Education & Scientific Research, Wasit University. https://doi.org/10.13140/RG.2.2.32083.73767
17. Tucillo D (2016) Activity-based costing in public services. Global Encyclopedia Public Adm Public Policy, Governance. https://doi.org/10.1007/978-3-319-31816-5_2318-1

Alteration in Autonomic Function Induced by Moderate Fluid Percussion Injury Model in Rats

Raphael Santos do Nascimento[1](✉), Fernando da Silva Fiorin[2,3], Luiz Fernando Freire Royes[3], Adair Roberto Soares Santos[2], and Jefferson Luiz Brum Marques[1]

[1] Institute of Biomedical Engineering, Federal University of Santa Catarina, Florianopolis, SC, Brazil
jefferson.marques@ufsc.br

[2] Laboratory of Neurobiology of Pain and Inflammation, Federal University of Santa Catarina, Florianopolis, SC, Brazil
adair.santos@ufsc.br

[3] Laboratory of Exercise Biochemistry, Federal University of Santa Maria, Santa Maria, RS, Brazil

Abstract. Traumatic brain injury (TBI) represents one of the most common social problems worldwide. This neurologic pathology often affects several body systemic organs, including the cardiovascular autonomic system. Changes in cardiac autonomic nervous system (ANS) regulation through the heart rate variability (HRV) analysis have been observed in all degrees of TBI in patients. On the other hand, little research has been addressed in TBI models using rats as a potential target for better explore therapeutic approaches in this pathology. In this sense, we investigate the relationship before and after an induced moderate fluid percussion injury (FPI) model in rats on the HRV features. The findings found in our work showed clearly a decreasing the sympathetic (LF), parasympathetic (HF), and overall HRV ($p < 0.01$ for mean NN) activities post-TBI when compared to baseline animals. In concomitant to this, the heart rate increased, such as showed by decreasing the mean NN. Therefore, this study indicates that the impacted severity moderate TBI in rats impairs some cardiac autonomic function parameters. However, further research is needed to better understand this pathophysiology.

Keywords: Traumatic brain injury · Fluid percussion injury · Autonomic nervous system · Heart rate variability · Rat

1 Introduction

Traumatic brain injury (TBI) is a devastating disease that affects approximately 10 million patients every year in the world. It is characterized by a set of immediate mechanical changes in the brain tissue and secondary damage [1]. The cardiovascular damage following TBI frequently constitutes one of the main causes of mortality, morbidity, as well

Y. Iano et al. (Eds.): BTSym 2020, SIST 233, pp. 300–307, 2021.
https://doi.org/10.1007/978-3-030-75680-2_34

as neurological impairment in children, sports athletes, civilian adult, military, and war veteran populations [2]. Although cardiovascular complications have been linked with all known degrees of injury (i.e., mild, moderate, and severe), it is understood that this imbalance cardiac occurs in a greater number when the injuries are of moderate and severe levels [3]. A body of clinical evidence indicating that the main impairment inside the abnormalities spectrum in cardiac complications includes the cardiac autonomic control changes [2, 4].

The autonomic nervous system (ANS) comprises the sympathetic and parasympathetic nervous system, in which it is responsible for regulation physiological, including the heart [5]. Therefore, perturbations of the ANS pathways following TBI may increase sympathetic activity and parasympathetic activity decreasing, leading the individual to sudden cardiac or ventricular tachyarrhythmias. Although, in milder cases of TBI, this process can recover naturally [6]. To investigate the sympathetic and parasympathetic activities of the ANS in TBI, the literature has utilized, extensively, the heart rate variability (HRV) method. This is defined as a standard non-invasive biomarker of cardiac autonomic modulation of heart rate, reflecting the sympathetic and parasympathetic nervous system activities under the heart's sinus node [7].

Regarding the HRV analysis following TBI, a clinical study recently conducted by Estévez-Báez and collaborators (2019) performed in 45 patients with a severe injury according to the classic Glasgow coma scale, shown that the overall HRV was decreased significantly [8]. In the same way, another study carried out in patients with mild, moderate, and severe TBI demonstrated an increased sympathetic activity reflected by both LF (normalized unit) and LF/HF parameters and decreasing the parasympathetic activity indicate by the HF (normalized unit) parameter. Additionally, it was also observed that the increasing severity of brain injury increasing, substantially, the ANS pathways (sympathetic and parasympathetic) impairment in patients [9]. These changes in the ANS's sympathetic and parasympathetic activities have been indicated to correlate with increasing the morbidity and mortality in moderate/severe TBI patients [6]. Although few exploratory studies are found in experimental TBI models with rats, previous research reported that rats inducted by a mild/moderate TBI model presented a decreasing heart rate [10].

The analysis of the HRV parameters in patients that surfing TBI has been presented in all degrees of injury mild, moderate, and severe [9]. Nonetheless, in TBI experimental models using rats, analysis of HRV have been little reported. In general, the difficulty is basically focused on the electrocardiogram (ECG) measurement on the animals, which in majority currently still require the use of surgical procedures, restraint of animals, or still anesthesia administration [11, 12]. For this reason, there is a difficulty in the neuroscience field and the pharmaceutical industries to develop solutions for the restoration of cardiovascular capacity after TBI due to the difficulty to mimic the cardiac physiology natural in these animals [13].

Therefore, considering the sympathetic and parasympathetic drives of the ANS following TBI, this study aimed to investigate the autonomic function balance in rats induced by the fluid percussion injury model. We analyzed the mean NN parameter and observed the ECG signal in the time-frequency domains, before and after the period the 24, 48, and 72 h of brain injury on the animals.

2 Materials and Methods

2.1 Animals

The experiments were performed on four adult male Wistar rats (400–500g, 200 days of age) provisioned from the Federal University of Santa Catarina. All animals were maintained in a controlled laboratory environment (12 h light-dark cycle, 50% relative humidity, and acclimatized room 22 ± 3 °C). They were housed in plastic cages (27 × 36 × 31 cm) to a maximum of 3 animals covered by shavings and allocated in ventilated racks. Animals obtained free access to water filtered [for heavy metals and fecal coliform (Jojaco® model J190, Capivari de Baixo, SC, Brazil)] and with ad libitium access to a standard laboratory diet (BioBase®, Águas Frias, SC, Brazil). All procedures carried out followed the Brazilian legislation regulated by the National Council for the Control of Animal Experimentation (CONCEA) and with the ethical principles in animal research established by the Brazilian Society of Science in Laboratory Animals. All experimental procedures were approved by the Committee in Care and Use of Experimental Animal Resources of the Federal University of Santa Catarina.

2.2 ECG Acquisition Protocol

Experimental procedures for ECG acquisition under the animals were carried out as previously described in a study [13]. Briefly, it was built a custom jacket under the measures thoracic of animals, allocating three electrodes in its internal surface to acquire the ECG signals. All animals were habituated in thirty days intercalary (approximately 1 h/day for each rat) in a transparent glass box (30 × 20 × 23), jointly with the jacket placed. With the help of a feed pellet, the animals were inducted inside of the jacket. After the thirty days of adaptation, the rats were conducted to the surgical procedures (craniectomy and TBI cannula implantation). Twenty four hours after the surgical procedures, we submitted all animals to the baseline ECG recording. The following day, animals were conducted to the application of the TBI model and subsequently sutured. Posteriorly, the animals were again submitted to ECG recording in a period of 24, 48, and 72 h after the TBI model application, identical to the baseline measurement protocol. Immediately after the ECG evaluation, all animals were euthanized according to the experimental protocol established.

2.3 TBI Model

Twenty four hours after the ECG recording baseline, the fluid percussion injury (FPI) was applied as previously reported, with small alterations [14, 15]. Animals were briefly prepared with anesthesia (ketamine 100 mg/Kg and xylazine 10 mg/Kg intraperitoneally) and subsequently transferred to a rodent stereotaxic device. A craniectomy of 3-mm diameter was performed parasagitally at lateral to the midline (3.0 mm) and posterior to the bregma (2.0 mm), with very attention to maintain the dura mater intact. A plastic injury tube was cemented (dental cement) on the dura exposed. After the hardening of dental cement, the cannula was filled with saline solution and closed with an adequate plastic cap. Posteriorly, the rats were removed from the stereotaxic apparatus and transferred to its plastic home cage. After 1 day, animals were anesthetized with isoflurane inhaled (1%), and the injury tube was attached to the FPI experimental device. This then produced a transient pressure fluid pulse, impacting the exposed dura on the animal. Pressure pulses were recorded on the skull using a pressure sensor (MS5803-14BA, Sparkfun Electronics Inc., Niwot, Colorado, USA) attached to a computer; the pressure pulse was then measured at the exact moment of the impact of the pendulum on the cylinder filled the saline for the FPI. The sensor recorded a pressure of 2.0 ± 0.2 atm at the moment of impact, indicating a moderate TBI in this experiment. Immediately after these procedures, all animals were anesthetized over isoflurane (1% inhaled), and the cannula with dental cement was removed. A surgical clamp was utilized at the location of the incision, with the animals re-anesthetized over anesthesia nose cone that was removed following the end of the procedure.

2.4 ECG Analysis

To ECG samples were measured using three electrodes positioned in lead I under the rats. These were coupled to an ECG amplifier employed in a hardware module produced in our laboratory [Institute of Biomedical Engineering of the Federal University of Santa Catarina (Florianópolis, Santa Catarina, Brazil)] [13]. The recorded raw data were exported as text files (.txt) in points of electrocardiograph amplitude and digitalized by a 12-bit analog-to-digital converter to a 500 Hz sampling frequency. Posterioursly, it was also digitally filtered by a band-pass (0.1–40 Hz) filter and stored for further HRV analysis. Before the beginning of the evaluations, animals were carefully scraped and allowed to acclimatize for a period of 20-min into the jacket. The assessment was performed during 8–10-min among all animals, where posteriorly a 5-min sample was visually inspected and chosen, eliminating the maximum of noise (ectopic beats, artifacts of movement, arrhythmias) for signal processing.

2.5 Heart Rate Variability

The Analysis of HRV features in the time-frequency domain was extracted using a custom-designed algorithm in Matlab (MathWorks Inc, Natick, MA, USA, version 9.4), where the signal variability (NN intervals) was detected, creating a time series with all NN intervals. For the HRV time-domain analysis, it was calculated the mean NN intervals (mean of all NN intervals) and the mean HR (heart rate), both reflecting the global magnitude of HRV. For frequency-domain analysis, we derived measures from NN intervals the fast Fourier transformation (FFT) algorithm, extracting in two power spectral bands: low frequency (LF, 0.1–1.0 Hz) and high frequency (HF, 1.0–3.5 Hz) [16]. LF component has been correlated to the sympathetic tone and vagal, whereas the HF to the parasympathetic tone. Both together (LF and HF) represent the HRV global [7].

2.6 Statistical Analysis

Statistical analysis was performed using the GraphPad Prism (GraphPad Software, Inc., San Diego, CA, USA, version 8.0) program. To evaluate whether HRV changed during the four measures made, a one-way analysis of variance (ANOVA) was used. Tukey post hoc test was applied for comparison between the measures. Data are presented as the mean ± standard deviation, with *$p < 0.05$ and **$p < 0.01$ for significant differences.

3 Results

In this present study, our data revealed that the TBI promotes changes in mean NN ($F[3, 12] = 6.063$; $p < 0.01$; Fig. 1A) parameter. In this same way, post hoc analysis demonstrated significant differences in the recording of 24 h ($p < 0.05$; Fig. 1A), 48 h ($p < 0.01$; Fig. 1A), and 72 h ($p < 0.05$; Fig. 1A) following TBI, when compared to the baseline recording. To confirm these effects, we randomly selected an animal and assessed the electrocardiographic tracing (a segment of 10s), jointly to the heart rate

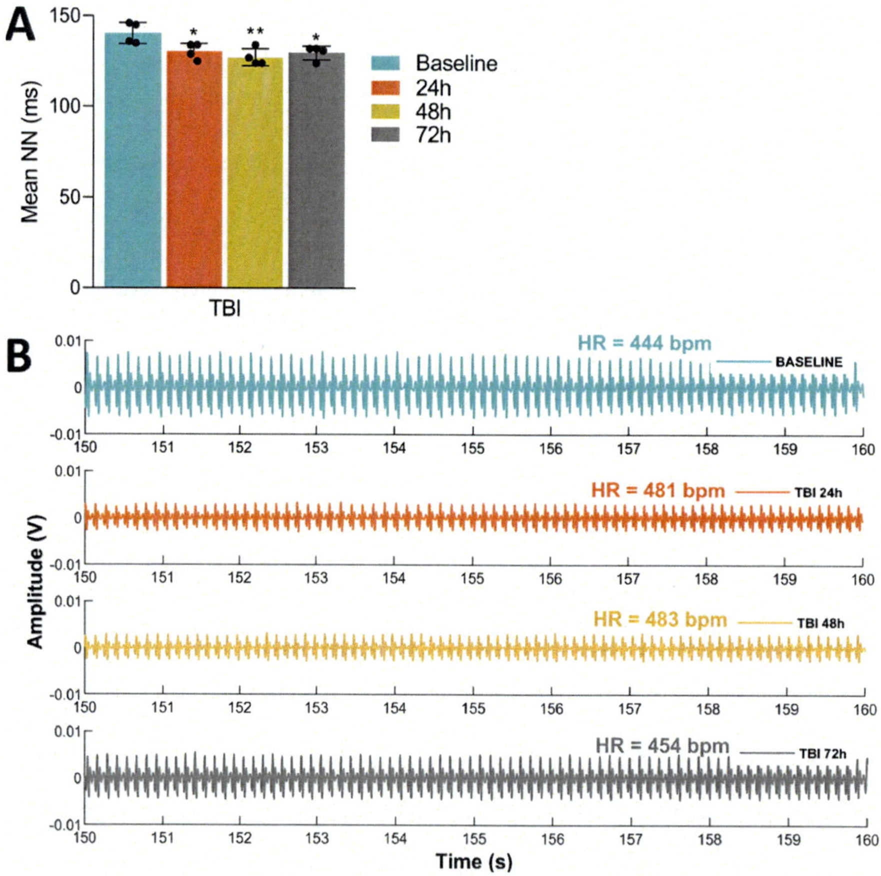

Fig. 1. Time-domain analysis before and after induced TBI model in rats. (A) Mean NN, mean of all NN intervals. Data were evaluated using the one-way analysis of variance (ANOVA). Tukey test was applied to comparing the measures. *$p < 0.05$ and **$p < 0.01$ compared to baseline recording. (B) A segment of 10s of ECG signal in an animal randomly chosen. HR, heart rate

(Fig. 1B). In this tracing, it was possible to observe a decrease of ECG signal amplitude in the recorded period in 24, 48, and 72 h compared to the baseline ECG signal. The increasing heart rate consequently accompanied this modification during these recorded periods (Fig. 1B).

To evaluation of HRV in the frequency domain, we also selected an animal randomly and estimated power spectral (5 min time series) (Fig. 2). In the full spectrum was seen that the sympathetic activity (indexed by LF band) was reduced in all recorded periods, whereas the parasympathetic activity (indexed by HF band) was reduced in 24 h and 48 h measures after brain injury. It is important to note still that in the 72 h recording, both the assessment of the ECG tracing and power spectral suggested a tendency towards recovery to the baseline values.

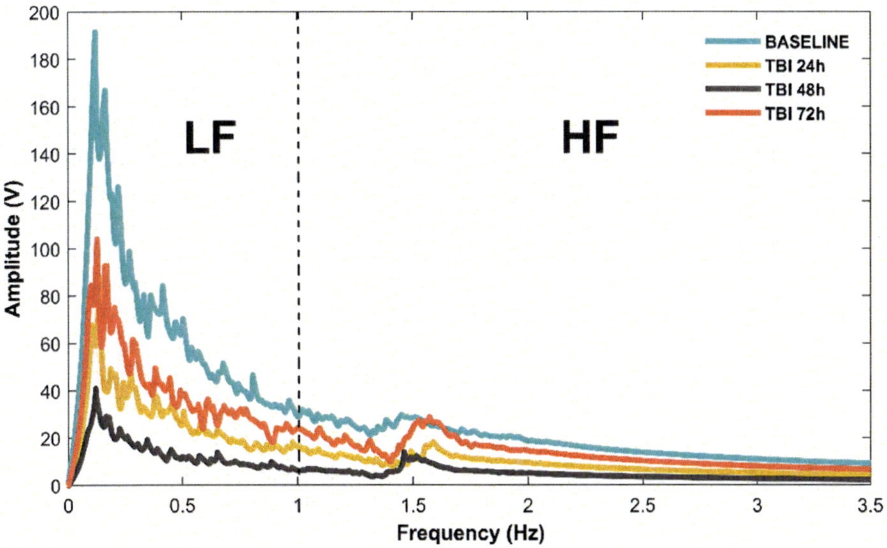

Fig. 2. Power spectral analysis in a random rat before and after induced TBI model. LF, low frequency and HF, high frequency are presented

4 Discussion

In this present work, moderate TBI in rats induced changes in the sympathetic and parasympathetic tone activities of ANS showed through the HRV features. Our study's findings revealed that the TBI promoted cardiovascular autonomic function impairment after twenty-four hours. Furthermore, HRV features following TBI triggered effects for both the time and frequency domain analysis.

The TBI causes damage to several systems of the body, including the cardiovascular autonomic nervous system. Global autonomic function decreasing is frequently found in severe TBI patients, affecting currently about 90% of patients who visited the intensive care units. This emerging panorama has been associated with a greater number of mortalities evidenced in the hospitals [17]. The ANS is linked directly with the hypothalamic-pituitary axis (HPA), where is concentered the center of the global body homeostasis modulation. Thus, a growing body of studies clinical and experimental elucidated that the TBI damage on the ANS can contribute to the impairment of several body organs, leading, consequently, to poor quality of life [18]. Based on this, the time-frequency HRV features analysis has been considered relevant to better understand the behavior of the autonomic nervous system activities after TBI, in order to search solutions to its prevention and treatment [9].

In this study, decreased mean NN was found in moderate TBI rats in comparison with baseline condition, pointing to low overall HRV. Corroborating with our results, studies performed in severity mild/severe TBI patients have shown an increased heart rate [3, 19]. Another study indicated that the mean NN parameter reduces according to the increasing degrees of injury [9]. Together, these data support the hypothesis that the overall HRV concomitant decreasing may increase the heart rate after TBI. It is still noteworthy that an overall HRV lower is seen as a risk factor for the incidence of morbidity and mortality [20]. Regarding the power spectral analysis, our data also demonstrated sympathetic (indexed by LF) and parasympathetic (indexed by HF) activities reduction after 48 h of TBI. Consolidating with our discovery, a study demonstrated that both the LF and HF band are reduced in the power spectral for elevated injury degrees in patients, unless in the presence of brain death [9]. It is important to emphasize that the mean NN, as well as the LF and HF band, presented recovery to the 72 h recording, indicating further therapeutic approaches for the restoration of regulation autonomic after TBI.

In this perspective, this work presented results in comparison with the found in clinical, providing subsidies important about the HRV tool for assessing the cardiac autonomic function in experimental models with rats. Besides our study have limitations, involving mainly a smaller number of animals and the evaluation of HRV measures few, this was conducted with an assessment of ANS modulation on the physiology natural of animals, in contrast to what occurs in previous studies explored.

5 Conclusion

In conclusion, considering all changes found in this work regarding the mean NN and the representations in the time-frequency of ECG signal, we suggest that the autonomic nervous system modulation was impaired following moderate TBI in rats. Additionally, we accredit that this study may be a precursor to understanding the ANS activities in TBI rats. Nonetheless, it reinforces the importance to the development of further research for the better integration between the clinical and experimental outcomes in TBI.

Acknowledgments. The authors gratefully acknowledge the student fellowships from the National Council for Scientific and Technological Development (CNPq) (to RSN) and the Coordination of Superior Level Staff Improvement (CAPES) (to FSF). ARSS, LFFR, and JLBM are grantees of CNPq research productivity fellowships.

References

1. Hyder AA, Wunderlich CA, Puvanachandra P, Gururaj G, Kobusingye OC (2007) The impact of traumatic brain injuries: a global perspective. NeuroRehabilitation 22:341–353
2. Gregory T, Smith M (2012) Cardiovascular complications of brain injury. BJA Contin Educ Anaesth Crit Care Pain 12:67–71
3. Baguley IANJ et al (2007) The incidence of dysautonomia and its relationship with autonomic arousal following traumatic brain injury. Brain Inj 21:1175–1181
4. Rapenne T et al (2001) Could heart rate variability predict outcome in patients with severe head injury? A pilot study. J Neurosurg Anesth 13:260–268
5. McCorry LK (2007) Physiology of the autonomic nervous system. Am J Pharm Educ 71:78
6. Hendén PL et al (2014) Can baroreflex sensitivity and heart rate variability predict late neurological outcome in patients with traumatic brain injury. J Neurosurg Anesth 26:50–59
7. Task Force of the European Society of Cardiology & The North American Society of Pacing and Electrophysiology (1996) Heart rate variability: Standards of measurement, physiological interpretation, and clinical use. Eur Hear. J 17:354–381
8. Estévez-Báez M et al (2019) Autonomic impairment of patients in coma with different Glasgow coma score assessed with heart rate variability. Brain Inj 33:496–516
9. Su C, Kuo TB, Kuo J, Lai H, Chen HI (2005) Sympathetic and parasympathetic activities evaluated by heart-rate variability in head injury of various severities *. Clin Neurophysiol 116:1273–1279
10. Umemoto Y, Chitravanshi VC (2017) Cardiovascular changes in urethane anesthetized Wistar rats after the traumatic brain injury. Res Reports 1:1–22
11. Fenske S et al (2016) Comprehensive multilevel in vivo and in vitro analysis of heart rate fluctuations in mice by ECG telemetry and electrophysiology. Nat Protoc 11:61–86
12. Mongue-Din H, Salmon A, Fiszman MY, Fromes Y (2007) Non-invasive restrained ECG recording in conscious small rodents: a new tool for cardiac electrical activity investigation. Pflugers Arch 454:165–171
13. Nascimento RS, Fiorin FS, Santos ARS, Royes LFF, Marques JLB (2020) A new approach for ECG recording in rats: an autonomic nervous system analysis. In: Proceedings of the 5th Brazilian Technology Symposium, Smart Innovation, Systems and Technologies, vol 202. Springer, Switzerland
14. Fiorin FS, Espírito Santo CC, Santos ARS, Fighera MR, Royes LFF (2018) Implication of surgical procedure in the induction of headache and generalized painful sensation in a fluid percussion injury model in rats. J Neurosci Methods 307:23–30
15. Fiorin FS et al (2020) Capsaicin-sensitive fibers mediate periorbital allodynia and activation of inflammatory cells after traumatic brain injury in rats: involvement of TRPV1 channels in post-traumatic headache. Neuropharmacology 176:108215
16. Stein AB et al (2007) Effects of anesthesia on echocardiographic assessment of left ventricular structure and function in rats. Basic Res Cardiol 102:28–41
17. Zygun DA et al (2006) Ventilator-associated pneumonia in severe traumatic brain injury. Neurocrit Care 5:108–114
18. Royes LFF, Gomez-Pinilla F (2019) Making sense of gut feelings in the traumatic brain injury pathogenesis. Neurosci Biobehav Rev 102:345–361
19. Pattoneri P et al (2005) Circadian blood pressure and heart rate changes in patients in a persistent vegetative state after traumatic brain injury. J Clin Hypertens 7:734–739
20. Thayer JF, Yamamoto SS, Brosschot JF (2010) The relationship of autonomic imbalance, heart rate variability and cardiovascular disease risk factors. Int J Cardiol 141:122–131

Analysis of Deforestation in Ucayali-Peru Using Satellite Imagery from Sentinel-2

Diego Velayarce[1], Manuel Alvarez[1], Diego Guevara[1], and Victor Murray[1,2,3(✉)]

[1] Department of Electrical Engineering, Universidad de Ingenieria y Tecnologia, Lima, Peru
{diego.velayarce,manuel.alvarez,diego.guevara}@utec.edu.pe, vmurray@ieee.org

[2] Department of Electrical and Computer Engineering, University of New Mexico, Albuquerque, NM, USA

[3] School of Engineering and Applied Sciences, Harvard University, Cambridge, MA, USA

Abstract. We propose a combination of methods for processing multitemporal multispectral satellite images to identify variations in vegetation indices for fixed regions across time to analyze deforestation in Peru. The images, taken from a location in Ucayali, Peru, are obtained from public repositories for Sentinel-2 satellite, which covers up to 13 spectral band layers with spatial resolutions of 10, 20, and 60 m and have a temporal resolution of 5 days with the same angle vision. We preprocessed the images for cloud and river removal and normalization to apply the normalized difference vegetation index (NDVI). All images were segmented with their corresponding model, leading to the analysis of the variance of this segmentation information across different time frames for a single location in which the normalized difference of time series (NDTS) was obtained for each year. Finally, a first approach was built for a linear regression model that could segment each frame into sectors, according to vegetation density, using the multispectral features as parameters. For the selected location, the results show an increment from about 4200 ha in 2018 to about 8200 in 2020.

Keywords: Multitemporal multispectral satellite images · NDVI · NDTS · Automatic deforestation detection · Sentinel-2

1 Introduction

Illegal logging in Peru's jungle region is happening at an all-time high rate, with up to 4 km^2 of forests disappearing per day [1]. Roughly 80% of all timber exported annually from Peru, as of 2017, stems from unregulated extraction [2]. Artificial deforestation, at rates like these, from a socio-economic and ecological perspective, is a significant threat to the sustainability of local communities [3]. Therefore, it is in the best interest of local public organizations and governments to find tools that enable them to take rapid measures against unauthorized logging. However, organized illegal logging is seldom as evident, as reported in [1]. Instead, it is often done in small sectors scattered across large

Y. Iano et al. (Eds.): BTSym 2020, SIST 233, pp. 308–316, 2021.
https://doi.org/10.1007/978-3-030-75680-2_35

regions of land, thus reducing the likelihood of detection starkly [3]. Fortunately, the increasing availability of high temporal-spatial resolution satellite imagery has proven to be significantly used to detect vegetation distribution changes inland. In particular, multispectral satellite cameras are useful due to the nature of the reflectance of vegetal material for radiation wavelengths outside of the visible spectrum [4]. Different digital image processing algorithms for segmentation and change detection of multispectral images are frequently used, for example, object-based segmentation [5], pixel-based index estimation, e.g., for normalized difference vegetation index (NDVI) [6], temporal gradient filtering, for video processing, and thresholding [3], pixel-based classification via support vector machines (SVM) regression [7], or wavelet transform analysis [8].

In this application, we preprocessed the images with atmospheric correction and normalization to remove clouds and possible rivers that could confuse deforested areas. We use a combination of NDVI, normalized difference of time series (NDTS) methods, and thresholding of grey scales images to obtain the percentage of non-vegetation found on a specific region of interest. Furthermore, we use a combination of temporal gradient filtering and thresholding with a pixel-based approach by classifying the spectral-spatial data using a regression model and finding an estimation of the time-domain change per each pixel in the classified and segmented images.

2 Methodology

In this section, we first describe the process of acquiring the images from the Sentinel Hub at the EO Browser. Next, we describe the image processing methods for preprocessing and segmentation of the satellite images. Then, we present normalized difference vegetation index calculations followed by the time-domain information processing methods.

2.1 Data Acquisition

The images for this study were retrieved from the Sentinel-Hub's EO Browser (available at apps.sentinel-hub.com/eo-browser), a private repository of satellite imagery that spans over 20 years of data from 15 satellite sources (as of 2020). The selected satellite, Sentinel-2, has three characteristics that are beneficial for the objectives of this project: high spatial resolution (10 m^2/pixel, 20 m^2/pixel, and 60 m^2/pixel for different bands), relative high temporal resolution (varies by latitude, five days between pictures with two satellites) and multispectral information (13 full bands) [9]. The information about each of the spectral bands retrieved from the satellite cameras for the Level-1C (L1C) is detailed in Table 1 [10].

The location with coordinates 8.37° S, 75.14° W, located in the Ucayali province of Peru, was selected (see Fig. 1), considering that it has rivers, small human communities, agricultural areas, and regular forest-only areas. Each image has a native resolution of 2500 × 1197 pixels with a projected resolution of 14m/pixel, spanning three years, with 13 spectral layers.

Besides the multispectral data, we have also acquired scene classification maps [10] corresponding to each location's frames. These will allow us to apply a linear regression method to create our classification maps that will only consider vegetation presence and no other factors, such as urbanization or water presence.

Table 1. Detailed information about employed spectral bands [10].

Band name	Name	Central wavelength (μm)	Camera
B01	Coastal aerosol	0.443	L1C
B02	Blue	0.490	L1C
B03	Green	0.560	L1C
B04	Red	0.665	L1C
B05	Vegetation Red Edge	0.705	L1C
B06	Vegetation Red Edge	0.740	L1C
B07	Vegetation Red Edge	0.783	L1C
B08	NIR	0.842	L1C
B08A	Vegetation Red Edge	0.865	L1C
B09	Water vapour	0.945	L1C
B10	SWIR-Cirrus	1.375	L1C
B11	SWIR	1.610	L1C
B12	SWIR	2.190	L1C

(a) (b) (c)

Fig. 1. Location: Ucayali (8.37° S 75.14° W), Peru. (a) 2020, (b) 2019, (c) 2018

2.2 Image Preprocessing

The images for the NDVI and NDTS analysis were normalized to have the same size and pixel count. After this, an atmospheric correction was applied to remove noise produced by clouds. Since regions have rivers, a threshold range was used to remove the river and have a final image with only vegetation and possible deforested areas.

For the regression model, we start by labeling and reordering the frames, according to date, using the visual labels' information in the upper-right corner of each frame. These images were converted to text using MATLAB's Optical Character Recognition (OCR) tool. Next, we select only the relevant sections of each frame for segmentation purposes. It is necessary to exclude these image sectors from the next procedures since the land in satellite images is densely covered by clouds. The implemented cloud recognition model (see Fig. 2) is based on SVM and is derived from the work in [11]. The final preprocessing step is normalization, which, for our data, was implemented as histogram equalization for each of the bands of each frame, procedure based on [3].

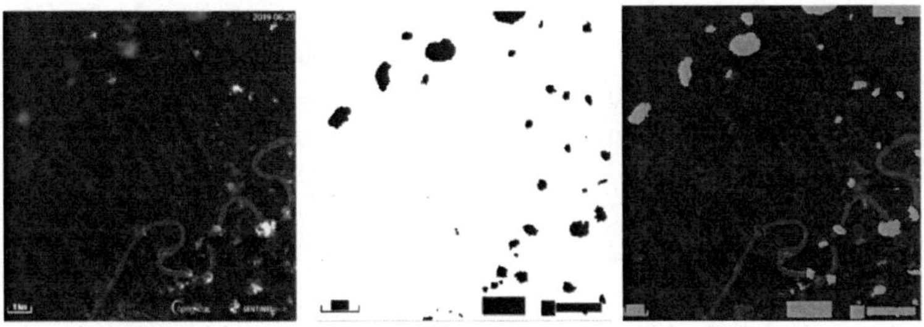

(a) (b) (c)

Fig. 2. Cloud and label masking. (a) Original Sentinel-2 image. (b) Computed mask to cover clouds and labels. (c) Resulting masked image.

2.3 Index Calculation

Images from the used satellites have a total of 10 to 13 bands, which can be downloaded separately from the Sentinel-Hub app. As discussed earlier, these bands provide a wide range of possible combinations that can give us indices related to agriculture, vegetation, and atmospheric penetration, among others (see [10] for more details). One specific combination of Sentinel-2 bands was selected for this analysis: bands 08 and 04. When combining these bands with a quotient of subtractions, we get a normalized difference vegetation index (NDVI) given by the combinations of the near-infrared (NIR) and red spectral bands (R) as

$$NDVI = \frac{NIR - R}{NIR + R}. \tag{1}$$

Once the NDVI of the region of interest (ROI) is acquired, vegetation information can be observed with a simple color map bar. Most importantly, the regions where there is no vegetation, i.e., deforested zones, are represented with a different color to those where vegetation density is high. With this information, it is possible to apply a threshold to the NDVI image to show only the deforested region. However, the optimum threshold can only be determined qualitatively by a trial-and-error approach.

2.4 Time-Domain Information Processing

Images for this work were taken in three different years: 2018, 2019, and 2020. For each zone, a different set of historical information was arranged, and the idea to compare the differences that occur across time was valid in the context of deforestation. Since the vegetation zone changes are visual indicators of deforestation on the acquired satellite images, a way to detect the variation and, more precisely, the absence of forest zones is of interest. For this purpose, the normalized difference of time series (NDTS) equation is used [12]:

$$NDTS = \frac{I_{t_2} - I_{t_1}}{I_{t_2} + I_{t_1}}. \tag{2}$$

Where I_{t_1} and I_{t_2} correspond to two adjacent time frames. By using (2), we find the difference between I_{t_1} and I_{t_2}, and this result is normalized. Given that the cloud-removal masks vary between adjacent time frames, the NDTS is instead calculated using pixels from adjacent time frames where the same pixel's class is known, i.e., not covered by a cloud.

2.5 Combination of NDVI and NDTS Method to Detect Possible Deforestation Zones

Once the NDVI image was obtained, rivers from some images taken were segmented as non-vegetation zones. A range threshold was applied to remove this from the general deforestation analysis. Thus, we obtain data only from vegetation indices and possible deforested zones. This threshold was tuned heuristically using a qualitative analysis. After a specific threshold was applied to every NDVI result, a binary image resulted in lower values representing the deforested zones. With the method applied with the appropriate threshold to every image, we decided to apply the enhanced NDTS method with certainty that clouds and rivers will not obscure the difference in the zones' vegetation.

The purpose of applying NDTS to the NDVI results is to reduce the error. NDVI shows the vegetation index for a specific location, but we do not know if the regions segmented as non-vegetation areas are deforestation. NDTS allows us to compute the changes between 2 images registered at different times. Therefore, if we apply NDTS over NDVI, we obtain the changes in non-vegetative areas in a fixed location. These changes include deforestation in separated areas and the expansion of non-vegetative areas.

2.6 Single-Frame Segmentation Using Multispectral Data

After preprocessing, the images were sorted within each satellite-camera group by a percentage of visible land, i.e., a percentage of the picture not covered by clouds. Then, the 15 least cloudy pictures of each group were chosen to calculate a linear regression coefficients matrix b that would be used to estimate the vegetation density from each frame's features, i.e., each spectra's values band.

The linear regression model is as $Xb = Y$, where X is the feature matrix, composed of n columns, where each column is the value of a spectral band layer, and m rows, where each row corresponds to a pixel on a selected frame, and Y is the class matrix, with m rows, corresponding to the pixels from the classification scene maps mentioned in Subsect. 2.1.

3 Results

We present the results for the location shown in Fig. 1. The results for NDVI and NDTS approaches were discussed in Sect. 2 and the combination of both algorithms. In this section, the numerical values for vegetation variances between two images using NDTS and the combination of NDTS and NDVI are detailed. In Table 2, we show the obtained non-vegetation total area and percentage for the binary image after the NDVI + NDTS method. The obtained values for the b coefficients for each band of the satellite, {B01, B02, B03, B04, B05, B06, B07, B08, B8A, B09, B10, B11, B12}, used to compute the classifier, were {8.994, −16.285, 18.741, 10.581, −17.010, 7.011, 0.582, −1.687, −3.728, 0.207, −4.348, −2.610, 8.518}, respectively.

Table 2. Non-vegetable zones percentages and areas calculated

Year	Non-vegetable percentage	Total area in hectares
2018	7.175	4208.59
2019	8.624	5058.27
2020	14.059	8246.23

In Fig. 3, we look at two true-color (RGB) images from one of the local maxima in the NDTS for the first quarter of the processed images.

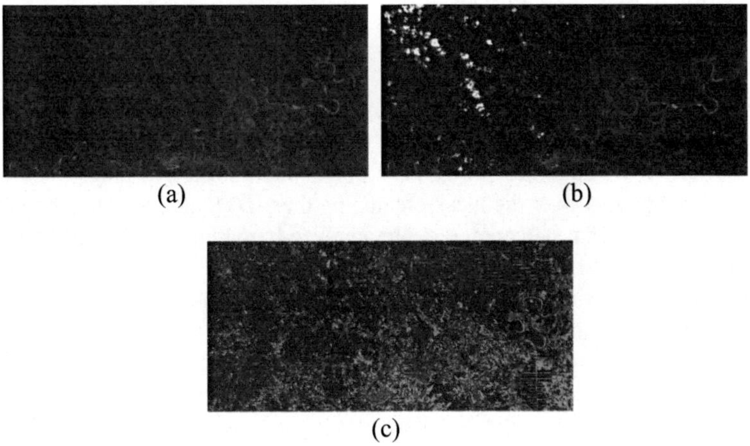

Fig. 3. RGB Images from Sentinel-2 from (a) 2020–08-28 (b) 2020–05-30. (c) NDTS result overlayed on the original image.

We show an NDVI analysis of an image from Fig. 4. First, in Fig. 4a, we show an image from August 28th, 2020. The NDVI analysis is shown in Fig. 4b, with the values normalized from 0 to 1, and the vegetation highlighted with a dark green tone. A threshold range was applied to remove information that corresponded to the rivers and clouds to obtain an image with vegetable area vs. non-vegetable area after computing the same algorithm in 2 previous years (see Fig. 4c).

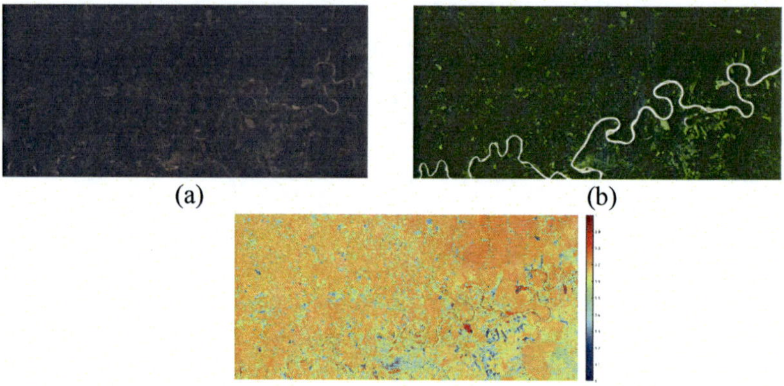

Fig. 4. NDVI analysis from August 28th, 2020. (a) Original image. (b) NDVI analysis of (a). (c) Segmented NDVI analysis of (b).

We can see how the method highlights these zones. With this final segmented NDVI image, the NDTS approach showed better qualitative results.

In 20 days of difference, deforestation zones appeared as small areas in the satellite image. The difference can be highlighted and calculated approximately by applying NDTS to these images directly, instead of applying it to the segmented images. The result is squared to enable the imaging of negative changes.

After analyzing the histogram of the NDTS, pixel intensity values lower than 0.004 can be considered noise or unimportant information. The result of these modifications is overlayed in the original image (see Fig. 3c). Considering that the resolution of satellite Sentinel-2 in the RGB band is 2500 × 1197 pixels and a projected pixel resolution of 14 m/pixel, we can work over the mask created by the NDTS analysis shown in Fig. 3c in order to calculate what percentage of the analyzed area has been deforested in the 20 days that differentiate them.

After spatially working with the time difference and recalling that each image is 2500 × 1197 pixels, with 14 m/pixel, equivalent to 58653 hectares (see Subsect. 2.1), we obtain the deforestation percentages of 7.175% for 2018, 8.624% for 2019, and 14.059% for 2020, approximately (see Table 2). The result shows that all of the new deforestation zones are segmented, but there is still noise caused by the clouds and their shadows. It is not easy to analyze the satellite images in the months from November to March. The high rain activity and clouds block the satellite images of the Amazon's surface, and this fact could seem like an excellent opportunity for informal loggers to increase their activity.

According to reports from the Ministry of the Environment of Peru [13], there were 25991 hectares deforested for 2018 in the whole region of Ucayali, in which the province of Padre de Abad was the most affected with deforestation. The images and the results we found are located in the Padre de Abad district, where we found calculated the deforested area, which correlates to the deforestation that took place in the region [13]. This study could help national organizations target specific zones and monitor monthly deforestation. This type of calculation could also be applied to other zones in the Jungle of Peru to have a national system to detect deforestation and send alerts automatically.

4 Conclusions

We have presented a method to analyze deforestation using multitemporal multispectral images from the satellite Sentinel-2. The method was applied to analyze the city of Ucayali, located in the Amazon rainforest in Peru. Deforestation is gradually increasing in Peru's jungle; the deforestation zones are small and sparse, i.e., not clustered, which is why high-definition spatial information is needed. The index approaches are considerably easier to implement but can only give qualitative information on vegetation distribution change or produce images for a human to interpret. However, computationally expensive, the proposed approach can give quantitative results thanks to processing multitemporal multispectral satellite images. As future work, a web application to monthly monitor deforestation would be implemented.

References

1. L. República (2018) El Perú pierde más de 427 hectáreas de bosques por día. https://larepublica.pe/sociedad/1381707-peru-pierde-427-hectareas-bosques-dia
2. British Broadcasting Corporation (2014) In pictures: Illegal logging in peru. https://www.bbc.com/news/world-latin-america-28926270
3. Shermeyer J, Haack B (2015) Remote sensing change detection methods to trackdeforestation and growth in threatened rainforests in Madre de Dios, Peru. J Appl Remote Sens 9(1):1–15
4. Huang C, Goward SN, Masek JG, Thomas N, Zhu Z, Vogelmann JE (2010) An automated approach for reconstructing recent forest disturbance history using dense landsat time series stacks. Remote Sens Environ 114(1):183–198
5. Ankayarkanni B, Leni ES (2014) Object based segmentation techniques for classification of satellite image. Indian J Comput Sci Eng 5(3):120–125
6. Motohka T, Shimada M, Uryu Y, Setiabudi B (2014) Using time series palsargamma nought mosaics for automatic detection of tropical deforestation: a test study in Riau, Indonesia. Remote Sens Environ 155:79–88
7. Koller R, Samimi C (2011) Deforestation in the miombo woodlands: a pixel-basedsemi-automated change detection method. Int J Remote Sens 32(22):7631–7649
8. Menaka E, Kumar SS, Bharathi M (2013) Change detection in deforestationusing high resolution satellite image with haar wavelet transforms. In: 2013 International Conference on Green High Performance Computing, pp 1–7
9. European Space Agency (2020) Sentinel-2. https://sentinel.esa.int/web/sentinel/missions/sentinel-2
10. Geogeek (2019) List of band combinations in sentinel 2. https://geogeek.xyz/list-of-band-combinations-in-sentinel-2a.html
11. Jeppesen JH, Jacobsen RH, Inceoglu F, Toftegaard TS (2019) A cloud detection algorithm for satellite imagery based on deep learning. Remote Sens Environ 229:247–259

12. Jensen JR (2015) Introductory Digital Image Processing: A Remote Sensing Perspective, 4th ed. Pearson Series in Geographic Information Science, May 2015
13. Sistema Nacional de Información Ambiental (2019) "Presencia del sectorambiente en ucayali," Ministerio del Ambiente (MINAM) del Peru, Technical report https://sinia.minam.gob.pe/download/file/fid/65889

The Method of Ontological Design of Safeguarding International Civil Aviation Against Acts of Unlawful Interference

Nikolai I. Plotnikov(✉)

Scientific Research Project Civil Aviation Institute «AviaManager», 630078 Novosibirsk, Russian Federation
am@aviam.org

Abstract. This work presents a method of ontological design, developing a complex of safeguarding international civil aviation against acts of unlawful interference. The complexity of international, legal, and political cooperatives countering terrorism presents the problem as almost insoluble. Terrorism manifests itself on a global and local scale. Overview of the hijacking in the XX Century and at the beginning of the XXI century is done. The analytical review was compiled based on many years of statistical research carried out over decades. Subject identification of acts of unlawful interference is studied. It is composed of a historical socio-political portrait of terrorism in the XX century. The historical content of the international regulatory framework is presented. Developing a complex of safeguarding is completed. The formulation of protection objectives is in the directions of the strategy for developing programs. Each area, goals, and plans are formulated in accordance with the methodology of strategic management. A normative description of the standard for countering aviation terror is compared with existing studies and analogs. The result of the implementation of this method is a new concept of protection against AUI event and a fundamentally different normative content, which differs from the ICAO standard.

Keywords: Aviation terrorism · Civil aviation · Acts of unlawful interference · Regulatory standards · Hijacking

1 Introduction

This paper presents a method of developing a complex of protection against aviation terrorism. Numerous studies of the phenomenon of terrorism and the growth of global threats show geopolitical, multicultural, religious, economic sources, and reasons for its manifestation. All new resonance acts of terrorism make the problem an absolute global priority. The complexity of international, legal, and political cooperatives countering terrorism presents the problem as almost insoluble. Terrorism manifests itself on a global and local scale. The most well-known practice of countering terrorism and creating security complexes in countries where the impact of terrorist acts occurs continuously and for decades, most often, refers to the experience in Israel. The manifestation of

Y. Iano et al. (Eds.): BTSym 2020, SIST 233, pp. 317–333, 2021.
https://doi.org/10.1007/978-3-030-75680-2_36

terrorism in aviation takes a special place. Unlawful interference (UI) in civil aviation (CA) activities, called hijacking, remains a vulnerable area of air transport and the global community. Acts of unlawful interference (AUI) cause a huge public resonance, violate international law, state sovereignty, and aggravate international relations. The uncertainty of AUI events obliges an increase in unpredictable investment in transport safety. This paper provides a historical overview and analysis of the main characteristics of terrorism and AUI.

The purpose of this study is to create a theoretical description of the subject of terrorism in order to develop a new defense approach. The result of the analysis is the compilation of a generalized social portrait of terrorism, on the basis of which global strategies to eliminate this phenomenon can be drawn up. The theoretical content can be the basis for the development of protection technologies. Efficiency is achieved based on the development of the protection circuit against AUI, which allows for a differentiated investment of defense measures. The development differs from the existing International Civil Aviation Organization (ICAO) standards in the acceptable completeness and accuracy of the content and allows for active and effective protection against AUI. The main conclusion is that it is necessary to revise the existing practice of protection from AUI and develop a new methodology for observing the subject area. The method of developing a complex of protection against aviation terrorism shows a new description of the subject of AUI and a normative description of the standard compared with existing studies and analogs.

2 Problem

The problem is characterized by the ambiguity of conceptual descriptions and substantive identification of terrorism, disagreements in the definitions of terrorism in different countries. For more than half a century in world practice, the concept of continuous scanning of people's flows and accumulations, baggage, and cargo has been used to detect threats. This concept is based on (1) constant monitoring of the entire flow, (2) technical devices, (3) due to the attention of security personnel in the visual viewing of monitors and in direct contact with objects. These flows and accumulations are not constant, continuous, and structureless; technical devices that are tuned to known possible threats have limits of technical solvability; human attention is fickle and subject to fatigue.

Any sophisticated invention of terrorists, such as plastic knives for cutting paper or transporting plastids under the clothes of suicide bombers, leads to new resonant terrorist attacks. The substantive vagueness of terrorism complicates the statistical recording, analysis, and planning of preventive measures. Operational protection measures, called preventive, are a control function in a semi-structured space. The normative and regulatory activities of defense against terrorism are predominantly reactive in nature. After the terrorist attacks, "increased" security measures are organized by temporarily increasing the number of employees, increasing attention within the framework of the same accepted methods and technologies for recognizing threats. After some time, everything gradually returns to the previous mode.

The security complex requires huge expenses, new methods, and technologies of protection. Thus, the set of problems consists of a) inadequacy of the identification of the object of terrorism, b) inadequacy of the observation method by semi-structured continuous scanning of threats, and c) the reactive nature of counter-terrorism strategies. The solution to the problem of global and local countering terrorism is possible both by combining international efforts and by an original initiative of approaches and the development of methods and mechanisms to ensure protection from terrorist influences, acts of unlawful interference. It is the second of the indicated approaches to resolving the problem of terrorist influences that determines the purpose of this work.

3 Subject of Terrorism

Studies of the nature and sources of violence and aggression in human behavior reveal many reasons: noise, pollution of the environment, television, instability of societies, etc. The incompleteness of knowledge is especially noticeable if to consider that the dependence of aggression in time is not revealed: a) an act of violence can be regular, periodic, and one-time in the life of an individual; b) violence more often does not coincide and is in no way connected with the intake of alcohol, drugs and other similar substances, which are the traditional culprits of inappropriate behavior; c) aggression can appear against the background of absolute mental health. The importance of studying terrorism is that the phenomenon of aggression and violence can manifest itself in a completely unpredictable way with global consequences and shocks.

The phenomenon of terrorism is considered as a way of influencing society through intimidation. In the term "terrorism", terror means horror, fear. Terrorism includes the use of violence and threats of destruction against people, groups, and states to fulfill personal goals. Terrorism is classified as an international crime, dangerous on a global scale, and requiring international countermeasures. The difference between terrorism and other types of violence lies in the following: it is always organized, specially prepared violence; it is the whipping up of maximum fear and uncertainty of the outcome for people; everyone is potential victims, terrorism has virtual targets for attacking anyone, anywhere, anytime, in any, least protected places and least protected people.

4 Overview of the Hijacking in the XX Century and at the Beginning of the XXI Century

4.1 General Overview

The term "air terrorism", as hijacking, means abduction of aircraft, air piracy, air banditry. The vulnerability of CA as an object and target of aviation terrorism has historically been predetermined. Aircraft hijacking poses minimal danger to a criminal and a great threat to passengers and crew lives. The consequences of terrorist attacks in aviation are very grave. The world's first seizure of the Fokker-7 by the Pervian revolutionaries took place in 1930 in order to drop leaflets over compatriots. During the period from 1930–67, there were 65 cases of hijacking of aircraft and two peaks of violence in 1948–51 and 1958–61. In 1970–79, there were more than 700 cases of attempted hijacking of aircraft, in which over 1,100 people died, more than 1,000 people were injured. Over the same period, over 10 thousand threats of seizure and sabotage were registered. In the USSR, from the first attempt in 1958 to 1979, 22 attempts were made, 6 hijackings of aircraft abroad, in which 10 people were killed and 7 people were wounded. Further, there is an escalation of terrorism more than five times. In 1990, 32 actions were carried out: 23 attempts and 9 cases of hijacking abroad. The organized hijacking of four planes by terrorists and the mass death of people on September 11, 2001, became a new unprecedented challenge to all of humanity [1].

Seizure and hijacking are regarded by a criminal as one of the most effective means of achieving the intended goal using minimal forces and means. Capture and hijacking can be carried out by a small group. The crime is committed individually, regardless of whether the perpetrator is a loner or an accomplice in a criminal group. The operating groups of criminals consider the seizure of the aircraft to be the easiest and cheapest way to gain popularity, release terrorists who are serving their sentences, and obtain a ransom. Escape to the state that provides them with asylum is of no small importance to them. In the event of a disaster, most of the passengers are killed, the terrorist remains anonymous.

4.2 Statistical Data Analysis

The analytical review was compiled on the basis of many years of statistical research carried out over decades. Detailed initial data are contained in works [1–12]. In the last century, it is the largest work, Frederic Dorey [3]. The initial data are processed by the author of this paper in monograph [13] in such a way that all known statistical indicators are converted into ratios (coefficients). Thus, it is obtained the quantitative parameters necessary for the subsequent analysis, which were not in the primary sources; for example, see data (5,7) in the rightmost column, Table 1.

Table 1. Attitude of seizures and victims of AUI 1930–1981 [3]

Continents	Seizures	Victims	b/a
	a	b	
Sum	116	656	5,7

Detailed analysis in several tables is presented in work [13]. An aggravation of terrorism accompanies international political transformation, the disintegration of countries, restructuring political and military blocs and alliances, and opening and closing borders. It is quite obvious that in the last third of the XX-th century and at the turn of the Millennium, there was an escalation of terrorism and a demonstration of even greater uncertainty and unpredictability of the phenomenon. The completed statistical survey of aviation terrorism can be supplemented by taking into account the acts of violence of the last decade, starting with the terrorist attack on September 11, 2001, in the United States. It should be recognized that the prevailing parameters of observations and analysis of statistical data constitute moderate opportunities for the development of normative decisions. However, the data is of reasonable completeness and accuracy, upon which protection strategies and technologies can be derived. This work's main conclusion is the need to revise the existing theory and practice of protection against AUI and develop a new methodology for observing the subject area. The survey of studies demonstrates the following analytical observation structure, Table 2.

Table 2. Statistical data analysis

Parameters	Conclusions
AUI amount	The largest number of AUI occurred on the American continents. The number of AUI has an indefinite character, without the connection of peaks with other events
Amount attempts and captures	Of the three acts, there are two captures per attempt. The capture chance is 0.71
Amount dead and wounded	The ratio of the number of wounded and dead, as well as the number of deaths to the total number of victims, is random
AUI amount and victims	There are about six victims per act: dead and wounded
AUI amount completed and unfinished	Completed captures could be completed or incomplete for some reason. The completed AUI is about half of the total number of acts

4.3 Domestic Air Terrorism

The introduction of a terrorist in the role of an air transport employee is due to the fact that in this role, the terrorist has the greatest capabilities to organize the delivery of threats and to commit a terrorist attack in many spaces. There have been six cases in the past 40 years (investigation: Air Crash Investigation Series, National Geographic). It is known who carried out the attack on September 11, 2001 - terrorists with professional flight training. In this case, service and passenger roles are mixed, which is an additional task of protection. The author of this work, being a captain of IL-14, is a witness of the event on September 26, 1976, in Novosibirsk as a result of a deliberate collision of the An-2 plane and crash with a five-story residential building. The cause was identified as the pilot's mental disorders associated with schizophrenia. The topic of internal domestic terrorism is touched upon here for the subsequent development of a method of protection.

5 Subject Identification of Acts of Unlawful Interference

5.1 Identifying the Phenomenon

The problem of protection from AUI is the difficulty of identifying the phenomenon of terrorism in various aspects. Historically, the regimes of violence and fear represent state structures: military, repressive, dictatorial, despotic, totalitarian, and revolutionary.

Until now, the attribution of state regimes and social movements to terrorism or progress is a question of the adepts' worldview. Examples include peasant uprisings, the terror of revolutions and civil wars, "Red brigades of Italy", "Irish Republican army", Peruvian "Sandoro Lumino", Spanish Basque group ETA, in Palestine FATAN, Japanese red army in Lebanon JAL, in France Action Directe, secret American army Acala, Kurdish People's Party in Iran RKK, the leftist trend in Turkey (Dev Sol), Muslim Brotherhood against Tourism in Egypt (Muslim Brotherhood), Palestinian group DELP-JC, pro-Iranian group (Hisbolla), Palestinian organization HAMAZ, international Islamist Sunni terrorist organization "Islamic State".

The identification of aviation terrorism has historically changed in terms and conceptual descriptions: armed seizure and hijacking of aircraft, unauthorized transportation of dangerous goods, dangerous hooliganism of passengers in flight, immigration, political blackmail, criminal blackmail, kidnapping, smuggling, hooliganism, murder, suicide. Including terms: for the purpose of obtaining insurance for relatives, drug trafficking, sabotage, shelling from the ground with portable missiles, flights in radioactive zones. Various names "air piracy", "air hijacking", "air terrorism" are used.

5.2 Object Identification

The complexity of object identification lies in the countless list of targets of attack, including centers of communication and public life: educational institutions and embassies, airlines and air terminals, airports and ticket agencies, trains and railway stations, metro and post office, hotels and libraries, churches and temples. The main objects are people: politicians, military leaders, diplomats, statesmen, business people, priests, heads of religious denominations, and especially children.

5.3 Causal Identification

The causal identification of terrorism can be formulated in a wide variety of descriptions: uneven distribution and distribution of world wealth and benefits among peoples; difference of cultures and views, national and ethnic roots; ideological and political origins; separatism and weak cooperation of the international community; the periodic resurgence of racism and nationalism; religious fanaticism and fundamentalism; neglect of criminal law; distrust of the institution of the state as a means of protecting a person; imitation and search for fame; lack of a common understanding of the phenomenon of terrorism.

5.4 Technological Aspect

The technological aspect of terrorism contains the use of the achievements of science, technological progress and inventions, education, the latest weapons and tools, missiles and radio-controlled devices, chemical, and biological drugs. The impact of cyberterrorism can be many times greater than physical threats: computer viruses, placing scary announcements, disabling networks and control systems, satellite communications with an aircraft.

5.5 Definitions Problem

There are many conceptual descriptions and definitions of the concept of terrorism [1, 7, 9]. The work [9] gives the following definition: "Terrorism is an international or domestic organizational and other activity aimed at creating a special organization and groups for committing murders and attempted murders, inflicting bodily harm, the use of violence and the capture of people as hostages, forcible imprisonment of a person, coupled with mockery of the person, the use of torture, blackmail; terrorism can be accompanied by the destruction and looting of buildings, small premises and other objects." The keywords in the definition of terrorism are "threat", "symbolic act of intimidation", and "violence". Subject identification complicates the conceptual description and definition of the phenomenon of terrorism. This paper examines a part of the subject area of terrorism related to air transport. Based on the aim of the research, it is composed of a historical socio-political portrait of terrorism.

5.6 Historical Socio-political Portrait of Terrorism of the XX–XXI Centuries

"A group who hijacked an aircraft for political reasons, which ended in a counter-terrorist operation, with the death and injury of people". The generalized portrait of terrorism is a global social group of organized terror against a huge variety of spheres of activity of targets of violence. The number of acts of terrorism is random, not determined by the targets of violence. Acts of violence have a significant likelihood of a completed result, with the number of victims one or two orders of magnitude higher than the number of terrorists. Terrorism spreads for geopolitical reasons in countries with contrasting social inequalities. Accordingly, global strategies can be formed of two types: a) counteraction and protection against terrorism; b) elimination of the causes of terrorism.

6 Safeguarding International Civil Aviation Against Acts of Unlawful Interference Regulatory Framework

6.1 Short Historical Content of the International Regulatory Framework

The normative work of international cooperation in civil aviation predates the phenomenon of aviation terrorism and is an understanding of the global nature of air transport. This is noted in the following international acts: 1919 - the Paris Civil Aviation Convention, the formation of the International Commission for Air Navigation ICAN; 1928 - Pan-American Convention on Air Navigation in Havana; 1968 - The Hague, Convention for the Prevention of Unlawful Seizure of Civil Aircraft, Document No. 8920. From that moment on, almost continuous international cooperation began to unite efforts against aviation terrorism. 1963, Tokyo, Convention on Offenses and Certain Other Acts Committed on Board the Aircraft. 1970, The Hague, Convention for the Suppression of Unlawful Seizure of Aircraft. 1971, Montreal, Convention for the Suppression of Unlawful Acts against the Safety of Civil Aviation. 1988, Montreal, Protocol for the Suppression of Unlawful Acts of Violence at Airports Serving International Civil Aviation, Supplementary to the 1971 Convention. 1991, Montreal, Convention on the Labeling of Plastics for the Purpose of Detection. At the Hague Conference in 1970, the term "illegal seizure of civilian aircraft" was adopted. Various sources cite the classification of 11 groups of acts of unlawful interference as discrimination against aviation security [1, 8, 9].

6.2 Scope and Content of the Concept Acts of Unlawful Interference

The common concept of "unlawful interference" in the documents [14–16] is established as a kind of generic concept "Aviation security": 1. a set of measures, as well as human and material resources, designed to protect civil aviation from acts of unlawful interference; 2. a state of protection of aviation from illegal interference in activities in the field of aviation. The concept of "act of unlawful interference" means unlawful actions, including terrorist acts, which may threaten the security of the transport complex. AUI is defined as "acts or attempts to commit acts that endanger the safety of civil aviation and air transport" and are divided into six groups:

1. Unlawful seizure of aircraft in flight.
2. Unlawful seizure of aircraft on the ground.
3. Taking hostages on board aircraft or at airfields.
4. Forced entry onboard the aircraft at the airport or at the location of the air navigation facility or service.
5. Placement on board the aircraft or at the airport of a weapon, a dangerous device, or material intended for criminal purposes.
6. Reporting false information that endangers the safety of the aircraft in flight and on the ground, passengers, crew members, ground personnel or the public, at the airport or at the location of facilities or services of civil aviation.

In groups 1–3, the criminals, under the threat of using a weapon or an explosive device against passengers and crew members, establish their control on the aircraft and put forward various demands: political asylum, changing the flight route, releasing accomplices from prisons, ransom for releasing hostages. In groups 4–5, bringing weapons or an explosive device onto the aircraft with the aim of capturing or detonating it in the air or on the ground; the laying of explosive devices inside or outside the passenger station, the use of weapons against passengers inside the building of the passenger station in places of their accumulation; carrying an explosive device or weapon on the body or in the hand luggage of a passenger; placing an explosive device in the passenger's baggage; laying an explosive device onboard the aircraft; laying an explosive device in mail, cargo, or onboard supplies. In group 6 of the reports of false information, the majority of the subjects of threats are not detected and are left unpunished.

Operational preventive measures include: inspection of aircraft before departure; screening of passengers, crew members, and carry-on baggage with them; inspection of checked baggage; inspection or other security controls of cargo, mail, and airborne supplies; special security measures when transporting convicted criminals, deportees, as well as when transporting weapons and armed persons who have the right to be on board the aircraft with weapons; the creation of restricted-access zones at airports, the introduction of an access regime for airport personnel and vehicles, as well as due diligence checks when hiring persons with access to these zones. Aircraft security at all stages of their operation and maintenance, adherence to access, and internal regulations at airports, checking personnel loyalty.

7 Developing a Complex of Safeguarding Against AUI

7.1 The Problem of Uncertainty of Safeguarding

Defending against a terrorist attack or preventing an event can be described as the mathematical task of recognizing images of the appearance of a terrorist with threat tools in the spaces that cause the greatest damage and victims. The complex of protection against AUI is considered in the spaces and areas of civil aviation responsibility, in crowded places, at airports, on an aircraft. This paper considers the problem of protection against the events of penetration of a terrorist and the commission of a terrorist attack on the aircraft. The task is fundamentally more complicated since the terrorist is in passengers' flow with a probability of about $P = 10^{-7}$. The terrorist is absent among the passengers if there was an intention to plant an explosive device and not to fly. The tools of the threat can be an explosive device, poisonous substances, flammable liquids, and other objects that are placed in the passenger service space in luggage, carry-on luggage, under clothes. The threat tools can be placed directly on the aircraft: from the service area of the airport, from the service area of the aerodrome. A terrorist may have options for behavior and intentions: to fly or place threats and remain outside the aircraft, survive or commit suicide, commit an act on the ground or in flight. The terrorist can be a passenger or an air transport worker. The description of the formalized conditions for blocking threats, the protection contour, and the AUI event's exclusion is the solution to the problem.

7.2 Resource Method of Ontological Design

The observed subject area is described by the space-temporal characteristics and properties of the physical world's quantities, which requires the use of mixed types of observations: qualitatively evaluated and quantitatively measured. The method of designing a complex of subject area observation is understood as heuristic identification and description of objects, one of which is defined as the observation base. Observation of each component requires selecting and defining the names of the categories and parameters of observation, the definition and formalization of the interconnections of the parameters, types of quantities, scales, and units of measurement and/or estimation. The resource method of protection against AUI events proposes to investigate objects in the category of relations: observation of a threatening object in the observation base, namely in the protected object's space - the aircraft. This technique concentrates on protection resources and is called information folding. Thus, the proposed method concept has two main elements: (1) protected object, (2) information resources folding [13].

7.3 Protection Complex: The First Information Resources Folding

Formal Task Statement. To formulate and solve the task, the formal apparatus of the symbolic designation of all protection complex elements is further applied. It is composed of the following content of the AUI subject area:

- E - AUI event;
- M - protected object (observation base);
- T - threatening object.

The E^{AUI} event occurs if a threatening object appears in space. The event does not occur $\overline{E}^{AUI}$ if the penetration of a threatening object into the aircraft is blocked by the defenses of the protected object M. The general condition of the problem is written (1):

$$\left\{ \begin{array}{c} M(T) \rightarrow f(E^{AUI}),\ \text{event occurs} \\ M(T) \rightarrow f'(\overline{E}^{AUI}),\ \text{event does not occur} \end{array} \right\} \tag{1}$$

Where f and f' are the functions of the initial and ultimate protection in solving the problem.

Objects are structured in subsets of values: parameters, consequences, observation scales, and protection objectives.

AUI Event Structure. AUI event is structured in the parameters of moral and material consequences E_i: human casualties E_H and/or material damage E_M. The consequences can be observed on the scales of relationships (2):

$$E^{AUI} = (E_H,\ E_M);\ e_i, i = e_h, e_m \tag{2}$$

Protected Object Structure. The protected object (M) and the set of its indicators and values M_i are structured in subsets: the space for placing threat objects M_S and delivery channels of threat objects $M_€$. In turn, the placement spaces M_S and a set of its indicators and values for M_j are structured in subsets: passenger m_{cp}, service m_{cc}, commercial m_{cc}, technical m_{ct}. Threat object placement spaces can be observed in order scales. Spaces are: passenger - salon, hand luggage shelves, toilet; service – cockpit crew, service compartments of the cabin crew - kitchen and others; commercial - compartments of baggage, mail, cargo; technical - external and internal compartments and hatches for operational maintenance of aircraft. Delivery channels $M_€$ are structured in subsets: passenger m_{cp}, commercial m_{cc}, technical m_{ct}. The delivery channels are: passenger - the flow of escorting passengers from the entrance to the airport, the entrance to the terminal to placement in the passenger cabin; commercial - moving baggage, mail, cargo to the aircraft; technical - the movement of service and technical transport in the area of the airport and airfield to the aircraft. Threat object delivery channels can be observed in interval scales.

Thus, the connection of the elements of the protected object is (3):

$$M = (M_S, M_C) : \left\{ \begin{array}{l} M_S, i = m_{sp}, m_{ss}, m_{sc}, m_{st} \\ M_C, i = m_{cp}, m_c, m_{ct}. \end{array} \right. \tag{3}$$

Threatening Object Structure. The terrorist alone and the threat tools only are not a threat object. Threatening object (T) and many of its indicators and values T_i are manifested in joint and simultaneous states of the threat subject (terrorist) $|T_S|$ and tools of threat $|T_D|$ (4):

$$T_i : (T_T|T_D) \rightarrow T \tag{4}$$

There are two options for creating a threat object: (1) a threatening object is created if the terrorist and the tools of threat are combined; (2) a threatening object becomes as tools of threat separately if it is placed in the space of the protected object via delivery channels. Thus, formulated the behavior of the threatening AUI object: For all M_{Si}, if the $M_{\in i}$ delivery channel resource is used, it is possible to create a threatening object based on the condition of the threatening object state. Record (5):

$$\exists(M_{Si}|M_{Ci}) \rightarrow (T_{Si}|T_{Di}) \rightarrow T_i \tag{5}$$

The terrorist T_S is structured in subsets of social roles: stp passenger, stc airline employee. A terrorist can be observed in the nominal scales. Tools of threat T_D are structured in the parameters: under the clothes of a terrorist-passenger t_{dp}; in hand luggage t_{dc}, in baggage t_{db}; in cargo and mail t_{dm}; in the space of the aircraft directly and/or from the service space of the airport and airfield t_{da}. Means of threat can be observed in order scales.

Thus, the relationship of elements of a threatening object is recorded (6):

$$T = (T_S, T_D) : \begin{cases} T_S, i = s_{tp}, s_{ts} \\ T_D, i = t_{tp}, t_{dc}, t_{db}, t_{da} \end{cases} \tag{6}$$

The presented concept of the complex of protection of the aircraft from AUI consists in the fact that the resources of object protection and blocking the channels of threat delivery to the aircraft must be equal to or exceed the resources of the threatening object (7):

$$\overline{E}^{aui} \subset (M_S \cup M_C) \geq (T_S \cup T_D)\# \tag{7}$$

In practical terms, changes can be made. For example, the motives of a terrorist can be considered in the roles of survival or suicide bomber. The latter requires a special version of threat recognition. Obviously, the threat parameter "cargo and mail" is also formed outside the airport space. Accordingly, this requires a separate task of recognizing the threat and describing the delivery channel, Table 3.

The developed protection package considers the most vulnerable part of the transport complex – the aircraft. When considering the threat of congestion of people and passengers in the airport areas, it is necessary to introduce appropriate data and other security complex structures. The parameters are structured in the indicators for each indicator are established in schools calculated values in units, and other data.

7.4 Protection Complex: The Second Information Resources Folding

A protection contour (PC) is a set of protection resources against the most plausible, possible, and probable combinations of threat subjects and threat tools ($T_{Si}T_{Di}$) in the placement spaces and delivery channels ($M_{Si}M_i$) of the protected object on the Car- tesian product of sets: {threat subjects} x {threat tools} x {delivery channels} x {placement spaces} (8):

$$(T_{Si} \text{ x } T_{Di}) \text{ x } (M_{Si} \text{ x } M_i) \tag{8}$$

Table 3. Subject area of the aircraft protection complex

Objects		Parameters	Symbol
1	3	3	4
Event, E^{AUI}	Victims	Moral	e_h
	Damage	Material	e_m
Threatening object |M|	Spaces for placing threatening objects, M_S	Passenger	m_{sp}
		Service	m_{ss}
		Commercial	m_{sc}
		Technical	m_{st}
	Threat delivery channels, M_C	Passenger	m_t
		Commercial	m_c
		Technical	m_p
Threatening object |T|	Terrorist, T_S	Passenger	t_{sp}
		Employee	t_{ss}
	Threat tools, T_D	Under clothes	t_{dp}
		Hand luggage	t_{dc}
		Luggage	t_{db}
		Cargo, mail	t_{dm}
		Aircraft	t_{da}

The PC is compiled by heuristic expertise as a tactical version of the implementation of the strategy for the defense of the aircraft from AUI. It is drawn up a three-color heatmap in a matrix of three rows and four columns of the aircraft protection contour. Using an expert method, it is drawn up the most probable conditions for organizing threats along the delivery and placement channels in the aircraft's space, Table 4.

Each cell of the matrix has the following designations: a name in the form of an alphanumeric number at the top of the cell; assessment in the "heat map" ("traffic light") scales in terms of risks: green - neglected, yellow - acceptable, red - marginal; assessment of the possibility of creating a threatening object ($T_{Si}T_{Di}$); quantitative assessment of a threatening object's size on a 10-point scale: 5–10 red, 2–4 yellow, 0–1 green. Table cells without ratings indicate negligible risk and the unlikely or impossibility of creating a threat object and AUI event. Completed columns indicate the likelihood or possibility of one or more types of creating a threatening object. Let us give examples of filling the cells of the protection contour.

- "3A": risk measure "red" "marginal", $(t_{sp}|t_{dp})$ x $(t_{sp}|t_{dc})$ the possibility of placing a threat under clothes and in hand luggage; quantitative assessment 10 - the largest value of a threatening object from the prevailing number of passengers.
- "3B": risk measure "red" "marginal", is formulated in natural language (NL) as follows: $(t_{ss}|t_{dp})$ "a service subject in the service space through the passenger de- livery

Table 4. Aircraft protection contour matrix

Threat delivery channels, M_C		Spaces for placing threatening objects, M_S			
		Passenger m_{sp}	Service m_{ss}	Commercial m_{sc}	Technical m_{st}
		A	B	C	D
Technical m_t	1	1A green $t_{ss}\|t_{da}$ 1	1B green $t_{ss}\|t_{da}$ 1	1C yellow $t_{ss}\|t_{db}$ 2	1D yellow $t_{ss}\|t_{dm}$ 4
Commercial m_c	2	2A green $t_{ss}\|t_{da}$ 2	2B yellow $t_{ss}\|t_{da}$ 2	2C red $t_{sp}\|t_{db}$; $t_{sp}\|t_{dm}$ 5	2D yellow $t_{ss}\|t_{da}$ 3
Passenger m_p	3	3A red $t_{sp}\|t_{dp}$; $t_{sp}\|t_{dc}$ 10	3B red $t_{ss}\|t_{dp}$; $t_{dp}\|t_{dc}$ 7	3C yellow $t_{ts}\|t_{dp}$ 0	3D green 0

channel underclothing is most likely to create a threatening object"; ($t_{ss}|t_{dp}$) "the creation of a threatening object is most likely by a passenger in the service space through the passenger delivery channel in hand luggage"; quantitative score 7, less than in "3A", since passenger access to service spaces is possible, but significantly limited.

- "1A" and "1B": risk measure "green" "neglected", the possibility of placing a threat through the passenger channel in the aircraft cabin and service compartments is possible during cleaning, the size of the threatening object is estimated as a minimum (1).
- "3D": risk measure "green", "neglected", the possibility of placing a threat through the passenger channel into the technical compartments of the aircraft is practically excluded, the value of the threatening object is zero.

Similarly, it is necessary to describe and fill in all the cells of the matrix. In the development of such matrices, the structure and the number of possible manifestations of a threatening object in the protection circuits can be described in a different configuration.

7.5 The Solution of the Task

The task of the PC is formulated as follows: "for all spaces of the placement of threatening objects along the delivery channels, the protection resources are greater than or equal to the possible manifestations of the threatening object." The task is written (9):

$$\exists \mathrm{M(T)} = \sum_{i=1}^{M} \mathrm{M_{Si}|M_{Ci}} \geq (\mathrm{T_{Si}|T_{Di}}) \rightarrow f'(\overline{E}^{AUI}) \tag{9}$$

In the probability space (Ω, F, P), the space of elementary outcomes (Ω) is the set of all possible elements of the AUI event. Estimating the probabilities (P) of events and their combinations presupposes the formation (F) of the σ-algebra of subsets of the set Ω generated by events from the set (M_S, M_D), that lead to the event (E^{AUI}) or exclude the event $\left(\overline{E}^{AUI}\right)$. The measure P($E^{AUI}$) is the probability heuristically estimated by experts. In the heuristic assessment, estimates of conditional probabilities are established: P($E^{AUI} \mid (M_i, M_j)$), based on a set of values of indicators derived by the soft estimation method using risk matrices compiled according to the principles of heatmap "traffic lights" models.

8 Design of a New Regulatory Framework of Safeguarding Against AUI

8.1 Safeguarding Strategy Development

Protection strategies contain a) strategic, b) regulatory, and c) technical developments and solutions. Let us present an example of drawing up a strategic plan and an example of a normative design of safeguarding CA against AUI. In the example of the development of a PC, it is carried out an assessment of the resources of the aircraft protection from AUI. Based on the qualitative and quantitative assessment of each cell of the matrix, it is possible to prioritize strategic tasks and decisions. Identification of a threatening object and inclusion in the PC is the task of economic feasibility of investing in security, Table 5.

The strategy assumes the effectiveness of protection in the following areas:

- Protection of the space of placement of objects of threat on the aircraft.
- Blocking channels for the delivery of threats to the aircraft.
- Recognition and blocking of the terrorist.
- Technical design of means of recognition and blocking of threats.

The formulation of protection objectives is a detailing of the directions of the strategy for developing programs. For each area, goals and plans are formulated in accordance with the methodology of strategic management used in this paper and developed in work [17].

8.2 New Safeguarding Standard Development

This is a proposal of a New Safeguarding Standard Development for ICAO Annex 17 to the Convention on International Civil Aviation Against AUI [16]. Chapter 4 of this document is summarized in the following sections.

CHAPTER 4. Preventive Security Measures
4.1 Objectives
4.2 Measures related to access control
4.3 Aircraft measures

Table 5. Strategic objectives of defense against AUI

Strategy name	Points	Task priorities
3A	10	Protection of the passenger spaces for the placement of objects of threat on the aircraft by blocking the passenger channels for the delivery of threats to the aircraft
3B	7	Protection of passenger spaces for placing threats on an aircraft by blocking service channels for the delivery of threats to aircraft
2C	5	Protection of commercial spaces for the deployment of threats on aircraft by blocking commercial channels for the delivery of threats to aircraft
1D	4	Protection of technical spaces for the placement of objects of threat on an aircraft by blocking technical channels for the delivery of threats to the aircraft
2D	3	Protection of technical spaces for the deployment of objects of threat on an aircraft by blocking technical channels for the delivery of threats to the aircraft
2A	2	Protection of the passenger spaces for the placement of objects of threat on the aircraft by blocking commercial channels for the delivery of threats to the aircraft
1C	2	Protection of service spaces for the placement of objects of threat on the aircraft by blocking the technical channels for the delivery of threats to the aircraft
1A	1	Threshold of feasibility of protection costs
1B	1	
3C	1	
3D	1	

4.4 Measures concerning passengers and their carry-on baggage
4.5 Measures concerning baggage carried in the hold
4.6 Measures concerning cargo, mail and other items
4.7 Measures concerning special categories of passengers

The existing content does not distinguish the concepts between protected and threatened objects. The concepts of delivery spaces and threat placement spaces are not clearly distinguished. There is no description and definition of what constitutes a threat. There is no content regarding the recognition and blocking of subjects and objects of threat. The main disadvantage of such content is the incompleteness of the development of the content of the observed subject area of the AUI, in the absence of the category of the observation base and the category of objects relations. Let us show the design of the new content of the document chapter.

CHAPTER 4. Measures to block the AUI threat in CA

4.1. Objectives. The tasks of blocking the AUI threat
4.2. Protected object
4.2.1. Tasks of protection and control of the threat placement space on an aircraft
4.2.1.1. Threat accommodation passenger spaces
4.2.1.2. Threat deployment service spaces
4.2.1.3. Commercial threat hosting spaces
4.2.1.4. Threat placement technical spaces
4.2.2. Tasks of protection and control of the threat delivery space
4.2.2.1. Threat delivery technical spaces
4.2.2.2. Commercial threat delivery spaces
4.2.2.3. Threat delivery passenger spaces
4.3. Threatening object
4.4. The tasks of recognizing and blocking threat subjects
4.4.1.1. Threat subject, passenger
4.4.1.2. Threat subject, employee
4.5. Threat objectives
4.5.1.1. Threat items placed under clothing
4.5.1.2. Threat items in carry-on baggage
4.5.1.3. Threat items in baggage
4.5.1.4. Threat items placed in cargo and mail
4.5.1.5. Airborne threats

9 Conclusion

To compare the structure of the document proposed, it is recommended the reader to look at the full content [16], which is used by the entire global CA community. In the newly developed content, it is established the definitions and terms of the AUI subject: protected object, threat, threatening object, subject of threat, delivery space, placement spaces. It is noted that the objects of threat differ from the abstract concepts of "threat" and "threatening object". The subject of threat is what must ultimately be found through the technical design of recognition and blocking tools. Only after its detection, the object is identified as a threatening object. The most important thing in the method is the choice of the observation base. The observer (defense manager) focuses on the protected object and not on the threat and terrorist. The detailed structuring of the AUI subject, according to the method, assumes different observation bases - time, groups, space. It seems that bringing this theoretical content to a standardized technology can lead to practical results of the method's applicability.

The method contains a new ontological model for developing standards and guidelines for joint actions of people on board in a state of extreme danger and ground flight operations and emergency services. The result of the implementation of this method is a new concept of protection against an AUI event, which differs from the ICAO standard [16] by the subject identification of terrorism, structured definitions of the concepts of danger and threats. The method contains the design of a protection complex against the AUI event and the reduction of this complex as a formal-logical protection contour. The protection contour is a pareto-optimal choice for countering threats through the channels of delivery of threats and the areas of penetration of the terrorist into the aircraft.

Practical significance: a) a radical change in organizational and technical protection measures that clearly define the delivery channels and the penetration space of threats into the protection contour, b) meeting public needs in air transportation, service providers - airlines, airports, and security services in active security measures and reducing the risks of emergency events of the AUI, c) saving lives, property and reducing the cost of protecting public facilities and aircraft by pareto-optimizing the structure of the antiterrorist protection complex.

References

1. Pidzhakov AY (2002) Legal regulation of the fight against international terrorism: Textbook. allowance. - SPb .: Ed. Center SPbGMTU. – p 262 (in Russian)
2. Cotron MJ (1989) The growing threat of terrorism. The futurist, July-August, pp 20–24
3. Dorey CF (1983) Aviation Security. Frederic C. Dorey– NY. – p 383
4. Jenkins BM (1987) The future course of international terrorism. The Futurist, July-August, pp 8–13
5. MacKenzie-Orr (1988) Aviation security in an age of terrorism. Flight Safety Digest 7(12):1–7
6. Shung-chai H (1987) Hijacking fall to 10-year low. FSF Flight Safety Digest. 6(9):10–13
7. Wilkinson P (1990) The future of terrorism. Futures. – October 1990, pp 493–504
8. DeJohn CA, Wolbrink AM, Larcher JG (2004) In-Flight Medical Incapacitation and Impairment of US Airline Pilots: 1993 to 1998. – DOT/FAA/AM-04/16 Final Report. – p 25
9. Karpets II (1979) International crimes. - M .: Legal Literature Publ. – p 264. (in Russian)
10. Ovodenko AA (2008) International terrorism: political analysis of risks and security strategies: in 3 volumes / SPb: Vol. 1: Globalization and security risks: trends in scientific analysis. - 493 p. Vol. 2: Models and strategies of international terrorism risk management. - 451 p. T. 3: International terrorism: Bibliography. Documents. Materials of socio-logical research – p 480 (in Russian)
11. Rundo AY (2011) Terminology problems in life safety. Safety in the technosphere. No 2. - S. 50–58. (in Russian)
12. Poyda AA (2002) Terrorism: Bibliographic index: 1975–2001. – Barnaul, p 21 (in Russian)
13. Plotnikov NI (2013) Resucy bezopasnosty transportnych kompleksov. Monographia. [Transport complex safety resources. Monography], Novosibirsk, AviaManager Publ, p 286 (in Russian)
14. ICAO (2016) Annex 19 to the Convention on International Civil Aviation. Safety Management. Second Edition, July 2016. - 999 Robert-Bourassa Boulevard, Montréal, Quebec, H3C 5H7 Canada, p 44
15. The Safety Management Manual (2018) (Doc 9859) AN/474, Fourth Edition, 999 Rob- ert-Bourassa Boulevard Montréal, Quebec H3C 5H7 Canada, p 149
16. ICAO (2011) Annex 17 to the Convention on International Civil Aviation, Security. Safeguarding International Civil Aviation Against Acts of Unlawful Interference. Ninth Edition March 2011. - 999 University Street, Montréal, Quebec, Canada H3C 5H, p 56
17. Plotnikov NI (2012) Consultant. Reorganization. Monograph. Ed. 2nd. - Novosibirsk, Avia-Manager Publ, p 592. (in Russian)

Design and Implementation of an ECG Recording System for *in Vivo* Experimentation in Rats

Raphael Santos do Nascimento[1](✉), Fernando da Silva Fiorin[2], and Jefferson Luiz Brum Marques[1]

[1] Institute of Biomedical Engineering, Federal University of Santa Catarina, Florianopolis, SC, Brazil
jefferson.marques@ufsc.br

[2] Laboratory of Neurobiology of Pain and Inflammation, Federal University of Santa Catarina, Florianopolis, SC, Brazil

Abstract. Rat electrophysiological measurements are widely used in experimental protocols to study heart diseases, drug assessment, and their interaction. However, commercially employed rat ECG recording systems demand a high cost for implantation due to the increased signal conditioning complexity and experimental arrangements compared to the ECG from a human subject. These issues preclude its greater use for experimental research in emerging countries. In this perspective, we designed and implemented a low-cost, simplified system for recording the cardiac electric activity in rats, meeting the minimum specifications for suitable data processing. The hardware developed in the system consisted of four blocks, including conditioning, conversion, transmission, and system power supply; the acquired signals are stored and processed on a personal computer. Our system demonstrated capacity for recording ECG signals successfully with low-noise and minimum requirements. Also, data analysis showed that with simple filtering was possible to identify the R peak, allowing the further analysis of ECG rhythmicity and morphology and calculation of heart rate variability indexes. Therefore, we concluded that this low-cost system proved to be adequate for ECG recording in rats. Nonetheless, we encouraged further work to better assessment in the different experimental protocols and other rodents.

Keywords: Electrophysiology · ECG recording · Low noise amplifier · Rat

1 Introduction

In recent decades, the recording and analysis of biomedical signals have been increasingly used in clinical medicine. This growth is mainly due to the rapid advances in the area of medicine and medical instrumentation, which has allowed us to better understand the acquired physiological signals [1]. In particular, the electrocardiogram (ECG) has become one of the more useful tools for the detection and evaluation of heart complications, both clinically and in experimental research animals. In this context, ECG signals

Y. Iano et al. (Eds.): BTSym 2020, SIST 233, pp. 334–341, 2021.
https://doi.org/10.1007/978-3-030-75680-2_37

from rats have been widely recorded in several experimental models for the extraction of information on cardiac electrical activity and physiology [2].

Although the ECG recording in humans is successful, in rats it is challenging yet. Such difficulties arise from significant attenuation on the signal due to the electrodes being positioned in different regions of the rat's body and a low ECG signal amplitude when compared with the human ECG recording [3]. Consequently, these factors require a more robust ECG recording system for the acquisition of an adequate signal from rats. Although the recent works suggested that the use of a telemetry system is the gold standard for electrophysiological data recording in rats, this requires a minimum body weight of animals, a high cost of implementation, and still an external power source, which is not viable for most research laboratories. Furthermore, a surgical procedure is necessary for the use of this technology, which may compromise the evaluation of recorded ECG data and animal well-being [4].

Therefore, the literature has reported that to an adequate ECG signal processing (i.e., ECG interval and morphology, ECG rhythm, and calculation of heart rate variability parameters), an ECG system must cover a sampling frequency typical or greater than 500 samples per second and with a minimum of 12-bit resolution for analog-to-digital data conversion [5–7]. Although, other studies argued that lower sampling frequencies (e.g., 250 and 100 Hz) might be used with mathematical interpolation [8, 9]. Nonetheless, most modern designed systems and, consequently, more expensive present specifications and computing power to sample at 1 kHz or more for data processing [10, 11].

In this study, we aimed to develop an integrated system for recording the cardiac electrophysiological activities in rats. The system was designed with a focus to reduce cost, smaller physic size, and considering the minimum parameters for suitable data processing in the research experimental settings.

2 Materials and Methods

2.1 Animal and Data Acquisition Method

For this study, we used one adult male Wistar rat (weighing about 400 g, 170 days of age) obtained from the Federal University of Santa Catarina animal house. The animal was maintained on laboratory-controlled conditions (water and food standard), housed in a plastic cage covered by shavings, being allocated in a ventilated rack with other animals. The experimental procedures for ECG acquisition on the animal anesthetized were performed as previously described in a study with some minor modifications [12]. Briefly, an inhaled isoflurane administration (2.5% mixed with 100% of oxygen) on the animal was done. Posteriourly, the rat was placed in a dorsal recumbence position, and an electrocardiograph gel (Ultra-Gel Electro, MultiGel Industry and Commerce, Import and Export of Products for Health., Silvianópolis, Minas Gerais, Brazil) was applied. Three disposable electrodes in the lead I configuration were positioned on the animal, and the ECG signals were recorded. Before the beginning of the recording, the rat was carefully scraped in the electrode positioning areas for gel application. Immediately after the recording end, the animal was euthanized following the experimental protocol settled. All procedures were conducted in compliance with the Brazilian legislation regulated by the National Council for the Control of Animal Experimentation (CONCEA). All procedures performed on the animal were also approved by the Committee in Care and Use of Experimental Animal Resources of the Federal University of Santa Catarina.

2.2 System Description

The hardware designed consisted basically of four central blocks: (1) signal conditioning, responsible for amplifying, filtering, and adjusting the signal; (2) data conversion, accountable by the data digitalization for further processing; (3) data transmission, which performed the data receiving and the transmission to storage in a computer, and (4) battery-power supply, responsible by powering all blocks integrated into the full system (Fig. 1).

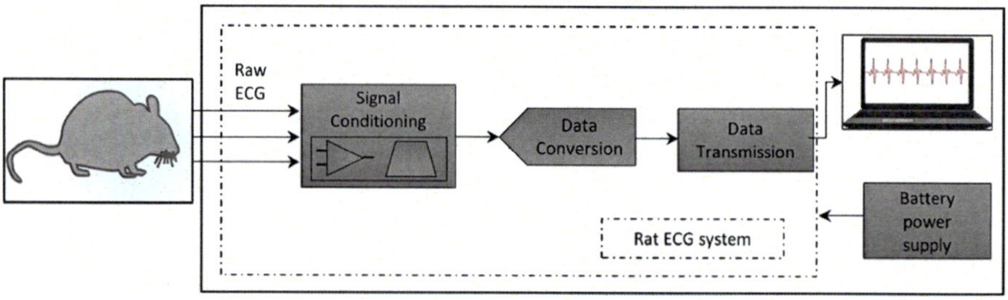

Fig. 1. General block diagram of rat ECG recording system design.

Signal Conditioning. For this study, we considered the use of lead acquisition cables of 30 cm (Herald Technologies Inc., Florence, South Carolina, USA) with push-button type connectors, isolating possible external electromagnetic interference. The raw signal collected (8 min) from the rat was received initially in an instrumentation amplifier (INA129, Texas Instruments Inc., Texas, USA), specific for the use of lower voltage (high CMRR = 120 dB and low-noise voltage density = 8 nV/$\sqrt{}$Hz) (Fig. 2). Additionally, this is an instrumentation amplifier of low-power and with high accuracy for medical instrumentation application. In this, we set the gain (G = 92,48) using the Eq. (1) provided in the datasheet, where the value of resistor Rg determined the gain (G) attributed to the signal. Subsequently, the signals were conditioned by two analog filters designed in compliance with the electrocardiograph features of rats, extracting only the signal frequency bandwidth of interest.

The filtering process consisted of a high-pass filter (first order, 0.1 Hz) with unit gain and a low-pass filter (Butterworth, fourth-order, 40 Hz), with a fixed gain in the first (G = 1.15) and the second (G = 2.235) stage. Posterioursly, it was implemented a gain adjustment circuit (G = 1 to 11) due to the low amplitude evidenced in the electrocardiogram of rats. For the implementation of the analog-to-digital converter, an appropriate level shifter circuit was constructed in order to ensure only positive voltage in the converter input. Finally, we also included a driven right leg circuit under the reference electrode to reduce possible noise of interference (e.g., common-mode), which the electrodes, the animal, as we as the wires could be susceptible to. All conditioner circuits designed in this block were implemented using operational amplifiers (LM324N, Texas Instruments Inc., Texas, USA).

$$G = \frac{49,4\ k\Omega}{Rg} \tag{1}$$

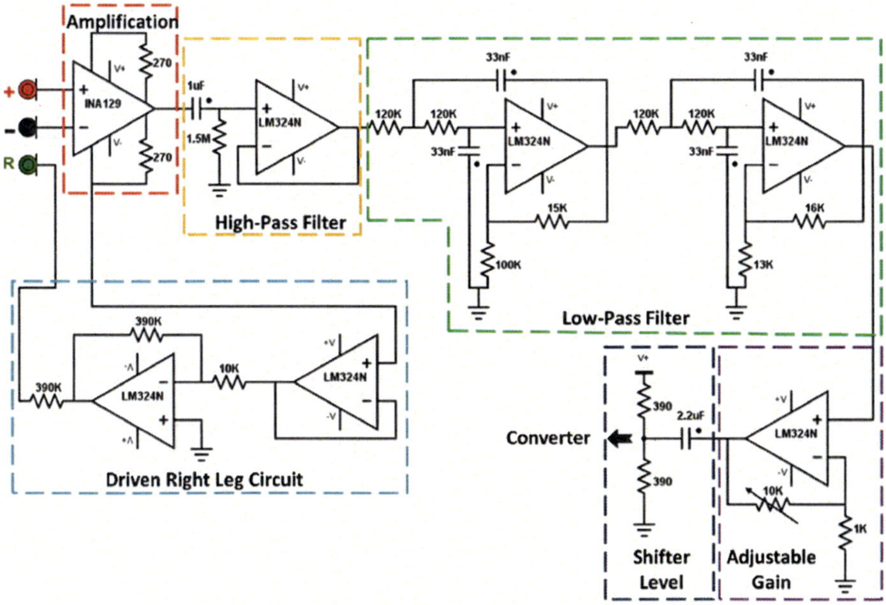

Fig. 2. General schematic circuit of the signal conditioning block.

Data Conversion. The signals were digitalized through a 12-bit analog-to-digital converter (ADS1015, Texas Instruments Inc., Texas, USA), which has a 3.3 kHz sampling rate constructed with sigma-delta topology (high precision conversion). This setting met the specifications of our system and was compatible with the data communication. In this way, the signals were converted and sent to the transmission block through the I2C communication protocol.

Data Transmission. To minimize our recording system's cost and size, we used a microcontroller Arduino (Arduino Nano board v.3.0., San Jose, California, USA) to transmit the recorded data. The data were read in a digital input in the Arduino platform and sent to a personal computer (Intel Core I3 [4.0 GB of RAM and 1TB of disk capacity]), where these data were stored for further analysis. The microcontroller provided data with voltage amplitude points in text files (.txt). We set the sampling rate in 500 samples per second, respecting the minimum sampling limits for the correct digital signal processing.

Battery-Power Supply. The full recording system's supply was carried out by a designed power supply circuit based on one 9 V battery (Fig. 3). The circuit was constructed with three integrated circuits, accompanied by passive elements (resistor and capacitor). Initially, we implemented a voltage converter (ICL7660, Renesas Technology Corporation., Koto, Tokyo, Japan) to receive the battery supply and converter from positive to negative. Posteriorly, it was used more two voltage regulators applied to the derived voltages (+9 V from the battery and −9 V from the ICL7660), for 5 V positive output (MAX666, Maxim Integrated Products, San Jose, California, USA) and a second with 5 V negative output (MAX664, Maxim Integrated Products, San Jose, California, USA), according to the supply specifications required by our system. The passive elements were added jointly with the integrated circuits as suggested by the fabricators, for reduction of noise and to better dynamic performance. Due to possible interference, which could be attributed to the signals (such as 60 Hz noise from powerline) and our system needing the symmetrical power supply, the use of the Arduino platform +5 V auxiliary power source was discarded.

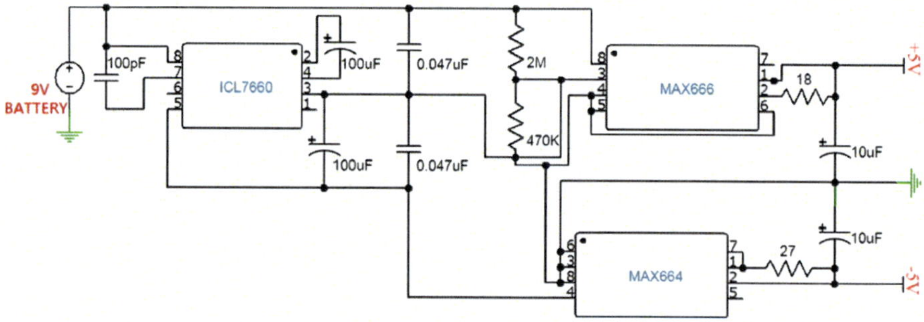

Fig. 3. Battery-power supply circuit design.

2.3 Data Analysis

The analysis of signals was done through a customized algorithm in Matlab (MathWorks Inc., Natick, MA, USA), where the signal variability was detected (R peaks), creating a time series with all R peaks. Moreover, we estimated the frequency spectrum on the data, using the Fast Fourier Transformation (FFT) with the specific bandwidth for rats (0–3.5 Hz) [13]. Before the analysis, the signals were digitally filtered by a band-pass filter (0.1–40 Hz). Although a longer recording of ECG had been collected, it was considered for processing and analyzing a sample of 5 min devoid of noise and artifacts.

3 Results and Discussion

In this study, we described with details an integrated system for ECG recording in rats. Our system was designed with a particular focus on low-cost and smaller size; however, implementing the minimum specifications for the possible further data processing. In this sense, all blocks designed in this system were integrated on a double-sided printed circuit board (110 × 70 × 1.6 mm) (Fig. 4a), which was posteriorly made into a customized polylactide box for full system packaging (125 × 80 × 100 mm) (Fig. 4b). Regarding the system features, this met a sampling frequency of 500 samples per second with a resolution of 12-bit, which is equivalent to several commercially employed systems in recent experimental research in rats [14, 15]. Our system also demonstrated low noise and interference, even in the absence of any shielding approaches used.

During the development of the system, we observed that the battery-supply and the instrumentation amplifier were essential for low noise recordings (in particular common-mode and 60 Hz of powerline). Nevertheless, it is noteworthy that external procedures such as the animal scraping and the electrocardiograph gel application also contributed significantly to the signal's improvement. Hereupon, our system was able to provide a raw rat ECG signal recording satisfactory, as observed in the data analysis (Fig. 5a). Although there is the presence of lower interference attributed to the raw signal, after the application of simple digital filtering, the unwanted noise frequencies were mostly eliminated (Fig. 5b).

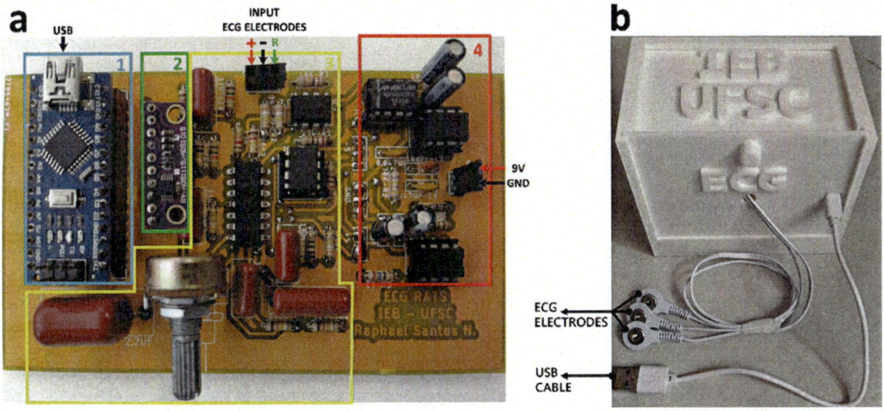

Fig. 4. ECG recording system. (A) Implemented printed circuit board (top view) and (B) Customized box for full system packaging.

From the filtered ECG signal was possible to detect the R peaks (Fig. 5c) and, consequently, estimate the frequency spectrum for the rat bandwidth (Fig. 5d). It is important to emphasize that the ECG analysis is commonly performed based on R peak detection and processing. Thus, for the detection of peak R, the rat ECG signal must be adequate for detection and processing. Moreover, the algorithm created must deal with the signal variation on time and under the different types of noise present on the rat ECG. Therefore when the R peak can be detected, the R-R intervals between consecutive heartbeats can also be calculated, allowing the ECG interval and morphology assessment, ECG rhythmicity, and computing the heart rate variability indexes [5].

The data show that the ECG system presented here to rat application provided a recording approaching harmless, safe, and at a lower cost, which can be easily employed with high repeatability. In this line of thinking, the present rat ECG recording system can be useful for investigating heart alterations in different experimental research designs.

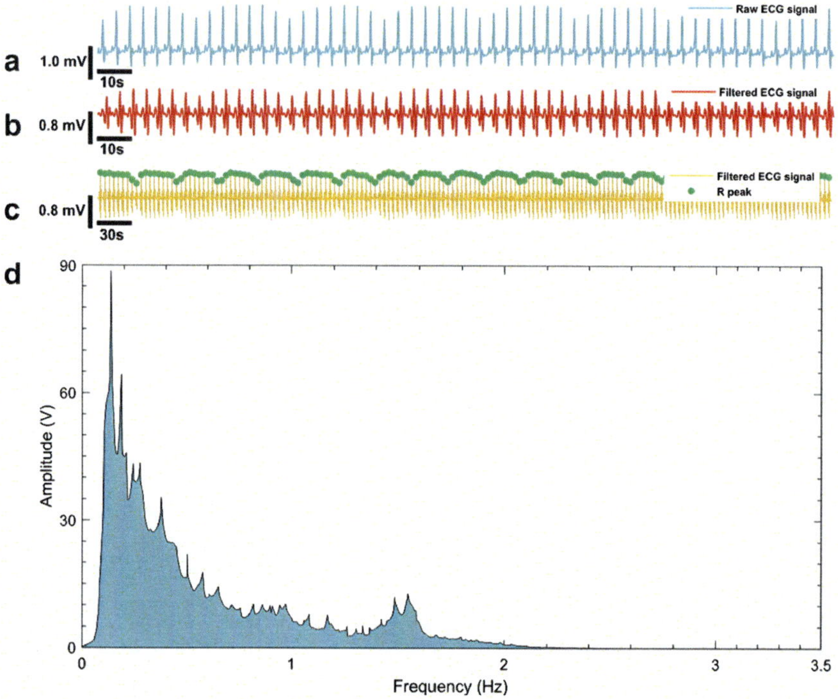

Fig. 5. Rat ECG signal representation. (A) Raw ECG signal, (B) ECG signal after digital filtering, (C) Detection of the R peaks, and (D) Frequency spectrum estimation of the signal.

4 Conclusion

In summary, we herein develop an ECG recording system for application in experimental research with rats, employing lower cost and simplified design. Moreover, the raw signal analysis allowed us to observe the full functionality and effectiveness of our system for cardiac electric activity recording in rats. Nonetheless, further studies using a greater number of animals are required to better illustrate this device's superiority and compatibility in the different experimental research settings and purposes.

Acknowledgments. The authors gratefully acknowledge the student fellowships from the National Council for Scientific and Technological Development (CNPq) (to RSN) and the Coordination of Superior Level Staff Improvement (CAPES) (to FSF). JLBM is a grantee of the CNPq research productivity fellowship.

References

1. Biswal D, Uvanesh K, Champaty B, Ray SS, Pal K (2014) Development of an ambulatory universal biopotential recording device. In: International Conference on Control, Instrumentation, Communication and Computational Technologies, pp 1124–1129. IEEE. https://doi.org/10.1109/ICCICCT.2014.6993130
2. Serhani MA, El Kassabi HT, Ismail H, Navaz AN (2020) ECG monitoring systems: review, architecture, processes, and key challenges. Sensors (Basel) 20:1796
3. Wang T-W, Lin S-F (2020) Non-contact capacitive sensing for ECG recording in small animals. Sci Technol
4. Niemeyer JE (2016) Telemetry for small animal physiology. Lab Anim 45:255–257
5. Task Force of the European Society of Cardiology & The North American Society of Pacing and Electrophysiology (1996) Heart rate variability: standards of measurement, physiological interpretation, and clinical use. Eur Heart J 17:354–381
6. Bolanos M, Nazeran H, Haltiwanger E (2006) Comparison of heart rate variability signal features derived from electrocardiography and photoplethysmography in healthy individuals. In: Engineering in medicine and biology society. IEEE, pp 4289–4294
7. Nazeran H (2006) Electrocardiography, computers in. In: Webster JG (ed) Encyclopedia of medical devices and instrumentation. Wiley, pp 34–52
8. Kwon O et al (2018) Electrocardiogram sampling frequency range acceptable for heart rate variability analysis. Heal Inf Res 24:198–206
9. Ziemssen T, Gasch J, Ruediger H (2008) Influence of ECG sampling frequency on spectral analysis of RR intervals and baroreflex sensitivity using the EUROBAVAR data set. J Clin Monit Comput 22:159–168
10. Arini PD, Liberczuk S, Mendieta JG, Santa Mariá M, Bertrán GC (2018) Electrocardiogram delineation in a Wistar rat experimental model. Comput Math Methods Med 2018:2185378
11. Pereira-Junior PP, Marocolo M, Rodrigues FP, Medei E, Nascimento JHM (2010) Noninvasive method for electrocardiogram recording in conscious rats: feasibility for heart rate variability analysis. An Acad Bras Cienc 82:431–437
12. Botelho AFM et al (2019) Non-invasive ECG recording and QT interval correction assessment in anesthetized rats and mice. Pesq Vet Bras 39:409–415
13. Nascimento RS, Fiorin FS, Santos ARS, Royes LFF, Marques JLB (2020) A new approach for ECG recording in rats: an autonomic nervous system analysis. In: Proceedings of the 5th Brazilian Technology Symposium, Smart Innovation, Systems and Technologies, vol 202 Springer
14. Jafarinezhad Z, Rafati A, Ketabchi F, Noorafshan A, Karbalay-Doust S (2019) Cardioprotective effects of curcumin and carvacrol in doxorubicin-treated rats: Stereological study. Food Sci Nutr 7:3581–3588
15. Kalkan F et al (2018) Protective and therapeutic effects of dexpanthenol on isoproterenol-induced cardiac damage in rats. J Cell Biochem 119:7479–7489

Sentiment Analysis of Song Lyrics Using Clustering

Miguel Vásquez-Leon and Willy Ugarte(✉)

Universidad Peruana de Ciencias Aplicadas (UPC), Prolongación Primavera 2390, Lima 15023, Peru
u201423494@upc.edu.pe, willy.ugarte@upc.pe

Abstract. Music often helps people to relax and have fun times. However, the search for this in a personalized way based on the feelings of the users is not present in the main search engines. The closest thing to this would be the playlists created arbitrarily by other users. For this reason, we propose a method that allows us to label artists representing different feelings according to their lyrics.

Keywords: NLP · Machine learning · Psychology · K-means

1 Introduction

Music is a universal language that is understood worldwide [1]. It crosses the borders of race, nationality, culture, and religion. We use music to accompany various moments of our day to day, to express our emotions; we could even say that music is the language of emotions. Currently, when trying to find music, you have to search by artist, soundtrack, playlist, or album. This is because search engines are not prepared to seek to adapt to the user's feelings [2]. The closest thing to this is user-created playlists on music streaming platforms [3], but these playlists lack tonality or lyric analysis, so the content is totally arbitrary. For what we consider ideal, carry out a true analysis of lyrics to find ideal songs according to the mood of the users. Additionally, we consider that this research is really relevant, given that music usually allows people to relax, be distracted, and even have fun moments doing different activities. So, in the global context that we are currently daring (global pandemic by Covid-19), where people have been forced to quarantine for months, it could help alleviate stress, boredom, or some type of people related problem[1]. This work passed and was approved by the faculty's Ethics Committee (and faculty name). Our contributions are as follows:

- We have carried out a lyrics analysis, thanks to natural language processing, in order to identify their positiveness and negativeness.

[1] "How Music Therapy Is Easing Anxiety and Isolation" - https://bit.ly/3o9iQ7o.

Y. Iano et al. (Eds.): BTSym 2020, SIST 233, pp. 342–350, 2021.
https://doi.org/10.1007/978-3-030-75680-2_38

– We have grouped these results into various groups in order to identify the predominant emotions in the songs matching with psychology studies.

The rest of the paper is organized as follows: Section 2 introduces the main concepts and methods related to our research. In Sect. 3, we explain in detail the experimentation that was carried out. After, Sect. 4 presents the works related to our approach. Finally, conclusions and future works are presented.

2 Methods

In this section, a brief introduction to the main concepts that we used is given.

2.1 Material

One of the main purposes of Machine Learning is to enable automatic learning for computers (i.e., without any human intervention).

Definition 1 (Clustering [4]). *Basically, a Clustering is a set of groups regrouping without intersecting all the data (point) objects.*

There are many proposals for computing clustering since there is no precise notion of a "cluster".

Definition 2 (K-Means [5]). *This is a clustering algorithm that regroups data objects into k clusters based on their similarities by minimizing the sum of the distances between each data object and its cluster centroid.*

K-Means has three steps:

1) *Initialization*: a number k of clusters must be chosen, then k centroids are randomly set.
2) *Assigning objects to centroids*: each data object is assigned to its nearest centroid.
3) *Update centroids*: each centroid position is updated to a new centroid that is the average of cluster objects.

Assignation and *Updating* are repeated until the centroids stabilize (i.e., they do not move anymore). K-means tackles an optimization problem by minimizing the quadratic distances sum of each data object to its respective centroid.

$$\min_{s} E(\mu_i) = \min_{s} \sum_{i=1} \sum_{x_j \in S_i} ||x_j - \mu_i||^2 \qquad (1)$$

The centroids are updated based on the averages of the elements present in the group.

Artificial intelligence (AI) is a field of study of computer science, which seeks to simulate the thinking of human beings processed by computer systems. It seeks to simulate the processes of the human being for tasks such as planning, movement, reasoning, learning, social intelligence, and even creativity.

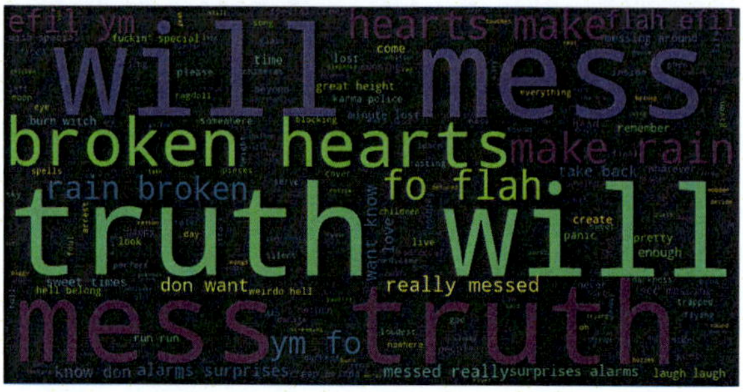

Fig. 1. Happy and sad words for Radiohead songs.

Definition 3 (Natural Language Processing (NLP) [6]**).** *In artificial intelligence, NLP is the branch that allows computers to understand, interpret, and manipulate human language. For this, the text is analyzed to group it, categorize it, and extract the structure and meaning of large volumes of words.*

NLP is usually used for content classification and text analysis.

Definition 4 (Sentiment Analysis [7]**).** *NLP, text analysis, and computational linguistics were used to identify, extract, and study affective states in text.*

A sentiment analysis was performed on lyrics of popular music artists[2] of various periods by analyzing the positivity, negativity, and neutrality present in the lyrics. Figure 1 shows the most relevant words for the band Radiohead.

2.2 Selection of Data

Our main goal is to be able to detect as many feelings as possible, independently from the musical genre or period. Therefore, we choose 100 artists and conduct an in-depth analysis of the prevailing sentiment according to their ten best-known songs. Additionally, we opted to choose English as the only language, given the great variety of English-spoken artists, and it straightens the analysis having a single language. For the selection of artists, we take as a basis the ranking of the 100 most important musicians of the last decade according to Billboard[3].

Then, to contract a little more current, we selected some names of the artists of the moment according to Billboard. We also selected some of the most renowned artists in history, according to Rolling Stone magazine[4] (e.g., the Beatles and Led Zeppelin). The aim was to cover the greatest possible variety.

Finally, we use the Genius API[5] to obtain the lyrics for each artist's songs from known databases. This API extracts all the lyrics from the database with the same structure, facilitating the pre-processing of the data.

[2] https://bit.ly/3dIU2gZ.
[3] Billboard - https://bit.ly/2ZqRs9Z.
[4] Rolling Stone Magazine - https://bit.ly/3iiCU5a.
[5] Genius API - https://bit.ly/2BUsF68.

2.3 Data Pre-processing

Even if the aforementioned API extracts lyrics data with the same structure, once this data is analyzed, it presents some inconsistencies (e.g., wrong words, inconsistent titles, ambiguity, etc.). We had to manually verify that each chosen song was from the chosen artist. For instance, when searching for "Queen", most of the results were from the band, but some were from "Beyoncé", so it ended up combining both when pulling the data. For this reason, we identify the artists having this conflict (e.g., Evanescense and Madonna), and replace them. We repeat the same process until no such conflict was found. For musical collaborations, as they represent one of the artists, this did not cause any problems.

2.4 Data Cleaning

As we mentioned before, it was very helpful that all the lyrics had the same structure. Since it allowed us to quickly identify that lines that began with square brackets "[" "]" or similar were usually used to indicate parts such as verses or choruses, so in the first place, we eliminated all these lines since they did not contribute to our data of study; we also removed the blank spaces. Like [8], we use the NLTK[6] to remove stopwords (i.e., self meaningless words).

2.5 Data Analysis

Thanks to NLTK, we could evaluate words (singularly or combined) to obtain three scores: positive, negative, and neutral.

Selection of Groups. As mentioned before, we seek to identify as many feelings as possible according to studies in psychology [9–11]. Various tests (the first with two groups, the second with four groups, and finally with eight groups) were conducted for this purpose. Assigning the artists to each group is made with their collected scores in the previous step. Thanks to K-means, any number of clusters can be computed based on these scores. After grouping the musicians, all the data from their songs were taken and grouped to be analyzed.

Match with Psychology. Once the lyrics were grouped, by using NLTK, we are able to identify the predominant words in each group, aiming to associate them with different emotions given various psychology studies. For the first case (see Fig. 2), there are only two groups[7]; we choose to divide it into the more general emotions from which the others stem: potential and negative. The second step is a little bit more complicated since it had to dig a little deeper to determine what emotions you would be looking for. Figure 3 shows the most important of

[6] https://www.nltk.org/.
[7] Emotions - https://bit.ly/3nqliq8.

them: *anger, fear, sadness, and joy* [12]. Finally, a well-known scheme in this field, called Plutchik's Wheel of Emotions, defines a wide variety of emotions but presents eight basic ones: aggressiveness, optimism, love, submission, awe, disapproval, remorse, and contempt.

Fig. 2. Positive and Negative emotions

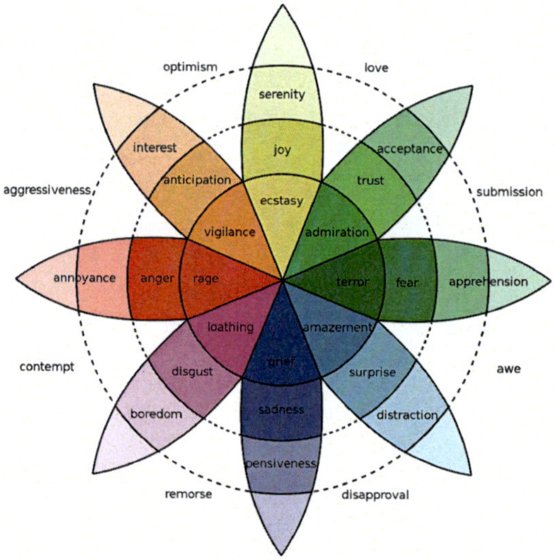

Fig. 3. 4 Basic Emotions and Plutchik wheel of emotion [12]

3 Results

Now, we present experiments to show the feasibility of our approach.

3.1 Experimental Protocol

By using the collected data (see Sect. 2.2), we could establish three scores for each song (positivity, negativity, and neutrality) for each song in the dataset thanks to NLTK. K-Means clustering is a fast, robust, and simple algorithm that gives reliable results when data sets are distinct. From Fig. 3, for this experiment, we propose three scenarios according to these scores: separating the songs into i) 2 groups, ii) 4 groups, and iii) 8 groups.

3.2 Analysis

At the end of the lyrics' processing with NLTK, three different scores were obtained: positive, neutral, and negative. However, this situation does not allow us to access adequate training data nor data labels, so it is not completely clear [5,13]. This led us to the K-means algorithm (see Definition 2), which is an unsupervised clustering technique. Since this algorithm generates groups based on centroids that are calculated based on the "chords" of each of the k groups defined to be able to label our data, these groups will turn out to be the (sentiment) labels.

This algorithm works by pre-selecting the value of K and that value defines the cluster values to be calculated. In general, there is no exact way to determine a value of K, but the most reliable way to achieve this is known as *the Elbow method* [11]; the idea is to run k-means clustering for a range of clusters k and for each value, computing the sum of squared distances from each point to its assigned center (distortions). This test led us to k values of 2, 4, and 8. Therefore, we analyzed each cluster as such for each k.

Scenario 1 (2 Clusters). By separating into only two groups, the result is evident since we can show that the main words of each of the groups allow us to make a clear distinction between positive and negative.

- The first group (see Fig. 4a) contains mainly words as: *"love"*, *"baby"*.
- The second group (see Fig. 4b) stands out for a variety of pejorative terms.

(a) Positive group (b) Negative group

Fig. 4. Scenario 1: Word clouds for 2 clusters

Scenario 2 (4 Clusters). For this scenario, we can see two positive groups and two negative ones, and they share the main words. However, if we look at the accompanying words, we can notice a subtle difference.

- Starting with the group where "joy" predominates (see Fig. 5a), it is accompanied by words such as *"night", "take", and "yeah"*.
- The second group (see Fig. 5b) presents words such as *"go", "leave", and "die"*, related to sadness.
- Contrarily, the third group (see Fig. 5c) contains words like *"never", "hell", and "alone"*, most of them related to negativity but going more on the side of fear.
- The other group (see Fig. 5d) has more predominant swear words like: *"bang bang", "hit", "gun"*, and other violent words that infer anger.

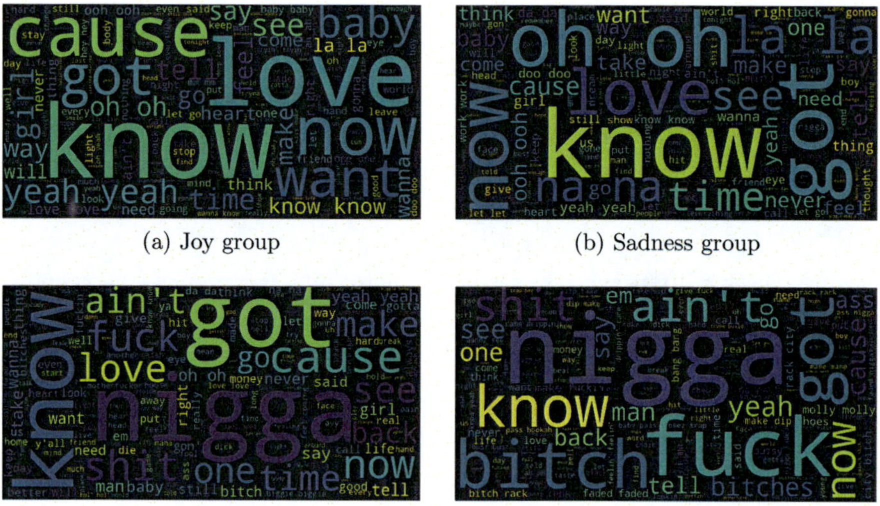

(a) Joy group (b) Sadness group (c) Fear group (d) Anger group

Fig. 5. Scenario 2: Word clouds for 4 clusters

Scenario 3 (8 Clusters). Finally, 8 groups that should fit with *aggressiveness, optimism, love, submission, awe, disapproval, remorse, and contempt.* In this case, making the match is very hard, mostly because there were no advanced studies in psychology and nothing ensured that all feelings were present. However, the anger group is easily identified at first glance since it's labeled identically as previously. Labeling the other seven groups require further psychology expertise.

4 Discussion

In [9], the authors try to directly address three problems: i) Neural networks need a large amount of data, ii) despite the advancement of the concept of natural language processing, there is still a lot of uncertainty regarding the use of word vectors and iii) it is not easy to find a simple and effective configuration for large amounts of data. To tackle the first problem, the effectiveness of various crowdsourcing alternatives is explored. A series of experiments were conducted with large text collections of various contents to address the second problem. Regarding the third problem, a series of experiments with convolutional neural layers and maximum grouping were carried out.

In [14], the authors claim that inferring the emotions caused by a musical piece turns out to be a complex task. Thus, a sentiment analysis method based on data mining is proposed, extracting various characteristics of musical pieces that enable to carry out a multi-label classification. They show that MLkNN was found to be the best algorithm for classifying songs into six different feeling groups. The main difference with our work is that we do not seek the construction of predictive models, and we focus more on clustering, using NLP to infer emotions.

5 Conclusions

In the present work, a sentiment analysis was carried out in order to classify a group of 100 artists, from different musical genres, based on a weighting of their ten best-known songs. It allows labeling them and recommending the best artists to listen to based on the user's mood. As shown in the results, we obtained a clear grouping in two and four mood states. However, trying to classify eight groups was complicated since more advanced knowledge of psychology is required. Finally, it is promising to do the same, but using the melodies directly; it is more laborious but should return better results.

References

1. Fitch WT, Popescu T (2019) The world in a song. Science 366(6468):944–945. https://doi.org/10.1126/science.aay2214
2. Bahraini A, Tilevich E (2019) Ask toscanini!: architecting a search engine for music scores beyond metadata. In: SAC, ACM. pp 796–803. https://doi.org/10.1145/3297280.3297356
3. Hadash S, Liang Y, Willemsen MC (2019) How playlist evaluation compares to track evaluations in music recommender systems. In: IntRS@RecSys. CEUR-WS, pp 1–9. http://ceur-ws.org/Vol-2450/paper1.pdf
4. Estivill-Castro V (2002) Why so many clustering algorithms: a position paper. SIGKDD Explor 4(1):65–75. https://doi.org/10.1145/568574.568575
5. Hartigan JA, Wong MA (1979) Algorithm as 136: a k-means clustering algorithm. JSTOR: Appl Stat 28(1):100–108. https://doi.org/10.2307/2346830

6. Gómez-Pérez, JM, Denaux R, García-Silva A (2020) A practical guide to hybrid natural language processing - combining neural models and knowledge graphs for NLP. Springer. https://doi.org/10.1007/978-3-030-44830-1
7. Rodríguez-Ibáñez M, Gimeno-Blanes FJ, Cuenca-Jiménez PM, Muñoz-Romero S, Soguero-Ruíz C, Rojo-Álvarez JL (2020) On the statistical and temporal dynamics of sentiment analysis. IEEE Access 8:87994–88013. https://doi.org/10.1109/ACCESS.2020.2987207
8. Chen XV, Tang TY (2018) Combining content and sentiment analysis on lyrics for a lightweight emotion-aware Chinese song recommendation system. In: ICMLC. ACM, pp 85–89. https://dl.acm.org/citation.cfm?id=3195148
9. Nielek R, Ciastek M, Kopec W (2017) Emotions make cities live: towards mapping emotions of older adults on urban space. In: WI. ACM, pp 1076–1079. https://doi.org/10.1145/3106426.3109041
10. Çano E (2018) Text-based sentiment analysis and music emotion recognition. CoRR abs/1810.03031. http://arxiv.org/abs/1810.03031
11. Thorndike RL (1953) Who belongs in the family. Psychometrika, 267–276. https://doi.org/10.1007/BF02289263
12. Jack RE, Garrod OG, Schyns PG (2014) Dynamic facial expressions of emotion transmit an evolving hierarchy of signals over time. Cur Biol 24:187–192. https://doi.org/10.1016/j.cub.2013.11.064
13. Kim DM, Kim K, Park K, Lee J, Lee K (2007) A music recommendation system with a dynamic k-means clustering algorithm. In: ICMLA. IEEE, pp 399–403. https://doi.org/10.1109/ICMLA.2007.97
14. Gómez LM, Navarro-Cáceres M (2017) Applying data mining for sentiment analysis in music. In: PAAMS. Springer, pp 198–205 (2017). https://doi.org/10.1007/978-3-319-61578-3_20

Recycling: A User-Friendly Oriented Mobile and Web Solution for Generators and Recyclers in the City of Lima

Sebastian Manrique(✉), Liz Eulogio, and Jimmy Armas

Universidad Peruana de Ciencias Aplicadas (UPC), Av. Primavera, 2399 Surco, Lima, Perú
u201421327@upc.edu.pe, jimmy.armas@upc.pe

Abstract. This paper proposes a solution composed of a mobile application and a web application. The design of the solution has been oriented to two groups of specific users, i.e., generators and recyclers, to integrate the processes in which both are participants through user-friendly interfaces. The formal waste pickers associations in the city of Lima work in precarious conditions with a complete lack of technological infrastructure to support their daily tasks. The proposed solution considers the needs of its users, allowing them to carry out a more effective waste segregation with the help of a simple barcode scanner integrated into an Android application. The solution also contributes to the recycler by helping him to collect and classify recyclable material. A web dashboard was also implemented for management and monitoring tasks. A validation in a test scenario was made for the generator, considering the state of emergency and policies of social isolation imposed by the Peruvian government and the municipality of Lima.

Keywords: Recycling · Mobile application · Mobile solution · Cloud computing · Java

1 Introduction

Nowadays, recycling is seen as one of the measures to combat pollution and its effects on the integrity of the environment and the beings that inhabit it. As such, environmental pollution represents one of the most important and severe problems in today's societies. The United Nations (UN) defines pollution as the introduction of substances or energies whose effects endanger human health, natural resources, and ecosystems [1]. Nations worldwide have been harmed in different areas such as economic, political, and social due to the impact of this problem. The UN states that pollution affects the working and recreational use of the environment and threatens the cultural values that many people attribute to the environment's diversity, both natural and human-made [1].

For this reason, recycling represents one of the best forms of participation in the fight against pollution for natural people from different communities. However, the data indicate that it is not an activity that is carried out as expected for the levels of waste generation in the world. According to the Organisation for Economic Co-operation and

Y. Iano et al. (Eds.): BTSym 2020, SIST 233, pp. 351–358, 2021.
https://doi.org/10.1007/978-3-030-75680-2_39

Development (OECD), in 2017, only 36% of the 675 million tons of waste generated worldwide was recycled, considering the countries that make up that organization [2]. In Peru, only 4% of the 900,000 tons of plastic generated each year is recycled [3]. There are many causes that lead to such low figures in terms of recycling, however, the main responsibility falls on the authorities in charge of the solid waste management process.

In recent years, the evolution of technologies such as the Internet of Things (IoT), RFID, cloud computing, mobile devices, and blockchain has enabled considerable progress in waste management and allowed the emergence of concepts such as Smart Waste Management (SWM). Therefore, activities such as recycling have presented multiple opportunities for the optimization or automation of this sub-process so that its effectiveness is increased and reflected in results.

In this paper, we hypothesize that, in the city of Lima, a solution is required for covering the specific needs of one or more groups of people in terms of the recycling process. Systems and solutions proposed by other authors who contribute to solid waste management in some specific contexts are evaluated. The technologies and other components used in these solutions are analyzed to confirm if the hypothesis is valid and if there is a specific gap to be filled in the field of research and contribution to the recycling process.

This paper is organized as follows. First, a state of the art is made based on some current solution proposals for recycling automation in order to discuss its most relevant characteristics. Then the proposed solution is explained, highlighting its characteristics and functionalities. A case study is then presented to validate the proposed solution in a conditioned environment for a real simulation taking into account one of the identified users. Finally, some conclusions and initiatives for future work are presented.

2 Literature Review

2.1 Solutions and Systems Literature Review

Anagnostopoulos [4] proposed a dynamic model for waste management through a series of infrastructure services for Smart Cities located on IoT. The model uses sensors, radio frequencies (RFID), and actuators for the monitoring process. In general, the set includes the following phases: planning and collection of waste when using routing solutions for trucks that control certain restrictions; transport to a collection center the type of waste to be transported; and, finally, the waste recycling phase for future reuse. The model was dynamically named for its ability to move in real-time to the parameters and planes of interest to waste collection while it is in operation.

Aazam provided a smart cloud-based solution for intelligent waste management consisting of a series of smart containers for each type of waste equipped with sensors that constantly monitor and update their status in the cloud, where different stakeholders are connected to receive information relevant to them [5]. The system not only acts for waste management but also for deciding the best collection route that follows the most economically effective path.

Saha [6] proposed a waste management solution based on providing intelligence to waste containers using a prototype IoT with sensors. It can read, collect, and transmit large volumes of data over the Internet. This information, applied to space-time contexts and processed by optimized algorithms, can be used to dynamically manage the collection mechanisms.

Chowdhury [7] proposed an intelligent waste management model as an improvement in a waste collection system that charges for waste generation. Through RFID, the model allows users to assign the exact cost based on the waste they have produced.

Wahab [8] proposed an intelligent system for recycling waste. The system automatically evaluates the type and amount of waste and provides a benefit in the form of points credited to a card for subsequent exchange of these for items or withdrawals through a bank network, simulating a virtual currency.

Thakker [9] presented a technological solution to the problem of waste containers not being cleaned in time. The system incorporates a monitor that collects an alarm and information for the agents that control that the container is ready to be handled.

2.2 Discussion and Conclusions of Reviewed Literature

From the above analysis, it can be deduced that it is necessary to provide models or technological solutions focused on the type of "citizen" waste generator. All the proposals reviewed cover broad and complex waste management systems from a technological perspective. Most of the proposals include IoT technologies and route optimization; others propose their own designs of intelligent containers for waste management. All proposals have been designed considering the municipal management of the environment in which they have been proposed. Although it is possible to take these cases as references to propose a solution suitable for the context of the city of Lima, we validate the hypothesis originally put forward in which it is stipulated that there are few solutions that focus on the end-users and the relationship between them.

The proposal will contemplate a waste management solution based on a mobile application that guides and encourages the user to segregate waste. The mobile solution must also be able to integrate the two main actors in the initial stages of the recycling process: segregation and collection; these users will be the generator and recycler. One of the features of the proposed solution includes the ability to track the waste from the generator's point of view in order to have greater traceability in the waste recycling chain, from the moment the waste is segregated and deposited by the generator until it reaches the hands of the recycler for its classification and final treatment. It has been decided to focus the scope of the solution, for the user-generator, to those who reside in condominiums or apartments because the characteristics of coexistence and administration of these places, together with the volume of the waste generation that occurs in them; it is attractive for the validation of the proposal.

3 Proposed Solution

The proposed solution comprises two mobile applications with functionalities designed exclusively for the compliance and integration of waste segregation and collection processes. Two groups of users have been defined for both applications: waste generators and recyclers.

The scope of the solution is focused, for the generators, on those that are waste in condominiums or apartments due to the volumes of waste generated in these places, in addition to the fact that the collection of waste in these places is carried out in specific points which facilitates a later transfer of the waste for its treatment. For the waste pickers, it has been decided to work with those formal associations that work together with the district municipalities in Lima.

As a previous step to designing the solution, a technological model divided into different layers was developed: application, processing, data analysis, and presentation, as we can see in Fig. 1.

Regarding the characteristics and technological tools used to develop the solution, both mobile applications were developed using the Java language. The operating system chosen for the deployment was Android. As part of the solution, a web interface was developed to monitor and systematize administration tasks such as creating, updating, and deleting records in user accounts and databases. The solution, both mobile and web, make use of storage and integration to a PostgreSQL relational database provided by the Heroku platform.

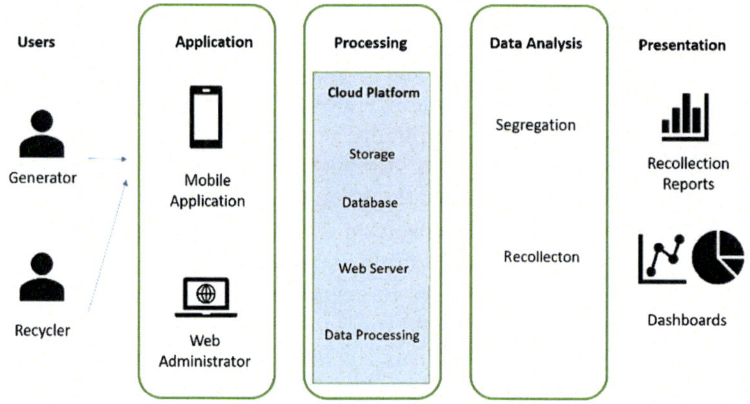

Fig. 1. Technological model of the proposed solution, which consists of a java frontend connected to a database backend.

Below, we detail the flow of basic tasks that each user can perform within their application, contrasting the integration between both processes.

3.1 Identification and Grouping of Recyclable Material

After logging in, a generator-type user must access the "Scan" functionality to perform a barcode reading printed on the label or packaging of the product to be recycled. This will provide relevant information to continue the flow of the application, as shown in Fig. 2.

In the first place, you will know if the material you want to recycle is considered, in the current scope of the solution, as it has been designed, considering plastic, paper, and cardboard packaging. Once this validation is done, you will obtain data such as the type of material, the weight of the package, etc. At that time, the user will have to make use, if desired, of the "My Bag" functionality to add the scanned package or packages to a unique virtual container, designed in a similar way to a shopping cart in e-commerce applications.

When the bag is full, the generating user will have to link it with a physical bag that will be previously provided. These physical bags will include a QR code printed with a unique identifier. To link a virtual exchange with a physical exchange, the user only needs to use the "Link exchange with QR" function in the exchange menu. Once this is done, the virtual exchange register will now include the unique identifier present in the QR code of the physical exchange. Finally, the user will place it in the container of his condominium or building.

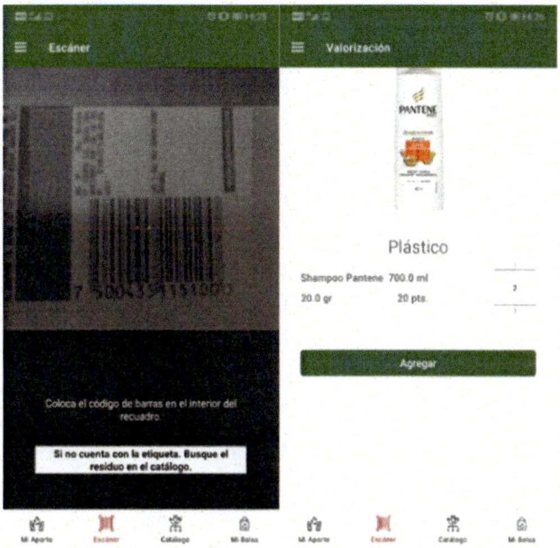

Fig. 2. Screenshots of the following features: scanner and valorization.

3.2 Content Collection and Validation

When the physical bag is already in the residential complex dumpster, the waste picker assigned to that location will make periodic visits as he considers for collection. When you decide to remove the bags from a dumpster, you must use your application and, after logging in, you will access the "Collection" functionality. You will be asked to read a QR code, which should be printed on each physical bag. When reading the code, you will be asked whether you decide to accept the bag or not. If you decide to accept the QR's unique identifier and, therefore, the virtual bag assigned to it, it will belong only to that recycler.

After accepting the desired number of bags, the recycler must check the contents of each bag. To do so, he will access the "Validate" function, which is accessible right after accepting a bag or also from the lower menu bar. Regardless of how to access this functionality, the recycler must scan the QR code of the bag whose contents he wants to validate. The system will evaluate whether the scanned code corresponds to a bag previously accepted by the recycler. If so, the user will be provided with a list of that virtual bag's contents and will have to make the corresponding comparison with the contents of the physical bag. If the user finds any discrepancy between both contents, he can send the set of observations found to the user who registered the bag.

4 Implementation

The application was installed in the cell phone of one of the neighbors of the condominium "Los Jardines"; in addition to this, a user's manual was previously provided with all the necessary instructions for the use of the application. As there is no real and robust database with all the records of products whose packaging is recyclable, test data previously agreed with the user based on his consumption habits was used. The user used the application for 4 consecutive days in which he generated different volumes of recyclable material (plastic beverage bottles). A cardboard container was installed in his home with a capacity to store a volume of 5 L; the purpose of this container was only to collect the bags after the user made the registration. When the container was installed, the user was also provided with 5 biodegradable 30-L capacity bags with unique QR codes printed on each one.

At the end of the 4-day period, the collection process of bags deposited by the user-generator was carried out. For this phase, in principle, it was planned to have the participation of a recycler member of the association ATIARRES, but, given the quarantine circumstances, it was necessary to have the voluntary collaboration of a family member of the generator. He was trained to play the role of a recycler and use the application, and, on the fourth day, he made the respective collection. The user-recycler collected 3 bags from the container and proceeded to validate the content.

During the previous phases, all the records made were stored correctly in the cloud-based database. Thanks to this, it was possible to generate reports in CSV format to demonstrate the potential of the solution as a tool for decision-making and analysis of recycling behavior.

As already mentioned above, the solution also includes a web interface for administration purposes. The user can manage the accounts of generators and recyclers to authorize account creations. In addition, considering that some position of the municipality uses this web interface, more information panels were included, more detailed than those present in mobile applications, as we can see in Fig. 3.

A total of 4 days passed during which the solution was tested in a simulated scenario under quarantine restrictions. This work passed and was approved by the institution's ethics committee.

Fig. 3. Information dashboard for monitoring the collection process.

5 Conclusions

In this paper, a solution was presented, consisting mainly of two mobile applications and a web dashboard designed for two groups of users. These groups correspond to the actors involved in the processes of segregation and collection of waste within the macro-process of recycling. The proposal focuses on integrating the two identified actors to facilitate the participation of each one and automate the activities of segregation and collection.

Future work concerns testing and validation processes for this user since, in a normal context, they carry out their work through collection routes between different areas of the city. It is necessary and mandatory to carry out a validation process that includes the mobility of waste pickers during their working hours.

References

1. United Nations (2017) United Nations Environment Programme Assembly on the Environment. https://papersmart.unon.org/resolution/uploads/k1708350s.pdf
2. Ramirez V (2019) World Recycling Day: OECD countries recycle only 36% of urban waste, lasexta, 17 May 2019. https://www.lasexta.com/noticias/sociedad/dia-mundial-reciclaje-paises-ocde-reciclan-solo-residuos-urbanos_201905175cde6d7f0cf27d3dafd455ec.html
3. RPP Noticias (2020) Peru still only recycles 4% of the 900,000 tons of plastic it discards. RPP Actualidad, 30 January 2020. https://rpp.pe/peru/actualidad/peru-solo-recicla-aun-el-4-de-las-900000-toneladas-de-plastico-que-desecha-noticia-1242755?ref=rpp
4. Anagnostopoulos T (2017) Challenges and opportunities of waste management in IoT-enabled smartcities: a survey. IEEE Trans Sustain Comput 2:275–289
5. Aazam M (2016) Cloud-based smart waste management for smart cities. In: 2016 IEEE 21st International Workshop on Computer Aided Modelling and Design of Communication Links and Networks (CAMAD), 23–25 October 2016
6. Saha H (2017) Waste management using Internet of Things (IoT). In: 2017 8th Annual Industrial Automation and Electromechanical Engineering Conference (IEMECON), 16–18 August

7. Chowdhury B (2007) RFID-based real-time smart waste management system. In: 2007 Australasian Telecommunication Networks and Applications Conference, 2–5 December 2007
8. Wahad M (2014) Smart recycle bin: a conceptual approach of smart waste management with integrated web based system. In: 2014 international conference on IT convergence and security (ICITCS), 28–30 October 2014
9. Thakker S (2015) Smart and wireless waste management. In: 2015 international conference on innovations in information, embedded and communication systems (ICIIECS), 19–20 March 2015

Cybersecurity and Privacy Capabilities Model for Data Management Against Cyber-Attacks in the Health Sector

Humberto Luis Salcedo Jara, Henry Bryan Pérez Navarro, and Jimmy Armas-Aguirre(✉)

Universidad Peruana de Ciencias Aplicadas (UPC), Lima 15023, Perú
{u201312945,u201516210,jimmy.armas}@upc.edu.pe

Abstract. In this paper, we propose a model of capabilities that identify the reliability degree of Cybersecurity and Privacy elements applied to the Health Sector. The increasing interactions between technology and the health sector have brought a new set of risks to be confronted, such as data breaches and cyber-attacks. However, in order to improve, a greater understanding of their current situation is needed. The proposal identifies the capability level for the organizations to know their maturity level comprehensively. This was achieved by selecting existing models, frameworks, and regulations, increasing their complexity, integrating their privacy and cybersecurity capabilities, and health data management. In this way, the proposal is supported by a tool prepared for outcome estimation and diagnosis. The model structure is organized into categories and subcategories, and the assessment is made according to the level of compliance with controls, for which five levels of maturity were defined: 1. Basic, 2. In Progress, 3. Defined, 4. Differentiated, 5. Continuous Improvement. The model was validated and proven in a private hospital in Lima, Peru. The preliminary results are related to the model application in the selected process. As a result, we found that the private hospital obtained a level of maturity of 2-In Progress. Based on this, we made some recommendations to improve the capacities of the assessed health provider. A comparison was made between the proposed model's results and the ones obtained through its root components. Said results were similar, thus proving that a coherent and comprehensive integration was achieved.

Keywords: Cybersecurity · Cyber threat intelligence · Maturity model · Dark web

1 Introduction

The advancement of digitalization in different sectors brings along with its new potential risks. We should not ask whether cyber-attacks and data breaches will happen, but when they will happen.

An IBM report indicates that, since 2014, the average total cost of data breaches globally across industries has increased by 12% from $3.5 million to $3.92 million in 2019 [1]. According to the latest IBM X-Force report, among the most attacked

Y. Iano et al. (Eds.): BTSym 2020, SIST 233, pp. 359–367, 2021.
https://doi.org/10.1007/978-3-030-75680-2_40

industries, health care represented 3% of all attacks in the top 10 industries [2]. Maturity models identify the degree of reliability or dependency that a company can place on a process to achieve desired goals or objectives. As a result, mature processes make it easier to know the current state of the organization, the monitoring, and compliance with regulations and standards [3].

Cybersecurity Maturity models integrate health elements holistically but lack an additional element, which is information privacy. In view of this, the model proposed in this article unites these three elements and allows for continuous assessment with a process improvement after each result is obtained. The model will allow organizations to identify their state using these three elements jointly, with a maturity scale to know which aspects are needed to promote improvements and thus reach the desired level. This process determines the state of the organization through compliance with controls that respond directly to good practices in cybersecurity, privacy, and health data management in a comprehensive manner.

This article is organized as follows. It begins with a brief discussion on cybersecurity and privacy, relating them to health data management. Subsequently, the development process is presented, which includes the analysis, design, and deployment of the solution. The article concludes with an explanation of the validation process, the metrics used, and a discussion of the preliminary results obtained through the corresponding case study.

2 Literature Review: Maturity Models

We carried out a systematic review and described the state of the art of cybersecurity, considering aspects of privacy in general and its integration to the Health Sector. The research questions provided a better understanding of Cybersecurity and Privacy jointly and the implications of both in the Health Sector.

Our research objective was to determine the classification and categories of relevant studies from a set of different case studies as well as the challenges and recommendations in Cybersecurity and Privacy and their implication in the Health Sector, including the relevant aspects affecting their integration. This was achieved through the use of the Scopus database and queries whose keywords were selected through the PICOC methodology, which is an anagram for Population, Intervention, Comparison, Outcome, Context. Our results were grouped in 4 significant approaches: Cybersecurity and the Health Sector [4, 5], Privacy and the Health Sector [6, 7], Cybersecurity and Privacy [8, 9], Cybersecurity – Privacy and the Health Sector [10, 11].

During the article search and selection process, we identified that none of them analyze Cybersecurity and Privacy in the Health Sector and the Management of Health Data comprehensively. In the same way, none proposes a unique solution that addresses all these aspects. The solutions found follow similar developments and shared approaches, which guarantee that these processes are reliable and valid. However, these maturity models, frameworks, methods, methodologies, and controls present a variety of evaluation and analysis criteria but no unique analysis to cover several areas. As a result, the analysis is not facilitated based on Cybersecurity and Privacy aspects in a comprehensive manner, the consequences of this union in the Health Sector, and the creation of diagnosis and continuous improvement processes.

3 Proposed Model of Capabilities

We developed the proposed Capability Model considering a previous research that provides a process to build and use a Maturity Model [12]. The proposed model is shown in Fig. 1.

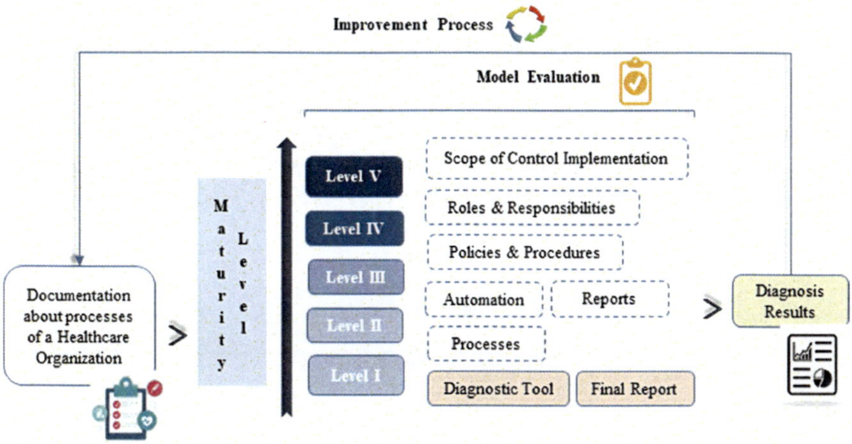

Fig. 1. Maturity model proposed

3.1 Purpose and Components

The proposed solution tries to evaluate these three perspectives: Cybersecurity, Privacy, and Management of Health Data comprehensively. The purpose of applying the model is to fully analyze the organization and thus to improve the general compliance, control, and organizational governance. To achieve this, the model uses controls that deal with these perspectives in a unified manner. These are used as compliance indicators according to the category and subcategory prepared and the control implementation level. Using the NIST Cybersecurity Framework (NIST CSF) and NIST Privacy Framework (NIST PF) controls structure as a reference, we built and arranged the controls in categories and subcategories. We also developed an additional grouping for specific Health Sector controls. We used NIST frameworks because they are highly compatible with the elements used for the construction of the model, as shown in Table 1. We decided to consider local regulations such as Law No. 29733 and No. 30024 because the case study is located in Peru. The final result is a capability maturity model that indicates the current state of the organization.

Table 1. Elements used to build the model

Approach	Name
Cybersecurity	NIST CSF [13], C2M2 v1.1[14], NIST 800–53 SP [15]
Privacy	AICPA/CICA Privacy Maturity Model [16], NIST PF [17], Personal Data Protection Law Regulations (Law No.29733) [18]
Health	Electronic Clinical Records National Registry Regulations (Law N° 30024) [19], Health Insurance Portability and Accountability Act - HIPAA [20]

Measuring Criteria. We built the criteria for assessing controls and defining the level of capabilities from the Enablers proposed by the fifth version of the Control Objectives for Information and Related Technology (COBIT 5) [21]. The criteria prepared are shown in the following Table 2.

Table 2. Measuring criteria

Criteria	Definition
Policies and procedures	Define expected behavior regarding the practice of daily management
Roles and responsibilities	Define key responsibilities to complete activities and make correct decisions
Scope of control implementation	Define the level of automation of control or practice
Reports	Define all the information produced and used by the organization
Processes	Define practices organized as a whole and activities to achieve goals. It allows achieving IT-related goals
Automatization	Defines the level of automation of control or practice

Categories. Controls are grouped following NIST's category structure. We added a new category named Legislate, which is defined as "Develop activities to measure the compliance with laws and regulations applied to cybersecurity and privacy in the health sector."

Subcategories. Additionally, within each category, subcategories, which contain the controls, are grouped. So that the model can perform a comprehensive and holistic assessment and diagnosis, 33 subcategories were defined.

Controls. Controls have been grouped in the subcategories previously defined. The objective is to use the capability levels obtained by control to define the level by subcategory and category with a simple average. Altogether, 172 controls were prepared, grouped into three elements: Cybersecurity, Privacy, and Health Data Management Regulations.

Artifacts. Diagnostic Tool: The tool developed provides a comprehensive vision of the current state of the organization regarding Cybersecurity, Privacy, and Health Data Management. The tool groups controls, which are evaluated by the level of compliance with regard to criteria; consequently, said results would reveal the level of maturity by category and subcategory.

Control Matrix: The prepared matrix contains the model controls, indicating whether they belong to Cybersecurity, Privacy, or Health Data Management. This will make it possible to identify the scope of each category and subcategory clearly, as well as to facilitate future updates of the model components.

Final Report: The model comprises a final report representing the current situation of the organization based on the diagnosis carried out. It has a maturity level in each category, the rationale for each of these levels, and recommendations to improve the result.

3.2 Specifications by Level of Capabilities

Level 1 – Basic. The organization has an unpredictable process with little control. There are no policies, procedures, and definitions of roles and responsibilities. Besides, there is no clear approach in implementing protection controls; the controls are all manual, and no reports are made.

Level 2 – In Progress. The organization has characteristic processes in place. The policies and procedures are limited and incomplete, and the roles are partially defined. Besides, the control implementation scope is limited, the controls are manual, and the reports are limited.

Level 3 – Defined. The organization has standardized and repeatable processes in place. The policies and procedures are integrated, defined, published, and the roles and responsibilities are determined and defined. Besides, the control implementation scope cover core assets, the controls are manual, and defined reports are made.

Level 4 – Differentiated. The organization has measured processes, controlled at a set frequency. The policies and procedures are published and reviewed continually, and the roles and responsibilities are reviewed at a set frequency. Besides, the control implementation scope is reviewed at a set frequency. There are manual and automatized controls, and the reports are reviewed and sent to the top management at a set frequency.

Level 5 – Continuous Improvement. The organization's processes are focused on continuous improvement and automatization. There are reviews and continuous improvement of the policies. Policy plans and improvements are in place, and there is a follow-up while the published policies are implemented. The roles and responsibilities are continuously reviewed. Besides, the control implementation scope is reviewed and improved continuously. The controls are automated and are improved continuously to achieve full coverage. The report requests are reviewed and updated continuously.

4 Case Study

4.1 Organization

The model was validated at the "A" Private Hospital. This private hospital is a service offered by the "B" Service Cooperative, and as this is the most profitable service of the cooperative, B needs to guarantee its proper operation. It is important to understand that developing clinics such as "A" need to achieve compliance with newer regulations with a reduced TI team, low budget, and a low prioritization in technology investment. Furthermore, said clinics are not clear about the first steps needed to attain constant improvement on the standardization and integration of their processes and, most importantly, their adaptation to their financial and technological capacities. The data used in this work were provided exclusively for academic purposes on the anonymity condition.

Our solution will help identify organizations' strengths and weaknesses similar to "A" and assist them in the preparation of strategies to improve and maintain the quality of the services they provide. We implemented the model on a chosen process from clinic "A," which was assessed and diagnosed with the tool. On the other hand, we validated the model through an assessment form prepared by the private hospital.

4.2 Implementation

It was done by analyzing the backup process of Clinic "A" with seven applicable controls from the model. The implementation steps were the following:

- Acquire Process Documentation: We submitted a formal request to the clinic in order to receive the documentation regarding their internal processes
- Scenario Selection: We selected a process that would allow us to evaluate multiple aspects of our model and compare the results to its root components
- Maturity Model Implementation: The model is used to evaluate the selected process, taking into consideration current controls and components
- Obtain Maturity Level: After finishing the implementation, we obtained the maturity level results

Based on the information collected, we could appreciate the value of an analysis that considers the aspects of Cybersecurity, Privacy, and the Management of Health Data comprehensively. The maturity model facilitated the control, monitoring, and compliance of controls, practices, and regulations. We found that the chosen scenario presented a maturity level of "in progress" (level 2). This indicates that "A" private hospital has aspects to improve. Additionally, we can observe the results at the level of the controls in Table 3.

The results indicate that the main need of the private hospital is to prepare and standardize processes since they were unable to reach capability level 3 in 4 controls. They also need to improve compliance with health regulations, as this is the only control in Level 1 – Basic.

Table 3. Implementation results

Category	Subcategory	Control	Level
Govern	Policies, processes, and governance procedures	GV.PO-1	2
	Policies, processes, and governance procedures	GV.PO-4	2
Govern	Review and monitoring	GV.MT-5	2
Protect	Data Security	PR.DSI-4	2
Protect	Data protection processes and procedures	PR.IPI-3	3
		PR.IPI-7	3
Legislate	Interoperability	LE.IN-3	1

4.3 Validation

A comparison between the controls used in our model and their sources in NIST Cybersecurity Framework, NIST Privacy Framework, and the Health Insurance Portability and Accountability Act (HIPAA) through our measuring criteria was made in order to confirm that a coherent integration between models was achieved. We would consider the integration successful if the results were within a 1 point margin of error. A few results of the said comparison can be seen in Tables 4, 5, and 6.

Table 4. Control comparison 1

ID	Source	Measuring criteria					
		P&P	R&R	CIS	Reports	Processes	Automatization
PR.IPI-3	Proposed model	4	3	4	3	2	3
PR.IP-4	NIST CSF	4	3	4	3	2	3

Table 5. Control comparison 2

ID	Source	Measuring criteria					
		P&P	R&R	CIS	Reports	Processes	Automatization
GV.PO-P4	Proposed Model	2	2	2	1	1	2
164.524(a)(1),(b)(1),(b)(2),(c)(2),(c)(3),(c)(4),(d)(1),(d)(3)	HIPAA	2	2	2	1	1	2

As seen in the table, the results obtained for each measuring criteria were the same for PR.IPI-3 and GV.PO-4 while there was a minor difference in the "Reports" criteria

Table 6. Control comparison 3

ID	Source	Measuring criteria					
		P&P	R&R	CIS	Reports	Processes	Automatization
PR-IPI-7	Proposed model	4	3	4	3	2	3
PR.IP-9	NIST CSF	4	3	4	2	2	3

for PR-IPI-7. This difference is within a margin of error of 1 point, and thus the result is still valid.

5 Conclusions

In this article, we proposed a capability model that integrates aspects of cybersecurity and privacy applied to the health sector. In this model, the level of capabilities is calculated by the level of compliance with controls. Additionally, a diagnostic tool was prepared to measure the level of capabilities of organizations in a comprehensive and holistic manner. The model was tested in a private hospital in Lima, Peru, where the level of capabilities was 2 of 5, which allowed us to identify important gaps and make recommendations such as the standardization of processes belonging to the categories used in the evaluation as well as the creation of new ones such as backup testing, response plans, and compliance evaluation.

On the other hand, we compared the results obtained through our proposed model and the results of its root components and found acceptable results within our defined error margin, thus proving that the integration was done coherently and comprehensively. Additionally, "A" suggested adding more controls focused on Peruvian regulations since they are more relevant to them. Such observation will be used as an opportunity to improve the model in future iterations.

References

1. Security I (2019) Cost of a Data Breach Report 2019. https://www.ibm.com/downloads/cas/ZBZLY7KL
2. IBM (2020) X-Force Threat. 49
3. ISACA Interactive Glossary & Term Translations | ISACA. (n.d.). https://www.isaca.org/resources/glossary
4. Ghobakhloo M (2019) Determinants of information and digital technology implementation for smart manufacturing. Int J Prod Res. https://doi.org/10.1080/00207543.2019.1630775
5. Le NT, Hoang DB (2017) Capability maturity model and metrics framework for cyber cloud security. Scalable Comput 18(4):277–290. https://doi.org/10.12694/scpe.v18i4.1329
6. Shen N, Sequeira L, Silver MP, Carter-Langford A, Strauss J, Wiljer D (2019) Patient privacy perspectives on health information exchange in a mental health context: Qual Study J Med Internet Res 21(11). https://doi.org/10.2196/13306

7. Kuo K-M, Talley PC, Cheng T-J (2019). Deterrence approach on the compliance with electronic medical records privacy policy: the moderating role of computer monitoring. BMC Med Inform Decis Making 19(1). https://doi.org/10.1186/s12911-019-0957-y
8. Khatoun R, Zeadally S (2017) Cybersecurity and privacy solutions in smart cities. IEEE Commun Mag 55(3):51–59. https://doi.org/10.1109/MCOM.2017.1600297CM
9. Mascetti S, Metoui N, Bettini C, Lanzi A (2018) EPIC: A methodology for evaluating privacy violation risk in cybersecurity systems EPIC: a methodology for evaluating privacy violation risk in cybersecurity systems. Trans Data Privacy 11. https://www.researchgate.net/publication/327529035
10. Zandona DJ, Thompson JM (2017) Going beyond compliance: a strategic framework for promoting information security in hospitals. Health Care Manager 36(4):364–371. https://doi.org/10.1097/HCM.0000000000000189
11. Jalali MS, Kaiser JP (2018) Cybersecurity in hospitals: a systematic, organizational perspective. J Med Internet Res 20(5). https://doi.org/10.2196/10059
12. Rose J (2013) Selecting, Using, and Creating Maturity Models: A Tool for Assurance and Consulting Engagements. July 1–27. https://www.iia.org.uk/media/358857/selecting__using__and_creating_maturity_models___a_tool_for_assurance_and_consulting_engagements.pdf
13. National Institute of Standards and Technolgy (2014) Framework for Improving Critical Infrastructure Cybersecurity. In Framework for Improving Critical Infrastructure Cybersecurity Note to Readers on the Update, pp 29–35. https://doi.org/10.6028/NIST.CSWP.04162018
14. U.S. Department of Energy (2014) Electricity subsector cybersecurity capability maturity model (Issue February). https://energy.gov/sites/prod/files/2014/02/f7/ES-C2M2-v1-1-Feb2014.pdf
15. Security and Privacy Controls for Federal Information Systems and Organizations (2013) https://doi.org/10.6028/NIST.SP.800-53r4
16. AICPA/CICA (2011) AICPA/CICA Privacy Maturity Model. Aicpa/Cica, March 1–42. https://www.kscpa.org/writable/files/AICPADocuments/10-229_aicpa_cica_privacy_maturity_model_finalebook.pdf
17. National Institute of Standards and Technolgy (2020) NIST Privacy Framework. 16 January 2020. https://doi.org/10.6028/NIST.CSWP.01162020
18. Empresa Peruana de Servicios Editoriales S.A. (2013) Reglamento de la Ley No 29733, Ley de Protección de Datos Personales. Diario Oficial El Peruano, 35, 15. www.minjus.gob.pe
19. Reglamento-de-la-Ley-N°-30024-Ley-que-crea-el-Registro-Nacional-de-Historias-Clínicas-Electrónicas. (n.d.)
20. U.S. Dept. of Labor Employee Benefits Security Administration (1996) HIPAA. Health Insurance Portability and Accountability Act of 1996. In: 104th Congress: Vol. Public Law. https://www.govtrack.us/congress/bills/104/hr3103/text
21. Ti G, Isaca Manuel G, Quintanilla Y, Erastus Mosha, MINTIC, Arte EDEL, Muñoz Periñán IL, Ulloa Villegas G, Copy P, Lanter D, Isaca S, Ocasio VM, Andres C, Saavedra S, MINTIC, Neira AL, Spohr JR, Weybrecht G (2012) COBIT 5 - Un Marco de Negocio para el Gobierno y la Gestión de las TI de la Empresa. In Guía de inspiración para la implementación de PRME: Segunda Edición: Aprender para Avanzar, vol 147, Issue 17. https://doi.org/10.9774/gleaf.9781783537846_16

Face Recognition for Criminal Identification

An Alert System in Suspects Scenarios

Bryan Orellana, Luiggi Álvarez, and Jimmy Armas-Aguirre(✉)

Universidad Peruana de Ciencias Aplicadas (UPC), Av. Primavera 2399 Surco, Lima, Perú
{u201316580,u201115670}@upc.edu.pe, jimmy.armas@upc.pe

Abstract. In this paper, we propose an alert system to identify suspects on the scene using mobile devices. This project's objective is to give the police officers a mobile application to receive alerts in real-time; it will help improve the current identification time by using facial recognition. This proposal is due to the increase in Peru's criminal acts that affects 27.3% of the population who has suffered more than one criminal act. Also, the current identification time is between 5 and 10 min. This alert system consists of 1. The collection and training of images of criminals; 2. IP video device configuration; 3. Sending data from the Python service to the application; 4. The process of receiving results in the application. The alert system was validated in a public area of the Rímac district in Lima, Peru. Preliminary results showed that the alert system reduces the current identification time by 91.46% and has an efficiency of 90% based on the tests performed.

Keywords: Face recognition · Video surveillance · Smart city · Security system · Cloud computing

1 Introduction

Currently, citizen security is considered one of the main problems affecting Peru. In September 2018, this problem reached 38.9%, occupying the second place of the main problems in our country according to the perception of the population [1]. Citizen security is not about reducing all criminal acts that may occur in our city but about having strategies to improve the citizens' quality of life. However, according to the National Security Plan published in 2019, the assigned budget for citizen security in the 2013–2018 period was scarce, making it difficult to carry out the planned activities [1].

Citizen security has been affected by the large number of criminal acts that are currently taking place and delinquent capture delay. In a report carried out between the years 2011 and 2018, we observe that 59% of the complaints registered in 2018 were for crimes related to theft or robbery in its various forms, being simple theft the crime with the largest number of complaints with 88,780 complaints at national level [2]. In the same year, the number of violent thefts increased by more than 151.59% compared to the year 2017. Regarding non-violent thefts, the increase was of 66.63% [3]. These numbers are reflected in the high victimization rates of citizens, where 12.4% of the population has been victim of more than one criminal act, and 27.3% has been the victim of a single

Y. Iano et al. (Eds.): BTSym 2020, SIST 233, pp. 368–377, 2021.
https://doi.org/10.1007/978-3-030-75680-2_41

crime [4]. Based on existing research and with the support of experts in citizen security, criminal identification time takes between 5 to 10 min.

According to the report of the National Institute of Statistics and Informatics (INEI) published in March 2020, in Lima, 65.8% of citizens feel insecure about walking on the streets at night, and the perception of insecurity continues to increase each year obtaining, as a result, 90.4% in this last survey. This is a consequence of the most common crimes that occur every day, such as money, wallets, and cell phones theft, with 18.4%. For this reason, the population considers not to report theft because 38.9% believe that it is a waste of time, and 16.5% do not know the offender [4].

Recent progress in technological solutions for citizen security has positive contributions to society, with apps that allow the number of crimes to be visualized through a heat map. In addition, they have facial recognition solutions to capture faces. However, they do not provide a quick identification alert.

Through this project, an alert system that is aligned with sustainable development goals (SDGs) agreed for 2030 by the United Nations (UN) in 2015 is offered, thus reducing all forms of violence [5]. That is why this solution allows optimizing the criminal identification process, obtaining a fast and effective attention time through facial recognition. This alert system consists of detecting and identifying the faces of criminals trained in the recognition model, allowing an identification alert to be sent to the web and mobile app in order to quickly receive data on criminals and their locations.

This paper is organized as follows. We will begin with the study of previous works related to the research. Then, we will describe the technological proposal and the case study. Finally, we will conclude the article with the results obtained in the case study.

2 Literature Review

Different factors were considered in the study stage of scientific articles. On one hand, an analysis of the factors related to the performance of the recognition service was performed. On the other hand, the factors that alter facial recognition were analyzed. In addition, we carried out a study of different models that have been used to develop video surveillance solutions, which use different video devices that will be detailed in this section.

2.1 Metrics to Measure the Performance of Cloud Recognition Service

It is very important to be able to measure the performance of the different recognition services that we have in the cloud. In Table 1, there are different measurement variables for the facial recognition service. This service is divided into facial detection and verification. Regarding metrics, precision is not always a number that tells us which service is the best, but its relevance could vary due to the scenario where these solutions are implemented. Processing time in Microsoft services is much faster in both cases, but the accuracy is less than Amazon services, which are 95% accurate, however, with a high processing time [6].

Table 1. Performance evaluation of face recognition service

Service	Metric	Amazon	Google	Microsoft
Face detection	Precision	95%	85%	88%
	Processing time	1080 ms	760 ms	250 ms
	Error ratio	0,12%	0,12%	0,50%
	Compound metric value	86%	77%	74%
Face Verification	Processing time	2500 ms	–	150 ms
	Compound metric value	76%	–	84%

2.2 Factors that Alter Recognition

There are different factors that can affect the effectiveness of the facial recognition service, as shown in the results of Table 2, which was performed with a data set known as (CAS-PEAL-R1) [7]. The results show us how different factors may affect the effectiveness, such as accessories or posture, which obtains one of the lowest identification rates. Microsoft and Amazon services present a high performance in the first three factors, but in the posture, we can see that Amazon shows a better acceptance rate [8].

Table 2. Factors that decrease the effectiveness of facial recognition

Factors	Identification rate (%)		
	Neurotechnology	Microsoft	Amazon
Accessories	92.61%	98.76%	98.76%
Expression	98.31%	99.57%	99.78%
Luminosity	63.42%	86.69%	83.95%
Posture	31.31%	74.47%	85.73%

2.3 Proposed Solutions for Video Surveillance

At the beginning of this project, research was done on solutions similar to the case study. This is related to facial recognition projects using video surveillance equipment. In this case, we find different models applied in the field of public security and also applied in smart cities (Table 3).

Table 3. Models applied to video surveillance

Models	References
Fog computing	[9–12]
Cloud computing	[13–16]
Edge computing	[17–19]
5G	[20, 21]
Microservices	[22]

2.4 Devices Used for Video Transmission

According to the research carried out, video devices play a fundamental role in facial recognition solutions. This is because they are the tools that generate the video transmission for capturing images (faces). For the present case study, the possibilities of image capture devices were considered, which will be shown below in Table 4.

Table 4. Video devices

Devices	References
Cameras	[11, 13, 17, 18, 20–24]
Drones	[9, 12, 15, 25]
Mobile	[26]
Webcam	[6, 27, 28]
Wearable	[16]

3 Face Recognition for Criminal Identification: Proposed Alert System

This proposal consists of an alert system based on the use of facial recognition to find criminals on the streets. The objective of the proposal is focused on optimizing the identification time of criminals through a mobile device that simulates a security camera, allowing an identification alert to be sent to the web / mobile app. This alert system is composed of different technologies that participate in stages of the process to obtain the final result that is the notification alert (Fig. 1).

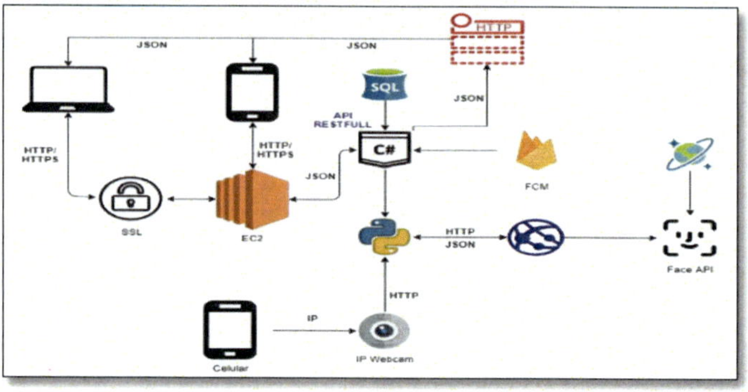

Fig. 1. Communication diagram

3.1 Collection and Training of Criminal Images

The image collection is based on specific requirements proposed by Microsoft Azure as a cognitive service in order to carry out a correct training. For this reason, we selected at least three images from a public dataset to imitate the images captured on the Document of National Identity (DNI). These images must be captured from the front. Once the images have been collected, they are stored in files with the label of their ID number per person. The recognition model training is done using Microsoft Azure Face API. That is why we proceeded to run the Python training script, where the Key of the cognitive service is associated, allowing communication with the cloud.

3.2 IP Video Device Configuration

The IP video device configuration was made through the IP Webcam app that was downloaded to an Android cell phone. This app allows us to have a view of a security camera from a monitoring center through the creation of a virtual server that takes the IP of the cellphone and assigns it Port 8080. It is important to indicate that we used a Samsung A50 cellphone to take the test videos. This video resolution must be 1080p or higher to get optimal results and avoid false positives. This IP is inserted in the Python script, allowing to capture frames in real-time, using the cell phone camera transmission as video input. These frames are the instantaneous images that are captured; in this case, it was configured to capture every 50 frames, allowing a faster time to take images from the camera.

3.3 Sending Data to the App from Python Service

As part of software development, two services are contemplated, one developed in Python and the other in C#. The Python service allows training and storing faces in the Microsoft Azure Face API service. This service provides a recognition ID, which is sent to the CosmosDB database through an SDK, storing it through collections and documents. Also, it communicates with the cell phone through its IP address and HTTP protocol. This allows the cell phone to send the request to the Face API while is transmitting in real-time and capturing faces through frames and then validate them with the face database that has been trained. This request will return a notification to the service, which works with Firebase Cloud Messaging (FCM) so that it can be communicated to the web-mobile platform developed in C # and hosted in the Amazon EC2 service.

3.4 Receiving Results in the Application

Once the alert notification has been received in the app, the offender's data and background are displayed. This allows the police or serenazgo to instantly see the information of the offender. Also, for each notification of the identified criminal, the criminological map is updated in real-time, showing the areas with the highest number of identified criminals through colors. In Fig. 2, we can see the list of identification alerts with some information such as the criminal name, the location, and the crime. Besides, the system has other functionalities that allow the police officers to register their assistant to an emergency and see the map to know the criminal location.

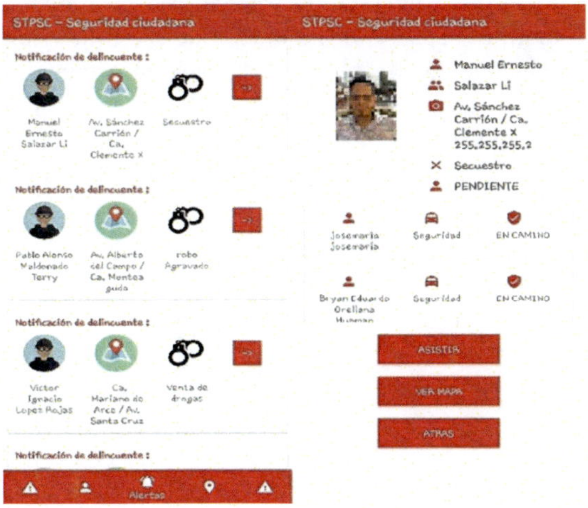

Fig. 2. Alerts view in mobile application

In Fig. 3, the system has a criminological map that changes its color depends on the quantity of camera identification. This functionality helps the police officers to know which zone is the most dangerous in the city and allow them to make strategies to fight crime.

Fig. 3. Alert heat map

4 Case Study

4.1 Organization

The validation of the proposed model was carried out in a public area in the Rimac district of Lima, Peru. In this validation, a real scenario of a monitoring center was simulated, allowing the transmission of the mobile device to be viewed in real-time. First, the height where the mobile device was placed was measured; it was positioned 5 m from the ground level to safeguard the integrity of the camera. This allows the direct focus to the hallway where the identification test was performed. The following factors were taken into account in this simulation: the current identification time, the type of connection (Wifi), the height of the camera, and the time (day or night). Based on these factors, it allowed us to measure the response time of identification of a criminal.

On the other hand, we performed a test to measure the effectiveness of the recognition system with a public dataset that contains a group of faces to train the model so that the system can recognize them by their National Identity Document (DNI).

4.2 Implementation

Target Recognition. After having performed the service training, the script developed in Python was executed to start capturing snapshots through the mobile device. This allows snapshots to be collected and sent to the cognitive service so they can analyze them and return the result of the person's face identification.

Alert Reception on Web/Mobile Application. The cognitive service sends all the results of the identifications made (name of the criminal and the camera that identified him) to the web and mobile app, allowing us to receive the information stored in an SQL database. Within the application, it is possible to view more information about the offender, such as background and personal data.

Collection and Processing Data. As the alerts of identifications increase, they are displayed on a graph called a heat map that allows recognizing the cameras where the highest number of identifications have been made. This makes it possible to keep tracking the areas where criminals are mostly involved and more likely commit a crime.

4.3 Results

Based on the tests performed in the case study, the following can be determined: In Table 5 and 6, it is evident that the alert system allows reducing the current identification time by 91.46% since the alert was received around 25.64 s on average after having identified the offender in the established attempts. This allows municipal authorities to quickly implement strategies to reduce criminal acts in districts based on an early warning.

Table 5. Identification time (day)

Current	Connection type	Height	Time of day	App (Seg)	Variance (%)
5 min	Wifi	5 m	Day	24.95	91.68
5 min	Wifi	5 m	Day	23.55	92.15
5 min	Wifi	5 m	Day	21.75	92.75
5 min	Wifi	5 m	Day	22.94	92.35
5 min	Wifi	5 m	Day	22.31	92.56

Table 6. Identification time (night)

Current	Connection type	Height	Time of day	App (Seg)	Variance (%)
5 min	Wifi	5 m	Night	28.50	90.50
5 min	Wifi	5 m	Night	26.90	91.03
5 min	Wifi	5 m	Night	27.56	90.81
5 min	Wifi	5 m	Night	29.59	90.17
5 min	Wifi	5 m	Night	28.32	90.56

On the other hand, it is shown that the alert system has a 90% efficiency based on the public dataset trained in the recognition model, in which 18 faces were correctly recognized. However, it is important to detail that the following factors altered the efficacy:

- A face was not recognized by the DNI (National identity document) that had been assigned (5%) since the training was carried out with an image format that did not correspond.
- A face was not recognized or detected (5%) since the trained image did not capture the complete face.

5 Conclusion

In this work, we proposed an alert system to optimize the criminal identification process. This alert system was tested in the Rimac district in Lima, Peru. The current time for identifying criminals was reduced by 91.46%; with this early warning alert, the authorities can have more tools to fight crime and carry out better strategies. All of this with the different functionalities that the application has, such as cameras location, heat map, criminal information, and the alert list.

This project can be part of different solutions that are carried out for the new smart city concepts. This solution could be integrated with other technologies that help obtain data and can be used to improve the monitoring and security of these cities.

It is important to indicate that there were limitations during the project due to the pandemic experienced by COVID-19. Therefore, the technologies used to capture images, which, in this project, were cellular, can be improved with specialized cameras for facial recognition. These improvements would benefit from getting an optimal identification and a much faster alert.

Finally, it is important to indicate that solutions with facial recognition are being used with greater concurrence in different cities, allowing good results in the sectors to be implemented, but these solutions must be accompanied by an adequate legal framework so that they are not affected the integrity of any citizen for the improper use of these. If we can achieve the above, this project can have a big impact on our society, as it focuses on improving the quality of life for all citizens.

References

1. Miniter (2019) Plan Nacional De Seguridad Ciudadana 2019-2023. CONASEC, Lima
2. Gestion (2019) Estos son los 10 delitos más denunciados en el Perú en el 2018. https://gestion.pe/fotogalerias/son-10-delitos-denunciados-peru-2018-269447-noticia/
3. Cruz R (2019) El Comercio. Se Reportan 18 Robos y Hurtos al Dia en el Cercado de Lima. https://elcomercio.pe/lima/seguridad/se-reportan-18-robos-y-hurtos-al-dia-en-el-cercado-de-lima-estoyalerta-noticia/?ref=ecr
4. INEI (2020) Estadisticas de Seguridad Ciudadana. Instituto Nacional de Estadistica e Informática, Lima
5. ONU (2015) Resolución aprobada por la Asamblea General el 25 de septiembre de 2015. Asamblea General de las Naciones Unidas, Nueva York
6. Xue Y, Zhang H, Ma H (2018) Performance evaluation of image and video cloud services. In: 2018 IEEE 20th international conference on high performance computing and communications; IEEE 16th international conference on smart city; IEEE 4th international conference on data science and systems (HPCC/SmartCity/DSS)
7. Gao W, Cao B, Shan S, Chen X, Zhou D, Zhang X, Zhao D (2008) The CAS-PEAL large-scale Chinese face database and baseline evaluations. IEEE Trans Syst Man Cybern-Part A: Syst Humans 38:149–161
8. Al-kawaz H, Clarke N, Furnell S, Li F, Alruban A (2019) Advanced facial recognition for digital forensics. In: Proceedings of the 17th European conference on information warfare and security
9. Haouari F, Faraj R, AlJa'am JM (2018) Fog computing potentials, applications, and challenges. In: 2018 International conference on computer and applications, ICCA 2018
10. Grambow M, Hasenburg J, Bermbach D (2018) Public video surveillance: using the fog to increase privacy. TU Berlin & Einstein Center Digital Future Mobile Cloud Computing Research Group

11. Nasir M, Muhammad K, Lloret J, Sangaiah AK, Sajjad M (2019) Fog computing enabled cost-effective distributed summarization of surveillance videos for smart cities. J Parallel Distrib Comput 126:161–170
12. Chen N, Chen Y, You Y, Ling H, Liang P, Zimmermann R (2016) Dynamic urban surveillance video stream processing using fog computing. In: Proceedings - 2016 IEEE 2nd international conference on multimedia big data, BigMM 2016
13. Amin AH, Ahmad NM, Ali AM (2016) Decentralized face recognition scheme for distributed video surveillance in IoT-cloud infrastructure. In: 2016 IEEE region 10 symposium (TENSYMP)
14. Chen T-S, Lin M-F, Chieuh T-C, Chang C-H, Tai W-H (2016) An intelligent surveillance video analysis service in cloud environment. In: 2015 international Carnahan conference on security technology (ICCST)
15. Koubaa A, Qureshi B (2018) DroneTrack: cloud-based real-time object tracking using unmanned. IEEE Access 6:13810–13824
16. Shaukat Z, Xiao MC, Aslam S, Farooq QU, Aiman S (2019) Facial recognition on cloud for android based wearable devices. In: International conference on applied human factors and ergonomics
17. Dautov R, Distefano S, Bruneo D, Longo F, Merlino G, Puliafito A, Buyya R (2018) Metropolitan intelligent surveillance systems for urban areas by harnessing IoT and edge computing paradigms. Wiley
18. Wang J, Pan J, Esposito F (2017) Elastic urban video surveillance system using edge computing. In: SmartIoT 2017 - proceedings of the workshop on smart internet of things
19. Prentice CT, Karakonstantis G (2018) Smart office system with face detection at the edge. In: 2018 IEEE smartworld, ubiquitous intelligence & computing, advanced & trusted computing, scalable computing & communications, cloud & big data computing, internet of people and smart city innovation (SmartWorld/SCALCOM/UIC/ATC/CBDCom/IOP/SCI)
20. Nizzi F, Pecorella T, Bertini M, Fantacci R, Bastianini M, Cerboni C, Buzzigoli A, Gattoni M, Fratini A (2018) Evaluation of IoT and videosurveillance applications in a 5G smart city: the Italian 5G experimentation in Prato. In: 2018 110th AEIT international annual conference, AEIT 2018
21. Hu H, Shan H, Zheng Z, Huang Z, Cai C, Wang C, Zhen X, Yu L, Zhang Z, Quek TQS (2018) Intelligent video surveillance based on mobile edge networks. In: 2018 IEEE international conference on communication systems (ICCS)
22. Nagothu D, Xu R., Nikouei SY, Chen Y (2019) A microservice-enabled architecture for smart surveillance using blockchain technology. In: 2018 IEEE international smart cities conference, ISC2 2018
23. Hitoshi I (2016) Video face recognition system enabling real-time surveillance. Special Issue on AI & Social Value Creation
24. Huang J-Y, Lee W-P (2016) A cloud-based video surveillance system for event recognition and face detection. In: Proceedings of the World Congress on Engineering, pp 99–104
25. Kim H-C, Lim C-S, Lee C-S, Choi J-H (2016) Introduction of real-time video surveillance system using UAV. J Commun 11:213–220
26. Hossain MS, Muhammad G (2015) Cloud-assisted speech and face recognition framework. In: 2015 IEEE 28th Canadian conference on electrical and computer engineering (CCECE)
27. Balla PB (2018) IoT based facial recognition security system. In: 2018 international conference on smart city and emerging technology (ICSCET)
28. Rana S, Deepu MP, Sivanantham S, Sivasankaran K (2016) Face detection system using FPGA. In: IC-GET 2015 - proceedings of 2015 online international conference on green engineering and technologies

Combined Model Based on Lean Healthcare and BPA to Reduce Waiting Times in Public Health Entities

Axel Zevallos-Aquije, Rosa Salas-Castro, Edgardo Carvallo-Munar, and Luis Cardenas-Rengifo(✉)

Universidad Peruana de Ciencias Aplicadas, Av. Prolongación Primavera 2390, Santiago de Surco, Lima, Perú
{u201310682,rosa.salas,edgardo.carvallo,luis.cardenas}@upc.edu.pe

Abstract. Economic penalties have been imposed when there are long waiting times in low-income health centers in Peru; this problem is a crucial cause of the deficient global assessment on the national health level. Within this context, a model is presented that combines diagnostic and solution tools based on applying the Lean Healthcare philosophy within the parameters of Good Storage Practices, demand forecasting, system simulation, and Gemba Kaizen formats. Among the results obtained, the average daily waiting times were reduced by 63.89% simultaneously, process times were decreased by 12.62%, and reject patients by 2%.

Keywords: Waiting times · System simulation · Demand forecast · Lean healthcare · Good storage practices · Gemba Kaizen

1 Introduction

The Peruvian health sector belongs to the group of countries that invest less in health. Also, management about investment is the most inefficient in South America. This severe crisis produces inefficiencies in terms of waiting times, infrastructure, and shortages. The latest research presented by the Association of Taxpayers of Peru, together with information from the World Health Organization, report a lack of investment and the incorrect management of resources as factors that create inefficiency in health services, which affect the health and economy of millions of Peruvians [1]. The national health system needs to find an improvement model to meet the area's needs and requirements; this context involves the limited economic, material, and personnel resources that the national state uses. The required model's objective should focus on reducing the waiting times as it was one of the primary factors of the mentioned problem, involving economic and social losses.

A proposal for improvement is presented through a combined model of diagnostic and solution tools based on applying the Lean Healthcare and BPA philosophy to focus on the health sector to reduce waiting times in the medical care process. To achieve the

Y. Iano et al. (Eds.): BTSym 2020, SIST 233, pp. 378–384, 2021.
https://doi.org/10.1007/978-3-030-75680-2_42

objective, the root causes of the diagnosed problems will be analyzed, and the value flow, to recognize the activities that are not producing value in their processes. This is followed by developing the appointment assignment improvement model based on a study of patient demand prognosis, which will be assessed by simulating the system capacity and minimizing queues. Consequently, Good Storage Practices are implemented, which leads to the proposal of flow standards in the workplace. Then, the processes' improvements are standardized, and Gemba Kaizen formats are developed, which will allow the procedures to be continuously improved. On the one hand, the importance of proposing a model using the combination of tools shown resides, in economic terms, in the penalties that amount to more than 12 million soles per year; on the other hand, the social importance lies in the excellent level of dissatisfaction created, since 40% patients in public entities have expressed approximately 40% level of discontentment that results in complaints [2]; whose primary root cause lies in reducing waiting and process times inpatient care in health centers.

Using the proposed solution is based on the need to improve public health services that do not have the resources to mitigate the problems presented. Unlike others, this model is based on the minimum amount of economic, material, and personnel resources. They are adapted in the context of the problematic national situation primarily created by the minimum amount of resources that the state invests in these entities.

2 State of the Art

Lean methodology has been used to improve service quality since its inception by Toyota in Japan. Its objective is to eliminate waste from processes that do not create value [3]. The Lean approach is to develop standardized work focusing on continuous process flow by identifying value at each step using support tools to find these wastes, eliminate them, and perform the process improvement [4]. In medical care—the first publications date from 2002—they show the need to use Lean techniques about quality improvement, patient safety, cost reduction, waiting time reduction, and staff satisfaction improvement [5].

For years, studies have been conducted about using simulation to provide solutions to health problems, specifically for bed planning in geriatric medicine. This case study integrates simulation with flow models demonstrating how there are changes in the admission rate, length of stay, and bed allocation (capacity), as an additional contribution. It highlights the need to consider the ability to maintain service efficiency [6]. Moreover, in 2006, models were already being applied where simulation was used to increase staffing effectiveness; this was concerning employers' arrival rate to optimize patients' timely care. This case focuses on optimization based on system capacity [7].

Focusing on the current period, some studies are based on capacity measurement. Analyses were performed based on practical capacity within an emergency service; this demonstrated that objective measures might be achieved inadequate in a health center using queuing theory and simulation [8]. Studies have been conducted on models that determine scheduling appointments based on queuing theory optimization models with a focus on the project. This case study was conducted at the West China hospital, one of the hospitals with the most significant single global access obtained due to the substantial relationship of appointment scheduling with queuing theory optimization models.

[9] Some studies analyze outpatient visits for the planning and reasonable allocation of resources (material, human) in medical care; here, the author uses the seasonal autoregressive integrated moving average (ARIMA) model in the daily time series exponential smoothing model. Consequently, better prediction performance was achieved with variance and average minimum errors. [10] The planning, design, and construction of medical information storage policies come in close association with the organization in which it is structured. Authors who establish these parameters propose ergonomic value as the basis for the models' development. Likewise, the considerations to be taken when designing this type of medical documentation storage are already established [11].

Lean Healthcare, as a variation of Lean, focuses on the field of health. Its recent applications involve using support tools, such as system simulation, to determine system capacity. Other authors have used the forecast of demand based on the ARIMA model to improve care capacity and enhancements when assigning medical appointments. Furthermore, the application of Good Storage Practices in the healthcare field has recently been used in improvements related to reducing time in care as a basis for improving procedures.

3 Contribution

3.1 Overview

A combined Lean Healthcare tools and Good Storage Practices model is presented to reduce waiting times through a patient allocation model and a newly restructured workflow area to increase patients' number per day. The proposed improvement model follows a scheme based on reducing queues through an optimal allocation model, applying Good Storage Practices, standardizing processes, and making a continuous improvement plan. In Fig. 1, the model achieved is defined.

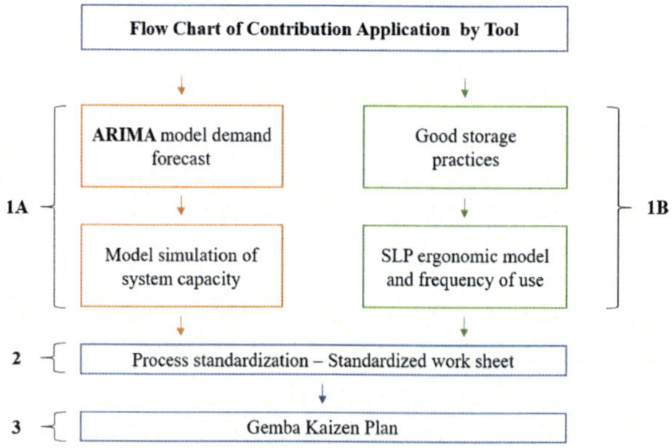

Fig. 1. Design of the improvement model.

3.2 Specific View

Phase 1A. Developing patient demand prediction in the medical area of study - Developing server system capacity (consultation table) from the physician's work minimizing the waiting queue to 0.

In this phase, a forecast of patient demand based on the ARIMA model is made to obtain a more accurate value for patients arriving at a medical consultation. Then, depending on the market, the model is adjusted according to the simulation to determine system capacity. The model is simulated based on the parameters obtained to attain the optimal minimum queue value in the health care system.

Phase 1B. Applying Good Storage Practices - Support methodology: Systematic Layout Planning. In this phase, the design and organization policies of the medical documentation storage are proposed.

Following this, a better design of the work area based on the Spaghetti Chart analysis is presented. The job's new design should be based on reducing movement and changing the medical document storage location.

Phase 2. Process standardization. In this phase, standardized work formats are proposed. The allocation parameters are suggested based on previous time-recording studies; these vary in patient allocation by critical and non-critical hours that also consider the patient care capacity evaluated in the Takt Times from the standardization process.

Phase 3. Gemba Kaizen Plan. As a final part of the model, and within the Lean Healthcare tools, Gemba Kaizen format models are always verified for continuous process improvement. These formats are compiled from the forms used by Mark Graban, a Lean Healthcare specialist researcher.

3.3 Indicator

The improvement achieved with the use of the combined model is measured through the following indicators:

- The percentage improvement in waiting time (iwt) is the average time the patient remains to be treated (scheduled time of appointment minus the time the doctor treated them).

$$\%iwt = \frac{waiting\ time\ before - waiting\ time\ after}{waiting\ time\ before} \times 100 \tag{1}$$

- The improvement in process time (ipt) is the average time it takes for the doctor to treat a patient (end of care time minus the start of patient care).

$$\%\text{ipt} = \frac{\text{process time before} - \text{process time after}}{\text{process time before}} \times 100 \tag{2}$$

- The percentage of rejected patients (rp) represents the rate of patients cited to be attended and who, due to delays in care times, have to be rescheduled.

$$\%\mathrm{rp} = \frac{\text{Number of rejected patients}}{\text{total number of patients}} \times 100 \tag{3}$$

4 Validation

For validation purposes, a pilot plan is used that has been conducted in a health center in a district of Lima, Peru; a specific process has been selected in which the external consultation process will be worked on, i.e., medical care. The phases of the model to be applied are divided into the following phases: phase 1 (which is applied at the same time as phase 2) focuses on the work capacity calculation; phase 2 restructures the job; following this, phase 3 standardizes the previous processes. In phase 2, the medical documentation warehouse has been designed and organized specifically in terms of ergonomics and medical documentation frequency. Also, the relocation of the warehouse in the workplace was based on reducing distances and re-circulations that have been taking place. Taking into account the following specifications:Storage dimension: 51 cm × 369 cm × 194 cm, Folder size: 31 cm × 26 cm × 39 cm and Ergonomic dimension: For 1.74 = Range: 68 cm–163 cm.

Following this, a better design of the work area was developed in this phase based on the elaborated Spaghetti Chart; Fig. 2 shows the workplace's new design based on the reduction of movement and change of place of the medical documentation warehouse.

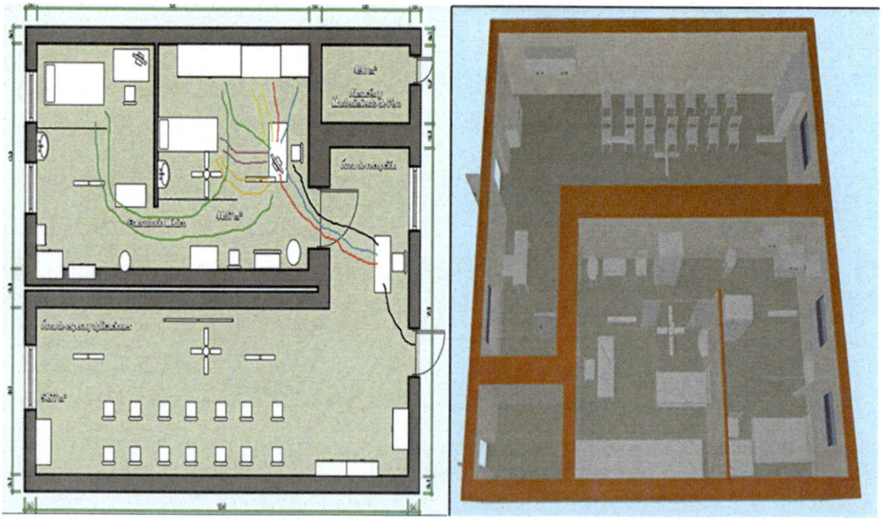

Fig. 2. Design of the improvement model.

Based on the implementation of the combined model phases on the established days, the following results were obtained in Table 1.

Table 1. Results obtained per indicator.

Indicators	Before applying the model (min)	After applying the model (min)	Improvement (%)
Waiting time	72	26	63.89
Process time	16.25	14.20	12.62
Rejected patients	10%	8%	2

Based on the initial diagnosis, values derived from the primary problem were obtained from the first day of application until days before applying the model. Daily waiting times reached an average matter of 72 min, and the average value acquired after using the 26-min model shows a reduction of 46 min.

As part of reducing waiting times, it is essential to minimize process times to achieve new free time and more patients treated within a day; the reduction value was 2.05 min based on the model's application. A decrease of 2% of the rejected patients was obtained, despite being small, the result with the best-obtained value of 8% is below the average established by the health entities. These three values represent the effectiveness of the model under study. Based on the scope of the applied project, it is necessary to consider that the labor policies imply a risk in terms of changes that could occur very frequently. Therefore it is required to incorporate future versions of the model techniques to mitigate these factors as techniques associated with risk management and change management.

5 Conclusions

A combined model of lean healthcare tools and best practice storage methodologies has been developed to reduce waiting times through a patient allocation model and a new restructured workspace flow to increase the number of patients served per day. The feasibility of using the combined model is shown by the favorable improvements based on reducing waiting times and process times that decreased by 63.89% and 12.62%, respectively, and a 2% reduction in patients rejected for care. Some factors can harm the model's objective; it is necessary to consider that the forecast model and the simulation model can present inconveniences with the expected results in a lean healthcare model. In a patient allocation model, it is necessary to consider the number of rejected patients. They can make claims that lead to economic penalties, which is the project's primary problem. It is worth considering that SUSALUD and INEI have identified other health sector issues that impact social value. These involve patient satisfaction that causes the national health system to be graded poorly in health systems' world ranking.

References

1. Sausa M (2017) Servicio de salud presenta más de 47 mil reclamos en lo que va del año
2. SUSALUD (2019) Registro de Sanciones
3. Ramaswamy R (2017) Gillings school of global public health

4. Rushton C, Robertson L, Taylor T, Taylor P, Button P, Alfred A (2016) Improving the service for patients receiving extracorporeal photopheresis using lean principles. Br J Nurs 25(16):917–921. https://doi.org/10.12968/bjon.2016.25.16.917
5. Costa LBM, Filho MG (2016) lean healthcare: review, classification, and analysis of literature. Prod Plan Control J 27(10):823–836. https://doi.org/10.1080/09537287.2016.1143131
6. Gorunescu F, McClean SI, Millard PH (2002) Using a queueing model to help plan bed allocation in a department of geriatric medicine. Health Care Manag Sci 5:307–312. https://doi.org/10.1023/A:1020342509099
7. Green L, Soares J, Giglio J, Green R (2006) Using queueing theory to increase the effectiveness of emergency department provider staffing. Acad Emer Med 13(1):61–68. https://doi.org/10.1197/j.aem.2005.07.034
8. Conislla MT, Saico SA, Le6n CC, Alvarez JM, Raymundo IC (2019) A lean healthcare model for increasing the availability of products in pharmaceutical SMEs. Paper presented at the ACM international conference proceeding series, pp 13–17. https://doi.org/10.1145/3364335.3364381
9. Calero, L, Maccasi, A, Raymundo, C (2020) Lean model of services for the improvement in the times of attention of the emergency areas of the health sector. https://doi.org/10.1007/978-3-030-25629-6_144
10. Luo L, Zhou Y, Han BT, Li J (2017) An optimization model to determine appointment scheduling window for an outpatient clinic with patient no-shows. Health Care Manag Sci 22(1):68–84. https://doi.org/10.1007/s10729-017-9421-7 Epub 2017 Oct 9. PMID: 28993965
11. Qiu Y, Song J, Liu Z (2016) A simulation optimization on the hierarchical health care delivery system patient flow based on multi-fidelity models. Int J Prod Res 54(21):6478–6493. https://doi.org/10.1080/00207543.2016.1197437

A Comparative Study of Deep Learning Techniques Aimed at Detection of Arrhythmias from ECG Signals

John Gómez, Alberto Quispe, and Guillermo Kemper(✉)

Universidad Peruana de Ciencias Aplicadas, Av. Prolongación Primavera 2390, Monterrico, Santiago de Surco, Lima, Perú
{u201315965,u201322186,pcelgkem}@upc.edu.pe

Abstract. This article presents a comparative study of the performance of Deep Learning techniques (LSTM, Dense, and CNN) applied to the detection and classification of arrhythmias in an ECG signal. The objective is to obtain the best prediction model for cardiac arrhythmias with each Deep Learning technique and compare their performances. The training was done with the Physionet MIT-BIH database, from which 108854 samples were taken from a total of 48 patients with and without proven arrhythmia. With this information, the three indicated neural networks were trained, and the performance comparison was carried out through the ROC curve. The one-dimensional convolutional neural network (1D-CNN) achieved the best performance. It was used to build an arrhythmia detection algorithm using one hidden convolutional layer and 3600 input samples at a time (10-s signal fragments). The obtained results were very satisfactory, achieving a precision of 92.07% for the 1D-CNN technique, 82.10% for LSTM, and 75% for Dense.

Keywords: CNN · 1D-CNN · Dense · LSTM · Comparison · Cardiac arrhythmia · ECG · Physionet · Deep learning

1 Introduction

A vital aspect of personalized health care is the continuous monitoring of a person's health. Each year, 17.3 million people deceased from any cardiovascular disease, representing 37% of total global deaths [1]. Many of these deaths could have been prevented with adequate patient follow-up.

Electrocardiography (ECG) is the most simple and accessible method today to diagnose diseases related to the heart and health in general since this method is non-invasive, and it is very easy to use. Visualization and analysis of this signal is the most widely used diagnostic method among physicians.

Nowadays, the growing use of portable biomedical devices makes possible the practical implementation of signal monitoring devices and the implementation of various intelligent algorithms that, through the continuous evaluation of some biomedical signals, can predict potential health hazards. This has encouraged recent studies to apply

Y. Iano et al. (Eds.): BTSym 2020, SIST 233, pp. 385–395, 2021.
https://doi.org/10.1007/978-3-030-75680-2_43

different Deep Learning techniques aimed at detecting anomalies in the ECG signal [2–4].

This work aims to compare various Deep Learning methods and identify the most reliable method for detecting cardiac arrhythmias from the ECG signal. We focus on comparing the performance of 3 different Deep Learning techniques: Dense Neural Network, CNN (Convolutional Neural Network), and LSTM (Long Short-Term Memory).

These techniques are currently the most widely used in state of the art to analyze and detect patterns in ECG signals and numerical sequences in general. The Dense Neural Networks are the simplest networks composed of layers of neurons whose activation function is non-linear [5]. The work of [3] proposes the use of this type of network for ECG analysis. In this study, an accuracy of 96.77% was achieved in the diagnosis. However, to achieve this result, the study applied the Discrete Wavelet Transform for signal decomposition besides using the neural network.

Convolutional neural networks are characterized by using kernels that apply the convolution operation over the signal and increase the signal depth [2, 4]. The study of MM Al Rahhal in [5] proposes, for example, the use of a one-dimensional CNN to determine arrhythmias. The results achieved show good efficiency in various scenarios for the classification of different types of arrhythmia. However, in this study, an additional generative module is used to convert the one-dimensional ECG signal into an image through convolution layers, which forces the use of two types of convolution at the same time (1D and 2D).

Finally, LSTM is used mainly for the analysis of numerical sequences [6]. Since the ECG is a signal in time, it is normal trying to use a Recurrent Neural Network (RNN) like LSTM. Unlike the other networks, this network has gates that keep in memory the past data information. In [7], LSTM is used to detect arrhythmias, achieving accuracy greater than 99%. However, the model only analyzes ECG signals of short duration (short fragments) and requires a complex model of compression of the ECG data to serve as input for the proposed network.

The three techniques work satisfactorily and have independently achieved excellent results; however, the works where these techniques are applied focus on using only one of them at a time and looking for the best combination of parameters to achieve the best results. A comparison is not made between the proposed techniques, with the purpose of determining the one that presents the best performance in detecting arrhythmias from an ECG signal. The present work focuses on identifying heart rhythm patterns with different networks in equal conditions (same training data and same validation data). For this purpose, records from the MIT-BIH Database were used to generate learning and prediction models using the results available in said database.

The results finally showed that the best performance was achieved by 1D-CNN, which will be described and discussed at the end of this manuscript.

2 Description of the Comparative Study

For the comparative study of Deep Learning algorithms, a classification and evaluation methodology based on Plawiak's proposal [8] was used. The block diagram of the process is shown in Fig. 1 and consists of five stages, which will be described below.

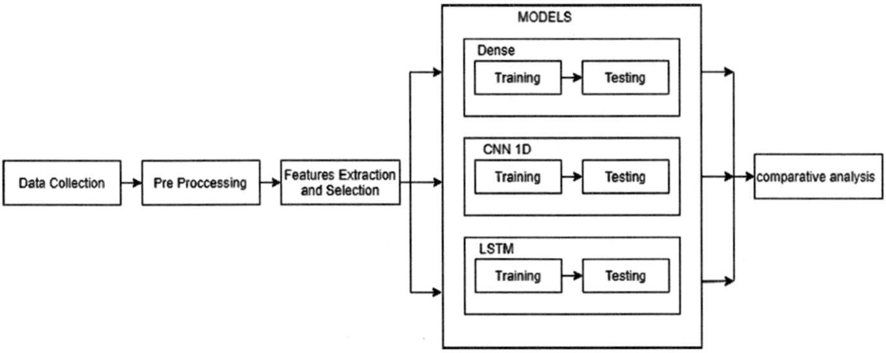

Fig. 1. Block diagram of the comparative study

2.1 Data Collection

The ECG signals used for the training and validation of the algorithms were acquired from the MIT-BIH arrhythmia database available from Physionet [9]. According to the information provided by Physionet, the signals were acquired at a sampling frequency of 360 Hz and a gain of 200 adu/mV. A total of 108854 signal fragments of a 3600 sample length each (corresponding to 10 s of signal) were extracted from a total of 48 patients. Of the 108854 fragments, 34229 corresponded to signals with arrhythmia, and 74625 corresponded to signals without arrhythmia. Figure 2 shows a fragment of a normal signal and a fragment of an arrhythmia signal. Note that the normal ECG signal has an almost uniform distribution of the peaks of the QRS complexes [9], while the arrhythmic signal has an abnormal variation around the peak marked in red.

2.2 Pre-processing

Initially, we define $x(n)$ as the ECG signal acquired from Physionet's MIT-BIH database. Then, the pre-processing of the signal consists of three stages.

Gain Reduction. The database information states that the signals were sampled with a device with gain $g = 200$. Therefore, to obtain the real values of the sensor, the gain of the signal present in the data was reduced using the following expression:

$$x1(n) = \frac{x(n)}{g} \tag{1}$$

Where $x1(n)$ is the ECG value after gain reduction, and g is the value of gain of the device on which the ECG signal was recorded.

Figure 3 shows a portion of the input signal and its output after the gain reduction.

Signal Segmentation. Next, $x1(n)$ is segmented into the 10-s fragments. Each fragment can be expressed as $x1_b(n)$, where b constitutes the fragment number of the signal.

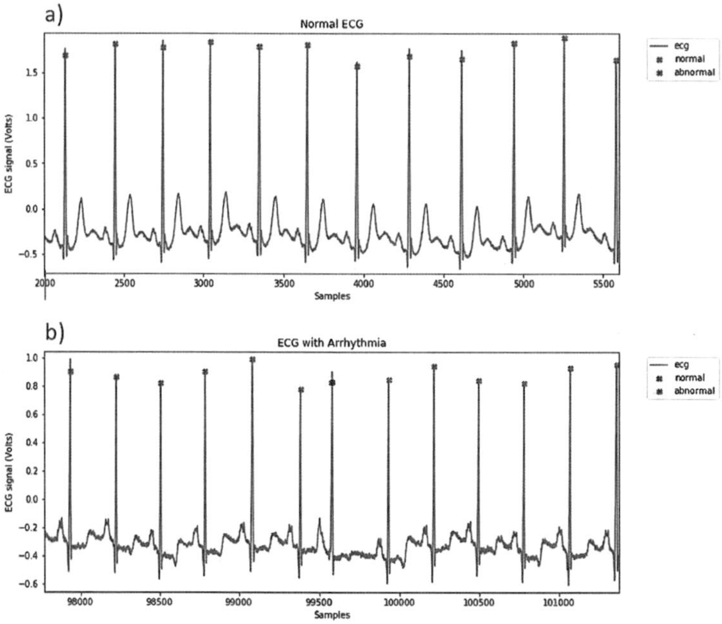

Fig. 2. a) Fragment of normal ECG and b) fragment of ECG with arrhythmia.

Rescaling. The signal was rescaled to the range from 0 to 1 to normalize all the ECG signal that will enter the network and reduce the effect of the DC component. The process is carried out block-by-block, and it can be expressed as:

$$x2_b(n) = \frac{x1_b(n) - x1_b(n_0)}{x1_b(n_1) - x1_b(n_0)} \tag{2}$$

Where:

$$x1_b(n) \geq x1_b(n_0) \forall n \tag{3}$$

$$x1_b(n) \leq x1_b(n_1) \forall n \tag{4}$$

The signal fragment after normalization is shown in Fig. 4.

2.3 Features Extraction and Selection

For the selection of the training data and the validation process, we performed two different tests for the division of the total data. We used all 108854 fragments belonging to the 48 patients in the first test, and we rearranged them randomly. From the total of fragments, 87083 random fragments were extracted for the training data and 21771 fragments for the validation data. We will call this data the test set 1.

In the second test, we divide the available data into patients. We use the data from 38 patients as the training data and the data from the remaining ten patients as the validation

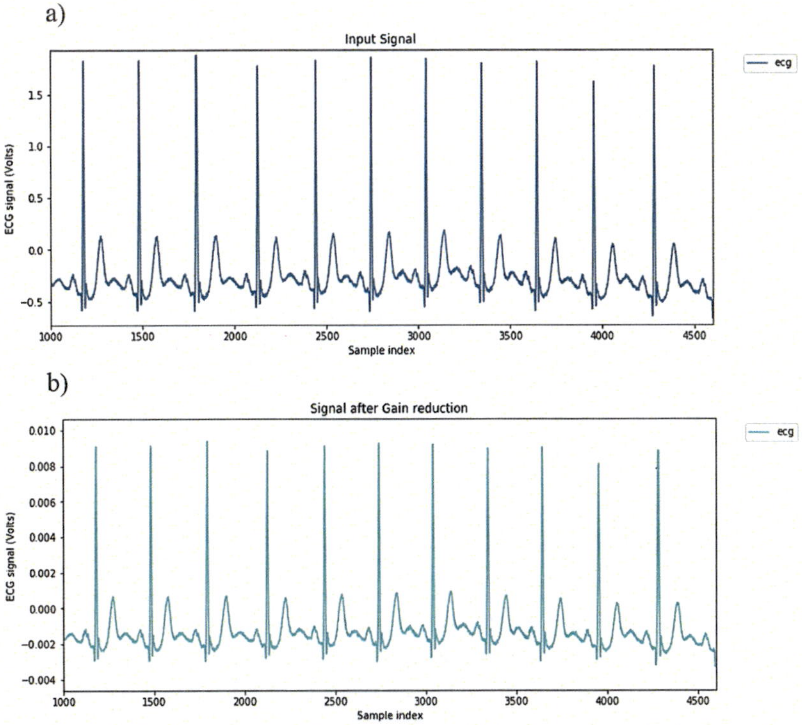

Fig. 3. a) Fragment of Input Signal and b) Signal after gain reduction.

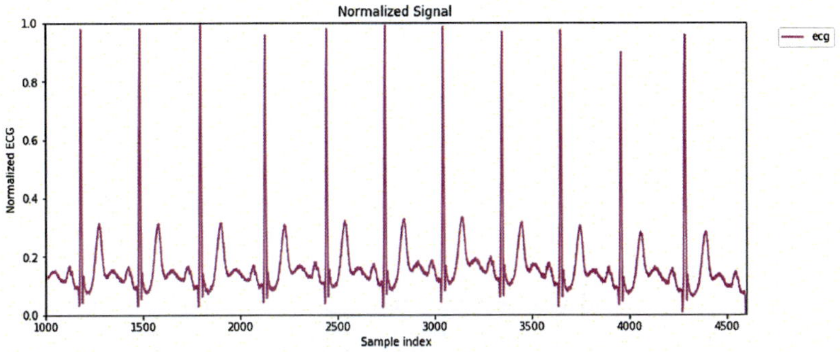

Fig. 4. Normalized signal fragment.

data. In total, 81861 fragments were used for training (75.2% of the total data) and 26993 signal fragments for validation. We will call this data the test set 2. Table 1 shows the distribution of the total fragments used for both sets.

Table 1. Distribution of fragments used to train and validate algorithms

Condition		Set 1		Set 2	
	Total fragments	Training	Validation	Training	Validation
Normal	74625	59665	14960	59009	15616
Arrythmia	34229	27418	6811	22852	11377
Total	108854	87083	21771	81861	26993

2.4 Proposed Models

We developed three different methods for arrhythmia detection and compared their performance. To standardize the conditions and make the comparison, the networks were designed with a single hidden layer of the corresponding type (except for dense) and a Dense output layer of one neuron. In all models, we use a Dropout layer of value 0.25. The dropout layer is a regularization technique used to prevent overfitting through the random elimination of some nodes [10].

NN Dense. In the case of NN Dense [11], the structure of the proposed model consists predominantly of dense-type layers, which is a simple perceptron network. To establish a fair comparison between the three models and maintain a level playing field, in this model, we use a simple structure of 2 densely connected hidden layers of 32 units each. With that number of hidden layers, we ensure the prevalence of dense layers in this model.

Since each input signal in the data set has a dimension of 3600 samples, we configure a first input layer with that same dimension. After the hidden layers, a dropout layer was added with rates of 0.25 to prevent overfitting. Finally, we use a dense layer of 1 neuron in the last layer of the network (binary output layer) in order to classify the ECG fragments into two types: with and without arrhythmia.

We use the training and validation data corresponding to test set 1 to train and validate the described network. The results showed an accuracy of 0.964 in the validation.

Then, for test 2, a network was trained with the same configuration. However, in this case, we used the training and validation data corresponding to test set 2. The results this time showed an accuracy of 0.750.

We noticed that the validation accuracy was greatly reduced in the second case, despite being the same network and using the same dataset. This might happen because in test set 1, data from the same patient can appear in both training sets and validation sets. This means that it is possible that, in the first case, relevant information from the patient signal was accidentally leaked into the data sets. Therefore, in this work, we divide our data by patients and use the test set 2 in the following tests. Table 2 presents the performance of the proposed Dense neural network.

Convolutional Neural Network (CNN). Convolutional neural networks are generally used to process images and any 2-dimensional matrix data; however, their use is also common for one-dimensional data processing. There are numerous studies where CNNs achieve excellent performance, especially for speech recognition and image processing.

Compared to other networks, CNN has few parameters to optimize, making it easier to work with [10].

We trained a one-dimensional CNN, which is widely used in the current literature for ECG signals [2, 5, 10]. The model structure consists of the layers of classical neural networks, but the structure of 1D-CNN is predominant. In the data set, each input signal has 3600 samples; therefore, the first input layer is of the same dimension. Only 1 CNN block was placed in the first hidden layer of the network to maintain a level playing field for comparison with the other models. In order to obtain the greatest amount of information and allow a greater extraction of characteristics in the signal (features maps), 128 convolution filters of size five were used.

To reduce the size of the obtained feature maps, we generated new feature maps using the maximum values of a specific region from the previous feature maps. We use the 1D MaxPool layer for this purpose. A flatten layer was used to adjust the output data of the convolutional layer and make it suitable for interacting with the next layer. After the flatten layer, a dropout layer was added at a rate of 0.25 to prevent overfitting. Finally, a dense layer of 1 neuron was placed at the output of the network (binary output layer), with the aim of classifying the ECG fragments into two types (with arrhythmia and normal).

Long Short-Term Memory (LSTM). For this work, we use a bidirectional Long Short-Term Memory (LSTM) network. An LSTM network is a kind of Recurrent Neural Network made up of memory blocks, memory cells, and gate units [7]. The advantage of the LSTM network is that, unlike other networks like Dense and CNN, it retains historical information of the signal for its processing. LSTM is suitable for the prediction of time series where there are repetitive patterns. The LSTM network used in this work consists of five layers. We used only one LSTM block of 32 units in the first layer of the network to maintain fairness for comparison with the other models. A flatten layer was used to fit the LSTM output data and make it suitable for interacting with subsequent dense layers. After the flatten layer, a 100-unit dense layer was added to get the most information from the previous layer's output. A dropout layer with rates of 0.25 was used to prevent overfitting. Finally, in order to automatically recognize the incoming ECG signals and classify the ECG fragments into two types (with arrhythmia and normal), a dense layer of 1 neuron was placed in the last layer of the network.

3 Results Comparison

The test set 2 was used in the 3 cases to evaluate the performance of the algorithms. The experimental validation results were carried out on a computer with an Intel (R) Core processor (TM) 1.99GHz i7-8550U. The three proposed networks were trained separately with the same training data from the mentioned dataset.

Figure 5 shows the training and validation performance graphs of the proposed networks. The AUC (Area Under the Receiver Operating Characteristic (ROC) Curve) was used as a performance parameter. This parameter is a performance indicator for diagnostic algorithms commonly used in Deep Learning [12]. Notice that the graphs show that the AUC increases as the number of patients trained for the 3 cases increases.

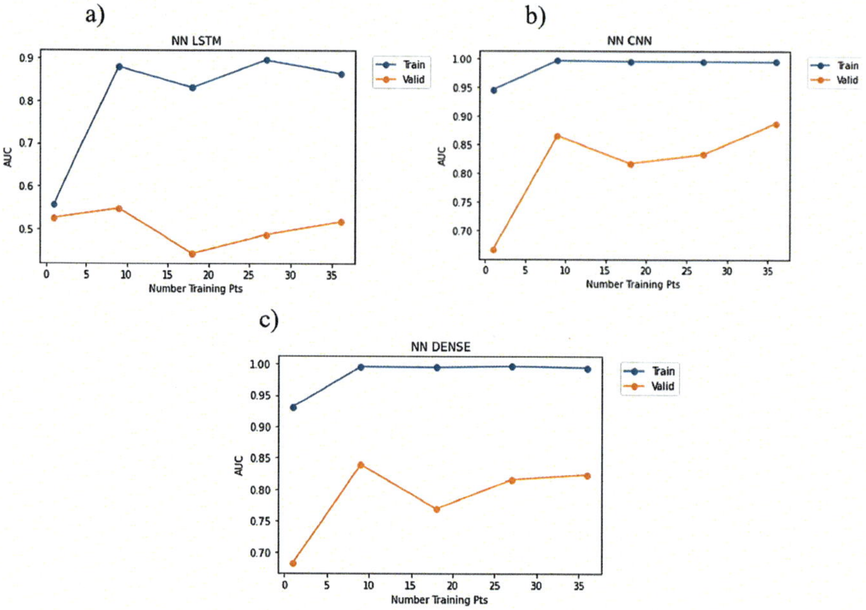

Fig. 5. Performance charts of a) LSTM, b) CNN, and c) Dense

For the 3 cases, the results of training achieved good precision. After ten epochs for the NN Dense, the training and validation stages achieved accuracy rates of 96.4% and 75%, respectively. For the LSTM network, the training and validation accuracy rates were 91.7% and 82.10%, respectively. Finally, accuracy rates of 96.6% were obtained in training and 92.07% in validation for CNN. These results are shown in Table 2.

Table 2. Accuracy in training and validation

Algorithm	Training accuracy (%)	Validation accuracy (%)
Dense	96.40	75.00
LSTM	91.70	82.10
1D-CNN	96.60	92.07

3.1 Evaluation Criteria

The standard evaluation criteria defined in [8], such as sensitivity (SEN), specificity (SPE), and general accuracy (ACC), were used to evaluate the performance of the proposed models.

The formulas used in the calculations for general precision (ACC), sensitivity (SEN), and specificity (SPE) are as shown below.

Accuracy [8]:

$$ACC = \frac{\left(\sum_{i=1}^{N} \frac{TP+TN}{TP+FP+TN+FN}\right)100}{N} \tag{5}$$

Sensitivity [8]:

$$SEN = \frac{\left(\sum_{i=1}^{N} \frac{TP}{TP+FN}\right)100}{N} \tag{6}$$

Specificity [8]:

$$SPE = \frac{\left(\sum_{i=1}^{N} \frac{TN}{FP+TN}\right)100}{N} \tag{7}$$

Where N is the total number of sets used for validation, TP True Positive, TN True Negative, FP is False Positive, and FN False Negative.

Table 3 shows the detail of the evaluation criteria calculated considering a decision threshold (cut-off value) equal to the simple average of the neural network's known outputs [12].

Table 3. Detail of the evaluation criteria used

Algorithm	SEN (%)	SPE (%)	Overall ACC (%)
Dense	88.20%	65.38%	75.00%
LSTM	78.40%	84.79%	82.10%
1D-CNN	91.30%	92.62%	92.07%

3.2 ROC Analysis

Once the three types of neural networks were trained and the standard evaluation criteria were calculated, we made a comparison of their performance based on the ROC analysis. ROC curve analysis is a method that provides tools to select optimal models and discard those that are not [12]. The ROC analysis is directly related to the cost/benefit analysis in making diagnostic decisions and shows the false positive rate versus the true positive rate for all possible decision thresholds.

In Fig. 6, the ROC curve of the three types of applied neural networks is shown.

We also evaluate the AUC-ROC (Area Under Curve ROC). For the calculation of the AUC, all possible decision thresholds (from 0 to 1) were considered [12].

The 1D convolutional neural network is the one with the highest AUC; for this reason, it can be stated that it has a better performance in the diagnosis of arrhythmias in an ECG signal, compared to LSTM and Dense.

It is also shown that AUC results in 1D-CNN improve rapidly as the number of trained patients increases. CNN achieves an AUC greater than 0.91 for a training data of 35 patients. We note that with a training data of 8 patients, results greater than 0.8 are obtained. These results are shown in Fig. 7.

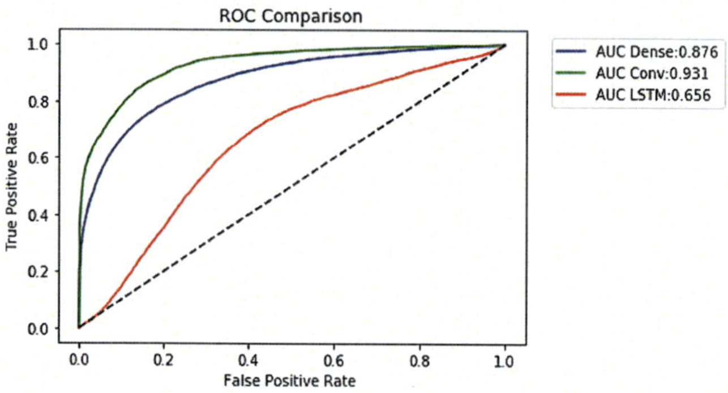

Fig. 6. ROC analysis of CNN, Dense, and LSTM neural networks

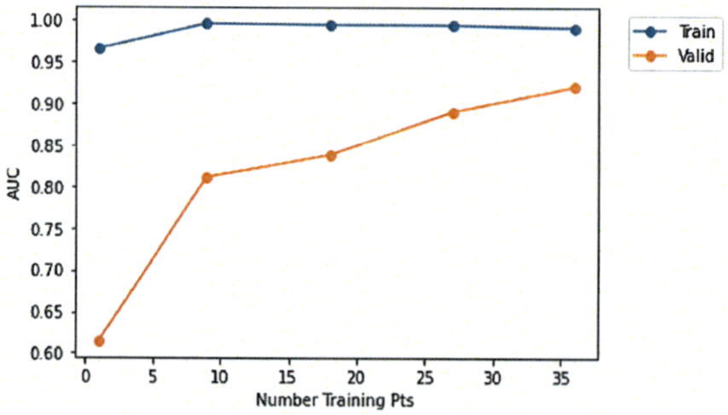

Fig. 7. 1D-CNN Learning curve.

4 Conclusions

The proposed one-dimensional convolutional neural network (1D-CNN) achieves better results, using the same amount of training data compared to the dense network and the LSTM. The achieved performance of the CNN network based on the AUC parameter is 0.931 compared to an AUC of 0.876 and 0.656 of Dense and LSTM, respectively.

The learning speed of (1D-CNN) is higher than the other models, considering the same size of training data.

The division of the training data by patients, instead of the random division by samples, positively influences the results and learning of the network, canceling the effect of overfitting generated in a division by samples.

References

1. American Heart Association (2017) Heart disease and stroke statistics. A report from the American Heart Association
2. Pranav R (2017) Cardiologist-level arrhythmia detection with convolutional neural networks. Cornell University. Computer vision and pattern recognition arXiv:1707.01836v1

3. Prasad GK, Sahambi JS (2003) Classification of ECG arrhythmias using multi-resolution analysis and neural networks. In: TENCON 2003. Conference on Convergent Technologies for Asia-Pacific Region, Bangalore, India, vol 1, pp 227–231. https://doi.org/10.1109/TENCON.2003.1273320
4. Pyakillya B, Kazachenko N, Mikhailovsky N (2017) Deep learning for ECG classification. In: Journal of physics: conference series, BigData conference (formerly international conference on big data and its applications) 15 September 2017, Moscow, Russian Federation, vol 913
5. Rahhal M, Bazi Y, Almubarak H, Alajlan N, Zuair M (2019) Dense convolutional networks with focal loss and image generation for electrocardiogram classification. IEEE Access 7:182225–182237. https://doi.org/10.1109/ACCESS.2019.2960116
6. Gamboa JCB (2017) Deep learning for time-series analysis. In: Cornell University, machine learning, seminar on collaborative intelligence in the TU Kaiserslautern, January 2016. arXiv:1701.01887
7. Yildirim O, Baloglu U, Tan Ru-San, Ciaccio E, Rajendra Acharya U (2019) A new approach for arrhythmia classification using deep coded features and LSTM networks. Comput Methods Programs Biomed 176:121–133. https://doi.org/10.1016/j.cmpb.2019.05.004 Epub 2019 May 10. PMID: 31200900
8. Plawiak P (2018) Novel methodology of cardiac health recognition based on ECG signals and evolutionary-neural system. Expert Syst Appl 92C:334–349. https://doi.org/10.1016/j.eswa.2017.09.022
9. Moody GB, Mark RG (2001) The impact of the MIT-BIH Arrhythmia Database. IEEE Eng Med Biol Mag 20(3):45–50. https://doi.org/10.1109/51.932724 PMID: 11446209
10. Yıldırım Ö, Pławiak P (2018) Arrhythmia detection using deep convolutional neural network with long duration ECG. Comput Biol Med 102:411–420. https://doi.org/10.1016/j.compbiomed.2018.09.009
11. Liu X, Yang D, Gamal AE (2017) Deep neural network architectures for modulation classification. In: Cornell University, machine learning, Asilomar conference on signals, systems, and computers, November 2017. arXiv:1712.00443
12. Bradley AP (1997) The use of the area under the ROC curve in the evaluation of machine learning algorithms. Pattern Recognit 30(7):1145–1159

Lean Manufacturing Model Using a Biotechnological Approach for Increasing Efficiency and Reducing Waste at a Small Plastic Production Company

Diego Samar-Tarazona, Alejandro Tapia-Landa(✉), Ernesto Altamirano-Flores, and Carlos Raymundo-Ibañez

Universidad Peruana de Ciencias Aplicadas (UPC), Lima, Perú
{u201512309,u201518029,pcinealt,carlos.raymundo}@upc.edu.pe

Abstract. This study analyzes the background and current problems experienced by companies in the plastics sector due to the ban on the manufacturing, distribution, and use of conventional bags. Simultaneously, the internal problems of manufacturing have been identified within the company, such as machine shutdown and inadequate use of personnel, thus generating manufacturing waste and downtime in plastic bag production. Therefore, a lean manufacturing model is proposed through a biotechnological approach at a small company operating within the plastics sector, where solution techniques such as 5S methodology, business process management, and biotechnology will be used to generate an efficient production of biodegradable bags, which is the primary outcome of this implementation and is as an alternative to conventional plastic. Hence, an innovative product will be created that guarantees economic, social, and environmental viability in the plastics sector businesses.

Keywords: Lean manufacturing · Manufacturing model · 5S methodology · Process management · Biotechnology

1 Introduction

World plastic production reached 348 million tons, with a 4% share of Latin America (2017) [1]. In Peru, the growth rate of the gross domestic product (GDP) is 4%, with a 4.1% increase in plastic manufacturing (2018) over the previous year [2]. The total supply of the Peruvian plastic industry (2018) was 41% from foreign companies and the rest from local companies. It is worth noting that, in the several applications of polymers in Peru, 20% are devoted to manufacturing flexible packaging, of which ~23% corresponds to plastic bags [3]. Currently, plastic companies are being affected by Law No. 30884, enacted by the Peruvian government, which prohibits the manufacturing, marketing, and use of conventional bags [4]. It should be noted that biodegradable bags are manufactured using natural resources, such as vegetable starch and glycerin,

Y. Iano et al. (Eds.): BTSym 2020, SIST 233, pp. 396–403, 2021.
https://doi.org/10.1007/978-3-030-75680-2_44

essential inputs for developing the innovative and sustainable product with a benefit for the environment and society.

In 2019, it was observed that, in the company, there was an operational inefficiency, which generated a high impact on its revenue. Therefore, a diagnosis was conducted where it was evident that the inefficiency in the company under study is generated because of two main causes: machine shutdowns, representing 61.54% of the impact, and waiting time observed while changing the filters in the extrusion machine of the production line. Here the operator must leave the production area and move toward the filter recovery area. Additionally, there is an inadequate use of personnel, at a 38.46% impact since they generate production waste and downtime due to a work environment that lacks cleanliness and order in the production line. Small- and medium-sized companies in the plastics sector require a manufacturing model with a biotechnological approach because recently several dispositions have been established, causing changes in the current manufacturing line. The data used in this work were provided exclusively for academic purposes on the anonymity condition.

This study aims to demonstrate the advantages of using the lean manufacturing methodology combined with biotechnology, which uses natural inputs, such as corn starch and glycerin, to provide support for the production of biodegradable bags as an added value to the business in the plastic sector.

2 Literature Review

2.1 Lean Manufacturing in a Manufacturing Model of Micro and Small Enterprises (MSE) in the Plastics Sector

Lean manufacturing tools support the implementation of a manufacturing model because they improve the coordination and effectiveness of operational processes. Among the most used ones, we can find the 5S methodology, a methodical application in the workplace that increases process performance, determining the adequate time needed in each activity through its five key principles: Seiri (Sort), Seiton (Set in order), Seiso (Shine), Seiketsu (Standardize), and Shitsuke (Sustain). Similarly, this tool provides an easy strategy and possesses great potential when developing a successful company. It is worth noting that awareness and communication with the company's employees at all levels are of vital importance for an optimal improvement [5, 6]. Another very important tool is Business Process Management (BPM) since its main goal is to improve business processes by applying both modeling and analysis to find ideas for solutions to reduce unnecessary costs and improve the level of service. To measure the improvements in the processes, it is necessary to use Key Performance Indicators (KPIs). Their choice must be associated with the fulfillment of the strategic goals of the organization [7, 8].

Companies that have used the lean manufacturing methodology have demonstrated excellent results due to its simplicity and effectiveness, achieving considerable reductions in the waste or changes generated by their operating processes. Likewise, through this application, operational efficiency in the organization is increased, evidencing a better optimization of personnel, machinery, and space, and tangible and quantifiable outcomes of great impact on the company.

2.2 Biotechnology in a Production Model in Micro and Small Enterprises (MSE) of the Plastic Sector

The plastic sector businesses are in search of improvement alternatives that allow them to have more efficient and cleaner processes. Many of these have decided to improve their production model so that their processes conform to the environmental guidelines established in recent years. The innovative technique that has revolutionized the global market due to the high expectations it has generated is biotechnology, a technique that requires the use of natural resources to create or modify products for specific uses. It is this opportunity to manufacture biodegradable plastics based on natural biopolymers, such as starch (corn, cassava, and potato), cellulose, and polylactic acid (PLA) [9, 10]. On the other hand, when a biotechnological procedure is implemented in the production model, cleaner processes are generated, unlike current processes. Likewise, there would be a better control of polluting waste and a decrease in energy consumption, guaranteeing the effectiveness of production processes and thus achieving increased benefits in the organization [11, 12].

By implementing a biotechnological procedure in their production model, businesses in the plastic sector would generate more efficient and cleaner processes. Therefore, this technique is beneficial to generate a replacement alternative to conventional plastic since biodegradable plastics will be produced by taking advantage of natural resources in order to improve the biodegradability of the plastic material.

3 Contribution

3.1 Rationale

The proposed model is developed for companies in the plastics sector that are currently being affected by various state regulations. Therefore, a lean manufacturing model will be implemented through a biotechnological approach, where starch (corn, yucca, and potato), cellulose, and polylactic acid (PLA), which are necessary for the production of biodegradable plastics, will be used as natural biopolymer alternatives [10] due to poor coordination when changing filters. For this reason, the application of lean manufacturing tools is critical to minimize waste and ensure increased efficiency in the organization [13]. Therefore, the 5S methodology will be implemented by using the five key principles: Sort, Set in Order, Shine, Standardize, and Sustain, which will improve the conditions in the work environment and generate value for all operational processes [7]. Another important tool that will be implemented is the BPM, which will help improve the business processes by applying modeling and analysis. It is important to find viable solutions to reduce unnecessary costs and improve the level of the service, generating a competitive advantage in the sector [12].

3.2 General View

3.3 Detailed Overview

This section describes the stages presented in Fig. 1.

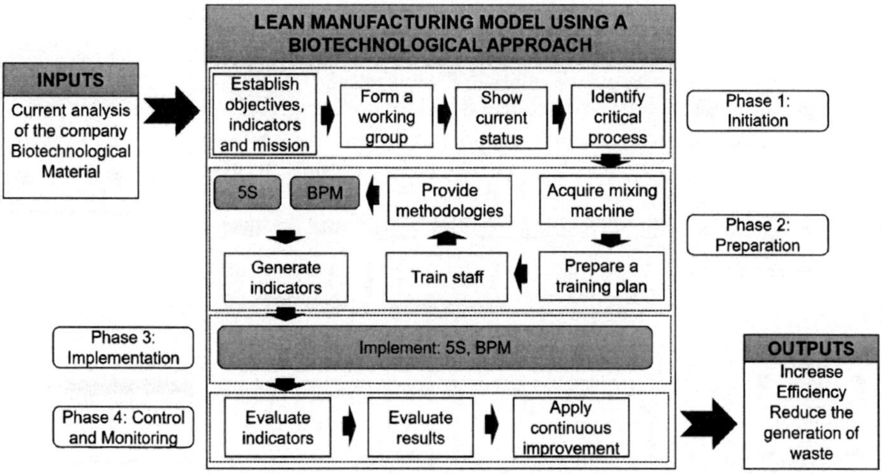

Fig. 1. Lean Manufacturing Model using a biotechnological approach.

Stage 1: Start. Set goals, indicators, and mission: The goals will be set to obtain efficiency in the manufacturing process above 85% and reduce the generation of waste in the critical processes by 2%. The goals will also be set as a mission for manufacturing a product based on biotechnological materials that are in accordance with the environmental guidelines established by the Peruvian State.

Create workgroups: A workgroup will be integrated according to the necessary requirements for the company, establishing hierarchies, procedures, and activities to be performed.

Assess current status: A general assessment of the current status of the company will be conducted, identifying flaws in the processes and possible solutions to be made as improvements.

Identify critical processes: A general inspection will be conducted in the company, which, according to the data obtained, reports an inefficiency in the production process, generated by the shutdown of machines with a 61.54% impact due to the waiting time for the change of filter in the extrusion machine of the production line and by the inadequate use of the personnel with a 38.46% of the impact, causing production waste in the extruded process and raw materials (RM) reception because of the presence of a work environment with a lack of order and cleanliness.

Stage 2: Preparation. Purchase mixing machine: A mixing machine will be purchased for the production line, as it is necessary to replace the crushing process with the mixing process to adapt the biotechnological approach of the manufacturing process.

Prepare a training plan: A training plan will be implemented, so the quality of the training material should be assessed, instructors or trainers specializing in biotechnology and manufacturing should be available, and the cooperation and motivation of the company's staff should be established, and it will also be necessary to define the workers who will receive the training.

Train operators: Training will be provided to the operators establishing the activities to be executed, the forms to be filled in, and the procedures to be taken into account, which will be done before, during, and after the training.

Implement methodologies: The advantages of applying methodologies such as 5S and BPM, such as reducing downtime and waste in the manufacturing process to ensure improvements in the company's efficiency, will be explained to the operators.

Stage 3: Implementation. The implemented methodologies are described below:

1. 5S methodology:

   - Seiri (Sort): All materials that contribute no value to the company will be removed. Also, obsolete and unnecessary tools for the manufacturing process should be separated, and an analysis will determine a feasible option between selling the equipment and performing an overhaul.
   - Seiton (Set in order): The areas for establishing an adequate order will be determined, the specific places for each tool will be identified, and the use of labels will allow an easy recognition linked to the manufacturing process. Additionally, after being used, it will be verified that the tools or pieces of equipment are placed in their assigned places.
   - Seiso (Shine): The manufacturing area will be cleaned, and then a preventive maintenance will be performed on all machines, establishing the actions and schedules for each activity. Also, a cleaning kit should be purchased for each operator.
   - Seiketsu (Standardize): All the information concerning the procedures that must exist in the work stations will be detailed in an exact way, as well as establishing the responsibilities for each operator. It is worth mentioning that information will be gathered to minimize unnecessary rounds in the manufacturing process.
   - Shitsuke (Sustain): A follow-up, order, and cleanliness of the organization will be performed, providing constant training to all the company's employees to maintain the 5S philosophy. Additionally, the records of the conformity indexes and the improvement of the implementation will be kept to verify if the work is being fulfilled efficiently.

2. Business Process Management:

Each manufacturing process should be constantly inspected, detailing the shift duration in hours of the operator's inspection and the use of the machine by the operator, and establishing the quantities of products under process. Also, detailing the cycle times and downtimes in minutes for each process and its corresponding downtime code. However, the possible flaws to be found in the manufacturing process are detailed, with the most common being machine shutdowns, process waste, improper machine starts, and poor machine maintenance. Similarly, the other flaws that may occur during the working day will be indicated.

Stage 4: Follow-up and Control. Assess indicators: The indicators obtained will be assessed to analyze if the goals were achieved with the expected baseline level.

Assess results: The results will be assessed by contrasting the current situation of the company with the improved situation, after the implementation of 5S and BPM methodologies, with the use of Arena software for the respective simulation. Simultaneously, possible additional improvements that can be applied in the company will be determined.

Apply continuous improvement: A continuous improvement will be applied in all the manufacturing process to guarantee that the actions performed are maintained and improved progressively.

4 Validation

4.1 Validation by Simulation

Initial Validation Scenario. To validate the proposed model, a simulation will be executed using the Arena software. Likewise, as inputs to the system, we have the statistical distributions of the times for each operating process, from the reception of RM to the final cutting and sealing process, as well as the specific number of machines and operators involved in each process. The restrictions in the simulation are the working hours (8 h/day), the capacity to produce in the manufacturing process (13 tons per month), and the generation of waste in the RM reception and extrusion processes (5% in each process).

Initial Diagnosis. The simulation of the current manufacturing process has revealed the flaws existing within the organization. Among them, the operational inefficiency in the whole process is caused by a delay in the change of filters of the extrusion machine, and an increase in the average standard time of the extrusion process has been observed by 22.43% (average standard time = 70 s). Also, this delay caused in the extrusion process generates queues or waiting times in the same process and in the previous process (crushing). Another flaw that is causing much impact on the manufacturing process is the lack of cleanliness and order in the RM reception and extrusion processes, causing discoordination when performing work activities.

The flaws identified in the current manufacturing process cause the assessment indicators not to be the most accurate for the organization. The simulation conducted yielded an operational efficiency of 80.27%, considered by the measurement indicators as regular, but the improvements should increase that result. Other results obtained were the generation of waste in the process of RM reception, which is estimated at 26.7 packages/month, and in the extrusion process, estimated at 23.5 packages/month. These are defective products considered as waste in the process. These indicators support improvements in the critical processes to obtain better results later.

Validation Design. Through the simulation performed, it was possible to reduce the queues or waiting times in the crushing and extrusion processes. Favorable results were obtained in the reduction of downtimes during the extrusion process, wherein the reduction was from ~85.7 to 75 s, representing a 12.49% reduction in the downtimes for this process.

The improvements made the operating efficiency increase from 80.27% to 90.26%, almost 10%. Additionally, the generation of waste in critical processes was reduced. We observed a downfall from 26.7 to 26.4 packages per month in the RM reception process, a 1.25% reduction. In the extrusion process, waste was reduced from 23.5 to 22.9 packages per month, a 2.56% reduction in the number of defective products considered waste.

5 Conclusions

Biotechnology is an innovative technique that can be applied in the business of the plastic sector. Through this procedure, a sustainable product can be manufactured according to the environmental guidelines established in recent years. This sustainable product called "biodegradable plastic" can become the best alternative to replace conventional plastic. Since it is manufactured from natural resources, such as starch and vegetable glycerin, it is easily biodegradable in any setting and environmental condition (~3 to 4 years).

With the implementation of the 5S tool, a greater availability of the materials required for performing the different manufacturing process activities is achieved. Likewise, this tool improves the conditions of the workplace where the collaborators develop their daily activities in an efficient and productive way.

With the application of BPM, operational efficiency is improved, showing a better optimization of personnel, machinery, and space, and tangible and quantifiable results of great impact in the organization.

The implementation of the biotechnological procedure in the organization will guarantee an increase in the efficiency of the manufacturing process by ~10%. Likewise, if this implementation is supported with an adequate preventive maintenance plan for the machines used in critical processes and training sessions for the workers involved, it will be possible to reduce 1.25% of the waste generated during the RM reception and 2.56% of the waste generated during the extrusion process, according to the results obtained in the simulation.

The economic assessment indicates that the project is viable, with the following profitability indicators: NPV of 663,904.89 > 0, IRR of 185% > 16%, with an opportunity cost of 16% and B/C ratio of 7.59 > 1.

References

1. Mundo PLAST: La Revista Profesional del Plástico y sus Tecnologías. https://mundoplast.com/produccion-mundial-plasticos-2017/
2. INEI: Panorama de la Economía Peruana 1950–2018. https://www.inei.gob.pe/media/MenuRecursivo/publicaciones_digitales/Est/Lib1654/libro.pdf
3. Sociedad Nacional de Industrias (SIN): Percepción Frente a los Proyectos de Ley sobre el Plástico de un Solo Uso. https://www.congreso.gob.pe/Docs/comisiones2018/Produccion/files/ppt/4ta_(2)sesion_presentaci%C3%B3n_ing._salazar_congreso_02-10-18.pdf
4. El Peruano D (2018) Ley que regula el plástico de un solo uso y los recipientes o envases descartables. https://busquedas.elperuano.pe/normaslegales/ley-que-regula-el-plastico-de-un-solo-uso-y-los-recipientes-ley-n-30884-1724734-1/

5. Sangani R, Kottur VKN (2019) Enhancement in productivity by integration of 5S methodology and time and motion study. In: Proceedings of international conference on intelligent manufacturing and automation. Springer, Singapore, pp 541–550
6. Agrahari RS, Dangle PA, Chandratre KV (2015) Implementation of 5S methodology in the small-scale industry: a case study. Int J Sci Technol Res 4(4):180–187
7. Mau M, Ramos R, Llontop J, Raymundo C (2019) Lean manufacturing production management model to increase the efficiency of the production process of a MSME company in the chemical sector
8. Van Der Aalst WM, La Rosa M, Santoro FM (2016) Don't forget to improve the process! Bus Process Manag 58(1):1–6
9. Rujnić-Sokele M, Pilipović A (2017) Challenges and opportunities of biodegradable plastics: a mini review. Waste Manag Res 35(2):132–140
10. Masmoudi F, Bessadok A, Dammak M, Jaziri M, Ammar E (2016) Biodegradable packaging materials conception based on starch and polylactic acid (PLA) reinforced with cellulose. Environ Sci Pollut Res 23(20):20904–20914
11. Paço A, Jacinto J, da Costa JP, Santos PSM, Vitorino R, Duarte AC, Rocha-Santos T (2018) Biotechnological tools for the effective management of plastics in the environment. Crit Rev Environ Sci Technol 49:1–32
12. Lin X, Negenborn RR, Lodewijks G (2018) Predictive quality-aware control for scheduling of potato starch production. Comput Electron Agric 150:266–278
13. Guillén K, Umasi K, Quispe G, Núñez V, Raymundo C, Moguerza JM (2018) Lean model for the optimization of plastic bag manufacturing in small and medium-sized companies in the plastics sector, vol 1, pp 40–45

A Diagnostic Model for Determining the Reasons Causing Low Quality Logistics Services in SME Logistics Operators

Edgardo Vega-Barros, Pamela Palomino-Ruiz, Claudia Leon-Chavarri(✉), and Pedro Bermudez-Sanabria

Universidad Peruana de Ciencias Aplicadas (UPC), Lima, Peru
{u201314299,u201512842,pcincleo,pcappber}@upc.edu.pe

Abstract. The outsourcing of logistics services has become an increasingly relevant practice for trading companies as it significantly improves their costs and productivity. However, service providers often fail to maintain standards that may guarantee adequate customer satisfaction levels. This is reflected in the logistics service quality issues faced by many Latin–American medium-sized logistics service providers. Within this context, the present research study aims to find a model that can diagnose the causes of these service quality issues in a timely manner. The proposed model uses operational mapping tools, simple information gathering, correlations, and control charts. In fact, the authors' motivation for this study lies in conducting more complete assessments and reducing undiagnosed spaces along with increasing knowledge in the use of diagnostic tools as a whole within the logistics services sector.

Keywords: Logistics problem diagnosis · 3PL industry · Warehouse operations

1 Introduction

The outsourcing of third-party logistics services (3PL) has become fundamental for reducing costs at trading companies [1] and are also known as logistics operators. These companies currently use logistics 4.0, which is a logistics that uses digital media as its main tools for the development of its activities within the supply chain [2]. This type of logistics has become a necessity for all companies; depending on their size, the technologies become more or less essential [3]. In addition, this logistics allows the reverse logistics of returns within the companies that provide service to be more fluid, and better control can be given [4]. However, merchandise returns are still a challenge today, especially in Latin American countries, despite the technologies used in the sector [5]. Companies that outsource their logistics services often manage to significantly reduce their storage and distribution costs while significantly increasing their productivity [6, 7]. Therefore, the service levels provided by logistics operators become critical to these operations. However, small- and medium-sized enterprises (SMEs) often exhibit operational and distribution problems, which negatively affects not only their performance but also the quality of logistics services provided [5, 6].

Y. Iano et al. (Eds.): BTSym 2020, SIST 233, pp. 404–411, 2021.
https://doi.org/10.1007/978-3-030-75680-2_45

Indeed, the quality of the logistics services provided by SME companies within the sector is rated at 58.25%, while the minimum quality expected by customers is at least 73.04% [6]. These poor logistics service quality ratings are frequently associated with merchandise storage and management issues [1, 6]. Also, the quality of the services provided by logistics operators is currently compromised, owing to several warehouse management triggers [1]. In other situations, the issues are grounded in poor merchandise distribution practices [8]. Furthermore, improper order processing creates major productivity hurdles as it is one of the most critical warehouse activities [9, 10].

Within this context, the present research study aims to develop a model that can jointly diagnose the different causes of these logistics service quality issues. It is expected to significantly reduce the problem as follows: Information normality tests will be used to ensure that we can work with reliable data. Then, control charts will be used to diagnose irregular processes. These processes were determined if they added value to the process or not with the value-added analysis tool. Finally, a correlation analysis will be carried out to determine the causes of the main irregular processes.

This research study will focus on a medium-sized Peruvian logistics operator while assessing studies by other authors that address similar issues and applying diagnostic tools that have been previously used in similar scenarios. This article will then discuss the contributions from the research study along with its rationale and the diagnostic tools used. Subsequently, the resulting percentage values will be listed. Finally, the last section of this article will present the conclusions drawn from the research study.

2 State of the Art

2.1 Problem

Often, diagnostic tools of 3PLs problems have been used in isolation, which leaves a gap where there is a percentage of unanalyzed problems [11]. This occurs because the problems not analyzed tend not to be detected by simple observation since no tools can be used in a linked way [12]. In addition, the use of a single tool most of the time is not enough to be able to issue a total diagnosis of a company.

An investigation carried out the diagnosis of the 3PL sector and showed that 8.77% of the problems of the quality of the service in the warehouses are not diagnosed on time, a smaller percentage that generates a great irregularity in service times and costs for both parties, thus causing many times the change of service provider for that of the competition, that is, loss of customers for 3PLs [7].

2.2 Model

Another research study used a mathematical model to reveal a 14.6% increase in travel times within warehouses. That study concluded that implementing layout changes may reduce travel times by 3.3 times [8]. Moreover, another study evidence that merchandise collection times at a given warehouse are 8% higher than the average and that this is caused by excessive travel distances and poor selection and segmentation. As part of this study, the authors conclude that the number of collectors used significantly influences performance [13]. The linear regression results from another research study reveal that 21% of dispatch errors are due to poor planning. Herein, the authors conclude that incomplete order incidents can be reduced through proper demand projections [14].

In a research study that used an algorithm-based model, a 30%–70% warehouse efficiency rate was obtained for the picking and packing process. [15]. Another algorithmic model determines that 38% of the delays experienced in collection processes are caused by inefficient warehouse layout designs [16]. In another study, the algorithm confirmed that order preparation times were 24% higher than the standard. This model is then deemed ideal for assessing times in conventional 2-block warehouses [17]. Finally, a linear programming model identified a 15% opportunity for improvement within the routing issues reported by the warehouse being assessed [18].

2.3 Method

A relevant research study used a method that revealed a 16% risk of delivery delays and an 8% risk of incorrect inventory, evidencing a correlation between them. That study concluded that the higher the correlation between these two activities, the higher the risk of occurrence [19]. In addition, another study managed to evidence that 23% of all warehouse issues reported at a particular location were caused by changes in their stocking keeping units, which generated procedure delays and rising costs [20]. In contrast, a different study applied the order quantity optimization method and identified a cost overrun issue in order processing. That particular study concludes that maintaining optimal quantities can reduce costs [21].

Another study used a simulation to assess initial operator performance, revealing that inefficient load times caused 60% of the processing time issues. That study also proved that simulations provide a more detailed scope for time-related issues [22].

3 Contribution

3.1 Rationale

This contribution is based on the fact that all current methods, models, and tools attempt to solve isolated diagnostic issues. In fact, it may even be considered that they only diagnose preselected specific issues. However, the contribution of this research study seeks to provide a more complete assessment to reduce undiagnosed spaces and increase awareness within the sector.

The proposed model comprises data normality tests, control charts, a value-added analysis, and a correlation analysis.

3.2 Model Details

Data Normality Tests. To begin the analysis, normality tests are performed using the collected data to increase the reliability of the data. These tests were performed for both the order processing and order data management processes. Then, two control charts were developed for non-compliant orders, one for each process under study. Both the picking and packing as well as the order data management processes were assessed at 95% confidence level using the Minitab software, under the following hypothesis.

Where:

Null Hypothesis (Ho): The datasets are normally distributed.
Alternative hypothesis: (H1): The datasets are not normally distributed.

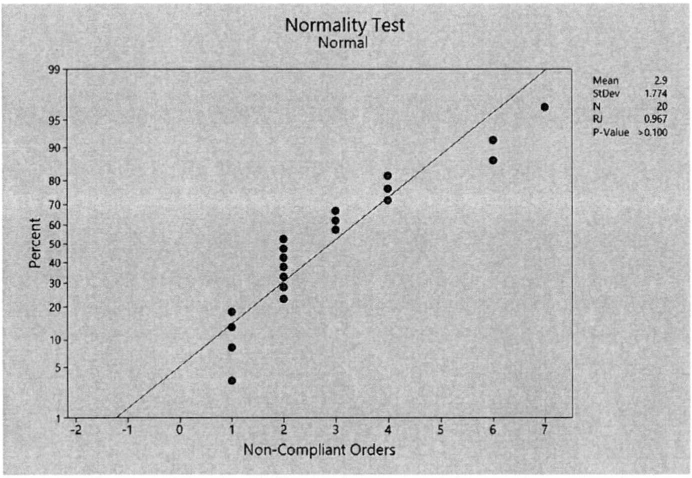

Fig. 1. Normality test-picking and packing

As shown in Fig. 1, both processes yielded a p-value of 0.1, which is >0.05, indicating that the datasets from both processes are normally distributed.

Control Charts. As the datasets are normally distributed, a P control chart is developed to determine the current status of the picking and packing process. A process assessment may then reveal how often orders are incorrectly dispatched using the historical records provided by the company, which evidence incidents that have occurred during this process.

In total, 20 samples of 26 orders each were selected, at a confidence level of 95% and a margin of error of 5%.

As shown in Fig. 2, Upper Boundary (UB) = 23%, Mean Proportional (P) = 11.15%, and Lower Boundary (LB) = 0%

The same P control chart was developed for the data management process, but reporting inventory and new merchandise data delivery incidents, where UB = 27.66%, P = 14.23%, and (LB) = 0.8%.

Both processes exhibit out of bound datasets, indicating non-compliant orders that fall outside the expected levels. In addition, 15% of the samples from both processes exhibit non-compliant orders, which deems both processes as out of control.

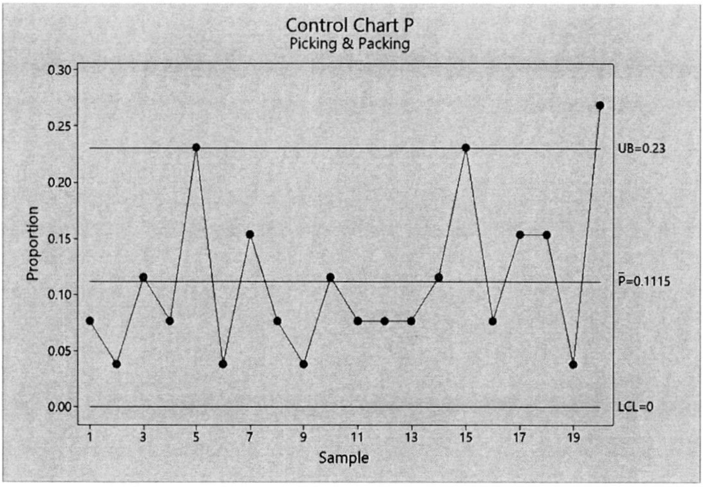

Fig. 2. P control chart – picking & packing

Value-Added Analysis. The order data management, delivery–storage and picking, and packing–dispatch processes were assessed to determine which activities added or did not add value to both the general process and the customers. To this end, an AVA matrix was prepared, using the previous times reported for each activity.

Table 1. Ava matrix - picking, packing, and dispatch process

Picking, Packing, and Dispatch process	Time (min)	VA	NNVA	NVA
Locate Merchandise	12		X	
Pick and Select	18		X	
Report Grouped Quantities	5			X
Move to Dispatch Area	8		X	
Check Documentation	2.82		X	
Load Merchandise	36	X		
Total	**81.82**	**36**	**40.82**	**5**
Percentage	**100%**	**44%**	**50%**	**6%**

In the picking and packing process, one activity was identified as a value-added (VA) activity to the customer. In addition, four activities did not add value to the customer but added value to the process (NNVA), and one activity did not add value to either the customer or the process (NVA). This information can be used to remove or group activities for improving overall process performance.

The same analysis was conducted for all other processes included herein, obtaining similar results to those displayed in Table 1.

Correlation Analysis. The defective merchandise identified in customer orders was due to three main reasons: merchandise damaged during forklift operations, unrecorded merchandise status, and poor merchandise stacking practices. To determine which of these causes exerted a greater influence on the defective merchandise, we correlated each cause to the defective merchandise identified using the Minitab software.

The Pearson's coefficients (r) provided by this software for merchandise damaged during forklift operations, unrecorded merchandise status, and poor stacking were $r = 0.902$, $r = 0.826$, and $r = -0.577$, respectively.

Therefore, both the damages caused during forklift operations and the operators that failed to properly record merchandise status as the merchandise ($r = 0.902$ and $r = 0.826$, respectively) were present when the merchandise was delivered to the warehouse, exhibiting a strong correlation with the defective merchandise included in the orders. Thus, the company must pay special attention to addressing these two causes.

4 Validation

Based on the correlations and important relationships established using these tools, we were able to statistically determine that the picking–packing process, the data management process, and the inefficient distribution of merchandise caused incorrect quantities in 22.7%, 28.3%, and 15.2% of all orders, respectively. In contrast, unrecorded merchandise statuses, merchandise damaged by forklift operations, and poor stacking practices included defective merchandise in 9.8%, 9.8%, and 6.3% of the orders. These data are presented in Table 2.

Table 2. Cause analysis results

	Incomplete orders	Excess orders	Orders with defective merchandise
Picking and Packing	13.3%	9.4%	–
Data Management	7.7%	20.6%	–
Unrecorded Merchandise Status	–	–	9.8%
Merchandise damaged by forklift operations	–	–	9.8%
Poor stacking practices	–	–	6.3%
Inefficient Merchandise Distribution	15.2%	–	–

Indicators Incomplete Orders = Orders with missing merchandise/Non-compliant Orders.

Excess Orders = Orders with excess merchandise/Non-compliant Orders.

Orders with defective merchandise = Orders with defective merchandise/Non-compliant orders.

5 Conclusions

Thus, the tools used herein identify 76.9% of the causes for low-quality logistics services while another 15.2% is identified via simple observation, and the remaining 7.5% is out of the scope of this study.

While poor stacking practices cause 6.3% of non-compliant orders, the correlation analysis reveals a weak correlation, implying that this cause might be circumstantial.

Therefore, this study proves that by using the information normality test tools, control charts, added value analysis, and correlation analysis together, a more complete diagnosis can be accomplished in the warehouse under study. In addition to the initial analysis, these tools can also be used to continuously monitor the activities of the company.

References

1. Alawneh F, Zhang G (2018) Dual-channel warehouse and inventory management with stochastic demand. Transp Res Part E Logist Transp Rev 112:84–106. https://doi.org/10.1016/j.tre.2017.12.012
2. Barreto L, Amaral A, Pereira T (2017) Industry 4.0 implications in logistics: an overview. Procedia Manuf 13:1245–1252. https://doi.org/10.1016/j.promfg.2017.09.045
3. Esmaeilian B, Sarkis J, Lewis K, Behdad S (2020) Blockchain for the future of sustainable supply chain management in Industry 4.0. Resour Conserv Recycl 163:105064. https://doi.org/10.1016/j.resconrec.2020.105064
4. Guo H, Zhang Y, Zhang C, Zhang Y, Han Z (2019) A multi-commodity location-inventory problem in a closed-loop supply chain with commercial product returns. Int J Prod Res 58:1–18. https://doi.org/10.1080/00207543.2019.1686186
5. Li C, Guo H, Zhang Y, Deng S, Wang Y (2018) An improved differential evolution algorithm for a multicommodity location-inventory problem with false failure returns. Complexity 2018. https://doi.org/10.1155/2018/1967398
6. Ali SS, Kaur R (2018) An analysis of satisfaction level of 3PL service users with the help of ACSI. Benchmarking 25(1):24–46. https://doi.org/10.1108/BIJ-10-2016-0163
7. Isaksson M, Hulthén H, Forslund H (2019) Environmentally sustainable logistics performance management process integration between buyers and 3PLs. Sustainability 11(11):3061. https://doi.org/10.3390/su11113061
8. Zhou L, Li Z, Shi N, Liu S, Xiong K (2019) Performance analysis of three intelligent algorithms on route selection of fishbone layout. Sustainability 11(4):1–17. https://doi.org/10.3390/su11041148
9. Calzavara M et al (2018) An integrated storage assignment method for manual order picking warehouses considering cost, workload and posture. Int J Prod Res 57:1–17. https://doi.org/10.1080/00207543.2018.1518609
10. Yoshitake H, Kamoshida R, Nagashima Y (2019) New automated guided vehicle system using real-time holonic scheduling for warehouse picking. IEEE Robot Autom Lett 4(2):1045–1052. https://doi.org/10.1109/LRA.2019.2894001

11. Yang B, Frimpong AAO, Yang Y, Liu L (2019) Application of the quick scan audit methodology in an industrial filter production process. Prod Plan Control 30:1–13. https://doi.org/10.1080/09537287.2019.1630684
12. Wo P, Dragan K (2019) Correlation between strength, hardness and the share of welding imperfections in an aw 5754 aluminium alloy weld executed with the tig method. J. Konbin 49(4):383–400. https://doi.org/10.2478/jok-2019-0092
13. Sad N (2019) Improving the order picking efficiency by optimising the orders' sequence 18:125–137
14. van Gils T, Caris A, Ramaekers K, Braekers K, de Koster RBM (2019) Designing efficient order picking systems: the effect of real-life features on the relationship among planning problems. Transp Res Part E Logist Transp Rev 125:47–73. https://doi.org/10.1016/j.tre.2019.02.010
15. Kim TY (2018) Improving warehouse labour efficiency by intentional forecast bias. https://doi.org/10.1108/IJPDLM-10-2017-0313
16. Çelik M, Süral H (2018) Order picking in parallel-aisle warehouses with multiple blocks: complexity and a graph theory- based heuristic. Int J Prod Res 57:1–19. https://doi.org/10.1080/00207543.2018.1489154
17. Hoser D, Cross AD, Design A (2018) A discrete cross aisle design model for order-picking warehouses. Eur J Oper Res Receiv 275(2):411–430. https://doi.org/10.1016/j.ejor.2018.11.037
18. Elbert RM, Franzke T, Glock CH, Grosse EH (2016) The effects of human behavior on the efficiency of routing policies in order picking: the case of route deviations. Comput Ind Eng 111:537–551. https://doi.org/10.1016/j.cie.2016.11.033
19. Tarei PK, Thakkar JJ, Nag B (2020) Benchmarking the relationship between supply chain risk mitigation strategies and practices: an integrated approach. Benchmarking 27(5):1683–1715. https://doi.org/10.1108/BIJ-12-2019-0523
20. Schwerdfeger S, Boysen N (2017) Order picking along a crane-supplied pick face: the SKU switching problem. Eur J Oper Res 260(2):534–545. https://doi.org/10.1016/j.ejor.2016.12.037
21. Republic S (2020) Using of optimizing methods in inventory management of the company, 9–16. https://doi.org/10.22306/al.v7i1.150
22. Zahra F, Moussa B, de Guio R, Dubois S, Rasovska I (2019) Study of an innovative method based on complementarity between ARIZ, lean management and discrete event simulation for solving warehousing problems. Comput Ind Eng 132:124–140

Productivity Improvement Model in Small and Medium Metal Extruding Companies, Applying Total Productive Maintenance, Six Sigma, and Process Standardization

Jorginhio Andoni Luciano-Apolinario, Ricardo William Meza-Flores, Claudia Carolina Leon-Chavarri(✉), and Pedro Martín Bermúdez Sanabria

Facultad de Ingeniería, Universidad Peruana de Ciencias Aplicadas (UPC), Prolongación Primavera 2390, Lima 15023, Perú
{u201515446,u201314295,pcincleo,pcappber}@upc.edu.pe

Abstract. Metallurgical companies must focus on improving efficiency and productivity in their production process because, through this approach, they can be competitive. It is important considering that metallurgy is a sector on which other manufacturing sectors such as construction, automotive, and electricity depend. This article proposes a productivity improvement model designed for small and medium enterprises (SMEs) in the metallurgical sector based on process standardization, Six Sigma, and total productive maintenance. The objective of the model is to reduce production losses, thereby increasing productivity; this was applied in a pilot plan model in which the reduction of the level was achieved at 2.5% and an improvement in productivity of 11%.

Keywords: Productivity system · Foundry · Metallurgy · Scrap reduction · Extrusion · Standardization · Total productive maintenance · Six Sigma

1 Introduction

Over the last 10 years, Peru has been a significant copper, zinc, gold, silver, lead, and molybdenum producer [1]. As such, it is considered a mining country in which the mining sector, along with the metallurgical sector, is a supplier to other manufacturing sectors, such as the metal-mechanic sector, which requires primary metal products [2]. Since 2013, the metallurgical sector's production volume of non-ferrous metals—one of its main products—has decreased [3]. This has had a negative impact on Peru's economic development, given that metallurgical mining activity represents a significant portion of the national GDP [4]. This decrease in production volume translates to the low productivity of micro and small businesses—a clear problem among SMEs in these sectors [5], where more than 70% of the country's employed people are housed.

Low productivity among SMEs of the metallurgical sector has led to a reduction in the productive efficiency in refined metal production processes and that of semi-products such as metal extrusions. In this context, the rate of defective products and the

Y. Iano et al. (Eds.): BTSym 2020, SIST 233, pp. 412–418, 2021.
https://doi.org/10.1007/978-3-030-75680-2_46

waste level of a manufacturing company that performs some metallurgical operations were reduced, resulting in productivity improvement through the use of the Six Sigma tool under the DMAIC methodology [6]. Further, process standardization is applied as the core of the model to eliminate processes that do not add value and, thus, to increase productivity at an automotive parts smelter [7]. Regarding Total Productive Maintenance (TPM), its implementation is shown in a metal manufacturing company in which it is achieved a waste reduction, a performance improvement, and, in this way, a productivity improvement is obtained [8].

In Peru, the metallurgical and mining sectors contribute to the country's economic development [9]. For this reason, this article presents a case study of a model proposed for improving productivity in the metallurgical sector. This model incorporates the three tools mentioned above. Thus, the proposal is aimed at aiding SMEs in this sector to improve their productivity. This article begins with the state of the art of each of the tools. Then, a case study of a waste reduction model is presented along with its validation and discussion. The last section comprises the conclusions of the study.

2 State of the Art

The metallurgical industry, which is a supplier to other industries such as the automotive, construction, energy, mechanical, and electrical sectors [10], includes metal smelting, whose production is growing worldwide [11]. In order to stay competitive and match the growth in demand, metallurgical companies have chosen to lower their operating costs since they are unable to reduce raw material costs [10]. Expenses are cut down by improving or redesigning the existing production system [12]. Process improvement is made by reducing or eliminating processes that do not add value through cultural and operational changes. This creates an efficient, effective operational model that reduces the operating costs of metal products [10]. Further, high production costs are incurred when a process is not optimized, leading to reprocessing and waste, which reduces companies' profit margins [13]. Therefore, metallurgical companies must be updated and must reconsider their business and production models to identify the processes that generate waste and eliminate those that are inefficient; further, they must keep the processes that facilitate their work within the framework of continuous improvement [10]. The goal of such models is to improve productivity [12]. In the framework of optimizing processes, this is achieved by the standardization of processes.

Process standardization in the manufacturing sector involves the identification of processes that do not add value through graphic tools such as the SWCT (standard work combination table). Improvements are then proposed; these include reducing waste such as unnecessary time, decreasing cycle times, reducing waste, and increasing production volume [8, 14]. Furthermore, standard procedures are determined through work informative sheets for the future learning of new employees [15].

Improved productivity in the manufacturing of metal bars such as steel is achieved by applying Six Sigma and using the DMAIC methodology. These tools reduce the level of rejected products, improve metal quality, and increase process performance [16]. This tool requires quantitative input data, known as quality metrics, for the analysis: process performance, waste percentage, sigma level, and DPMO [7].

3 Contribution

3.1 Foundations

The proposed model is focused on reducing waste. It includes the process standardization tool, which identifies processes that do not add value based on implementation stages and increases production volume [8]. It is based on the DMAIC Six Sigma methodology, which allows controlling the metal extrusion process based on quantitative variables and, thus, to reduce the waste percentage [17].

3.2 Proposed Model

Using the tools mentioned above, the continuous improvement process allows a productivity improvement model, carried out by reducing waste in a metallurgical company that produces metal extrusions. The model shows a continuous flow of information (Fig. 1), which enables constant improvement, guaranteeing competitive production.

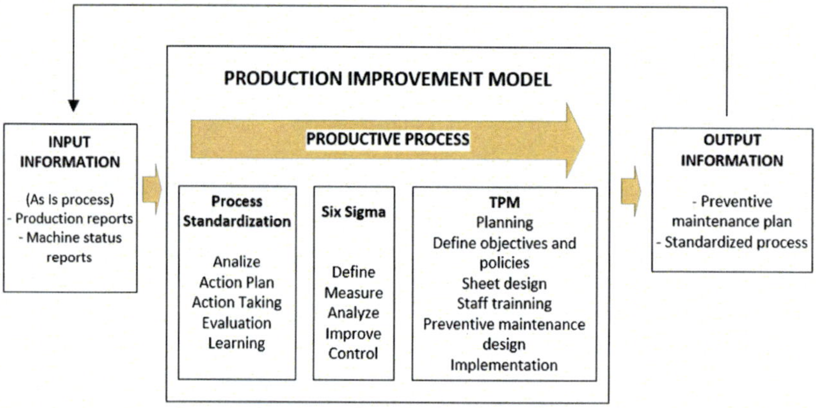

Fig. 1. Model of the proposed solution

3.3 Components of the Proposed Model

Process Standardization. A tool that evaluates processes, implements improvements, and standardizes them.

- Analyze: Analyze records and identify improvement processes using SWCT;
- Action plan: Develop an action plan for the identified improvements;
- Action taking: Implement improvements;
- Evaluation: Analyze and evaluate results;
- Learning: Standardize results.

Six Sigma. A tool that controls product quality and, thus, is capable of reducing the waste level.

- Define: Define the problem and objectives;
- Measure: Measure the initial quality level;
- Analyze: Create a flowchart of the process and determine the problem;
- Improve: Create a countermeasure matrixand implement improvements;
- Control: Standardize the improvements made.

TPM. A tool to resolve machine breakdowns in production.

- Planning: Determine performance indicators of machinery;
- Define objectives and policies: Define the policies and objectives regarding preventive maintenance;
- Sheet design: Design formats to record breakdowns, routine inspections, and data on the machinery;
- Staff training: Train staff;
- Preventive maintenance design: Design preventive maintenance registries;
- Implementation: Implement preventive maintenance through a pilot plan.

The model requires an input or flow of information necessary to apply the three tools of the model. Similarly, the information output shows the results of the model, which subsequently serves as information for a new development cycle for the model.

3.4 Process

This section presents a flowchart that shows the procedures that are carried out during project development (Fig. 2). This is important in implementing the productivity improvement model.

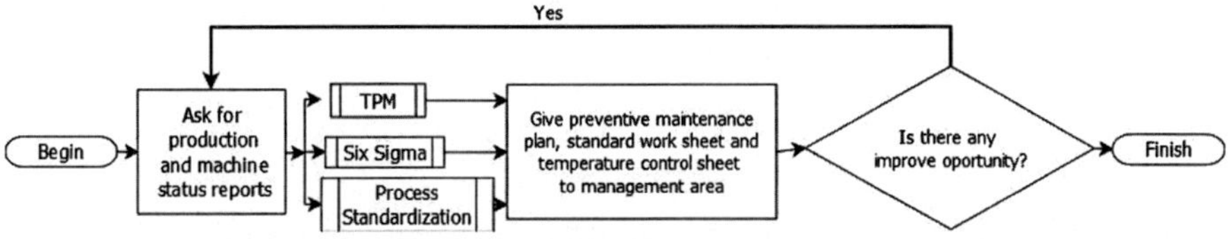

Fig. 2. Project flowchart

3.5 Indicators

Waste Percentage. This ratio provides insight into the amount of waste that is generated per kilogram of metal produced. It is the main indicator in the implementation of process standardization and Six Sigma over the two production process stages.

$$Percentage\ of\ scrap = \frac{Scrap(kg)}{Metal\ produced(kg)} \tag{1}$$

Productivity. This indicator will be applied to the production process, considering the two stages, based on the raw material's production factor.

$$Productivity = \frac{Extrudable(kg)}{Raw\ material(kg)} \quad (2)$$

OEE. This indicator provides insight into the condition and efficiency of machinery used in production. It is the primary indicator for the TPM.

$$OEE = Availability \times Performance \times Quality \quad (3)$$

Sigma Level. This indicator shows the quality level under the framework of applying Six Sigma. Its value is obtained through the Minitab program.

4 Validation

4.1 Case Study

This case study was carried out in an SME in the metallurgical sector producing extrusions and metallic pieces, located in the city of Lima. The data used in this work were provided exclusively for academic purposes on the anonymity condition.

Prior to the implementation of the pilot plan, the productivity of the company was 69.63%, a low level compared to the standard productivity level, which is 92.6%. The company under study has two productive processes: extrusion and smelting.

In the smelting process, an average waste of 9.85% was calculated. This value is much higher than the standard waste value in smelting, which is 3.5%. This difference represents a negative balance for the company because waste and productivity are inversely proportional. There was an average waste of 16.2% in the extrusion process, which also represents a negative balance for the company because the percentage of standard waste for this process is 10%.

The company's primary problem was identified as low productivity. To this end, an evaluation was carried out to determine the root causes of the main problem.

4.2 Assessment

For the main problem identified, a root cause analysis was carried out (Table 1).

The most important causes were selected according to the incidence rate. The following is the analysis carried out for each of the causes:

Improper Temperature of Pieces. Considering the temperature at which the metallic pieces were treated and the waste generated at each working temperature evaluated, a strong correlation (90%) between the waste and temperature variables was determined.

Breakdown of the Hydraulic Press. A hydraulic press that was used in the extrusion process broke down. The OEE analysis was used to determine the current performance condition of these machines. The standard OEE level is 60%, which indicates that the company falls below the required level.

Table 1. Determination of root causes

Root causes	Incidence percentage
Inadequate temperature	35.22%
Hydraulic press fault	25.34%
Leak in the casting process	16.52%
Off-center billets	15.52%
Extrusion needle failure	7.72%

Spill in the Casting Process. Spills of molten metal occur in the casting process. This process' activities are performed by operators who did not demonstrate similar performance, which is an indicator of a lack of standardized work procedures. The objective was to demonstrate that there is variability in operator performance, which causes different quantities of waste.

4.3 Results

For the root causes identified, the tools that were previously developed were used (Table 2).

Table 2. Outcomes of the pilot plan

Problem	Indicator	Initial condition	Ideal condition	Final condition
Low productivity in extrudable production	Productivity	69.93%	92.6%	81.13%
	% Average Waste in the Casting Process	9.73%	4%	7.64%
	% Waste per Inappropriate °C	23.31%	10%	11.52%
	Hydraulic Press OEE	41%	60%	60.33%

5 Conclusions

The study's objective was to design a model that would increase the productivity of the company under study. This was achieved by using the Six Sigma tools, process standardization, and TPM. Together, these tools increased the company's productivity to 81.13%, compared to the initial value of 69.93%.

The waste percentage in the casting and extrusion processes was reduced by 2.09% and 11.79%, respectively. The machinery's performance was optimized based on the use of inspection and supervision formats proposed in preventive maintenance, and the company's OEE increased by approximately 20%.

The percentage of waste reduction due to inadequate temperature was subject to the conditions of the furnace in activity, whose limitation is antiquity.

It is recommended for subsequent projects to consider the age of the machinery and apply the respective control and monitoring.

Implementing the Six Sigma tools and process standardization made it possible to reduce the waste generated in the extrusion and casting processes.

However, the results obtained were not the expected values, which resulted from limitations such as workspace and operation of the machinery used.

References

1. Dirección General de Minería (2018). https://www.minem.gob.pe
2. Servicio Nacional de Geología y Mineria (2019). https://www.sernageomin.cl
3. Instituto Peruano de Economía (2017). https://www.ipe.org.pe/portal/wp-content/uploads/2018/01/IPE-Junio-2017-El-Valor-Agregado-de-la-Miner%C3%ADa.pdf
4. Ministerio de la Producción (2018). https://www.produce.gob.pe
5. Ministerio de la Producción (2018). https://ogeiee.produce.gob.pe
6. Comisión Económica para América Latina y el Caribe (2018). https://repositorio.cepal.org/bitstream/handle/11362/43540/66/cap10_Desarrollo_e_igualdad_es.pdf
7. Kumar D, Kaushish D (2015) Scrap reduction in a piston manufacturing industry : an analysis using six sigma and DMAIC methodology. IUP J Oper Manag 14(2):7–25
8. Mor RS, Bhardwaj A, Singh S, Anish S (2019) Productivity gains through standardization-of-work in a manufacturing company. J Manuf Technol Manag 30(6):899–919
9. OSINERGMIN (2007). https://www.osinergmin.gob.pe
10. Adis T (2016) Sustainable cost reduction, 55(4):846–848
11. Soiński MS, Kordas P, Skurka K (2016) Trends in the production of castings in the world and in Poland in the XXI century. Arch Foundry Eng 16(2):5–10
12. Daroń M, Górska M, Dunay A (2018) Improvement of the production system functioning with using quality management tools in the metallurgical enterprise. In: MATEC Web Conference, vol 183
13. Allwood JM, Cullen JM, Waugh R (2010) Options for achieving a 50% cut in industrial carbon emissions by 2050. Environ Sci Technol 6(44):1888–1894
14. Halim NHA et al (2015) Standardized work in TPS production line. J Teknol 76(6):73–78
15. Herlambang A, Hidayati J (2019) Strategy and policy for increasing work productivity of operators in the steel industry through work improvement with lean method (journal review). IOP Conf Ser: Mat Sci Eng 505:012074. https://doi.org/10.1088/1757-899X/505/1/012074
16. Naeem K, Ullah M, Tariq A, Maqsood S, Akhtar R, Nawaz R, Hussain I (2016) Optimization of steel bar manufacturing process using six Sigma. Chin J Mech Eng (Engl Ed) 29(2):332–341. https://doi.org/10.3901/CJME.2015.1225.155
17. Girmanová L, Šolc M, Kliment J, Divoková A, Mikloš V (2017) Application of six sigma using DMAIC methodology in the process of product quality control in metallurgical operation. Acta Technologica Agriculturae 20(4):104–109. https://doi.org/10.1515/ata-2017-0020

Reducing Waste in Fast-Food Restaurants

A Demand-Driven MRP-Based Collaborative Model

Gianella Carbajal-Roman, Cesar Lopez-Vela, Gino Viacava-Campos(✉), and Juan Quiroz-Flores

Ingenieria Industrial, Universidad Peruana de Ciencias Aplicadas, Prolongación Primavera 2390, Lima 15023, Perú

{u201319864,u201316666,pcingvia,pciijqui}@upc.edu.pe

Abstract. Medium- and large-sized businesses in the fast-food sector face problems in aligning their operational strategies to their business strategies, generating missed opportunities in sales, unforeseen rising costs, and high inventories. In this regard, this study proposes a mixed model using resource planning and improvement stabilization tools and applies it to an actual case study through a 3-month pilot plan. The combined use of these tools will increase service levels and can be replicated in companies managing several lines of business. As a result of its application, a 1.9% reduction in stock breaks was achieved, the gap between actual and projected consumption decreased to 5.9%, and losses against sales costs were reduced to 5%.

Keywords: Demand forecasting · Integrated business planning · Performance supply · Fast-food chain · Service level

1 Introduction

At the global level, restaurants in the food-and-beverage sector face constant challenges due to concepts such as offer-related issues and the need to uphold their service-level standards. In a study found in the literature [1], the fast-food industry is characterized as one of the highest growing segments since its inception in the Peruvian market. To date, this sector contributes 1.53% to the national GDP [2], and due to its association with dynamic tourism, it constantly generates more jobs to cover consumer demands [3]. Indeed, publications that detail how these companies are constantly looking to improve their market position reveal that, in the first lines, they can increase sales by expanding the scope of their consumer network and offering more menu variety. However, other studies show that 50% of the establishments close within three months [4]. Among the reasons for these untimely closings, 60% are due to poor service, 25% due to poor management, and the other 15% due to limited menu options [5]. As it may be observed, these companies experience business-management deficiencies that hinder their overall growth. Within this context, these companies require to better align their strategies to improve their service level.

Hence, in the current scenario and for the aforementioned reasons, the fast-food sector is missing sales opportunities and facing unexpected rising costs and high inventories.

Y. Iano et al. (Eds.): BTSym 2020, SIST 233, pp. 419–426, 2021.
https://doi.org/10.1007/978-3-030-75680-2_47

In a study [6], the authors reported planning problems due to an inadequate demand projection model, poor warehouse coordination, and an inefficient distribution system. This caused certain supplies to perish or not be available for placement in stores. In another study [7], some new products, ideas, or behaviors quickly crossed an apparent threshold or "inflection point" and became established, while many others did not. When explaining why internal company inefficiencies play a crucial role when competing against other companies, the authors argued that these inefficiencies are often clearly identified within the company; however, companies fail to address them in a timely manner because, at first, they are not considered a crucial issue for the organization.

On the contrary, other studies on fresh-food supply chains only propose solutions that address some of the issues identified. In other words, they only focus on demand management [8] or on managing inventory levels and leftover risks [9]. Instead, in today's highly volatile and variable service environment where planning scenarios are more complex, a more widespread strategy, including careful planning, scheduling, and execution synchronization against material consumption, is required. In this sense, the scientific papers reviewed contain little information on using mixed models, which means that further research is needed to provide a meaningful contribution to the scientific community under the current context. For this reason, this study aims to solve the problems identified in this sector through a change model pilot plan conducted as an actual case study using demand planning tools and lean manufacturing methodology.

This paper describes different cases focused on the same business sector and problem. In addition, the model proposed for increasing service levels at a fast-food restaurant are explained in detail. Finally, the contributions of this study are listed, and the effectiveness of the proposed model is discussed. The data used in this work were provided exclusively for academic purposes on the anonymity condition.

2 State of the Art

The selected papers have been classified according to a specific typology.

2.1 Theoretical Papers – Demand Management

The literature has shown how demand management models have adapted to the volatile and dynamic environment over the decades. In a study [10], the authors detail demand forecast calculations; another [11] seeks to satisfactorily integrate supply with demand. Finally, there are other studies [12, 13] wherein the methodology integrates supply and demand and finance, and later validating this integration in a case study. In this sense, these studies considered that the work environment must be considered in the implementation of the proposed solution.

2.2 Solution-Based Papers

In this section, the studies that provided tools to solve the aforementioned problems are discussed. A study [14] engaged the entire supply chain coupled with specific strategies to obtain greater supply efficiency. Another study [15] looked for more accurate results to predict the demand. The studies [16, 17] obtained operational results by assessing and improving the dynamic variables of the production chain. Another study [18] focused on finding results in the warehouse and validating as much as possible the inventory of fresh products and their batches at a systematic level. As such, there are multiple tools that aim at solving the problem; however, the most complete tool is the integrated business planning (IBP) as it provides solutions for the entire chain, ranging from warehouse (RFID) to demand (forecast).

2.3 Indicator-Based Papers

In this section, the indicators used by the selected studies for their respective solution proposal is discussed. In the first paper [19], the authors listed demand projections, error estimates, and days in inventory as their indicators. Here, they could reduce merchandise costs and inventory levels by 4% and 6%, respectively. In the other papers [20–23], the corresponding authors considered demand projections, reduction of days, stock breaks and fulfillments, sales volumes, supply fill rates, storage costs, and financial flow from inventories as measurement factors. Overall, we can conclude that IBP is the model that not only exhibits the largest use of indicators but also provides a better market control parameter than the rest. In fact, the improvements it has provided to the organizations are supported by numerical results.

3 Contribution

The proposed model (Fig. 1) is developed based on the DDMRP [24] management model, adopting a qualitative system that generates accurate information through semistructured interviews, documents, and archive records. This output is essential for triangulating planning, scheduling, and execution against material consumption. As this model is supported by other tools such as demand forecasting [15], this paper provides a guide for balancing demand over supply, thus allowing for combined manufacturing and supply footprint decision-making. The model is also supported by IBP planning [25], where the planning and tactical processes are merged to align the management of important marketing events for a business with its demand planning, as well as by distribution resource planning (DRP) [11], which validates whether the distribution system directly impacts expected cost savings under an integrated strategy. The model includes a lean manufacturing approach for sustaining improvements using its just-in-time (JIT) tool [26], where specific innovation enablers are identified in material and production management. The combined usage of these approaches is expected to improve the service levels in the company.

As a first stage, a pre-application assessment is required, i.e., an evaluation of the supply chain to determine critical processes, existing problems, and the factors that are affecting the activities. Accordingly, a qualitative analysis was conducted through brainstorming and quantitative analysis using the Vester matrix to identify the central problem, i.e., inefficient resource planning. In this manner, we were able to determine that the problem spans multiple areas that can be supplied under projected consumption. In addition, the information collected by the planning team can be reused for determining material and supply as well as storage and picking requirements and distributing the sales orders required by restaurants.

Then, in the preliminary stage, a management commitment is established, demonstrating the analysis from the previous stage. For this reason, new rules and policies are required to understand the limits and resources available to secure project continuity. The owners for the different project activities are appointed, forming a committee. For the successful completion of this stage, employees in the planning, sales, finance, and distribution areas are trained, sharing key project concepts and activities. Consequently, the global indicators for the project were determined by assessing the impacts generated and are listed in Table 1.

Table 1. Global Indicators

Variables	Indicators	Value
Inefficient resource planning	Valued loss/Cost of sale	6.91%
	% Stock break (warehouse)	2.81%

The third project stage, known as "launching," starts with the application of demand forecasting. Here, we determine demand trends and the forecast model applicable to the case study (decreasing seasonal trend and additive decomposition model). Next, for the deployment of IBP planning, the supply system is developed and integrated with the MRP and production plan. At this moment, the monitoring of the innovations and promotions matrix starts. Finally, for the DRP, the supply system of restaurants is assessed to develop a distribution dashboard.

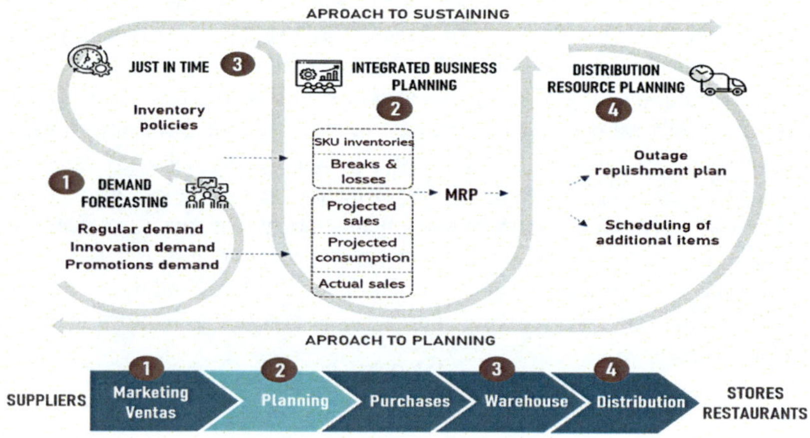

Fig. 1. Proposed model. Self elaboration.

In the fourth stage, called "improvement stabilization," the assessment starts by identifying viable activities within the sales, planning, and distribution processes. After this, flowcharts are documented, and new policies are defined. This combined model is assessed through indicators, as detailed in Table 2.

Table 2. Support indicators

Variables	Indicators	Value
Inadequate scheduling of materials and supplies	Number of times without inventory (restaurant)	71 times
	% Actual vs. projected consumption	11.6%
Scheduling of uncoordinated spaces	Inventory level	56 days
Lack of projection of shared demands	Actual Sales vs. projected sales	35.68%
Inefficient distribution system	Allocated vs. actual freight budget	4.3%
	% Additional freight charges on total freight	12.5%

In the final stage, the application results, in addition to the aforementioned indicators, are assessed and reported to company management. Together with the project team, the committee determines the adjustment of the indicators and the new work plans, continuing with the evaluation to improve the results and foster continuous improvement.

4 Validation

The improvement design is presented to the supply and sales management teams with the main purpose of developing it through a pilot plan within the selected business line, detailing the benefits that the application of the designed lean manufacturing tools would bring to the business processes of the company. With approval from management and the support of the different process teams, we started developing each of the steps required for the implementation of all the proposed tools. The piloted tools were demand forecasting, IBP, DRP, JIT supply, and working methodology standardization over a period of four months.

First, stock breaks were reduced to 2.01%. This change is critical because it is associated with service levels. Then, losses were reduced to 5.64%; this favors company profitability. Even so, the gap between the actual and projected consumption narrowed to 5.9%, and the average number of times a restaurant ran out of inventory was 58 times. Similarly, the gap between actual and projected sales was reduced by 5.55%. Subsequently, the inventory level decreased from 56 days to 38 days. Finally, additional freight expenses represented 10.81% of the total freight expenses.

The proposed model aimed to manage indicators without allowing them to become critical. However, due to human factors, the goal was not 100% achieved. With this model, no indicator is completely out of control, with two indicators being completely managed, two managed but critical, and four managed but not critical, as it may be observed in Table 3.

Table 3. Comparison of performance indicators before and after the application of the model

Indicator	As – is (monthly)	To - be	February
Valued loss/cost of sale	6.91%	5%	5.64%
% Break stock (warehouse)	2.81%	1.9%	2.01%
Number of times without inventory (restaurant)	71 times	50 times	58 times
% Actual vs. Projected consumption	11.6%	8%	5.9%
Inventory level	56 days	34 days	38 days
Projected vs. Real sales	14.5%	7.5%	8.95%
Allocated vs. Actual freight budget	4.3%	2.5%	2.75%
% Additional freight charges on the total freight	12.5%	9%	10.81%

The main reason why some indicators remained critical despite being managed is that human resources still need to adapt to the changes implemented. As this happens, the proposed tools will yield better results in the medium- and long-term.

5 Conclusions

The collaborative model application yielded an average reduction of 18% and an FR equivalent of 2.01% over a period of four months. In addition, the sales cost was reduced by 0.77% for the business unit in which the pilot plan was conducted. This saved the company S/. 3,406 for the selected restaurant. We estimate that, if the model is applied to all the restaurants in this business unit, at least S/. 376,397 in savings can be obtained. When compared against the initial analysis (S/. 4,279,760), these savings represent 8.79%. Similarly, we have other economic indicators, such as an NPV of 217 thousand, an IRR of 571%, and a payback projected in 9 months, that also validate the effectiveness of this proposal. However, we could not implement all the tools in all the restaurants due to cost and time constraints. Despite this, improvements were observed in these processes, thus decreasing sales costs while increasing the monthly gross margin of the company. This model is focused on resource planning and continuous process improvement, which allows us to integrate innovation and promotion demands with the MRP plan to reduce the gap between the projected and actual sales and, in turn, increase service levels. Finally, through the application of the JIT tools, both the distribution process and its replenishment schedules were reviewed and monitored. It should be noted that, to guarantee a successful model implementation, the engagement and commitment of the entire team is required.

References

1. Gestión (2018) https://gestion.pe/economia/mercados/fast-food-mueve-us-585millones-comida-sana-le-pone-freno-241080
2. Gestión (2018) Fast food mueve US$ 585 millones. https://gestion.pe/economia/mercados/fast-foodmueve-us-585-millones-comida-sana-le-pone-freno-241080
3. Instituto Nacional de Estadistica e Informatica (2017) Boletin estadistico del sector servicios. INEI Editores, Lima. https://www.inei.gob.pe/media/MenuRecursivo/boletines/boletinestadistico-del-sector-servicios-n-12-diciembre-2017.pdf
4. Instituto Nacional de Estadistica e Informatica (2018) https://www.inei.gob.pe/estadisticas/indice-tematico/productobruto-interno-por-departamentos-9089/
5. PerúRetail (2017) Análisis del sector de comida rápida en Lima. https://www.peru-retail.com/analisis-sector-comidarapida-lima/
6. Amit Sinha WP (2016) Matching supply with demand in supply chain management education (Adecuación entre oferta y demanda en la gestión de la cadena de suministro). Int J Logist Manag
7. Ann Vereecke KV (2018) Mind the gap – assessing maturity of demand planning, a cornerstone of S&OP (Tenga en cuenta la brecha: evaluar la madurez de la planificación de la demanda, una piedra angular de S&OP). Int J Oper Prod Manag
8. Aris Syntetos ZB (2016) Supply chain forecasting: theory, practice, their gap and the future (Predicción de la cadena de suministro: teoría, práctica, su brecha y el futuro). Eur J Oper Res, 252:1–26
9. Arteaga JD (2018) IBP El camino hacia la excelencia operativa en la planificación de la cadena de suministros. Universidad Católica de Córdoba, Argentina
10. Bhattacharyya S (2015) Improving inventory demand forecasting by using the sales pipeline: a case study. Inst Bus Forecast Plan 33:7
11. Carvajalino EA (2018) Modelo de planeación logística de una productora de lácteos en un proceso ibp (integrated business planning). https://repository.unimilitar.edu.co/handle/10654/17762
12. Coker J, Helo P (2016) Demand-supply balancing in manufacturing operations (Equilibrio entre oferta y demanda en las operaciones de fabricación). Benchmark: Int J
13. Contreras Juárez A, Atziry Zuniga C, Martínez Flores J, Sánchez Partida D (2016) Análisis de series de tiempo en el pronóstico de la demanda de almacenamiento de productos perecederos
14. Dreyer Heidi Carin KK-P (2018) Proposals for enhancing tactical planning in grocery retailing with S&OP (Propuestas para mejorar la planificación táctica en la venta minorista de comestibles con S&OP). Int J Phys Distrib Logist Manag
15. Helo JC (2016) Demand-supply balancing in manufacturing operations. Int J Process Manag Benchmark 23:564–583
16. James N (2017) Supply Chain Digitalization— Delivering Sustainable Cross-Functional Change (Digitalización de la cadena de suministro: entregando un cambio multifuncional sostenible). Int J Forecast
17. Jiang J, Rim Suk-Chul (2017) Strategic wip inventory positioning for make-to-order production with stochastic processing times. Math Prob Eng 2017:1–7
18. Jonsson JA (2018) Big data in spare parts supply chains. Int J Phys Distrib Logist Manag 48:524–544
19. Jonsson P (2016) Future of supply chain planning: closing the gaps between practice and promise. Int J Phys Distrib Logist Manag 46:62–81
20. Kortabarria A, Apaolaza U, Lizarralde A, Amorrortu I (2018) Material management without forecasting: from MRP to demand driven MRP. J Ind Eng Manag 11(4):632

21. Li D, Wang X (2016) Dynamic supply chain decisions based on networked sensor data: an application in the chilled food retail chain
22. Mark Moon PA (2015) From sales & operations planning to business integration (De la planificación de ventas y operaciones a la integración empresarial). Foresight: Int J Appl Forecast
23. Martin ME (2017) Planificación de la cadena de suministros: desde el S&OP hasta el IBP. https://www.esan.edu.pe/conexion/actualidad/2017/10/20/laplanificacion-de-la-cadena-de-suministros/
24. Annika Alftan RK (2015) Centralised grocery supply chain planning: improved exception management (Planificación centralizada de la cadena de suministro de comestibles: mejora de la gestión de excepciones). Int J Phys
25. Sorensen D (2016). Beyond S&OP and IBP to enterprise planning and performance management (Más allá de S&OP y IBP para la planificación empresarial y la gestión del rendimiento). Foresight: Int J Appl Forecast
26. Jiang J, Rim Suk-Chul (2017) Strategic WIP inventory positioning for make-to-order production with stochastic processing times. Math Prob Eng 2017:1–7

Waste Reduction Model Design in Textile Industry

A Lean Manufacturing Approach

Sebastián Torres-Luna, Javier Valdivia-Ríos, Iliana Macassi-Jáuregui, Edgar Ramos Palomino, Gino Viacava-Campos(✉), and Claudia León-Chavarri

Ingeniería Industrial, Universidad Peruana de Ciencias Aplicadas, Prolongación Primavera 2390, Lima 15023, Perú

{u201516587,u201416539,pcadlmac,pcineram,pcingvia, pcincleo}@upc.edu.pe

Abstract. Lean manufacturing is a methodology that focuses on continuous improvement and waste elimination. This methodology is supported by its tools, namely, lean tools, to identify and eliminate waste. This article presents a model for reducing waste in the Peruvian textile industry. The Peruvian textile manufacturing industry is characterized by its manual or semi-manual operation, differentiating itself from foreign competition, which tends to use automated machinery. Such foreign competition forces the local industry to maximize its operations. In a case study of a company, waste identified was reprocessing, with an index of 21.63%. To reduce this waste, tools identified in the literature, such as lean tools, 5S, work-station design, and process management, are used for reducing rework to 11.58%
.

Keywords: Lean tools · 5S · Clothing industry · Workstation design · Waste · Process management

1 Introduction

The textile industry is one of the three pillars of Peru's economic growth and is separated into two groups: the textile industry and the textile–clothing industry [1]. The United States is the main buyer of textile exports, accounting for 49.3% of the share in 2017. At the national level, textile exports showed a deficit between 2013 and 2017, resulting in a decrease in the US clothing suppliers ranking from position 20 to 23 [2]. Figure 1 shows the industry supply chain. Based on the aforementioned statistics, this industry must improve its efficiency to be competitive in the foreign market.

It is important to identify and reduce waste in manufacturing processes. In the last few decades, different lean practices have been established to eliminate waste in manufacturing processes [3].

The implementation of these practices has yielded reduced time and costs in the manufacturing sector [4]. The lean manufacturing approach focuses on reducing activities that do not add value to the process [5]. This article identifies the problems of this sector and their root cause and proposes a solution model based on the lean concepts reported in the literature.

Y. Iano et al. (Eds.): BTSym 2020, SIST 233, pp. 427–434, 2021.
https://doi.org/10.1007/978-3-030-75680-2_48

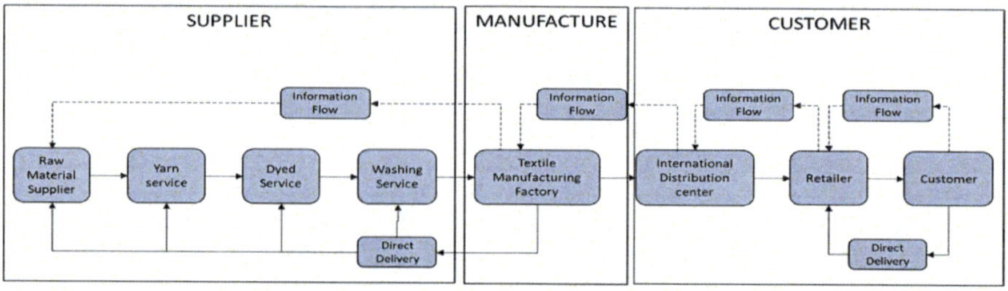

Fig. 1. Textile apparel supply chain

2 Literature Review

A literature study is conducted for tools along with their impact on manufacturing industries. For this purpose, the literature in international journals was researched using the keywords listed in this article.

2.1 Lean Manufacturing

The lean concept focuses on eliminating the activities that do not add value to the process and affect the production cost, which is considered an operational improvement in the manufacturing sector [5]. The implementation of lean manufacturing in small- and medium-sized companies uses lean tools as indicators of improvement [6]. Lean and other improvement tools, such as Visual Control, are used to help organizations identify areas that show opportunities to reduce waste and improve productivity [7]. The advantages of lean manufacturing include a decrease in raw material waste, cycle volatility, and production and delivery times, and an increase in performance and infrastructure and capacity utilization [8]. In manufacturing industries, the lean methodology application represents a simple and economical approach because it yields advantageous results by facilitating daily activities for both employees and companies [9].

To correctly implement a lean system, special attention must be paid to the human factor because it largely depends on the support of employees and other members of the organization [10].

2.2 Lean Tools

Lean tools support the implementation of lean manufacturing theory in companies [6]. These tools identify waste and then reduce it. Kaizen is one of the tools used to identify opportunities that can be improved. The function of this tool is to involve each worker of the company in the work to achieve continuous improvement [6, 7]. The stages for Kaizen implementation include 1) the selection of improvement opportunity, 2) quantification and subdivision, 3) root cause analysis, 4) level of performance and design, 5) design and program solutions, and 6) solution implementation.

5S is a lean tool for organizing and cleaning the work area with the aim of eliminating waste and improving the productivity area [11]. It is considered the foundation of "lean house" and is used in research to yield sustainable performance in companies [12].

Alternatively, Visual Management is a tool used in conjunction with 5S to help workers understand the changes implemented [11]. Studies have reported that, to use these two tools, the key is organizational commitment and discipline [13]. Combining both of these techniques produces a reduction in search times within the company, which are caused by lost or misplaced items in the inventory [14].

2.3 Wastes

Wastage is any activity in the production process that does not add any value to the process. Waste is classified into seven categories: overproduction, defects, transport, inventory, overprocessing, delay, and unnecessary movements [3, 14, 15]. Several studies have used similar tools for waste identification and reduction, and lean tools are the most popularly used because they allow establishing a streamlined and high-quality system that manufactures finished products, which are intended to satisfy customer demand [16]. Tools such as 5S and Poka-Yoke are used to avoid product defects [17, 18]. Other tools such as Kaizen and VSM are used to identify waste in the production process by identifying its origin [8, 19, 20].

2.4 Workstation Design

Efficiently designed workstations are essential for a business to allow flexibility and effective mass production. The station design, oriented toward the user and activity requirements, allows the company to increase its production indicators and quality levels [21]. This tool complements the use of Visual Management and 5S by incorporating key changes in the workstations to reduce waste originated at the station.

2.5 4-Point System

The 4-point inspection system is the most common and widely used fabric quality inspection system for all types of fabrics in the textile and clothing industry [22]. The 4-point system, also called the point classification system by the American Apparel Manufacturers Association (AAMA), is widely used by garment producers and the United States Department of Defense to determine the fabric quality, with the backing of the American Society for Quality Control (ASQC) [23].

2.6 Framework

The proposed design is based on continuous improvement and waste reduction of the lean philosophy. By using lean tools, it will be possible to identify and eliminate waste in the manufacturing process. Tools such as 5S and workstation design can aid the optimization of the work area. The key to implementing 5S is the commitment of all workers in the area; hence, spreading awareness is necessary before implementing the tool [13].

Using Kaizen, opportunities for improvement in the manufacturing process are identified [6]. By supporting Kaizen with tools such as Poka–Yoke and Visual Management, the improvement opportunity is controlled and managed. Then, using process management and the fourth step of 5S, Seiketsu, all changes and improvements incorporated are standardized. Finally, VSM and Shitsuke are used to analyze the identified values. The model design is shown in Fig. 2.

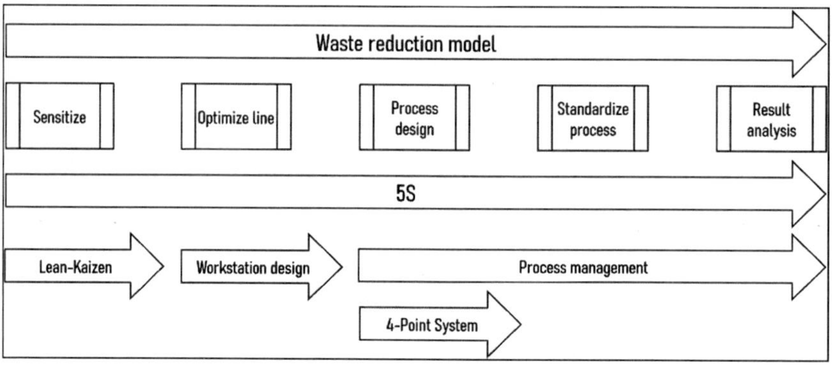

Fig. 2. Solution model design.

3 Case Study

3.1 Background

The Peruvian clothing textile industry is characterized by manual and artisan work. According to a study of a sector conducted by the Ministry of Production, the processes identified are design, cutting, sewing, finishing, and packaging. The company used for this case study implements the same processes. Using VSM, a reprocessing rate of 21.63% was identified in the T-shirt manufacturing process in this company. This rework was attributed to errors such as insufficient preventive maintenance, nonstandard-cleaning process, ignorance of sewing methods, and lighting below the recommended range; these results were obtained on the basis of a cause-effect analysis. The data used in this work were provided exclusively for academic purposes on the anonymity condition.

3.2 Model Implementation

The model involves five phases based on continuous improvement lean philosophy in conjunction with lean tools to reduce waste and optimize company processes.

Phase 1: Raise awareness and sensibility among the organization's workers. For this purpose, it is necessary to perform an initial audit, accompanied by a survey of personnel to understand their knowledge of the lean and 5S philosophy [15]; the audit results are shown in Fig. 3. The results indicated that the current knowledge of issues related to standardized work was the one best handled by workers. Moreover, it was identified that the workers had less knowledge about the discipline and classification tools. Additionally, unlike the workers' perception, the area had a lower level than expected.

Awareness spread should be performed, considering the survey results. The lowest values obtained were classification and discipline, which are the main concepts that must

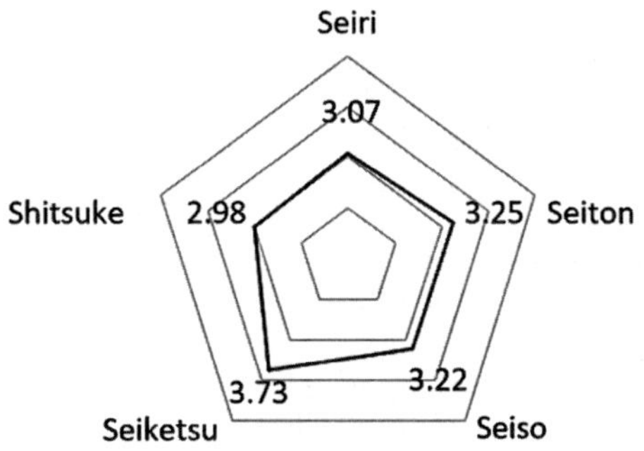

Fig. 3. Preliminary survey result

be considered. The awareness-raising approach includes the following concepts: What is the improvement cycle? What is lean Kaizen? How does 5S work? How and why are tools classified? What are the benefits of the 5S for the staff? Moreover, the concept of discipline was covered by emphasizing the importance of everyone performing their functions, providing information to understand the impact they produce on the work performance of their colleagues.

Phase 2: Optimize production line. It includes improving the manufacturing area. This phase begins by classifying the work tools of the 5S using the Seiri method. During the information gathering, it was observed that 3.59% of all tools had to be discarded, and 25.15% had to be stored because they were rarely used.

Next, the workstation design method was used to optimize the workstations using a lighting study. It was identified that the station reached a maximum of 722 lx, with 1000 lx or more being recommended for manual work. To achieve this, the source lumen level was raised from 1,100 to 2,026 by implementing LED spotlights.

The third task of phase 2 includes the use of the Seiton, in which everything identified as unnecessary in Seiri is discarded. The remaining tools are classified using frequency, leaving everything that is used one or more times a day in the stations; the rest is located in the supervisors' stations.

Phase 3: Optimize process. Using the five principles of Seiso, the station-cleaning process is created. For this process, four key activities were considered: open the machinery, use an air valve to remove dust, close the machinery, and clean the workstation using benzene (Fig. 4). These four key activities are aimed at reducing the presence of dust on the garment. To implement it correctly, those responsible for the process and their starting point were identified.

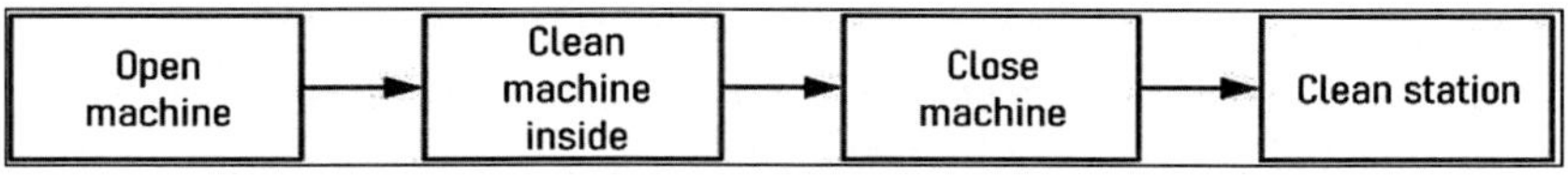

Fig. 4. Cleaning method

With the support of the 4-Point system, a raw material review process was created for the fabrics entering the cutting process. This method allows classifying the error types and measuring the size and frequency of these errors to decide the fabric quality according to the number of errors per 100 m^2.

Phase 4: Changes standardization.

- Documentation: The process standardization stage begins by documenting all changes made in phase 2, which are included in the 5S manual. The processes created are documented in procedures and instructions following the principles of process management.
- Training: It includes training all workers in the clothing area in their procedures and instructions in addition to cutting and finishing workers in the materials inspection procedure. It also includes explaining the use of the formats and the correct manner of conducting these activities.
- Monitoring: After training, a process-monitoring period is implemented to identify errors or inconveniences when conducting work.

Results analysis: At the end of the monitoring period, the final 5S audit is performed to identify the improvements achieved by the implementation of the model in the clothing area, in addition to performing the final audit of the implemented processes. Moreover, this audit must verify the indicators presented in Table 1.

Table 1. Audit indicators

Indicator	Current index
Station-cleaning compliance index	42%
Rejected fabrics index	NA
Index of reprocessed garments	21.63%
5S audit average	52.38%

4 Results

The implementation of lean manufacturing philosophy, the use of lean tools, and waste management lead to achieving improvements in the case under study, such as the reduction in reprocesses, which indicate an improvement in the quality and productivity of the company's processes and products. This model can be implemented in other companies in the textile industry, and its application is more appropriate in manual and semi-manual confections. The implementation results are listed in Table 2. In Table 3, the improvement in the audit indicators is provided.

Table 2. Improvement by error type

Error type	Before	Improved	% Improvement
Oil	0	0	0.00%
Production	53.37	41.77	21.74%
Contaminated	20.1	10.4	48.26%
Dust	99.6	40.5	59.34%

Table 3. Resulting indicators

Indicator	Before	Improvement
Station-cleaning compliance index	42%	94%
Reprocessed Garments Index	21.63%	11.58%
5S audit average	52.38%	66.67%

5 Conclusions

The current analysis shown using this model allows us to observe that the main problem affecting the company under study is reprocessing because they affect productivity by spending time correcting errors made in production. The use of lean tools (VSM, Kaizen, 5S, and Visual Management) was suggested to solve this issue. Such an application and the use of these techniques yield several benefits to the production system.

The implementation of lean manufacturing philosophy, the use of lean tools, and waste management yield improvements in the case under study, such as a reduction in reprocesses, which indicate an improvement in the quality and productivity of the company's processes and products. This model can be implemented in other companies in the textile industry because its application is more appropriate in manual and semi-manual confections.

Using the proposed model, production waste was reduced from 21.63% to 11.58%, indicating an improvement of 46.46%. This finding suggests that the efficiency of the workers increased, allowing the completion of the customers' daily demand quota. This model is applicable to other companies in the textile sector at the national level, considering that this type of problem is frequent. Further, its implementation benefits include simple application steps, and it does not require high investments in terms of new machinery or the recruitment of specialized personnel. In conclusion, these tools improve the production processes based on the results of the case study.

References

1. Ministry of Production (2017) Textile and Clothing Industry: Sector Research Study, pp 1–157
2. ADEX: ADEX Data Trade. https://www.adexdatatrade.com/

3. Xiong G, Shang X, Xiong G, Nyberg TR (2019) A kind of lean approach for removing wastes from non-manufacturing process with various facilities. IEEE/CAA J Autom Sin 6(1):307–315
4. Raj D, Ma Y, Gam H, Banning J (2017) Implementation of lean production and environmental sustainability in the Indian apparel manufacturing industry: a way to reach the triple bottom line. Int J Fash Des Technol Educ 10(3):254–264
5. Sabiya K, Shilpa M, Appaiah S (2019) Scrap reduction in TMT reinforced bar production by the application of lean techniques. Int J Recent Technol Eng 8(2):3483–3487
6. Dresch A, Veit DR, de Lima PN, Lacerda DP, Collatto DC (2019) Inducing Brazilian manufacturing SMEs productivity with Lean tools. Int J Product Perform Manag 68(1):69–87
7. Ahuja I, Singh H (2018) Evaluating the effectiveness of 5S implementation practices in Indian manufacturing industry. Int J Prod Qual Manag 25(4):506
8. Kumar S, Dhingra A, Singh B (2018) Lean-Kaizen implementation: a roadmap for identifying continuous improvement opportunities in Indian small and medium sized enterprise. J Eng Des Technol 16(1):143–160
9. Pombal T, Ferreira LP, Sá JC, Pereira MT, Silva FJG (2019) Implementation of lean methodologies in the management of consumable materials in the maintenance workshops of an industrial company. Procedia Manuf 38:975–982
10. Vukadinovic S, Macuzic I, Djapan M, Milosevic M (2018) Early management of human factors in lean industrial systems. Saf Sci 119:392–398
11. Choomlucksana J, Ongsaranakorn M, Suksabai P (2015) Improving the productivity of sheet metal stamping subassembly area using the application of lean manufacturing principles. Procedia Manuf 2:102–107
12. Khan SA, Kaviani MA, Galli BJ, Ishtiaq P (2019) Application of continuous improvement techniques to improve organization performance. Int J Lean Six Sigma 10(2):542–565
13. Manfredsson P (2016) Textile management enabled by lean thinking: a case study of textile SMEs. Prod Plan Control 27(7–8):541–549
14. Schonberger RJ (2019) The disintegration of lean manufacturing and lean management. Bus Horiz 62(3):359–371
15. Dinis-Carvalho J, Guimaraes L, Sousa RM, Leao CP (2018) Waste identification diagram and value stream mapping. Int J Lean Six Sigma 10(3):767–783
16. Goshime Y, Kitaw D et al (2018) Lean manufacturing as a vehicle for improving productivity and customer satisfaction. Int J Lean Six Sigma 10(2):691–714
17. Oliveira J, Sá JC, Fernandes A (2017) Continuous improvement through 'Lean Tools': an application in a mechanical company. Procedia Manuf 13:1082–1089
18. Dhiravidamani P, Ramkumar AS, Ponnambalam SG (2017) Implementation of lean manufacturing and lean audit system in an auto parts manufacturing industry – an industrial case study. Int J Comput Integr Manuf 31(6):579–594
19. Behnam D, Ayough A et al (2018) Value stream mapping approach and analytical network process to identify and prioritize production system's Mudas (case study: natural fibre clothing manufacturing company). J Text Inst 109(1):64–72
20. Azyan Z, Pulakanam V, Pons D (2017) Success factors and barriers to implementing lean in the printing industry: a case study and theoretical framework. J Manuf Technol Manag 28(4):458–484
21. Teixeira M, Salonitis K (2017) Lean assessment tool for workstation design of assembly lines. Procedia CIRP 60:386–391
22. Said S, Nahar N, Hossain S, Sajib SH (2018) Amendment of finished knitted fabric quality by reducing the intensity of defects and improvement techniques. Am J Mater Sci 8(1):6–14
23. Chandurkar P, Kakde M, Patil C (2016) Minimization of defects in knitted fabric. Int J Text Eng Process 2(3):13–18

A Production Management-Based Lean Manufacturing Model for Removing Waste and Increasing Productivity in the Sewing Area of a Small Textile Company

Stefanny Zamora-Gonzales, Jose Galvez-Bazalar, and Juan Quiroz-Flores(✉)

Ingenieria Industrial, Universidad Peruana de Ciencias Aplicadas, Lima, Peru
{u201516862,u201422811,pciijqui}@upc.edu.pe

Abstract. Small- and medium-sized companies within the textile sector face serious problems in their production flows, mostly because of the absence of standardized production systems and adequate resource planning. Therefore, a diagnostic analysis is conducted on an SME textile company to evaluate and define the deficiencies and factors affecting their productivity. Thus, in this study, we develop a model that contributes to waste elimination in the production flow, improves resource planning, and provides improved working conditions. The proposed model is developed by applying lean manufacturing tools and a material supply planning system through a continuous improvement approach based on the Kaizen philosophy.

Keywords: Productivity · Lean Manufacturing · Textile Industry · Kaizen · MRP

1 Introduction

Currently, small and medium-sized enterprises (SMEs) are essential sources for improving productivity, ensuring national development, and enhancing international trade. 96.5% of all Peruvian businesses are SMEs, which play a very important role in the national economy because they provide employment to 75% of the economically active population (EAP) [1]. The textile sector is the largest manufacturing industry in the country, with 16.5% of the companies dedicated to clothing production [2]. Within the textile sector, there has been a growth rate of more than 15% because of the considerable international demand and the strategy of adopting Peruvian products [3]. However, the considerable demand for clothing has resulted in the participation of 67.1% informal companies within the textile sector, which do not comply with the regulations implemented by SUNAFIL [4]. Similarly, there is 65% non-compliance with the occupational health and safety regulations within the textile sector [5], affecting workers and causing deficiencies in processes. To correct this situation, companies use lean manufacturing tools, which seek to identify and eliminate faults and improve quality, time, and production [6].

Y. Iano et al. (Eds.): BTSym 2020, SIST 233, pp. 435–442, 2021.
https://doi.org/10.1007/978-3-030-75680-2_49

This study aims to develop the lean manufacturing approach based on the 5S, Jidoka, process standardization, SLP, and MRP-I tools to reduce operating costs, increase productivity, and generate a high gross margin.

The subsequent sections will present the development of the proposed model for increasing productivity based on a review of the success stories in literature. Thus, the data collection and the assessment of the current situation will be detailed to suggest an appropriate proposal based on the reported conditions. Finally, the investigation results will be evaluated.

2 State of the Art

Among the tools that will be used is the MRP-I; in this tool, it is important to manage material scheduling such that no production stoppages are generated due to material shortage [7, 8]. Similarly, MRP-I converts the master production schedule (MPS) into a detailed schedule to ensure the availability of raw materials and components for the final products at the appropriate time and in appropriate quantities, with the main objectives of providing management with important details, such as knowing when to order or how much to order from suppliers, and generating new orders and rescheduling orders. The aforementioned tasks must be performed successfully to meet the ever-changing needs of customers and complete the entire manufacturing process [9, 10].

The Jidoka tool can be observed within the lean manufacturing philosophy. According to previous studies [11, 12], the usage of this tool ensures that each process can self-manage its own quality by searching for the root cause. In this sense, the interaction and cooperation between humans and automation is necessary. However, the systematic layout planning (SLP) tool can be used to organize a workplace in the plant using two areas with logical and high-frequency relations with each other; further, it makes the material flows faster and reduces cost [13, 14]. Similarly, the 5S tool is used to eliminate waste and improve efficiency and performance. In this regard, previous studies [15, 16] have indicated the importance of using this tool to reduce the accident risk and improve the worker efficiency and production times.

The Kaizen methodology is based on waste elimination from the production system and operations; this view has been shared by some authors [17, 18] who have commented that it is a way of thinking and organizing everything, from the way in which one works and the way in which a team works together. In addition, lean manufacturing is about eliminating waste or efforts that do not add value to the company. In fact, eliminating or reducing this waste is an ongoing battle. A company can better react to the needs of its customers by continuously focusing on waste reduction; however, to increase its effectiveness, this approach must be supplemented by the Kaizen methodology, which uses a continuous improvement approach [19].

3 Contribution

3.1 Rationale

The companies in the textile sector have a high market share. However, majority of the problems associated with the field of sewing can be attributed to the fact that this field includes several activities and that it requires more time. Therefore, according to the literature review and the factors related to the central problem, the combination of lean manufacturing tools, i.e., process standardization, Jidoka, SLP, 5S, and the proposal of a material supply plan, was defined as the proposed solution. All of them have been validated to improve the productivity and efficiency of the production chain.

3.2 Proposed Model

In this study, we develop a model that combines the application of lean manufacturing tools under a continuous improvement approach. This model is developed to increase the productivity of the sewing area through waste elimination and adequate planning of resources. Figure 1 shows the approach involved in the proposed improvement model, which includes three stages in which the lean manufacturing tools are developed and a supply plan is prepared. Each execution phase has been planned under a logical sequence to improve the production flow. Similarly, each proposed tool is based on the solution of critical factors identified within the process such as the high index of defective products, unequal workload, high transfer times, low operational efficiency, and the number of unfulfilled production orders.

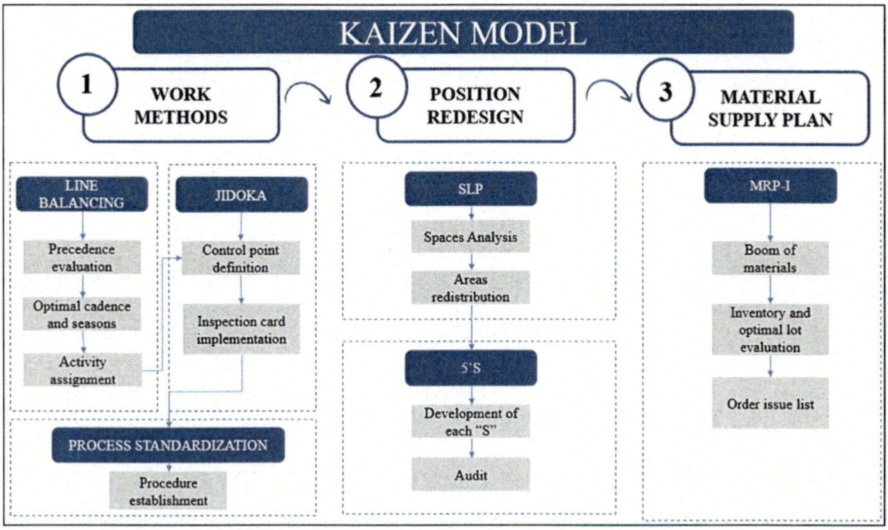

Fig. 1. Proposed improvement model

Thus, the proposed improvement model is developed based on the Kaizen philosophy, which exhibits a continuous improvement approach.

3.3 Model Details

3.3.1 Phase 1

The first phase of the proposed model balances the production line, which has the main objective of balancing the manufacturing line by optimizing the variables that affect productivity. When balancing the manufacturing line in this study, each preceding activity and the times used for the development of the said activities are considered. Based on this, a process standardization is proposed to unify all the procedures of the new production system. Similarly, the Jidoka tool is developed, which complements the approach of the new production system, by considering the definition of the inspection points within the sewing line based on which inspection cards will be assigned. The main objective of this tool is to reduce defective products and achieve quality control inspections at each assigned point [12].

3.3.2 Phase 2

The development of the second phase is based on the restructuring of job positions. First, the SLP tool proposes a redistribution model, which considers the analysis of spaces in each area and the improvement in the fluidity of the production flow. Further, this tool proposes to redistribute the areas in such a way that the conditions of each job position improve and the operational effort is reduced [14]. After making the aforementioned changes, we proceed with the development of the 5S tool to create a culture of order and cleanliness for improving the productivity of each operator by complying with the ergonomic principles.

3.3.3 Phase 3

The final phase of the proposed improvement model involves the development of an MRP material planning model, which aims to define the exact moment at which a purchase order will be issued.

In this stage of development, the demand forecast, the supplier lead time, and the inventory amount in stock are considered to reduce the orders for unsupplied production number and increase the productivity of the company being investigated [10].

3.4 Proposed Process

The proposed improvement model, which is shown in Fig. 1, has a focus on increasing productivity, improving working conditions and planning materials; however, it will have an impact on quality management, supply management, and occupational health and safety management. This proposal began after having analyzed the current one of the company under study; Likewise, a comparison was made with the literature reviewed where it showed the experience in other companies. Everything described above has made it possible to select the appropriate tools for the application of a more productive, efficient and innovative model compared to those that currently exist. After that, we proceeded with the training of personnel so that the implementation of the proposal can be carried out, achieving an increase in productivity and a reduction in operating costs. It should be noted that the proposal is focused on planning materials, eliminating activities or factors that do not generate value in the production chain, as well as an approach to continuous improvement based on the Kaizen methodology.

3.5 Model Indicators

In the following points, indicators are proposed for the evaluation of the development of the proposed model, whose main purpose is to objectively measure the evolution of the production management system. Likewise, it allows evaluating the performance of the organization under study against its goals, objectives and responsibilities.

3.5.1 Productivity

Objective: Increase the multifactorial productivity of the company to 0.13 und/soles

$$\frac{Produced\ units\ (unt)}{MO + MP + MQ + General\ Expenses\ (soles)} \tag{1}$$

This can be used to validate whether the globally applied tools have been effective and whether the production costs have been reduced.

3.5.2 Index of Defective Products

Objective: Reduce the defective product rate by 16%.

$$\frac{Defective\ units\ (unt)}{Produced\ units\ (unt)} \times 100 \tag{2}$$

Its estimation is necessary because it will be possible to verify whether the control points were accurately located.

3.5.3 Efficiency

Objective: Increase the efficiency of the sewing line by 8%.

$$\frac{Real\ production\ (unt)}{Expected\ production\ (unt)} \times 100 \tag{3}$$

The evaluation of this indicator is necessary because it is used to measure the level of execution of the process and its performance.

4 Validation

4.1 Scenario Description

This research presents an innovative model as a proposal for improvement, which will be developed and validated through a pilot implementation system. The model was implemented at a small textile company in Lima, Peru. Which approved the use of private information and was approved by the institution's ethics committee. A Lean Manufacturing model is developed which is based on production management under a waste elimination approach, which will be validated through a pilot model. Said model is developed under an optimal sample size, which has been evaluated based on the population sample size, obtaining a result of 269 samples. The implementation of the pilot model is developed during a month of production, where the 3 phases of the proposed model are implemented in order to increase the productivity of the company under study.

4.2 Initial Assessment

The company under study belongs to the textile sector, whose standard product is the production of polo shirts, which represent 52.5% of annual sales with a turnover of 75,349.33 soles. Likewise, there has been a reduction of 13.69% in income during the last year, which led to an analysis of productivity where a result of 0.104 und / soles was obtained. Said productivity is below the optimal indicator in similar processes; For this reason, an analysis was carried out applying the Value Stream Mapping tool in order to identify the critical factors and determine the central problem.

4.3 Validation Design

Stage 1 - Working methods: The first phase includes the development of a new production system model based on line balancing and the implementation of inspection points. Control points has been strategically defined to ensure continuous control of the sewing process and reduce the number of defective products. The implementation of this phase contributes to the development of a uniform process, which will later be standardized, thereby achieving a 10.19% reduction in the defective product index.

Stage 2 - Redistribution of Job Positions: Phase 2 of this pilot model restructures jobs. First, plant redistribution was conducted based on the system layout planning tool. Thus, the transfer times were reduced by 3.13% with respect to the cycle time, reducing the effort of the operators and improving the production flow. Further, the 5S methodology was implemented, which is mainly used to improve the working conditions based on the principles of order and cleanliness. The pilot implementation of this tool contributed to improve the ergonomic conditions of the work stations, achieving a 35% reduction in non-compliance with ergonomic principles. Thus, based on the combination and complementation of the aforementioned tools, the operational efficiency index was observed to increase by 10.2%.

Stage 3 - Material Supply Plan: In the pilot implementation of the third stage, an MRP supply planning model was developed based on demand planning and the issued production orders. Herein, we focused on the development of a planning model, which allows to determine the exact date of issuance of purchase or production order for each material. Thus, an improvement was obtained with respect to the number of orders for stock-out production (a reduction of 11.91%), indicating that stock-out products were reduced and the production flow was improved.

At the end of the implementation of the 3 stages of the model, there is a new production system which allows the production flow to be continuous and to reduce stoppages due to defective products and shortages of materials. Also, the conditions in the jobs improve, which generates that the operational efficiency increases. With all this, it is important that monthly internal audits are carried out, in order to evaluate and collect observations in such a way that it can be constantly improved.

The main objective of the model developed in this study is to increase productivity based on waste elimination by applying lean manufacturing tools and focusing on continuous improvement. Therefore, a compliance assessment is conducted based on the indicators presented in Table 1. This assessment will allow us to determine if each factor of the proposed model contributes favorably to the fulfillment of the main objective.

Table 1. Comparison of the indicators

Indicator	Current value	Target value	Final result
Productivity	0.104 und/soles	0.13 und/soles	0.13 und/soles
Defective products	31.19%	15%	21%
Efficiency	68.30%	75%	78.50%
Standard sewing time	7.63 min/polo	6.5 min/polo	6.97 min/polo
Effort	40,425 kg.m	34,500 kg.m	33,245 kg.m

5 Conclusions

This article investigates the relationship and impact of Lean Manufacturing tools and the sourcing plan on production management. For this reason, the elaboration of the model managed to increase productivity to 0.13 und / soles, which shows an increase of 20% in the main objective. On the other hand, through the implementation of control points in the production chain of the sewing area, the rate of defective products is reduced to 21%, which indicates that the Jidoka methodology is building an autonomous defect control system. Likewise, it was possible to increase operating efficiency by 10.7% through the improvement of ergonomic conditions and the redistribution of certain areas; however, there were areas that could not be redistributed due to limitations imposed by production machines. Finally, the sewing time was reduced with respect to the cycle time, achieving an index of 38.72%, which indicates that the line balancing contributed to the elimination of activities that do not generate value to the production chain, thereby proposed to standardize the processes obtaining an increase of 22 units in the number of production per day, which shows that the productive flow has improved.

References

1. Zuniga J. Panoramica-Las Pymes en la Alianza del Pacífico. https://www.panoramical.eu/columnas/las-pymes-en-la-alianza-del-pacifico/

2. Universia (2019) Fundación Universia, 27 Marzo 2019. https://noticias.universia.edu.pe/practicasempleo/noticia/2019/03/27/1164222/aumentadopymes-peru.html
3. INEI (2019) Demanda internacional
4. INEI (2016) Características de las Empresas del Emporio Comercial de Gamarra
5. SUNAFIL (2018) Manual de Seguridad y Salud en el Trabajo
6. Correa F (2007) Manufactura Esbelta- Principales herramientas. Handle: RePEc:cey:panora 2(2):85–112
7. Mourtzis D, Doukas M, Lalas C, Papakostas N (2015) Cloud-based integrated shop-floor planning and control of manufacturing operations for mass customisation. Procedia CIRP 33:9–16
8. Damand D, Ammar O, Lepori E, Barth M (2013) Analysis method of the relations between MRP parameter and performance indicator based on a literature review. IFAC Proc Vol 46(9):377–382. https://doi.org/10.3182/20130619-3-RU-3018.00606
9. Kiran D (2019) Manufacturing resource. In: Production planning and control, vol 31, pp 441–455. https://doi.org/10.1016/C2018-0-03856-6
10. Díaz-Madroñero M (2015) Material requirement planning under fuzzy lead times. IFAC-PapersOnLine 48(3):242–247. https://doi.org/10.1016/j.ifacol.2015.06.088
11. Grisales N (2017) Objetivos y decisiones estratégicas operacionales como apoyo al lean manufacturing. Suma de Negocios 8(18):106–114. https://doi.org/10.1016/j.sumneg.2017.11.005
12. Romero D (2019) Rethinking Jidoka Systems under Automation & Learning Perspectives. IFAC PapersOnLine 52:899–903. https://doi.org/10.1016/j.ifacol.2019.11.309
13. Kiran DR (2019) Chapter 19 - systematic layout planning. In: Production planning and control, pp 279–292. https://doi.org/10.1016/C2018-0-03856-6
14. Németh I (2019) Layout configuration, maintenance planning and simulation of AGV based robotic assembly systems. IFAC-PapersOnLine 52(13):1626–1631. https://doi.org/10.1016/j.ifacol.2019.11.433
15. Susie T (2019) The application of the Toyota production system LEAN 5S methodology in the operating room setting, 54:53–59. https://doi.org/10.1016/j.cnur.2018.10.008
16. Neves P (2018) Implementing lean tools in the manufacturing process of trimmings products. Procedia Manuf 17:696–704. https://doi.org/10.1016/j.promfg.2018.10.119
17. Garza-Reyes J (2018) The effect of lean methods and tools on the environmental performance of manufacturing organisations. Int J Prod Econ 200:170–180. https://doi.org/10.1016/j.ijpe.2018.03.030
18. Maarof MG (2016) A review of contributing factors and challenges in implementing Kaizen in small and medium enterprises. In: 7th international economics & business management, vol 35, pp 522–531. https://doi.org/10.1016/S2212-5671(16)00065-4
19. Ortiz C (2010) Kaizen vs. Lean: distinct but related. Metal Finish 108(1):50–51. https://doi.org/10.1016/S0026-0576(10)80011-X

Lean Process Optimization Model for Improving Processing Times and Increasing Service Levels Using a Deming Approach in a Fishing Net Textile Company

Anthuane Carrillo-Corzo, Erick Tarazona-Gonzales, Juan Quiroz-Flores(✉), and Gino Viacava-Campos

Ingenieria Industrial, Universidad Peruana de Ciencias Aplicadas, Prolongación Primavera 2390, Lima 15023, Peru
{u201510435,u201313028,pciijqui,pcingvia}@upc.edu.pe

Abstract. Over the years, the textile industry has become increasingly important, both domestically and internationally, owing to its great contribution to the economy. Therefore, with prior planning of resources and times, companies now need to increase each of their processes' efficiency to achieve the maximum potential in the shortest period of time. Based on this, order processing times become more relevant because they are critical for fulfilling this objective. Companies want to prevent cost overruns and sanctions. Therefore, they seek to implement improvements that may provide greater sustainability and reliability to both their production processes and end customers. For these reasons, this study proposes the application of Lean tools such as SMED, Kanban, 5S, and standardization based on the Deming Continuous Improvement approach. The main objective is to increase On Time In Full (OTIF) service levels from 56% to 63% by increasing the number of orders delivered on time. The case study is conducted in a fishing net textile company. This study is motivated by the lack of in-depth studies executed in this significant subsector, which exhibits a great development potential.

Keywords: Fishing nets · Service level · Deming cycle · Lean · Standardization

1 Introduction

In the past few years, the growth of the textile industry has been significant. The sector accounts for 3.5% of the national GDP, which is equivalent to $ 151,248.82 million US dollars [1]. Moreover, 61.2% of manufacturing companies have conducted innovation activities, reinforcing their desire to improve production processes and efficiency indexes [2]. Another subsector of the textile industry that has exhibited a significant performance is the production of ropes, twine, and threads. In fact, this sector has grown 56.7% in recent years based on the dormant demand linked to fishing, an economic activity with a great potential value for the economy [3]. Nevertheless, the problem that manufacturing companies face is inefficient processes, which encourages their need to improve their

Y. Iano et al. (Eds.): BTSym 2020, SIST 233, pp. 443–451, 2021.
https://doi.org/10.1007/978-3-030-75680-2_50

production processes, from delivering inputs to the ultimate marketing of their finished products in the shortest possible times [4].

The proposal centers on improving service level indicators for a fishing net textile company by facilitating the on-time delivery of orders based on a Lean Deming model. This model will optimize setup times, improve machine availability, and reduce failures and Work in Progress (WIP) products. Four guidelines are also listed for the Plan, Do, Check, and Act stages of the Deming model.

2 State of the Art

2.1 Optimization of Processes and Times

For process time optimization and, in turn, for process improvement, several Lean Manufacturing tools may be used. For example, some studies successfully used SMED to decrease production changes by 43%, thus reducing internal activities by 15% [5]. Furthermore, standardization increases productivity by 16%, whereas Kanban improves inventories by 30%, which means that it reduces excess inventories [6].

2.2 Human Factor

The integration of the human factor contributes to improving production line processes because a trained and motivated employee with adequate knowledge on the correct performance of their functions and supported by an appropriate procedure can reduce project improvement costs from 30% to 50%. Companies must assess not only the strong skills but also the soft skills of workers to improve the performance of their functions by correctly manipulating and creating sequences of activities to improve process efficiency [7, 8].

2.3 Lean and Deming

By combining Lean and Deming, specific solutions can be implemented for production process problems at a company. Moreover, the Deming approach develops a structured and planned method of not only revealing problems but also ensuring constant and sustainable improvement. According to J. Mohd and S. Mojib, the combination with Lean allows reducing bottleneck times by 56%, which is a 44% gain for the wood manufacturing industry [9].

3 Contribution

3.1 Proposed Model

Figure 1 illustrates the model proposed with different components, which jointly generate a novel contribution to the fishing net textile industry. This model seeks to improve service levels at a manufacturing company by applying the Deming methodology based on Continuous Improvement and by applying Lean Manufacturing tools.

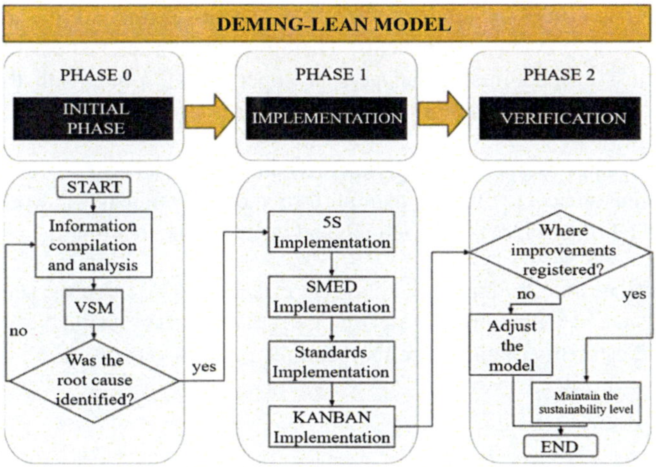

Fig. 1. Proposed model

3.2 Model Details

Phase 0: Initial Phase. Phase 0 includes the compilation of information and the analysis of the current situation in which the main problem and its causes are identified. The VSM tool is applied to map the process to reduce or waste, decrease great delivery times, standardize its procedures, and stabilize its processes [10].

Phase 1: Implementation. Phase 1 includes the tools presented below.

5S. The 5S tool will be presented first to the project team, and then the current situation will be assessed for later actions. All personnel involved will be trained. Subsequently, to proceed with the execution of the action plan, visual management formats will be implemented. Finally, activities will be constantly monitored through audits.

SMED. Activities that aim at changing production processes are classified as internal and external. The current situation will be assessed by recording times after implementing the production changes. Correspondingly, internal and external activities will be categorized. All potential internal activities will be converted into external activities to continue with their optimization. Finally, the process will be standardized, and the staff will receive training on the improvements applied to the production processes.

Process Standardization. This tool is used to standardize activities and processes by applying OPL and SOP forms.

Kanban. To implement Kanban, information will be compiled, current variables will be assessed, and the reorder point and Kanban inventory number will be estimated. The next step is to design the Kanban board and cards. Thereafter, all staff will receive training on the Kanban principles to establish a work culture.

Visual Management and Communication. This tool aims to improve the training sessions conducted on previous components. Therefore, the parties responsible for communication must be appointed (managers or supervisors), along with the types of communications issued (graphic/printed, digital, or face to face) and their planning.

Skills and Knowledge Matrix. By applying this matrix, the goal is to provide staff members with training on how to improve their performance and efficiency at work. For these purposes, critical requirements for each job and task must be identified.

Phase 2: Verification. In Phase 2, audits were executed to measure the success of the tools implemented in the previous Phase. Later, to attain project sustainability, corrective actions and adjustments to guarantee the continuous improvement of the model were taken.

3.3 Model Indicators

Indicators will be used in this study to measure the actual success rate of the model implemented. Different indicators and metrics used for these assessments are listed in Table 1.

Table 1. Model indicators

Project Indicators	
OTIF	$\text{OTIF} = \frac{\text{\#Orders delivered on time and in full}}{\text{Total number of orders}}$
SETUP	This indicator records production change times
AVAILABILITY (%)	$\%\text{Availability} = \frac{\text{Operating time}}{\text{Time for that a machine is available}}$
WIP	This indicator measures products that have entered theproduction process but are not yet for selling because they have not been finished

4 Validation

4.1 Scenario Description

The model was implemented at a fishing net textile company in Callao, Lima, Peru. This study focuses on the production line of anchovy trawl nets. This line corresponds to 60% of total sales. One of the most significant problems reported by the company is late orders, which affects the company's revenue because late orders are usually subject to penalties and fines, which change depending on the client. The current OTIF indicator is set at 56% of the order amount. This indicator represents $781,500 in annual penalties and fines for the company. The data used in this work were provided exclusively for academic purposes on the anonymity condition.

4.2 Initial Diagnosis

Based on the diagnosis, the following results were obtained:

- 76% of all orders in a year are delivered late.
- The average OTIF in 2019 was 56%.
- The main production line is anchovy trawl nets (60%).

4.3 Validation Design

Phase 0: Initial phase. In Phase 0, information was compiled for the anchovy trawl net production line by applying a Value Stream Mapping (VSM) diagram. The main waste, bottlenecks, weaving, revision, and remeshing activities were identified in the diagram. Moreover, production process deficiencies were identified, and the main problem was assessed. It was observed that the low service levels within the anchovy trawl net production line were the main issue. When the root causes of this problem were identified, the biggest impact was caused by the lack of equipment availability at 48%. The two main reasons why equipment may be unavailable are noncompliance with a preventive maintenance plan and the excessive amount of time required from SETUP to production processes. Another cause identified was the large number of WIP at 41%. This situation arises from production failures, which were set at 70%, due to poor production changes.

Phase 1: Implementation. The aforementioned components were implemented in Phase 1.

5S. The 5S tool was implemented to foster the creation of an organizational culture of discipline in terms of order and cleanliness, starting with the most critical processes (weaving, reviewing, and remeshing).

This tool will support the application of additional Continuous Improvement components, which were identified in the design. The reason for this is that it also allows staff members to identify themselves and commit to their equipment and workplace. In the implementation, audits were executed to control improvements that were initially developed, and critical points were emphasized (Table 2).

Table 2. Results of the audit process

5S Audit				
Concept	1st	2nd	3rd	4th
Results	1.87	2.50	2.72	3.52

SMED. The SMED tool was implemented to decrease the SETUP time by optimizing activities and standardizing the knotted net weaving production process. At the beginning of this stage, both operational and administrative staff members were trained as deemed necessary. Moreover, to avoid mechanical or electrical breakdowns in production change processes, machines were inspected. To ensure better results from the SMED tool, activities were classified into groups (Table 3). Therefore, better controls could be kept after implementation.

Table 3. Analysis and grouping of activities

#	SITUATION	AS IS		GOALS	
	Activities	# of people	Time period (min)	# of people	Time period (min)
1	Unload materials	1	40	-	0
2	Change of warp	2	132	3	85
3	Finish the weaving of previous product	1	37	2	15
4	Preparation of material (reels)	1	76	1	0
5	Conditioning of the warp	2	215	3	175
6	Mechanical adjustments	2	320	2	157
	Total Time		820 min		432 min

After making improvements in December, weekly SETUP times were monitored in January and February (Table 4), which revealed significant time reductions. Increasing the availability of weaving area machines is one of the principal objectives of the SMED tool.

Table 4. % Machine availability

	S1	S2	S3	S4	S5	S6	S7	S8	S9
Availability	75%	76%	83%	78%	77%	80%	81%	85%	75%

Process Standardization. The standardization tool is centered on streamlining activities in the production change process. Formal documents were used and aligned with company tasks. An OPL for production changes was developed and posted in strategic areas to help employees correctly follow instructions. Furthermore, a standard operating procedure was developed with detailed process activities.

This application aimed to reduce the number of defective products (trawl nets). The result is presented in Table 5.

Table 5. Results of failures

Monthly results	Jan - 20	Feb - 20
Failures/cloths	45	42

Kanban. Kanban is used to create an effective and efficient method for production flow control. Information was compiled to generate the results required to ensure the fidelity of results in simulations. The results are shown in Table 6.

Table 6. Kanban variables assessed

Kanban Variables	Value
Demand (M)	7439 fathoms/month
Demand (d)	247.96 fathoms/day
Time	28.3 min/u
Safe Stock Levels	18%
# of changes	2
Kanban inventories	38 nets
Reordering point	1 net

Using this new lead time, the OTIF was calculated considering the largest number of orders processed in the least possible time. The number of fulfilled orders increased from 65 to 72 per month, which set the OTIF indicator at 63%. In the WIP analysis, the percentage of increase was compared to the base safety stock. The 45,060 kg of in-process inventory reported in December decreased to 34,396 kg, reflecting a 31% improvement. The results obtained was OTIF: 63%, Work in Process – WIP: 31%

Visual Management. To reinforce the tools implemented, visual management will be applied, especially in the form of murals, such as the 5S board. Each OPL developed for SMED will also be placed at strategic locations.

Skills and Knowledge Matrix. Additionally, the implementation of the skills and knowledge matrix was conducted.

The previous step was to evaluate workers on the skills required by the position. In the present case, weaving skills were needed.

If the operating personnel present a minimum score in this skill, they will need to retrain.

Phase 2: Verification. In relation to initial indicator results, corrective actions were taken for the model proposed.

4.4 Final Indicator Results

A pilot implementation of 5S, SMED, process standardization tools, and the simulation method was conducted with the Kanban tool (Table 7). The implementation method corresponding to the model has been developed.

Table 7. Final indicators

Concept	Indicators	AS IS	GOAL	Measurement Results
MODEL	OTIF	56%	70%	63%
SMED	Setup time	540 min	432 min	452 min
	Availability of machines	70%	80%	80%
Kanban	WIP (%increase)	46%	25%	31%

5 Conclusions

In conclusion, in this project, through the SMED tool and the application of standards, it was possible to increase the availability of the machine to 80%, also related to the level of products from inventories in the process (WIP), which was reduced by 31%. As a result, the service level increased through the OTIF indicator to 63%, representing 72 monthly orders delivered on time for the present case study.

Over time, the sustainability of the improvements relies on the human factor. Along with training, the implementation of novel engineering techniques would increase process efficiency.

We recommend further studies focusing on the fishing net textile sector because it is an unexplored source of information and latent yields. This sector has the potential to become one of the main resources for the national and international fishing and textile industries.

References

1. Ministerio de la Producción (2017) Anuario Estadístico Pesquero y Acuícola 2017. Impacto monetario del sector textil
2. Ministerio de la Producción (2015) Resultados de la Encuesta Nacional de Innovación en la Manufactura
3. Ministerio de Producción (2017) Crecimiento por industria
4. Narayan B (2018) Productivity growth and efficiency change: comparing manufacturing – and service based firms in India. Econ Model 7:447–457
5. Azadeh A, Zarrin M (2016) An intelligent framework for productivity assessment and analysis of human resource from resilience engineering, motivational factors, HSE and ergonomics perspectives. Saf Sci 89:55–71

6. Pombal T, Pinto L, Sá J, Pereira M, Silva F (2019) Implementation of lean methodologies in the management of consumable materials in the maintenance workshops of an industrial company. Procedia Manuf 38:975–982
7. Garza J, Kumar V, Chaiki S, Hua Tan K (2018) The effect of lean methods and tools on the environmental performance of manufacturing organizations. Int J Prod Econ 200:170–180
8. Fruggiero F, Riemma S, Ouazene Y, Macchiaroli R, Guglielmi V (2016) Incoporating the human factor within manufacturing dynamics. IFACPapersOnLine 49–12:1691–1696
9. Mohd J, Mojib S (2015) Production line analysis via value stream mapping: a lean manufacturing process of color industry. Procedia Manuf 2:610
10. Helleno A, Isaias de Moraes A, Tadeu A (2017) Integrating sustainability indicators and lean manufacturing to assess manufacturing processes: application case studies in Brazilian industry. J Clean Prod 153:405–216

The Method of Soft Computing of Pilot Reliability Depending on Age

Nikolai I. Plotnikov(✉)

Scientific Research Project Civil Aviation Institute «AviaManager», 630078 Novosibirsk, Russian Federation
am@aviam.org

Abstract. This work is a statistical and analytical modeling of the individual reliability of a person depending on age. The soft computing method is used. Descriptions of statistical, experimental studies and results are given. Studies of flight safety civil aviation are executed. It is used heatmaps for calculating of levels of dependability of a person's age. Soft Computing (SC), statistical and expert methods for calculating a person's properties are used. SC concept is proposed. The author of this work gives SC formal definition output. New quantitative dependences of the frequency of aviation accidents (AA) events on the pilot's age in non-integer continuous values for years are obtained. An integral indicator of the total pilot's flight time of hours was established. The results' novelty is to establish the accounting of indicators for time periods: flight task (day), month, 90 days, six months, year. This structure and the established values of indicators allow for continuous monitoring and rationing of operational dependability of airline flight resources to ensure flight safety. The task of calculating the values of aviation accident probabilities depending on the pilot's age by years is set and solved.

Keywords: Soft computing · Reliability · Dependability · Civil aviation · Flight safety · Pilot · Age

1 Introduction

This paper presents a method of modeling and soft computing resources of individual reliability of a person depending on age. The practical area of research is the activity of pilots of civil aviation (CA), statistical analysis of the dependence of aviation accidents (AA) on pilots' age. The source data of the world AA statistics is used. The problems of low-frequency characteristics of statistical data necessary for reliable analysis and calculations are investigated. An approach to solving the problem of interval analysis of statistical data series is proposed. The proposed interval design gives the best representation of age continuity due to the displacement of the average calculated values. The AA probability is calculated as a variation series for each year and age group: the ratio of the number of AA to the total or annual flight hours of the corresponding age group of pilots. Changes in the pilot's reliability resources are represented by a non-linear function called the "U-profile" of reliability, through which calculations are made. It is possible to quantify AA per age values by years in fuzzy probability intervals of the U-profile.

Y. Iano et al. (Eds.): BTSym 2020, SIST 233, pp. 452–462, 2021.
https://doi.org/10.1007/978-3-030-75680-2_51

2 Problem

Until now, theory and methods that could take into account the differences in the characteristics of specialists remain uncertain, which leads to the use of untenable standards of professional training and labor rationing in world CA [1]. The content of the problem of analyzing and calculating the age dependence of AA is as follows. The AA event is extremely rare, making it impossible to establish a link with age for a particular individual and decide about admission to work. This can be done based on the analysis of large statistical studies over several decades. However, due to the high labor intensity and cost, such statistical researches are extremely rare.

Statistics and analyses in studies [2–10] made up a period of up to 10 years with the number of subjects more than 3000 people. However, these results also did not allow to create continuous statistical series of the required power.

There are contradictions in the construction of statistical series. Complete information for detecting weak effects from age is related to the continuity of the series. Statistical frequencies are piecemeal when establishing continuity. A statistical set is a probability that can be determined by differences between groups and depends on the parameters: a) the size of the difference between groups (the size effect); b) the composition of samples, that is, the number of groups compared; c) the level of statistical significance, from 5%. Studies have allowed the calculation of probability through differences in age groups. To overcome this disadvantage, interval data aggregation is performed. The greater the aggregation, the greater the power of the statistical set. However, aggregating data in intervals reduces the number of indicators to analyze. As shown in [8], the method of aggregation of large intervals of statistical data increases their frequency power for probabilistic calculations but reduces the accuracy and possibility of calculating the values of AA risks for a specific age of the pilot. In other words, when aiming for the statistical significance of the 5% threshold ($p < 0.05$), aggregating data into age groups reduces the power of analysis, which is necessary to detect differences in AA dependencies on the age of the pilot between groups. Grouping also reduces the accuracy of results: any specific detection will be assigned to a group rather than a specific person age. However, the technique of aggregating data intervals is a compromise and is used to increase the frequency power of the data.

The total content of the problem include: a) the need to investigate rare events up low-frequency characteristics; b) mixing statistical sampling types of pilots, airlines, transportation, markets: commercial, non-commercial, regular, irregular; c) all available databases, including the most quoted NTSB and the FAA, details not structured; d) the grouping of samples is based on the type the pilot's testimony and the lack of accounting data on the types of transactions and markets. The cause of problems: 1) non-numeric nature of the object (person); 2) limited methods of mathematics; 3) rarity of AA events; 4) insufficient frequency of statistical analyses and numerical fuzziness of data; 5) extreme cost of statistical and expert research.

3 Method

3.1 Soft Computing

The largest part of the values of the properties and states of organizations, social groups, and people is determined by the methods, which are used in conjunction with the heatmap risk model, the "traffic light model". It is used heatmaps for calculating of levels of dependability of a person's age. Soft computing (SC), statistical and expert methods for calculating a person's properties are used. SC concept is proposed by the founder of the theory of fuzzy sets L. Zadeh, (1994) [11]. The author of this work gives SC formal definition output in the following statement. All definitions are introduced with symbols (Def-1).

Def-1. Soft computing is a set of methods: hard estimation (HE), soft measurement (SM), soft estimation (SE) in metric and non-metric scales with the ability to simultaneously process quantitative numerical and qualitative linguistic data.

The presented interpretation of SC can be considered as the notion of the power (1):

$$SC : \Omega = <\mu_{SM}, \mu_{HE}, \mu_{SE}> \tag{1}$$

Where fuzzy (soft) measures are indicators of data inaccuracy, Ω - finite universal set.

The powers of these measures are minimal for single-point sets. The most accurate values of the objects are measures μ_{SM}, less accurate μ_{HE}, and the greatest inaccuracy measures have μ_{SE}, in the limiting case (2):

$$\mu_{SE} \cong \Omega \tag{2}$$

Thus, insufficient statistical research densities are overcome by a developed method that gives acceptable quantitative data analysis results.

3.2 Task Statement

The most important methodological issues are the time coverage of the statistical sample, which is especially relevant for observing rare events and the representation of age as an indicator of continuous independent interval variable. It is used the following methods for compiling statistical series: a) discrete discontinuous interval series: 20, 25,..., 55, 60 years; b) continuous non-intersecting interval series: 20–24, 25–29,..., 50–54, 55–60 years; c) a continuous intersecting interval series: 20–24, 23–26,..., 56–59, 60–63. The proposed interval design gives the best representation of age continuity due to the displacement of the average calculated value. However, with this method, the characteristics of a single pilot may be present at different intervals. The best results are obtained by simultaneously analyzing different data structures, which gives a comparison of the convergence of results.

4 Modeling

4.1 Age as an Indicator of a Continuous Variable

To solve the problem, age is structured as an indicator of a continuous variable. The task is formulated as follows: perform calculations of the pilot's reliability resources in the likelihood metric in probability indicators, using stratification analysis techniques.

Sequence: 1) determine the degree of dependence of AA on the age of the pilot; 2) determine the degree of dependence of AA in the age group over 60 years. To obtain valid evidence, it is particularly important: a) the number and variety of available studies on this topic; b) the subject sample of different categories of pilots; c) the time intervals of the research sample; d) the use of age as an independent variable; e) determining the types of events to include in the numerator of the calculation; e) determining the number of flight hours to include in the denominator of the calculation of flight risks; g) determining acceptable statistical methods of analysis.

4.2 Statistical Data

The initial data is contained in [3, 9]. The sample for analysis was made up of 3,306 pilots aged 45–54, of which 1,632 pilots were involved in 66 AA over the 11-year period 1987–1997 (49.4% of the total sample), of which 105 pilots died. These ages are structured into two groups of 45–49 and 50–54 years. The combination of age and flight hours: <5 000, 5000–9999, 10000–14999, $\geq$ 15000. In this sample of pilots with a total flight time of 12.9 million hours, 5.1 accidents per million flight hours occurred. The average age is 48.5 years, and the average flight time is 9749 h. Statistically distinguishable results of the analysis are the following indicators: age 50–54 has a risk index of 22% higher; transitions from flight intervals of < 5000 h to the interval of 5000–10000 h and especially > 10000 h are accompanied by a noticeable effect of reducing the relative risk of AA. Statistical calculations are based on the number of AA divided by the number of flight hours. The AA probability is calculated as a variation series for each year and age group: the ratio of the number of AA to the total or annual flight hours of the corresponding age group of pilots, Table 1, Fig. 1.

Table 1. AA dependence on age in years [9]

Age, years	20	25	30	35	40	45	50	55	60	65
AA per 1000 pilots	32	15	10	8	6	7	8	7	6	5

The age and AA dependence analysis shows that by the age of 35 and with a flight of 10,000 h, the pilot reaches a "safety plateau", the best reliability. In the first two years of operation, when developing new equipment, the share of AA assigned to the pilot is 80%; after 10 years, it is about 30%. To prepare for statistical modeling at the search stage, these AA values per flight hour are distributed according to the pilots' characteristics. Differences in AA values are estimated in relative risks of 95% confidence intervals.

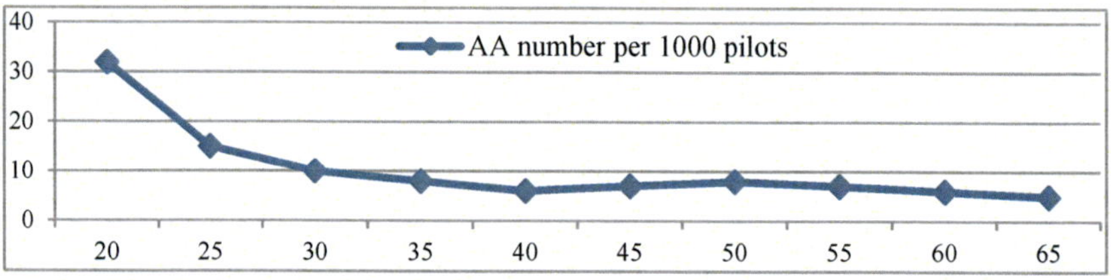

Fig. 1. AA dependence on age in years

The effects of potential interaction between independent variables have been studied by stratification analysis. At the modeling stage, the number of flight hours preceding the AA is analyzed as a function of independent variables using a proportional hazard model. The model is based on the assumption that the accumulated flight hours preceding the AA has an exponential distribution so that the relative risk associated with this independent variable has a constant value over time. In this analysis, the pilot's age of 45 years is used as an independent variable in time to enter the model.

4.3 Solving the Problem via the U-Profile of Dependability

The solution is to calculate the values of AA depending on the pilot's age using the "U-profile" of reliability. Studies of AA and age dependence have one common drawback: weak frequency capacities of the analyzed data for representing numerical calculations in metric scales of measurement. Most studies lead to conclusions about the difficulty of distinguishing the statistical dependence of accidents on age since this indicator "dissolved" other indicators, primarily flight experience, calculated by flying hours. Changes in pilot reliability resources are represented by a non-linear function called the U-profile (3):

$$U(D) = f(A, TFH) \tag{3}$$

Where D is the dependability, A is the pilot's age, TFH is the pilot's total flight hours.

Three enlarged intervals are aggregated in the U-profile. (1) from the beginning of the pilot's career to the mature age of 35–45 years, the flight time is 5000–10000 h. During this period, there is a multiple (up to an order of magnitude) increase in dependability and reduction of AA risks. In this left part of the "U-profile", the number of AA has a steep character of acceptable clarity up to about 35–40 years. (2) the average part of the profile is stabilized at the minimum number of AA in the interval of 35–55 years. (3) the right side of the profile is flat, with an increase in AA at the age of 55–65 years. This statement is true for large samples of statistical analysis: a period of more than 10 years, age groups in intervals of at least 10 years, and a noticeable frequency of observed events. In the intervals of age after 50 years with a flight time of more than 10,000 h, the increase in dependability is stabilized. At the turn of 60 years of age, there is a decrease in dependability, which is statistically difficult to verify but confirmed in non-metric evaluation scales.

To solve the problem, the selected statistical series is assumed to be conditionally continuous and converted with equal values of data for years in intervals as a continuous function. The calculation method aims to establish the function and formula for the dependence of two variables: finding the maximum or minimum of a continuous statistical series in the interrelated change in the values of AA and the age of the pilot.

To implement the method, we use the initial data [3]: integer values of five-year intervals AA per age of pilots the largest known statistical analysis of the sample, time period, and number of subjects.

Calculation task: to set non-integer indicators values of a continuous series of AA per age of pilots. The method is implemented in the following sequence.

4.4 Assigning the Status of an Independent Variable

Procedures:

- The main axiom of the plausibility of statistical research data is postulated, in which the character of AA per age variables definitely correspond to the U-profile;
- The original data is converted and approximated using the Ordinary Least Squares (OLS) method to several classical functions in MS Office Excel: linear, interval, interval continuous, exponential, logarithmic, polynomial.

4.5 Estimation of the Likelihood Approximation

Procedures:

- Alternate expert estimation of the likelihood (probability) of classical functions for the greatest AA per age correspondence to the U-profile;
- Selecting one function from them and assigning the status to a variable in the statistical series under study;
- Splitting the selected approximating function into intervals corresponding to the levels of the risk matrix;
- Determination of thresholds and probability intervals for quantitative calculations of AA per age values;
- Estimation of statistical series values in the nominal scales and order in a three-dimensional risk matrix.

It is important to note that the method does not use calculations using probability theory operations. The calculations and numerical values performed are carried out in likelihood measures, which are then reduced to probabilistic indicators.

Thus, with these procedures in the fuzzy probability intervals of the U-profile, the possibility of quantitative calculations of AA per age values by years is achieved.

5 Performance of Calculation

In accordance with the task, a table with numerical values of the interval function AA from age for five-year periods with the corresponding calculation formulas was compiled. The series is converted with equal data values for years in intervals as a continuous function, Table 2, Fig. 2 (Figs. 3, 4, 5, 6 and 7).

Table 2. Selection of approximating functions

Function	Formula	Reliability
Linear	$y = -1{,}9394x + 21{,}067$	$R^2 = 0{,}5256$
Interval	By table formulas (AA/age calculation) [11]	
Interval continuous	$y = 0{,}0226x^2 - 1{,}4356x + 26{,}408$	$R^2 = 0{,}8668$
Exponential	$y = 20{,}025e^{-0{,}15x}$	$R^2 = 0{,}6867$
Logarithmic	$y = -9{,}842\ln(x) + 25{,}265$	$R^2 = 0{,}7934$
Power function	$y = 25{,}471x^{-0{,}705}$	$R^2 = 0{,}8903$
Polynomial	$y = 0{,}5568x^2 - 8{,}0644x + 33{,}3167$	$R^2 = 0{,}8029$

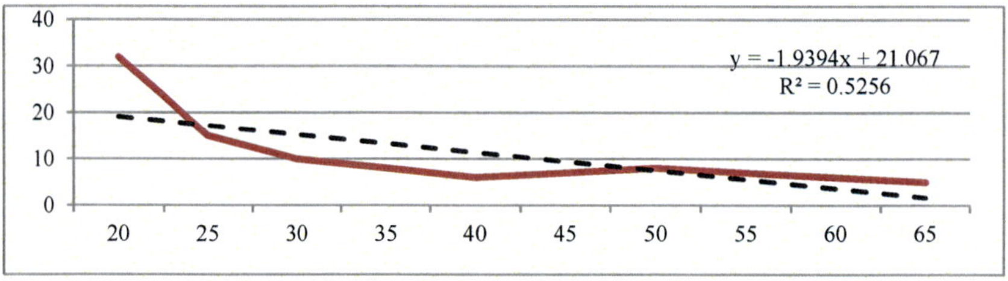

Fig. 2. Linear function

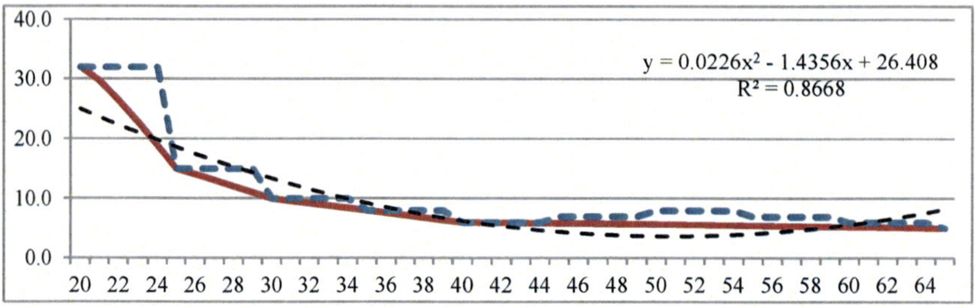

Fig. 3. Interval continuous function

Formulas are obtained by converting data to various classical functional distributions by querying in MS Office Excel the correspondence of a statistical data series to any classical function. Next, it is built a graph of the investigated function of the source data and a graph of the classical function with the formula. Graphs of the linear, interval, interval continuous, exponential, logarithmic, polynomial functions with corresponding formulas and coefficient values are compiled to establish the likelihood of AA per age correspondence to the U-profile,

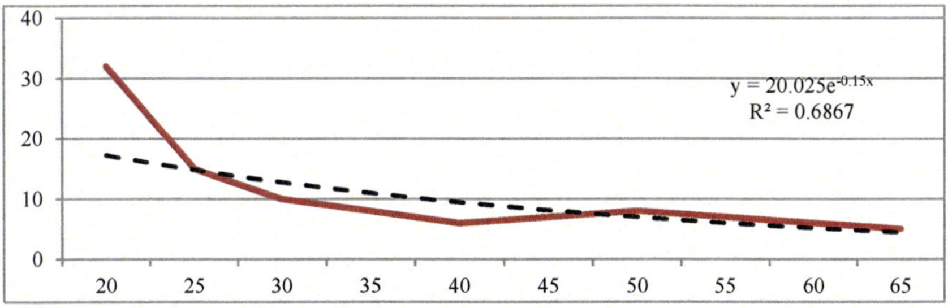

Fig. 4. Exponential function

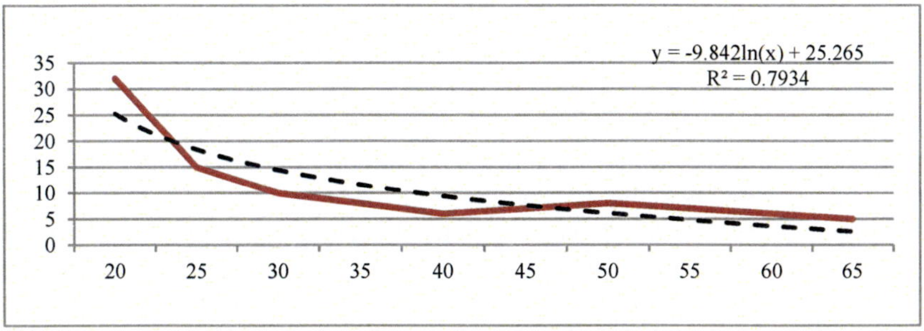

Fig. 5. Logarithmic function

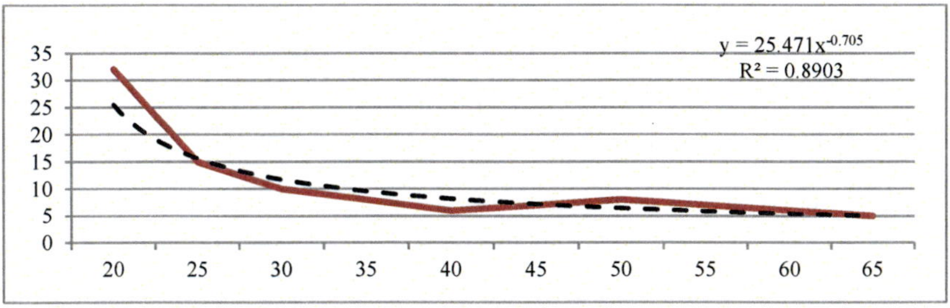

Fig. 6. Power function

Seven calculation formulas, the approximation confidence value, and graphs are presented in the papers [11, 12]. Determination of the approximation function for calculating the probability of risks AA per age is carried out by expert evaluation according to the following conditions:

1. The likelihood of compliance with the U-profile;
2. The accuracy of the approximation.

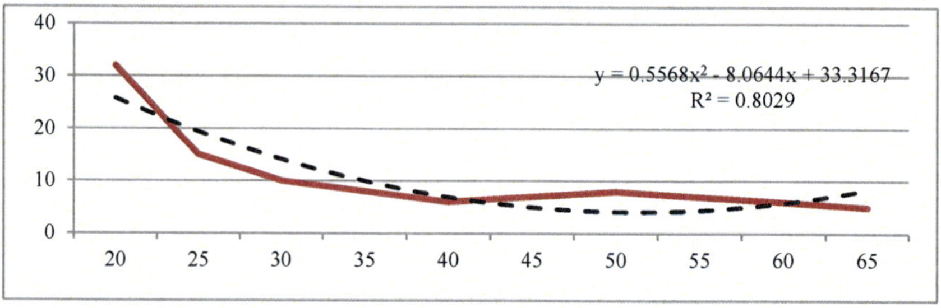

Fig. 7. Polynomial function

We see that the power function $R^2 = 0{,}8903$ is the most satisfying condition (2) in terms of numerical values, followed by the interval continuous $R^2 = 0{,}8668$.

However, in the pilot age range of 50–65 years, these functions under condition (1) do not satisfy the U-profile likelihood axiom, which was previously postulated. Since the given axiom is accepted in the priority procedure, we establish that condition 1 is most satisfied by the polynomial function $R^2 = 0{,}8029$.

Determination of the AA probability intervals and the normative space of the individual dependability of the pilot's age is carried out in the following sequence.

The studied statistical analysis is performed on the space of five-year intervals of the age of pilots 20–65 years. Values of interval indicators are used to output the following definitions.

Def-1. Values of a statistical series of age indicators <20, 35, 45, 55, 60, 65> we call the thresholds for the values of individual dependability of the pilot's resources by age.

Def-2. The distances of neighboring values of the values of indicators are called the intervals of values of individual dependability of the pilot's resources by age.

The space intervals of the age structure in the range of one AA of the order of probability $10^{-5} \geq P > 10^{-6}$. To output the reference probability values, there is a table for the polynomial function by calculating the probability of the AA 1000 pilots on the original data when the number of flights 3800, corresponding to 10 years of pilot experience. Number of AA per 3800 flights = log10(A (i)/3800000). This value is selected based on the U-profile transition to the minimum AA per age indicators [8].

6 Results

New quantitative dependences of the frequency of AA events on the pilot's age in non-integer continuous values for years are obtained. Assessment in the risk matrix is performed by dividing the values of the pilot's age indicator into levels corresponding to the three-level traffic light model (heatmap): 1 (green), 2 (yellow), 3 (red). The space of values of the polynomial function $y = 0{,}5568x^2 - 8{,}0644x + 33{,}3167$; $R^2 = 0{,}8029$ is estimated by the nature (likelihood) of the AA number curve and the corresponding AA probabilities is structured into intervals:

- (3 – red, $10^{-5,2} > P > 10^{-5,4}$);
- [2 – yellow, $10^{-5,6} > P > 10^{-5,8}$];
- (1 – green, $10^{-5,9} > P > 10^{-6,0}$).

Thus, heatmap risk levels, age intervals, and probability values are recorded:

- 3 - red [20–30];
- 2 - yellow (35–40);
- 1 - green (45–55);
- 2 - yellow (60–65];
- 3 - red [>65].

It is noted that two interval ranges are set for level 2 – yellow. Accordingly, it is marked the boundaries of the values of indicators that are evaluated by the SE method with round brackets and mark them with square brackets using the HE method [13].

The result of calculations is the conversion of piecewise-defined integer interval statistics AA per age into non-integer values of a continuous statistical series of the polynomial function $R^2 = 0{,}8029$.

An integral indicator of the total pilot's flight time of hours was established. Pilots have the highest professional dependability in terms of flying for more than 10,000 h.

The novelty of the results is to establish the accounting of indicators for time periods: flight task (day), month, 90 days, six months, year. This structure and the established values of indicators allow for continuous monitoring and rationing of operational dependability of an airline flight resources to ensure flight safety. Some fragments and a detailed description of the developed theory, method, experiments, and expert assessments are presented in the papers [12, 13].

7 Conclusion

This paper describes a statistical analysis of AA that establishes the relationship between the age and experience of pilots of flight crews by flight hours. The analyses are based on large statistical studies of more than 3,000 people over several decades to produce continuous statistical series of the required capacity. Calculation work aims to establish fuzzy intervals of pilot experience indicators that correspond to the rules of calculation by definition SC. This makes it possible to calculate the dependability of the pilot in individual values of age with an assessment of risk levels for making management decisions. It is reliably established that the decrease in individual dependability and the increase in the risks of AA events becomes noticeable after 55 years and certain after 60 years of age of the pilot. It was found that the frequency of AA in the age of over 60 years is twice higher than in the general sample of statistical analyses before this age. The thresholds and intervals of the normative space of individual dependability of the pilot by age in the values of a statistical series of indicators are determined. Results are obtained in qualitative methods for calculating flight safety and risk states of airlines. Indicators for air transport production risk management standards and decision-making tools were obtained. Calculated indicators of pilot dependability values are a model for developing the airline's strategy, quantitative assessments of flight specialists, standardizing professional activities, and managing training costs.

References

1. Kane MR (1990) Air Transportation. Kendall Hunt Publishing Company, USA, Iowa. 500 p
2. Aarons RN (1987) Safety statistics for prudent pilot. Bus. Commer. Aviat. 61(1):54–57
3. Broach D, Joseph KM, Schroeder DJ (2003) Pilot age and accident rates report 3: an analysis of professional air transport pilot accident rates by age. Civil Aeromedical Institute, Human Resources Research Division, FAA. 63 p
4. CAST: Commercial Aviation Safety Team (2004) Process for Conducting Joint Implementation Measurement and Data Analysis Teams (JIMDATs), DRAFT, June 2004
5. Hancock PA, Chignell MH (1988) Mental workload dynamics in adaptive interface design. IEEE Trans. Syst. Man Cybern. 18(4):647–658
6. Massie DL, Campbell KL, Williams AF (1995) Traffic accident involvement rates by driver age and gender. Accid. Anal. Prevent. 27:73–87
7. Mortimer RG (2000) Differences in accidents of professional pilots and non-professional pilots as a means of identifying training needs. National Aviation Safety Data Analysis Center. Introduction to the databases. FAA. https://www.nasdac.faa.gov/internet/fw_learn.htm
8. Salthouse T (1990) Influence of experience on age differences in cognitive functioning. Hum. Factors 32:551–569
9. Salvatore S, Stearns M, Huntley S Jr, Mengert P (1986) Air transport pilot involvement in general aviation accidents. Ergonomics 29(11):1455–1467
10. Shaw RR (1985) Airline Safety 1950–2000. Aircraft 1:32–34
11. Zadeh LA (1994) Fuzzy logic, neural networks, and soft computing. Commun. ACM 37(3):77–84
12. Plotnikov NI (2013) Resucy pilota. Nadezhnost. [Pilot resources. Dependability.], Novosibirsk, Russia, AviaManager Publ. 264 p. (in Russian)
13. Plotnikov NI (2012) Automated system for predicting and preventing accidents at the organization and production of air transport. The intermediate stage № 4: «Adaptation of the developed algorithms and software AS». – Scientific- and-technical report «2010-218-02-068», № State registration 01201150118 from 12.01.2011, № 194, Ulyanovsk, Section 3, pp. 154–238; Applications: I, K, L, M, N., pp. 1048–1258. 1340 p. (in Russian)

Melanoma Classification Based on Three Different Very Deep Neural Networks

Pablo Minango[1,3](✉), Yuzo Iano[1], Alex M. R. Ruelas[1], Gabriel Gomes de Oliveira[1], Maria Cecilia Luna Alvarado[1], Juan Minango[3], Rangel Arthur[2], and Diego A. Pajuelo Castro[1]

[1] School of Electrical and Computer Engineering, University of Campinas, Campinas, Brazil
yuzo@decom.fee.unicamp.br, a151594@dac.unicamp.br

[2] School of Technology, University of Campinas, Campinas, Brazil
rangel@ft.unicamp.br

[3] Instituto de Fomento al Talento Humano. IFTH - SENESCYT, Quito, Ecuador

Abstract. The present study proposes a methodology for classifying dermoscopic skin lesions images with melanoma and non-melanoma issues. Earlier diagnosis of melanoma gives the patient an estimate of a 5-year survival rate of 95%. Our main objective is to develop a computational tool that assists medical specialists in the task of recognizing melanoma in its initial stages. For this reason, our system is intended to generalize a prediction model, applying image processing techniques and very deep neural networks.

Keywords: Melanoma classification · Convolutional networks · Very deep neural network

1 Introduction

According to the World Health Organization (WHO), melanoma skin cancer has increased in the last decade due principally to the depleted ozone, which caused the loss of the atmosphere's protective filter function, and more UV radiation passes directly to people [1]. Melanoma is one of the most dangerous skin cancer because it spreads quickly over other organs if not treated in its early stages [2].

Melanoma is produced by melanocytes, which are cells that give skin its color. As we mentioned, UV radiation is one of the main factors to developed melanoma because the energy can have enough energy to eliminate an electron-ionize from an atom or a molecule with the principal purpose of damage the DNA in cells, which in turn may lead to skin cancer [3].

Melanoma is detected by a visual examination of a medical specialist, and it is called the "ABCDE" criterion [4]. The medic evaluates the skin lesion characteristics: Assimetry, Border, Color, Diameter, and Evolution. For this reason, a computer medical assistant plays an important role to aid in a complementary diagnosis in order to help specialists make an early decision. When melanoma is

Y. Iano et al. (Eds.): BTSym 2020, SIST 233, pp. 463–476, 2021.
https://doi.org/10.1007/978-3-030-75680-2_52

diagnosed in its initial stages, there is an estimate of 5-year survival rate of 95% [5].

This paper is organized as follows: Sect. 2 presents the model and problem formulation of the system. Section 3 shows the design of a very deep neural network. Section 4 describes the experiments and discussion. Finally, the concluding remarks of the work are given.

2 System Model and Problem Formulation

In this section, we explain the mathematical model of the melanoma classification task. Let $X^i = (X^1, X^2, .., X^T), X^i \in R^{AxB}$ be the group of image samples and $y^i \in \{0, 1\}$ be the label of each instance 'i', where 1 represents melanoma and 0 non-melanoma. The resolution of the images are AxB pixels.

From the signal analysis perspective, the task is finding the optimal representation from the signal of interest, i.e., melanoma or not melanoma. Draw out the following mathematical representation $X = Ac$, where X is the signal of interest, A is a representation matrix, and c is the coefficient vector, also known as *feature vector*. Based on this analysis, we would have to assume the system linear; however, we will allow a certain degree of non-linearity with the commitment to provide a model with a high degree of accuracy. To do this, it is decided to divide the classification task into two stages: the feature engineering stage and the decision stage.

2.1 Feature Engineering Stage

This stage considers the application of digital image processing methods, which aim to extract feature vectors that allow optimal clustering of the two classes that try predicting the final mathematical model.

This stage is modeled as a multi-layer cascade system. A layer is defined as the core building block, in which a few digital image processing operations, e.g., linear and non-linear spatial filtering methods, are applied to a certain input signal. A basic operational unit is modeled as "Rectified-Correlations on a Sphere" (RECOS) [6].

Let $x = (1, x_1, x_2, ..., x_N)$ be the input image patch plus a bias term, whose coefficients are arranged by the neighborhood pixels around a certain center point within the image represented as a matrix symbolically, and $a_k^l \in R^{N+1}, k = 1, 2, .., k, ., K)$ be the K anchor vectors, so-called as *kernel* or *filter mask* of the respective layer 'l', $l = (1, 2, ..., L)$. Remark that the bias weight is included in each a_k^l vector. It is defined an scalar output z expressed in (1) as the dot product of vectors x and a_k^l:

$$z = x^T a_k^l \tag{1}$$

This value is associated with the degree of similarity that exists between two vectors within a sphere. Likewise, the negative value of y must be prevented by applying nonlinear clipping function f^{Rec} [6], e.g., mostly used the ReLU function, and it is expressed in (2):

$$z_r = f^{Rec}(x^T a_k^l) \tag{2}$$

The value of N is set as the dimension of an anchor vector a_k^l, thereby the dimension of the resulted matrix depends on the chosen displacement, so-called *stride*, applied to the correlation process to an entire image. For instance, if the stride is two, the resulting matrix dimension would be half of the input image dimension.

The RECOS transform unit B_l for a layer 'l' encompasses the correlation process and the element-wise rectification function operating on the corresponding input x_{l-1}. It is expressed as:

$$B_l = R \circ A_l, \ where \ A_l{}^T = [a_1^l, a_2^l, .., a_K^l] \tag{3}$$

Let $A_l \in R^{KxN}$ be the feature maps and R is a matrix that represents the element-wise rectification function for a layer 'l'.

Finally, the mathematical expression of the feature vector $x_f \in R^2$ is expressed in (4). This 2-dimensional vector represents the output of the 'L−1' layer. Its dimension refers to the number of classes to predict; we consider two classes. Clearly, B_L is equal to x_f.

$$x_f = B_{L-1} \cdots B_l \cdots B_1 x \tag{4}$$

2.2 Decision Stage

At this stage, the logistic regression model [7] is used for the prediction model. This model is thought to formulate the system's decision-making mathematically, whose main task is to classify two classes: melanoma or not melanoma.

Let $\hat{y}^i$ be the prediction of the outcome y^i for a instance 'i'. Hence, the logistic regression model prediction is expressed in (5):

$$\hat{y}^i = \begin{cases} 0, & \text{if } p < 0.5 \ (non - melanoma) \\ 1, & \text{if } p \geq 0.5 \ (melanoma) \end{cases} \tag{5}$$

Consider p as the estimated probability of a specific class from the extraction of the feature vector x_f. Equation (6) shows the mathematical relationship between them.

$$p = \sigma(x_f) \tag{6}$$

The function $\sigma(\cdot)$, expressed in (7), is a **sigmoid function** that outputs a number between 0 and 1 performing transition between both values linearly. This function is denoted in (7).

$$\sigma(h) = \frac{1}{1 + e^{-h}} \tag{7}$$

2.3 Problem Formulation

Once we have defined the modeling of the system, the type of problem to formulate is a parameter estimation problem. For this, a cost function 'J)' (log loss) is defined in (8). The model parameters are represented by a, which corresponds a_k^l, representing the K anchor vectors for a specific layer 'l'.

$$J(a) = -\frac{1}{T}\sum_{i=1}^{T}[y^i \log \hat{p}^i + (1 - y^i) \log (1 - \hat{p}^i)] \tag{8}$$

Finally, the problem becomes a cost function minimization problem, as expressed in (9) since it is of the utmost importance to minimize the prediction error, for each input instance 'i', with arguments that depend on the anchor vectors a.

$$\min_a J(a) \tag{9}$$

3 Design of a Very Deep Neural Network

In this section, we describe the overall design of three different very deep neural networks. The proposed system sketches a convolutional neural network (CNN), which follows the RECOS modeling. A brief introduction to the CNN concept will be described below, as well as the conceptual explanation of the three different very deep neural networks used in this work.

3.1 Convolutional Neural Network

Convolution Neural Network is a particular category of neural networks for understanding image content or detecting an object within an image or even a region within an image, drawn out the main ideas of a *neocognitron* [8] inspired by the experiments of scientists David H. Hubel and Torsten Wiesel [9]. It has shown extraordinary work in image segmentation, classification, and detection related tasks [10,11].

CNN-based models are a considerable breakthrough in the field of image processing and computer vision because it required a bare minimum of image processing compared to the other prevalent techniques of the time. The successful achievement of CNN's has captured the attention of academia and the industry companies for investigating new CNN designs.

CNN models usually consist of convolutional layers, hidden layers, pooling layers, and fully connected layers. It contains a trainable filter mask per layer. These are known as *receptive field*, which symbolically represents the anchor vectors, that is scanned over a layer below it to compute an output feature map. The output map is then sub-sampled using sum or max pooling, the latter being more common in order to reduce sensitivity in the upper layers. This process is varied up to some point when the features become specific to the problem at hand. Therefore, CNN is a deep neural network that increasingly learns more and more abstract dense features that can later be used as feature detectors for recognition problems.

3.2 AlexNet

AlexNet was the first deep CNN model for object detection and image classification. In 2012, it won the ImageNet Large Scale Visual Recognition Challenge (ILSVRC) by achieving a 15.3% in top-5 test error rate [12]. The main difference with previous CNN architectures, e.g., LeNet [13], is its network depth, which allows a significant amount of adjustable parameters. The network architecture contains eight layers from which five are convolutional and three fully connected (FC) as shown in Fig. 1.

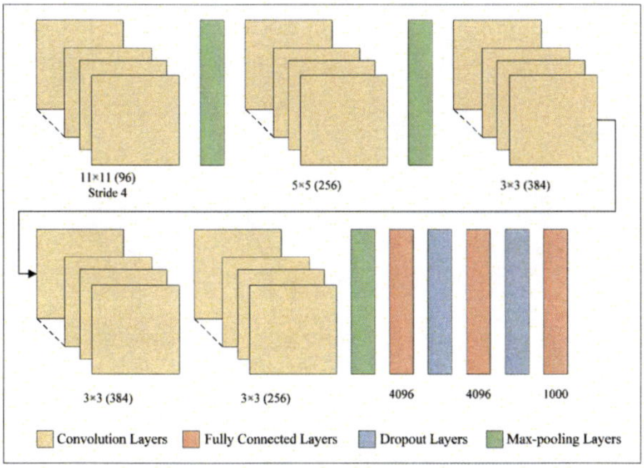

Fig. 1. The AlexNet architecture [14].

The first two FC layers are used with dropout to reduce over-fitting allowing the network to perform better generalizations to previously unknown inputs. The final FC layer is fed to a 1000-way SoftMax [14]. The first layer filters the input image with 96 kernels size $11 \times 11 \times 3$ and performs convolution and max pooling operations. In the second layer, the same operations are performed with 256 kernels of size $5 \times 5 \times 48$ over the output of the first convolutional layer. The third, fourth, and fifth convolutional use 3×3 filters with 384, 384, and 256 kernels respectively [12].

This architecture has multiple applications in recent deep learning developments. A variation of AlexNet is used in Korean character recognition showing an average of 90.12% success rate over commercial OCR programs [15]. Similarly, CNN's are widely used in biometrics like palmprint recognition [16] and finger vein validation systems [17]. Here AlexNet is implemented, replacing common computer vision combinations of descriptors and algorithms, showing much better results in feature extraction and, therefore, accuracy. A few papers have discussed the implementation of AlexNet architecture for disease leaf classification [18,19]. Other applications of AlexNet take place in the medical field to diagnose benign and malignant thyroid lesions [20] and breast cancer [21].

3.3 VGG16 and VGG19

VGGnet is one of the most popular CNN models since its introduction in 2014. It uses small-sized convolutional kernels, unlike other CNNs. It strictly uses 3×3 convolution kernels with 2×2 max pooling layers [22]. The fact that VGGnet uses small-sized filters results in very deep networks. There are five different VGG architectures that differ only in their depth. The most popular configurations are D and E. VGGnet won second place in the ILSVRC-2014 challenge with a 7.3% test error using an ensemble of 7 models implementing variations of C, D, and E configurations. After the submission, they manage to reduce it to 6.8% using an ensemble of 2 models, configurations D and E. VGG16, i.e., D configuration, has 13 convolutional layers and 3 FC layers, a total of 16 layers [23], as shown in Fig. 2.

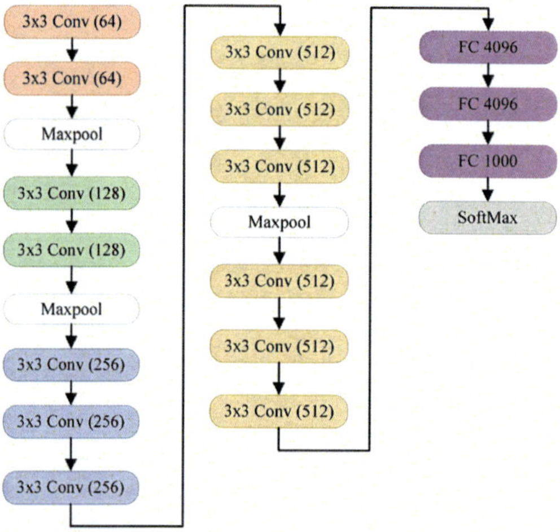

Fig. 2. VGG-16 architecture [23].

On the other hand, VGG19 (see Fig. 3), i.e., E configuration, has a total of 19 layers from which 16 are convolutional and 3 FC layers. This configuration is the most commonly used because in [23] it is evident that the classification error notably decreases as the depth of the networks increases from 11 to 19 layers.

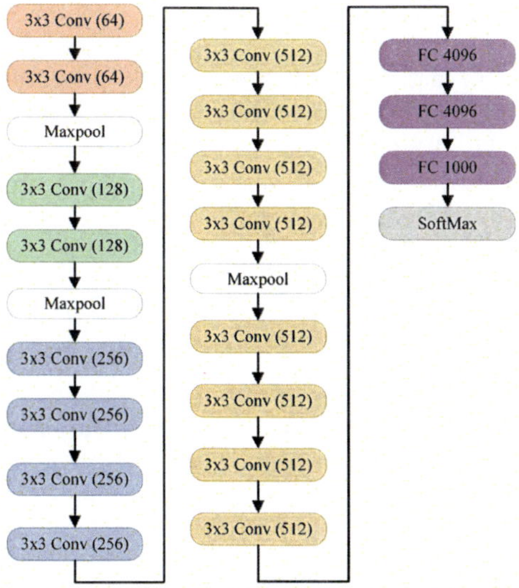

Fig. 3. VGG-19 architecture [23].

VGG16 and VGG19 have been implemented in multiple deep learning applications. In [24], the architecture VGG-19 is used to predict melanoma thickness since the prognosis of patients depends on it. In [25], a system based on VGG19 is used for musculoskeletal abnormality detection in four types of study: elbow, wrist, finger, and humerus. Also, VGG16 has huge acceptance in the industry where applications for PCB error inspection are being developed [26]. Moreover, because of its performance VGGnet is widely use in multiple fields for tasks like pathological brain image categorization [27], tobacco leaf pests classification [28], traffic sign detection [29], etc.

3.4 Related Works

The use of CNNs captured the attention of many researchers that work in different problems like segmentation, detection, and classification of skin cancer due to the high-level ability to extract features through the semantic and texture, and low-level features from edges and shapes automatically [30]. Kawahara et al. [31] proposed the classification of ten types of skin lesions through a multi-tract CNN. Ebtihal et al. [32] used support vector machines to proposed a classification method to identify melanoma. Skin lesion classification using CNN's was implemented by Esteva et al. [33] and Ulzii et al. [34]. Mahbod et al. [35] show a fully automatic computerized method for skin lesion classification using transfer-learning through AlexNet, VGG16, and ResNet-18, as a deep feature generator.

3.5 Gradient Descent Method

Once the architecture of the three different very deep neural networks is set, a generic algorithm is proposed to resolve the unconstrained optimization problem in (9) based on supervised learning by the application of the back-propagation algorithm. The batch gradient descent method is the chosen one, and the partial derivatives of the cost function with regards the model parameters a (anchor vectors) are expressed in (10).

$$\frac{\partial J}{\partial a_k^l} = \frac{1}{T/t} \sum_{i=1}^{T/t} (\hat{y}^i - y^i) x_k^l \tag{10}$$

Being the batch processing method considered, the calculation of the cost function and the partial derivatives are referenced on a portion of the total number of instances to train. Hence, the variable t is being referred to.

4 Experiments and Discussion

In this section, we experiment with the three different architectures defined above and evaluate their performance for the classification task in terms of the accuracy and loss metric of the prediction model. We adopt the Keras packet [36] as a deep learning framework, and the Matplotlib package [37] for visualizing the results. Also, the ISIC dataset [38] is considered for the training process.

In order to validate the prediction mode, the cross-validation technique is performed. This technique split the dataset into three different subsets: training set, validation set, and test set. Over the total images, represented as $T = 9044$, we choose 8244 (4122 represents melanoma and the remaining images the non-melanoma) for the training process, 600 images for the validation set (300 for each class), and finally, 200 images in the test set (100 for each class). In order to improve our prior work [39], the training set increased, and each class maintained the same quantity of images to solve the imbalance problem into the classes with the purpose to do not affect the final results.

The motivation of the following experiments attempts to analyze the repercussion on the performance of different arrangements of the multi-layer cascade system. Lastly, to optimize and evaluate the learning process, we compared the performance of three state-of-the-art optimizers: Adam, RMSprop, and SGD [10].

4.1 AlexNet Experiments

We configure a random network initialization for the Alexnet experiment. Table 1 shows the performance of the three optimizers in which the Adam optimizer has a better performance in the validation and test set than the other optimizers.

Hence, Fig. 4a shows the accuracy values by each epoch of training for the Adam optimizer. We observe that the training and validation curve increase

Table 1. Alexnet - Accuracy and loss comparisons of the different types of optimizer into the validation and test sets

		Adam	RMSprop	SGD
Validation	Acc (%)	93.17	92.50	93.00
	Loss	0.0683	0.0750	0.0700
Test	Acc (%)	92.50	91.50	92.33
	Loss	0.0750	0.0850	0.0767

together until epoch 20, when the two curves begin to separate. For this case, it is necessary to check the log loss function shown in Fig. 4b in order to avoid over-fitting. For this reason, the early stopping function was applied to stop the training.

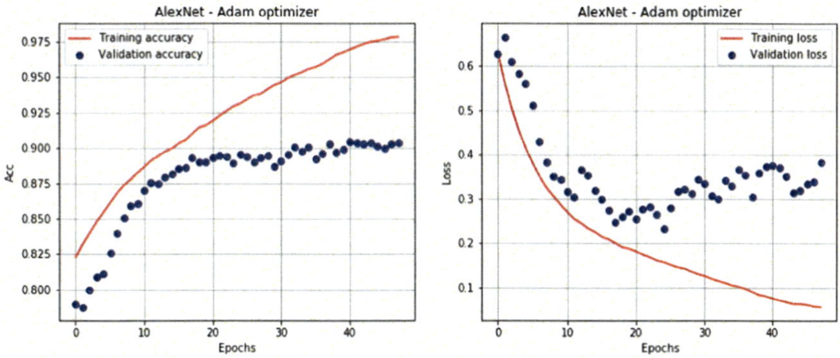

Fig. 4. Performance of AlexNet architecture with the Adam optimizer. (a) Accuracy, (b) Loss

4.2 VGG16 Experiments

We configured the network Imagenet-weights based initialization [40] for the VGG16 experiment. Table 2 gives the performance of the accuracy and the loss metric of the validation and the test set. We analyze that the better optimizer at the moment to evaluate the performance in the validation set is the Adam optimizer, but if we analyze the results over the test set, we see that RMSprop is the best optimizer at the moment to classify new images.

We observe that the behavior of accuracy of the training and the validation curves, shown in Fig. 5a, starts to converge around epoch 10, but to prevent a possible over-fitting, we also observe the log function, shown in Fig. 5b. The training loss presents its values decaying over the epochs, but the validation

Table 2. VGG16 - Accuracy and loss comparisons of the different types of optimizer into the validation and test sets

		Adam	RMSprop	SGD
Validation	Acc (%)	93.50	92.50	92.00
	Loss	0.0650	0.0750	0.0800
Test	Acc (%)	90.83	91.00	89.33
	Loss	0.0917	0.0900	0.1067

loss has a variation of the error values between 0.20 and 0.30. This may happen when the validation set does not have enough information to evaluate the learning. Also, we appreciate around epoch 23 an increase of the error value prognosticating the over-fitting of the prediction model with the dataset. For this reason, we employed early stopping, which allows evaluating the learning stage in each epoch and saves the better values of the accuracy and the loss.

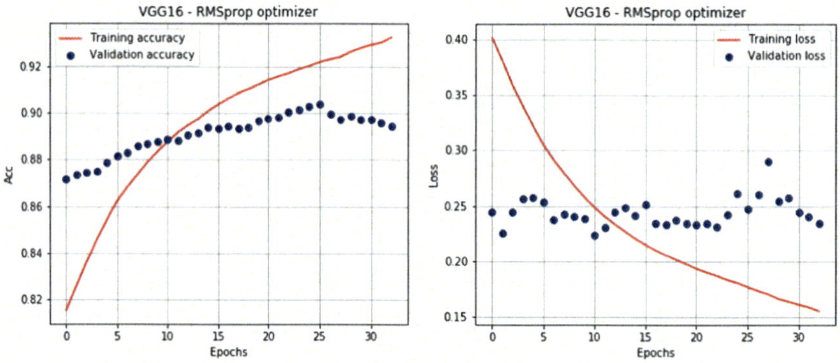

Fig. 5. Performance of VGG16 architecture with the RMSprop optimizer. (a) Accuracy, (b) Loss

4.3 VGG19 Experiments

We configured the network Imagenet-weights based initialization [40] for the VGG19 experiment. According to Table 3, the Adam optimizer presents better performance results in the validation set, but the RMSprop achieves 91.17% of accuracy in the test set.

Similarly, the figures for the performance of accuracy and loss were smoothed to appreciate the learning curves. Figure 6a shows the accuracy of the training and the validation set where we appreciate that it is better than the last architecture tested because when the two curves converge around epoch 11, the validation curve shows values very similar to the training set, thereby performing an excellent generalization of the prediction model.

Table 3. VGG19 - Accuracy and loss comparisons of the different types of optimizer into the validation and test sets.

		Adam	RMSprop	SGD
Validation	Acc (%)	94.00	93.00	88.00
	Loss	0.0600	0.0700	0.1200
Test	Acc (%)	91.00	91.17	87.00
	Loss	0.0900	0.0883	0.1300

Figure 6b shows the loss function where both training and validation curves show a similar decrease. However, around epoch 21 the early stopping command is applied.

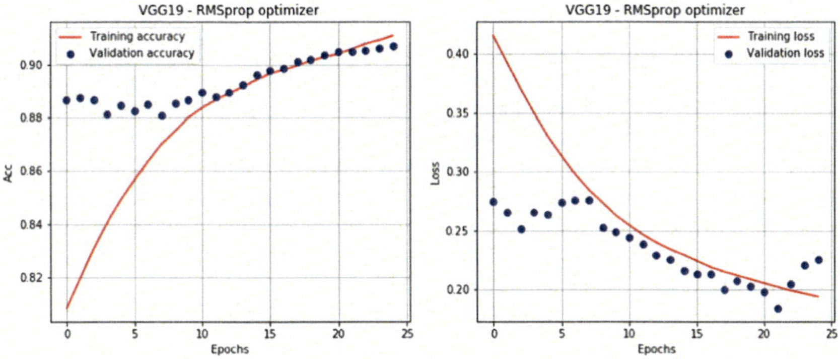

Fig. 6. Performance of VGG19 architecture with the RMSprop optimizer. (a) Accuracy, (b) Loss

5 Conclusion

In this paper, we propose a method for melanoma or non-melanoma classification based on the RECOS model. To seek the generalization of the proposed prediction model, three different very deep neural networks were tested to analyze the impact of many layers in cascaded. The AlexNet, VGG16, and VGG19 architectures are selected for this purpose. Experimental results indicate that the model generalization with the VGG19 was better than the other architectures. However, the stacking of much larger did not improve the accuracy of the system for the three architectures.

References

1. World Health Organization (2017) Skin Cancers. https://www.who.int/uv/faq/skincancer/en/index1.html

2. Melanoma - the skin cancer foundation (2018). https://www.skincancer.org/skin-cancer-information/melanoma/
3. American Cancer Society (2017) Ultraviolet (UV) Radiation. https://www.cancer.org/cancer/cancer-causes/radiation-exposure/uv-radiation.html
4. Nachbar F, Stolz W, Merkle T, Cognetta AB, Vogt T, Landthaler M, Bilek P, Braun-Falco O, Plewig G (1994) The ABCD rule of dermatoscopy. J Am Acad Dermatol 30(4):551–559
5. American Cancer Society (2017) What is melanoma skin cancer?: What is melanoma? https://www.cancer.org/cancer/melanoma-skin-cancer/about/what-is-melanoma.html
6. Kuo CCJ (2016) Understanding convolutional neural networks with a mathematical model. J Vis Commun Image Represent 41:406–413
7. Hosmer DW Jr (2013) Applied logistic regression. Wiley, Hoboken
8. Fukushima K (1980) Neocognitron: a self-organizing neural network model for a mechanism of pattern recognition unaffected by shift in position. Biol Cybern 36(4):193–202
9. Hubel DH, Wiesel TN (1962) Receptive fields, binocular interaction and functional architecture in the cat's visual cortex. J Physiol 160(1):106–154
10. Shrestha A, Mahmood A (2019) Review of deep learning algorithms and architectures. IEEE Access 7:53040–53065
11. Khan A, Sohail A, Zahoora U, Qureshi AS (2019) A survey of the recent architectures of deep convolutional neural networks. arXiv preprint arXiv:1901.06032
12. Krizhevsky A, Sutskever I, Hinton GE (2017) ImageNet classification with deep convolutional neural networks. Commun ACM 60(6):84–90 https://dl.acm.org/doi/10.1145/3065386
13. LeCun Y, Bottou L, Bengio Y, Haffner P (1998) Gradient-based learning applied to document recognition. Proc IEEE 86(11):2278–2323
14. Khan S, Rahmani H, Shah SAA, Bennamoun M (2018) A guide to convolutional neural networks for computer vision. Synth Lect Comput Vis 8(1):1–207
15. Lee SG, Sung Y, Kim YG, Cha EY (2018) Variations of AlexNet and GoogLeNet to improve Korean character recognition performance. J Inf Process Syst 14(1):205–217
16. Gong W, Zhang X, Deng B, Xu X (September 2019) Palmprint recognition based on convolutional neural network-AlexNet. In: Proceedings of the 2019 federated conference on computer science and information systems, FedCSIS 2019. Institute of Electrical and Electronics Engineers Inc., pp 313–316
17. Fairuz S, Habaebi MH, Elsheikh EMA (November 2018) Finger Vein identification based on transfer learning of AlexNet. In: Proceedings of the 2018 7th international conference on computer and communication engineering, ICCCE 2018. Institute of Electrical and Electronics Engineers Inc., pp 465–469
18. Arya S, Singh, R (February 2020) A comparative study of CNN and AlexNet for detection of disease in potato and mango leaf. Institute of Electrical and Electronics Engineers (IEEE), pp 1–6
19. Nachtigall LG, Araujo RM, Nachtigall GR (January 2017) Classification of apple tree disorders using convolutional neural networks. Institute of Electrical and Electronics Engineers (IEEE), pp 472–476
20. Sun J, Sun T, Yuan Y, Zhang X, Shi Y, Lin Y (January 2019) Automatic diagnosis of thyroid ultrasound image based on FCN-AlexNet and transfer learning. In: International conference on digital signal processing, DSP, vol 2018-November. Institute of Electrical and Electronics Engineers Inc

21. Titoriya A, Sachdeva S (November 2019) Breast cancer histopathology image classification using AlexNet. In: 2019 4th international conference on information systems and computer networks (ISCON). IEEE, pp 708–712
22. Venkatesan R, Li B (2017) Convolutional neural networks in visual computing: a concise guide
23. Simonyan K, Zisserman A (2015) Very deep convolutional networks for large-scale image recognition. In: 3rd international conference on learning representations, ICLR 2015 - conference track proceedings, pp 1–14
24. Jaworek-Korjakowska J, Kleczek, P, Gorgon M (June 2019) Melanoma thickness prediction based on convolutional neural network with VGG-19 model transfer learning. In: 2019 IEEE/CVF conference on computer vision and pattern recognition workshops (CVPRW). IEEE, pp 2748–2756
25. Mondol TC, Iqbal H, Hashem MM (September 2019) Deep CNN-based ensemble CADx model for musculoskeletal abnormality detection from radiographs. In: 2019 5th international conference on advances in electrical engineering, ICAEE 2019. Institute of Electrical and Electronics Engineers Inc., pp 392–397
26. Li YT, Guo JI (August 2018) A VGG-16 based faster RCNN model for PCB error inspection in industrial AOI applications. In: 2018 IEEE International Conference on Consumer Electronics-Taiwan, ICCE-TW 2018. Institute of Electrical and Electronics Engineers Inc
27. Kaur T, Gandhi TK (March 2020) Automated brain image classification based on VGG-16 and transfer learning. Institute of Electrical and Electronics Engineers (IEEE), pp 94–98
28. Swasono DI, Tjandrasa H, Fathicah C (July 2019) Classification of tobacco leaf pests using VGG16 transfer learning. In: Proceedings of 2019 international conference on information and communication technology and systems, ICTS 2019. Institute of Electrical and Electronics Engineers Inc., pp 176–181
29. Qiao K, Gu H, Liu J, Liu P (December 2017) Optimization of traffic sign detection and classification based on faster R-CNN. In: Proceedings - 2017 international conference on computer technology, electronics and communication, ICCTEC 2017. Institute of Electrical and Electronics Engineers Inc., pp 608–611
30. Xie F, Fan H, Li Y, Jiang Z, Meng R, Bovik A (2017) Melanoma classification on dermoscopy images using a neural network ensemble model. IEEE Trans Med Imaging 36(3):849–858
31. Kawahara J, Hamarneh, G (October 2016) Multi-resolution-tract CNN with hybrid pretrained and skin-lesion trained layers
32. Al-Mansour EA, Jaffar MA (2016) Classification of dermoscopic skin cancer images using color and hybrid texture features
33. Esteva A, Kuprel B, Novoa R, Ko J, Swetter S, Blau H, Thrun S (2017) Dermatologist-level classification of skin cancer with deep neural networks. Nature 542:115–118
34. Dorj UO, Lee KK, Choi JY, Lee M (2018) The skin cancer classification using deep convolutional neural network. Multimed Tools Appl 77:9909–9924
35. Mahbod A, Schaefer G, Wang C, Ecker R, Ellinge I (2019) Skin lesion classification using hybrid deep neural networks. In: ICASSP 2019 - 2019 IEEE international conference on acoustics, speech and signal processing (ICASSP), pp 1229–1233
36. Keras (2015) https://keras.io
37. Hunter JD (2007) Matplotlib: a 2D graphics environment. Comput Sci Eng 9(3):90–95
38. Public skin lesion image archive (2018). https://www.isic-archive.com

39. Minango P, Iano Y, Monteiro A, França R, Oliveira G (2019) Automatic skin lesions classification from dermoscopic images employing deep learning. Set Int J Broadcast Eng Set IJBE 5:108–115
40. Deng J, Dong W, Socher R, Li LJ, Li K, Fei-Fei L (2009) ImageNet: a large-scale hierarchical image database. In: CVPR 2009

A Multi-criteria Modelling for Ranking CO_2 Emitting G20 Countries from the Kaya Identity and Their Impacts on Elderly Health

Leila Matos Abreu[1], Henrique Rego Monteiro da Hora[1](✉), João José Assis Rangel[1], Milton Erthal Jr[1], Navid Razmjooy[2], Vania Vieira Estrela[3], Thierry Oscar Edoh[4], Gabriel Gomes de Oliveira[5], and Yuzo Iano[5]

[1] Instituto Federal de Educação Ciência e Tecnologia Fluminense, Campos dos Goytacazes, RJ, Brazil
joao.rangel@ucam-campos.br
[2] Department of Electrical Engineering, Tafresh University, Tafresh, Iran
[3] Department of Telecommunications, Fluminense Federal University, Niteroi, RJ, Brazil
[4] Department of Applied Software Engineering, Technical University of Munich, Munich, Germany
edoh@in.tum.de
[5] School of Electrical and Computer Engineering, State University of Campinas, Campinas, SP, Brazil
oliveiragomesgabriel@live.com, yuzo@decom.fee.unicamp.br

Abstract. Understanding the factors driving CO_2 emissions and their spatial behavior is vital to lessen natural and anthropogenic emissions. Lessen CO_2 emissions can impact elderly health, especially those living in affected regions. The constant search for sustainable development motivated the elaboration of this work, whose purpose is to build a ranking of the most sustainable countries concerning CO_2 emissions, according to the criteria established in the Kaya identity indicator. Furthermore, this study aims to push to enact policy for improving the population health, elderly ones. For the study, the economic block G20 was selected due to its representativeness in the world scenario, with the application of the multi-criteria method PROMETHEE II for the country order, whose data were obtained from the International Energy Agency. As a result, the five best positions in the ranking are represented by countries with excellent economic projections and energy efficiency according to the criteria.

Keywords: CO_2 emissions · Kaya identity · G20 · PROMETHEE II · Green design · Smart cities · Pollution surveillance

1 Introduction

The issues regarding sustainable development, which intends to achieve a balance between economic, social, and environmental aspects, became relevant to humanity

Y. Iano et al. (Eds.): BTSym 2020, SIST 233, pp. 477–487, 2021.
https://doi.org/10.1007/978-3-030-75680-2_53

from the reflexes of its anthropic actions. With industrialization and the advance of the economy, one of the main hurdles for the planet's sustainability is Greenhouse Gases, particularly carbon dioxide. However, recent and harsher environmental and economic programs have been implemented, aiming to reduce these emissions. Current studies are demonstrating a decoupling trend between various countries' CO_2 emissions and their GDP. This behavior is observed when pressure or environmental impact is lower than their economic strength, pointing to economic growth in developed countries without a compelling growth in their CO_2 emissions [1].

The G20, or Group of Twenty, assembled on September 25, 1999, in the face of the 1990's financial crises, is an international forum of economic and financial cooperation between developed and emergent countries, aiming to establish a broader dialogue about the global economy. These countries' integration aims to promote sustainable growth coupled with economic stability through financial programs and regulations that aim to reduce the risk of crisis in an international scenario [2].

This group consists of the European Union and the nineteen major world economies: Argentina, Australia, Brazil, Canada, China, France, Germany, India, Indonesia, Italy, Japan, Mexico, Russia, Saudi Arabia, South Africa, South Korea, Turkey, United Kingdom and the United States of America. Together, the member nations of G20 represent approximately 90% of the world GDP, 80% of international trade, two-thirds of the world population, and 84% of the greenhouse gases emissions, granting them significant input on the financial system management and global economy [3].

The Multi-Criteria Decision Making Methods (MCDMs) consist of methodological formulations or theories, with well-defined axiomatic structures that can be utilized to elaborate a decision model aiming to solve a problem [4]. The Kaya identity modeling provides robust utility by allowing the calculation of emissions based on populational growth, economic growth, energetic consumption by GDP per unit, and CO_2 emissions by energy per unit predictions [5]. This tool has been utilized to project greenhouse gases future emissions in various climatic projections.

In this context, this article ranks the most sustainable countries regarding their CO_2 emissions and according to the criteria addressed by the Kaya identity indicator, with analyses of economic, populational, energetic, and carbon dioxide emissions criteria. The MCDM model used was the Preference Ranking Organization Method for Enrichment Evaluation (PROMETHEE II) to ordinate the member nations of G20 due to their representativeness in the global scenario. 3D GIS modeling practices integrating geospatial methodologies, e.g., photogrammetry and laser scanning, help to intensify the effectiveness of environmental documentation and preservation. This struggle brings up technological approaches that eventually will lead to better administration of CO_2. The present study will be employed in emission inventories, along with maps, remote sensing techniques, and a modification of the Kaya identity in a further study about municipalities. The authors will need to combine these components, incorporating the applicable criteria that follow.

Section 2 discourses about the Kaya identity methodology. Section 3 shows some experimental results and discusses them. Section 4 concludes this paper.

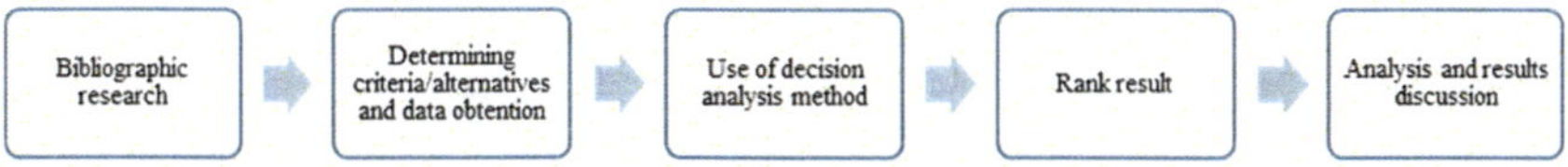

Fig. 1. Research steps.

Table 1. Evaluation criteria

	Criteria	Description	Weight
C1	CO_2 emissions	This criterion will evaluate the amount of carbon dioxide emissions by fossil fuel. The country with the least amount of emissions will be considered the most sustainable. It is a minimization criterion, that is, the least the better	1.0
C2	GDP per capita	This criterion will evaluate the economic growth concerning the population. The country with the best relation will be considered the most sustainable. It is maximization criteria, i.e., the bigger, the better	1.0
C3	Energetic intensity	This criterion is represented by the ratio between the total amount of energy generated by a country and its GDP. It shows the degree of efficiency between the energy utilization and the country's wealth. The country with the best efficiency will be considered the most sustainable. It is a minimization criterion, that is, the lower, the better	1.0
C4	Carbon intensity	This criterion is represented by the ratio between emissions and total energy consumption, evaluating the energetic efficiency. The country with lower results will be considered the most sustainable. It is a minimization criterion, that is, the lower, the better	1.0

2 Methodology

This article details some steps to build a rank of the most sustainable countries in light of established criteria. The research was organized and structured with the following sequential steps presented in Fig. 1.

With the objectives determined for the application of the decision analysis method, we selected as alternatives member nations from G20, and the criteria were determined according to the Kaya identity indicator. Data were obtained in the public database from IEA – International Energy Agency, for the year 2016 [6, 7].

Kaya identity is an extension of the IPAT identity, in which population (P), affluence (A), and technology (T) impact the environment (I). Affluence is related to the consumption or income/product per capita [8]. The Kaya identity [9] is a mathematical decomposition used to establish a relationship between carbon dioxide emissions produced by human activity and four important factors: demographic, economic, energy intensity, and carbon intensity, as seen in (1) [10].

$$C = P \times \frac{PIB}{P} \times \frac{E}{PIB} \times \frac{C}{E} \tag{1}$$

C is the carbon dioxide emission rate (Mt CO_2); P is the population by millions of people; PIB/P is the ratio per capita of the GDP (1 billion dollars in 2010 per person);

and the E/PIB ratio is the energetic intensity, the generation of primary energy (Mtoe) per GDP unit (toe/a thousand dollars in 2010), with toe – tonne of oil equivalent; and the C/E ratio is the carbon intensity, the carbon emissions per energy unit generated (tCO_2/toe).

Table 2. Payment matrix

Criterions/ alternatives	CO_2 emission (min)	GDP per capita (max)	Energy intensity (min)	Carbon intensity (min)
Argentina	191	10.11	0.19	2.22
Australia	392	60.88	0.09	3.02
Brazil	417	10.81	0.13	1.46
Canada	541	50.78	0.15	1.93
European Union	1470	29.26	0.10	2.06
France	293	41.96	0.09	1.20
Germany	732	46.12	0.08	2.36
India	2077	1.86	0.35	2.41
Indonesia	455	3.98	0.22	1.98
Italy	326	34.11	0.07	2.16
Japan	1147	47.66	0.07	2.69
Korea	589	25.61	0.22	2.09
Mexico	445	10.32	0.15	2.41
China	9057	6.89	0.31	3.06
Russia	1439	11.31	0.45	1.97
Saudi Arabia	527	21.59	0.30	2.51
South Africa	414	7.50	0.33	2.96
Turkey	339	14.38	0.12	2.47
United Kingdom	371	41.79	0.06	2.07
United States	4833	52.38	0.13	2.23

This research utilized the LTI method, an MCDM of the PROMETHEE family, in particular, developed by PROMETHEE II [11], which establishes a complete order between alternatives destined to sorting problems. The software we utilized for the application was Visual PROMETHEE Academic Edition.

As established, the model consists of twenty alternatives and four criteria. Table 1 presents the evaluation criteria with their respective descriptions and weights. There is no weight distinction between criteria. Table 2 shows the data from the payment matrix of the alternatives - the member nations of G20 - with a group of four criteria, the quotient of the determining factors of Kaya identity. This study considered four of the seven criteria from the indicator, not analyzing population, GDP, and primary energy in isolation, considering that these criteria are part of the ratios analyzed in the indicator, such as GDP per capita, energetic intensity, and carbon intensity. Only carbon emissions were utilized individually, considering they are the central part of the problem. The method calculation consists of 4 steps [12].

Firstly, the difference between the performances (δ_{ik}) of alternative x_i with alternative x_k regarding criteria j, and the relative preference function (P) of each j criteria is determined. Relative preference is given by Eq. (2).

$$P(x_i, x_k) = P_j\big(u_j(x_i) - u_j(x_k)\big) = P_j(\delta_{ik}) \tag{2}$$

The difference between each pair of alternatives is denoted by Eq. (3).

$$\delta_{ik} = u_j(x_i) - u_j(x_k) \tag{3}$$

Next, the preference index (S_{ik}) of the alternative x_i is compared to the other alternatives x_k, as seen in Eq. (4). w_j is the weight of each criterion, that is, the importance given to the criteria.

$$S_{ik} = \frac{\sum_j w_j \cdot P(\delta_{ik})}{\sum_j w_j} \tag{4}$$

The alternative ranking is done based on the decreasing order of their respective liquid flows, in which the more prominent the flow, the better is the alternative performance, and, if the liquid flows are the same, the alternatives are neutral between themselves.

The third step represents the overflow. Equation (5) has the positive overflow, or outflow, which represents the preference intensity of an alternative over all other alternatives. Equation (6) has the negative overflow, or inflow, which represents the preference intensity of all other alternatives over one of the alternatives. The bigger the Φ^+, the better the alternative, and the lower the Φ^-, the better the alternative.

$$\Phi_i^+ = \sum_k S_{ik} \tag{5}$$

$$\Phi_{\mathrm{i}}^- = \sum_k S_{ki} \tag{6}$$

The fourth and last step allows us to obtain the general order of the alternative, utilizing the overflow or liquid flow given in Eq. (7), representing the results between power and weakness of the alternative.

$$\Phi(a) = \Phi^+(a) - \Phi^-(a) \tag{7}$$

3 Results and Discussion

The PROMETHEE II approach has produced a rank of the most sustainable countries in liquid flow decreasing order, that is, from the most sustainable to least sustainable

Table 3. Alternatives rank and PROMETHEE II liquid flow

Rank	Alternatives	ϕ
1	France	0.7113
2	United Kingdom	0.5875
3	Italy	0.5049
4	Canada	0.3285
5	Australia	0.2680
6	Brazil	0.2348
7	Germany	0.1773
8	Japan	0.0921
9	Turkey	0.0830
10	European Union	0.0631
11	Argentina	0.0542
12	United States	0.0244
13	Korea	−0.0755
14	Indonesia	−0.1462
15	Mexico	−0.1902
16	Russian	−0.3029
17	Saudi Arabia	−0.3188
18	South Africa	−0.4874
19	India	−0.7481
20	China	−0.8600

according to the established criteria, as seen in Table 3. Taking the spotlight in the first position, France is considered the most sustainable country according to the model applied, with a liquid flow of 0.7113. The twentieth and last position on the rank is China, which is considered the least sustainable country with a liquid flow of −0.8600.

Figure 2 represents a graphic of dispersion with an analysis of the 20 countries rank. The axis-ordinates represents the rank of the world's major economies, and in the axis-of-abscissas, the sustainability rank was obtained by the PROMETHEE model. Amongst the twenty alternatives, France obtained the best results, with the best performance in all criteria, taking the spotlight with the least CO_2 emissions and least carbon intensity. It is possible to observe in the graph that France is the sixth major world economy and the most sustainable country. The United Kingdom and Italy followed France with the best energetic intensity, both countries with positive performance in all criteria. On the other hand, China provides the worst performance in all criteria, with the most carbon dioxide emissions and carbon intensity. China takes second place in the GDP rank and the twentieth and last position in the rank of sustainable countries. India and Saudi Arabia also obtained negative performances in all criteria, emphasizing India as the worst economic relation per habitant. Analyzing the economy, the United States of America has the world's largest economy by GDP, and it is the second major carbon dioxide emitter, only behind China, but occupies the twelfth place in the sustainability rank due to its GDP per capita relation.

Regarding the CO_2 emission criteria, the most relevant to this study, apart from the countries already mentioned, the countries with the best performance in the order of liquid flow were Australia, Brazil, Turkey, Argentina, Indonesia, Mexico, and South Africa. Although they emit fewer greenhouse gases in the atmosphere, they all present strengths and weaknesses in the remaining criteria. Argentina had the most sustainability relevance regarding emissions, with the lowest emissions in the group, and Australia stands out in the GDP per capita relation. On the other hand, the countries with negative performance regarding CO_2 were Canada, Germany, Japan, the European Union, South Korea, and Russia. They presented positive and negative behaviors in the remaining criteria, even though they are the biggest offenders regarding emissions, with Russia having the worst energetic intensity. For the European Union, 28 countries are considered, excluding the ones with double participation on G20, like Germany, France, Italy, and the United Kingdom.

Besides that, the country with the lowest GDP, between the 20 economies, is South Africa, occupying the 18th spot on the sustainability rank, as we can see in Fig. 2. Argentina is the country with the lowest amount of CO_2 emission amongst the 20 countries, but occupies the 11th spot in the flow rank, considering its weak performance regarding its GDP per capita and energetic intensities, taking the second to last position amongst the twenty economies.

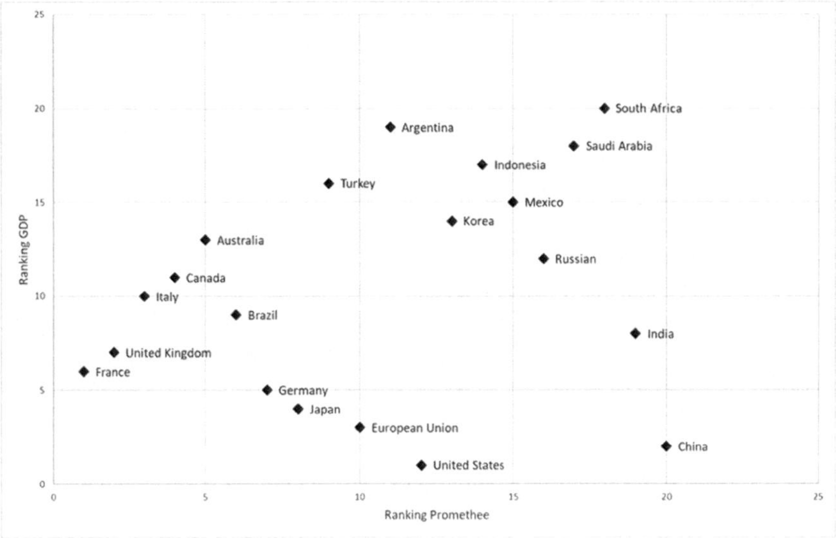

Fig. 2. Dispersion graphic between the GDP rank and the PROMETHEE II rank.

Kaya identity allows us to analyze the development of these indicators, being an essential tool to implement actions that contribute to the reduction or limitation of Greenhouse Gases emission more intensely and less expensively, aiming to revert the effects of the populational and economic growth.

Analyzing the model results with the intensity indicator, we can verify the ascending or descending behavior of the curves. This behavior indicates the relative situation of a country in a specific moment concerning the base-year of 1990 (index = 1). The period analyzed ranges from 1990 to 2016, with the data available on the IEA database. Below, in Fig. 3, we present the behavior of countries that occupied the first and last positions of our rank with the Kaya identity model.

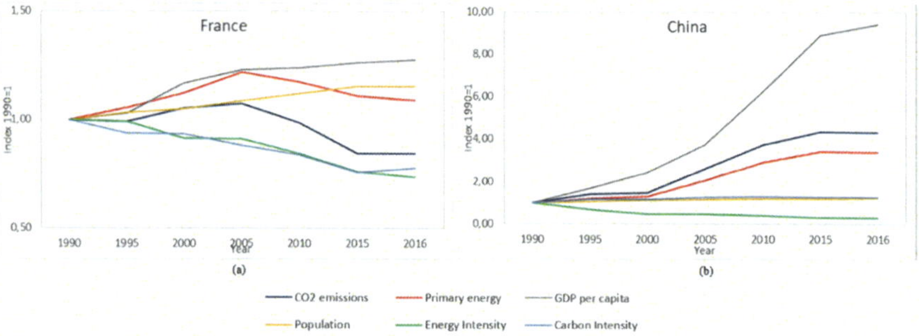

Fig. 3. Kaya indicators development for France and China.

France in 2016 was one of the countries with the least amount of CO_2 emissions, regarding the twenty analyzed countries, second only to Argentina, with 1.12% of the emissions. Hence, we can observe in Fig. 3 (a) the reduction in emissions, carbon, and energetic intensities throughout the years, maintaining its economic growth. On the other hand, with its relevant economic development and industrialization, China was responsible in 2016 for 34.76% of carbon dioxide emissions, taking the lead on harmful emissions amongst all countries in the world. This increases all analyzed factors, as shown in Fig. 3 (b), especially their GDP per capita and carbon intensities. In both cases, populational growth is constant.

Next, two analyses are presented. Figure 4 shows a correlation between the amount of carbon dioxide emitted by GDP with the amount of carbon dioxide per capita from 1990 to 2016, demonstrating a retraction from French emissions and a strong growth in emissions from China. Figure 5 shows a comparison of the two energy matrices from both countries, regarding their electric energy generation in GWh, from different sources. Nuclear energy in France is responsible for 72.49% of its energetic matrix, and with growing investments in technologies and clean energy, 17.97% belongs to renewable energies and 9.54% to fossil fuels and other sources. However, China utilizes 71.13% of its energetic matrix based on fossil fuels and 25.26% for renewable energies, with a growing trend for the latter, in order to mitigate the environmental effects. Only 3.61% belongs to nuclear energy and other sources.

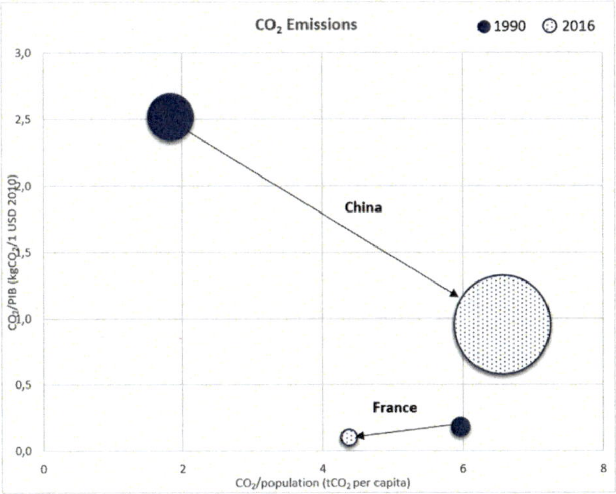

Fig. 4. CO_2 emissions intensity

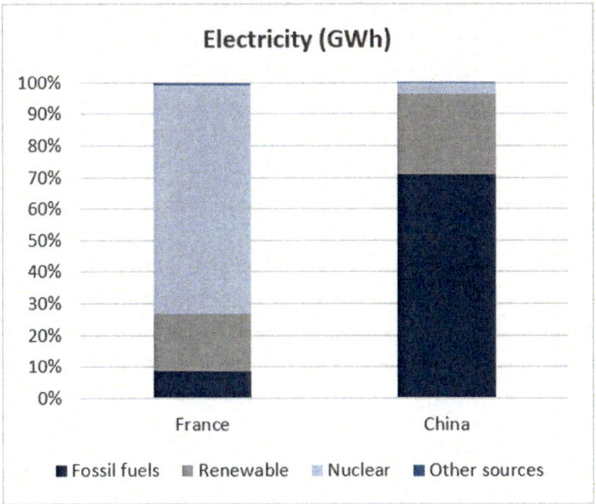

Fig. 5. Energetic matrix.

The economic growth of a country implies the increase of energy consumption and pressure growth in the environment. The efficiency of the energy sector and the substitution of fossil fuels to renewable energy is a necessary measure.

The environment is suffering from economic growth. The population's health is, in turn, severely impacted. Health conditions like chronic respiration diseases and asthma are the consequences. Elderly people are more affected by such a condition. Therefore, this tool will help choose a living place for elderly people and prevent them from any health complications.

4 Conclusion

Analyzing the aspects that result in CO_2 emissions and their spatial comportment are crucial tasks to minimize both the natural as well as anthropogenic emissions.

This study aimed to rank the G20 member nations from most sustainable to least sustainable, regarding their CO_2 emissions according to economic, populational, energetic, and CO_2 emissions criteria of the Kaya identity indicator. The objective was achieved by using the PROMETHEE II method, showing France in the first place and China in the last position.

With this result, we can verify that France adopts techniques that assure them the first place in sustainability amongst the world's largest economies, with an energetic matrix mostly consisting of nuclear energy and potential growth of renewable energies, along with politics and programs that promote environmental impacts reduction in climate change. These practices are part of the French project "Energy Transition for the Green Growth," which aims to make the country a world leader in sustainable development, reducing Greenhouse Gases emissions, diversifying their energetic model, and increasing the development of sustainable energies [13].

On the other hand, China, being the world's largest population and with notable economic growth, faces the big challenge of maintaining its economic development sustainably. One of the things that most contribute to the emissions of CO_2 is the use of coal in its energetic matrix, considering that China is the world's biggest coal producer and the world's biggest coal importer. However, China has the most number of installations for renewable energies, with almost 30% of all renewable energies such as solar, hydroelectric, and eolic installations being in China according to the Key World Energy Statistics report [7].

Finally, this study can provide a tool to ordinate results like these and analysis that aim to contribute to sustainability, such as determining better energetic sources, in the face of economic, environmental, and social aspects. In future applications, the authors intend to use this type of analysis to map and quantify urban carbon emissions using the Cyber-Physical System paradigm and the Internet of Things [14–16].

References

1. Wang Q, Li R, Liao H (2016) Toward decoupling: growing GDP without growing carbon emissions. Environ Sci Technol 50(21):11435–11436. https://doi.org/10.1021/acs.est.6b05150
2. G20 (2018). https://g20.org/en/
3. Itamaraty (2018). https://www.itamaraty.gov.br/pt-BR/politica-externa/diplomacia-economica-comercial-e-financeira/15586-brasil-g20
4. Almeida AT (2013) Processo de decisão nas organizações: construindo modelos de decisão multicritério. Atlas, São Paulo
5. Budzianowski WM (2013) Modelling of CO_2 content in the atmosphere until 2300: influence of energy intensity of gross domestic product and carbon intensity of energy. Int J Global Warm 5(1):1. https://doi.org/10.1504/IJGW.2013.051468
6. IEA - International Energy Agency (2018). https://www.iea.org/statistics/
7. Key World Energy Statistics (2018) (p. 51). https://www.iea.org/statistics/kwes/

8. Ehrlich PR, Holdren JP (1971) Impact of population growth. Sci New Ser 171:1212–1217
9. Kaya Y (1989) Impact of carbon dioxide emission control on GNP growth: interpretation of proposed scenarios. Paper presented at the IPCC Energy and Industry Subgroup, Response Strategies Working Group, Paris, France
10. Tavakoli A (2018) A journey among top ten emitter country, decomposition of "Kaya Identity." Sustain Cities Soc 38:254–264. https://doi.org/10.1016/j.scs.2017.12.040
11. Brans JP, Vincke PA (1985) Preference ranking organisation method: (the PROMETHEE method for multiple criteria decision-making). Manag Sci 31(6):647–656
12. Brans JP, Vincke P, Mareschal B (1986) How to select and how to rank projects: the PROMETHEE method. Eur J Oper Res 24(2):228–238. https://doi.org/10.1016/0377-2217(86)90044-5
13. French Government (2018). https://www.gouvernement.fr/en/energy-transition
14. Estrela VV, Saotome O, Loschi HJ, Hemanth DJ, Farfan W, Aroma RJ et al (2018) Emergency response cyber-physical framework for landslide avoidance with sustainable electronics. Technologies 6:42. https://doi.org/10.3390/technologies6020042
15. Estrela VV, Monteiro ACB, França RP, Iano Y, Khelassi A, Razmjooy N (2019) Health 4.0: applications, management, technologies and review. Med Tech J 2(4):262–276. https://doi.org/10.26415/2572-004X-vol2iss1p262-276. https://medtech.ichsmt.org/index.php/MTJ/article/view/205
16. Razmjooy N, Estrela VV (2019) Applications of Image Processing and Soft Computing Systems in Agriculture. IGI Global. https://doi.org/10.4018/978-1-5225-8027-0

Digital Garbage Bin Monitoring System (DGBMS)

A Smart Garbage Monitoring and Management Cyber-Physical System

Thiagarajan Yogamoorthi[1](✉), Vania Vieira Estrela[2], Thierry Oscar Edoh[3], Navid Razmjooy[4], Abdeldjalil Khelassi[5], Henrique Rego Monteiro da Hora[6], Gabriel Gomes de Oliveira[7], Gabriel Caumo Vaz[7], and Yuzo Iano[7]

[1] Department of Electrical and Electronics Engineering, Christ College of Engineering and Technology, Puducherry, India

[2] Department of Telecommunications, Fluminense Federal University, Niteroi, RJ, Brazil

[3] Department of Applied Software Engineering, Technical University of Munich, Munich, Germany
edoh@in.tum.de

[4] Department of Electrical Engineering, Tafresh University, Tafresh, Iran

[5] University of Abou Bekr Belkaid Tlemcen, Tlemcen, Algeria

[6] Instituto Federal de Educação Ciência e Tecnologia Fluminense, Campos dos Goytacazes, RJ, Brazil

[7] School of Electrical and Computer Engineering, State University of Campinas, Campinas, SP, Brazil
yuzo@decom.fee.unicamp.br

Abstract. In day-to-day life, all could witness the odor of scattered garbage and unkempt garbage bins because there is no proper time interval schedules and routines, which results in garbage overflow. It also creates hygienic problems, land pollution, and landscape unpleasantness. This scenario demands a system that observes the garbage bin status and provides evidence to the authorities administrating the collection intervals for emptying the bins. The so-called Digital Garbage Bin Monitoring System (DGBMS) is a smart garbage management system relying on the Internet of Things (IoT) technology that can solve this problem. This system monitors the garbage excess and the moisture in the garbage bin. In this project, a sensor network detects the garbage level and the garbage moisture in the dustbin via an Infrared (IR) sensor and then sends evidence to the decision-making party through an IP address. The UNO ARDUINO board interfaces the sensors with the IP address and handles the actuators. An RFID channel constantly monitors the necessary garbage information regarding the different locations of the wastebaskets. Developing countries, especially the rural areas, are severely concerned with these issues that can cause severe health problems among the population. Implementing this solution could contribute to addressing many health-related issues in these countries.

Keywords: Humanitarian engineering · Green design · Smart cities · Infrared sensor · Waste management · Radio Frequency Identification (RFID) · Sensor networks · Internet of Things · Cyber-physical system

Y. Iano et al. (Eds.): BTSym 2020, SIST 233, pp. 488–497, 2021.
https://doi.org/10.1007/978-3-030-75680-2_54

1 Introduction

Despite the existence of more advanced technologies in developing countries, public awareness of waste management remains low, which calls for citizen consciousness to safeguard a clean and healthy environment. During the last few years, the operational cost for solid waste management has increased gradually, and lots of solutions relied on setting the proper cleaning intervals to solve this problem. Nevertheless, visually checking the status of the bin is a challenging job. This paper uses a framework employing a Sensor Network that transmits data from the trash can to the control station through a gateway and then stores them in a corporation or institution database [1, 2].

In Coimbatore, the garbage bins from both public and private spaces can overflow. These unhygienic conditions may provoke several diseases, even with a regular garbage collection schedule, which varies according to the target population. This is quite frequent in festivals seasons or some functions, where people generate lots of garbage material in a specific area. This project aims to avoid such situations.

Section 2 presents a literary survey on the topic of waste management. The necessary design requirements are discussed in Sect. 3, while Sect. 4 depicts the materials and the methodology used. The proposed Digital Garbage Bin Monitoring System (DGBMS) is described in Sect. 5. Future works are shown up in Sect. 6, and the conclusions are presented in Sect. 7.

2 Literature Survey

A suitable site selection policy for solid waste disposal using a wireless remote control and Geographical Information System (GIS) comes into view in [3]. Landfill siting is a complicated and challenging practice that entails an evaluation of many different criteria. This work employs analyses for a multi-criteria decision on geospatial data for the best choice of landfill sites. Its two-stage analysis forms a Spatial Decision Support System (SDSS) intended for Solid Waste Management (SWM) in fast-growing urban regions.

An optimal location and a nearness distance estimation policies for Municipal Solid Waste (MSW) collection bin using GIS appear in [4]. Urban areas in India demand the development of better MSW management plans. Moreover, to simplify the MSW management, improved MSW initial conditions related to factors such as collection and transportation parameters can speed up convergence to an optimal solution. GIS appraises and analyzes the environmental and economic effectiveness functionals, which depend on the optimal positions of a new trash can concerning present bin vicinity, road networks, and population densities.

The investigation in [5] resulted in a GIS-based optimal garbage collection course for MSW with the best storage and collection plan using a Geologic Information System in one of the wards. The district map is digitized and geo-referenced, supporting the recognition of collection bins followed by the MSW collection route optimization for degradable and recyclable wastes individually. The optimized vehicles' route for recyclables is 82.5% smaller than the corresponding result for degradable wastes.

An MSW collection route optimization model using the ArcGIS network analyst is discussed in [6]. The Network Analyst estimates interrelations among the dynamic aspects, like network traffic alterations in the area undergoing studies to find optimized solutions. All the required dynamic parameters for creating an initial scenario can be defined or modified, which generates alternative scenarios and leads to several solutions. As a final point, the optimal solution results for the involved functional parameters such as the shortest distance path, road interconnection networks as well as social/environmental repercussions.

An application of GIS in SWM for Coimbatore city, India, appears in [7]. Urban SWM is one of the most tedious ecological problems public authorities face in developing countries. Rapid urbanization, together with cumulative industrial, commercial, and economic development, amplified the generation of various types of waste. The trash generation rates stem from socio-economic development, the degree of industrialization, and climate. Usually, the higher the economic growth and the percentage of the urban population involved, the more significant is the amount of solid garbage produced. GIS benefits planners by providing visualization of the real solid waste circumstances and assisting route analysis through mapping.

3 Problem-Solving

The emerging Cyber-Physical System (CPS) computing paradigm builds on the novel technology developments [8–10]. As depicted in Fig. 1, where sensors and actuators handle the physical world and the subsequent records go to the cyber world to be evaluated to infer the environment state and produce the corresponding outputs to optimize and control it through actuators. The Internet of Things (IoT) interconnects smart objects containing self-configuring abilities intermediate the physical and the cyber domains with secure and energy-efficient information exchange. The union of CPS with IoT

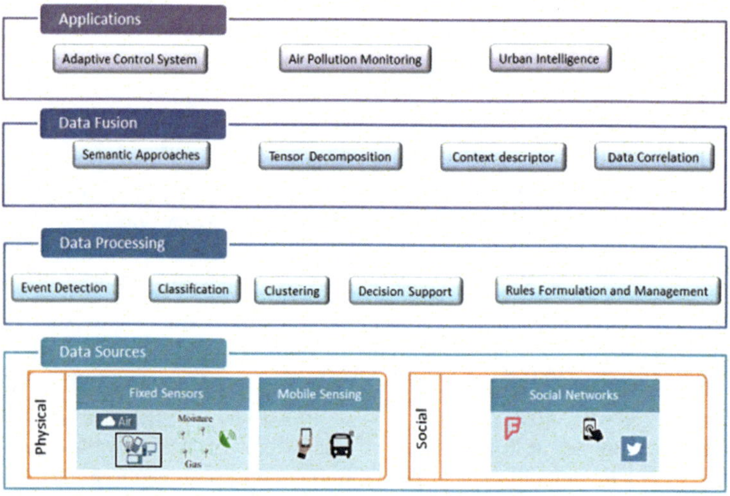

Fig. 1. Digital Garbage Bin Monitoring System (DGBMS) modeled as a Cyber-Physical System (CPS).

influences the physical world observations, detected by smart objects with the computational distributed processing resources. It enables modeling and reasoning of the physical events with efficient communication and data processing for efficient sensing and actuation.

The variety of monitoring devices, as depicted in Fig. 2, encompasses Wireless Sensor Networks (WSNs) for environment probing as well as actuators such as alarms and lights.

Fig. 2. Controlling unit for the Wireless Sensor and Actuator Network (WSAN) using Arduino UNO.

The trash cans overflow quite often, and identifying the people responsible for this so that they can further dump the domestic waste in the right bin can help a lot. Such a framework prevents lousy odors, as well as stopping toxicity in the unhygienic surrounding areas. It turns the city unpleasant, which is a way to disseminate the air pollution and too many harmful diseases in the area.

This project aims at the placement of bins in the right place while finding an optimal route until the waste dump yard. This process is cumbersome when done visually and manually. The distribution of bins relies on the population density and the waste generated in the areas. The goal is to preserve the areas clean, which leads to seeking the optimal route for the garbage collecting vehicles. The corresponding Digital Garbage Bin Monitoring System (DGBMS) can be modeled as a CPS that comprises all stages From Fig. 1 [8, 9, 11].

4 Materials and Methodology

4.1 Arduino Uno

The DGBMS has as its fundamental instrumentation level component the Wireless Sensor and Actuator Network (WSAN), which can be implemented using an Arduino Uno. Such a module is an open-source PC standalone programmable board, with communications interfaces for constructing programmed devices and intelligent objects that can detect and control real-world issues. The equipment data sheets are provided under a Creative Commons Attribution-Share license (Fig. 2). The Atmel 8-bitAVRRISC microcontroller has the following features: (i) a 32 kB ISP flash memory; (ii) a 1 kB EEPROM; (iii) a 2 kB SRAM; (iv) 23 general-purpose I/O lines; (v) 32 general-purpose registers; (vi) 3 timer/counters with comparison modes; (vii) internal and external interrupts; (viii) a serial programmable USART; (ix) a byte-oriented 2-wire serial interface; (x) an SPI serial port; (xi) a 6-channel 10-bit A/D converter; (xii) a watchdog timer with an internal oscillator that can be programmed; and (xiii) 5 power saving modes controlled by software.

4.2 IR Sensor

Most thermal radiation discharged by things close to room temperature is IR. Like all EMR, IR passes on brilliant vitality and acts both like a wave just as a photon (quantum particle) (it would be ideal looking at Fig. 3).

Fig. 3. IR sensor

Molecules emanate or assimilate IR radiation after changing their dipole moment, making a helpful phenomenon to think about the energy states of molecules for the best possible range of frequencies. IR spectroscopy takes a glance at the absorption and transmission of photons in the IR interval.

4.3 Gas Sensor

A gas sensor is a transducer that can recognize gas leaks or other emanations. They can interface with a control framework to naturally close down a procedure and sound an alert to flag the leak area, so individuals and animals can sidestep any harm. These transducers sense burnable and harmful gases and oxygen consumption and can be found in Fig. 4.

Fig. 4. A gas sensor (MQ-2)

4.4 Passive Infrared Sensor

A Passive Infrared (PIR or PID) sensor captures infrared (IR) emanation from objects in its Field of View (FoV) as exemplified by Fig. 5. Quite often, PIR-based motion detectors sense the movement of animals, people, or other entities to automatically activate alarms.

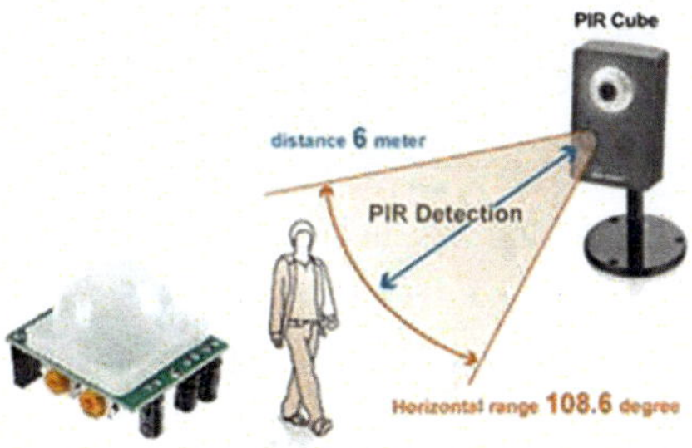

Fig. 5. Passive Infrared Sensor (PIR)

4.5 Wi-Fi Module

This onboard processing unit is autonomous with storage capability that allows integration with the sensors and application-specific actuators through its GPIOs with minimal development effort and runtime loading. Its features decrease the need for external circuitry, comprising the front- end module within a minimal PCB area.

The ESP8266 supports APSD for VoIP and Bluetooth with a self-calibrated RF and without peripheral RF parts. The ESP8266 has instructions to work as an Internet of Things (IoT) module.

4.6 Ethernet Shield

The W5100 is a single-chip network interface with a built-in 10/100Mbs Ethernet controller (refer to Fig. 6). An on-chip system provides high performance and a highly stable low-cost Internet connection without an Operating System (OS), and it is able to handle IEEE 802.3 10 BASE-T and 802.3u 100 BASE-TX.

The W5100 contains a full TCP/IP protocol stack relying on hardware. It has the Ethernet Media Access Control (MAC) layer in addition to the physical layer (PHY). This market-tested hardware implements the TCP/IP stack for the following protocols: TCP, IPV4, UDP, ICMP, IGMP, ARP, and PPoE. Furthermore, the W5100 provides 16 KB of memory for data transmission using socket programming.

The W5100 interfacing is three-fold: (i) the direct parallel bus, (ii) the indirect parallel bus, and (iii) the SPI bus. They can be addressed like external memory, and they come with the W5100 Ethernet library to connect to a microcontroller.

Fig. 6. Ethernet shield

5 Digital Garbage Bin Monitoring System (DGBMS)

5.1 Garbage System

The DGBMS WSAN node with sensors and actuators appears in Fig. 7. It can receive the input power from solar panels via a charge controller to store the energy in the battery. The +5V supply sustains the Arduino controller. This system used three different configuration IR sensors to detect the garbage level in the bin, and a moisture sensor detects the unhygienic and poisonous gases. If any toxic is produced, then the PIR sensor detects, and a message is sent to the corporation via IP address to warn humans, which can be used as voice playback. If the waste does not go to the garbage bin, then Wi-Fi will be called upon, which is controlled by a relay.

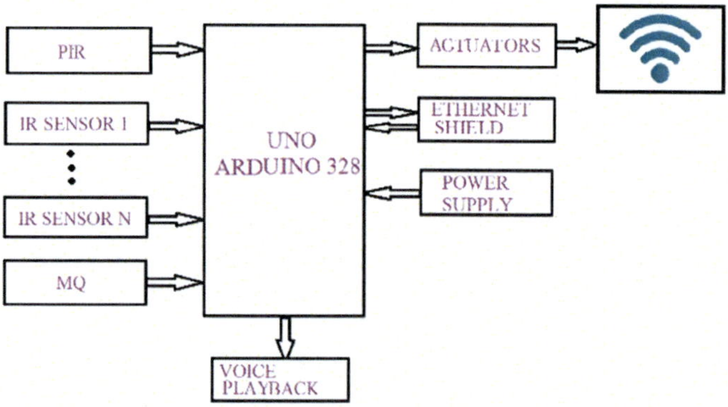

Fig. 7. Block diagram of the garbage system

5.2 Transport System

The proposed transport system will be automated with the manual driver to receive the signal from places via Bluetooth [3], as seen in Fig. 8. Then which route is a better way to collect the garbage is calculated. The relay used to the alert information for the filled bins is the nearest place.

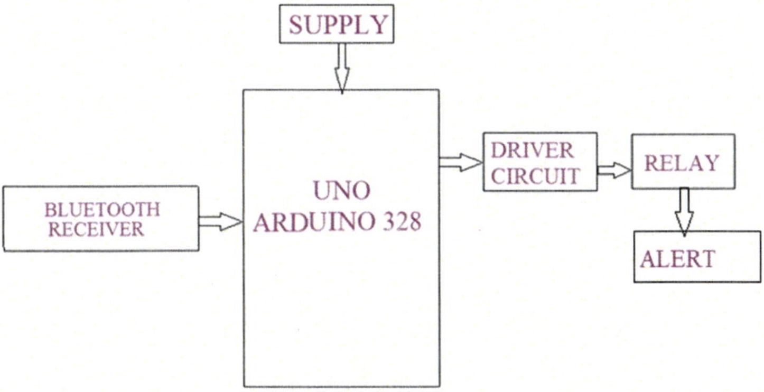

Fig. 8. Block diagram for the transport system

6 Usage in Resources Poor Countries

An open research question is how to adapt and launch this system in resource and technology poor countries as well as how to integrate it into the national health care system to prevent the population from any eventual diseases.

Many developing countries do not own adequate IoT infrastructure like TTN (the things network) to transfer data. Since mobile phone communication is well developed, it will be worth exploring how to base the entire communication on mobile communication.

Beyond the communication issues, the system must ensure that people efficiently use the bins, no waste should be deposited on the floor. It is also important to judiciously place the bins so that the elderly can easily access them.

We recommend conducting a future study to investigate the impact of this system on the population well-being and health, particularly in poor regions, where people are less aware of the environment.

7 Conclusion

Garbage administration is a huge concern for any public or private facility that immediately needs to act upon it. This framework cautions when the dust is placed in the dustbin and can employ voice playback via sensors. At the point when the wastebasket is full or flooded, the IR sensor recognizes its trash level or if some wet rubbish is tossed into the dustbin employing a moisture sensor or some offensive smell just as poisonous gases are created when the gas sensor with a fake nose acknowledges the data. These sensors feed the microcontroller through web convention for information transmission over long separations. The Arduino controller is utilized to send the message to cleaning specialists who clean the dustbin. It is a real-time monitoring framework with the dustbin status showed on a PC-based GUI. The message is sent to the specific transport framework. At that point, a central government agency gathers and cleans the wastebaskets in this structure. This framework is utilized in school grounds, workplaces, and numerous spots where it is appropriate, and these days more urban communities are becoming smart cities. The outcomes from this investigation recommend that geospatial approaches, together with IoT innovation, improve the basic administration process for determining suitable trash receptacles and, later, the relating directing of their residues to the landfill locales lessening the expense and time to the waste administration legitimately.

A data-centric view is recommended to accomplish the CPS for urban and rural areas because it helps create abstractions of the high-dimensional and high-volume data produced and traded by the heterogeneous devices for further analyses and fuse analytical and perceptive data [9, 12–18]. As the physical world information is mostly dependent on its nearby environment, crowd-sensing infers relevant knowledge from external data such as locations, time, and nature of ongoing events.

The distribution of sensors and allocation of actuators involves optimization problems that may require real-time processing as well as decision-making that requires extensive computational loads. Such tasks can be accelerated by employing computational intelligence [8, 19–24].

References

1. Mahajan K, Chitode JS (2014) Waste bin monitoring system using integrated technology. Int J Innov Res Sci Eng Technol 3(7):14953–14957
2. Pavithra (2014) Smart trash system: an application using Zigbee. Int J Sci Technol 1(8)
3. Patil MV, Gajbhiye SM (2015) A review on internet of things based garbage bins detection systems. Int Sci Res (IJSR) 1699–1702. ISSN (Online) 2319-7064
4. Nithya R, Velumani A (2011) Optimal location and proximity distance of municipal solid waste collection bin using GIS: a case study of Coimbatore City. WSEAS Trans Env Dev 8:107–119
5. Velumani A (2013) GIS based optimal collection routing model for municipal solid waste: case study in Singanallur, India. Int J Eng Innov Tech (IJEIT) 3(5). ISSN 2277-3754
6. Bhambulkar AV (2008) Municipal solid waste collection routes optimized with Arc GIS network analyst. Int J Adv Eng Sci Tech 11(1):202–207
7. Shoba B, Rasappan K (2013) Application of GIS in solid waste management for Coimbatore city. Int J Sci Res Publ 3(10). ISSN 2250-3153
8. Estrela VV, Hemanth J, Saotome O, Grata EGH, Izario DRF (2019) Emergency response cyber-physical system for flood prevention with sustainable electronics. In: Iano Y, Arthur R, Saotome O, Estrela VV, Loschi HJ (eds) Proceedings of the 3rd Brazilian technical symposium BTSym 2017. Springer, Cham
9. Anantharam P, Barnaghi P, Thirunarayan K, Sheth A (2015) Extracting city traffic events from social streams. ACM Trans Intell Syst Technol 6:1–27
10. Estrela VV, Monteiro ACB, França RP, Iano Y, Khelassi A, Razmjooy N (2019) Health 4.0: applications, management technologies and review. Med Tech J 2(4):262–276
11. De S, Zhou Y, Abad IL, Moessner K (2017) Cyber–physical–social frameworks for urban big data systems: a survey. Appl Sci 7. https://doi.org/10.3390/app7101017

12. Digiesi S, Facchini F, Mossa G, Mummolo G, Verriello RA (2015) Cyber-based DSS for a low carbon integrated waste management system in a smart city. In: IFAC 2015, vol 48, pp 2356–2361
13. Zhou Y, De S, Moessner K (2016) Real world city event extraction from Twitter data streams. Proc Comput Sci 98:443–448
14. McAfee A, Brynjolfsson E (2012) Big data: the management revolution. Harv B Rev 90(10):60–66
15. Miller S (2014) Collaborative approaches needed to close the big data skills gap. J Org Des 3(1):26–30
16. Schadt EE (2012) The changing privacy landscape in the era of big data. Mol Syst Biol 8(1):1–3
17. Schouten P (2013) Big data in health care solving provider revenue leakage with advanced analytics. Healthc Financ Manag 67(2):40–42
18. Yang B, Castell N, Pei J, Du Y, Gebremedhin A, Kirkevold, Ø (2016) Towards crowd-sourced air quality and physical activity monitoring by a low-cost mobile platform. In: Proceedings of the international Conference on smart homes and health telematics, 25–27, Wuhan, China, pp 451–463
19. Hemanth DJ, Estrela VV (2017) Deep learning for image processing applications, advances in parallel computing. IOS Press. ISBN 978-1-61499-821-1 (print). 978-1-61499-822-8 (online)
20. Aviso KB, Mayol AP, Promentilla MAB, Santos JR, Tan RR, Ubando AT, Yu KDS (2018) Allocating human resources in organizations operating under crisis conditions: A fuzzy input-output optimization modeling framework. Resour Conserv Recycl 128:250–258
21. Estrela VV, Saotome O, Loschi HJ, Hemanth DJ, Farfan WS, Aroma RJ, Saravanan C, Grata EGH (2018) Emergency response cyber-physical framework for landslide avoidance with sustainable electronics. Technologies 6:42
22. Teo TA, Cho K-H (2016) BIM-oriented indoor network model for indoor and outdoor combined route planning. Adv Eng Inform 30:268–282
23. Razmjooy N, Mousavi BS, Khalilpour M, Hosseini H (2014) Automatic selection and fusion of color spaces for image thresholding. Sig Image Video Process 8(4):603–614
24. Mousavi B, Somayeh F, Razmjooy N, Soleymani F (2014) Semantic image classification by genetic algorithm using optimised fuzzy system based on Zernike moments. Sig Image Video Process 8(5):831–842

Safety Management Applied to Smart Cities Design

Telmo Cardoso Lustosa(✉), Yuzo Iano, Gabriel Gomes de Oliveira, Gabriel Caumo Vaz, and Valéria Sueli Reis

State University of Campinas, São Paulo, Brazil

Abstract. According to the study group 5 of ITU-T [1], smart cities are characterized as follows: "A smart sustainable city (SSC) is an innovative city that uses information and communication technologies (ICTs) and other means to improve quality of life, efficiency of urban operation and services, and competitiveness, while ensuring that it meets the needs of present and future generations with respect to economic, social and environmental aspects." Given this definition, we will present in this work part of ICT infrastructure safety management, some real cases that show the threats to the ICT networks, and the current legislation concerning the subject.

Keywords: Smart cities · Management · Security · Infrastructure of cities · Technology

1 Introduction

Writing about Smart Sustainable Cities (SSC) means treating almost all the topics in the technological, social, economic, and environmental areas. Figure 1 brings an overview of SSCs and the role of information and communication technologies.

In October 2014, the Study Group 5 of ITU-T (Industry Standardization of the International Telecommunication Union) established the following definition of a Smart Sustainable City: "A smart sustainable city (SSC) is an innovative city that uses information and communication technologies (ICTs) and other means to improve quality of life, efficiency of urban operation and services, and competitiveness, while ensuring that it meets the needs of present and future generations with respect to economic, social and environmental aspects" [1]. Therefore, within this universe, we will address the part of ICT infrastructure security management.

At first, it is important to clarify that there is a difference between prevention and city safety concerning ICT network security management. The security of the city is an assignment of the civil defense office, municipal police, environmental protection, and others as established by the City Hall. Network security management is an assignment of ICT specialists.

In the following sections, we will detail the concepts and tasks related to network security.

Y. Iano et al. (Eds.): BTSym 2020, SIST 233, pp. 498–510, 2021.
https://doi.org/10.1007/978-3-030-75680-2_55

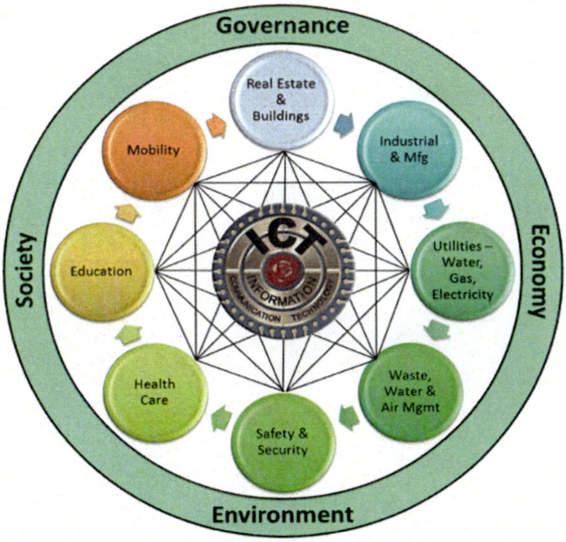

Fig. 1. An overview of SSCs and the role of information and communication technologies

2 Theoretical Development

From the 90s, ITU-T, through the series M3000 recommendations, established the concept of TMN (Telecommunication Management Network) and the functional areas of management, as shown in Fig. 2. Its objective was to define standardized interfaces to communicate independent management networks and solve the problematic Operation, Administration, Maintenance, and Provisioning (OAM&P) of telecommunications networks and services in an open and multivendor environment [2].

Though these recommendations, at the time, were developed for communications networks, they apply to smart cities ICT projects, as presented by the author in his master's thesis [3].

About these functional areas, described in the following subsections, we point out that in the past, there were many security problems and equipment and networks were poor in intelligent features of faults, performance, configuration, and accounting.

Today, exactly the opposite happens; the devices have a high level of intelligence to meet the various demands, and security is the biggest challenge for any business and public agencies in all countries.

2.1 Performance Management

This is the set of activities that ensure the equipment, networks, and services operate with the specified quality. Performance management functionalities include monitoring performance parameters, using measurements, trend analysis, monitoring network utilization, and application controls.

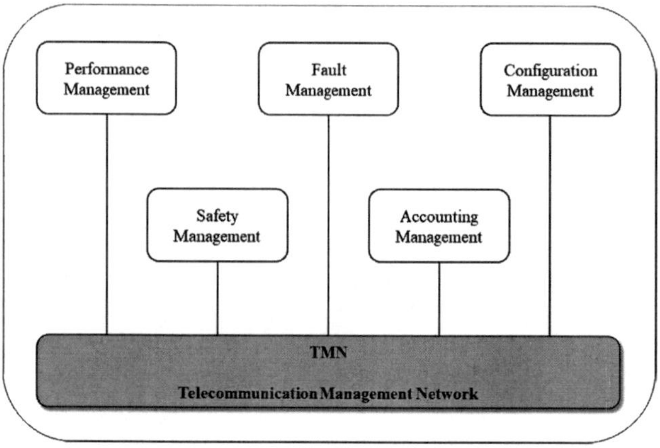

Fig. 2. ITU-T M 3400 recommendation

2.2 Fault Management

This is the set of activities that can detect, isolate, and correct the abnormal operation of the equipment. The fault management features include alarm monitoring, event analysis, presentation of alarms, recovery management, administration, testing, and failure analysis tickets.

2.3 Configuration Management

This management defines the functions to exercise control over identifying, collecting, and changing the data related to equipment, networks, and services through the ICT infrastructure. It exercises functions that modify the resources and facilities through the control of a database.

It meets the following functions: planning and network engineering; installation; planning and changing services; supply; control over the status quo.

2.4 Safety Management

According to pre-established security policies, this is the set of activities that allow the prevention, detection, and remediation actions that could compromise the use and dissemination of information in the TMN interfaces.

It should include the following security mechanisms: authentication - ability to provide/verify the identification of a communication entity (user and/or application); Control access to management resources - ability to check the authorization of access to network resources, also including the decision if a communication entity user/application is allowed to connect to a given destination or to perform certain operations.

Information needed for security management must be stored and maintained in a specific database SMIB (Security Management Information Base).

2.5 Accounting Management

It is the management that enables the use of measures of ICT services networks and the determination of costs to provide the services and customer costs for the use of services. It also supports the pricing of services.

Accounting Management includes the following function groups: use of measures of services; support for setting prices and tariffs; collection and financial control; corporate control.

In this case, there are some differences between a telecommunications services company, which aims to profit, and the city that provides public services for free. However, the city must meet its costs.

3 ICT and Security Platform

Figure 3 illustrates our proposed ICT platform for IC design. In the author's master's thesis [3], it is possible to find a more detailed description of each layer.

Notice that safety should be a concern from the physical layer, referring to sensors, to the application layer.

We will see in the next section that hacker attacks occur in each of these layers. It is not enough, for example, to create security features in public service systems, just at the application layer.

4 Examples of Cases of Hacker Attacks

As examples of the importance of security management nowadays, the following subsections present some data reported in press vehicles. In each presented case, it is important to observe the growing access people have to the connected world and, on the other hand, the growing risk to which they are exposed when sharing their data.

The companies also have to worry about their systems and data security without compromising their clients' experience, and it is true to the cities as well because, when they develop an ICT platform, it is unavoidable to send sensitive data to external networks where malicious people could steal them.

It is also essential to prevent external access to the local networks, which could compromise the correct operation of the public services.

4.1 Fraud via SS7 Signaling Network

Even with security threats, people usually think that the communication made through encrypted services will remain private and protected forever. However, it is possible to steal a mobile identity and pass as the user without the user's consent. It can give access to messages, calls, and supposedly secure applications [4].

The problem, called SS7 spoofing, is a security breach in the Signaling System 7 (SS7), an international standard that defines how network elements shall exchange information on a signaling network. This standard was developed in 1975 and has never been updated to protect the advances in mobile technology or the rise of cybercrime [4].

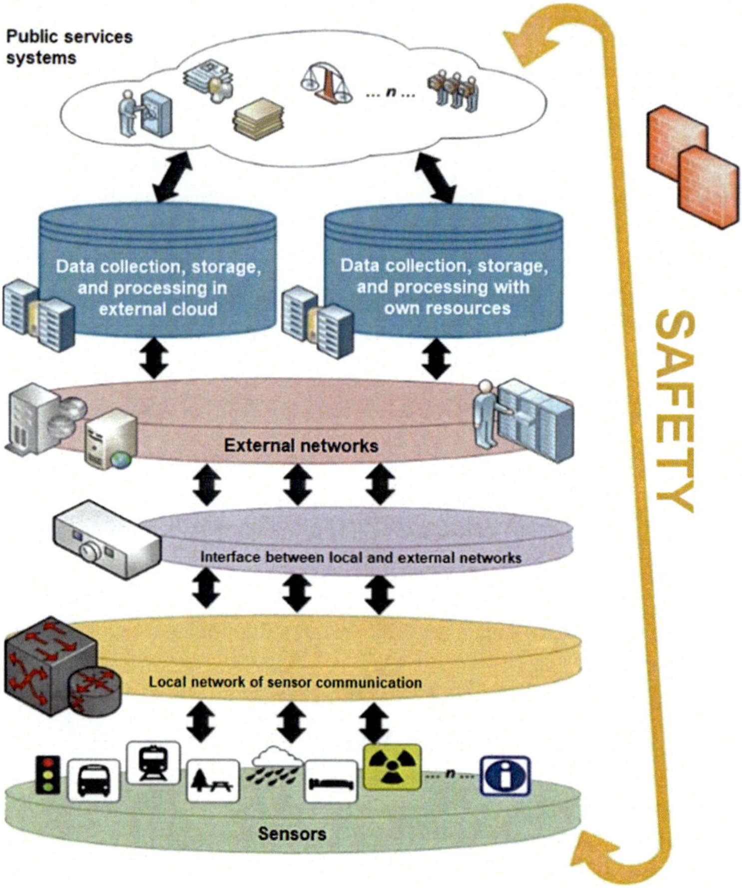

Fig. 3. ICT and security for IC platform

The SS7 network was, in the beginning, designed for closed communications, but it was used in Public Telephone Networks as well. With the growing number and complexity of interfaces, its intrinsic vulnerabilities arose because it was not designed for this purpose. Besides, there is an increasing interdependence and interconnectivity between SS7 networks and the Internet, which results in more vulnerabilities [5].

SS7 is used worldwide and allows the operators to exchange the necessary information to transmit messages, calls, and text (SMS) and ensure correct billing. Although it is a key technology to keep the world connected, there are vulnerabilities of an SS7 network that allow a fraudster without sophisticated equipment to generate SS7 packets and determine the location of the subscriber, tap Call and obtain personal data [4].

With this access level, fraudsters do not need to crack an encrypted application because the most usual tool to identify a user is through SMS messages, and a fraudster who is exploiting an SS7 vulnerability can intercept the messages and create a second user account without the user knowing [4].

4.2 Rio 2016

According to the data collected by [6], 92% (136,034) of all unique devices authenticated by the network access control platform (Cisco Identity Services Engine) were connected through the 3,536 Wi-Fi hotspots provided by Cisco for the Games, and the remaining devices (8%) were connected using a wired connection.

Of all devices connected to the Wi-Fi network during the Games, 10% belonged to the Olympic Family and athletes, 35% were accredited media, and 55% of the Rio 2016 workforce. The internet traffic in the Games network was 492 terabytes, and the use peaked at 5.4 Gbps.

A total of 4.2 million security events were recorded, and 731,607 attempted Distributed Denial of Service (DDoS) attacks were blocked by the Cisco security infrastructure implemented for the Games.

In the protection of all official public websites and mobile applications of Rio 2016, Cisco solutions and other security partners of Rio 2016 detected 40 million security events, blocked 23 million attempted attacks, and mitigated 223 DDoS attacks.

The "CISCO Cybersecurity Report," released in July 2018 [7], has other very interesting data that confirms the need for ongoing concern about hacker attacks. It is important to stress that the data concerning the companies can be applied to the municipalities' systems because, in practice, they suffer or will suffer, from the implementation of the ICT, with the same threats.

4.3 Check Point Warned Cyber-Attacks

According to [8], in 2018, Check Point, an Israeli digital security company, could prevent more than 100 million cyber-attacks. The company spent on cybersecurity 11% more than in the previous year worldwide. The data were presented at the CPX event held in São Paulo to partner companies and Check Point customers.

However, according to the company's CMO, Peter Alexander, the world undergoes a significant change concerning the attacks since they currently can reach entire cities, large corporations and deeply hurt the economy of countries or companies.

4.4 China Inserted Microchips into Servers

In 2018, [9] revealed an attack executed by Chinese spies, whose objective was creating a stealth doorway into networks by introducing microchips into motherboards that would be used in servers. It is estimated that these servers could have compromised the network security of almost 30 companies and institutions, including banks, big techs, and the Department of Defense of the USA.

With these chips in the servers, it was possible to insert malicious codes into the operating system, which could, for example, by-pass authentication protocols and steal encryption keys. Basically, any information inside the server could be exposed on the Internet.

Some weeks after the article, the motherboard manufacturer's CEO announced that a third-party company audited the hardware, and no evidence of malicious chips had been found [10].

4.5 Huawei Refutes Accusations

As reported by [11], in the MWC19 (Mobile World Congress 2019), the chairman of Huawei, Guo Ping, affirmed that Huawei devices do not have backdoors. His declaration came after the USA government's charges against the company, which resulted in tension between the USA and China. The executive also defended the 5G of the company and accused the USA Cloud Act, a federal law enacted in 2018, of allowing their own entities to access cross-border data.

The CEO of Huawei suggested that the Network Equipment Security Assurance Scheme (NESAS), an initiative of the GSM Association (GSMA), the 3rd Generation Partnership Project (3GPP), and the International Telecommunication Union (UIT), should be extended to other countries, to standardize the security of mobile networks.

The NESAS aims to provide a security baseline to ensure that network equipment satisfies a list of security requirements. It would be used with other mechanisms to ensure a network is secure [12].

4.6 Movie "Privacy Hacked"

This film addresses several episodes of the USA presidential election of 2016, the BREXIT referendum in England, elections in many countries, and the external and fake news interference. Furthermore, it shows the beginning of a great discussion in the USA about security and legislation on the subject [13].

It evidences the use of the data shared by people in social media and how severe its consequences are, with the power of influencing people's decisions in questions of collective interest.

5 Safety Legislation

The cases presented in Sect. 4 are just some examples of hacker attacks that daily hinder the various ICT services. Local governments are subject to these risks when offering the various public services, and they must keep citizens' privacy in accordance with the legislation presented in this section.

The first concern a project and IC operation manager should have is the legislation dealing with personal data security. Any unauthorized information leakage can cause serious legal problems to city officials.

Following, we present a summary of the national legislation on the subject.

5.1 Law No. 12,527, of November 18th, 2011

This law, known as Transparency in Public Administration Law, deals with access to information in public agencies. It ensures the population's right to know the source and destination of public resources and determine the entities that are subject to this law.

It states that all the public agencies listed in its first and second articles must grant access to any information requested by a citizen unless the information is totally or partially confidential [14].

This law is important because it specifies the data that must be considered of public interest, i.e., those data that must remain public even in an environment where privacy and security are imperative.

5.2 Law No. 12,965, of April 23rd, 2014

This is called the law of Internet Civil Mark. Its Section II, Article 10, states: "The guard and the availability of records of connection and access to internet applications addressed by this law, as well as personal data and the content of private communications, must meet the preservation of intimacy, private life, honor, and image of the parties directly or indirectly involved" [15].

5.3 Law No. 13709 of August 14th, 2018

This General Law of Protection of Personal Data disposes about the protection of personal data and amends Law No. 12,965, of April 23rd, 2014 (Internet Civil Mark).

President Michel Temer signed this law with several vetoes to the original text sent by the Senate. The main change was the removal of articles 55 to 59, which determined the creation of a National Data Protection Authority [16].

Without the authority, experts from the Brazilian Institute of Consumer Defense (IDEC) [17] fear that the result is a law without oversight on its compliance, i.e., there would be a lack of administrative structure to monitor illegal practices concerning the treatment of personal data.

5.4 Provisional Measure 869/18, of December 28th, 2018

The Government published in the Official Diary of the Union (DOU) this provisional measure establishing the National Data Protection Authority (ANPD) [18].

The ANPD will be composed of a board made up of five directors to be stated by the President. The board members, whose term shall be four years, should have a "high concept in the specialty field for which they will be named."

It will also include a National Council of Personal Data and Privacy Protection composed of 23 members from the Executive Power, the Senate, the House of Representatives, the National Council of Justice, the National Council of the Public Ministry, the Internet Steering Committee in Brazil, scientific, technological and innovation institutions, civil society organizations with proven performance in personal data protection, and entities representing the business sector related to personal data processing area.

5.5 Law No. 13,853, of July 8th, 2019

After approval of Provisional Measure 869/18 in the House of Representatives and Senate, it was converted in the law 13,853/19 [19] and sanctioned by President Jair Bolsonaro. With its sanction, the creation of the National Data Protection Authority (ANPD) was formalized.

It was assigned to the ANPD transient nature of federal public administration agency linked to the Presidency of the Republic. It can be transformed into autarchy after two years at the Executive Power discretion.

Despite the administrative link between ANPD and the Presidency, the law ensures its technical and operative autonomy.

Concerning the internal organization of ANPD, it should follow this structure:

- A Board of Directors (maximum steering body).
- A National Council of Personal Data and Privacy Protection.
- A Magistrate.
- An Ombudsman.
- A proper legal advice Authority.
- Administrative and specialized units required to implement the provisions of the law.

It is essential to study this law by all those who work or will work directly or indirectly with the IC.

5.6 General Data Protection Regulation (GDPR)

The GDPR, 2016, is a regulation of European law about personal data privacy and protection, applicable to all individuals in the European Union (EU) and European Economic Area (EEA). It also regulates the exportation of personal data outside the EU and EEA. The GDPR aims to give citizens and residents ways to control their personal data and unify the European regulatory framework [20].

It was the first document worldwide about personal data protection. Although limited to within the EU, it serves as a reference for studies on the subject. Besides, any company outside the EU must fulfill this regulation to do business in the EEA.

5.7 ISO/IEC 27000:2018

This is a document whose objective is to provide an overview of Information Security Management Systems (ISMS) [21]. It applies to all types and sizes of organizations, and the concepts presented within can be extended to the municipalities that are developing ICTs platforms.

It is an essential document for all who are working with IC projects. Concerning the design of smart cities, it is especially important for the municipal managers who will be responsible for the safe access to ICT networks and public systems.

5.8 National School of Cyber Defense (ENaDCiber)

The ENaDCiber was inaugurated on February 2nd, 2019, at Fort Marechal Rondon, in Brasilia. It reports to the Cyber Defense Command and aims to promote and disseminate the necessary skills for Cyber Defense concerning the National Defense. It also contributes to the research, development, operation, and management of the subject [22].

The school contributes to the training and specialization of human resources, both civil and military, for the work in the cyber sector under the Army responsibility. The courses and internships are available for the military of the armed forces as well as academia.

According to the authorities in Cyber Defense, the most essential, in the context of cyber defense, is the human material, but there is a lack of specialized personnel in the area. The importance of the school is evidenced because of this need for human resources [22].

Concerning the development of ICTs in the cities, the municipal departments responsible for the IC design and implementation should contact the National School of Cyber Defense to qualify their technical staff.

5.9 Safety Technical Group

In August 2019, the Federal Government created Technical Groups of Broadcasting Critical Infrastructures Security and Telecommunication Critical Infrastructures Security, both of the House of Foreign Affairs and National Defense, whose objective is studying and proposing the implementation of measures concerning the communications critical infrastructures security [23].

Both the Technical Groups aim to survey threats, vulnerabilities, and control measures related to the respective sectors (broadcasting or telecommunication). The groups must meet at least eight times during the year, and, at the end of the works, they must submit to the House of Foreign Affairs and National Defense a report containing their conclusions [23].

5.10 Strategic Plan Campinas Smart City (PECCI)

The PECCI is a municipal level plan concerning the city of Campinas and has a chapter exclusively dedicated to the security subject.

This chapter highlights the growing use of technological devices in the urban space and emphasizes these devices' ability to collect the citizens' data, monitor their activities, and identify them [24]. There is a concern about the protection of privacy and personal data of individuals.

Bringing these issues to the smart city subject, the privacy of citizens must be ensured, and the Government must accomplish this task through the observance of good practices and the current legislation and standards.

6 Results and Conclusion

This paper aimed to present technical requirements, real cases, and a legal overview of Information and Communication Technologies and highlight the importance of safety management in smart city projects.

In the real cases presented, it was possible to observe some critical issues. The first is a communication standard used for almost 50 years without updates and corrections of known vulnerabilities, which worsens with the growing number of connections and complexity of services.

Following this line of reasoning, it is essential to emphasize that the number of connected devices and the demand for services on ICTs platforms are noticeably increasing and, with them, there is a growing number of security events, from attacks to the services providers, which compromise the infrastructure of servers and networks, to private data stealing, which can affect any entity, i.e., personal data, companies systems, or public services systems.

Another issue is the concept that data is power. From this idea arise the cases of espionage in hardware, software, and network levels. This work presented cases in which state agencies have been affected by malicious components in their equipment, companies have been accused of sending cross-border data to foreign governments, and people data have been used to influence elections.

All these cases show how important technical knowledge is for the development of ICT platforms in smart cities because it is unavoidable sending sensitive data through the Internet, and when it is done, both the data and the local network are exposed to the risks mentioned above. On the other hand, just technical knowledge is not enough because laws and standards rule personal data security and must be obeyed.

From a legal perspective, this work presented the Brazilian laws that rule the privacy and data safety on the Internet. The set of national laws in Sect. 5 work together determining what is public or confidential data concerning public agencies, defining the rights of the citizens when they use the Internet, holding companies and agencies responsible for the users' data, and creating the regulatory agency responsible for ensuring the compliance with legislation.

When the local legislation and standards do not specify how to act in a specific situation, it is recommended to follow international references such as the General Data Protection Regulation, which was the first legal document about personal data protection and is a reference to other laws worldwide, and the ISO/IEC standard, which gives the direction when there is a lack in the legislation.

The Government created a school to disseminate Cyber Defense skills and improve the country's capability to defend itself from cyber-attacks as those mentioned above. It also created the Safety Technical Groups to identify threats, vulnerabilities, and control measures at the national level.

Finally, the last document presented in this work is the strategic plan for Campinas becoming a smart city. As shown in the document, the cities must have a plan that follows all the legislation, recommendations, and good practices.

Nowadays, information safety is an issue that must be faced at the personal, business, and public levels. Personal because people are increasingly sharing their data on social media without concerning about who can access them. Business because the companies have many sensitive and confidential data, and it becomes more serious when the company works for public agencies. Public because the Government has many sensitive data of the citizens.

When a smart city is designed, safety management is crucial because the services must operate correctly without compromising the citizens' data, and this is a challenge in a world where the systems are exposed to so many risks. Section 3 showed the layers involved in an ICT platform, and all of them are susceptible to cyber-attacks. That means the project manager must be prepared to deal with safety issues and ensure privacy and data security for both the municipal agencies and the citizens. The development team must be composed of qualified staff with technical and legal knowledge to keep services and data integrity.

References

1. ITU-T Focus Group on Smart Sustainable Cities (2014) Smart sustainable cities: an analysis of definitions. https://www.itu.int/en/ITU-T/focusgroups/ssc/Documents/Approved_Deliverables/TR-Definitions.docx
2. Glitho RH, Hayes S (1995) Telecommunications management network: vision vs. reality. IEEE Commun Mag. ISBN 0163-6804-95-$04.00
3. Lustosa TC (2017) Proposal of methodology for projects of integrated management network and telecommunications services in smarts cities. 1 recurso online (142 p). Dissertação (mestrado) - Universidade Estadual de Campinas, Faculdade de Engenharia Elétrica e de Computação, Campinas, SP. https://www.repositorio.unicamp.br/handle/REPOSIP/325062
4. Biavaschi B (2016) Alerta de fraude: explorando a rede de sinalização SS7 para cometer desvios. J Adm. https://administradores.com.br/artigos/alerta-de-fraude-explorando-a-rede-de-sinalizacao-ss7-para-cometer-desvios
5. Lorentz G, Moore T, Manes G, Hale J, Shenoi S (2001) Securing SS7 telecommunications networks. In: Proceedings of the 2001 IEEE workshop on information assurance and security. United States Military Academy, West Point, NY, 5–6 June 2001. ISBN 0-7803-9814-9/$10.00
6. Lustosa TC, Iano Y, Loschi HJ (2016) O recorde das TIC's nos Jogos Olímpicos do Rio 2016; BTSym 2016 Proceedings. ISSN 2447-8326. V.1
7. CISCO (2018) Pequenas e poderosas. Como as empresas de pequeno e médio porte podem fortalecer suas defesas contra as ameaças atuais. https://www.cisco.com/c/dam/global/pt_br/products/PDF/cisco-2018-smb-report.pdf
8. Butcher I (2019) Check Point preveniu mais de 100 milhões de ataques cibernéticos em 2018. Time Mobile. https://www.mobiletime.com.br/noticias/05/06/2019/check-point-preveniu-mais-de-100-milhoes-de-ataques-ciberneticos-em-2018/
9. Robertson J, Riley M (2018) The Big Hack: How China Used a Tiny Chip to Infiltrate U.S. Companies. Bloomberg. https://www.bloomberg.com/news/features/2018-10-04/the-big-hack-how-china-used-a-tiny-chip-to-infiltrate-america-s-top-companies
10. Mobile Time (2018) Investigação afirma que não existe chip malicioso em placas-mãe da Super Micro. https://www.mobiletime.com.br/noticias/11/12/2018/investigacao-afirma-que-nao-existe-chip-malicioso-em-placas-mae-da-super-micro/
11. Amaral B (2019) Huawei rebate acusações e ironiza governo norte americano. Teletime. https://teletime.com.br/26/02/2019/huawei-rebate-acusacoes-e-ironiza-governo-norte-americano/
12. GSM Association (2020) Network Equipment Security Assurance Scheme – Overview. Version 1.1. https://www.gsma.com/security/wp-content/uploads/2020/09/FS.13-NESAS-Overview-v1.1.pdf
13. Netflix (2019) Privacy hacked. https://www.netflix.com/br/title/80117542
14. Brazilian Presidency of the Republic (2011) Lei Nº 12.527, de 18 de novembro de 2011. https://www.planalto.gov.br/ccivil_03/_ato2011-2014/2011/lei/l12527.htm
15. Brazilian Presidency of the Republic (2014) Lei Nº 12.935, de 23 de abril de 2014. https://www.planalto.gov.br/ccivil_03/_ato2011-2014/2014/lei/l12965.htm
16. Brazilian Presidency of the Republic (2018) Lei Nº 13.709, de 14 de Agosto de 2018. https://www.planalto.gov.br/ccivil_03/_ato2015-2018/2018/lei/L13709.htm

17. Brazilian Consumer Protection Institute (2018) Governo sanciona Lei de Proteção de Dados Pessoais. https://idec.org.br/noticia/governo-sanciona-lei-de-protecao-de-dados-pessoais
18. Brazilian Presidency of the Republic (2018) Medida Provisória Nº 869, de 27 de dezembro de 2018. https://www.planalto.gov.br/ccivil_03/_ato2015-2018/2018/Mpv/mpv869.htm#:~:text=MPV%20869&text=Altera%20a%20Lei%20n%C2%BA%2013.709,que%20lhe%20confere%20o%20art.
19. Brazilian Presidency of the Republic (2019) Lei Nº 13.853, de 8 de julho de 2019. https://www.planalto.gov.br/ccivil_03/_Ato2019-2022/2019/Lei/L13853.htm
20. Official Journal of the European Union (2019) L119. https://eur-lex.europa.eu/legal-content/PT/TXT/PDF/?uri = OJ: L: 2016: 119: FULL & from = EN
21. International Organization for Standardization (2018) ISO/IEC 27000:2018 – Information technology – Security techniques – Information security management systems – Overview and vocabulary
22. Brazilian Ministry of Defense (2019) Escola Nacional de Defesa Cibernética é inaugurada em Brasília. https://www.gov.br/defesa/pt-br/assuntos/noticias/ultimas-noticias/escola-nacional-de-defesa-cibernetica-e-inaugurada-em-brasilia
23. Teletime (2019) Governo cria GTs para estudar segurança de infraestruturas de comunicações. https://teletime.com.br/23/08/2019/governo-cria-gts-para-estudar-seguranca-de-infraestruturas-de-comunicacoes/?utm_source=akna&utm_medium=email&utm_campaign=TELETIME+News+-+23%2F08%2F20193A34+23%
24. City Hall of Campinas (2019) Plano Estratégico Campinas Cidade Inteligente (PECCI). https://www.campinas.sp.gov.br/arquivos/desenvolvimento-economico/pecc-2019-2029.pdf

Emerging Trends in Systems Engineering Mathematics and Physical Sciences

5G - Active Antenna Applications

Daniel Izario[1(✉)], Yuzo Iano[1], João Brancalhone[2], Karine Izario[3], Gabriel Gomes[1], and Diego Pajuelo[1]

[1] University of Campinas, Campinas, SP, Brazil
yuzo@decom.fee.unicamp.br
[2] National Telecommunications Institute, Santa Rita do Sapucai, MG, Brazil
[3] Educational Anhanguera, Indaiatuba, SP, Brazil

Abstract. Active antennas are devices used in diverse applications, among them the mobile devices in which it operates in the range of microwave and millimeter waves. Therefore, they are useful to the modern 5G systems, in which there is the use of antennas based on the integration of active and irradiation elements, developed with compact technologies, presenting smaller dimensions than the traditional passive antennas. They can be projected in order to present a high directivity either, allowing the operation in a wider bandwidth.

Keywords: 5G · Active antennas · Integrated active antenna · Applications

1 Introduction

An active antenna is made of an irradiator along with electronic appliances able to process radio frequency signals according to established requirements. Studied for more than years, they were initially built using electronic valves in the active parts; later, they have changed into semiconductors-based components, all of them attached to a laminated printed circuit board. The semiconductors have low power consumption, low weight, and a more complete assemblage, better fitting to the current market. Therefore, the active antennas have an important potential along with 5G technologies, which increasingly requires more efficient and compact devices, including low power consumption [1].

The associated functions may include resonant circuits, filter circuits, duplexing functions, mixers, and amplifiers, all acting together in order to assure the efficacy of the essential activities of emitting and capturing magnetic energy. Generally, active antennas are categorized according to the involved actions related to its constitution. The first category is made of an irradiator along with an oscillator [2], the other including an amplifier and irradiator element [3] and the third one, an antenna associated with a frequency converter [4], mainly to the application in communications systems with the use of high-frequency waves inside the microwave range as well as the general use in the range of millimeter waves. It is possible, such as in the case of transceivers, the use of the three previously mentioned options together [5].

An active antenna can operate isolated or as part of a network element, generally used to establish a control about the irradiation diagram as well as the control of other

Y. Iano et al. (Eds.): BTSym 2020, SIST 233, pp. 513–519, 2021.
https://doi.org/10.1007/978-3-030-75680-2_56

features. It is also possible, if carefully performed, the development of active antennas containing small dimensions radiators and able to operate in a larger swath width than the passive models [6]. Due to its small dimensions, the use of integrated active antennas suits especially mobile application terminals as well as on big urban centers in which big passive antennas cannot be deployed in its buildings. Furthermore, some active antenna models are used as radiators, demanding control or commutation in the irradiation diagram as well as in signal combiners, high-performance power amplifiers, among others.

2 Methodology

During the writing of the methodology, several comparisons are raised for analysis: in the first, the modes of operation of the active antennas.

2.1 Active Antenna Integrated with an Oscillator

The first model uses the oscillator's outbound step acting as a radiating element and built with special diodes such as the Gunn [7] and the IMPATT - Impact Avalanche and Transit Time ionizing diode, which, among other similar versions [8, 9], may also present negative resistance. As a matter of curiosity, it is also possible to build an oscillator deploying special transistors to operate in high frequencies, with a positive feedback supply.

Using negative resistance diodes, the systems can assure bigger power ratings in the range of millimeter waves and grant a low level of noise. They present, however, low efficiency in the energy conversion between DC current to radiofrequency energy.

When used as an irradiation element, the antenna curiously integrates the oscillator circuit having its resonant elements, and its dimensions are determined according to the required frequency and oscillation. However, the entry impedance is part of the oscillator's load, this specification being part of the active circuit project and the desired features in the radiating element, requiring a tuning element to achieve the properly desired performance to the whole set. In this case, simple printed antennas with rectangular or rounded shapes are inadequate, as their good performance is restricted to the narrow range. In this case, one of the suggested tuning elements is the varicap diode, able to vary its capacitance according to the reverse voltage applied in its terminals [10].

The structures in which the printed or printed radiated irradiation elements are employed, adaptable to active circuits, have many different aspects than the antennas exclusively built with passive elements. In the case of using transistors, the circuit should provide some tolerance towards inbound and outbound impedance and the expected transistor gain. Its performance suffers more important influences in the antenna's features, factors such as its dimensions in relation to the wavelength as well, the used dielectric, among others.

Since the irradiator is supposedly built in small dimensions, it will also be small compared to the wavelength, offering little gain. The action of the active circuit can partially compensate for the little gain. Besides the gain related aspects, the small dimensions radiator present restrictions relative to its inbound impedance [11]. In this way, the active antenna uses a convenient impedance transform system in order to assure the efficiency of the complete system.

An active antenna connected into an oscillator built with a dynamic negative resistance diode, as in Fig. 1, acts as the transmitting element, with a limited application to the power rating of the active element. However, it can be among the elements of antennas, and, with an adequate project, assure an increasing of the radiating power, even in the case of using millimeter-waves [12].

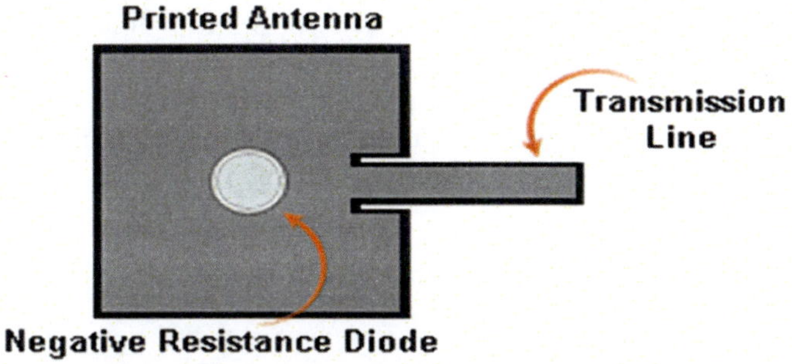

Fig. 1. Example of a printed and integrated active antenna with an oscillator (negative dynamic resistance).

In printed antennas with a transistorized oscillator, it is suggested using frequency stabilization techniques, such as the ones used in the PLL - Phase-Locked Loop. An example, a MOSFET - Metal Oxide Semiconductor Field Effect Transistor-based amplifier, using a ribbon microline as the feedback link. Its length loads and impedance adaptation circuits, as well as the performance of the active antenna, are responsible for the necessary positive feedback.

Another example, now using a transistorized oscillator, is based on the use of two separated antennas, one acting as a transmitter, the other as a receiver. The oscillator must then use the PLL, synchronized with the inbound signal, which can be the inbound signal applied in the range [13]. The feedback supply is performed through a ribbon microline directional coupler [14].

2.2 Active Antenna Integrated with an Amplifier

In the transmitting mode, the irradiation element acts as the system terminal outbound, and, in the receiver, this element is connected as the inbound access. The processing signal has very small amplitude in the receiver, requiring the use of techniques granting acceptable gains and low noise rate [15, 16].

Associated amplifiers and passive irradiation elements allow increasing the gains in both the swath width and the range, also somehow controlling the outbound noise. Using arrangements, it is also possible to achieve greater radiating power along with more accurate control of the irradiation diagram. In this case, multiple studies experienced the distribution of elements in planar networks, creating a grating of radiating elements to very high frequencies, including the millimeter waves.

2.3 Active Antenna Integrated with a Frequency Converter

The active antenna integrated to a frequency converter is used for reception, properly to the operation in frequencies over the range of microwave as well as the millimeter wave [17]. Hence, the high-frequency signal capture element is connected into a mixing, which receives the inbound signal from a local oscillator. Compact structures also suggest integrating the local oscillator into the same block as the antenna and the mixer.

However, the previously mentioned scenario is not mandatory, as the capture element can also capture the signal from an internal source. In this case, both the local and the expected outbound signal are attached to the element, which will mix them. It is also possible to develop both the oscillating and mixing functions together in the same device. The justification is the ability of any non-linear component of mixing signals, allowing, in that way, assembling, in a single circuit, the local oscillator and the mixing, using the non-linearity of the used devices [17].

2.4 Combined Active Antennas for Increased Power

This mentioned application in which there is a combination denominated oscillators cascade is very convenient, but it bears some concerns. First, synchronizing all the different oscillators in the same frequency but with strict phase control is needed. Having a planar and compact structure, the antennas are built along the same hypothetic axis. It is then proposed a transversal network with the maximum irradiation in a perpendicular direction towards its axis, requiring to its excitations a strict phase alignment [11]. The synchronism command may come from an external source or internally throw oscillators. In this scenario, a network architecture example is a range of MOSFET built oscillators and either the allowance of external synchronization.

The mutual synchronism among oscillators bears, as an advantage, the lack of an external circuit specifically for this purpose. This procedure is known as ILC - Internal Locked Synchronism or as II - Internal Injection [18]. Usually, this kind of synchronism is built using the emitted field from one element or throws the use of different transmission lines connecting the oscillators. There is the schematic of a structure with four antennas with the possibility of synchronism through the mutual coupling method among elements. It presents as a disadvantage the inability of very easy control of the coupling between the elements.

As maximum irradiation in the normal direction to its axis is necessary for a transversal linear network, all the elements must be excited with proper phase synchronism. Then, a different discrepancy between the excitations led to changes in the direction of the range's maximum issuance, besides multiple consequences about other network parameters [11].

There is a stronger coupling between the network elements using parts of the transmission line. In this case, it is necessary to use sections of half wavelength in the fundamental frequency. Hence, in the second harmonic of this frequency, there are separations of one wavelength as well as frequency discrepant and phase combined excitations in the radiating elements in the established fundamental frequency. The combination of phase excitations with a frequency in the double of the fundamental will result in an increase of its power [19].

This structure has the disadvantage of allowing the excitation in higher modes. The greater the number of elements, the greater the number of excitations in the range, making the formation of phase synchronized excitations more difficult. In a structure made of many elements, even with this presented disadvantage, it is also possible to reach a proper arrangement, leading to proper gains and a controlled radiation diagram.

2.5 Active Antennas in Network Formation

It is possible to control the excitation phase of the inbound signals in the active antennas network, then controlling the irradiation diagram. To this purpose, along with the amplifiers and the radiating elements, lagers are used, when possible, also integrated into the same assembly. Instead of lagers, it is possible to apply the internal injection locking method, establishing a progressive phase lag. To this option, it is possible to adopt either the weak or the strong coupling. It is also possible to keep the phase fixed and quickly modify the excitation frequency of an element in relation to others [20]. Consequently, it results in changes in the wavelength and the phase of multiple excitation elements, as in Fig. 2.

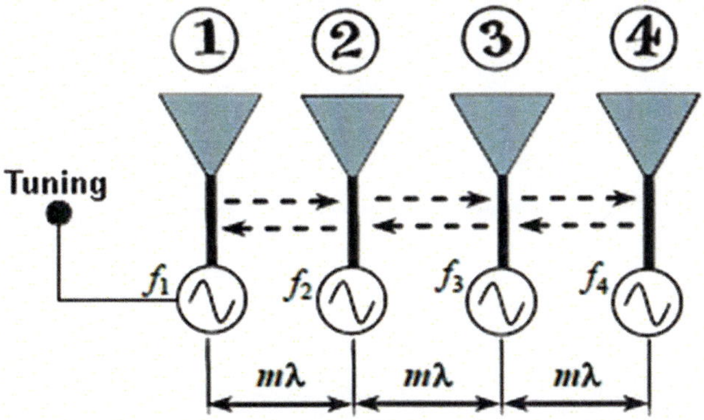

Fig. 2. Formation of a network of active antennas with control of the irradiation diagram.

In Fig. 2, the separation between the radiator elements is fixed according to the functions of multiple parameters, including the multiple coupling represented by the arrowheads. There is no more need for lagers with this solution, allowing the frequency set to continuously vary into the beam's maximum radiation direction. It creates, then, a very efficient system of controlling the radiation diagram [21].

3 Results

In the last updates of the global market demands, multiple models of active antennas are indicated, following the silicon technology procedure. Thus, a model was developed over a plane with 2,400 mm^2 (60 mm × 40 mm) with the total area occupied, considering the area occupied by active and irradiation elements, of 12.4 mm × 24 mm, as in Fig. 3.

As the laminate material, it was used the FR4 of low loss, having a thickness of 0.2 mm and 4.4 of dielectric constant, as determination of PCMCIA - Personal Computer Memory Card International Association. It also contains a ground plane in order to reduce the mutual influence between the different electronic control devices irradiated fields [6]. Active circuits are high stability VCO - Voltage Controlled Oscillator, providing total gains of 4,6 dBi in a central frequency of 2.44 GHz.

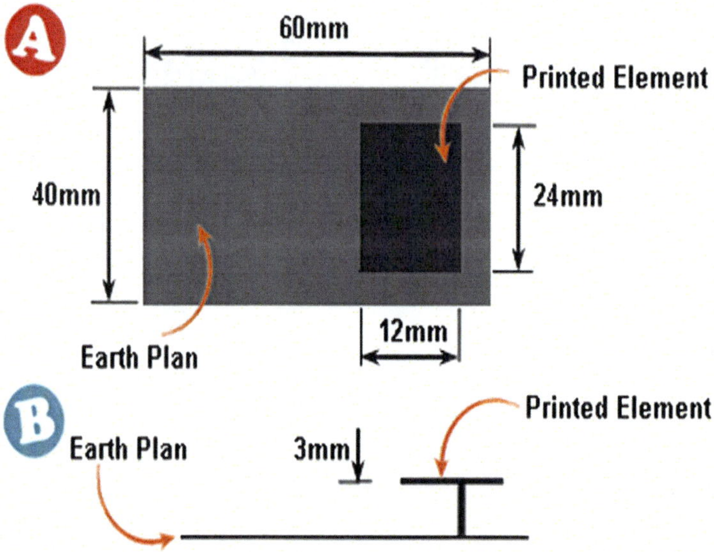

Fig. 3. Example of an active antenna developed for certain applications. (a) Top view. (b) Seen from the side.

The set presented 11 dB as the front to end ratio and a very good impedance matching, presenting 23 dB of return loss. Considering that, generally, the operation is in a frequency range with losses bigger than 10 dB. It was finalized with a limit between 2.39 GHz e 2.52 GHz.

4 Conclusion

Some principles used in the development and deployment of active antennas were established in order to join the development of the 5G technology integrating circuits able to generate and process electrical signs to the irradiative elements, reaching several objectives and positive aspects to the antennas. That said, it creates a development's possibility of new models with small dimensions, in which proper impedance matching and compensating method are performed in order to widen the operation's frequency range.

With an adequate and properly elaborated project, it is possible to achieve a low noise performance as well as the suppression of interferences when operating as a receiving antenna. The modern technology of semiconductor-based integrated circuits also enables the performance of microwave operations, in which the active antenna can process carriers in a very high frequency, modulated in a very high transmission rate. Due to its allowance of integration with diverse elements, it turns up as a compatible solution to the mobile applications used in the 5G, requiring small dimensions communication terminals and rates of tens of megabits per second.

References

1. Copeland JR, Robertson WJ, Verstraete RG (1964) Antennafier arrays. IEEE Trans Antennas Propagat 12(3):227–233
2. Thomas HJ, Fudge DL, Morris G (1981) Gunn source integrated with microstrip patch. Microw RF 24(2):87–91
3. Roberts B, Razban T, Papiernick A (1992) Compact amplifier integration in square patch antenna. Electron Lett 28(19):1808–1810
4. Stephen KD, Camilleri N, Itoh T (1984) A quasi-optical polarization duplexed balanced mixer characterized by isotropic conversion loss. IEEE Trans Micro Theory Tech 31(2):97–102
5. Birkeland J, Itoh T (1989) FET-based planar circuits for quasi-optical sources and transceivers. IEEE Trans Microw Theory Tech 33(9):1452–1459
6. Bilotti F, Urbani F, Vegni L (2006) Design of an active integrated antenna for a PCMCIA card. Progr Electromag Res 61:253–270
7. Gunn JB (1963) Microwave oscillations of current in III-IV semiconductors. Solid State Commun 1(9):88–91
8. Makintosh IW (1975) Applications of microwave solid state devices. In: Howe MJ, Morgan DV (eds) Microwave devices. Wiley, London
9. Camilleri N, Bayaraktaroglu B (1988) Monolithic millimeter-wave IMPATT oscillator and active antenna. IEEE Trans Microw Theory Tech 36(12):1670–1676
10. Navarro JA, Hummer KA, Chang K (1991) Active integrated antenna elements. Microw J 34(1):115–126
11. Balanis CA (2009) Teoria de antenas. Trad. de J. R. Souza. LTC, Rio de Janeiro
12. Popovic ZB, Weikle RM, Kim M, Rutledge DB (1991) A 100-MESFET planar grid oscillator. IEEE Trans Microw Theory Tech 39(2):193–200
13. Birkeland J, Itoh T (1992) An FET oscillator element for spatially injection locked aarays. IEEE Int Microw Symp Dig 3:1535–1538
14. Collin RE (1991) Foundations for microwave engineering, 2nd edn. McGraw-Hill, New York
15. Robert B, Razban T, Papiernick A (1992) Compact amplifier integrated in square patch antenna. Electron Lett 28(19):1808–1810
16. Hall PS (1993) Analysis of radiation from active microstrip antennas. Electron Lett 29(1):127–129
17. Hwang VD, Itoh T (1988) Quasi-optical HEMT and MESFET self-oscillating mixers. IEEE Trans Microw Theory Tech 36(12):1701–1705
18. Stephan KD (1986) Inter-injection-locked oscillator for power combining and phased arrays. IEEE Trans Microw Theory Tech 33(11):1017–1025
19. Mortazawi A, Foltz HD, Itoh T (1992) A periodic second harmonic spatial power-combining oscillator. IEEE Trans Microw Theory Tech 40(5):851–856
20. David CW (2018) All-silicon active antennas for high performance 5G/SATCOM terminals. Microw J 61:44–60
21. Ribeiro JAJ (2012) Engenharia de antenas: fundamentos, projetos e aplicações. Érica, São Paulo

Micro-generation of Electricity Through Photovoltaic Conversion

Daniel Izario[1](✉), Yuzo Iano[1], João Brancalhone[2], Karine Izario[3], Gabriel Gomes[1], and Diego Pajuelo[1]

[1] University of Campinas, Campinas, SP, Brazil
yuzo@decom.fee.unicamp.br
[2] National Telecommunications Institute, Santa Rita do Sapucai, MG, Brazil
[3] Educational Anhanguera, Indaiatuba, SP, Brazil

Abstract. The current Brazilian and global panorama and perspective about electricity microgeneration throw photovoltaic conversion, explaining and comparing important approached points such: The Brazilian potential in generating solar energy; the solar energy production worldwide; photovoltaic energy conversion and the viability of implementation, focusing on the cost-benefit as well as its future perspectives, along with the visions of students and specialists in the Energy Engineer field.

Keywords: Energy conversion · Energy engineering · Energy potential · Implementation feasibility · Photovoltaic · Energy · Solar energy

1 Introduction

The energetic demand due to the technological, industrial, and demographic expansion can have an important part of its supply from the photovoltaic sources instead of being mostly or even sometimes entirely based on the crude oil, as in the current scenario. Using the solar irradiation, photovoltaic cells, with different performances and presenting different dimensions, can last for circa 25 years [1, 2].

Since Edmund Bequerel discovered the photovoltaic effect in 1839, the use of solar energy has been exponentially growing. In 1954, a state of art cell used to present an area of 2 square centimeters with a generation power of 5 mW and an efficiency of 5% [3, 4]. Lately, after 50 years of research, a hundred million photovoltaic cells were available in the market to customers with an efficiency of 16%. In 2012, photovoltaic modules used to present an efficiency of 20% on sunny days. In the present scenario, the maximum efficiency is 25%, but some cells, developed in a test performed in 2017, presented an efficiency of 44.5% [5, 6].

Besides the explanation from Bequerel, along with the explanation from Albert Einstein about the photoelectric effect, awarding the physics Nobel prize in 1921 as well as the semiconductors physical band theory [7, 8], the technological breakthrough was assured to the photovoltaic panels, as 90% of the photovoltaic cells contains silicon transistors as its materials, through the doping and purification [9, 10].

Y. Iano et al. (Eds.): BTSym 2020, SIST 233, pp. 520–527, 2021.
https://doi.org/10.1007/978-3-030-75680-2_57

In a nutshell, the efficiency increment as well as the decrement of costs of the photovoltaic panels in figures of 70% in Europe between 2000 and 2011, gives Brazil a chance of better exploitation of its huge solar power available in its territory, able to supply an average of 5 Wh/day or 1825 kWh/year [11].

Comparing to Brazil, China, Japan, Germany, the USA, and Italy, however, have a small presented solar power, but lead the use of this alternative, sustainable, and widely available technology [11, 12].

2 Methodology

Considering the emitted solar radiation, just a part reaches the Earth's plane since much of the sunbeams are absorbed when crossing the atmosphere. However, all other indirect forms of energy, such as the oceans, fossil fuel, biomass, among other means of energy, are just possible towards the energy given from the sun [13].

In order to receive the sun's heat, the Earth has two main motions: rotation and translation. The day-night cycle is the consequence of the rotation within its own axis. The different stations of the year are the consequences of the elliptical movement towards the sun's axis. Towards this celestial body, there is the minimum distance, called perihelion approximately on the January 3rd, and the maximum distance, called aphelion on the July 4th, in which, through an effect generated by the distance, generate meaningful variations, but determinate through the latitude towards the equator's line [14]. Consequently, there are multiple kinds of weather in each of the world's regions with their temporal positions, determining the receiving amount of solar irradiation, also influenced by the ocean currents and winds, which transport the tropic's heating to the poles [15].

In the Brazilian southern hemisphere, for example, one of the most important uses of photovoltaic energy is through the photovoltaic panels, focused to the north direction with an inclination angle close to the local latitude, which enables the photon, originated from the sun, to convert into electricity through the electricity conversion principle [15, 16]. Concerning the global potential of solar energy production and humanity consumption, the annual proportion of the available natural resources in the year 2019 is shown in Fig. 1.

According to Fig. 1, the energetic potential is much bigger than the energy consumption. Oil, one of the most used sources of energy, is the less abundant source, followed by natural gas, mineral coal, uranium, and the sun, which enables photosynthesis, as well as hydraulic and eolian conversions. It must also be observed that this resource can provide 7.500 times the figure of 450EJ, which is the total consumption of primary energy in the world in the period of a year. This presented scenario propitiates a wide field of actions regarding the use of photovoltaic conversion as a source of energy [17].

Describing the Brazilian potential of solar energy use, it is possible to highlight the Northeast region, containing the highest levels of solar radiation in the whole country, a potential bigger than the German and the Japanese together, distributed in a 1.558.000 km^2 area. The maximum potential value achieved in the central part of Bahia, though, with figures of 6.5 $kWh/m^2/day$, result of the low rainfall, stable cloudiness weather through the year, as well as high solar incidence.

From a global perspective, China, Japan, the USA, and Germany are the main solar energy producers. They produce respectively 77,9 GW, 42,9 GW, 42,8 GW e 41,1 GW

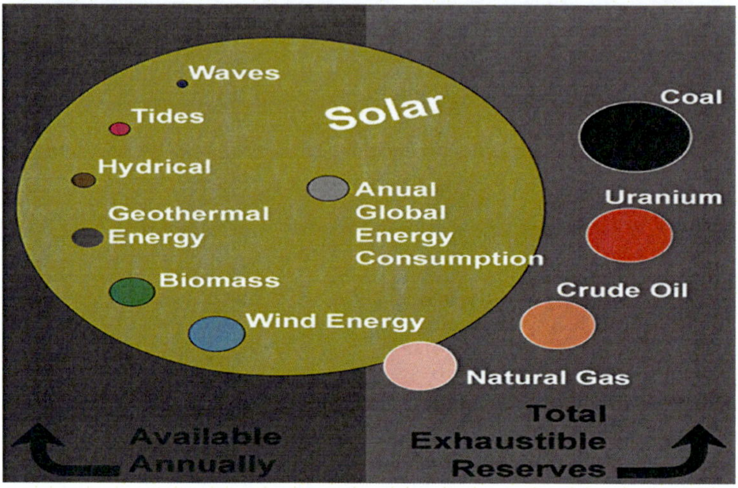

Fig. 1. Total Earth energy resources.

of solar energy, corresponding to the respective percentages of 25%, 14%, 13,8% and 13,4%. Besides the considerable smaller area of Japan and Germany compared to those others, those countries are the only ones which a considerable part of its electric matrix come from photovoltaic solar energy, with rates of about 7%, compared to figures of 1,11% in China and 1,37% in USA [17].

In the German case, this high participation percentage of the photovoltaic energy in the energetic matrix is mainly due to the governmental benefits, especially considering its low solar incidence index. In the USA, there is also a deduction in the incoming taxes for people who buy solar panels [5, 15].

Governmental policies in the residential and commercial scope to the solar energy generation can already be identified in the country, leading to bigger competitiveness than the consumption fee established by the ANEEL [13]. Since 2006, the decentralization of solar energy generation was pointed as strategical to develop the segment in Brazil, expecting the contribution of the use diffusion of the photovoltaic systems connected to the electrical grid to contribute to the expansion of this market [5, 9].

The first steps of the government in this way, however, come later with the implementation of the REN 482/2012 and, more recently, the ProGD in 2015 [17]. However, the use of roofs in the urbanized regions in order to install FV panels known as solar rooftop, which allowed the expansion of solar energy in Germany [8], along with governmental benefits, can be an important measure in order to better develop this kind of market in Brazil.

In the Brazilian scenario, differently than most of the other mentioned countries, this kind of energy generation was mainly used in isolated systems, mostly in areas without electrical supply. One example of it is the program "*Luz para todos*" ("Light for everyone", in English) through the PRODEEM - *Programa de Desenvolvimento Energético de Estados e Municípios*, in English, "Energetic Development Program to States and Municipalities", where, since the 2000s, alternative sources, including the photovoltaic among them, were installed in households without power supply, mainly to the use of small appliances and water pumps.

The use of solar energy in integrated power systems was regulated in Brazil in 2012, the first calls for its use being made in 2011, through companies specialized in this kind of central. As a result, in 2013, there were already 5MW in integrated photovoltaic systems installed. The Brazilian performance, even very discreet to international standards (only 0.1 of the Brazilian energetic matrix come from photovoltaic sources) presented an increase, between 2015 and 2016, of 267,5% of the installed solar energy, which represents a 66,5% increase in the generation of this kind of energy [12].

Considering the high costs of appliances, as well as the little governmental efforts, solar energy is not very often considered viable energy in the Brazilian scenario, the most efficient way of using it being the called "hybrid energy". In this system, there is the photovoltaic cell and thermic energy generation, using solar power to heat the water used in taps and showers. Considering that a big percentage of power consumption in Brazilian households is with the use of electrical showers and that central water heating systems are uncommon, it is a very efficient provision.

Presenting a very high social inequality compared to other main solar energy producers, it is important to note that the energy consumption in Brazil is extremely unequal, mainly linked to the different social classes, as in wealthy families some appliances connected to the grid such as air conditioner and heaters are much more common.

In many needy households, there are few appliances, presenting a very low consumption. There are either many different fees between the states as well as very different levels of solar radiation. With the presented panorama, the viability will be widely varied in different regions of Brazil. In Curitiba, considering the low solar irradiation, the power consumption from the electrical showers was normally very high, in the way that just a water heating system would bring an important effort, even considering the very low solar radiation as the heating system is very efficient. In cities such as Macapá and Boa Vista, there are not many presented advantages with high solar power radiation as the energy fees are much lower. Considering the high temperatures, the hot water demand is lower [5, 17].

3 Future Perspectives

Nonrenewable energy sources such as oil, coal, natural gas, and nuclear energy represent figures of around 80% of the global electric energy matrix, just the other 20% corresponding to the renewable sources. In Brazil, however, the electrical matrix is one of the countries with mostly renewable matrix in the world, responsible for approximately 87,8% of the electricity generation, mainly because of the high hydroelectrically potential, corresponding to more than 63,7% of the available electrical power [17, 18], as shown in Fig. 2 and Fig. 3.

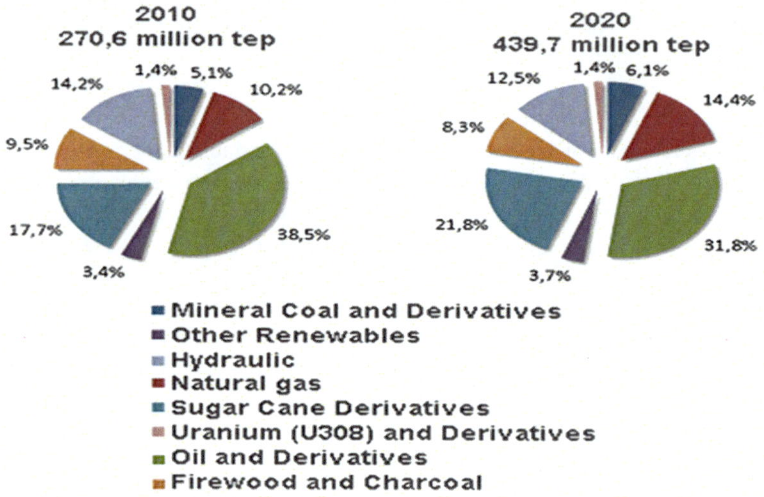

Fig. 2. Electricity generation - Brazil without contracted import, comparison between 2010 and 2020.

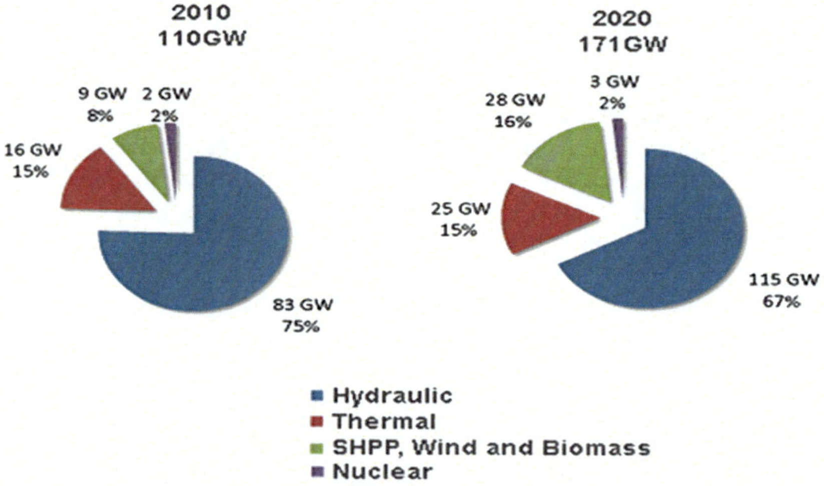

Fig. 3. Electricity generation - Brazil without contracted import, comparison between 2010 and 2020.

In Fig. 2, it is highlighted the low use of solar energy in Brazil since the country has favorable conditions for its development, offering high levels of solar radiation as well as high-quality quartz reserves, which allow the production of high purity silicon, considering that it is a product used in cells and solar modules, high aggregated values products.

The world photovoltaic market has an accumulated installed capacity of 402,5 GW (+98 GW in relation to 2017), growing continuously, which has made Brazil enter the world's power row of photovoltaic solar power [18, 19].

The use of solar energy is a world trend for facts such as low cost and being ecologically correct and renewable, bringing an important market differential regarding social responsibility. A bigger role in renewable energy sources in the energetic matrix is boosting worldwide, multiple countries adopting public policies to achieve bigger energetic safety and sustainability. An interesting bet is the use of sustainable or photovoltaic roofs. Visually like normal roofs, most of these tiles contain four photovoltaic cells, the installation performed under the roof, producing a reasonable amount of energy: an installed area of 45 m^2 of tiles can supply a power of approximately 3 kW [18].

A complete or partial roof made with photovoltaic tiles can supply around the total power amount required for a residency. The photovoltaic tiles usually are more expensive than the conventional photocells, even its installation being typically performed like any other roof: an innovation in the energetic sector, awarding international highlights. With better and more natural aesthetic aspects, the photovoltaic ceramic tile was created to overcome problems from the traditional photovoltaic cells. A breakthrough in many countries, it will certainly be, soon, a tendency in Brazil [18, 19]. Figure 4 shows a roof made up of photovoltaic ceramic tiles.

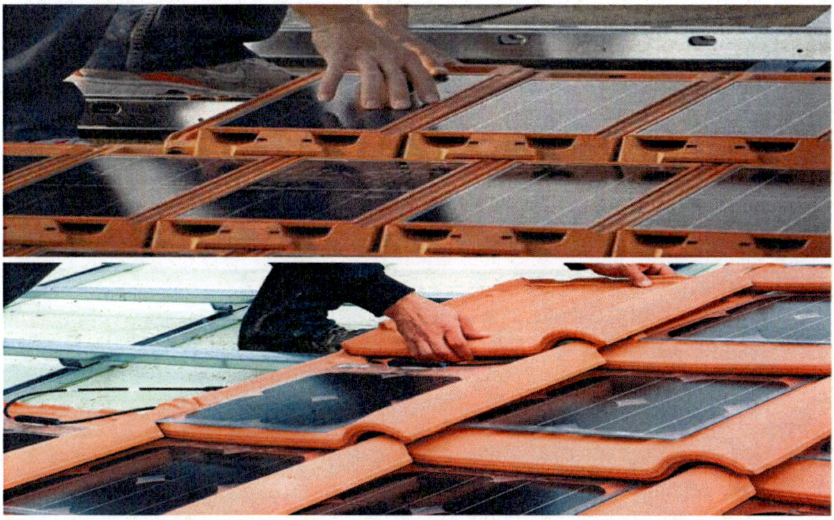

Fig. 4. Photovoltaic roof of the Tegola Solare model.

4 Conclusion

In a nutshell, solar energy fulfills, potentially, an important role in the participation of alternative energy sources in the world, considering its abundance and its wide availability on the Earth's surface. Its international tendencies are typically dedicated to its new breakthroughs, such as the photovoltaic energy installed in public and private buildings, the photovoltaic energy installed in public and private buildings, the energy storage technologies becoming massive and more applied into grid-connected systems as well as stationary storage and the global electric demand management.

That said, this kind of energy has meaningful importance in the Energy Engineering field, where everything is continuously renewed, always looking for new technological solutions besides inserting and using the photovoltaic energy, which is widely available on the whole planet, being essential and cheap to the society as a whole, though.

References

1. Torres RC (2012) Energia solar fotovoltaica como fonte alternativa de geração de energia. SP: São Carlos, Universidade de São Paulo
2. Equipe eCycle, Poderia toda a energia do mundo ser obtida pela radiação solar? https://www.ecycle.com.br/component/content/article/37-tecnologia-a-favor/3095-poderia-toda-a-energia-do-mundo-ser-obtida-por-energia-solar.html
3. Zilles R (2012) Energia fotovoltaica. SP: São Paulo, USP. https://www.iee.usp.br/sites/default/files/biblioteca/producao/2012/Livros/zillesenergiasolar.pdf
4. Vallera AM, Brito MC (2006) Meio século de história fotovoltaica. Faculdade de Ciências da Universidade de Lisboa, Lisboa: Campo Grande. https://solar.fc.ul.pt/gazeta2006.pdf
5. Ambiente Energia (2017) Nova célula solar tem eficiência de 44,5%, a maior do mundo. Ambiente Energia. https://www.ambienteenergia.com.br/index.php/2017/07/nova-celula-solar-tem-eficiencia-de-445-maior-mundo/32187#.W8UnuyyJLIU
6. Jones J (2017) Nova célula solar fotovoltaica promete a maior eficiência de sempre. https://www.portal-energia.com/celula-solar-fotovoltaica-maior-eficiencia
7. Santos CA (2008) As Digitais de Einstein em Nosso Cotidiano. Universidade Estadual do Rio Grande do Sul. https://cienciahoje.org.br/coluna/as-digitais-de-einstein-em-nosso-cotidiano
8. Moreira IC. Einstein e seus Trabalhos de 1905 e 1915. UFRJ. https://www.educacaopublica.rj.gov.br/biblioteca/fisica/0004.html
9. Santos CA (2004) O Prêmio Nobel de Einstein. https://www.if.ufrgs.br/einstein/efeitofotoeletricopremionobel.html
10. Hely F, Swart JW (2014) Energia solar fotovoltaica de terceira geração. www.ieee.org.br/wp-content/uploads/2014/05/energia-solar-fotovoltaica-terceira-geracao.pdf
11. Castro RD (2015) Energia solar térmica e fotovoltaica em residências: estudo comparativo em diversas localidades do Brasil. SP: Campinas. https://taurus.unicamp.br/bitstream/REPOSIP/265829/1/Castro_RodolfoDamasiode_M.pdf
12. Júnior ACOM, Júnior JCL, Júnior HXS (2017) Potencial de Geração de Energia Solar: O Brasil e o Mundo. IEEE. https://www.meioambientepocos.com.br/anais-2017/trabalhos/390.%20POTENCIAL%20DE%20GERA%C3%87%C3%83O%20DE%20ENERGIA%20SOLAR%20O%20BRASIL%20E%20O%20MUNDO.pdf
13. ANEEL (2017) Energia Solar. https://www2.aneel.gov.br/aplicacoes/atlas/pdf/03-energia_solar(3).pdf
14. UFPR. Radiação Solar e Terrestre. Balanço De Calor. https://fisica.ufpr.br/grimm/aposmeteo/cap2/cap2intro.html
15. UFPR. Movimentos da Terra, estações. https://fisica.ufpr.br/grimm/aposmeteo/cap2/cap2-1.html
16. Hargreaves C (2015) The Worldwide Distribution of Solar Resources by Nat. Act on Climate. https://www.wrforum.org/wp-content/uploads/2015/10/SS1-Hargreaves.pdf

17. Marques FMR. Perspectivas para a Energia Solar no Brasil. https://crasp.gov.br/wp/wp-content/uploads/Artigo-Perspectivas-para-a-energia-solar-no-Brasil.pdf
18. Cennatech. Telhas solares fotovoltaicas uma aposta no futuro. https://www.cennatech.com.br/telhas-solares-fotovoltaicas-uma-aposta-no-futuro
19. Portal da Energia. Telhas fotovoltaicas: conheça a autogeração discreta. https://portaldaenergia.com/telhas-fotovoltaicas

A Practical and Precise Method for Heating Calculus in Agitated Jacketed Vessels with Half-Pipe Coil

Maria Thereza de Moraes Gomes Rosa[1](✉), Leandro Paulino Júnior[2], Natan Mastrocollo Mota[2], Eric Henrique Ferreira[2], Luana Spósito Valamede[1], and Daniela Helena Pelegrine Guimarães[3]

[1] Mackenzie Presbyterian University, Campinas, Brazil
gomes.mtms@gmail.com
[2] São Francisco University, Campinas, Brazil
[3] São Paulo University, Lorena, Brazil
dhguima@usp.br

Abstract. Agitated jacketed vessels are frequently used in the food, pharmaceutical, and chemical industries. This study presents a computational system for calculating the time required for heating. The calculations were performed analytically and without any assistance of specific thermal area software, and it was based entirely on literature concepts and equations of heat transfer. The agitated jacketed vessel used in this study has a circular cross-sectional area with a torispherical dome. The half-pipe coil (Schedule Tube 10S – 3″ of stainless steel) was used as a heat-transfer device. A mean heat-transfer rate was obtained to estimate the heating time using an integral function. The average rate of the heat-transfer result was 808 kW with a heating time of 55 min.

Keywords: Agitation · Coil · Heat transfer

1 Introduction

The agitated jacketed vessels are important process equipment capable of mixing a fluid that can promote physical and chemical transformations in many types of products. Pieces of equipment are easily found in pharmaceutical, food, chemical, petrochemical, and metallurgical industries and refineries [1]. The agitated jacketed vessel is compounded by a cylindrical tank with a lid, and it is equipped with agitation systems, heat transfer systems, and control process instruments [2]. According to Torré et al. [3], the tank geometry, the impeller type, and the number and type of baffles can vary depending on the purpose of the operation.

The optimization of energy used in an industrial plant is very important to aggregate value to the final product at a low cost to the organization. The agitation plays an important role in these vessels because the fluid mixing reduces non-uniformities by eliminating gradients of concentration and temperature, characterized by heat and mass transport

Y. Iano et al. (Eds.): BTSym 2020, SIST 233, pp. 528–536, 2021.
https://doi.org/10.1007/978-3-030-75680-2_58

[4]. Mixing and heat transfer in agitated vessels are very important operations. Nassar & Mehrotra [5] demonstrates some of the basic principles and calculations of heat transfer under agitation.

Most industries require the use of these agitated jacketed vessels in their product processing. These industries produce without control over the steam flow conditions and without precise knowledge about the physical properties of the product and the correct agitation level. However, the reduction of energy consumption by time process and activation potency minimization is of utmost importance to decrease the operational costs [6].

This study aims to apply the classic theory of heat transmission to estimate the global heat-transfer coefficient and the heating time in an agitated jacketed vessel. To promote this analysis, a half-pipe coil was used as a thermal exchange mechanism. The calculations were made analytically and without any help of specific thermal area software, and they were completely based on literature concepts and equations of heat transfer. Therefore, this study is relevant to key pieces of industrial equipment and some organizations that do not have specific thermal software because of the high acquisition price, maintenance, and licensing costs.

2 Methodology

The calculations present in this study do not have any artifice that originated from specific software that promotes the thermal calculations in agitated jacketed vessels. There were totally estimated and based on equations and concepts found in the literature of heat transfer. Figure 1 shows the process flowchart to obtain thermal sizing and heating time in an agitated jacketed vessel.

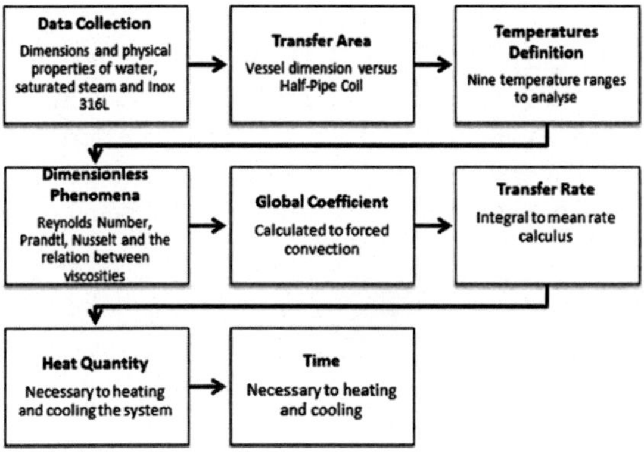

Fig. 1. Process flow chart of the process to calculate the heating time in an agitated jacketed vessel with the use of a half-pipe coil.

The agitated jacketed vessel used in this study has a circular cross-sectional area with an internal diameter of 1,480 m, with a torispherical dome. It is supported by legs and built with stainless steel 316 L (thermal conductivity of 13.4 W/m.K and specific heat of 468 J/kg.K), and the agitation is sustained by background bearings with baffled.

The vessel has external half-pipe coils utilized like heat-transfer dispositive as well as like stone wool thermal coating and many mouthpieces to access the process fluids. The agitated jacket vessel volume is 6 m^3, and the vessel wall thickness is 0,008 m. Figure 2 presents the technical design of the studied equipment.

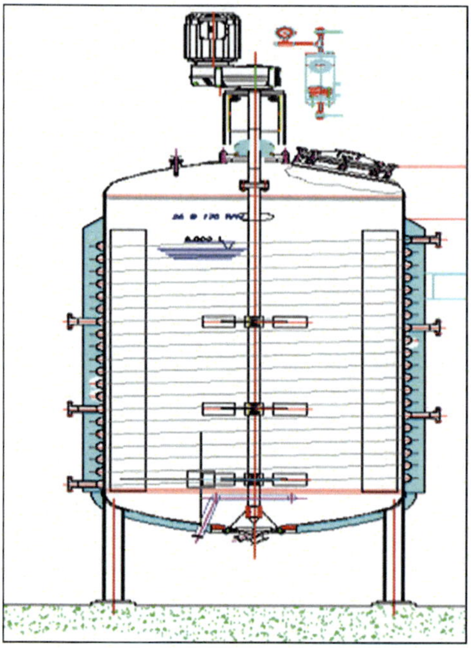

Fig. 2. Technical design of the agitated vessel.

The half-pipe coil (Schedule Tube 10S – 3″ of stainless steel) was used as a heat-transfer device. The coil wrapped on the vessel had 37 turns with a pitch of 116 mm between them. The dimensionless data of the coil (0.089 m external diameter, 0.083 m internal diameter, 0.003 m wall thickness, and 6.450 kg/m linear weight) were provided by the tubes Schedule fabricator (FEITAL Group). The total area jacketed by half-pipe coils was obtained by Eq. 1.

$$A = \left(\sqrt{(\pi.D_T)^2 + P^2}\right).N_{esp}.D_{int} \tag{1}$$

Where: A is the total area that was jacketed by half-pipe coils (m^2); *DT* is the tank diameter (*m*); P is the pitch between the coils (*m*); *Nesp* is the number of coils (un.); *Dint* is the tube internal diameter (*m*).

The total area of heat exchange between half-pipe coils and the tank was obtained by multiplication of length of the tube and tube internal diameter. The length of the tube was calculated using the Pythagorean Theorem to have better accuracy of the heat exchange area due to the configuration in the spiral. According to Fig. 3, the precise length of the tube was obtained considering the circumference length of the tank, the pitch between the coils, and the number of coils.

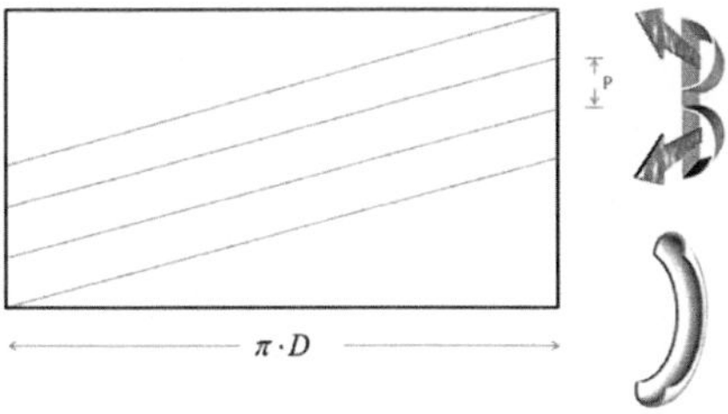

Fig. 3. Estimation of the total area of heat exchange.

Saturated steam with 3.5 bar of pressure and 148 °C of temperature was used as a heating source. The thermal conduction of the working fluid that, in this case, is condensed water, was evaluated in nine temperature ranges, with the initial temperature 23 °C and the last temperature 122 °C. The nine temperature ranges were determined with a logarithmic function for a better description of the dynamic temperature behavior in relation to time. The logarithmic function that ruled the conduct of nine temperature study points was established through those equation of points (1 to 9) in the function of the temperature gradient (23 °C to 122 °C). The values of density, viscosity, specific heat, and thermal conductivities were obtained from Incropera et al. [7] and Van Wylen et al. [8].

To determine the heat-transfer rate (Eq. 2), it was necessary to obtain the global thermal exchange coefficient (U) (Eq. 3). The equipment in this study is used in unit batch processes, and frequent cleanings are made between these processes to avoid fouling. The fluids used have low viscosity and do not support the generation of deposits on the walls. Therefore (ff_i) and (ff_j) factors in Eq. 3 can be disregarded. The saturated steam is considered one of the best heating fluids because of if high heat-transfer power. It is known that one of the heat-transfer coefficients is generally lower than the other; only the first coefficient governs and dominates the heat-transfer [7]. Saturated steam has a much higher wall heat transfer coefficient. In this sense, the convection coefficient of the coil side has a high value of 10.000 (W/m^2. K), according to Singh & Heldman [9], causing the annulment of the term.

$$q = U.A.\Delta T \tag{2}$$

$$\frac{1}{U} = \frac{1}{h_i} + ff_i + \frac{e}{k} + ff_j + \frac{1}{h_j} \tag{3}$$

Where: U is the overall coefficient of heat transfer (W); A is the contact area (m^2); ΔT is the difference between hot and cold fluid temperatures (K); hi is the internal convective coefficient (W/m^2.K); ff_i is the internal fouling coefficient (W/m^2.K); e is the tank wall thickness (m); k is the thermal conductivity (W/m.°C); ff_j is the external fouling coefficient (W/m^2.K); h_j is the external convective coefficient (W/m^2.°C).

The convection heat-transfer coefficient of the internal side (h_i) was determined using the Nusselt Number, defined as the dimensionless numbers Reynolds and Prandtl, according to Rosa et al. [10]. The Nusselt number can be defined as an increase (or an improvement) in the rate of heat transfer by convection over the conductive mode. For example, if Nu = 1, there is no improvement in the heat transfer rate due to convection. However, if Nu = 5, the convective heat transfer rate due to fluid movement is five times the heat transfer rate per conduction [9].

According to Dream et al. [4], the Reynolds Number in agitated jacketed vessels is determined using Eq. 4.

$$\Re = \frac{D_{IMP}.N.\rho_f}{\mu_f} \quad (4)$$

Where: DIMP is the impeller diameter (m); ρ_f is the fluid density (kg/m^3); μ_f is the fluid viscosity (Pa.s); N is the impeller rotation (rps).

The value obtained for Reynolds showed that the flow regime is turbulent. With the information about the type of agitation and the use of baffled, the utilization of Eq. 5 was more effective for this study, according to Dream et al. [4].

Equations 5 and 6 were utilized to calculate the Prandtl and Nusselt number, respectively.

$$Pr = \frac{c_p.\mu}{k} \quad (5)$$

Where: Pr is the Prandtl number; cp is the specific heat (J/kg.K); μ is the viscosity (Pa.s); k is the thermal conductivity (W/m.K).

The thermodynamic properties (density, specific heat, viscosity, and thermal conductivity) were obtained from tabulated data, from which empirical equations were developed.

In order to develop a mean heat transfer rate that unifies all the results, a curve was set that describes all transfer rates as a function of the temperature gradient. The proposed calculation was obtained through nine temperature ranges, and all results obtained from such study intervals are implicit in the consequent heat transfer rate results. A curve that describes all transfer rates in the function of the temperature gradient was adjusted.

$$qx = -0.029T^2 - 5.1T + 1362 \quad (6)$$

Where: q is the heat transfer rate (W); T is the temperature (°C).

The value of the heat-transfer rate average was obtained according to Eq. 7, using the area equality below both functions of Fig. 4 (curve and straight).

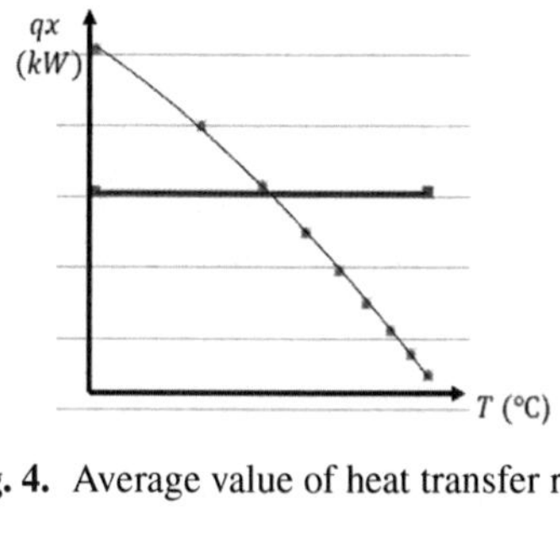

Fig. 4. Average value of heat transfer rate.

The time was estimated related to the sensitive heat and the heat-transfer rate average.

$$\grave{q} = \frac{\int_{ti}^{tf} qx.dx}{(t_f - t_i)} \quad (7)$$

Where: $\bar{q}$ is the average heat transfer (W); t_i is the initial temperature (°C); t_f is the final temperature (°C).

3 Results and Discussion

The research developed in this study aims to analyze the heat-transfer in agitated jacketed vessels in the function of the temperature increase during the heating process system. Given that several factors, which include the heating calculus, vary with the temperature variation, the analysis in more than one point turns the calculus more precise and proves more clearly the results behavior point by point. Inside of equipment where the heat-transfer happens, when the heat capacity rate of hot fluid is much bigger than a cold fluid, the wall temperature becomes constant. In practice, when the steam in condensation is used, the heat capacity tends to infinite. Incropera et al. [7] also say that the temperature of a hot fluid will stay constant in all the heat-change equipment, while the temperature of a cold fluid will increase, as well as demonstrated in Fig. 5.

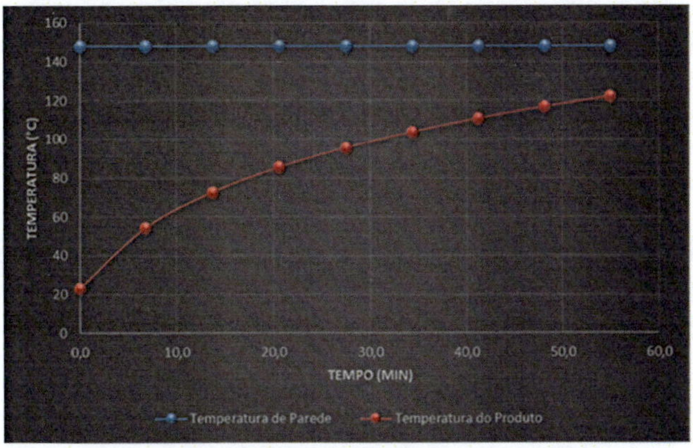

Fig. 5. Temperature behavior in function of time

The logarithmic function, which ruled the conduct of 9 temperature study points, was established through the equation of points (1 to 9) in the temperature gradient function (23 °C to 122 °C). The values obtained are shown in Table 1. Table 1 also shows the results gotten to the dimensionless numbers and the convective heat-transfer coefficient.

Arfelli [6] affirms that when it works with processes that occur a non-linear temperature variation, the arithmetic average will not be usual once that the values do not come next to the ideal ones. In this case, most authors calculate the temperature logarithmic average, which consists of a function of the flow conduct that happens on the studied process [10].

Table 1. Dimensionless numbers and convective heat-transfer coefficient (hi)

Temperature range	T (°C)	Re	Pr	Viscosities relations	Nu	hi (W/ m^2K)
T1	23.0	213,047	6.87	5.08	3,173	1,303
T2	54.2	445,102	3.05	2.40	3,580	1,567
T3	72.5	568,022	2.31	1.86	3,713	1,668
T4	85.5	650,512	1.98	1.62	3,787	1,725
T5	95.5	711,917	1.79	1.47	3,836	1,761
T6	103.7	760,437	1.66	1.36	3,873	1,787
T7	110.7	800,305	1.57	1.29	3,901	1,805
T8	116.7	833,981	1.50	1.23	3,925	1,820
T9	122.0	863,020	1.44	1.18	3,944	1,830

Due to the decrease of density and viscosity, the friction between fluid and impeller, which spin 80 rpm, will be reduced, so the turbulence will increase considerably, as shown in Table 1. The Reynolds values increase with increasing temperature. The Prandtl number tends to decrease because this dimensionless number relates to the thicknesses of thermal and hydrodynamic boundary layers. With the temperature increase, the thermal layer stands out from the velocity layer, which is responsible for heat exchange. The relation among viscosities relates to the fluid viscosity on the wall in certain moments. Once the wall viscosity is constant, and the fluid viscosity decreases in function of the temperature increase, this dimensionless number tends to decrease. The Nusselt number defines the convection heat-transfer, and this number increases with the temperature because the amount of heat for the transfer needs to be highest as the temperature increases. The equation to this calculus was determined empirically, and it considers the agitation inside the reactor and the type of reactor. In this study, the equation to the calculus of Nusselt number (Eq. 8) was obtained through an equation that came from the practical tests made by Dream et al. [4].

$$Nu = 0,36(Re)^{0,67}(Pr)^{0,33}\left(\frac{\mu}{\mu_w}\right)^{0,14} \tag{8}$$

Where: Nu is the Nusselt number; Re is the Reynolds number; Pr is the Prandtl number; μ is the dynamic viscosity (Pa.s); μ_w is the dynamic viscosity of the fluid at the wall temperature.

In order to obtain the equations, it was necessary to perform many empirical tests, besides the type of agitator, the use of 70 baffled, the Reynolds number, and the heat-transfer advice. This calculus is so relative that some authors choose to use or not the relation among viscosities, changing their values, and with this dimensionless number becomes estimative. The convective coefficient is obtained with the relation of Nusselt, conductivity, and internal diameter. The results increase as the Nusselt number. Table 2 shows the results of the global coefficient and heat-transfer rate obtained in this study.

In Table 2, the energy values that need to be transferred by time unit decreased once the temperature variation also decreases with the course of the process. Figure 5

Table 2. Global coefficient and the heat-transfer rate (kW).

Temperature (°C)	Global coefficient (W/ m^2K)	Transference rate (kW)
T1 – 23.0	693	1,218
T2 – 54.2	749	1,002
T3 – 72.5	771	831
T4 – 85.5	783	699
T5 – 95.5	791	592
T6 – 103.7	796	503
T7 – 110.7	799	426
T8 – 116.7	802	359
T9 – 122.0	804	299

shows better this behavior, in which the heating fluid temperature is constant, and the temperature curve of fluid gets closer to the heating fluid temperature to achieve the equilibrium. After the calculation and results introduced in Table 2, it was necessary to calculate an average of the heat-transfer rate results. In order to maintain the calculus precision, the mean was obtained through the integral area of the points in the temperature limits of the process, according to Fig. 6.

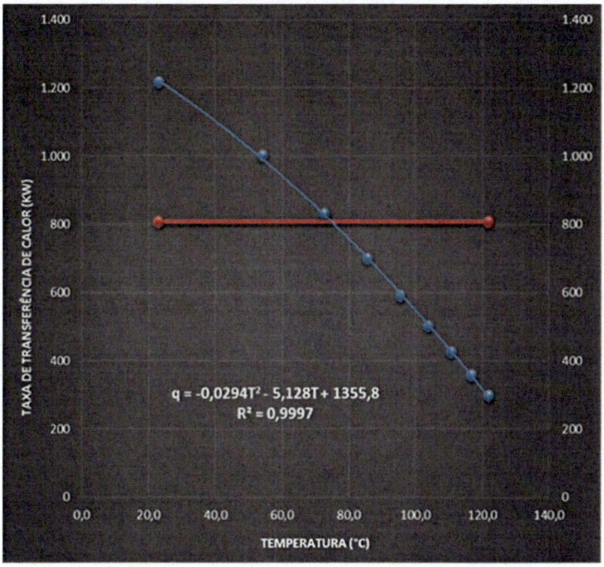

Fig. 6. Average index of the heat-transfer rate

In order to get the curve equation, thereafter, the integral was applied to determine the curve area. This area matched with the area of the constant function, finding Eq. 7 to obtain the average rate of heat-transfer.

The average rate of the heat-transfer result was 808 kW. Once it is known the average rate of energy to be transferred to the system, it was calculated the total energy necessary to heat the process fluid and the metallic mass of equipment (vessel + coil) until the final process temperature, which is the sensible heat. In this method, we utilized the relation between the energy necessary to heat the system and the heat-transfer rate, and, with this, the estimated heating time of 55 min was obtained.

4 Conclusion

From the results, it can be concluded that the obtained time of 55 min was satisfactory, taking into consideration the other studies already made. Also, the values that were gotten are precise, once all phenomena resulting from the temperature increase and their behaviors were analyzed. Finally, the average of the heat rate (obtained by using an integral) was the most precise way found so that there was no discrepancy in the results.

References

1. Katsuki LK, Razuk PC (2001) Software para cálculo da geometria de tanques verticais e horizontais. Revista Brasileira de Agrocomputação 1(2):31–36
2. Joaquim Junior CF, Cekinski E, Nunhez JR, Urenha LC (2007) Agitação e mistura na indústria. LTC Editora, Rio de Janeiro
3. Torré JP, Fletcher DF, Lasuye T, Xuereb C (2007) An experimental and computational study of the vortex shape in a partially baffled agitated vessel. Chem Eng Sci 62(7):1915–1926
4. Dream RF, Engineers PLG, Chopey NP (1999) Heat transfer in agitated jacketed vessels. Chem Eng 106(1):90–96
5. Nassar NN, Mehrotra AK (2011) Design of a laboratory experiment on heat transfer in an agitated vessel. Educ Chem Eng 6:e83–e89
6. Arfelli SL (2009) Projeto mecânico e análise térmica de tanques cilíndricos verticais com agitação e superfície de troca de calor. Dissertação de mestrado - Universidade Estadual Paulista, Faculdade de engenharia, Bauru
7. Incropera FP, Dewitt DP, Bergman TL, Lavine AS (2008) Fundamentos de transferência de calor e massa. 6 ed LTC Editora, Rio de Janeiro
8. Van Wylen G, Sonntag RE, Borgnakke C (1995) Fundamentos da termodinâmica Clássica. 4 ed Edgard Blucher Ltda
9. Singh RP, Heldman D (2013) Introduction to food engineering. Academic Press, Cambridge
10. Rosa MTMSG, Pelegrine DHG, Arce PFC (2019) Experimental measurements and simulation with artificial neural networks of the fouling phenomena of natural proteins. Int J Heat Mass Transf 129:1075–1085

Unscented Kalman Filter and Gauss-Hermite Kalman Filter for Range-Bearing Target Tracking

Gabriel Barragán[1(✉)], Saba Infante[1,2], and Aracelis Hernández[2]

[1] University Yachay Tech, Urcuquí, Ecuador
{gabriel.barragan,sinfante}@yachaytech.edu.ec
[2] University of Carabobo, Valencia, Venezuela
arhernan@uc.edu.ve

Abstract. Radar systems are used for estimating the position and velocity of aircraft and ships from range and bearing measurements. In this article, we consider two motion models for a civilian aircraft: a constant velocity model and a coordinated turn model. Both kinematics models are written in state-space terms, where the velocity perturbations are modeled by a white noise acceleration model. In order to estimate the aircraft states given noisy measurements obtained from sensor outputs, we follow a Bayesian statistical approach that calculates the estimators of the unknown states. In our simulation, we recreate an air traffic control scenario and implement two nonlinear filtering algorithms in order to perform target tracking of the aircraft. The nonlinear Bayesian-based filters for this target tracking problem are the unscented Kalman filter and the Gauss-Hermite Kalman filter. Finally, the performance of both nonlinear filters is evaluated with a performance metric, root mean squared error, in Monte-Carlo runs.

Keywords: Target tracking · Unscented Kalman filter · Gauss-Hermite Kalman filter

1 Introduction

Since target tracking allows to estimate position, velocity, and/or acceleration of remote objects from information of detection systems (infrared sensors, active or passive sensors, radio frequency sensors, etc.), researchers and practitioners have developed tracking techniques for a wide range of systems such as air traffic surveillance systems, spacecraft control, and military underwater systems [1–3]. A challenging task in tracking community and air traffic services is the real-time tracking of multiple flights in a determined zone. This is a particular case of tracking that requires knowing aircraft states (location, attitude, velocity, and heading) from recovered flight data. In the literature, alternative approaches have been proposed for the target tracking problem of an aircraft, such as models for bearing-Doppler measurements [1] and bearing-only measurements [2]. An important problem in aircraft tracking is the uncertainty in both process model (aerodynamics) and measurement model (available data) due to noise that

Y. Iano et al. (Eds.): BTSym 2020, SIST 233, pp. 537–544, 2021.
https://doi.org/10.1007/978-3-030-75680-2_59

adds redundant information or errors in the models. In this sense, filtering algorithms are useful to reduce noise and estimate the parameters of an aircraft. In the case of a linear dynamical system with white Gaussian noise processes, the Kalman filter can be used to calculate the minimum mean squared error (MMSE) estimates of the unknown states. However, the Kalman filter has a deficient performance for estimating the states of non-linear dynamical systems with non-Gaussian noise processes [4]. These nonlinearities along with non-Gaussianity assumption have motivated the development and the design of nonlinear filtering algorithms such as the derivative-based extended Kalman filter (EKF) [2, 4], the unscented Kalman filter (UKF) [5, 6], cubature Kalman filter (CKF) [3], ensemble Kalman filter (EnKF) [7, 8], particle filter (PF) [2], and Gauss-Hermite Kalman filter (GHKF) [2, 4, 9, 10]. This article contributes to the tracking of an aircraft, which is described by a dynamical system with two kinematics motion models (constant velocity model and coordinated turn model), and range and bearing measurements. For this purpose, we follow a Bayesian statistical approach that computes state estimates using recursive predictor-corrector nonlinear filtering algorithms (UKF and GHKF). The remaining of this paper is given as follows: Sect. 2 introduces the problem of state estimation and the dynamics of a civilian aircraft; In Sect. 3, we give the background of Bayesian estimation theory and also the main aspects of the UKF and the GHKF; In Sect. 4, we present the results of the performance of the nonlinear filters in an air traffic control scenario; Finally, we conclude the paper in Sect. 5.

2 Problem Description

In order to have a general framework for the proposed filtering algorithms, we consider the discrete-time state-space model for a dynamical system given by the following pair of difference equations [11, 12].

$$x_k = f(x_{k-1}) + v_{k-1}, \text{ *state equation*} \tag{1}$$

And

$$y_k = h(x_k) + w_k, \text{ *observation equation*} \tag{2}$$

Where $x_k \in R^{n_x}$ and $y_k \in R^{n_y}$ are the unknown states and the measurement of the unknown states at time step k, respectively. While f and h are known functions that describe the dynamics of the state equation and the observation equation, respectively. The sequences $\{v_k\}$ and $\{w_k\}$ are considered as zero-mean Gaussian noises with their respective covariances Q_k and R_k.

In classical target tracking, the states of a moving object (position, velocity, and acceleration) are expressed in Cartesian coordinates. Furthermore, we consider two models that describe a civilian aircraft's motion: constant velocity model and coordinated turn model [1].

2.1 Constant Velocity Model

For straight-line segments, an aircraft follows a constant velocity (CV) model in R^2. This motion model can be expressed as a state equation, also known as process equation, where the noise source is due to perturbations accelerations [1].

$$x_k = \begin{pmatrix} 1 & \Delta & 0 & 0 \\ 0 & 1 & 0 & 0 \\ 0 & 0 & 1 & \Delta \\ 0 & 0 & 0 & 1 \end{pmatrix} x_{k-1} + \begin{pmatrix} \frac{\Delta^2}{2} & 0 \\ \Delta & 0 \\ 0 & \frac{\Delta^2}{2} \\ 0 & \Delta \end{pmatrix} v_{k-1} \tag{3}$$

Where Δ is the sampling time interval between two consecutive states, $x = \left[\xi,\ \dot{\xi},\ \eta,\ \dot{\eta}\right]^T$ is the state vector with $(\xi,\ \eta)$ for the horizontal coordinates, i.e., x and y axis, and $(\dot{\xi},\ \dot{\eta})$ represent their respective velocity components. The term $v_k \sim N(0, Q_k)$ corresponds to the process equation's errors caused by turbulence, small acceleration, and other modeling errors.

2.2 Coordinated Turn Model

The turn of an aircraft usually follows a pattern known as coordinated turn (CT), where the turn rate is denoted by Ω. As in the CV model, we consider a model with perturbation accelerations as follows [1, 3]:

$$x_k = \begin{pmatrix} 1 & sin(\Omega\Delta) & 0 & -\frac{1-cos(\Omega\Delta)}{\Omega} & 0 \\ 0 & cos(\Omega\Delta) & 0 & -sin(\Omega\Delta) & 0 \\ 0 & \frac{1-cos(\Omega\Delta)}{\Omega} & 1 & \frac{sin(\Omega\Delta)}{\Omega} & 0 \\ 0 & sin(\Omega\Delta) & 0 & cos(\Omega\Delta) & 0 \\ 0 & 0 & 0 & 0 & 1 \end{pmatrix} x_{k-1} + \begin{pmatrix} \frac{\Delta^2}{2} & 0 & 0 \\ \Delta & 0 & 0 \\ 0 & \frac{\Delta^2}{2} & 0 \\ 0 & \Delta & 0 \\ 0 & 0 & \Delta \end{pmatrix} v_{k-1} \tag{4}$$

In this model, we consider an augmenting state vector $x = \left[\xi,\ \dot{\xi},\ \eta,\ \dot{\eta},\ \Omega\right]^T$. It is important to notice that the noise sequence in the CT model has a different dimension than the noise sequence in the CV model.

2.3 Measurement Model

In this article, we consider a measurement model that tracks an aircraft's motion in polar coordinates. This is common in static passive sensors that measure range, $\rho(\xi, \eta)$ and bearing (azimuth), $\theta(\xi, \eta)$ [3]. The corresponding observation equation of this model is given by

$$y_k = \begin{pmatrix} \rho(\xi_k, \eta_k) \\ \theta(\xi_k, \eta_k) \end{pmatrix} + w_k = \begin{pmatrix} \sqrt{(\xi_k - \xi_r)^2 + (\eta_k - \eta_r)^2} \\ \tan^{-1}\left(\frac{\eta_k - \eta_r}{\xi_k - \xi_r}\right) \end{pmatrix} + w_k \tag{5}$$

Where $w_k \sim N(0, R_k)$, and (ξ_r, η_r) is the position of the radar in the Cartesian plane.

3 Bayesian Filtering Theory

The main objective of Bayesian estimation methods is to calculate the posterior density $p(x_k|y_{1:k})$ from the predictive density $p(x_k|x_{k-1})$, and the measurement likelihood $p(y_k|x_k)$. In order to carry out state estimation, Bayesian-based methods incorporate prior knowledge of the system and also the information of the measurement model that determines the relationship between unknown states and measurements. The calculation of the posterior density can be done by a recursive filtering algorithm known as Bayesian filter. One of this filter's main characteristics is its two-step predictor-corrector recursion: time update and measurement update [2, 4].

1. *Time update.* The predictive density, given the state equation (1), can be calculated by the Chapman-Kolmogorov equation as follows

$$p(x_k|y_{1:k-1}) = \int p(x_{k-1}|y_{1:k-1})p(x_k|x_{k-1})dx_{k-1} \tag{6}$$

2. *Measurement update.* Given a new measurement y_k and the Eq. (2), the posterior density is updated with the Bayes' rule as follows

$$p(x_k|y_{1:k}) = \frac{p(x_k|y_{1:k-1})p(y_k|x_k)}{\int p(x_k|y_{1:k-1})p(y_k|x_k)dx_k} \tag{7}$$

A good estimation of the Bayesian filter is obtained if we assume that both predictive density and likelihood density are Gaussian. Under this Gaussianity assumption, we have that the posterior density is also Gaussian [2]. Therefore, the recursive Bayesian filter is reduced to operate with means and covariances of the probability densities. If the state equation and observation equation are linear, the Kalman filter provides the optimal solution to the filtering problem [4]. However, in most real-life problems, the dynamical systems have nonlinear state and observation equations and/or non-Gaussian noises [2, 4]. In this last case, the UKF and the GHKF give better-filtered estimates than the Kalman filter.

3.1 Unscented Kalman Filter

The unscented Kalman filter is an algorithm that performs a weighted statistical linearization through a deterministic sampling technique called unscented transform (UT). UT is a technique that generates a set of deterministic points called sigma points, which are propagated through the nonlinearities of the state-space model. The goal of this filtering algorithm is to capture the mean and covariance of a target probability density [2]. In order to illustrate the UT, let $x \in R^{n_x}$ be a random variable normally distributed with mean μ, and covariance P (i.e., $x \sim N(\mu, P)$), then the UT expected value approximation and the UT covariance approximation of a function $g(X)$ are given by

$$E[g(x)] \approx \sum_{i=0}^{2n_x} W_i^m Y^i, \quad Cov[g(x)] \approx \sum_{i=0}^{2n_x} \cdot W_i^c \left(Y^i - \mu\right)^T \left(Y^i - \mu\right)^T \tag{8}$$

Where $Y^i = g\left(X^i\right)$, and the set of $2n_x + 1$ sigma points X^i are given by

$$\begin{aligned} x^o &= \mu \\ x^i &= \mu + \left[\sqrt{(n_x+\lambda)P}\right]_i,\ i = 1,\ldots,n_x \\ x^i &= \mu - \left[\sqrt{(n_x+\lambda)P}\right]_i,\ i = n+1,\ldots,2n_x \end{aligned} \tag{9}$$

We denote $\sqrt{A}$ as square root matrix of matrix $\sqrt{A}$ (i.e. $\sqrt{A}\left(\sqrt{A}\right)^T = A$), and $[A]_i$ as the i−th column of matrix A. The UT constant weights W_i^m and W_i^c are obtained as follows

$$\begin{aligned} W_0^m &= \tfrac{\lambda}{n_x+\lambda}\quad W_0^c = \tfrac{\lambda}{n_x+\lambda} + (1-\alpha^2+\beta) \\ W_i^m &= W_i^c = \tfrac{1}{2(n_x+\lambda)},\ i = 1,\ldots,2n_x \end{aligned} \tag{10}$$

Where λ, α and β are parameters defined in [6].

A characteristic of this filter is that it captures the mean and covariance of the posterior density to the third order for any nonlinearity [4]. An advantage of the UKF over the EKF is that the former does not need any Jacobian or partial derivatives, which are computationally expensive and also unstable, for the estimation of filtered estimates [2, 4, 6]. For more details of the UKF, see [2, 4–6].

3.2 Gauss-Hermite Kalman Filter

The Gauss-Hermite Kalman filter (GHKF) is a nonlinear filtering algorithm based on the Gauss-Hermite quadrature. This filter approximates the Bayesian filter's integrals with the help of a set of quadrature points ξ and their respective weights W using a Gauss-Hermite scheme [9, 10]. For illustration purposes, let us consider a random variable $x \in R^{n_x}$ such that $x \sim N(\mu, P)$, then the expected value of function $g(x)$ is approximated by

$$E\left[g(x)\right] = \int_{R^{n_x}} g(x)N(\mu,P)dx \approx \sum\nolimits_{i_1,\ldots,i_{n_x}=1}^{p} W_{i_1,\ldots,i_{n_x}} g\left(\mu + \sqrt{P}\xi^{(i_1,\ldots,i_{n_x})}\right) \tag{11}$$

Where $\xi^{(i_1,\ldots,i_{n_x})} = \left[\xi^{i_1}\cdots\xi^{i_{n_x}}\right]^T$, and $W_{i_1,\ldots,i_{n_x}} = W_{i_1}\times\cdots\times W_{i_{n_x}}$. In order to use these $p-$order quadrature points ξ^i, it is required to extract the roots of a Hermite polynomial of order p. For more details and extensive references on implementing the GHKF and numerical methods for computing quadrature points and weights, see [2, 4, 9, 10, 12].

An advantage of the GHKF over the EKF is that the former only needs to compute a set of quadrature points and propagate them through the nonlinearities. Furthermore, GHKF can be applied in filtering problems with correlated or non-additive Gaussian process and measurement noises. However, a disadvantage of this filter is the computational cost when it is used in filtering problem with state vectors of higher dimension (the GHKF requires to calculate p^{n_x} quadrature points in each recursion) [4]. Another problem of this filter occurs when the GHKF is applied for a long period of time; in this case, the round-off errors accumulate and may produce divergent filtered estimates [2, 4].

4 Simulation

In this section, we illustrate the performance of the UKF and the GHKF on an aircraft traffic control (ATC) scenario. We consider that the motion of a civilian aircraft in the horizontal plane is described by Eqs. (3) and (4), and the measurement model is given by Eq. (5).

For this target tracking problem, we set $\Delta = 5\,\text{s}$, and the entire tracking time to be 600 s. The actual initial estimate is

$$x_0 = \left[10000\,\text{m},\ 125\,\text{ms}^{-1},\ 10000\,\text{m},\ 0\,\text{ms}^{-1},\ 0\,\text{rad s}^{-1}\right]^T$$

In order to initialize both filtering algorithms, the initial state estimate is chosen as $\hat{x}_0 \sim N(x_0, P_0)$, where the initial covariance, P_0, is taken as

$$P_0 = diag\left(100\,\text{m}^2,\ 10\,\text{m}^2\text{s}^{-2},\ 100\,\text{m}^2,\ 10\,\text{m}^2\text{s}^{-2},\ 1e^{-5}\,\text{rad}^2\text{s}^{-2}\right)$$

The process covariance is assumed as $Q = diag(q_1 M,\ q_1 M,\ q_2 \Delta)$ where

$$M = \begin{pmatrix} \frac{\Delta^3}{3} & \frac{\Delta^2}{2} \\ \frac{\Delta^2}{2} & \Delta \end{pmatrix} \tag{12}$$

With intensity parameters $q_1 = 0.1$ and $q_2 = 0.00175$.

The measurement covariance is set to be $R = diag\left(r_\rho^2,\ r_\theta^2\right)$ with measurement noise parameters $r_\rho^2 = 110$ and $r_\theta^2 = 0.0017$. The sensor is located at $(\xi_r,\ \eta_r)^T = (750\,\text{m},\ 1750\,\text{m})^T$.

In order to compare the performance of the UKF and the GHKF of order $p = 3$, we analyze the filtered estimates of each one in $M = 200$ Monte-Carlo runs. For this ATC scenario, we use the root mean squared error (RMSE). The RMSE is a performance metric used to evaluate the accuracy of a proposed estimation method. The lower the RMSE value of the filtered estimate, the better the accuracy of the method. In this sense, we define the $RMSE_k$ in position as follows

$$RMSE_k = \sqrt{\frac{1}{M} \sum_{n=1}^{M} \left(\left(\xi_k^n - \hat{\xi}_k^n \right)^2 + \left(\eta_k^n - \hat{\eta}_k^n \right)^2 \right)} \tag{13}$$

Where $\left(\hat{\xi}_k^n,\ \hat{\eta}_k^n\right)^T$ is the filtered estimates of $\left(\xi_k^n,\ \eta_k^n\right)^T$ at time k of the n–th Monte-Carlo run.

In Fig. 1, we depict the tracking scenario and the filtered estimates of both UKF and GHKF during a particular simulation run. In this figure, it is shown that the aircraft has maneuvered in three-time intervals: $\Omega = 1°$ for 90 s, $\Omega = -3°$ for 30 s, and $\Omega = -1°$ for 90 s.

In Fig. 2, we observe the RMSE in position for 200 Monte-Carlo runs of UKF and GHKF. It can be seen that the RMSE in position is lesser in the UKF than in the 3–th order GHKF.

We consider that the low accuracy of the GHKF is due to the round-off errors of this algorithm. In addition, we remark that the GHKF needs to calculate a larger quantity of sample points ($p^{n_x} = 3^5 = 243$) in comparison to the UKF ($2n_x + 1 = 2(5) + 1 = 11$) at each time step.

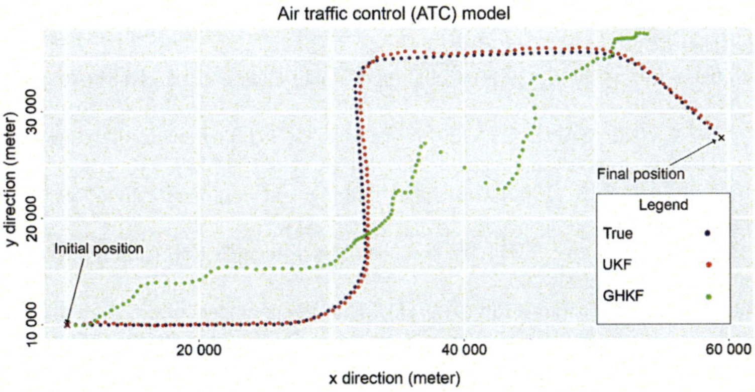

Fig. 1. Filtered estimates of the UKF and the GHKF.

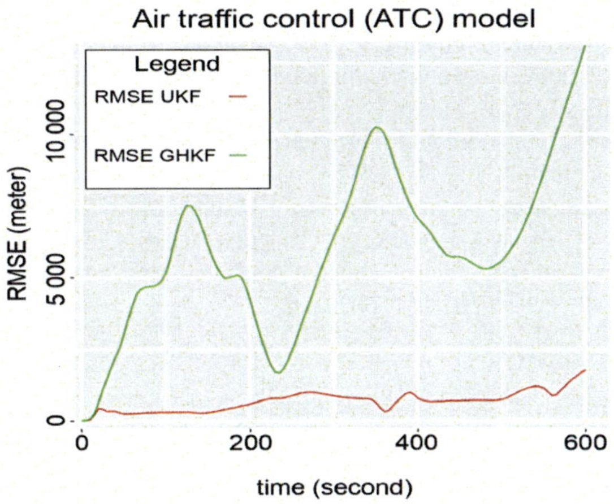

Fig. 2. RMSE in position for 200 Monte-Carlo runs of UKF and GHKF

5 Conclusions

In this article, we have provided a real application of the UKF and the GHKF in an air traffic control scenario. In this target tracking problem, we have considered the dynamics of an aircraft in the horizontal plane given by the constant velocity model and coordinated turn model, and also the measurement model with range and bearing observations. Finally, we have compared the performance of both filtering algorithms using the RMSE, and we have concluded that the UKF produced better-filtered estimates than the GHKF filtered estimates.

References

1. Bar-Shalom Y, Li XR, Kirubarajan T (2004) Estimation with applications to tracking and navigation: theory algorithms and software. Wiley, Hoboken

2. Särkkä S (2013) Bayesian filtering and smoothing, vol 3. Cambridge University Press, Cambridge
3. Arasaratnam I, Haykin S (2009) Cubature Kalman filters. IEEE Trans Autom Control 54(6):1254–1269
4. Afshari HH, Gadsden SA, Habibi S (2017) Gaussian filters for parameter and state estimation: a general review of theory and recent trends. Sig Process 135:218–238
5. Julier SJ, Uhlmann JK (2004) Unscented filtering and nonlinear estimation. Proc IEEE 92(3):401–422
6. Wan EA, Van Der Merwe R, Haykin S (2001) The unscented Kalman filter. Kalman Filter Neural Netw 5(2007):221–280
7. Sánchez L, Infante S, Griffin V, Rey D (2016) Spatio-temporal dynamic model and parallelized ensemble Kalman filter for precipitation data. Braz J Probab Stat 30:653–675
8. Soto J, Infante S (2019) Ensemble Kalman filter and extended Kalman filter for state-parameter dual estimation in mixed effects models defined by a stochastic differential equation. In: International conference on 'knowledge society: technology, sustainability and educational innovation'. Springer, pp 285–300
9. Arasaratnam I, Haykin S, Elliott RJ (2007) Discrete-time nonlinear filtering algorithms using Gauss-Hermite quadrature. Proc IEEE 95(5):953–977
10. Ito K, Xiong K (2000) Gaussian filters for nonlinear filtering problems. IEEE Trans Autom Control 45(5):910–927
11. Infante S, Sanchez L, Hernandez A (2020) Stochastic models to estimate population dynamics. Stat Optimiz Inf Comput 8(1):136–152
12. Barragán G (2020) State and parameter estimation in stochastic dynamical system. Bachelor's thesis. Escuela de Ciencias Matemáticas y Computacionales, Universidad de Investigación de Tecnología Experimental Yachay, Urcuquí, Ecuador. https://repositorio.yachaytech.edu.ec/bitstream/123456789/137/1/ECMC0020.pdf

Amoxicillin Determination by Colorimetric Methods: An Experience Report Using the DMPD and NCS Methods

Mariana Cardoso Barros Ribeiro[1], Rodrigo Fernando dos Santos Salazar[2], and André Luis de Castro Peixoto[1](✉)

[1] Instituto Federal de Educação, Ciência e Tecnologia de São Paulo (IFSP), Av. Dr. Enio Pires de Camargo 2971, Capivari, SP 13360-000, Brazil
alcpeixoto@ifsp.edu.br

[2] Mondial Laboratório de Produtos Químicos Ltda (MONDIAL LAB). 3275, Cruz Alta, RS 98045-138, Brazil

Abstract. In general, pharmaceutical compounds are difficult to degrade (persistent compounds) in conventional and advanced treatment systems. In parallel, to ensure these compounds' monitoring, analytical and detection methods must minimize or neutralize the interferences in the sample matrix. In this sense, the present study characterizes it as being diagnostic and descriptive in that, from a conceptual approach of the fundamentals of analytical chemistry and chemical reaction mechanisms, understand the advantages and limitations of the methods of determination of amoxicillin in samples containing components photocatalytic oxidation residues. One method makes use of the N-N-dimethyl-p-phenylenediamine reagent (DMPD) and the other of the N-chlorosuccinimide reagent (NCS). Even in the presence of several constituents of the medicine and of transformation products generated by its oxidation, the methods discussed here presented themselves as selective.

Keywords: Amoxicillin · N-N-dimethyl-p-phenylenediamine (DMPD) · N-chlorosuccinimide (NCS)

1 Spectrophotometry: Fundamentals, Applications, and State-of-the-Art

To apply UV-visible spectrophotometric techniques for component analysis in aqueous media correctly and reliably, it is necessary to understand the principles behind this technique and guidelines on the best laboratory operational practices to obtain reliable spectral data [1, 2]. In this way, the understanding of the fundamentals that permeate the proper use of spectrophotometric techniques takes into account the basic concepts: 1) the interaction of light with solutions containing chromophores and other absorption and dispersion processes; 2) the calibration requirements to ensure adequate accuracy and precision and consequent obtaining reliable spectral data results; 3) good skills

Y. Iano et al. (Eds.): BTSym 2020, SIST 233, pp. 545–552, 2021.
https://doi.org/10.1007/978-3-030-75680-2_60

with analytical chemistry to correct application of certified reference standards and analytically validated procedures [1, 3, 4].

Regarding the requirements of analytical nature and solution chemistry, it is necessary for the analyst who will use spectrophotometry to have tacit knowledge of the requirements that make up the analytical sequence that will guarantee the integrity and validity of the data obtained [1, 4, 5]. The analytical sequence can also be called the analytical frequency and has the following steps: 1) problem definition; 2) selection of the method; 3) sampling with a sample collection that is representative; 4) preparation of samples that, for spectrophotometric techniques, require digestion or solubilization of samples in a liquid medium; 5) use of spectrophotometric buffers or chemical correctors, when necessary, to eliminate or minimize interference; 6) execution of the analytical measure and calibration of the equipment signal with a set of standards with defined, stable and well-known concentrations, since the spectrophotometric techniques are not absolute analytical techniques; 7) calculation of results and statistical validation and; 8) issuing the opinion and making a decision with the result issued [5, 6]. Among the steps that constitute the analytical frequency, the steps of sampling and sample preparation are the most critical for spectrophotometric determination in comparison to the other steps that constitute, since they are more subject to contamination and factors that lead to systematic errors that end up compromising the reliability of the results [7, 8]. In this sense, the importance of the integrity and security of the data has made it fundamental to the requirements that advocate good laboratory practices (GLP), serving as a starting point for laboratories to comply with the different regulatory requirements that certify the quality of laboratory tests such as the ISO 17025 Guide, the last update occurred in the year 2017 [1, 7].

In any case, with the advance in analytical procedures involving sampling and sample preparation and, in parallel, technological advances have allowed the development of this technique in terms of detection systems and strategies [7]. Consequently, there is a consolidation of the use of UV-Vis spectrophotometry as a simple, sensitive, reliable, and low-cost technique, which allows the determination of low concentrations of compounds and the use of minimal amounts of samples. Besides, the versatility and portability of the equipment and strategies developed with the use of UV-Vis have also contributed to the permanent use and growth over the years when compared to other spectroanalytical techniques [8]. In this sense, several fields, such as agriculture, food, clinical, pharmaceutical, and environmental, have allowed the evolution of detection in UV-Vis spectrophotometry to continue over the next years by the development of more robust, portable systems with ample capacity for applications in different environments to meet different specificities [3].

2 Analytical Methods to Amoxicillin Determination

Among the chemical compounds belonging to the penicillin group, amoxicillin is the only one that presents the phenolic group. Several spectrophotometric methods explore this difference in the functional group. In commercial formulations, amoxicillin is commonly used with potassium clavulanate. This substance aims to inhibit the production of β-lactamase. In addition to starch and sugars, pharmaceutical formulations of amoxicillin contain excipients such as dyes and flavors. In the studies of our research group, a pharmaceutical formulation was used in the form of a powder for an aqueous suspension containing the following excipients declared by the manufacturer: sodium benzoate, sodium citrate dihydrate, cherry aroma, strawberry aroma, silicon dioxide, xanthan gum, red erythrosine disodium, and sucrose.

It is necessary to know the pharmaceutical formulation minimally to ensure that the methodology to be applied in the quantification of amoxicillin is selective. It means that the overlapping of absorption signals by the different substances present in the sample must be avoided. Besides, using the artifice of chemical reactions that generate a colored chemical substance with the amoxicillin molecule, one must consider that the reaction must be kinetically favorable and that the colored product is stable. These characteristics (selectivity, favorable kinetics, and stability) are important to be considered in a laboratory routine.

We opted to analyze the theoretical and practical aspects of two spectrophotometric methodologies to analyze amoxicillin molecules in the present work. The first methodology evaluated involved the use of the chemical reagent N-N dimethyl-p-phenylenediamine (DMPD). Due to several difficulties encountered with the handling of the DMPD reagent, a second methodology was sought. Thus, the research group carried out some tests with the N-chlorosuccinimide reagent (NCS). The proposed use of these methodologies is part of the context of the study of photooxidative degradation of amoxicillin in an aqueous medium. This manuscript will consider the qualitative and quantitative aspects of both methodologies for amoxicillin analysis and the advantages and limitations of both methodologies within our work proposal. Social and environmental aspects are interconnected and must be dealt with jointly. For example, toxicity aspects of chemical reagents are essential for the safety of personnel involved in the laboratory routine and the environmental safety of waste generated after using these same reagents.

In academic studies involving effluents treatment, the main advantage of using UV-Vis spectrophotometric methodologies is the low cost of analysis. In general, spectrophotometers have a lower purchase cost when compared, for example, with high-performance liquid chromatography equipment (HPLC). Even considering the simplest configuration of a GC equipment, which can also be used to quantify amoxicillin, purchasing a spectrophotometer is still more advantageous from a financial point of view. The financial factor is relevant to academic institutions from developing countries. Considering the use of consumables, in general, spectrophotometric methods have an advantage over chromatographic methods from a financial point of view. In the visible region, the spectrophotometric methods use glass cuvettes, distilled water, and analytical grade reagents. Chromatographic methods, on the other hand, require chromatographic columns (*e.g.*, C8, C18, DB-5), guard columns, in addition to organic solvents with a higher degree of purity. All of this, without considering the maintenance costs of the analytical equipment, which is more expensive with chromatographs.

2.1 A Case Report Employing DMPD Methodology

This methodology was the first to be used in our academic research laboratory. The analysis method consists of the reaction, in an alkaline medium, between amoxicillin, N,N-dimethyl-p-phenylenediamine (DMPD), and potassium hexacyanoferrate(III). According to Al-Abachi et al. [9], by oxidation with potassium hexacyan-ferrate(III), DMPD loses two electrons and a proton, forming an electrophilic intermediate. This intermediate, II, undergoes electrophilic substitution by the phenolic group of the AMX molecule, forming a blue-colored product (Fig. 1).

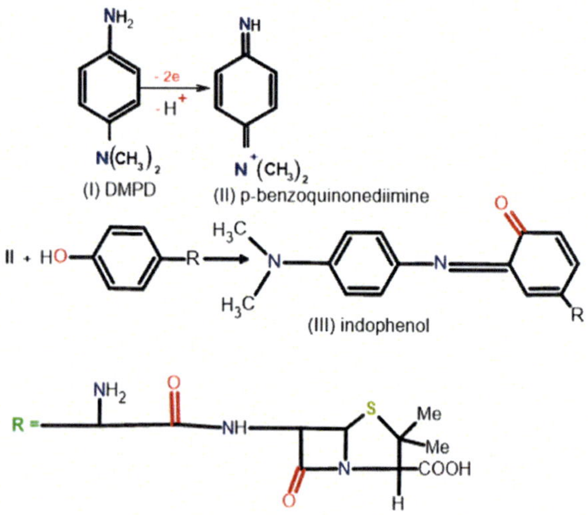

Fig. 1. Reaction mechanism involved in the analysis of amoxicillin with the DMPD reagent.

The analyses were performed in the laboratory with a room temperature stabilized at 20 °C. The samples had their pH values corrected to ~7. In 10.00 mL volumetric flasks, 0.50 mL of ammonium hydroxide (50 mM), 0.80 mL of potassium hexacyanoferrate(III) (10 mM), 0.80 mL of DMPD (5 mM), and 2.00 mL of sample were added. The volume was completed with ultrapure water. After 15 min of reaction, the absorbance reading in the spectrophotometer (660 nm) was performed. The blank was carried out with all reagents except AMX.

Amoxicillin trihydrate solutions were prepared to perform the calibration curve ranging from 2 to 40 μg mL^{-1} (Eq. 1), in which:

$$[\text{amoxicillin}] = 53,406 * \text{abs} - 0,1815 \tag{1}$$

$$R^2 = 0,9954$$

The term [amoxicillin] refers to the concentration of amoxicillin (μg mL^{-1}) in the sample, and the term *abs* refers to its absorbance.

The chemical instability of the DMPD reagent is what makes the laboratory work routine more challenging. Due to its sensitivity to light and oxygen, it is necessary to immediately use the DMPD solution after its preparation. Thus, during the analyses, there was a control of luminosity, turning off laboratory lights and making analyses always near sunset time. The volumetric flasks were wrapped in aluminum foil until reading on the spectrophotometer. Ideally, the use of amber-type volumetric flasks is a better choice.

Regarding storage, the DMPD reagent was kept under an inert atmosphere (N_2) according to the manufacturer's technical support instructions. It was performed by injecting the inert gas into the reagent bottle and immediately sealing it. An additional precaution was taken by storing the bottle with an inert atmosphere in a glass desiccator and under low pressure.

Even paying attention to all the details discussed above, it was necessary to carry out the analyses quickly and with another student's help (Fig. 2). The entire experimental procedure was time-controlled. As an unstable reagent, the DMPD solution was discarded after use. Figure 2a illustrates the blank and the solutions ready for absorbance reading in the spectrophotometer (bluish color) at 15 min of reaction. Figure 2b shows one of these solutions after 24 h of preparation.

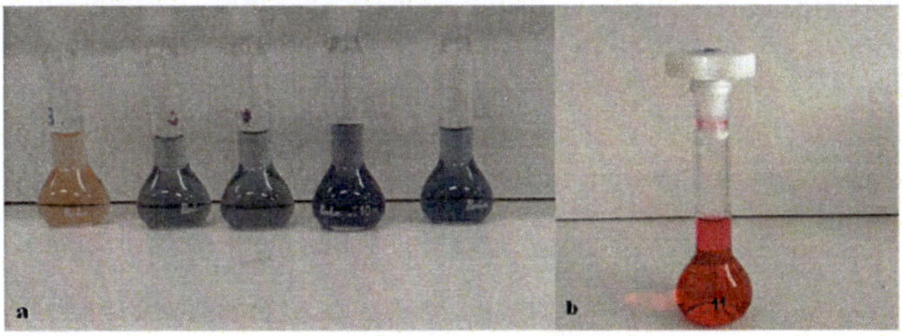

Fig. 2. Volumetric flasks prepared with DMPD methodology: (a) using the freshly prepared DMPD solution; (b) using the DMPD solution after 24 h of preparation.

In addition to being an unstable reagent, DMPD is also quickly impregnated in different materials (Fig. 3). Figure 3a shows a silicone hose used to generate a vacuum in the desiccator (storage of the DMPD under low pressure). Figure 3b shows the silica gel from that same desiccator, impregnated with the DMPD reagent. Figure 3c shows a Falcon-type tube used in the preparation of the DMPD solution.

In glassware (*e.g.*, glass cuvettes), it was possible to remove traces of DMPD with some cleaning steps. First, this cleaning procedure was done with a solution of sodium alcoholate 10% w/v. Then, neutral soap with an ultrasonic bath was used. Finally, the abundant use of water was made.

Fig. 3. Laboratory materials impregnated with the DMPD reagent: (a) silicone hose; (b) silica gel; (c) Falcon-type tube made of polypropylene.

2.2 A Case Report Employing NCS Methodology

Due to the characteristics of DMPD, as discussed, it was decided to seek a second methodology for the AMX quantification, a methodology that would be more practical for a laboratory routine. After a stage of bibliographic execution, it was used Saleh's methodology [10]. Saleh proposed two spectrophotometric methods for amoxicillin analysis that stand out for being simple to perform, fast and selective [10]. In these methodologies, the amoxicillin phenolic group is oxidized with (i) N-bromosuccinimide (NBS) and (ii) N-chlorosuccinimide (NCS) in an alkaline medium, developing a colored solution (light yellow), as shown in Fig. 4. In the method developed by Saleh [10], different types of samples (syrup, tablets, and capsules) containing amoxicillin were tested. With commercial formulations, solutions containing amoxicillin with concentrations varying between 1 and 200 $\mu g\ mL^{-1}$ were prepared.

Fig. 4. Volumetric flasks prepared with DMPD methodology: blank solution in the left; with amoxicillin 26 $\mu g\ mL^{-1}$ in the right.

The N-bromosuccinimide reagent (NBS) has high toxicity to the aquatic environment, presenting half-maximal effective concentration (EC_{50}) of 0.65 mg L^{-1} in 48 h for *Daphnia magna* and other aquatic invertebrates. Its biodegradability is not fast, but it is also not bioaccumulative. It has acute toxicity by inhalation and can damage the respiratory tract. Contact with the skin causes irritation and even burns. Contact with the eyes can cause severe eye damage. Its half lethal dose is median, with LD_{50} >2000 mg kg^{-1} [11].

However, the NCS reagent is less harmful to the environment and does not develop acute and/or chronic toxicity to organisms in the aquatic environment. It is also less harmful to the skin compared to NBS. Its low toxicity, plus the fact that it is up to 20% cheaper than NBS, made us choose NCS in our routine analyses.

Saleh tested eight different solvents in his study [10], namely: water, methanol, ethanol, propan-2-ol, propan-1-ol, acetone, dimethyl sulfoxide (DMSO), and dimethyl fumarate (DMF) [10]. Among these solvents, methanol showed the highest color intensity (0.460). Although Saleh concluded that methanol is the best analytical option for quantifying amoxicillin in pharmaceutical formulations [10], we consider propan-1-ol solvent in our laboratory routine. Methanol has higher toxicity and volatility than the propan-1-ol solvent. However, because it has a lethal dose for rats of 143 mg kg^{-1}, much lower than for propan-1-ol (LD_{50} 3670 mg kg^{-1}), and due to its higher vapor pressure, we opted to use propan-1-ol, which was the second-best solvent (0.435).

The analyses were carried out in a laboratory with a controlled room temperature at 20ºC. The samples had their pH values corrected to ~7. In 10.00 mL glassware flasks, 1.00 mL of sample, 1 mL of sodium hydroxide (0.1 mM), 1 ml of NCS (5%), and propan-1-ol were added to complete the flask volume. Then, the absorbance reading was performed in the spectrophotometer. The blank was made with all reagents except AMX. Each sample was promptly read because of the solvent volatility. With the use of methanol, the issue of solvent volatility is more critical. The calibration curve prepared from the analytical standard of amoxicillin trihydrate (1 to 10 μg mL^{-1}) is shown by Eq. (2):

$$[\text{amoxicillin}] = 76.955 * \text{abs} + 0.1815 \tag{2}$$

$$R^2 = 0.9990$$

The term [amoxicillin] refers to the concentration of amoxicillin (μg mL^{-1}) in the sample, and the term *abs* refers to its absorbance.

Using this methodology, the laboratory routine became more practical. While the DMPD method required two people's work, the NCS method could be performed by only one student. The NCS method is less susceptible to errors and also less time-consuming. Although both methodologies discussed here use reagents that are not stable for a long time, NCS is more stable than DMPD, and its solution can be used for about 1–2 h. This time is sufficient for the preparation and measurements of many samples containing amoxicillin. However, it is emphasized that DMPD and NCS solutions should not be stored, and the preparation of these solutions should be done close to the moment of analysis.

3 Final Consideration

Spectrophotometric methods are versatile, selective, and cost-effective in laboratory routines. Depending on the analytical objectives, UV-Vis spectrophotometry can be as useful as liquid or gas chromatography, with the advantage of less investment in the purchase and maintenance of the equipment. Both the DMPD method and the NCS method have a low detection limit using cells of 1 cm of optical path. However, this detection limit can be reduced with cells with a larger optical path or even using pre-concentration of samples. The discussion pointed out throughout the text, we indicate the use of the method with the NCS reagent and the propan-1-ol solvent. Thus, there is a sensitive method, with relative stability of the primary reagent, and good characteristics of solvent vapor pressure, cost, and low toxicity. Therefore, factors of technique, cost, as well as environmental and social issues were considered. The NCS method was successfully applied to AMX photooxidative degradation. Such oxidative processes were based on AMX pharmaceutical formulation, in a complex aqueous matrix and its transformation products.

Acknowledgments. Grants #2018/05698-0 and #2019/09991-6, São Paulo Research Foundation (FAPESP).

References

1. Burgess C (2017) The basis for good spectrophotometric UV–visible measurements. In: Burgess C, Thomas O (eds) UV-visible spectrophotometry of water and wastewater, 2nd edn. Elsevier, Amsterdam, pp 1–35
2. Thomas O, Brogat M (2017) Organic constituents. In: Burgess C, Thomas O (eds) UV-visible spectrophotometry of water and wastewater, 2nd edn. Elsevier, Amsterdam, pp 73–138
3. Pérez-Lemus N, López-Serna R, Pérez-Elvira SI, Bernardo E (2019) Analytical methodologies for the determination of pharmaceuticals and personal care products (PPCPs) in sewage sludge: a critical review. Analytica Chimica Acta 1083:19–40
4. de Oliveira MA, Porto BL, Bastos CDA, Sabarense CM, Vaz FA, Neves LN, Duarte LM, Campos NDS, Chellini PR, da Silva PH, de Sousa RA (2016) Analysis of amino acids, proteins, carbohydrates and lipids in food by capillary electromigration methods: a review. Anal Methods 8(18):3649–3680
5. Krug FJ (2008) Métodos de preparo de amostras: fundamentos sobre preparo de amostras orgânicas e inorgânicas para análise elementar, 1st edn. EdiUSP, Piracicaba
6. Thomas O, Causse J (2017) From spectra to qualitative and quantitative results. In: Burgess C, Thomas O (eds) UV-visible spectrophotometry of water and wastewater, 2nd edn. Elsevier, Amsterdam, pp 37–72
7. ABNT (2017) NBR ISO/IEC 17025 Requisitos Gerais para Competência de Laboratórios de Ensaio e Calibração. ABNT Editora, Rio de Janeiro
8. Passos MLC, Saraiva M, Lúcia MFS (2019) Detection in UV-visible spectrophotometry: detectors, detection systems, and detection strategies. Measurement 135:896–904
9. Al-Abachi MQ, Haddi H, Al-Abachi AM (2005) Spectrophotometric determination of amoxicillin by reaction with N, N-dimethyl-p-phenylenediamine and potassium haxacyanoferrate (III). Analytica Chemica Acta 554(1–2):184–189
10. Saleh GA (1996) Two selective spectrophotometric methods for the determination of amoxicillin and cefadroxil. The Analyst 121(5):641–645
11. MERCK. https://www.merckmillipore.com

Detection of Simultaneous Faults Using State Observer Methodology

Camilo Rocha Araújo(✉) and Gilberto Pechoto de Melo

Department of Mechanical Engineering, UNESP - Ilha Solteira, Av. Brasil Centro, 56, Ilha Solteira, SP CEP 15385-000, Brazil
{camilo.rocha,gilberto.pechoto}@unesp.br

Abstract. Due to the big interest of industries in predictive maintenance, the development of techniques to diagnose failures has been gradually acquiring a relevant role in the current industrial scenario. For this reason, this paper aims to evaluate a fault diagnosis methodology based on the state observer theory, considering simultaneous failures. In order to validate this methodology, a system of vibrating tables was built to perform experimental tests, which consist of inputting two failure modes in a mechanical structure: mass loss and stiffness loss. The system is, then, excited by a shaker, and its response is measured. Applying the measured signals (the input and output) in the mathematical system model, the state observer methodology was, thus, executed to estimate the observable system states. The estimated state corresponding to the measured state is compared to the measured signal, resulting in a fault metric that can be used for failures characterization.

Keywords: Simultaneous failures · State observer methodology

1 Introduction

The fault identification in mechanical structures is widely studied, and, thus, several methodologies for failure detection have been developed in the last years [1–3]. Amidst those methodologies, the methodology based on state observer is among the ones that stand out [4–6], allowing detection, localization, and quantification of faults using the estimate of at least one observable state of the system that is compared to its respective measured state [7], i.e., if it is possible to measure and estimate at least one state of the system, then, the signals (measured and estimated) are compared with each other, resulting in the creation of a fault metric that can be analyzed to identify failures in the system [8]. This methodology has been applied lately to characterize different failure types. The last work published was developed by Leão et al. [9]. They applied state observer methodology to identify a transversal crack in a horizontal rotating shaft, obtaining good results in his simulations.

However, so far, nobody has applied this methodology to detect simultaneous faults. For this reason, this paper aims to evaluate the state observer methodology to identify two fault modes occurring at the same time. As it is an unprecedented investigation, a

Y. Iano et al. (Eds.): BTSym 2020, SIST 233, pp. 553–560, 2021.
https://doi.org/10.1007/978-3-030-75680-2_61

mechanical system simple enough to be modeled as a mass-spring-damper system, with four degrees of freedom, was considered. Also, the fault modes taken into account were mass loss and stiffness loss.

2 Mathematical Model

The mechanical system to be modeled can be seen in Fig. 1, which consists of a structure built of three aluminum plates (modeled like concentrated masses) connected to some steel blades (modeled like stiffness elements) and a vibration absorber (modeled like a concentrated mass and a stiffness element) coupled in the upper plate. The structure is excited by a shaker coupled in the bottom plate. The approximated model can be seen in Fig. 2.

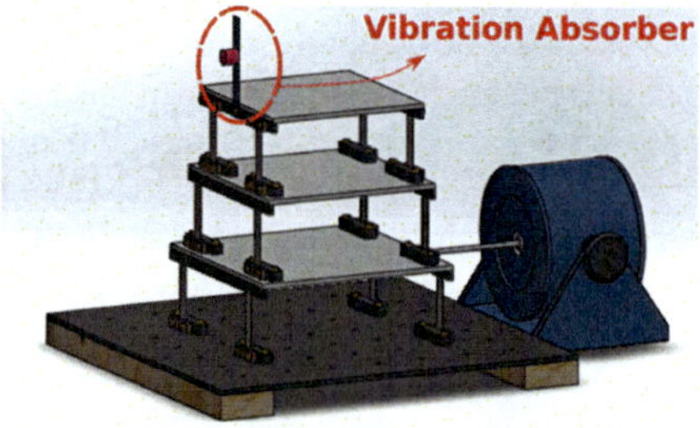

Fig. 1. Mechanical system.

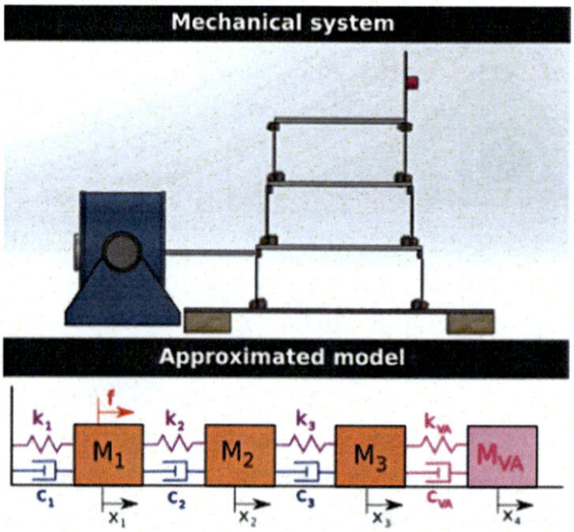

Fig. 2. Mechanical system and approximated model.

After applying Newton's Second Law in the approximated model, the following equation of motion is obtained:

$$[M]\ddot{X}(t) + [D]\dot{X}(t) + [K]X(t) = F(t) \tag{1}$$

Where: $[M]$, $[D]$ and $[K]$ are, respectively, mass, damping, and stiffness matrices. The vectors $X(t)$ and $F(t)$ contain the displacements and the inputs of the system, respectively.

$$X(t) = \left[x_1(t)\ x_2(t)\ x_3(t)\ x_4(t)\right]^T \tag{2}$$

$$F(t) = \left[f(t)\ 0\ 0\ 0\right]^T \tag{3}$$

Equation 1 can be written in state space, resulting in the following equation:

$$\begin{aligned} \dot{X}(t) &= [A]X(t) + [B]F(t) \\ y(t) &= [C]X(t) \end{aligned} \tag{4}$$

Where: $[A]$, $[B]$ and $[C]$ are, respectively, dynamic, input and output matrices of the system; $y(t)$ is the output of the system. The vector $X(t)$ is redefined as a state vector (contains displacements and velocities of the system).

$$A = \begin{bmatrix} [O] & [I] \\ [M]^{-1}[K] & [M]^{-1}[D] \end{bmatrix}; B = \begin{bmatrix} [O] \\ [M]^{-1} \end{bmatrix}; C = \left[0\,0\,0\,0\,0\,0\,1\,0\right] \tag{5}$$

$$C = \left[x_1(t)\ x_2(t)\ x_3(t)\ x_4(t)\ \dot{x}_1(t)\ \dot{x}_2(t)\ \dot{x}_3(t)\ \dot{x}_4(t)\right] \tag{6}$$

Where: $[O] \epsilon IR^{nxn}$ is a null matrix, and $[i] \epsilon IR^{nxn}$ is an identity matrix and is the number of states of the system ($n = 8$). In the $[C]$ matrix, just the seventh element is equal to 1, and it indicates that just the seventh state ($\dot{x}_3$) is available to be measured, so, $y(t) = \dot{x}_3$.

3 State Observer Methodology

The state observer was built from the mathematical model of the mechanical system, whose states were interested in estimating. As an input, this model needs at least one output and all inputs measured from the mechanical system.

Starting with Eq. (4), it is possible to build a state observer defined by the following equation:

$$\begin{aligned} \dot{\grave{X}}(t) &= [A]\grave{x}(t) + [B]F(t) + [L]\big(y(t) - \grave{y}(t)\big) \\ \grave{y}(t) &= [C]\grave{X}(t) \end{aligned} \tag{7}$$

Where: $\grave{x}$ denotes the estimated states and $\grave{y}$ corresponds to the state observer estimated output, which can be compared with the mechanical system's output ($\grave{y}$). $[L]$ is the gain of the observer.

The gain $[L]$ can always be calculated if the system is observable, i.e., if the following condition is satisfied:

$$rank(M_o) = rank\left(\left[C\ CA \ldots CA^{n-1}\right]^T\right) = n \tag{8}$$

Where: M_o is called the observability matrix.

For the considered system, condition 8 is satisfied. Then, $[L]$ can be calculated so that estimation error (Eq. 9) tends to zero as fast as possible.

$$e(t) = y(t) - \grave{y}(t) \tag{9}$$

Based on the estimation error, a fault metric (*fm*) is defined as a quantitative parameter, which can be used to identify the failure.

$$fm = \frac{1}{RMS(y(t)) - RMS(\grave{y}(t))} \tag{10}$$

If $fm \to \infty$, it means that the estimated $y(t)$ and measured $\grave{y}(t)$ states are almost equal, i.e., the estimation error tends to zero, and else, the state observer is not able to estimate the states correctly.

Thus, a state observer is designed based on the plant without failures. Then, the system is excited with a sine wave by the shaker, and the velocity of the third plate is measured (seventh state). If there is no fault, the estimate works well, and the estimation error tends to zero, i.e., $fm \to \infty$. Otherwise, if there is some failure, the estimate does not work well, and the estimation error does not tend to zero because the mechanical system has undergone changes due to the failure. These changes have not been updated in the mathematical model.

When a failure is detected, the parameters of the state observer are modified until the estimate works well. It is possible to know which fault mode is occurring when seeing the system parameters that make the estimation error tends to zero.

4 Experiment

The experimental assembly can be seen in Fig. 3.

The data acquisition system used was the Quanser Q8 - USB Data Acquisition Device (letter E in Fig. 3), which is a portable and affordable data acquisition system for real-time measurement and control. "*The Quanser Q8-USB is a single-point I/O, eight-channel data acquisition device that delivers reliable real-time performance via a USB interface*" [10].

The mechanical properties are shown in Table 1:

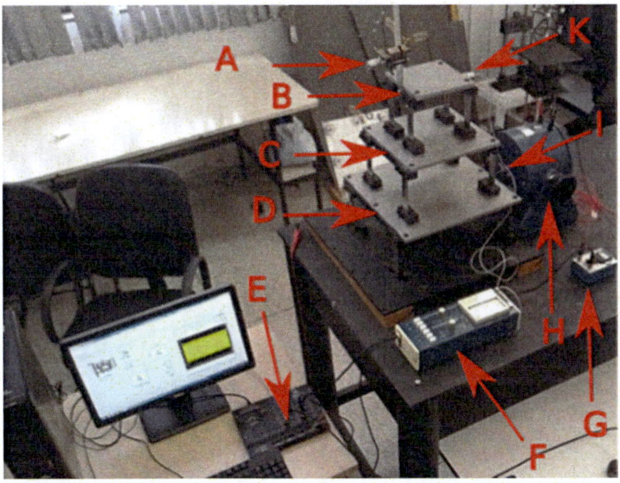

Fig. 3. Setup: (A): Vibration absorber mass, (B) Third plate, (C) Second plate, (D) First plate, (E) Q8 - USB Data Acquisition Device, (F) Analogue integrator, (G) Conditioner and amplifier, (H) Shaker, (I) Dynamic load cell and (K) Accelerometer.

Table 1. Mechanical parameters.

m_1	m_2	m_3	m_{VA}	c_1	c_2	c_3	c_{VA}	k_1	k_2	k_3	k_{VA}
[kg]				$\left[\frac{\mathrm{Ns}}{\mathrm{m}}\right]$				$\left[10^3 \frac{\mathrm{N}}{\mathrm{m}}\right]$			
6.70	4.73	1.96	0.54	13.7	16.5	6.3	0.4	58.6	60.0	39.5	1.6

The experiment consisted of exciting the first plate with a sinusoidal force signal and then measuring the acceleration of the third plate using an accelerometer. Finally, the acceleration signal goes through a digital integrator, resulting in a velocity signal.

First, the experiment was run without any faults. Then, two failure modes were introduced in the system: mass loss and stiffness loss.

Mass loss was introduced by withdrawing the mass of the absorber vibration.

Each column connecting the second and the third plates is made of a set of eight steel blades. So, to introduce stiffness loss, half of the blades are withdrawn from the second floor.

5 Results

For the case without any faults, the seventh state estimation (third plate's velocity) worked well, as it is possible to see in Fig. 4.

When two failure modes were introduced in the structure, the estimation did not work well, as it is possible to see in Fig. 5.

Looking at the scenario in Fig. 5: once a failure is detected, the observer state's parameters are changed until the estimation starts working well again. In this case, it

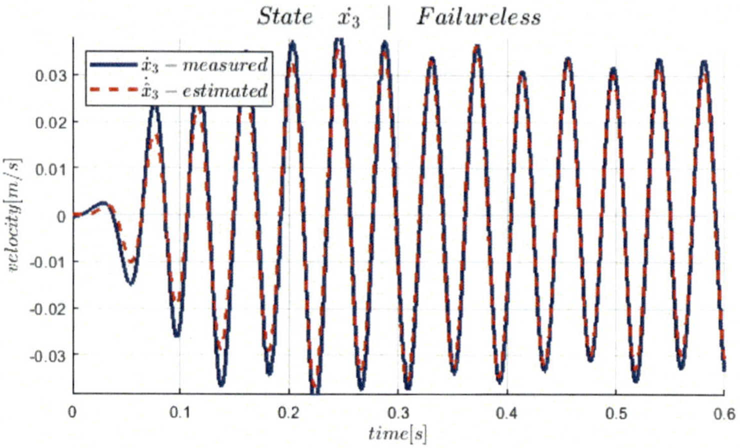

Fig. 4. Structure without any faults - the seventh state measured and estimated.

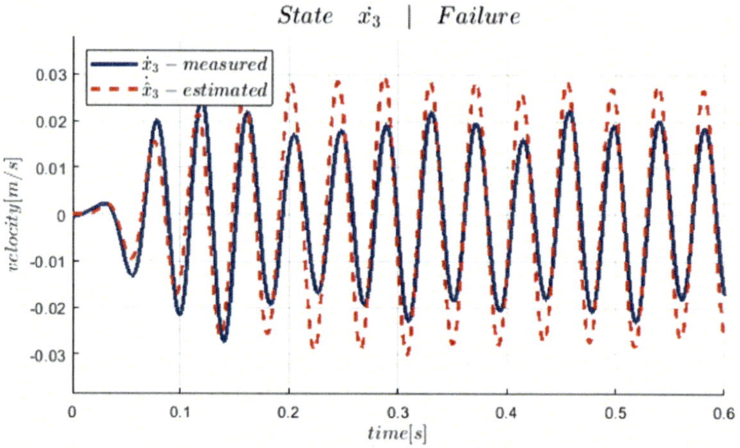

Fig. 5. Structure with two failure modes (mass loss and stiffness loss) - seventh state measured and estimated.

is considered the simplifying hypothesis that the failure modes are mass loss in the vibration absorber and stiffness loss on the second floor. These parameters are changed in the state observer model in order to obtain a good estimate, i.e., $fm \rightarrow \infty$. This can be seen in Fig. 6.

From Fig. 6, the stiffness loss is 24.4, and the mass loss is 61.6. So, the new values of k_2 and m_{VA} are, respectively, $(1 - 0.244)k_2 = 47.7\,\text{kN/m}$ and $(1 - 0.616)m_{VA} = 0.21\,\text{kg}$. Replacing it in the state observer model, the estimate works well again, as can be seen in Fig. 7.

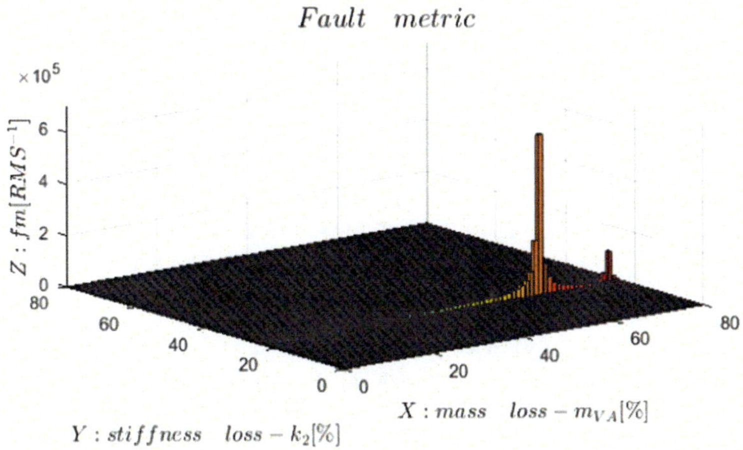

Fig. 6. Fault metric - *fm*.

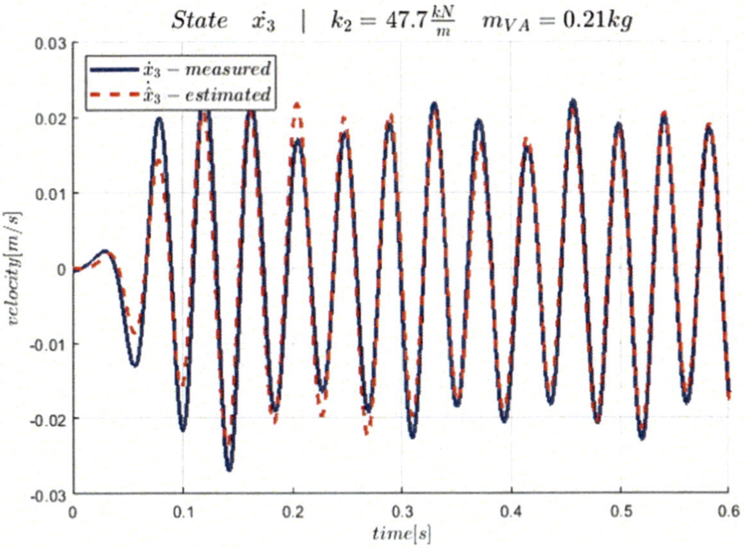

Fig. 7. Identified failure modes.

6 Conclusion

The state observer methodology displayed good results, and, after its implementation, it was possible to identify two fault modes occurring simultaneously: mass loss and stiffness loss of a mechanical structure modeled like a mass-spring-damper system with four degrees of freedom. Therefore, we verified that the state observer methodology could be used to identify two simultaneous failure modes in a simple mechanical system, which is an innovation in terms of the use of this methodology. For future works, a new challenge is to investigate whether it is possible to identify simultaneous faults in a complex system, like a dynamics rotor, for example.

References

1. Xu J, Xu L (2017) Integrated System Health Management. Academic Press, pp 55–99. ISBN 9780128122075. https://doi.org/10.1016/B978-0-12-812207-5.00002-X
2. Zhao R, Yan R, Chen Z, Mao K, Wang P, Gao RG (2019) Deep learning and its applications to machine health monitoring. Mech Syst Sig Process 115:213–237. 0888–3270. https://doi.org/10.1016/j.ymssp.2018.05.050
3. Oliveira LR, Melo GP (2018) Crack detection and dynamic analysis of a cracked rotor with soft bearings using different methods of solution. In: Cavalca K, Weber H (eds) Proceedings of the 10th International Conference on Rotor Dynamics – IFToMM. IFToMM. Mechanisms and Machine Science, vol 61. Springer, Cham. https://doi.org/10.1007/978-3-319-99268-6_1
4. Bohn C, Magnor O, Schultalbers M (2005) State observer-based analysis of crankshaft speed measurements with application to misfire detection. In: International Conference on Control and Automation, Budapest, vol 1, pp 239–244. https://doi.org/10.1109/ICCA.2005.1528124
5. Oliveira LR, Melo GP (2016) Experimental determination of stiffness and damping in rotating systems using metaheuristic hybrid optimization and state observers. J Braz Soc Mech Sci Eng 38:59–66. https://doi.org/10.1007/s40430-015-0413-6
6. Cavallini Jr. AA, Melo GP, Franco VR, Gonsalez CG, Lopes V (2009) A new technique for structural monitoring based on state observers in modal domain. In: 20th International Congress of Mechanical Engineering, COBEM 2009, November 2009, Gramado, RS, Brazil
7. Melo GP, Morais TS (2007) Fault detection using state observers with an unknown input, identified by orthogonal functions and PI observers. Brazilian Society of Mechanical Sciences and Engineering, pp 358–373. ISBN 978–85–85769–30–7
8. Melo GP, Araujo M (2016) Using fault models in state observers methodology for crack detection in continuous systems. Revista Sul-Americana de Engenharia Estrutural. https://doi.org/10.5335/rsee.v13i2.573
9. Leão LS, Cavalini AA, Morais TS, Melo GP, Steffen V (2019) Fault detection in rotating machinery by using the modal state observer approach. J Sound Vib 458:123–142. https://doi.org/10.1016/j.jsv.2019.06.022
10. QUANSER. Q8 – USB Data Acquisition Device. https://www.quanser.com/products/q8-usb-data-acquisition-device/

Digital Twin as a Tool to Select CT Scan Parameters

Crhistian R. Baldo[1](✉), Thiago L. Fernandes[2], Gustavo D. Donatelli[2], and Wim Dewulf[3]

[1] Federal University of ABC, Santo André, SP 09210-580, Brazil
crhistian.baldo@ufabc.edu.br
[2] CERTI Foundation, Florianópolis, SC 88040-970, Brazil
{thf,gd}@certi.org.br
[3] KU Leuven, Celestijnenlaan 300, 3001 Leuven, Belgium
wim.dewulf@kuleuven.be

Abstract. Digital manufacturing is one of the key drivers associated with the idea of flexible mass customization, which has been boosted in recent years by the substantial advances in additive manufacturing techniques. Complex free-form geometries, irregular shapes, internal and reentrant features, and variable thicknesses can be reproduced using the CAD model as a geometrical template. The dimensional quality control of additively manufactured parts, on the other hand, requires a measurement technique capable of sampling the high data density of those structures, including the hidden ones. Computed Tomography (CT) is the technology of choice to inspect additively manufactured parts. CT, however, features a complex measurement chain and the measurement accuracy relies on selecting proper scan parameters. Moreover, CT systems for dimensional metrology are expensive assets to buy and maintain. In this regard, the creation and use of the digital twin of a CT system to support these decision-making processes is presented in this work, as well as the correlation of the digital-based results with those obtained with the physical twin.

Keywords: CT metrology · Physical and digital twins · Scan parameters

1 Introduction

Digital manufacturing is one of the key drivers behind the concept of mass customization, which has been considerably boosted in recent years by the advances in additive manufacturing technologies, to give just one example. Broadly speaking, additive manufacturing processes make objects by adding layer-upon-layer of material; as a consequence, complex free-form geometries, irregular shapes, internal and reentrant features, and variable thicknesses, to name but a few, can be replicated based on the geometrical template obtained from 3D model data; additive processes, on the other hand, have shown some limitations in terms of production rate, mechanical behavior, surface finish, and dimensional accuracy [1, 2]. For the dimensional quality control of additively manufactured parts, the measurement technology should be capable of sampling high

Y. Iano et al. (Eds.): BTSym 2020, SIST 233, pp. 561–569, 2021.
https://doi.org/10.1007/978-3-030-75680-2_62

point coordinate data density of those structures, including the hidden ones [3]. In this regard, Computed Tomography (CT) is the technique of choice to inspect parts obtained by additive processes and respond to the inherent needs of mass customization [4–6].

CT is capable of holistically capturing part features, with unprecedented capability of measuring the internal ones; therefore, CT as a measurement technique can contribute a tremendous value to the new manufacturing paradigms. However, CT features a very complex measurement chain and the accuracy of CT measurements relies on, but not only, the choice of proper settings of the CT system, such as (x-ray tube) voltage and current, (detector) integration time, image averaging and gain, (rotational stage) number of radiographs, (object) magnification and orientation, (image and data processing) edge detection and fitting algorithm; which need a skilled CT metrologist who can make the proper tuning of all them [7]. In addition, CT systems for dimensional metrology are expensive assets to buy and maintain; therefore, before considering buying a CT system, it is important to check if it can be kept busy enough to justify the expense, e.g. if it will be suitable to measure several part models [8].

In this sense, analytical simulation tools have been developed and used [9–12] to, e.g., assist the metrologist with the choice of proper parameters for a given measuring task, thus supporting the preparation of written measurement protocols, as well as assist the company's decision-makers with the selection of a CT system for a given product range. With this in mind, the digital twin, i.e., the virtual replica of the actual physical process, of the CT system needs to be created using consistent input data. The digital twin can be then used to virtually generate the 2D radiographic images (CT projections) of the object and reconstruct the volumetric matrix (CT image), from which the outputs of the measurement task can be predicted and, if required, optimized (by adjusting the CT settings and rerunning the virtual experiment) prior to measuring the object on the physical twin. This paper presents the creation of the digital twin of a CT measurement process based on data obtained from manufacturers' specifications, standards, and empirical tests (Sect. 2), the CT chosen settings to scan the object in the digital environment and their correlation with the physical twin (Sect. 3), the visual findings (Sect. 4), and the concluding remarks (Sect. 5).

2 Materials and Methods

Dimensional metrology using CT can benefit from using digital twins to support the selection of optimized scan parameters. By doing so, the variables under the control of the metrologist can be previously tuned in the simulation platform, leaving just some minor adjustments to be performed on the CT system. In this work, the computer tool for quantitative description of radiographic testing – aRTist, version 2.10.0, by the German Federal Institute of Materials Research and Testing (BAM); was chosen to create the digital twin of a CT system optimized for dimensional measurements. The virtual scene allows the metrologist to set parameters for the x-ray source, detector, and object.

2.1 Description of the Physical Twin

In the simulation platform, the digital twin of the first generation of the Carl Zeiss IMT METROTOM 1500, which integrates the CERTI Foundation's coordinate metrology infrastructure, was created. The CT system comprises a proprietary kinematic system, a microfocus x-ray source by VISCOM (model XT 9225-DED), and a flat panel detector by PerkinElmer (model XRD 1620 AN14). Figure 1 shows the CT system and the three main components: x-ray source, rotational stage, and detector.

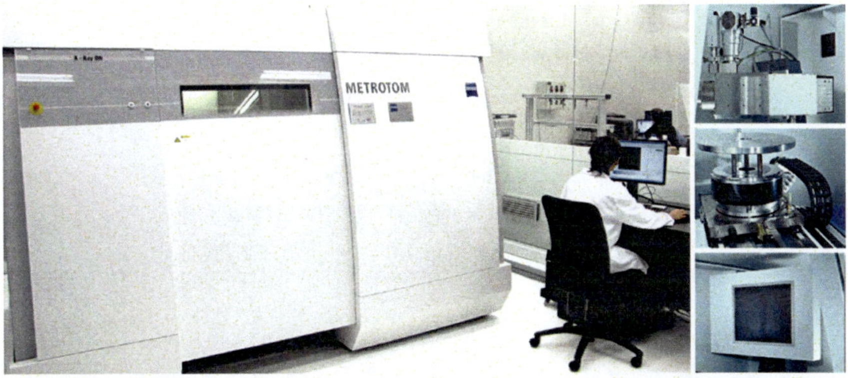

Fig. 1. Photo of the CT system Carl Zeiss IMT METROTOM 1500 modeled in the simulation tool and the three main components: x-ray source, rotational stage, and detector.

2.2 Description of the Digital Twin

Most of the specifications used to create the digital twin were extracted from the manufacturers' documentation, summarized in Table 1; yet, some details on the imaging performance of an x-ray detector were obtained from standards (ASTM E2597, IEC 62220-1) and some experimental results from others were used to input more realistic data on the focal spot size [10]. Basically, the x-ray source and detector were modeled in the digital environment with the information listed in Table 1, using, respectively, the 'spectrum calculator' and 'detector calculator' tools. The positioning system, which is software-corrected, was regarded as having negligible errors in this work.

2.3 Description of the Test Object

The chosen test object to embrace the virtual CT measurement process was a part that forms the drive system of a car window lift mechanism, made of acetal photopolymer (nominal density $\rho = 1.42$ g/cm^3) and produced by injection molding. Some views of the object can be seen in Fig. 2. Attributes such as size, shape, and material make it a very likely candidate for CT. This is because objects made of plastic require rather less radiation energy than, for example, objects made of light-weight metals; thus, they are much less prone to be impacted by beam hardening and scattering artifacts, which may significantly affect the measurement accuracy. It is worthy of notice that the object was scanned as well on the physical twin in order to compare both measurement setups.

Table 1. Technical specifications of the components of the CT system as per manufacturers.

Component	Parameter	Value/Attribute/Type
x-ray source	Acceleration voltage	(30 to 225) kV
	Filament current	(10 to 3000) μA
	Target material	Tungsten/1.0 mm thick
	Intrinsic filtering	Aluminum/0.3 mm thick
Flat panel detector	Total area	409.6 mm × 409.6 mm
	Pixel matrix	1024 × 1024 @ 400 μm
	Radiation energy	40 keV to 15 meV
	Scintillator material	Gd_2O_2S:Tb/208 μm thick
	ADC resolution	16 bits/65536 Gy levels

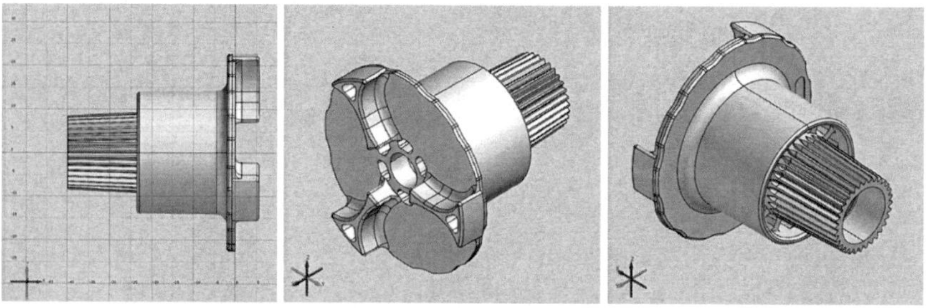

Fig. 2. Views of the test object made of POM chosen to compose the virtual measurement setup.

3 Experimental Setup

In this section, an overview of the measurement conditions for both virtual and real cases is provided. In both cases, the variables under the control of the metrologist were tuned observing the lessons learned from previous researches of the authors, e.g., [13–16]. The comparison between the two environments is also drawn to illustrate how well the virtual measurement setup mirrors that considered for the physical twin of the CT measurement process, allowing us to make inferences about the accuracy of the virtual environment for the case under study.

3.1 Measurement Setup of the Digital Twin

In the digital twin, the object's virtual model, described by STL geometries, was loaded and coarsely positioned between the x-ray source and detector. Taking some good CT measurement practices into account, the virtual model was aligned in such a way to ensure first that the maximum distance to be traveled by x-rays and the variation of penetration depth during object rotation are attenuated, and second that the surfaces to be scanned are inclined at least 10° to the rotation axis. The virtual model was then finely positioned along the magnification axis to project it using the detector's maximum possible area (thus improving the spatial resolution). Next, the voltage was set high enough to avoid complete beam absorption, and the detector integration time adjusted to a convenient value; the current was tuned to enhance image contrast and brightness; and the number of projections was chosen as approximately the number of pixels covered by the object's shadow in the projection. In addition, before each simulation run, flat-field correction (available in the simulator) was applied to remove the shading in the image due to the different path lengths of the x-ray cone beam.

In Fig. 3, screenshots of the virtual environment can be seen, which exhibit the source panel (left), detector panel (center), measurement scene (top-right), and CT scan panel (bottom-right). Table 2 lists the parameters selected for the virtual experimental setup. After sampling the informed number of projections per complete rotation of the object in the virtual scene, the volumetric matrix was created using the reconstruction module embedded in the simulator, which uses the conventional filtered back-projection method. As for the edge detection, the standard 'iso-50%' threshold value was applied globally (a reasonable choice as the object is assumed to be made of a single homogeneous material) within the software 'VG Studio MAX, version 3.3.2 64 bit'. Then, a CT image can be viewed on the screen and used to check part integrity, perform dimensional measurements, and compare the actual part with the CAD model.

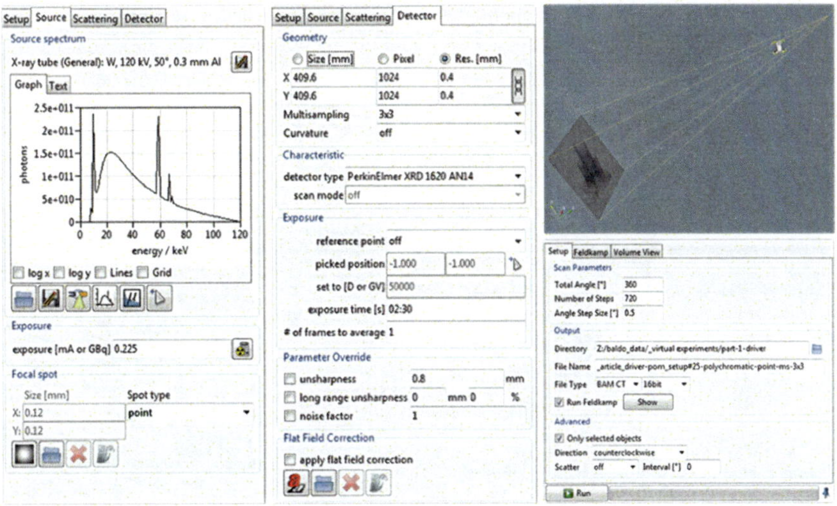

Fig. 3. Screenshots of the virtual environment illustrating the source panel (left), detector panel (center), measurement scene (top-right), and CT scan panel (bottom-right).

3.2 Measurement Setup of the Physical Twin

The same rationale behind the digital CT system setup was considered for the object's orientation and position, x-ray tube and detector settings, and image acquisition setup.

Some of the values shown in Table 2 are, however, approximated or nominal, as the actual values could either change within limits during scanning (e.g., voltage, current), differ from the nominal (e.g., due to inaccurate correction of the positioning error), or not be measured (e.g., object's orientation relative to the rotation axis). The reconstruction was performed on the software of the CT system with the filtered back-projection method. The edge detection strategy was identical to the virtual setup.

Table 2. Selected parameters on the physical and digital twins for scanning the test object.

Component	Parameter	Virtual setup	Real setup
x-ray source	Acceleration voltage	120 kV	110 kV
	Filament current	225 μA	225 μA
Flat panel detector	No. of pixels (H/V)	1024	1024
	Integration time	2 s	2 s
	Unsharpness	0.8 mm	0.8 mm
Test object setup	Source-to-object distance	250 mm	250 mm
	Geometrical magnification	6x	6x
	Rotation about x-, y-, z-axes	0°/20°/−110°	0°/20°/−110°
	No. of projections	720	720

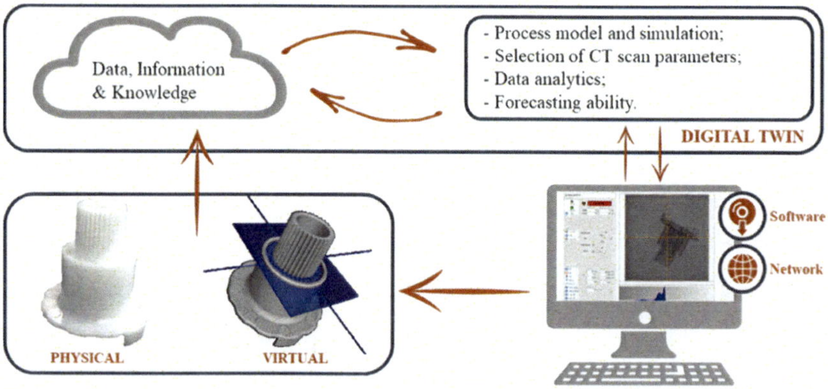

Fig. 4. Interaction between the physical and digital twins in the broad context of the measurement process realization.

3.3 Comparison of Measurement Setups

As shown in Table 2, most of the nominal input values were identical in both setups. Only the voltage needed to be set slightly higher in the digital twin in order to ensure a good signal-to-noise ratio. From the CT metrologist viewpoint, which would make use of the virtual environment outputs to preset the CT system, the close correlation between the physical and digital setups is a good indication that the metrologist could profit from the virtual environment to predefine the measurement setup before moving to the CT system and validating it. This allows us to infer that the digital twin created can be regarded as not only a digital but also a functional twin of the physical system, that the effective use of the CT infrastructure can be augmented, and the return on investment can be optimized. From the company's decision-makers' view, the use of the digital twin, provided by a third-party consultancy, could help choose the most suitable CT system to encompass the company's product mix. Therefore, the measurement process realization, which includes all activities associated with the entire measurement lifecycle, from equipment definition to measurement execution [17], could be performed in a more structured manner, as depicted in Fig. 4.

4 Qualitative Analyses

In addition to the comparison just presented, qualitative analyses could be carried out to observe how changes in a given input parameter affect the quality of the CT image. This 'what-if' analysis can be used to check the effect of using a detector with higher/lower resolution, the effect of increasing the integration time (thus reducing tube current and spot size), and the effect of varying the capacitance of the photodiodes (thus changing the detector responsivity). In this work, the qualitative effect of changing the gain factor of the photodiodes was studied as the metrologist may face situations in which the current needs to be reduced due to inadequate focal spot size or the scanning time needs to be squeezed for throughput purposes.

By doubling the gain, either the current or the integration time can be halved at the cost of increasing the image noise. The effect of varying the detector gain can be perceived on the tomograms in Fig. 5; from left to right, the tomograms for a noise factor equal to (a) 1x - default case, (b) 10x, (c) 20x, (d) 100x. The image definition is visually worsened from a factor of 20x; this could impair the measurement accuracy.

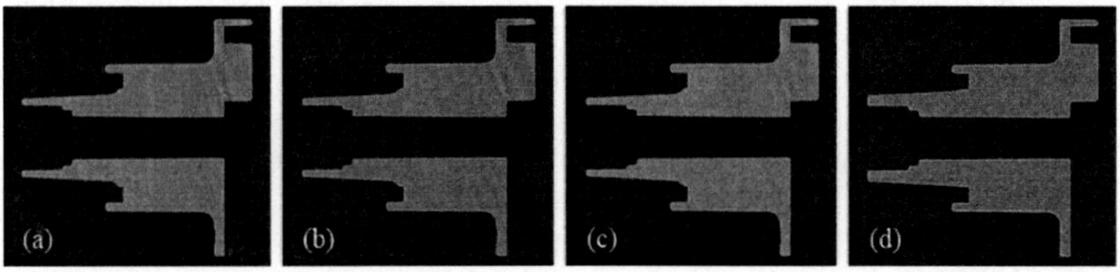

Fig. 5. Tomograms obtained in the virtual environment with noise factor set to: (a) 1x - default, (b) 10x, (c) 20x, and (d) 100x; mimicking changes in the capacitance of the photodiodes.

5 Concluding Remarks

This paper presented the creation of the digital twin of a CT system using data obtained from manufacturers' specifications, standards, and experiments. For the measuring task under analysis, most of the CT settings defined in the digital twin could be used as well for the real measurements. Therefore, the CT metrologist could benefit from the virtual environment to find a proper object orientation (that reduces penetration depth) and position along the magnification axis (that improves spatial resolution) and predefine the various CT settings. In this situation, just a few minor adjustments would be required on the physical twin, minimizing the x-ray tube parts' wear and downtime. As for the high acquisition and ownership costs, the use of digital twins could be thought of as a tool to assist decision-makers with the bill of requirements that would need to be fulfilled by the CT system, including it in the methodology outlined in [8]. In both cases, it is worth noting that simulation time, including the reconstruction time, was of the order of 75 min. This time can be reduced by lowering the mesh accuracy of the virtual model of the test object. The time needed to create the digital twin, on the other hand, highly depends on the availability of information to model the CT components.

Future works will focus on expanding the scope of this work by addressing the effect of different CT variables on the dimensional content (e.g., diameters, distances, form errors) of the test object used here and other test objects with dissimilar geometries and materials. The following influence factors are to be considered in this quantitative sensitivity analysis using the simulation software: (related to the x-ray source) spectrum and spot size, (related to the x-ray detector) noise and unsharpness, and (related to the interaction of x-ray and matter) beam hardening and scattering. The same CT system will be used in this new set of experiments, and the dimensional content of the test objects will be measured as well on a coordinate measuring machine in order to evaluate how well the virtual environment predicts the actual measurement values.

Acknowledgments. The first author gratefully thanks the financial support of São Paulo Research Foundation (FAPESP), grant #2018/24757-7.

References

1. Jiménez M et al (2019) Additive manufacturing technologies: an overview about 3D printing methods and future prospects. Complexity 2019:1–30
2. Jared BH et al (2017) Additive manufacturing: Toward holistic design. Scripta Materialia 135:141–147
3. CCL strategy document 2018–2028. Consultative Committee for Length (CCL). https://www.bipm.org/utils/en/pdf/CCL-strategy-document.pdf
4. De Chiffre L et al (2014) Industrial applications of computed tomography. CIRP Ann—Manuf Technol 63(5):655–677
5. Imkamp D et al (2016) Challenges and trends in manufacturing measurement technology - the "Industrie 4.0." J. Sens Sens Syst 5:325–335
6. Villarraga-Gómez H et al (2019) X-ray computed tomography: from medical imaging to dimensional metrology. Precis Eng 60:544–569
7. Menoncin M et al (2019) New training concept on industrial computed tomography for dimensional metrology. In: NDT.net 24(3), Conference on ICT, Padova, IT
8. Baldo C et al (2017) Drawing up an industrial CT system purchase specification: a Brazilian case study. In: NDT.net 22(3), Conference on ICT, Leuven, BE

9. Reisinger S et al (2011) Simulation-based planning of optimal conditions for industrial computed tomography. In: NDT.net 16(11), International Symposium on DIR and CT, Berlin, DE
10. Hiller J et al (2012) A computer simulation platform for estimation of measurement uncertainties in dimensional x-ray CT. Measurement 45:2166–2182
11. Helmecke E et al (2016) Numerical measurement uncertainty determination for dimensional measurements of microparts with CT. In: NDT.net 21(2), Conference on ICT, Wels, AT
12. Wohlgemuth F et al (2020) Convergence behaviour of numerical uncertainty evaluation using a virtual metrological CT system. In: NDT.net 25(2), Conference on ICT, Wels, AT
13. Baldo C et al (2015) Experimental investigation of CT dimensional capability using modular test artefacts. In: Proceedings of the IMEKO World Congress, Prague, CZ
14. Baldo C et al (2016) Experimental assessment of CT dimensional performance using modular test pieces made of POM and aluminum. In: NDT.net 21(2), Conference on ICT, Wels, AT
15. Arenhart FA et al (2016) Experimental investigation of the influence factors on the structural resolution for measurements with CT systems. In: NDT.net 21(2), Conference on ICT, Wels, AT
16. Fernandes T et al (2016) Experimental evaluation of the uncertainty associated with the result of feature-of-size measurements through CT. J Phys: Conf Ser 733:012056
17. Baldo C et al (2019) CT measurement realization process: the need for advanced correction methods. In: NDT.net 24(3), Conference on ICT, Padova, IT

Estimating the Relevance of Different Frequency Peaks of Undamped Systems Through Recursive Algorithm

A Power Harmonics Approach

André Carneiro Rocha[1(✉)], Bruno Luís Soares de Lima[2], and Ricardo Janes[2]

[1] WEG Drives & Controls, São Paulo, SP 04077-023, Brazil
andrerocha@weg.net
[2] Mackenzie Presbyterian University, São Paulo, SP 01302-907, Brazil
{bruno.lima,ricardo.janes}@mackenzie.br

Abstract. Nowadays, many applications need to trace and reconstruct input signals as the presence of undesired frequency components can be not only quite challenging to avoid, but harmful to various costly equipment in the industry. As a quality of energy case study for the present work, the growing presence of variable frequency drivers (VFD) in the industry, as well as many other non-linear loads, can indirectly cause serious damage to equipment installed downstream the power source due to harmonic components generated by semiconductors or fast switching operations. Should those components be correctly neglected by a passive filter or, more likely, on recent applications, an algorithm-based filter, the system's input signal could be reconstructed with a degree of accuracy based on how many frequency components are taken into consideration. The present work aims to implement a recursive algorithm on MATLAB capable of receiving the samples of a signal from an undamped system and sort the different magnitude peaks according to their frequency values (i.e., frequency peaks) through the use of Fast Fourier Transform (FFT). The higher the magnitude of a given frequency component, the greater is the distortion it can cause to the original signal. Finally, the algorithm must reconstruct the signal based on the desired accuracy for the process and estimate each peak's impact on the original system, making it possible to analyze the process from a performance or accuracy trade-off point of view.

Keywords: Undamped system · MATLAB · FFT · Frequency peaks · Recursive algorithm · Power harmonics

1 Introduction

Over the past few decades, the increasing use of electronics, and thus semiconductors devices, was a watershed for an infinite variety of fields. Medical equipment, computer-aided design (CAD) or manufacturing (CAM), communications, digital signal processing, and many other applications could be easily hindered without the latest contributions of Electronic Engineering. Even Moore's Law was recently put to the test in front of the

Y. Iano et al. (Eds.): BTSym 2020, SIST 233, pp. 570–577, 2021.
https://doi.org/10.1007/978-3-030-75680-2_63

exponential growth of technologies and possibilities within increasingly smaller components [1]. However, great advances not only lead to better and valuable processes but also to challenging conditions as well.

Each field of study has its own challenge. In power electronics, fast-switching devices at higher frequencies, such as insulated-gate bipolar transistors (IGBT), which are widely used on variable frequency drivers (VFD) to generate pulse width modulation (PWM) waveforms and hence controlling three-phase motors, can produce power harmonics [2]. Those are an inconvenience to any electrical system as they can cause onerous damage to sensible equipment on the same network or compromise the overall quality of the energy due to distortions [3, 4].

As a case study in the present work, the influence of power harmonics on an electrical system will be used to explain and validate the practical implementation of the proposed algorithm, since the electrical power source signal itself represents an undamped system signal, preferably a sinusoid in an ideal alternating current network.

The power harmonics hereby referred to are the Total Harmonic Distortion (THD) that is the ratio of total harmonics value to the fundamental frequency value, considering an electrical system imbued with power harmonics, regardless of the cause. It is increased along with the increase of non-linear loads in the system [5].

A common VFD topology is depicted in Fig. 1.

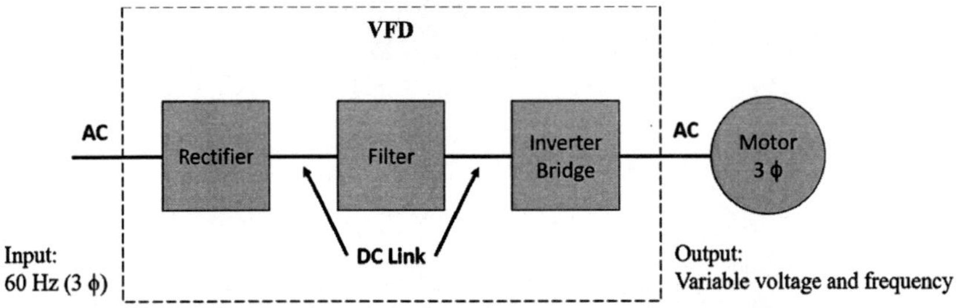

Fig. 1. Common VFD topology as a block diagram to illustrate the conversion from AC power source to DC voltage and then back to AC after the inverter bridge (IGBTs).

Some VFD manufacturers develop useful ways to mitigate THD interference. One method that has been used for a few years by now is to place an inductor on the DC link, before the output inverter (IGBTs), to operate as a passive filter. However, it is not an absolute resource as the harmonic distortion stress on the inductor will diminish its life span [6, 7]. Moreover, output reactors can also be used and prove to be a powerful ally not only against THD but to operate as a dv/dt filter as well [8].

The method proposed here can be used for almost any system that needs to predict the input signal with a variable degree of accuracy (regarding to the application), for instance, a mechanical system with four degrees of freedom that are strongly affected by vibrations, as explored by As'arry et al. [9], or noise-canceling, as investigated by Ozawa et al. [10], and many other applications.

In the present work, an electrical network with harmonic distortions caused by VFD was used as a didactic yet reliable case study to investigate and prove the recursive algorithm's operation.

2 Materials and Methods

2.1 Algorithm Block Diagram

The development is done entirely in MATLAB environment. In Fig. 2, it is illustrated the block diagram for the proposed algorithm.

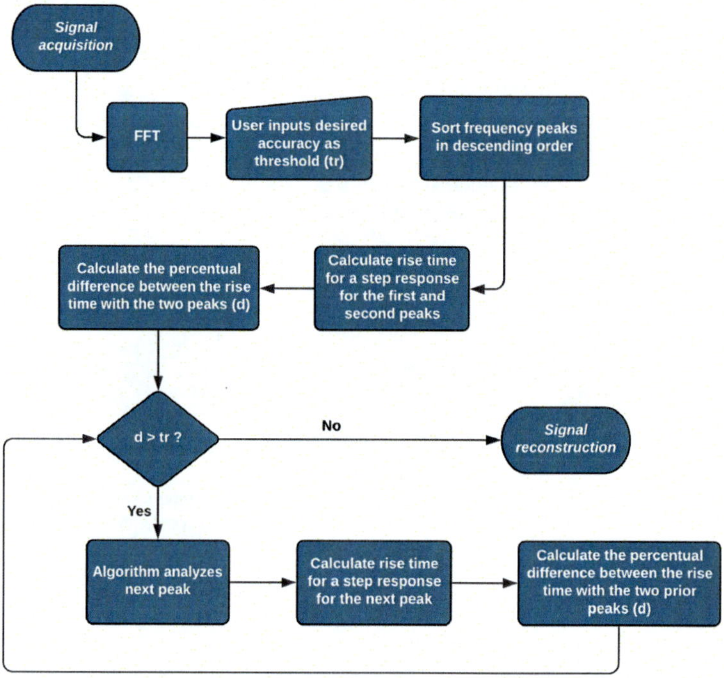

Fig. 2. Block diagram for the proposed algorithm.

Signal Acquisition and Data Processing. First of all, the algorithm must acquire the undamped system signal. It can be an online monitoring through the use of sensors or offline from data collection. The frequency spectrum will be achieved through Fast Fourier Transform (FFT), highlighting the magnitude peaks and their respective frequency values in Hz. The user inputs the desired accuracy for the process, i.e., the threshold that has to be attended. The more iterations, the higher will be the accuracy and processing load to hardware. Up to now, the system is capable of sorting the magnitude of the frequency peaks in a descending order, storing it in a two-dimensional data array.

Recursive Loop. According to the accuracy informed by the user, it will be settled the number of frequency peaks to be analyzed as well as the number of recursive iterations.

Step Response Analysis and Signal Reconstruction. Each time the algorithm iterates, the undamped system's function is updated with the impact of the next peak and it tests the function for a unitary step input. A proportional-integral-derivative (PID) compensator is needed for compelling the response to damp after closing the control loop with unitary feedback. This stage is completely described in Sects. 2.3 and 2.4, where the mathematical modeling is elucidated. Finally, the undamped signal can be reconstructed exclusively with the influence of selected peaks.

2.2 Magnitude Vectorization Using FFT

The process of vectorizing the magnitude peak for each frequency is performed after the algorithm applies FFT on acquired signal or received data.

For instance, a sinusoid input with peak value of 10 V and 60 Hz is summed to its fifth and seventh harmonic components, with magnitudes of 5 and 3 V, respectively. It is an easy task to evidence the three major frequency peaks, but it becomes a burden when the signal is corrupted with some random noise, like when White Gaussian Noise comes into play [11]. The analysis with the presence of random noise is far important for the purpose of the present work as it provides a free way to the applicability of the algorithm on non-periodic signals.

After detecting a set of different frequency peaks, the algorithm now sorts their magnitudes in descending order, along with its corresponding frequency value, in a two-dimensional array. The number of the elements in this array is determined by the accuracy set by the user and hence the number of recursive iterations.

2.3 Mathematical Modeling

In possession of the data array, the system can use it to test the response for a unitary step in order to estimate the relevance of the several peaks on the reconstruction of the original signal, one by one. Since the object of study of the present work is the analysis of undamped systems, the standard transfer function times the half of the amplitude for that frequency component can be given by (1).

$$T_{amp}(s) = \frac{\omega_n^2}{s^2+\omega_n^2} \times \frac{Amp_n}{2} \tag{1}$$

That is the method the algorithm analyzes a single frequency peak. However, it will be n frequency peaks as the accuracy grows with the number of iterations. So, it is convenient to use a function that sums each term and updates itself through the use of recursion and returns the transfer function that will be tested on the next iteration for the unitary step input. It is possible to reach the general equation of the proposed algorithm, as seen in (2), where n is the number of recursive iterations.

$$T_{general}(s) = \sum_{i=1}^{n} \left[\frac{\omega_{0i}^2}{s^2+\omega_{0i}^2} \times \frac{Amp_i}{2} \right] \tag{2}$$

2.4 Step Response and Signal Reconstruction

To be possible to test an undamped system for a unitary step, a compensator in closed loop is needed. Due to the enormous versatility of PID controllers, along with the fact that it is widely present in the industry [12] and has a great variety of libraries for programming and prototyping purposes. The function given by (2) will be tested for a unitary step along with a PID compensator in a closed-loop system.

More than one parameter can be used to analyze the impact of a frequency peak to the whole system, e.g. rise time, settling time, and overshoot. In the present work, the parameter chosen was the rise time to be compared at each iteration. The accuracy set in previous stages will determine the percentual threshold that has to be compared to the first iteration for breaking the recursive loop. If the accuracy is set to higher values, the number of iterations and the array needed to store the data will be greater due to the aforementioned condition. All of the step response curves, one for each iteration, will be put together in a single plot.

Finally, the algorithm is capable of reconstructing the system signal taking only the analyzed peaks into consideration and plots it afterward for the user to verify.

3 Results and Discussion

To validate the algorithm and prove the technique proposed in the present work, a sinusoid of 220 V RMS was used, with an approximate peak value of 311 V and a frequency of 60 Hz. The power source signal is corrupted, not only with high-level harmonic components but also with random White Gaussian Noise.

As it is possible to verify the original signal in Fig. 3a and its frequency spectrum in Fig. 3b, the FFT of the power source signal shows many frequency peaks of various amplitudes. The highest peak, however, is clearly the pure input signal itself, at 60 Hz. Right after the frequency spectrum, there is the plot with the seven highest magnitude peaks in descending order in Fig. 4. As a proof of concept, the accuracy, 'tr', was set by a value of 3%. The plot shown in Fig. 4 results from the unitary step input applied to each one of the seven iterations.

As expected, there are huge differences between the iterations due to each peak's relevance in terms of magnitude and frequency. The major difference is spotted at the second iteration (green curve), when the step response presents a great distance from the first iteration (red curve), whereas the fifth, sixth, and seventh peaks do not seem to change, significantly, the step response curve as shown in Fig. 5.

To better compare the progress at each iteration, it is possible to arrange the observed data into a table. Therefore, Table 1 summarizes each peak's rise time and shows the percentual difference from the current peak to the major peak, i.e., the first iteration. As the accuracy was chosen in terms of percentage, the absolute value (seconds) is useful to verify how much of the maximum peak the system could discard and classify as noise. Each iteration is indexed into the array in descending order, from the greatest amplitude peak (iteration 1) to the smallest (iteration 7).

The column "Percentage" in Table 1 shows the percentual difference at each iteration. Once the accuracy was set to 3%, the algorithm discarded the sixth and onwards

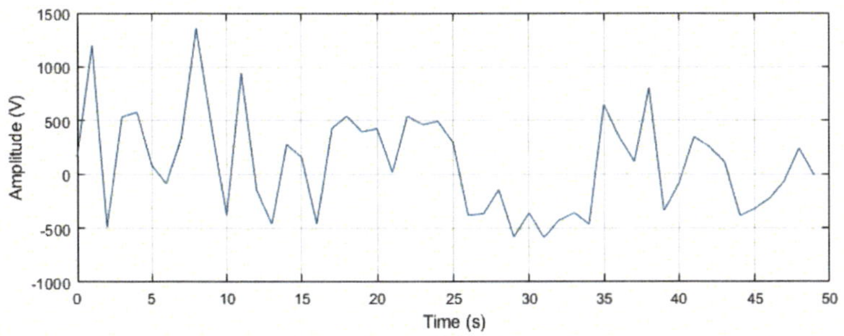

(a)

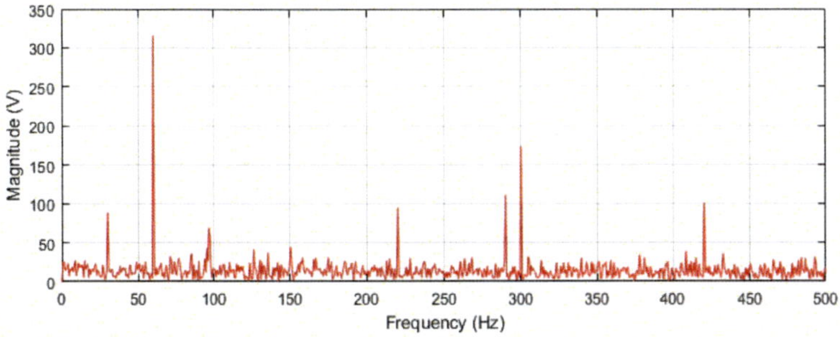

(b)

Fig. 3. a) The input signal in the time domain. b) The frequency spectrum of the input signal.

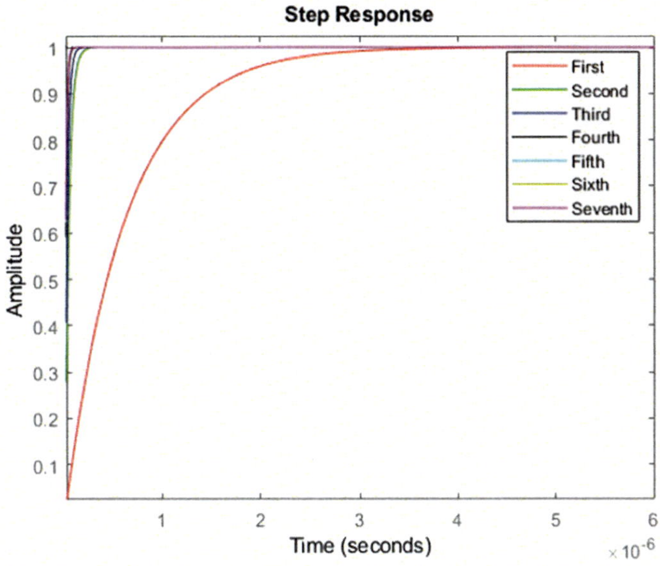

Fig. 4. Step response curve for each iteration.

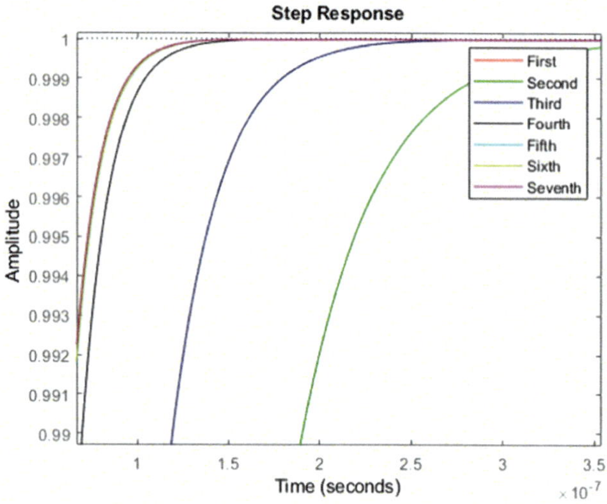

Fig. 5. Zoomed step response curves.

iterations, as they do not show any changes to the percentage parameter. However, the sixth and seventh iterations are present in the table just for exemplification purposes.

Table 1. Acquired results for unitary step response for each iteration.

Iteration number	Amplitude (V)	Frequency (Hz)	Rise Time (10E-7 s)	Percentage (%)
1	311.13	60	13.970	-
2	174.17	300	0.93518	6.7
3	111.89	290	0.59304	4.2
4	100.74	420	0.49186	3.5
5	95.403	220	0.31145	2.2
6	89.034	30	0.31101	2.2
7	69.580	97	0.30790	2.2

4 Conclusion and Future Scope

In the present work, a method to estimate the relevance of frequency peaks of undamped systems was developed. The algorithm herein could isolate the highest peaks of the frequency spectrum and sort them in a two-dimensional vector. Thus, it was possible to test each iteration to a unitary step input on a closed-loop PID compensator in order to plot a graph containing every curve for better visualization and comparison. The results proved to be reliable whilst the percentage setting for accuracy returned a satisfactory output.

Though the application of power harmonics is a case study with relatively low perturbance and chaotic patterns compared to many other application possibilities, it turned out to be a convenient and didactic concept proof. For further investigations, the algorithm could be used for noise and vibration canceling, energy quality analysis applied to renewable sources, perturbance-resistant self-driving machines and so on.

References

1. Williams RS (2017) What's Next? [The end of Moore's law]. Comput Sci Eng 19(2):7–13. https://doi.org/10.1109/MCSE.2017.31
2. Suresh Kumar V, et al (2006) Harmonic distortion in a modern cement industry-a case study. In: 2006 IEEE power India conference, New Delhi.https://doi.org/10.1109/POWERI.2006.1632630
3. Nelson JP (2004) A better understanding of harmonic distortion in the petrochemical industry. IEEE Trans Ind Appl 40:220–231. https://doi.org/10.1109/TIA.2003.821802
4. Kumar VS, et al (2007) Optimal estimation of harmonics in power system using intelligent computing techniques. In: 2007 international joint conference on neural networks, Orlando, FL, 2007, pp 142–147. https://doi.org/10.1109/IJCNN.2007.4370945
5. Thakur P (2020) Load distribution and VFD topology selection for harmonic mitigation in an optimal way. IEEE Trans Ind Appl. 56(1):48–56. https://doi.org/10.1109/TIA.2019.2946111
6. WEG Drives & Controls. Induction motors fed by PWM frequency inverters, Jaraguá do Sul, SC, Brazil, November 2018, code 50029350, revision 30
7. Ramesh P et al (2019) Analysis of high frequency impact on the DC link inductor of VFD under the influence of shunt active filter. In: Innovations in power and advanced computing technologies (i-PACT), Vellore, India. https://doi.org/10.1109/i-PACT44901.2019.8960226
8. Bruha M et al (2020) High speed electrical drives – perspective of VFD manufacturer. E3S Web Conf https://doi.org/10.1051/e3sconf/202017801006
9. As'arry A et al (2016) Suppression of Parkinson's hand-like tremor using fuzzy-PID controller. In: 2016 IEEE international conference on automatic control and intelligent systems (I2CACIS), Selangor, pp 1–5. https://doi.org/10.1109/I2CACIS.2016.7921041
10. Ozawa K et al (2019) Suppression of two noise sources in the same direction by instantaneous spectral subtraction. In: 2019 IEEE 8th global conference on consumer electronics (GCCE), Osaka, Japan, pp 745–746. https://doi.org/10.1109/GCCE46687.2019.9015244
11. Singh SK et al (2015) Power system harmonic parameter estimation using Bilinear Recursive Least Square (BRLS) algorithm. Int J Electr Power Energy Syst 67:10. https://doi.org/10.1016/j.ijepes.2014.11.006
12. Knospe C (2006) PID control. IEEE Control Syst Mag 26(1):30–31. https://doi.org/10.1109/MCS.2006.1580151

Soft Computing Method in Events Risks Matrices

Nikolai I. Plotnikov(✉)

Scientific Research Project Civil Aviation Institute «AviaManager», 630078 Novosibirsk, Russian Federation
am@aviam.org

Abstract. The existing normative definition of risk, based on probability theory, is a particular definition of risks and does not correspond to the variety of measures for calculating life risks. Calculating risks in other fuzzy measures: likelihood, possibility, confidence, necessity, opens up new possibilities to assign numerical values to objects and complex events. This paper presents a method of design and soft computing in risk matrices, which instead of a single outcome measure called a consequence, represents samples of several qualifiers, evaluation areas, parameters, and semantic rows of natural language. For the first time, the risk matrix definition is presented, a template for designing risk matrices and samples for use in practice are developed. The purpose of risk matrices is to make it possible to assign numerical values to the properties of evaluated objects through expert subjective judgments. The randomness measure absorbs a wide class of soft measures likelihood, necessity, confidence, conviction, possibility, probability. The measure of the magnitude of the outcome combines the concepts of severity, damages, losses, and victims. The method is aimed at direct application of expert assessments of any objects and their components.

Keywords: Soft computing · Estimation · Matrix · Risk · Emergency · Event

1 Introduction

There are numerous concepts of risk in studies of the risks of vital activity: one-dimensional, multidimensional, dependent and not dependent on a person (natural disaster); individual: motivational, active, passive, voluntary, forced; social: personal, group, business, organizational, deterministic (systematic), random (stochastic), accepted, acceptable, voluntary, conscious, perceived, controlled, estimated, regulated, controlled, neglected, acceptable, unacceptable, tolerable, residual, processed. The subject of risk is studied in the human sciences, philosophy, psychology, technosphere activity, economics, and finance as the expectation of negative activity outcomes.

Event management is dominated with approaches to measuring, calculating risks, developing national and global security programs, and managing risks in emergency situations. There are three approaches to study the subject of risk: the measurement of risk, the socio-cultural evaluation, and the psychometric evaluation. Risk measurement

Y. Iano et al. (Eds.): BTSym 2020, SIST 233, pp. 578–588, 2021.
https://doi.org/10.1007/978-3-030-75680-2_64

focuses on how to transform data on damage, casualties, financial losses, and how risk is influenced. The socio-cultural evaluation looks at the impact of group and cultural variables on risk. Risk psychometry establishes the emotive responses of people to risk situations that form risk judgments [1–3].

In the understanding of the author, socio-cultural and psychometric risks are assessed qualitative risks as opposed to measurable quantitative risks. It is shown in the diagram. Figure 1.

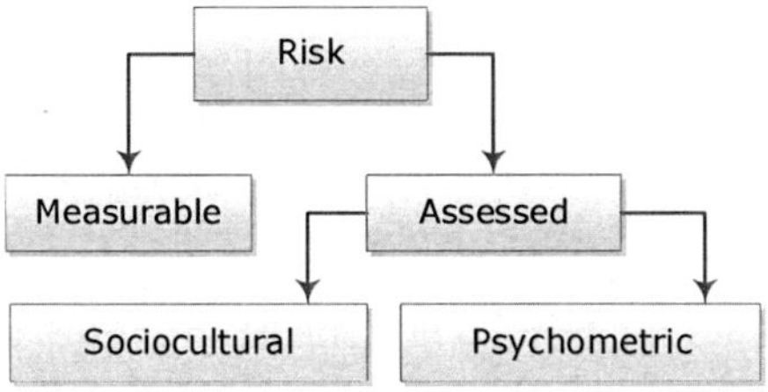

Fig. 1. The paradigm for the study of life risks

The risk assessment approach focuses on matrix risk estimation. Out of all the diversity of studying the risks of life activity, this work considers two most important aspects of the problem: (1) identification and the validity of calculating risks, (2) the validity of the approach to assessing risk matrices in the instrumentation. This paper describes a method of design and soft computing (SC) in risk matrices.

2 Problem

The problems of subject risk are as follows: a) the unclear origin of the term "risk", b) numerous incompatible classifications and definitions of risks, c) groundlessness of the concept of risk measurement, d) groundlessness of risk identification through uncertainty, e) groundlessness of risk identification as a negative consequence through a combination probability and damage.

2.1 Identification and Calculating Risk

Concept of risk measurement and the regulatory standards [3–7] risk is defined as follows: risk (a) is the product of multiplication (b) the probability, and (c) the consequences (impact, damage) of an event (1):

$$\begin{gathered}\left[\overleftrightarrow{R} = \mathbf{P} \cdot \mathbf{D}\right] \\ \left[Risk = Probability \times Damage\right]\end{gathered} \tag{1}$$

The scientific validity of the shown formula raises many questions: (1) Risk constitutes any content of human life and nature, where outcomes coexist that can be assessed as neutral, negative, and positive. However, the risk in the formula is determined only in the negative understanding of the consequences as damage. (2) Risk contains the expectation of an event in the form of probability, and probabilistic measures are used for the calculation. (3) Events take place in time. The frequency of events is set as an average per unit of time, usually a year. This calculation is mathematically incorrect, especially with uneven distribution of events throughout the year. (4) The ratio (quantification) of consequences and likelihood is reduced to the mathematical multiplication. However, it is impossible to perform a calculation using this formula since the variable consequences also have a probabilistic random nature. The ratio can be summative, multiplicative, unspecified, non-linear. In any case, calculating risks through the product of consequence and probability can be a particular option.

So, the problem of the normative description and standards of risks used by the world community is as follows: why in the formula (1) only probability is used, and no other measures, damage, not positive and neutral outcomes, multiplication, and not others ratio. **The scientific basis for the risk formula remains debatable and controversial.**

The consequence of these problems is the fact that, in practice, the calculation of risks by the ratio of probability and consequences becomes uncertain. If there are a number of consequences and the probabilities for different outcomes differ, then the total risk is determined by the sum of the mathematical multiplication (2):

$$R = \sum_{i=1}^{n} P_i \cdot D_i \tag{2}$$

This description of risks is met in finance and insurance. They use risks as prime numbers and can be compared. To make decisions, insurance companies do not add or multiply the consequences and probabilities of events but evaluate them exclusively separately, especially when evaluating very large assets. When establishing criteria for what to assess risks, it is common to establish a value threshold against which outcomes and event rates are compared. In conclusion, they subjectively tend to prefer decisions of evaluating events of the type “large consequences with low probability”.

2.2 Assessing Risk in Matrices

Problems of assessing risk in matrices are discussed for a very long time in the scientific literature [2, 8–13]. When compiling risk matrices, a fundamental mistake is made when assigning linear numerical values for the matrix’s rows and columns. These numbers are not simple scalars. For example, in the matrix [$7 \times 7 = 49$], the multiplication value in the upper right cell may not be correct because of the probability density functions and the consequences are unknown. When using risk matrices, events are assessed in terms of consequences: negative (red), moderate (yellow), positive (green). However, only the outcome of damage appears in the basic definition of risk. Multidimensional assessments in weak scales are the reason for the long-debated problem of “weak consistency” in strict risk matrix assessment [8, 10]. Therefore, the vast majority of practiced risk management methods cannot be recognized as scientifically justified.

3 Method

This paper presents the development of a new matrix method for event risk soft assessment. The method's basis is the concept of the richness of natural language (NL) compared with any formal models used for calculating the risks of complex events. The application of the method is demonstrated and shows experimental solutions of practical tasks.

Fuzzy modeling in the theory of fuzzy sets (TFS) solves the problem of describing the properties of objects, their structure and parameters by applying classical optimization methods and metaheuristic algorithms. The difficulties and problems of applying each of these methods are associated with high computational costs of maintaining fuzzy inference. In the scientific literature, there are claims that computational tasks using TFS methods are redundant and cumbersome. The development of a new soft computing approach (SC) in this paper is considered a non-identical TFS. Probably, this circumstance is the reason that the founder of TFS in 1954 (L. Zadeh) proposed the concept of SC 40 years later (1994) [14]. Since that, this concept is developed by many authors, for example [15]. The author does not know the formal definitions of SC in the literature, so the following definition (Def.-i) is proposed.

Def. 1 Soft computing (SC) defines as a set: hard estimation (HE), soft measurement (SM), soft estimation (SE) in metric and non-metric scales with the ability to simultaneously process quantitative numerical and qualitative linguistic data.

4 The Aim of the Research

The aim of the research is to present a new method of design and soft computing in risk matrices. The task and method of calculating risks in matrix assessments are set out in the following content: a) theoretically justify and present a method for developing matrices of risk SE using fuzzy measures, not only probability measures but also including a wide class of NL assessment concepts; b) develop a definition of the concept of risk matrix; c) develop a design template and samples of risk matrices; d) conduct an experiment of risk soft assessment (SA) and use in practice. The method description assumes a well-founded structured use of NL concepts, which display the parameters and areas of the events in the semantic series of increasing and decreasing estimated events.

5 Method Development

5.1 Risk Matrix Design Template

In this paper, it is proposed to use the following tools that lead to the estimation of the risk value to create event estimation matrix content (3)

$$RISK\ MATRIX : [word,\ color,\ letter,\ digit,\ interval,\ bitdepth] = risk\ value \quad (3)$$

It is presented the development of a risk matrix design template. Since there are no definitions related to risk matrices and their elements in the literature, it is suggested the following definitions (Def-i):

Def. 2 Risk matrix: $|x_{i,j}|$ is a relational table of the (m-n) category consisting of m-rows of names of indicators of the event randomness measure and n-columns of names of indicators of the event outcome measure.

Def. 3 Risk indicator: is the quantified value of |mi, mj| matrix cells, which displays the value of the estimated state of the object.

Def. 4 Risk value: is a numeric value, color indication, and name of the risk measure that displays the object status.

Risk assessment in this matrix is carried out on a nominal scale as a combination of the random value of the outcome and the event's randomness. The symbolic scores in the matrix are the sum of the event outcome value, called severity, and frequency combined with opportunity, Table 1.

Table 1. Risk matrix design template

Content	Event randomness measure	Issue magnitude measure
Word	Possibility, probability, plausibility,...	Outcome, consequence, severity,...
Color	Green, yellow, red,...	Green, yellow, red, ...
Letter	Alphabet: a, b, c, d, e, ...	
Digit	Natural number: 1, 2, 3, 4, 5,...	
Interval	Continuous in the definition area [0,1]	
Bit depth	$2 \times 2, 3 \times 3, 4 \times 4, 5 \times 5, \ldots, mn$	

The event is observed in the randomness of the outcome and in the measure of the magnitude of the outcome. The purpose of risk matrices is to make it possible to assign numerical values to the properties of the evaluated objects through expert subjective judgments. In this work method, the randomness measure absorbs a wide class of soft measures - likelihood, necessity, confidence, conviction, possibility, probability. The measure of the magnitude of the outcome combines the concepts of severity, impact, damages, losses, victims.

Estimations of values can be performed on the data of the numerical domain for determining pairs of fuzzy qualifiers [0, 1]. Such calculus is called HE. The regions can be expanded on sets of semantic series by a sequence of NL words, which are a chain of increasing and decreasing properties of an object.

To solve the tasks of the SE, it is used the nominal scales of fuzzy qualifiers. To build scales, it is suggested compiling semantic rows of words that set the values of event parameters. Each word is assigned a numerical value of the chosen heuristic school.

Def. 5 An indicator located in the middle of a semantic row has a value of the norm or *normal activity.*

The meaning of the words of the definition area (evaluation) is made up in order of strengthening (weakening) of the evaluated property of the object with the corresponding number of increasing (decreasing) numeric values assigned to each qualifier. When creating a table, it is necessary to heuristically align the left and right borders of the definition areas based on the principle of strengthening (weakening) the evaluated object property.

5.2 Establishment of the Parameters Quantification

The solution to the problem of calculating risks is to establish a method of quantification of the parameters. The cell displays one value of the estimated value of risks in natural numbers or in the names of the NL. The method prescribes a sequence in the concepts of NL: a) the search and establishment of an acceptable parameter, b) the establishment of the concept of the domain of definition (estimation), c) the expansion of the domain into a semantic series of words of increasing-decreasing of the estimated properties, d) assigning a numerical value to each of the words. It is possible to set various quantification parameters. In assessing life activity risks, categories are used intuitively and rationally - concepts that have the greatest generalization of meaning. In Table 2, the most used parameters of quantification and areas of HE of event risks are presented: distance, frequency, duration, magnitude, impact, speed.

Table 2. Events risk hard estimation

Quantification parameters	Hard estimation area	
	Left border	Right border
Distances	[near,	far]
Frequency	[always,	never]
Duration	[instant,	forever]
Value	[a lot,	a little]
Impact	[strong,	weak]
Speed	[fast,	slow]

Below are samples of semantic series of words NL, corresponding to the quantification parameters and samples of matrices: frequency, duration of events, and truisms.

5.3 Evaluating the Frequency of Events

The area of "frequency" [always, never] is expanded by a semantic series of SE object properties: [always, very often, often, regularly, constantly, periodically, rarely, extremely rarely, never], as shown in Table 3.

It is introduced the notation [never, always] and the abbreviation [new, alps] for the convenience of subsequent evaluation procedures by symbolic descriptions. Let's

Table 3. Events frequency

Qualifiers [als, nev]	Ω	
Always	als	10
Very often	vfn	9
Often	ofn	8
Constantly	cst	7
Regularly	rg	6
Periodically	pd	5
Sometimes	st	4
Rare	rar	3
Extremely rare	exr	2
Never	nev	1

introduce numerical scale values for expert procedures and quantitative calculations. Since the semantic series we have compiled includes 10 words, you can enter a 10-point scale.

5.4 Events Time Duration Estimation

A semantic row is compiled for the time duration qualifier. Assessments using qualifiers of particular duration overtime should be set in accordance with the goal Table 4.

Table 4. Events duration

Qualifiers [exs, fev]	Ω	
Extremely, instantly	exs	10
Instantly	ins	9
Very fast	verf	8
Fast	fst	7
Soon	sn	6
Moderately	mod	5
Slow	Sl	4
Long	Lg	3
Very long	vlg	2
Forever	fev	1

5.5 Events Truisms Estimation

The words called truisms have the greatest blurring. It is more difficult to compose a semantic row of truisms, Table 5.

Table 5. Events truisms estimation

Truism qualifier [abs, lst]	Ω	
Absolutely	abs	7
Totally	tot	6
Most	mst	5
Virtually	vrt	4
Definitely	def	3
Quite	qt	2
The least	lst	1

To make a row of 5, 7, or 10 words, it is reasonable to choose the same numerical value scale. Truisms are used for evaluation separately and in combination with other qualifiers. For example, truism + frequency qualifier: "most frequently"; truism + duration qualifier: "almost instantly". A seven-word row for truism qualifiers is composed. The advantage of the above matrices is simplicity; the disadvantage is that you can evaluate a property by one of the states displayed by the meaning of NL row. To overcome this disadvantage, a multidimensional profile of the matrix is developed in this paper.

5.6 Complex Matrices Design Development

The soft risk assessment matrix is designed for managing complex events. The number of subsets of fuzzy spaces in the set of fuzzy measures is six. This makes up a numerical scale [1 - the greatest fuzziness, 6 - the least fuzziness]. Weights of estimates are indicated by a double-digit, separated by a hyphen; the first digit has an estimate of the object's structural properties; the second digit evaluates the measures of clarity. The presented matrix has a bit size (6 × 10) and is intended for expert assessment of any events' risks. The matrix is intended for evaluating events in terms of duration randomness [exs, fev] and fuzziness of the plausibility-conviction measures [pl, con]. The resulting score is a measure of the outcome, denoted as a fuzzy measure of the first kind, and written (4):

$$R : <\mu_{[exs,fev]}, \mu_{[pl,con]}> \quad (4)$$

It is supposed that the (6 × 10) bit rate is offered as the highest for expert use. The size of the matrix and the number of cells can be smaller by choice for evaluating a specific complex event.

Def. 6 This symbolic expression is called the profile of the matrix of SE ob-ject values in the selected research area.

Risk assessment using this profile is an alternative probability measure compared to the accepted metric. Unlike analogs, the advantage of the matrix is the resolution of the problem of "weak consistency": the exclusion of mismatch of color and numeric values of risks.

6 Event Risk Soft Assessment Example

Let's create a scenario of practical soft assessment method (SAM) application of the risks of the complex event "Readiness of the Olympiad". No event was existent. Event risk SA example is a thought experiment. Choose a three-color matrix metric: the lower-right cells are red, the upper-left cells are green, and the intermediate cells are yellow. The duration qualifiers are reasonably reduced to six, excluding the top two and bottom two. The extreme right and extreme left measures can be excluded from fuzzy measures, Table 6.

Table 6. Multidimensional profile of matrix events risk soft computing

Duration (μ[exs, fev])	Fuzzy measures μ[pl, con]					
	Plausibility	Possibility	Probability	Necessity	Belief	Conviction
	1	2	3	4	5	6
10 Instantly exs	pl-exs 10-1 green	pos-exs 10-2 green	prob-exs 10-3 green	nes-exs 10-4 green	bel-exs 10-5 green	con-exs 10-6 green
9 soon quickly ins	pl-ins 9-1 green	pos-ins 9-2 green	prob-ins 9-3 green	nes-ins 9-4 green	bel-ins 9-5 yellow	con-ins 9-6 yellow
8 very quickly verf	pl-verf 8-1 green	pos-verf 8-2 green	prob-verf 8-3 green	nes-verf 8-4 yellow	bel-verf 8-5 yellow	con-verf 8-6 yellow
7 quickly fst	pl-fst 7-1 green	pos-fst 7-2 green	prob-fst 7-3 yellow	nes-fst 7-4 yellow	bel-fst 7-5 yellow	con-fst 7-6 yellow
6 soon sn	pl-sn 6-1 green	pos-sn 6-2 yellow	**prob-sn** **6-3** **yellow**	nes-sn 6-4 yellow	bel-sn 6-5 yellow	con-sn 6-6 red
5 moderately mod	pl-mod 5-1 green	pos-mod 5-2 yellow	prob-mod 5-3 yellow	nes-mod 5-4 yellow	bel-mod 5-5 red	con-mod 5-6 red
4 slowly sl	pl-sl 4-1 yellow	pos-sl 4-2 yellow	prob-sl 4-3 yellow	nes-sl 4-4 red	bel-sl 4-5 red	con-sl 4-6 red
3 long time lg	pl-sl 3-1 yellow	pos-sl 3-2 yellow	prob-sl 3-3 red	nes-sl 3-4 red	bel-sl 3-5 red	con-sl 3-6 red
2 very long time vlg	pl-vlg 2-1 yellow	pos-vlg 2-2 red	prob-vlg 2-3 red	nes-vlg 2-4 red	bel-vlg 2-5 red	con-vlg 2-6 red
1 forever fev	pl-fev 1-1 red	pos-fev 1-2 red	prob-fev 1-3 red	nes-fev 1-4 red	bel-fev 1-5 red	con-fev 1-6 red

As a result, a reduced 4 × 6 matrix can be used for this practical task. This reduction of the matrix is caused by our common-sense judgments that extreme estimates are improbable since they correspond to the failure of the Olympiad project. Let's assume that after the calculation, a total expert assessment of the event's risks is obtained (5):

$$\text{RISK} :< (\text{prob}-\text{soon}), (6-3)\left[\text{yellow}\right] > \equiv \{\text{plausible}-\text{realistic}\}, \qquad (5)$$

Where (prob – soon) is the measure of event randomness of expert values in terms of NL, (6–3) is weight values in duration qualifiers, [yellow] is a color estimation of the event according to the heat map method.

General risk assessment in terms of NL can only be carried out in one measure of randomness and considered as {plausible – realistic}. HE matrices require only highly qualified experts. In contrast, the presented matrix solves the possibility of participation of experts of different qualifications. In this case, the assessments can be placed in different cells, which can be combined by brainstorming.

7 Results

A formal definition of the concept of soft computing is presented. Method of design and soft computing of event management in risk matrices is proposed. A risk matrix design template is proposed. Risk matrix, risk indicator, risk value definitions are developed. Definitions allow to formulate the concepts of normal activity in numerical, non-numeric, and color indicators. Examples of evaluating various event parameters are presented: distances, frequency, duration, value, impact, speed. The complex matrix is to be used for expert assessment of risks of any events. Risk assessment using this profile is an alternative to the accepted probability measure metric. In contrast to the analogs, the advantage of the proposed matrix is the resolution of the problem of "weak consistency" of the matrices: the elimination of mismatch of color and numeric values [8, 10].

8 Conclusion

The purpose of risk matrices soft computing is to make it possible to assign numerical values to the properties of evaluated objects through expert subjective judgments. The randomness measure absorbs a wide class of soft measures. The measure of the magnitude of the outcome combines the concepts of severity, damages, losses, and victims. The method is aimed at the direct application of expert assessments of any objects and their components.

This practice is used by insurance corporations to evaluate very complex and large events of great value. They use commercial databases of damage values without calculating the randomness of events in any fuzzy measures, probabilistic or possible due to the high complexity of calculations. To evaluate the measure of the outcome value, in accordance with the method, it is necessary to create a number of semantic qualifiers of the value and perform an evaluation similar to the previous one.

Symbols proposed and used are normative guidelines for practical application. The matrix template allows to select a set of tools to indicate the randomness and magnitude of risks. The content provides a sample design of risk assessment matrices using the developed template using the soft computing method. The details of the method were previously drafted in the author's works [16, 17].

References

1. Loewenstein GF, Weber EU, Hsee CK, Welch E (2001) Risk as feelings. Psychol Bull 127(2):267–286
2. Slovic P, Fischhoff B, Lichtenstein S (1979) Rating the risks. Environment 21(3):14–20, 36–39
3. Slovic P (1987) Perception of risk. Science 236(4799):280–285
4. ICAO (2016) Annex 19 to the Convention on International Civil Aviation. Safety Management, Second Edition, July 2016. - 999 Robert-Bourassa Boulevard, Montréal, Quebec, Canada H3C 5H7, 44 p
5. The Safety Management Manual (SMM) (Doc 9859) AN/474 (2018) Fourth Edition, 999 Robert-Bourassa Boulevard Montréal, Quebec H3C 5H7Canada, 149 p
6. Gost R (2002) 51897-2002 Risk management. Terms and Definitions. Gosstandart of Russia, Moscow, 8 p. (in Russian)
7. Gost R (2006) 51898-2002 Safety aspects. Rules for inclusion in standards. Standartinform, Moscow, 8 p
8. Ale B, Burnap P, Slater D (2012) Risk Matrix Basics. Confidential – Draft for publication, 1 March 2012. https://www.cambrensis.org/wp-content/uploads/2012/08/RiskMatrices-The-Basics3-0.pdf
9. Cox LA Jr (2009) Cox's risk matrix theorem and its implications for project risk management. https://eight2late.wordpress.com/2009/07/01/cox%E2%80%99s-risk-matrix-theorem-and-its-implications-for-project-risk-management/
10. Cox LA Jr (2008) What's wrong with risk matrices? Risk Anal 28(2):497–512
11. Slovic P (2002) Perception of risk posed by extreme events. Decision Research and University of Oregon Elke U. Weber Columbia University and Wissenschaftskolleg zu Berlin. This paper was prepared for discussion at the conference "Risk Management strategies in an Uncertain World," Palisades, New York, 12–13 April 2002. https://www.ldeo.columbia.edu/chrr/documents/meetings/roundtable/white_papers/slovic_wp.pdf/
12. Sjöberg L (2003) Risk perception is not what it seems: the psychometric paradigm revisited. In: Andersson K (ed) Stockholm VALDOR conference, pp. 14–29
13. Sjöberg L (2002) Perceived information technology risks and attitudes. https://swoba.hhs.se/hastba/papers/hastba2002_005.pdf
14. Zadeh L (1994) Fuzzy logic, neural networks, and soft computing. Commun ACM 37(3):77–84
15. Hodashinsky IA (2007) Soft methods of evaluation: monograph. TUSUR, Tomsk, 152 p. (in Russian)
16. Plotnikov NI (2018) The development of the subject domain observation complex for management purposes. In: 2018 14th international scientific-technical conference APEIE – 44894, Part 1, vol 1. NSTU, pp 268–272
17. Plotnikov NI (2013) Resucy bezopasnosty transportnych kompleksov. Monographia. [Transport complex safety resources. Monography], Novosibirsk, Russia, AviaManager Publ. 286 p. (in Russian)

Pothole Identification in Flexible Pavement Using Unmanned Aerial Vehicles (UAVs)

Joaquin Humberto Aquino Rocha(✉), Nahúm Gamalier Cayo Chileno, Silvia Natalia Azurduy Rodriguez, Alex Isaac Arrázola Brañez, José Gabriel Terán Camacho, and Héctor Luis Sánchez Miranda

Universidad Privada del Valle, Tiquipaya, Bolivia
jaquinor@univalle.edu

Abstract. Unmanned Aerial Vehicles (UAVs) have more and more applications in Civil Engineering. Recent studies show the facilities and advantages of using this technology in infrastructure inspection. However, there are uncertainties about its actual scope. The objective of this study is to identify potholes in flexible pavement with UAV. The methodology consisted of selecting a damaged flexible pavement, where defects were measured both manually and digitally, the last through the UAV processed images. Data were collected at three flight altitudes (20 m, 30 m, and 40 m) and six hours (8:00 a.m., 10:00 a.m., noon, 2:00 p.m., 4:00 p.m., and 6:00 p.m.). Subsequently, the error between the field and digital measurements was calculated. The results show that the lower the flight altitude, the fewer errors occur. The morning hours are presented with fewer errors than the afternoon hours. 6:00 p.m. was the time with the highest error and standard deviation. The maximum error was 6,85%. Although the use of UAV allows easy and fast detection of potholes, other aspects such as the environment and obstacles in the inspected place must still be considered since they can generate problems in the identification and measurement of deterioration.

Keywords: Inspection time · Flight altitude · Error

1 Introduction

Highway inspection activities include the evaluation of both physical and functional aspects [1]. The inspection work must be extensive and complete to determine the pavement state in order to propose the repair type, estimate budgets, and, through this way, provide adequate service levels for vehicular traffic [2].

Traditionally, highway inspection activities are based on detecting problems visually. This procedure requires a great deal of time and work [3]. Traffic interruption is necessary, generating limitations in access to infrastructure (road or avenue) and inspection time. This produces delay and, therefore, higher cost. Additionally, visual inspections require professionals with extensive experience and skill. On certain occasions, inspectors risk their lives when exposed to dangerous conditions, such as direct exposure to vehicular traffic, considering the roads are not properly closed. Finally, quantitative data is rarely

Y. Iano et al. (Eds.): BTSym 2020, SIST 233, pp. 589–597, 2021.
https://doi.org/10.1007/978-3-030-75680-2_65

available, so, in general, the information reported is subjective, irrelevant, and limited [1, 4].

To counteract the drawbacks of visual inspection, different technologies have been proposed and developed to detect pavement defects, and their characteristic is the acquisition of more precise data, which allows the quantitative analysis of the pavement [5]. The use of Unmanned Aerial Vehicles (UAVs) and image processing techniques are among the most recent applications in pavement inspection [6, 7].

UAV's growing use is motivated by its advantages, such as flexibility, relatively low cost compared to other technologies, easy drive, remote control, information from the Earth's surface, and the rapid acquisition of data and high-resolution images [2, 8, 9]. These factors help in the early problems' detection. This aspect favors timely maintenance, avoiding major consequences, both economic and physical [10].

Despite the favorable results reported in the literature, it has been shown that the flight altitude and the specific conditions of the study site influence the detection of defects in the pavement. Limitations include spatial resolution and correct damage identification [5, 11]. This article's main objective is to identify deterioration in flexible pavement with UAV, specifically potholes, considering different flight altitudes and inspection schedules.

2 Materials and Methods

The present study was carried out through three defined stages. The first corresponds to the recognition and selection of the area to be studied. The second stage comprises the acquisition of field data, digitally with UAV and manually, considering the last as the real measurement. In the third stage, the data was processed. The digital images followed the procedure of adding photo, aligning, point cloud construction, mosaic model, digital elevation model (DEM), and, finally, orthomosaic, where the measurements were made. Subsequently, the real measurement in the area was compared with the UAV processed data, obtaining the absolute error in percentage (%).

The selected area is located at km 5 of Petrolera Avenue in the city of Cochabamba - Bolivia. The section is characterized by a deteriorated state of flexible pavement (presence of abundant potholes) and the little vehicular traffic on weekends, period in which the study was carried out. The deterioration of flexible pavement is due to many factors such as traffic, misuse, rain (bad drainage system), thermal expansion, lack of inspection and maintenance, among others [12]. Figure 1 shows the section location studied and the potholes analyzed in the investigation.

The DJI Phantom 4 Pro[1] was used for taking images. This equipment is used in different investigations due to its advantages such as simplicity when operating, configuration, availability in the market, and economy in relation to other technologies [3, 13, 14].

[1] DJI Phantom 4 Pro technical characteristics at https://dl.djicdn.com/downloads/phantom_4_pro/20170719/Phantom_4_Pro_Pro_Plus_User_Manual_ES.pdf, last accessed 2020/11/24.

DroneDeploy program was used to create the flight routes, which allows calculating the area and time necessary for taking images. In addition, different parameters can be modified: altitude, avoiding obstacles, structural mode, etc. With the use of this program, UAV can be operated remotely and automatically, both on the mission and during the takeoff and landing.

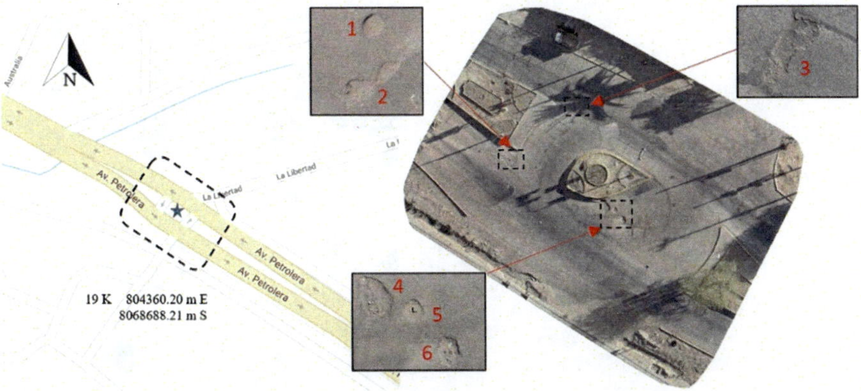

Fig. 1. Location of the selected area and studied defects.

An overlap of 85%, longitudinal and transverse, and an imaging speed of 0,3 m/s were used. For the present study, three flight altitudes (20 m, 30 m, and 40 m) and different times of imaging (8:00 a.m., 10:00 a.m., noon, 2:00 p.m., 4:00 p.m., and 6:00 p.m.) were considered in order to evaluate the influence of these variables on the measurement precision of the potholes area when compared to the actual measurement.

To process the images, Pix4D Mapper Enterprise[2] software was used. This program is used in research to generate an image alignment, DEM reconstruction, and generation of the orthomosaic [3, 13]. Potholes were measured using Global Mapper 20[3] software. The measurement tool was used to draw an arbitrary polygon (pothole) on the previously generated orthomosaic. The software performs a vector analysis to obtain the area and perimeter of the identified pothole. The procedure described is shown in Fig. 2. Field measurement was also performed to calculate the error (%) between both measurements. Saad and Tahar [10] proposed the error calculation method used.

The characteristics of the used computer were: MSI GS65 Stealth Thin 8RF, 8th generation Intel Core i7 processor, 2.20 GHz, and 16 GB of RAM memory.

---

[2] Pix4Dmapper Pix4D characteristics at https://www.pix4d.com/product/pix4dmapper-photogrammetrysoftware, last accessed 2020/11/24.

[3] Global Mapper characteristics at https://www.bluemarblegeo.com/products/global-mapper.php, last accessed 2020/11/24.

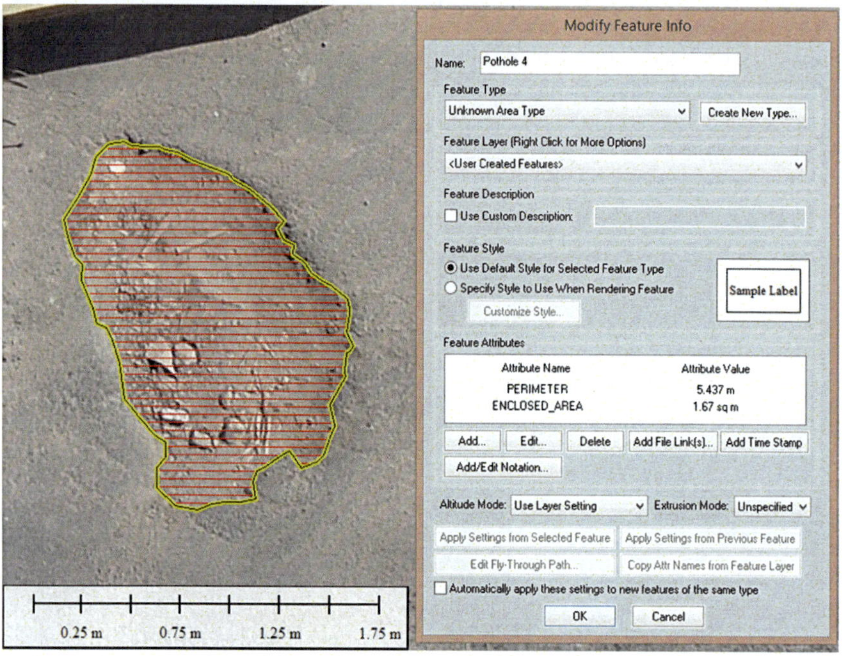

Fig. 2. Areas measurement with Global Mapper 20.

3 Results and Discussion

Figure 3 presents the average error considering the altitude. The average is obtained from the six hours studied for each of the altitudes; this procedure was performed for each defect. The flight altitude of 20 m presents the smallest errors for the 6 defects; the average of all data at this point is 1,42%, and the standard deviation is ±0,85%. For 30 m, the average is 2,37%, and the standard deviation is ±0,90%. Finally, for 40 m, the highest error and standard deviation of all the defects analyzed is presented, an average of 3,82% and a standard deviation of ±1,61%. This is comparable with the results reported by Saad and Tahar [10], where the higher the flight altitude, the higher the error rate.

To observe this behavior in detail, Fig. 4 shows the box-and-whisker plot for the flight altitude error of all the data. For the three flight altitudes analyzed, there is a wide dispersion of data; however, the error ranges are different. The smallest range is for the flight altitude of 20 m, from 0,23% to 3,68%; for 30 m, the values vary from 0,94% to 4,43%, and the flight altitude of 40 m presents the largest range, 1,13% to 6,85%, and the last value was the maximum error found in the investigation.

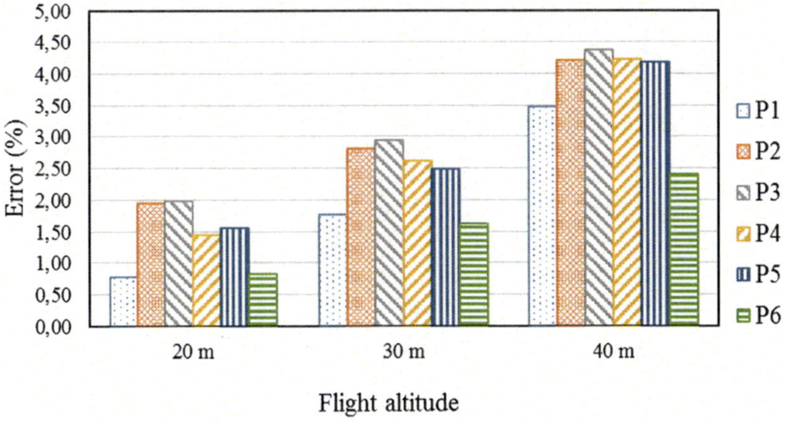

Fig. 3. Average error by flight altitude.

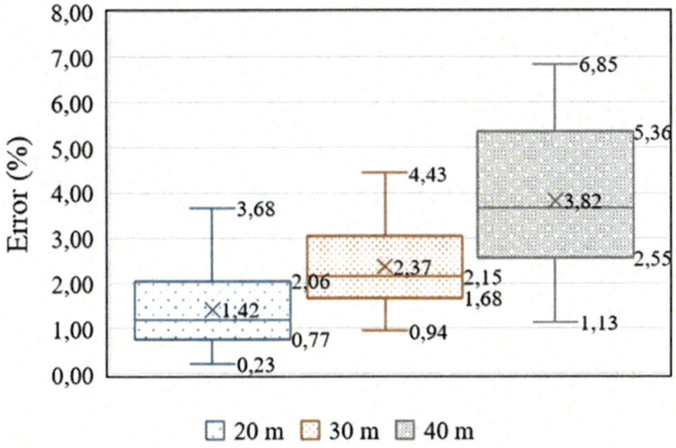

Fig. 4. Box-and-whisker plot for altitude error data.

In a qualitative results analysis, Fig. 5 shows Pothole 6 for the three flight altitudes analyzed. Visually, the image at a flight altitude of 20 m has better resolution and detail. In comparison, the one with the lowest resolution is the image taken at a flight altitude of 40 m. This is due to the number of pixels; the lower the flight altitude, the higher the image resolution, which generates fewer errors [3].

Regarding the time of data collection, Fig. 6 shows the default average error for the analyzed hours. This value is calculated by averaging the error of the three flight altitudes for each hour. The data are scattered; however, it is observed that the morning hours have less error than the afternoon hours. The error for 6:00 p.m. is the largest for all cases, and the error for 8:00 a.m. is the smallest except for Pothole 5.

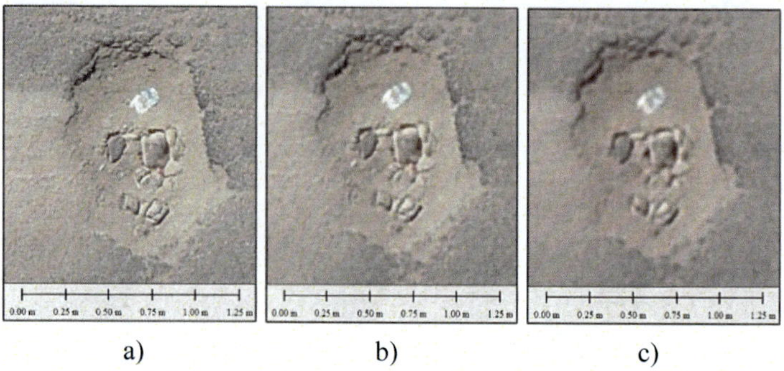

Fig. 5. Pothole 6 at 2:00 p.m. at different flight altitudes: a) 20 m, b) 30 m, and c) 40 m.

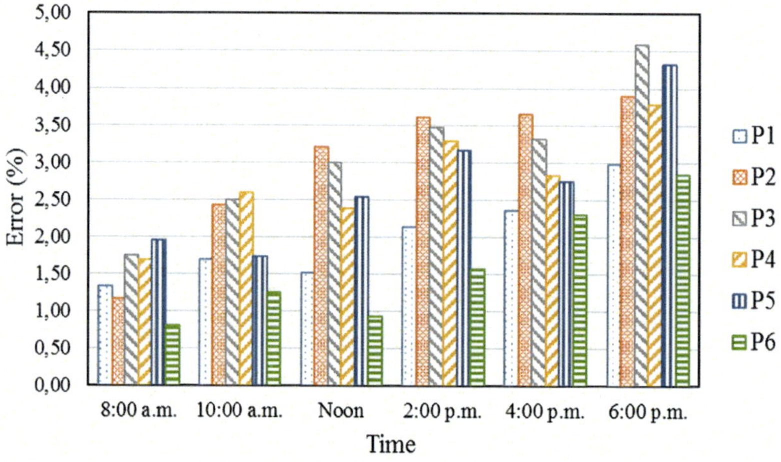

Fig. 6. Average error per hour.

The mean error for 8:00 a.m. for all defects is 1,45%, with a standard deviation of ±0,79%, being the smallest average error. At 10:00 a.m., the mean error is 2,03% and standard deviation is ±0,78%. At noon, the average error is 2,26% and a standard deviation of ±0,74%, being the lowest of all cases. For 2:00 p.m., the mean error is 2,87%, and the standard deviation is ±0,95%. At 4:00 p.m. the mean error is 2,67% and standard deviation ±0,90%. The highest mean error and standard deviation are presented for 6:00 p.m. with values of 3,74% and ±1,12%, respectively.

The whisker-and-box plot for all the error data, considering the test hours, is presented in Fig. 7. The morning and noon hours present less data dispersion compared to the hours of the afternoon. The lowest data range is for 8:00 a.m., from 0,23% to 3,21%; followed by noon with a variation of 3,18% (0,60% to 3,78%). The largest range of data is for 4:00 p.m., 0,58% to 6,58%, closely followed by the 6:00 p.m. time, from 0,95% to 6,85%.

Figure 8 shows Pothole 1 and 2 at a flight altitude of 20 m for the six hours studied. In Fig. 8a (8:00 a.m.), it is observed that the sun position creates a shadow on the defects, causing interference; however, the shape of the defects can be determined. At 10:00 a.m.

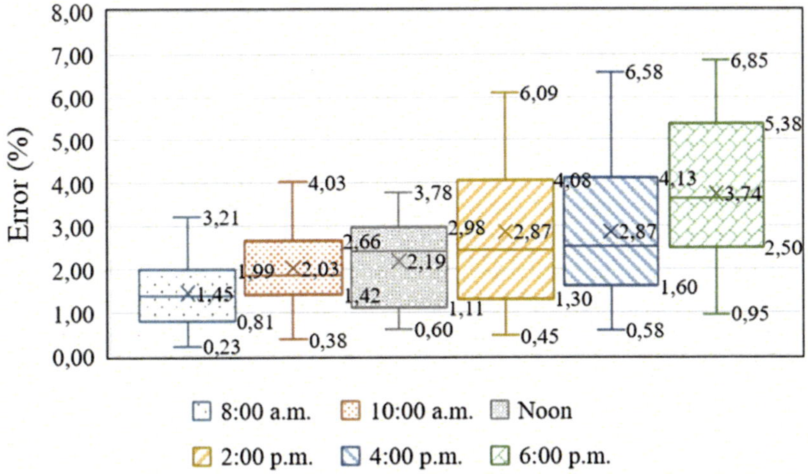

Fig. 7. Box-and-whisker plot for error data per hour.

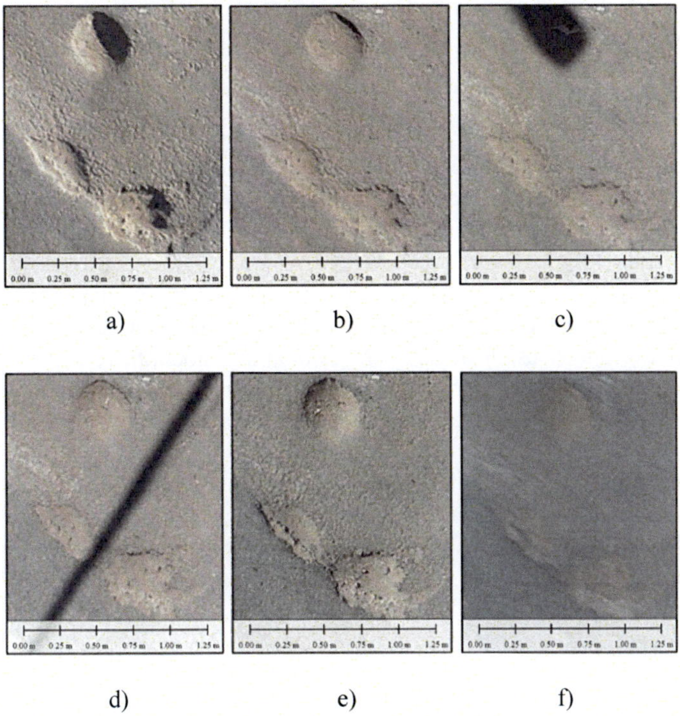

Fig. 8. Pothole 1 and 2 at a flight altitude of 20 m and different times: a) 8:00 a.m., b) 10:00 a.m., c) noon, d) 2:00 p.m., e) 4:00 p.m. and f) 6:00 p.m.

and noon, the defects are clearly identified, except for Pothole 1 at noon, the presence of the shadow of a post prevents their complete visualization. In the afternoon, at 2:00 p.m., the potholes are visualized only with a small obstacle in Pothole 2. At 4:00 p.m., as in the morning hours, the solar position generates shadows on the defects. Finally, at 6:00 p.m., the defects visibility is difficult and the image loses quality due to the lack of sunlight.

Pothole identification and measurement are influenced by the environment, sun position, and ambient conditions. In the first case, there may be obstacles that prevent the identification of defects during the day. Sun position casts shadows from both the environment and the defect itself, which can lead to inaccuracies when measuring potholes. Environmental conditions such as strong winds and cloudiness affect the results in data collection and the limitation of sunlight, respectively.

4 Conclusions

In the present investigation, an experimental study was carried out on defect measurement and identification precision with UAV. Different flight altitudes and times were considered for data collection. Inspection and measurement of defects are possible quickly and accurately when compared to the traditional (visual) method. The maximum error found was 6,85%, considering the most unfavorable conditions, maximum flight altitude, and the last time of the test.

The UAV operator's ability and the person in charge of analyzing and measuring the defects with the program must also be considered so as not to have a significant error in the data. For UAV use in the identification of potholes, environmental conditions and the environment must also be considered in order to avoid errors in data collection and analysis of results.

Regarding the flight altitude, the errors are minor for 20 m and maximum for 40 m; this is mainly due to the image's resolution. About the inspection schedule, the results are similar in average error; however, from 8:00 a.m. to noon, there is less variability of the data compared to the afternoon hours. At 6:00 p.m., there is the highest mean error and standard deviation of the data, indicating an inappropriate time for the pothole inspection.

It was possible to verify that the UAV inspection method is efficient and more effective (e.g., less time, safer, greater accessibility to the site, and less personnel), compared to the visual inspection method in the field, thus becoming a convenient alternative, for inspection of defects.

References

1. Agnisarman S, Lopes S, Chalil Madathil K, Piratla K, Gramopadhye A (2019) A survey of automation-enabled human-in-the-loop systems for infrastructure visual inspection. Autom Constr 97:52–76. https://doi.org/10.1016/j.autcon.2018.10.019
2. Romero-Chambi E, Villarroel-Quezada S, Atencio E, Muñoz-La Rivera F (2020) Analysis of optimal flight parameters of unmanned aerial vehicles (UAVs) for detecting potholes in pavements. Appl Sci 10(12):4157. https://doi.org/10.3390/app10124157

3. Tan Y, Li Y (2019) UAV photogrammetry-based 3D road distress detection. ISPRS Int J Geo-Inf 8(9):409. https://doi.org/10.3390/ijgi8090409
4. Ellenberg A, Kontsos A, Moon F, Bartoli I (2016) Bridge related damage quantification using unmanned aerial vehicle imagery. Struct Control Health Monit 23(9):1168–1179. https://doi.org/10.1002/stc.1831
5. Ragnoli A, De Blasiis M, Di Benedetto A (2018) Pavement distress detection methods: a review. Infrastructures 3(4):58. https://doi.org/10.3390/infrastructures3040058
6. Roberts R, Inzerillo L, Di Mino G (2020) Exploiting low-cost 3D imagery for the purposes of detecting and analyzing pavement distresses. Infrastructures 5(1):6. https://doi.org/10.3390/infrastructures5010006
7. Kim S, Irizarry J, Kanfer R (2020) Multilevel goal model for decision-making in UAS visual inspections in construction and infrastructure projects. J Manag Eng 36(4):04020036. https://doi.org/10.1061/(asce)me.1943-5479.0000803
8. Máthé K, Buşoniu L (2020) Vision and control for UAVs: a survey of general methods and of inexpensive platforms for infrastructure inspection. Sensors 15(7):14887–14916. https://doi.org/10.3390/s150714887
9. Inzerillo L, Di Mino G, Roberts R (2018) Image-based 3D reconstruction using traditional and UAV datasets for analysis of road pavement distress. Autom Constr 96:457–469. https://doi.org/10.1016/j.autcon.2018.10.010
10. Saad A, Tahar K (2019) Identification of rut and pothole by using multirotor unmanned aerial vehicle (UAV). Measurement 137:647–654. https://doi.org/10.1016/j.measurement.2019.01.093
11. Zhang S, Lippitt C, Bogus S, Neville P (2016) Characterizing pavement surface distress conditions with hyper-spatial resolution natural color aerial photography. Remote Sens 8(5):392. https://doi.org/10.3390/rs8050392
12. Bridle R, Porter J (2002) The motorway achievement: frontiers of knowledge and practice. Thomas Telford, United Kingdom
13. Ferrer-González E, Agüera-Vega F, Carvajal-Ramírez F, Martínez-Carricondo P (2020) UAV photogrammetry accuracy assessment for corridor mapping based on the number and distribution of ground control points. Remote Sens 12(15):2447. https://doi.org/10.3390/rs12152447
14. Burdziakowski P (2020) Increasing the geometrical and interpretation quality of unmanned aerial vehicle photogrammetry products using super-resolution algorithms. Remote Sens 12(5):810. https://doi.org/10.3390/rs12050810

Multi-agent-Based Approach for Complex Industrial Process Modeling

Kenza Redjimi(✉) and Mohammed Redjimi

Faculty of Science, Department of Computer Science, LICUS Laboratory, University 20 August 1955, 21000 Skikda, Algeria
{k.redjimi,m.redjimi}@univ-skikda.dz

Abstract. A large number of components that interact with each other characterize complex systems. The modeling of such systems is generally based on a fine division of sub-assemblies of components of less complexity. Each component can be individually modeled and then simulated. This article deals with a top-down modeling method using multi-agent systems. The overall system is considered as a set of interacting agents. The behavior of the global system emerges from the behavior of its parts. This approach was applied to the liquefied natural gas treatment process (decarbonation) at an Algeria gas plant. The results of the simulation can be used to train recruits in this factory.

Keywords: Multi-Agent System (MAS) · Industrial system modeling and simulation · Agent/Group/Role (AGR) Model · Aalaadin · Madkit

1 Introduction

Today, industrial systems employ advanced technologies for the development of complex industrial processes. The vast majority of these systems can be considered complex in the sense that they are composed of a large number of interacting elements. To develop such systems, to predict their behaviors before beginning to build them, or for educational training, these systems are modeled and undergo a set of tests.

The use of computer tools for modeling offers the advantage of building a computer model that is very faithful to the considered system, a set of its parts, then to judge the behavior of the obtained model by injecting it with a set of data as inputs. This makes it possible to handle, observe, and improve understanding of its behavior and allows avoiding any possible errors concerning its future evolutions. The shutdown of certain large so-called critical systems can cause material and financial damages. For these systems, the use of models representing them is of substantial help. In fact, instead of stopping the operation of all or part of a plant, it suffices to consider the models, which represent them. This is the case, for example, with industrial boilers, the shutdown of which can cause the shutdown of a whole part of the plant, which depends on it.

The notion of system is at the center of several fields and disciplines. A system is characterized by a set of communicating and interacting elements. A system receives inputs and delivers results. Various systems can be envisaged, such as industrial systems,

Y. Iano et al. (Eds.): BTSym 2020, SIST 233, pp. 598–608, 2021.
https://doi.org/10.1007/978-3-030-75680-2_66

economic systems, political systems, etc. A metric often used to characterize a system concerns its complexity.

The modeling of systems has been the subject of many studies and researches considering its important impact on the phenomena implemented. P. Fishwich's [1] considers modeling as a set of metaphors and analogies to better understand any phenomena. For C. Oussalah's [2], modeling is a transformation between a system and a model to give a simplified and observable representation of its structure and behavior.

The experimentation of the model can be considered as its simulation. Practically, a simulation is the execution phase of the model [3]. Given a set of pertinent data as input of the model, this phase consists of evolving the model and observing and analyzing the obtained results. So, works can be observed as modeling of actual cases in the laboratory.

Simulation of complex systems can take hours or even days and consume a lot of CPU time. This is why modelers have recourse to methods of reduction of complexity. Among these methods, there are parallel simulation [4], decomposition of the phenomenon in several entities of fewer complexities [5], and the use of fast and parallel CPUs [6].

The search for high fidelity, precision, and adaptation requires modeling and simulation data processing tools to be chosen with great care. Their use in the field of modeling and simulation is very widespread, ensuring excellent qualities for which the improvements are continuous with reasonable costs.

Several modeling approaches, methods, and tools have been and continue to be used in the literature. Petri nets (PN) [7] are among the most efficient and widely used tools. Despite their 62 years of existence, they continue to be relevant today.

These networks, despite their simplicity, make it possible to represent the dynamics of systems. Several variations and extensions of these networks have been developed, such as Colored PN (CPN) and Timed PN [8, 9].

The UML statechart diagrams are other powerful semi-formal tools that can represent the behavior of systems [10]. At a higher level, multi-agent systems (MAS) are modeling tools that offer the advantage of a fine decomposition of a real complex system into a set of its components. The behavior of the global system emerges from that of the behaviors of its entities. The notion of agent designates an entity whose functioning is autonomous and which is capable of carrying out a set of tasks [11, 12]. These systems are used more and more in computer systems because of their flexibility, simplicity, effectiveness, and possibility of extension. Moreover, a human observer can easily understand the description given by MAS.

The rest of this paper is organized as follows: Sect. 2 presents the multi-agent systems-based approach. Section 3 is dedicated to the proposed approach. Section 4 introduces a use case of industrial system simulation, and finally, Sect. 5 concludes this paper.

2 The Multi-agent System (MAS)

Multi-agent systems (MAS) provide high-level tools for modeling distributed systems. Nowadays, their application has become very widespread, and they have been applied with success in various domains, especially those based on artificial intelligence. A set of autonomous agents cooperate and collaborate in the execution of a common objective. From the behavior of all these agents emerges the overall behavior of the system [11–20].

2.1 Agent Definition

There is no common agreement about a definition of the term agent. In general, an agent is an entity (hardware or software), which is different from an ordinary program in the sense that it is endowed with autonomy for functioning in environments shared by other agents [12, 16]. An agent is able to evolve in an environment through the acquisition of information emitted by the latter and by acting on it. An agent's architecture varies from a simple structure formed of production rules often linked to very complex structures where the agent can be endowed with advanced cognition and reasoning mechanisms, referring to it as an associative and social character [21].

3 The Proposed Approach

The work concerned with this present theme is related to the development of industrial process simulators in the hydrocarbon sector. This involves providing educational support to technical staff so that they can understand the different stages of the process.

The proposed approach is a top-down approach, which starts, therefore, from the overall process and then divides it up according to the fundamental units (see Fig. 1) after having identified them.

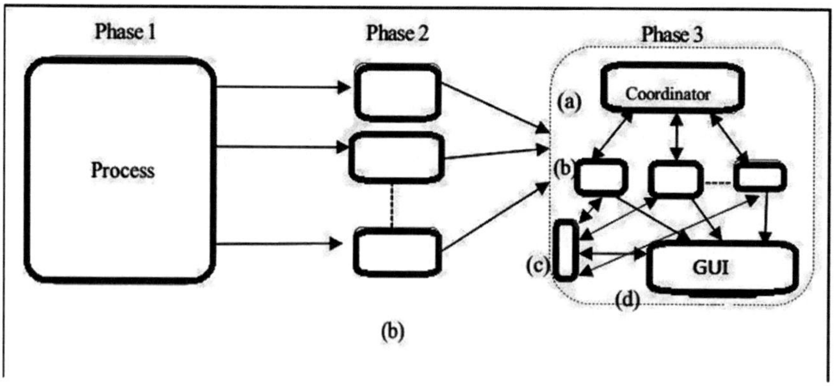

Fig. 1. Process breakdown.

The process cutting procedure is carried out as follows:

1. The process is studied in detail. The principle of its operation is clearly established.
2. The different units making up the process are identified according to the detailed technical diagram of the process (b).
3. Each unit is isolated to be individually modeled, and the obtained units are assembled in a global diagram.

Three additional control and communication units are added to the model:

1. A coordinator whose role is to control and manage the execution of the whole system.
2. An input interface through which the user can act on the system.
3. An interface for visualizing the progress of the process according to the inputs sent by the user.

3.1 Tools for the System Design

For the design of this kind of simulators, a solution using multi-agent systems is very appropriate. An agent represents each unit. The same goes for the coordinator, the input interface, as well as the display interface.

Nowadays, there are several multi-agent systems-based approaches well suited to the modeling and simulation of systems. Likewise, a wide variety of multi-agent platforms is available. Among the existing methodologies, the AGR (Agent-Group-Role) model gives very good results in terms of modularization of systems.

In the AGR model, an agent can play one or more roles in one or more groups. A group is made up of at least one agent. An agent who does not belong to a group can request to join this group. Several agents can play the same role. An agent can communicate with another only if the two agents are in the same group. The interactions between agents are founded on the messages exchange principle and allow coordination, collaboration, and synchronization between agents.

The AALAADIN meta-model is an organizational model for the modeling of multi-agent systems based on AGR concepts. AALAADIN provides a set of conceptual tools for the implementation of applications based on multi-agent systems. AALAADIN was developed at the computer science, robotics, and micro-electronics laboratory, Montpellier (LIRMM) [22–24].

3.2 The MADKIT Platform

The Multi-Agent Development Kit platform (MadKit) is a free software for the design and the simulation of Multi-Agent Systems (MAS) [25–29]. MadKit is a set of packages of Java classes. It implements a reduced size agent kernel, various libraries of messages, probes, and agents. In addition, Madkit includes a useful graphical development environment and many other agent handling systems. Madkit implements different tools for the management of agents, groups, and roles according to the AGR model. This platform is based on the AALAADIN meta-model.

4 Use Case

Pure natural gas (NG) is a fuel unfit for consumption due to the presence of impurities, particularly water and carbon dioxide (H2O and CO_2), which reduces its calorific value. Liquefaction of NG requires cooling to about −165 °C. Knowing that the freezing temperature of the water is at 0° C and that of the CO_2 at −20 °C, this will clog the pipes of the machines during the liquefaction process. For these reasons, NG must be treated (decarbonized and dehydrated) before the liquefaction operation. Decarbonation is the extraction of carbon dioxide CO_2 from natural gas, and dehydration consists of extracting water from it. There are several industrial ways to execute the decarbonation. [30] presents the process adopted in this work.

4.1 The Natural Gas Decarbonation Process

In this use case, there are two operations for the decarbonation process: i) the washing operation, which is intended to extract the CO_2 from the NG with an MEA ($C_2H_5ONH_2$), and the regeneration phase, where the results from the washing operation enter a regeneration tower so that they release CO2 and then restore the MEA solution. Figures 2 and 3 illustrate the dedicated equipment.

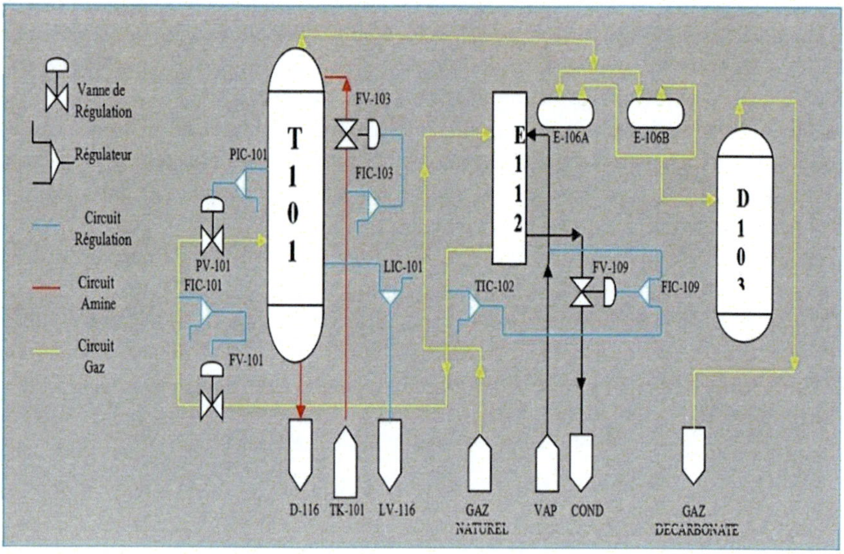

Fig. 2. Technical drawing of the washing process.

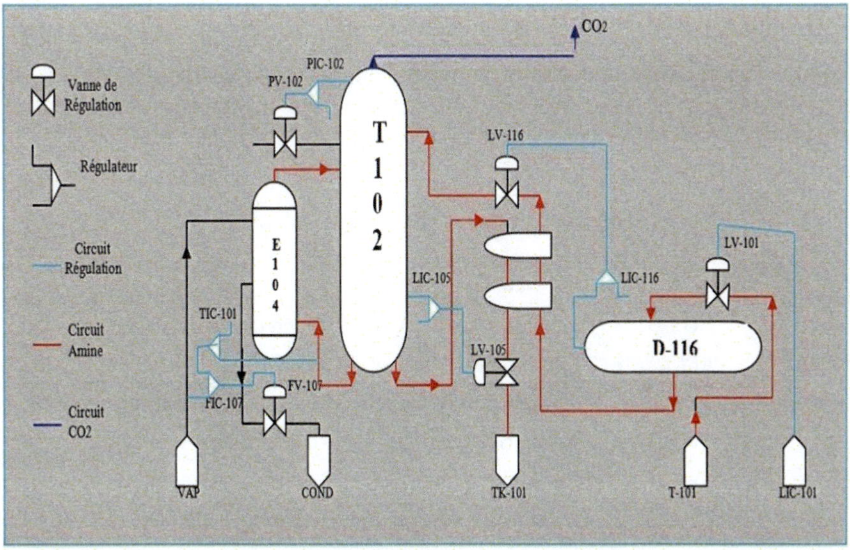

Fig. 3. Technical drawing of the regeneration process.

According to the breakdown system phases presented in Fig. 1, at first and after studying and analyzing the whole system as presented in Figs. 2 and 3, the following four points are highlighted:

- **The system ordered:** it is the decarbonation process. It contains a set of organs (T-101, T-102, D-116, E-112, E-104…) which carry out the various chemical reactions to achieve the decarbonation of the Natural Gas.
- **The control system:** this is the distributed control system for the natural gas decarbonation process. It is formed by a set of autonomous, modular, and distributed instruments. They are intended for the control and regulation of the process.

These instruments are divided into two sections:

- **The washing section:** it groups the regulation valves (FV-109, FV-103, FV-101, PV-101), which are regulated by a control system, which contains the different regulation loops.
- **The regeneration section:** also contains a set of control valves (FV-107, PV-102, LV-116, LV-105, LV-101), which are regulated by the same control system. To carry out this adjustment, the control system is based on the set of instructions.

4.2 The Proposed AGR Model

To build an interactive simulator for the above-mentioned industrial gas process, the three following groups, as depicted in Table 1, have been distinguished. Note that this point corresponds to phase 2 of the proposed simulation approach:

Table 1. Obtained agents, groups, and roles.

Id_Group	Id_Agent	Designation	Role
Perturbation	GUS	Graphical User System	To interact with the system
Washing	APM	Agent Process Manager	To acquire the Instructions
	APM	Agent Process Manager	To regulate the washing group
	FV-103	Flow Valve Agent	To control the Amine flow
	FV-101	Flow Valve Agent	To check the natural gas flow
	PV-101	Pressure Valve Agent	To check the pressure of T-101
	FV-109	Flow Valve Agent	To check the steam flow of E-102
Regeneration	APM	Agent Process Manager	To set the regeneration group
	FV-107	Flow Valve Agent	To check steam flow E-104
	PV-102	Pressure Valve Agent	To check the pressure T-102
	LV-101	Level Valve Agent	To check the level of T-101
	LV-116	Level Valve Agent	To check the level of D-116
	LV-105	Level Valve Agent	To check the level of T-102

This decomposition was obtained after a meticulous study of the NG decarbonation process as well as the material structure and the arrangement of the interactions between the different material components (pressure, level, flow control, etc.). This article will not go into the precise details of how the system works.

The APM agent ensures the time synchronization between the different units. This particular agent (APM) serves as a coordinator in the system.

4.3 The AALAADIN Model

The entire model with links between all the components (agents) of the system is depicted in Fig. 4 below. This diagram does not describe the roles of each agent. This step corresponds to phase 3 of the proposed demarche.

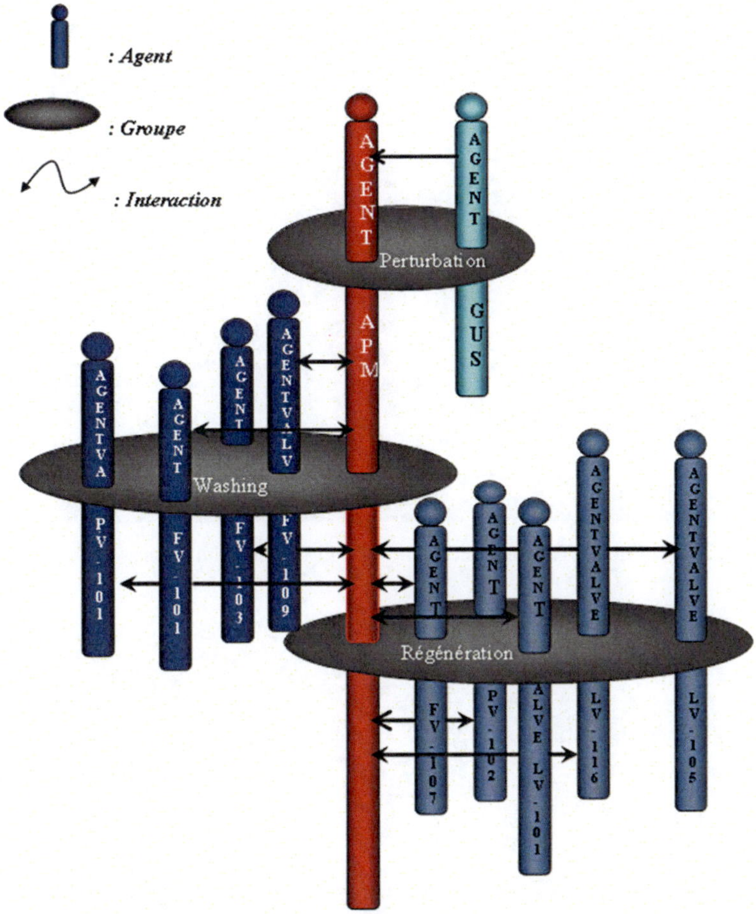

Fig. 4. The AALAADIN model of the simulator.

4.4 The System Implementation

The simulator was developed by using the Mad-Kit agent platform based on AALAADIN model. The two main interfaces of the project are presented in Figs. 5 and 6. These two interfaces concern the washing section and the interface of the regeneration section. They allow the user to drive the washing and the regeneration operations and supply the system with data inputs.

4.5 Discussion

This work aims to implement the MAS mechanisms of cooperation, collaboration, and synchronization between agents. These tools are used to design and to realize industrial simulators. The use case presented and implemented here, which can be very useful as pedagogical support, consolidates and validates the proposed approach.

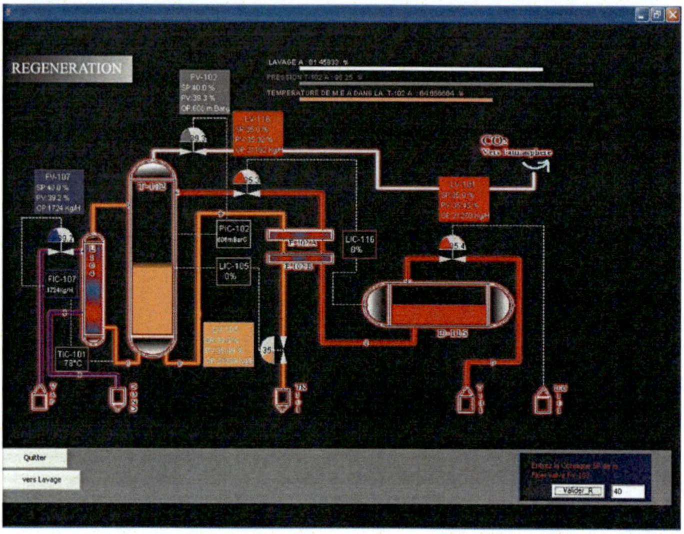

Fig. 5. The interface of the washing section simulator.

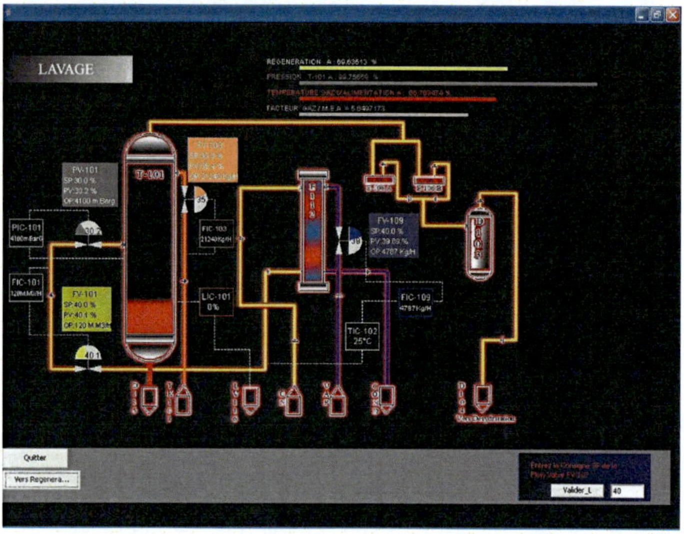

Fig. 6. The interface of the regeneration section simulator.

5 Conclusion

The concerns of this work are to introduce some notions of the design of industrial simulators. This work responds to a real need of providing a tool allowing young beginners in the company to understand the evolution of an industrial process. In this case, it is the process of liquefied natural gas decarbonation. The equipment allowing this procedure is modeled using agents. The coordination and cooperation between these agents allow the exchange of information and the evolution of the process. This simulator is currently operational, and its use allows significant financial and material gains.

All the components of the system are modeled using the agent paradigm. The interaction between these agents ensures the evolving of the system, and the global behavior emerges from the local agent's execution. The use of consistent and coherent tools (for example, AALAADIN and MADKIT) is of great interest in obtaining efficient software products, as is the case here. In future work, we project to extend this approach to distributed systems modeling.

References

1. Fishwick P (1995) Simulation model design and execution: building digital worlds
2. Oussalah C (1988) Modèles hiérarchisés multi-vues pour le support de raisonnement dans les domaines techniques. Technical report
3. Fishwick P, Paul A (1997) Computer simulation: growth through extension. Trans Soc Comput Simul Int 14(1):13–23
4. Mubarak M, Carothers CD, Ross RB, Carns P (2016) Enabling parallel simulation of large-scale HPC network systems. IEEE Trans Parallel Distrib Syst 28(1):87–100. https://doi.org/10.1109/TPDS.2016.2543725
5. Wainer G, Liu Q, Jafer S (2016) Parallel simulation of DEVS and Cell-DEVS models in PCD++. In: Discrete-event modeling and simulation, pp 223–270, 20115630
6. Martınez-del-Amor MA, Riscos-Núnez A, Pérez-Jiménez MJ (2017) A survey of parallel simulation of P systems with GPUs. Bull Int. MC Soc 55–68
7. Juranic J, Pavkovic N, Naumann T, Marjanovic D (2017) Modelling the design parameters dynamics with Petri nets. In: DS 87-2 Proceedings of the 21st international conference on engineering design (ICED 17) Vol 2: design processes, design organisation and management, Vancouver, Canada, pp 091–100
8. Jensen K, Kristensen LM (2009) Coloured petri nets. Springer, Heidelberg. https://doi.org/10.1007/b95112
9. Jensen K, Kristensen LM, Wells L (2007) Coloured Petri Nets and CPN Tools for modelling and validation of concurrent systems. Int J Softw Tools Technol Transfer 9(3–4):213–254. https://doi.org/10.1007/s10009-007-0038-x
10. Sunitha EV, Samuel P (2019) Automatic code generation from UML state chart diagrams. IEEE Access 7:8591–8608. https://doi.org/10.1109/ACCESS.2018.2890791
11. Mess M, Guerrits B (2018) Multi-agent Systems. Lecture notes in logistics, pp 611–636. https://doi.org/10.1007/978-3-319-92447-2-27
12. Ferber J (1995) Les systèmes multi-agents: vers une intelligence collective. Informatique, intelligence Artificielle, Intereditions Paris
13. Chen F, Ren W (2019) On the control of multi-agent systems: a survey. Found Trends Syst Control 6(4):339–499
14. Weiss G (1999) Multiagent systems – a modern approach to distribute artificial intelligence. MIT Press, Cambridge
15. Wang B, Wang J, Zhang B, Li X (2016) Global cooperative control framework for multiagent systems subject to actuator saturation with industrial applications. IEEE Trans Syst Man Cybern Syst 47(7):1270–1283
16. Ferber J (1999) Multi-agent systems. An introduction to distributed artificial intelligence. Addison-Wesley, Boston
17. Shoham Y (1993) Agent-oriented programming. Artif Intell 60(1):51–92

18. Franklin S, Gasser A (1997) Is it an agent, or just a program? A taxonomy for autonomous agents. In: Muller, Wooldridge, Jennings (eds) Intelligent agents III. Agent theories, architectures, and languages. Springer, pp 21–35
19. Wooldridge M, Jennings NR (1995) Agent theories, architectures, and languages. In: Wooldridge, Jennings (eds) Intelligent agents. Springer, pp 1–22
20. Russell S, Norvig P (2010) Artificial Intelligence: A Modern Approach. Pearson, London
21. Ye P, Wang S, Wang F (2018) A general cognitive architecture for agent-based modeling in artificial societies. IEEE Trans Comput Soc Syst 5(1):176–185. https://doi.org/10.1109/TCSS.2017.2777602
22. Ferber J, Gutknecht O, Michel F (2004) From agents to organizations, an organizational view of multi-agent systems. In: Agent oriented software engineering (AOSE) IV, vol 2935, pp 214–230
23. Souidi MEH, Songhao P, Guo L, Lin C (2016) Multi-agent cooperation pursuit based on an extension of AALAADIN organizational model. J Exp Theoret Artif Intell 28(6):1075–1088. https://doi.org/10.1080/0952813X.2015.1056241
24. Russell SJ (1997) Rationality and intelligence. Artif Intell 94:57–77
25. Ferber J, Gutknecht O (1998) Aalaadin: a meta-model for the analysis and design of organizations in multi-agent systems. In: Demazeau Y (ed) ICMAS (International conference on multi-agent systems), Paris. IEEE Press, pp 128–135
26. Chebout MS, Mokhati F, Badri M, Babahenini MC (2019) Monitoring open multi-agent systems: an aspect-oriented programming based approach. Multi-agent Grid Syst 15(2):155–177. https://doi.org/10.3233/MGS-190307
27. https://mansour.saber.free.fr/termadkit/site/madkit/doc/devguide/devguide.html
28. Gutknecht O, Ferber J (2000) The MadKit agent platform architecture. Laboratoire d'Informatique de Robotique et de Microélectronique de Montpellier
29. https://www.madkit.net/madkit
30. https://patents.google.com/patent/WO2014199036A2/en

DLP 3D Printer with Innovative Recoating System

Italo Leite de Camargo[1,2(✉)], Rogério Erbereli[1], João Fiore Parreira Lovo[1], and Carlos Alberto Fortulan[1]

[1] Department of Mechanical Engineering, EESC/USP, Trabalhador São Carlense, 400, São Carlos, SP, Brazil
{rogerio.erbereli,joao.lovo,fortulan}@usp.br

[2] Institute of Education, Science and Technology of São Paulo, IFSP, Primeiro de Maio, 500, Itaquaquecetuba, SP, Brazil
italo.camargo@ifsp.edu.br

Abstract. Digital Light Processing (DLP) is an additive manufacturing process capable of printing parts with tiny complex features made of polymer, composites, and ceramic materials. Research related to this topic is mainly about the development of new raw materials and the characterization of the manufactured parts, and thus there are few studies related to the mechanical design of equipment for this application. Spreading constant and homogeneous layers is a key factor in the performance of this process. This paper describes the development of a simple and innovative recoating system for DLP 3D Printers composed of two blades with sequential and distinct functions: spreading and leveling. A DLP top-down test bench was built and proved to be able to work even with challenging high solid loading materials as ceramic slurries. The concept involves saving and reusing raw material in a fast and economical process enabling small-scale production suitable for research and development laboratories and small businesses.

Keywords: Additive manufacturing · Digital Light Processing · Stereolithography

1 Introduction

Vat photopolymerization (VP) is the pioneer additive manufacturing process in which liquid photopolymer is selectively cured by polymerization activated by a light source [1, 2], and it still offers the best relationship between resolution and surface quality [3]. In addition, Digital Light Processing (DLP) is a type of VP in which an entire layer is formed at the same time, projecting light from the sections, being faster than processes based on scanning the region to be polymerized [2, 4, 5].

Moreover, VP processes can be divided according to their building direction. This way, in the bottom-up process, the light falls from bellow through the transparent bottom of the vat and, after curing each layer, the building platform rises the equivalent of one layer thickness, promoting the detachment between the piece and the bottom and

Y. Iano et al. (Eds.): BTSym 2020, SIST 233, pp. 609–616, 2021.
https://doi.org/10.1007/978-3-030-75680-2_67

allowing an uncured layer to be placed [3, 4]. The main disadvantage of this process is that the reported periodic detachment introduces stresses in the part being printed, a problem that is aggravated in high viscosity processes, as in ceramic VP, which uses slurries with high solid concentration [3, 6].

On the other hand, the light falls from above in the top-down process. Thus, the platform moves downward whenever a layer is cured [3, 4]. In this process, the main challenge is the recoating, which is the process of depositing a new layer of uncured material between the previously cured layer and the light source. A successful recoating process establishes a fresh layer of thickness exactly equal to the desired one within a short time. The effects related to the viscosity and surface tension properties of the raw material are obstacles to this process. Incorrect layer thickness will impair the accuracy of the printed parts [3, 4, 7].

Although there are some recoating systems used in the bottom-up approach, such as the one from Cerafab500 (Lithoz) [8] or the system patented by Mauritsio [9], this approach usually does not require a specific apparatus for this purpose, since the movement of the platform upwards will already cause the resin to fill the space spontaneously.

In contrast, a recoating system is indicated for the top-down approach to promote uniform layers with constant thickness [3, 4, 10], presenting different solutions such as dip and scrape, deposit and scrape, Zephyr blade, curtain recoating, etc. [11] Dip and scrape is the most widely recoating method. It has been used since the first generation of SLA-250 (3D System) [11] with its patent filed in 1989 [12]. This method includes a "deep dip", in which the part is dipped into the resin vat and then raised until it is a layer thickness from the top of the tank [13]. A blade then moves from one side to another, removing excess, spreading the photopolymerizable material, and creating a new layer [11–13].

Several inventions have been used to replace the "deep dip". Instead, the platform descends the equivalent of layer thickness, and the material is deposited and spread. For example, Optoform equipment, designed to handle high viscosity pastes [14], uses a special scraper [15] to spread material from an adjacent tank whose level is controlled by a piston [11]. Moreover, Acculas SI-C1000 (D-MEC Co.Ltd) deposits material with a dispenser before spreading it with a blade [16].

In addition, the Zephyr System (also known as active blade, application bar, or inverted U recoater) is used in Promaker 600 (Prodways) and current 3D System printers. In this system, a vacuum blade (symmetric hollow blade with a resin reservoir in the center) applies and sweeps a thin layer of resin across the part [11, 13, 17].

On all systems previously mentioned, the scraper speed must be restricted to minimize disturbances, and waiting time (z-wait) is required after spreading until disturbances subside and the resin surface is level. Considering a part of thousands of layers, it considerably increases the build time and can even make the process unfeasible [17].

Furthermore, in the liquid curtain system, a deposition head is moved from one side to another while the raw material is pumped from a reservoir, forming a stable curtain of liquid and depositing a new layer [18]. Unlike the recoating mechanisms previously presented, it is a contactless system; therefore, there is no mechanical contact between the scraper and the resin, preventing loads from being exerted on the previously built layers. However, this system is indicated only to high volume applications as Mammoth

Printer (Materialise) [11]. Lastly, there are other patented recoating systems with no known commercial application such as coating brush mechanism [19], counter-rotating rollers [20], thin-film recoater [21], etc.

Most commercial DLP equipment was not designed to handle high viscosity, as is the case with ceramic raw materials (suspensions with a high content of solid load) and the few capable of producing technological ceramics with good mechanical properties are costly or time-consuming [22]. In this work, an innovative low-cost recoating system was developed and a DLP top-down test bench was built and its performance was tested using commercial resins and own formulated ceramic slurry.

2 Materials and Methods

A top-down DLP test bench was designed and assembled and a schematic is shown in Fig. 1. The decision process applied to this project [23] considered inputs (state of art and process requirements), business context (considering a medium-size company capable of performing traditional manufacturing operations), and environment context (that points to the need for a low-cost machine), similar to the one previously discussed [24]. A commercial slide projector (InFocus), which contains a mercury-vapor lamp (2200 lumens) and whose UV filter was removed, was chosen as a light source. A projection lens was placed between the projector and the building platform to provide an image of the building's size (60 × 45 mm^2). We selected a linear guide actuator (KR33A, THK) driven by a stepper motor with a torque of 1 N.m (NEMA 23) to move the building platform. The top-down approach required a recoating system (not shown in Fig. 1), whose blade-based design will be discussed in the "Results and Discussion" section. Creation Workshop Software was chosen to process the STL files, creating light projections, and controlling the platform's movement and of the recoating system. Designing an economical and compact piece of equipment able to handle high viscosity photopolymerizable materials was the major challenge.

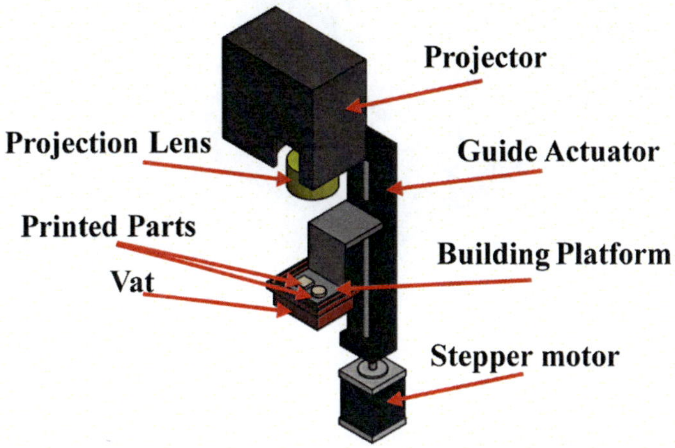

Fig. 1. Schematic diagram of the top-down DLP test bench

2.1 Experimental Validation Procedure

A commercial resin (Aqua Blue, Anycubic) and zirconia slurries with viscosity exceeding 2000 mPa.s (made of curable resin, photoinitiator, ceramic loading, and additives) were printed to evaluate performance. The layer thickness adopted was 100 μm.

3 Results and Discussion

3.1 Recoating System

Figure 2 is a sketch of the designed Recoating system. The system has two blades instead of one (as is common with existing equipment). The first blade has the function of spreading the resin or photopolymerizable suspension (happening in the forward movement), while the second has the function of leveling and thickness calibration (in the backward movement). Therefore, the two blades can have different shapes and materials. The exchange of the blade that will be active happens through the change of positioning that occurs mechanically at the end of the stroke. The position is locked by a mechanism in which a sphere is pressed into a housing by a spring (Fig. 3). The blade holder's linear forward and backward movement is created by rectified shafts and bushings, driven by a stepper motor with a torque of 0.5 N.m (Nema 17), and transmitted by pulleys and timing belts. The built DLP printer prototype and the recoating system are shown in Figs. 4a and 4b, respectively.

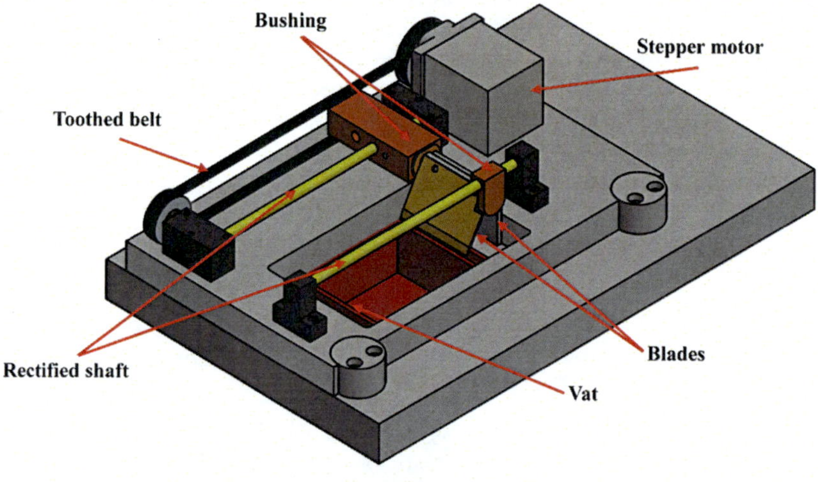

Fig. 2. Sketch of the designed recoating system.

3.2 Validation

Pieces were printed with commercial resin (Figs. 5a and 5b) and photopolymerizable ceramic slurries (Figs. 6 and 7). The developed system proved to be suitable for DLP additive manufacturing of polymeric and ceramic materials.

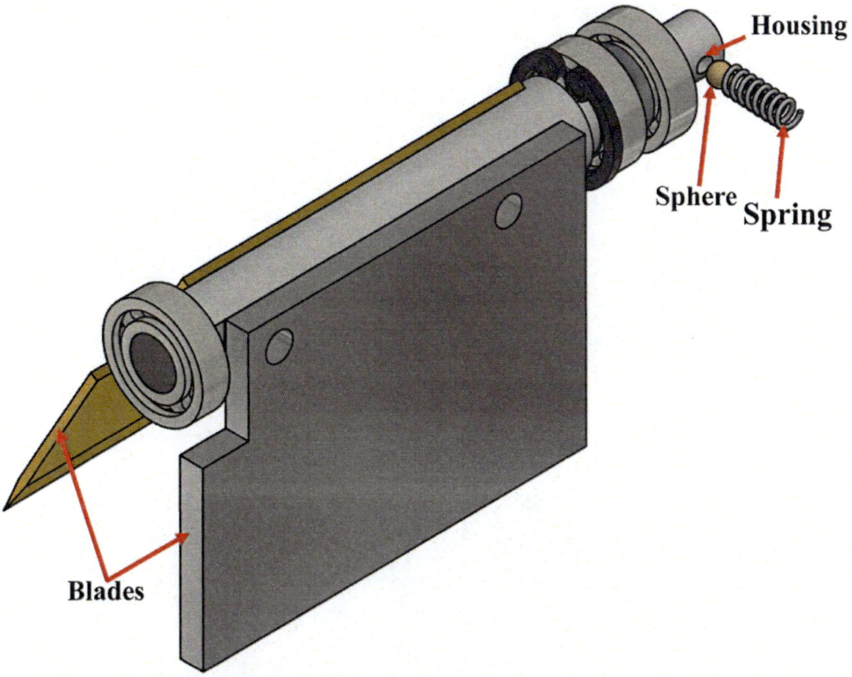

Fig. 3. Sketch of the designed mechanism that alternates the active blade

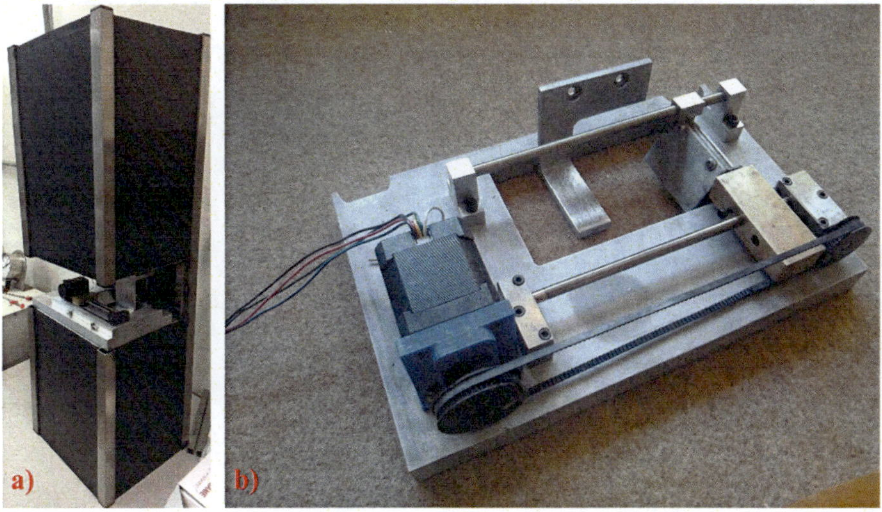

Fig. 4. DLP printer. a) Built prototype. b) Recoating system.

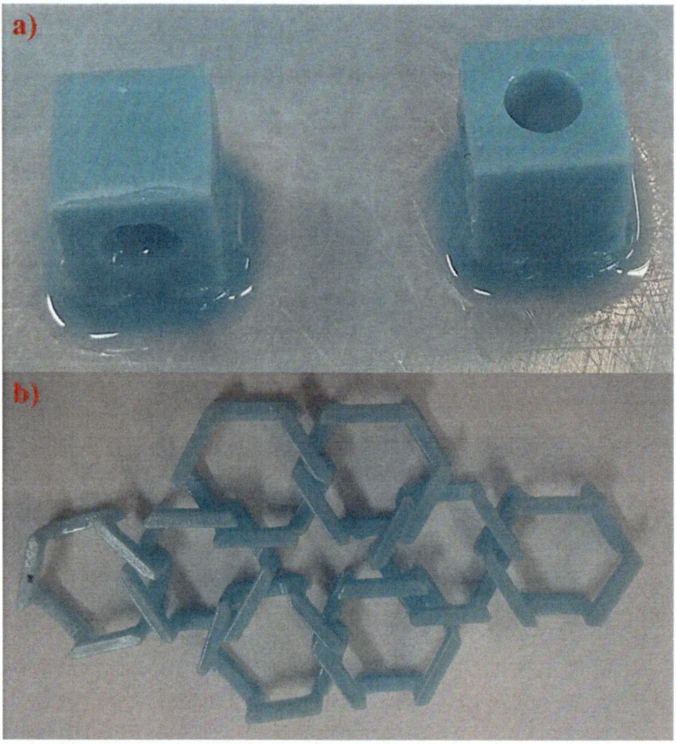

Fig. 5. Printed parts. a) Cube with holes. b) Textile structure based on the design of Correa el al. [25]

Fig. 6. Green ceramic printed gear.

Fig. 7. Magnified view of the layers of a sintered printed part.

4 Conclusion

In this work, an innovative recoating system for DLP 3D Printer was designed and a test bench was built. The two blades apparatus allowed an economical design made by traditional manufacturing operations. The tests performed indicate that the developed system is capable of handling different materials, even the challenging high solid loading ceramic slurries. The influence geometry and materials of the blades as well as the spreading speed may be the subject of future work.

Funding. This study was financed in part by the Coordenação de Aperfeiçoamento de Pessoal de Nível Superior—Brasil (CAPES)—Finance Code 001.

References

1. ISO (2015) IO for S. ISO/ASTM 52900:2015(en), Additive manufacturing—General principles—Terminology. Geneva
2. Gibson I, Rosen D, Stucker B (2015) Additive manufacturing technologies: 3D printing, rapid prototyping, and direct digital manufacturing. 2nd edn. Springer Science, New York, pp 1–498
3. Santoliquido O, Colombo P, Ortona A (2019) Additive manufacturing of ceramic components by digital light processing: a comparison between the "bottom-up" and the "top-down" approaches. J Eur Ceram Soc [Internet] 39(6):2140–2148. https://doi.org/10.1016/j.jeurceramsoc.2019.01.044
4. Ahrens CH (2017) Processos de AM por fotopolimerização em Cuba. In: Volpato N (ed) Manufatura Aditiva Tecnologias e Aplicações da Impressão 3D, 1st edn. Blucher, pp 129–144
5. Lian Q, Yang F, Xin H, Li D (2017) Oxygen-controlled bottom-up mask-projection stereolithography for ceramic 3D printing. Ceram Int [Internet] 43(17):14956–14961. https://doi.org/10.1016/j.ceramint.2017.08.014
6. Dumene R, Earle G, Williams C (2018) Characterization of additively manufactured cellular alumina dielectric structures. IEEE Trans Dielectr Electr Insul 25(6):2236–2240. https://doi.org/10.1109/TDEI.2018.007444
7. Venuvinod PK, Ma W (2004) Rapid prototyping. Laser-based and other technologies. vol. 23, Automotive Engineer (London). Springer US
8. Schwentenwein M, Homa J (2015) Additive manufacturing of dense alumina ceramics. Int J Appl Ceram Technol 12(1):1–7. https://doi.org/10.1111/ijac.12319

9. Mauritsio KE (2012) Stereolithographic machine for making of 3D object and method of stereolithography implemented by said machine (RU2562494C2)
10. Lovo JFP, de Camargo IL, Erbereli R, Morais MM, Fortulan CA (2020) Vat photopolymerization additive manufacturing resins: analysis and case study. Mater Res 23(4):e20200010. https://doi.org/10.1590/1980-5373-MR-2020-0010
11. Hafkamp T, Van Baars G, De Jager B, Etman P (2017) A trade-off analysis of recoating methods for vat photopolymerization of ceramics. In: Proceedings of the 28th annual international solid freeform fabrication symposium [Internet], pp 687–711. https://sffsymposium.engr.utexas.edu/sites/default/files/2017/Manuscripts/ATradeOffAnalysisofRecoatingMethodsforVat.pdf
12. Almquist TA, Hull CW, Modrek B, Serkuczewski AR, Jacobs PF, Lewis CW et al (1989) Recoating of stereolithographic layers (EP0361847B1)
13. Chua CK, Leong KF, Lim CS (2010) Rapid prototyping: principles and applications, 3rd edn. World Scientific
14. Doreau F, Chaput C, Chartier T (2000) Stereolithography for manufacturing ceramic parts. Adv Eng Mater 2(8):493–496. https://doi.org/10.1002/1527-2648(200008)2:8<493::AID-ADEM493>3.0.CO;2-C
15. Allanic A-L, Schaeffer P (2000) Fast three-dimensional modeling method and device (US6764636B1)
16. Miyamoto Y, Chen W, Kirihara S (2007) Smart processing of 3D micro ceramic devices by CAD/CAM micro-stereolithography and sintering. Trans JWRI 36(2):57
17. Pham DT, Ji C (2003) A study of recoating in stereolithography. Proc Inst Mech Eng Part C J Mech Eng Sci 217(1):105–118. https://doi.org/10.1243/095440603762554659
18. Renap K, Vancraen W (1997) Method and device for making a three-dimensional object from a hardenable liquid medium (EP0928242B1)
19. Saito N, Hayano S (1991) High precision photosolidification modelling device (EP0484182A1)
20. Almquist TA, Hull CW, Thayer JS, Leyden RN, Jacobs PF, Smalley DR (1996) Reenduction rapide d'objets tridimensionnels formes sur une base de coupe transversale (WO1996023647A2)
21. Hull CW, Cohen AL, Spence SL, Lewis CW (1994) Apparatus and related method for forming a substantially flat stereolithographic working surface (US5447822A)
22. Diptanshu M, Ma C (2019) Vat photopolymerization 3D printing of ceramics: effects of fine powder. Manuf Lett 21(July):20–23. https://doi.org/10.1016/j.mfglet.2019.07.001
23. National Research Council Committee on Theoretical Foundations for Decision Making in Engineering Design (2001) Theoretical foundations for decision making in engineering design. National Academy Press
24. de Camargo IL, Erbereli R, Lovo JFP, Fortulan CA (2019) Planetary mill with friction wheels transmission aided by an additional degree of freedom. Machines 7(2):33. https://doi.org/10.3390/machines7020033
25. Corrêa M, Dedini FG, Silva BMS, Centro AYSD (2019) Estruturas Têxteis Fabricadas Por Manufatura Aditiva: Análise Comparativa. In: 10° Congresso Brasileiro de Engenharia de Fabricação, São Carlos, SP, Brasil

Corrosion of AISI 316L Stainless Steel Pipe in a Complex Ammoniacal Medium

Felipe de Aquino Lima(✉), Ana Paula Neiva de Moura Santos, and Dalila Moreira da Silveira

Department of Chemical Engineering, Federal University of São João del Rei, Ouro Branco, Brazil
dalilams@ufsj.edu.br

Abstract. AISI 316L stainless steels are widely used in industrial pipes due to their resistance to corrosion and oxidation. However, some medium may favor incrustation and corrosion. In this scenario, the objective of the present research was to evaluate the deterioration of an AISI 316L steel tube for 12 months of industrial use. The medium contains 26,8011.00 mg/L of phosphate, 9,457.00 mg/L of chloride, 6.69 mg/L of sulfide, 6.08 mg/L of ammonia, and 1.88 mg/L of cyanide at a temperature of 200 °C and pressure of 12 kg/cm^2. After the period of industrial use, the tube was analyzed by scanning electron microscopy, dispersive energy spectroscopy, and optical microscopy. The chemical composition of the incrustation on the tube surface was verified by X-ray diffraction and X-ray fluorescence spectrometry. It was verified that sulfur compounds, such as sulfide, were the main corrosive agents on the tube, and cyanide accelerated this process. The medium's characteristics, such as the presence of chloride and phosphate salts, temperature, and pH (5.05), increased the corrosion rate on the tube, consuming about 30 cm of its length. It was observed the occurrence of stress corrosion, added to these compounds, caused sensitization of the material and weakening of metallic bonding.

Keywords: Corrosion · Tube · Distillation · Ammonia · Incrustation

1 Introduction

In the industrial project implementation, pipe costs represent, on average, 20% of the total cost of installation and 45% of the equipment costs. In this scenario, the industry faces the challenge of the choice of piping material, since this material must be resistant to the operational conditions and the environment they will be submitted to guarantee the mechanical integrity of the pipes and the feasibility of this application [1].

Austenitic stainless steels, such as AISI 316L, are widely used in industrial plants due to their excellent resistance to corrosive process at low temperatures and high resistance to oxidation and fatigue at high temperatures. However, its low hardness and low wear resistance can limit these applications [2–4].

Therefore, it is necessary to evaluate the process's operating conditions, such as temperature, pressure, and chemical compounds present in the medium, before deciding

Y. Iano et al. (Eds.): BTSym 2020, SIST 233, pp. 617–626, 2021.
https://doi.org/10.1007/978-3-030-75680-2_68

to use this type of steel. Some authors identified that the presence of compounds such as sulfides and chlorides, associated with temperatures in the range of 60 to 90 °C caused corrosion failure in AISI 316 L steel tube [5].

In a treatment plant of coke oven gas, the most aggressive compounds to the pipes are impurities, such as ammonia, sulfide, cyanide, and chloride [6]. In some parts of the process, when the concentration of these gases increases, the medium is more susceptible to corrosion.

During the treatment of coke oven gas, it is possible to selectively absorb ammonia using phosphoric acid and to distill this solution to obtain anhydrous ammonia. This product is intended for the production of dyes, pharmaceutical products, naphthalene, insecticides, also destined for the leather and concrete industry [7, 8].

Dias and collaborators [9] studied the corrosion in three types of stainless steels (AISI 304, 316 L e 321) used in a steelmaking process. Electrolytic tests were carried out using ammonia liquor and a chloride solution containing 2.5% (w/w) of NaCl. The electrolytic tests showed that the presence of molybdenum in the chemical composition of AISI 316 L steel did not affect the resistance to the corrosive action of the medium, being evidenced in the NaCl tests.

In another study about the corrosion in the heat-affected zone (HAZ) of AISI 316L stainless steel, caused by heavy crude oil, the samples heat-treated at 200 °C and 300 °C showed corrosion by iron sulfide products. This shows that media containing sulfide at this temperature can induce corrosion in this type of steel [10].

The presence of cyanide in the medium intensifies the corrosion process. This compound reacts with the iron sulfide and exposes the metal surface, leaving it unprotected and susceptible to a new corrosive attack. The iron oxides present in the medium reacts with the formed ferrocyanide complex and produces ferric ferrocyanide, which shows intense blue coloration, easily identified [11].

Thus, although there are studies that evaluate the resistance of stainless steels in various media [12–17], the evaluation under real operating conditions needs to be explored. In this scenario, this work aimed to study the corrosion mechanism and its products that affect the tube used in a distillation column of ammonia and phosphoric acid in a real industry.

2 Materials and Methods

2.1 Tube and Medium Characteristics

The tube used in this work was allocated in a real steel industry for 12 months, specifically in a distillation column to recover ammonia and phosphoric acid from an ammonia absorber, according to the schematic representation in Fig. 1. The operating pressure of the column was 12 kg/cm^2. The chemical composition of the steel before the operation is shown in Table 1. Table 2 shows the chemical composition of the medium in which the tube was submerged at a temperature of 200 °C. The characterization followed the methodology available in the Standard Methods for the Examination of Water and Wastewater [18].

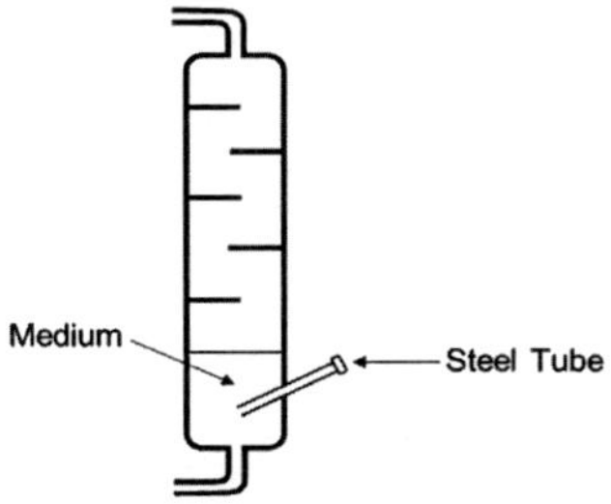

Fig. 1. Schematic representation of the distillation column showing the location of the tube.

Table 1. Nominal chemical composition of AISI 316L stainless steel before the use

Elements	C	Mn	Si	P	S	Cr	Ni	Mo	HV	HB
wt (%)	0.03	2.00	0.75	0.045	0.030	17.00	14.00	3.00	185	200

Table 2. Chemical composition of the medium in which the tube was submerged

Parameter	Unit	Method [12]	Concentration (mg/L)
Total ammonia	mg/L	4500 - B	6.08
Cyanide	mg/L	4500 CN^- - D	1.88
Calcium Hardness	mg/L	3500 Ca^- - B	5.18
Phenol	mg/L	5530 - C	1.43
Total iron	mg/L	3500 Fe iron - B	52.00
Oils and greases	mg/L	5520 - D	67.20
pH	–	4500	5.05
Total sulfur	mg/L	4500 - E	0.10
Silicon Dioxide	mg/L	4500 - C	12.00
Suspended solids	mg/L	2540 - B	215.00
Sulfide	mg/L	4500 - B	6.69
Chloride	mg/L	4500 Cl^- - D	9,457.00
Phosphate	mg/L	4500 P - C	26,8011.00

2.2 Analysis of AISI 316L Stainless Steel Tube

After 12 months of industrial use, tube samples were submitted to analysis of Optical Microscopy and Scanning Electron Microscopy (SEM) coupled to the Spectroscopy System Energy Scattering X-rays (EDS). The micrographs were obtained using a Clamex Optical Microscope. SEM/EDS analyses were carried out using a Quanta 400 FEI microscope equipped with an EDS system Bruker XFlash 4030, and the magnification used was 120x. Tube samples without incrustations were submitted to Vickers Microhardness (HV) and Brinell Microhardness (HB) analyses according to ASTM A370 [19] and as chemical composition by X-Ray Fluorescence Spectrometry (Shimadzu XRF 1800 spectrometer).

2.3 Corrosion Deposits Analysis

Corrosion deposits were removed from the tube surface and analyzed by X-ray diffraction (XRD) and X-Ray Fluorescence Spectrometry (XRF) to quantify inorganic compounds. The XRD analysis was carried out in a Rigaku diffractometer (Miniflex 600) operating with CuKα rad radiation by the powder method. The goniometer velocity was 2°/min to detect the crystalline phase at a range between 5° and 80° (2θ). For XRF analysis, the sample was oven-dried and calcined at 875 °C for 4 h. After calcination, the sample was fused with potassium carbonate, digested with hydrochloric acid, and then analyzed on a Shimadzu XRF 1800 spectrometer.

3 Results and Discussion

This work used an AISI 316L stainless steel tube in a steam pipe from a distillation process to recover ammonia and phosphoric acid. Figure 2 shows the tube before and after its installation in the real process.

Fig. 2. AISI 316L stainless steel tube (a) new (b) after 12 months of industrial use.

Figure 2(b) displays the visual aspect of the tube, showing that the tube was deteriorated when compared with the new one (Fig. 2(a)). It can be seen that the length of the tube was reduced around 30 cm due to the exposition of ammonia, cyanide, sulfide, and chloride in an acid medium (pH of 5.05) subjected to high pressures and temperatures.

Table 3 shows the chemical composition of steel and Vickers and Brinell Microhardness analysis after 12 months of industrial use. The analysis of Microhardness resulted in the same value, 165 HV, HB, different from its nominal value given in Materials and Methods. According to ABNT NBR-5601, the maximum value for this steel type is 217 HV [20], showing that the corrosive process did not significantly affect this parameter.

Changes in composition in relation to the nominal values of the new tube were observed. As AISI 316L austenitic stainless steel contains molybdenum, Rokosz and

Table 3. Chemical composition of AISI 316L stainless steel after 12 months of industrial use.

Elements	C	Mn	Si	P	S	Cr	Ni	Mo	HV	HB
wt (%)	0.02	1.40	0.56	0.035	0.003	16.86	10.23	2.03	165	165

collaborators [17] demonstrated that in the presence of phosphate and sulfate medium, a stable surface layer of iron, chromium, and molybdenum compounds is formed. However, a very corrosive medium can access the entire thickness of the material, damaging it. The decrease in carbon (from 0.03 to 0.02%), molybdenum (from 3.00 to 2.03%), and chromium (17.00 to 16.86%) in samples after industrial use in relation to the initial values can be associated with the formation of this passive layer, which due to the corrosivity of the medium was not enough to prevent degradation of the pipe.

Figure 3 shows the electronic images of the most deteriorated region of the tube obtained by SEM. In the demarcated area, in Fig. 3(a), the crack propagation can be observed due to the sensitization of steel by the corrosive environment, such as by sulfides and chlorides, together with the high pressure of the steam and the distillation column [12, 13]. Table 4 presents the semiquantitative analysis of the spectra obtained by EDS of the areas numbered in Fig. 3(b), through which it was possible to identify the presence of corrosion product consisting basically of oxygen (O), phosphorus (P), and sulfur (S) elements.

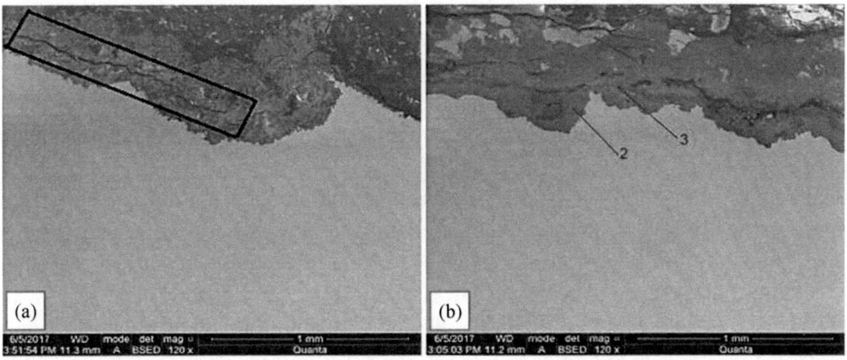

Fig. 3. Electronic image with an approximation of 120x of the most deteriorated region (tip) presents in the tube, where (a) shows the demarcated area of crack propagation and (b) shows the points where spectra were obtained by EDS.

Table 4. Semiquantitative medium chemical composition obtained by EDS.

Elements	O	P	Ni	Cr	Fe	S	Ca	Mn
wt (%)	37.12	24.68	10.91	10.66	10.31	5.64	0.56	0.13

The micrograph of the tube, presented in Fig. 4, showed that the basic structure is constituted by austenitic matrix and little occurrence of ferrite. There were no structural abnormalities that could be associated with the tube corrosion; nonetheless, it is observed that the operating conditions to which the tube was submitted may have weakened this region. The small change in nickel and manganese content in Table 3 corroborates these data, as this material plays an important role in the stability of the austenitic structure at high temperatures [21–23].

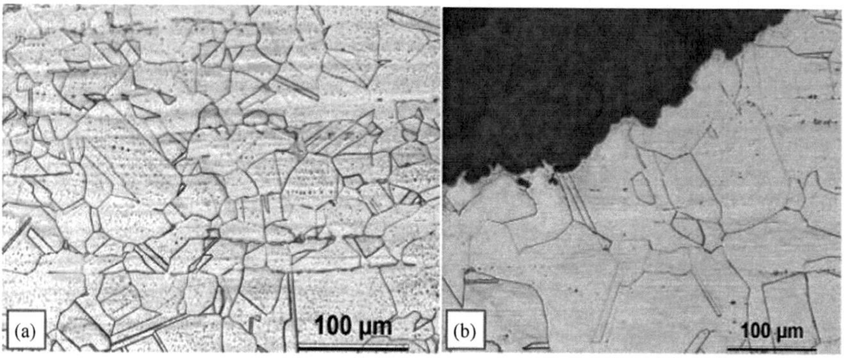

Fig. 4. Micrograph of the tube after 12 months of industrial use, where (a) shows the basic structure present in the tube sample and (b) shows the microstructure in the corrosion region.

Figure 3 and 4 showed that the structure of the tip of the tube deteriorated. These images showed how intense the degradation of stainless steel was due to the contact with the compounds of sulfides, chlorides, cyanides, phosphates, and ammonia. It is observed that the material has a generalized corrosion, with many cracks and deformations on the exposed surface.

The composition of the incrustations was confirmed in XRD and XRF. The diffractogram is shown in Fig. 5, and the chemical composition of XRF is shown in Table 5. X-ray diffractogram of the corrosion deposits collected from the tube surface shown in Fig. 5 pictures the presence of characteristic peaks of ferric ferrocyanide (17.37°, 24.71°, and 35.16°), that responsible for the Prussian blue pigment found on the tube surface; of Pyrite FeS_2 (28.96°) and hematite Fe_2O_3 (36.04°). The presence of ferric ferrocyanide shows the influence of the cyanide ion on the corrosion process and in the deposit formation, since the Prussian blue coloration was also found in regions less deteriorated by corrosion [24]. The presence of hematite was also verified by XRF, as presented in Table 5, corroborating with the XDR analysis. As presented in Table 5 and Fig. 5, it was possible to verify that the column's operating temperature favored the formation of hematite in relation to other oxides, such as magnetite and goethite, probably due to the reactions occurring between the stainless steel and the compounds present in the column.

The identification of other oxides in the XRF analyses, which were not previously identified by XRD, is due to the sample's preparation, since, in the XRF analyses, it was previously calcined, which caused the elimination of organic compounds and the obtaining the oxide phase.

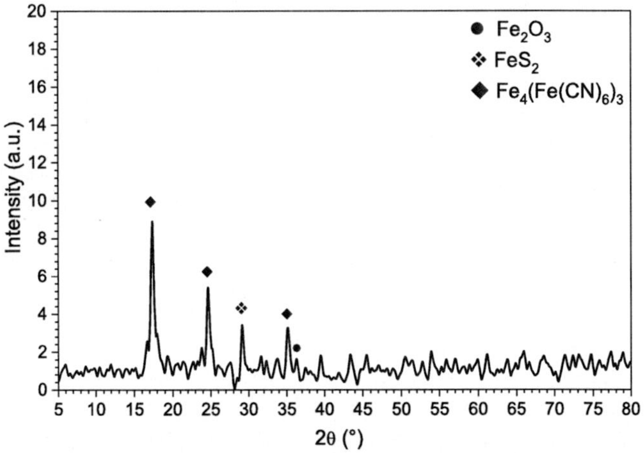

Fig. 5. X-ray Diffractogram of the tube incrustations after 12 months of industrial use.

Table 5. Medium chemical composition of deposits collected from the tube surface.

Oxide (%)	P_2O_5	Fe_2O_3	Na_2O	SiO_2	MgO
	59.27	43.52	6.28	0.74	0.25

The data obtained with the XRD and XRF corroborate with an increase in the composition of sulfur, phosphorus, and oxygen obtained by the EDS analyses. Also, the composition of the incrustations contains compounds that were possibly formed by contact with the medium, occurring reactions that led to the appearance of elements of the medium that were not previously found in the tube. The presence of sulfide compounds in the medium is aggressive due to their reduction to H2S, which is indicated as the main corrosion agent by sulfation [25]. This process is intensified in temperatures between 200 and 500 °C. The reaction of this compound with the tube surface can favor the increase of the corrosion rate through the formation of a sulfide layer. The free cyanide present in the medium can destabilize this layer, while sulfur inhibits the reaction of recombination of molecular hydrogen, allowing its permeation in the steel's crystal lattice [26–28].

Two effects can be observed when occurs the hydrogen permeation in the metal structure and their accumulation in the sites of the crystal lattice: the sensitization of the mechanical properties and the weakening of the metallic bonds, followed by the cracks nucleation, which can be accelerated in the presence of cyanide [29, 30].

The joint action of the stresses caused by the fluid at this operating pressure and the medium's composition may have conditioned stress corrosion and wakening by hydrogen. In the tube shows in Fig. 6, it is possible to observe that the deterioration intensity was greater in the area near the point that was completely submerged in the medium, as identified by the arrow.

Fig. 6. Visual analyses of the tube with the arrow indicating the direction of the steam flow.

It was possible to see that the stainless steel in the presence of corrosive compounds in a distillation column influenced the corrosion process, in which the reactions generated their bluish and greenish appearance. These reactions were influenced by the high temperature and pressure that the tube was put through. The use of this material in these conditions is not indicated, mainly due to the constant changes that must be carried out, exposing people to the risk of coming into contact with this distillation tower's contents, such as ammonia and phosphoric acid.

As a suggestion to future work, it is interesting to explore ways to increase the useful life of steam injection tubes inserted in corrosive media, such as other types of materials or metal coatings to protect the material. It should also be used a higher number of samples to obtain more statistically significant results.

4 Conclusions

The corrosion process in AISI 316L stainless steel tube was evaluated and associated with the occurrence of stress corrosion and hydrogen weakening. In summary, it was observed that in a real process, the main cause of corrosion was due to the high operating pressure, temperature, and the presence of sulfides, phosphates, and cyanides in the medium. The composition of the corrosion products was mainly phosphorus, oxygen, and sulfur. The presence of cyanide intensifies the corrosion process and led to the formation of ferric ferrocyanide, a compound associated with Prussian blue coloration observed. In this context, this work aims to contribute to the industrial field, showing that the AISI 316L stainless steel can be a corrosion target in the distillation column of ammonia. In the end, further works can be developed by exploring ways to avoid the corrosive process to diminish costs with equipment in the industry.

References

1. Telles PCS (2012) Tubulações Industriais: Materiais, Projetos e Montagem. Editora LTC, Rio de Janeiro

2. Arifvianto B, Suyitno M (2012) Effects of surface mechanical attrition treatment (SMAT) on a rough surface of AISI 316L stainless steel. Appl Surface Sci 258:4538–4543
3. Li Y, Wang Z, Wang L (2014) Surface properties of nitrided layer on AISI 316L austenitic stainless steel produced by high temperature plasma nitriding in short time. Appl Surface Sci 298:243–250
4. Pham MS, Solenthaler C, Janssens KGF, Holdsworth SR (2011) Dislocation structure evolution and its effects on cyclic deformation response of AISI 316L stainless steel. Mater Sci Eng: A 528:3261–3269
5. Adair ST, Attwood PA (2014) In-service stress corrosion cracking of AISI 316L stainless steel in an H2S environment. Corros Eng Sci Technol 49:396–400
6. Vieira WP (2004) Estudo de corrosão em tubulações de gás de coqueria [Doctor thesis]. Federal University of Espírito Santo, Vitória
7. Busca G, Pistarino C (2003) Abatement of ammonia and amines from waste gases: a summary. J Loss Prev Process Ind 16:157–163
8. Ferreira OS (2019) Modelos de Negócios e Sustentabilidade Ambiental - a ecologia industrial na siderurgia no Brasil. Braz J Bus 1:669–681
9. Dias FMS, Oliveira AS, Carneiro FP, Cândido LC (2008) Estudo da corrosão de aços inoxidáveis do tipo AISI: 304, 316 L e 321 em meios amoniacais de uma usina siderúrgica empregando-se ensaios eletroquímicos. In: Proceedings of 63° Congresso Anual da ABM. Associação Brasileira de Metalurgia e Materiais, São Paulo, pp 2354–2361
10. Silva CC, Sant'ana HB, Farias JP (2009) Evaluation of AISI 316L stainless steel welded plates in heavy petroleum environment. Mater Des 30:1581–1587
11. Silva PR, Ponte HA, Ponte MJJS, Kaminari NMS (2011) Development of a new electrochemical methodology at carbon steel/Na_2S system for corrosion monitoring in oil refineries. J Appl Electrochem 41:317–320
12. Hsu JP, Wang D, Kahn H, Ernst F, Michal GM, Heuer AH (2013) Fatigue crack growth in interstitially hardened AISI 316L stainless steel. Int J Fatigue 47:100–105
13. Kim YH, Kim DG, Sung JH, Kim IS, Ko DE, Kang NH, Hong HU, Park JH, Lee HW (2001) Influences of Cr/Ni equivalent ratios of filler wires on pitting corrosion and ductility-dip cracking of AISI 316L weld metals. Met Mater Int 17:151–155
14. Ferreira EA, Noce RD, Fugivara CS, Benedetti AV (2013) Influence of ethanol, acidity and chloride concentration on the corrosion resistance of AISI 316L stainless steel. J Braz Chem Soc 24:397–405
15. Hoseinieh SM, Shahrabi T (2017) Influence of ionic species on scaling and corrosion performance of AISI 316L rotating disk electrodes in artificial seawater. Desalination 409:32–46
16. Mejía-Caballero I, Palomar-Pardavé M, Trinidad JM, Romero-Romo M, Pasten-Borja RP, Lartundo-Rojas L, López-García C, Campos-Silva I (2015) Corrosion behavior of AISI 316 L borided and non-borided steels immersed in a simulated body fluid solution. Surface Coat Technol 280:384–395
17. Rokosz K, Lahtinen J, Hryniewicz T, Rzadkiewicz S (2015) XPS depth profiling analysis of passive surface layers formed on austenitic AISI 304L and AISI 316L SS after high-current-density electropolishing. Surface Coat Technol 276:516–520
18. American Public Health Association (2012) Standard methods for examination of water and wastewater. Washington DC
19. American Society for Testing and Materials (2008) Standard Test Methods and Definitions for Mechanical Testing of Steel Products: A370. West Conshohocken (2008)
20. Associação Brasileira de Normas Técnicas (2011) NBR-5601: Aços inoxidáveis - Classificação por composição química. 2ª ed
21. Revie RW, Uhlig HH (2008) Corrosion and corrosion control. Wiley Interscience, Hoboken

22. Potgieter JH, Olubambi PA, Cornish L, Machio CN, Sherif ESM (2008) Influence of nickel additions on the corrosion behaviour of low nitrogen 22% Cr series duplex stainless steels. Corros Sci 50:2572–2579
23. Bautista A, Blanco G, Velasco F (2006) Corrosion behaviour of low-nickel austenitic stainless steels reinforcements: a comparative study in simulated pore solutions. Cem Concr Res 36:1922–1930
24. Cabrera-Sierra R, Sosa E, Oropeza MT, González I (2002) Electrochemical study on carbon steel corrosion process in alkaline sour media. Electrochimica Acta 47:2149–2158
25. Beidokhti B, Dolati A, Koukabi AH (2009) Effects of alloying elements and microstructure on the susceptibility of the welded HSLA steel to hydrogen-induced cracking and sulfide stress cracking. Mater Sci Eng A 507:167–173
26. Vieira MM, Gomes JACP, Baptista W, Joia CJBM (2002) Efeito do pH na corrosão e permeação de hidrogênio em aço carbono em presença de água ácida desaerada. In: Proceedings of 6° COTEQ - Conferência Sobre Tecnologia de Equipamentos & 22° CONBRASCORR - Congresso Brasileiro de Corrosão. Instituto Brasileiro de Petróleo, Gás e Biocombustíveis, Rio de Janeiro
27. Gentil V (2012) Corrosão, 6th edn. LTC, Rio de Janeiro
28. Jambo HCM, Fófano S (2009) Corrosão: Fundamentos, Monitoração e Controle. Ciência Moderna Ltda, Rio de Janeiro
29. Vianna CS, Chicot D, Lesage J, Miranda PEV (2004) Difusão do Hidrogênio em Martensita. Revista Matéria 9(1):01–11
30. Tiwari GP, Bose A, Chakravartty JK, Wadekar SL, Totlani MK, Arya RN, Fotedar RN (2000) A study of internal hydrogen embrittlement of steels. Mater Sci Eng A 286:269–281

Software Development of Water Plugins as a Complement for Automatic Detection of Wetlands Based on the Digital Terrain Model

Herondino dos Santos Filho(✉)

Universidade Federal do Amapá, Rod. Juscelino Kubitscheck, Km 02, Caixa Postal 261, Macapá, AP 68902-280, Brazil

Abstract. The identification of Wetlands has been a purpose for governments around the world since the Ramsar convention. These areas are ecosystems of great environmental value, owning at the national and state scope the protection of their natural resources through their sustainable use. They are considered ecosystems of great importance for the environment because they present a diversity of flora and fauna; furthermore, they are responsible for supplying aquifers, improving water quality, retaining organic carbon, regulating biogeochemical cycles, and also more controlling flooding and climate regulation. The definition of these limits is extremely important for their conservation and dialogue with the populations that depend on them. Water plugin is a complement created for the automatic identification of Wetlands. The development of this software is the main contribution of this work. It allows the automatic delimitation of the wetlands from the digital terrain model, associated with models of classical statistics clusters, together with a local or global perspective. As an example, this work presents two case studies: the wetland surrounding the Fazendinha Environmental Protection Area and also the wetland called Açaí's "ressaca", both located in the city of Macapá, capital of the state of Amapá. The Kappa index was used in their evaluation, showing excellent results, that is, values ranging from 0.8 to 1.0.

Keywords: Automatic detection · Wetlands · Digital terrain model · Limits

1 Introduction

Wetlands belong to one of the ecosystems most affected and threatened with destruction by man. The Ramsar international convention, conducted in the city with the same name in February 1971, sought an intergovernmental treaty with the initial objective of protecting the aquatic habitats important for the conservation of migratory birds. Then, its concern with the Wetlands increased to promote its conservation and sustainable use, as well as the welfare of the populations that depended on it [1].

In 1993, Brazil signed the RAMSAR Agreement, but it was only enacted by Decree No. 1,905 of May 16, 1996, which presupposes a national policy for the intelligent management and protection of wetlands and their biodiversity [2].

Y. Iano et al. (Eds.): BTSym 2020, SIST 233, pp. 627–636, 2021.
https://doi.org/10.1007/978-3-030-75680-2_69

They are considered ecosystems of great importance for the environment because they present a diversity of flora and fauna; furthermore, they are responsible for supplying aquifers, improving water quality, retaining organic carbon, regulating biogeochemical cycles, and also controlling flooding and climate regulation [3].

The search for the definition of its limits has been the subject of recent research, mainly with the use of radar images, which may be the best alternative for emergency cartography of flooded areas that do not have cartography [4].

Among the images generated by Synthetic Aperture Radar Interferometry techniques - inSAR, the digital terrain model (DTM) is originated by computational mathematical representation directly from the altimetric quota model to the grey-level values model.

For [4], among all the topographic variables analyzed, topographic high points have the greatest potential for automatic delineation of wetlands.

The creation of the Water Plugin has, as its main element, the use of statistical models to classify a set of numerical data, the digital terrain model, for the detection of wetlands.

Among the different statistical methods used to estimate grouped data, there are: the Sturge's formula, Scott's normal reference function, the Freedman-Diaconnis estimator, and the Square Root estimator (Square root), between others [5]. These methods used to find the optimal value for the number of classes in a histogram will be used in an experimental approach to finding Wetlands.

Developed in the Phyton language of the Qgis, the experimental extension Water Plugin has as its main objective the creation of representative models of the Wetlands using statistical models.

In this work, we present two case studies in which this plugin was used. The detection of the limits of the surroundings of the Environmental Protection Area - APA Fazendinha [6] and also of the wetland called Açaí's "Ressaca" [7], both inserted in the metropolitan area of the city of Macapá-AP, presented an excellent result for the Kappa index.

This work's organization follows in an orderly manner with presentation of the following topics: materials and methods used, the development of the Water Plugin software, case studies, and, finally, the conclusions reached.

2 Materials and Methods

To achieve the proposed objective, the main tools were used: version 2.28 Qgis (Palmas) and the Plugin Builder extension.

QGIS is a software developed by the international scientific community with free access and a General Public License (GNU) character. Gary Sherman initiated it in 2002, and then this project was developed with the Open Source Geospatial Foundation (OSGeo) in 2007. Its first version was updated in 2009, and there is continuous development by a community of scientists around the world [8]. This tool allows the creation of extensions in its programming through the aggregation of add-ons called plugins.

Plugin Builder is one of the add-ons used to create a template in Qgis as a starting point for the development of other plugins, that is, it allows you to generate an initial structure.

In building the water plugin, an experimental approach was used, following the steps for its development, as shown in the diagram in Fig. 1.

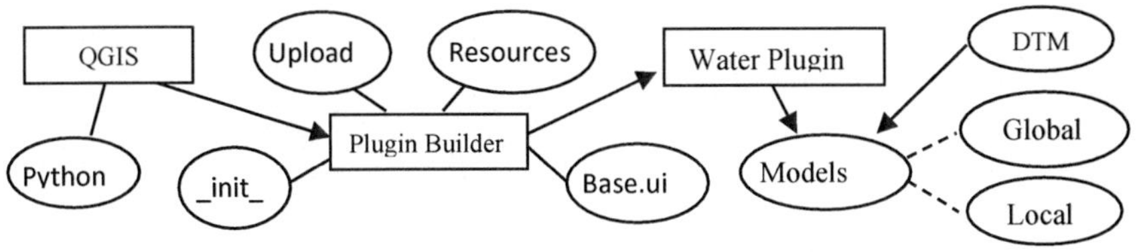

Fig. 1. Water plugin development steps

The main files generated by the Plugin Builder are: __init__, plugin_upload.py, resources.py, water_area.py, water_area_dialog.py, water_area_dialog_base.ui where the first three are to load the plugin in Qgis and the last to generate the plugin interface with the user and the application's algorithms.

Among the libraries to be used we have: PyQt4 used to create the user interface; GDAL used for the geospatial data format, as in the extraction of raster information or for the creation of raster and its interaction in Qgis; Numpy, which is a library for scientific computing that gathered greater support in dealing with matrices and vectors.

The main algorithms used for the development of the intended application have the procedures adapted for the construction of functions, statistical models, image subdivisions, and blend operations.

The generic format of a function is given by [9]. A block always starts with "def" to define the function. All parameters and input arguments must be placed within the parentheses when defining the function. The code block always comes after the colon (:). Using the "return" command will exit the function.

The models were created as functions in accordance with the definitions and arguments presented below.

2.1 The Sturges Model

Authors in [10] published a model to find the optimal class (k) of a frequency histogram. The count of the umpteenth (nth) term is given by the binomial coefficient $\binom{k-1}{i}$, $i = 0, 1, \ldots, k-1$. As k increases, this ideal histogram approaches the shape of normal density [11]. The total sample size is specified by:

$$n = \sum\nolimits_{i=0}^{k-1} \binom{k-1}{i} = (1+1)^{k-1} = 2^{k-1} \quad (1)$$

Therefore, the class number when constructing a histogram from normal data is given by the formula:

$$k = 1 + \log_2 n \quad (2)$$

In this way, Sturges established the optimal number of k intervals that can also be represented by the expression:

$$k = 1 + 3,322 * \log_{10} n \quad (3)$$

According to [11], the Sturges model has probably survived so long because of the fact that for moderate values n (<200), it gives results similar to alternative models, and therefore produces reasonable results. However, in our case, it does not work for a large number of "n". So, it suggests other models like Scott's or Freedman's and Diaconnis's.

2.2 Scott Function

This model was proposed by [12] and is based on the calibration of the normal distribution with the variance. It is given by the equation:

$$k = 3.49\sigma n^{-1/3} \tag{4}$$

Where σ is the standard deviation of the sample.

Although Gaussian density forms the basis of this equation, this proposition is not as strong as a parametric Gaussian assumption, that is, the use of this equation in non-Gaussian data will not result in a model that looks Gaussian [12].

2.3 Freedman and Diaconnis Estimator

Freedman's and Diaconnis' estimator is a modified version of Scott's function, where standard deviation and factor 3.49 are replaced by 2 interquartile (IQ), leaving the formula less sensitive for the interval. This model was proposed by [13] and is given by the following equation:

$$k = 2(IQ)n^{-1/3} \tag{5}$$

Where IQ is the sample of the interquartile width.

2.4 The Gonzales and Woods Model (Potency)

Authors in [14] for the digital image creation process, show the conversion of the detected continuous data to the digital format, that is, for the digitization of the amplitude values, called quantization, of a number n of discrete greys levels allowed for each pixel, is given by the equation:

$$n = 2^k \tag{6}$$

Which can be presented as:

$$k = log_2 n \tag{7}$$

The approximation with Sturge's formula draws attention.

2.5 Square Root

Square Root is a very common method that is used for its speed and simplicity [5]. In a distribution, the number of classes can be determined as the square root of the number of observations in a data set, as shown in the following equation:

$$k = \sqrt{n} \tag{8}$$

2.6 2n Square Root

This model is a variation of Square Root being given by the equation:

$$k = \sqrt{2n} \tag{9}$$

2.7 Fifth Root of n to the Third

It is also based on Square Root since, in the form of fractional power, this model is written 0.6 = 3/5 close to 0.5 = 1/2 and can be expressed as:

$$k = \sqrt[5]{n^3} \tag{10}$$

Two perspectives will be adopted for the models used: a Global one where the DTM image will be taken at a scale of 1: 25,000 and another Local where a window will be used whose size can be scaled.

3 Development of a Water Plugin

Construction of the complement had three main stages: the elaboration of the interface, the development of its functioning, and the realization of experiments through the use of cases.

3.1 Interface

Plugin interface has four main fields: the layer selection, the output file, the proposed model, and the model's perspective, as shown in Fig. 2.

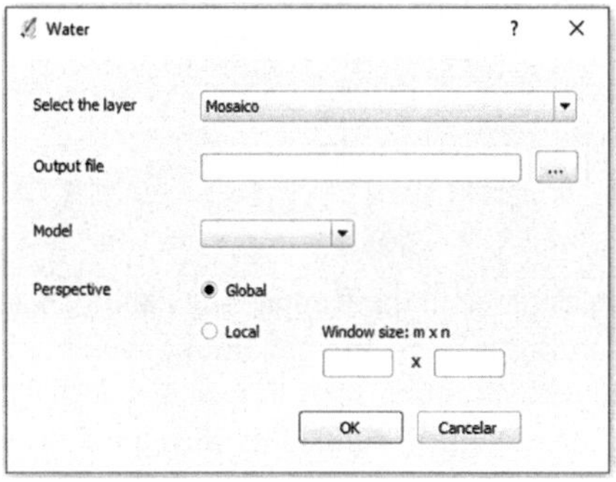

Fig. 2. Interface water plugin

Layer selection: this field is used to select the DTM layer to which the model is applied.

Output file: it is the result of a binary image in the format (.TIF) of the application of a model in the DTM layer.

Model: allows the option for the generation of the following models: Sturges, Scott, Freedman and Diaconnis, Square Root, Fifth Root of n the Third, and the Potency presented by [14].

Perspective: this field has the option of taking the total image (Global) or dividing it into smaller parts by means of the size of the adopted window, in which it was called Local.

3.2 Water Plugin Operation

The Digital Terrain Model's data treatment was accomplished by processing its representative image using the models developed in the Water Plugin extension, as shown in the flowchart in Fig. 3.

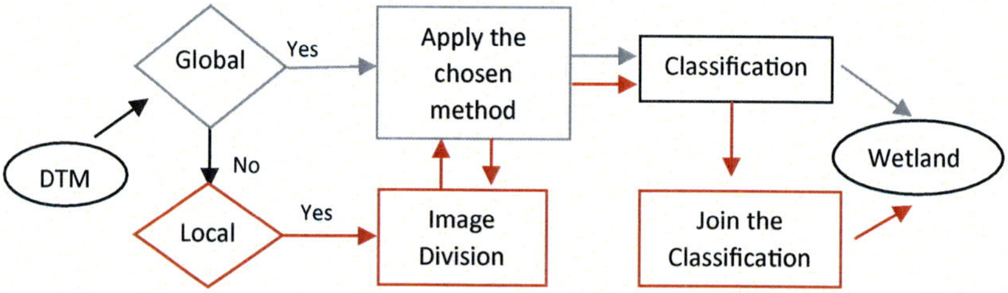

Fig. 3. Operation of the water plugin

The operation of the algorithm begins with the choice of its perspective: global or local. If global, the entire image is used, and the desired model is applied through its selection, which, after activation, performs the classification showing the wetlands. If the option is Local, the window size for processing must be established. The image is divided, and the chosen model is applied, which, after classification, is made to join and then present the desired wetlands.

4 Cases Study

Two case studies were presented, and the Kappa index and general precision were used for their evaluation. According to [15], the general precision (*Po*) is derived from the error matrix and is calculated summing the number of correctly classified values and dividing by the total number of values, defined by the equation:

$$Po = \frac{\sum_{i=1}^{M} n_{ii}}{N} \tag{11}$$

Where nii are elements of the main diagonal (number of pixels correctly classified), M is the number of classes present in the error matrix and N is the total number of samples. The Kappa index (Kp) is calculated using the ratio given by [15]:

$$Kp = \frac{Po - Pc}{1 - Pc} \tag{12}$$

Where in Pc is the proportion of units that agree by chance given by:

$$Pc = \frac{\sum_{i=1}^{M} n_{i+} n_{j+}}{N^2} \tag{13}$$

Where $n_{i+} = \sum_{j=1}^{M} n_{ij}$, that is, the total queue for a category, and $n_{j+} = \sum_{i=1}^{M} n_{ij}$, that is, is the column total for a given category.

The acceptance range for the Kappa index is shown in Table 1 [15].

Table 1. Acceptance range

Kappa Index	Characteristics
K ≤ 0.0	Too bad
0.0 < K ≤ 0.2	Bad
0.2 < K ≤ 0.4	Reasonable
0.4 < K ≤ 0.6	Good
0.6 < K < 0.8	Very good
K ≥ 0.8	Great

4.1 Açaí's Ressaca – Macapá/AP

Locally, in Macapá, a Wetland is known by the Portuguese term "ressaca". This case study depicted the wetland north of the metropolitan area of the city of Macapá with a minimum area of about 200 hectares. Among the statistical instruments used, the comparison with [16] resulted in the Kappa index values shown in Table 2 (*local perspective and **global perspective). Rounding to the decimal part, an excellent Kappa index was obtained according to [15].

A relevant factor is that the reference [16] used optical images from China-Brazil Earth Resources Satellite (CBERS-2B) with a resolution of 2.5 m, while those used by DTM came from radar waves in the P band (72 cm in length), which is also at the same resolution. And yet, these waves pass through the treetops, take information directly from the ground, taking advantage of optical images for the detection of wetland. The general precision in classification obtained by [16] was 98.97 and its Kappa index of 89.89.

Table 2. General precision and Kappa index

Model	General precision (%)	Kappa
Potency*	96.962035	0.789293
Square root*	96.948949	0.789098
Square root**	96.859395	0.777457
Sturges**	96.597888	0.772214
Potency**	96.888576	0.771335
Sturges*	96.272027	0.765123
Freedman & Diaconnis*	96.188566	0.760630
Fifth root of n to third*	95.904415	0.750537
Scott*	95.589233	0.739460
2n Square Root*	95.540623	0.737030
Fifth root of n to third**	95.538312	0.735136
2n Square Root**	95.322023	0.726705
Scott**	94.577635	0.698972
Freedman & Diaconnis**	93.575517	0.664023
[17]	91.856529	0.318499

4.2 Fazendinha's APA and Surroundings at the Macapá/AP - Brazil

The Environmental Protection Area (APA) of the Fazendinha district is located within the urban perimeter, in the South Zone of the city of Macapá. In this experiment, in addition to the spatial exploratory analysis, the general precision and the Kappa index in relation to the best result (local square root) were used to evaluate other models. Moreover, it was compared with [17], which used a mosaic of high-resolution images and MDE data from the Shuttle Radar Topography Mission (SRTM) in its interpretation. The results are shown in Table 3 below:

Table 3. General precision and Kappa index

Model	General precision (%)	Kappa
Potency*	99.022433	0.980203
Scott**	97.890527	0.957552
Fifth root of n to third*	96.420532	0.927123
Potency**	96.356286	0.926909
Square root**	95.222430	0.904470
Sturges*	94.512793	0.887837
Scott*	94.528931	0.888280
2n square root*	94.141979	0.880165
Sturges**	93.032344	0.861176
Fifth root of n to third**	91.863290	0.838268
2n square root**	91.079652	0.822973
[17]	38.900282	0.244143

All models presented excellent results when compared to the best result found, analyzed using high-resolution orthoimage (20 cm) of the Digital and Continuous Cartographic Base of Amapá used as a terrestrial truth.

5 Conclusions

The implementation of the statistical models in the Water Plugin allowed the generation of 14 detection models, 7 from a global perspective and the other 7 from a more local approach. When compared to models found in the literature, these models allowed the analysis of results that showed a better representation of the detected wetlands.

Comparing the results of the analysis of the Açai Wetland, the best values found were those obtained in decreasing order: Fifth root of N to third power (R5N3*), Sturges (St*), Square root (RN*), Potency (P*) and Scott (S*), all from a local perspective. The RN* and P* models have good results for the minimum values, and the others for the maximum ones.

In addition, the Kappa index also showed the best results for the Square Root and Potency models, achieving excellent results. However, it should be mentioned that all the other models were between 0.6 and 0.8, which, according to [15], are very good results.

For the wetland around the Fazendinha environmental protection area, the best results for the minimum water level increase were those observed for the local models: Square root (RN*) and Potency (P*). For the values of a maximum elevation of the water level, the following models were found: Fifth root of N to third power (R5N3*) local and Scott Global (Sc**).

In this work, the best results found were obtained using the local perspective method. In both case studies, the Square Root and Potency models stood out for a minimum value. In the rounding to its decimal part, it achieved excellent according to [15]. For maximum values, the Sturges (*), Scott (**), and Fifth root of n to third power (*) and 2n square root (*) models had good results.

It is worth mentioning that the results show similarities in all other models. Furthermore, larger experiments are necessary for its appreciation.

References

1. Ministério de Meio Ambiente (MMA). Áreas úmidas – conversão Ramsar. Governo Brasileiro. https://www.mma.gov.br/biodiversidade/biodiversidade-aquatica/zonas-umidas-convencao-de-ramsar.html
2. Brasil. Presidência da República, Casa Civil, Subchefia para assuntos Jurídicos. Decreto nº 1905, de 16 de maio de 1996. https://www.mma.gov.br/biodiversidade/biodiversidade-aquatica/zonas-umidas-convencao-de-ramsar.html
3. Gomes CS, Magalhães Junior AP (2018) Sistemas de Classificação de Áreas úmidas no Brasil e no mundo: panorama atual e importância de critérios hidrogeomorfológicos. Geo UERJ, Rio de Janeiro 33:34519
4. Muñoz VA (2014) Mapeamento de planícies de inundação a partir de dados SRTM: contribuição para análise da exposição a inundação na bacia do rio Itajaí, Tese (Doutorado em Sensoriamento Remoto) - Instituto Nacional de Pesquisas Espaciais (INPE), São José dos Campos, p 75

5. Cakmak UM, Cuhadaroglu M (2018) Mastering Numerical Computing with Numpy. Packt, Birmingham, Mumbai, pp 69–73
6. Santos Filho H (2020) Detecção automática de Área Úmida com o uso do Water Plugin: Apa da Fazendinha e seu entorno, Macapá, Brasil [Unpublished manuscript]. Departamento do Meio Ambiente, Universidade Federal do Amapá
7. Santos Filho H (2020) Detecção automática de Área Úmida com o uso do complemento experimental para o Qgis Water Plugin: Ressaca do Açaí, Macapá, AP, Brasil [Unpublished manuscript]. Departamento de Meio Ambiente, Universidade Federal do Amapá – UNIFAP
8. Menke K, Smith R, Pirelli L, Hoesen JV (2016) Mastering QGIS: go beyond the basics and unleash the full power QGIS with practical, step-by-step examples, 2nd edn. Pack Publishing Ltd, Birmingham-Mumbai
9. Fangohr H (2015) Introduction to Python for Computational Science and Engineering, Faculty of Engineering and the Environment University of Southampton, p 68
10. Sturges HA (1926) The choice of a class interval. J Am Stat Assoc 21(153):65–66. https://doi.org/10.1080/01621459.1926.10502161
11. Hyndman RJ (1995) The Problem with Sturges rule for constructing histogram, July. https://robjhyndman.com/papers/sturges.pdf
12. Scott DW (1979) On optimal and data-based histograms. Biometrika 66(3):608. https://biocurious.com/files/histo-bin-size-scott.pdf
13. Freedman D, Diaconis P (1981) On the histogram as a density estimator: L2 theory. Z. Wahrscheinlichkeitstheorie verw. Gebiete 57:454. https://doi.org/10.1007/BF01025868
14. Gonzalez RC, Woods RE (2003) Processamento de Imagens Digitais. Edgard Blücher Ltda, São Paulo
15. Moreira MA (2005) Fundamentos do Sensoriamento Remoto e Metodologias de Aplicação/Maurício Alves Moreira. 3a. Edição. ed. Viçosa: UFV, v. I
16. Santos Filho H (2011) Mapeamento e classificação das áreas de ressaca na região metropolitana de Macapá-AP utilizando imagens do satélite CBERS-2B. 2011. 125 f. Dissertação (Mestrado) - Curso de Modelagem Matemática e Computacional, Diretoria de Pesquisa e Pós-Graduação, Centro Federal de Educação Tecnológica de Minas Gerais (CEFET/MG), Belo Horizonte
17. Santos VF (2016) Dinâmica de inundação em áreas úmidas costeiras: Zona urbana de Macapá e Santana, costa amazônica Amapá. PRACS 9(3):121–144

Effect of Wire Spacing and Air Velocity on the Electrostatic Precipitation of Nanoparticles

Felipe de Aquino Lima(✉) and Vádila Giovana Guerra

Department of Chemical Engineering, Federal University of São Carlos, São Carlos, Brazil
vadila@ufscar.br

Abstract. Although electrostatic precipitators are commonly used in the removal of particulate matter, the use of this equipment in the removal of nanoparticles needs to be further elucidated. Thus, this study's objective was to evaluate the use of a single-stage, wire-plate electrostatic precipitator in the collection of sodium chloride nanoparticles using aerosol velocities of 1.9, 2.9, and 3.9 cm/s and voltages of 8.4, 8.8, and 9.2 kV. Three discharge electrodes (wires) were used and spaced 4, 6, and 12 cm apart for evaluation. It was evidenced that the increase in voltage favors the collection of nanoparticles, with efficiencies of 99.9% for the voltage of 9.2 kV. The increase in wire spacing increased the current inside the equipment and improved the collection efficiency, with greater variations in the voltage of 8.8 kV. The prevalence of electric forces at low gas velocities led the particle collection under these conditions.

Keywords: Electrostatic precipitation · Nanoparticles · Environmental control · Atmospheric pollution · Wire spacing

1 Introduction

The presence of ultrafine particles in the air has become a problem with increased emissions and the difficulty in characterizing geographically or over time [1, 2]. In general, inhalation is one of the most dangerous access routes of nanoparticles. Studies show that, after inhaled, nanoparticles can reach the gas exchange region, cross the air-blood barrier, access the bloodstream, and redistribute in other parts of the body, such as different organs and secondary tissues, causing systemic health effects [3, 4]. They can also be a risk factor for other diseases such as Parkinson's disease, Huntington's disease, Alzheimer's disease, and primary brain tumors [5].

Among the equipment that can be used in its collection, electrostatic precipitators (ESP) stand out for their versatility, high efficiency in a wide range of particle size (up to 99.9%), the possibility of being operated at high temperatures, cleaning can be performed using high volumetric rates simultaneously with low pressure drops, low maintenance requirements, and, if used dry operation, allows the direct recovery of materials [6, 7]. In any case, this technique needs to be evaluated in order to obtain such efficiency on the nanoscale.

Y. Iano et al. (Eds.): BTSym 2020, SIST 233, pp. 637–646, 2021.
https://doi.org/10.1007/978-3-030-75680-2_70

In this scenario, it is encouraged to look for new ways to optimize the electrostatic precipitation process to remove nanoparticles. In addition to operational parameters, this optimization involves the evaluation of the influences of geometric parameters on the collection of nanoparticles, such as wire spacing.

Few studies evaluate the influence of this parameter on the collection of nanoparticles by electrostatic precipitation. Lu [8] evaluated the collection of particles between 0.2 and 15 μm, and there was an increase in collection efficiency up to 100 mm, after which it decreased with increasing spacing due to small spacings not providing sufficient time to form a stable field. El Dein and Usama [9] concluded that the increase in spacing negatively affects the precipitator's performance due to the decrease in the resulting electric field along the field lines, with a consequent decrease in corona current under the same applied voltage. However, with very small spacings occurs the phenomenon of electrostatic shielding, decreasing the current inside the equipment. Kasdi [10] also verified the effect of electrostatic shielding with numerical simulations, especially in the central wires.

In this way, this work intended to verify this parameter's influence along with air velocities and voltages. Also, from the analysis of the ratio between the electrohydrodynamic number and the Reynolds number, the relationship between the primary and secondary flow and its effect will be evaluated and its effect on nanoparticles' collection efficiency by a wire-plate ESP. Therefore, this study can contribute to the technology of electrostatic precipitation of nanoparticulate.

2 Materials and Methods

Figure 1 presents a scheme of the experimental system used in the tests. In the main line, the air was fed through a compressor (Schulz, model MSV 12/175) and taken to a purification system composed of a purification filter (TSI, model 3074B) for removal of impurities. At the same time, a particle atomizer generator (TSI, model 3079) injected compressed air to atomize the aqueous solution of NaCl 0.1 /L. The aerosol (air + nanoparticles) was sent to a diffusion dryer (TSI, model 3062) to remove the water particles surrounding the particulate. This current was introduced to the main line where, after mixing, it passed through a neutralizing source of Kr-85 (TSI, model 3054), upstream of the electrostatic precipitator, in order to remove any residual charges.

The aerosol went to the electrostatic precipitator, which was connected to a high voltage source (Spellman, model SL 1200) to generate the corona effect. A three-way valve collected samples now upstream and now downstream of this equipment. The rest of the chain passed through a rotameter (Gilmont, model D7200-D7299) connected at the main line's end. The sample went to a neutralizing source of Am-241, following in the sequence for the measurement of the concentration by the Scanning Mobility Particle Sizer (TSI, model 3936), composed of an Electrostatic Classifier (model 3080) attached to a Differential Mobility Analyzer (DMA) and an Ultrafine Particle Counter (UCPC, model 3776), both sent data to a microcomputer.

The electrostatic precipitator used in experiments was a single-stage wire-plate type made of acrylic, according to Fig. 2. The equipment has 2 collecting plates (30 cm long and 10 cm high) spaced 4 cm apart. Three discharge electrodes with a diameter of 0.4 mm

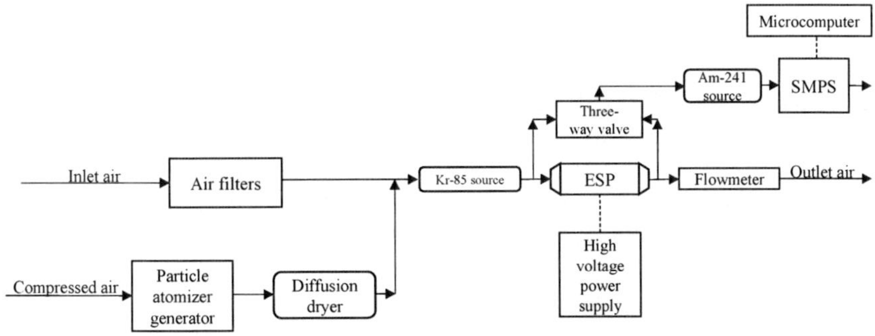

Fig. 1. Scheme of the experimental system.

were used with wire spacing of 4 cm, 6 cm, and 12 cm. The collection efficiency was obtained for three gas velocities of 1.9 cm/s (V1), 2.9 cm/s (V2) and 3.9 cm/s (V3) and three applied voltages of 8.4 kV (C1), 8.8 kV (C2), and 9.2 kV (C3). Air humidity was measured and kept below 20% for all tests. Each experiment was carried out in triplicate.

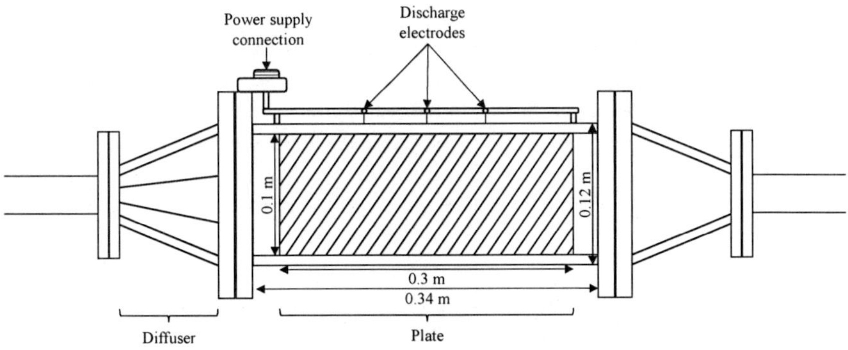

Fig. 2. Scheme of the ESP used in the tests

The experimental grade collection efficiency (η) was calculated from the relation between the particle concentration in the inlet (C_i) and the outlet (C_o) of the ESP for each diameter of the particle size distribution (6.15 nm to 241.4 nm), as in (1). The overall collection efficiency for each test was also calculated by (1) with the SMPS's total concentration.

$$\eta\,(\%) = 100(C_i - C_o)/C_i \tag{1}$$

To relate the primary and secondary flow, the ratio between the electrohydrodynamic number (N_{EHD}) and Reynolds squared number (Re^2), obtained by (2) and (3), respectively, was used. Thus, it was possible to determine the predominant phenomenon in electrostatic precipitation and the influence of electrical force on the gas's inertial and viscous forces [11].

$$N_{EHD} = I L^3 / \left(v_f^2\, r_f\, Z_i\, A\right) \tag{2}$$

$$\mathrm{Re} = \mathrm{u\,L}/\mathrm{v_f} \tag{3}$$

In these equations, I is the electric current, L is the characteristic length, r_f is the gas density, Z_i is the ionic mobility of the ions (commonly for gaseous ions under normal conditions is 2.10E10^{-4} m^2/Vs), v_f is kinematic viscosity, A is the surface area of the collecting plates, and u is the gas velocity. The characteristic length considered in the calculations of the N_{EHD}/Re^2 was the distance between the plates (L = 0.04 m; A = 0.06 m^2).

3 Results and Discussion

The inlet concentration in number of particles of the electrostatic precipitator was in the order of 10^5 particles/cm^3 for all tests of this work. This concentration is in agreement with Heitbrink [12] as the average particle emissions in this size range at several points. Also, most particles had a median close to 70 nm and similar distributions, with a geometric standard deviation around 1.72. In this size range, the diffusion charge is predominant and related to the Brownian motion, which intensifies with the decrease in the particles' size [13].

Figure 3 presents the current-voltage curves, at each velocity, for the three-wire spacings used in this work. The increase of the current with the wire spacing is observed, mainly for voltages higher than 14 kV. For lower voltages, the differences in current were more subtle. This phenomenon was associated with the reduction of electrostatic shielding, i.e., a decrease in the field lines' interference was caused by the increase in spacing [6, 14]. Moreover, due to the progressive increase in the current, by the decrease in corona suppression, it can be inferred that the greatest spacing still occurs before the maximum current point, from which the total specific current decreases [6]. The opposite behavior was obtained in other studies [8, 9], in which the increase in wire spacing led to the consequent decrease of the corona current under the same applied voltage. However, these studies did not evaluate only nanoparticles, a fact that can have great influence since the charging for the nanometric size range occurs by diffusion and for micrometric size by field.

It can also be verified that the increase in gas velocity from 1.9 cm/s to 2.9 cm/s increased the corona current, while the increase in gas velocity from 2.9 cm/s to 3.9 cm/s decreased the corona current. The reduction with increased velocity is reported in the literature [15, 16]. However, at the lowest velocities, this did not occur and was associated with the residence time. As reported by Oliveira and Guerra [17], it is possible that there was enough residence time for the particles to be charged and collected by the plates, so that the time was longer than the required to collect the particles and a large part of the ions did not participate in the charging.

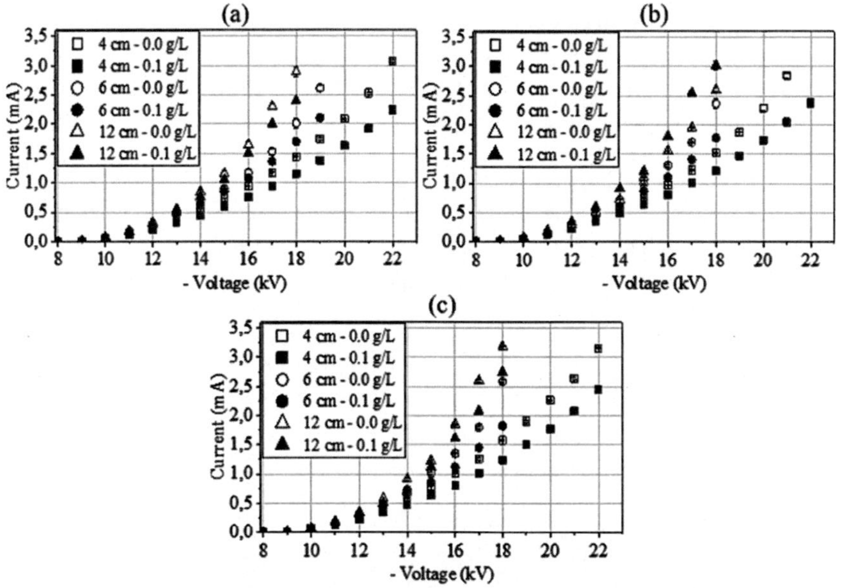

Fig. 3. Current-voltage curves for tests with velocity of: (a) 1.9 cm/s; (b) 2.9 cm/s; (c) 3.9 cm/s.

To verify the influence of primary or secondary flow in the process, the valor of N_{EHD}/Re^2 was calculated according to (2) and (3) and presented in Table 1. It should be noted that for electrostatic wire plate-type precipitators, values much higher than 1 indicate a strong influence of secondary flow or ionic wind [18]. In all tests performed, it was found that the electric force is high enough in relation to the other forces, i.e., a predominance of electric force influence on the flow pattern in relation to the viscous and inertial forces. Also, it was found that the increase in gas velocity provided a decrease in the values of this ratio. This phenomenon's occurrence was expected because the increase in velocity, even at low velocities, provides an increase in the number of Reynolds and decreases the influence of ionic wind [19]. In addition, a considerable increase occurred with the increase in the applied voltage due to the larger electric field between the discharge and collection electrodes, causing a greater formation of vortex and turbulence due to the influence of the electric field on the flow pattern [20, 21].

Analyzing the behavior of N_{EHD}/Re^2 in relation to the wire spacing, it is verifiable that, at the same velocity, there was an increase between the spacings of 4 cm and 6 cm, but a decrease in relation to the spacing of 6 cm and 12 cm. This behavior can be explained by the range of voltages used in this work (8.2 to 9.2 kV), necessary to be able to analyze the behavior of the parameters. In this range, it is observed that the current values are quite similar (Fig. 3). Moreover, as observed in Table 1, the overall efficiency for values above 9.2 kV would be greater than 99% and, if higher voltages were applied, the small changes could not be associated only with the influence of the parameters but also with experimental errors.

From the overall efficiency given in Table 1, it is verified that the increase in voltage applied to the electrostatic precipitator provided a higher collection efficiency, with

Table 1. Overall efficiency and N_{EHD}/Re^2 ratios obtained in the tests

Tests	Overall efficiency (%)	N_{EHD}/Re^2
C1/V1 – 4 cm	40.58 ± 1.23	14.66
C2/V1 – 4 cm	65.16 ± 3.01	29.31
C3/V1 – 4 cm	99.79 ± 0.19	146.56
C1/V2 – 4 cm	31.43 ± 0.69	9.44
C2/V2 – 4 cm	54.01 ± 6.25	15.73
C3/V2 – 4 cm	99.87 ± 0.07	62.91
C1/V3 – 4 cm	30.55 ± 0.55	8.69
C2/V3 – 4 cm	52.45 ± 1.93	10.44
C3/V3 – 4 cm	99.04 ± 0.43	34.78
C1/V1 – 6 cm	28.58 ± 0.78	21.98
C2/V1 – 6 cm	88.43 ± 9.21	58.63
C3/V1 – 6 cm	98.43 ± 1.12	172.22
C1/V2 – 6 cm	29.28 ± 0.27	6.29
C2/V2 – 6 cm	98.56 ± 0.52	15.73
C3/V2 – 6 cm	99.56 ± 0.01	56.62
C1/V3 – 6 cm	43.63 ± 3.54	3.48
C2/V3 – 6 cm	67.80 ± 3.27	6.96
C3/V3 – 6 cm	99.46 ± 0.07	173.93
C1/V1 – 12 cm	99.01 ± 0.12	37.33
C2/V1 – 12 cm	98.22 ± 1.66	51.29
C3/V1 – 12 cm	99.99 ± 0.00	146.56
C1/V2 – 12 cm	32.44 ± 0.66	3.15
C2/V2 – 12 cm	95.07 ± 1.68	44.15
C3/V2 – 12 cm	99.99 ± 0.00	153.47
C1/V3 – 12 cm	27.52 ± 0.85	1.74
C2/V3 – 12 cm	67.12 ± 10.61	6.08
C3/V3 – 12 cm	99.98 ± 0.00	127.83

efficiency above 98% in all tests with a voltage of 9.2 kV, evidencing that the electrical field created is already strong enough to guarantee efficient ionization of the air and charging of the particles. In general, the corona discharge, created by the discharge electrodes, carries the particles suspended in the air, distorts its trajectory with the electric field induced towards the collection electrode, and intercepts them in the collecting plate [22]. As the applied electric field increases, in addition to favoring the charging of particles, there is also an increase in electrical force. Because of this, particles tend to be collected more quickly.

As expected, the increase in gas velocity negatively affected the collection efficiency, as a result of the reduction of particle residence time, mainly for the voltages of 8.2 and 8.8 kV. This occurs because, inside the equipment, the particles are subject to viscous and electrical forces. The increase of gas velocity reduces the probability of the charged particle to move towards the collection electrode and be collected before it passes through the ESP, carried by the gas flow [6, 15].

In addition, the increase in wire spacing provided a greater overall collection efficiency in the conditions applied in this study. As observed by the current-voltage curves, the increase in spacing increased the electrical current inside the equipment. The most significant differences were observed mainly for the voltage of 8.8 kV. At this voltage, near the beginning of the corona, it proves to be a region of great instability (as observed by most tests with high standard deviation). However, it is not so weak as to not be able to ionizer a large number of particles, but it is also not strong enough to ensure an efficiency of 99%. However, in some tests, such as tests with a velocity of 2.9 cm/s and a voltage of 8.8 kV, the increase in spacing from 6 to 12 cm has not increased considerably the efficiency (around 67%).

For the applied voltage of 8.4 kV, all overall efficiency was low for any condition used, ranging from 27 to 43%. It was not observed a defined behavior with the increase of the wire spacing. However, the test using a gas velocity of 1.9 cm/s and spacing of 12 cm showed an efficiency of 99.01 $\pm$ 0.12%, probably by combining low velocities with the increase of the electric current in this spacing, in relation to that of 4 and 6 cm. However, the use of spacings of 12 cm and voltage of 9.2 kV is advantageous from the economic point of view, reducing the expenses with the amount of wires used in an industrial process and maintaining an efficiency above 99%.

The results of grade efficiency for the different voltages and wire spacings used in this study are evidenced in Fig. 4. The behavior of the grade efficiency curve is similar at different velocities, only with the reduction of the collection efficiency, so it was chosen to present only the curves for 1.9 cm/s.

The tendencies of the curves also demonstrate the effect of the electric force on the collection efficiency. In fact, the greatest ratios between the electrohydrodynamic number and the number of Reynolds occur at the highest voltage used. In all tests, there is a decrease in efficiency for particles greater than 120 nm. This was attributed to the intensity of the applied voltage not being strong enough to promote particle charging, except for 9.2 kV. This decrease can be associated with low electrical mobility and low number of charges acquired by particles in this range [17].

In addition, different points of maximum efficiency were observed when varying the spacing of the wires, especially in the lower voltages. As shown in Fig. 4(a), when applying a voltage of 8.4 kV, a maximum efficiency point occurs around 25 nm for the spacings of 6 and 12 cm, while this phenomenon occurs around 75 nm for the spacing of 4 cm. The influence of electrical forces in relation to viscous (evident by N_{EHD}/Re^2) may be associated with this phenomenon. In fact, the increase in N_{EHD}/Re^2 (14.44, 21.98, and 37.33 respectively to, 4, 6, and 12 cm) demonstrates the greater tendency of the formed electric field to be strong enough to collect particles greater than 120 nm and smaller, below the value that occurs the maximum efficiency point.

Additionally, it was verified an anomalous tendency of the test's grade efficiency curve with voltages of 8.8 kV (Fig. 4(b)), which presented several peaks of efficiency. For the spacing of 4 cm, this occurs in the particle diameters around 38, 56, and 108 nm, while for the spacing of 6 cm, this occurs around 20, 56, and 90 nm. The tests with the spacing of 12 cm presented only one peak at this velocity, around 20 nm. It is possible that the local vortexes formed due to the interaction between the ion flow and the gas stream, associated with the effect of particle size on the electrical mobility and on the number of charges acquired in its volume, have led these results in this configuration of operational conditions [23].

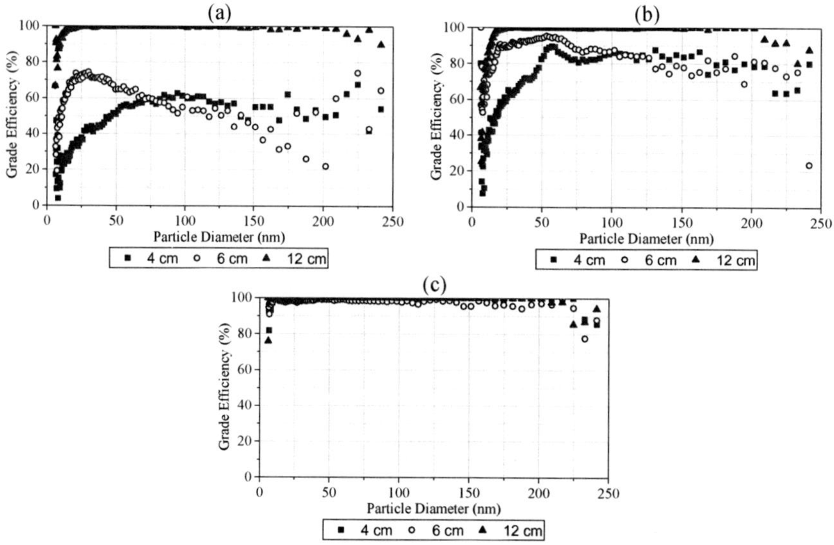

Fig. 4. Grade efficiency for the different wire spacing of the voltages of: (a) 8.4 kV; (b) 8.8 kV; (c) 9.2 kV.

In any case, at the highest applied voltage (Fig. 4(c)), high efficiency was obtained for most particle diameters evaluated in this study. A fall in efficiency is observed for particles smaller than 10 nm and particles greater than 220 nm. This behavior is associated with a small number of particles in these ranges; as evidenced earlier, the median in all tests is around 60 nm. Moreover, this drop occurs mainly for the spacings of 4 and 12 cm, and, as evidenced in Table 1, the electrohydrodynamic number for the spacing of 6 cm was higher for this speed and voltage.

Studies involving the influence of wire spacing in the efficiency of ESP collection generally evaluate microparticles or a size range from nano to microparticles [8–10]. Thus, the investigation only in the nanometric range is necessary since the particles' charging occurs differently to larger particles. This allows choosing wires spacings (such as the 12 cm) that would provide efficiency as good as using smaller spacings. Also, studies involving the effect of velocity and, consequently, residence time present variations in particulate concentration [24, 25]. In fact, Oliveira and Guerra [17] demonstrated that a more appropriate way to perform this analysis is to maintain a similar concentration in all tests, which was performed in this study. Moreover, the effect of the ionic wind on the primary flow is usually not taken into account in these analyses and, as evidenced, the influence of electrical forces in relation to viscous and inertias has a great impact on the efficiency of electrostatic precipitators [19–21]. Therefore, the present results demonstrate that in the analysis of wire spacing, voltage and velocity should be taken into account factors such as the residence time and the electro-fluid dynamics in order to evaluate the effect of these parameters.

As a suggestion for future work, it is interesting a more comprehensive evaluation of geometric parameters along with higher aerosol velocities, as well as statistically analyze the influence of wire spacing. In addition, evaluate other geometric parameters such as the numbers or types of wires and spacing between plates.

4 Conclusions

Electrostatic precipitation has the potential for the removal of nanoparticles, with efficiencies above 99% for voltages of 9.2 kV. The increase in voltage provided a higher efficiency of removal of nanoparticles, while the increase in velocity negatively affected the collection. The ionic wind's influence on the gas stream was preponderant for low gas velocities and increases with increasing wire spacing. Also, the increase in wire spacing provided a greater efficiency of collection in the conditions applied in this study, indicating that the use of spacings of 12 cm can be advantageous to reduce the expenses with the number of wires used in an industrial process and maintain an efficiency above 99%.

References

1. Li N, Georas S, Alexis N, Fritz P, Xia T, Williams MA, Horner E, Nel A (2016) A work group report on ultrafine particles (American Academy of Allergy, Asthma & Immunology): why ambient ultrafine and engineered nanoparticles should receive special attention for possible adverse health outcomes in human subjects. J Allergy Clin Immunol 138(2):386–396
2. Weichenthal S, Dufresne A, Infante-Rivard C, Joseph L (2008) Determinants of ultrafine particle exposures in transportation environments: findings of an 8-month survey conducted in Montréal Canada. J Expo Sci Environ Epidemiol 18(6):551–563
3. Adams RA, Potter S, Bérubé K, Higgins TP, Jones TP, Evans SA (2019) Prolonged systemic inflammation and damage to the vascular endothelium following intratracheal instillation of air pollution nanoparticles in rats. Clin Hemorh Microcirc 72:1–10
4. Gwinn MR, Vallyathan V (2006) Nanoparticles: health effects - pros and cons. Environ Health Perspect 114(12):1818–1825
5. Win-Shwe TT, Fujimaki H (2011) Nanoparticles and neurotoxicity. Int J Mol Sci 12(9):6267–6280
6. Parker KR (2003) Electrical operation of electrostatic precipitators, 1st edn. Institution of Engineering and Technology, London

7. Lu Q, Yang Z, Zheng C, Li X, Zhao C, Xu X, Gao X, Luo Z, Ni M, Cen K (2016) Numerical simulation on the fine particle charging and transport behaviors in a wire-plate electrostatic precipitator. Adv Powder Technol 27(5):1905–1911
8. Lu C, Yi C, Yi R, Liu S (2017) Analysis of the operating parameters of a vortex electrostatic precipitator. Plasma Sci Technol 19(2), 025504
9. El Dein AZ, Usama K (2014) Experimental and simulation study of V-I characteristics of wire-plate electrostatic precipitators under clean air conditions. Arab J Sci Eng 39(5):4037–4045
10. Kasdi A (2016) Computation and measurement of corona current density and V-I characteristics in wires-to-plates electrostatic precipitator. J Electrostat 81:1–8
11. IEEE (2003) Recommended international standard for dimensionless parameters used in electrohydrodynamics. IEEE Trans Dielectr Electr Insul 10(1):3–6
12. Heitbrink WA, Evans DE, Peters TM, Slavin TJ (2007) Characterization and mapping of very fine particles in an engine machining and assembly facility. J Occup Environ Hyg 4(5):341–351
13. Alonso M, Alguacil FJ (2002) Electrostatic precipitation of ultrafine particles enhanced by simultaneous diffusional deposition on wire screens. J Air Waste Manag Assoc 52(11):1342–1347
14. White HJ (1963) Industrial electrostatic precipitation. Addison-Wesley, Reading
15. Hinds WC (1998) Aerosol technology: properties, behavior, and measurement of airborne particles, 2nd edn. Wiley, New York
16. Dean JA (1999) Lange's handbook of chemistry, 15th edn. McGraw-Hill, New York
17. Oliveira AE, Guerra VG (2018) Influence of particle concentration and residence time on the efficiency of nanoparticulate collection by electrostatic precipitation. J Electrostat 96:1–9
18. Podliński J, Dekowski J, Mizeraczyk J, Brocilo D, Chang JS (2006) Electrohydrodynamic gas flow in a positive polarity wire-plate electrostatic precipitator and the related dust particle collection efficiency. J Electrostat 64(3–4):259–262
19. Fujishima H, Morita Y, Okubo M, Yamamoto T (2006) Numerical simulation of three-dimensional electrohydrodynamics of spikedelectrode electrostatic precipitators. IEEE Trans Dielectr Electr Insul 13(1):160–167
20. Chun YN, Chang JS, Berezin AA, Mizeraczyk J (2007) Numerical modeling of near corona wire electrohydrodynamic flow in a wire-plate electrostatic precipitator. IEEE Trans Dielectr Electr Insul 14(1):119–214
21. Ning Z, Podlinski J, Shen X, Li S, Wang S, Ping H, Yan K (2016) Electrode geometry optimization in wire-plate electrostatic precipitator and its impact on collection efficiency. J Electrostat 80:76–84
22. He Z, Dass ETM (2018) Correlation of design parameters with performance for electrostatic precipitator. Part II. Design of experiment based on 3D FEM simulation. Appl Math Model 57:656–669
23. Oliveira AE, Guerra VG (2018) Effect of ionic flow on the electrostatic precipitationof nanoparticles. In: Proceedings of the 2018 Brazilian technology symposium, vol 1. LCV - Laboratory of Visual Communications, Campinas
24. Liu Y, Hu B, Zhou L, Jiang Y, Yang L (2016) Improving the removal of fine particles with an electrostatic precipitator by chemical agglomeration. Energy Fuels 30(10):8441–8447
25. Zukeran A, Sawano H, Yasumoto K (2019) Collection characteristic of nanoparticles emitted from a diesel engine with residual fuel oil and light fuel oil in an electrostatic precipitator. Energies 12(17):3321

Production of Organic Acids by Batch Fermentations

L. C. Fardelone[1](✉), G. C. Silveira[1], T. S. Bella de Jesus[1], Y. P. David[1], G. P. Valença[1], and P. J. S. Moran[2]

[1] School of Chemical Engineering, University of Campinas, Campinas, Brazil
gustavo@feq.unicamp.br

[2] Chemistry Institute, University of Campinas, Campinas, Brazil
moran@unicamp.br

Abstract. The global organic acid market is expected to reach USD 36.86 Billion in 2026. This study aimed to produce citric, itaconic, and gluconic acids by using a biotechnological process in a lightly stirred bioreactor with fungus from the genus *Aspergillus spp*. These fermentations processes provided good yields, 22.6%, 99.9%, and 24.8% (48 h) for citric, gluconic, and itaconic acids, respectively.

Keywords: Organic acid · Citric acid · Gluconic acid · Itaconic acid · Fermentation process

1 Introduction

Organic acids represent an important group of bio-based chemical products because they contain functional groups, exhibit a high transformation capacity, and can be produced by fermentation or direct transformation processes - enzymatic or synthetic catalysis. Citric, gluconic, and itaconic acids (Fig. 1) and their derivatives are widely used in the food, chemical, and pharmaceutical industries [1–12].

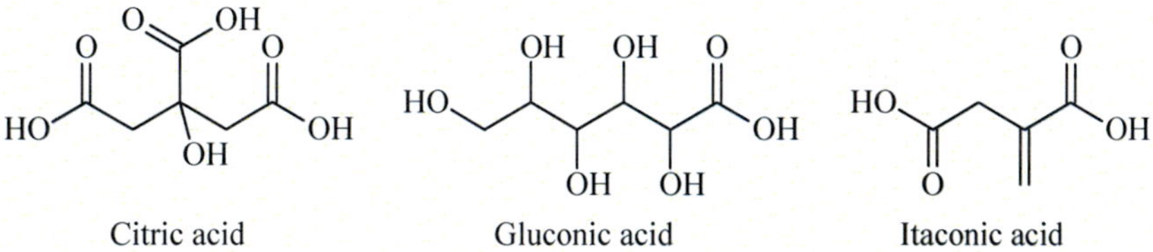

Fig. 1. Structural formulas of citric, gluconic, and itaconic acids

The global organic acid market was valued at USD 19.91 billion in 2018 and is expected to reach USD 36.86 billion by 2026, at a CAGR of 8.0%. The market is growing due to the wide-scale use of organic acids in the production of several products [13].

Y. Iano et al. (Eds.): BTSym 2020, SIST 233, pp. 647–653, 2021.
https://doi.org/10.1007/978-3-030-75680-2_71

The search for environmentally friendly processes is constantly growing, and to avoid the use of petroleum derivatives, these acids can be produced by living cells and manufactured via large-scale biotechnological routes, using selected strains of filamentous fungi of the genus Aspergillus spp. [14].

The citric acid is produced during the Krebs Cycle by using glucose or sucrose as carbon source for the Aspergillus niger. In the cytosol, the glucose is metabolized in an acid, denominated as pyruvate. A molecule of pyruvate is oxidized with the formation of acetyl-CoA by the mitochondrial pyruvate dehydrogenase complex, and another one is carboxylated to oxaloacetate in the cytosol by pyruvate carboxylase. The oxaloacetate is transported into the mitochondrion and condensed with acetyl-CoA to form citrate by the action of the enzyme citrate synthase [15]. The same microorganism produces the gluconic acid, but directly in the extracellular medium, the glucose is oxidized in a two-step reaction in gluconic acid by the action of glucose oxidase. The itaconic acid production by Aspergillus terreus follows a parallel route to the production of citric acid by Aspergillus niger, adding two steps: the dehydration of citrate by aconitase, forming cis-aconite and decarboxylation by cis-aconite decarboxylase to form itaconic acid [15].

Applications of these acids are extensive. Citric acid is used as an acidulant agent, antioxidant, buffering, chelating, among others. The itaconic acid is widely applied in the industry for the synthesis of resins, bioplastics, polyesters, absorbent material, and in the preparation of active compounds used in agriculture and medicines. The gluconic acid has applications as an additive in cement to increase strength and water resistance, to soften the sour taste in food products, and different salts of gluconic acid are used in the pharmaceutical industry to treat deficiencies, as well as in calcium and iron supplements [1, 2, 16–22].

In this work, we reported citric, gluconic and itaconic acids production by using Aspergillus niger and Aspergillus terreus in fed-batch fermentation with glucose as carbon source, which provided us with a greater process.

2 Methodology

2.1 General Process

Fermentations were conducted using 3 g of fungi cells in a batch regime with 0.220 L of culture medium in a 0.5 L lightly stirred bioreactor for 48 h. The bioreactor used was an Infors HT, Multifors model with online control of production parameters, such as temperature, pH, aeration, and agitation. All the culture media were sterilized in an autoclave at 121 °C for 15–20 min.

The concentrations of organic acids produced and glucose consumed were determined by HPLC with a refractive index detector, coupled to an Aminex HPX-87H column (300 mm × 7.8 mm, Bio-Rad) and using 5 mmol/L mobile phase H2SO_4, with flow 0.6 mL/min at 50 °C. Quantifications of products and sugars were performed from calibration curves using an external standard.

Microorganism. Aspergillus niger and Aspegillus terreus are being used for this process, which has high cell mass and organic acids production. These strains of Aspergillus niger and Aspergillus terreus are founded in our laboratory and are stored in the form of MCB (Master Cell Bank) and WCB (Working Cell Bank), respectively, and were prepared in accordance with international precepts for good production and laboratory practices [23–25].

Preparation of Culture Media. For cell expansion (inoculums and pre-inoculums) of Aspergillus niger and Aspergillus terreus strains, it was used a culture medium with industrial components previously established with a composition of yeast extract, peptone, and 20 g/L of glucose. To obtain a high cell density (OD) baffled flask is being used, which promotes a better shaker mix.

Analytical Methodology. The final concentration of organic acids and residual glucose was determined by High-Performance Liquid Chromatography (HPLC) with a refractive index detector coupled to an Aminex HPX-87H column (300 mm × 7.8 mm, Bio-Rad) and uses 5 mmol/L mobile phase $H2SO_4$, with 0.6 mL/min flow at 50 °C. Product and residual carbohydrate quantification are performed from calibration curves using external standards for organic acids and glucose.

3 Results and Discussion

Fermented batch processes were carried out for citric, gluconic, and itaconic acids production by using fungi living cells, and the results are shown in Table 1.

Table 1. Citric, itaconic, and gluconic acids production by using *Aspergillus niger* and *Aspergillus terreus* fungi with glucose as a carbon source

Acid	Fungi	Glucose (g)	Acids (g)	Productivity (g/Lh)	Yield (%)
Citric	*Aspergillus niger* ATCC 11414	7.31	1.76	0.22	22.6
Gluconic	*Aspergillus niger* ATCC 9029	3.06	2.64	0.38	> 99.9
Itaconic	*Aspergillus terreus* ATCC 10020	6.74	1.21	0.13	24.8

For citric acid, the results obtained by using Aspergillus niger ATCC 11414 were 1.76 g, with glucose consumption of 7.31 g, yield of 22.6%, and a productivity of 0.22 g/Lh. Aspergillus niger ATCC 9029 furnished 3.06 g gluconic acid in quantitative yield and with total glucose consumption, with productivity of 0.38 g/Lh. Aspergillus terreus ATCC 10020 furnished 1.21 g of itaconic acid and providing 24.8% yield, with a productivity of 0.13 g/Lh with 6.74 g of glucose consumption.

In Fig. 2, it was shown a typical profile of glucose consumption and citric (a), gluconic (b), and itaconic (c) acids production.

In submerged culture, two forms of filamentous morphology are used, pellets and free filaments, depending on conditions of culture media and the genotype of the strain

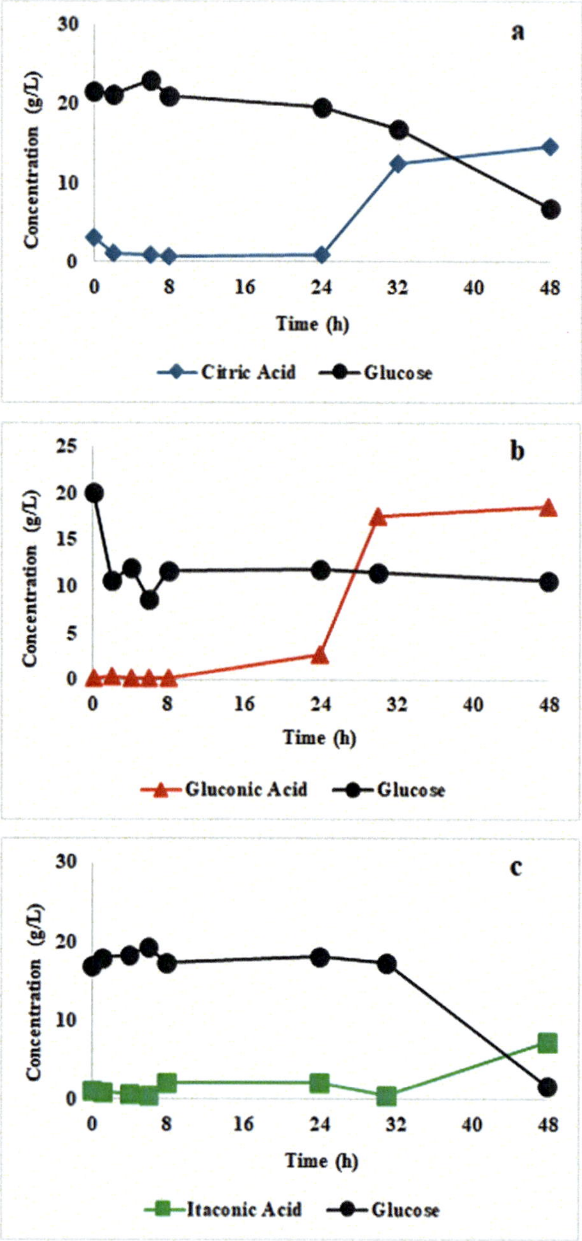

Fig. 2. Carbohydrate concentration and citric (a), gluconic (b), and itaconic (c) acids production profiles in batch fermentation

[6]. In this work it was used a culture media that provides a suspended cell culture, which depending on the strain utilized, can reach lower yields, however, reduces the production costs, decreasing the steps needed to purify the product.

Industrial recovery of these organic acids can be achieved by precipitation or adsorption by using ion-exchange resin. Precipitation requires the removal of the micelles from the fungus and fermentation medium by filtration, treated with active carbon and passed in ionic resins exchangers, concentrated by evaporation in a vacuum for crystallization [6, 26].

In fermentative processes using aerobic fungi, aeration, and agitation are crucial parameters that must be analyzed, especially during scale-up studies. The aeration of the growing microorganism must be carried out equally in the entire work volume of the bioreactor and needs to be enough for its maintenance. Another benefit of a good system of aeration and agitation is to diminish the size of mycelial aggregates, making the access to cells easier to oxygen [27–30]. The online bioreactor used provides a good system of aeration and agitation.

Another parameter extremely important to microorganism development is the pH of the culture media. When the pH is too acid, it can inhibit the microorganism's metabolism [31]. It can be observed that during the production process of citric and itaconic acids, the concentration of acid in the culture medium may have caused an inhibition of the microorganism's metabolism and substrate absorption. This condition can also be avoided using an extraction system coupled to the bioreactor, with continuously removing the product from the medium, increasing the production.

In Fig. 3, it was shown a picture of the bioreactor during a batch fermentation process.

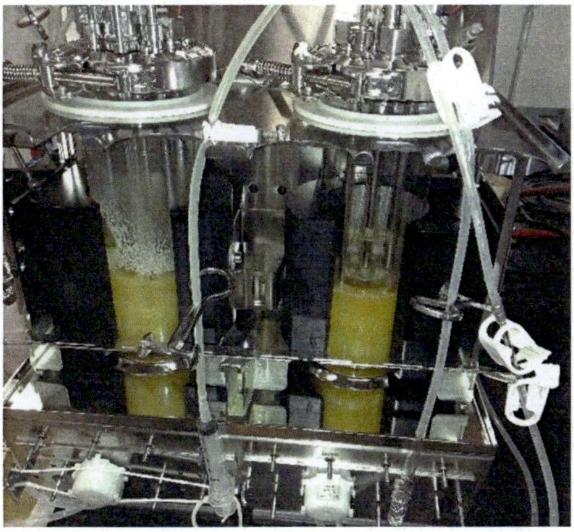

Fig. 3. Picture of the batch fermentation process using an Infors HT Bioreactor during 48 h

4 Conclusions

The results showed that the production of gluconic acid by Aspergillus niger during batch fermentation processes can achieve excellent yields. It was possible to observe that for the production of citric and itaconic acids, the concentration of acid in the culture medium may have caused an inhibition of the microorganisms production, which can be avoided by performing an acid extraction system coupled to the fermentative process, and the Research Group is already studying techniques.

Acknowledgments. The authors thank Higher Education Personnel Improvement Coordination (CAPES, 88882.329694/2019–01) and National Council for Scientific and Technological Development (CNPq) for financial support.

References

1. Connor JM (2007) The citric acid industry. In: Global price fixin, studies in industrial organization book series – SIOR, 26:113–116
2. Zaman NK, Law JY, Chai PV, Rohani R, Mohammad AW (2017) Recovery of organic acids from fermentations broth using nanofiltration technologies: a review. J Phys Sci 28:85–109
3. Cañete-Rodríguez AM, Santos-Dueñas IM, Jiménez-Hornero JE, Ehrenreich A, Liebl AW, García-García I (2016) Gluconic acid: properties, production methods and applications - an excellent opportunity for agro-industrial by-products and waste bio-valorization. Process Biochem 51:1891–1903
4. Naraian R, Kumari S (2017) Microbial production of organic acids. In: Gupta VJ, Treichel H, Shanpaval VO, Oliveira LA, Tuohy MG (eds) Microbial functional foods and nutraceuticals. John Wiley & Sons, Hoboken, pp 93–121
5. Fardelone LC, Bella de Jesus TS, da Silveira GC, David YP, Valença GP, Moran PJS (2020) Production of organic acids by biotechnology. Environmental sustainability and development in organizations: challenges and new strategies. In: Poveda AC, Martínez CI (eds) science publishers (An Imprint of CRC Press/Taylor & Francis Group). in press
6. Pal P, Kumar R, Banerjee S (2016) Manufacture of gluconic acid: a review towards process intensification for green production. Chem Eng Process: Process Intensification 104:160–171
7. Bafana R, Pandey RA (2018) New approaches for itaconic acid production: bottlenecks and possible remedies. Critical Rev Biotechnol 38:68–82
8. Saha BC (2017) Emerging biotechnologies for production of itaconic acid and its applications as a platform chemical. J Ind Microbiol Biotechnol 44:303–315
9. Robert T, Friebel S (2016) Itaconic acid - a versatile building block for renewable polyesters with enhanced functionality. Green Chem 18:2922–2934
10. Kuenz A, Krull S (2018) Biotechnological production of itaconic acid-things you have to know. Appl Microbiol Biotechnol 102:3901–3914
11. Duarte JC, Valença GP, Moran PJS, Rodrigues JAR (2015) Microbial production of propionic and succinic acid from sorbitol using *Propionibacterium acidipropionici*. AMB Express 5:13
12. Weastra SRO (2013) Determination of market potential for selected platform chemicals: itaconic acid, succinic acid, 2, 5-furandicarboxylic acid. BioConSepT. https://www.bioconsept.eu/wp-content/uploads/BioConSepT_Marketpotential-forselected-platformchemicals_report1.pdf

13. Reports and Data (2019) Organic Acids Market Analysis, By Product Type (Formic Acid, Citric Acid, Others), By Source Type (Biomass, Chemical Synthesis, Others), By Application (Food & Beverages Industry, Pharmaceutical Industry, Others), Forecasts To 2027
14. Kirimura K, Honda Y, Hattori T (2011) Gluconic and Itaconic Acids. In: My M (ed) Bio-Based Chemicals. Elsevier B.V, Tokyo, pp 143–147
15. Magnuson JK, Lasure LL (2004) Organic acid production by filamentous fungi. In: Tkacz JS, Lange L (edsd) Advances in fungal biotechnology for industry, agriculture and medicine. Springer US, Boston, pp 307–340
16. Show PL, Oladele KO, Siew OY, Zakry FAA, Lan JC-W, Ling TC (2015) Overview of citric acid production from *Aspergillus niger*. Front Life Sci 8:271–283
17. Wang B, Li H, Zhu L, Tan F, Li Y, Zhang L, Ding Z, Shi G (2017) High-efficient production of citric acid by *Aspergillus niger* from high concentration of substrate based on the staged-addition glucoamylase strategy. Bioprocess Biosyst Eng 40:891–899
18. Krull S, Hevekerl A, Kuenz A, Prüße U (2017) Process development of itaconic acid production by a natural wild type strain of *Aspergillus terreus* to reach industrially relevant final titers. Appl Microbiol Biotechnol 101:4063–4072
19. El-Imam AA, Du C (2014) Fermentative itaconic acid production. J Biodivers. Bioprospecting Dev 1:119
20. Ahmed AS, Farag SS, Hassan IA, Botros HW (2015) Production of gluconic acid by using some irradiated microorganisms. J Radiation Res Appl Sci 8:374–380
21. Singh R, Mittal A, Kumar M, Mehta PK (2017) Organic acids: an overview on microbial production. Int J Adv Biotechnol Res 8:104–111
22. Papagianni M, Mattey M (2006) Morphological development of *Aspergillus niger* in submerged citric acid fermentation as a function of the spore inoculum level. Application of neural network and cluster analysis for characterization of mycelial morphology. Microbial Cell Factories 5:3
23. Coecke S, Balls M, Bowe G, Davis J, Gstraunthaler G, Hartung T, Hay R, Merten OW, Price A, Schechtman L, Stacey G (2005) Guidance on Good Cell Culture Practice. ATLA 33:261–287. Available in the Best Practices (GCCP) on the ESACT-UK (2005). The UK Society for Cell Culture Biotechnology website. https://www.esactuk.org.uk
24. Stacey G (2004) Fundamental issues for cell-line banks in biotechnology and regulatory affairs. In: Fuller, B.J., Lane, N., Benson, E.E. (eds) Life in the frozen state. CRC Press LLC, Boca Raton, Florida, pp 437–452
25. Health Products and Food Branch Inspectorate (2002) Annex 2 to the current edition of the good manufacturing practices guidelines schedule drugs (biological drugs) - GUI-0027
26. Hilbod NJ, Schab F (2013) Bio-based building blocks, case study of a large-scale manufacturing process. Chimica Oggi 31:28–32
27. Bakri Y, Mekaeel A, Korieh A (2011) Influence of agitation speeds and aeration rates on the Xylanase activity of *Aspergillus niger* SS7. Braz Archiv Biol Technol 54:659–654
28. Jafari AR, Sarrafzadeh MH, Alemzadeh I, Vosoughi M (2007) Effect of stirrer speed and aeration rate on the production of glucose oxidase by *Aspergillus niger*. J Biol Sci 7:270–275
29. Nemestóthy N, Bakonyi P, Komáromy P, Bélafi-Bakó K (2019) Evaluating aeration and stirring effects to improve itaconic acid production from glucose using *Aspergillus terreus*. Biotechnol Lett 41:1383–1389
30. Shin WS, Lee D, Kim S, Jeong YS, Chun GT (2013) Application of scale-up criterion of constant oxygen mass transfer coefficient (kLa) for production of itaconic acid in a 50 L pilot-scale fermentor by fungal cells of *Aspergillus terreus*. J Microbiol Biotechnol 23:1445–1453
31. Riscaldati E, Moresi M, Petruccioli M, Federici F (2000) Effect of pH and stirring rate on itaconate production by *Aspergillus terreus*. J Biotechnol 83:219–230

L2 Cache Robust Partitioning in Multicore Processors

Thiago Silva de Oliveira Duarte(✉) and Osamu Saotome

Instituto Tecnológico de Aeronáutica, São José dos Campos-SP, Brazil
osaotome@ita.br

Abstract. Automotive and, especially, the aerospace industry requires reducing the size, weight, and power (SWaP) of the embedded electronics. Additionally, both industries demand an increment of the embedded electronics' power capability and low data transfer latency. Multi-core processors' (MCP) advent allows integrating multiple applications in the same System on Chip (SoC) in high-performance computers. The usage of MCP processor in safety-critical systems such as Flight Control Computers and Autonomous Driving depends on solving determinism issues related to the processor's resource contention. The usage of a resource from one core can impact the other cores, preventing the delivery of the algorithm output at a specific time, and this failure can lead to a catastrophic event. This paper introduces an analysis of interference paths to map the possible resource contention in MCP and presents an approach to mitigate the low-level cache (LLC) contention using the page coloring technique.

Keywords: Multicore · L2 cache · Bus contention

1 Introduction

Safety-critical software development employs fault-tolerant designs to achieve system requirement goals. Usually, using a single-core computer (SCP), each processor has its cache memory, main memory, and I/O isolated from other computers. The physical limits of the semiconductor electronics due to the heat prevent the increase of the SCP's computational capability. The multi-core architecture provides a means to expand the processing power by adding more cores to a system on chip (SOC). One application can be split into the threads, executing each thread in different cores, or two different applications can run in parallel [1].

The parallel execution of algorithms does not mean that the cores are running isolated because, in MCP architectures, some processors' resources are shared, as shown in Fig. 1. However, the sharing of the processor resources like interconnection, memories, cache, and I/O, can lead to the contention of them and, consequently, timing anomalies or non-determinism [2].

Although, the usage of MCP provides a means to develop integrated architectures using a high-performance computer. In one SoC, two different applications run in parallel, reducing the number of Line Replaceable Units (LRU) or Electronic Control Units

Y. Iano et al. (Eds.): BTSym 2020, SIST 233, pp. 654–661, 2021.
https://doi.org/10.1007/978-3-030-75680-2_72

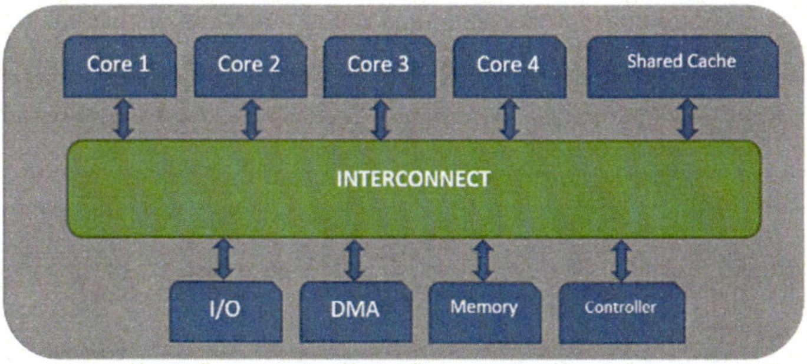

Fig. 1. MCP interconnections.

(ECU), consequently the cabling and power consumption. The certification authorities, especially in the aerospace industry, have provided documentation and guidance to the industry to help in the safety-critical systems certification. The CAST-32 is one complement of the DO-178C with objectives to accomplish in terms of development to consider a software architecture over an MCP certifiable. It requires mapping the MCP used configuration, listing the resources shared between the cores, identifying the approach to prevent the bus contention, the mitigation when possible, and the most crucial objective, the methods to implement the partitioning of those resources [3].

This work introduces the interference path analysis and a robust partitioning of the low-level cache (LLC) in an MCP with four cores. The cache coloring technique is implemented using an ARMv8-A53 processor and tested using five algorithms found in the autonomous system, aerospace industry, and IoT to prove the technique's efficiency.

The organization is structured as follows: in section II, we describe the interference analysis in a multi-core architecture and the impacts on system-level and the low-level cache contention and the page coloring technique to mitigate the contention. Section III describes the experiment to evidentiate the LLC interference and the us'e of the cache coloring and the measured Worst-Case Execution Time (WCET). Section IV presents the conclusion and future work.

2 Background

2.1 Interference Path Analysis

The interference path analysis is one of the recommendations of the CAST-32. This analysis performs the identification of all the processor shared resources, such as:

- Shared Memory;
- Shared Cache;
- Shared I/O; or
- And interconnection.

The next step of this analysis is to classify the interference as acceptable, tolerable, or unacceptable. The acceptable and tolerable contentions will not lead to a delay that can impact the system determinism. The unacceptable requires a plan of mitigation to prevent catastrophic events [4].

Figure 2 shows one example of an interference path analysis in a dual-core system and the possible bus contention when accessing the processor resources like memory and I/O. This figure also shows the Direct memory access (DMA) controller that also works in parallel with the cores interfering with the core execution. The prevention can be done through a hardware configuration, controlling the bandwidth or priority access to physical resource or software architecture.

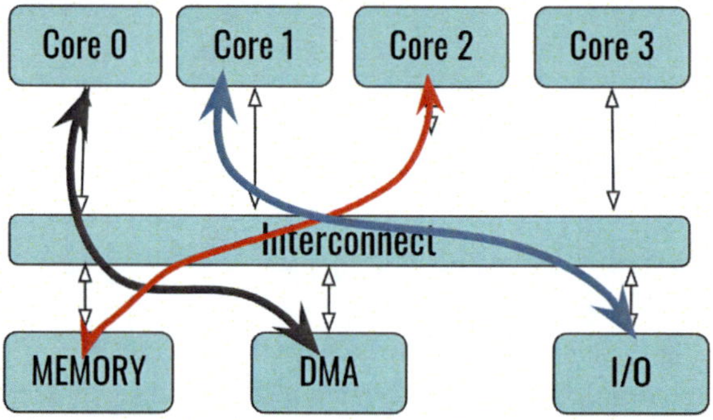

Fig. 2. Interference path analysis example.

In some architectures, the interference path might not be avoidable due to the physical connection between the cores and the resources, like the shared LLC cache. In this case, the mitigation shall provide a reduction of the contention using the time partitioning or other mechanism to minimize the impact of mutual access.

2.2 Cache Coloring

The cache memory is an essential component of modern processors, interposed between the processor and the main memory. However, the cache memory also adds a non-determinism in terms of timing. When the instructions or data need to be used by the microprocessor and those are already in the cache, also called cache hit, it minimizes the time to read them. On the other hand, when data or instruction are not in the memory cache, also called cache miss, the data's fetch in the main memory might increase the execution time.

The cache state is a challenge for the WCET tools and methods because it demands a prediction of the previous processing stage and the cache state. A context switch or an interruption can lead to a cache miss and an additional execution time to update the main memory and load the new block of data to the cache memory [5].

A significant number of MCP architectures provide not only a L1 cache but also a shared cache L2 interconnecting all the cores, as exemplified in Fig. 3.

This second level of cache must be listed as interference path and demand a way to minimize the cores' intra-interference, excluding or minimizing the cache miss in LLC raised by the neighbor cores, i.e., creating a cache partitioning. The usage of

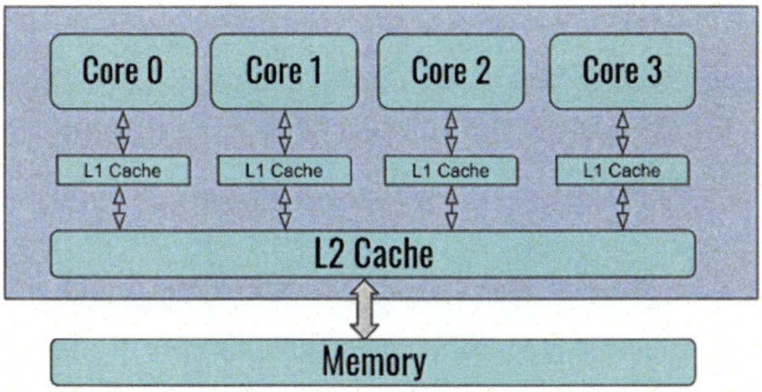

Fig. 3. Example of dual core processor with a shared L2 cache memory.

shared cache is even discouraged on safety-critical system to make the WCET feasible in safety-critical systems [6].

Cache coloring technique uses the set-associative caches' behavior to allow the cores to access only the memory associated with the cache blocks reserved for them.

The coloring technique aims at making a shared cache behave like a private one. Cache coloring aims at reducing interference in caches by ensuring that data and instructions used by software running on different cores will be stored in different sets [7].

Figure 4 shows how the memory is segmented in pages that are associated with the segment in the memory cache.

The page coloring creates a set of memory regions, also named colors, that can be segregated, preventing the same LLC index to be reused by different cores. The memory

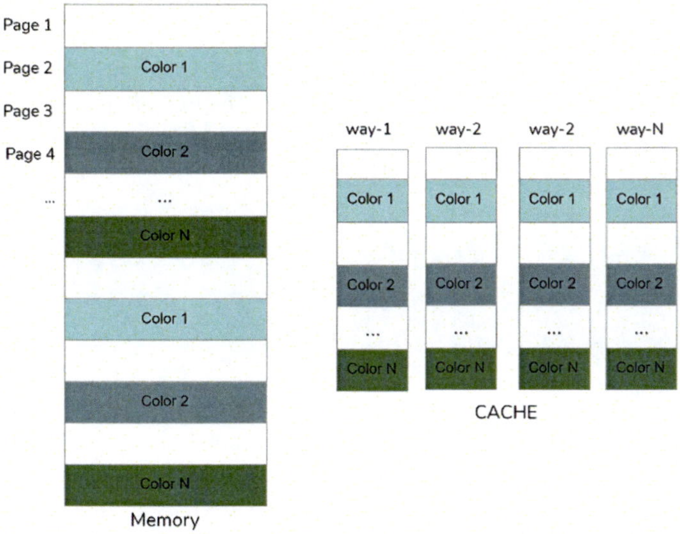

Fig. 4. Page coloring [8].

management unit (MMU) is used to map the virtual address in the physical address related to a specific color.

3 Experimental Result

This work develops an environment to prove the cache interference and the efficiency of the page coloring technique, using a quad-core processor running in three of the four core applications accessing the memory interruptedly, forcing the contention of the L2 cache, and in the core 0, the execution of algorithm to measure the interference and the WCET improvement. The quad-core processor used in this experiment is the A53, a high-efficiency processor that implements the ARMv8-A64 bits architecture present in the single-board computer Raspberry Pi 3B.

In the Raspberry Pi 3B, the core zero is the firmware entry point after the boot provided by Graphics Processing Unit (GPU). The core zero works as an initiator, executing an ELF (Executable and Linkable Format) binary containing the ELF binary of the four cores. The initiator ELF binary is also responsible for performing the boot loading and the clock frequency, the individual core configuration as stack size, the execution level, and the loading of the ELF binary of the four cores in their memory position.

Splitting the L2 cache in eight colors, the core zero's ELF binary implements five applications with a focus on mathematics operations, autonomous flight, and a pathfinding algorithm:

- FFT over an array of 1024 positions;
- Matrix multiplication;
- A* Search Algorithm;
- Quadcopter Control Law; and
- Voice Activity Detection.

Figure 5 shows the allocation of the algorithms. The execution of the algorithms follows an asynchronous multiprocessing architecture (AMP). It means that each core runs a separate algorithm. There is no core controlling or synchronizing the others.

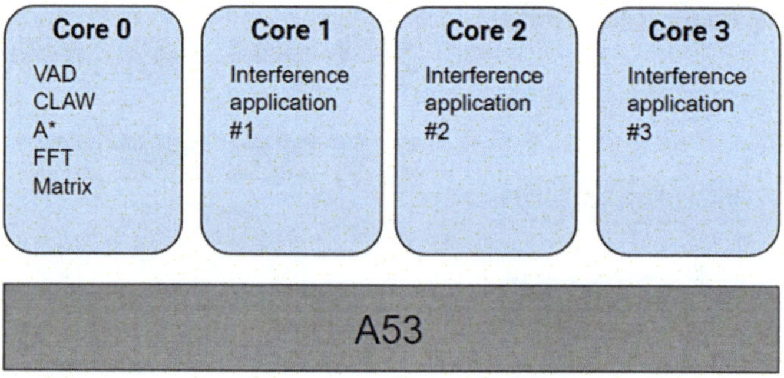

Fig. 5. Asynchronous multiprocessing architecture (AMP)

The core one, two, and three are responsible for executing the same algorithm disturbing the core zero, interfering in the L2 cache as follows:

```
Algorithm 1: Auxiliary core algorithm
Index = 0;
While True do
  Buffer1[Index] = Buffer2[Index] + Buffer3[Index];
  Buffer2[Index] = Buffer1[Index] + Buffer3[Index];
  Buffer3[Index] = Buffer1[Index] + Buffer2[Index];
  Index = Index + 1;
  if Index == 4096 then
    Index = 0;
  end
end
```

The mix of those algorithms intends to show how applications with different criticality levels can suffer interference in the L2 cache. This interference can result in a higher WCET that can drive a delay in response to safety-critical applications, such as control law or the response of an IoT device, with an impact on user experience. Figure 6 shows a comparison of the WCET between the three scenarios:

- Scenario 1: Solo execution of core 0;
- Scenario 2: Four cores running in parallel without any cache partitioning;
- Scenario 3: Four cores running in parallel with cache coloring.

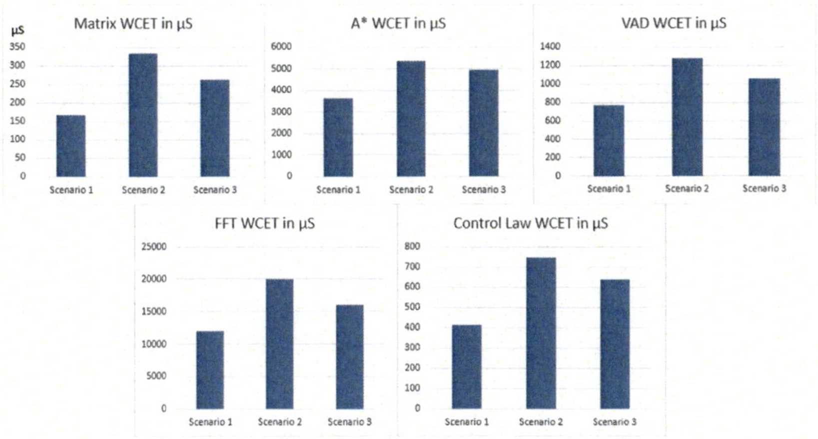

Fig. 6. Comparison of the 3 scenarios.

In the first scenario, the core zero executes the five algorithms solely, creating a baseline to compare the other cores' timing interference. The execution of the four cores in parallel shows explicitly how the interference affects the algorithms under tests. As follows, the results indicate a performance until 94% worse than the solo execution of core zero. It must be considered not only the L2 cache interference but also the contention of the main memory.

The use of cache coloring in the third scenario shows that the technique by itself does not establish the WCET to the same level as scenario one but presents a significant improvement from scenario two. The WCET of the third scenario is 55% worse compared with the first scenario and 39% better in comparison to scenario two.

4 Conclusion and Future Work

This work shows the interference path analysis in a quad-core processor, highlighting that the shared cache is one resource that demands mitigation to ensure determinism. From a shared cache point of view, this work aims to provide determinism, creating a partitioning of the L2 cache using the page coloring technique.

The three test scenarios highlight the impact of the cache contention in the WCET and how the page coloring improves the WCET and consequently the determinism. When comparing the first scenario, a solo execution of one core, and the third scenario, the page coloring technique usage, it is still possible to see some other three cores' interference over the core 0. The configuration with the paging coloring technique is still 50% worse than the core zero's solo execution. The contention of other resources like the main memory explains the bigger WCET even after the cache's partitioning.

The results also prove that the page coloring technique is a suitable solution to the cache contention in safety-critical systems. From the programmer's point of view, the segmentation of the memory is transparent. No code is needed to enable it on the application. Usually, hypervisors already protect the memory of the cores using the MMU [9].

As a drawback, this technique does not allow the mapping of the memory in blocks smaller than the page size. It can bring a waste of memory in an embedded system with a small amount of memory [10]. The mapping of the shared memory used to address the I/O or perform the inter-core communication can also be an issue. In this case, cache partitioning might exclude the I/O memory area, or only one core might be responsible for managing the access to the peripherals. The use of scratchpad memory (SPM) and cache coloring is also an option that could offer a time predictable and consistent access to shared memory.

References

1. Kinnan L (2009) Use of multi-core processors in avionics systems and its potential impact on implementation and certification, p 1.E.4-1. https://doi.org/10.1109/DASC.2009.5347560
2. Fuchsen R (2010) How to address certification for multi-core based IMA platforms: Current status and potential solutions. In: 29th digital avionics systems conference, pp 5.E.3-1-5.E.3-11
3. Agirre I, Abella J, Azkarate-askasua M, Cazorla FJ (2017) On the tailoring of CAST-32A certification guidance to real COTS multi-core architectures. In: 2017 12th IEEE international symposium on industrial embedded systems (SIES) (2017), pp 1–8

4. Boniol F, Pagetti C, Sensfelder N (2019) Identification of multi-core interference. In 2019 IEEE 19th international symposium on high assurancesystems engineering (HASE), pp 98–106
5. Sheikh SZ, Pasha MA (2020) Energy-efficient realtime scheduling on multicores: a novel approach to model cache contention. ACM Trans. Embed. Comput. Syst. 19:4 (2020). Article 28 (July 2020), 25 pages. https://doi.org/10.1145/3399413
6. Domingues RP, Bezerra JDM, Hirata CM (2015) Design recommendations for the development of airborne multi-core systems. In: 2015 IEEE/AIAA 34th digital avionics systems conference (DASC), Prague, pp 1–32. https://doi.org/10.1109/DASC.2015.731162
7. Mandalapu S (2016) White Paper on Issues Associated with Interference Applied to Multicore Processors
8. Zhang X, Dwarkadas S, Shen K (2009) Towards practical page coloring-based multi-core cache management, pp 89–102. https://doi.org/10.1145/1519065.1519076
9. Dasari D, Nelis V (2012). An analysis of the impact of bus contention on the WCET in multicores. In: 2012 IEEE 14th international conference on high performance computing and communications
10. Martins J, Tavares A, Solieri M, Bertogna M, Pinto S (2020) Bao: a lightweight static partitioning hypervisor for modern multi-core embedded systems. InNG-RES@HiPEAC

Increase in Bearing Capacity in Subgrade Composed of Low Plasticity Clays Using Stabilization with Fiberglass Powder

Cynthia Carhuapoma[1,2(✉)], Jaime Tito[1,2], Manuel Silvera[1,2], and Fernando Campos[1,2]

[1] Peruvian University of Applied Sciences, Av. Prolongación Primavera 2390, Lima, Peru
{u910754,u201416765,manuel.silvera,pccifcam}@upc.edu.pe
[2] Springer Heidelberg, Tiergartenstr. 17, 69121 Heidelberg, Germany

Abstract. Low plasticity clays are characterized by having a high moisture content, low cohesion, and a poor capacity to resist shear stress between soil particles. The behavior of these clays in subgrade for pavements is low performance due to their low resistance capacity to imparted cyclical load. Considering these limitations, subgrade soils must be stabilized to improve their bearing capacity, which must be able to withstand the loads imparted throughout the pavement structure. A sustainable stabilization alternative is the use of recycled fiberglass dust, which improves the performance of the mechanical properties of resistance to the cutting of subgrade soils. This research aims to provide a stabilization proposal applied to supersaturated clay soils with a high percentage of humidity, which reduces the water content between particles and increases the pavement subgrade's bearing capacity. For this, stabilization was carried out in clay soils of low plasticity incorporating recycled fiberglass powder in proportions of 6%, 8%, 10%, and 12% by dry weight of material. To validate the stabilization proposal, the California Bearing Ratio Test (CBR) was performed. Laboratory results showed that CBR increased from 3.7% to 13%. Finally, the Bearing capacity of soils improved by 250% under stabilized conditions.

Keywords: Bearing capacity · Soil stabilization · Expansive soils · Fiberglass dust

1 Introduction

The presence of clay soils with low plasticity in saturated subgrade conditions represents a great limitation in the useful life of the roads and affects the performance of all types of pavement. This type of soil has a Bearing Capacity (CBR) value of 3.7%, which makes it inadequate to support the upper layers of the pavement structure. Also, physical characteristics, such as low cohesion between clay fines, low degree of dry density 1.64 g/cm^3, and high moisture content of 36.14% can trigger scenarios where the pavement structure can collapse [1]. In this sense, the need to use a stabilization technique in soils with the features mentioned above has a very important role because

Y. Iano et al. (Eds.): BTSym 2020, SIST 233, pp. 662–669, 2021.
https://doi.org/10.1007/978-3-030-75680-2_73

it helps increase the bearing capacity of the land and better distribute the efforts, which, in turn, reduces deformations and deflections in the pavement layers.

The solutions to stabilize a soil with low plasticity area of the chemical type with lime, cement, sodium or magnesium chloride and of the physical type whit geosynthetics. The effectiveness of each one varies according to the physical properties of the soil [2]. Besides, these types of stabilizations on certain occasions tend to be expensive solutions and not easily accessible due to legal and supervisory restrictions for their purchase. Likewise, the increase in the CBR value in subgrade is variable for clays with low plasticity since it depends on the degree of compaction and moisture they present [3].

In this context, the present research is focused on proposing a sustainable method that allows increasing the bearing capacity of low plasticity clays under conditions of high saturation, using as a stabilizing agent a material from the recycling of construction materials, such as the application of fiberglass powdered. This material has a high compressive strength of 1060 Mpa, a moderate tensile strength of 69 Mpa, and a high hardness value of 7 [3]. These mechanical properties allow to increase the bearing capacity of the subgrade and reduce the presence of fatigue failures, rutting, cracks, and settlements in the different layers of the pavement, prolonging the useful life and allowing the development of the functionality of long term transport routes [4].

2 State of Art

Different investigations have implemented fine soil stabilization methodologies in order to increase the bearing capacity of pavement subgrade. One of the materials with potential for recycling and efficiency in stabilizing fine soils is glass dust from industrial and construction waste. Its application seeks to increase the CBR of clay soils [5]. As this value is higher, it allows increasing cohesion, which in turn reduces volumetric changes and the percentage of humidity in subgrade soils. The clay soil particles combine with the SiO_2, Na_2O_3, and Al_2O_3 ions made up of the glass dust, generating a chemical reaction that develops physical and chemical changes in the soil particles [4]. The increase in strength occurs only up to a certain optimum percentage of glass powder does not cause any significant effect on the floor [6].

Next, different studies that were carried out based on clay soil stabilization methodologies are presented. In research conducted in India, it was proposed to stabilize a lower plasticity clay soil using white cement [7]. The dosages made were 2%, 5%, and 7%. The natural soil sample improved its value from 24.8% to 95% when 7% of white cement was added regarding the CBR test. However, the humidity percentage has been reduced by 18.1% [7]. Other research also carried out in India used recycled glass powder with the objective of improving the physical and mechanical properties of the soil [8]. The proposed stabilization increased the bearing capacity and the CBR value of cohesive soils, with a value of 0.5 kg/cm^2 and 1.56%, respectively. For this research, soil and glass powder dosages of 2%, 4%, 6%, 8%, and 10% of dry material were made. The combination that presented the best results was that of cohesive soil with the addition of 10% of glass powder, increasing its resistance to shear forces represented by the CBR test with a value of 10.4% [8].

3 Methodology

This section describes the procedure developed for the chemical mechanical stabilization of clay soils with low plasticity under constant saturation conditions by adding recycled fiberglass dust. Figure 1 briefly shows the stabilization method of this investigation. The tests carried out to obtain physical and mechanical values of the stabilized soil follow the recommendations of the American Society for Testing and Materials regulations (ASTM) [9–12].

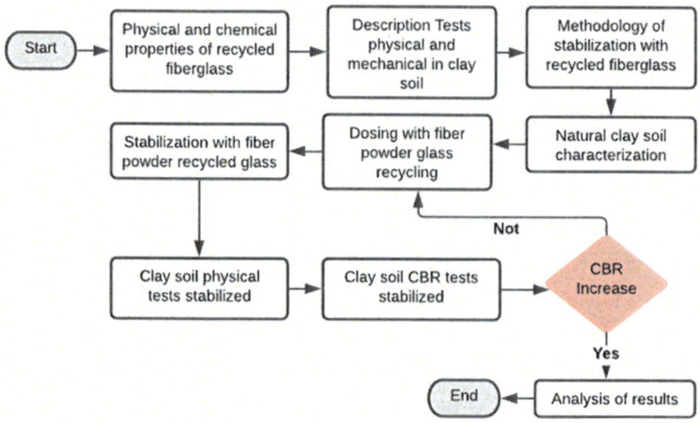

Fig. 1. Investigation methodology

3.1 Fiberglass Powder

The fiberglass dust used in this research comes from waste from building and road construction activities. The material was exposed to grinding with a granulometry that passes 100% the 200 mesh, being a very fine material and conducive to easy mixing with clay. Figure 2 shows the ground fiberglass powder. Likewise, Table 1 presents data on the chemical composition and mechanical properties of the recycled fiberglass powder [7].

Fig. 2. Fiberglass in recycled (left) and ground (right) state.

Table 1. Chemical composition and mechanical properties of recycled fiberglass.

Chemical composition	Unit	Value
Silicon oxide (SIO_2)	%	76.0
Sodium carbonate (NA_2O_3)	%	11.6
Aluminum oxide (AL_2O_3)	%	11.5
Other non-heavy minerals	%	0.9
Mechanical properties	Unit	Value
Specific weight	g/cm^3	2.9
Compressive strength	Mpa	1080
Tensile strength	Mpa	69.0
Hardness	–	7

3.2 Physical and Mechanical Tests of Clay Soils

Granulometric Analysis by Sieving (ASTM D422). The natural soil sample of the subgrade was characterized to determine the presence of the number of percentages of fines according to their size on the granulometric scale [9].

Limits of Attemberg (ASTM D4318). The Liquid Limit and Plastic Limit Tests were carried out to determine the moisture content and plasticity of the soil [10].

Modified Proctor (ASTM D1557). The clay sample was compacted by applying 25 blows with a 10 lb pneumatic hammer to determine the optimum moisture content and the maximum dry density [11].

California Bearing Ratio (ASTM D1883). The CBR Test was run to determine the bearing capacity and qualify of the stabilized clay soil. This test consists of compacting the soil sample in a normalized mold, immersing them in water, and applying a load with a standardized piston [12].

4 Stabilization with Recycled Fiberglass Powder

This section describes the physical properties of low plasticity clay soil and fiberglass powder. Likewise, the process of mechanical stabilization of clay soil is explained.

4.1 Natural Clay Soil

Table 2 shows the results obtained from the physical tests of Granulometry, Limits of Attemberg, Moisture Content, and Mechanics of modified Proctor and CBR carried out on the natural sample of clay soil with low plasticity.

Table 2. Low plasticity natural clay soil physical properties.

Physical properties of soil under study			
Soil parameters	Unit	Value	Analytical method
Classification AASHTO	–	A-6 (4)	ASTM D6913
Classification SUCS	–	CL	ASTM D6913
Liquid limit	%	34.0	ASTM D4318
Plastic limit	%	23.0	ASTM D4318
Plasticity index	%	11.0	ASTM D4318
Moisture content	%	16.5	ASTM D4643
Maximum dry density	g/cm^3	1.731	ASTM D1557
OCH	%	17.2	ASTM D1557
Soil density	g/cm^3	1.627	ASTM D1557
CBR	%	3.7	ASTM D1883

4.2 Dosage of the Clay Soil Sample to Be Stabilized

The experimental plan Will be developed by carrying out tests on different samples of natural clays dosed with 6%, 8%, 10%, and 12%. The physical and mechanical tests follow the guidelines established by the Federal Highway Administration (FHWA) [13]. Figure 3 shows the characterization process of the natural clay soil sample dosed in the laboratory.

Fig. 3. Characterization of a clay soil sample dosed in the laboratory.

5 Validation and Results

The tests carried out and the results obtained, which validate the clay soil stabilization methodology, are described in this section.

5.1 Influence of Fiberglass Dust on LL, LP e IP

The variation of the Liquid Limit (LL), Plastic Limit (PL), and Plasticity Index (PI) with the application of different percentages (6%, 8%, 10%, and 12%) of fiberglass powder is presented in Table 3. Therefore, with the addition of fiberglass powder, the LL is reduced from 34% to 19% and the LP from 23% to 12%. This is due to the high content of silica in fiberglass dust, which develops permeability between soil particles.

Table 3. Variation of the liquid limit, plastic limit, and plasticity index

Fiberglass powder added (%)	LL (%)	PL (%)	PI (%)
0	34	23	11
6	28.2	19.8	8.4
8	25.6	17.7	7.9
10	22.9	15.4	7.5
12	19	12	7

5.2 Influence of Fiberglass Dust on Maximum Dry Density and Optimal Moisture Content

Table 4 shows the variation of the Maximum Dry Density (MDD) and the optimum moisture content (OMC) of the clay soil and of the mixtures with different percentages of fiberglass dust (6%, 8%, 10%, and 12%). The MDD of the natural soil was initially 1.727 g/cm^3 and increased to 1.897 g/cm^3 with the addition of 12% glass fiber powder. However, OMC decreased from 17.2% to 14.1%. The increase in MDD is mainly due to the increase in specific gravity, and the decrease in OMC was due to the fact that the fiberglass dust reduces the pore index of the soil, achieving better densification of the subgrade soil and.

Table 4. Variation of Maximum dry density and Optimum moisture content.

Fiberglass powder added (%)	Maximum dry density (g/cm)	Optimal moisture content (%)
0	1.727	17.2
6	1.759	16.7
8	1.813	16.3
10	1.872	15.6
12	1.897	14.1

5.3 Influence of Fiberglass Dust on CBR

The CBR test was performed based on standard load penetrations of 2.5 mm and 5 mm in accordance with ASTM D1883 [12]. The CBR value of the natural clay soil obtained was 3.7%, initially; however, when it was stabilized with 12% of fiberglass powder, a CBR value of 13% was obtained. Consequently, the largest percentage of the value of CBR was 250%. Also, its initial classification changed from very poor soil to very good soil, according to AASHTO [14]. The influence of the different fiberglass powder dosages applied to samples of natural low plasticity clay soils is presented in Table 5.

Table 5. Variation of the CBR value in clay soil samples stabilized with fiberglass powder

Dosage	0%	6%	8%	10%	12%
Penetration 2.5 mm	3.7	7.7	9.4	12.1	13
Penetration 5 mm	4.2	9.3	11.8	14.5	15.4

The increase in the CBR value is mainly due to the fiberglass's high silica content, which allows reducing the moisture content and increasing the cohesion between particles of the soil and fiberglass mixture, which causes it to rise bearing capacity and improve the quality of the subgrade soil. The graphical representation of the variation of fiberglass powder is shown in Fig. 4.

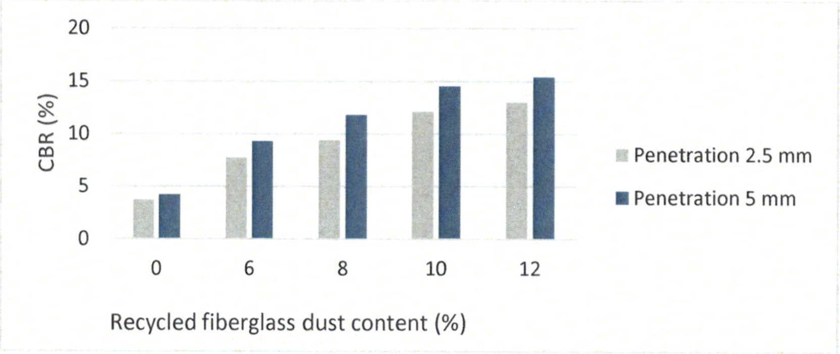

Fig. 4. Variation of the CBR value in natural soil sample and samples stabilized with fiberglass powder.

6 Conclusions

This research concluded, based on the result obtained, that recycled fiberglass powder improves clay soil properties. Using this material for stabilization not only helps to reduce the environmental effect but also for economic solutions as this material is locally and cheaply available. The addition of fiberglass powder reduced the LL from 34% to 19% and the PL from 23% to 12%. This indicates that fiberglass dust reduces the soil's liquid retention capacity due to the high silica content of fiberglass dust.

From the test results of Modified Proctor, it could be determined that the addition of fiberglass powder has a minimal influence due to the fact that the MDD increased from 1.727 g/cm^3 to 1.897 g/cm^3, and the OMC decreased from 17.2% to 14.1%. The CBR value of the clay soil was initially 3.7%; this was increased by 250% with the addition of 12% of fiberglass powder, obtaining a final CBR value of 13%. The results potentially improve the strength and quality of clay subgrade. This research shows that the addition of fiberglass dust can reduce the thickness of the structural layers of pavement due to the increase in the CBR value.

References

1. Meysam P, Mohammad M, Vahid T (2018) Experimental investigation of using a recycled glass powder- based geopolymer to improve the mechanical behavior of clay soils. Constr Build Mater 2(170):302–313
2. Gowtham S, Naveenkumar A, Rankithkumar R (2018) Stabilization of clay soil by using glass and plastic waste powder. Int J Eng Tech 4(2):2395–1303
3. Olufowbi J, Ogundoju A (2014) Clay soil stabilisation using powdered glass. J Eng Sci Technol 9(5):541–542
4. Jinu Rose B, Joseph Jolly K, Juny Mareena S, Thomas M (2017) Effect of glass powder on engineering properties of clayey soil. Int J Eng Res Technol 6:2278–0181
5. Syed Aaqib J, Sudipta C (2020) Effects of waste glass powder on subgrade soil improvement. World Sci News 144:30–42
6. Benny J, Jolly J (2017) Effect of glass powder on engineering properties of clayey soil. Int J Eng Res Technol 6(5):2278–0181
7. AL- Homidy A, Dahim A, Abd El Aaal A (2016) Improvement of geotechnical properties of dabkha soil utilizing cement kiln dust. J Rock Mech Geotech Eng 16:1674–7755
8. Canakci H, Al-Kali A, Celik F (2016) Stabilization of clay with waste soda lime glass powder. Proc Eng 161:600–605
9. American Society for Testing and Materials (2007) Standard Test Method for Particle- Size Analysis of Soils (ASTM D 422). https://www.astm.org/Standards/D422.htm
10. American Society for Testing and Materials (2017) Standard Test Methods for Liquid Limit, Plastic Limit, and Plasticity Index of Soils (ASTM D 4318). https://www.astm.org/Standards/D4318.htm
11. American Society for Testing and Materials (2012) Standard Test Methods for Laboratory Compaction Characteristics of Soil Using Modified Effort (ASTM D 1557). https://www.astm.org/Standards/D1557.htm
12. American Society for Testing and Materials (2016) California Bearing Ratio (ASTM D 1883). https://www.astm.org/Standards/D1883.htm
13. Federal Highway Administration (2019) Materials Management & Performance (FHWA). https://www.fhwa.dot.gov/pavement/materials/
14. American Association of state Highway Transportation Officials (1998) AASHTO Guide for Design of Pavement Structures, 4th edn. Washington, DC

Design of a Warm Mix Asphalt (WMA) with Addition of Recycled Rubber for the Reduction of Permanent Deformation

Lucero Sandoval(✉), Luis Marceliano, Manuel Silvera, and Fernando Campos

Universidad Peruana de Ciencias Aplicadas UPC, Av. Gral. Salaverry 2255, San Isidro 15076, Lima, Peru
{u201722938,u201615515,manuel.silvera,pccifcam}@upc.edu.pe

Abstract. In recent years, the growth of the automotive fleet in Lima has caused its main roads to experience fatigue failures in the asphalt layer due to permanent deformations that affect the pavement's useful life. Warm Mix Asphalt meets the sufficient requirements to improve the structural behavior of flexible pavements. They have similar or superior behavior to conventional asphalt mixtures since they reach higher compaction densities. This allows a reduction in the aging of the binder, which works at lower mixing and compaction temperatures. The use of rubber in the warm mix not only improves the strength of the asphalt mixes but also improves durability, rutting, and fatigue. The laboratory results, the product of the softening point, and penetration tests show that the Warm Mix Asphalt with 18% recycled rubber reduces the deformations caused by fatigue by 50% when experiencing a cyclic load of 650N and 1300N, which represents felt the traffic congestion. In conclusion, it can be said that the Warm Mix Asphalt with the addition of recycled rubber improves the permanent deformations of the pavements, as a result of a higher viscosity in the mixture and a greater amount of air, which gives greater rigidity to the asphalt binder.

Keywords: Warm Mix Asphalt · Permanent deformation · Recycled rubber

1 Introduction

In this research, a Warm Mix Asphalt (WMA) with the addition of recycled rubber (RR) will be evaluated to reduce permanent deformations in pavements in the city of Lima. Since the last decade, the automotive fleet has been on a growing rise; this causes road structures to reach and even exceed their design limits, generating premature damage in the life of pavements [1].

The damages that we can find, for the most part, are the product of permanent deformations. They can be developed in the different layers that make up the pavement structure, even at the asphalt layer level. Because of this, the development of an asphalt pavement project must be complete. Under this premise, not only the quality of the materials should be analyzed; it is necessary to take into account the controls in the

Y. Iano et al. (Eds.): BTSym 2020, SIST 233, pp. 670–678, 2021.
https://doi.org/10.1007/978-3-030-75680-2_74

asphalt mixture, the structural design of the pavement, the study of traffic, the type of soil of the subgrade, the drainage conditions, the weather, etc. [2].

Warm mix asphalt, using different techniques, can reduce mixing and compaction temperatures without significantly altering its mechanical properties. The use of WMA represents a reduction in fuel consumption and polluting emissions to improve environmental quality and reduce production costs [3]. In addition, it enables the production temperatures of traditional mixtures to be reduced. WMA helps to reduce the working temperature of asphalt by up to 30 °C [4].

The main benefits of improving the performance of bituminous mixtures are the reduction of the mixture's sensitivity to temperature, the high elastic performance of asphalt, the reduction of the oxidation effect, and the improvement in the performance of the mixture in response to phenomena related to fatigue. The latter is reflected in the cracking and plastic deformation that pavements suffer throughout their useful life [5]. The voids percentage of warm mixes is lower than that of a hot mix; this makes the mix much more compact, and, as a result, density values very close to the design density are obtained; these are classified as high-performance asphalt mixtures [6, 7].

It is public knowledge that the addition of polymers produces a more resistant mixture against the joint action of the load by vehicular congestion and the temperatures applied to the conventional pavement. The use of recycled rubber in WMA has allowed improving its weak mechanical properties; this mixture has allowed a decrease in permanent deformability and a greater resistance to traffic stresses due to its elastic properties [8].

This research's structure is presented below: The following item presents a detailed analysis of the concepts of WMA and permanent deformations. In state of art, the methodology of all the research is presented. In the fourth section, the WMA plus rubber addition proposal is validated with the results obtained from the Resilience Modulus tests. This specific test will allow us to obtain the modulus of elasticity of the proposed pavement. Finally, the results of the investigation are presented at the end of this article.

2 State of Art

Warm mix asphalt represents an ecological solution for road engineering; this is achieved by lowering the asphalt production temperature, which causes a reduction in energy consumption. The energy reduction leads to a reduction in CO_2 emissions [9]. This characteristic in the temperature of the mixture allows higher percentages of recycled materials, such as rubber. Likewise, a longer time can be achieved in the transport of the mixture and an opening for roads' operation in less time [10].

There are comparative studies between warm and hot mix asphalt, which show a similar behavior between both techniques. It has been validated that there are important characteristics in warm mix asphalts that improve the rigidity of the mat designed with this type of mix, the most important being the reduction of voids content [11]. This allows the flexible pavement behavior to be like a rigid pavement, which would prolong the useful life of the road [12]. Permanent deformations can occur due to different factors such as the construction process, construction materials, and others. Latest research has shown that vehicular speed greatly influences fatigue deformations on pavements [13].

3 Methodology

For this type of experimental research, within the methodology, as seen in Fig. 1, laboratory tests will be used. These include the Marshall design and modulus of elasticity tests. Both are complemented using formulas obtained from the literature studied for the calculation of permanent deformation asphalt mixtures: hot and warm with the addition of rubber.

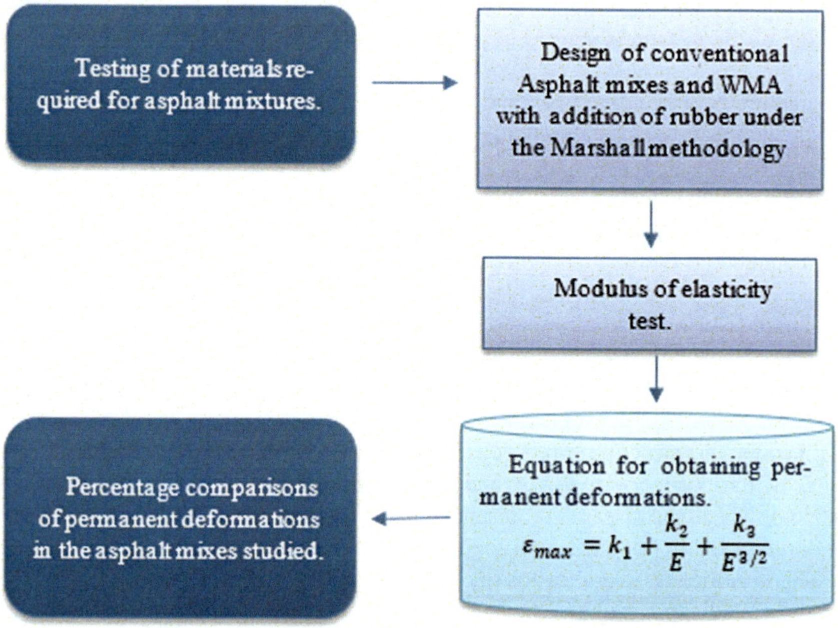

Fig. 1. Methodology used in the research work

3.1 Material Test

The materials used in this research were designed with MAC-2 aggregates, according to the Peruvian technical standard. The type of asphalt cement is 60–70 times the annual average temperature of the city of Lima, according to the indications of the MTC [14].

The additive used was ITERLOW-T; this allows the production of asphalt mixtures between 125 °C–130 °C. The Marshall tests were given in 0.3% and 0.5% of additive to determine the adequate percentage of ITERLOW-T.

3.2 Marshall Test

The Marshall design was made by taking into account the considerations of the MTC EG -2013 technical standard. In this design, an optimum percentage of asphalt of 5.9%

was obtained after 75 hits. Finally, a void percentage of 4.3% was determined; this characteristic will be improved in the Marshall design of the warm asphalt mix.

Regarding the warm asphalt mix design, the hot mix design has been taken as a reference. The additive content has been considered in 0.3% and 0.5% for the dosage, the best percentage being 0.5% because it complies with all the concepts required in the Peruvian technical standard.

3.3 Test of Modulus of Elasticity

The modulus of elasticity was estimated in the laboratory through the UTM Universal Press and at 35 °C. As for the hot asphalt mix, the value of the contact load was 65N, and the sample was exposed to a cyclic load of 650N. The warm asphalt mix was also tested with a contact load of 130N and was exposed to a cyclic load of 1300N.

3.4 Formula to Obtain Permanent Deformations

The calculation of permanent deformations was carried out using the formula proposed by Schmidt and Santucci [15].

$$\varepsilon_{max} = k_1 + \frac{k_2}{E} + \frac{k_3}{E^{\frac{3}{2}}} \tag{1}$$

Where ε_{max} is the maximum deformation, and E is the modulus of elasticity.

With the elaboration of 360 samples for the tests, the values of $k1 = 6.24 \times 10^{-6}$, $k2 = 0.48$ and $k3 = 15.79$ were obtained.

3.5 Percentage Comparisons of Permanent Deformations in the Asphalt Mixtures

In the comparative analysis carried out between the hot and warm mixtures, the evolution of the percentage differences in the number of voids and the maximum deformations allowed by the Peruvian technical standard was taken into consideration.

4 Validation

The research validation model focused on the comparison of two samples of asphalt mixes. Two samples were tested, one of them was a WMA with the addition of 18% recycled rubber, and the other sample was a conventional blend of HMA. This validation process is described and supported below.

4.1 Characteristics of the Asphalt Mixes

In this investigation, the addition of the ITERLOW-T additive was used in percentages of 0.3 and 0.5, both with the same initial characteristics and with 18% recycled rubber in its composition. After comparing both samples, the WMA with 0.3% + 18% RR was defined as the sample with the best characteristics since it complies with the EG 2013 standard, as shown in Table 1. Regarding the volume of asphalt mix, it was observed that the WMA at 0.3% had a result of 18.5%. This value is lower than that obtained with the WMA of 0.5%. This means a reduction in materials, cost, and energy production. In addition, the value of the percentage of voids obtained from the optimal mixture was 4.7%. According to the regulations, the value of the void percentage must fluctuate between 3–5%. The percentage of voids filled with asphalt was 74.77%, a value like the hot mix. In addition, the stability/flow parameter was estimated at 3225.8 kg/cm, a value that remains within the admissible range of 1700–4000 kg/cm. On the other hand, the percentage of rubber was assumed because research has been carried out to support the use of this amount of recycled rubber [11].

Table 1. Marshall design features.

Design parameters		Summary of results			Specification EG 2013
% WMA	%	0.3	0.5	0	
Asphalt cement	%	5.9	5.9	5.9	
Unit weight	kg/m3	2.433	2.411	2.475	
Empty	%	4.667	5.537	4.3	3–5
VMA (Minimum)	%	18.500	19.244	17	14
V. LL.C.A.	%	74.773	71.229	74	
Dust/Asphalt	%	0.817	0.817	0.7	0.6–1.3
Flow	mm	14.333	14.667	13.8	8–14
Stability (Minimum)	kN	1154.400	1328.773	1007.5	8,15
Stability/flow	kg/cm	3225.849	3627.520	2928.7	1700–4000

4.2 Modulus Elasticity Test (E)

A sample of HMA PEN 60/70 and a WMA 0.3% + 18% RR were tested at a specific temperature of 35 °C, with void percentages of 4.5% and 4.7%, respectively, as shown in Table 2. Both samples were subjected to five pulsations, of which the mean, standard deviation, and the coefficient of voids (% CV) were obtained. With the HMA sample, the mean Modulus of Elasticity of 825 Mpa could be determined.

The tests carried out to calculate the modulus of elasticity, with 5 pulsations, in the WMA 0.3% + 18% RR obtained an average of 1462 Mpa, as shown in Table 3. Also, the modulus obtained from the proposed mix is 43.57% higher than the hot mix asphalt.

4.3 Calculation of Permanent Deformations

Using the equations of Schmidt and Santucci [15], it was possible to calculate the permanent deformations for each pulse obtained in the test of modulus of elasticity; in these calculations, it was possible to observe that the difference in the deformations between

Table 2. Results of the modulus of elasticity test for the hot mix asphalt.

Test results	Pulse 1	Pulse 2	Pulse 3	Pulse 4	Pulse 5	Average	Stand. dev.	% CV
Resilient module (Mpa)	793	789	814	873	857	825	33.97	4.12
Peak load force (N)	648	656	653	655	644	651.2	4.5	0.7
Settlement force	74	66	64	72	84	72	7.04	9.5
Total, recoverable deformation	2.67	2.62	2.57	2.46	2.47	2.56	0.08	3.21
Recoverable horizontal deformation # 1 (μm))	1.49	1.41	1.39	1.32	1.25	1.37	0.08	5.86
HORIZONTAL RECUPERABLE #2 (μm)	1.18	1.21	1.18	1.13	1.22	1.18	0.03	2.54

Table 3. Results of the modulus of elasticity test for the warm asphalt sample with the addition of rubber.

Test results	Pulse 1	Pulse 2	Pulse 3	Pulse 4	Pulse 5	Average	Stand. dev.	% CV
Resilient module (Mpa)	1406	1408	1475	1545	1479	1462	51.74	3.54
Peak load force (N)	1290	1299	1283	1286	1293	1290	5.6	1.5
Settlement force	123	118	142	136	139	131.6	9.4	8.3
Total, recoverable deformation	3.46	3.72	3.67	3.82	3.87	3.75	0.14	2.81
Recoverable horizontal Deformation # 1 (μm))	1.53	1.62	1.76	1.74	1.99	1.76	0.17	9.83
Horizontal recuperable #2 (μm)	2.03	2.15	2.05	2.11	1.83	2.03	0.1	5.02

both types of mixtures is higher in approximately 50%, as shown in Table 4, which allows us to recognize the added value of rubber in terms of viscosity and adherence of the mixture [14].

Table 4. Calculation of permanent deformations.

Test	E (HMA) (Mpa)	Emax (mm)	E (WMA + RR) (Mpa)	Emax (mm)	Δ (%)
Pulse 1	793	1.319E−03	1406	6.17E−04	53.207
Pulse 2	789	1.327E−03	1408	6.46E−04	51.320
Pulse 3	814	1.276E−03	1475	6.10E−04	52.156
Pulse 4	873	1.168E−03	1545	5.77E−04	50.615
Pulse 5	857	1.196E−03	1479	6.08E−04	49.119
Average	825	1.254E−03	1462	6.17E−04	50.812

In the calculation of permanent deformations, it is observed that the warm asphalt mix with the addition of rubber at 18% presents smaller deformations. The number of repetitions of equivalent axes of a representative sample was taken as reference, which simulates a congested urban intersection in Lima. This was done to calculate the allowable deformation, according to RD N °10-2014-MT /14, and to be able to verify compliance with the regulations.

$$D_{adm} = (1.15/N)^{0.25} \tag{2}$$

Where N: Number of repetitions of equivalent axes in millions.

The value assigned for N was 309 million equivalent shafts, and an admissible deformation of 7.81E−03 mm was obtained. It was observed that the values obtained from the warm mix asphalt differ from the admissible deformation, compared to the results from the hot mix asphalt. This validates a better performance of the proposed warm mix.

5 Conclusions

The warm mix asphalt with addition of rubber shows a better behavior against the calculated permanent deformations. This percentage difference has a value greater than approximately 50%. In pulse No. 4, the highest modulus of elasticity was obtained in terms of WMA and HMA, giving 1445 MPa and 873 MPa, respectively. These modules represent maximum deformations of 1.168E−03 mm and 5.77E−04 mm. With these results, it is shown that the best elastic behavior is possessed by WMA + 18% RR, from which the lowest maximum deformation value is obtained.

In a conventional situation, in Lima, an admissible deformation of 7.81E−0.3 mm could be reached according to the Peruvian Technical Standard. The proposal of WMA + 18% RR presents a maximum deformation of 6.46E−04 mm; this value represents 8.27% of what is admissible in the regulations.

The percentage of voids of the WMA + 18% RR was 4.67%, a value higher than that of the HMA by 0.36%. However, the difference in the modulus of elasticity obtained was 77.21%, which validates the theory that warm mixtures have a more rigid behavior due to their percentage of voids.

Acknowledgments. The authors of this research are grateful to the group Achirana Ingenieros y Contratistas. Without their support and follow-up, this research would not have been possible.

References

1. Lopera C (2017) Vulnerabilidad de la infraestructura vial ante incremento del parque automotor en la ciudad de Cajamarca. Master's Thesis. Universidad Nacional de Cajamarca. Cajamarca. Perú
2. Rico A, Téllez R, Gárnica P (1998) Pavimentos Flexibles. Problemática, metodologías de diseño y tendencias. Secretaría de comunicaciones y transportes. Instituto Mexicano de Transporte. México D.F. México
3. Almeida-Costa A, Benta A (2016) Estudio de impacto económico y ambiental del asfalto de mezcla caliente en comparación con el asfalto de mezcla caliente, J Clean Prod
4. Vaitkus A, Čygas D, Laurinavičius A, Perveneckas Z (2009) Analysis and evaluation of possibilities for the use of warm mix asphalt in Lithuania. Baltic J Road Bridge Eng 4(2):80–86. https://doi.org/10.3846/1822-427x.2009.4.80-86
5. Flores R, Rojas J (2019) Comportamiento de las mezclas asfálticas tibias adicionando tensoactivos y sus beneficios respecto a las mezclas asfálticas en caliente. Specialty Thesis. Universidad Ricardo Palma. Lima. Perú
6. Lopera C (2011) Diseño y producción de mezclas asfálticas tibias a partir de la mezcla de asfalto y aceite crudo de palma. Master's Thesis. Universidad Nacional de Colombia. Medellín. Colombia
7. Benítez Barreto A (2015) Evaluación de la resistencia de una mezcla asfáltica tibia, sustituyendo en su fabricación, parte del agregado pétreo, por escoria de alto horno Universidad Católica de Colombia. Bogotá D.C. Colombia
8. Maguiña W (2019) Caucho reciclado de llantas en la mezcla de Asfalto a Compresión para mejorar las Propiedades Mecánicas. Master's Thesis. Universidad Ricardo Palma. Lima. Perú
9. Lopera C (2011) Diseño y Producción de Mezclas asfálticas tibias, a partir de la mezcla de asfalto y aceite crudo de palma. Thesis. Universidad Nacional de Colombia. Bogotá D.C. Colombia
10. López CH, Thenoux G, Sandoval G, Armijos V, Ramírez A, Guisado F, Moreno E (2017) Estudio de mezclas asfálticas templadas con emulsión super- estabilizada. Revista de Ingeniería de construcción 32(1):57–64
11. Lemus EA, Mosquera JL, Virviescas XA (2018) Estudio de mezclas asfálticas densas con adición de materiales alternativos. Specialty Thesis. Universidad católica de Colombia. Bogotá D.C. Colombia
12. Castro JE (2017) Evaluación de la capacidad estructural de una mezcla asfáltica tibia MDC-19 utilizando aditivos modificadores de viscosidad. Specialty Thesis. Universidad católica de Colombia. Bogotá D.C. Colombia

13. Rondón HA, Reyes FA, Urazán CF (2013) Efecto de la disminución de la velocidad vehicular en la durabilidad de la capa asfáltica. Revista chilena de ingeniería. 21(1):139–146
14. MTC- Ministerio de Transportes y Comunicaciones (2014) Manual de carreteras. Suelos, Geología, Geotecnia y Pavimentos. Sección suelos y pavimentos. Lima, Perú
15. Schmidt RJ, Santucci LE (1969) El efecto de las propiedades del asfalto en el agrietamiento por fatiga del concreto asfáltico en el proyecto Zaca-Wigmoretest. Asociación de Tecnólogos de Pavimentación Asfáltica 38(1):39–64

Analysis of Replacement of Luminaires in Public Lighting System

Eder Carlos Fernandes(✉), Ederaldo Luíz Ratz, Joice Marin, Luiz Ariovaldo Fabri Junior, Marli de Freitas Gomes Hernandez, and Rangel Arthur

School of Technology, State University of Campinas, Limeira, SP, Brazil
ederaldo@ratz.com.br, rangel@ft.unicamp.br

Abstract. The impact on the consumption of energy destined for public lighting generates a high index cost. Even more, there are significant financial impacts by street lamps that do not have a very high useful life. Therefore, high amounts of electrical energy are needed to keep the streets lit up. New technologies for this concept in order can allow a significant economy in public lighting networks. The LED lamps have a longer useful life and also lower electrical energy consumption and a better luminance. The replacement of conventional lamps for LEDs can save a lot for the city, and the money saved in public lighting can be used in health and education. This article describes a case study in Sumaré, a small Brazilian city, where a simulation is performed to replace the public lighting systems. The financial impact of replacing the entire conventional lighting system with a lighting system with LED luminaires is studied and discussed. The simulation shows the return on investment in replacing a conventional system with a more modern one and its energy consumption over time.

Keywords: Public lighting · LED lighting · Energy efficiency · Energy consumption · Sustainable energy

1 Introduction

Energy efficiency is an issue not only in Brazil but also worldwide, such as the European Union's concern for energy efficiency issues, according to the European Parliament's Directive 2012/27/EU [1, 2].

Energy efficiency has been gaining strength due to the energy destruction crisis, global warming, which has been very accentuated lately, and other environmental issues. So, not only in lighting but also in construction concepts and very overloaded with the concept of energy efficiency, it is considered a high-performance project [3].

Efficiency is defined as the ability to perform tasks or jobs effectively and with minimal waste; productivity.

Energy efficiency is a concept that determines the level of economy of energy resources in the various consumer devices [4]. It is a technical-economic study area aiming to provide the best energy and water consumption, with reduction of related

Y. Iano et al. (Eds.): BTSym 2020, SIST 233, pp. 679–686, 2021.
https://doi.org/10.1007/978-3-030-75680-2_75

operating costs, minimization of contingencies in the supply of these inputs, and introducing elements and instruments necessary for the organization's energy management. The main purposes are to reduce consumption by taking measures such as replacing lighting devices with more efficient ones, use of automation systems, lighting only for specific needs, electric quantities as harmonics and power factor to the characteristics of the operation in question, and substitution of energy input as electricity by natural light [5].

Due to socioeconomic factors, street lighting is a fundamental role as an inhibitor of violence and a mechanism for the development of low-income communities, who live daily with the lack of urban infrastructure [6, 7].

According to government reports, public lighting energy consumption corresponds to approximately 4.5% of the national demand and 3.3% of the country's total electricity consumption, which is equivalent to 3GWh and consumption about 9.7 billion KWh/year [8, 9].

According to reports from the government of the state of São Paulo - Brazil, in June 2020, the total electricity consumption in the state of São Paulo corresponds to 9,398,697 MWh, from which the public lighting of the state corresponds to 262,901 MWh [10]. More details are shown in Table 1.

Table 1. Consumption by end-use sector (MWh) [10]

Consolidated state of São Paulo June/2020		
Class	Consumptions in MWh	%
Residential	3.313.729	35,26
Industrial	3.206.183	34,11
Commercial	1.704.954	18,14
Rural	284.784	3,03
Public powers	171.136	2,80
Public lighting	262.901	1,82
Public Service	447.696	4,76
Own use	7.314	0,08
Total	**9.398.697**	**100**

In this scenario, we need to prevent the increase in demand for electric energy for public lighting. Adding this to the high costs of alternative power sources, there is a growing need from federal government, states, and municipalities to adopt consumption management policies to achieve rationalization of the use and, consequently, energy efficiency.

Aiming to attend to this demand, the Brazilian government, with the support of the Ministry of Mines and Energy (MME) through Eletrobras Agency, implemented The National Program for Efficient Public Lighting (Reluz) [11–13]. The primary purpose is to promote the development of energy efficiency in public lighting systems. For this objective, this program encourages the use of more energy-efficient technologies, energy conservation principles, and accordance with the current laws, regulations, and technical standards [14].

This paper is a case study in which real data were obtained from quantities of public lighting points in the city of Sumaré, located in the State of São Paulo, Brazil. By simulation, considering the market values to exchange the components of illumination obtained in public bids. The cost and return time analysis is performed for the implementation of these actions.

The article will be organized as follows. Section 2 discusses the main research on LED lighting and its benefits over the years.

Section 3 presents a comparison of conventional lighting with LED lighting.

Section 4 discusses the methodology of this article.

Section 5 comments about the studies and simulations present in this article and, finally, the conclusion.

2 Lighting

In recent years, the research on devices and methods of lighting with greater energy efficiency significantly grew up, [15] mentions the comparison of different luminaires. It is noticed significant changes in lighting technology and semiconductor based on solid-state lighting (SSL). It is associated with the electronic toy indicator lamp. In the future, solid-state lighting will be combined with LEDs (light-emitting diodes). It has been shown to be more efficient than incandescent lamps and almost all visible wavelengths, where a white light source of 150–200 lm/W [16].

With this discovery, consumption in the world of electricity in lighting in general, such as public, residential, building lighting, etc., will decrease by more than 50% of the total consumption of electricity [16].

Recently LEDs have attracted a lot of attention and interest because the luminance in this case is very high with less consumption. A good example is a 3 W LED lamp that replaces a 40 W halogen lamp [15]. A LED lamp has an extremely long life and can emit colored light if using colored filters, the light generated by the LED lamps is highly directional.

It is perceived that LED technology is becoming a satisfactory and economical solution due to its low energy consumption, long service life, i.e., longer operating time, less environmental impact [17], decreasing cost and investments [18], and many other benefits and advantages.

In recent years, the efficiency of LED lamps has improved by about 25 lm/W to more than 160 lm/W, according to Pattison et al. [19]. A LED lighting system performs better than high-pressure sodium luminaires. This performance stands out in economy and energy quality, a reduction in energy efficiency index of approximately 40% compared to the high-pressure sodium luminaire.

Juntunen [20] produced data collected from an intelligent LED lighting pilot installation along an illuminated route in a residential area in Helsinki. They developed a system of controls tracking user movement and location along the street with infrared sensors. They state that the ideal behavior of lighting is one that can save energy without reducing the safety of users of the route during night hours.

3 Conventional Lighting x Led Illumination

The conventional street lighting approach refers to environments where incandescent, fluorescent, mercury vapors, sodium, and metallic lamps are used. Incandescent lamps have been gone from the Brazilian market since 2016 [21] as a measure to minimize the waste in the consumption of electric energy. This kind of lamp has a tungsten filament through which the electric current circulates, causing it to heat to the point of incandescence and, consequently, releasing light but showing a very low energy efficiency [22].

In the process of evolution, the fluorescent lamp was created. In this kind of lamp, light emission is carried out by the action of the electric current, which generates radiation through a tube containing a mixture of gases and mercury vapor. This tube is phosphor-coated, converting the radiation into visible light. In this way, the fluorescent lamp has a high energy efficiency by emitting more electromagnetic energy in the form of light than heat.

For conventional lighting of industries, shops, and public roads, mercury, sodium, and metallic vapor lamps are the most common kind of lamps on the market. Mercury vapor lamps emit a bluish-white light, have a long average lifespan, and show a poor color rendering index. Sodium vapor lamps are known for showing a yellowish/orange light and are seen in most public street lighting in Brazil. Another characteristic of this type of lamp is the low color reproduction index, impairing drivers and pedestrians' visibility.

An evolution of mercury vapor lamps is the metal vapor one, which shows a higher color reproduction index than the sodium and mercury vapor lamps and also has a long life.

With the development of lighting technologies, LED lamps stand out because they are easily accessible and cheaper than metal vapor lamps, sodium vapor lamps and mercury vapor lamps. When energized, the LED lamp emits visible light with low heat emission with energy loss. Quality LED lamps have a life expectancy of about 100,000 h, which is equivalent to an average of 12 years if used 24 h a day, 7 days a week. This type of lighting offers many differentials in relation to conventional lighting, such as: [5].

- Higher electrical and light efficiency;
- Longer service life;
- Better flexibility (light beam);
- Improvement of Color Temperature;
- Lower heat emission;
- Greater stability and operational safety;
- Higher color rendering index;
- Improvement of applications.

4 Methods

The study presented here is based on a process of simulating the total cost of replacing conventional lamps in the city of Sumaré/SP with lamps with LED technology and analyzing the value achieved with energy savings each month and the amortization time of the necessary investment.

The data used in this work were provided by the Municipal Works Secretariat in the city of Sumaré/SP, which consents to the disclosure of data by the authors of this work, which reported the number of public lighting points currently installed, as well as the installed power. Table 2 shows the lighting points installed in the city of Sumaré, in the interior of São Paulo.

Table 2. Summary table of lighting points and Powers

Report of public lighting points of the municipality of Sumaré – table summary					
Lamp type	Quantities	Power of the lamp (W)	Wattage reactor (W)	Power of the reactor (W)	Power installed (KW)
Mercury Steam	*4.799*	*80*	11	91	436,71
Mercury Steam	*6.484*	*125*	15	140	907,76
Sodium vapor	*1.452*	*70*	15	85	123,42
Sodium vapor	*3.285*	*100*	18	118	387,63
Sodium vapor	*1.125*	*150*	26	176	198,00
Sodium vapor	*1.622*	*250*	37	287	465,51
Metallic Steam	*800*	*400*	38	438	350,40
Metallic Steam	*500*	*250*	30	280	140,00
Total Municipality				**Total**	**3.009,43**
				Value of KW/h	**0,488**
				Total monthly amount R$	**1.468.603,30**

The values related to the total cost of the LED lamp set together with the costs of the change services were obtained in consultation with public announcements from other city halls that carried out similar services, being about R$ 200.00 per set. The costs of the energy in kWh (kilowatt-hour) without considering taxes, according to the National Tariff Ranking available on the ANEEL website, is about R$ 0.488 [23].

5 Study and Simulation

Table 3 presents the result of the calculation of power and consumption costs, considering the replacement of conventional lamps with LED equivalents ones [13].

Considering the same energy costs in kW/h (0.488) for both cases scenarios, the study shows that it is possible to save monthly amounts of R$ 871,798.82, which corresponds to 59% of the amount currently paid. The values are seen in the graph in Fig. 1.

According to the price survey carried out in similar public bids, the average cost for replacing the luminaires, lamps, and reactors, all the new equipment included, is about R$ 200.00.

Table 3. Total power with equivalent LED lamps

Summary chart – total power with LED lamps			
Lamp type	Quantities	Power of the lamp (W)	Power installed (KW)
Mercury Steam	*4.799*	*30*	143,97
Mercury Steam	*6.484*	*60*	389,04
Sodium vapor	*1.452*	*30*	43,56
Sodium vapor	*3.285*	*50*	164,25
Sodium vapor	*1.125*	*60*	67,50
Sodium vapor	*1.622*	*120*	194,64
Metallic Steam	*800*	*200*	160,00
Metallic Steam	*500*	*120*	60,00
Total Municipality		**Total**	**1.222,96**
		Value of KW/h	**0,488**
		Total monthly amount R$	**596.804,48**

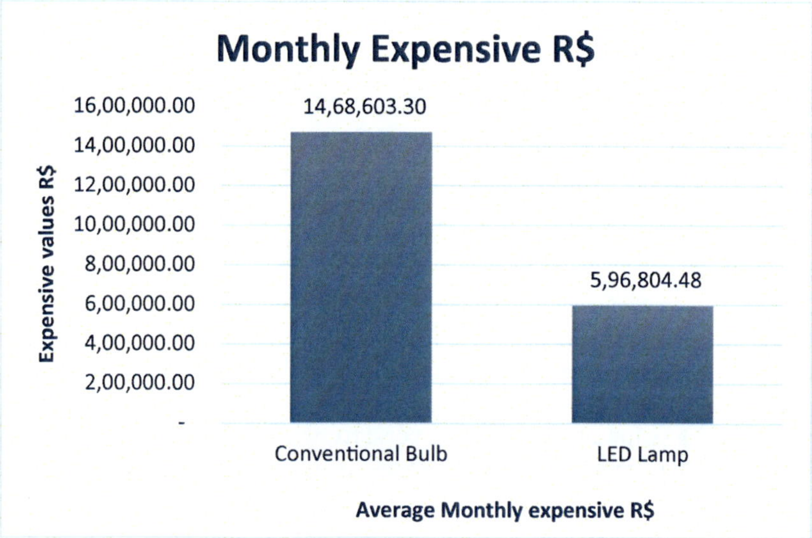

Fig. 1. Conventional monthly expenditure x LED

In this study, the number of luminaires to be replaced will be 20,067, corresponding to an investment of about R$ 4,013,400.00.

Based on this calculation, it can be observed in Fig. 2, that after the fifth month from the change of lamp system, the amortization of the investment would take place and financial savings would begin. At the end of a 12-month cycle, it would be possible to achieve savings of R $ 6,448,185.84.

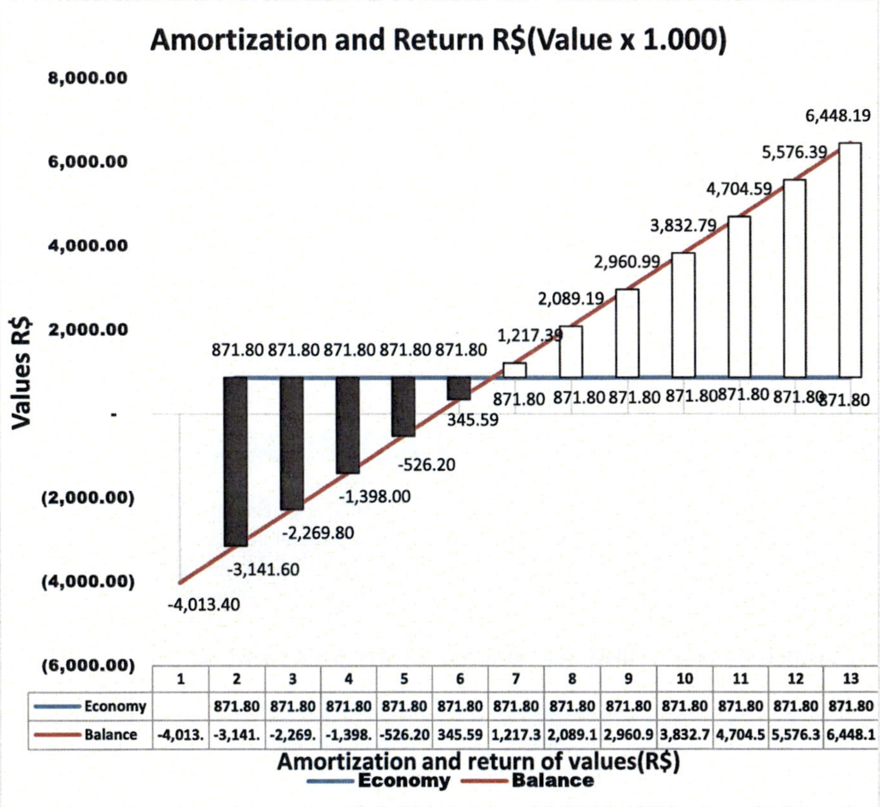

Fig. 2. Amortization of investiment

6 Conclusion

This work's motivation was a solution to the energy economy by replacing conventional lamps with LED lamps in public lighting.

This research has demonstrated that the replacing can bring many benefits such as improved visual comfort and safety of the environment by eliminating the disposal of lamps with pollutants and, above all, the financial question.

In the case study, monthly expenses with lamps with metallic steam, sodium vapor, and mercury vapor are R$ 1,458,603.30, while LED lamps have a monthly cost of R$ 596,804.48.

Another important aspect to consider is that this economy would also allow the expansion of the public lighting system without the need to increase the capacity of electricity generation.

References

1. Dolega W (2014) Selected aspects of energy efficiency in Poland. In: 9-th international conference on deregulated electricity market issues in South-Eastern Europe DEMSEE 2014,

Nicosia, Cyprus, 25–26 September 2014. Section Renewable energy sources and energy efficiency I, full paper 3, pp 1–5
2. Dolega W (2018) Energy efficiency—case study Poland. In: 2018 innovative materials and technologies in electrical engineering (i-MITEL), Sulecin, pp 1–6. https://doi.org/10.1109/IMITEL.2018.8370463
3. Lerum V (2008) Edifício de alto desempenho. Wiley, Hoboken
4. Menkes M (2004) Eficiência energética, políticas e sustentabilidade. UnB, Brasília
5. AALOK: Iluminação convencional X Iluminação LED. https://aalok.com.br/blog/iluminacao-convencional-x-iluminacao-led
6. HTL – Instalação e Manutenção Elétrica, Tabela de equivalência entre lâmpadas LED e lâmpadas convencionais. https://www.htlbrasil.com/Arquitetura/PDF/tabela-de-equivalencia-de-iluminacao-e-economia-de-energia-eletrica.pdf
7. Fidalgo F, Ferreira G, Tiryaki GF (2007) Iluminação Pública em Salvador: Gestão, Eficiência e o Papel do Agente Regulador. In: V Congresso Brasileiro de Regulação, 2007, Recife. V Congresso Brasileiro de Regulação
8. Dantas CAP, Junior FK (2015) Iluminação Pública e eficiência energética. https://www.osetoreletrico.com.br/iluminacao-publica-e-eficiencia-energetica
9. Empresa de Pesquisa Energética (2017) Anuário Estatístico de Energia Elétrica de 2017 ano base de 2016. Ministério de Minas e Energia. Governo Federal do Brasil
10. São Paulo. Resumo Executivo, dados de produção e consumo de energia elétrica. https://dadosenergeticos.energia.sp.gov.br/portalcev2/intranet/BiblioVirtual/eletrica/Resumo_Executivo_EE.pdf
11. Eletrobras. Procel – iluminação pública. https://www.eletrobras.com/ELB/data/Pages/LUMIS0389BBA8PTBRIE.htm
12. G1 (2015) Lâmpada incandescente de 60 watts deixa de ser vendida; Inmetro fiscaliza. https://g1.globo.com/economia/seu-dinheiro/noticia/2015/07/lampada-incandescente-de-60-watts-deixa-de-ser-vendida-inmetro-fiscaliza.html
13. Eletrobraas/Procel Reluz (2004). Iluminação Pública Eficiente. Rio de Janeiro - RJ Guia Técnico
14. PROCEL. Legislação. https://www.procelinfo.com.br/main.asp?TeamID={0B311C4C-D52D-4159-9842-FEA7AD2E25D5}
15. Bonda P, Sosnowchik K (2007) Sustainable commercial interiors. John Wiley, Hoboken
16. Tsao JY (2004) Solid state lighting: lamps, chips and materials for tomorrow. IEEE Circ Dev 20(3):28–37
17. Campisi D, Gitto S, Morea D (2018) Economic feasibility of energy efficiency improvements in street lighting systems in Rome. J Clean Prod 175:190–198
18. Witjes S, Lozano R (2016) Toward a more circular economy: proposing a framework linking sustainable public procurement and sustainable business models. Resour Conserv Recycl 112:37–44
19. Pattison PM, Hansen M, Tsao JY (2018) Eficácia da iluminação LED: status e direções. CR Phys 19(3):134–145
20. Juntunen E, Sarjanoja EM, Eskeli J, Pihlajaniemi H, Österlund T (2018) Smart and dynamic route lighting control based on movement tracking. Build Environ 142:472–483
21. Eletrobrás/Procel (1998) Gestão Energética Municipal. Brasília-DF Guia Técnico
22. de Araujo Moreira V (1999) Iluminação Elétrica. Editora Edgard Blücher Ltda - Primeira Edição
23. ANEEL, Ranking das Tarifa. https://www.aneel.gov.br/ranking-das-tarifas

EMC Issues in Grid-Connected Photovoltaic Systems

The Brazilian Regulatory and Standardization Scenario

Leonardo dos Santos[1(✉)], Yuzo Iano[1], Hermes Loschi[2,3,4], Douglas Nascimento[2,3], Navid Razmjooy[5], Euclides Chuma[1], and Carlos Bertolassi[1]

[1] Department of Communications (DECOM), School of Electrical and Computer Engineering (FEEC), University of Campinas (UNICAMP), Campinas, Brazil
{leonardoncsantos,euclides.chuma}@ieee.org, yuzo@decom.fee.unicamp.br, carlos.bertolassi@br.abb.com

[2] Institute of Automatics, Electronics and Electrical Engineering, University of Zielona Gora, Zielona Gora, Poland
{eng.hermes.loschi,eng.douglas.a}@ieee.org

[3] Power Electronic and EMC (PE) Group, Faculty of Electrical Engineering, Mathematics and Computer Science (EEMCS), University of Twente, Enschede, Netherlands

[4] Department of Electrical and Electronic Engineering, School of Electrical and Electronic Engineering, University of Nottingham, Nottingham, England

[5] Department of Electrical and Control Engineering, Tafresh University, Tafresh, Iran

Abstract. Exponential growth of photovoltaic systems connected to the grid has been observed over the last decade in Brazil concurrently with concerns about the power quality and operational compliance of related equipment. In the past decades, there has been much research and debate regarding the definition of electromagnetic compatibility requirements. Since Brazil has both government and private entities responsible for the product certification and homologation process, it is important to discuss and analyze these entities' roles from an electromagnetic compatibility viewpoint in photovoltaic systems connected to the network. Thus, it is our aim to amplify current discussions on the Brazilian scenario. This work provides, first, an analysis of the origin and propagation of conducted and radiated electromagnetic interference in grid-connected photovoltaic systems, highlighting such concepts as symmetrical and asymmetrical noise and unintended antenna effects. Finally, the standardization framework is investigated based on the Brazilian scenario and current standards.

Keywords: Electromagnetic compatibility · Grid-connected photovoltaic systems · Electromagnetic interference · EMC standards

1 Introduction

Traditionally, there are concerns regarding electromagnetic compatibility (EMC) in the various types of photovoltaic power generation systems, given that connection of various

Y. Iano et al. (Eds.): BTSym 2020, SIST 233, pp. 687–694, 2021.
https://doi.org/10.1007/978-3-030-75680-2_76

items of electronic equipment to the network may give rise to disturbances not only in distribution lines, as seen in [1], but also electronic devices in the vicinity, as addressed in [2].

In recent years, the investigation of conducted and radiated electromagnetic interference (EMI) from photovoltaic power generation systems has become more active in Brazil [3]. According to the Brazilian Solar Photovoltaic Energy Association (ABSOLAR) [4], the cumulative operational capacity of distributed photovoltaic energy generation has recently reached 3GW, driven by the exponential growth of grid-connected installations in the last decade. Moreover, with the advent of smart cities, IoT (Internet of Things), and cloud computing, electromagnetic environments (EME) have become even more present, with the coexistence of a wide range of systems and wireless communication technologies [5].

In [6], it is stated that the EMC with grid-connected photovoltaic systems might be achieved based on an understanding of the EME. Attempts should be made to identify and characterize EMI sources, electrical equipment items that are susceptible to emitting or receiving EMI, and relevant non-electrical conditions from an EMC viewpoint, e.g., distance, humidity, and triboelectricity [7]. Such an understanding of the EME, together with the EMI requirements proposed in international standards, is normally necessary to ensure that systems are safe, correct, and reliable.

However, as suggested [8], the identification of possible violations may show a lack of EMC standards specific to the various items of equipment that comprise grid-connected photovoltaic systems, principally the photovoltaic (PV) inverter. Usually, such standards as CISPR 11, EN 55011, and EN 55022 are used for PV inverters, i.e., these devices are treated as household appliances [9] or as items of information technology equipment (ITE) in CISPR 22 [10].

Therefore, aiming to expand the current discussion regarding the Brazilian regulatory and standardization scenario, this work provides an overview of the EMI effect and investigates the scope of EMC's Brazilian regulation in grid-connected photovoltaic systems.

2 Electromagnetic Environment

As also observed in [6, 11, 12], Fig. 1 illustrates a PV inverter which, although essential to the installation, may give rise to EMI emissions since it operates in a non-linear manner based on switching and high-frequency clock signals [13, 14]. As a result, high *dv/dt* and *di/dt* occur, injecting conducted disturbances on both the DC and AC sides, comprising residential loads and the distribution network [1, 6, 9].

As indicated by [3, 11], the conducted EMI circulates via cables and elements, creating a parasitic reactance condition in the form of ground loops. According to [9, 15, 16], these loops could also make the grid-connected photovoltaic system operate as unwanted antennas in the EME, irradiating electromagnetic fields via radiofrequency waves (Fig. 1).

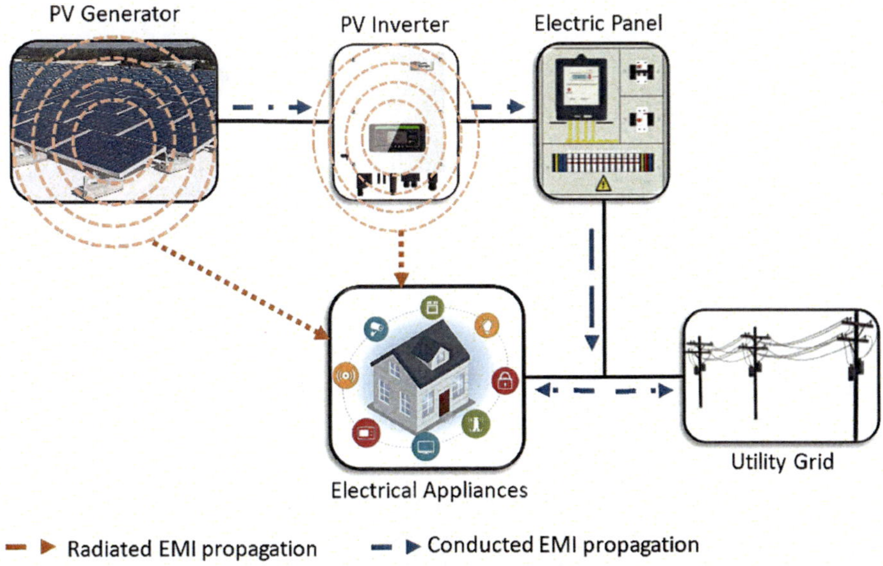

Fig. 1. Possible propagation paths for conducted and radiated EMI.

3 Electromagnetic Interference

Normally, EMI in the grid-connected photovoltaic system occurs in a conducted or radiated manner, such that propagation of one may generate the other, based on indirect emissions, as seen in [6, 17]. As observed by [8, 9], these disturbances are often divided into two types, according to frequency range: i) 150 kHz–30 MHz, in which conducted EMIs are normally addressed; and ii) 30 MHz–200 MHz, in which radiated EMI phenomena are constantly observed.

3.1 Symmetrical and Asymmetrical Conducted EMI

A conducted EMI analysis is normally made based on common and differential mode interference, as noted by [11]. According to [18], common-mode (CM) or asymmetrical interference originates from PV inverter switching processes, propagating oscillating voltage levels via DC cables and returns as ground currents from parasitic capacitance couplings. This is due to the fact that the cables behave electrically as undesirable inductors and the grounded modules as undesirable capacitors.

In general, the CM interference path is considered to be from the PV inverter to the load via circuit grounding. Furthermore, as indicated by [11, 17], there is an indication that the EMI couplings may occur with the photovoltaic solar modules interface in empirical proof processes. In this context, Fig. 2 illustrate the possible paths for CM current and floating electrostatic capacitance (C_{CM}) for a typical grid-connected photovoltaic system with a single-phase PV inverter.

Normally, as illustrated in Fig. 2, the transistors switches are high-frequency EMI sources due to their rapid switching process, generating CM voltages, represented as

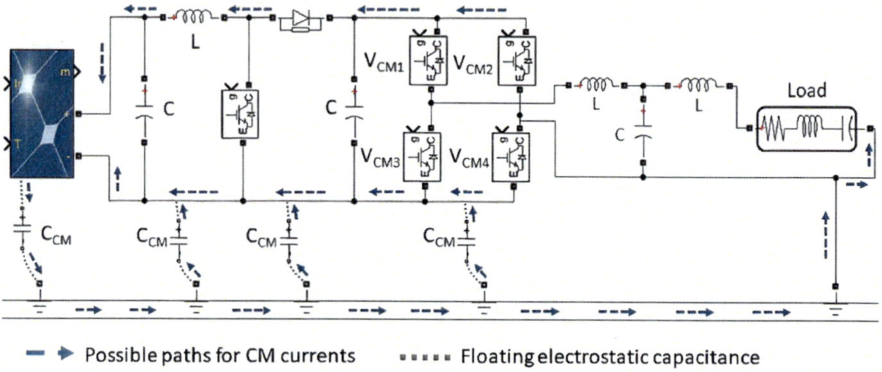

Fig. 2. CM currents possible paths for a typical grid-connected photovoltaic system with a single-phase PV inverter.

V_{CM}, [19]. Once the V_{CM} starts to flow through a parasitic ground, it may return by a capacitive coupling, generating various loop currents [16–19].

On the other hand, studies [3, 9–11, 15–19] indicate differential mode (DM) currents propagate in opposite, symmetrical directions, i.e., from the photovoltaic solar modules to the PV inverter (Fig. 3). The parasitic elements as series-equivalent inductance and resistance (X_{DM}), illustrated in the Fig. 3, are also sources of conducted EMI. As a result, the X_{DM} may provide EMI through DC cabling [20].

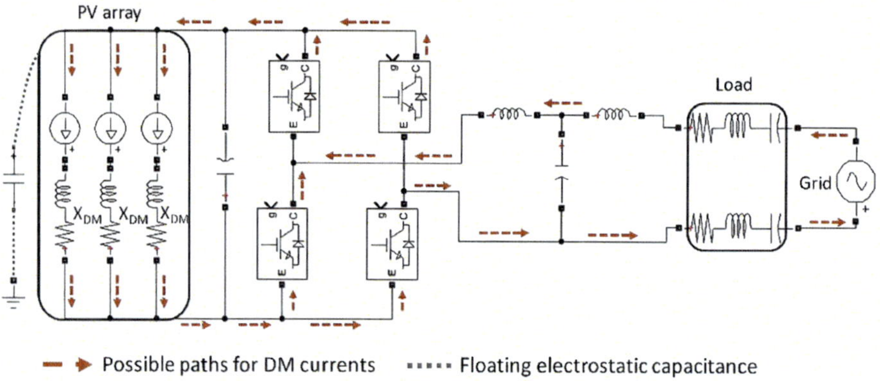

Fig. 3. DM currents paths propagation for a typical grid-connected photovoltaic system with a single-phase PV inverter.

Figure 3 illustrates the DM currents generated by photovoltaic solar modules that may flow through the AC side, propagating through the load and even to the grid [20]. However, as suggested [21], an EMI filter may filter the DM currents, traditionally dominant in high-frequency operations, if connected with a PV inverter.

3.2 Unintended Antenna Effects

Devices that unintentionally radiate electromagnetic energy are normally identified as active unintended antennas, e.g., cables conducting alternating current or openings/slots in the screens of devices that conduct radiofrequency (RF) currents [3]. On the other hand, devices that emit some type of EMI, in an unwanted manner, in the transmission and propagation of an electromagnetic signal are known as passive unintended antennas.

In the context of a grid-connected photovoltaic system, the modules are normally compared to active unintended antennas since the CM and DM emissions are generated with the interface of photovoltaic solar modules [15, 16]. Furthermore, as indicated in [17, 18], the DC side of a grid-connected photovoltaic system may operate as a resonant circuit, normally characterized in the EME as unintended inductors, and the photovoltaic solar modules as unintended capacitors, due to such parasitic elements as distributed longitudinal inductance and distributed transversal capacitance in DC cabling. Therefore, as proposed by [3, 18, 22], a cable-generator setup is highly desirable, as resonant frequencies generally reduce ground impedance on eliminating reactive impedances.

4 Brazilian EMC Standardization Framework

Although IEC standards regulate the production of photovoltaic equipment in Brazil, installers and consumers should be aware of additional Brazilian EMC standards. The National Telecommunications Agency (ANATEL) is responsible for mandatory regulation of telecommunications services in Brazil. Also, the Brazilian Technical Standards Association (ABNT) provides equipment certification based on international standards, aiming to unify Brazilian requirements. At the same time, research centers and institutions may conduct equipment tests and conformity certification, which will later be accredited by the National Institute of Metrology, Quality, and Technology (INMETRO), recognized as the Official Brazilian Accreditation Agency.

Over the last decade until mid-2017, ANATEL's standard 442 was used as one of the main references in conformance and EMC tests in order to regulate product certification based on IEC 61000-3-4, CISPR 11, and CISPR 22. In the context of grid-connected photovoltaic systems, ANATEL standard 442 was frequently used to measure the conducted EMI noise. Furthermore, for the measurement of radiated EMI noise, the CISPR 22 standard was used.

In 2018, ANATEL resolution 1120 was created to replace ANATEL standard 442. In addition, the ANATEL resolution 1120 establishes measurement and test method requirements for power supply equipment as sources, converters, inverters, and rectifiers, although it does not deal directly with the context of photovoltaic power generating systems.

Therefore, the main concept on which the Brazilian EMC standardization scenario is based assumes that ANATEL establishes the EMC requirements applicable to equipment subject to mandatory certification and in the absence of specific requirements. However, as indicated by [3], there are empirical cases in which photovoltaic power generating systems exceed the national limits imposed, which could be related to the fact that compulsory national standards do not specifically cover the EMC requirements and test methods for this kind of application.

Analogously, as indicated by [23], although INMETRO has recently submitted several quality tests and possible tests for conducted and radiated noise for photovoltaic inverters, there is no certification applicable for these items of equipment in its scope of accreditation. On the other hand, with regard to Brazilian voluntary certification, some standards have been considered by ABNT regarding EMC for photovoltaic equipment (Table 1).

Table 1. EMC standards applied in the certification of photovoltaic equipment in Brazil.

Standards	Description	Equipment covered	Tests performed
CISPR 22	Establishment of limits for radiofrequency disturbances as well as requirements for measurement methods and immunity of ITE equipment	Industrial, scientific, and medical (ISM) and ITE equipment	Conducted and radiated emissions
CISPR 11	Application of EMI limits for ISM equipment, requirements for photovoltaic semiconductor power converters (SPCs), and the introduction of a fully-anechoic room (FAR) for EMC measurements	ISM and photovoltaic PCEs	Radiated emissions; Conducted emissions input power
ANATEL 1120	Homologation of EMC certification based on requirements for EMI emissions, immunity, and resistibility. Application of test methods for converters, inverters, and rectifiers regarding EMC measurement	Telecommunication products	EMI conducted and radiated emissions; Immunity tests to electrical surges and noise; Resistibility measurement
IEC 62920	Provides classification and definition of PCEs and establish EMC requirements regarding emission limits and immunity for PCEs. Test conditions and measurement methods in relation to photovoltaic PCEs are also presented	Photovoltaic power conversion equipment (PCE)	Conducted and radiated emissions; Immunity tests to electrostatic discharge, surges, and transients; Direct/Indirect application of discharges to PCEs

5 Conclusions and Trends

This work provides an overview of the EMI effect and investigates the scope of the Brazilian regulation on EMC in grid-connected photovoltaic systems. As the standards that deal with EMC issues for the grid-connected photovoltaic system (Table 1) are voluntary, we can assume that the regulatory framework could be one of the causes that explains why there is still some irregular equipment sold in Brazil.

In this sense, although voluntary certification can add competitive value to a product, it generally increases its selling price, making the need for certification an unlikely proposal on the part of the supplier. Therefore, it is recommended that Brazilian standardization institutions, such as ANATEL, establish mandatory EMC requirements, especially with regard to the PV inverter.

In addition, it is reasonable to consider that this standard could also facilitate the implementation of technologies related to the concept of Smart Cities in Brazil since reducing the emission of EMI from grid-connected photovoltaic system installations would provide EMC, which can lead to improved cost-benefit for the EME in the long term.

References

1. Shadare E, Sadiku MNO, Musa SM (2017) Electromagnetic compatibility issues in critical smart grid infrastructure. In: IEEE electromagnetic compatibility magazine, vol 6, no 4, pp 63–70. Fourth Quarter. https://doi.org/10.1109/MEMC.0.8272283
2. Zhu B, Leskarac D, Lu J, Wishart M (2016) Electromagnetic interference investigation of solar PV system for microgrid structure. In: 2016 Asia-Pacific international symposium on electromagnetic compatibility (APEMC), Shenzhen, pp 456–459. https://doi.org/10.1109/APEMC.2016.7522767
3. Loschi HJ, Ferreira LAS, Nascimento DA, Cardoso PER, Carvalho SRM, Conte FD (2018) EMC evaluation of off-grid and grid-tied photovoltaic systems for the Brazilian scenario. J Clean Energy Technol 6(2):125–133
4. ABSOLAR: Geração Solar Distribuída Atinge Marca de 3 GW de Potência Instalada No Brasil, Public, June 2020. https://www.absolar.org.br/noticia/noticias-externas/geracao-solar-distribuida-atinge-marca-de-3-gw-de-potencia-instalada-no-brasil.html
5. Cormier K (2018) EMI intrudes on a wireless world. EE-Eval Eng 57(5):10–13
6. Jiraprasertwong J, Jettanasen C (2016) Electromagnetic interference in photovoltaic system and mitigation of conducted noise at DC side. In: 2016 IEEE region 10 conference (TENCON), Singapore, pp 915–920. https://doi.org/10.1109/TENCON.2016.7848138
7. Baxley RJ, Altman JT (2017) Efficient localization of transmitters within complex electromagnetic environments. US Patent n. 9,551,781, 24 January 2017
8. Stenumgaard P, Linder S (2019) Interference impact from solar-panel systems on air traffic control communications. In: 2019 international symposium on electromagnetic compatibility - EMC EUROPE, Barcelona, Spain, pp 1044–1048. https://doi.org/10.1109/EMCEurope.2019.8871512
9. Kubík Z, Skála J (2019) Electromagnetic interference from DC/DC converter of photovoltaic system. In: 2019 international conference on applied electronics (AE), Pilsen, Czech Republic, pp 1–4. https://doi.org/10.23919/AE.2019.8867026
10. Hegarty T (2018) An overview of conducted EMI specifications for power supplies. Texas Instruments white paper SLYY136
11. Miloudi M, Bendaoud A, Miloudi H (2017) Common and differential modes of conducted electromagnetic interference in switching power converters. Rev Roum Sci Technol–Électrotechn Et Énerg 62(3):246–251
12. Subramaniam U, Bhaskar SM, J.Almakhles D, Padmanaban S, Leonowicz Z (2019) Investigations on EMI mitigation techniques: intent to reduce grid-tied PV inverter common mode current and voltage. Energies 12:3395. https://doi.org/10.3390/en12173395
13. Loschi H, Lezynski P, Smolenski R, Nascimento D, Sleszynski W (2020) FPGA-based system for electromagnetic interference evaluation in random modulated DC/DC converters. Energies 13:2389. https://doi.org/10.3390/en13092389
14. Loschi H, Smolenski R, Lezynski P, Nascimento D, Demidova G (2020) Aggregated conducted electromagnetic interference generated by DC/DC converters with deterministic and random modulation. Energies 13:3698. https://doi.org/10.3390/en13143698
15. Zhang Y, Wang S, Chu Y (2019) Investigation of radiated electromagnetic interference for an isolated high-frequency DC–DC Power converter With power cables. IEEE Trans Power Electron 34(10):9632–9643. https://doi.org/10.1109/TPEL.2019.2892706
16. Prajapati M, See KY (2019) Extraction of equivalent noise source model from photovoltaic systems. IEEE Trans Electromagnet Compat 61(3):903–910.https://doi.org/10.1109/TEMC.2018.2836665
17. Di Piazza MC, Viola F, Vitale G (2018) Evaluation of ground currents in a PV system with high frequency modeling. Int J Renew Energy Res 8(3):1770–1778

18. Yu C, Xu H, Liu C, Wang Q, Zhang X (2019) Modeling and analysis of common-mode resonance in multi-parallel PV string inverters. IEEE Trans Energy Convers 34(1):446–454. https://doi.org/10.1109/TEC.2018.2877911
19. Vázquez N, Vázquez J, Váquero J, Hernández C, Vázquez E, Osorio R (2017) Integrating two stages as a common-mode transformerless photovoltaic converter. IEEE Trans Ind Electron 64(9):7498–7507. https://doi.org/10.1109/TIE.2017.2682014
20. Shuhao D et al (2016) Optimized hybrid modulation strategy for AC bypass transformerless single-phase photovoltaic inverters. J Power Electron 16(6):2129–2138. https://doi.org/10.6113/JPE.2016.16.6.2129
21. Liu Y, Lai CM (2018) LCL filter design with EMI noise consideration for grid-connected inverter. Energies 11:1646. https://doi.org/10.3390/en11071646
22. Bensaad D, Hadjadj A, Ales A, Valtchev S (2018) Identification of the common mode impedance of a microgrid DC-DC buck converter in normal service and under insulation fault. In: 2018 IEEE 18th international power electronics and motion control conference (PEMC), Budapest, pp 1079–1084. https://doi.org/10.1109/EPEPEMC.2018.8521856
23. Almeida Neto JCS (2017) Avaliação de conformidade de inversores para micro e mini geração fotovoltaica: a implantação da NBR 16150 e NBR IEC 62116. Dissertação de Mestrado, Instituto de Energia e Ambiente, Universidade de São Paulo, São Paulo. https://doi.org/10.11606/D.106.2017.tde-06122017-102326.Recuperadoem 2020-09-26, dewww.teses.usp.br

Steady-State Simulation of a Gas Turbine Operating with Sewage Sludge Biogas

M. Tvrzská de Gouvêa[1](✉), C. D. O. Maciel[1], J. P. Caly[1], and M. T. M. G. Rosa[2]

[1] Universidade Presbiteriana Mackenzie, Rua da Consolação 896, prédio 6, Escola de Engenharia, São Paulo, SP 01302-907, Brazil

[2] Centro de Ciências e Tecnologia, Universidade Presbiteriana Mackenzie, Av. Brasil 1220, Campinas, SP 13073-148, Brazil

Abstract. In this paper, a steady-state phenomenological rigorous model to simulate a gas turbine's operation that uses biogas from a sewage treatment plant is presented. The model makes use of isentropic efficiencies and operating curves. The model's parameters were obtained so that the model reproduces the nominal operating point of the gas turbine presented by França Junior [1] that uses biogas produced from sewage sludge. Simulations were further performed in order to analyze the effect of disturbances on the operation of the turbine. Perturbations in ambient air conditions and biogas feed were applied, and the simulator successfully solved the simulation problems. Although the gas turbine exhibited a nonlinear behavior, simulation results showed smooth responses. The simulator enables one to understand how each perturbing variable affects the operation of the turbine. Hence, the simulator can be used not only to predict the operations under different operating scenarios but also as a training tool for the operators of sewage treatment plants.

Keywords: Gas turbine · Sewage sludge biogas · Simulation

1 Introduction

Obtaining energy and electricity from renewable sources is a demand of the XXI century. Although Brazil is a country that traditionally uses renewable sources to produce energy, the kinds of sources are being changed because of environmental impacts and availability. In the period between 2016 and 2017, the contribution of hydropower plants has decreased by 2.9% in the energetic balance, and the usage of thermoelectric power plants has increased to 25.3% [2]. Still, 67.6% of the fuels used in Brazilian thermoelectric power plants are not from renewable sources despite the great biomass availability in Brazil. So, there is a need of increasing the usage of biomass as a fuel source. Particularly, there is an interest in using biogas produced in sewage treatment plants as a source of fuel. Contrary to what is observed in Germany, this biogas is burned in flares in the sewage wastewater treatment plants (SWTP) in Brazil [3]. Two main causes contribute to this waste of an energetic power source, namely, the high investment needed to build

Y. Iano et al. (Eds.): BTSym 2020, SIST 233, pp. 695–702, 2021.
https://doi.org/10.1007/978-3-030-75680-2_77

a power plant to produce energy from the biogas and the fear of operating a power plant that exhibits higher operational risks than a SWTP.

Among others, França Junior [1] showed the economical viability of producing energy in SWTPs using a gas turbine in a combined cycle. However, few pieces of information from the literature can be retrieved in regard to what affects the operation of a sewage sludge biogas power plant. Recently, simulation studies based on commercial simulators have been appearing [4, 5]. However, either constant efficiencies are used in the process model or confidential equations. In this paper, we present a steady-state model of a biogas gas turbine that can be used to explore the effects of changing operating variables and disturbances in the gas turbine's operation. We aim to contribute with pieces of information and an open simulator that may help to enable changes to the current scenario of wasting this important energy source, particularly in Brazil.

2 Description of the Mathematical Model

For modeling purposes, the sewage sludge biogas gas turbine is divided, as is the common practice, into compressors, combustion chamber, and turbine (see Fig. 1).

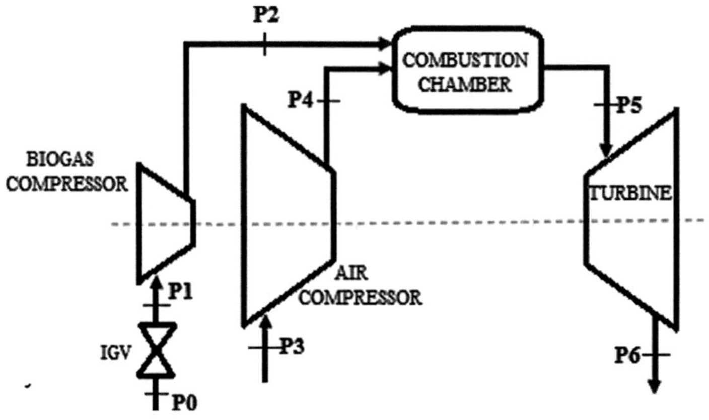

Fig. 1. Modeling components of the gas turbine.

Two compressors are considered in this work: the air compressor and the biogas compressor. The shafts of both compressors and the turbine are the same, so that, part of the mechanical work produced by the turbine is used to run the compressors in normal operation. The internal streams of the gas turbine and the inlet guide vane of the biogas compressor are also indicated in Fig. 1. The inlet guide vane is represented by IGV in Fig. 1.

The main modeling assumptions include: ideal gas behavior, steady-state operation, negligible friction loss in the combustion chamber, and kinetic and potential energies are considered much smaller than internal energies. Heat capacities of the ideal gas were taken from Koretsky [6] and enthalpy of formation from Himmelblau & Riggs [7]. Air properties, including relative humidity and water vapor pressure, are modeled accordingly to ASHRAE standards [8]. Moreover, the model presented here was adjusted to meet the heat and mass balances of the nominal point of the gas turbine considered by França Junior [1].

The molar mass balances for each specimen present in the compressors, combustion chamber, and turbine are given in Eq. (1). The chemical compounds considered in this paper are presented in Table 1, where the compositions of dry air and biogas are also shown. The latter are those considered by França Junior [1], and the former ones were taken from ASHRAE [8]. The molar flow rates of each stream k, $\dot{n}_k$, in Eq. (1), can be related to the mass balances by the molecular weight of the stream. In Eq. (1), $x_{k,i}$ represents the molar fraction of ith compound in k^{th} stream, $\alpha_{j,i}$ is the stoichiometric coefficient of specimen i in the jth reaction. Positive values are considered for the products, and negative values for the reactants. The reaction rate is given by $\dot{n}_{R,j}$.

$$0 = \left(\dot{n}_1 x_{1,i} + \dot{n}_3 x_{3,i}\right) - \dot{n}_5 x_{5,i} - \sum_{j=1}^{3} \alpha_{j,i} \dot{n}_{R,j} \tag{1}$$

Table 1. Composition of the inlet streams of the gas turbine.

Chemical species	Molecular weight (g/mol)	Molar fraction of each chemical compound	
		Biogas	Dry air
Oxygen (O_2)	31.999	0.0089	0.2094350
Nitrogen (N_2)	28.013	0.1320	0.7808180
Nitrous oxide (N_2O)	44.013	0	0.0000003
Methane (CH_4)	16.042	0.8080	0.0000052
Carbon dioxide (CO_2)	44.010	0.0407	0.0003850
Carbon monoxide (CO)	28.010	0	0.0000002
Hydrogen (H_2)	2.0160	0	0.0000005
Argon (Ar)	39.948	0	0.0093320
Hydrogen sulfide (H_2S)	34.100	0.0006	0
Water (H_2O)	18.015	0.0098	0

The combustion reactions considered in this paper are the following ones:

$$CH_4 + 2O_2 \rightarrow CO_2 + 2H_2O$$

$$H_2S + \frac{3}{2}O_2 \rightarrow H_2O + SO_2$$

$$H_2 + \frac{1}{2}O_2 \rightarrow H_2O$$

A conversion model was considered to model the rate of reactions. Hence, the reaction rate of each reaction is given by Eq. (2), where X_i the conversion of reactant i. Since

França Junior [1] presented no composition data for the exit stream of the gas turbine, we assumed that combustion reactions of H_2S and H_2 ran to completion. At the nominal point, the conversion of methane was assumed to be 97.68%. That value was obtained assuming adiabatic operation of the combustion chamber and adjusting the model to meet the exit temperature of the combustion chamber presented by França Junior [1].

$$\dot{n}_{R,i} = X_i\left(\dot{n}_1 x_{1,i} + \dot{n}_3 x_{3,i}\right) \ ; \ i = CH_4, H_sS, H_2 \tag{2}$$

The energy balance for each jth machine is given by Eq. (3). The parameter β_j is 1 for the compressors and −1 for the turbine. The following values for the product of the global heat exchange coefficient with heat exchange area $\left(U_jA_j\right)$ were considered, respectively, for the air compressor, biogas compressor, and turbine: 33.54W/m^{2}K, 293.6 W/m^{2}K, and 916 W/m^{2}K. Those values were obtained by adjusting our model to the heat and mass balances provided by França Junior [1]. Equation (3) can also be used to obtain the reversible work. For that purpose, the machine work $\dot{W}_j$ is exchanged with the reversible work $\dot{W}_{R,j}$, and the exit specific enthalpy h_{i+2} is replaced with the exit specific enthalpy h^*_{i+1} of the reversible process. The ith mass rate entering the machine is given by $\dot{m}_i$, T_i and T_{i+1} are the entry and exit temperatures of the machine and T_∞ is the ambient temperature.

$$0 = \dot{m}_i(h_i - h_{i+1}) + \beta_j \dot{W}_j - U_jA_j\left(\frac{T_i + T_{i+1}}{2} - T_\infty\right) \tag{3}$$

In order to characterize the reversible exit states of the machines, the entropy balance considering the reversible operation for each machine is used as given by Eq. (4), where s_i is the specific entropy of the ith stream and s^*_{i+1} is the specific entropy at the exit stream $i+1$ of each machine operating at maximum efficiency, i.e., performing a reversible work. In Eq. (4), $\dot{Q}_j$ represents the heat exchange with external air of the jth machine and is equal to $U_jA_j\left(\frac{T_i+T_{i+1}}{2} - T_\infty\right)$, which appears in the energy balance given by Eq. (3).

$$0 = \dot{m}_i\left(s_i - s^*_{i+1}\right) - \frac{\dot{Q}_j}{T_\infty} \tag{4}$$

The real and the reversible works are related through the isentropic efficiency η_j of the jth machine as generically shown in Eq. (5). The parameters a and b are +1 and −1 for the compressors and −1 and +1 for the turbine.

$$\eta_j = \dot{W}^a_{R,j} \dot{W}^b_j \tag{5}$$

Compressor isentropic efficiencies are usually related to the inlet volumetric flow rate by a quadratic model. In this work, the biogas isentropic efficiency was considered to be a constant value equal to 0.761, and the air compressor isentropic efficiency was modeled accordingly to Eq. (6), where the nominal inlet mass flowrate $\dot{m}_{3,nom}$ into the compressor is 12.71 kg/s. Turbine isentropic efficiency models usually take into account not only the flow rate but also the shaft speed. In this work, a constant value of 0.632 was assumed for the gas turbine. This was done due to the lack of information regarding the turbine in França Junior's work [1].

$$\eta_{air} = 0.5237 - 0.55\left(\frac{\dot{m}_3}{\dot{m}_{3,nom}}\right) - 0.28\left(\frac{\dot{m}_3}{\dot{m}_{3,nom}}\right)^2 \tag{6}$$

Phenomenological compressor flow rate models are usually quite complex [9], and, therefore, the most common approach is to use operating curves provided by the manufacturers, which typically present how the polytropic head is related to the volumetric inlet flowrate [10]. The polytropic head curve can be transformed into the difference between the exit specific enthalpy of the reversible compression h^*_{i+1} and the inlet specific enthalpy h_i. This is the approach used in this work. The generic operating curve of both compressors is presented in Eq. (7).

$$h^*_{i+1} - h_i = h_m\left(230 - 0.012\dot{m}_c^4 + 0.53\dot{m}_c^3 - 9.5\dot{m}_c^2 + 67\dot{m}_c\right) \tag{7}$$

For the air compressor, $h_m = 1$ and $\dot{m}_c$ is the mass flow rate through the compressor. For the biogas compressor, $h_m = 0.8$, and $\dot{m}_c$ is given by Eq. (8).

$$\dot{m}_c = 6.355 + \frac{14.9}{0.405}(\dot{m}_1 - 0.135) \tag{8}$$

It is a common practice to relate the flow rate through the turbine with the pressure difference [9, 11]. The Stodola equation is largely used for that purpose and is presented in Eq. (9) as used in this work. All variables in Eq. (9) are in International System Units. In Eq. (9), ρ_5 is the mass density of stream 5, P_5 and P_6 are, respectively, the pressures of streams 5 and 6.

$$\dot{m}_5 = 1.667\sqrt{\frac{\rho_5\left(P_5^2 - P_6^2\right)}{P_5}} \tag{9}$$

The IGV of the biogas compressor was modeled as a valve and was modeled accordingly to Eq. (10), where all variables are in International System Units.

$$\dot{m}_5 = 1.973\sqrt{\rho_o(P_o - P_1)} \tag{10}$$

3 Simulation Results

The gas turbine's nominal operating point considered here is the one considered by França Junior [1] and is presented in Table 2. It is characterized by the values of temperature, pressure, and mass flowrate for each stream. Ambient air is fed into the compressor, and the air to biogas mass ratio is considered to be 47.07 at the nominal operating point. Besides those specifications, dry air was considered to enter the gas turbine [1]. In Table 2, we also present the differences between consecutive streams enthalpies evaluated by França Junior [1] and by our model. The relative errors are very small, showing that the model presented in this paper is able to reproduce the nominal operation of the gas turbine presented by França Junior [1]. It is noteworthy to say that the proposed model's parameters were adjusted so that the temperatures and pressures of the nominal operating point of the model exactly met the data from França Junior [1]. The model was simulated using Matlab® R2006b. For solving the system of nonlinear algebraic equations, the SQP optimization method was selected using *fmincon* function. A constant objective function was used for that purpose. The optimization problem was formulated to have 12 decision variables.

Table 2. Heat and mass balance data for the nominal operating point of the gas turbine.

Stream	$\dot{m}$ (kg/s)	P (bar)	T (K)	$\Delta\hat{h}_{\text{mod}}$ (kJ/kg)	$\Delta\hat{h}_{FJ}$ (kJ/kg)	Percentage error (%)
P0	0.27	1.013	308	579.01	570.09	2
P2	0.27	15.00	571			
P3	12.71	1.013	298	399.96	399.84	0.03
P4	12.71	15.00	682			
P5	12.98	15.00	1272	−568.03	−580.55	2
P6	12.98	1.059	789			

In Table 3, the heat losses for each machine are presented, as well as each machine's work. Hence, the net power produced by the gas turbine in this paper is 1054kW, while the gas turbine from França Junior [1] produced 1063 kW. So the prediction error is quite small (0.8%) and, thus, we can consider that the proposed model reproduces quite well the nominal operating point presented in [1].

Table 3. Heat losses and machine work at nominal operating points.

Machine	Heat loss (kW)	Power (kW)
Air compressor	137.2	5221
Biogas compressor	13.84	179
Turbine	91.93	6454

In order to evaluate the behavior of the gas turbine, some disturbances were applied. To simulate the effect of disturbances, it was assumed that the heat exchange coefficients used to predict heat losses remain constant. It is known that inlet temperatures, the humidity of the air, and the amount of methane affect the operation of the gas turbine. Therefore, simulations were performed, where disturbances to each of those variables were applied. Simulation results are shown in Table 4. When ambient temperature decreases, the net power produced by the turbine increases. This occurs because the density of the air into the compressor increases, improving its efficiency and reducing the volumetric flow rate under constant mass rate operation. Increasing the humidity ratio above 10% causes the decrease of the temperature at the exit of the combustion changer, which lowers the net power produced by the turbine. Augmenting the humidity ratio up to 10% shows little effect on the operation of the gas turbine. When the amount of methane is increased in the biogas stream under constant air to biogas ratio, more net power was produced by the turbine as expected. This occurs because air is kept in excess. Note that the proposed model was able to deal with large variations in all the disturbance variables, showing a robust numerical behavior.

Table 4. Heat and mass balance data for the nominal operating point of the gas turbine.

Perturbed variable	Perturbation amplitude	T_5 (°C)	T_6 (°C)	P_4 (bar)	$\dot{W}_{t,net}$ (kW)
T_3	+8 °C	1283	795.1	14.74	1020
	+4 °C	1277	792.1	14.87	1041
	−4 °C	1266	785.9	15.13	1085
	−8 °C	1261	782.8	15.26	1108
Air relative humidity	10%	1271	788.6	14.97	1054
	20%	1270	788.1	14.95	1047
	50%	1266	786.78	14.88	1025
T_0	+4 °C	1272	789.18	15.00	1063
	−4 °C	1272	788.9	15.00	1062
$x_{0,\ CH4}$	0.02	1276	791.5	15.01	1089
	– 0.02	1268	786.4	14.98	1036

Observing the results presented in Table 4, we can see that the gas turbine exhibits a nonlinear behavior. Finally, it is noteworthy to say that the compressors were operated far from the surge condition.

4 Conclusions

Simulation results show that the proposed model reproduces the operating point of the gas turbine of a sewage treatment plant presented by França Junior [1]. The proposed model makes use of general compressor operating curves, which can be tuned to a specific compressor and the isentropic efficiency of the air compressor is dependent on the airflow rate. All equations were explicitly shown as used. The model was easily and rapidly solved using Matlab® R2006b. Moreover, the proposed model takes into account process disturbance like ambient conditions and was able to properly predict the effect of each disturbance applied to the process model. Hence, the model can be used to understand, plan, and optimize the operation of the biogas turbine. Although the gas turbine exhibited a nonlinear behavior, the responses were smooth. By simulating the gas turbine's behavior, it is possible to train operators, making them not fear operating a gas turbine in a sewage treatment plant.

References

1. França Junior AT (2008) Análise do aproveitamento energético do biogás produzido numa estação de tratamento de esgoto. Master thesis. Faculdade de Ilha Solteira da UNESP
2. EPE: Balanço Energético Nacional. https://epe.gov.br/pt/publicacoes-dados-abertos/publicacoes/balanco-energetico-nacional-2018
3. Moreira HC (2016) Iniciativas do Projeto Brasil – Alemanha de fomento ao aproveitamento energético do biogás no Brasil. In: 6th Encontro Nacional das Águas, ABCONSINDCON, São Paulo
4. Henry S, Baltrusaitis J, Luyben WL (2021) Dynamic simulation and control of a combustion turbine process for biogas derived methane. Comput Chem Eng 144:9. (4january – in press)
5. León E, Martín M (2016) Optimal production of power in a combined cycle from manure based biogas. Energy Convers Manage 114:89–99
6. Koretsky MD (2007) Termodinâmica para engenharia química, 1st edn. LTC, Rio de Janeiro
7. Himmelbalu DM, Riggs JB (2004) Basic principles and palculations in chemical engineering. Pearson
8. ASHRAE (2017) Handbook of fundamentals. ASHRAE Inc.
9. Thomas P (1999) Simulation of industrial process for control engineers. Butterworth-Heinemann
10. Stephenson G (2011) Integrate compressor performance maps into process simulation. Chem Eng Progress 107(6):42–47
11. Alobaid F, Starklaff R, Pfeiffer S, Karner K, Epple B, Kim HG (2015) A comparative study of different dynamic process simulation codes for combined cycle power plants – part A: part loads and off-design operation. Fuel 153:692–706

Parallel Adaline for Active Power Filter

Roger Chuquipiondo[1], Christian Flores Vega[2], and Julien Noel[2](✉)

[1] Universidad Tecnologica del Peru, Lima, Peru
rchuquipiondo@utp.edu.pe
[2] Universidad de Ingenieria y Tecnologia – UTEC, Lima, Peru
{cflores,cflores}@utec.edu.pe

Abstract. Due to the rise of electronics and electrical devices into the electrical grid, the network's power quality has become sensitive. Harmonics parameters in emerging power systems have to be controlled in order to give reliable power/energy to the grid. Adaptive power filter is one of the reliable solutions to compensate harmonics and the unwanted components of the power system. For this effect, we have designed a control strategy for a balanced load in a three-phase four-wire system with an adaptive neural network algorithm. The results obtained for the current harmonic compensation demonstrate a reduction of the total current harmonic distortion below 5%. Those include the reduction in the neutral current and the equilibrium phase currents.

Keywords: Adaline · Harmonics · Power quality · Fryze theory · Adaptive power filter

1 Introduction

Nowadays, power systems distortion, faults, and harmonics affect the power quality of any systems and have a significant impact on smart grid and renewable energy production as sudden changes in energy production and consumption threaten the grid stability and security [1–5]. Harmonic distortion in power system voltage and current waveforms has increased rapidly because of the load that is electronically controlled. Many of the power system loads are dynamic in nature, which implies the time-varying amplitude of current waveforms. In order to provide quality to the delivered power, harmonics parameters such as magnitude and phase are essential to design filters to reduce or eliminate the effects of the harmonics in a power system [6, 7].

Active power filter (APF), a static power converter that generates currents or voltages to compensate the unwanted components of the power system, is a solution to eliminate harmonics, as it does have advantages over passive filters. In AC distribution systems, APF improves electric power quality through harmonic mitigation, correction of voltage sags and swell, and balancing the unbalanced line currents. According to the IEEE 519 standard [8], it is suggested that harmonics estimation should be adaptive in nature and has to be robust against noise and transients. In this work, we propose the design of a control strategy to obtain the reference current in a three-phase four-wire system based on the development of an adaptive neural network in Matlab/Simulink. This work shows

Y. Iano et al. (Eds.): BTSym 2020, SIST 233, pp. 703–712, 2021.
https://doi.org/10.1007/978-3-030-75680-2_78

improvement in comparison to [9] in term of the PWM and APF control design, including the Adaptive Neural Network that estimates the reference compensation currents; to [10] with a reduction of the Total Harmonic Distortion (THD); to [11] in improving the ADALINE control system and the THD of the 3-phase 4 wire simulated system; and to [12] in the compensation by extracting harmonic currents and realize the selective compensation of the APF. To estimate the voltage and current signals, an adaptive neural system algorithm with a Combined Least Mean Square (CLMS) was proposed to update the weight vector and obtain the reference current. A three-phase electrical system in star connection was evaluated to test the ability of the designed adaptive control. The APF simulation results obtained demonstrate the reduction of the total current harmonic distortion below 5%, including the reduction in the neutral current and the equilibrium phase currents.

2 Design of the APF System

With both IEEE standards [8, 13], the concept of power quality has generated plenty of interest as electromagnetic phenomena affect power systems and power quality due to the problems of the load's instability into the network.

Harmonics distortions in power systems have been reduced with passive and active compensation. Passive compensation does not seek to eliminate the harmonic currents of the system but adapt them to the electrical system elements so that equipment and tools are less sensitive to the presence of harmonics. Meanwhile, active compensation uses systems such as transformers, passive or active filters to suppress or counteract harmonics. APF such as shunt active power filter, series active power filter, hybrid filters, and unified power quality conditioner have been used, studied, and presented widely as a solution to reduce and eliminate harmonics from the electrical network.

Different technical tools for the analysis of harmonics have been used to detect, estimate, and eliminate the presence of noise in distorted signals. For example, shunt APF has been used to assure the current control in the AC side of the APF and voltage or current control in the DC side of the APF [14–16]. As systems become more complex, different approaches involve a combination of subsystems for harmonics estimations and assure the removal of those [10, 12, 17–21].

To evaluate the performance of the APF, three types of loads can be defined: (1) non-linear balanced load, (2) linear unbalanced load, and (3) non-linear unbalanced load. Various methods are used to calculate non-active load, such as instantaneous reactive power compensation (PQ), DFT, SRF, Fryze [22], Hosseini, etc. The design implemented here has considered a three-phase ideal network where the non-linear balanced load consists of an uncontrolled three-phase bridge rectifier connected to every single phase with a series RL load where $R = 10\ \Omega$ and $L = 1$ mH (see Fig. 1).

Nonetheless, the APF performance depends on harmonic detection methods to generate current references, current control method, and dynamic characteristics of the converter circuit.

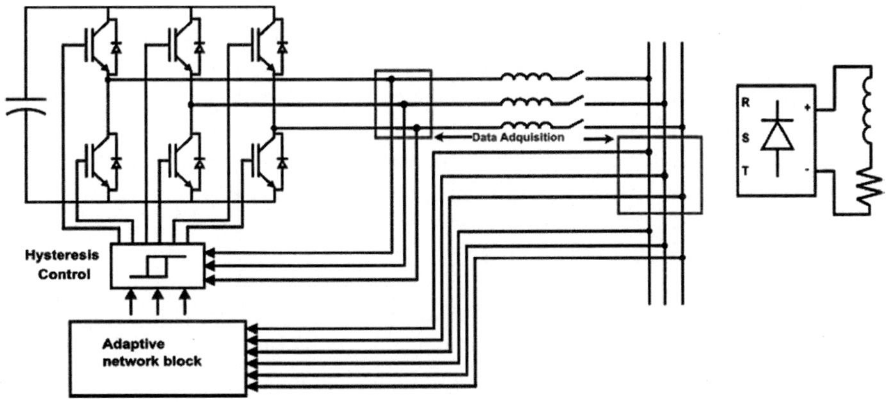

Fig. 1. Designed APF structure.

3 Harmonic Estimation Using Parallel Least Mean Square (LMS)

3.1 Adaptive Neural Network

A periodic discrete signal contains a fundamental angular frequency and N harmonics. Thus, this signal can be represented by Fourier Series:

$$y(n) = \sum_{n=1,\ldots,N} [A_n \cos(nw_r t) + B_n \sin(nw_r t)] \tag{1}$$

Where A_n and B_n are the amplitude of each N harmonics:

$$y(n) = W(n)^T x(n) \tag{2}$$

Where:

$$W(n) = [A_1 B_1 \ldots A_N B_N]^T \tag{3}$$

And:

$$x(n) = [\sin(w_r n), \cos(w_r n), \sin(2w_r n)\cos(2w_r n), \ldots \sin(Mw_r n), \cos(Mw_r n)]^T \tag{4}$$

The network topology and the weights update the adaptive neural network (see Fig. 2). The inputs x(n) of the ADALINE provide information to obtain the unknown parameters W(n). The input x(n) is multiplied with the weighting W[n], so the partial outputs are summed to estimate the signal $y_{est}(n)$. The adaptive neural network updates the weights W(n) in order to obtain the $y_{est}(n)$ for each time n. The error signal e(n) is the difference between the desired y(n) and the estimate $y_{est}(n)$ [23].

$$e(n) = y(n) - y_{est}(n) \tag{5}$$

$$e(n) = y(n) - \sum_{n=1,\ldots,N} [A_n \cos(nw_r t) + B_n \sin(nw_r t)] \tag{6}$$

$$e(n) = y(n) - (W(n))^T x(n) \quad (7)$$

After the initial random estimation recommended in [24], an ADALINE network uses an adaptive algorithm to update the weight vector [25] W(n), estimate $y_{est}(n)$ to the desired y(n). Several authors [26, 27] used conventional steepest descent method based weight up-gradation rule called modified Widrow–Hoff delta rule [27]:

$$w(n+1) = w(n) + \frac{ue(n)p(n)}{x^T(n)p(n)} \quad (8)$$

Where u is the learning parameter and p the input vector of dimension n. Traditional ADALINE networks are reported to be incompatible for real-time power system signal analysis [28]; therefore, we are using a convex combination of two adaptive filters in order to achieve faster convergence and self-synchronizing property [29].

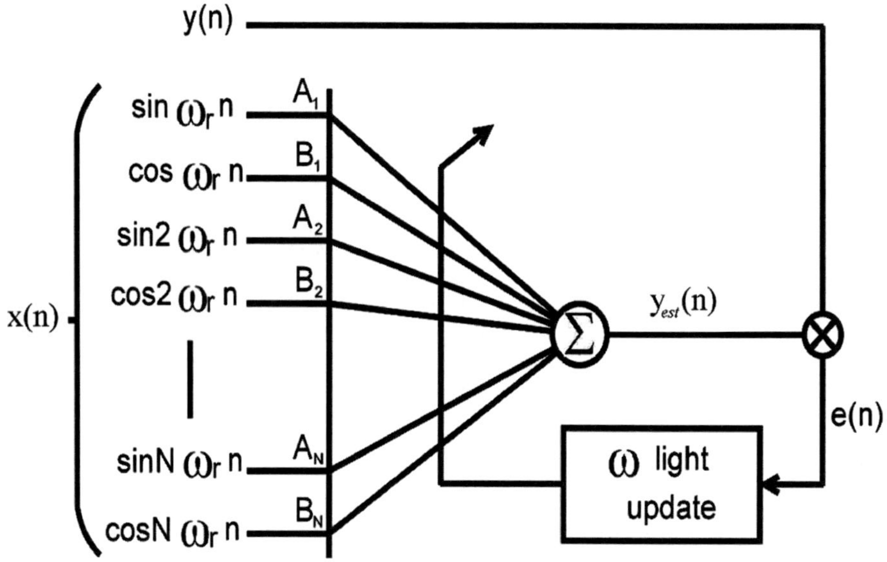

Fig. 2. Architecture of adaptive linear neuron.

3.2 Harmonic Estimation Using Combined Least Mean Square (CLMS)

This method is a combination of two adaptive LMS filters; one uses a high step size $\mu 1$ and the other for little adjustment $\mu 2$. Initially, both filters use their own adaptation rule.

$$w_i(n+1) = w_i(n) + \mu_i e_i(n)x(n), i = 1, 2 \quad (9)$$

Both filters are combined in a linear convex manner:

$$w(n) = \lambda(n)w_1(n) + (1 - \lambda(n))w_2(n) \quad (10)$$

The output of both filters is (see Fig. 3):

$$y(n) = \lambda(n)y_1(n) + (1 - \lambda(n))y_2(n) \tag{11}$$

Where λ(n) is a parameter that depends on the new variable a(n). This relation is defined as follows:

$$\lambda(n) = sgm[a(n)] = \frac{1}{1+e^{-a(n)}}, 0 < n(n) < 1 \tag{12}$$

The combination scheme can be seen in a new filter between w_1 and w_2 where the rule for adaptation of $a(n)$ is:

$$a(n+1) = a(n) - \frac{\mu_a}{2}\frac{\partial^2 e^2(n)}{\partial a(n)} \tag{13}$$

Otherwise, it can be written:

$$a(n+1) = a(n) + \mu_a e(n)[(w_1(n) - w_2(n))T]u(n)\lambda(n)[1 - \lambda(n)] \tag{14}$$

Where μ_a is the step-wise that tunes the update parameter of $a(n)$.

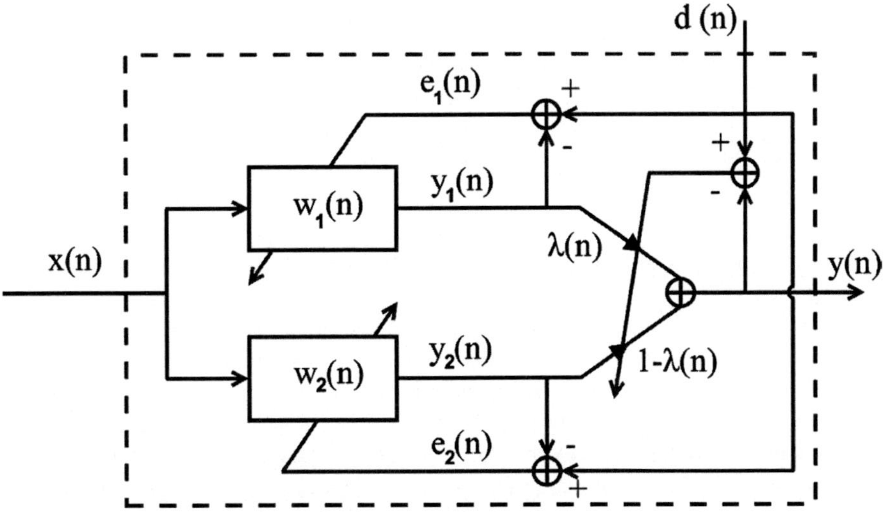

Fig. 3. Adaptive convex combination of two ADALINE.

3.3 Estimation of the Current Compensation for a Tree-Phase System

When estimating the current phase compensation for a one-phase system, this one arose using an adaptive neuron to obtain the current of the fundamental frequency as described in [9]:

$$i_s = i_1 + i_c \tag{15}$$

Where i_c is the current of the compensation signal, is the current of the power source, i_1 is the current of the fundamental where $i_1 = A_1 cos(wt) + B_1 sin(wt)$ where A_1 and B_1 are the weight on the neuronal component. Fryze multi-phase circuits respond to the currents, voltages instantaneous, and *RMS* values. Using the current from the power source, the active current, the reactive current, and the harmonic current; then, the difference between the power source and the active current obtained the compensation current [30]. Adapting the Fryze formula for a three-phase system with four wires, the active current for each balanced phase to reduce the neutral current and compensate the reactive power as demonstrated in [9] is:

$$ip = \frac{\sum_{i=L1,L2,L3} V_{i,11} V_{i,11} + V_{i,11} V_{i,11}}{\sum_{i=L1,L2,L3} V_{i;11}^2 + V_{i;12}^2} \left[V_{L1,11} cos(wt) + V_{L1,12} sin(wt) \right] \tag{16}$$

4 Analysis of the APF: Balanced Case

The balanced case configuration in Fig. 1 is composed of a three-phase uncontrolled rectifier with resistive-inductive load, where the APF is connected in parallel with the load to enable the harmonics reduction. Figure 4 displays the voltage and current waves that the source brings to Phase Uncontrolled Rectifiers. At a first analysis, it shows that the non-linear load distorts the current wave because it injects harmonics into the electrical network.

Before accomplishing the compensation, the APF must be able to estimate the compensated reference signal. This signal is obtained through the online training of the adaptive neural system. First, the adaptive neuronal estimation block estimates the voltage and current signal in the load, obtaining the values of the amplitudes of the corresponding fundamental signal. Next, the estimation block obtains the active current through the Fryze theory. The reference compensated signal is then obtained from the subtraction of the current in the load and the active current (see Fig. 5). After obtaining the reference compensation current, it is compared with the current compensated signal, and, by means of the hysteresis band current control, the compensated signal is then injected into the electric network, thus reducing the harmonic distortion generated by the non-linear load. The compensated current signal is then injected into the load.

The THD magnitude spectrum of the voltage and current wave before and after the APF compensation is resumed in Table 1. The control algorithm and the APF implemented show that the total harmonic distortion reduced considerably. In tension, the THD was reduced from 4.20 to 0.56; in current, the THD was reduced from 22.71 to 2.30 and is below the 5% range recommended by IEEE 519 standard.

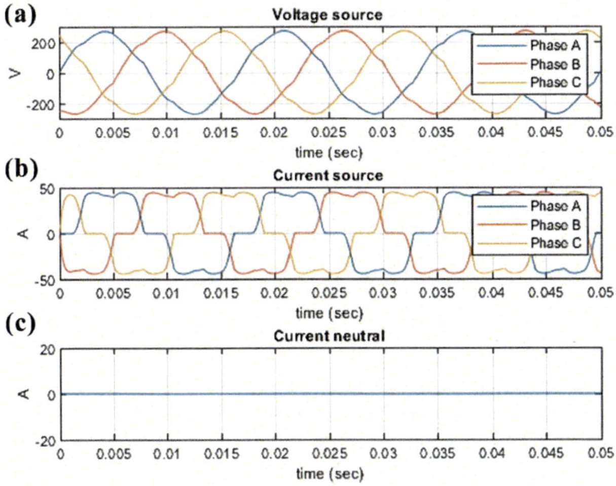

Fig. 4. a) voltage, b) current, and c) neutral signal from the source.

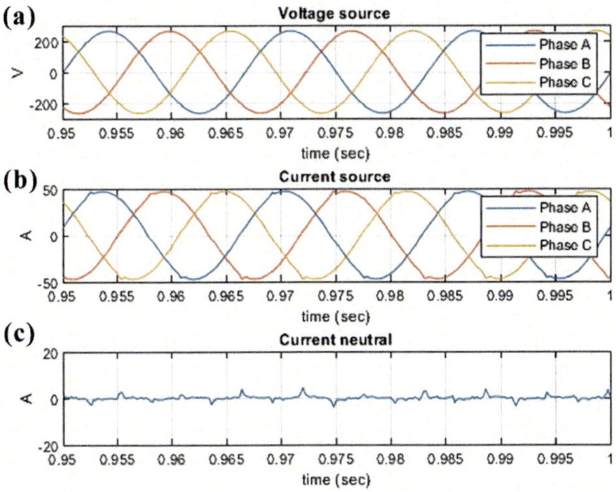

Fig. 5. a) estimated current in the load, b) active current, and c) compensated current.

Table 1. Total harmonics distortion before and after compensation.

Phase	THD (V)		THD (I)	
	Before	After	Before	After
A	4.21	0.54	22.71	2.24
B	4.18	0.56	22.71	2.30
C	4.21	0.56	22.71	2.29

5 Conclusion

In the last decade, the network's power quality has become more sensitive due to the effects of the harmonics of the electronics and electric components of the power systems. In order to give a reliable solution to those effects, with have implemented an adaptive power filter and its control strategy to a balanced load to a three-phase four-wire system.

The control of the proposed adaptive power filter based on neural networks (explained in Sects. 3 and 4) improved the delivered power quality due to the elimination of generated harmonics by the non-linear loads, meanwhile reducing the harmonic distortion, the reduction of the neutral current, and the equilibrium of the phase currents. The results obtained from the estimation of the reference current are based on the application of the LMS standardized combination algorithm since it presents better stability and faster convergence in the update of the neuron weights. The total current harmonic distortion demonstrated great results. In tension it was reduced from 4.20 to 0.56 and in current from 22.71 to 2.30, and is below the 5% range recommended by IEEE 519 standard meanwhile using the adaptive power filter.

References

1. Escudero R, Noel J, Elizondo J, Kirtley J (2017) Microgrid fault detection based on wavelet transformation and park's vector approach. Electr Power Syst Res 15(2):401–410
2. Espinoza F, Mar M, Ramirez E, Noel J (2016) Control of real power in asynchronous machine using a SCADA system in a smart grid. In: ANDESCON 2016 IEEE. IEEE, pp 1–4
3. Blancas J Noel J (2018) Short-term load forecasting using fuzzy logic. In: 2018 IEEE PES transmission distribution conference and exhibition Latin America (TD-LA), pp 1–5
4. Cardenas H, Zhang L, Noel J (2017) Improvement on energy trade capacity for asynchronous power system by application of hybrid multi-infeed direct current transmission system. In: 2017 IEEE URUCON, pp 1–4
5. Arias Velasquez RM, Romero Ramos YL, Meldrum I, Dıaz C, Noel J (2020) Dynamic model for transmission lines maximum disconnection time on wind farm. Ain Shams Eng J
6. Noel J (2014) Design of an 10 kva–45 a automatic static transfer switch module. In: Power Electronics, Machines and Drives (PEMD 2014). In: 7th IET international conference on, IET, pp 1–4
7. Cardenas H, Zhang L, Noel J (2018) Modeling, simulation and application of modular multilevel converter in hybrid high voltage direct current transmission system. In: 2018 IEEE PES transmission distribution conference and exhibition - Latin America (TD-LA), pp 1–5

8. IEEE Std 519-2014 (2014) IEEE Recommended Practice and Requirements for Harmonic Control in Electric Power Systems, pp 1–75
9. Vazquez J, Salmeron P (2003) Active power filter control using neural network technologies. IEEE Proc Electr Power Appl 150(2):139–145
10. Panda G, Ray PK, Puhan PS, Dash SK (2013) Novel schemes used for estimation of power system harmonics and their elimination in a three-phase distribution system. Int J Electr Power Energy Syst 53:842–856
11. El Shatshat R, Kazerani M, Salama M (2002) Power quality improvement in 3-phase 3-wire distribution systems using modular active power filter. Electr Power Syst Res 61(3):185–194
12. Merabet L, Saad S, Abdeslam DO, Omeiri A (2013) A comparative study of harmonic currents extraction by simulation and implementation. Int J Electr Power Energy Syst 53:507–514
13. IEEE draft recommended practice for monitoring electric power quality (2018) IEEE P1159/D5, December 2018, pp 1–109
14. Mingxuan D, Zhihao H, Wu J, Xu D, Ke H (2012) Study on three-leg-based three-phase four-wire shunt active power filter. In: proceedings of the 7th international power electronics and motion control conference, vol 1. IEEE, pp 552–556
15. Vardar K, Akpınar E, Surgevil T (2009) Evaluation of reference current extraction methods for DSP implementation in active power filters. Electr Power Syst Res 79(10):1342–1352
16. El-Mamlouk WM, Mostafa HE, El-Sharkawy MA (2011) Active power filter controller for harmonic suppression in industrial distribution system. Ain Shams Eng J 2(3–4):161–172
17. Sharma N, Gupta R, Sharma D, Swarnkar A, Gupta N (2010) Estimation and elimination of harmonics in power system using modified FFT with variable learning of ADALINE. In: proceedings of the 2010 electric power quality and supply reliability conference. IEEE, pp 79–86
18. Flieller D, Abdeslam DO, Wira P, Merckle J (2009) Distortions identification and compensation based on artificial neural networks using symmetrical components of the voltages and the currents. Electr Power Syst Res 79(7):1145–1154
19. Keshawala RP, Pandya SN, Patel SK, Suthar DP (2018) Comparative analysis of 3-phase 3-wire shunt active power filters using various control strategies. In: 2018 IEEMA engineer infinite conference (eTechNxT). IEEE, pp 1–6
20. Sejpal JA, Ayalani M (2018) Comparison of 3-phase 3-wireshunt active power filters through current control topologies. In: 2018 international conference on current trends towards converging technologies (ICCTCT). IEEE, pp 1–6
21. Chauhan SK, Shah MC, Tiwari RR, Tekwani P (2013) Analysis, design and digital implementation of a shunt active power filter with different schemes of reference current generation. IET Power Electron 7(3):627–639
22. Dadawala J, Shivani S, Kamani P (2014) Review on shunt active power filter for three phase four wire system. Int J Eng Dev Res 2:419–424
23. Li S, Wang J, Wang T, Du H, Duan Y, Gao H, Zhao B (2018) The bidirectional ac/dc power converter with capability of suppressing the harmonic current in hybrid micro grid. In: 2018 Asian conference on energy, power and transportation electrification (ACEPT). IEEE, pp 1–6
24. Czarnecki LS, Pearce SS (2010) CPC-based comparison of compensation goals in systems with nonsinusoidal voltages and currents. In: 2010 international school on nonsinusoidal currents and compensation. IEEE, pp 27–36
25. Yuan C, Yigang H, Zhong L, Jinping W (2015) Improved FBD reactive power and harmonic current detecting method based on voltage sequence decom-position. In: 2015 12th IEEE international conference on electronic measurement & instruments (ICEMI), vol 2. IEEE, pp 567–572
26. Dash PK, Panda SK, Mishra B, Swain DP (1997) Fast estimation of voltage and current phasors in power networks using an adaptive neural network. IEEE Trans Power Syst 12(4):1494–1499

27. Chang GW, Chen CI, Huang BC, Liang QW (2008) A comparative study of two weights updating approaches used in ADALINE for harmonics tracking. In: 2008 13th international conference on harmonics and quality of power, pp 1–5
28. Joorabian M, Mortazavi S, Khayyami A (2009) Harmonic estimation in a power system using a novel hybrid least squares-ADALINE algorithm. Electr Power Syst Res 79(1):107–116
29. Dash P, Panda S, Liew A, Mishra B, Jena R (1998) A new approach to monitoring electric power quality. Electr Power Syst Res 46(1):11–20
30. Thahab RT, Asumadu JA (2017) Current decomposition based on a double/multiple synochrouns reference frames and fryze-buchholz-depenbrock theory for a non-islanded microgrid with a finite control set-model predictive controller: a comparative approach. In: 2017 IEEE 3rd international future energy electronics conference and ECCE Asia (IFEEC 2017-ECCEAsia). IEEE, pp 356–363

Control and Monitoring System of Marine Aquarium

Renan Tadeu Baldini de Brito, Thyago Netto Baltazar, and Filipe Ieda Fazanaro(✉)

Federal University of ABC (UFABC), Santo André, São Paulo, Brazil
filipe.fazanaro@ufabc.edu.br

Abstract. The complexity and high maintenance cost of marine aquariums imply various problems to the aquarists. Many parameters as temperature, density, pH level, and lightness must be controlled to avoid abrupt changes, which could cause the death of the animals and/or the development of contagious diseases that condemn the entire tank. Therefore, it is necessary to measure such parameters and eventually adjust them, which could become a complex task due to the number of processes, equipment, and chemical kits needed. In this context, this work aims to develop a unified system to control and monitor the parameters of the marine aquariums and, moreover, employing an app to smartphones, which is capable of informing the actual state of each parameter to the aquarist, reducing the spent time to the maintenance of the aquarium, making the work easier to be done by the aquarist and ensuring a more stable environment.

Keywords: Control · Monitoring · Marine aquarium · Parameters · App

1 Introduction

1.1 Contextualization

Historically, evidence shows that the first objects that could be considered aquariums appeared in Ancient Egypt, but only during the 20th century the aquarium has become popular, especially due to the air transport diffusion which allowed the specimens to be transported from their original ecosystem to any other place [1, 2]. The marine aquarium is not only connected to leisure, but also to food stock, therapeutic treatments and environmental education incentive [3, 4]. On the other hand, marine aquariums require to control multiple parameters, making its maintenance a laborious task. In this scenario, the automation of the marine aquarium's maintenance could be interesting, which could help the aquarist as it allows a better control of each parameter and integrates all the control process. In this way, one of the most recent advances was the Arduino platform and its subsequent ESP32 [5].

Y. Iano et al. (Eds.): BTSym 2020, SIST 233, pp. 713–722, 2021.
https://doi.org/10.1007/978-3-030-75680-2_79

1.2 The Marine Aquarium and the Automation Process Considerations

Commonly, the marine aquariums can be classified by its capacity - in liters - and the kind of specimens contained. For example, a 70 L "reef only" aquarium designates a tank with 70 L of capacity and contains only corals, rocks, little crustacean, and some other specimens according to the interest of the aquarist, without any specimen of fish. Another example is the classification "fish only", which designates an aquarium without corals, containing only fishes, crustaceans, rocks, and other little animals [6].

Considering the several parameters associated to a marine aquarium, the water is, obviously, the most important as it can be obtained by natural sources or be prepared by different means. For instance, the producers of marine synthetic salt determine the use of 30 to 35 g of salt per liter to obtain the ideal density of 1,024 kg/L, which can be measured with a densimeter or a refractometer. It is critically important to the maintenance of life in the marine aquarium that partial water changes occur every 30 days, changing approximately 20% of the aquarium's total capacity in liters or changes between 10 to 15% of the capacity if this procedure occurs every 15 days. These changes are responsible for reducing undesirable parameters, as ammonia, nitrites, organic matter, low oxygenation, and cyanobacteria occurrence [6], and assures the renovation of the water, helping the restoration of important parameters of the environment, as the magnesium and the calcium levels, the pH and the kH. Furthermore, the aquarium must contain the substrate, which is composed of sand, aragonite (milled shells), dolomite, Halimeda, and other compounds that contribute to the aquarium's maturation (the balance of the marine biology). The periodical water changes are usually sufficient to process the ammonia and nitrites produced after the maturation, but, before this stage, porous rocks are important to help this process [6–8].

Another important parameter is the aquarium's temperature, which must stay, typically, between 24 °C to 27 °C, due to the fact that if it is over 27 °C, the corals can be overheated, and bellow 24 °C, the fishes can perish. These values can be modified according to the specimens contained in the aquarium. Usually, the water warming is made by thermostats considering approximately 1 W per liter, and the cooling is made by forced ventilation. However, the aquarium's salinity and the humidity produce improper conditions for the fans' operation, drastically reducing its lifetime [6].

Another variable of great importance is the pH, which must be in the range between 8.0 and 8.4, ideally being equal to 8.2. As previously noted, these values can be adjusted according to the specimens within the aquarium. A correctly built aquarium will hardly present problems with the level of acidity as soon as all the elements used to build the environment conduces to a chemical buffered alkaline balance. There is still the possibility of adding buffers to help balance the pH, especially to aquariums in the cycling period [6, 9].

In addition to the previously discussed variables, one can consider the lighting of the aquarium, which provides good aesthetics and, more importantly, is associated with biological functions, being essential to the development of the corals and regulating the animals' biological cycle. For this reason, white and blue lighting sources are commonly used, which can work in cycles that simulate day and night, avoiding abrupt changes in light [6].

Finally, there is the process of water filtration, which occurs in the called "sump", a compartment that is usually located on the back or under the aquarium when it is considered, respectively, small and big aquariums. In this compartment, the water passes by a filter process with many steps, which can be summarized as follows: (1) filtering by bubbling through an equipment called "skimmer", (2) mechanic filtering through perlon (an acrylic wool that catches the macro impurities), (3) chemical filtering using active coal and (4) biologic filtering by proper ceramics means. Many aquarists also use additional filtering equipment as a UV sterilizer, aiming to achieve better results.

In this context, it is immediate to realize that even in the small ones, the maintenance of a marine aquarium is not a simple task, requiring a lot of time and dedication. Thereby, it is interesting to automate the system maintenance as the aquarium operation could be improved, reducing the effort needed and make the process easier. Therefore, this work aims to propose and develop an automatic system maintenance for marine aquariums based on an ESP32 Platform that controls the sensors (temperature, pH, infrared to the density, level), actuates through the Peltier chip and pumps, and shows the status of each measure to the aquarist on a mobile app, enabling full control of the aquarium on the aquarist's hands.

This work is organized as follows. Section 2 describes the methodology of the execution of the project. Section 3 presents the results obtained from the tests using the developed prototype and its functionalities. And, finally, in Sect. 4, conclusions related to the discussions presented here are made.

2 Methodology

This work's development was based on the ESP32 platform, especially because of its built-in WIFI module and the number of analogic inputs available. The ESP32 has a Tensilica Xtensa 32-bit LX6 dual-core microprocessor, which works with a typical clock of 160 MHz. The block diagram in Fig. 1 illustrates the automation system proposed and how the ESP32 controls the main sensors and modules used, which are described in the following section.

2.1 Density Sensor Module

The first module considered here is a device capable of measuring the water density and translating its measure to a digital signal. It was developed a system using a floating densimeter installed inside a PVC pipe which was connected directly to the aquarium in a way that was possible to obtain a water column at the same level of the tank and, also, sampling the water accurately. This module is illustrated in Fig. 2 where it can be seen that at the top end of the densimeter, it was attached a plastic resistant component, as slight as possible, with its diameter approximately 1mm smaller than the intern pipe diameter. This pipe was sealed with a cover of the same material. Initially, an ultrasonic sensor was attached to this cover. However, after initial tests, the measures obtained were not accurate and, because of this, this sensor was replaced by an infrared one.

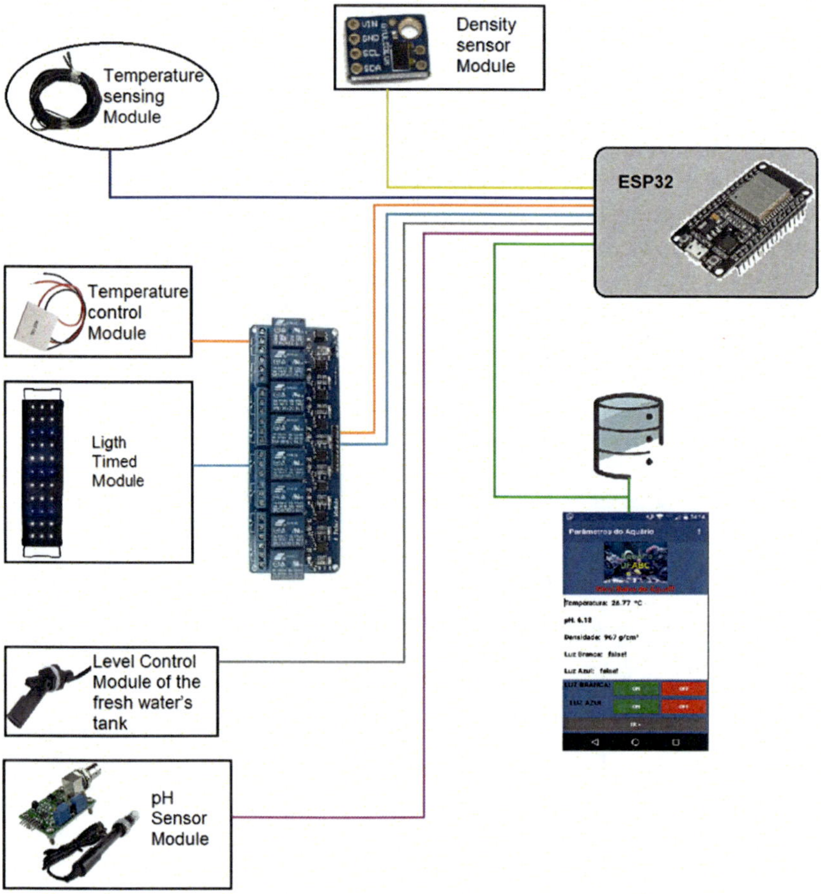

Fig. 1. Project blocks diagram.

2.2 Temperature Sensing and Control Module

To measure the temperature, it was used a 10k NTC sensor [10], which is a thermal sensitive sensor with a negative temperature coefficient. In other words, its resistance decreases with the increase of the temperature. It was developed a library based on its documentation to realize the measure of the temperature following the Steinhart-Hart equation [11], shown in Eq. (1), where T is the temperature, A, B and C, coefficients obtained by the Steinhart-Hart matrix equation, presented in Eq. (2), and R is the resistance measured by the NTC on that temperature. On the Steinhart-Hart matrix equation, R1, R2, and R3 are resistances obtained experimentally on three different operating points and T1, T2, and T3 the respective temperatures.

$$\frac{1}{T} = A + B + lnR + Cln^3R \quad (1)$$

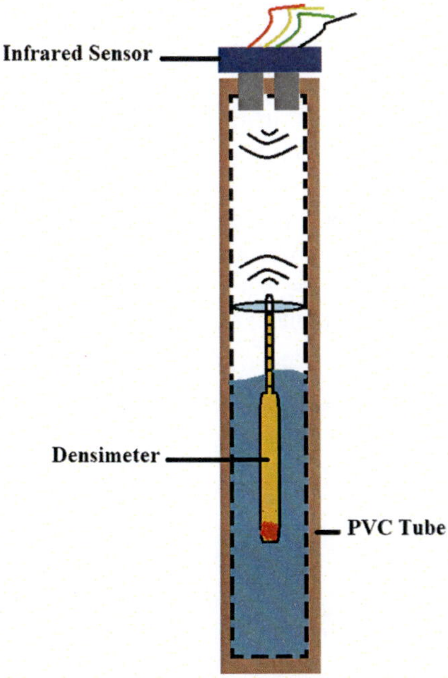

Fig. 2. Assembly schema of the density sensor.

$$\begin{bmatrix} 1, \ ln\, R_1, \ ln^3\, R_1 \\ 1, \ ln\, R_2, \ ln^3\, R_2 \\ 1, \ ln\, R_3, \ ln^3\, R_3 \end{bmatrix} \begin{bmatrix} A \\ B \\ C \end{bmatrix} = \begin{bmatrix} \frac{1}{T_1} \\ \frac{1}{T_2} \\ \frac{1}{T_3} \end{bmatrix} \tag{2}$$

In Table 1, the chosen three points of temperature and the respective resistances measured are shown.

Table 1. Operating points for the calculation of Steinhart-Hart coefficients.

T(ºC)	R (KΩ)
9,2	19,11
17,8	13,42
59,2	3,30
T (K)	R (Ω)
282,35	1911
290,95	1342
332,35	3300

Based on the values presented in Table 1, after solving the matrix in Eq. (2) [11], the coefficients A, B, and C got the values: 2.8386514 × 10-4, 3.48586482 × 10-4, and – 1.86357303 × 10-7, respectively.

The developed code processes the thermistor resistance measurement considering a fixed reference resistor, calculates the logarithm of this value, and obtains the temperature employing the Steinhart-Hart equation. Furthermore, the code implements the temperature system's control (a relay module, two water pumps, and a Peltier module installed between heat sinks) to maintain the temperature between the limits chosen by the user. Basically, the system leads the water through the appropriate heat sink, cooling or warming it before leading back to the tank.

2.3 Level Control Module of the Fresh Water Tank

It was employed a level sensor, a magnetic float type proper to Arduino [12], that activates a red LED when the water level inside the tank is low. Normally, this level is approximately equal to 15% of the tank.

2.4 Timed Functions

Some functionalities must be controlled considering the time, e.g., an action must be done every day at 11 P.M. or change the light's color at 6 A.M. and, again, at 6 P.M. To do so, it was used the NTP (Network Time Protocol), that updates the local date and time through the webserver.

2.5 pH Sensor Module

The pH reading is carried out by the PH-4502C sensor of "DiyMore" [13]. It is an analogic sensor with supply voltage of 5 V and working current between 5 and 10 mA (this value varies according to the pH level). The range of detection covers the entire pH scale, i.e., it can measure pH variations from 0 to 14. The reading electrode is connected, through a BNC connector, to a module with a proper circuit. This circuit has the function to amplify and linearize the sensor's output because the pH follows a logarithmic scale.

The calibration of the pH sensor requires the use of three different aqueous solutions: an acid, a neutral, and an alkaline.

The initial measurements of the sensor had a significant oscillation behavior. The manufacturer informs that the reading stabilization time is approximately equal to 2 min after the sensor is immersed in the solution. However, even after this period, the sensor's output voltage shows a periodic oscillation of approximately 0.3 V. This oscillation ("ripple voltage") was possibly caused by the 5 V power supply used in the prototype. Therefore, the solution adopted uses a digital filter with Gaussian average. The use of the moving average causes a delay of approximately 60 s to stabilize the pH reading. This would imply in a delay of the same magnitude associated to control this parameter. Another possible solution would be based on the use of a capacitive filter to correct the ripple voltage.

This filtering procedure can cause a delay, which is related to the sampling. In this work, it was executed 100 readings for each value, resulting in a delay of approximately 60 s. Fortunately, this value does not affect the overall behavior of the system.

2.6 The Smartphone Application

The Firebase platform, from Google, was used to implement the smartphone application. The Firebase is a digital tool created to assist the development of web and mobile applications. Its real-time database employs a proper communication protocol and allows the storage and the synchronization between the user and the device through a NoSQL Data Bank hosted on the cloud [14].

To use the Firebase, initially, it was needed to create a Google account and access the console through the "Spark Plan" package. It was then created a project called "Aquario UFABC" which was necessary to generate an access key and a host address to write and read data.

The updated data were synchronized through all of the connected devices in seconds. During the development of the application for the aquarium automation, it was detected a delay of approximately 8 s. Besides that, the data remains available even if the application is offline, being updated immediately when the re-connection with the network occurs.

The communication between the Firebase and the ESP32 can be realized by installing a specific library called "FirebaseESP32.h" and then the definition of the variables "host" and "Senha Fire".

The platform chosen to the development of the application was the Kodular [15]. This online tool commonly used for the development of Android applications allows the programming to be realized simply and intuitively through logic blocks. The Kodular offers a large variety of functions allowing the creation of many applications, making it a very didactic and interesting tool. Moreover, the Kodular has specifics functionalities that allow the flow of data with the Firebase, which is advantageous due to the use of this database on this project. Kodular offers two creation aspects, i.e., the application could be developed using a programming blocks interface or a graphic interface.

3 Results

The project's development was based on modular aspects, i.e. each functionality was projected, developed, and tested, aiming to verify if it is "running" as expected. This enables better analysis of each module and simplifies the correction of eventual bugs.

The first tests were analyzed by the serial plotter of the Arduino IDE. The density module was tested by printing the variable "Distance" that was the distance captured by the infrared sensor in centimeters and the variable "Densidade" that shows the calculation's result to convert the distance into density. The equation takes the relation between the distance (in cm) and a known value of density (1000, which means freshwater density) and makes an interpolation to present the values to other distances measured.

The temperature's module calculates the temperature based on the Stein-Hart equation, with the calculated coefficients. The initial tests were analyzed by the serial plotter, as realized during the density testes, and prints on the screen the values of voltage (measured on the NTC pin and calculated by the code for the fixed resistor used as reference) using a digital filter (following the same discussion associated to the pH sensor), the current and resistance calculated on the code and the temperature calculated using the resistance and the Stein-Hart equation.

The other modules use proper libraries associated with the sensor used and print the values on the screen. It was adopted that only the level sensor's module used a LED to show the sensor's state.

The application developed demonstrated full functionality and shown the results of the measures clearly. Moreover, the application has a direct connection with the ESP32 and the possibility of future improvements. The complete integrated prototype presented a delay approximately equal to 8 s, which is acceptable to the scenario. The screen of the application can be observed in Fig. 3 and the warning signals implemented can be seen in Fig. 4.

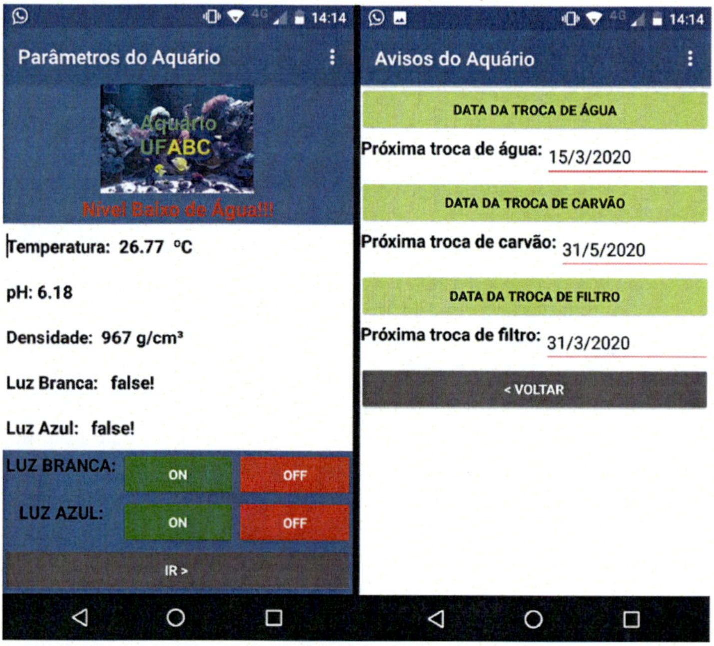

Fig. 3. App's screen.

The final code of the project and the related files are available on GitHub[1]. It was recorded a demonstration of the system working, which is available on YouTube[2].

[1] GitHub (short) link: https://cutt.ly/Df8FuJN.

[2] YouTube link: https://youtu.be/yhSx_I89C_8.

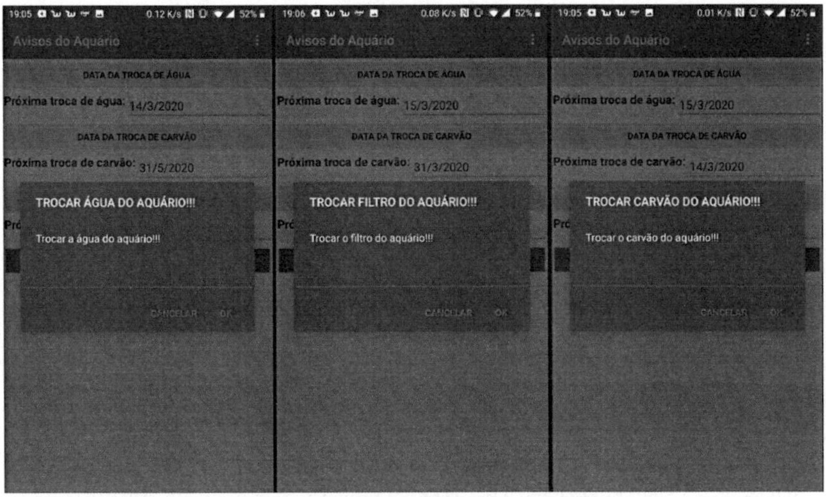

Fig. 4. Warning signs implemented.

4 Conclusions

The project of the Marine Aquarium Control and Monitoring System represents a solution for the development of artificial aquatic ecosystems, making it possible, through the modification of different parameters (temperature range, pH, salinity, etc.) through programming, to adapt the aquarium to the more different organisms. In this way, it is ideal for different simulated aquatic environments, making it feasible for applications ranging from hobbyists to researchers.

Among the main contributions of the project, the monitoring of the main reference parameters to maintain the aquarium's biology can be highlighted through centralization and automation through an IoT application; which makes the practice of fishkeeping accessible to anyone, regardless of their prior knowledge on this topic.

Like all technology, improvement is essential, always looking for the most modern. In this way, the developed project opens a horizon for many improvements, resulting in a plausible product for the related market.

A possible future improvement would be implementing a control system for some parameters that are currently only read by the developed module. For example, it could be possible to control the pH using acidic and alkaline solutions. These solutions would be in separate reservoirs, which would be dosed according to the desired pH range based on the pH readings over time. When this sensor indicates a value outside the desired referential, it would be automatically administered a solution in order to correct the pH value, always keeping it within a desirable range, generating an effect similar to the pH buffering. Therefore, implementing a closed-loop control system. The same principle could be applied to all other aquarium parameters, as is already done for the temperature parameter. Another possible improvement would be the development of a database, which stores the history of the values for all parameters of the aquarium over time, allowing the user to periodically obtain reports and graphs that could demonstrate the behavior of the aquarium variables over time.

References

1. Trevisol F, Mori MR, Da Silva MJD (2014) Sistema de controle e monitoração dos fatores determinísticos para o desenvolvimento de macrófitas aquáticas em aquários, Trabalho de Conclusão de Curso de graduação do Curso Superior de Tecnologia em Automação Industrial, Universidade Tecnológica Federal do Paraná, Curitiba
2. Ribeiro FAS, Lima MT, Fernandes CJBK (2010) Panorama do mercado de organismos aquáticos ornamentais. Boletim ABLimno
3. Aragão S (2011) A utilização de aquários e terrários como ferramenta de ensino: um olhar pelo viés da experimentação. Departamento de Ciências Biológicas, Universidade Estadual de Feira de Santana. https://www2.uefs.br/semic/upload/2011/2011xv-025tha053-220.pdf
4. Gomes A et al (2018) Aquário na sala de fisioterapia com crianças portadoras de necessidades especiais. In: 9º Salão Internacional de Ensino, Pesquisa e Extensão, vol 9, no 7
5. Monk S (2017) Programação com Arduino: começando com sketches, 2nd edn. Bookman, Porto Alegre
6. Guia de Montagem de um Aquário Marinho. https://www.aquariossobrinho.com/guia-montagem-aquario-marinho
7. O que é Rocha Viva? https://oaquashow.wordpress.com/reefs/o-que-e-rocha-viva
8. Como reduzir nitritos. https://www.comidadecorais.com/como-reduzir-nitritos
9. Como consertar e manter o pH do seu aquário. https://www.aquariossobrinho.com/post/consertarph
10. Cantherm, Pearl-Shaped Precision NTC Thermistor for Temperature Measurement. https://img.filipeflop.com/files/download/Datasheet_termistor_mf52_1.pdf
11. ILX Lightwave Corporation, Application Note: Thermistor Calibration and the Steinhart-Hart Equation. https://www.newport.com/medias/sys_master/images/images/h67/hc1/8797049487390/AN04-Thermistor-Calibration-and-Steinhart-Hart.pdf
12. Sensor de Nível de Água com Bóia Horizontal. https://www.usinainfo.com.br/sensor-de-nivel-arduino/sensor-de-nivel-de-agua-com-boia-horizontal-2580.html
13. How to use a PH probe and sensor. https://www.usinainfo.com.br/index.php?controller=attachment&id_attachment=553
14. Firebase. https://firebase.google.com/?hl=pt
15. Kodular. https://www.kodular.io

Comparative Analysis of Multiphase Flow in a T Type Micro Junction

Gabriel M. Guimarães(✉), Kevin L. Pinto, and Raquel J. Lobosco

Federal University of Rio de Janeiro, Rio de Janeiro, Brazil
rlobosco@macae.ufrj.br

Abstract. This scientific research investigates the behavior of two incompressible fluids in a micro-T-junction channel by means of a computational simulation through the physical properties of water and air and the discretization of conservation equations for the dispersed phases. It was possible to represent different regimes of droplets formation in the computational domain. OpenFOAM version 4.0.1 software is an open platform used in the academic environment for fluid dynamic simulations and was chosen to perform the flow's numerical simulations due to its great acceptance in the scientific environment . The InterFOAM solver using the VOF method (Volume Of Fluid) developed for computational simulations of multi-phase flows was adopted in the simulations studied here. This study develops a comparative analysis of the droplet's formation behavior at the injection time and evaluates the flow aspect of the dispersed phase through the continuous phase in the main channel. From the variations in the flow rates, it was possible to verify the flow phases' dependence in determining the flow regime. This research was developed based on a literature review and evaluates the bubble injection shape and droplet coalescence in a T-junction microchannel. The results were compared with the literature data.

Keywords: Multiphase · Micro-fluidic · OpenFoam

1 Introduction

The multi-phase flow behavior has several characteristics depending on the conditions in which they are applied. In observing the flow of two incompressible fluids, with a constant density, in a microchannel, different flow regimes can be predicted depending on the speed rate, as showed in Fig. 1.

T-shaped joints are widely used for industrial cases of injection of a fluid and are frequently used to study the formation of drops in microchannels [1]. In this configuration, the continuous phase flows through the main channel, and the dispersed phase is injected, causing the formation of the droplets.

In a multi-phase microchannel flow, the fluid transport regime is characterized by the number of capillarities, shear stress, flow rate, and viscosity rate [2]. Among the various profiles already observed for multi-phase flows in microchannels in type T joints, five stand out, of which they are considered main as described below.

Y. Iano et al. (Eds.): BTSym 2020, SIST 233, pp. 723–731, 2021.
https://doi.org/10.1007/978-3-030-75680-2_80

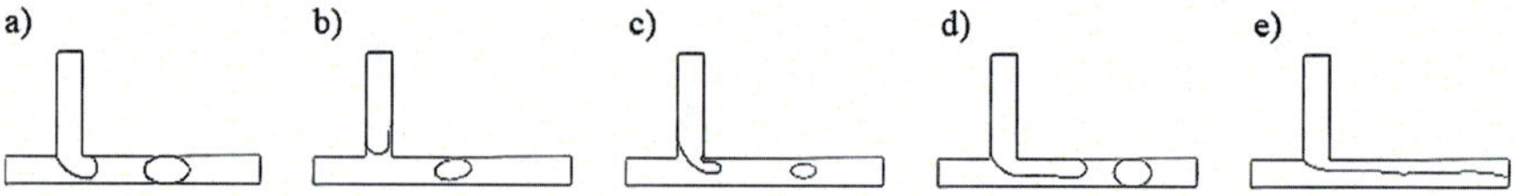

Fig. 1. Different types of flow observed. They are classified into five main types (a) Squeezing, (b) Transition, (c) Dripping, (d) Jetting and (e) Parallel Flow. Illustration of the numerical simulation at the OpenFoam software.

1. In the Squeezing regime, the drop fills the entire section of the main channel. The dispersed phase blocks the continuous phase in the channel for a brief moment, causing the drop before it detaches to gain a length, giving it a flow format.
2. In the Transition regime, the dispersed phase partially blocks the channel, forming drops with a flow rate, but with a shorter length than the Squeezing drops.
3. In the Dripping regime, the drops are spherical in shape and do not fill the entire channel. The point of separation of the drops is closer to the point of junction of the channels.
4. In the Jetting regime, part of the dispersed phase takes over the main channel with an elongated shape, during which the two immiscible fluids flow in parallel. The droplets detach themselves away from the junction point.
5. The Parallel flow profile has a particularity because it does not generate drops. The two fluids start to flow simultaneously in parallel through the main channel without any mixing of the fluids [3].

2 Modeling

The parameters used to perform the numerical analysis in the OpenFOAM open software version 4.0.1 is described below. The numerical results attend to be mass and momentum conservation for two incompressible flows.

2.1 Geometry of the Flow Channel

To validate the numerical results of the computational model, a geometry similar to the study carried out by S. Arias [4], as shown in Fig. 2. The geometry differs only in the length of the main channel, which in the study carried out here, is twice as long as the one studied by S. Arias, totaling 20 mm, in order to make it possible to analyze the development of drops in a computational domain.

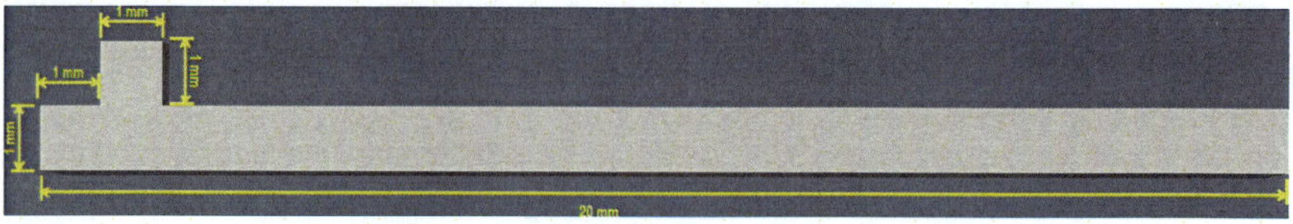

Fig. 2. T-junction dimensions of the geometry.

The T-type joint has two inlet and one outlet faces. The phases have identical input characteristics differentiated only by the fluid injection direction. The liquid phase has a horizontal entrance while the gas enters in the vertical direction. As in microchannel studies, the gravitational forces are considered null and the phase input direction doesn't affect the final results [5].

The mesh was defined to distribute the entire geometry in cells of the same size and shape, aiming to separate all geometry in symmetrical cubic unit cells with a 0.1 mm edge in a hexahedral mesh. The mesh characteristics are very important for capturing the interface in a multi-phase flow because of the interface capturing numerical challenges. Figure 3 illustrates the inlet boundary conditions. The gas inlet is at the green surface, and the liquid inlet is at the red face. The cell mesh is illustrated in Fig. 4.

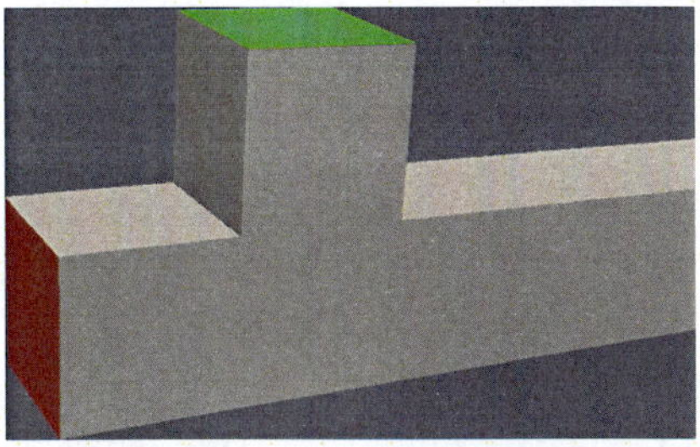

Fig. 3. Entry Faces - red for water and green for air.

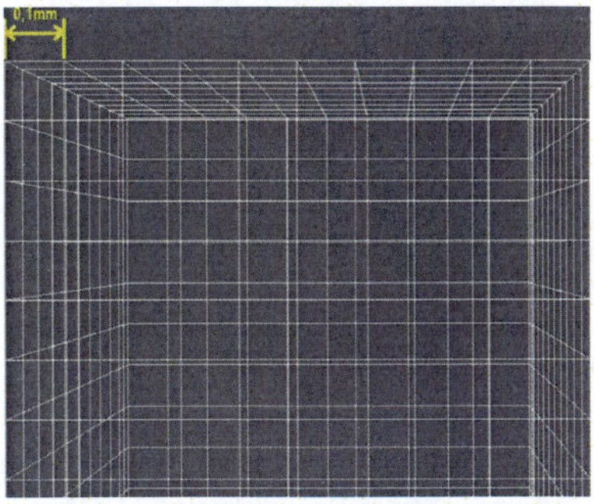

Fig. 4. Unit cell detail.

2.2 Boundary Conditions

Some boundary conditions are defined in order to simplify and make flow analysis possible.

To perform the study through a computational analysis, the OpenFOAM solver called interFoam was used. That has the multi-phase simulation capabilities for two incompressible fluids. This solver performs the studies following the methodology of interfacial treatment between volumetric fractions (VOF; Volume of fluids). Regardless of the application in OpenFOAM, the fluid volume method has been shown to be quite effective and robust in the treatment of multi-phase flows in several platforms [6].

All the possible parameters to be set as constants were thus made, with the water and air properties, so that the analyzes depended only on the variation of the fluid inlet speed to change the flow profiles.

The geometry faces are defined in four different groups where the boundary conditions are set. Each of those groups has a different characteristic, and they are defined as Wall, Gas inlet, Liquid inlet, and Outlet, as shown in Figs. 2 and 3.

The group of wall surfaces makes the flow an internal flow. They were defined with the adhesion principle and without slip.

Input1 face is defined as a continuous velocity flow, and the called face input2 has characteristics similar to input1. The only difference is the normal direction and is defined as a fixed velocity for the gas.

Finally, the outlet face, towards which the flow of the phases is directed. This face has interaction with the external environment, making it necessary to consider external atmospheric pressure.

2.3 Flow Speed Variation

Once the parameters mentioned above are fixed, the flow properties will depend only on the variations of the phases' input velocities. Based on studies of the same case, a range of speeds was adopted in which it is expected to obtain comparative results [4].

At first, we seek to analyze the formation of two distinct flow profiles, the dripped and the slug. For this, speeds between 0.106 m/s and 0.531 m/s were adopted for the continuous phase, and, for the dispersed phase, a variation between 0.08 m/s and 0.34 m/s.

For the second part of the analysis, three speeds were tested for each phase. Continuous phase with 0.531 m/s, 0.318 m/s and 0.106 m/s and dispersed phase of 0.08 m/s, 0.24 m/s and 0.34 m/s in order to treat with these data a drop formation map and analysis of the frequency of drop formation.

3 Results and Discussions

The results were carried out to determine the development between the main flow profiles analyzed in experimental studies of a similar analysis [4]. Three speeds were defined for the continuous phase, and, for each of them, a combination with three different speeds for the dispersed phase was conducted. These analyzes generated a total of nine analyzed cases. In addition, two other cases for specific velocities were conducted in order to compare the flow profiles found with the one carried out by [4].

3.1 Droplet Formation

The characteristics of the dispersed phase injection process happen following the sequence, as shown in Fig. 5.

The images show the progress of the droplet formation in the Slug flow regime. The characteristics of the bubble formation was recently defined by [2].

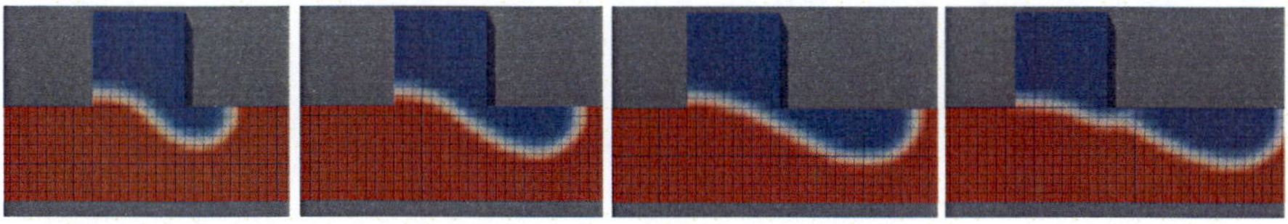

Fig. 5. Droplet formation

3.2 Flow Profiles

In other to evaluate the flow regimes, numerous simulations were carried out to characterize the drip and slug flow profiles. The inlet boundary conditions were varied out in the intervals between the maximum and minimum gas and Liquid velocity (VG and VL, respectively), as shown in Table 1.

Dripping Flow. The dripped flow occurred with speeds VL = 0.531 m/s and VG = 0.08 m/s generating the following flow profile, as shown in Fig. 6. This profile is defined by the semi-spherical shape that the drop presents, the high frequency of drop formation and the partial filling of the main channel. The dispersed phase takes the upper part of the channel and, as noted, the release time of the drops is 5 ms.

Comparing the Arias [4] experiment, the Dripping flow happens with the same speed values of phases.

Slug Flow. Slug Flow Occurs with Velocities VL = 0.350 m/s and VG = 0.242 m/s. This flow profile is characterized by the elongated shape that the drop takes, less spacing between drops, and partial obstruction of the main channel, as shown in Fig. 7. Comparing with the study by Arias and D. Legendre [4] to obtain the same drop profile, it was necessary to vary the velocity of the dispersed phase.

3.3 Droplets Formation Frequency

As a means of standardizing the data reading, flows were analyzed from the moment the first drop reaches the end of the main channel. Nine simulations were carried out combining three speeds for the continuous phase (0.106 m/s, 0.318 m/s and 0.531 m/s) and another three for the dispersed phase (0.08 m/s, 0.24 m/s and 0.34 m/s). The period T in seconds read for each case is described in Table 2.

The data from Table 2 of frequencies for the simulations performed in OpenFOAM were plotted in the graph shown in Fig. 8 along with the lines taken from the study by Arias and D. Legendre [4] that refer to the frequencies of drop formation in controlled experiments in the laboratory. Most of the data obtained in the simulations were satisfactory in relation to the laboratory study. Except when the velocity of the dispersed phase of 0.24 m/s is simulated combined with the velocities of the dispersed phase at 0.318 and 0.106, the frequencies found were slightly above the expected.

Despite the divergence in the frequencies of two of the speed combinations, the droplet formation pattern in the simulations responded well to the result obtained in the laboratory by [4].

Table 1. Interval between maximum and minimum phase speeds.

-	Máx.	Min.
V_L (m/s)	0.531	0.318
V_G (m/s)	0.242	0.068

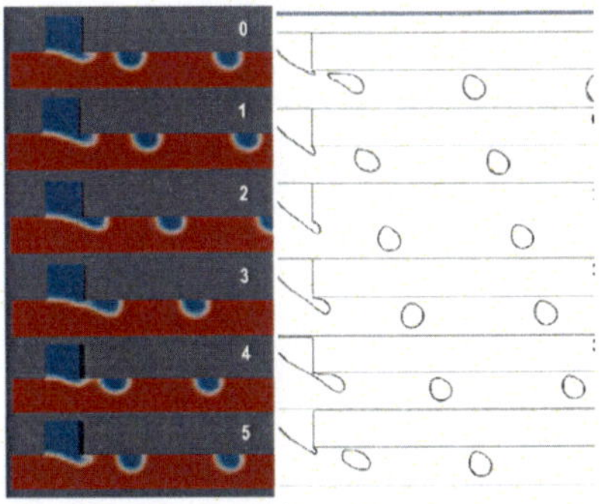

Fig. 6. Comparison between simulation performed in OpenFOAM 4.0.1 and bubbles shape experiments performed by Arias and D. Legendre [4].

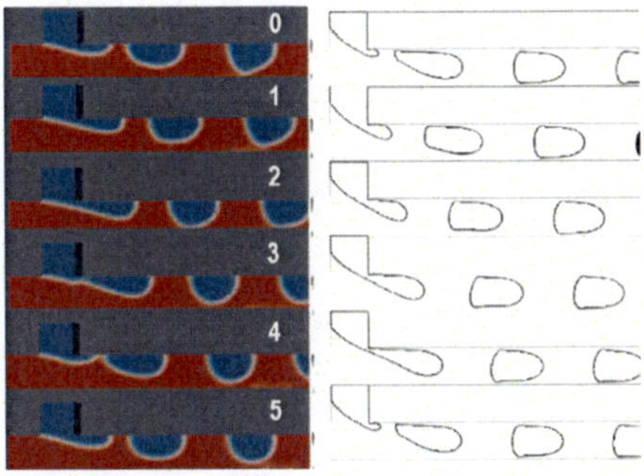

Fig. 7. Comparison between simulation performed in OpenFOAM 4.0.1 and the bubbles profile performed by Arias and D. Legendre [4].

Table 2. Droplet generation frequency.

f(Hz)	$V_G = 0.08$ m/s	$V_G = 0.08$ m/s	$V_G = 0.08$ m/s
$V_L = 0.106$ m/s	43.48	333.33	100
$V_L = 0.106$ m/s	111.11	200	200
$V_L = 0.106$ m/s	0.003	0.0025	0.003

The frequencies obtained in the flow profiles analyzed show that there is a well-defined pattern of droplet formation for each flow rate of the continuous phase, that the increased injection speed of the dispersed phase makes the pattern of droplet formation stable, and the frequency of formation will be the same for higher injection speeds of the dispersed phase.

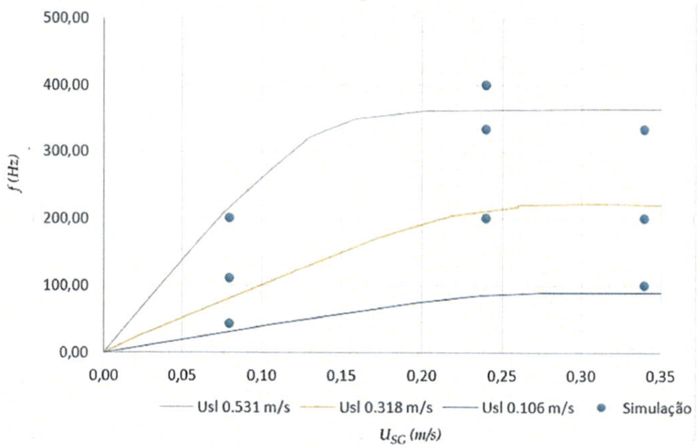

Fig. 8. Comparison between frequencies simulated in OpenFOAM and graph generated in laboratory analysis. Source: Arias and D. Legendre [4].

4 Conclusions

The purpose of this scientific research was to perform in OpenFOAM the multi-phase flow in a microchannel at T junction and to evaluate the aspects of the drip and Slug flow profiles based on the laboratory analysis performed by [4].

Through the computer simulations, it was possible to obtain a satisfactory result for the bubble formation profiles.

As it is a flow of micro fluids, the parameters of initial conditions, boundary conditions, geometric properties, fluid, and flow properties must be well defined in order to minimize numerical errors and simulation failure.

The InterFOAM solver used the Volume of Fluid method (VOF) and showed capabilities to reproduce the interfacial effects in the tested regimes in a microscale.

New analyses still need to be carried out to better understand the bubble coalescence phenomena.

References

1. Leman M (2015) Microfluidique en gouttes à l'échelle femtolitrique. L'Université Pierre et Marie Curie, Paris
2. Malekzadeh S, Roohi E (2015) Investigation of different droplet formation regimes in a t-junction microchannel using the vof technique in openfoam. Springer, Media Dordrecht, Mashhad, Iran, p 13
3. Guillot P, Colin A (2005) Stability of parallel flows in a microchannel after a t junction. The American Physical Society, Pessac, p 4
4. Arias S, Legendre D, González-Cinca R (2012) Numerical simulation of bubble generation in a t-junction. Institut de Mécanique des Fluides de Toulouse (IMFT), Tolouse, França, p 60

5. Qian D, Lawal A (2010) Numerical study on gas and liquid slugs for taylor flow in a t-junction microchannel. Chermical Engineering Science, Hoboken, USA, p. 17, 2006.Author, F.: Contribution title. In: 9th international proceedings on proceedings, Publisher, Location, pp 1–2
6. Deshpande SS, Anumolu L, Trujillo MF (2012) Evaluating the performance of the two-phase flow solver interfoam. Computational Science & Discovery, Madison, p 37

Effect of Granulometry on the Physicochemical and Sensory Properties of Green Banana Peel Flour-Based Snacks for Two Treatments

Lizbeth Cruzado-Muñoz[1](✉), Deysi Valdiviezo-Quipuscoa[1], and Guillermo Linares Lujan[2]

[1] Universidad Privada del Norte (UPN), Trujillo, Perú
[2] National University of Trujillo, Trujillo, Perú
glinares@unitru.edu.pe

Abstract. Agroindustry generates a high production of waste, including the banana shell that can be used for flour extraction and the production of snacks. In this context, the objective of the study was to evaluate the effect of granulometry on the physicochemical and sensory properties of green banana peel flour-based snacks for two treatments in order to determine the acceptability of the product. 4 samples were prepared (AC-100, AC-140, A-100, A-140), with AC being the treatment of flour extraction with metabisulphite at 0.01% plus citric acid 0.5%. To the technique of extraction of flour in water, the coding of 100 and 140 are the mesh number used for standardization. The results indicate that granulometry influences both the humidity and color of the snack; the lower the granulometry, the higher the percentage of humidity, higher values of L* and b*, but lower values of a*. The sensory analysis determined that treatments with higher granulometry have a darker color and mild crispness, the most acceptable sample being that of treatment A-140.

Keywords: Snack · Banana peel flour · Granulometry · Sensory properties of snacks

1 Introduction

Through the implementation of unit operations, the raw materials used in agribusiness are transformed to generate added value; however, it is one of the main environmental problems by the high production of waste [1]. The level of waste we produce globally has accelerated in recent decades [2]. A World Bank study indicates that 2,010 million tonnes of solid waste are generated worldwide each year, and 33% of them are not treated [3].

Today, the company demands a high consumption of bananas, located as the fourth most important cultivation in the world, is considered a basic and export product, becoming a source of employment and income in many developing countries [4]. Guerrero et al. [5] have estimated that, from a usable area of 38,604 hectares, they would generate

Y. Iano et al. (Eds.): BTSym 2020, SIST 233, pp. 732–739, 2021.
https://doi.org/10.1007/978-3-030-75680-2_81

190,102 tonnes per year of discarded banana (fresh matter) and 198,602 of lignocellulosic biomass (dry matter). The trend of banana consumption has evolved in different versions, such as fresh for direct consumption, dehydrated, and frozen [6]. In addition, locally, it is offered for the production of chifles or as flour [7], generating a gran amount of shells, a residue that is not used.

Research reports that not only bananas have nutrients needed for humans but also by-products generated as shell, which has been shown to have a high nutrient content, specifically proteins, lipids, fiber, and antioxidant-based compounds [8]. The shell represents approximately 30% of the fruit weight [9]. These wastes can well be used to fortify food or create new ones in order to facilitate food accessibility.

Banana peel has a high dietary fiber (DF) content (50 g/100 g) [9], which can be used in the food industry; the high intake of DF has been linked to various physiological effects, and its consumption prevents different diseases such as colon cancer, diabetes, cardiovascular diseases, among others [10].

Dietary fiber includes those components of plant foods that cannot be degraded by man's digestive enzymes and that are fundamentally in the cell wall, whose current interest begins from the epidemiological association between a diet rich in fiber and a lower incidence of chronic non-communicable diseases, including diabetes mellitus [11].

Dietary fiber incorporated into enteral nutrition formulas has beneficial effects on gastrointestinal function and glucose metabolism in patients with diarrhea, constipation, inflammatory bowel diseases, short bowel syndrome, and glucose metabolism [12].

One of the optimal presentations to take advantage of the DF present in the banana shell is in flour, which, through unitary operations, are transformed into various products such as snacks and cereals that constitute part of the food worldwide [13].

Currently, there is a wide variety of snack products on the market, both sweet and savory [14]. On the other hand, for the formulation of snacks, Hives [15] mentions that banana peel flour has a higher content of raw protein and dietary fiber than banana pulp flour and compound flours. In addition, according to Dussan [16], granulometry in flour is a key factor in achieving the desired physical properties in the final product. A flour with high granulometry uniformity promotes a better sensory quality of texture, taste, and appearance to the final product.

The mention of the green banana is a state of maturity of the same. In his research, Mazzeo [17] obtained as a result that green banana flour has a higher yield and superior quality to products with similar characteristics, being able to substitute other traditional flours for industrialization purposes.

Due to the environmental problems generated by the high production of waste in agribusiness, it is justified to find procedures that allow the use of banana residues for the extraction of flour and its application to develop snacks, a useful product with added value for its beneficial components.

In this sense, the objective of this research is to evaluate the effect of granulometry on the physicochemical and sensory properties of green banana peel flour-based snacks for two treatments in order to determine the acceptability of the product.

2 Materials and Methodology

2.1 Obtaining Banana Peel Flour

Bellaco Hartón banana *(Musa paradisiacal Var. Harton)*, the procedure proposed by Djioua [18], was followed, consisting of washing the fruit and disinfecting in a sodium hypochlorite solution at 100 ppm for 10 min. Subsequently, the banana was peeled, and we separated the shell to apply a metabisulphite treatment to 0.01% plus citric acid at 0.5% (AC-sample 1) and a sample in water (A-sample 2), where they were submerged for a period of 5 min [19], the shell was then dried in the GRT-CT-C industrial dehydrator, at 60 °C for 5 h [20]. Then, a grinding and a sieving were performed to obtain particle sizes between 105 to 149 microns; the procedure described by Bedolla, [21], which consists of weighing 100 g of the sample and placing on a series of meshes number 140 and 100, was stacked in increasing order of mesh size. It was waved for 10 min using a TYLER sieve model RX-29[1]; the amount of sample retained in the meshes was expressed as a percentage. Once the flour samples have been obtained (AC-100, AC-140, A-100, A-140), moisture was determined according to Mexican Standard NMX-F-428-1982 [22].

2.2 Preparation of the Snack

It was incorporated into a 30% share of Bellaco Hartón banana (AC-100, AC-140, A-100, A-140), 30% cornmeal, and 40% wheat flour for each treatment proposed by Garcia [19]. The dough was molded with a thickness of 2 mm and 3 cm in diameter and then went into a frying process with soybean oil at 180 °C for a period of 10 s [23].

2.3 Physicochemical Properties

Moisture Determination. For moisture determination, Mexican Standard NMX-F-428-1982 [22] was followed. It consists of weighing 0. 5 g of the sample and insert into the thermobalance plate OHAUS MB45[2].

Color Determination. To determine the color, it was made with the Mexican standard NMX-F-116-SCFI-2012 [24]. A spectrophotometer (Konica-Minolta) was used. Color is determined by the CIE (International Lighting Commission) color scales, evaluating L* (lightness), a* (green or red), and b*(blue or yellow). Calibration with black and white standards must be performed before color measurement.

Sensory Analysis. Sensory analysis was carried with an evaluation sheet to 30 panelists without training; these were chosen based on random choice through convenience sampling [25]. Furthermore, informed consent was obtained from all individual participants included in the study. To determine for the overall acceptability of the snacks, the characteristics evaluated were color, oiliness, and crispness.

---

[1] https://www.equiposylaboratorio.com/portal/userfiles/RX29.pdf.

[2] https://www.totalweight.com/pdf/ANALIZADOR_HUMEDAD_OHAUS_MB.pdf.

For color analysis, the intensity level present on the snack surface was assessed on a scale of 1 to 6, where the limit is very pale, pale, slightly pale, slightly dark, dark, and very dark, respectively. For the degree of crispness, it was evaluated on a scale of 1 to 6, the energy with which crack-crunch is made when the snack is bitten, where the limit goes from very mild, mild, moderately mild, moderately intense, intense, and very intense, respectively. The feeling of oiliness in the snack was evaluated on a scale of 1 to 6, whose classifications are none, mild, moderately mild, moderately intense, intense, and very intense. For the level of acceptability of the product, it was used a scale of 1 to 6, where the levels are very upset to me, disgusts me, disgusts me moderately, I like moderately, I like it, and I like it a lot, respectively.

2.4 Data Analysis

A variance analysis (ANOVA) was performed for the determination of moisture, ash, colorimetry, and sensory analysis. All statistical analyses were carried out with a 95% confidence level. The Statgraphics Centurion software[3] was used to process the data. The confidence interval has been established by consensus at 95%, which indicates that, within the given range, there is the real value of a parameter with 95% certainty [26].

3 Results and Discussions

3.1 Physicochemical Properties

Moisture Determination. Table 1 shows that the humidity percentage of snacks would be between 2.61 and 4.84%. According to Sota [27], the percentage of humidity allowed for snack products should not exceed 7%, so the Peruvian Technical Standard (NTP) indicates that the percentage of moisture set for a fried snack is 3% maximum and, for an extruded one, the maximum is 6% [28]. Based on the results obtained, it can be said that all snacks comply with the established.

Table 1. Percentage of moisture from snacks to different treatments and granulometries

Snacks	Humidity %
AC-100	$3{,}087 \pm 0.660^a$
A-100	$2{,}607 \pm 0.777^a$
AC-140	$3{,}847 \pm 0.636^a$
A-140	$4{,}837 \pm 1{,}305^b$

Lowercase letters indicate whether samples are homogeneous with 95%confidence.

[3] https://www.statgraphics.com/.

It is also observed that the treatments with the highest percentage of humidity are those with the lowest granulometry (A-140 and AC-140). These results agree with Bravo & Ortega [29], who claim that granulometrics are related to the diffusivity of water in particles, that is, to the ability of flour to absorb water; as flour granulometry decreases, the percentage of moisture is higher because particles with smaller sizes absorb proportionally more water quickly than larger particles, because there is a larger surface area to interact with water molecules. Therefore, the uniformity of granulometry favors the distribution of water in the mass.

Color Determination. With regard to color, Table 2 shows that for parameter L*, it was found that higher granulometry treatments (AC-100 and A-100) had a lower lightness and a significant difference from lower granulometry treatments. Values of a* have a significant difference from treatment and granulometry, with the highest value being A-100, i.e., the sample began to acquire red hues. The values of b* are higher for lower granulometry treatments (AC-140 and A-140), and the samples had yellow hues. Similar research by Picon [30] states that thick granulometry rice snacks had low lightness values (L*) and higher values of a* relative to those with lower granulometry. García [31] mentions that the speed of color formation in frying varies depending on the type of oil, food, and frying temperature. The results showed that the highest lightness value obtained was 54, of a scale whose maximum value is 100. The lightness value obtained was reached in a time of 111 s of frying at a temperature of 167°.

Table 2. Instrumental color parameters in the different treatments of snacks

Color parameters	Treatments			
	AC-100	A-100	AC-140	A-140
L*	48.01 ± 0.269^{a}	$45{,}645 \pm 0.233^{a}$	$54 \pm 1{,}117^{b}$	$53{,}275 \pm 0.219^{c}$
a*	$4{,}385 \pm 0.049^{a}$	4.52 ± 0.113^{b}	$3{,}685 \pm 0.078^{c}$	$4{,}175 \pm 0.021^{d}$
b*	11.78 ± 0.269^{a}	$9{,}495 \pm 0.035^{a}$	14.23 ± 0.212^{b}	$14{,}225 \pm 0.205^{c}$

Lowercase letters indicate whether samples are homogeneous with 95%confidence.

3.2 Sensory Analysis

Table 3 shows the sensory evaluation results; with respect to the color, they were obtained between "Slightly dark" and "dark", with a significant difference between treatment and granulometry. Treatments of lower granulometry obtained red hues, while in the treatments of greater granulometry, brown tones were evident. These results are consistent with Picon [30], where he mentions that his snacks of greater granulometry reached a dark color because the thick granulometry causes a greater expansion of the snack, therefore greater heating surface generating the caramelization of sugars and dark color.

Table 3. Result of sensory assessment of snacks

Attributes	Treatments			
	AC-100	A-100	AC-140	A-140
Color	5.9 ± 0.305[a]	5,967 ± 0.183[b]	4,333 ± 0.479[c]	5,667 ± 0.479[c]
Crispness	1.3 ± 0.466[a]	1.3 ± 0.535[a]	2,167 ± 0,950[b]	2,033 ± 0.765[b]
Acetosity	2.2 ± 0.484[a]	1,533 ± 0.507[b]	1,567 ± 0.504[c]	1,233 ± 0.430[b]

Lowercase letters indicate whether samples are homogeneous with 95% confidence.

With regard to crispness, there is a significant difference in treatments; those with lower granulometry had mild crispness, while those with higher granulometry very mild. According to Salinas, the flour with the highest granulometry produces a softer textured product with uniform empty spaces and friability in the mouth [32]. In the case of oiliness, no significant differences were obtained regarding granulometry, and most treatments had no oiliness except for AC-100 treatment that had the presence of mild oiliness.

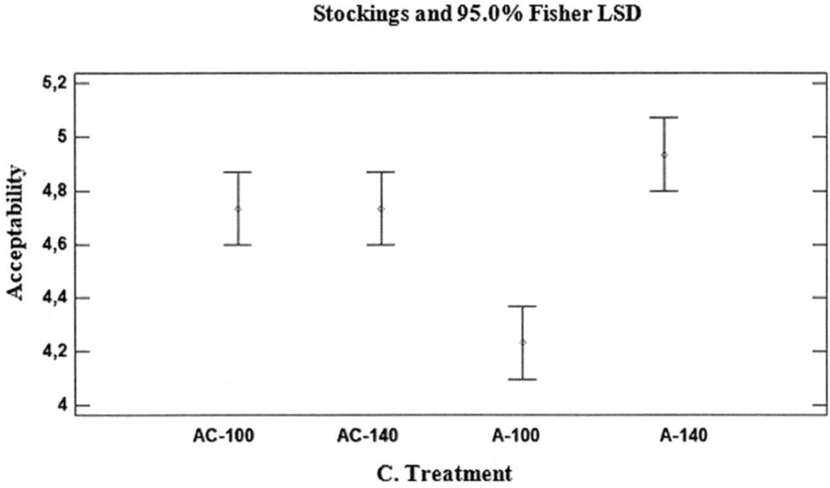

Fig. 1. Stocking chart for the overall acceptability of snacks

Figure 1 shows the mean graph with a 95% confidence level indicating that there is homogeneity between AC-100, AC-140, A-140, and A-100 treatments. In addition, it is observed that the level of acceptability of the snacks is between "I like moderately" and "Like", the most acceptability being the treatment A-140 because it has a less dark color, greater crispness, and a level of oiliness almost null, parameter that differentiates it from AC-100 and AC-140 treatments.

4 Conclusions

The addition of green banana shell flour in different treatments and granulometry affects the physicochemical and sensory properties of the snack, in addition to the thickness and frying time. Granulometry influences both the humidity and color of the snack; the lower the granulometry, the higher the percentage of humidity, higher values of L* and b*, but lower values of a*. The sensory analysis determined that higher granulometry treatments have a darker color and mild crispness, while, concerning oiliness, there is no significant difference. Finally, the A-140 treatment gained better acceptability of snacks. However, for future work, it is recommended to analyze other physicochemical properties such as texture.

References

1. Almeida A, Albertini S, Fernandes G, Pecini R, Fillet M (2008) Agroindustry use of solid waste from minimally processed melon. Food Sci Technol 28(3):733–737
2. BBC (2018) Global Garbage Crisis: 3 shocking figures about America's role. News Mundo
3. World Bank (2019) Living with garbage: the future we don't want
4. Ministry of Agriculture and Animal Resources (2014) Trends in production and trade and banana production and trade in the international and domestic market. Lima, Peru
5. Guerrero A, Aguado P, Sánchez J, Curt M (2016) GIS-based assessment of banana residual biomass potential for ethanol production and power generation: a case study. Valuation Waste Biomass 7(2):405–415
6. Gestion (2016) Peru could become the leading supplier of organic bananas in USA Lima, Peru
7. Management (2015) Peruvian banana and banana production would grow 2.6% in 2015. Lima, Peru
8. Soto V (2010) Quantification of total starch and resistant starch in green platano flour (Musa Cavendishii) and green banana (Musa Paradisiaca). Boliv J Chem 27(2):94–99
9. Blasco G (2014) Functional properties of banana (Musa Sp). Rev Med UV, 14(2):23–26
10. Matos R, Chambilla E (2010) Importance of diethetic fiber, its functional properties in human food and in the food industry. J Res Food Sci Technol 1(1):4–17
11. Zapata M, Hoet A, Simonini D (2013) Evaluation of dietary fiber consumption in patients with type 2 diabetes who attend a care center in the city of Rosario. Spanish J Hum Nutr Diet 17(3):95–101
12. Valenzuela A, Maiz A (2006) The role of dietary fiber in enteral nutrition. Chil J Nutr 33(2):342–311
13. Boxes J (2015) Effect of rice flour granulometry on the volume and alveolate of gluten- free bread. In: Latin American food _CONICET, pp 64–68
14. Bonifacio B, Destinobles C, Buendía E, Fretelli J, Rojas J (2018) Snacks based on products native to Peru. USIL. Lima, Peru
15. Hives M (2018) Preparation of pulp flour and green banana shell *Clon Harton* common for the formulation of a mixture of banana-based arepas flour: corn.Venezuela
16. Dussan S (2019) Granulometry, functional properties and color properties of quinoa and chontaduro flours. Technol Inf 30(5):3–10
17. Mazzeo M (2010) Industrial use of harvest and post-harvest residues of bananas in the department of Caldas. Eng Educ 5(9):128–139
18. Djioua T (2009) Improving the storage of minimally processed mangoes. Postharvest Biol Technol 52(2):221–225

19. Garcia M (2007) (2012) Potential of the green male banana for processing. Ibero-Am J Educ Res Dev 3(5):20–30
20. Mendez A, Aguilar K (2013) Use of male banana peel (Musa paradisiacal L.) as a substitute for wheat flour for the production of two baking products. Lacandonia, 7(2):73–76
21. Bedolla S (1984) Characteristics of US and Mexican instant maize flours for tortilla and snack preparation. CFW 29:732–735
22. Industrial Microlab (2011) Thermobalance method
23. Alvis A (2016) Loss of moisture oil absorption during frying of banana slices (paradisiacal muse L.). Biotechnol Agric Agribus Sector 14(2):119–124
24. Turegano C (2012) Norma Mexicana nmx-f-116-scfi-2012 Food-oils and vegetable or animal fats-color determination-test method. Mexico
25. Otzen T, Manterola C (2017) Sampling techniques on a study population. Int J Morphol 35(1):227–232
26. Sota B (2012) Determination of adequate humidity in the proportions of cañihua (chenopodium pallidicante aellen) and corn (zea mays) expaded by extrusion. Institutional Repository UNA, pp 43–47
27. Candia R (2005) Confidence intervals. Med J Chile 133(9):1111–1115
28. NTP (2011) Peruvian Technical Standards NTP 209.229-1984, pp 4–7
29. Bravo E, Ortega J (2017) Effect of granulometry and formulation on the quality of a rice-based extruded snack (Oryza sativa l.), quinoa (Chenopodium quinoa willd.) and degressed chia cake (Salvia hispánica l.). Universidad Nacional del Santa, Peru
30. Picon J (2014) Study of the influence of the type of flour on the development of gluten-free cookies, Spain
31. García M, Ferrero C, Bertola N, Martino M, Zaritzky N (2002) Edible coatings from cellulose derivatives to reduce oil absorption in fried products. Innov Food Sci Emerg Technol 3(4):391–397
32. Salinas S (2014) Development of a snack based on biofortified bean flours Honduras Nutritivo (Phaseolus vulgaris) and nixtamalized maize (Zeamays). Honduras

Thermal Stability of Anthocyanins in Grape Skin Extracts from Red Winemaking Residues

María Fernanda Merino-Miñano[1], Gladys Esmeralda Luján-Herrera[1], and Ricardo Vejarano[2(✉)]

[1] Ingeniería Agroindustrial, Universidad Privada del Norte, Trujillo, Peru
[2] Dirección de Investigación y Desarrollo, Universidad Privada del Norte, Trujillo, Peru
ricardo.vejarano@upn.edu.pe

Abstract. The stability of anthocyanins in extracts from red grape skins after applying dehydration temperatures of 40 °C and 150 °C was evaluated. Two fractions of red grape skins: unfermented fresh grape skins (F) and skins from grapes after winemaking (V) were used. Extracts of both types of grape skins were prepared, and the total content of anthocyanins (by high-resolution liquid chromatography) and antioxidant capacity and color parameters (by UV-vis spectrophotometry) were measured. Anthocyanins in extracts from grape skins dehydrated at 40 °C (F40 and V40) were more stable, with a total content of anthocyanins of 99.86 and 63.45 mg L^{-1}, respectively. Besides, the extracts of grape skins F40 showed the highest antioxidant capacity (AC), obtaining 100% of inhibition of the radical DPPH with 1.0 g of these skins; meanwhile, the same amount of the other grape skins showed a percentage of inhibition lower than 46.01% . According to the color analysis, the extracts F40 and V40 showed the highest values of color intensity (CI of 1.21 and 0.57, respectively), as well as the highest percentage of red color (56% and 60%, respectively). In contrast, the extracts F150 and V150 showed the highest percentage of yellow color (76% and 51%, respectively), indicating a thermal degradation of anthocyanins to their colorless forms. According to the results, a greater stability at high temperatures was obtained in the fermented grape skins, which would be related to the synthesis of the most stable anthocyanin derivatives during the fermentative process.

Keywords: Grape residues · Anthocyanin stability · Antioxidant capacity

1 Introduction

In addition to providing nutrients, functional foods can confer physiological benefits and help reduce the incidence of disease [1] due to their content of bioactive compounds, including anthocyanins, phytosterols, probiotics, fiber or fatty acids, and whose consumption has been increasing considerably in recent years. These bioactive compounds have demonstrated antioxidant, anti-inflammatory, anticancer, antimicrobial, among others activities [2], helping to prevent the prevalence of morbidity and mortality.

Anthocyanins are important phenolic compounds in food products, in which, besides conferring the typical red color, these bioactive compounds can help prevent several

Y. Iano et al. (Eds.): BTSym 2020, SIST 233, pp. 740–749, 2021.
https://doi.org/10.1007/978-3-030-75680-2_82

diseases related to oxidative stress. These compounds act by capturing electrons that are not paired and generating less reactive species. Other mechanisms involved include the interruption of autoxidation reactions, deactivation of singlet oxygen, synergy with other antioxidant compounds, activation of antioxidant enzymes, action against nitrosative stress, and oxidative enzymes, among others [2, 3].

Anthocyanins can be obtained from different raw materials such as red grape, blueberry, elderberry, purple corn, purple sweet potato, etc. An important source is the residual materials generated after the transformation of these raw materials into their derived products. One of the most important anthocyanins sources is the red grape, especially when used to elaborate red wine. In Peru, a considerable increase in the *per capita* consumption has been observed, especially in the production of red wine, in the last years [4], whose disposal of residues is becoming a key issue.

In the red wine, the phenolic compounds are extracted from the skins and seeds during the fermentative process, in which the sugar from the grapes is converted into alcohol, and it acts as a solvent of the phenolic compounds, mainly anthocyanins [5].

The winemaking process (conversion of the grape into wine) generates considerable amounts of solid residues, called grape pomace. In many cases, the treatment and disposal of these winemaking residues involve high costs [6]. Nevertheless, the grape pomace contains phenolic compounds because of their incomplete extraction into the grape-must/wine during the winemaking process. It is estimated that it is composed of approximately 25% seeds, 25% stems, and 50% skins [7]. All of them could be reused, taking advantage of its high phenolic content, for example, the grape skins, whose anthocyanins content can reach up to 830 mg per 100 g fresh skin [8].

Several studies have focused on finding new applications to recycle and/or valorize these winery residues as an additional income source for wineries. Existing literature reports applications such as (a) extraction of bioactive compounds [9, 10], soil fertilizers [11], recovery of tartaric acid [12], oil extraction [13], production of hydrolytic enzymes [14], production of bioethanol by solid-state fermentation [15], manufacturing of lightweight bricks [6], biochar [10], among other applications.

From the perspective of specific use for foodstuff development, residual skins from red grape have been applied as a source of antioxidants in snack formulations [16]. Another alternative could be the elaboration of infusions as previously developed from residues of fruits with a high content of anthocyanins [17].

Additionally, it is important to consider the applied treatments in order to reduce the degradation of the bioactive compounds. In that sense, the study aimed at determining the effect of two dehydration temperatures on the stability of anthocyanins of grape skins, based on their total content of anthocyanins, antioxidant capacity (AC), and color parameters, as a basis for the future development of an infusion of residual red grape skins.

2 Materials and Methods

2.1 Treatment of Grape Skins

Red grapes (*Vitis vinifera* cv. Gross Collman) were purchased from Cascas (La Libertad, Peru). The grapes were separated into two fractions: intact grape berries and grape berries for winemaking. In the case of intact berries, the skins were removed with a scalpel. For its part, the other fraction was used to elaborate red wine, and the residual skins were recovered at the final of the fermentative process.

Both types of grape skins: unfermented fresh grape skins (F) and skins from grapes after winemaking (V), were dehydrated in a UN55 PLUS oven (Memmert GmbH Co. KG, Germany) at temperatures of 40 °C (72 h, to preserve the anthocyanins and confer varietal aromas) and at 150 °C (30 min, to evaluate the anthocyanins stability and confer toasted character).

The dehydrated skins were ground in a knife mill to obtain a homogeneous powder.

2.2 Preparation of Extracts

Extracts were prepared with a solution of water:methanol:formic acid (48.5:50:1.5) [18]. For the extraction, 1.0 g of each grape skin powder was macerated in 30 mL of extractant solution. These suspensions were stirred for 30 min at 300 rpm. After that, the extracts were separated and made up to 50 mL. These extracts were centrifuged at 4000 rpm for 10 min at 4 °C, and the supernatant fractions were reserved.

2.3 Analysis of Anthocyanins

The anthocyanins were analyzed using a high-resolution liquid chromatograph series 1100 coupled with a diode array detector (HPLC-DAD) (Agilent Technologies, Palo Alto, United States) as described by [19]. For separation, two solvents were used as mobile phase: eluent A based on water (95%) and formic acid (5%), and eluent B based on methanol (95%) and formic acid (5%). Both eluents at a flow of 0.8 mL min^{-1} through a reverse-phase Kinetex C18 column (Phenomenex, Torrance, United States). 100 μL of each extract was filtered (cellulose membrane of 0.45 μm, Tecknokroma, Barcelona, Spain) and injected into the HPLC chromatograph. The anthocyanins were identified based on their retention times and by comparing the UV-vis and mass spectra with the data from the literature [20]. To quantify, malvidin-3-*O*-glucoside (Extrasynthèse, Genay, France) was used as an external standard.

2.4 Antioxidant Capacity

Antioxidant capacity (AC) analysis was performed on the basis of the DPPH assay [21]. A solution of 50 mg L^{-1} DPPH in 80% (v/v) methanol was prepared. The extracts of both skin fractions (F and V) were used. Each extract was diluted with 80% (v/v) methanol up to 10 mL, as shown in Table 1:

Table 1. Dilutions of extracts of both F and V skin fractions.

Dilution *(mL of extract / mL 80% methanol)*	0.5/9.5	1.0/9.0	2.5/7.5	4.0/6.0	5.0/5.0
Equivalent of dehydrated grape skin *(g L^{-1})*	1.0	2.0	5.0	8.0	10.0

Each dilution was mixed with the DPPH solution (ratio of 1/10) and incubated for 30 min, after which the absorbance at 520 nm was measured using a UV-vis 8453 spectrophotometer (Agilent Technologies, Palo Alto, USA).

The antioxidant capacity (AC), i.e., the ability of phenolic compounds to neutralize/inhibit the oxidizing radical DPPH, was expressed as the percentage of inhibition or antioxidant efficiency Q (%) of each extract, according to Eq. 1 [22]:

$$Q(\%) = \frac{A_{sample}}{A_{DPPH}} \times 100 \quad (1)$$

Where, A_{sample}: sample in contact with DPPH, and A_{DPPH}: pure DPPH agent.

2.5 Color Parameters

The color intensity (CI) and the tonality were determined using a UV-vis 8453 spectrophotometer (Agilent Technologies, Palo Alto, USA) following the Glories procedure [23]. The absorbance was measured with a 1 mm optical path length cell at 420, 520, and 620 nm. In addition, the percentages of yellow, red, and blue of each extract were also determined [24].

2.6 Statistical Analysis

IBM SPSS Statistics 25 software (IBM Corp., Armonk, United States) was used to perform the analysis of variance and the least significant difference test (Fisher's LSD test, $p < 0.05$).

3 Results and Discussion

3.1 Analysis of Anthocyanins

Figure 1 shows the HPLC chromatograms of the extracts, with a greater number of retention peaks (RT) in the extracts F40 and V40, corresponding to the grape skins dehydrated at 40 °C, while in the extracts of the grape skins dehydrated at 150 °C (F150 and V150) a lower number of peaks is observed, which would indicate degradation of anthocyanins at high temperature [25, 26].

When comparing the extracts of the grape skins dehydrated at 150 °C (F150 and V150), the greater intensity is observed in the peaks of the extract of the fermented grape skins (V150), which indicates greater stability due to the synthesis of anthocyanin derivatives, mainly vinylphenolic pyranoanthocyanins and vitisins, more stable due to the robustness of the heteroaromatic ring in their structures [27].

Table 2 shows the total anthocyanin content (mg L^{-1}), with significant differences between all treatments (Fisher's LSD test), with higher content in the following order F40 > V40 > V150 > F150.

The extract V40 has a lower anthocyanin content than the extract F40, despite the fact that both types of grape skins were dehydrated at the same temperature (40 °C). The lower anthocyanin content in the extract V40 may be a consequence of the fermentative

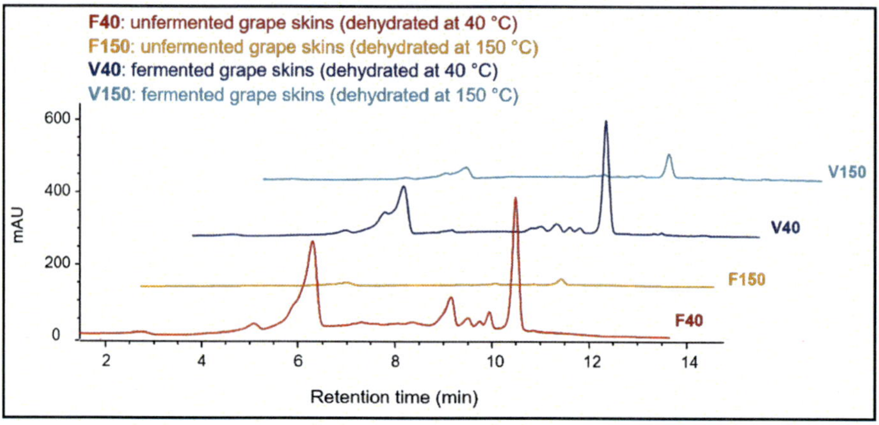

Fig. 1. HPLC chromatograms for separation of anthocyanins in the extracts of red grape skins.

Table 2. Anthocyanins content in extracts of both F and V fractions of red grape skins, expressed as mg L^{-1} (mean ± standard deviation of three independent replicates).

Anthocyanin	RT (min)	Unfermented grape skins		Fermented grape skins	
		F40	F150	V40	V150
Delphinidin-3-*O*-glucoside (D3G)	2.73	1.33 ± 0.03 a	-	0.48 ± 0.08 b	0.18 ± 0.02 c
Cyanidin-3-*O*-glucoside (Cy3G)	5.07	2.16 ± 0.07 a	-	0.71 ± 0.06 b	0.33 ± 0.06 c
Peonidin-3-*O*-glucoside (Pn3G)	5.90	5.16 ± 0.14 b	-	5.54 ± 0.08 a	1.32 ± 0.04 c
Malvidin-3-*O*-glucoside (M3G)	6.28	40.49 ± 0.47 a	0.91 ± 0.09 d	20.75 ± 1.08 b	5.10 ± 0.10 c
Malvidin-3-*O*-glucoside-vinyl adduct (vitisin B; VITB)	7.15	-	-	0.18 ± 0.02 a	-
Malvidin-3-*O*-glucoside-pyruvate adduct (vitisin A; VITA)	7.23	-	-	0.14 ± 0.04 a	-
Delphinidin-3-*O*-(6"-acetylglucoside) (D3G6Ac)	7.27	0.56 ± 0.02 a	-	0.31 ± 0.06 b	-
Petunidin-3-*O*-(6"-acetylglucoside) (Pt3G6Ac)	8.32	0.56 ± 0.05 a	-	-	-
Peonidin-3-*O*-(6"-acetylglucoside) (Pn3G6Ac)	8.95	1.77 ± 0.06 a	-	0.75 ± 0.01 b	0.33 ± 0.02 c
Malvidin-3-*O*-(6"-acetylglucoside) (M3G6Ac)	9.13	7.34 ± 0.04 a	0.29 ± 0.05 d	1.34 ± 0.06 b	0.56 ± 0.02 c
Cyanidin-3-*O*-(6"-*p*-coumaroylglucoside) (Cy3GCm)	9.46	2.04 ± 0.06 a	-	2.00 ± 0.12 a	-
Petunidin-3-*O*-(6"-*p*-coumaroylglucoside) (Pt3GCm)	9.71	0.57 ± 0.01 b	-	0.68 ± 0.04 a	0.20 ± 0.01 c
Peonidin-3-*O*-(6"-*p*-coumaroylglucoside) (Pn3GCm)	9.92	2.70 ± 0.10 a	0.11 ± 0.07 d	0.70 ± 0.01 b	0.33 ± 0.05 c
Malvidin-3-*O*-(6"-*p*-coumaroylglucoside) (M3GCm)	10.46	35.20 ± 2.28 a	1.71 ± 0.20 d	29.50 ± 1.05 b	5.65 ± 0.25 c
Malvidin-3-*O*-glucoside-4-vinylphenol (M3GVPh)	11.45	-	-	0.13 ± 0.02 a	-
Malvidin-3-*O*-glucoside-4-vinylguaiacol (M3GVG)	11.61	-	-	0.24 ± 0.04 a	0.10 ± 0.01 b
Total		**99.86 ± 2.34 a**	**3.01 ± 0.11 d**	**63.45 ± 1.93 b**	**14.11 ± 0.24 c**

Values in each row with the same small letter do not differ significantly by the Fisher's LSD test.

process, mainly due to the adsorption of anthocyanins in the yeast cell walls [28] and to the expression of the anthocyanin-β-glucosidase activity, which causes anthocyanin hydrolysis [29].

In contrast, in the extract of the fermented grape skins dehydrated at 150 °C (V150), a higher total content of anthocyanins (14.11 mg L^{-1}) was obtained, compared to the extract F150 (from fresh unfermented grape skins dehydrated at 150 °C) (3.01 mg L^{-1}), in accordance with results of Fig. 1. This indicates that in the skins that have undergone the fermentative process (V150), anthocyanin derivatives were synthesized, mainly vinylphenolic pyranoanthocyanins and vitisins, with greater stability than their anthocyanin precursors [27].

3.2 Antioxidant Capacity

Table 3 shows the values of the percentage of inhibition Q (%) of the oxidant radical DPPH (Molyneux, 2004) at different concentrations of each extract. The highest value of Q was obtained with the extract of fresh unfermented grape skins dehydrated at 40 °C (F40), in which 1.0 g of these skins was enough to completely neutralize the oxidant radical DPPH (Q = 100%). While dehydration of the same grape skins at 150 °C (F150) decreased their AC, being necessary 5.0 g to obtain the same effect (Q = 100%), which indicates the loss of AC at high temperatures [25, 26], in agreement with its lower anthocyanin content (Table 2).

A similar behavior showed the extracts of the fermented grape skins, where the dehydration process at 150 °C (V150) decreased their ability to inhibit the radical DPPH with respect to the grape skins dehydrated at 40 °C (V40).

When comparing the extracts from skins dehydrated at 40 °C, 2.0 g of the skins V40 would be necessary to totally inhibit the radical DPPH (Q = 100%), compared to 1.0 g required of the grape skins F40. This indicates a lower anthocyanin content in the grape skins V40 (Table 2) due to the fact that part of the anthocyanins has been transferred to grape-must/wine from grape skin during the fermentative process [8, 30].

Table 3. Inhibition of DPPH by extracts of red grape skins, expressed as Q (%).

Equivalent of dehydrated grape skin (g/L)	Unfermented grape skins		Fermented grape skins	
	F40	F150	V40	V150
1	**100.00**	46.01	37.89	37.89
2	100.00	85.39	**100.00**	71.18
5	100.00	**100.00**	100.00	**100.00**
8	100.00	100.00	100.00	100.00
10	100.00	100.00	100.00	100.00

3.3 Color Parameters

The color parameters represent the content of coloring matter in the sample. The extract from fresh unfermented grape skins dehydrated at 40 °C (F40) showed the highest color intensity (CI = 1.21) (Table 4), in addition to a high percentage of red (56%) (Fig. 2), which is the component that characterizes the flavillium cation, the most colored form of anthocyanins [31], in accordance with its high content of anthocyanins (Table 2). On the contrary, the extract F150 showed the lowest value of CI (0.40) and the lowest percentage of red (20%) and blue (4%). This extract also showed the lowest content of anthocyanins (Table 2).

In addition, the highest value of tonality (3.82) was shown by the extract F150 (Table 4), in concordance with the highest percentage of yellow (76%) (Fig. 2). Tonality represents the ratio of yellowish pigments over the red ones, so its value directly relates to the percentage of yellow (Fig. 2).

Table 4. Color intensity (CI) and tonality in extracts of red grape skins.

Extract	CI	Tonality
F40	1.21 ± 0.00 a	0.69 ± 0.00 c
F150	0.40 ± 0.00 d	3.82 ± 0.01 a
V40	0.57 ± 0.00 b	0.56 ± 0.00 d
V150	0.48 ± 0.00 c	1.18 ± 0.00 b

Values in each column with the same small letter do not differ significantly by the Fisher's LSD test.

The high percentage of yellow in extracts F150 and V150 (Fig. 2: 76% and 51%, respectively) is related to the degradation of anthocyanins, as a consequence of the loss of glycosidating sugar bound at position 3 of the anthocyanidin molecule, producing the opening of the ring, with the subsequent conversion to its colorless forms [25].

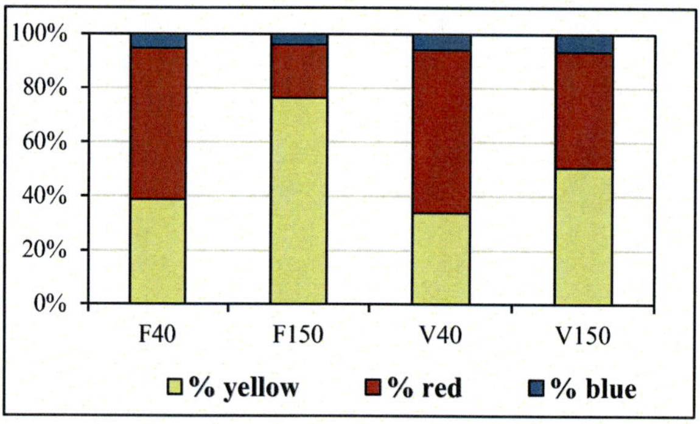

Fig. 2. Percentage of yellow, red, and blue in extracts of red grape skins.

Additionally, the high percentage of yellow indicates the presence of pigments derived from the thermal degradation of the grape skin constituents, mainly phenolic compounds and lignocellulosic material. Exposure of lignocellulosic material to high temperatures releases several compounds, for example, furanic aldehydes [32]. These aldehydes react with malvidin-3-*O*-glucoside (M3G) with the consequent synthesis of M3G-furfural, M3G-methylfurfural, among other anthocyanin-furanic aldehyde adducts [33]. These compounds can contribute to the yellowish appearance (Fig. 2).

Regarding the effect of high temperature on the color parameters, the extract V150 showed a higher CI than the extract F150 (Table 4: 0.48 and 0.40, respectively), which was also reflected in the higher percentage of red (Fig. 2: 43% and 20%, respectively) and blue (Fig. 2: 7% and 4%, respectively). These results agree with the higher content of anthocyanins in the extract V150 (Table 2), i.e., the presence of more stable anthocyanin derivatives, produced during the fermentative process, confer to the extract greater protection against degradation [27].

4 Conclusions

The stability of anthocyanins in extracts from red grape skins after applying dehydration temperatures of 40 °C and 150 °C was studied. In view of the results, anthocyanins from grape skins dehydrated at 40 °C showed greater stability, with a total content of anthocyanins up to 99.86 mg L^{-1} and the highest antioxidant capacity (up to 100% of inhibition of radical DPPH), as well as the best color parameters. Likewise, anthocyanins from fermented grape skins (winemaking residues) dehydrated at 150 °C showed greater stability than unfermented fresh grape skins dehydrated at the same temperature, indicating the synthesis of most stable anthocyanin derivatives during the fermentative process, which confer greater protection against degradation. Finally, our results suggest the potential of red grape residues as a source of antioxidant compounds, with potential application in the development of new food products with functional character, contributing at the same time to recycle and/or valorize these winery residues.

Acknowledgment. This study was supported by the Universidad Privada del Norte (projects UPN- 20171003 and UPN-20191001). The authors thank the personnel from the Chemistry and Food Technology Department in the Universidad Politécnica de Madrid (Spain) for excellent technical assistance in the HPLC analysis.

References

1. United States Department of Agriculture. https://www.ars.usda.gov/SP2UserFiles/Place/00000000/NPS/FinalFunctionalFoodsPDFReadVersion6-25-10.pdf
2. Vejarano R, Gil-Calderón A, Díaz-Silva V, León-Vargas F (2019) Improvement of the bioactive profile in wines and its incidence on human health: Technological strategies. In: advances in grape and wine biotechnology. IntechOpen, London. https://doi.org/10.5772/intechopen.85861
3. Peng Y, Zhang H, Liu R, Mine Y, McCallum J, Kirby C, Tsao R (2016) Antioxidant and anti-inflammatory activities of pyranoanthocyanins and other polyphenols from staghorn sumac (*Rhus hirta* L.) in Caco-2 cell models. J Funct Foods 20:139–147. https://doi.org/10.1016/j.jff.2015.10.026
4. International Organisation of Vine and Wine. https://www.oiv.int/es/statistiques/?year=2016&countryCode=PER
5. Waterhouse AL (2002) Wine phenolics. Ann N Y Acad Sci 957:21–36. https://doi.org/10.1111/j.1749-6632.2002.tb02903.x
6. Taurino R, Ferretti D, Cattani L, Bozzoli F, Bondioli F (2019) Lightweight clay bricks manufactured by using locally available wine industry waste. J Build Eng 26:100892. https://doi.org/10.1016/j.jobe.2019.100892

7. Jin B, Kelly JM (2009) Wine industry residues. In: biotechnology for agro-industrial residues utilisation: utilisation of agro-residues. Springer, Dordrecht, pp 293–311. https://doi.org/10.1007/978-1-4020-9942-7_15
8. Romero I (2008) Extracción de compuestos fenólicos de la uva al vino. Papel de los enzimas de maceración. Ph.D. thesis, Departamento de Tecnología de Alimentos, Nutrición y Bromatología, Universidad de Murcia, España
9. Castro-López C, Rojas R, Sánchez-Alejo EJ, Niño-Medina G, Martínez-Ávila GCG (2016) Phenolic compound recovery from grape fruit and by-products: an overview of extraction methods. In: grape and wine biotechnology. IntechOpen, London. https://doi.org/10.5772/64821
10. Sette P, Fernandez A, Soria J, Rodriguez R, Salvatori D, Mazza G (2020) Integral valorization of fruit waste from wine and cider industries. J Clean Prod 242:118486. https://doi.org/10.1016/j.jclepro.2019.118486
11. Arvanitoyannis IS, Ladas D, Mavromatis A (2006) Potential uses and applications of treated wine waste: a review. Int J Food Sci Technol 41(5):475–487. https://doi.org/10.1111/j.1365-2621.2005.01111.x
12. Yalcin D, Ozcalik O, Altiok E, Bayraktar O (2008) Characterization and recovery of tartaric acid from wastes of wine and grape juice industries. J Therm Anal Calorim 94(3):767–771. https://doi.org/10.1007/s10973-008-9345-z
13. Fiori L (2007) Grape seed oil supercritical extraction kinetic and solubility data: critical approach and modeling. J Supercrit Fluids 43(1):43–54. https://doi.org/10.1016/j.supflu.2007.04.009
14. Díaz AB, De Ory I, Caro I, Blandino A (2009) Production of hydrolytic enzymes from grape pomace and orange peels mixed substrate fermentation by *Aspergillus awamori*. In: chemical engineering transactions, vol 17. The Italian Association of Chemical Engineering, Milano, pp 1143–1148. https://doi.org/10.3303/CET0917191
15. Li C, Rodríguez LF, Khanna M, Spaulding AD, Lin T, Eckhoff SR (2010) An engineering and economic evaluation of quick germ-quick fiber process for dry-grind ethanol facilities: model description and documentation. Bioresour Technol 101(14):5275–5281. https://doi.org/10.1016/j.biortech.2010.01.139
16. Bender ABB, De Mello Luvielmo M, Loureiro BB, Speroni CS, Boligon AA, Da Silva LP, Penna N (2016) Obtention and characterization of grape skin flour and its use in an extruded snack. Brazilian J Food Technol 19:e2016010. https://doi.org/10.1590/1981-6723.1016
17. da Silva JK, Batista ÂG, Cazarin CBB, Dionisio AP, Brito ES, Marostica-Junior MR (2015) A new functional tea from Brazilian berry: bioactive profile of extract aqueous of jaboticaba peel. In: Anais do Simpósio Latino Americano de Ciências de Alimentos. Universidade Estadual de Campinas, Campinas, p 35482. https://proceedings.science/slaca/slaca-2015/papers/a-new-functional-tea-from-brazilian-berry--bioactive-profile-of-extract-aqueous-of-jaboticaba-peel
18. Morata A (2004) Influencia de la maduración antociánica de la uva y de la biotecnología fermentativa en color, aroma y estructura de los vinos tintos. Ph.D. thesis, Departamento de Tecnología de Alimentos, Universidad Politécnica de Madrid, España
19. Del Fresno JM, Morata A, Loira I, Escott C, Suárez-Lepe JA (2020) Evolution of the phenolic fraction and aromatic profile of red wines aged in oak barrels. ACS Omega 5(13):7235–7243. https://doi.org/10.1021/acsomega.9b03854
20. Monagas M, Núñez V, Bartolomé B, Gómez-Cordovés C (2003) Anthocyanin-derived pigments in Graciano, Tempranillo, and Cabernet Sauvignon wines produced in Spain. Am J Enol Vitic 54(3):163–169. https://www.ajevonline.org/content/54/3/163
21. Floegel A, Kim DO, Chung SJ, Koo SI, Chun OK (2011) Comparison of ABTS/DPPH assays to measure antioxidant capacity in popular antioxidant-rich US foods. J Food Compos Anal 24(7):1043–1048. https://doi.org/10.1016/j.jfca.2011.01.008

22. Molyneux P (2004) The use of the stable free radical diphenylpicryl-hydrazyl (DPPH) for estimating antioxidant activity. Songklanakarin J Sci Technol 26(2):211–219. http://rdo.psu.ac.th/sjstweb/journal/26-2/07-DPPH.pdf
23. Glories Y (1984) La couleur des vins rouges. 2e partie : mesure, origine et interpretation. OENO One 18(4):253–271. https://doi.org/10.20870/oeno-one.1984.18.4.1744
24. Morata A, Benito S, Loira I, Palomero F, González MC, Suárez-Lepe JA (2012) Formation of pyranoanthocyanins by *Schizosaccharomyces pombe* during the fermentation of red must. Int J Food Microbiol 159(1):47–53. https://doi.org/10.1016/j.ijfoodmicro.2012.08.007
25. Garzón GA (2008) Anthocyanins as natural colorants and bioactive compounds. A review. Acta Biol Colomb 13(3):27–36. http://www.scielo.org.co/scielo.php?script=sci_abstract&pid=S0120-548X2008000300002
26. Saucier C (2010) How do wine polyphenols evolve during wine ageing? Cerevisia 35(1):11–15. https://doi.org/10.1016/j.cervis.2010.05.002
27. Morata A, Bañuelos MA, Tesfaye W, Loira I, Palomero F, Benito S, Callejo MJ, Villa A, González MC, Suárez-Lepe JA (2015) Electron beam irradiation of wine grapes: effect on microbial populations, phenol extraction and wine quality. Food Bioprocess Technol 8(9):1845–1853. https://doi.org/10.1007/s11947-015-1540-x
28. Loira I, Vejarano R, Morata A, Ricardo-Da-Silva JM, Laureano O, González MC, Suárez-Lepe JA (2013) Effect of Saccharomyces strains on the quality of red wines aged on lees. Food Chem 139(1–4):1044–1051. https://doi.org/10.1016/j.foodchem.2013.01.020
29. Wightman JD, Wrolstad RE (1996) β-glucosidase activity in juice-processing enzymes based on anthocyanin analysis. J Food Sci 61(3):544–548. https://doi.org/10.1111/j.1365-2621.1996.tb13153.x
30. Morata A, Loira I, Vejarano R, Bañuelos MA, Sanz PD, Otero L, Suárez-Lepe JA (2015) Grape processing by high hydrostatic pressure: effect on microbial populations, phenol extraction and wine quality. Food Bioprocess Technol 8(2):277–286. https://doi.org/10.1007/s11947-014-1405-8
31. Giusti MM, Wrolstad RE (2001) Characterization and measurement of anthocyanins by UV-visible spectroscopy. Curr Protoc Food Anal Chem 00(1): F1.2.1–F1.2.13. https://doi.org/10.1002/0471142913.faf0102s00
32. Vejarano R, Morata A, Loira I, González MC, Suárez-Lepe JA (2013) Theoretical considerations about usage of metabolic inhibitors as possible alternative to reduce alcohol content of wines from hot areas. Eur Food Res Technol 237(3):281–290. https://doi.org/10.1007/s00217-013-1992-z
33. Sousa A, Mateus N, Silva AMS, Vivas N, Nonier MF, Pianet I, De Freitas V (2010) Isolation and structural characterization of anthocyanin-furfuryl pigments. J Agric Food Chem 58(9):5664–5669. https://doi.org/10.1021/jf1000327

Study of the Resonance Frequency Variation in UHF RFID Tags by Changing the Internal Dimensions of an Inlay

Rafael Soleiman Franco[1](✉), Edson Tafeli Carneiro dos Santos[2], and Cristiano Akamine[1]

[1] Mackenzie Presbyterian University (PPGEEC), São Paulo, Brazil
rafasole@uol.com.br, cristiano.akamine@mackenzie.br
[2] Mackenzie Presbyterian University (EE), São Paulo, Brazil
edson.santos@mackenzie.br

Abstract. This paper aims to establish an analysis of the impact in the resonance frequency of a Radio Frequency Identification (RFID) inlay operating in Ultra High Frequency (UHF) by changing its internal dimensions. Obtaining the resonance frequency in terms of the S_{11} parameter is crucial in order to design RFID labels that meet specific requirements. The use of a Computer Aided Design (CAD) tool was considered to design the RFID inlay, while the simulations were performed using the MWS-CST Studio Student Edition software. Satisfactory results were met in order of verifying central frequency shifting as long as the internal dimensions of the inlay have been adjusted in two rounds of simulations. Additionally, it was possible to demonstrate a gain greater than 2 dB and 30 MHz bandwidth, showing that the tested conditions can be seen as a good starting point for future works to advance the studies in the context herein presented. The RFID tags designed in this paper benchmarks an antenna widespread in the RFID market. Meanwhile, the theoretical basis obtained by observing the antenna theory and analyzing the results has been used to contribute to all the process of building an RFID tag and stimulate the expansion of RFID technology worldwide.

Keywords: RFID · Tags · UHF · Resonance frequency · Radio frequency identification

1 Introduction

The Radio Frequency Identification (RFID) had its peak of popularization in the early 2000s; since then, this technology has been undergoing several improvements that ended up with the first interactions around the concept of the Internet of Things (IoT).

The success obtained in the most diverse market applications encourages the use of this technology in different scenarios and unexplored conditions. Thus, the popularization of technology has been growing since its first implementations, and, with that, investments in research and development in this area [1]. The use of RFID is quite diverse and is applicable for different market segments; among the most well-known

Y. Iano et al. (Eds.): BTSym 2020, SIST 233, pp. 750–758, 2021.
https://doi.org/10.1007/978-3-030-75680-2_83

applications are those that related to the retail and apparel segment, where RFID tags are applied to clothing items and respond to the infrastructure of readers and the software installed in a logistic distribution center or inside the stores.

Generally, a passive RFID tag has three main components: Integrated Circuit (IC), an antenna, also known as an inlay, and a substrate. The set operates by backscattering the Radio Frequency (RF) signals from a base station known as RFID Reader, which transmits a modulated signal and converts the voltage at the antenna terminals into Direct Current (DC), the energy is then transmitted from the reader to the tag that energizes the IC and sends back its information contained therein, thus changing the complex input impedance of the system [1].

One of the most important parameters to consider during an RFID tag development is the match between the antenna and the IC; usually, RFID tags are designed to answer the IC specifications; otherwise, it would be cost-wise unfeasible to develop a new IC for every particular application.

There are a few authors publishing different ways to develop an RFID inlay focusing on a specific market, including biotechnology [2], healthcare [3], RTLS (Real-Time Location System) [4, 5].

The regulations for UHF RFID frequencies depends on the countries or regions where the technology is being used; for example, in Europe, the local regulations permit the use of 866 MHz–869 MHz band, North and South America operate in 902 MHz–928 MHz, Japan and some Asian countries, from 950 MHz to 956 MHz. In addition, regulatory bodies are associated with the governments of the regions where communications are regulated; in Brazil, for example, ANATEL (National Telecommunications Agency, in Portuguese, Agência Nacional de Telecomunicações) manages and regulates the use of the frequency spectrum in telecommunications, and in Europe, the ETSI (European Telecommunications Standards Institute) [6–8].

This article aims to demonstrate the displacement of the resonance frequency in terms of parameter S_{11} in relation to the variation of the internal dimensions of an RFID inlay developed from a reference model used in this segment. The results obtained in this article aim to propose new constructive approaches in the scenario of RFID inlays in order to contribute to the viability of the technology in conditions that require specific characteristics.

This article is divided into four fundamental groups in which Sect. 1 presents a historical review of RFID technology, and Sect. 2 refers to the theoretical approach addressed as a baseline for the construction proposed here. Section 3 presents the development process applied to the methodology used in the simulations, as well as the premises and starting points that were taken into account in this article, while the conclusions and results can be found in Sect. 4, which includes a consolidation of the simulations performed and results obtained.

2 Theoretical Characteristics of Impedance Matching

In RFID projects, reading range, also known as reading distance, is given by the maximum distance that the RFID reader is able to decode the information contained in an RFID tag; this parameter, in many cases, defines the use of a certain label at the expense of others [1].

Some environmental conditions directly affect the efficiency of the reading range; most of these conditions are correlated with the resonance frequency matching and the bandwidth in which the system is propagating the RF waves, in addition, the orientation of the label, the immersion or contact with ionic and/or metallic environments are also amidst the conditions that induce RF distortions in the UHF band [9].

One of the most important parameters to consider when developing an RFID tag is the match between the antenna's impedances and the IC of an RFID tag. RFID tags are generally designed to meet the specifications of a market IC since the cost and complexity of developing a new IC for every single use case would be impracticable.

As in most antenna calculations, the reading range can be approximated by the Friis equation in free space, as shown in (1) where λ is the wavelength, Pt is the power transmitted by the reader, Gt is the gain of the transmission antenna, Gr is the gain of the receiving antenna, Pth is the power threshold necessary to supply the RFID inlay IC and τ is the transmission coefficient expressed by (2) [10].

By the qualitative analysis of Eq. (2) in which the impedance of the IC is given by $Zc = Rc + jXc$ and the antenna impedance by $Za = Ra + jXa$, it is clearly possible to verify that the maximum reading distance is obtained when the modules of Xa and Xc has the same values. The authors of [8] illustrate, on page 2, the contours of the correlation between Zc and Za.

The sensitivity of the RFID reader is typically higher compared to the sensitivity of the tag. In addition, the reading distance is determined by the thresholds of the responsiveness of the tag. It is important to notice that the reading distance is mainly influenced by the means of propagation and also by the type of material on which the label is operating [8].

$$r = \frac{\lambda}{4\pi}\sqrt{\frac{PtGtGr\tau}{Pth}} \tag{1}$$

$$\tau = \frac{4RcRa}{|Zc + Za|^2}, 0 \leq \tau \leq 1 \tag{2}$$

Therefore, according to [10], it is possible to establish a correlation between the expected gain and the transmission coefficient. This type of analysis is essential to understand the compromise between gain, impedance, and bandwidth of an RFID tag.

3 Simulations and Results

3.1 Methodology

The simulations were carried out through practical experiments to verify the behavior of a RFID tag; thus, the development of the antenna was performed using the software MWS-CST Studio Student version that allows to design and simulate an antenna that meets the proposed requirements.

The antenna's development phase consists essentially of obtaining the best electromagnetic coupling ratio between IC and inlay. Thus, variations in the layout and dimensions of these inlays significantly alter the frequency response and the power radiated by the RFID tag, permitting to adjust the correlation between gain and resonance frequency to meet the proposed requirements.

Important factors such as the substrate on which the inlay will be deposited, the adhesive layers, the protective screening, and so the synthetic paper often used to print visual data were not considered in the experiments. Strictly speaking, only the direct response of the inlay and the interferences of the environment in which it was observed were considered.

The interferences of the environment in which the RFID tag is inserted are very relevant in the study of the behavior of RFID tags and can limit the choice of one type of tag over others as well as the communication protocol that governs the communication of data between an RFID tag and the reader infrastructure in which it communicates. The protocol used for RFID UHF communication is ISO 18006-C, which aims to standardize a series of requirements that, in addition to the frequency spectrum regulations of each country and/or region, enable the use of technology worldwide.

The results presented in this article were obtained through two simulation rounds that consisted of checking the parameter S_{11}, and then the realized gain. The first round of simulations was performed with the original measurements of the inlay built from the reference model, while the second round was performed after some adjustments in the internal dimensions of the RFID inlay. The results and particularities of each of them will be described in this article.

The use S_{11} parameter was due to the fact that this parameter serves as a comparative reference for the central frequency of the antenna under development and its respective geometric adjustments. On the other hand, the realized gain is a very relevant measure that shows the relationship between the radiated power and the directivity of an antenna, since the energy radiated in a given direction must obtain a minimum efficiency in order to guarantee the information transmitted by the antennas meets the requirements from a given use case [11].

3.2 Reference Design

The starting point for the construction of the RFID inlay built for the simulations described in this article was based on the ALN-9770 model manufactured by the Alien Technology company, also known as "Bat" inlay [12].

The reason to choose this reference model is due to the fact that it is a very widespread label in the market, and it has a good efficiency in several environmental conditions, such as close to metallic environments and close immerse to ionic moisture [12].

As well as the reference of the contours and the layout of the inlay, the maximum dimensions defined for the model used in the simulations will also be based on the Alien "bat" label, that is, 83 mm × 32 mm [12]. In order to obtain a constructive safety margin, it was established that the maximum dimensions for the label under development in this article are 100 mm × 40 mm.

Two other parameters were established as constructive starting points for the design proposed in this article: the minimum reading sensitivity of −20 dBm and the Parasitic Capacitance at −0.21 pF; both values were taken from a commonly used IC in RFID applications for general purpose, the Monza R6 IC from Impinj manufacturer [13].

Taking into account the premises described, Fig. 1 illustrates the inlay used for the simulations presented in this article.

Fig. 1. Inlay under development used S_{11} and realized gain measurements.

3.3 Results Obtained

The starting point established for the dimensions defined in the antenna under development was based on the internal measurements of the Alien "bat" tag. Figure 2 presents the dimensions and the references that list the main measures that were taken into account during the simulations proposed here, and Table 1 presents the values of each dimension for the first round of simulations.

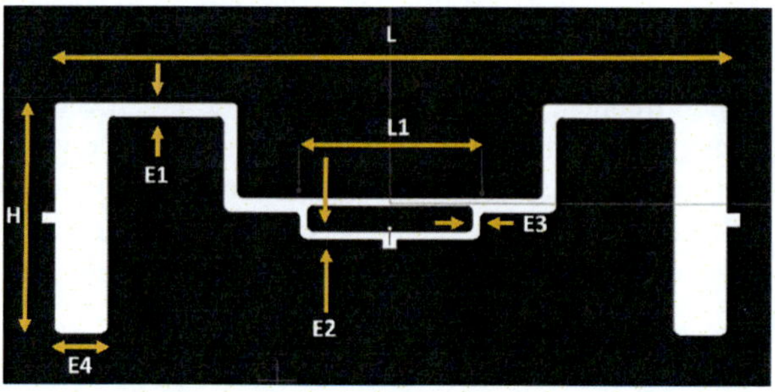

Fig. 2. Dimensions of main internal dimensions of the RFID inlay under development.

Table 1. Dimensions for designed inlay parameters in the first round of simulations

Dimensions	L	L1	E3	E4	H	E1	E2
Measure (mm)	98	12	0.9	16	34	0.9	0.9

Once the dimensions of each of the measures presented in Table 1 are defined, the goal of this first analysis is to understand what is the resonance frequency established by parameter S_{11} for this scenario. It is worth mentioning that the tests did not consider the application of the label on any type of surface; that is, the results obtained include the inlay applied only on the substrate in PEC.

Figure 3 illustrates the first simulation scenario considering the measures presented in Table 1. The first simulation round showed that the bandwidth for S_{11} at -10 dB varies between 925 MHz–939 MHz, and the S_{11} resonance occurs at 930 MHz at −22 dB.

Also, considering the measures established in the first round of simulations, the realized gain, as shown in Fig. 4, reached its maximum of 1.98 dB at 930 MHz, that is, at the same resonance frequency seen in S_{11}. Two complementary measures at 900 MHz and 915 MHz were taken of −2.07 dB and 0.49 dB, respectively, showing a positive trend of gain as it approaches the resonance frequency.

Mathematically, when analyzing the gain of 2 dB as a baseline for establishing the bandwidth for this configuration, a range of 30 MHz bandwidth could be seen.

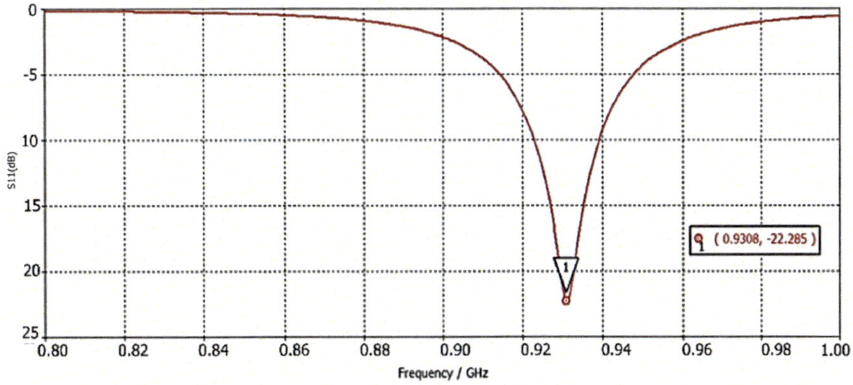

Fig. 3. Resonance frequency in S11 for the measurements established in Table 1.

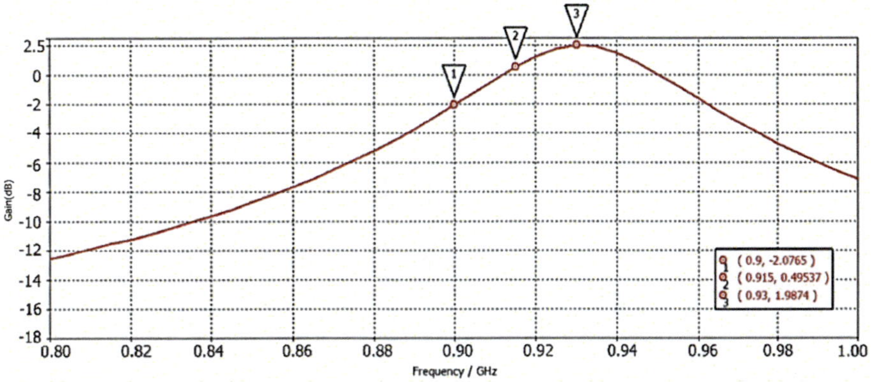

Fig. 4. Realized gain in dB for the measures established in Table 1.

The second round of simulations was carried out similarly to the previous condition, this time adjusting the values of the internal dimensions of the inlay under development for the measures corresponding to Table 2. For this new scenario, the dimensions E3 and E2 were changed, from 0.9 mm in the first round of simulations to 1.1 mm, and from 0.9 mm to 1.2 mm, respectively.

The region comprised of the E3 and E2 dimensions is known as the inductive loop; this area is very sensitive to changes in thickness and reflects directly in the S_{11} resonance frequency since it is in this region where the magnetic coupling occurs. As can be seen in Sect. 2, changes in the imaginary components of the input impedances can change the impedance matching between the IC and the inlay.

In order to enable a comparative analysis of the results obtained in the two rounds of simulations, for the second round, the same simulations were carried out in terms of the resonance frequency in S_{11} and the realized gain, as shown in Fig. 5 and Fig. 6 respectively.

Table 2. Dimensions for designed *inlay* parameters in the second round of simulations

Dimensions	L	L1	E3	E4	H	E1	E2
Measure (mm)	98	12	1.1	16	34	0.9	1.2

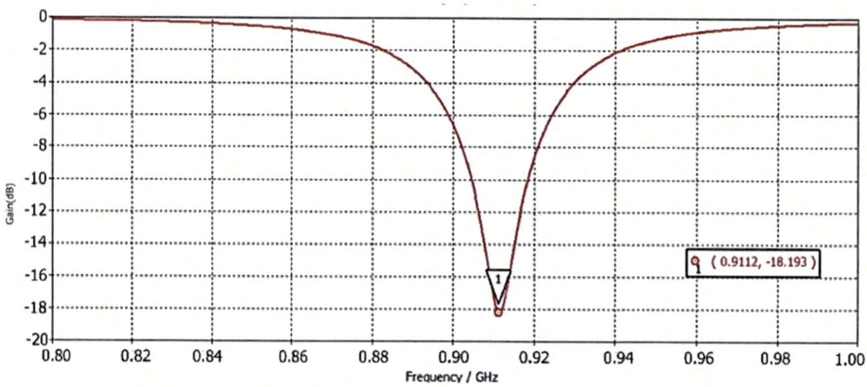

Fig. 5. Resonance frequency in S_{11} for the measures established in Table 2.

The second round of measurements showed that the resonance frequency in S_{11} is at 911 MHz and -18 dB and that a -10 dB range for S_{11} varies from 908 MHz–918 MHz, as shown in Fig. 5.

Still, for the second round of measurements, Fig. 6 shows the realized gain reaching its maximum point at 915 MHz and 2.0 dB; in this case, it has been shifted by around 4 MHz from the S_{11} resonance frequency. The two adjacent measures at 900 MHz and 930 MHz showed realized gain at 0.48 dB and 0.46 dB, respectively.

Establishing the same baseline reference at 2 dB of gain seen in the first round of simulations, the bandwidth was 8 MHz in this scenario.

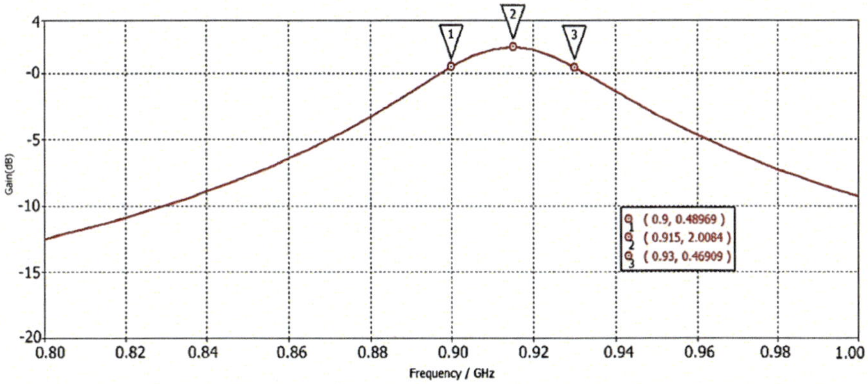

Fig. 6. Realized gain in dB for the measures established in Table 2.

4 Conclusions

The results from the simulations analyzed two different scenarios where there were adjustments on the internal inlay dimensions. Both scenarios have shown that there is a compromise between the measures of internal dimensions of an RFID inlay and its resonance frequencies as well as its realized gains.

In summary, the dimensional adjustments analyzed in Table 1 and Table 2 indicated that the resonance frequency shifted from the range close to 930 MHz to roughly 915 MHz. However, despite the average 2 dB maximum point of the realized gain in the two simulated situations, there was a significant loss in bandwidth in the second round of simulations, varying from 30 MHz to 8 MHz; such variation occurs due to the variation in the capacitive components of the input impedances of the RFID inlay.

Finally, it was possible to demonstrate that modifications of the dimensions in the thicknesses and lengths of the RFID inlay can be a good factor in order to get the desired frequencies and realized gains in a UHF RFID tag design.

However, further studies presented in this article can be driven in terms of knowing how the interaction of these dimensional adjustments affects or complements one another in order to obtain a gain that could be sustained by a larger bandwidth.

References

1. Nikitin PV, Rao KVS (2006) Theory and measurement of backscattering from rfid tags. IEEE Antennas Propag Mag 48(6):212–218
2. Chrysler A, Furse C (2016) Biocompatible, implantable UHF RFID antenna made from conductive ink. In: 2016 IEEE international symposium on antennas and propagation (APSURSI), Fajardo, pp 467–468
3. Amendola S, Lodato R, Manzari S, Occhiuzzi C, Marrocco G (2014) RFID technology for IoT-based personal healthcare in smart spaces. IEEE Internet of Things J 1(2):144–152
4. Ciftler BS, Kadri A, Güvenc İ (2017) IoT localization for bistatic passive UHF RFID systems with 3-D radiation pattern. IEEE Internet of Things J 4(4):905–916
5. Yu H, Chen J, Hsiang T (2015) Design and implementation of a real-time object location system based on passive RFID tags. IEEE Sens J 15(9):5015–5023

6. GS1 (2020) Regulatory status for using RFID in the EPC Gen2 (860 to 960 MHz) band of the UHF spectrum. [S.l.]
7. Agência Nacional de Telecomunicações. https://www.anatel.gov.br/institucional
8. European Telecommunications Standards Institute. https://www.etsi.org
9. Balanis CA (2005) Antenna radiation efficiency. In: Antenna theory: analysis and design, 3rd edn. Wiley-Interscience, New Jersey, pp 85–87
10. Rao KVS, Nikitin PV, Lam SF (2005) Antenna design for uhf rfid tags: a review and a practical application. IEEE Trans Antennas Propag 53(1):3870–3876
11. Balanis CA (2005) Gain. In: Antenna theory: analysis and design, 3rd edn. Wiley-Interscience, New Jersey, pp 65–69
12. ALIEN (2016) ALN-9770 "Bat" Battery Inlay. [S.l.]
13. IMPINJ (2020) Monza R6 Tag Chip Datasheet. [S.l.]

Flydubai 981: An Analysis of the Accident Through the Quantitative Loss of Control Criteria

Vinicius Birindelli Caracik(✉), João Paulo Macedo, and Jorge Henrique Bidinotto

Universidade de São Paulo, São Carlos, SP, Brazil
{vinicius.caracik,joao.macedo}@usp.br, jhbidi@sc.usp.br

Abstract. Aeronautical accidents and related casualties occur since the beginning of aviation. In the past decade, loss of control in-flight (LOC-I) occurrences were responsible for several aviation-related deaths. LOC-I accidents are highly complex events in which contributing factors of various natures combine. Quantitative approaches can significantly contribute to the phenomenon's characterization, providing new information and perspectives for investigation teams. In the present work, the Quantitative Loss of Control Criteria (QLC) is applied to FDB981 accident data in order to validate the quantitative approach to flight parameters for investigation purposes. Through the use of the quantitative approach, the aircraft dynamics are observed from the perspective of the deviation from normal flight conditions, which, ultimately, the QLC classifies as a LOC-I. It follows that the application of the QLC criteria achieves the objective of providing a different approach to air crash investigation procedures while determining the loss of control.

Keywords: Loss of control in flight · Quantitative Loss of Control Criteria · Aircraft accident investigation · Aviation safety

1 Introduction

Since the beginning of commercial aviation, especially in the jet era, aeronautical accidents have been prevented by the increasing knowledge and safety recommendations resulting from accident investigations. This philosophy helped the fatal accident rate per million flight drop to 0.11 in 2019 in comparison to a rate of more than 11 in 1959, according to Airbus (see [1], p. 13).

The technology incorporated in the most recent aircraft generation allows the addressing of recurrent issues. One example is the development of the Ground Proximity Warning Systems (GPWS) over the 1980s and 1990s, specifically oriented to prevent Controlled Flight Into Terrain (CFIT) accidents. The system, widely spread into commercial aviation, definitely helped to enhance aviation safety records.

More recently, loss of control in-flight (LOC-I) became the accident associated with the highest number of fatalities. According to Airbus (see [1], p. 24), from 1999 to 2019,

Y. Iano et al. (Eds.): BTSym 2020, SIST 233, pp. 759–765, 2021.
https://doi.org/10.1007/978-3-030-75680-2_84

LOC-I alone accounted for 33% of fatal accidents. For such a reason, and being CFIT gradually mitigated, the industry, allied to regulators and operators in the Commercial Aviation Safety Team (CAST), turned its' attention to LOC-I occurrences. The study of several occurrences classified as loss of control resulted in many safety enhancement recommendations. For example, the development of flight envelope and control protection technologies (see [2], p. 32), today widely used in fourth-generation aircraft. One other recommendation of CAST's report is an investigation about how to quantitatively characterize a LOC-I event (see [2], p. 22). In 2004, Wilborn and Foster (see [3]), representing private and governamental institutions, respectively, developed a quantitative approach to determine through the analysis of flight data the occurrence of a loss of control condition. The approach, named Quantitative Loss of Control Criteria (QLC), is used to investigate a real accident in this paper. Other studies, such as Macedo and Bidinotto (see [4]), relate this quantitative approach to pilots' assessment on the controllability, with the Cooper-Harper scale (see [5]), of aircraft in different flight conditions.

According to the International Air Transport Association (see [6], p. 48–49), between 2015 and 2019, from the 292 accidents in commercial aviation, 22 were classified as LOC-I accidents, being responsible for 69.8\% of the total deaths. This period has recent examples to be analyzed via the criteria described by Wilborn and Foster (see [3]). Flydubai's flight 981, which crashed in Rostov-on-Don, Russia, is an example and was selected as the object of study.

2 Methodology

2.1 Quantitative Loss of Control Criteria

In an effort to better understand the loss of control phenomenon and ultimately reduce related casualties, private and governamental institutions worked in the development of a quantitative complement to the qualitative description of an event classified as LOC-I. Wilborn and Foster (see [3]) studied the accidents displayed in CAST's report (see [2]) concerning loss of control occurrences to determine which flight parameters better describe LOC-I situations. Their research proposes that plotting different flight parameters against each other, instead of the traditional timeline study of the events, may provide a different insight over the facts for investigators.

According to Wilborn and Foster (see [3], p. 3), the primary causes related to LOC-I events are identified as stall, sideslip-induced rolls, rolls from other causes, pilot induced oscillations (PIO), and yaw, alongside with overspeed as an important factor in the majority of CAST LOC-I events. Thus, the flight parameters which best relate to these causes and that should be assessed in order to flag a loss of control condition are promptly provided by the flight data recorder (FDR); these are:

- Angle of attack (α);
- Sideslip angle (β);
- Pitch attitude (θ);
- Roll attitude (φ);
- Equivalent airspeed (V_E);
- Normal load factor (n);

- Pitch control (δ_e or δ_c);
- Pitch rate (q);
- Roll control (δ_a or δ_w);
- Roll rate (p).

The QLC methodology consists of the evaluation of five envelopes that capture the relationship between the listed parameters.

Adverse Aerodynamics Envelope (AA). This envelope maps the normalized sideslip angle (β_{NORM}) against the normalized angle of attack (α_{NORM}). The attack angle is normalized by the angle of attack corresponding to the stall warning activation (α_{SW}), while the sideslip angle for a non-crabbed approach in maximum demonstrated crosswind (β_{MDXW}) serves as the β normalization parameter. In the present paper, however, due to the lack of data, crabbed approach data is used to determine β_{MDXW}.

The α and β limits reflect the usual operational limits encountered by pilots in normal conditions, including emergency maneuvers predicted in the manuals. Hence, this envelope indicates stall, sideslip-induced stall, and large sideslip angles.

Unusual Attitude Envelope (UA). The unusual attitude envelope maps roll attitude against pitch attitude comparing the data to the limits accepted by the industry as "usual attitude", which are $-45° \leq \varphi \leq +45°$ and $-10° \leq \theta \leq +25°$.

The envelope indicates sideslip-induced roll, roll from other causes, and nose-high or nose-low attitudes. The axis in which the upset condition was initiated also is important to determine the correct recovery technique.

Structural Integrity Envelope (SI). The structural integrity envelope considers a function of the normalized speed (V_{NORM}) and the normal load factor (n), both depending on the aircraft's flap configuration. This envelope considers stall warning speed (V_{SW}), the maximum operating speed (V_{MO}), maximum flaps extended speed (V_{FE}), indicating insufficient or excessive airspeed, excessive structural load, and accelerated stall. The limits of n are -1.0 g's to $+2.5$ g's for flaps-up operations (shown in full lines in Fig. 1) and 0.0 g's to $+2.0$ g's for flaps-down operations (shown in dashed lines in Fig. 1), as required per Federal Aviation Regulations Part 25 for structural loads. The normalization is given by Eqs. (1) for flaps-up and (2) for flaps-down configurations, respectively.

$$V_{NORM} = \frac{V_E - V_{SW}}{V_{MO} - V_{SW}} \tag{1}$$

$$V_{NORM} = \frac{V_E - V_{SW}}{V_{FE} - V_{SW}} \tag{2}$$

Dynamic Pitch Control Envelope (DPC). The DPC maps the percentage of control deflection against the dynamic pitch attitude (θ') of the aircraft, being the latter the sum of the current pitch attitude with the pitch rate in one second, as shown in the Eq. (3).

$$\theta' = \theta + \theta[1] \tag{3}$$

This enables the assessment of the control usage compared to the aircraft current and trend states, allowing the determination of how the pilot's commands influenced the aircraft's attitude dynamics. For example, an aircraft with a nose-down attitude with a positive pitch rate has a reduced dynamic pitch attitude and better chances of being in normal conditions according. This envelope's limits are similar to the pitch limits used in the UA envelope with tapered corners from the upper and lower limits to full negative and full positive deflection percentages, respectively. Stabilizer mistrim, mechanical failures, longitudinal PIO, and inappropriate pitch control inputs typically appear in this envelope.

Dynamic Roll Control Envelope (DRC). The DRC is analog to the DPC but considers the lateral axis. It maps the dynamic roll attitude (φ'), calculated through the Eq. (4), against the control authority in the roll axis, in percentage.

$$\varphi' = \varphi + \varphi[1] \tag{4}$$

It indicates rolls from sideslip, wake vortex encounters, structural failures, lateral PIO, or inappropriate roll control inputs.

Loss of Control Determination. Wilborn and Foster (see [3], p. 5–9) applied the QLC to different flight data, such as flight test maneuvers, in order to determine how they confirm the loss of control. The criteria were developed considering specific flight test maneuvers, usually considered as marginally controllable or even as loss of control. This way, according to Wilborn and Foster (see [3]), the excursion of three or more envelopes configure the loss of control of a commercial aircraft.

Critical Window. According to Wilborn and Foster (see [3], p. 9), the evaluation of the envelopes excursion in a time-history format may be used to further characterize the events, assessing the interval between each excursion and the order in which envelopes are excursed. The critical window is the time between the initial upset condition, given by the first excursion, and the loss of control. It shows the time the crew had to deal with an upset condition until the control was lost. The order in which the envelopes excursions occur may help investigators to determine the chain of events that lead to the accident (see [3], p. 10).

2.2 Analysis of the Flydubai's Accident

On March 19th, 2016, Flydubai's Boeing 737-8KN, callsign A6-FDN, was performing a two-way trip from Dubai International Airport to Rostov-on-Don International Airport. According to the Interstate Aviation Committee's (IAC) final accident report (see [7]), during the approach in Rostov-on-Don, the airborne wind shear detection and alert system predicted the occurrence of wind shear ahead, sounding a predictive wind shear warning. The crew decided to abort the approach, and after performing a wind shear escape manoeuvre, waited in a holding pattern for safe weather conditions to perform a second try.

After waiting for almost two hours, the second approach was initiated. Due to overspeed, the crew decided to perform another go-around, however unsuccessfully. The Pilot In-Command (PIC) increased the engines thrust to takeoff/go-around (TO/GA) power and tried to maintain the nose down, using the pitch trim. Allied with irregular inputs on the control column, the aircraft started to pitch-down quickly, resulting in negative load factor conditions.

After altitude oscillations, the Boeing 737 pitched down, entering in a rapid descent. Although the normal procedure would be to pitch-up the aircraft, the report (see [7]) shows that the inputs were to pitch-down even more. Ultimately, the aircraft crashed on the runway at almost 340 knots, with a final nose-down attitude of 50° and 60° banked to the left. The accident resulted in the death of all 62 occupants.

2.3 Flight Data

In order to determine the contributing factors to the accident, flight recorders play a crucial role in the investigation. The Cabin Voice Recorder (CVR) contains the last two hours of sounds recorded in the cockpit, allowing investigators to determine what happened inside the cockpit. The Flight Data Recorder (FDR) contains flight-related data that allows the reconstruction of the moments before the accident. Both were found at the crash site and investigated by IAC. To which concerns the present paper, the analysis of A6-FDN's flight used FDR extracted data, available in the final accident report (see [7]).

In addition to the available FDR data, there is a simulation performed by the manufacturer to determine the aircraft's recovery capabilities, using the parameters recorded by the FDR. The relevant simulation data, which was published by IAC (see [7]), consisted of the last 93 s of the flight. Due to the 737 engineering model constraints, the simulation was limited up to three seconds before the impact, thus, assessing 90 of the 93 mentioned seconds.

Since the final events of that flight, such as the end of the second approach, the beginning of the second go-around, and the moment of the crash, occur in this 93 s period, it was selected as the analysis window of this paper.

3 Results and Discussion

The results of the application of the QLC to A6-FDN's flight data are shown in Fig. 1[1].

According to the methodology proposed by the QLC, all five envelopes were crossed. Indeed, as indicated by the black asterisks, at 00:41:35 UTC a total of three envelopes are extrapolated, a sufficient condition for the method to classify the event as a LOC-I. Furthermore, the control of the aircraft was not recovered after it was lost since at least three envelopes were being exceeded simultaneously until the moment of impact.

The first excursion occurred 50.8 s before the impact, while the third excursion occurred 13.7 s before the crash, determining a critical window of 37.1 s. Moreover, after the third excursion, the subsequent envelope crossings occurred much more rapidly: 1.4 s between the third and the fourth and 2.0 s between the fourth and the fifth.

Regarding the order of excursions, it followed the sequence AA-SI-DPC-UA-DRC. Going back into the facts pointed by the IAC (see [7]), due to overspeed at the flap 30

[1] The plot lines are coloured according to the timeline of the accident, being the time indications displayed in the Universal Coordinated Time (UTC) standard.

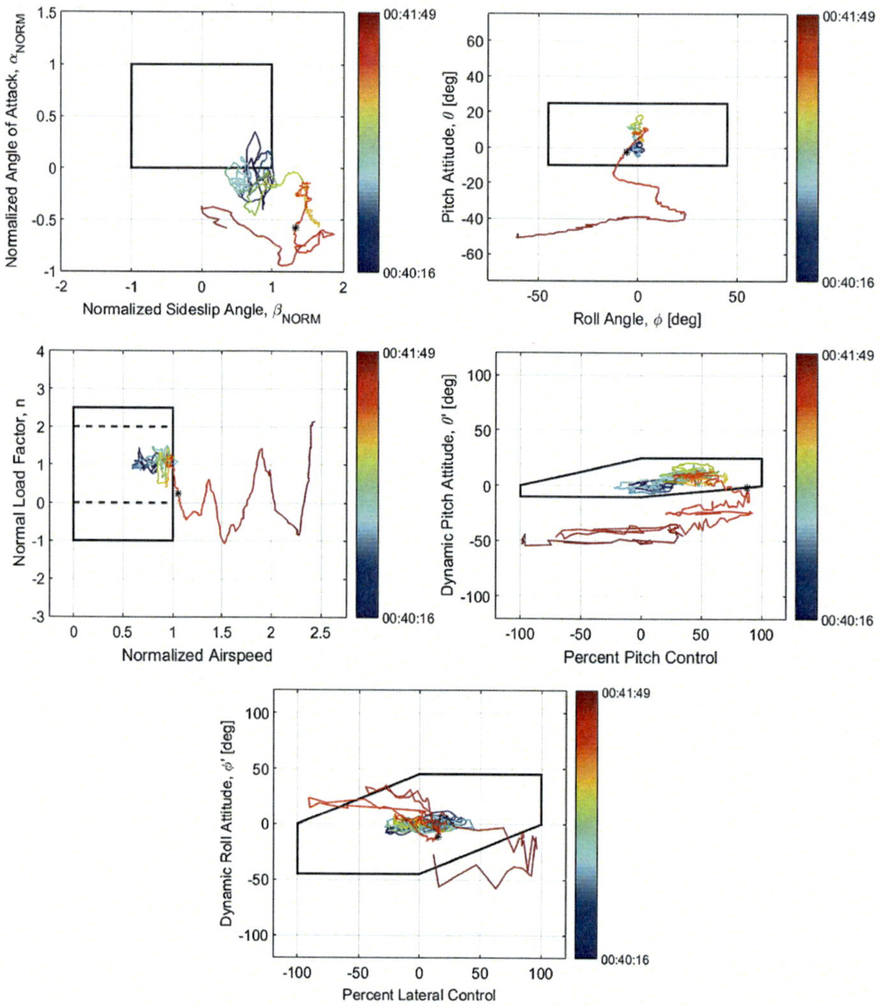

Fig. 1. The five QLC envelopes for FDB981's flight data. From left to right, top to bottom: AA, UA, SI, DPC, DRC. Source: Produced by the authors based on A6-FDN's flight data.

configuration, the crew initiated a second go-around. The results shown in Fig. 1 show that the initial upset condition came from an insufficient angle of attack (first excursion – AA) that, allied with the reconfiguration of the aircraft to flaps 15, impacted on the capability to maintain the correct go-around profile. With the engine thrust increased to TO/GA power, the crew experienced overspeed (second excursion – SI) and a pitch-up moment, leading the PIC to use the pitch trim switch to trim the nose down. Since the PIC maintained the switch pressed for 12 s while pushing the control column, the following pitch down movement of the aircraft was aggressive and with a significant pitch rate, resulting in a negative load factor condition. According to IAC (see [7], p. 158), a negative load condition can easily cause pilot disorientation, especially for those not trained for these circumstances. After the peak in altitude and the following disorientation, the PIC pushed the control column, pitching the aircraft down in addition to the pitch trim inputs earlier provided (third excursion – DPC). These three reported envelope crossings already flag a LOC-I condition to the QLC. On the continuity of the event, the aircraft descended rapidly into an unusual pitch attitude (fourth excursion – UA) and, in a condition of complete spatial disorientation (see [7], p. 155) of the PIC, the aircraft was rolled to the left with large control wheel inputs (fifth excursion – DRC).

4 Conclusion

Flydubai's A6-FDN crashed after a loss of control in-flight. After a failed second attempt to approach Rostov-on-Don's International Airport, the PIC started a second go-around with an incorrect procedure. The control was lost, resulting in the loss of 62 lives.

Although LOC-I fatal accident rates are gradually diminishing, it is still a very complex event. Several studies have been done in order to better understand the phenomenon. Wilborn and Foster (see [3]) developed quantitative criteria to determine the occurrence of loss of control of commercial aircraft through flight data analysis.

From the application of the method to a real case, Flydubai's flight 981, it follows that the criteria constituted an adequate tool to assess LOC-I since it confirms the loss of control using the FDR data. Although it correctly assessed LOC-I occurrence in this particular case, more events must be analyzed via the criteria.

References

1. Airbus (2020) A statistical analysis of commercial aviation accidents 1958–2019. Technical report, Airbus, pp 1–31. https://www.airbus.com/content/dam/corporate-topics/publications/safety-first/Statistical-Analysis-of-Comercial-Aviation-Accidents-1958-2019.pdf
2. CAST (2000) JSAT Loss of control - CAST approved final report - Results and analysis. Commercial Aviation Safety Team, pp 1–246. https://www.cast-safety.org/pdf/jsat_loss-control.pdf
3. Wilborn JE, Foster JV (2004) Defining commercial transport loss-of-control: a quantitative approach. In: AIAA atmospheric flight mechanics conference and exhibit, american institute of aeronautics and astronautics, Providence, USA, pp 1–11. https://doi.org/10.2514/6.2004-4811
4. Macedo JP, Bidinotto JH, Bromfield MA (2020) Loss of control in-flight: comparing qualitative pilot opinion with quantitative flight data. In: AIAA AVIATION Forum, American Institute of Aeronautics and Astronautics, pp 1–23. https://doi.org/10.2514/6.2020-2911
5. Cooper GE, Harper RP (1969) The use of pilot rating in the evaluation of aircraft handling qualities. NASA Technical Note D-5153, NASA, pp 1–60. https://ntrs.nasa.gov/search.jsp?R=19690013177
6. IATA (2020) Safety report 2019. Technical report, International Air Transport Association, pp 1–260. https://www.iata.org/en/publications/safety-report
7. IAC (2019) Boeing 737-8KN A6-FDN fatal accident - final report. Interstate Aviation Committee, pp 1–175. https://www.mak-iac.org/upload/iblock/3d1/report_a6-fdn_eng.pdf

Arduino-Based Prototype for Measuring Chlorinated Solution Volumes in Water Treatment Systems

Marco Alfredo Sánchez Peña(✉), Segundo Dobbertin Sánchez, and Rosa López Martos

Environmental Engineering, Universidad Privada del Norte, Av. Vía de Evitamiento Norte s/n cuadra 15, Cajamarca, Perú
marco.sanchez@upn.pe, {segundo.dobbertin,rosa.lopez}@upn.edu.pe

Abstract. Poor drinking water quality in rural areas is still a common issue in Latin American countries, mainly because of the flaws in drinking water systems. For example, in Peru, drinking water chlorination systems exhibit a large number of deficiencies from the dosing process to the monitoring of chlorine volumes in storage tanks. In fact, as the water tanks are often far from communities, chlorine solution levels are hardly ever properly monitored. Therefore, the system proposed in this paper will allow users to monitor the real-time chlorinated solution volume levels in drinking water systems, thus, rendering their operations effective. The design of the prototype relies on an intelligent two-way system comprising of a sender and a receiver. Both parts are Arduino-based and communicate with each other via a long-range wireless transmission protocol. In addition, the system features low energy consumption and uses cloud connectivity for storing chlorination process information.

Keywords: Arduino · Chlorination · Water treatment system · Intelligent systems · Ultrasound

1 Introduction

Water is the most important liquid for all living forms. Over the years, water has been at the core of sustainable and socioeconomic development, energy, food production, healthy ecosystems, and, primarily, human survival itself [1]. Conversely, poor sanitation and pollution are directly related to the transmission of diseases such as hepatitis A, cholera, diarrhea, typhoid fever, and poliomyelitis.

Today, the importance of using technological tools is undisputable in all fields. According to WHO (World Health Organization), at least 1.8 billion people worldwide use a drinking water source contaminated with feces. Fortunately, technology has been able to offer solutions for this kind of issue [2].

Thus, throughout these years, one of the technological tools such as Arduino has been the protagonist of thousands of technological and educational projects [3]. In fact, to this

Y. Iano et al. (Eds.): BTSym 2020, SIST 233, pp. 766–773, 2021.
https://doi.org/10.1007/978-3-030-75680-2_85

date, Arduino continues to provide open-source hardware and software for developing new projects with fast-evolving current technologies [4].

Indeed, several projects attempting to solve water problems using technological tools have been developed over the years, such as the "Wireless Intelligent Water Level Instrument Device" developed by Wang Zhongming, He Yunhua, and Liu Wei [5]. This device adopts the low-power-consumption wireless transmission module. It does not need an external antenna, it is convenient to install and debug, and it offers a long service life. It can be debugged and configured by keys or remotely, which is convenient for the maintenance and updating of the device [5].

Another example is the "Wireless Intelligent Inspection Well Liquid Level Monitoring Terminal" project developed by Li Bizhong, Zhao Enguo, Shi Zhaoyang, Liu Renjie, Zhao He, Huang Jinyuan, Wang Chuang, Hu Gangquan, and Lin Weicong [6]. This terminal comprises a pressure-type liquid level sensor, a first-stage water immersion sensor, a second-stage water immersion sensor, and a monitoring terminal body. The pressure-type liquid level sensor is installed in the middle area of a drainage pipeline [6].

Likewise, the "Internet-of-Things-based Automatic Monitoring and Transmitting Device for Precipitation Well Water Levels" project, developed by Zhang Yalong [7], performs a regular automatic water level monitoring function to accurately and stably monitor water levels while improving the limitations associated with using traditional monitoring, thus, avoiding operations personnel from conducting on-site measurements [7].

The main features of an Arduino-based intelligent system are its versatility, easy implementation, and low cost [8], which makes it a viable option for application in different technological fields, such as autonomous navigation [9], intelligent system to avoid obstacles[10], photovoltaic panel data collection systems [11, 12], intelligent lighting systems [13], and data collection used in laboratory work [14].

This prototype was developed for the implementation of an intelligent system that monitors chlorinated solution tank levels for the water treatment systems in Peruvian rural areas and provides real-time solution volume information to the users and operators. Likewise, it was only tested in the laboratory; it was not implemented due to the pandemic covid-19. However, government measures are expected to allow free transit for implementation.

2 Methodology

2.1 General System Design

The volume measuring system uses an ultrasound sensor controlled by an Arduino Nano board, powered by an independent solar battery. This system captures the data required to calculate volume levels and transmits the information signal through a long-range antenna. This is the only device that transmits the data collected. The receiving device, powered by a direct current line, receives the information through a long-range antenna. This antenna is connected to an Arduino Mega board, which processes the information, performs the required calculations, and obtains the current chlorine solution volume level inside the tank. Once this volume level has been estimated, it is displayed on an

LCD screen for immediate viewing. Then, the device interfaces with the server via the "ThingSpeak" platform, where it uploads this information to the cloud through a Wi-Fi module so that it can be available anywhere and anytime for monitoring (Fig. 1). The representation of the system topology is in Fig. 2.

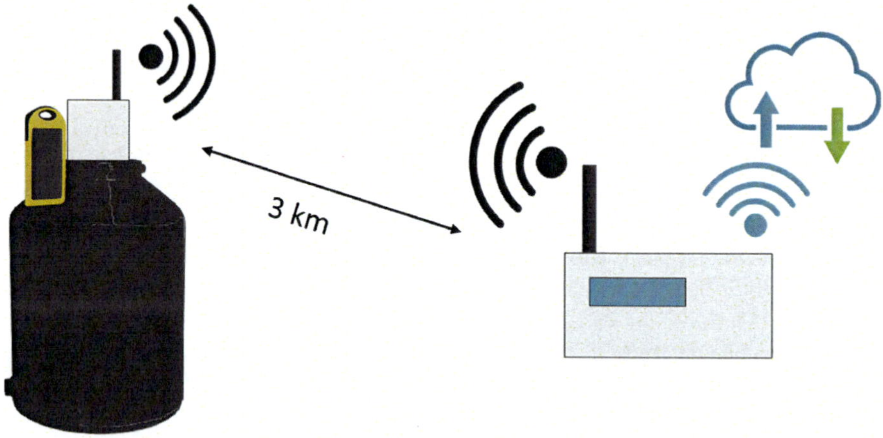

Fig. 1. Volume measuring system structure, signal sender, and receiver

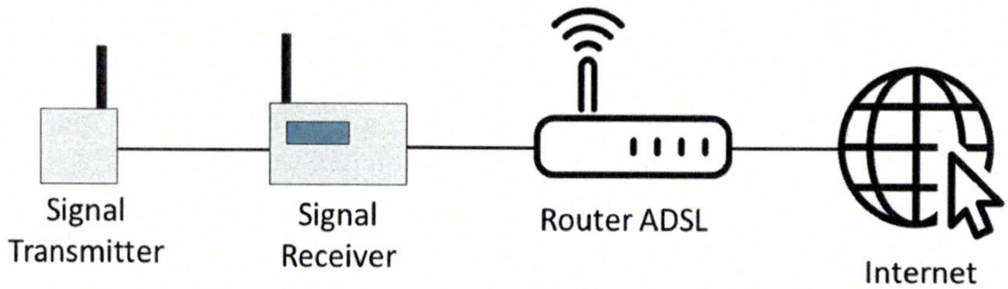

Fig. 2. Communication topology of the chlorine solution volume measurement system prototype.

2.2 System Hardware Design

Transmitter Design. To remain energetically convenient, the sender uses an HC-SR04 ultrasound sensor, which is not only an affordable but also a less energy-consuming distance sensor. The system was encapsulated in a custom-designed housing in a 3D printer. The sensor did not present any problems in the test environment, so the same behavior is expected in field tests. It operates at 5 V and 15 mA and uses ultrasound at a frequency of 40 kHz to determine the distance of an object located at approximately 2 to 450 cm [15]. Arduino Nano board, an 18.5 mm × 43.2 mm development board that uses a 5 V ATmega328P microcontroller by Atmel [16], controls this sensor. Likewise, this board controls the LoRa E32-TTL-100 SX1278 device, which is a long-range and low-energy consumption wireless transmission protocol developed by Semtech. This device uses an SMA antenna, operates at 2.3 to 5.5 V with a maximum effective range of 3 km, has an intensity of 2 dB, an information transfer rate of 2.4 Kbps, and working frequency ranging from 410 MHz to 441 MHz, with an encoding rate of 4/5 and a spreading factor (SF) of 12, set by default, no tests were performed with other SF. [17]. This long-range device transmits the information obtained by the ultrasound sensor over the air, and the entire system is powered by a 5 V, 3000 mA solar battery protected by a weatherproof housing (Fig. 3a and b). The environmental conditions of the emitter's work in terms of relative humidity in the area are 35 to 50% daily average.

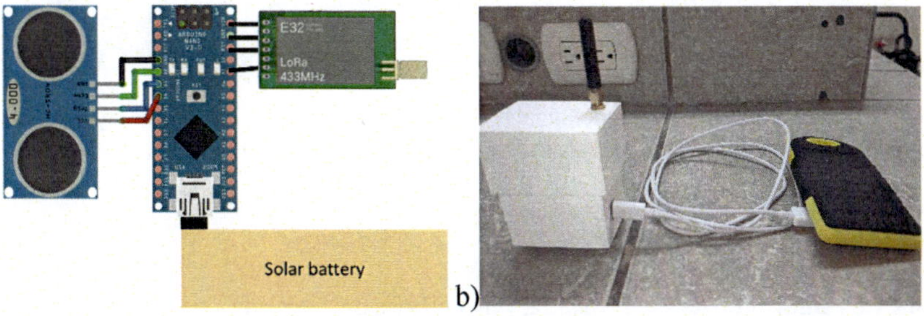

a) b)

Fig. 3. a) Diagram of the long-range signal transmitter device, b) long-range signal transmitter device

Receiver Design. The receiver consists of two parts. First is based on a long-range LoRa E32-TTL-100 SX1278 transmission and reception device [17], exactly like the sender device. This device is also powered and controlled by Arduino Mega board, a 101.52 mm × 53.3 mm development board that uses a 5 V ATmega 2560 microcontroller [18]. The information received is processed by the Arduino Mega board, which estimates the current solution volume within the tank and immediately displays this value on a standard 16 × 2 LCD screen. The second part consists of a Wi-Fi ESP 8266-S01 (ESP-01) module that uses an ultra-low-power UART-WiFi chip designed for mobile devices and Internet of Things (IoT) applications as it features a rich hardware interface that supports UART, I2 C, PWM, GPIO, and ADC and operates at 3.0 to 3.6 V [19]. This module uploads the resulting volume to the cloud and stores it in "ThingSpeak," which uses default security protocols, an open-source IoT application programming interface (API) used to store and retrieve data over the Internet through the HTTP and MQTT protocols (Fig. 4a and b). The entire receiver is directly powered by a 9 V, 2 A charger. At this stage, no specific security layers have been designed in the data transmission of the prototype.

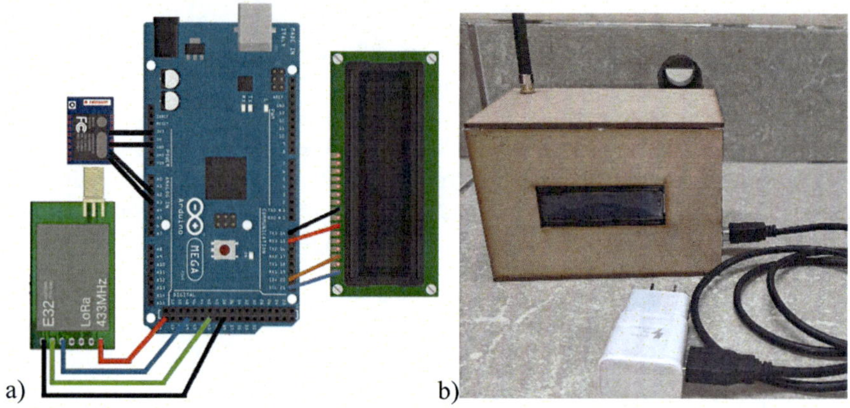

Fig. 4. a) Diagram of the long-range signal reception device, responsible for displaying the volume information and uploading this information to the cloud, b) long-range signal reception device.

2.3 System Software Design

Sender Design. The sender is located vertically on the tank lid. This way, the transducers are pointing perpendicularly toward the surface of the chlorinated solution. One of the transducers sends an ultrasound pulse every 10 μs that bounces off the solution, and the echo produced is captured by another transducer. The distance between the transducers and the chlorinated solution is proportional to the time taken by the sound wave to reach the fluid and the echo to bounce off the surface. Considering the sound speed of 340 m/s and converting it to the required units, we obtain 59 cm/μs. As the total distance traveled is short, we can safely consider this speed as constant. Therefore, using the speed and travel time values, we obtain Eq. (1), which provides the distance between the sensor and the solution surface. This value is transmitted and then used by the receiver to calculate the current solution volume level. The entire process is shown in Fig. 5.

$$h_1 = \frac{t}{59}[\text{cm}] \quad (1)$$

Where h_1 is the distance and t the time the sound takes to travel from the sending transducer, bounce off the surface, and reach the receiving transducer.

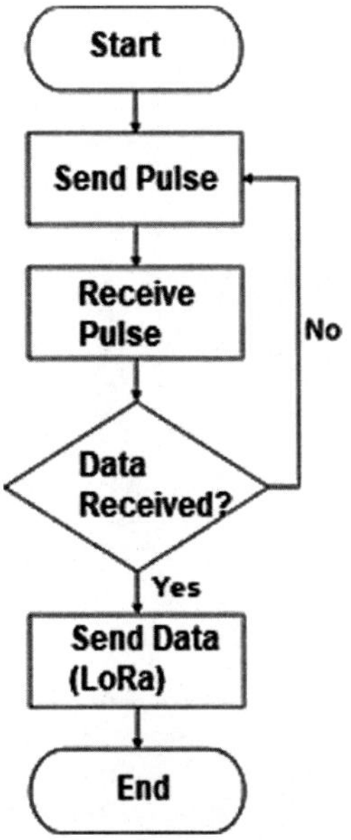

Fig. 5. Sender flow chart

Receiver Design. Upon receiving the distance value from the ultrasound sensor to the surface of the fluid in the tank, the receiver uses this information to estimate the current solution volume levels. This estimation assumes that all the tanks of a given capacity have standard measurements according to the manufacturer. Hence, based on tank dimensions, such as height, diameter, and the maximum solution storage height, the current volume level can be determined as specified below. First, we calculate the maximum height to which the chlorine solution may reach:

$$h_2 = H - h_1 \tag{2}$$

Where H is the total height of the tank, h_2 is the maximum height of the solution, and h_1 is the distance from the ultrasound sensor to the solution surface.

With the diameter D of the tank, we can calculate the volume of the stored content using Eq. (3):

$$V = \pi * \frac{D^2}{4} * h_2[\mathrm{cm}^3] \tag{3}$$

The estimated volume is then converted into liters, and the resulting value is displayed on the LCD screen. Likewise, the Arduino Mega board uploads this information to the cloud using the ESP8266-S01 (ESP-01). This entire process is shown in Fig. 6.

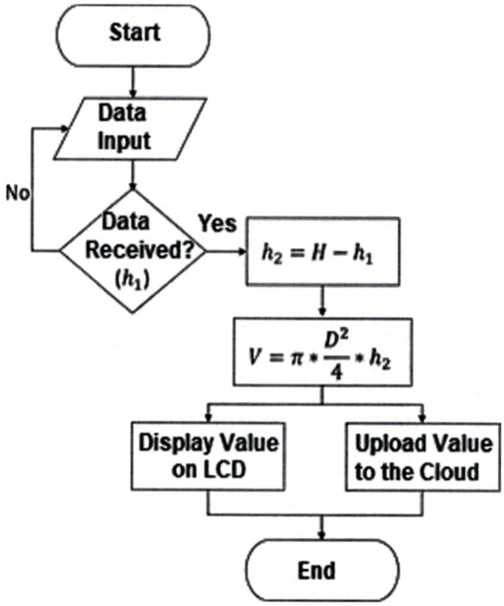

Fig. 6. Receiver flow chart.

3 Conclusions

As part of this project, the authors developed and tested a prototype of an intelligent system based on Arduino using free software and libraries available on the Github platform that monitors the volume levels of chlorine solution in water treatment tanks in rural Peru. This prototype will allow rural communities to optimize their water chlorination processes and reduce the risk of disease due to microbiological contamination of the water. In addition, the system achieved an efficient communication between the emitter and the receiver, which allowed the real volumes of chlorinated solution to be displayed in real-time on the receiver's screen for the correct operation of the user.

References

1. United Nations. https://www.un.org/es/sections/issues-depth/water/index.html
2. Armenta MH (2017) Tecnología, la solución para nuestros problemas con el agua. https://www.forbes.com.mx/tecnologia-la-solucion-para-nuestros-problemas-con-el-agua
3. Bisták P (2019) Arduino support for personalized learning of control theory basics. ScienceDirect 52(27):217–221. https://doi.org/10.1016/j.ifacol.2019.12.759
4. Arduino A (2018) Aprendiendo Arduino. https://aprendiendoarduino.wordpress.com/2018/05/06/arduino-en-la-educacion
5. Zhongming W, Yunhua H, Wei L (2019) Wireless intelligent water level instrument device. Espace net
6. Bizhong L, Enguo Z, Zhaoyang S, Renjie L, He Z, Jinyuan H, Chuang W, Gangquan H, Weicong L (2019) Wireless intelligent inspection well liquid level monitoring terminal, Espacenet

7. Yalong Z (2019) Internet-of-things-based automatic monitoring and transmitting device for water level of precipitation well, Espacenet
8. Gerardo MS, Zach LG, Jonathan D (2019) A low-cost Arduino-based datalogger with cellular modem and FTP communication for irrigation water use monitoring to enable access to CropManage. ScienceDirect 6:e00066
9. Emilian Oltean S (2019) Mobile robot platform with Arduino Uno and Raspberry Pi for autonomous navigation. ScienceDirect, 572–577
10. Feng Li Z, Fan Li X, Jian Yang Y, Xiao J, Wen X (2020) Intelligent tracking obstacle avoidance wheel robot based on arduino. ScienceDirect 166:274–278
11. Vestenicky M, Matuska S, Hudec R (2019) Simple method of photovoltaic panel power characteristic measurement based on Arduino hardware platform. ScienceDirect, 579–585
12. Ulloaa C, Nuñeza JM, Suáreza A, Lin C (2017) Design and development of a PV-T test bench based on Arduino. ScienceDirect, 71–75
13. Li Z, Li J, Li X, Yang Y, Xiao J, Xu B (2020) Design of office intelligent lighting system based on Arduino. ScienceDirect, 134–138
14. Vallejo W, Diaz-Uribe C, Fajardo C (2020) Do-it-yourself methodology for calorimeter construction based in Arduino data acquisition device for introductory chemical laboratories. ScienceDirect 6(3):e03591
15. ElecFreaks. https://cdn.sparkfun.com/datasheets/Sensors/Proximity/HCSR04.pdf
16. Arduino. https://www.arduino.cc/en/uploads/Main/ArduinoNanoManual23.pdf
17. Alldatasheet. https://pdf1.alldatasheet.com/datasheet-pdf/view/800241/SEMTECH/SX1278.html
18. Arduino. https://store.arduino.cc/usa/mega-2560-r3
19. Alldatasheet. https://pdf1.alldatasheet.com/datasheet-pdf/view/1179098/ETC2/ESP-01.html

Fault Detection in Transmission Towers Using State Observers

Gilberto Pechoto de Melo(✉), Lucas Ferreira Bertão, and Camilo Rocha Araújo

Department of Mechanical Engineering, UNESP - Ilha Solteira, Av. Brasil Centro, 56, Ilha Solteira, SP CEP 15385-000, Brazil
{gilberto.pechoto,camilo.rocha}@unesp.br

Abstract. Over the years, the demand for tools that facilitate the analysis of static and dynamic systems only increases, and the fidelity of the results obtained with these tools is also another point of interest. The present work aims to analyze towers of transmission lines used by companies responsible for producing and selling electricity. The analyses made were based on the fact that each element of the tower behaves like a beam element and the finite element method was used for the development of mathematical models, with the structures being subjected to general and random forces simulating winds. The results showed that simple and small forces cause structures to suffer from vibrations of large amplitudes resulting in accidents, since the most important are the existing frequencies, leading them to resonances. It was also used the method of state observers to reconstruct unmeasured or difficult to access states and detect possible failures and/or crack propagation. The open/closed crack model was adopted and the irregularity caused by the crack was observed in the local change in stiffness and tension.

Keywords: Transmission towers · Finite element method · State observers

1 Introduction

The first transmission line registered in Brazil was built around 1883 in the city of Diamantina, Minas Gerais, by the National Electric Energy Agency [1]. Studies on structures were not so in-depth at that time, making the projects more vulnerable to technical errors due to theoretical ignorance or even overlooking some important detail [2]. The occurrence of several accidents over the years was inevitable. Over time these accidents could be predicted and, in many cases, avoided with the advancement of technologies that involved numerical simulation, which allows the reproduction of a computational model very similar to the real model. Numerical simulations combined with physics and mathematics resulted in powerful software capable of performing analyses, whether static or dynamic, for example, Ansys® and GNU Octave.

Transmission Towers, like any other structure or mechanical system, are not entirely free from failures just because their designs have been subjected to extensive analysis on a computer. Several factors interfere and end up causing failures in structures.

Y. Iano et al. (Eds.): BTSym 2020, SIST 233, pp. 774–781, 2021.
https://doi.org/10.1007/978-3-030-75680-2_86

In the case of Transmission Towers, the main causes of failure, in addition to unforeseen structural behaviors, are external agents: rains, winds, temperature fluctuations, among others. Because of this, several techniques for detecting and locating faults have been improved, avoiding abrupt machine stops and preventing collapses in structures. Although the numerous theoretical analysis tools have reached such a level of sophistication, there are still difficulties in diagnosing problems that result in failures, such as the inaccuracy of mathematical models combined with the difficulty, or even the impossibility, of measuring relevant variables of the system. It is in this context that the State Observers fit. They consist of elaborating a model for the system or structure in question to compare the estimated output signal with the output signal measured using instruments and still manage to reconstruct the unmeasured states or in places of difficult access [3].

2 Methodology

There are countless cases of transmission lines that come to the ground due to external agents and/or design flaws. Today, Brazil's low energy production, associated with the fact that transmission lines are not carrying energy, can result in widespread chaos due to power outages. The development of mathematical models that simulate real systems' behavior is of fundamental importance in dynamic analysis. The Finite Element method, together with the Lagrange equations, can provide very satisfactory mathematical models, and the State Observers methodology can reconstruct unmeasured states, detect possible irregularities and monitor the growth of possible cracks in the structures. In an analysis of dynamic systems, the existence of forces and, consequently, displacements that vary over time must be considered. The number of degrees of freedom of a mechanical system is of fundamental importance, since it references the number of independent coordinates necessary to fully describe a system's physical state. In the case of Finite Elements, to determine the elemental Mass, Stiffness, and Damping Matrices, the number of degrees of freedom per node must be considered. Element Stiffness Matrix Each element to be discretized has its coordinates in the local and global systems. Figure 1 shows the element in its local coordinates, considering 6 degrees of freedom per node, that is, 3 displacements and 3 rotations. As such, the element matrix will be 12×12 [4].

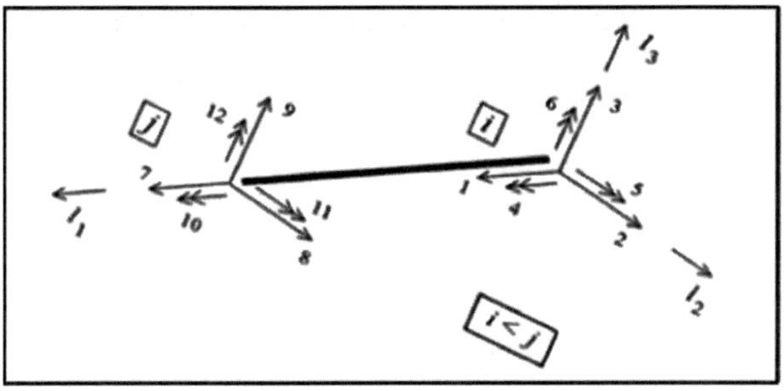

Fig. 1. Element with 6 degrees of freedom per node (local coordinate system).

In the analysis that was carried out, proportional damping was considered:

$$[C] = \alpha[K] + \beta[M] \tag{1}$$

Where [C] is the Damping matrix, [K] is the Stiffness matrix, [M] is the Mass Matrix, and α and β are constant.

2.1 State Observers

Many control systems are based on the supposition that the full state vector is available for direct measurement, but in practice, not always all the variables are available, and those unavailable must be estimated. In such a way, an observer can be built to estimate them. The schematic is as follows:

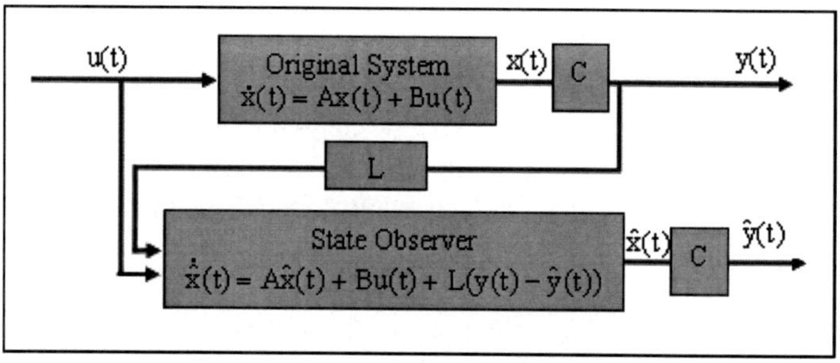

Fig. 2. Scheme of a state observer.

Basically, the observer is a copy of the original system, having the same input and almost the same differential equation. An extra term compares the actually measured output y(t) to the estimated output $\hat{y}(t)$.

Control systems using state observers can reconstruct the non-measured states or estimate the values of difficult access points in the system. However, the necessary condition for this reconstruction is that all the states should be observable [5].

Consider a linear time-invariant system described by:

$$\{\dot{x}(t)\} = [A]\{x(t)\} + [B]\{u(t)\} \text{ and } \{y(t)\} = [C_{me}]\{x(t)\} + [D]\{u(t)\} \tag{2}$$

Where: $[A] \in R^{nxn}$ is the dynamical matrix; $[B] \in R^{nxp}$ Is the input matrix; $[C_{me}] \in R^{kxn}$ Is the measure matrix; $[D] \in R^{kxp}$ Is the matrix of the direct inputs; n is the order of the system, p the inputs {u(t)}, k the outputs {y(t)}.

One of the advantages of this representation type is that the state vector {x(t)} contains enough information to completely summarize the system's last behavior, and the future behavior is governed by a simple differential equation of first order. A state observer for the system of Fig. 2 is defined by:

$$\begin{aligned}\{\dot{\bar{x}}(t)\} &= [A]\{\bar{x}(t)\} + [B]\{u(t)\} + [L](\{y(t)\} - \{\bar{y}(t)\}) \text{ and} \\ \{\bar{y}(t)\} &= [C_{me}]\{\bar{x}(t)\}\end{aligned} \tag{3}$$

Where: [L] is the observer matrix; $\{\overline{y}(t)\}$ is the output of the observer; $\{\overline{x}(t)\}$ is the states vector of the observer.

In addition, there is the possibility of several parameters failing or failing at the same time. The solution to circumvent this problem is the assembly of robust observers to all parameters that may suffer malfunctions or failures in order to achieve a more effective control. Figure 3 illustrates an observation system assembly that has an input (excitation force vector) and an output (response vector).

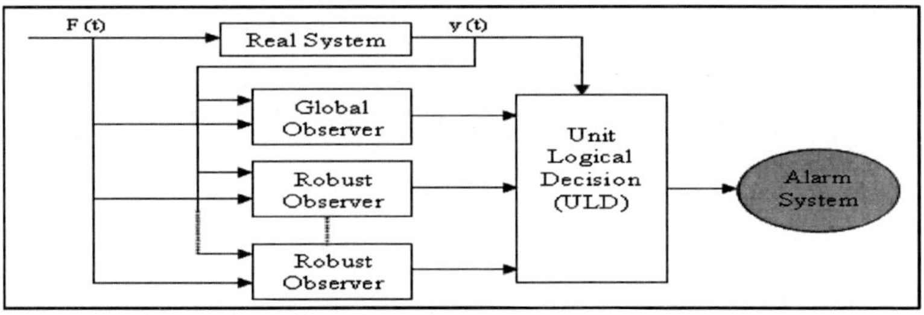

Fig. 3. Scheme of a state observer.

Still in Fig. 3, there is a logical decision unit responsible for collecting the data from the assembled observers and comparing it with the system's response to locate the changes.

2.2 Fracture Mechanics and Transmission Towers

In order to establish some brief comments on the topic, it can be said that fracture mechanics is the area of engineering that studies the causes of the appearance and propagation of failures in structures that cause the decrease of the material's resistance and may lead to the fault. This branch of study uses, for example, analytical methods of solid mechanics to calculate the driving force in a crack and experimental methods to calculate material resistance. Aeronautical engineer Alan Arnold Griffith developed the study in this area during the First World War. Their motivation was two contradictory facts: the fact that a glass block requires 100 MPa of applied voltage to fail (break) and a molecular bond needs 10000 MPa. The mechanics of solids then appear to reconcile such conflicting observations [6].

The fundamental role of a transmission tower is to accommodate transmission lines as safely and effectively as possible. The energy that travels through the transmission lines can be transported underground or under overload. Overloaded energy transport is generally the best option due to ease of maintenance and installation. In this case, the transmission lines are accommodated in metal structures of high altitudes, the transmission towers. Transmission towers are generally installed in planned and open locations and away from any inhabited place to ensure safety. Such towers are structures composed of bars combined in order to give strength and stability to the final structure. In theory, these safety measures associated with others should encompass any and all types of predictions regarding accidents that could occur in the towers, but this is not the case.

As in several other cases, the winds' action was also analyzed and standardized by the Brazilian Association of Technical Standards - ABNT. In its set of regulations, there is the standard NBR 6123, called Forces due to wind in buildings, which informs the steps to deal with the winds' action in different types of structures and buildings. Established in 1987, NBR 6123 nevertheless failed to prevent structural accidents from occurring. At the national level, it is possible to mention, as an example, the fall of towers that occurred in the south of the state of Rio de Janeiro, reaching some cities such as Resende, Itatiaia, and Porto Real [7].

A simplified mathematical model with a real tower configuration was used, using beam elements that support bending and twisting moments, since the bar element that only accepts axial effort does not reflect the real behavior. The model chosen was a self-supporting tower used by CESP in the state of São Paulo that conducts 450 kV transmission lines (Fig. 4).

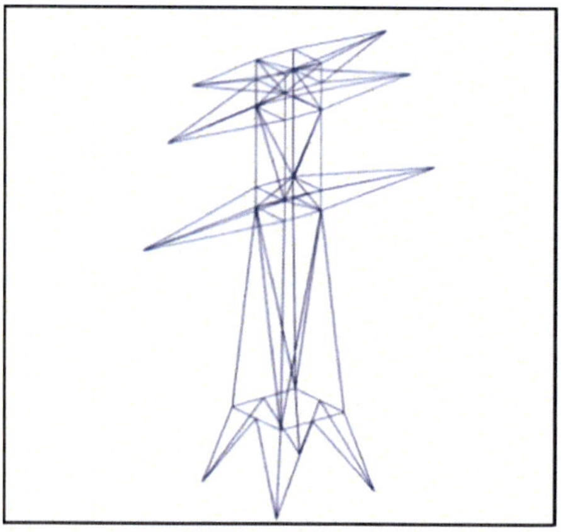

Fig. 4. Transmission tower model used in the study.

3 Fault Diagnosis via State Observers (Computational Simulation)

Detecting possible failures, a 20% change in stiffness was made in one of the most stressed elements (a previous stress calculation was performed). An analysis was made with the tower, assuming that it has all the cables connected to it and connected to the other towers.

With the conditions determined, the routine was performed and the data obtained. The tables below show the response signal of the most tensioned elements when excited at a frequency of 10 Hz, just above the third natural frequency - 7.1 Hz - and the inverse of the module of the signals' RMS values of each element for each percentage failure analyzed. The inserted failure was 5% in the stiffness matrix of the element that suffered the greatest stress for a uniform lateral loading of F = 100 kN, that is, the degrees of freedom in the x - horizontal direction - received a force vector so that the direction to the tower and not "coming out" of it (see Fig. 5). The incidence of lateral wind in the tower's structure caused a greater tension in the elements located near its base. Figure 5 shows some of these elements.

The most stressed elements are: 15 (20.75 MPa), 23 (19.60 MPa), 4 (15.61 MPa), 5 (15.26 MPa) and 3 (14.14 MPa).

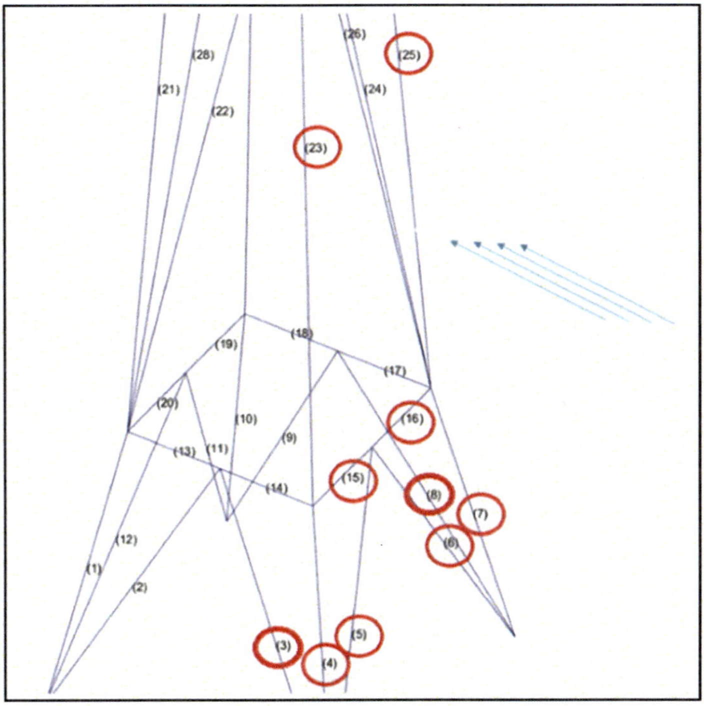

Fig. 5. The most stressed elements.

The following tables show the RMS differences between the Real System without Faults and The Global Observers. For each of the 3 cases in each figure above, the RMS difference from one signal to the other highlights the differences between them, which images sometimes failed to highlight. When no system failure was imposed (decreased rigidity), the RMS difference between the Global Observer and the Real System was of the order of E-20 (they are practically coincident). When any irregularity or loss of rigidity occurs, these curves start to distance, increasing the RMS difference regardless of percentage or location. When robust Observers curves are fitted to the percentage of failures and are compared with the Global Observer without failures, they are logically not coincident. This coincidence will start to happen when the real system starts to fail and gets close to the robust observer at the specific failure percentage. In this case, a 5% loss of rigidity loss was imposed on Element 15.

Table 1. RMS values - RMS difference. Real System with Failure x Robust Observer percentage to rigidity percentage (highlight for element 15 with 5% Failure).

	5% failure	10% failure	15% failure	20% failure
Element 15	2.475233E-16	1.028203E-6	2.131661E-6	3.306646E-6
Element 23	7.835963E-8	4.598887E-7	8.551390E-7	1.251953E-6
Element 4	1.093152E-8	7.627374E-8	1.661209E-7	2.579256E-7
Element 5	1.030128E-9	9.171479E-9	1.941660E-8	2.971915E-8
Element 3	-1.493890E-7	-3.947483E-7	-6.451633E-7	-9.006294E-7

4 Results and Conclusions

The model under analysis consists of a transmission tower embedded in the 4 points where it touches the ground, being hit by a constant wind current (with a natural frequency set at 10Hz, just above the third natural frequency - 7.1 Hz). The first two natural frequencies obtained were very close (around 5Hz), which shows the almost symmetry of the tower - avoided by the presence of the side arms that serve as guides for the line cables. From the third frequency, it is possible to notice the appearance of vibrating modes that twist the tower's structure. The justification for the third mode to be highlighted was as follows: the low frequency that it presents - 7.1 Hz - can cause it to be excited with relatively light gusts and, for being the first twisting mode of the tower, can compromise the structure by exploiting any pre-existing failure. The natural frequencies were determined: 1st - 2.5 Hz; 2nd - 2.8 Hz; 3rd - 7.1 Hz; 4th - 13.2 Hz; 5th - 15.1 Hz; 6th - 18.8 Hz; 7th - 26.7 Hz; 8th - 30.5 Hz; It was possible to observe that at the beginning of the excitation, the tower presented greater amplitude of oscillation (greater amplitude approached 5×10^{-4}) until, after almost 2 s, the movement entered in regime decreasing the amplitude of the movement.

It can be said that the short time that the tower structure takes to reach a stationary oscillation regime gives us signs of some characteristics that can be inferred from the structure under analysis. One of them is about structural stiffness (object of calculation in the computational routine using a proportional method). Structures with less rigidity are more susceptible to oscillate for a longer time when under the action of some force (whether oscillatory or not), thus taking more time to enter into regime. As in this case, the force in question is of harmonic nature. If it establishes itself for a long time in the structure without ceasing, it is possible that the occurrence of failures caused by fatigue due to the number of cycles that will perform it.

As it is a force that simulates a constant gust of wind, and that winds do not remain constant to the point of causing some type of fatigue failure at once, it can be said that if the structure is going to suffer from this type of failure, it will be after a long period. As it was predetermined that the fault should be present in one of the five most stressed elements, it was plausible that it was located in the most stressed element. The fault identification was then possible to be verified in Table 1, element 15, with a 5% loss of stiffness. Figure 6 - Module of the RMS Difference between the Response Signals.

Regarding the calculations and analyses carried out before the implementation of the state observer method, it can be said that they contributed significantly with respect to the input matrices for the observers' calculations. These matrices of stiffness, mass, and damping were obtained by the finite element method.

The failure detection was made by comparing the RMS values of each element of the structure's response signals before it was affected by the failure and after the insertion of the altered stiffness, which demonstrates the great capacity of the method.

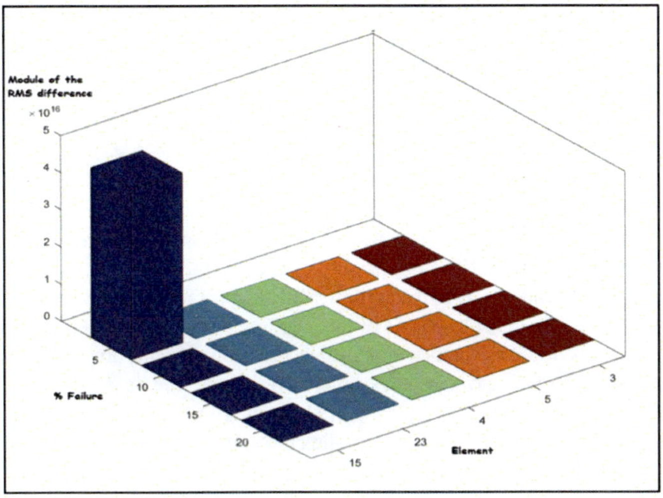

Fig. 6. Module of the inverse RMS difference between the response signals.

References

1. Guedes CFB (2010) Políticas públicas de estímulo à P&D: uma avaliação dos resultados do programa regulado pela Agência Nacional de Energia Elétrica – ANEEL, Dissertação Mestrado – UnB, Brasília
2. Behzad M, Ghadami A, Maghsoodi A, Hale JM (2013) Vibration based algorithm for crack detecion on catilever beam containing two different types of cracks. J Sound Vibr 332(24):6312–6320
3. Cavalini Jr. AA (2009) Monitoramento da integridade estrutural de sistemas mecânicos via observador de estado modal. Dissertação Mestrado – UNESP, Ilha Solteira, 128 p
4. Dawe DJ (1984). Matrix and finite element displacement analysis of structures, vol 11. Oxford University Press, USA, pp 411–487
5. Morais TS, Leão LS, Cavalini AA, Steffen V Jr, Melo GP (2019) Fault detection in rotating machinery by using the modal state observer approach. J Sound Vibr 1:1–10
6. Rao NP, Knight GMS, Lakshmanan N, Iyer NR (2012) Studies on failure of transmission line towers in testing. Eng Fail Anal 35:55–70
7. Fu X, Li HN, Wang J (2019) Failure analysis of a transmission tower subjected to combined wind and rainfall excitations. Struct Design Tall Spec Build 28:350–362

Up- and Down-Projection Unit in a Dual Branch Scenario for Single Image Super Resolution

Miguel Bozer da Silva(✉) and Ricardo Suyama

Centro de Engenharia, Modelagem e Ciências Sociais Aplicadas, Universidade Federal do ABC, Avenida dos Estados, 5001 - Bairro Santa Terezinha, Santo André, SP, Brazil
ricardo.suyama@ufabc.edu.br

Abstract. High-Resolution images are used in many fields, such as medicine, astronomy, security, and monitoring. However, in many cases, digital cameras usually do not have the desired resolution. Thus, the Super Resolution field has emerged in order to assist this need. The most used methods nowadays are based on only a single image super-resolution (SISR), and the Convolutional neural networks (CNN) are considered one of the potential methods for this task. In this paper, we analyze a dual branch model that uses the up- and down-projection units proposed by [1], and it was possible to verify that the dual branch model can outperform the single branch model. Using more up- and down-projection units can lead to a better performance in the dual branch scenario, and our model outperforms one of the proposed models of [2].

Keywords: Super-resolution · Convolutional neural networks · Deep learning · Up and down-projection Units

1 Introduction

Images can have their quality degraded due to several processes, such as undersampling, noise, and relative movements [3]. In these cases, Super-Resolution (SR) techniques can be applied to the low resolution (LR) images in order to recover the high resolution (HR) image. SR is very useful in many areas, such as surveillance [4], medical imaging [5], video applications [6], generation of depth maps for 3-D applications [7], astronomical images [8], just to mention a few.

SR methods may explore several LR images to recover the HR image [9]. But in most scenarios, we have a single LR image, and, in this scenario, another approach arises, named Single Image Super-Resolution (SISR) [10], which include interpolation-based methods, reconstruction-based methods, such as the iterative back-projection algorithm and methods that learn how to construct a HR from LR images based on training images.

Some methods following this latter approach assume that small image patches in LR and HR forms manifolds, and this can be used to explore the similarities of the local geometry between those patches in the training samples. One such technique use the locally linear embedding (LLE) algorithm [11], which can capture the local geometry of the pair of HR and LR images patches [12].

Y. Iano et al. (Eds.): BTSym 2020, SIST 233, pp. 782–789, 2021.
https://doi.org/10.1007/978-3-030-75680-2_87

Other methods explore deep learning algorithms for the SISR problem. The work of [2] has shown that the process of the reconstruction of the HR image can be viewed with a Convolutional Neural Network (CNN) based approach. This inspired many articles for the SISR using CNN, such as [5, 7, 13–15].

In this context, Haris et al. [1] proposed the Deep Back-Projection Networks (DBPN), which is based on two units named up- and down-projection units (UDPU). Compared to other state-of-art super-resolution models, the DBPN leads to better results in Peak Signal Noise Ratio (PSNR) and structural similarity index metrics (SSIM). This model is very promising because it uses fewer parameters than other state-of-art super resolution models and should be analyzed in different scenarios [3].

In this paper, exploring the ideas presented in [1], a new method based on Convolutional Neural Networks is proposed for SISR. The idea is to explore a dual branch architecture, each branch employing UDPU units. In the first branch, the LR is upsampled by transposed convolution, and in the second one, we used bicubic interpolation.

2 Related Works

In recent years, CNN based structures have been widely applied in the SISR problem. Following the concept of the SRCNN, Kim et al. [16] proposed a model that uses residual learning, named Very Deep Super Resolution (VDSR). In this model, the output of the shallow layers is connected to the output of other layers, even those that are close to the output of the network.

Shi et al. [17] proposed a different approach, named Efficient Sub-Pixel Convolutional Network (ESPCN), which uses a convolutional layer to upsample and obtain the HR image, which allows it to have a few learnable parameters.

Ledig et al. [18] proposed a deep residual network with skip-connections (SR-ResNet). This method used the residual networks and was able to recover images with a higher scale factor. Inspired by this model, Lim et al. [19] improved the SRResNet and propose the Enhanced Deep Super-Resolution network (EDSR). The changes create a new model more compact than the SRResNet.

Most structures usually use a sequence of convolutional layers to extract the feature maps and recover the HR image. Haris et al. [1] proposed a different approach by creating the iterative up- and down-sampling layers named Deep Back-Projection Networks (DBPN). In this method, the feature maps extracted from the LR image are sequentially up and down-sampled in the model. Thus, each up or down-sampling unit provides an error used to enable the self-correction of the proposed model features.

Other works use two branches of convolutional layers in order to obtain different pieces of information of the LR image. One example of this approach is the Differential Convolutional Network (DCN), proposed in [20]. This module produces two outputs, one of them is used in the next convolutional layer, and the other one is used in the other branch of the proposed model, which are then combined to generate the HR image.

3 Proposed Work

In this paper, we propose a structure for SISR based upon [1] and [20]. Our model is a dual branch network that uses the UDPU proposed by [1]. In one of the branches, we use the bicubic interpolation to upsample the image to the HR space, and in the other one, we use the transposed convolution for this task. These upsampled feature maps in the HR space are sequentially processed by n UDPU. Furthermore, we analyze our structure in a scenario with more UDPU and with only one branch in order to analyze how this changes the results.

3.1 Up- and Down-Projection Units (UDPU)

The UDPU [1] is shown in Fig. 1. The up-projection unit upsamples the feature maps while the down-projection units downsample it. The up-projection unit receive the previous LR feature map (L_{T-1}) produced by a down-projection unit or the input of the model and create an intermediate HR feature map (H_0). With this intermediate HR feature map, the model attempts to map it back to LR (L_0). Then the residual difference between L_{T-1} and L_0 are upsampled to the HR space (H_1). H_1 is added to H_0 producing the output of this unit H_T. The same idea is used in the down-projection units. But this unit has the objective to generate an output in the LR space (L_T) by receiving the HR feature map (H_T) produced by another up-projection unit or by an image in the HR space.

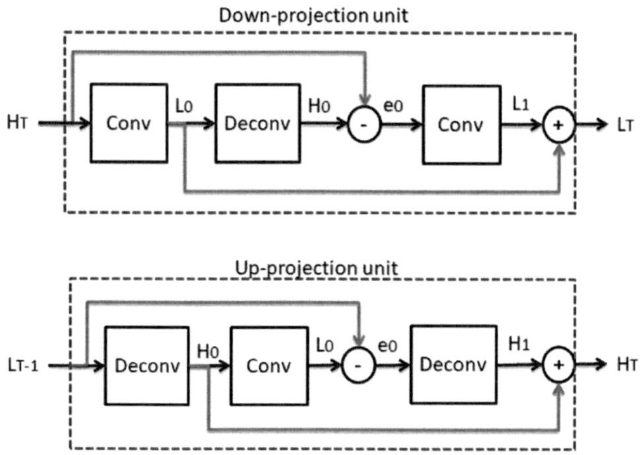

Fig. 1. Up- and down-projection units (UDPU)

3.2 Proposed Architecture

The proposed architecture is shown in Fig. 2. The input of the network is a LR image (I_{LR}) and the first operation in this image is the feature extraction. This module is composed of three convolutional layers with $n_1 = 128$, $n_2 = 32$, and $n_3 = 1$ filters, respectively, and a kernel size of 3×3 in the first layer and 1×1 in the other two layers. The output of the feature extraction feeds both branches of the model. In the first one, the upsample is made by transposed convolution, and in the second one it is used the bicubic interpolation. The transposed convolution has 2 filters, with a kernel size of 8×8. The stride depends on the scale factor used for the upsampling: for ×2 a stride of 2 × 2 is used, for ×3 the stride is 3 ×3, and for ×4 the stride is 4 × 4.

After the upsampling layers, the feature maps are fed to then down-projection and up-projection units, where n is the number of UDPU used. The parameters of all convolutions and transposed convolutions of both projection units are as follows: the number of filters is 2, the kernel size 8 × 8, and the stride is the same as mentioned before.

The DCN network [20] combined shallow feature maps generated by each DCN unit in another branch in the network. Inspired by that, every HR feature map created by each UDPU are concatenated and fed to the reconstruction layer, which will produce the HR image (I_{HR}). The reconstruction layer is a convolutional layer with 1 filter, kernel size of 3, and stride of 1 × 1. Every activation layer of the network, except the reconstruction one, uses Parametric Rectifier Linear Units (PReLU).

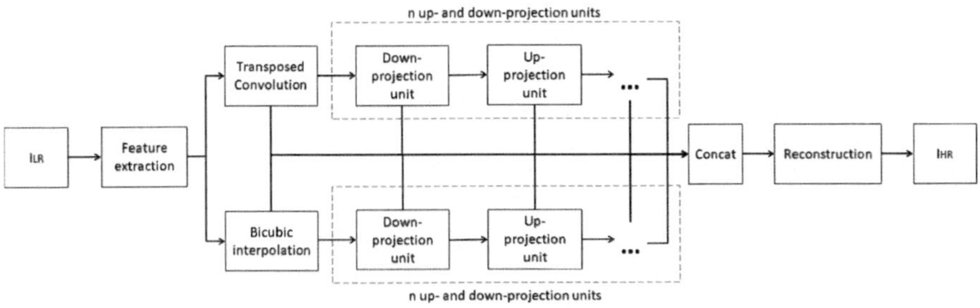

Fig. 2. Proposed architecture

4 Experiments

In this section, the training phase details are first introduced, such as the size of the images, datasets, and the characteristics of the computer used. Then the results are presented and discussed.

4.1 Dataset and Preparing the Images

The dataset used in the training phase was T91 [21], which consists of 91 HR images. The LR images were created from the 91 HR images by first applying a blur operator and then cropping the resulting image. The size of the images was also reduced in order to generate a bigger dataset: in the dataset, the size of the pair of images in HR and LR respectively was set to 32 × 32 and 16 × 16 for the scale factor of ×2, 33 × 33, and 11 × 11 for the scale factor of ×3, and 32 × 32 and 8 × 8 for the scale factor of ×4. It was used a stride of 14 in the cropping process. After this step, the final size of the HR images was reduced to 30 × 30 for the scale factor of ×2 and ×4, and 31 × 31 for the scale factor of ×3. This is necessary because the reconstruction layer of the network, Fig. 2, does not include zero-padding in order to avoid the border effect as described in [2]. However, it should be mentioned that it is possible to include the zero-padding step in the test phase. The validation and test set was composed by Set5 dataset [22], and the same preparation for the training dataset was applied in this dataset.

4.2 Training Details

ADAM algorithm was used in the training phase with the following parameters: $\varepsilon = 10^{-4}$(learning rate), $\beta_1 = 0.9$ and $\beta_1 = 0.999$. The training phase used a batch size of 20, and it was performed in 775 epochs.

The proposed architecture is scale specific, so it was made three different pieces of training for the scale factors of ×2, ×3, and ×4. We analyze our structure in three scenarios; first, we analyze the situation with just one branch in order to verify whether the two branch scenario can outperform a structure with only one branch. For this, we train a structure without the bicubic interpolation branch, and we use $n = 1$. The second proposed scenario verifies how the number of UDPU can improve the network's performance in terms of PSNR and SSIM. For this task, another training was made with $n = 4$. The third scenario compares our structure with one of the proposed structures of SRCNN [2] (the one with three layers:$n_1 = 64$, $n_2 = 32$ and $n_3 = 1$ and kernels sizes of 9 × 9, 3 × 3, and 5 × 5). For this task, another train was made for this structure.

The experiments were performed on an NVIDIA GEFORCE 940M, 8GB RAM, and we used the Keras library to create the models and perform the training and tests.

4.3 Results

The comparison with the proposed model with two branches and the one branch model can be viewed in Table 1 (one branch $n = 1$ and two branches $n = 1$). In this comparison,

Table 1. Comparations of PSNR and SSIM on dataset Set5 for scale factor ×2, ×3, and ×4 of each trained model and bicubic interpolation

Scale	Image	Bicubic interpolation		SRCNN [2]		Ours one branch n = 1		Ours two branches n = 1		Ours two branches n = 4	
		PSNR	SSIM	PSNR	SSIM	PSNR	SSIM	PSNR	SSIM	PSNR	SSIM
×2	Baby	34.945	0.936	36.753	0.955	36.806	0.955	**36.846**	**0.957**	36.831	0.956
	Bird	34.532	0.962	39.064	0.978	38.773	0.962	39.334	0.980	**39.794**	**0.983**
	Head	33.015	0.828	34.109	0.864	33.973	0.864	34.128	0.865	**34.189**	**0.868**
	Butterfly	25.399	0.895	31.415	0.955	31.132	0.951	31.232	0.957	**31.767**	**0.960**
	Woman	30.121	0.933	33.728	0.960	33.602	0.933	33.815	0.962	**33.959**	**0.963**
	Average	31.602	0.911	35.014	0.942	34.857	0.933	35.071	0.944	**35.308**	**0.946**
×3	Baby	32.815	0.899	33.404	0.904	33.437	0.909	33.509	0.910	**33.515**	**0.911**
	Bird	31.684	0.926	33.735	0.946	33.706	0.945	33.945	0.946	**34.063**	**0.948**
	Head	31.557	0.777	32.072	0.793	32.072	0.794	32.100	0.794	**32.120**	**0.796**
	Butterfly	23.082	0.822	26.316	0.900	26.156	0.901	26.554	0.906	**27.079**	**0.915**
	Woman	27.664	0.889	29.532	0.917	29.540	0.916	29.927	0.922	**30.112**	**0.924**
	Average	29.360	0.862	31.012	0.892	30.982	0.893	31.207	0.896	**31.378**	**0.899**
×4	Baby	30.722	0.852	31.066	**0.904**	31.165	0.855	**31.237**	0.860	31.160	0.857
	Bird	29.316	0.876	30.125	0.890	30.228	0.888	30.245	**0.892**	**30.304**	0.891
	Head	30.253	0.728	30.545	0.734	30.535	0.736	**30.675**	0.738	30.620	**0.739**
	Butterfly	21.130	0.740	23.011	0.834	22.957	0.831	23.017	0.836	**23.350**	**0.849**
	Woman	25.576	0.835	26.380	0.859	26.502	0.857	26.309	0.860	**26.829**	**0.867**
	Average	27.399	0.806	28.225	**0.844**	28.277	0.833	28.296	0.837	**28.453**	0.840

it is possible to verify that the model with two branches outperformed the model with one branch in almost every image in the Set5 when we compare the PSNR and SSIM of the test results. When we check the average of each scale factor, the two branches model outperformed the one branch model in every scale factor. When analyzing the second scenario, training a dual branch model with $n = 4$ we can see that this model outperformed every other model trained, in PSNR and SSIM, in almost every image in the Set5 and it has the highest average in all scale factors, except for the SSIM when compared with ×4 of SRCNN. With this result, we can view that adding more UDPU units can improve the results in terms of PSNR and SSIM, and our model leads to better PSNR and SSIM when compared to the trained SRCNN structure and the bicubic interpolation.

For visual comparison, a few images are shown in Fig. 3. We choose three different images from Set5 for each scale factor. Also, we displayed the image with the bicubic interpolation in order to show the difference between a traditional method and our method. It is possible to see that the model with $n = 4$ can lead to a slight improvement in the edges of the images, especially when we compare the images of the butterfly and the woman. Thus, the model with $n = 4$ reconstructs the wings of the butterfly with slightly more details than the other models. When we check ×2 images, we can not see any visual differences between the three scenarios of the proposed model.

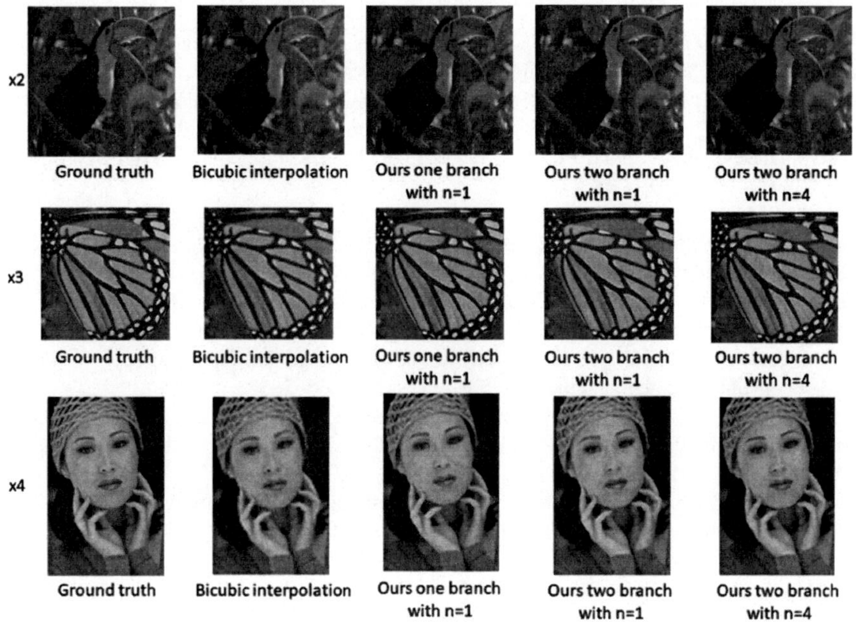

Fig. 3. Reconstructed images of the Set5 for each scale factor for the proposed model and bicubic interpolation

5 Conclusions and Future Work

In this study, a super-resolution model based on convolutional neural networks with the Up- and Down- Projection Units has been studied in order to analyze the performance of the dual branch scenario. We could verify that the dual branch scenario can outperform a model with only one branch, and using four up- and down-projection units lead the model to a better performance in PSNR and SSIM. Furthermore, our model leads to better performance in PSNR and SSIM when compared to one proposed structure of SRCNN. In future works, we shall investigate the model's performance by adding more filters in the up- and down-projection units, increasing the kernel size of the model, and checking the performance of this network in the RGB color space.

Acknowledgment. This work was funded in part by the Coordenação de Aperfeçoamento de Pessoal de Nível Superior - Brasil (CAPES) - Finance Code 001 - and the National Council for Scientific and Technological Development – CNPq.

References

1. Haris M, Shakhnarovich G, Ukita N (2018) Deep back-projection networks for super-resolution. In: 2018 IEEE/CVF conference on computer vision and pattern recognition, pp 1664–1673
2. Dong C, Loy CC, He K, Tang X (2016) Image super-resolution using deep convolutional networks. IEEE Trans Pattern Anal Mach Intell 38(2):295–307
3. Wang Z, Chen J, Hoi SC (2019) Deep learning for image super-resolution: a survey, p 23
4. Zhang L, Zhang H, Shen H, Li P (2010) A super-resolution reconstruction algorithm for surveillance images. Signal Process 90(3):848–859. https://doi.org/10.1016/j.sigpro.2009.09.002
5. Zhang S, Liang G, Pan S, Zheng L (2019) A fast medical image super resolution method based on deep learning network. IEEE Access 7:12319–12327. https://doi.org/10.1109/ACCESS.2018.2871626
6. Huang Y, Wang W, Wang L (2018) Video super-resolution via bidirectional recur-rent convolutional networks. IEEE Trans Pattern Anal Mach Intell 40(4):1015–1028
7. Ni M, Lei J, Cong R, Zheng K, Peng B, Fan X (2017) Color-guided depth map super resolution using convolutional neural network. IEEE Access 5:26666–26672
8. Guo R, Shi X, Zhu Y, Yu T (2018) Super-resolution reconstruction of astronomical images using time-scale adaptive normalized convolution. Chin J Aeronaut 31(8):1752–1763. https://doi.org/10.1016/j.cja.2018.06.002
9. Nasrollahi K, Moeslund TB (2014) Super-resolution: a comprehensive survey. Mach Vis Appl 1423–1468. https://doi.org/10.1007/s00138-014-0623-4
10. Yue L, Shen H, Li J, Yuan Q, Zhang H, Zhang L (2016) Image super-resolution: the techniques, applications, and future. Signal Process 389–408. https://doi.org/10.1016/j.sigpro.2016.05.002
11. Saul L, Roweis S (2001) An introduction to locally linear embedding. J Mach Learn Res 7
12. Chang H, Yeung D-Y, Xiong Y (2004) Super-resolution through neighbor embedding. In: Proceedings of the 2004 IEEE computer society conference on computer vision and pattern recognition, CVPR 2004, vol 1, p I
13. Zhao Y, Li G, Xie W, Jia W, Min H, Liu X (2018) Gun: gradual upsampling network for single image super-resolution. IEEE Access 6:39,363–39,374

14. Sun Y, Zhang W, Gu H, Liu C, Hong S, Xu W, Yang J, Gui G (2019) Convolutional neural network based models for improving super-resolution imaging. IEEE Access 7:43042–43051. https://doi.org/10.1109/ACCESS.2019.2908501
15. Liu P, Hong Y, Liu Y (2019) A novel multi-scale adaptive convolutional network for single image super-resolution. IEEE Access 7:45191–45200. https://doi.org/10.1109/ACCESS.2019.2908003
16. Kim J, Lee JK, Lee KM (2016) Accurate image super-resolution using very deep convolutional networks. In: 2016 IEEE conference on computer vision and pattern recognition (CVPR), pp 1646–1654
17. Shi W, Caballero J, Huszár F, Totz J, Aitken AP, Bishop R, Rueckert D, Wang Z (2016) Real-time single image and video super-resolution using an efficient sub-pixel convolutional neural network. In: 2016 IEEE conference on computer vision and pattern recognition (CVPR), pp 1874–1883
18. Ledig C, Theis L, Huszár F, Caballero J, Cunningham A, Acosta A, Aitken A, Tejani A, Totz J, Wang Z, Shi W (2017) Photo-realistic single image super-resolution using a generative adversarial network. In: 2017 IEEE conference on computer vision and pattern recognition (CVPR), pp 105–114
19. Lim B, Son S, Kim H, Nah S, Lee KM (2017) Enhanced deep residual networks for single image super-resolution. In: 2017 IEEE conference on computer vision and pattern recognition workshops (CVPRW), pp 1132–1140
20. Liu P, Hong Y, Liu Y (2019) Deep differential convolutional network for single imagesuper-resolution. IEEE Access 7:37555–37564
21. Yang J, Wright J, Huang TS, Ma Y (2010) Image super-resolution via sparse representation. IEEE Trans Image Process 19(11):2861–2873
22. Bevilacqua M, Roumy A, Guillemot C, Alberi Morel ML (2012) Low-complexity single-image super-resolution based on nonnegative neighbor embedding. In: Proceedings of the British machine vision conference. BMVA Press, pp 135.1–135.10. https://doi.org/10.5244/C.26.135

Simulation of Damage Location Algorithm for Structural Health Monitoring System

Vanessa Botinelly[1(✉)], Crístian Müller[2], Osamu Saotome[1], and Renato Machado[2]

[1] Technological Institute of Aeronautics - ITA, São José dos Campos, São Paulo, Brazil
{vanbot,osaotome}@ita.br
[2] Federal University of PAMPA – UNIPAMPA, Alegrete, RS, Brazil
cristianmuller@unipampa.edu.br, rmachado@ita.br

Abstract. This article proposes a novel algorithm to detect structural damages by simulating the propagation of guided waves on an isotropic structure. To properly configure the transducer network, Damage Index and Triangulation methods, extended to Multilateration methods, are used. In this way, magnitudes are assigned to each method transducer and, among them, according to the interference caused by the damage and, among them, those three with the highest output value are selected to perform the triangulation to determine the damage location, in addition to locating it in the aluminum plate. Simulation test results show the validity of the proposed model.

Keywords: Structural health monitoring · Structural damage · Crack detection · Lamb waves · Damage index · Triangulation

1 Introduction

Most research for structural safety purposes focuses on Structural Health Monitoring (SHM) techniques [1]. In the aeronautical industry, these techniques are used to investigate new technologies to improve aircraft structures' safety and efficiency. The damage is described by researchers Farrar and Worden as any change in a system that impairs its performance [2]. Thus, to prevent accidents from degradation and failure, modern airplanes are equipped with many sensors and sensing systems to monitor and provide information about the operation of each flight system [3].

The evolution of SHM is due to advances in sensor technology and signal processing [4]. Therefore, for the constant assessment of the integrity of the materials and, consequently, structural state to detect anomalies, means have been sought to improve the systems in terms of measurement accuracy and precision in results [5] as well as in the ability to sustain the safe and efficient operation of these vehicles by the early detection of damage or degradation of critical components. A system capable of carrying out SHM facilitates the detection of damages, minimizes the risk of failures, downtime, operating and maintenance costs, in addition to being an area of technical and scientific interest [6–8].

Y. Iano et al. (Eds.): BTSym 2020, SIST 233, pp. 790–800, 2021.
https://doi.org/10.1007/978-3-030-75680-2_88

Aircraft structures are made of both composite and metallic materials, where their maintenance programs include a Non-Destructive Inspection (NDI) plan and a repair strategy [4]. However, their service conditions and failures are planned without precision. Therefore, composite materials still have some shortcomings when compared to isotropic materials due to limited experience in the field, guaranteed performance, and lack of failure prediction [7, 9]. SHM technology can be used to remedy the problems described above in addition to detecting failures and being able to indicate early physical irregularity to the structure.

The SHM system is based on a sensor network installed in the host structure, a diagnostic hardware with monitoring principles, and software with anomaly identification algorithms [5, 10]. The main resources used to monitor and detect structural damage are given by the type of sensing technology used. SHM can be divided into two types: passive SHM systems use sensors only, as they continuously monitor the structure, and the sensors will pick up the acoustic waves from any impact on the structure; in active SHM systems, the structure is excited by an actuator and sensed by the sensors, so according to the changes in the signals captured by the sensors, the location of the damage can be determined [11]. For these types of systems, it is worth mentioning the Optical Fiber (OF) sensors or Bragg Fiber Grids (BFGs), piezoelectric materials, air/vacuum galleries, thermography, nanomaterials, and eddy current film sensors [10, 12, 13].

Among these, the most used in aeronautical structures are piezoelectric materials, as they present low cost and high efficiency [14]. As a result, piezoelectric transducers (PZTs) have been used as actuators and sensors for Ultrasonic Guided Waves (UGW). In this way, PZTs are widely used for SHM applications, mainly for the detection of problems in isotropic plates. Therefore, this is a promising technique due to its propagation capacity and sensitivity to changes in structures [1, 15–18]. To analyze the data acquired by the sensor network, several SHM methods were tested and validated on plates to determine the location of structural damage [14–16, 19–24].

The pioneers in research for structural defects used the Lamb waves in the non-destructive assessment to detect damage, as this wave is a UGW that propagates in thin plates [25]. Thereby, in [26] carried out experiments using isotropic structures to locate acoustic sources based on the triangulation method.

In addition, the authors in [16] used methods to locate cracks in plate-like structures using piezoelectric transducers, the triangulation method and a multilateration algorithm to obtain more accurate solutions for the fault positioning.

Based on this state-of-art, this work proposes simulation methods to determine the position of a fault by estimating the distances of the transducers in a plate-like structure. The PZT sensor network configuration follows the square format with twelve transducers, where the algorithm uses the Damage Index to expose which transducers are most affected by the existence and proximity of the damage. The triangulation method will be applied to estimate the irregularity position with double verification by means of loci in the shape of circumferences and ellipses. The project is scalable for both the size of the network and the structure.

2 Development of Detection Strategy

For the development of an automated system to locate damage in aluminum plates, the technique that omnidirectional waves spread from the failure is explored. The parameterization developed by [16] is used for isotropic structures. The development of this system depends on analytical formulations to adjust the signals that represent the propagation of waves in the structure. The Lamb waves propagate through the damaged specimen, and the irregularity spreads these waves omnidirectionally.

Using the developed technique, the damage will be located by two determination stages that make use of the Time of Flight (ToF) of the Lamb waves in the structure. The first stage will be choosing the transducers by the Damage Index (DI). And the second stage will be the application of the triangulation method in these transducers to obtain an estimate of the defect location.

2.1 Damage Index

The Damage Index (DI) is based on the variation of the signals' amplitudes in the current structure with a defect in relation to the signals in the primitive structure. DI is considered the simplest damage identification algorithm. The signal is analyzed by a formula to produce a scalar value. This value is used as a quantitative measure of whether or not the specimen is damaged [9, 27]. In this work, the metric was based on [4], where a DI was proposed according to the maximum residual signal envelope energy for each actuator-sensor path. Next, a comparison of the signature signs with the current state is performed, where the resulting sign $R_{ij}(t)$ is the difference between the initial state $S_{ij}(t)$ and the current state $P_{ij}(t)$ in the ij path, given by:

$$R_{ij}(t) = S_{ij}(t) - P_{ij}(t) \tag{1}$$

Where S_{ij}, P_{ij} and R_{ij} are two-dimensional signs whose values at some point in time are specified by two parts, a real part and an imaginary part, given by:

$$X(t) = x(t) + i\tilde{x}(t) \tag{2}$$

Where $\tilde{x}(t)$ is related to $x(t)$ by the Hilbert Transform [28]. Using a traditional representation of the analytical signal in its trigonometric form, its instantaneous amplitude (envelope, magnitude, modulus) can be determined by:

$$A(t) = \pm|X(t)| = \pm|x(t) + i\tilde{x}(t)| = \pm\sqrt{x^2(t) + \tilde{x}^2(t)} \tag{3}$$

Where $H(x(t))$ or $\tilde{x}(t)$ is the Hilbert transform of the sign, given by [16, 28]: $H[x(t)] = \tilde{x}(t) = \frac{1}{\pi}\int_{-\infty}^{\infty}\frac{x(\tau)}{t-\tau}d\tau$. The resulting envelope of R_{ij} is the difference between the initial state envelope $S_{ij}(t)$ and the current state envelope $P_{ij}(t)$ in the path ij given by:

$$E(R_{ij}) = \left|R_{ij} - iH(R_{ij})\right| \tag{4}$$

From Eq. (4), it is concluded that DI is the maximum amplitude of the resulting signal envelope for each path ij, as follows:

$$DI_{ij} = max(E(R_{ij})) \tag{5}$$

Equation (5) shows how to calculate the DI values; however, it is necessary to apply two thresholds to characterize the failure's existence. The first threshold applied is 'DI Threshold-Mean', whose average values of DI_{ij} indicate the existence of the damage. However, in order to highlight which path ij the anomaly is near, the second threshold is used, that is, the standard deviation, called 'DI Threshold-Mean + Standard Deviation'. When the DI_{ij} value is high, this means that the defect is closer to the path of the corresponding actuator-sensor [4, 9]. The network of N transducers has a i actuator that transmits the signal, and the reception is done by $N-1$ sensors j sequentially. The signal is repeated until all sensors have acted as an actuator. In this action, it is observed which transducers were most affected by the damage, applying the average of N DIs for each i actuator, given by:

$$DI_i = \frac{1}{N}\sum_{i=1}^{N}\sum_{j=1}^{N} E\big(R_{ij}(t)\big) \tag{6}$$

After verifying the existence of the damage, it is necessary to determine its location. This process begins with the selection of transducers using the following methods: The first transducer i_1 is selected by the highest value of DI_i found. Then, a restriction applies to neighboring transducers, $i_1 - 1$ and $i_1 + 1$, which cannot be chosen. From there, the next i_2 transducer is selected by the second-highest value of the remaining DI_i. With the two chosen transducers, the same restriction applies, but for the N paths of each selected i actuator. With this, the third transducer is selected by the highest DI_{ij} value of the ij actuator-sensor path, once, the number of three transducers is the minimum number required for triangulation-based methods.

2.2 Triangulation

The triangulation method will be used to provide estimates of the damage location. This will be done by inserting the locus around the transducers selected by the DI. This locus covers every location that the defect may be. The development of this locus is based on the system's ToFs.

In this work, the locus for the triangulation method with both ellipses and circumferences was developed. The loci are constructed based on the ToFs of the incident waves of the actuator and reflected from the damage. The size of the locus is determined so that the irregularity is defined along its length. The estimated damage location is found by the coordinates' average on the structure's points with the highest number of loci that represent the detection estimate [16].

Triangulation with Ellipses. The ellipses triangulation method is used to provide initial estimates of the failure position with respect to the actuator and the sensor. The ellipse-shaped locus will represent the possible locations. Any point in the outline of the ellipse is represented with possible damage.

Using ellipse relations, the PZTs are placed in the focal position. The distance between the PZT in direct sight is $d\_i - j = 2c$. The distance between the PZT going through the damage is $d_{i-D-j} = 2a$. For a horizontal ellipse, the height of the ellipse is given by $b = \sqrt{a^2 - c^2}$. Based on this, it was possible to develop the ellipse for each ij path.

Triangulation with Circumferences. From the estimates obtained by triangulation with ellipses, it is possible to estimate the irregularity position in relation to the actuator. For each PZT, there will be a circular shaped locus that represents the possible anomaly sites. This proposal was developed using the relation, where d_{i-D-j} is the distance between the *PZTi* until the damage and after the damage until the *PZTj*, thus

$$d_{i-D-j} = d_{i-D} + d_{j-D} \tag{7}$$

Using three PZT to do the triangulation. For example, when PZT_1 is an actuator, PZT_2 and PZT_3 are sensors, and so on. To develop the circumference, it is necessary to find the distance between the actuator and damage by:

$$d_{1-D} = (d_{1-D-2} + d_{1-D-3} - d_{2-D-3})/2 \tag{8}$$

$$d_{2-D} = d_{1-D-2} - d_{1-D} \tag{9}$$

$$d_{3-D} = d_{1-D-3} - d_{1-D} \tag{10}$$

3 Simulations

In this section, the proposed method is evaluated the performance of the proposed algorithm to locate damage in an aluminum plate through simulations using the MATLAB software. For this, a scenario was considered corresponding to a square aluminum plate with a length of 40 cm. The permanent installation of PZT in the structure increases the sensitivity to defects. And the structure was filled with twelve PZT distributed according to Fig. 1. Considering that if there is any damage in the area of effective detection, it will be located [16].

The positive-defined, compact support signals obey the formulation of *Morlet wavelet* signals, as shown in Fig. 2. For them, the ToFs of these signals are the information of interest in this simulation.

In [16], the calibration of the speed and displacement of the Lamb waves was developed. The relationship between the ToF and the distance between actuator and sensor was used for these calculations, even though the theoretical speed of the Lamb wave can be calculated from the Rayleight-Lamb equations. An important factor that must be taken into account is the delay in instrumentation, which causes uncertainties in distance measurements. Therefore, it is necessary to first perform a calibration, the particular result of which is:

$$d = 5296.1 \cdot t - 0.0077323 \tag{11}$$

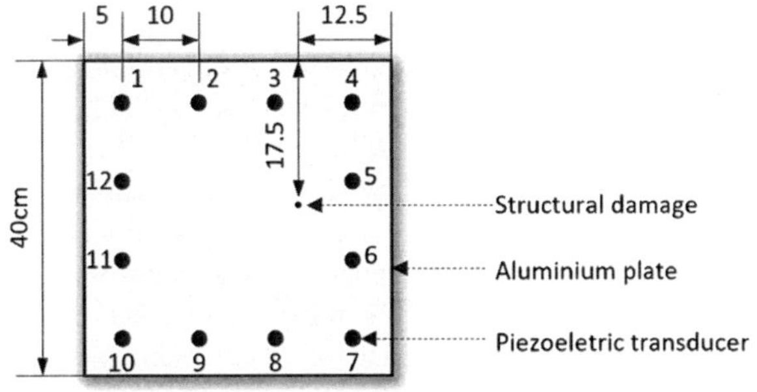

Fig. 1. Structure Diagram based on [16] with twelve piezoelectric transducers and one structural damage.

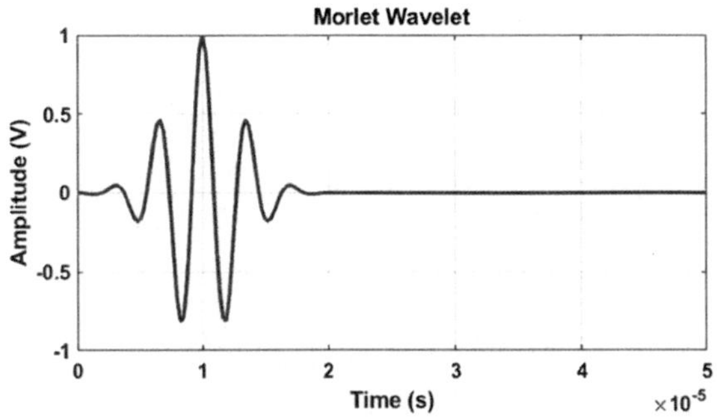

Fig. 2. Morlet Wavelet, the exciting pulse.

The first measurement is performed with the specimen intact to check the initial condition, this measurement is also called the signal signature. The measurements after this are called the current signal, according to your answers, you can monitor the health of the structure. These changes are highlighted by subtracting the signature from the current signal. As a result, there is the instant of arrival of the current signal.

When the system is selected to operate in *pitch-catch* mode, the PZTs are activated in sequence; however, the PZT that is functioning as an actuator and will not function as a sensor. Therefore, one PZT is responsible for transmitting and the others for receiving Lamb waves. After the twelve PZTs acted as an actuator, the previous steps start, such as checking for damage, selecting the three sensors, and later applying triangulation.

4 Results and Discussions

In these experiments, the system is selected to operate in *pitch-catch* mode. The transducer network has the PZTs activated in sequence. However, the PZT that is functioning as an actuator will not function as a sensor. Therefore, a PZT is responsible for transmitting and receiving Lamb waves. After the twelve PZTs acted as actuators, the steps mentioned above, such as the verification of the existence of defect and selection of the three sensors, and subsequent application of the triangulation, start.

The diagnostic signal illustrated in Fig. 2 was developed according to the Lamb wave in the S0 symmetrical mode, seeking to provide stronger reflections of the damage and a signal propagation through the less dispersive structure. The frequency used was 380 kHz because of the wavelength and the blind detection zones [16].

The first simulation was carried out with the mentioned frequency to minimize the effects of the attenuation and to capture the structure data without any anomaly. After recording the signature data, the damage of a punctual character was inserted, as if it were a mass added to the specimen to simulate the spreading due to the changes that occurred in the place that would represent what caused the damage. A simulated fault was inserted in the position (27.5, 22.5), shown in Fig. 1. The current state data is captured according to the position of the mass.

It was adopted that the signs of the $i-j$ and $j-i$ paths of the system are identical, even knowing that, in a real aluminum plate, there are small variations in the paths. By comparing the data of the current and the subsequent status, it is possible to start the application of the Damage Index described above. The data were obtained and released from the PZTs. The average of the DIs for each actuator can be seen in Fig. 3. These values were obtained by Eq. (4) to expose which actuator was the most affected by the damage.

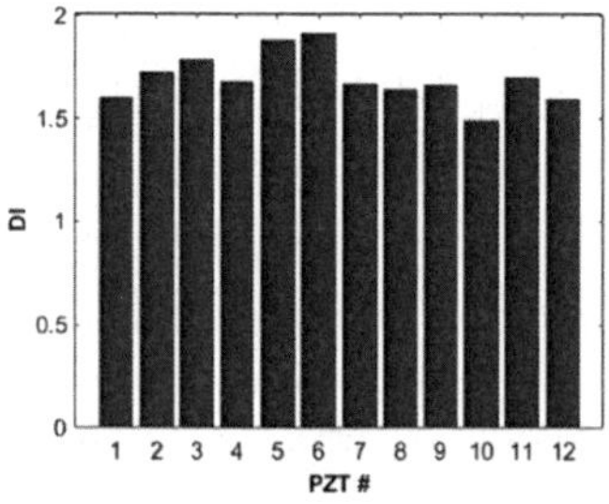

Fig. 3. DI for each actuator in the PZT network, according to the proximity of actuator to damage, the higher the DI value.

To select the transducers that will participate in the triangulation, it is observed that the $PZT6$ has the highest value of DI, and for this reason, it is the first element of the PZT, by the aforementioned restriction rules, where neighbors cannot be selected. Therefore, $PZT5$ and $PZT7$ are disregarded, so $PZT3$ is indicated as the second PZT of the triangulation. Consequently, $PZT2$ and $PZT4$ are disregarded in the next steps as well. The third PZT is chosen by the maximum amplitude of the resulting signal envelope for each path obtained by Eq. (4). Figure 4 illustrates the DIs for each path from selected actuators and as applied restrictions. The yellow bars are the PZTs mentioned, either because they were elected or eliminated by the restriction. When there is no DI value, it means that this PZT is working as an actuator; therefore, it is not activated as a sensor. The blue bars are the DIs available for selecting the third PZT. Therefore, PZT8 presents the highest value of the remaining DIs; its residual signal is maximum for the 6–8 path. Thus, it completes the minimum quantity of three PZTs for an estimate of location based on triangulation.

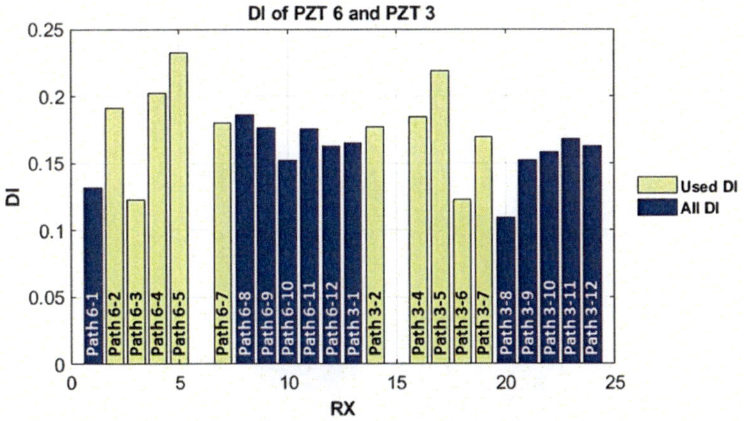

Fig. 4. DI for each path ij. The first columns are associated with the first PZT selected for triangulation, in this case actuator 6; and the remaining twelve columns are associated with the second chosen PZT, actuator 3. In blue are 'All DI' available to choose the third PZT of the triangulation, and in yellow are the 'Used DI' columns and eliminated by the previous steps.

The next diagnostic step is to execute the triangulation method using a PZT {6, 3, 8} network. At this time, the only parameter used is ToF. Time of Flight is converted to the distance covered by the signal in the structure by applying Eq. (9), seeking to estimate the locus for this method. The first result of the damage position estimation was based on triangulation with ellipses in Fig. 5a. To increase the detection's accuracy, triangulation is performed with the circumferences shown in Fig. 5b.

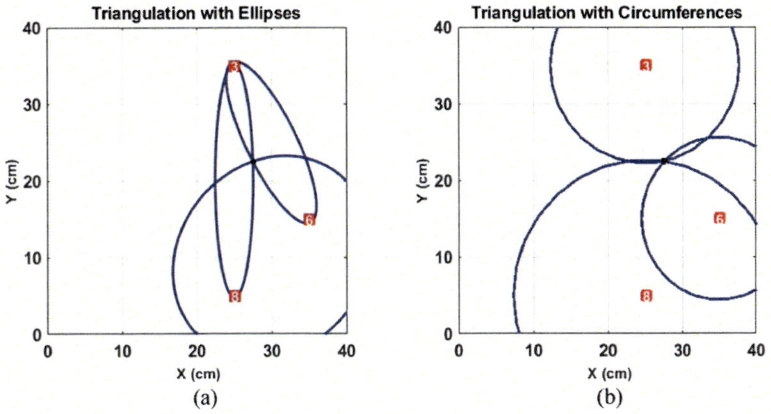

Fig. 5. Triangulation method using three sensors placed at positions 3, 6, and 8 with projection of: (a) elliptical locus and (b) circular locus.

Figure 6 explains how the position in the network is extracted using triangulation. By mapping the structure and dividing it into pixels, the value of each pixel can be obtained using the sum of the Gaussian distribution described in [15]. The highlighted spectrum represents the magnitude of the spatial spectrum of each P (x, y) mapped point. Each circumference has its contribution in this magnitude, and the peak of this spectrum represents the damage point located by the presented triangulation algorithm.

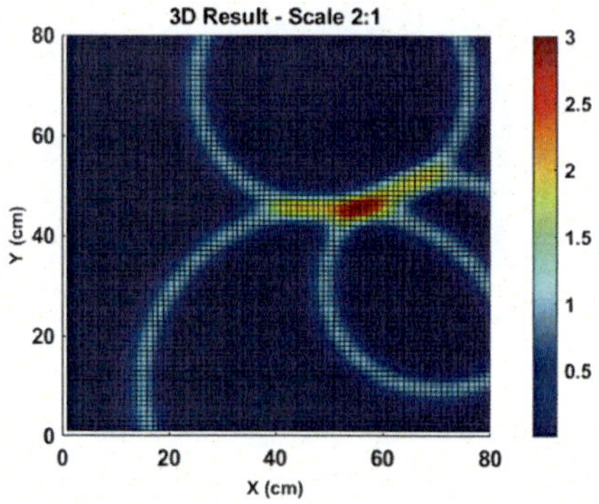

Fig. 6. Estimation result of damage position based on the sum of the Gaussian distribution using the sensing paths of three PZTs.

After using the contextualized logic in the previous chapters and implementing it in a Matlab simulation environment, it was possible to verify that the proposed algorithm was effective in detecting anomalies in the simulated structure in such a way as to be compatible with the location of the actual damage, as shown in Fig. 7.

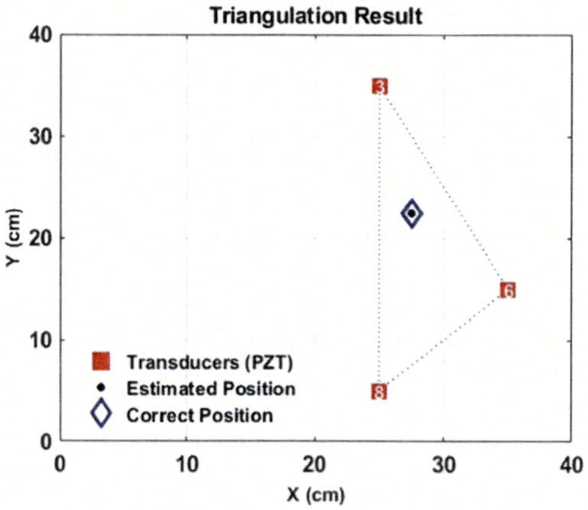

Fig. 7. Result of the triangulation exposing the position estimated by the algorithm and the correct position of the damage. The transducer network is indicated by the red square and its numbering.

5 Conclusion

A model to simulate the propagation of guided waves on an isotropic structure for detecting structural damage was presented. Through simulations in the Matlab software, the proposed algorithm was tested with a network of twelve transducers and structural irregularity. The simulations revealed that the Damage Index is not able to locate the damage, but it clearly exposes which transducers reflect the most changes in their signals according to the proximity of the problem. These transducers were chosen by the aforementioned DI strategy and, consequently, by selecting the PZTs with the largest residues in their path and using this network to detect damage based on triangulation. The result shows the importance of selecting the transducers to obtain precision in the damage detection algorithm. Triangulation method using ToF to determine the location of the anomaly caused an efficiency when used with both ellipses and circumference in its development, as well as a chance of errors, as it has one more way to obtain information and generate a more reliable estimate for the damage position. The advantage of the strategy is that it is easily applicable and scalable for larger structures. The estimated position is in accordance with the location of the simulated defect, and the results show that the damage detection method based on triangulation is effective in estimating the damage position in an aluminum plate.

References

1. Hua J, Lin J, Zeng L (2015) Struct Health Monit 14(1):20
2. Farrar CR, Worden K (2007) Philos Trans R Soc A: Math Phys Eng Sci 365(1851):303. https://doi.org/10.1098/rsta.2006.1928
3. Merhav S (1998) Aerospace sensor systems and applications. Springer Science & Business Media
4. Khodaei ZS, Aliabadi M (2016) Materials 9(9):790
5. Mei H, Haider M, Joseph R, Migot A, Giurgiutiu V (2019) Sensors 19(2):383. https://doi.org/10.3390/s19020383
6. Ross R (2016) Structural health monitoring (SHM) in aerospace structures, Edited by Yuan FG. Woodhead Publishing, pp. 3–31. https://doi.org/10.1016/B978-0-08-100148-6.00001-9
7. Carrino S, Maezzoli A, Scarselli G (2020) Materials Today: Proceedings. https://doi.org/10.1016/j.matpr.2019.12.048
8. Lin M, Qing X, Kumar A, Beard SJ (2001) Smart structures and materials 2001: industrial and commercial applications of smart structures technologies, vol 4332, Edited by McGowan AMR. International Society for Optics and Photonics (SPIE), pp 98–106. https://doi.org/10.1117/12.429646
9. De Luca A, Perfetto D, De Fenza A, Petrone G, Caputo F (2020) Theor Appl Fract Mech 105:102408
10. Qing X, Li W, Wang Y, Sun H (2019) Sensors 19(3):545. https://doi.org/10.3390/s19030545
11. Fu H, Khodaei ZS, Aliabadi MHF (2019) IEEE Internet of Things J 6(1):1183. https://doi.org/10.1109/JIOT.2018.2867722
12. Alvarez-Montoya J, Carvajal-Castrillon A, Sierra-Perez J (2020) Mech Syst Signal Process 136:106526. https://doi.org/10.1016/j.ymssp.2019.106526
13. Salmanpour M, Khodaei ZS, Aliabadi M (2017) J Intell Mater Syst Struct 28(5):604
14. Rebillat M, Mechbal N (2020) Struct Health Monit 19(1):305. https://doi.org/10.1177/1475921719843453
15. Migot A, Bhuiyan Y, Giurgiutiu V (2019) J Intell Mater Syst Struct 30(4):618. https://doi.org/10.1177/1045389X18818775
16. Zou F, Rao J, Aliabadi M (2018) NDT E Int 94:1. https://doi.org/10.1016/j.ndteint.2017.11.002

17. Li J, Khodaei ZS, Aliabadi M (2019) Int J Mech Sci 155:235. https://doi.org/10.1016/j.ijmecsci.2019.02.045
18. Li L, Haider MF, Mei H, Giurgiutiu V, Xia Y (2020) Struct Health Monit 19(2):357. https://doi.org/10.1177/1475921719848149
19. Wang Y, Luo Y, Qiu L (2020) IEEE Sens J 20(1):102. https://doi.org/10.1109/JSEN.2019.2941533
20. Martins T, Infante V, Sousa L, Fonseca A, Antunes P, Moura A, Serrano B (2020) Int J Fatigue 132:105348. https://doi.org/10.1016/j.ijfatigue.2019.105348
21. Sundararaman S, Adams DE, Rigas EJ (2005) Struct Health Monit 4(2):171
22. Li F, Meng G, Kageyama K, Su Z, Ye L (2009) J Intell Mater Syst Struct 20(10):1147. https://doi.org/10.1177/1045389X09102562
23. Kundu T, Das S, Martin SA, Jata KV (2008) Ultrasonics 48(3):193. https://doi.org/10.1016/j.ultras.2007.12.001
24. Zhao X, Gao H, Zhang G, Ayhan B, Yan F, Kwan C, Rose JL (2007) Smart Mater Struct 16(4):1208
25. Worlton D (1961) J Appl Phys 32(6):967
26. Tobias A (1976) Non-Destructive Test 9(1):9. https://doi.org/10.1016/0029-1021(76)90027-X
27. Giurgiutiu V (2014) Structural health monitoring with piezoelectric wafer active sensors, 2nd edn. Edited by Giurgiutiu V. Academic Press, Oxford. https://doi.org/10.1016/B978-0-12-418691-0.00001-0
28. Feldman M (2011) Mech Syst Signal Process 25(3):735. https://doi.org/10.1016/j.ymssp.2010.07.018

Performance Analysis of LoRa and Zigbee for Application in Industry 4.0

Renan R. Mendes, Rafael M. Silva, Carlos E. Capovilla, and Ivan R. S. Casella(✉)

Federal University of ABC, Santo Andre, SP, Brazil
{renan.mendes,carlos.capovilla,ivan.casella}@ufabc.edu.br,
r.montoia@aluno.ufabc.edu.br

Abstract. The Industry 4.0 concept has been progressively consolidated worldwide and much of its success is due to the use of LPWA communication technologies. Among them, LoRa and Zigbee have stood out since, in addition to low power consumption and long-range, they are low cost and use spread spectrum technique to increase the system's robustness to interferences. In this context, this work presents a study of these technologies' performance for indoor and outdoor environments to analyze the feasibility of their use in Industry 4.0.

Keywords: Industry 4.0 · IoT · LPWA · LoRa · Zigbee

1 Introduction

In recent years, the world industry has undergone a real revolution. The concept of Industry 4.0 has been deployed to offer Internet of Things (IoT) technologies for the monitoring and control of industrial processes [1]. One reason for its rapid growth was the adoption of wireless LPWA (Low Power Wide Area) communication technologies. Among them, the LoRa (Long Range) and Zigbee technologies have shown to be very promising since they offer devices with affordable prices and of easy acquisition, reduced deployment costs and good immunity to interference (spread spectrum techniques), in addition to low consumption (e.g., in the order of 10 years) and long-range (e.g., few km) [2–4].

The performance of these technologies in different radio propagation environments, such as those found in industrial plants, is currently a very important issue for the successful implementation of the Industry 4.0 concept and, as a result, several works have been published recently in the scientific literature [1, 5–7].

In this context, this work aims to present a comparative study between LoRa and Zigbee technologies based on experimental measurements carried out in both indoor and outdoor environments to verify their feasibility in Industry 4.0. The analysis will use the measured values of RSSI (Received Signal Strength Indication), PER (Packet Error Rate), and the estimated maximum range as performance references.

In addition to the introduction, this text is divided into four additional sections. Section 2 presents a brief description of the evaluated technologies. Section 3 shows the

Y. Iano et al. (Eds.): BTSym 2020, SIST 233, pp. 801–810, 2021.
https://doi.org/10.1007/978-3-030-75680-2_89

measurement tests. Section 4 shows a radiopropagation model analysis. Finally, Sect. 5 concludes the work.

2 Technologies Overview

LoRa uses Chirp Spread Spectrum (CSS) and has been specified to operate on some specific ISM (Industrial, Scientific, and Medical) bands. In this work, the 915 MHz band standard was chosen as it uses the frequency band adopted in Brazil. LoRa has a very interesting feature: it allows the adjustment of some important parameters of the system, such as SF (Spreading Factor), BW (Bandwidth), and CR (Coding Rate), which can vary, respectively, from 7 to 12, 125 to 500 kHz, and 4/5 to 4/8 [8]. According to these parameters, the data rate can vary from 0.3 to 50 kbps.

On the other hand, Zigbee employs Direct Sequence Spread Spectrum (DSSS). It has been standardized to operate on three different ISM bands, each with very different characteristics [9]. In this work, the 2.4 GHz standard will be employed, as it is the most used standard in Brazil. It employs quasi-orthogonal modulation based on OQPSK (Offset Quaternary Phase Shift Keying) and can achieve data rates in the order of 250 kbps [6].

3 Performance Analysis

This work aims to evaluate and compare the performance of LoRa and Zigbee technologies from the point of view of quality (RSSI and PER) and transmission range. Due to industrial plants' characteristics, the performance of the systems will be analyzed for both indoor and outdoor environments.

For LoRa performance analysis, two ESP-LoRa devices based on the SX1276[1] chip operating at 915 MHz and positioned 2.0 m from the ground were used, one acting as a transmitter (TX) and the other as a receiver (RX). The research was based on the eight possible LoRa operating configurations presented in Table 1 for the different possible combinations of SF, BW, and the transmission power (P_{TX}) [1].

For Zigbee analysis, two Xbee-Pro devices (TX and RX) operating at 2.4 GHz were used. Each device was connected to an embedded Arduino Mega system for the control and transmission of packets and, as in the LoRa tests, both were positioned at 2.0 m from the ground.

3.1 Indoor Environment

The indoor environment tests were carried out in the basement of the A building of the Federal University of ABC (UFABC), Santo André. The performance of LoRa and Zigbee was analyzed from the measurements of the RSSI and PER parameters. During

[1] https://semtech.com/products/wireless-rf/lora-transceivers/sx1276.

the measurements, the TX was kept fixed and the RX was moved from 10 to 10 m until reaching 100 m. It is important to note that the systems were exposed to different interferences on the site, such as, for example, cochannel interference from WiFi (Wireless Fidelity) networks in operation. Specifically for LoRa, measurements were performed for all configurations presented in Table 1.

Table 1. LoRa configurations.

Configurations	SF	BW (kHz)	P_{TX} (dBm)
Config 1	7	125	14
Config 2	12	125	14
Config 3	7	125	20
Config 4	12	125	20
Config 5	7	500	14
Config 6	12	500	14
Config 7	7	500	20
Config 8	12	500	20

The indoor environment where the tests were performed is presented in Fig. 1.

Fig. 1. Indoor environment.

Figure 2 presents the RSSI variation as a function of d for the analyzed indoor environment. It can be seen that for LoRa, the measured values of RSSI was between –60 and –96 dBm. For Zigbee, it can be seen that up to 50 m, the RSSI was higher than –50 dBm, while, above 50 m, it was higher than –65 dBm.

In turn, Fig. 3 presents the results of PER as a function of *d* for Lora (Config 1 and Config 6) and Zigbee in the analyzed indoor environment. In the tests, three different measurement points were defined (10, 50, and 100 m) and, at each point, 10.000 packets were transmitted. It can be seen that up to 100 m, the LoRa presented a maximum PER of 2.2% for Config 1 and 1.8% for Config 6. For the Zigbee, the maximum PER was 0.4% between 10 m and 50 m. However, for 100 m, there was a significant increase in PER, which was around 22.5%.

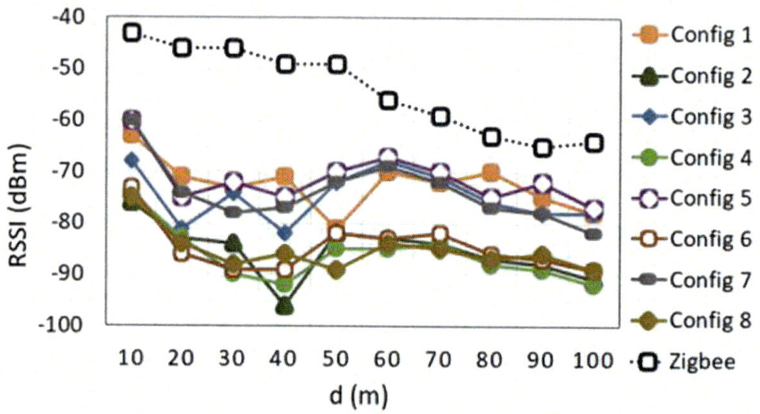

Fig. 2. RSSI as a function of *d* in the indoor environment.

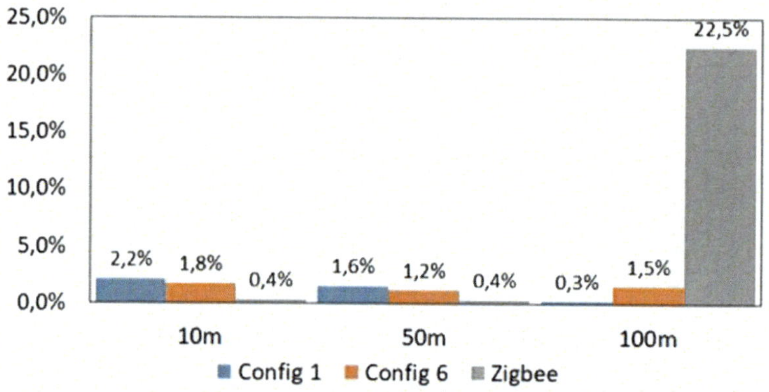

Fig. 3. PER as a function of *d* in the indoor environment.

3.2 Outdoor Environment

The outdoor measurements were carried out at the Faculty of Technology (FATEC), São Bernardo do Campo. Figure 4 shows the environment where the tests were performed. During the measurements, the TX was kept fixed and the RX was placed at 100, 200, 250, and 300 m away from TX, both at a height of 2.0 m from the ground. For simplicity, LoRa measurements were performed only for Config 1 and Config 6. Figure 5 shows the measurement points (a) and the terrain's elevation profile (b). It can be seen that the maximum slope between TX and RX is of 22.0 m.

Fig. 4. Outdoor environment.

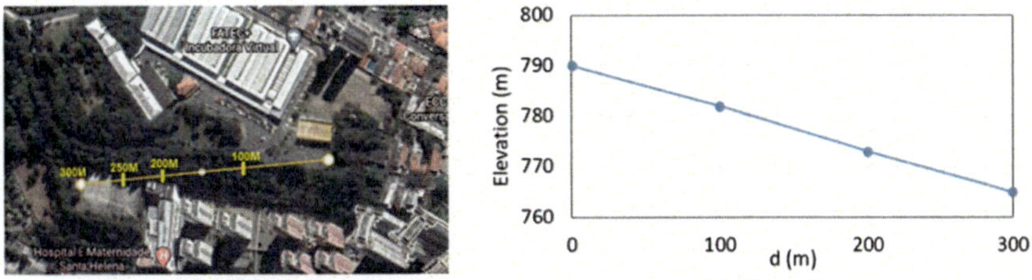

(a) Distances (b) Elevation

Fig. 5. Measurement points and terrain elevation of the outdoor environment.

Figure 6 presents the RSSI as a function of *d* for the analyzed outdoor environment. For LoRa, it can be seen that the RSSI varied from –80 to –110 dBm for Config 1 and from –85 to –110 dBm for Config 6, for distances between 50 and 300 m. As the reception sensitivity of LoRa devices is –148 dBm, the system operated properly within the analyzed range (making it possible to cover even greater distances, by the margin shown). In the case of Zigbee, the RSSI varied from –57 to –83 dBm, between 50 and 250 m. However, it was not possible to measure the RSSI in 300 m, since all packets were received with errors, indicating the level of the received signal was below the reception sensitivity of –102 dBm of Zigbee devices.

On the other hand, Fig. 7 presents the results of PER as a function of *d* for Lora (Config 1 and Config 6) and Zigbee in the analyzed outdoor environment. For Lora, it can be seen that up to 300 m, Config 1 presented a maximum PER of 3.0%, while Config 6 showed a maximum PER of 2.5%. For the Zigbee, the PER at 200 m was 34.71%, a value worse than 6.17%, measured at 250 m. This is mainly due to the environment's dynamic characteristics, such as the circulation of people and vehicles around the tests. However, at 300 m, Zigbee stopped working properly and all data packets were received with errors (corresponding to a PER of 100%).

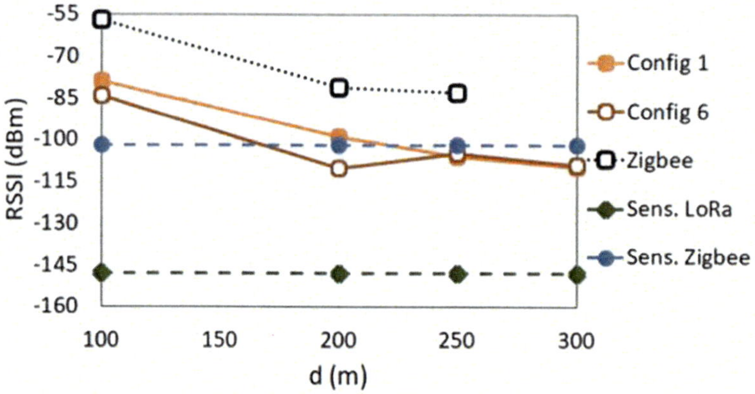

Fig. 6. RSSI as a function of d in the outdoor environment.

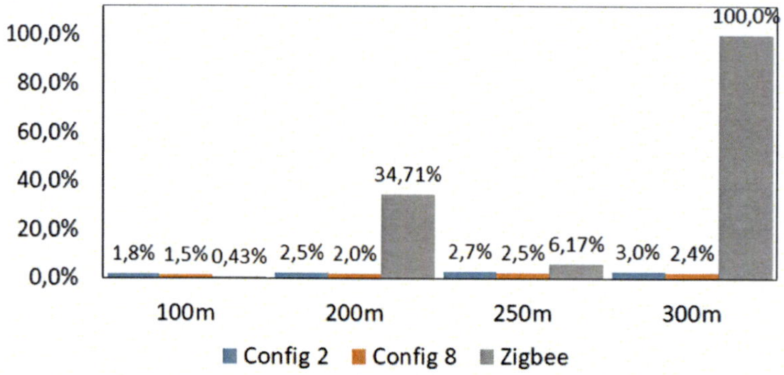

Fig. 7. PER as a function of d in the outdoor environment.

4 Shadowing Propagation Model

To estimate the mean received power (P_{RX}) as a function of d, the Shadowing Propagation Model (SPM) was used due to its flexibility, simplicity, and possibility of application when practical measures are taken. This model is based on the modified free space model and uses a log-normal random variable X_{SH} to represent the shadowing effect of the environment, which can be quantified by the value of its standard deviation σ_{SH}, expressed in dB [6, 10]. The P_{RX} for the SPM can be represented by:

$$P_{RX}(d) = P_{RX}(d_0) - 10 \cdot \gamma \cdot \log_{10}\left(\frac{d}{d_0}\right) + X_{SH} \tag{1}$$

Where: d_0 is the reference distance, $P_{RX}(d_0)$ is the mean power received at d_0, $P_{RX}(d)$ is the mean power received in d, and γ is the propagation coefficient of the environment.

Table 2 shows the values of γ and σ_{SH}, determined by a logarithmic regression method [6], for LoRa (for all configurations analyzed) and Zigbee frequency bands in the indoor environment.

Table 2. Obtained γ and σ_{SH} of the indoor environment.

Configurations	γ	σ_{SH}
Config 1	1.35	3.86
Config 2	1.35	4.38
Config 3	0.9	5.10
Config 4	1.74	4.07
Config 5	1.57	4.54
Config 6	1.60	4.60
Config 7	1.97	4.86
Config 8	1.45	3.13
Zigbee	1.8	3.9

Figure 8 shows the measured P_{RX} and the P_{RX} obtained through SPM for the analyzed indoor environment. In order to make the LoRa analysis clearer, without loss of generality, only Config 1 curves were presented in the figure. Analyzing the results, it is possible to verify that the P_{RX} shows some variations along d, mainly caused by cancellations of multipath components of the order of multiples of half the wavelength of the transmitted signal. For Zigbee, it can be seen that the measured P_{RX} also varies due to the cancellations of the multipath components. It is important to note that, due to the physical limitation of the environment (i.e., 100 m), it was not possible to obtain an estimate of γ and σ_{SH} for distances greater than the critical distance (D_c) [6]. In turn, Fig. 9 shows the random effect of the shadowing component on the P_{RX} curve obtained by SPM in the indoor environment for LoRa (Config 1) and Zigbee. The measured P_{RX} curve is also shown as a reference.

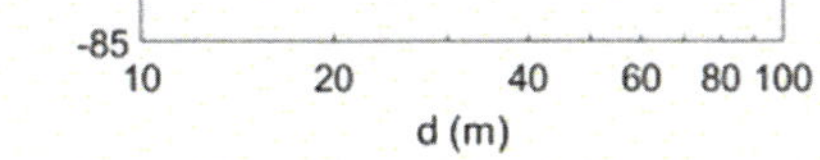

(a) LoRa (Config 1)

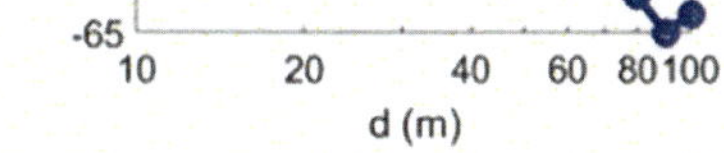

(b) Zigbee

Fig. 8. P_{RX} as a function of d for the indoor environment.

Table 3 presents the obtained values of γ and σ_{SH} for Lora (Config 1 and Config 6) and Zigbee in the analyzed outdoor environment. It can be seen that high values of γ were obtained due to the characteristics of the analyzed environment (e.g., slope between

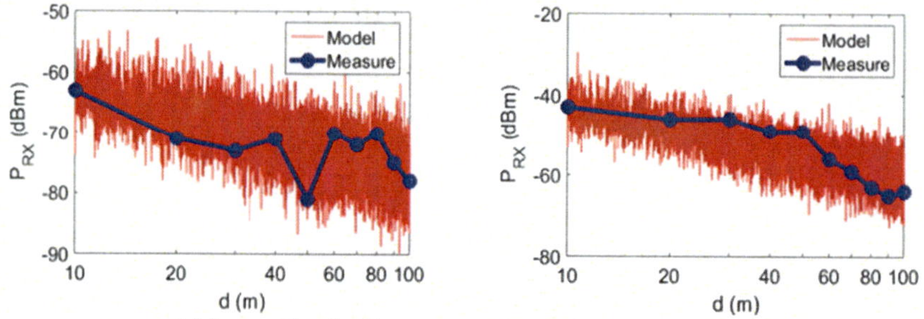

(a) LoRa (Config 1) (b) Zigbee

Fig. 9. P_{RX} as a function of d for the indoor environment, considering the shadowing effect.

TX and RX, path obstructions) and the measurements use some points with distances greater than D_c (e.g., after D_c, γ change to a higher value due to ground reflections) [6].

Table 3. Obtained γ and σ_{SH} of the outdoor environment.

Configurations	γ	σ_{SH}
Config 1	6.62	0.44
Config 6	5.8	4.73
Zigbee	7.06	2

Figure 10 shows the values of the measured P_{RX} and the P_{RX} obtained by the SPM for the outdoor environment. For simplicity, only the LoRa results for Config 1 are shown. It can be seen that a good fit was obtained between the model used and the measured values for both systems. On the other hand, Fig. 11 shows the effect of shadowing on the P_{RX} curve obtained by the SPM for LoRa (Config 1) and Zigbee in the outdoor environment. The measured P_{RX} curve is also shown.

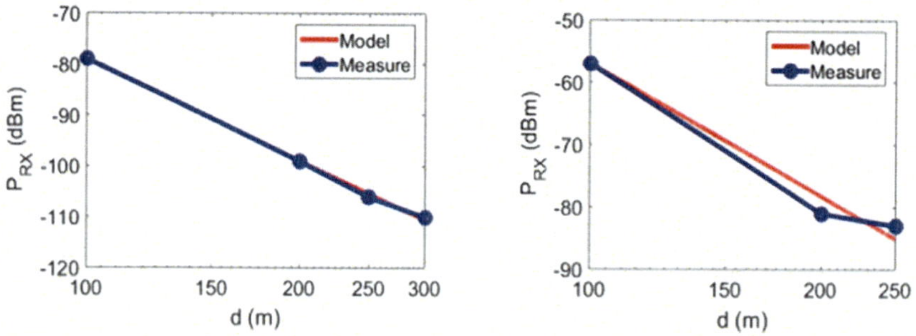

(a) LoRa (Config 1) (b) Zigbee

Fig. 10. P_{RX} as a function of d for the outdoor environment.

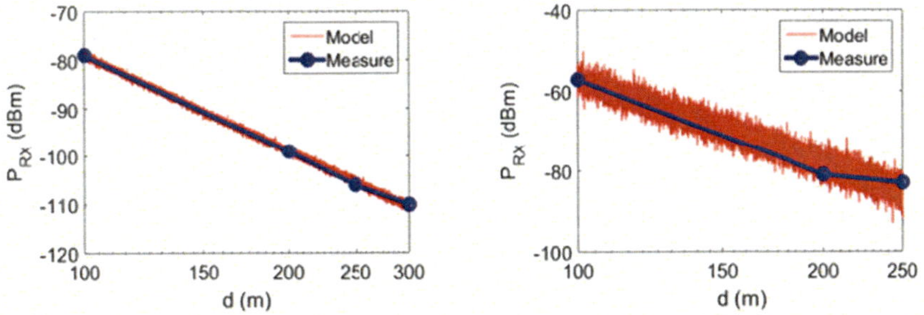

(a) LoRa (Config 1) (b) Zigbee

Fig. 11. P_{RX} as a function of d for the outdoor environment, considering the shadowing effect.

An estimate of the analyzed systems' maximum range can be obtained (according to the reception sensitivity) by extrapolating the P_{RX} curve obtained from the SPM. In this way, it can be seen that the maximum range of LoRa (Config 1) is 1.1 km, while the range of Zigbee is limited to 433.9 m.

5 Conclusions

The results obtained indicate that the evaluated Zigbee system has a better performance than the LoRa system used for a range of up to 50 m for indoor and outdoor environments. However, in a more in-depth analysis, also considering coverage above 50 m, the LoRa proved to be the most suitable choice for low rate, low consumption, and long-range applications. Operating in the 915 MHz frequency band, which is less congested than the 2.4 GHz band, and having a better reception sensitivity, the LoRa offers wider coverage with better signal quality, which is ideal for use in Industry 4.0.

References

1. Zhou Q, Zheng K, Hou L, Xing J, Xu R (2019) Design and implementation of open LoRa for IoT. IEEE Access 7:100,649–100,657
2. Adame T, Bel A, Bellalta B (2019) Increasing LPWAN scalability by means of concurrent multiband IoT technologies: an industry 4.0 use case. IEEE Access 7:46990–47010
3. Aheleroff S, Xu X, Lu Y, Aristizabal M, Velásquez PJ, Joa B, Valencia Y (2020) IoT–enabled smart appliances under industry 4.0: a case study. Adv Eng Inform 43:101,043
4. Sisinni E, Ferrari P, Fernandes Carvalho D, Rinaldi S, Marco P, Flammini A, Depari A (2020) LoRaWan range extender for industrial IoT. IEEE Trans Ind Inform 16(8):5607–5616
5. El Chall R, Lahoud S, El Helou M (2019) LoRaWAN network: radio propagation models and performance evaluation in various environments in Lebanon. IEEE Internet Things J 6(2):2366–2378
6. Casella IRS, Pereira AL (2016) A bioinspired propagation model for brazilian digital TV system based on MLP and RBF networks. IEEE Latin Am Trans 14:3941–3948
7. Filsoof R, Bodine A, Gill B, Makonin S, Nicholson R (2014) Transmitting patient vitals over a reliable zigbee mesh network. In: 2014 IEEE Canada International Humanitarian Technology Conference, IHTC 2014, pp 1–5

8. Croce D, Gucciardo M, Mangione S, Santaromita G, Tinnirello I (2018) Impact of LoRa imperfect orthogonality: analysis of link-level performance. IEEE Commun Lett 22(4):796–799
9. Raposo D, Rodrigues A, Sinche S, Sá Silva J, Boavida F (2018) Industrial IoT monitoring: technologies and architecture proposal. Sensors 18(10):3568
10. Cheung KW, Sau JHM, Murch RD (1998) A new empirical model for indoor propagation prediction. IEEE Trans Veh Technol 47(3):996–1001

Thermographic Evaluation of Organic Photovoltaic Cells Under Real Working Conditions

Josué Marcos de Moura Cardoso[1](✉), Lia Toledo Moreira Mota[2], Claudia Cotrim Pezzuto[2], Valeria Cristina dos Santos Silva[2], Gabriel Oliveira Gomes[1], and Yuzo Iano[1]

[1] Universidade Estadual de Campinas, Campinas, SP, Brasil
oliveiragomesgabriel@ieee.org, yuzo@unicamp.br
[2] Pontifícia Universidade Católica de Campinas, Campinas, SP, Brasil

Abstract. Photovoltaic technologies are fundamental resources for creating a sustainable environment. Since the emergence of the first polycrystalline silicon photovoltaic cells (pc-Si), the very evolution of science has enabled the development of these conductors with organic compounds today. However, these organic photovoltaic cells have not yet spread to areas such as civil construction, for example. The main advantages of these cells are flexibility and easy handling/implementation, and the aesthetics of the built environment can also be favored. This paper intends to evaluate the thermal performance of organic cells from the images captured by an infrared camera. The results showed that climatic and environmental conditions have a significant role in the performance of these cells, and that they have an almost instantaneous cooling capacity, based on the reduction of the solar incidence in these cells. As per the captured thermographic images, it is observed that the organic PV cells operate at temperatures higher than those determined in norms and standardizations performed with a focus on the performance of photovoltaic modules.

Keywords: Organic photovoltaic cells · Operating temperature · Infrared camera

1 Introduction

The improvement of the technologies and chemical compounds adopted for the elaboration of photovoltaic (PV) cells resulted in cells with flexible and adaptable structures. Silicon (Si) is the main element adopted for making photovoltaic cells. Initially, cells with monocrystalline silicon (m-Si) and polycrystalline (pc-Si) were developed, characterizing the first generation cells. Later, amorphous Silicon (a-Si) brought thin films as PV modules already in the second generation of cells [1–3]. Years later, other elements were also inserted in the manufacture of cells such as the compounds of Gallium Arsenide (GaAs), Cadmium Telluride (CdTe), among others. The third generation contemplates organic compounds as raw material, being Anthracene (C14H10), Tetracene

Y. Iano et al. (Eds.): BTSym 2020, SIST 233, pp. 811–823, 2021.
https://doi.org/10.1007/978-3-030-75680-2_90

(C18H12) and Pentacene (C22H14) some of the elements present in these PV cells [2, 4, 5].

Despite the development of photovoltaic technologies, PV modules have their efficiency impaired by a many factors, as well as the malfunction of the internal components of the equipment, failure in the internal resistances of the electronic circuits [3, 6, 7], in addition to factors such as dust and dirt in general [2, 8]. Therefore, they are subject to internal equipment manufacturing failures that affect their output power (P_{OUT}) [9, 10]. Environmental and climatic factors also contribute to the efficiency of the PV cells, such as air temperature (T_a), cell operating temperature (T_{OP}), wind speed (S) and solar radiation (I). Those factors have been the most discussed in the literature [11–13]. Other environmental parameters such as relative humidity, wind speed and wind direction also contribute to the energy efficiency of the modules [8, 14].

The standards ASTME-1036, ASTME1036–15 and IEC1215 adopt the standard test conditions STC (Standard Tests Conditions) as parameters to be used in laboratory tests that contain photovoltaic cells. Therefore, they consider the following values: $I = 1000\ W/m^2$ for solar irradiance; $T_{OP} = 25\ °C$ in the PV cells; and 1.5 air mass (AM). Another standard that supports the standardization of tests performed on PV modules is the NOCT (Nominal Operating Cell Temperature) standard. This standard includes the variables of solar irradiance (I), wind speed (S), air mass (AM) and T_{OP} delimited in the values of 800 W/m^2, 1 m/s, 1.5 and 20 °C, respectively [13–17]. In Brazil, for example, the standards NBR11876/EB2176 and NBR12137 bring the same values mentioned in the STC and NOCT standards applied to photovoltaic panels [3, 18, 19]. However, under real working conditions, PV modules operate under different values. The Table 1 below presents some compounds applied to the manufacture of photovoltaic cells and their respective energy conversion efficiencies.

Table 1. Efficiency in the conversion of electrical energy considering the type of photovoltaic cell [20]

Photovoltaic cells types		Efficiency (%)
Silicon	mc-Si	25 ± 0,5
	pc-Si	20,4 ± 0,5
	Thin films on reusable metal/glass polymer structures	20,1 ± 0,4
Boron (13) and Nitrogen(16)	GaAs (thin film)	28,8 ± 0,9
	GaAs (polycrystalline)	18,4 ± 0,5
	InP (monocrystalline)	22,1 ± 0,7
Chalcogen Compounds	CdTe (thin film)	18,3 ± 0,5
Amorphous silicon/nanocrystalline	Amorphous (a-Si) (thin film)	10,1 ± 0,3
	Nanocrystalline (nc-Si)	10,1 ± 0,2
Multi-junction Cells	InGaP/GaAs/InGaAs	37,7 ± 1,2
Organ cells (thin films)		10,7 ± 0,3

2 Previous Work

The National Renewable Energy Laboratory (NREL) takes an approach considering, mainly, the chemical elements used to manufacture PV cells. The separation consists of: multi-junction cells; crystalline Si cells; cells consisting of GaAs with unitary junction; thin films and, finally, emerging technologies [5, 20, 21]. In other perspectives, photovoltaic cells are classified into generations, where their peculiarities and particularities are considered for the grouping of each technology [1–3], as per Table 2 presents some characteristics of the three types of PV cells for the first three generations, as well as each type of PV cell is illustrated by this table.

In the third generation of PV cells, the devices are designed to better optimize the energy contained in photons (solar radiation). Much of the energy contained in the sun's rays is not converted into electricity. As a result of the effects of the equipment itself and the meteorological and local conditions, solar energy ends up being transformed into heat. One of the ways to optimize the process of capturing and converting this energy is to use multi-junction cells. Therefore, these cells are made up of sub-layers of cells with more elaborate energy conducting elements, as shown in the Fig. 1. In this type of multi-function, the cells are superimposed on each other, improving the use of the solar spectrum and decreasing the effect of excessive heating in this type of equipment [4, 20, 22, 23].

The T_{OP} of the PV cells contributes directly to the thermal performance of the PV module. This temperature is related to the structural aspects of the equipment, since each part associated with the panel contributes to the heating of the equipment, in general [1, 24, 25]. The heat flow inherent to the direct contact established by the sealing and protection layers in a PV module. The heat distribution in which each part of the module (glass, encapsulating material, teddlar, aluminum cladding, etc.) absorbs and transmits heat corroborates so that T_{OP} also rises. Therefore, climatic conditions, electrical properties and installation conditions of the modules are other important factors when evaluating the behavior of T_{OP} [3, 7, 21, 25].

Also, the ambient temperature (Ta) plays a fundamental role in the performance of PV modules [7, 27]. It acts mainly on semiconductor cells that have their functioning affected by the elevation of this temperature [12, 28, 29]. The operating temperature of the cells (TOP) obeys an inversely proportional relation to Ta, therefore, when the temperature rises, the tension drops. Another effect caused by the increase in Ta is a minimal, but noticeable, change in the short circuit current for low voltage values. Thus, the power of the maximum point, results in significantly lower values [9, 16, 20, 30]. The Fig. 2 shows the graphic of electric voltage and electric current at different values of ambient temperatures.

The T_{OP} can be described according to Eq. (1) whose value is associated with the ambient temperature at the installation site of the PV module (T_a) and the temperature variation of the elements that make up the modules (ΔT) [31].

$$T_{OP} = T_a + \Delta T \tag{1}$$

The TOP of PV cells varies according to the three thermal transfer phenomena: conduction, convection and radiation. Convection is caused by the ambient temperature

Table 2. First, second and third generation photovoltaic cell types [20, 22, 23, 26]

PV CELL GENERATION	CHARACTERISTICS	EXEMPLIFICATIVE FIGURE
FIRST	First type of PV cells launched on the market, basically characterized by the presence of mc-Si and pc-Si cells, and also GaAs (Gallium Arsenide). This type of technology still dominates the photovoltaic capture and conversion market in the world (around 90% of the entire generation in the world consists of silicon PV modules and panels, essentially);	pc-Si  mc-Si 
SECOND	It has a-Si as an essential element and stands out for its constitution through thin films. The most common chemical elements in this generation are CdTe (Cadmium telluride) and $CuInGaSe_2$ Thin films have shown a slight growth over the past decades and today they participate with around 10% of the energy production by solar panels in the world;	a-Si 
THIRD	These are the most recent technologies associated with the use of solar radiation. This generation includes cells made using organic compounds. They are sensitized by dyes, with low efficiency, and also adopt multi-function compounds with high utilization of PV conversion, however, they have a high commercial price.	Organic cell 

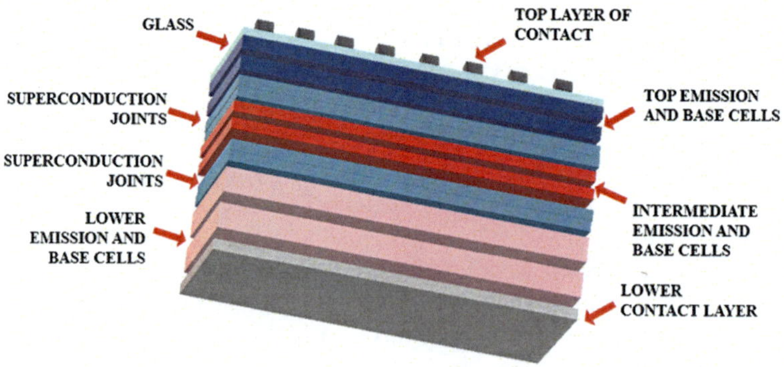

Fig. 1. Layers of components present in a photovoltaic module with organic cells [4]

(Ta), since an additive effect is forced convection, resulting from the mass flow of air that has a cooling effect. This effect is mainly influenced by the installation characteristics of the photovoltaic system (free standing, flat roof, inclined roof, integrated facade or roof). In contrast, the conduction phenomenon expresses heat transfers between the different layers of the PV modules (glass, Ethylene Vinyl Acetate - EVA, cell, teddlar). The heat transfers from the radiation caused by the incidence of solar radiation is the most influential factor in the temperature variation of the photovoltaic cells. The TOP of the PV module increases 4.4 °C when the solar radiation increases from 800 to 900 W/m^2, rising to 1 °C when the Ta increases 1 °C, adding, still, 0.5 °C when the humidity increases from 50% to 60%. Conversely, the TOP drops 3.3 °C when the wind speed accelerates from 1 m/s to 2 m/s [6, 7, 10].

In practical terms, the PV modules operate in different situations and that differ from the conditions outlined by the standardized test conditions (STC) where the T_a must be at 25 °C. In cases of real work applications, the cells reach expressive values of temperature, because of that, it was necessary to establish new standards for the behavior of these compounds when being subjected to the characteristic environmental and climatic conditions [17, 19, 21]. In this perspective, the increase in the T_{OP} of the PV cells causes two effects on the electrical parameters of the PV module. The voltage produced by the photovoltaic effect (V) decreases and the current (I) rises proportionately. As a result, the potential of electric power generation (P_{OUT}) of the equipment is impaired, since there is a decrease in the maximum power point of the modules, where the electrical conversion degrades [20, 25, 27–29].

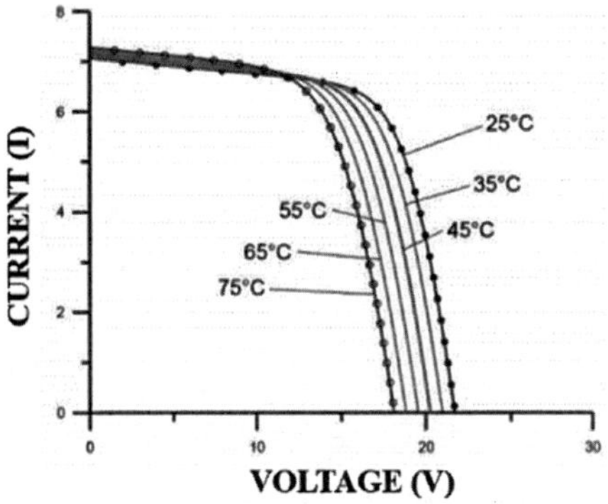

Fig. 2. Voltage and electric current from photovoltaic cells for different ambient temperature values [19]

Thermography images helps to see the heat distribution in PV modules. It is a non-intrusive and effective means that allows readings on the thermal range that these modules have under operating conditions. The environmental and climatic configuration at the site, as well as the physical characteristics of each photovoltaic panel, also interfere in studies involving thermographic machines as a technique for monitoring surface temperature. The effects inherent to the process of solar radiation, reflection, refraction and diffraction are important factors that will change the emissivity of the material [32, 33].

Thermographic cameras provide module thermal information based on the intensity of infrared radiation in the PV module. This is a punctual and temporal analysis, it makes it easy and practical to collect the data of interest, where the upper part of the modules is used as part of the studies. These cameras record a considerable amount of heat points at which the surface temperature can be estimated from each captured point. As a result, a thermal gradient of the panel is generated in great detail. Thus, when considering the local climate as an integral agent in the TOP analyzes of the PV modules, with each variation of the environmental factors it is possible to verify this performance in the images captured by the thermographic cameras [19].

As per proposed by [32] the authors produced a work evaluating the thermal behavior of PV modules in an open environment with the help of thermographic machines. Based on the intrinsic relationship between the operating temperature of the PV modules and their efficiency, the authors develop a methodology for acquiring data captured by a Flir Infrared Camera. Thus, in order to analyze the change in the actual energy production by the module and to track the maximum power point with the temperature in real time, it is important to determine the exact temperature of the photovoltaic module in external operating conditions. The authors' approach consisted of a non-invasive and more accurate technique to measure the module's temperature under operating conditions. The experimental installation was established with outdoor and infrared images of the PV module. The images were subsequently processed quantitatively to calculate the module temperature. In evaluating the results, the output data were compared with values calculated with the aid of well-established relationships available in the literature with different temperatures calculated using existing techniques.

The infrared camera detects the heat emissions by infrared radiation emitted by the bodies, the that makes it possible to interpret its temperature [34]. Thus, the process is different from that of conventional cameras, because unlike these, the thermal imager does not need the reflection of light on the object under study, the signal comes from the vibratory movement of the molecules that make up the materials of that object [19, 33]. Thus, thermal radiation can be measured at different levels over the electromagnetic spectrum, and most infrared temperature measurements are made in the region between 0.2 to 20 μm. The thermal imagers use detectors, which through an optical system for converting infrared radiation into an electrical signal, and complex signal processing, to build the thermal image [35], allow reading the temperature of objects.

Therefore, in this work, the thermographic images from an infrared photographic camera will be analyzed to visualize and identify the TOP of the PV cells made of organic material.

3 Methodology

In order to add new generations of photovoltaic cells to the work, a PV module composed of organic cells was also used in the tests carried out. The Fig. 3 shows the PV module consisting of organic cells used for bench tests. This PV module is produced by the manufacturer [36]. It is a strip of organic PV cells, with flexible structure, fixed with semitransparent adhesive on the front and back. This equipment works as a "photovoltaic adhesive" with versatile implementation. It can be applied to rods, blinds, window frames, glass, aircraft wings, cabinets, etc. The Fig. 4 shows the adaptation made with this PV module to perform the tests. The organic PV cells were glued to an EVA surface. The physical and electrical characteristics of this module are described in Table 3 where V_{OC} (open circuit voltage), I_{SH} (short circuit current), V_{OUT} (output voltage) and I_{OUT} (output current).

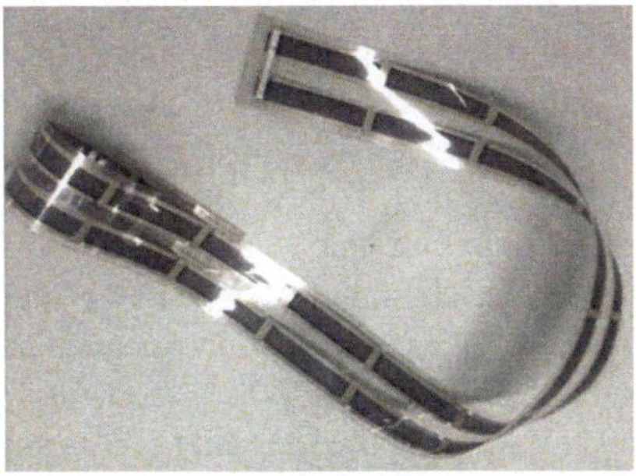

Fig. 3. PV module with organic cells [11]

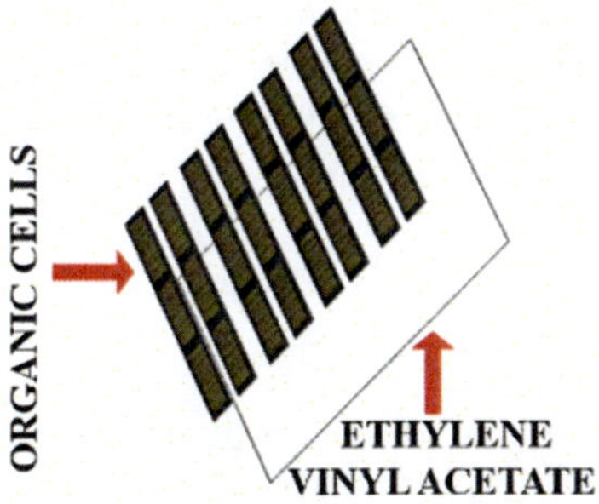

Fig. 4. PV panel with organic cells applied over EVA [11]

Table 3. Electrical and physical characteristics of PV module with organic cells

V_{OC}	20 V
I_{SH}	50 mA
V_{OUT}	1 V
I_{OUT}	0,5 mA
Number of cells	24
Size	30 × 30 cm

In this paper, the organic cell strip was adapted by cutting the PV cells into groups of 6 units, composing 4 groups in total (24 cells) (Fig. 5). Despite contact with suppliers, the chemical composition of the PV cells in this module has not been made available. The 24 organic PV cells were connected in series, just as the original tape was connected. However, during the photovoltaic tape handling stage, as it is a highly sensitive equipment, one of the groups was damaged. This resulted in three functioning groups, connected in series. Even so, the 4 groups were maintained during the performance of the tests precisely to approximate the methods linked to each test carried out by approaching the PV module. During the validation steps of this PV module, each groups did not exceed 1V of electrical voltage. Another point to be highlighted is the lack of bibliographic framework dealing with photovoltaic modules with organic cells. Specifically, for the module developed in the methodology of this work, the foundations used for the PV modules with pc-Si cells were used, in an attempt to approximate the structure of the modules in their physical composition.

The Fig. 6 shows the equipment of the manufacturer FLUKE model Ti100 that was supplied by the Thermal Comfort Laboratory at PUC-Campinas. This equipment is characterized by its resistance and reduced weight, easy handling and adjustment, with effective focus and shooting capabilities. The system for focusing the machine is automatic. This thermal imager assists in the sampling of temperature points distributed on any surface. To ensure the effective data collection with the Ti100 machine, the distances between the equipment and the bench were kept the same for capturing data from each specimen. This description of each test performed contains this information. All images captured by the Ti100 machine can be saved in file extensions compatible with image reading applications. Thus, at the end of each test, the images were downloaded to the base node of the monitoring network for checks and analyzes. The FLUKE thermographic machine has a rechargeable battery and expandable memory thanks to the memory card slot [37].

Fig. 5. PV module with organic cells adapted to collect thermographic images [22]

Fig. 6. Thermographic camera model Ti100 (FLUKE) used to capture the images [22]

4 Results

The tests were performed on 12/12/18 from noon and the thermographic images were taken 75 cm away from the PV module (latitude: -22,8861/longitude: -47,0673 – PUC-Campinas). Figure 7 (A) shows the heat distribution in the adapted PV module, where it can be seen that the cell temperature reached 51.8 °C. The Fig. 7 (B) shows the T_{OP} of the organic PV cells at 1:37 pm (Brasília time) with average temperature in 40, 7 °C. As shown in Fig. 7 (C) the T_{OP} of organic photovoltaic cells can reach 51.8 °C. Figure 7 (D) shows the T_{OP} of these cells at 4:34 pm, where the average temperature was 45.2 °C.

Near sunset, the T_{OP} of organic cells can be seen in Fig. 7 (E). Without the incidence of direct solar radiation (I) the cells showed considerable cooling (when compared with the previous thermographic images). Thus, the average temperature at 5:27 pm was 38.2 °C. At the end of the test day, Fig. 7 (F) was captured with the thermographic camera to check the cooling of the PV module. As can be seen by this figure, at 7:58 pm the average temperature in the equipment was 30,2 °C.

Table 4. T_{OP} average of organic cells during the tests.

TIME (12/12/2018)	T_{OP} AVERAGE (°C)
12:23 am	51.1
1:37 pm	40.7
2:53 pm	51.8
6:34 pm	45.2
5:27 pm	38.2
7:58 pm	30.2

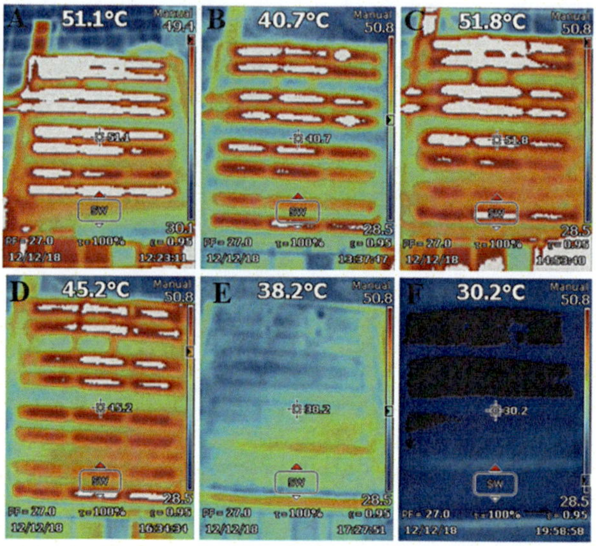

Fig. 7. Thermographic images of the PV module composed of organic cells under real working conditions.

Table 4 above shows the compilation of data collected by the thermographic camera. Note that at 2:53 pm the TOP reached its peak (51.8 °C), and it is possible to observe that the average of the TOP, in general, was 50.8 °C. The Fig. 7 (F) was captured an hour and a half after the end of the tests to validate the TOP in the PV module. Despite the absence of solar radiation, the average TOP at that time was 30.2 °C. The cooling capacity of the PV modules is also associated with their location (roofs, photovoltaic plants, etc.).

5 Discussion

As per the adaptation made on the tape with the organic solar cells, it was possible to validate the heat distribution in the adapted PV module. In this paper the objective was to validate the T_{OP} of organic cells under real working conditions, and the data collection took place on 12/12/2018. During the tests it was observed that this photovoltaic technology showed rapid cooling, being extremely sensitive to solar radiation (I) and to ambient temperature (T_a). It should be noted the low electrical performance of these cells, which are even capable of producing energy from artificial lights. In this context, in areas such as civil construction, where many professionals are still resistant to the insertion of PV modules in their projects for aesthetic reasons, the adaptability and flexibility made possible by the PV technology studied in this paper may be factors that contribute to the dissemination of organic cells in the built environment, for example.

6 Conclusion

The focus of this paper was to register, using thermographic images, the T_{OP} behavior of organic PV cells under real working conditions. In fact, it has been proven that organic cells operate at temperatures much higher than those proposed by STC and NOCT, and in STC standardization T_{OP} is not one of the factors considered for the preparation of tests containing PV modules. The temperature (either ambient or cell operation) directly interferes with the P_{OUT} of this equipment. The higher the temperature, the less the photons can convert into electrical energy. The organic PV cells play a fundamental role in the dissemination of renewable energy in the built environment. These cells can adapt to windows, blinds, roofs, doors, walls, among other building elements.

References

1. Falcão VD (2005) Fabricação de Células Solares de CdS/CdTe. 2005, 120 f. Master's Dissertation. Curso de Mestrado em Ciência dos Materiais do Instituto Militar de Engenharia. Rio de Janeiro
2. Mcevoy A, Castaner L, Markvart T (2012) Solar cells: Materials, Manufacture and Operation, 2nd edn. Academic Press, Cambridge
3. Ortega LLM (2013) Conversão Fotovoltaica: Comparação de Modelos de Desempenho. 2013, 116 f. Master's Dissertation. Programa de Pós-Graduação em Metrologia para Qualidade e Inovação, Pontifícia Universidade do Rio de Janeiro. Rio de Janeiro
4. Avrutin V, Izyumskaya N, Morkoç H (2011) Semiconductor Solar Cells: Recent Progress in Terrestrial Applications. Superlatives and Microstructures, vol 49, no 4, p 337–364
5. Neves GM (2016) Influência do Espectro da Radiação Solar em Módulos Fotovoltaicos. Master's Dissertation, Instituto Nacional de Pesquisas Espaciais. São José dos Campo. São Paulo
6. Tina GM, Abate R (2008) Experimental Verification of Thermal Behavior of Photovoltaic Modules. In: The 14th IEEE Mediterranean Electrotechnical Conference, Ajaccio, France, p 579–584, May 2008
7. Tina GM, Scrofani S (2008) Electrical and Thermal Model for PV Module Temperature Evaluation. In: The 14th IEEE Mediterranean Electrotechnical Conference, Ajaccio, France, p 584–590, May 2008
8. Mekhilef S, Saidur R, Kamalisarvestani M (2012) Effect of dust, humidity and air velocity on efficiency of photovoltaics cells. Renew Sustain Energy Rev 12:2920–2925
9. Bühler AJ (2007) Determinação de Parâmetros Fotovoltaicos a partir de Ensaios de Curva Característica sem Iluminação. 2007, 89 f. Master's Dissertation. Programa de Pós-Graduação em Engenharia Mecânica, Escola de Engenharia Universidade Federal do Rio Grande do Sul, Porto Alegre
10. Ramos CM (2006) Procedimentos para Caracterização e Qualificação de Módulos Fotovoltaicos. 2006, 140 f. Master's Dissertation, Programa de Interinidades de Pós-Graduação em Energia. Universidade de São Paulo. São Paulo
11. Armstrong S, Hurley WGA (2010) Thermal Model for Photovoltaic Panels Under Varying Atmospheric Conditions. Applied Thermal Engineering, vol 30, no 11–12, p 1488–1495
12. Cantor GAR (2017) Influência dos Fatores Climáticos no Desempenho de Módulos Fotovoltaicos em Regiões de Clima Tropical. 176 f. Master's Dissertation. Programa de Pós-Graduação em Energias Renováveis. Universidade Federal da Paraíba, João Pessoa

13. Sahri A, Toumi S, Ramenah H, Tanougast C (2013) temperature effects on the power production based photovoltaic cells. In: 25th International Conference on Microelectronics (ICM), Beirut, Lebanon
14. al-Nimr MA, al-Ammari WA, (2016) A novel hybrid PV-distillation system. Solar Energy 135:874–883
15. U.S. Department of Energy Office of Scientific and Technical Information. American Society for Testing and Materials (1985) "E 1036 Standard Methods of Testing Electrical Performance of non-Concentrator Terrestrial Photovoltaic Modules and Arrays Using Reference Cells". https://cvtg.net/app/download/630523004/ASTM+C1036.pdf
16. U.S. Department of Energy Office of Scientific and Technical Information. American Society for Testing and Materials (1997) "E 927–91 Standard Specifications for Solar Simulation for Terrestrial PV Testing". https://www.astm.org/DATABASE.CART/HISTORICAL/E927-91R97.htm
17. Buday MS (2011) Measuring Irradiance, Temperature and Angle of Incidence Effects on Photovoltaic Modules in Auburn Hills, Michigan. Master's Practicum, Univesity of Michigan: Ann Arbor 1–26, 64 f
18. Dhar S, Sridhar R, Avasthy V (2012) Modeling and Simulation of Photovoltaic Arrays. In: National Power System Conference IIT-BHU, Varanasi, India
19. Trinuruk P, Sorapipatana C, Chenvidhya D (2009) Estimating operating cell temperature of BIPV modules in Thailand. Renew Energy 34(11):2515–2523
20. Laboratório de Modelagem e Estudos de Recursos Renováveis de Energia; Centro de Ciência do Sistema Terrestre; Instituto Nacional de Pesquisas Espaciais. Atlas Brasileiro de Energia Solar. São José dos Campos, São Paulo. 2ª edição. https://labren.ccst.inpe.br/atlas_2017.html
21. Santolin EA (2016) Aplicação de Sensores à Fibra Óptica no Monitoramento Térmico de Painel Fotovoltaico. 105 f. Master's Dissertation. Programa de Pós-graduação em Engenharia Elétrica. Universidade Tecnológica Federal do Paraná. Pato Branco
22. Pinho JT, Galdino MA (2014) "Manual de Engenharia para Sistemas Fotovoltaicos". Rio de Janeiro: CEPEL–CRESESB. 529 p. https://www.cresesb.cepel.br/publicacoes/download/Manual_de_Engenharia_FV_2014.pdf
23. U.S. International Energy Agency (2014) Technology Roadmap – Solar Photovoltaic Energy
24. Cardoso JMM (2019) Sistemas BAPVS e BIPVS: Avaliação Do Desempenho Da Integração E Aplicação De Módulos Fotovoltaicos A Elementos Construtivos. 2019. 270 f. Master's Dissertation. Programa de Pós-Graduação em Sistemas de Infraestrutura Urbana. Pontifícia Universidade de Campinas
25. Cardoso JMM, Mota LTM, Pezzuto CC, Silva VCS, Soeira M, Gomes GO, Iano Y (2019) Sensors network to evaluate the performance of photovoltaic modules in BAPV systems. In: Brazilian Technology Symposium, 2019, Campinas. Brazilian Technology Symposium BTSYM 2019
26. Mattei M, Notton G, Cristofari C, Muselli M, Poggi P (2006) Calculation of the polycrystalline PV module temperature using a simple method of energy balance. Renew Energy 31:553–567
27. Zhou JC, Zhang Z, Liu HJ, Yi Q (2017) Temperature distribution and back sheet role of polycrystalline silicon photovoltaics modules. Appl Therm Eng 111:1296–1303
28. Centro de Pesquisas de Energia Elétrica Centro de Referência Para as Energias Solar e Eólica Sérgio de Salvo Brito, Ministério de Minas e Energia (2006) Energia Solar – Princípios e Aplicações. https://www.cresesb.cepel.br/download/tutorial/tutorial_solar_2006.pdf
29. Prieb, CWM (2002) Desenvolvimento de um Sistema de Ensaio de Módulos Fotovoltaicos. 67 f. Master's Dissertation. Programa de Pós-Graduação em Engenharia Mecânica. Universidade Federal do Rio Grande do Sul. Porto Alegre
30. Beltrão REA (2008) Efeito da Temperatura na Geração de Energia de Módulos Fotovoltaicos Submetidos a Condições Climáticas Distintas: Estudo de Caso para as Localidades de Recife

e Araripina. 171 p. Master's Dissertation. Programa de Pós-Graduação em Tecnologias Energéticas e Nucleares, Universidade Federal de Pernambuco, Recife
31. Dubey S, Sarvaiya JN, Seshadri B (2013) Temperature dependent photovoltaic (PV) efficiency and its effect on PV production in the world - a review. Energy Procedia 33:311–321
32. Nordmann LCT, Clavadetscher L (2003) Understanding temperature effects on PV system performance. In: Proc. Third World Conference on Photovoltaic Energy Conversion, Osaka, Japan, p 2243–2246
33. Wilcox JR (2013) Solar Cell Temperature Dependent Efficiency and Very High Temperature Efficiency Limits. 304 p. Master's Dissertation. Faculty of Purdue University, West Lafayette, Indiana, United States
34. Irshad; Jaffrey ZA, Haque A, (2018) Temperature measurement of solar module in outdoor operating conditions using thermal imaging. Infrared Phys Technol 92:134–138
35. Silva BPA (2017) Metodologia para Determinação de Temperatura em Painéis Fotovoltaicos por Meio de Termográfica Quantitativa. 2017, 60 f. Master's Dissertation. Programa de Pós-Graduação em Engenharia Mecânica. Universidade Federal de Minas Gerais. Belo Horizonte
36. KSC - Kennedy Space Center (2012) Thermography Technique AT-9. Florida, EUA
37. LAND Instruments International (2004) A Basic Guide to Thermography. Publication No: Training ThermographyV1/1104. Newtown EUA

Near-Ground Propagation Model in an Archaeological Park in Cusco for Low Power Wireless Sensor Network

Yhon Lezama[1(✉)], Jinmi Lezama[2], Cesar Briso[3], and Jorge Arizaca[1]

[1] Universidad Nacional de San Antonio Abad del Cusco, Cusco, Perú
jorge.arizaca@unsaac.edu.pe
[2] Universidad Nacional Tecnológica de Lima Sur, Lima, Perú
[3] Universidad Politécnica de Madrid, Etsi Sistemas Telecomunicación, Madrid, España
cesar.briso@upm.es

Abstract. In the present work, we develop the propagation channel modeling for point-to-point (PtP) communication network between devices using LoRa technology within an archaeological park in Cusco, Peru, at 920 MHz, due to the propagation model is an essential component in the design and deployment of the wireless sensor network (WSN). The signal indicator received from the nodes is obtained, while the transmitter moves at a constant speed in different linear routes, under three outstanding environments of the archaeological park, being the CAPELLANPAMPA esplanade, the Inca walls, and the WATAYCARCEL platforms. By means of an exhaustive analysis of the influence of different heights of the receiving antenna and the factors of each terrain, the use of the moving average filter with a window length of 40λ is determined in order to separate the fades, then interpolate the average received power during the deployment of the PtP network within each scenario. This study emphasizes what happened within the scenarios trying to explain each one through the models developed. Each model is validated and chosen based on the value of the correlation coefficient between the measured data and the developed model.

Keywords: LoRa · Received power · Channel model · Archaeological park

1 Introduction

Nowadays Cusco is the historical capital of Peru, so it has several archaeological parks, and one of the most important is the archaeological park of Chinchero; it has been affected over the years and by the presence of people that has been increasing, so it needs new ideas for prevention and maintenance of these archaeological parks.

Technological advances allow using electronic devices for different applications, especially in areas such as telecommunications, agriculture, logistics, smart networks, geolocation, security, and others [1].

Y. Iano et al. (Eds.): BTSym 2020, SIST 233, pp. 824–832, 2021.
https://doi.org/10.1007/978-3-030-75680-2_91

One example is the systems to localized people in different places, such as GPS systems and wireless sensor networks. Some activities of localization systems can be developed using the wireless sensor network to avoid the degradation of the archaeological park of Chinchero.

Recent advances in wireless and electronic communications have made it possible for these multifunctional sensors to be low-cost, low-power, and small, making their use massive for different applications. Moreover, localization technics are based on the trilateration algorithm of RSS value and time of arrival signal (ToA) [2]. The precision of these technics depends on signal sensitivity and also on the channel propagation model.

Due to sensor networks communicate wirelessly, these wireless links are not reliable because their behavior varies with time and space. Therefore, modeling of the propagation channel between the transmitter and receiver node will be an important factor to ensure the proper functioning and deployment of a network; this, in turn, will depend mainly on the type of terrain, the type of dispersers, and the foliage of the environment [3].

The signal received is usually difficult to describe because it undergoes different phenomena such as diffraction, reflection, and dispersion, so it reaches the receiver by multiple paths. For this reason, mathematical models are used, where a wireless channel model is a set of mathematical expressions and algorithms. This is used to represent the phenomena of signal propagation within a specific environment [4].

The document is organized as follows. In Sect. 2, some works related to this research are briefly described. In Sect. 3, the data collection methodology is described, including the measurement equipment and the scenarios in more detail. In Sect. 4, the results of the measurements in the developed environments are presented and analyzed, that is to say, the esplanade with a wide line of sight, the walls and the platforms that constitute the most relevant scenarios within the archaeological park of Chinchero, and finally, the conclusions are presented.

2 Related Work

In most investigations, the sensor nodes are located on the ground a few centimeters from the ground. In [5], a near-ground propagation channel was studied in the frequency for 800 MHz and 1000 MHz, and 11 sites were measured with an antenna height of 15 cm. In [6], propagation channel measurements were made in 3 different scenarios at the frequency of 1925 MHz with an antenna height of about 20 cm above the ground; values of received power at spaces of 5 m up to a total distance of 40 m were obtained; there are 128 points, and each point has 300 data obtained, which are averaged to perform a linear regression and obtain the equation model for these scenarios, also the free space model and the two-ray model.

In [7], the authors make a measurement bell in a flat environment, university yards, and a lawn park, with a working frequency of 868 MHz; for this case, random distances were taken every 0.5 m up to 3 m., then it varies to 1m up to 10 m and finally every 2 m up to the total distance, all this in a straight line. Many of the models used today have antenna heights greater than 10 m. In [8], the authors give some characteristics of the sensor networks, for example, the small heights of the sensor nodes, the directivity of the antennas, the low transmitting power, and the stationary topology. This paper extracts the propagation models for sensor networks in outdoor scenarios from several publications, being these the variable soil reflectivity model, the two-ray model, the free space model, the log-normal double-slope model, and the free space model in outdoor scenario.

In [9], the propagation signal of the mobile radio is analyzed, and the data is recorded with an RF receiver for 2500 s. The multipath signal is shown both in dBm and in Watts power. The moving average filter is used to separate small-scale variations from large-scale variations, indicating that this filter window value varies between tens of wavelengths. It also analyzes these variations statistically through the probability function and the cumulative distribution function.

3 Measurement Setup

3.1 Hardware Setup

The measurement system basically consists of two TTGO T-Beam wireless nodes, one as a transmitter and the other as a receiver. A computer was connected to the receiver node to show the received data packets and store the Received Signal Strength (RSS) value. For the communication between the receiver and the laptop, Arduino's Integrated Development Environment (IDE) was used, as well as an external application to store this data.

Both nodes are designed for low power WSN applications and work in the free industrial, scientific, and medical (ISM) frequency bands at 868 MHz (Europe) and 915 MHz (USA). Detailed information about the TTGO T-Beam module is given in Table 1. The W1063 antenna is also used [10].

Table 1. LoRa signal parameters [11]

Parameter	Settings
Transceiver chip	Semtech LoRa SX1276
Bandwidth	500 kHz
Center frequency	920 MHz
Transmission speed	5470 bps
Data rate	DR3 37.5 kbps
Power transmission	+20 dBm
Transmit current	120 mA @ + 20 dBm

3.2 Measurements Scenarios

The archaeological park of Chinchero is located to the northeast of the city of Cusco, at a distance of 30 km; it is at an average altitude of 3766 m.a.s.l. Three different scenarios were selected within the archaeological park, which can be described as an esplanade, Inca walls, and platforms. The esplanade named CAPELLANPAMPA, that used for fulled palatial and ceremonial functions, located to the north of the main square, gives great extension, is in the highest zone, with respect to the platforms, it is composed by an average area of 9656.30 m^2 and a perimeter of 430.87 m, of rectangular form. It has a slightly wavy floor covered with natural grass of a height of 1 cm [12]. It is showed in Fig. 1(a).

On the other hand, in the Inca walls, which have a type of sedimentary rigging, the stones used are limestone and andesite; their shape is tetrangular, among the most common rectangular and trapezoidal, with profiles generally flat and convex [13], these have an average height of 2.1 m, with a length of up to 80 m long. They are showed in Fig. 1(b).

Finally, the platforms named WATAYCARCEL, located to the northeast of the esplanade, which are longitudinally staggered constructions with collateral walls, with a number of 14 structures, which were made for agricultural purposes and some of the support, has rigging type polygonal rustic and cyclopean fine finish, composed of limestone and joined with mud mortar [12]. Approximately, they get to have an average height of 2.1 m, lengths of up to 85 m long, and width of the floor of 3 m; it also has a slightly wavy floor covered with grass of minimum height. It is showed in Fig. 1(c).

In this scenario, the transmitter and receiver in some measurements came to have non-line-of-sight (NLOS) communication.

(a) Esplanade (b) Inca Walls (c) Platforms

Fig. 1. Measurements sites

3.3 Measurement Methods

The transmitting antenna was placed at the height of 140 cm, and the receiving antenna was variable in sizes of 20 cm, 50 cm, 80 cm, and 120 cm above the ground, in order to maintain harmony and aesthetics within the archaeological park; it was developed in a cold-dry climate, with light wind currents.

There were no people within the archaeological park, so the variations are due exclusively to the environment; the constant speed at which the transmitter moves were 0.4 m/s.

For our first scenario, the receiver was placed in the upper center of the concourse, and then routes were drawn in a straight line as shown in Fig. 2(a); for each route, two trips were made to obtain better statistical fidelity. About 4200 data were obtained per route. On the other hand, measurements on the Inca walls were made on straight routes, according to the length of each Inca wall, the transmitter was placed 0.5 m from the wall, obtaining 4100 data per wall.

For the third scenario, measurements were obtained at each platform, obtaining 500 data for each of them; the receiver was placed in the highest area and then vary floor to floor the position of the transmitter, obtaining a height variation of approximately 40 m. Figure 2(b).

The height of the receiving antenna was varied in sizes of 20 cm, 50 cm, 80 cm, and 120 cm. The data is stored in a notebook, in format CSV for further processing in the MATLAB software, to extract the fading and propagation loss exponent.

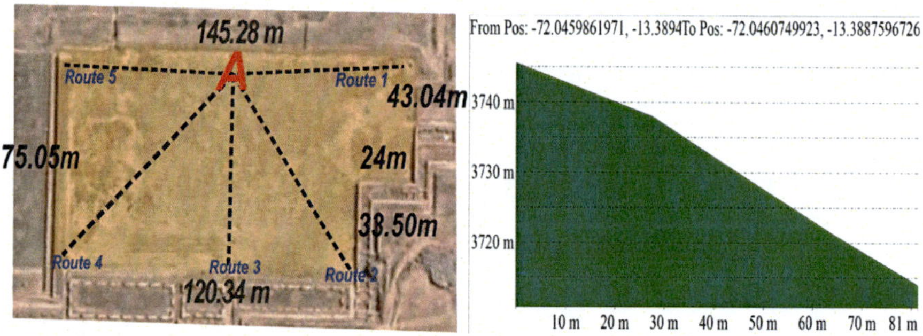

(a) Esplanade (b) Terrain Profile- Platforms

Fig. 2. Measurements routes

4 Measurement Results and Analysis

For this research, the distribution of tasks presented in Fig. 3 is taken into account for the correct analysis of measurement data.

4.1 Extraction of Loss Propagation Exponent and Shadowing Fading

After obtaining the data from the measurement campaign in the three scenarios, an analysis of these was carried out. The received power data is passed through a moving average filter with a window length of 40λ [14] in order to mitigate the small scale fade variation; the fade filtering follows the following reasoning: the filter is applied to the experimental samples set, see Fig. 4(a), obtaining the small scale fade, which is subtracted from the experimental measurements obtaining the large scale fade, Fig. 4(b).

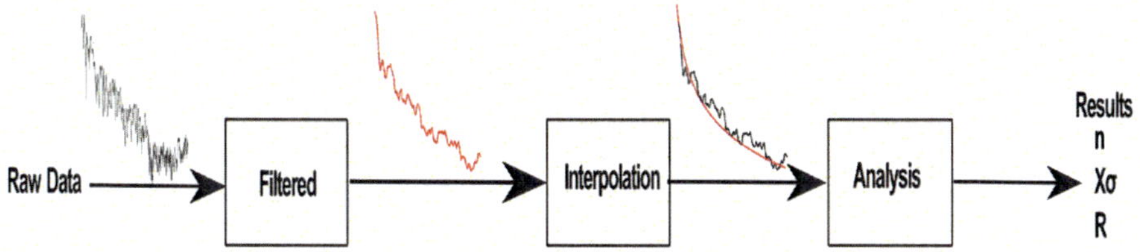

Fig. 3. Block diagram

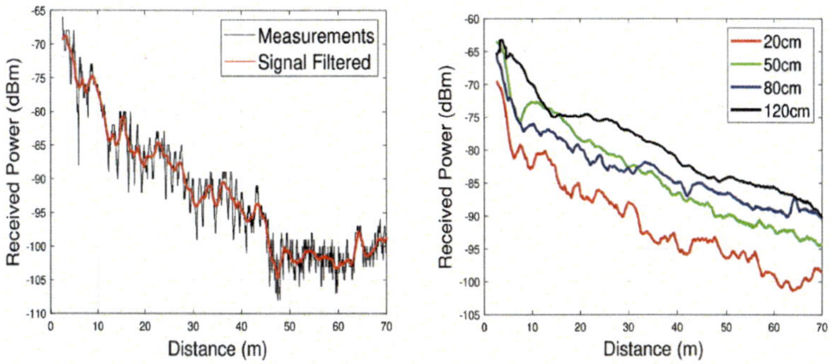

(a) Filtering of the Received Signal (b) Signal Filtered Route 3

Fig. 4. Signal filtered

With the filtered received power data, the propagation loss exponent resembles the logarithmic distance propagation loss model, where the propagation loss exponent represents the change in propagation loss along with the distance. This exponent and the shadow fading are extracted from the measured results using the following expression Eq. 1.

$$\mathrm{Pr}(d) = Pr_0 + 10n\ \log_{10}(d) + X_\sigma[dB] \tag{1}$$

Where d is the distance between Tx and Rx in meters, $\mathrm{Pr}(d)$ is the propagation loss (small-scale fading already filtered) expressed in dBm, Pr_0 is the average propagation loss at the reference distance d_0, n is the propagation loss exponent, and X_σ is a zero-mean log-normally distributed random variable with σ standard deviation in dB, expressing the shadow fading. By using the least squares criterion, the propagation loss exponent n can be obtained. The two-slope model is adopted to adjust the data of the propagation loss in the Inca Walls, using the following Eq. 2 [14].

$$\mathrm{Pr}(d) = \begin{cases} \mathrm{Pr}(d_0) + 10\mathrm{n}_1\ \log_{10}\left(\frac{\mathrm{d}}{\mathrm{d}_0}\right) + \mathrm{X}_{\sigma 1}(\mathrm{d} \le \mathrm{d}_\mathrm{b}) \\ \mathrm{Pr}(d_\mathrm{b}) + 10\mathrm{n}_2\ \log_{10}\left(\frac{\mathrm{d}}{\mathrm{d}_\mathrm{b}}\right) + \mathrm{X}_{\sigma 2}(\mathrm{d} > \mathrm{d}_\mathrm{b}) \end{cases} \tag{2}$$

Figure 5 shows an example of the adjustment to the minimum square of n in the esplanade at 80 cm (Fig. 5(a)), double slope model for the wall at 50 cm (Fig. 5(b)), and the one slope model for platforms at 20 cm (Fig. 5(c)). The values of the shadow fade can be calculated by simple subtraction of the above formula with the filtered data.

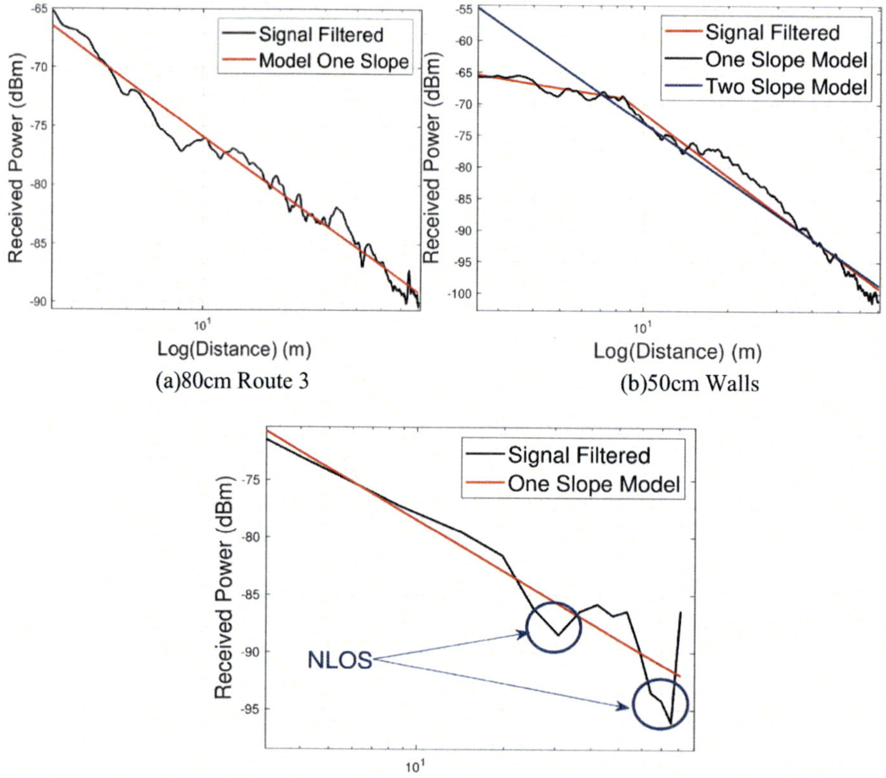

(a)80cm Route 3 (b)50cm Walls

(c)20cm Platforms

Fig. 5. Models of the scenarios

The value of the propagation loss exponent, the reference power, and its standard deviation are obtained and detailed in Table 2, for Esplanade and Platforms, and in Table 3 for Inca Walls.

Table 2. Large-scale fading characteristics of esplanade and platforms.

One Slope								
Height (cm)	Esplanade				Platforms			
	Pr_0(dBm)	n	X_σ(dBm)	R^2(%)	Pr_0(dBm)	n	X_σ(dBm)	R^2(%)
20	– 53.18	2.52	1.43	0.97	– 63.41	1.75	2.59	0.83
50	– 50.83	2.22	1.96	0.94	– 57.51	1.77	2.69	0.82
80	– 52.62	2.01	1.66	0.94	– 66.18	1.70	3.55	0.80
120	– 53.95	1.83	1.63	0.93	– 57.27	1.89	3.50	0.79

Table 3. Large-scale fading characteristics of inca walls.

Height (cm)	One Slope				Two Slope					
	Pr_0(dBm)	n	X_σ(dBm)	R^2(%)	Pr_1(dBm)	n_1	Pr_2(dBm)	n_2	X_σ(dBm)	R^2(%)
20	- 55.20	2.92	3.58	0.90	- 67.22	1.07	- 75.36	4.15	1.84	0.97
50	- 53.36	2.44	2.33	0.93	- 63.52	0.90	- 69.16	3.30	1.30	0.98
80	- 54.25	2.29	2.76	0.90	- 67.47	0.64	- 71.77	3.17	1.60	0.96
120	- 53.50	2.30	2.97	0.89	- 64.56	0.95	- 72.04	3.05	1.35	0.98

5 Conclusions

To accurately determine the characteristics of the environment, we measure the RSS values at different heights near the ground so as not to affect the structural harmony within the archaeological park. The moving average filter is used to eliminate small-scale fading. Thus, we obtain the filtered signal to which we establish statistical models of propagation losses and analyze the influence of the height of the receiving antenna and terrain factors. The linear regression in the obtained data indicates that the model based on the logarithmic distance is still viable to model the received power in places close to the ground and for this scenario of the archaeological park since the values found are close to 2, the value of the exponent in the free space, but it is not the case in the Inca walls scenario, where there is a strong reflection in the walls; it is here where the double slope model shows a higher accuracy, obtaining values of R higher than the one slope model.

In addition, it is observed that the exponent of propagation loss decreases with respect to the height of the antenna, while in the platform scenario, this exponent is less than 2 since it has a wide line of sight, in addition to especially influencing the height of the antenna, that is, for each platform the height of the antenna increases by 2.1 m., reaching a height difference of 40 m at the last measurement point.

Finally, it is concluded that a proper study and analysis of the deployment and operation of a sensor network must take into account a suitable propagation model with parameters adjusted for each specific scenario.

Acknowledgment. This work was supported by Fondo de Desarrollo Científico y Tecnol6gico (FONDECYT) and Banco Mundial.

References

1. Swami A, Zhao Q, Hong YW, Tong L (2007) Wireless sensor networks: signal processing and communications perspectives. Wiley, Hoboken
2. Zekavat R, Buehrer RM (2011) Handbook of position location: theory, practice and advances. Wiley, Hoboken
3. Stallings W (2009) Wireless communications & networks. Pearson Education India, Chennai
4. Otero CE, Haber R, Peter AM, AlSayyari A, Kostanic I (2014) A wireless sensor networks analytics system for predicting performance in on-demand deployments. IEEE Syst J 9(4):1344–53

5. Sohrabi K, Manriquez B, Pottie GJ (1999) Near ground wideband measurement in 800–1000 mhz. In: IEEE 49th vehicular technology conference
6. Alsayyari A, Kostanic I, Otero CE, Aldosary A (2017) An empirical path loss model for wireless sensor network deployment in a dense tree environment. In: 2017 IEEE sensors applications symposium (SAS) IEEE. pp 1–6
7. Martinez-Sala A, Molina-Garcia-Pardo JM, Egea-Ldpez E, Vales-Alonso J, Juan-Llacer L, Garcia-Haro J (2005) An accurate radio channel model for wireless sensor networks simulation. J Commun Netw 7(4):401–7
8. Kurt S, Tavli B (2013) Propagation model alternatives for outdoor wireless sensor networks. In: 2013 IFIP wireless days (WD) IEEE. pp 1–3
9. Lavanya V, Rao GS, Bidikar B (2016) Fast fading mobile channel modeling for wireless communication. In: International conference on computational modeling and security (CMS 2016) procedia computer science, vol 85, pp 777–781
10. Pulse Part Number W1063. https://pulselarsenantennas.com/wp-content/uploads/2017/01/W1063.pdf
11. TTGO-T-Beam-Car-Tracker. https://github.com/tekk/TTGO-T-Beam-Car-Tracker
12. Huallani Banda R (2008) Informe final de investigación arqueológica parque arqueológico de Chinchero sector Wataycarcel, Dirección Desconcentrada de Cultura Cusco, pp 17–19
13. Vilca Hacho F (2016) Recuperación del Monumento Prehispánico del Sector Q´entepata del Parque Arqueológico de Chinchero Urubamba Cusco, Dirección Desconcentrada de Cultura, Cusco, Perú, Informe anual de obra y preliquidación 2015
14. Rappaport TS (1996) Wireless communications: principles and practice, 2nd edn. Prentice Hall PTR, Hoboken

Assessment of Geothermal Potential and Estimation of Ground Settlements Following the Implementation of a Thermo-active Piles System in Southern Peru

Anel Canturin(✉), Yanell Casabona, and Gary Duran

Civil Engineering, Peruvian University of Applied Sciences, Santiago de Surco, Prolongación Primavera 2390, Lima, Peru
{u201610264,u201619378,gary.duran}@upc.edu.pe

Abstract. Thermo-active piles constitute a type of sustainable energy because these foundations extract heat stored in the ground through intern pipes for climatize buildings. This technology uses geothermal heat pumps (GHP) that make it possible to dispense with conventional external heating and cooling during winter and summer periods. The implementation of this system requires previous studies to know the geological, geotechnical, thermal, and hydraulic properties of the ground that guarantee the geothermal potential of the soil at a certain depth. In Peru, the existence of hydrothermal sites would make it possible to implement the technique. Despite the presence of these resources, there is a widespread lack of knowledge of the local soils characterization. Therefore, it was necessary to start with research aimed at the behavior of geothermal structures for their future application. In this sense, an analysis of the potential of three areas located in southern Peru (Arequipa, Cusco, and Tacna) was carried out to estimate the maximum energy that can be extracted from the subsoil. Besides, the effect of thermal loads is analyzed using a programming code that will facilitate the calculation of settlements in thermo-active piles. In this way, the research proposes a map of potential and a code of free access that facilitates the understanding of thermomechanical behavior in this type of foundation.

Keywords: Thermoactive pile · Heat transfer · Settlements · Thermal load · Building energy

1 Introduction

The use of geothermal energy in shallowness (heterothermal zone) represents an innovative technology, where the soil is used as a source of heat for the supply of heating and/or cooling of buildings. This type of technology is known as a borehole heat exchanger system and has been studied and applied for several years, being the first application in the 1980s [1]. The main advantage of these thermo-active foundations, in addition to their structural function, is the use of geothermal energy through their heat exchangers. The

Y. Iano et al. (Eds.): BTSym 2020, SIST 233, pp. 833–844, 2021.
https://doi.org/10.1007/978-3-030-75680-2_92

heat transfer process generates stresses within the energy pile, generating a thermal axial deformation in the soil. The calculation of these thermal effects represents a complex theme for initiating a thermo-mechanical analysis since it is necessary to define input parameters referring to building loads and thermal properties of the soil (thermal conductivity, heat capacity, thermal expansion, among others). The thermal parameters must be obtained by carrying out laboratory or in situ tests to define these values. However, there is a lack of information regarding geothermal resources in Peru and the application of the technique, which represents a strong limitation to obtain a characterization of local soils and an adequate analysis of the thermo-mechanical behavior between the soil-structure interaction.

About the utilization of ground thermal energy to heat and cool several types of building, thermos-active foundation systems are an innovative and sustainable alternative against ground-source heat pumps and other conventional systems [2]. Thermal piles have an embedded system closed-coils absorber pipe, which adds a new function as heat exchanger elements. Also, the higher thermal capacity of concrete than grout or rock makes this system a better option than a borehole heat exchanger because of lower installation and maintenance cost and less land area. Therefore, these elements are environmentally sustainable because of their high thermal productivity. Ground temperature, annual fluctuation, and variation with depth are critical parameters to energy piles design calculation.

On the other hand, soil thermal properties as heat capacity and thermal conductivity define the heat flux propagation in soil and the variation of stresses on thermal piles [3]. This variation of stress results from constant temperature changes that occur within the structural elements after their operation. Studies of calibration of numerical models using field data estimate that the temperature within an energy pile varies by up to 20% in the core during heating [4]. Moreover, studies focusing on the calculation of displacement under thermo-mechanical effects argue that the accumulation of non-reversible displacements arises during the process transition phase, and a maximum displacement increase of 8% was estimated during the expansion generated in the pilot [5].

In the present investigation, the geothermal potential of three cities in southern Peru was determined. These areas under study have different meteorological and geotechnical parameters, which define the amount of geothermal energy that will be possible to extract from the terrain. In this way, it is proposed to show the results of each zone in a geothermal potential map, allowing quantifying the heat stored in the soil and its possible use. Besides, based on the proposed map, a thermo-mechanical analysis on thermo-active foundations was carried out using a programming code drawn up to calculate the effects arising from the heat transfer.

2 Thermo-active Foundations

2.1 System Components

The thermo-active foundations form a closed geothermal pump system (GHP), which aims to implement the air conditioning of buildings through the use of geothermal resources. This system consists of three basic components: the primary circuit, the heat pump, and the secondary circuit. The primary circuit made up of a set of closed pipes installed within the structural elements in contact with the earth (piles, walls, surface foundations, among others), where a heat-carrier fluid is pumped, exchanging the building's energy with the ground. The secondary circuit is a closed secondary pipe network responsible for heating, which is embedded in the floors and walls of the structure. Finally, both circuits are connected through a heat pump consisting of an expansion-compression circuit responsible for transmitting heat from a low-temperature spotlight to a higher temperature hot spot [6].

2.2 Application Feasibility in Peru

The implementation of heat pump systems requires that the soil has some geothermal characteristics to ensure the proper functioning of the technique in the future, such as the presence of hot rock bodies or thermal deposits. This is because the conditions of that soil ensure a potential to be extracted. However, thermo-active foundations are characterized by being regarded as low enthalpy systems, which require low-temperature levels to operate. This facilitates the implementation of the technique in the heterothermic zone, considered as the first meters of the surface, where the temperature of the soil varies depending on the time and depth. Also, in Peru, there are researches of the Geological Mining and Metallurgical Institute (INGEMMET) that show the presence of thermal deposits in various parts of Peru.

3 Thermo-mechanical Analysis Based on Soil Potential

3.1 Soil Temperature Estimation

Soil temperature is not constant as a result of the heat exchange that is transferred through solar radiation, rainfall, and other atmospheric effects. In this sense, the temperature of the soil varies depending on the time and depth expressed through a sinusoidal function (1), due to changes in ambient temperature and its spread in the environment. To do this, we must draw the temperature records of the environment according to the meteorological stations found in the National Service of Meteorology and Hydrology of Peru (SENAMHI).

$$T(z,t) = T_m + A_0 e^{-\frac{z}{d}}\left[sen\left(\omega t - \frac{z}{d}\right)\right] \tag{1}$$

Where T_m [°C] is the average temperature, A_0 [°C] is the difference between the maximum and minimum temperature, z [m] indicates depth, the d factor quantifies the damping of the annual fluctuation, ω [rad/days] is annual radial frequency, t [days] denotes time, and T [°C] is soil temperature.

3.2 Thermal Characterization of Soil

The thermal properties of the soil are based on geotechnical conditions and in-situ tests such as SPT, Thermal Response Test, Needle Test, among others. However, for the present investigation, the studies by Pahud [7] and Bourneo-Webb et al. [8] were consulted to know about the typical values of thermal conductivity, heat capacity, and thermal expansion according to the type of soil. In addition, correlations of different authors, such as Peck, Hanson, and Thornburn [9], Brinkgreve et al. [10], and Jaky [11], were used to determine the friction angle, dilatance angle, and lateral coefficient of lands at rest.

3.3 Estimation of Energy Potential in Soils

The estimation of the maximum energy potential is done by recognizing soil conditions (lithological units), which will allow us to obtain a range of representative values for the maximum amount of heat transfer.

3.4 Estimation of Thermal Loads

The expansion and contraction of a thermo-active pile develop around the null point, which is located at an NP depth where the sum of the mobile friction along the top plus the reaction of the structure is equal to the sum of the friction mobilized along the bottom plus the reaction at the base [12]. To determine the thermal deformations, the following expression shall be used.

$$\epsilon_t = \epsilon_L = \alpha \cdot \Delta T \tag{2}$$

Where ϵ_L is the free thermal axial strain, α [°C^{-1}] is the coefficient of thermal expansion of the material, and ΔT [°C] indicate temperature variation.

For the assessment of thermal loads, an iterative procedure is applied, which allows us to obtain the deformation observed by the method described below:

1. An initial value is chosen for the observed deformation. Initially, the pilot is supposed to be completely free to move. The first displacement is determined from Eq. (2).
2. Calculation of thermal displacement from an initial displacement other than zero and a state of deformation induced by mechanical load is performed. Using the t-z curve, the first set of mobilized reaction voltages is obtained.
3. The axial tension is calculated in the pile (σ_t), which corresponds to the sum of all external forces divided by the pile section.
4. The blocked thermal deformation (ϵ_r) is calculated from the division of the axial tension on the material module in the pile.
5. Finally, the observed deformation (ϵ_0), corresponding to blocked deformation minus free deformation, is calculated. If it is different from the initial value, steps 2 to 5 should be repeated with the new set of observed deformations until the value converges.

3.5 Thermo-mechanical Analysis in a Thermoactive Pile

The thermo-mechanical analysis represents the calculation of the effect generated by mechanic loads from the superstructure, and thermal loads, generated by the heat transfer. The loads transferred at the shaft and base of the pile from mechanical and thermal applications shall be decomposed into n_1, n_2, and n_3 parts. The pile area will also be divided into different sub-areas, where each part will receive a certain baseload (3) and shaft load (4), which will generate forces and tensions in a new system with local coordinates and be determined by Mindlin equations [13]. The load coming from the building is symbolized as the load P [tons] applied to the foundation, and the influence it has on any point on the ground is called B with coordinates in x, y, and z (Fig. 1).

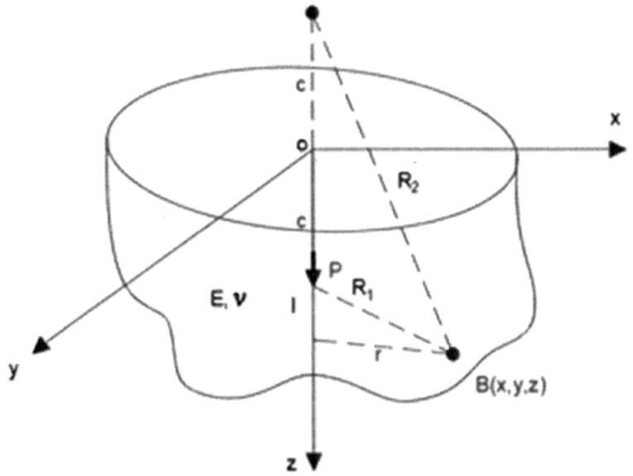

Fig. 1. System in three dimensions of analysis with the Mindlin equations (Adapted from Force at a point in the interior of a semi-infinite solid, by Mindlin, 1953).

$$P_{i,j} = \frac{P_b}{n_1 \cdot n_2} \tag{3}$$

$$P_{i,k} = \frac{(D_2 - D_1)}{2n_3}\left[2f_1 - \frac{2k-1}{n_3} - (f_1 - f_2)\right] \tag{4}$$

Where P_b [tons] is baseload on pile, n_1 denote the number of circumference divisions, n_2 is the number of base radius divisions. The forces $P_{i,j}$ [tons] and $P_{i,k}$ [tons] correspond to each subarea discretized of point and shaft, respectively. In the shaft area case, D_2 and D_1 indicate the initial and final depth of the stresses. Also, n_3 denote the division of shaft load.

Based on this reduction, the influence of each element at a given point in the soil is analyzed to overlap the effects and obtain the total settlements (5) using the Aoki-Lopes formulation [14].

$$w = \epsilon_z \Delta_z = \frac{1}{E}\left[\sigma_z - \mu\left(\sigma_x + \sigma_y\right)\right]\Delta_z \tag{5}$$

Where ϵ_z is the strain soil, Δ_z [meters] is the layer thickness and E [kg/cm^2] is the Young Module. The x, y, and z denote the directions of the stresses.

3.6 Numerical Implementation

The numerical method applied has been encoded using the Python software. This code allows the analytical calculation of settlements produced under thermo-mechanical loads.

Among the considerations of this tool there is the ease of performing an analysis on a stratified soil where the specific properties of the soil for each layer must be defined. The geometric characteristics of the pilot and the stratigraphic profile should also be entered as input data. In addition, the thermal parameters of the pile material, considered as constants in relation to depth, are necessary.

4 Case Study

This section will analyze two areas chosen for the analysis of surface thermal power of the soil, through which an energy potential map was developed. Furthermore, this result allows us to analyze the application of a group of thermo-active piles in that town, with the aim of estimating settlement during the heating and cooling processes.

4.1 Geothermal Potential Analysis

This process was developed based on the standard values: VDI4640 Part 2 – Thermal use of the subsoil (German Standard 2001) [15] and taking into account the lithologies present in the national geological maps of Peru, 1:500000, carried out by INGEMMET [16], a specific table has been developed for these lithological units, assigning a range of values to the thermal capacity of surface extraction (Table 1). The physical, thermal and hydraulic properties of each lithological unit have also been considered for classification.

Table 1. Groups of lithologies established on the basis of thermal characteristics of the rock and corresponding thermal potential formation

Rock type (lithologies)	Heat collection capacity (W/m)
Gravel, sand (dry)	<25
Clay and wet marls	35–50
Basalt	40–65
Limestones and dolomites (massive)	55–70
Sandstone, gravel, sand with water	65–80
Granite	68–85
Metamorphic rocks	70–85
Gravel and sand with great water circulation	80–100

After complementing both information, the geothermal potential map was prepared for the Tacna (Fig. 2), Arequipa (Fig. 3), and Cusco (Fig. 4) quadrangle. The presence of different lithologies in each area makes it possible to increase geothermal resources in southern Peru. It is important to note that it will depend on the depth requested by the foundation design to take advantage of the heat from the soil.

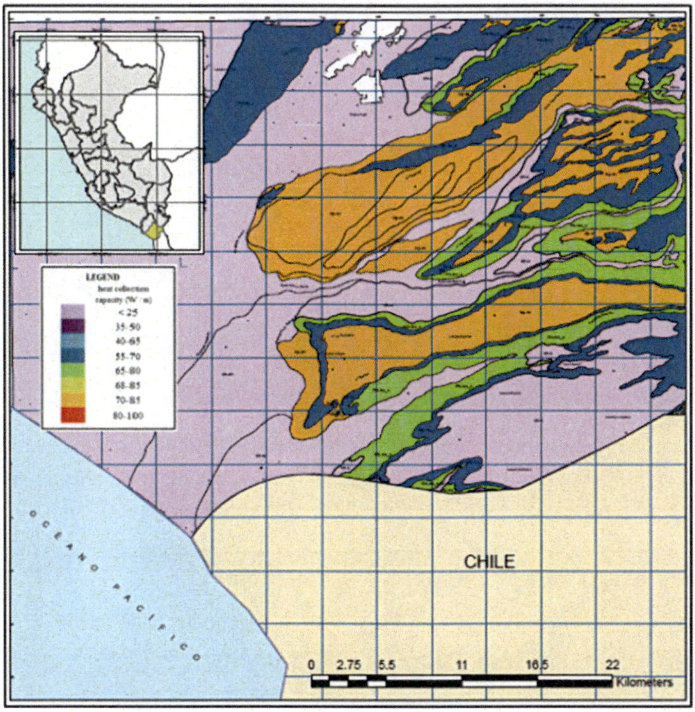

Fig. 2. Geothermal potential map of Tacna.

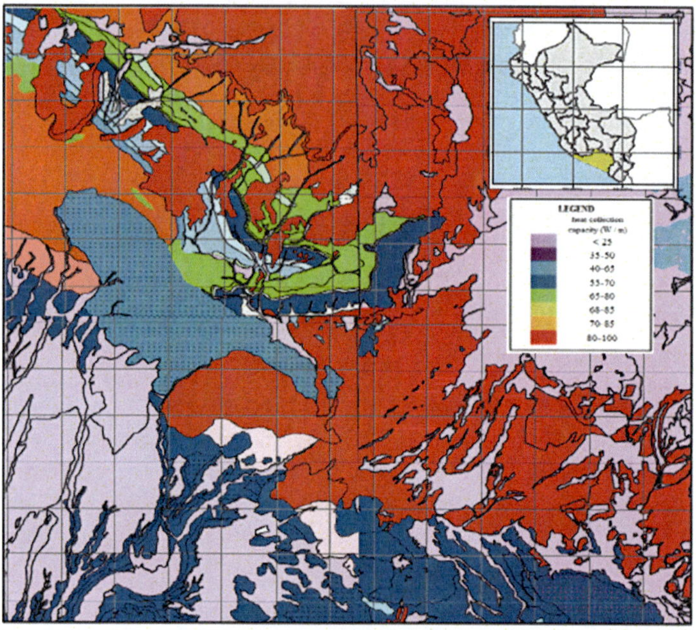

Fig. 3. Geothermal potential map of Arequipa.

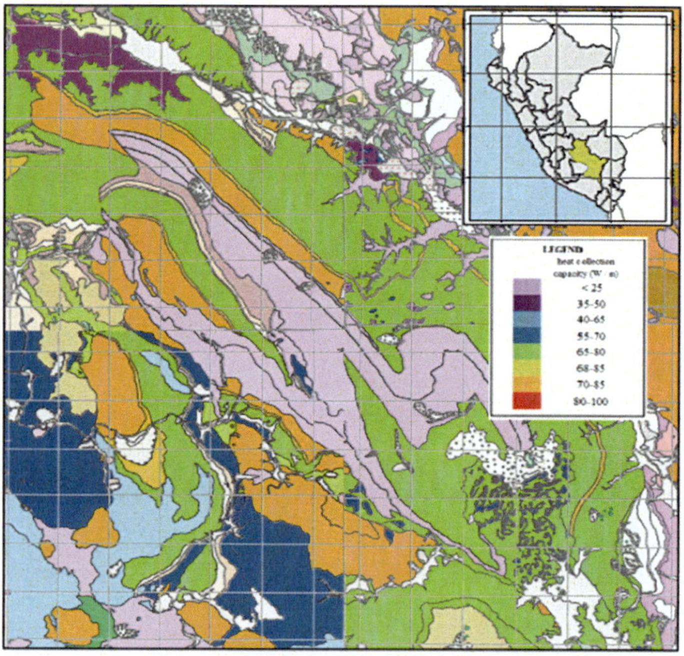

Fig. 4. Geothermal potential map of Cusco.

4.2 Thermo-mechanical Analysis

The analysis of soil behavior through thermo-mechanical analysis requires information on soil characteristics, such as defining mechanical and thermal properties and stratigraphy. The geometry of the structural elements to be applied is also necessary.

In this context, the soil of Arequipa is made up of pink tuffs of the Sencca Volcanic Formation, which has a SUCS (Unified System of Soil Classification) of SM (Silty sand). The soil has a medium to fine granulometry, poorly graded, sporadic presence of gravels; in some cases, gravels of volcanic nature (pumice and lava). On the other hand, the soil of Cusco is a rocky stratum altered from the surface to a depth of 17.7 m, classified as a soil type GP (graves poorly graded) with the sporadic presence of clays. Also, the foundation designed for each pilot conforms to the requirements of each stratigraphy, as can be seen in Table 2.

Table 2. Information about pile geometry

Description	Arequipa	Cusco	Units
Number of piles	9	9	piles
Diameter	0.85	1.00	m
Length	15.00	8.50	m
Separation	2.55	3.00	m

Determination of Soil Properties. The mechanical and thermal parameters determined for soils under study are shown in Table 3 and Table 4.

Table 3. Mechanic properties of materials.

Location	Material	γ (kN/m^3)	N	E' (Mpa)	v'	c' (kPa)	φ'(°)	Ψ(°)	K_0
Arequipa	Stratum 1	16.26	0.32	29.42	0.35	0	30.3	0.33	0.49
Cusco	Stratum 1	26.00	0.78	117.68	0.30	19.61	32.0	2.00	0.47
	Concrete	24	–	3000	0.20	–	–	–	–

Table 4. Thermal properties of material.

Location	Material	k(m/s)	ρ(kg/m^3)	C_S(J/kg/°C)	λ(W/m/°C)	α(10^{-6}/°C)
Arequipa	Stratum 1	4.61×10^{-5}	1670	801	2.4	12
Cusco	Stratum 1	4.61×10^{-5}	2679.82	846	3.57	12
	Concret	–	2440	890	1.60	9.9

For the calculation of soil-produced displacements, the methodology allows us to analyze in three dimensions, making it easier to calculate the effect at any point in the soil you want to analyze. Also, according to the programming developed, the center on the surface of the central pilot is considered as the starting point (0; 0; 0).

In the following cases, the results will be presented for the soils analyzed in Arequipa and Cusco, where St represents the immediate settlement calculated under mechanical loads and St + Sth represents the settlement under thermomechanical loads during heating and cooling processes.

Case 1: Arequipa Soil. The settlement results obtained after applying the thermo-active foundation in this soil are shown in Table 5.

Table 5. Settlements obtained under thermo-mechanical loads in heating and cooling processes in Arequipa.

X (cm)	Y (cm)	Z (cm)	St (cm)	St + Sth (cm) Heating	St + Sth (cm) Cooling	Observations
500	542.5	1900	1.2014	1.2405	1.1628	Perimeter
797.5	585	1900	1.2673	1.3083	1.2269	Perimeter
1052.5	585	1900	1.5819	1.6321	1.5325	Perimeter
500	797.5	1900	1.3611	1.4048	1.3179	Perimeter
797.5	797.5	1900	1.5371	1.5860	1.4890	Center
1095	797.5	1900	1.7733	1.8291	1.7184	Perimeter
500	1052.5	1900	1.4694	1.5163	1.4232	Perimeter
797.5	1010	1900	1.8444	1.9023	1.7875	Perimeter
1095	1052.5	1900	2.3459	2.4183	2.2746	Perimeter
– 202.5	744.868	1900	0.4026	0.4143	0.3910	Ground settlement
– 97.2368	218.553	1900	0.3934	0.4048	0.3821	Ground settlement
8.02632	639.605	1900	0.3936	0.4051	0.3823	Ground settlement
113.289	429.079	1900	0.3719	0.3827	0.3613	Ground settlement
1481.71	744.868	1900	0.7451	0.7667	0.7237	Ground settlement
1586.97	1060.66	1900	0.9353	0.9625	0.9085	Ground settlement
1797.5	955.395	1900	0.6514	0.6703	0.6327	Ground settlement

Case 2: Cusco Soil. The settlement results obtained after apply-ing the thermo-active foundation in this soil are shown in Table 6.

Table 6. Settlements obtained under thermo-mechanical loads in heating and cooling processes in Cusco.

X (cm)	Y (cm)	Z (cm)	St (cm)	St + Sth (cm) Heating	St + Sth (cm) Cooling	Observations
500	550	1000	0.0597	0.0698	0.0498	Perimeter
850	600	1000	0.0698	0.0812	0.0585	Perimeter
1150	600	1000	0.0870	0.1007	0.0734	Center
500	850	1000	0.0676	0.0787	0.0566	Perimeter
850	850	1000	0.0863	0.0999	0.0728	Perimeter
1200	850	1000	0.1011	0.1167	0.0857	Perimeter
500	1150	1000	0.0747	0.0868	0.0628	Perimeter
850	1100	1000	0.1111	0.1280	0.0943	Perimeter
1200	1150	1000	0.3278	0.3736	0.2824	Perimeter
− 202.5	744.868	1000	0.0122	0.0138	0.0105	Ground settlement
− 97.2368	218.553	1000	0.0185	0.0209	0.0160	Ground settlement
8.0263	639.605	1000	0.0207	0.0235	0.0180	Ground settlement
113.289	429.079	1000	0.0148	0.0168	0.0129	Ground settlement
1481.71	744.868	1000	0.0592	0.0671	0.0514	Ground settlement
1586.97	1060.66	1000	0.0481	0.0545	0.0418	Ground settlement
1797.5	955.395	1000	0.0363	0.0412	0.0315	Ground settlement

5 Conclusions

The estimation of energy potential from the analysis of the lithology presented by national geological maps of Peru has made possible the production of an energy potential map, which allows us to design proposals for thermo-active foundations for future application. In addition, the quadrangle of the Arequipa presents the highest geothermal potential due to the presence of welded tuffs. This material is a type of igneous rock originated by the southern volcanic axis that exists in Peru. In this way, the geothermal potential maps presented provide a reference for the use of heat from the ground in strategic areas of each city such as Arequipa, Cusco, and Tacna.

The application of two proposals for a foundation with the objective of calculating the soil settlement under thermo-mechanical loads was also considered. This process has the objective of knowing the thermal influence by ensuring the complex interactions between the thermal storage and the mechanical behavior of the geostructure. From this procedure, it was concluded that heating induces compression in the pilot generating an increase in the settlement, while cooling generates a tensile tension in the pilot.

In thermo-active structures, care should be taken with the additional forces generated within the energy pilot, caused by temperature variations and soil-structure interaction. From the analysis developed, it was obtained that the thermal influence is in a range of 0.05 to 0.15 mm.

References

1. Brandl H (2006) Energy foundations and other thermo-active ground structures. Géotechnique 56:81–122
2. Huerta L, Krarti M (2015) Foundation heat transfer analysis for buildings with thermal piles. Energy Convers Manag 89:449–457
3. Suryatriyastuti M, Mroueh H, Burlon S (2012) Understanding the temperature-induced mechanical behaviour of energy pile foundations. Renew Sustain Energy Rev 15:3344–3354
4. Caulk R, Ghazanfari E, McCartney J (2016) Parameterization of a calibrated geothermal energy pile model. Geomech Energy Environ 5:1–15
5. Coletto A, Sterpi D (2016) Structural and geotechnical effects of thermal loads in energy walls. Procedia Eng 158:224–229
6. Sauer HJ, Howel RH (1983) Heat pump systems, 1st edn. Wiley, Hoboken
7. Pahud D (2002) Geothermal energy and heat storage. Laboratorio di Energia, Ecologia ed Economía, Italy
8. Bourne-Webb P, Burlon S, Javed S, Kürten S, Loveridge F (2016) Analysis and design methods for energy geostructures. Renew Sustain Energy Rev 65:402–419
9. Peck RB, Hanson WE, Thornburn TH (1974) Foundation engineering, 2nd edn. Wiley, Hoboken
10. Brinkgreve RBJ, Engin E, Swolfs WM (2017) PLAXIS 2D material models manual. Netherlands
11. Jaky J (1944) The coefficient of earth pressure at rest. J Soc Hung Architects Eng 78:355–358
12. Peron H, Knellwolf C, Laloui L (2011) A method for the geotechnical design of heat exchanger piles. In: American society of civil engineers geo-frontiers congress 2011, Geo-Frontiers. pp 470–479
13. Mindlin R (1936) Force at a point in the interior of a semi-infinite solid. Physis 7:195–202
14. Aoki N (1975) Estimating stresses and settlements due to deep foundations by the theory of elasticity. In: Pan American congress of soil mechanics and foundation engineering, Argentina, 1, pp 377–386
15. Verein Deutscher Ingeniure Richtlinien 4640 (2001) Thermal use of the ground. Ground source heat pump systems. VDI 4640, Part 2, Beuth Verlag GmbH, Berlin
16. Geological, Mining and metallurgical institute. National geological chart of perú. https://www.geocatminapp.ingemmet.gob.pe/complementos/Descargas/Mapas/publicaciones/cgn_50_000/CG_100mil.html

Towards an Energy-Efficient Approximate Computer Implementation

Marcelo Jara[1](✉), Rodolfo Azevedo[2], and Lucas Wanner[2]

[1] Comp. Systems Lab, LSC-IC, University of Campinas, Campinas, SP, Brazil
mjara.perez@ic.unicamp.br
[2] Institute of Computing (IC), University of Campinas, Campinas, SP, Brazil
rodolfo@ic.unicamp.br, lucas@ic.unicamp.br

Abstract. Under the continuously growing demand for energy-efficiency and high-performance of present computer and communication systems, the approximate computing paradigm emerges as an effective option to satisfy design requirements feasible at different hierarchy levels. In this paper, the implementation design process of an approximate core processor is introduced. This hybrid precise/approximate processor is based on an open-source RISC-V architecture, aiming to characterize the impact and evaluate the overhead of approximate arithmetic circuits, when integrated under the core hierarchy, after completion of physical design.

Keywords: Approximate Computing · ASIC physical design · Low-power

1 Introduction

The continuous and ever-increasing demand for high-performance and energy efficiency in computing and communication systems is leading to fundamental changes in the ASIC/SoC design methodologies [1].

Therefore high-density SoCs are exceeding the fundamental thermal and power density criteria, already reaching critical limits when running at full capability.

Under this scenario, the Approximate Computer (AC) paradigm emerges as one of the more relevant methodology options to achieve the ultimate low-power and high-performance SoC design targets.

The relevant point of AC is to allow a controlled degradation in performance or accuracy of some results produced by the hardware/software system, but without producing a relevant impact on the correct operation of the whole integrated system. In such a way, some applications are less sensitive to some precision loss, as for instance, those ones based on deep learning or neural-based algorithms, computer vision, and multimedia.

Sometimes, accuracy can be dynamically adjusted in a controlled way, depending on the particular application; however, this is only possible at expense

Y. Iano et al. (Eds.): BTSym 2020, SIST 233, pp. 845–853, 2021.
https://doi.org/10.1007/978-3-030-75680-2_93

to include more custom approximate hardware. The same overhead effect is present when accurate and approximate computing structures simultaneously coexist on-chip.

This work aims the design and characterization of an approximate processor, based on the idea of including precise and approximate arithmetical functional units (AFUs) on the same RISC-V core [2]. The corresponding AFUs can be activated by enabling/disabling AC modules dynamically at execution time. Another option is to set always active the approximate and precise units and their associated datapaths. The desired results could be, then, captured from specific registers.

The processor version presented is based on an open-source RISC-V core rv12 architecture proposed by *RoaLogic*$^{(TM)}$ [3], implemented from a SystemVerilog RTL description, synthesized targeting a Cadence CMOS 45nm technology [4]. Several AFUs from EvoLibs [5] were also integrated in an hybrid exact/approximate ULA. Our workflow included several experimental implementation versions, pursuing a highly reliable characterization of the Virtual Prototype through a feasible physical design flow. The impact and final overhead of the arithmetic approximate units on the final core processor implementation were then evaluated and characterized.

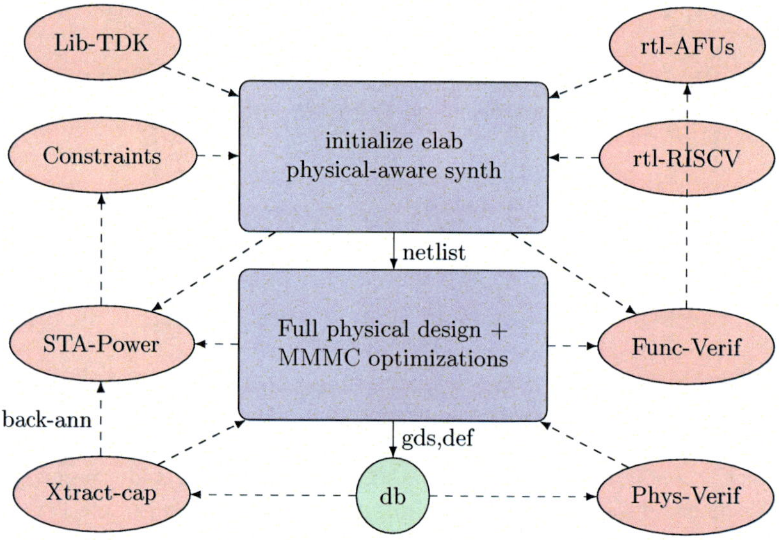

Fig. 1. Workflow Approximate Functional Units (AFUs) + RISC-V (rv12)

In Fig. 1 the complete workflow for (AFUs) Adders, Multipliers and core (RISC-V) is represented, all internal processes (elaboration, synthesis, physical design, and MMMC: multi-mode, multi-corner optimization) are repeated until convergence by fine adjusting of flow, constraints and optimization efforts.

In an initial stage (Phase-1), a set of 1,000 AFUs (adders and multipliers) from evoLibs [5] were simulated, synthesized, and physically implemented in a virtual chip prototype. The AFUs were functionally verified at both RTL and post-synthesis (gate-level) stages, and its errors and distance from exact results were quantified by Eq. 1. This phase aimed to evaluate and choose the best AFU candidates to be integrated under the rv12 core hierarchy.

The following stage (Phase-2) was focused on the RISC-C rv12 core design. In this context, the workflow was initiated at the micro-architecture design level with the SystemVerilog coding and specifically by evaluating the most relevant HDL parameters available, with direct impact on the synthesis results. All these processes were executed with the original rv12 core (using precise arithmetic units) and with the AFUs chosen in the previous phase, integrated under its hierarchy. The results from such procedures are discussed in the next sections.

2 A Summary of Approximate Computing Proposals

A collection of reviews, surveys, and proposals about AC can be found in the literature, e.g., [6–11]. Some systems applying approximate computing techniques and using related methodologies under several approaches have been proposed, covering from a pure software solution to combined hardware/software or purely hardware-centric way. Some of such proposals are summarized in the following parts of this section.

In [9], a complete survey describing several techniques and AC strategies for diverse abstraction levels and scalability, ranging from circuit devices to software solutions and IDE software platforms, can be found. Tables were organized showing the different strategies applied at each abstraction level in classified entries identifying the technique (e.g., error injection, memoization, inexact or faulty hardware, voltage scaling) and its corresponding references. Each technique is shortly but very clearly described across the paper.

Some AC methodologies are discussed [10] aiming communication issues between computers. The article explores the idea of using AC in order to increase the efficiency of communication data transfer between interconnected machines in order to improve efficiency, lowering data rate and in such way to decrease power and overhead communication. Some of the reduction techniques include using an approximation for data compression and GPUs to reduce bandwidth in off-chip memory access, shared data and error-tolerant relaxed synchronization (e.g., for producer-consumer tasks), acceleration (speculative, shared), and combination possibilities of some of such techniques. About 10 to 20% of power is reported to be consumed by the network links in High-Performance Computer (HPC) systems and particular parallel architectures, like Network-on-Chip (NoCs) and other massive parallel models. But even when a lower fraction of time is spent in communication, this can be potentially the primary source of bottleneck as the number of processors increases for an application requiring high parallel processing. In this way, AC is presented as a good solution to reduce the communication overhead between processors and routing devices.

In [8], several sub/near-threshold circuits are presented aiming ultra-low-power applications. It is also considered to mitigate the impact of Process-Voltage-Temperature (PVT) variability, two design techniques aiming to generate inexact circuits are explored: Probabilistic Pruning and Speculative Adders.

In [11], by another hardware approach, an energy-efficient design is proposed with the design space explored from higher levels modeling (HLS), aiming approximate computing accelerator, based on a dynamically reconfigurable micro-architecture module, continuously monitoring the workload distribution.

3 The Approximate Arithmetical Functional Units

This section presents a description of the IP models (Intellectual Property blocks) used for the approximate arithmetic units, with approximate adders and multipliers circuits integrated as ALU components of the RISC-V rv12 architecture modified for simultaneous fixed and approximate computer operation. The used EvoLib database includes a big set of more than 1000 arithmetic 8-bit units [5]. The library also includes 16 and 32-bit adders and multipliers, which were preliminary included in our workflow. However, after carefully quality-results analysis, comparing synthesis and corresponding functional results, only the 8-bit units were finally considered to be included in our Phase-2 flow. The 8-bit AFUs were used as basic units to compose 32-bit adders and multipliers.

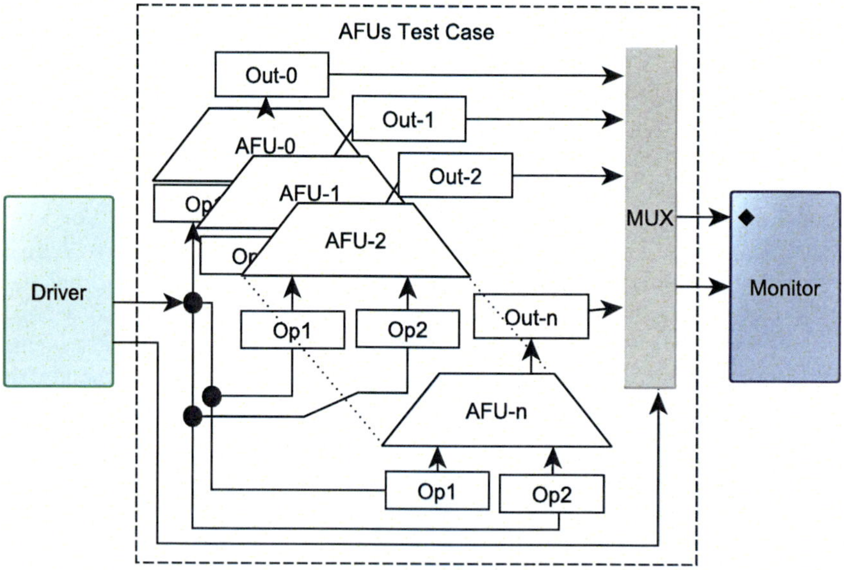

Fig. 2. AFU adder/multiplier testbench

To functionally validate the set of AFUs, a verification environment was created for both adders and multipliers to execute all simulation processes at both RTL and gate-level (after synthesis).

The functional scheme built is represented in Fig. 2, with a conventional driver-monitor structure sending stimulus and captures responses from the DUT (Device-Under-Test), a $(n - bit)$ AFU instance.

The AFU Testbench and Test Case were described in RTL SystemVerilog (SV), with flexibility to instance both an individual AFUs, a selected group, or all the AFUs from the evoLib (AFU-0 to AFU-n) simultaneously. A custom top instance was created to integrate the approximate adders and the corresponding multipliers. Each top instance was stimulated by a custom SV testbench, which included basic UVM (Universal Verification Methodology) features for ad-hoc proving and automatized input vector generation and results capture. After functional verification results analysis, the complete implementation flow (Fig. 1) is executed from the top-level AFU Test Case instance described in a top-down approach.

The synthesized/optimized netlist is functionally verified at gate-level (optimized netlist), instanced in the same testbench structure.

3.1 Approximate Adders

In this section, the adder arithmetic results are discussed, only the 8-bit modules (504 units) from the EvoLib library version were considered to continue the Phase-2, and the best-optimized unit was chosen to build a 32-bit version, also integrated under a top-level hierarchy (see Fig. 2) for verification. For the RISC-V processor, approximate functional units of 32-bit input data (A, B operands) were generated based on such 8-bit module versions, which presented minimum power consumption after physical-aware synthesis and optimization procedures completion. Its results are presented below.

Table 1. Add8 performance after logic and physical-aware synthesis

Timing model	Opt. effort	Type	Cells/area - power(nW)		
WLM	0	Approx	3	15	4466
		Exact	8	40	9803
Phys-1	1	Approx	3	15	8115
		Exact	8	45	22679
Phys-2	2	Approx	3	15	3329
		Exact	31	73	15665

In Table 1, the first column (left) represents the timing model used for synthesis (WLM, for logic synth., and Phys-1, Phys-2 for physical-aware). The 2nd column represents the effort level of the EDA tool to execute the optimization steps using physical timing models, where basic physical place and routing procedures are launched. The physical-aware synthesis should represent a more realistic design, where physical constraints should impact the feasibility. The 3rd column represents the type of arithmetical unit (approximate vs. exact/precise). The 4th and 5th show the total number of cells (from the technology library) and its associated area, measured in um^2. The last column shows the power in nW (NanoWatts) units, evaluated after synthesis is finished and optimized.

3.2 Approximate Multipliers

The integration and results data of the Approximate multiplier modules and its results are discussed in this section. The complete EvoLib version includes 505 AFU multiplier units, all IPs integrated under the approx. multiplier 8-bit top hierarchy. The AFU multiplier module reporting the minimum power after physical-aware synthesis is represented in the table below. This AFU was the module chosen to be the baseline for the approximate 32-bit multiplier integrated under the ALU RISC-V version.

Table 2. Mult8 performance after logic and physical-aware synthesis

Timing model	Opt. effort	Type	Cells/area - power(nW)		
WLM	0	Approx	57	140	26841
		Exact	120	364	84076
Phys-1	1	Approx	57	229	85034
		Exact	134	561	231245
Phys-2	2	Approx	57	181	38671
		Exact	254	682	127222

The results in Table 2 are presented in the same format as Table 1, i.e., for logic and physical-aware synthesis. It is important to mention that physical timing models include data extracted from both cells and metal tracks used for routing. This is the reason why logic synthesis data extracted from WLM models are not really representative of final physical silicon implementation and could mask real results, especially if high-impact physical constraints potentially exist (congestion, etc.); therefore, physical-aware synthesis should be used whenever is possible.

3.3 Quantifying Errors

Several metrics are proposed to evaluate the Error rate in approximate arithmetic circuits; the chosen option is normally based on the methodology that best represents a particular application. The focus of the work here described is to evaluate the feasibility and constraints for physical implementation of an

Approximate Computing custom processor. Therefore there is no such particular application targeted for this proposal. The metrics recommended in [5] was adopted, named MRE (Mean Relative Error), to quantify errors for (a + b) or (a × b) operations, defined by the expression below:

$$fMRE = \frac{1}{2^{Ni}} \sum_{i=1}^{N} Dist_i \tag{1}$$

where :

$$Dist_i = \frac{O_{Exact}^{(i)} - O_{Approx}^{(i)}}{max(1, O_{Exact}^{(i)})} \tag{2}$$

The O_{Approx}, O_{Exact} terms correspond to the decimal representation of the i-th circuit. And 2^{Ni} represents the total number of primary a + b or a × b inputs (input vectors), meaning each a, b operand width is Ni divided by 2 bits. In this way, for the 8-bit operand width, $2^{Ni} = 65536$.

For the Approximate AFUs described in Tables 1 and Table 2, the MRE evaluated for each operation is: - Adder8-bit: 0.07 - Multiplier8-bit: 12

For both AFU cases, the golden reference is 1 (i.e., exact/precise adder/multiplier).

4 Implementation Results - The Approximate RISC-V

After building and integration of 32-bit AFUs (from chosen 8-bit units), with the original rv12 RISC-V processor, the design flow in Fig. 1 was executed again, starting with the parameterized RTL, following to physical-aware synthesis and the complete place + route processes, including clock tree and all the needed optimization steps in MMMC mode, for the more relevant PVT (Process-Voltage-Temperature) conditions available by the PDK library. The results are presented below; both synthesis and Place and Route (PR) were run at 400 MHz. The original netlist used to initiate the floorplan/powerplan, and the following place and route processes correspond to a netlist generated by a physical-aware synthesis with physical effort level classified as Phys-1 (medium), but with high optimization effort. The MMMC option was activated only at the physical design phase in order to avoid unnecessary overhead.

Table 3. RISC-V performance after physical virtual prototyping

PR start:physical-aware synthesis netlist - high opt. effort						
Precision type	Gates count	Cells number	Area umt2	WNS nsec	Power nanoW	Density %
Approximate	299289	82653	307070	−0.5870	3.7M	58.3
Exact	292947	80112	300563	+0.0880	2.9M	57.0

The data in Table 3 corresponds to a final version with clock-gating capabilities enabled, WC (Worst Case) PVT conditions with Voltage = 1.08 V and

Temperature = 125C, using regular Vt std cells from TDK library, the OCV (On-Chip Variation), enabled, which considers some variability of the semiconductor processes. WNS is Worst Negative Slack (timing). Density is the occupation rate.

5 Conclusion

This proposal aims to design a processor where its physical implementation feasibility and energy efficiency are the primary targets, while performance and silicon area minimization continue as a part of a multi-objective design. The early results show that there is low overhead when approximate structures for arithmetic operations are included for fixed-point operations. But STA also indicates timing needs some critical paths fixing at micro-architecture level. There is no particular application aimed at this point, although this architecture should be a good candidate for some applications like deep/machine learning accelerators and low power portables, like wireless sensor networks or other similar IoT-related devices. In such direction, further activities need more development, mainly involving deeper functional verification at RTL, micro-architecture, and post-layout power and performance analysis. A set of singular AFUs could be selected and fine-tuned to some applications, identifying the best units for optimized energy and acceptable functional degradation for each particular workload.

References

1. ITRS International Roadmap Committee (Several working Groups Key Contributors to ITRS 2.0) (2015) The International Technology Roadmap Semiconductors 2.0; Executive Report, Ed
2. Waterman A, Lee Y, Patterson D, Asanovic K (2014) The RISC-V Instruction Set Manual, Vol. I: User-Level ISA, vers 2.0. Technical Report No. UCB/EECS-2014-54, Electrical Engineering And Computer Science, University of California at Berkeley, May 6
3. Roa Logic BV (2018) RV12 RISC-V 32/64-bit CPU core; Datasheet (v1.3), 01-Feb
4. Cadence Design Systems (2016) Reference Manual Generic 45 nm Salicide 1.0/1.8 V.; Process Design Kit and Rules Deck (PRD), Rev.5.0
5. Mrazek V, Hrbacek R, Vasicek Z, Sekanina L (2017) EvoApprox8b: library of approximate adders and multipliers for circuit design and benchmarking of approximation methods. IEEE Xplore, 15 May 2017, pub. in Design, Automation & Test in Eu. Conf. & Exhib., DATE
6. Xu S, Carrion Shaffer B (2017) Exposing Approximate Computing Optimizations at Different Levels: From Behavioral to Gate -Level; IEEE Transactions on Very Large Scale Integration (VLSI) Systems, Vol 25, No 11, November
7. Shafique M, Hafiz R, Rehman S, Henkel J (2016) INVITED paper: Cross-Layer Approximate Computing: From Logic to Architectures; 53nd ACM/EDAC/IEEE Design Automation Conference
8. Schlachter J, Camus V, Enz C (2015) Near/Sub-Threshold Circuits and Approximate Computing: the Perfect Combination for Ultra-Low-Power Systems; IEEE Computer Society Annual Symposium on VLSI

9. Mittal S (2015) A survey of techniques for approximate computing; ACM Comput Surv, Vol a, No b, Article 1, Pub
10. Betzel F, Khatamifard K, Suresh H, Lilja D, Sartori J, Karpuzcu U (2018) Approximate communication: techniques for reducing bottlenecks in large-scale parallel systems; ACM Comput Surv 51,1, Article 1, Jan
11. Xu S, Schafer B (2017) Approximate reconfigurable hardware accelerator: adapting the micro-architecture to dynamic workloads; IEEE 35th int. Conf. on Computer Design, ICCD Boston, MA

Terahertz Imaging and Machine Learning in the Classification of Coffee Beans

Patricia Uceda(✉), Hideaki Yoshida, and Pedro Castillo

Universidad Privada del Norte, Perú Environmental Engineering, Avenue Vía de Evitamiento Norte s/n cuadra 15, Cajamarca, Perú
{patricia.uceda,pedro.castillo}@upn.edu.pe, N00162805@upn.pe

Abstract. The geographical origin of coffee beans represents an effect on the attributes and quality of the product due to the different soil and weather conditions for a specific location. Therefore, the development of methods for rapid classification and authentication of coffee beans based on their geographical origin is essential. This research was done with the purpose of determining the capacity of coffee (*Coffea arabica*) varieties classification with the use of Terahertz (THz) imaging and machine learning. THz images of coffee beans samples from 3 different geographical origins were acquired with a time-domain spectrometer and then used to measure the classification performance of methods such as neural networks, random forests, and support vector machines. The results obtained reached an accuracy up to 91.2%, which showed that the use of THz imaging and machine learning is an effective method for the non-destructive analysis of coffee variables and classification based on geographical origin.

Keywords: THz spectroscopy · Supervised learning · Chemometrics · Coffee post-harvest

1 Introduction

Worldwide coffee production shows a tendency of growth since the last decade, producing 113,6 million bags between the year 2000 and 2001 to 158,9 million bags between 2017 and 2018 [1].

Link et al. [2] explain the importance of proving the geographic and genotypic origin of coffee varieties in the coffee industry using reliable methods, given that these farming characteristics give special attributes and qualities that raise their value on the market. Therefore, it is essential to use faster and low-cost analytic methods for differentiation and certification of coffee composition and quality.

Recent studies have proven the potential of mid and near-infrared spectroscopy (MIR/NIR) in roasted coffee quality analysis using calibration models and reference samples [3] as well as its application in discriminating coffee between varieties arabica and robusta [4].

Given this background and the need for non-destructive analysis technology development as well as efficient methods for classification and identification of coffee by

Y. Iano et al. (Eds.): BTSym 2020, SIST 233, pp. 854–861, 2021.
https://doi.org/10.1007/978-3-030-75680-2_94

these technologies in the coffee industry, the present research study worked in studying the viability of using THz spectroscopy and machine learning for discriminating coffee bean varieties.

2 Background

Liu et al. [5] worked in the classification of transgenic rice seeds from their non-transgenic counterparts with the use of THz spectroscopy and chemometric methods without sample pre-treatment. To achieve this, principal component analysis, least-square support vector machines, and random forest were used over the frequency domain THz spectra with first and second derivative and standard normal variate transformation pre-treatments. An accuracy of 97.67% in the prediction set was achieved, so the study concluded that the use of THz spectroscopy together with chemometrics has potential use in simple, rapid, and non-destructive identification of transgenic rice seeds in situ without sample pre-treatment.

Link et al. [2] used a Radial Basis Function (RBF) network to classify the geographical and genotypic origin of arabica coffee. In their study, spectra obtained in the Fourier Transform Infrared (FTIR) were analyzed and used to develop sequential optimization methods and set averaging for the selection of parameters and generalization of the neural network, respectively. The optimized RBF achieved 100% correct geographic classification and 94.44% correct genotypic classification; therefore, it was concluded that the combination of RBF networks and FTIR spectroscopy is a fast and effective technique for the classification of arabica coffee samples.

Bona et al. [6] developed methods for the geographical classification of different arabica coffee genotypes with the use of Support Vector Machines (SVM). To achieve this, spectra were collected in the range of Near Infrared (NIRS) and Mid Infrared (FIRS), and for the data analysis, a SVM was built using radial basis as kernel function, and parameters of the support vector machine were optimized using a genetic algorithm. The results obtained 100% sensitivity and specificity in the NIRS classification tasks, so it was concluded that infrared spectroscopy is a viable method for the certification of geographical origin with reduced pre-treatment of samples and without the production of chemical wastes.

The authors Monteiro et al. [7] compared Proton Transfer Reaction Mass Spectroscopy (PTR-MS) with Near Infrared Spectroscopy (NIRS) in the authentication of the geographical and agricultural origin of Brazilian coffee. Partial Least Squares - Discriminant Analysis (PLS-DA) was applied and provided results up to 89% of prediction success, and the study concluded that PTR-MS and PLS-DA techniques form an alternative for rapid analysis and differentiation of geographical origin of Brazilian coffee.

The authors Craig et al. [8] evaluated and compared the performance of FTIR and Near Infrared (NIR) spectroscopy for the quantification of defective beans in roasted coffee. To achieve this, quantitative models with partial least squares regression were developed. The results obtained correlation coefficients that ranged from 0.032 to 0.891 and mean square errors from 0.026 to 0.953 in the validation for FTIR and NIR models. After a comparison between the two techniques, the study concluded that NIR provided more robust models.

3 Method

For this research, mathematical and statistical analysis were performed using the Scikit-learn library for the Python programming language. The proposed methodology is resumed in Fig. 1.

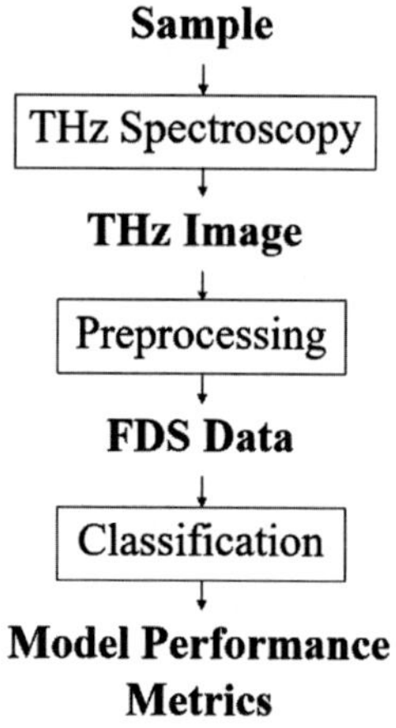

Fig. 1. Schematic structure of the method

3.1 Coffee Samples

For this research, arabica coffee beans of 3 different geographic origins were used. A total of 300 coffee beans were used after the selection of 100 for each type. This selection process was manually done to use similar height beans to facilitate the focusing of the THz beam over the samples and remove defective beans from the samples.

The number of samples used for this study was based on the research done in previous THz application publications by authors Liu et al. [5] and FTIR classification of arabica coffee by Link et al. [2].

3.2 Terahertz Spectroscopy

THz imaging was done by using a time-domain spectrometer, ranging from 0.02 to 8 THz. Coffee beans of each type were arranged in 10 by 10 order per image acquired, totaling a number of 3 THz images with 100 coffee bean samples acquired for each type. An example of the arranged coffee beans THz image acquired with the time domain spectrometer is shown in Fig. 2.

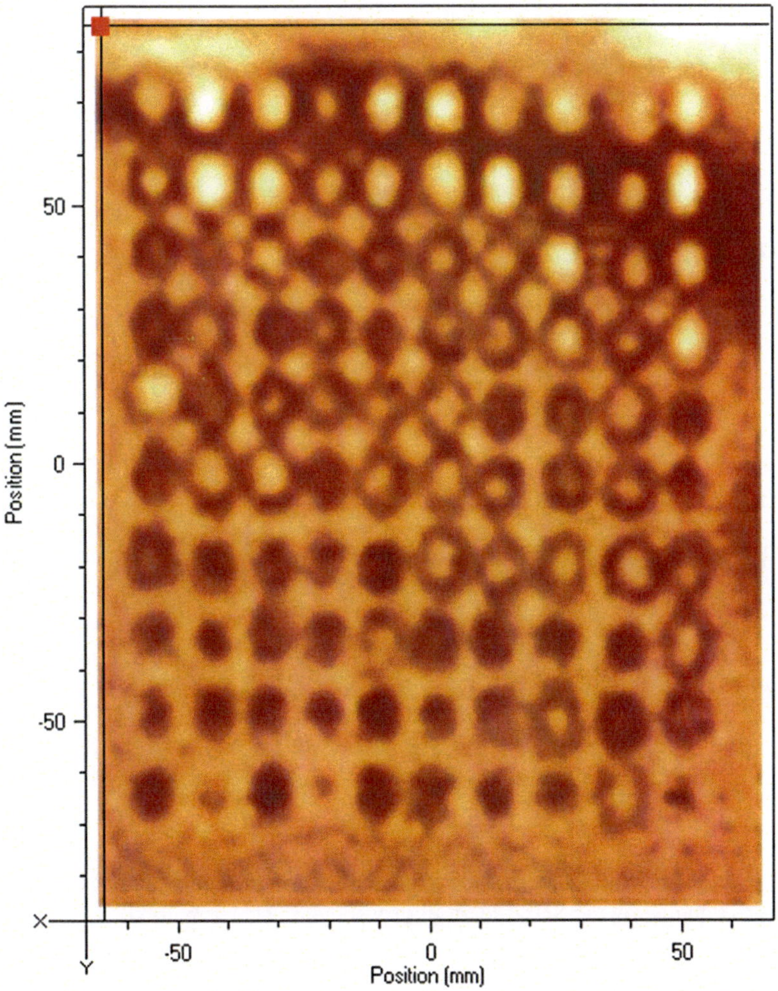

Fig. 2. THz image of coffee samples

3.3 Preprocessing

Once the Time Domain Spectra (TDS) of the coffee samples were acquired, the data was transformed into NxMx2048 arrays, the first two dimensions representing the two-dimensional space for the image pixels and the third one being the electrical field captured for each optical delay time point.

Next, a Fast Fourier Transform (FFT) algorithm [9] was applied to the data matrix to obtain the Frequency Domain Spectra (FDS) of the samples.

Figure 3 and Fig. 4 show examples of TDS and FDS acquired from the coffee samples.

To acquire the data for classification, points of interest for each individual sample containing the full 2048 optical delay time points were extracted and sorted for the training and testing sets.

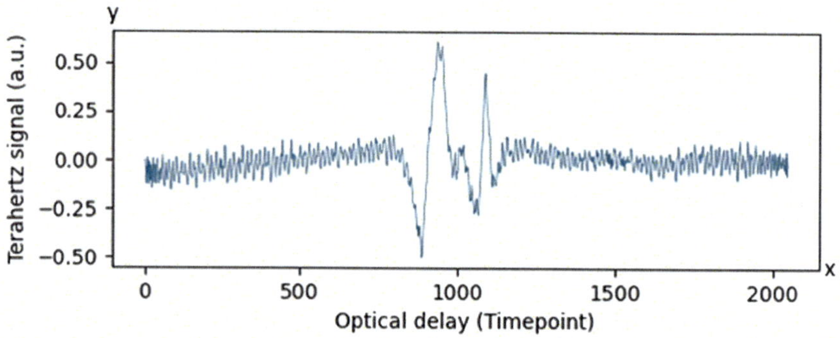

Fig. 3. Time-domain waveform of the coffee sample.

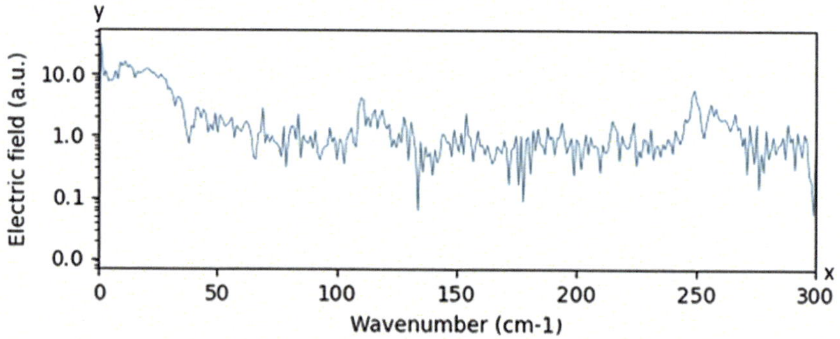

Fig. 4. Frequency domain waveform after FFT.

Feature standardization was applied by removal of the mean and scaling to unit variance, as detailed by the Scikit-learn library standard scaler method.

3.4 Neural Network

For the neural network model training, the most relevant parameters used were as follow:

Activation function for hidden layer: Logistic sigmoid function.
Weight optimization solver: Limited-memory Broyden-Fletcher-Goldfarb-Shanno algorithm.
Regularization term: 1e-5
Learning rate: Constant learning rate of 1e-2
Number of iterations: 900.

3.5 Support Vector Machine

For the support vector machine model training, the most relevant parameters used were as follow:

Kernel: Linear.
Error term regularization: 0.1.
Stopping criterion tolerance: 1e-3.

3.6 Random Forest

For the random forest model training, the most relevant parameters used were as follow:

Number of trees: 100.
Number of features: Square root of the number of features.
Minimum number of samples required to split an internal node: 2.
Minimum number of samples for leaf node: 1.
Bootstrapping: Enabled.

All learning models were trained with the same data set of samples to compare the performance between models.

4 Results

Classification task results, such as true positive, false positive, true negative, and false negative from each machine learning model, were registered and then used to calculate the classification performance metrics of accuracy, F-score, precision, and recall.

The average performance results of the machine learning models showed that random forest achieved higher accuracy (91.2%), F-score (86.7%), precision (86.7%), and recall (86.8%), as listed in Table 1.

Detailed results of accuracy and F-score for each sample testing set are shown in Table 2 and Table 3.

Table 1. Average measures result per model

Metric	Neural network	Support vector machine	Random forest
Accuracy	0.839	0.817	0.912
F-score	0.757	0.720	0.867
Precision	0.769	0.719	0.867
Recall	0.758	0.724	0.868

Table 2. Accuracy results per model

Test set	Neural network	Support vector machine	Random forest
1	0.777	0.791	0.875
2	0.806	0.769	0.879
3	0.934	0.890	0.982
Sum	2.517	2.450	2.736
Average	0.839	0.817	0.912

Table 3. F-score results per model

Test set	Neural network	Support vector machine	Random forest
1	0.697	0.671	0.815
2	0.675	0.640	0.811
3	0.901	0.848	0.973
Sum	2.273	2.159	2.599
Average	0.758	0.720	0.866

5 Discussion

From Table 1, the values of the average machine learning model performance reached values above 80% accuracy and 70% F-score, precision, and recall for neural network and support vector machine, while random forest achieved 91% accuracy and above 86% F-Score, precision, and recall. These results evidence that there is a clear difference between different models, coinciding with the research of Luna et al. [10], where machine learning models for classification of clonal varieties of coffee were implemented.

In the results shown in Table 2, the average accuracy was 83% for neural network, 82% for support vector machine, and 91% for random forest. This shows a slight improvement of 5.57% in the accuracy for the neural network shown in the work of Liu et al. [5], where neural network, support vector machine, and random forest techniques were implemented for classification of rice seeds with THz spectroscopy. This could be due to the chemometric variations of different compounds between these products giving a better performance in the training processes of neural networks for coffee bean classification.

Finally, in the results of Table 3, the average F-score was 76% for neural network, 72% for support vector machine, and 86% for random forest. Likewise, in the work of Monteiro et al. [7], where chemometric methods and spectroscopy were used for the authentication of the geographical origin of coffee, similar results to this study were obtained, providing insight and evidencing that these chemometric methods represent a viable way for non-destructive analysis of coffee.

6 Conclusions

This study demonstrated a significant difference of chemometric methods over coffee variety identification capabilities, both in the accuracy dimension and F-score dimensions used as measures for the evaluation of the classification task performances. As a synthesis, THz imaging and machine learning models represent a viable way for non-destructive analysis of coffee beans as shown preliminary in the presented research, showing the potential application for other food-related products in the agricultural industry field.

For future works, the development of scalable modules for the execution of machine learning python scripts and statistical data analysis of the classification results as well as optimized file systems for the handling of THz images and FTT algorithms will be required to move forward with a practical application of these methods in the coffee industry.

Acknowledgments. P. Uceda and H. Yoshida acknowledge the financial support from Project Concytec – The World Bank "Mejoramiento y Ampliación de los Servicios del Sistema Nacional de Ciencia Tecnología e Innovación Tecnológica" 8682-PE through Fondecyt [contract no 006–2018].

References

1. Ministerio de Agricultura y Riego (2019) Plan Nacional de Acción del Café Peruano 2019–2030
2. Link JV, Lemes ALG, Marquetti I, Scholz M, Bona E (2014) Geographical and genotypic classification of arabica coffee using fourier transform infrared spectroscopy and radial-basis function networks. Chemometr Intell Lab Syst 135:150–156
3. Barbin DF, Felicio A, Sun D-W, Nixdorf SL, Hirooka EY (2014) Application of infrared spectral techniques on quality and compositional attributes of coffee: an overview. Food Res Int 61:23–32
4. Bertone E, Venturello A, Giraudo A, Pellegrino G, Geobaldo F (2016) Simultaneous determination by NIR spectroscopy of the roasting degree and arabica/robusta ratio in roasted and ground coffee. Food Control 59:683–689
5. Liu W, Liu C, Hu X, Yang J, Zheng L (2016) Application of terahertz spectroscopy imaging for discrimination of transgenic rice seeds with chemometrics. Food Chem 210:415–421
6. Bona E, Marquetti I, Link JV, Makimori GYF, Arca V, Lemes ALG, Ferreira JMG, Scholz M, Valderrama P, Poppi RJ (2017) Support vector machines in tandem with infrared spectroscopy for geographical classification of green arabica coffee. LWT - Food Sci Technol 76:330–336
7. Monteiro P, Santos J, Alvarenga V, Pasini C, Koot A, Boerrigter-Eenling R, van Ruth S, Georgouli K, Koidis A, Granato D (2018) Comparison between proton transfer reaction mass spectrometry and near infrared spectroscopy for the authentication of Brazilian coffee: a preliminary chemometric study. Food Control 91:276–283
8. Craig A, Franca A, Oliveira L, Irudayaraj J, Ileleji K (2015) Fourier transform infrared spectroscopy and near infrared spectroscopy for the quantification of defects in roasted coffees. Talanta 134:379–386
9. van Drongelen W (2018) Continuous, discrete, and fast fourier transform. In: Signal processing for neuroscientists, Academic Press, pp 103–118
10. Luna A, Silva A, Silva C, Lima I, Gois J (2019) Chemometric methods for classification of clonal varieties of green coffee using Raman spectroscopy and direct sample analysis. J Food Compos Anal 76:44–50

Prevention and Control of Ravines and Gullies to Consolidate Green Economy Models

Caiubi Emanuel Souza Kuhn[1,2](✉) and Fabio Augusto Gomes Vieira Reis[2]

[1] Federal University of Mato Grosso, Cuiabá, Mato Grosso, Brazil
[2] São Paulo State University (UNESP), Campus Rio Claro, São Paulo State, Brazil
fabio.reis@unesp.br

Abstract. Spatial planning is essential to make the transition to a green economy. Among the various elements related to this theme, soil stands out for its social, environmental, and economic importance. For example, it is essential to contribute to carbon sequestration, food production, the maintenance of biodiversity, and the preservation of water resources. Soils are not renewed on the human time scale, so accelerated erosion becomes a severe public management problem. This paper presents an analysis of the importance of actions to prevent ravines and gullies, in the scope of territorial, urban, and rural development, for the consolidation of green economy initiatives in the State of Mato Grosso, Brazil. For this, a bibliographic review was carried out to identify the main causes and possible initiatives that may contribute to this transition. The results demonstrate the need to consolidate government initiatives to a) carry out territorial planning, considering the susceptibility to erosion; b) to identify and monitor sensitive areas; c) to promote the prevention of erosion on roads and green roads; d) develop erosion prevention and control actions in agricultural areas and in urban areas. The proposed actions make it possible to carry out territorial management focused on soil preservation and the prevention of accelerated erosion, considering the promotion of good practices and the development of prevention and mitigation.

Keywords: Green Economy · Ravines and gullies · Sustainable development

1 Introduction

Ravines and gullies are linear erosions, forming open and unstable channels, with depths ranging from a few to many meters, and a long evolution that can be characterized by several stages of development [1]. These processes can remove and transport large volumes of soil in a short period of time [2] and are found in all climates, except polar, predominating in temperate climates, but are frequent in tropical climates [3]. Anthropic action is considered a fundamental factor for the development of ravines and gullies, and the evolution of these processes is conditioned by the historical evolution of land use and management practices [3]. Erosion control and correct territorial planning become essential in achieving a balance between agricultural production and conservation [4].

In Europe, ravines are found under forests that have already reached their final stage of development, which were formed during the Bronze Age (Poland), Roman period

Y. Iano et al. (Eds.): BTSym 2020, SIST 233, pp. 862–869, 2021.
https://doi.org/10.1007/978-3-030-75680-2_95

(Belgium), medieval times (Germany, Poland, and Central Spain), or more recent periods (for example, Slovakia, Romania, and Italy) [5]. The evolution of linear erosions can take hundreds to thousands of years, growing 10 to 50 times the initial size and developing geomorphic characteristics that mark the record of human-environment interaction [1]. The annual evolution of these processes depends on several factors, but it can exceed the extension of 150 m [6]. The conditions for triggering the erosion process may be the result of several factors such as sudden storms, changes in land use, degradation of vegetation, overgrazing, intensified agricultural activities, and changes in runoff due to the construction of roads and urban centers, associated with predispositions related to geological and geomorphological characteristics of the terrain [7, 8].

Among the direct impacts caused by ravines and gullies, there are: the destruction of houses, bridges, and infrastructure; loss of soil productivity; reduction of carbon stock; silting of lakes and rivers with a consequent reduction in the useful life of reservoirs created for supply or generation of electricity; rapid drainage of terrestrial aquifers resulting in the lowering of the water table and subsequent drying of the wells; temporary or permanent loss of work capacity; and in rare cases causing deaths. Among the indirect impacts, there are: the increase in the cost of agriculture in the affected places, or even the abandonment of the areas; and the increase in the risk of flooding in regions downstream from the basin due to the silting up of the rivers [9–13].

Erosion prevention and control, by means of public policies and measures of correct territorial ordering and recovery of degraded areas, are fundamental to prevent these processes from arising or advancing to become a permanent part of the landscape, reducing the socioeconomic and environmental functions of the territory affected [14]. Erosion control is related to underground and superficial water flow regulation, contributes to climate regulation by expanding carbon sequestration, and favors biological diversity reaching a balanced environment [15]. All these factors are important for consolidating several goals related to sustainable development objectives and for the creation or consolidation of green economic models.

The State of Mato Grosso since 2016 is part of the initiative Partnership for Action on Green Economy (PAGE), which is a program developed since 2013 by five UN agencies, namely Environment Program (UNEP), International Labor Organization (ILO), Development Program (UNDP), Industrial Development Organization (UNIDO) and Institute For Training And Research (UNIT). The partnership's objective is to assist in the transition from the economic model to sustainable forms that are oriented according to the objectives of sustainable development, resource-efficient, socially inclusive, and with low carbon. During the elaboration of the program's logical matrix in the State of Mato Grosso, territorial ordering was chosen as one of the four priority lines by the program's steering committee. The State of Mato Grosso is located on the country's agricultural expansion frontier and, in recent decades, has received significant investments in infrastructure, in addition to expanding its grain production, consolidating itself as a national leader in agribusiness. However, it also took the lead in relation to the number of natural disasters caused by ravines and gullies in the country. Therefore, it is necessary to think of strategies for territorial ordering capable of minimizing the impacts of the recent human occupation in the Cerrado, Amazon, and Wetland biomes. So, this work presents a theoretical analysis of the importance of actions to prevent gullies and ravines for the development of a territorial order focused on soil preservation and good practices for the consolidation of green economy initiatives.

2 Methodology

The work was developed in two stages: a bibliographic review based on a bibliometric survey and a discussion of the data considering the challenges for a transition to a green economy. The bibliographic review was carried out through the portal of Capes journals, where access to the Scopus database was obtained. Combinations of keywords were used for research, such as a) or gully ravines; b) natural hazards or natural disasters and ravines or gully; and c) ravines or gully and green economy. After reading the material, the information related to good practices and the impacts of ravines and gullies were systematized and analyzed.

3 Results and Discussion

In the literature, few works relate the ravines or gullies with the green economy. One of the few existing studies addressed linear erosions in Ethiopia and found, through field observations, that soil loss (1,080,782.6m^3) and biodiversity loss are the main physical impacts caused by erosion processes. The author related the possibility of developing a green economy during the rehabilitation of areas affected by ravines [16]. However, when analyzing gullies and ravines with natural disasters, the number of articles available is much larger. Through this bibliography, it is possible to identify challenges and good practices that are linked to the management of linear erosions and methods and actions that can be related to the green economy, which are related to a) carry out territorial planning, considering the susceptibility to erosion; b) to identify and monitor areas susceptible to gullies and ravines and to contain erosion processes in urban and rural areas, and c) promoting the prevention of erosion on green roads and transport.

3.1 Carry Out Territorial Planning, Considering the Susceptibility to Erosion

The occupation of the Brazilian territory was accompanied by environmental problems, such as ravines and gullies, which were developed because of the incorrect use of the soil and the lack of territorial planning. There are records of widespread problems with erosion and silting up in drainages, triggered by the rapid occupation of the Planalto Occidental Paulista in the 1920s [17]. Erosion problems are also reported in the municipalities of Altamira and Brasil Novo, in the state of Pará [18]. In the state of Mato Grosso, studies analyzed the environmental quality and identified problems related to gullies and ravines in the municipality of Jaciara [19].

The link between land-use change and the development of linear erosions is widely accepted in the literature. However, it is necessary to identify which governmental measures can assist in the construction of sustainable development. The analysis of aerospace images, through remote sensing, for the development of a territorial ordering is one of the tools with great potential to help identify areas with high sensitivity. For example, ASTER's automatic classification of optical images has been successfully used to identify ravines in portions of the Brazilian Cerrado in the State of Minas Gerais [20]. Other authors used machine learning to identify erosive processes, also demonstrating the potential of these technological tools to identify areas with a high occurrence of erosive processes [2]. Multivariate analysis, applied through the decision tree technique, was able to identify patterns that distinguish typologies from linear erosions [21].

Geological-geotechnical cartography remains one of the best alternatives to understanding the physical environment's limitations and avoiding natural disasters and environmental problems. Factors such as the absence of vegetation, geological characteristics, soil type, hydrogeological conditions, channel order, and the terrain's curvature are related to the development of gullies [3, 14]. The construction of maps of susceptibility to erosion contributes to decision making, considering the trend of space, and allowing public managers to guide investments and develop policies with local communities [2].

The existence of geological, geomorphological maps, soil maps, among others, with a good scale of detail, in addition to high-definition orbital images, are fundamental for the development of territorial analyzes capable of providing detailed studies that can predict the behavior of the environment and thus allow the correct territorial ordering.

3.2 Identify and Monitor Susceptible Areas in Urban and Rural Areas

The erosion susceptibility maps mentioned in the previous topic allow identifying the appropriate place and the type of measures necessary to prevent linear erosion [22]. In addition, other measures, such as monitoring the sedimentary load of reservoirs, can help to identify hydrographic basins that should be prioritized [23].

The monitoring of areas after forest fires is important because this process favors the triggering of new juices and gullies [23]. Several authors have studied measures to control ravines and gullies. They can be separated into three different main classes: a) structural, b) vegetative, and c) structural/vegetative [15]. Other authors detail the classification of recovery measures into: a) ecological (e.g., shrub barriers); b) agricultural (e.g., soil preparation); c) mechanics (e.g., road maintenance); d) structural macro-drainage (e.g., dams); e) structural micro drainage (e.g., galleries); and f) bioengineering (e.g., gabions with vegetation) [24].

Actions can be carried out within the scope of the hydrographic basin, with emphasis on improving the hydrographic basin to regulate runoff; the control of rainwater along with the head of the ravine; the diversion of surface water above the ravine area; stabilize the ravine by means of structural measures and revegetation; flatten the land in the vicinity of erosion, and undertake measures to recover the area [1]. Although many techniques effectively prevent and control gullies, they are rarely adopted by farmers, as many of them do not have a rapid benefit in terms of increased productivity, and the absence of incentives aggravates the situation [25].

Some studies have obtained promising results regarding the cost-benefit of erosion containment techniques, as is the case with the use of bioengineering [26] and biological geotextiles [27]. The promotion of research in this field may be a government measure that is able to provide the discovery of solutions with high cost-benefit adapted to different scenarios. In addition, it is necessary to train and structure state agencies to monitor the development of erosion processes, as well as to carry out the necessary actions to contain these erosions.

Erosions have a major impact on the field. Among the reasons that lead to the abandonment of lands, there is the impact of ravines and gullies, especially those that have great depths, due to low productivity, the cost of restoration, and the cost of labor [28]. There is a record of areas that were not viable for agricultural activity due to progressive erosion, but the author points out that the place can be used for other activities such as fishing and ecotourism if the erosion processes are controlled [29]. In addition to the reduction of land available for cultivation, erosion processes lead to crop losses and increase the workload in the fields [23]. Although the control of linear erosions can be expensive, in the long run, the cost-benefit ratio is usually positive [15]. For the authors, the control measures applied in each area must consider the monetary aspects and the ecosystem services developed in these areas. Among the existing benefits, there are the restoration of productive areas and the sequestration of carbon and stock of organic matter in these productive areas [30].

In Brazil, measures such as no-till farming already contribute to reducing erosion. In places where erosion processes are already underway, it is necessary to apply containment methods, as well as the construction of dams to contain the sediments released by ravines and gullies. This latter technique is widely used in the Loess Plateau in China, where more than 100,000 dams have been built in 50 years [31]. Brazil needs to adopt a soil and water conservation policy related to a sustainable model of agriculture. In between the scenarios studied by the authors, the advance on native areas can increase erosion in the country; on the other hand, the restoration of degraded areas, especially by livestock, can reduce the impact of soil erosion [32].

Accelerated erosion in the urban environment causes the destruction of urban residences and structures. Small ravines and gullies can be easily contained with simple measures such as the diversion of water flow and the rehabilitation of the affected area. However, when the erosion dimensions are large, a greater engineering effort is required [6]. In poorer neighborhoods, where control measures are not developed, the impact can be even greater, increasing the cost involved in carrying out the intervention [33].

Erosive processes do not arise immediately. Between the beginning of the occupation of the area, with the construction of the first houses, and the development of large ravines and gullies, there is a time-lapse of years [34]. Among the causes for the development of erosive processes are extreme rainfall events, uncontrolled urbanization, the lack of land management, combined with environmental weaknesses [13].

The implementation of measures to prevent erosion associated with the analysis of the physical environment, through geotechnical maps, hydrological studies related to the water connectivity of the land and the local water cycles and management of the land to guarantee places for rainwater infiltration, some measures can contribute to avoiding erosive processes [34–36]. These measures require the development of information with good precision and the carrying out of case studies considering local conditions. Many Brazilian municipalities do not have technical teams or resources to carry out these activities, so the partnership with research institutions and the support of state and national agencies are some of the alternatives.

3.3 Foster Erosion Prevention on Roads and Green Roads and Transport

Roads are one of the most common causes for the development of ravines and gullies. It was identified that in areas where there were no roads, the presence of ravines was low, whereas unpaved roads were associated with erosive processes, while paved roads show a lesser relationship with the development of linear erosions [37]. There is a clear influence of roads on the rate of change of the head of ravines and gullies [8]. Other studies indicate the development of nozzles on unpaved roads, demonstrating that paved roads provided an increase in the runoff by 146–227%, with peak discharge between 178–575%, but with a reduction in the sediment load in 90–94% [36]. The results of the authors' work demonstrated that soil erosion on unpaved roads without vegetation was 3.09 times greater than the projected roads on which grasses were used [38].

However, erosion problems are not restricted to just this transportation mode, with the risk of forming nozzles in linear projects such as railways [1]. Linear erosions were recorded along the Malha Paulista (SP) railroad [21]. The impacts of road obstructions due to the development of linear erosions cause economic, social, and environmental impacts related to transport paralysis, accidents, contamination of the soil and water resources, among others.

4 Conclusion

Although few studies addressing the prevention and control of ravines and gullies and their relationship with the green economy are scarce, the review made it possible to identify several good practices and actions that can help the State of Mato Grosso and other units of the federation to adopt measures related to spatial planning considering the best way to use soil ensuring its preservation and full provision of ecosystem services. The proposed measures allow to reduce environmental and social impacts and to increase economic gains in the medium and soon term.

References

1. Dagar JC (2018) Ravines: formation, extent, classfication, evolution and measures of prevention and control. In JC Dagar, AK Singh (eds.) ravine lands: greening for livelihood and environmental security. https://doi.org/10.1007/978-981-10-8043-2_2
2. Arabameri A, Chen W, Loche M, Zhao X, Li Y, Lombardo L, Cerda A, Pradhan B, Bui DT (2020) Comparison of machine learning models for gully erosion susceptibility mapping. Geosci Front 11(5):1609–1620
3. Castillo C, Gomez JA (2016) A century of gully erosion research: urgency, complexity and study approaches. Earth Sci Rev 160:300–319. https://doi.org/10.1016/j.earscirev.2016.07.009
4. Bhat SA, Dar MUD, Meena RS (2019) Soil erosion and management strategies. In Meena R., Kumar S., Bohra J., Jat M. (eds) Sustainable management of soil and environment. Springer, Singapore. https://doi.org/10.1007/978-981-13-8832-3_3
5. Zhang W, Liu Y (2019) Research on visual interpretation and spatial distribution pattern of the erosion gully in luoyugou watershed of China. Environ Nat Resour Res 9(3):23–31
6. Hudec PP, Simpson F, Akpokodje EG, Umenweke MO (2005) Anthropogenic contribution to gully initiation and propagation in southeastern Nigeria. GSA Rev Eng Geol 16:149–158

7. Nosko R, Maliariková M, Brziak A, Kubáň M (2019) Formation of gully erosion in the Myjava region. Slovak J Civ Eng 27(3):63–72
8. Xu Q, Kou P, Wang C, Yunus AP, Shua XJ (2019) Evaluation of gully head retreat and rates based on high-resolution satellite images in the loess region of China. Environ Earth Sci. 78:465
9. Argüello GL, Dasso CM, Sanabria JA (2006) Effects of intense rainfalls and their recurrence: case study in corralito ravine, province of córdoba. Argent Quatern Int 158(1):140–146
10. Balzerek H, Fricke W, Heinrich J, Moldenhauer KM (2003) Man-made flood disaster in the savanna town of Gombe/NE Nigeria. The natural hazard of gully erosion caused by urbanization dynamics and their peri-urban footprints. Erdkunde **57**(2), 94–109
11. Costea M, Tăuşan I (2016) Land degradation due to erosion in public perception. Case study: secaşul mare river basin settlements (Transylvanian Depression, Romania). Environmental monitoring and assessment 188(4)
12. Daba S, Rieger W, Strauss P (2003) Assessment of gully erosion in eastern Ethiopia using photogrammetric techniques. Catena 50:273–291
13. Busnelli J, Neder LDV, Sayago JM (2006) Temporal dynamics of soil erosion and rainfall erosivity as geoindicators of land degradation in Northwestern Argentina. Quatern Int 158(1):147–161
14. Real LSC, Crestana S, Ferreira RRM, Sígolo JB, Rodrigues VGS (2020) Proposition for a new classification of gully erosion using multifractal and lacunarity analysis: a complex of gullies in the palmital stream watershed, Minas Gerais (Brazil). Catena 186, 104377
15. Romero-Díaz A, Díaz-Pereira E, De Vente J (2019) Ecosystem services provision by gully control. Rev Geogr Res Lett. 45(1):333–366
16. Hassen G, Bantider A (2020) Assessment of drivers and dynamics of gully erosion in case of tabota koromo and koromo danshe watersheds. South Cent Ethiop Geoenvironmental Disasters 7(1):5
17. Bezerra MÂ, Etchebehere ML de C, Saad AR, Married F da C (2009) Geoenvironmental analysis of the region of marília, sp: susceptibility to erosive processes in view of the area's historical occupation. São Paulo, UNESP, Geosciences, **28**(4) 425–440
18. Costa DF, de Alvarez W, P, Herrera JA, (2020) Erosive susceptibility in watershed in Amazonia: landscape exploration and burial of springs in the jaurucu sub-basin in Brasil Novo. Pará. Rev Geográfica Venez 61(1):88–98
19. Wantzen KM, Siqueira A, Da Cunha CN, De Sá M, De FP (2006) Stream-valley systems of the Brazilian Cerrado: Impact assessment and conservation scheme. Aquatic Conserv: Mar Freshw Ecosyst 16:713–732
20. Vrieling A, Rodrigues SC, Bartholomeus H, Sterk G (2007) Automatic identification of erosion gullies with ASTER imagery in the Brazilian cerrados. Int J Remote Sens 28(12):2723–2738
21. Souza N.da C, Pitombo C, Cunha AL, Larocca APC, Almeida Filho GS, (2017) Classification model of gully erosion process along of railway line through decision Tree algorithm and geotechnology. Bol de Ciên Geodésicas 23(1):72–86
22. Akgün A, Türk N (2011) Mapping erosion susceptibility by a multivariate statistical method: a case study from the Ayvalik region. NW Turk Comput Geosci 37(9):1515–1524
23. Martínez-Murillo JF, López-Vicente M (2018) Effect of salvage logging and check dams on simulated hydrological connectivity in a burned area. Land Degrad Dev 29(3):701–712
24. Rotta CMS, Zuquette LV (2014) Erosion feature reclamation in urban areas: typical unsuccessful examples from Brazil. Environ Earth Sci 72:535–555. https://doi.org/10.1007/s12665-013-2974-y
25. Valentin C, Poesen J, Li Y (2005) Gully erosion: Impacts, factors and control. Catena 63:132–153

26. De Brito Galvão TC, Pereira AR, Coelho AT, Pereira PR, Coelho JFT (2011) Straw blankets sewn with recycled plastic threads for erosion and urban sediments contro. Geotech Geol Eng 29:49–55. https://doi.org/10.1007/s10706-010-9371-z
27. Guerra AJT, Bezerra JFR, Fullen MA, Mendonça JKS, Jorge MCO (2015) The effects of biological geotextiles on gully stabilization in São Luís, Brazil. Nat Hazards 75:2625–2636. https://doi.org/10.1007/s11069-014-1449-0
28. Pani P (2016) Controlling gully erosion: an analysis of land reclamation processes in Chambal Valley. India. Dev Pract 26(8):1047–1059
29. Abdo MTVN, Vieira SR, Martins ALM, Silveira LCP (2013) Gully erosion stabilization in a highly erodible kandiustalf soil at pindorama, São Paulo State, Brazil. Ecological Restoration 31(3)
30. Somasundaram J, Parandiyal AK, Jha P, Kala S, Ali S (2018) Ravines: prospective zone for carbon sequestration. ravine lands: greening for livelihood and environmental security. In JC Dagar, AK Singh (eds.), Ravine lands: greening for livelihood and environmental security
31. Xiang-zhou X, Hong-wu Z, Ouyang Z (2004) Development of check-dam systems in gullies on the Loess Plateau, China. Environ Sci Policy 7:79–86
32. Merten GH, Minella JPG (2013) The expansion of Brazilian agriculture: Soil erosion scenarios. Int Soil Water Conserv Res 1(3):37–48
33. Kayembe KW, Wolff EM (2015) Contribution of the geographic approach to the study of human factors involved in the intraurban gully erosion in Kinshasa (DR Congo) Geo-Eco-Trop, 39(1), 119–138
34. Imwangana FM, Vandecasteele I, Pierre PT (2015) The origin and control of mega-gullies in Kinshasa (DR Congo). Catena 125:38–49
35. Albuquerque AO, Carvalho Júnior OA, Guimarães RF, Gomes RAT, Hermuche PM (2020) Assessment of gully development using geomorphic change detection between pre- and post-urbanization scenarios. Environ Earth Sci 79:232. https://doi.org/10.1007/s12665-020-08958-9
36. Gudino-Elizondo N, Biggs TW, Bingner RL, Yuan Y, Langendoen EJ, Taniguchi KT, Kretzschmar T, Taguas EV, Liden D (2018) Modeling ephemeral gully erosion from unpaved urban roads: equifinality and implications for scenario analysis. Geosciences 8:137. https://doi.org/10.3390/geosciences8040137
37. Selkimäki M, González-Olabarria JR (2017) Assessing gully erosion occurrence in forest lands in catalonia (spain) land degradation and development, 28(2), 616–627
38. Cao CS, Li C, Gao W, Chen Y, Yan M (2006) Impact of planting grass on terrene roads to avoid soil erosion. Landscape Urban Plann 78:205–216

Classification of Daily-Life Grasping Activities sEMG Fractal Dimension

Elmer Escandón(✉) and Christian Flores

Universidad de Ingeniería y Tecnología – UTEC, Lima, Peru
{elmer.escandon,cflores}@utec.edu.pe

Abstract. Applications in pattern recognition and feature extraction for hand tasks are widely applied in prosthesis design through superficial electromyographic signals (sEMG) characterization. Novel applications still require higher classification accuracies and inter-subject invariability. Moreover, as machine learning techniques are implemented in a prosthesis, higher interest is focused on the training data, considering real-life variables as muscle fatigue and continuous data collection. This paper presents the detection of three different grasping action groups using two electrodes positioned in the extensor and flexor digitorum from a benchmark database with acquired real-life signals. Higuchi's Fractal Dimension feature extraction technique is applied to determine a feature vector as training input data. Consequently, the training algorithm with a Support Vector Machine (SVM) technique for two kernel functions: linear and radial. Results indicate accuracies of 97.2%, 92.2%, 89.7% for two, three, and four task grasping actions with a Radial Basis Function kernel, respectively.

Keywords: Higuchi's fractal dimension · Support Vector Machines · sEMG · Task classification · Daily-life grasping activities

1 Introduction

Prosthesis designs require effective and outperforming results in its tasks classification techniques to operate according to the end-user [1]; however, these methods are still a matter of study due to the difficulties presented in real-life conditions, where muscles may fatigue, and people perform the tasks continuously. Additionally, the created algorithms require independence for inter-subject variations, regularly a problem for adequate task classification. In that way, new implementations need to obtain higher classification accuracies for real-life situations and perform satisfactorily task recognition without further modifications [2].

Electric biosignals have been studied to detect hand and arm movements. Mostly, non-invasive techniques through superficial electrodes are preferred due to their rapid and low-risk procedure. These external electromyographic signals (sEMG) were characterized using time, frequency, and fractal domains to provide insight for classification in muscle-related activities [3]. Consequently, Root-Mean Square (RMS), Discrete Wavelength Transform (DWT), and additional time-domain characteristics have been applied

Y. Iano et al. (Eds.): BTSym 2020, SIST 233, pp. 870–877, 2021.
https://doi.org/10.1007/978-3-030-75680-2_96

to several classification techniques with an inter-subject analysis, resulting in 83% accuracy with a linear Support Vector Machine (SVM) algorithm as the highest result for the second database [4]. Similarly, the spectrogram of the signals applying the Fast Fourier Transform (FFT) with Principal Component Analysis (PCA) was classified using SVM; in consequence, the proposed method achieved 76% of classification accuracy in basic hand movements [5]. Classification techniques were also used in daily-life activities only through time-domain features as RMS for several muscles, leading to desirable results of 97% accuracy with a real-time response for robotic hand prosthesis [6].

On the other hand, the fractal features have been demonstrated to correlate with the forearm's sEMG signals. Fractal characteristics associated with conduction velocity in the muscles were evaluated on both young and elder subjects to identify fatigue indicators. Hence, the Fractal Dimension identifies differences in change rate with statistical significance detecting motor unit synchronization [7]. Besides, the Grassberger and Procaccia technique to obtain the fractal dimension is applied to recognize pronation and supination movements in the arm, providing a small distance among data points and low variance; as a consequence, two identifiable regions are detected for each action [8]. The Higuchi's fractal dimension and the Maximum Fractal Length were features applied in a twin SVM. Therefore, the proposed framework achieved 82% to 95% and 78% to 91% with different algorithm modifications [9].

Thus, this paper's primary goal is to utilize Higuchi's Fractal Dimension characteristic with an SVM classification technique using an online database, which provides a benchmark standard for real-life signal conditions. Additionally, the sEMG signals from only two electrodes/channels of 15 subjects were selected to identify their relationship with three different group tasks; consequently, the combined feature vector serves as input data for the SVM algorithm. This work is presented as an intra-subject analysis extension using clustering and Higuchi's Fractal Dimension [10].

The paper is organized as follows. Section 2 shows the database information and electrode selection, the proposed feature extraction, and the classification algorithm. Then, Sect. 3 presents the results obtained. Finally, the conclusions and future work of the paper are exposed in Sect. 4.

2 Methodology

2.1 Benchmark Database

The Ninapro N°2 is an available online database created to study the application of robotic hand prosthesis through a non-invasive procedure. This dataset provides the sEMG signals for 40 intact subjects performing 49 different hand tasks divided into Exercises based on the type of movement and muscle contractions. Therefore, Exercise B has eight static contractions with nine basic wrist movements; Exercise C, 23 tasks between grasping objects and functional activity; finally, Exercise D, nine force patterns movements of the fingers [4]. Moreover, the signals from the database provide information from 12 wireless electrodes with a sampling rate of 2 KHz with a baseline noise fewer than 750 nV RMS, where eight electrodes are positioned equidistantly around the forearm and one in each muscle of the extensor digitorum, flexor digitorum, triceps brachii, and biceps brachii [4].

Since the database objective is to emulate real-life conditions, each subject performed the tasks continuously; consequently, the signals were recorded in one take. Each task was performed six times with an ideal rest time of 3 or 6 s between repetitions or task change. Additionally, a relabeling method [11] is applied due to the mismatch between the actual task duration and the video reference during the experiments.

2.2 Preprocessing

A 250 ms window is applied due to the maximum time reaction response in prosthesis control [12], while the overlapping time variate depending on the length of data points in every repetition to the number of segments among task repetitions.

2.3 Fractal Dimension

The Higuchi's fractal dimension is obtained for every segment of time X with N size. According to [13], consider small portions of the X signal can be modeled as X_m^k, where k indicates the number of steps and m, the index of the initial position. Consequently, the length of the curve $L_m(k)$ for any X_m^k is calculated using

$$L_m(k) = \frac{1}{k}\left[\left(\sum_{i=1}^{\frac{N-m}{k}} |X(m+ik) - X(m-(i-1)k)|\right)\frac{N-1}{\frac{N-m}{k}}\right] \quad (1)$$

This value is defined as the sum of the absolute difference of two consecutive data points for every i, while the truncated value $N - m$ divided by k steps represent the summation limit. Lastly, the proportional relation defined as $L_m \propto k^{-D}$, the slope of the $\log(L_m(k)) - \log(k)$ characterize the fractal dimension value. To determine the adequate fractal dimension of a signal, the K_{max} parameter must be selected. In this case, the value of 6 [14] is chosen according to the highest score in our previous work [10].

2.4 Support Vector Machines

The Support Vector Machines (SVM) for classification introduced by Boser et al. [15] is a supervised machine learning algorithm to classify information based on margin distance maximization among training patterns. From a training data $(x_1, y_1), \cdots, (x_m, y_m)$ with m samples, where each input sample has d dimensions corresponding to the number of features, assume that a hyperplane function is defined to separate the classes as Eq. (2).

$$f(x) = \boldsymbol{\omega}^T \phi(\boldsymbol{x}) + \boldsymbol{b} \quad (2)$$

It divides the positive and negative sample data x_m to identify class difference with the minimum value of the parameter $\boldsymbol{\omega}$ with a bias $\boldsymbol{b}$, besides the function $\phi(\boldsymbol{x})$ maps the sample $\boldsymbol{x}$ into a higher dimensional space [16].

In that way, the minimum value of $\boldsymbol{\omega}$ is required to flat the decision function in the high dimensional space. As a consequence, the following optimization problem is defined.

$$\min_{\omega,b,\xi} \frac{1}{2}\omega^T\omega + C\sum\nolimits_{i=1}^{m} \xi_i$$
$$s.t. y_i\left(\omega^T\phi(x_i) + b\right) \geq 1 - \xi_i \quad (3)$$
$$\xi_i \geq 0, i = 1, \ldots, m$$

Where ξ_i is the deviation parameter from the target result y_i and C is the regularization parameter. The presented equation is solved by applying the Lagrangian multipliers, represented by α_i, and quadratic programming to find the optimal value of ω, resulting in Eq. (4).

$$\omega = \sum\nolimits_{i=1}^{m} y_i\alpha_i\phi(x_i). \quad (4)$$

Finally, the normalized decision function from –1 to 1 is presented in the following equation.

$$sgn\left(\omega^T\phi(x) + b\right) = sgn\left(\sum\nolimits_{i=1}^{m} y_i\alpha_i K(x_i, x) + b\right) \quad (5)$$

The $\boldsymbol{K}$ equals to $\phi(x)^T\phi(x)$ as the Kernel function [17]. In this paper, the Linear and Radial Basis Function (radial) kernels are selected. The `LIBSVM` library from scikit-learn [18] with its parameters as default is used to classify the C tasks group classes in a *one vs. rest* (OVR) approach. Consequently, a 6-fold cross-validation technique with a stratified sampling method is applied; in specific, five samples for each fold task were training data while the left one as test data.

2.5 Proposed Structure

The proposed method is structured as follows. Firstly, the sEMG signals are acquired for four tasks and two electrodes. The selected electrodes are positioned in the flexor and extensor digitorum (see Fig. 1), which are related to the four non-opposite fingers' movements during flexion and extension. Likewise, the tasks have a relation with the previously mentioned finger movements, specifically in grasping.

In that way, the data is structured in three task groups, where the main goal is to classify two, three, and four tasks, as shown in Table 1. Consequently, Higuchi's fractal dimension is applied in every repetition segment to obtain a 25 new feature vector with the overtime fractal dimension. From that, the initial signal can be represented as its similarity pattern. The extracted vectors are then matched to each task, creating the training and test data classes for the Support Vector Machine training. Lastly, the classification accuracy is obtained using an average of each accuracy from a 6-fold cross-validation technique with stratified sampling, particularly five samples for each class are training and one as a test.

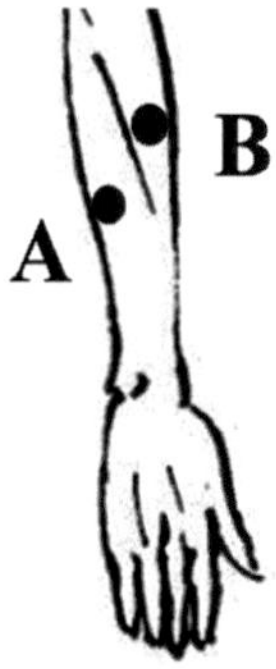

Fig. 1. Electrode positioning: (A) Extensor digitorum and (B) Flexor digitorum.

Table 1. Tasks groups and descriptions for the proposed structure.

NinaPro DB2 tasks	Description	Belongs to a task group
T1	Thumbs up	GT3
T6	Fingers flexed together in a fist	GT2, GT3
T5	Abduction of all fingers	GT1, GT2, GT3
T17	Wrist extension with a closed hand	GT1, GT2, GT3

3 Results

The results were obtained from 15 intact subjects, which demonstrated separability in a previous study [10]. Consequently, the SVM classification from two electrodes' signals was applied in three group tasks, as shown in Table 2, using the linear and radial kernel. Furthermore, Table 2 indicates the accuracy of the 6-fold cross-validation for every subject, kernel, and group task.

The GT1 has average accuracies of 95% and 100% classification for 60% of the subjects using the linear kernel. Additionally, all the subjects' average is 97% and 66.6% for the linear and radial kernel, respectively. Moreover, the radial kernel improved the accuracy for subjects S16 and S25 while it reduced the standard deviation from the average by 2.1%.

Secondly, the GT2 for the linear kernel presents a 90.3% average accuracy, and the lowest accuracy of 61.1% for the subject S25. On the other hand, the radial kernel shows an average accuracy of 92.2%, due to the improvement of 4 subjects, including S25. Both Kernels have the same percentage of residents with 100% classification, whereas the standard deviation was reduced from 9.6% to 7.3% in the linear and radial kernel, respectively.

Thirdly, the GT3 presents average accuracies below 90% and 20% of 100% classification in both kernels. The linear kernel has an average of 84.2% accuracy and its lowest value of 54.2% in the subject S22. On the other hand, the radial kernel presents

Table 2. Accuracy for task classification using SVM and Higuchi's fractal dimension

	GT1		GT2		GT3	
Subject	Linear	Radial	Linear	Radial	Linear	Radial
2	100.0	100.0	100.0	100.0	100.0	100.0
3	100.0	100.0	100.0	100.0	100.0	100.0
8	100.0	100.0	88.9	88.9	62.5	66.7
14	91.7	91.7	94.4	94.4	83.3	87.5
15	100.0	100.0	88.9	94.4	87.5	95.8
16	83.3	91.7	88.9	94.4	91.7	95.8
17	100.0	100.0	83.3	83.3	70.8	87.5
19	100.0	100.0	94.4	88.9	95.8	91.7
21	91.7	91.7	88.9	88.9	87.5	91.7
22	100.0	100.0	83.3	94.4	66.7	91.7
25	83.3	100.0	61.1	72.2	54.2	62.5
27	91.7	91.7	94.4	94.4	91.7	95.8
30	100.0	100.0	100.0	100.0	100.0	100.0
33	91.7	91.7	88.9	88.9	83.3	83.3
40	100.0	100.0	100.0	100.0	87.5	95.8
Avg	95.6	97.2	90.3	92.2	84.2	89.7
Std. dev	6.0	3.9	9.6	7.3	13.8	11.0

a precision of 89.7% and the lowest accuracy of 62.5% in the same subject. The radial kernel maintains a lower standard deviation while improving the results for ten subjects.

The intra-subject analysis shows the adequate classification of two, three, and four tasks with two electrodes. In specific, the two-task identification proved outstanding accuracies with an average higher than 95% in both Kernels. Additionally, the GT2 also maintains higher precision, even though its standard deviation has increased. Although GT3 presents an adequate classification accuracy, both kernels' standard deviation demonstrates higher dissimilarities in the results. Overall, the performance does not exceed 80% precision in any of the kernels with different group tasks, where the radial kernel shows the highest accuracies and lowest standard deviations. In comparison to [5], the radial kernel exhibits the best performance compared to other kernels for task classification with sEMG.

The effect of the selected muscles maintains a relationship with the tasks performed [19]; therefore, the electrodes' selection allows the adequate identification of the group tasks. The intra-subject analysis reveals that the algorithm was unperformed in subjects 8 and 25 when classifying three and four tasks. The previous results obtained from a clustering approach [10] in Table 2 presents the selected cluster selection method versus the expected result. Since all the subject's hand classification was applied only for two tasks (GT1), the expected cluster selection must be only two clusters of the data. Despite the different metrics used between our previous and current work, this proposed approach presented a higher indicator.

4 Conclusions

The classification of three group tasks using SVM was applied using the Higuchi's Fractal Dimension in a benchmark online database for 15 intact subjects using two electrodes positioned in the forearm. The signals retrieved were acquired considering real-life situations with daily life tasks related to grasping. Consequently, Higuchi's Fractal Dimension algorithm was applied into small segments of the signal repetition. The SVM algorithm utilized two kernel functions: radial basis and linear. Lastly, the accuracy results were obtained using 6-fold cross-validation. As a result, the radial kernel presents higher average accuracy compared to the linear kernel, where the two and three tasks show a precision higher than 90%. On the other hand, the four-task classification has 89.7% precision and a standard deviation of 11.0%. The intra-subject classification indicates that two electrodes enable adequate identification for two, three, and four tasks related to those muscles' movement, deliberately selected due to its correspondence in the hand actions.

Since the database provides signals from 40 subjects and 12 electrodes, further information could be extracted to obtain further insight into other muscles' relation to task detection. Moreover, future work includes more tasks and electrodes related to those tasks; thus, a more versatile and useful selection can be implemented.

References

1. Calado A, Soares F, Matos D (2019) A review on commercially available anthropomorphic myoelectric prosthetic hands, pattern-recognition-based microcontrollers and sEMG sensors used for prosthetic control. In: 2019 IEEE International conference on autonomous robot systems and competitions (ICARSC), IEEE, pp 1–6
2. Iqbal N, Subramaniam K, Asmi S (2018) A review on upper-limb myoelectric prosthetic control. IETE J Res 64(6):740–752. https://doi.org/10.1080/03772063.2017.1381047
3. Phinyomark A, Quaine F, Charbonnier S, Serviere C, Tarpin-Bernard F, Laurillau Y (2013) EMG feature evaluation for improving myoelectric pattern recognition robustness. Expert Syst Appl 40(12):4832–4840
4. Atzori M, Gijsberts A, Castellini C, Caputo B, Hager AGM, Elsig S, Giatsidis G, Bassetto F, Müller H (2014) Electromyography data for non-invasive naturally-controlled robotic hand prostheses. Sci Data 1(1):1–13
5. Zhai X, Jelfs B, Chan RH, Tin C (2016) Short latency hand movement classification based on surface EMG spectrogram with PCA. In: 2016 38th Annual international conference of the IEEE engineering in medicine and biology society (EMBC), IEEE, pp 327–330
6. Castellini C, Fiorilla AE, Sandini G (2009) Multi-subject/daily-life activity EMG-based control of mechanical hands. J Neuroeng Rehabil 6(1):1–11
7. Boccia G, Dardanello D, Beretta-Piccoli M, Cescon C, Coratella G, Rinaldo N, Barbero M, Lanza M, Schena F, Rainoldi A (2015) Muscle fiber conduction velocity and fractal dimension of EMG during fatiguing contraction of young and elderly active men. Physiol Meas 37(1):162

8. Xiao H, Zhi-zhong W, Xiao-mei R (2005) Classification of surface EMG signal with fractal dimension. J Zhejiang Univ Sci B 6(8):844–848
9. Arjunan SP, Kumar DK, Naik GR (2010) A machine learning based method for classification of fractal features of forearm sEMG using twin support vector machines. In: 2010 Annual international conference of the IEEE engineering in medicine and biology, IEEE, pp 4821–4824
10. Escandón E, Flores C (2020) Clustering of sEMG signals on real-life activities using fractal dimension and self-organizing maps. In: 2020 IEEE engi-neering international research conference (EIRCON), pp 1–4. https://doi.org/10.1109/EIRCON51178.2020.9253761
11. Kuzborskij I, Gijsberts A, Caputo B (2012) On the challenge of classifying 52 hand movements from surface electromyography. In: 2012 Annual international conference of the IEEE engineering in medicine and biology society, IEEE, pp 4931–4937
12. Smith LH, Hargrove LJ, Lock BA, Kuiken TA (2010) Determining the optimal window length for pattern recognition-based myoelectric control: balancing the competing effects of classification error and controller delay. IEEE Trans Neural Syst Rehabil Eng 19(2):186–192
13. Higuchi T (1988) Approach to an irregular time series on the basis of the fractal theory. Physica D 31(2):277–283
14. Accardo A, Affinito M, Carrozzi M, Bouquet F (1997) Use of the fractal dimension for the analysis of electroencephalographic time series. Biol Cybern 77(5):339–350
15. Boser BE, Guyon IM, Vapnik VN (1992) A training algorithm for optimal margin classifiers. In: Proceedings of the fifth annual workshop on computational learning theory, pp 144–152
16. Ameri A, Kamavuako EN, Scheme EJ, Englehart KB, Parker PA (2014) Support vector regression for improved real-time, simultaneous myoelectric control. IEEE Trans Neural Syst Rehabil Eng 22(6):1198–1209
17. Chang CC, Lin CJ (2011) LIBSVM: A library for support vector machines. ACM Trans Intell Syst Tech (TIST) 2(3):1–2
18. Pedregosa F, Varoquaux G, Gramfort A, Michel V, Thirion B, Grisel O, Blondel M, Prettenhofer P, Weiss R, Dubourg V, Vanderplas J, Passos A, Cournapeau D, Brucher M, Perrot M, Duchesnay E (2011) Scikit-learn: machine Learning in Python. J Mach Learn Res 12:2825–2830
19. Brorsson S (2008) Biomechanical studies of finger extension function. Analysis with a new force measuring device and ultrasound examination in rheumatoid arthritis and healthy muscles. Institute of Clincial Sciences. Department of Orthopaedics

Recognition of Hand-Towards-Face Movements Prototype to Fight COVID-19 Using an IMU Sensor and Deep Learning Model Embedded in a Single-Board Computer

Alvaro Martín Aspilcueta Narvaez(✉)

Aubo, Lima, Perú
alvaro.aspilcueta.n@uni.edu.pe

Abstract. People touch their faces an average of 16 to 23 times per h; this unconscious habit is very difficult to eradicate and allows the SARS-CoV-2 virus to enter the body after a contaminated surface was touched with hands, causing COVID-19 disease that could become a serious illness for older people with underlying medical problems. To help combat this behavior, a Recognition of hand-towards-face movements prototype has been proposed by using an IMU Sensor and Deep Learning Model embedded in Single-Board Computer. It works using linear and angular accelerations from an MPU 6050 IMU sensor and a Raspberry Pi Model 3B+ that collects data and executes a TensorFlow Lite model that determines whether a person is going to touch their face or not to avoid unintentional infections.

Keywords: HAR · COVID-19 · TinyML

1 Introduction

COVID-19 is an infectious disease caused by SARS-CoV-2. You can get the infection by touching your eyes, mouth, or nose with your hands after a contaminated surface was touched. According to [1] and [2] studies, people touched their faces an average of 16 and 23 times per h, respectively. One of the best ways to keep yourself safe is not touching your face as much as possible and avoiding unintentional infections. For this reason, it was developed a Recognition of hand-towards-face movements prototype to fight COVID-19 by using an IMU Sensor and Deep Learning Model embedded in Single-Board Computer.

2 Proposal

2.1 Data Collection

Hardware. An MPU-6050 IMU sensor with an accelerometer and gyroscope sensitivity configuration of $\pm$ 2g and $\pm$ 250 degrees/second, respectively, was used. A Raspberry

Y. Iano et al. (Eds.): BTSym 2020, SIST 233, pp. 878–885, 2021.
https://doi.org/10.1007/978-3-030-75680-2_97

Pi Model 3B+ with a 1.4 GHz 64-bit quad-core processor was used to develop the prototype faster. Jumpers were used to connect the MPU-6050 and Raspberry Pi, as shown in Fig. 1. The MPU6050 IMU sensor was placed on the wrist of the dominant hand (right hand in this case). The orientation of the Y+ axis points to the hand, as shown in Fig. 2.

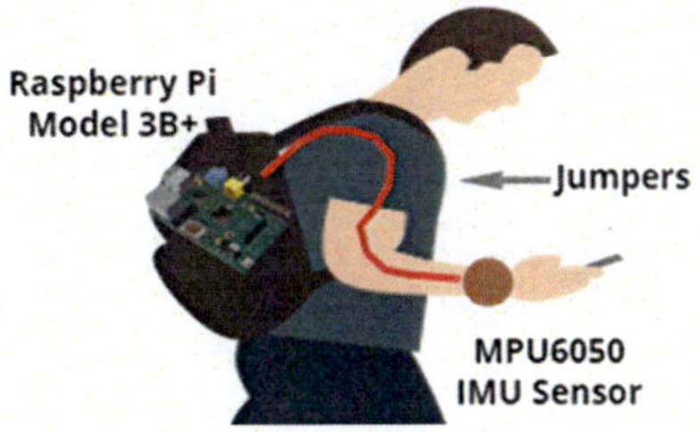

Fig. 1. Location of hardware on the person's body.

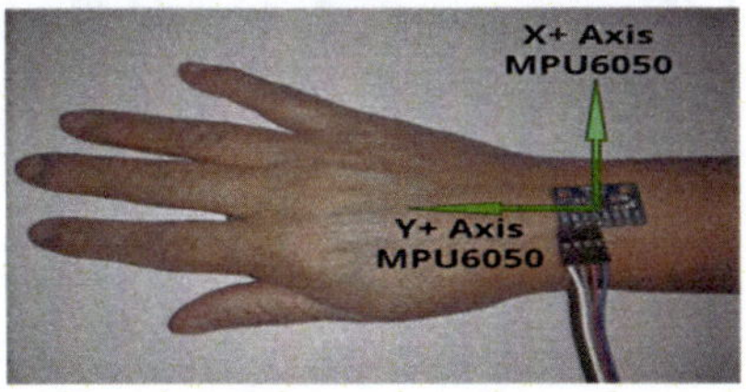

Fig. 2. Location of hardware on the wrist of the dominant hand.

Software. The operative system running on Raspberry Pi is the Raspbian OS, the programming language used was Python 3.7 and TensorFlow Lite to reduce the size of the model.

Methodology for Measurement of User's Movements. It was created a program on Raspberry to collect the data and save it as a.csv file. Six typical positions of the hand towards face movements were recorded and labeled as "1", "2", "3", "4", "5", and "6", as shown in Fig. 3, and "7", that is related to other movements different from the labels "1" to "6", could be considered as random movements. It is important to realize that the model will be trained with the data collected from just one user due to the quarantine, so it is a specific model. The steps to collect each of these datasets are:

- Put the dominant hand in the starting position (in the mouse, touchpad, etc.)
- Move the dominant hand from the starting position to your face (eyes, nose, mouth) until you touch them, then stop by 3–5 s.
- After waiting for 3–5 s, move your dominant hand to the starting position, wait another 3–5 s.
- Repeat after more than 20,000 data points were collected to make data balanced. In this case, the program showed the number of data points recorded on a screen.

The reason to wait 3–5 s in each movement is to differentiate them better when the data is analyzed.

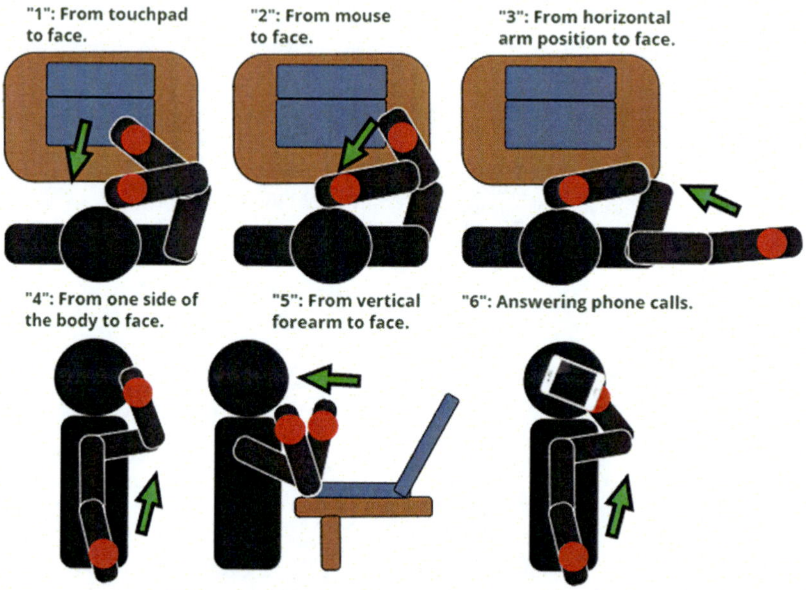

Fig. 3. User's movements recorded for the experiment (red circle represents the IMU sensor).

2.2 Data Preprocessing

Normalization. In this study, missing values were not presented. Standardization with mean and standard deviation was applied to data, as shown in Fig. 4. This helps understand the dynamics of data, label it, and avoid bias.

Window's Length. The complete movement towards the face lasts an average of 0.85 s, represented by the blue box in Fig. 4. The inference model has to detect this movement before it is completed since the objective of the prototype is to avoid touching your face with your hands. So, the length of the window is considered as 0.5 s. It contains ten samples per window due to a sample rate of 20 Hz.

Offset of the Time Window. There are, basically, two approaches according to [3]: FNOW (Fully non-overlapping temporal window) and SNOW (Semi non-overlapping temporal window). In this study, a SNOW approach was used with a 50% overlap because the movement duration is relatively short. So, the movement may be partially collected. Besides, 50% is not either a low value that has to trigger the inference model frequently.

Extraction of Movements for Training. When the data collection is finished, there are seven.csv files, each of them representing one of the movements labeled as "1" to "7". As

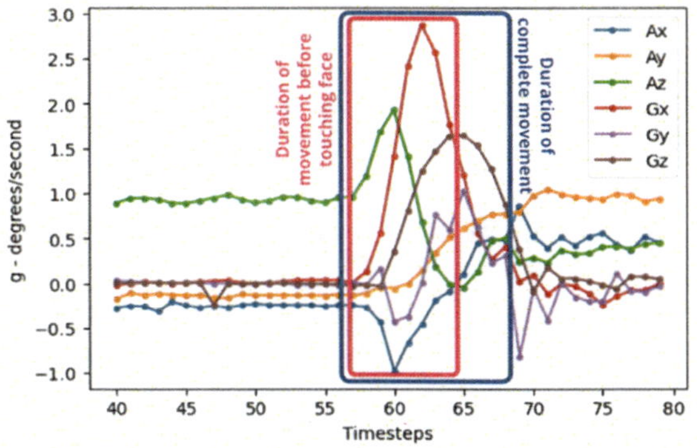

Fig. 4. Data from Linear accelerations (Ax, Ay, Az) on g units and Angular accelerations (Gx, Gy, Gz) on degrees/second units corresponding to the touchpad towards face movement.

shown in Fig. 5, it will be extracted from the normalized data corresponding to label "1" in this case. The purple boxes that contain the data for the movement from the touchpad to the face will be used to train the model; the blue boxes correspond when there is a pause in the movement, and the orange boxes, corresponding to the movement from the face to the touchpad, will not be used. All this extraction was visually and manually done for each label. In the case of label "7", all random movement data are useful to train the model.

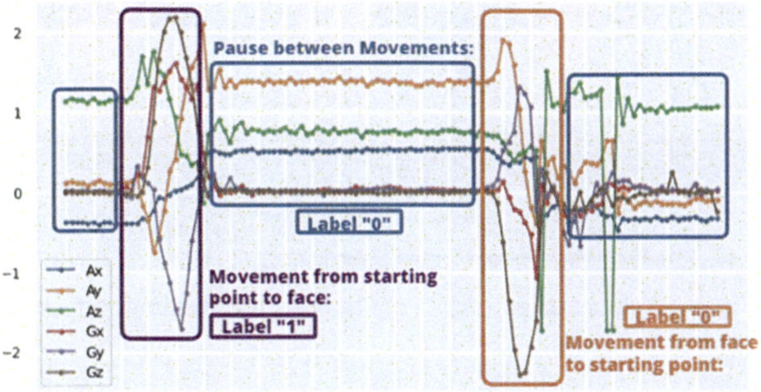

Fig. 5. Components of the normalized data corresponding to label "1".

Creation of the Input Dataset for the Model. After data was extracted for each class, it was split into Training (70%), Validation (20%), and Testing (10%) approximately. Finally, the datasets of training and validation of each class were combined in a unified dataset for training and validation containing all classes. Then, in the process of segmenting this dataset in time, windows features of shape (window's length, number of features: Ax, Ay, Az, Gx, Gy, Gz) = (10, 6).

- "1" or "The face was not touched": data extracted from labels "6" and "7".

• "2" or "The face was touched": data extracted from labels "1" to "5".

Data Augmentation. After the process of creating the input dataset, data was augmented following the subsequent criteria (this was not applied to label "7"):

• Scaling features data by multiplying by 0.95, 0.96, 0.97, 0.98, 0.99, 1.01, 1.02, 1.03, 1.04, 1.05.
• 25 samples were created by multiplying each element of the (10, 6) window feature by a random sample that is uniformly distributed over the half-open interval [0.85, 0.95).
• Another 25 samples were created by multiplying each element of the (10, 6) window feature by a random sample that is uniformly distributed over the half-open interval [1.05, 1.30).

Any value within the given interval is equally likely to be drawn. So, for each sample, 60 were generated. This was made to create a more robust and precise model besides compensating the imbalance because there were more samples from label "7" (random movements). Before, it was considered to multiply each value by a random distribution with mean = 1 and standard deviation = 0.2, but the disadvantage of this method was that, as values could be greater or less than 1 for a sample, this deformed the original shape of the curves as can be seen in Fig. 6.

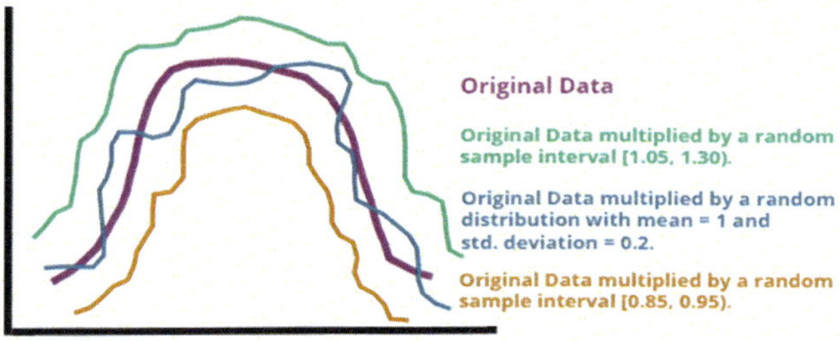

Fig. 6. Augmentation of data multiplying it by random values.

2.3 Creation of Model

Feature Extraction. There are two options: Hand-crafted feature extraction (time domain, frequency domain, or others) or Deep Learning feature extraction (automatically extracts appropriate features from raw data [4], and exploits its sequential nature [5], in a data-driven way within the network [6]). In this study Deep Learning approach was used due to its advantages; although it has a disadvantage related to the size of the model and its parameters, this is solved by using TensorFlow Lite, which is, according to [7], an "open source deep learning framework for on-device inference". It has useful characteristics to work with this kind of device: lightweight, low-latency, improved power consumption, and no internet connection required.

Architecture of the Best Deep Learning Model. It is described below:

- 1D Convolutional Layer: filters = 16, kernel size = 3, relu activation function.
- 1D Convolutional Layer: filters = 8, kernel size = 3, relu activation function.
- Dropout Layer: rate = 20%.
- 1D Max Pooling Layer: pool size = 2.
- Long Short Term Memory Layer: units = 16, relu activation function.
- Dense Layer: units = 16, relu activation function.
- Dense Layer: units = 1, sigmoid activation function.

Adam optimizer and Binary Cross Entropy Loss were used; the metric to monitor was Binary Accuracy with Early Stopping with patience set to 3. The steps per epoch used were 318; validation steps were 85; batch size was 256.

3 Results and Discussion

In order to obtain the performance and other metrics for the developed model, we evaluated it on the Testing dataset, that is, the data that the model has not used for its training. Therefore, the model has not seen that data before. The classification metrics like Accuracy, Precision, Recall, and F1-Score can be seen in Table 1. The Confusion Matrix is shown in Fig. 7.

Table 1. Classification metrics evaluated on Testing dataset.

Class or label	Accuracy	Precision	Recall	F1-Score
"1" or "The face was not touched"	0.962	0.979	0.928	0.953
"2" or "The face was touched"	0.962	0.950	0.986	0.968

The F1-Score, the most significant metric in classification, is greater than 95% (see Table 1). From the Confusion Matrix in Fig. 7, there is more than 98% of probability that the model detects when someone will touch the own face when this happens. On the other hand, the probability that the model detects when someone makes a movement different from touching their face is 92% when this happens, but there is a 7% chance that the model detects that someone touches the own face when it was a random movement. This can be improved if more data of the user's movements are recorded so that the model gets more data to train and gets more robust. Another way to improve it is not using random movements as a label for classification; instead, it could be used an open classification approach so that the model learns to ignore data that do not correspond to its labels. Finally, if the random movements were better distributed, the accuracy would improve because it looks like some samples of random movements appear in training and validation datasets but not in the testing dataset. So this decreases the accuracy of this dataset. The last step is to embed this model into a Raspberry Pi.

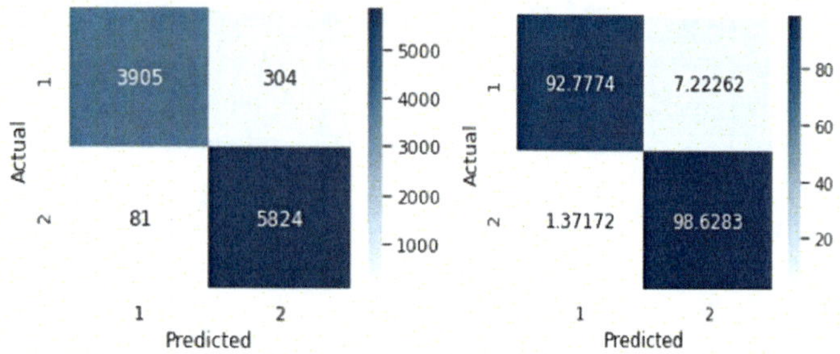

Fig. 7. Confusion matrix of testing dataset. Absolute values (left) and relative values (right).

Exportation of Model and Embedding in the Raspberry. After the model was created and saved, it is converted to the TensorFlow Lite format, exported, and downloaded. A program was created for Raspberry to read the data and transform it under the following logic (see Fig. 8): as the window's length is 10 samples and the overlap is 50%, that means that 10 values are read, formatted with the shape the model needs, the model is invoked, the model makes the inference, return the results, and an alert is emitted based on the results: hand-towards-face movement detected or not. Then, in the next loop, 5 values are read and appended to the 10 values read before, and the first 5 values are erased, so these 10 values follow the process mentioned before. In this way, this allows the process to be efficient since the data recorded is not saved but is processed in streaming. As expected, the model running on Raspberry Pi works well. The time the model takes to make the inference is 0.01 s; this allows a good user's experience when it is used.

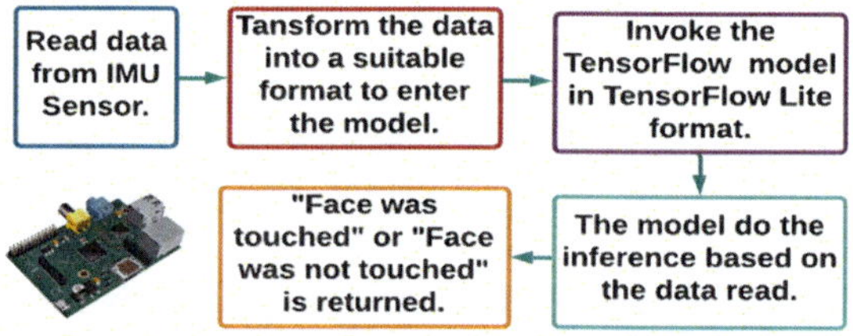

Fig. 8. Raspberry Pi pipeline.

4 Conclusions and Future Work

In this study, it has been explored if hand-towards-face movements could be detected before it happens to fight COVID-19 by developing a prototype based on an IMU Sensor and Deep Learning Model embedded in a Single-Board Computer (SBC). Many approaches have been tried to solve this problem, such as the data treatment as an Open Classification problem so that random movements were unnecessary because they are practically infinite, but it could not be done. After that, it was determined that it was more convenient to use an Standardization instead of MinMax scaling because it reduces the variance and slows the convergence. The data has to be labeled twice to capture just the most notable parts of the movements because otherwise, this creates confusion to the algorithm. After that, other architectures were tried, like Time Distributed 1D Convolutions that apply a layer to every temporal slice of an input, also ConvLSTM was tried, but both architectures could not be converted to a TensorFlow Lite model by now. So, normal 1D Convolutions had to be used. Also, random movements were difficult to treat because, at the time to shuffle the data, it is difficult to get a balanced distribution of examples in Training, Validation, and Testing. For this reason, although, in training and validation, the model did it well, in testing, its performance decreased because it is possible that not all the examples in testing were seen in training and validation. Change the way to augment data was crucial to improve the performance of the model. Finally, it was a decisive decision to, instead of trying a multiclass classification for every label, try it with a binary classification "The face was touched" and "The face was not touched", because that improved the performance of the model.

As future work, the use of an open classification approach to reject unknown classes would be useful, as well as embed the program into a microcontroller like ESP32. Also, the use of Transfer Learning or Train the model with more people data to build the final model because the data collected samples were only taken by the author. Develop a program that distinguishes when the data do not change to not trigger the model each time. Automate this step and facilitate the development of the model.

References

1. Nicas M, Best D (2008) A study quantifying the hand-to-face contact rate and its potential application to predicting respiratory tract infection. J Occup Environ Hyg 5(6):347–352. https://doi.org/10.1080/15459620802003896
2. Kwok YL, Gralton J, McLaws ML (2015) Face touching: a frequent habit that has implications for hand hygiene. American J Infect Control 43(2):112–114. https://doi.org/10.1016/j.ajic.2014.10.015
3. Singh SP, Lay-Ekuakille A, Gangwar D, Sharma MK, Gupta S (2020) Deep ConvLSTM with self-attention for human activity decoding using wearables. In: arXiv preprint arXiv:2005.00698
4. Xia K, Huang J, Wang H (2020) LSTM-CNN Architecture for Human Activity Recognition. IEEE Access 8:56855–56866
5. Agarwal P, Alam M (2020) A lightweight deep learning model for human activity recognition on edge devices. Procedia Comput Sci 167:2364–2373
6. Jalal A, Quaid MA, Sidduqi MA (2019) A Triaxial acceleration-based human motion detection for ambient smart home system. In: IBCAST 2019 16th International bhurban conference on applied sciences and technology, pp. 353–358
7. TensorFlow Lite. https://www.tensorflow.org/lite/

A Dual PRF IR-UWB Pulse Generator Implemented in 130nm CMOS Process for Data Communication Systems

Filipe Ferreira Caetano[1(✉)], Osamu Saotome[1], Marcus Henrique Victor Júnior[1], and Luiz Carlos Moreira[2]

[1] Instituto Tecnológico de Aeronáutica (ITA), São José Dos Campos, SP, Brazil
{filipeffc,osaotome,marcus}@ita.br
[2] Universidade Católica de Santos, Santos, SP, Brazil
lcm@unisantos.br

Abstract. This paper presents an Impulse-Radio Ultra-Wide Band (IR-UWB) pulse generator circuit implemented in 130 nm Complementary Metal Oxide Semiconductor (CMOS) technology. The pulse generator was implemented with a Negative Edge Combiner (NEC) connected to an RLC filter to produce a monocycle per clock. This monocycle can be generated by dual Pulse Repetition Frequency (PRF) using a Voltage Control Oscillator (VCO) connected to a Voltage Selector (VS) circuit. Its measured peak-to-peak output voltage was 132 mVpp on a 50 Ω load for a central frequency of 4.8 GHz, and a supply voltage of 1.2 V. The size of the main chip occupies a small area of 214 μm × 234 μm (without pads).

Keywords: IR-UWB pulse generator · Negative edge combiner circuit · Shunt-capacitance · Microelectronics · CMOS

1 Introduction

The first transmissions of the UWB pulse occurred in 1901 when Guglielmo Marconi first transmitted the Morse Code across the Atlantic ocean [1]. Since then, the technology has time crossed until the present decade, when IR-UWB circuits have shown several applications. The main areas of application are the automotive industry, such as Radio Detection And Ranging – (RADAR) [2] and medical imaging [3], and aircraft structural monitoring [4]. CMOS technology contributes extremely to the development of IR-UWB wireless technology due to the many advantages of transmitting and receiving data over a broad spectrum. Additionally, the IR-UWB systems have low-cost, low complexity, low power consumption design, and can operate at high data rates [5, 6]. There is a high demand for high-speed data transmission due to the Internet of Things (IoT) market growth [7]. IoT's main features include small devices, such as pervasive systems, smart homes and sensor networks, and a wide variety of wireless systems. Thus, we propose radio frequency (RF) circuits with IR-UWB technology characteristics as a possible solution for this demand.

Y. Iano et al. (Eds.): BTSym 2020, SIST 233, pp. 886–893, 2021.
https://doi.org/10.1007/978-3-030-75680-2_98

Today the world is globalized, and communication enables the internet with several countries in real-time. For this, there is a demand to implement more adaptive systems to work with different transmission rates. Thus, the proposed new pulse generator can operate with two alternate PRF's in communication systems working with variable transmission rates. The central frequency at the output can be tuned based on the target spectrum profile. The paper is organized as follows: Sect. 2 shows the IR-UWB pulse generator design's details in the 130 nm CMOS process with dual PRF; In Sect. 3, we will present the proposed layout simulations, followed by the conclusions in Sect. 4.

2 Full Pulse Generator Block

The complete block diagram of the pulse generator consists of the five blocks to generate a monocycle at the output, as shown in Fig. 1. All blocks have been implemented at the process 130 nm CMOS technology. The first block is a Voltage Selector Block (VSB) and is connected at the input of the Voltage Control Oscillator (VCO) to control pulse width, as shown in Fig. 1(a). This circuit is implemented with two voltage inputs and a Vbias voltage to select input voltage (V1 and V2). The second block is a VCO composed of two sub-blocks: 1) a current mirror using topology P and N; 2) a current starved inverter. In this design, the current mirror controls the current starved inverter chain [8], as shown in Fig. 1(b, c). The third block consists of a shunt-capacitor inverter delay chain to control central operating frequency, as shown in Fig. 1(d); the fourth block is a NEC block with nMOS transistors to generate a current through an inductor [2], as shown in Fig. 1(e); the fifth block is a RLC filter to generates a monocycle at the output, as shown in Fig. 1(f). The circuit is polarized with 1.2 V. These blocks are explained in detail below:

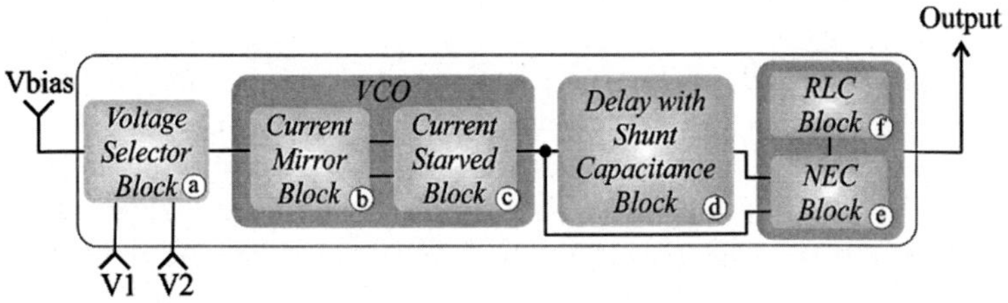

Fig. 1. Complete block diagram of the pulse generator circuit: (a) Voltage selector block, (b) and (c) VCO implemented with current mirrors circuits P and N types and current starved Circuits, (d) Delay block using shunt-capacitance topology, (e) NEC block and (f) RLC filter.

2.1 Voltage Selector Block (VSB)

The VSB was implemented to select the voltage V1 or V2 at the input of the VCO block. This voltage defines PRF operational required, so, when Vbias = 0 V, the supply V2 is selected to connect to the input of VCO. Similarly, when Vbias = 1.2 V, the supply V1 is selected to connect to the input of VCO, as shown in Fig. 2. This block was implemented with two nMOS transistors and an inverter circuit. This circuit has two different modes of operation: 1) The first mode of operation occurs when a logic level 0 is connected to the Vbias input. In this condition, transistor Mn1 is "on", and transistor Mn2 is "off"; 2) The second mode of operation occurs when a logic level "1" is connected to the Vbias input. In this condition, transistor Mn1 is "off" and transistor Mn2 is "on". In this form, the VCO block can operate with two operating frequencies and, consequently, with two PRFs.

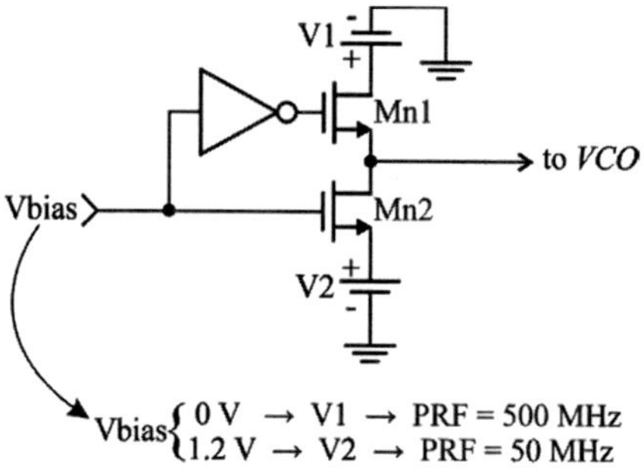

Fig. 2. Voltage selector circuit.

2.2 Voltage Control Oscillator (VCO) Block

The block of the VCO circuit is composed of three sub-blocks: 1) Current mirror circuits type P and N, as shown in Fig. 3(a); 2) A ring oscillator employing the current starved topology, as shown in Fig. 3(b); 3) A buffer composed of the two inverters connected in series, as shown in Fig. 3(c). The VCO can generate a pulse width depending on the variation of V1 and V2.

The VCO proposed in this paper is implemented with current mirrors type P and N. In this design, the VCO block, as implemented, produces just two operating frequencies. For example, the 50 MHz frequency occurs at the output when a voltage V1 is connected at the input of VCO. However, when voltage V2 is connected at the input of VCO, a frequency of 500 MHz is generated at the output. The frequency of the oscillator is given by Eq. (1):

$$f_{osc} = \frac{Id}{N \cdot C_{tot} \cdot Vdd} \tag{1}$$

Where Id is a current generates at the current mirror; N is a number of inverter blocks at VCO; total capacitance C_{tot} of the node is the sum of input capacitance C_{in} and output C_{out} of the inverter; Vdd is the 1.2 V supply voltage of the 130 nm CMOS process. From this equation, we can optimize the W (width) values of the transistors in the current mirror circuit.

2.3 Delay Block Using Shunt-Capacitance Topology

The pulse generator proposed in this project was implemented with an NEC block and delay circuits using the Shunt-Capacitance topology [9, 10], as shown in Fig. 4.

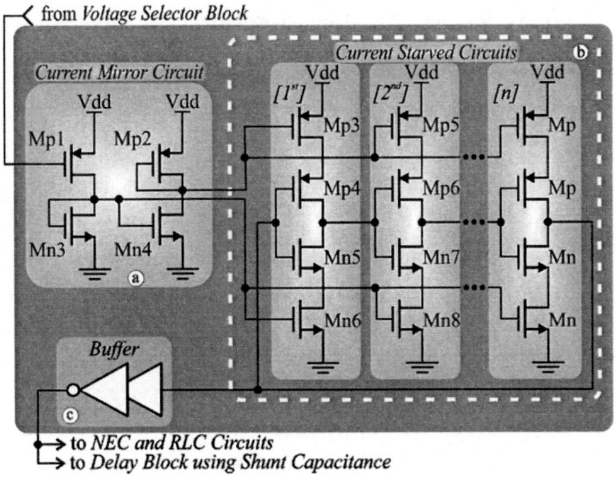

Fig. 3. Voltage control oscillator (VCO) Circuit: (a) Current mirror circuit; (b) Current starved inverter; (c) Buffers.

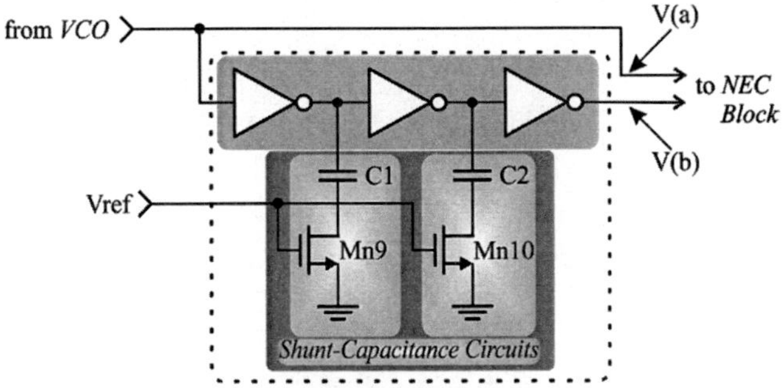

Fig. 4. Delay block using shunt-capacitance topology.

This topology is known among designers and is used to control the capacitor's charge time, which consequently controls the pulse period. The delay block was implemented with a chain of inverters using transistor nMOS (Mn9 and Mn10) connected in series with a capacitor. The voltage Vref at the gate transistor controls the flow of the current of transistors Mn, as shown in Fig. 4. Thus, we can vary the capacitance as a function of the voltage. This configuration is efficient and can control the capacitor's charge time, which consequently controls the pulse delay connected to the transistor port of the NEC block.

2.4 Negative Edge Combiner (NEC) and RLC Blocks

The NEC and RLC blocks generate a monocycle at the output of the transmitter. This circuit uses a phase difference waveform between the signals at inputs V (a) generated by the VCO and V (b) generated by the delay block (shunt-capacitance inverter) [10], as shown in Fig. 5. The NEC block contains two nMOS transistors (Mn11 and Mn12) connected in series. Figure 5 shows the transistor Mn11 connected at the output of the transmitter circuit. The source of Mn12 transistor is connected to ground. This circuit can control of current flux of the inductor and generates a monocycle pulse. The RLC block uses three passives components such as planar inductor square implement on metal layer 6 and 5; a poly resistor and a capacitor using Metal-Insulator Capacitor (MIM). The inductor, in this configuration, is connected between Vdd and the drain of the Mn11 transistor, as shown below. The output impedance is a capacitor connected in series with a 50 Ω resistor, and both are connected to the drain of the transistor circuit. The NEC was implemented to generate three states, as follows: the first state occurs before the clock edge, when logic signals are V(a) = 1 and V(b) = 0, consequently, transistor Mn11 is on, and Mn12 is off. Therefore, there not a voltage across the inductor, and NEC block is high impedance condition; the second state occurs at each edge clock transition, when the signal V (a) falls from "1" to "0" and V (b) rises from "0" to "1". In this condition, both transistors (Mn10 and Mn20) are on, and a monocycle is generated at the output of circuit with opposite polarity to the original applied voltage third state, is like to first state and occurs after edge clock transition, when logic signals V(a) = 0 and V(b) = 1. Thus, there is no voltage across the inductor, and NEC block is in high-impedance condition.

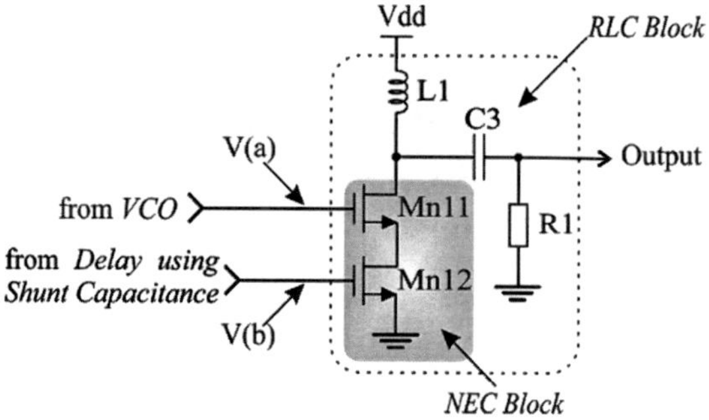

Fig. 5. Negative edge-combiner and RLC locks.

3 Simulation and Transmitter Layout Generation

This section will present the simulations results of the pulse generator IR-UWB transmitter implemented in the 130 nm CMOS process. We simulated the circuit after the post layout and parasitic extraction. Furthermore, all simulations are considered default temperature (27°C), 1.2 V power supply. The peak-to-peak output voltage is 132 mVpp on a 50 Ω load for a central frequency of 4.8 GHz, and a supply voltage of 1.2 V. The capacitor was implemented with 4pF using a Metal-Insulator-Metal (MIM) using metal layer 5 and 6 to avoid noise and other parasitic effects. In this design, the inductor was implemented with 2 nH and a series resistance of 11.6 Ω. At the output, it was implemented a 50 Ω resistor using poly resistances and a ring guard with substrate biased at ground potential to avoid noise under of substrate. The peak current in the inductor is

2 mA in just one state, as shown above. In the other states, the peak current is 0, and the transistors are turned off. The peak-to-peak voltage at the output of the monocycle was ~ 132 mVpp, as shown in Fig. 6(a). Figure 6(b, d) shows the waveforms of PRFs of 50 and 500 MHz. Thus, both voltages at the input of VCO generate a PRF of 20 ns and others of 2 ns, respectively. Figure 6(c) shows –48 dBm@4.8 GHz the PSD at the output of the circuit by monocycle.

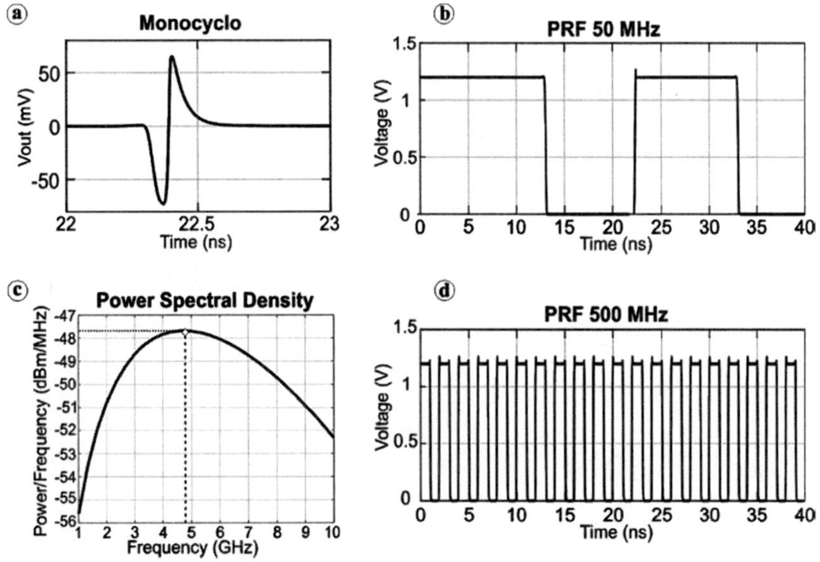

Fig. 6. Waveforms of the pulse generator circuit: (a) Monocycle; (b) PRF = 50MHz; (c) Simulated output spectra of the pulse generator circuit; (d) PRF = 500MHz.

The layout of the designed pulse generator circuit without PAD's is shown in Fig. 7. It has occupied an area of 214 µm × 234 µm. The average power consumption of the power supply is 3.85 mW. Table 1 provides a comparative summary of our works and two other references.

Table 1. Comparative summary of IR-UWB pulse generators and transmitters. Legend: EC- edge combiner; Filtered EC; NA - not available.

Results	This work	Ref [5]	Ref [6]
CMOS (nm)	130	130	130
Supply Voltage(V)	1.2	1.2	1.2
Pulse (GHz)	4.8	4	3.1–10.6
Method	Filtered EC	EC	Filtered EC
PRF (MHz)	50 and 500*	100	100
Power Supply (mW)	3.85	14.7	3.8
Area (mm^2)	0.05	0.11	0.54

*Dual PRFs; **Without PAD's.

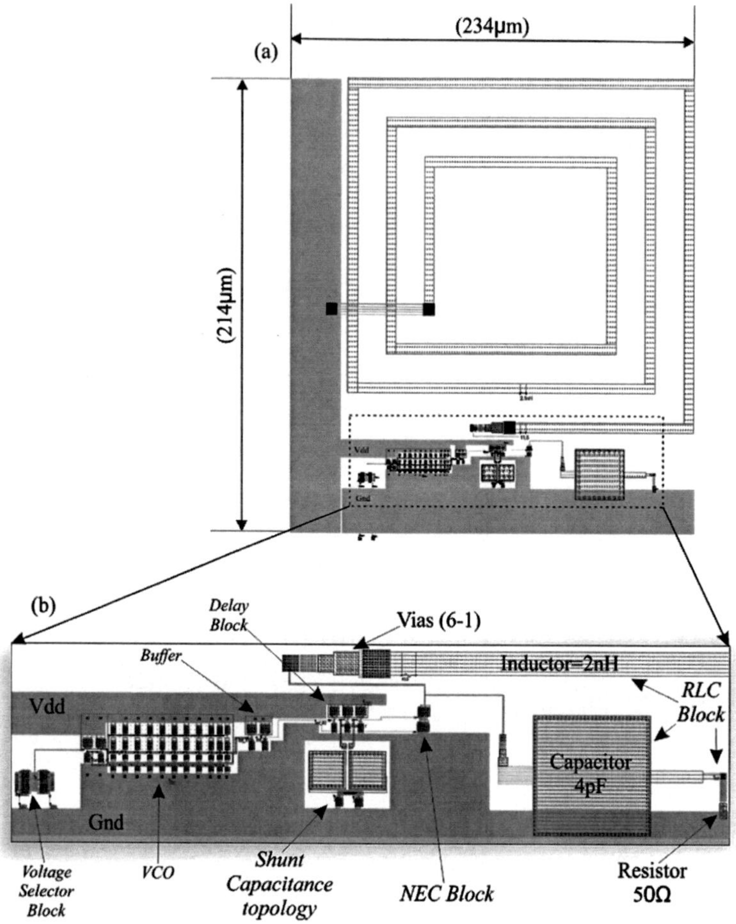

Fig. 7. Complete layout of proposed IR-UWB pulse generator without PADs: (a) Compact layout; (b) Final layout.

4 Conclusion

An efficient pulse generator implemented with a NEC block using 130 nm CMOS technology was presented. This circuit generates a monocycle with a frequency of 4.8 GHz and amplitude of 132 mVpp with PRFmin of 2 ns (500 MHz) and PRFmax of 20 ns (50 MHz). The design without the PADs and the inductor occupies an area of only 214 um × 234 um. The transmitter has a small and mainly small area and low complexity. Thus, this circuit can be used in other domains that employ wireless communication. Finally, the pulse generator using dual PRF can significantly lead to communication systems working with variable transmission rates.

References

1. Win MZ, Dardari D, Molish AF, Wiesbeck W, Zhang J (2009) History and applications of UWB. In: Proceedings of the IEEE, vol 97, no 2, February 2009
2. Oncu A, Badalawa BBMW, Wang T, Fujishima M (2007) 22–29GHz Ultra-WideBand CMOS pulse generator for collision avoidance short range vehicular radar sensors. In: Proceedings of the Asia and South Pacific Design Automation Conference, pp 94–95
3. Mahbub SKI (2015) A low power impulse-radio ultra-wide band (IR-UWB). Transmitter for biomedical applications. CMOC Symposium
4. Yuan F (2016) Structural health monitoring (SHM) in aerospace structures. Phased array techniques for damage detection in aerospace structures. Woodhead Publishing, pp 285–301
5. Vauche R, Muhr E, Fourquin O, Bourdel S, Gaubert J et al (2017) A 100 MHz PRF IR-UWB CMOS transceiver with pulse shaping capabilities and peak voltage detector. IEEE Trans Circ Syst I: Regul Pap 64(6):1612–1625
6. Bourdel S, Bachelet Y, Gaubert J, Vauche R, Fourquin O, Dehaese N, Barthelemy H (2010) A 9-pJ/Pulse 1.42-Vpp OOK CMOS UWB pulse generator for the 3.1–10.6-GHz FCC band. IEEE Trans Microw Theory Tech 58(1):65–73
7. Ling RWC, Gupta A, Vashistha A, Sharma M, Law CL (2018) High precision UWB-IR indoor positioning system for IoT applications. In: IEEE 4th World Forum on Internet of Things (WF-IoT)
8. Norimatsu T, Fujiwara R, Kokubo M, Miyazaki M, Maeki A, Ogata Y, Kobayashi S, Koshizuka N, Sakamura K (2007) A UWB-IR transmitter with digitally controlled pulse generation. IEEE J Solid-State Circuits 42:1300–1309
9. Jovanović GS, Stojčev MK (2005) Linear current starved delay element. In: XL International Scientific Conference on Information Communication and Energy Systems and Technologies (ICEST), pp 59–62. July
10. Maymandi-Nejad M, Sachdev M (2003) A digitally programmable delay element: design and analysis. In: Transactions on Very Large-Scale Integration (VLSI) Systems. IEEE, pp 871–878

Effect of the Use of Iron Base Nanostructures Supported on Rice Husk Ash Residue Applied in the Making of Cell Concrete Blocks

M. T. Nunes(✉), F. S. Rodrigues, and J. Boita

Laboratório de Síntese E Caracterização de Nanomateriais (LSCNano), Universidade Federal de Santa Maria, Cachoeira do Sul, Brasil
jocenir.boita@ufsm.br

Abstract. The use of agricultural waste has become a necessity due to the high environmental cost, as in the case of rice husk ash (RHA), produced by its indiscriminate burning, as well as, we have the need to replace some components in concrete, such as cement and its aggregates due to its scarcity in nature. Thus, the nanoparticles (NPs), due to their characteristics, such as the increase of the surface area and quantum effects, can bring interesting properties to the production of cell concrete blocks (CCB), besides allowing the use of residues in its synthesis. The study has the object of using RHA as a support material for NPs based on iron oxide, producing CCB's with the 2:1 trait and characterizing the stages of making the blocks. In addition to making a comparative when elaborating a block made with cement and ash of rice husk, in the same proportion, aiming to analyze the function that NPs play in the block with respect to mechanical resistance. Observing at the end of the study that the CCB obtained from NPs supported in RHA has better mechanical strength, since the CCB obtained without the use of NPs showed a 64.4% lower resistance, proving the essential function that the NPs based iron oxide supports In RHA play in the manufacture and strength of the blocks.

Keywords: Rice husk ash · Nanoparticles · Cell concrete blocks

1 Introduction

The large volume of rice husk ash (RHA) produced during the burning process in the rice processing industry for thermal energy generation becomes worrying [1, 2]. Therefore, when discarded improperly, it promotes pollution in the soil since it presents a significant amount of residual carbon and alkalinizing oxides [3].

RHA's silica content, almost always above 90%, when burned with temperature control, allows obtaining a microstructure of the amorphous type [4, 5]. However, it is estimated that most RHA production has secondary applications due to burning without temperature control, which gives it some degree of crystallinity [6]. An alternative for the application of RHA is in concrete, of which in 1924, the first study with the use of rice husk ash in concrete was patented. After that, other researchers began to use RHA derived from uncontrolled burning [7].

Y. Iano et al. (Eds.): BTSym 2020, SIST 233, pp. 894–904, 2021.
https://doi.org/10.1007/978-3-030-75680-2_99

In this same scenario, we need to replace some concrete components, such as sand and its aggregates, due to its high cost and scarcity in nature. This substitution can be accomplished by using waste, where the use of these tailings can improve the properties of concrete, as in the case of cell concrete blocks (CCB), which has its resistance impaired due to its low density.

Foamed cellular concrete is obtained by the introduction of air bubbles into the mortar, from which the foaming agent can be any chemical capable of producing stable air bubbles inside the mortar and remaining intact, according to ABNT NBR 12645:1992 [8].

The cellular concrete presents great advantages in its application because its structure of closed pores ensures an excellent degree of thermal and acoustic insulation, good resistance to water absorption, and good resistance to fire, characteristics conferred by its completely inorganic composition. However, all these attributes are obtained in detriment of the material's mechanical strength, which decreases with increasing porosity of the same [9, 10]. In this sense, the use of nanomaterials becomes a possibility in this branch of engineering.

The nanomaterials have very peculiar characteristics differing significantly from other materials, mainly by the increase of the surface area and also by the quantum effects [11, 12]. The nanomaterials are all those composed or agglomerates of particles having a size between 1 and 100 nm [13].

The purpose of the study is to use RHA as a support material for iron oxide nanoparticles. Therefore, using the nanoparticles (NPs) supported the construction of foamed cellular concrete blocks, where the residue used as a support material comes in a complete replacement of the conventional sand. In addition to performing a comparative in the elaboration of a block made with cement and ash of rice husk ash, in the same proportion, aiming to analyze the function that the nanoparticles play in the block with respect to mechanical resistance.

2 Material and Methods

The production processes of NPs and CCB were conducted at the Laboratory of Synthesis and Characterization of Nanomaterials (LSCNano), at the Federal University of Santa Maria (UFSM), Campus Cachoeira do Sul. In order to obtain the cellular concrete blocks from iron oxide nanoparticles supported on rice husk ash, we must divide it into three main steps: the preparation of the nanoparticles, the formulation of the foam material, and, finally, the construction of the blocks of cellular concrete.

The pre-treated rice husk ash was homogenized, dried in an oven at 80°C for 15 min, and only the material passed through the 200 Mesh sieve was used as support material for the NPs.

The iron oxide NPs supported with RHA were obtained by the polyol method in an aqueous medium, the iron oxide being synthesized via a hydrothermal route, of which its components are described in Table 1.

For the use of the residue as support material for the NPs, 2 g of rice husk ash was used, which was composed mainly of SiO_2. The synthesis remained unstable for 2 h 30 min, where after this time, it acquired its stability. It was then placed in a muffle furnace at

Table 1. Synthesis components of iron oxide nanoparticles.

Agent	Function	Quatity	Dilution in H_2O
$Fe_2(SO_4)_3.5H_2O$	Metal Precursor	1,3 Mol	4 ml
$C_6H_5O_7Na_32H_2O$	Stabilizing	0,5 Mol	4 ml
$C_6H_8O_6$	Reducer	1,1 Mol	8 ml
C_6H_9 $(NO)_n$	Encapsulator	1,8 Mol	4 ml
Source: Patent BR 102016006952–1 [14]			

350°C for 30 min, resulting in a solid material. Soon after, the sample was macerated with the aid of a mortar until a powder with uniform granulometry was obtained.

The air incorporated inside the CCB occurs by the use of a foamy substance, type detergent foam. The preparation of a new synthesis of NPs, according to the indications already mentioned in Table 1, producing 20 ml, without the support material, was necessary for the composition of the foamed material that will form the CCB. Adding 1.5 ml of concentrated detergent (sodium lauryl ether sulfate), which mechanically agitates the system for 5 min, generates a dark and stable coloring foam.

The purpose of this study was to analyze the resistance of CCB's made in the ratio of 2:1 (Table 2), one consisting of cement and nanoparticles, the other of rice husk ash and cement, with the purpose of analyzing the function that the nanoparticles play in the block with respect to mechanical strength.

Table 2. Relation of the quantity of CCB components in the proportion 2:1.

Type of CCB (2:1)	Mass of cement	Mass of "aggregate"	Foam	Water
Cement:NPs	133 g	67 g of NPs supported in RHA	20 ml of NPs + 1.5 ml of concentrated detergent	100 ml
Cement:RHA	133 g	67 g of RHA	20 ml of water + 1.5 ml of concentrated detergent	100 ml

After obtaining the NPs supported in RHA, it consists of realizing the mixture of the solid components. In this study, Portland cement CP IV-32 was used.

After that, it is necessary to mix the foamed solid components together with the addition of water, little by little, yielding a flowable cementitious paste.

Therefore, the cementitious paste obtained for the CCB's should be placed in non-corrosive metal molds with a threadable base with their inner sidewall covered by polymeric material. In this way, all molds have an internal diameter of 50 ± 0.2 mm and a height of 100 ± 0.2 mm.

The drying process of the blocks was conducted under normal conditions of temperature and pressure during the 28 d period, where, at 14 days after the date of manufacture, the blocks were removed from their respective molds, where they followed the drying process until the 28th day of age.

The X-ray diffraction measurements were performed at the Analytical Center of UFSM, in a Mini Flex of the Rigaku brand, in the range of 5°–100°, with Cu Kα anode (17.5 mA, 40 kV and $\lambda = 1.54$ Å). The diffraction peaks were indexed using the Crystallographica Search-Match software and the crystallographic database International Centre for Diffraction Data (ICDD), which consists of a database with a large number of crystalline materials already cataloged [15].

The XPS was measured at the National Nanotechnology Laboratory (LNNano) at the National Center for Energy and Materials Research (CNPEM). Measurements were performed in conventional mode on a Thermo Scientific model K-Alpha. For data treatment purposes, the energy was corrected based on the peak of C 1s (284.5 eV) and measured with energy of 1486.6 eV through Al kα anode. Data processing was performed using CASA XPS.

The density of the dry mass of CCB's (kg/m^3), that is, the concentration of matter by volume, was obtained from the ratio of the mass of CCB (kg) by its volume (m^3).

The rupture of the specimens, to determine the compressive strength, was performed with the aid of the Rílen accessory in the Alfred J. Amsler and Co. Hydraulic Press equipment. Before carrying out the test, the specimen was capped with sulfur.

3 Results and Discussion

Figure 1 shows, in the upper part, the measurements obtained for the rice husk ash and, in the lower part, the comparison, according to the crystallographic database. Thus, it is possible to identify the silica (SiO_2) in the cristobalite form (PDF 76–937). This occurs when there is excess amorphous silica in a thermal reaction at approximately 1260°C. It crystallizes by acquiring the form of cristobalite. However, we have this naturally through the residue of rice husk ash [16].

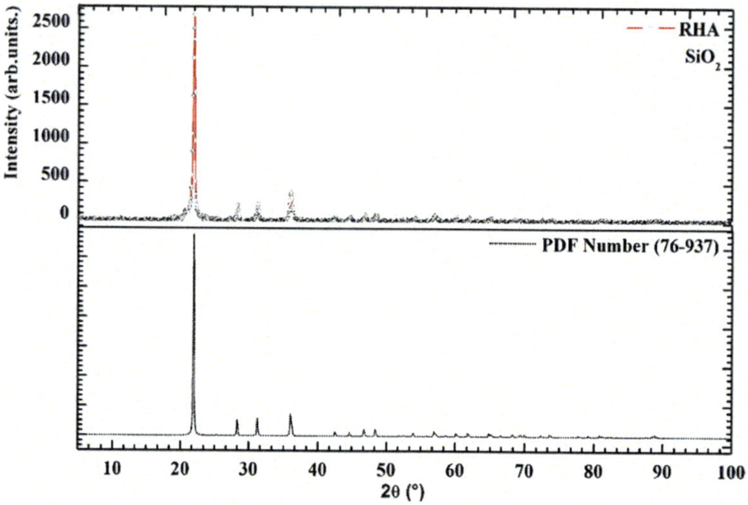

Fig. 1. X-ray diffraction analysis of the rice husk ash residue compared to the Cristobalite pattern.

The XRD measurement performed for the NPs without support material shows only a bottom (Background) highlighted between 15–35°, without a definition of the diffraction peaks, similar to an amorphous material. Research shows that the widening or blurring of the diffraction peaks may be directly related to the structure's size, revealing signs of nanoscale [17, 18]. However, for this reason, X-ray diffraction analysis becomes complex, with evidence of hematite NPs as compared to the standard (PDF 65–390), shown in the image below (Fig. 2).

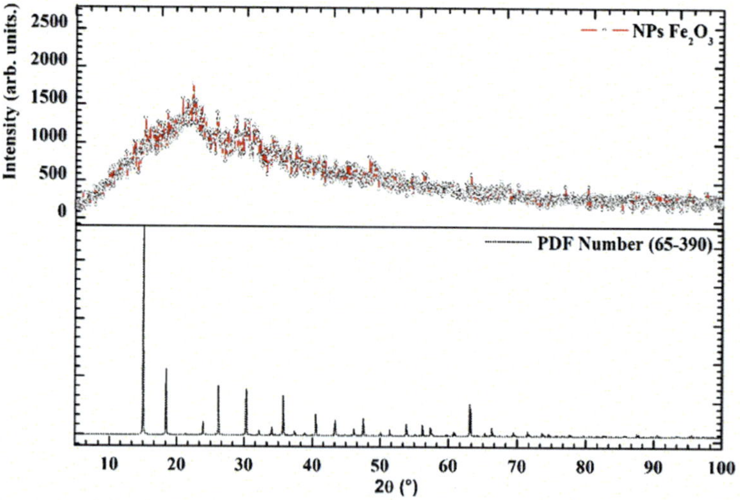

Fig. 2. X-ray diffraction analysis of Fe_2O_3 compared to the Hematite standard.

Figure 3 shows the XRD result for the iron NPs supported on rice husk ash, where it shows evidence that the RHA residue acted as support material for the synthesis of NPs, because in the lower part of the said figure, we have the combination of hematite data (PDF 65–390) represented in blue. In black, the cristobalite (PDF 76–937), providing an assertion of obtaining NPs of Fe_2O_3/RHA, which is represented in the upper part of the figure.

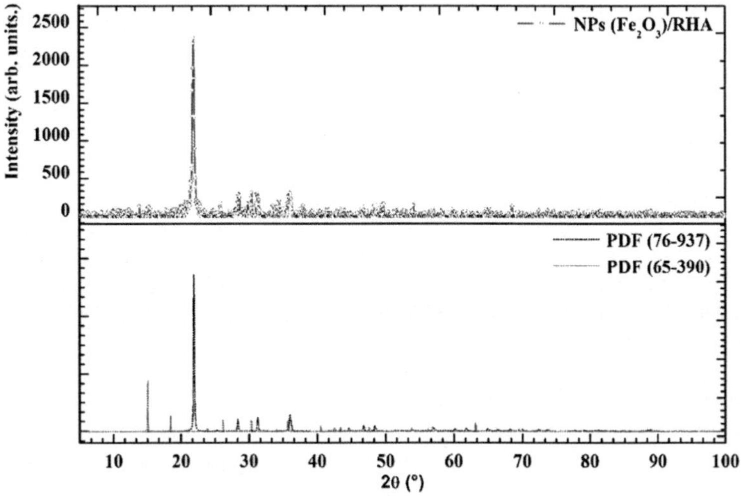

Fig. 3. X-ray diffraction analysis of Fe_2O_3 NPs supported with RHA compared to the combination of Hematite and Cristobalite.

The CCB made with NPs supported in RHA also had its crystalline arrangement observed through the X-ray diffraction. In this way, the constitution of different components was verified since, by including the cement to the mass that will originate the block, we have the addition of various chemical components, such as, for example, calcium sulfate, calcium carbonate, among others. Thus, the chemical and crystalline composition of the block changes when the iron NPs supported in rice hull ash are maintained, but the characteristics of cristobalite peaks predominate, as can be observed in Fig. 4.

Also, in Fig. 4, in the part corresponding to the reading of the CCB sample, the presence of iron oxide is observed, maintaining its undefined curves and peaks, characteristic of nanoparticles. However, when comparing with the standards established by the ICDD database, we observed the presence of the magnetite (Fe_3O_4) through the PDF 88–315 register instead of hematite (Fe_2O_3), as observed in the supported NPs.

In some studies with the use of magnetite in concrete blocks, it has been found that concrete containing only magnetite ore concentrate, cement, and water has considerably better radiation protection properties than ordinary concrete [19].

Analysis of XPS results was conducted in a manner similar to XRD, divided into three steps that will be discussed in the sequence.

The chemical characterization of the iron NPs is shown in Fig. 5, through the XPS analysis. Where (a) is exposed to the wide-energy scanning for iron nanoparticles without

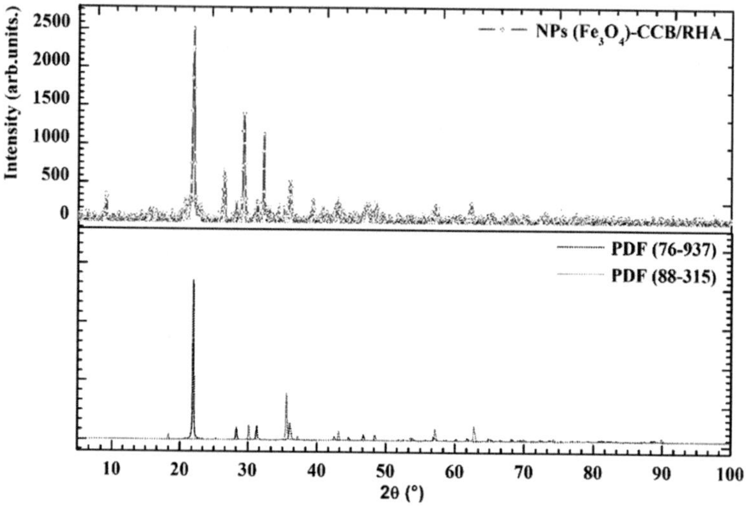

Fig. 4. X-ray diffraction analysis of the cellular concrete block obtained from Fe_2O_3 NPs supported with RHA compared to the combination of the Magnetite and Cristobalite standards.

support material, in which the pattern is shown in the lower part of the figure, allowing the comparison with that established for the sample. In (b), the scanning of the local region of Fe 2p 3/2 is presented, as well as, in the lower part, this pattern is obtained.

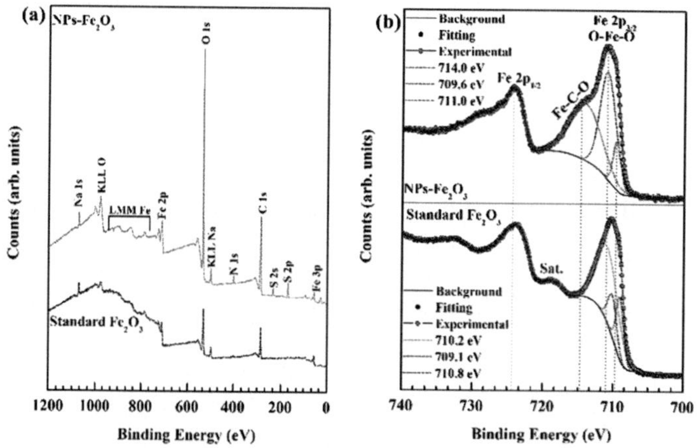

Fig. 5. (a) Spectrum with wide sweep in energy for the iron nanoparticles without support material with all the identified regions, (b) Scanning of the local region of Fe 2p 3/2.

When examining the XPS data for iron nanoparticles (Fig. 5), we noticed the great similarity of the measurement performed with the Fe_2O_3 standard in the sweep of wide energy, maintaining the main peaks. Therefore, in (b), it is possible to identify O-Fe-O and Fe-CO bonds in the satellite peak region of the standard sample since the standard measure is obtained in a controlled manner since the NPs underwent a treatment process, but the binding energy remains close to the standard obtained for the Fe 2p 3/2 region of the hematite.

The XPS data for iron nanoparticles supported on rice husk ash Fig. 6.

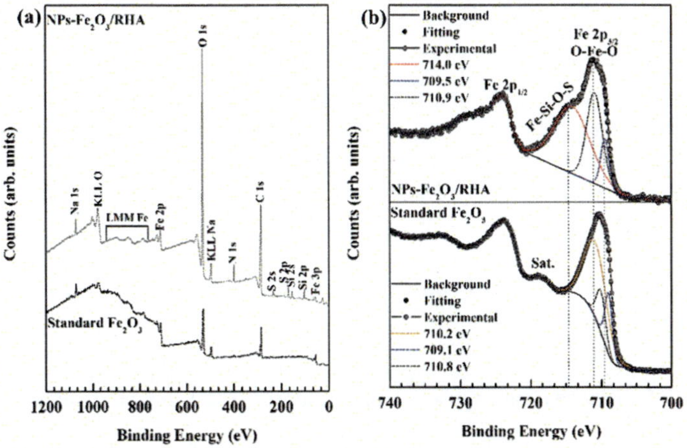

Fig. 6. (a) Spectrum with wide sweep in energy for the iron nanoparticles supported in RHA with all identified regions, (b) Scanning of the local region of Fe 2p 3/2.

It can be seen in Fig. 6a that iron NPs with rice husk ash maintain the hematite pattern with similar tops. In Fig. 6b, the Fe-Si-O-S bond is detected by identifying the Fe 2p 3/2 region. The carbon present in the binding in the unsupported NPs is altered by silicon in the NPs supported on rice husk ash, thus showing that the nanoparticle synthesis adhered to the RHA as carrier material.

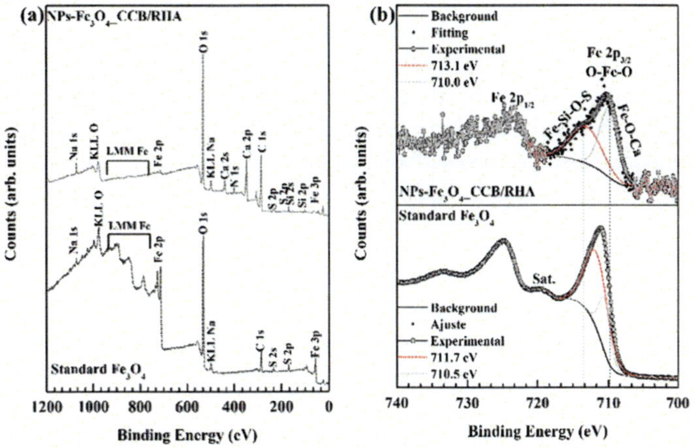

Fig. 7. (a) Spectrum with wide scanning energy for the CCb obtained from NPs supported in RHA with all identified regions, (b) Scanning of the local region of Fe 2p 3/2.

The XPS data for the cell concrete blocks, in Fig. 7a, shows the sweep in wide energy, allowing to identify the modification in the oxidation state of the iron. This is now presented as magnetite (Fe_3O_4), as already observed during the investigation of the crystalline structure of the cellular concrete block by XRD measurements.

In Fig. 7b, when investigating the region of Fe 2p 3/2, the correspondence to the magnetite pattern is observed, observing in a generalized way its peaks and binding energies. However, when analyzing the sample of the block in said region, we can distinguish Fe-Si-O-S, O-Fe-O, and Fe-Ca-O bonds. Of which, probably the calcium that appears in this last connection comes from the cement since this one owns a great quantity of calcium carbonate and sulfate. We also have a decrease in the count, consequence related to the process of confection of the cellular concrete block.

Therefore, XPS results correlate with those of XRD, confirming the veracity in the chemical and crystallographic analysis of the confection stages of the cellular concrete blocks analyzed.

Table 3 presents the results of density and compressive strength for CCB's obtained with and without the use of NPs.

Table 3. Density and compressive strength results for the cellular concrete blocks.

Trace	Density (kg/m^3)	Compressive strength (MPa)
Cement:NPs	916,23	2,11
Cement:CCA	478,53	0,75

The block without NPs presented a density of 478.53 kg/m^3, becoming perceptible to low density by the large number of pores formed, according to Fig. 7b. However, this showed a compressive strength of only 0.75 MPa, revealing a resistance of approximately 64.4% lower than the block made with nanoparticles of the same trace. This shows the importance of the use of the NPs supported in rice husk ash for the production of cellular concrete blocks, directly influencing the resistance, because, for this same trace 2:1 (cement: NPs), the value of 2.11 MPa.

In this way, the results show that the NPs have a crucial function in the characteristics related to the resistance and, especially, the sand substitution in the blocks, which brings a great economic and also environmental advantage.

4 Final Considerations

It is concluded from the results obtained that it was possible to use the rice husk ash residue as support material for iron oxide nanoparticles, as well as through the results, the NPs have evidence of nanometric material.

Through the XPS and XRD data, it was possible to observe the chemical and crystallographic analyzes of the stages of making the block, highlighting along with the formation the Fe-Si-OS bonds resulting from the incorporation of RHA in the synthesis and, at the end, in the CCB the Fe-Ca-O bond, with calcium originating from the cement used to produce the block. It was also possible to identify, during synthesis, the formation of hematite that had its shape changed to magnetite in the cellular concrete block produced, showing the possibility of using this material in civil construction due to its high resistance.

Regarding the compressive strength, it was observed that the block NPs: cement presented the best performance, presenting a resistance 64.4% greater than the CCB made without NPs. This proves the essential function of nanoparticles based on iron oxide and rice husk ash for cellular concrete blocks' resistance.

Therefore, the block made with NPs, besides having excellent resistance, has as a differential the use of rice husk ash residue. In this way, the production of CCB's with NPs supported in RHA provides a great economic advantage and also to the environment, enabling a block with innovative technology, sustainable and ecologically correct.

References

1. Ferraro RM, Nanni A, Vempati RK, Matta F (2010) Carbon neutral off-white rice husk ash as a partial white cement replacement. J Mater Civ Eng 22:1078–1083
2. Zhang ZY, Gonzalez AM, Davies EGR, Liu Y (2012) Agricultural wastes. Water Environ Res 84:1386–1406
3. Da Silva Teixeira JB, Islabão GO, Vahl LC, Guidotti R, Kath AH, De Sousa Oliveira R (2019) High rice husk ash doses applied to the soil can lead to phosphorus water contamination. Environ Sci Pollut Res 26:6399–6408
4. Dey KP, Ghosh S, Naskar MK (2013) Organic template-free synthesis of ZSM-5 zeolite particles using rice husk ash as silica source. Ceram Int 39:2153–2157
5. Sarangi M, Nayak P, Tiwari TN (2011) Effect of temperature on nano-crystalline silica and carbon composites obtained from rice-husk ash. Compos B: Eng 42:1994–1998
6. Zerbino R, Giaccio G, Isaia GC (2011) Concrete incorporating rice-husk ash without processing. Constr Build Mater 25:371–378
7. Liu X, Chen X, Yang L, Chen H, Tian Y, Wang Z (2016) A review on recent advances in the comprehensive application of rice husk ash. Res Chem Intermed 42:893–913
8. Brazilian Association of Technical Standards - ABNT NBR 12645 (1992) Execution of foamed cellular concrete walls molded on site – Procedure
9. Ramamurthy K, Nambiar EKK, Ranjani GNIS (2009) A classification of studies on properties of foam concrete. Cem Concr Res 37:388–396
10. Kolias S, Georgiou C (2005) The effect of paste volume and of water content on the strength and water absorption of concrete. Cem Concr Res 27:211–216
11. Kang JG, Sohn Y (2012) Interfacial nature of Ag nanoparticles supported on TiO_2 photocatalysts. J Mater Sci 47:824–832
12. Zhang QL, Du LC, Weng YX, Wang L, Chen HY, Li JQ (2004) Particle-size-dependent distribution of carboxylate adsorption sites on TiO_2 nanoparticle surfaces: insights into the surface modification of nanostructured TiO_2 electrodes. J Phys Chem B 108:15077–15083
13. Valsami-Jones E, Lynch I (2018) How safe are nanomaterials? Sci 350(6259):388–389
14. Boita J, Rodrigues FS, Nunes MT Processo de obtenção de nanocompósito - BR 102016006952–1
15. Gates-Rector S, Blanton T (2019) The powder diffraction file: a quality materials characterization database. Powder Diffr 34(4):352–360
16. Zhang S, Tie S, Zhang F (2018) Cristobalite formation from the thermal treatment of amorphous silica fume recovered from the metallurgical silicon industry. Micro Nano Lett 13:1465–1468

17. Boita J, Nicolao L, Alves MCM, Morais J (2014) Observing Pt nanoparticle formation at the atomic level during polyol synthesis. Phys Chem Chem Phys 16:17640–17647
18. Boita J, Nicolao L, Alves MCM, Morais J (2017) Controlled growth of metallic copper nanoparticles. New J Chem 41:14478–14485
19. Creutz E, Downes K (1949) Magnetite concrete for radiation shielding. J Appl Phys 20:1236–1240

Fuzzy PID Control System Analysis for a Wind Turbine Maximum Power Point Tracking Using FAST and Matlab Simulink

Eduardo Muñoz[1](✉), Edy Ayala[1], Nicolás Pozo[1], and Silvio Simani[2]

[1] Universidad Politécnica Salesiana, Calle Vieja 12-30 and Elia Liut Ave., Cuenca, Ecuador
{emunozp2,apozoa}@est.ups.edu.ec, eayala@ups.edu.ec
[2] Ferrara University, Via Savonarola, 9, 44121 Ferrara, FE, Italy
silvio.simani@unife.it

Abstract. The present research proposes a control method applied to a Wind Power Generation Systems (WEGS) for Maximum Power Point Tracking (MPPT) technique based on a Mamdani Fuzzy observer and complemented by a Proportional and Integral (PI) controller using the Direct Speed Control (DSC). This approach allows commanding the rotor side reference current i_{qr} through the variations of the electromagnetic torque and active electrical power in a Doubly Fed Induction Generator (DFIG) model. Consequently, the speed of the generator is controlled to ensure obtaining a rapid response of the maximum Power Coefficient (C_p). This strategy's construction starts with direct measurements of the electrical and mechanical variables using computational tools such as FAST and Matlab-Simulink for simulations of a 1.5MW wind turbine model. This DSC strategy presents a rapid performance regarding the tracking of the maximum C_p considering the wind turbine dynamics. This approach contrasts with a traditional PI controller improving the extracted power to analyze the proposed procedure.

Keywords: DFIG · DSC · FAST · Fuzzy · PI · Power coefficient · MPPT · Wind turbine

1 Introduction

Energy production efficiency has been a topic of interest for developing technologies in renewable energies. For instance, wind energy has led to research focused on the subsystems immersed in wind turbines to improve the global system's response. This fact is evidenced in the increase of the effectiveness in generating electrical energy; hence, minimizing faults and power losses assisted by control strategies with better performance. The issue of extracting the maximum available energy from the incident wind is a crucial factor to guarantee maximum efficiency, and it can be evidenced in the estimation of the C_p based on mechanical and electric losses.

Significant research has been carried out for MPPT control through the monitoring of the maximum extracted power. For instance, a nonlinear control strategy using the adaptive backstepping method is described in [1]. In [2], a perturb and observe scheme is

Y. Iano et al. (Eds.): BTSym 2020, SIST 233, pp. 905–917, 2021.
https://doi.org/10.1007/978-3-030-75680-2_100

developed for MPPT by employing a state machine applied to low-inertia and low-output-power wind turbines. In other research works, the use of direct torque control (DTC) is also proposed to evaluate the flow and electromagnetic torque as explained in [3–5]; and in [6], a DTC with an Artificial Neural Network (ANN) control is recommended. Other types of studies are raised in [7], where the authors refer to research approaches in three control techniques such as power signal feedback (PSF), tip-speed ratio (TSR), and hill-climb searching (HCS), widely used and applied to wind turbines with squirrel cage induction generators (SCIG), DFIG, and generators of type permanent magnet synchronous (PMSG).

Fuzzy controllers applied to MPPT have demonstrated significant results in different works. A Fuzzy Takagi-Sugeno-Kang (TSK) controller for peak tracking power is exposed in [8]; and in [9], vector control in a DFIG based on the use of a Fuzzy controller of electrical variables is explained. In other researches, the Fuzzy controller has been combined with various control techniques as evidenced in [10], which presents a Fuzzy slope HCS control method to vary the search step in the HCS and improve the system response. In [11], a Variable Universe Fuzzy Controller (VUFC) is described for the system powers correct control.

The combination of Fuzzy and Proportional, Integral, and Derivative (PID) controllers is quite common. To be specific, a Fuzzy-PI controller focused on torque control is proposed in [12]; and in [13], maximizing power transfer with a similar approach is described. Other control techniques focus the attention on various parameters such as those analyzed in [14].

In this work, within the context of MPPT, the use of a Fuzzy controller is proposed under the DSC method complemented by a PI to regulate the reference current i_{qr}. This aims to command the triggers of the electronic switches of the back-to-back converter operating at the rotor side. Additionally, the mechanical and electrical variables of the wind turbine have been modeled to act on the electromagnetic torque, power, and reference current i_{qr}. Thus, the generator speed is regulated according to the dynamics of the system in response to the wind entry at each instant of time. Consequently, with the Fuzzy controller, a more appropriate C_p tracking – with greater accuracy and speed of stabilization compared to a classic PI controller – is achieved. The FAST simulation software from NREL [15] and the Matlab/Simulink [16] were used to model the wind turbine, thus verifying the effectiveness of the developed proposal.

In this sense, the proposed work is supported by a description, in section two, of the aerodynamic model of a wind turbine with a 1.5MW DFIG, including the two-mass mechanical drivetrain under physical representation of the transmission stages. In section three, the mathematical description of the DFIG for the simulation model is presented. In section four, the integration of the FUZZY and PID controllers into the DSC observer-based method is extended. Section five presents the simulation results and the comparison between the DSC MPPT control and the PI control. Finally, section six compiles the conclusions drawn from the present work.

2 Wind Turbine Model

Traditional wind turbines have a generic structure, mainly made up of blades, that capture the wind and allow the system to function, connected to the low-speed shaft (LSS), and this is followed by the gearbox for the transmission of movement towards the high-speed shaft (HSS) until the machine which is usually linked to the electrical grid [17].

2.1 Aerodynamic Model

The available energy that is finally extracted by the wind turbine, expressed in (1), is directly affected by the wind speed and factors related to its construction and characteristics. Inertia is significant for the system dynamic modeling and all the parameters that estimate the output power, as explained in [8].

$$P = \frac{1}{2}\rho C_p(\lambda, \beta)A{V_w}^3 \tag{1}$$

$$\lambda = \frac{w_r R}{V_w} \tag{2}$$

Where the mechanical power of the rotor side is P (watts), ρ is the air density (kg/m^3), $A = \pi \cdot \mathrm{R}^2$ is the swept area (m^2), R is the geometric length until the blade tip (m), V_w is the wind speed (m/s), C_p is the rotor power coefficient, w_r is the turbine angular velocity (rad/s), β is the pitch angle of the blades (deg), and λ is the TSR, which denotes a relationship that considers two speeds: at the end of the blades and the wind input [8].

The C_p indicates a reliable measure of the efficiency in the extraction of power, reflected as the fraction or percentage of the incident wind kinetic energy on the blades, that is then converted by the turbine into useful electrical energy; and it depends on both the pitch angle β, according to the blade model, and the TSR [8, 18].

During MPPT, as the desired situation, the TSR reaches its optimum value, λ_{opt}. In the face of wind variations, the rotor speed must be controlled in some way to keep track of the maximum operating point given by the relationship C_p and TSR [19]. The wind turbine torque is calculated from (3), which is the base for the DSC technique.

$$T_r = \frac{P}{w_r} \tag{3}$$

The wind turbine operates, conforming to the value of incident wind speed, within three zones: the zone of non-operation, the MPPT control zone, and the pitch control zone [1]. This paper is performed on zone 2 or MPPT to maximize the power extracted, within the operating limits, at different wind speeds, in which w_r is controlled on a torque basis. The pitch orientation is always fixed at its minimum value.

2.2 Mechanical Description

The mechanical representation of a generic wind turbine to achieve a more precise response of its dynamic behavior against wind fluctuations can be approximated as a two-mass powertrain model [20] as it is detailed in Fig. 1. By applying Newton's second law, and assuming the inertias in the gearbox, as well as the global damping and rigidity coefficients of the rotor and the generator to be negligible from the movement descriptions, the dynamic equations of the mechanical system, are extracted [21]:

$$N = \frac{\omega_g}{\omega_r} = \frac{\dot{\omega}_g}{\dot{\omega}_r} = \frac{T_{ls}}{T_{hs}} \tag{4}$$

$$\begin{cases} J_r\dot{\omega}_r = T_r - T_{ls} \\ J_g\dot{\omega}_g = -T_{em} + T_{hs} \end{cases} \tag{5}$$

Where ω_r and ω_g are the angular speeds (rad/s) and, $\dot{\omega}_r$ and $\dot{\omega}_g$ are the angular accelerations (rad/s^2) of the rotor and the generator, respectively; J_r and J_g are the rotor and generator moments of inertia (kg· m^2); T_r, T_{em}, T_{ls} and T_{hs} are the turbine aerodynamic torque, the electromagnetic torque of the generator, and the torques defined at the LSS and HSS (N·m), respectively, and N is the gearbox ratio that joins the LSS and the HSS.

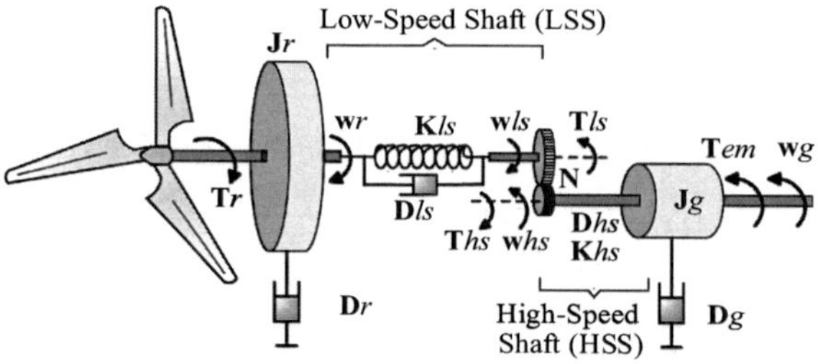

Fig. 1. Two mass mechanical structure representation of the wind turbine.

Relating Eq. (5) to the gearbox ratio N in Eq. (4) and dividing by the angular acceleration of the generator, a simplified expression in (6) is obtained.

$$J_r + N^2 J_g = \frac{T_r - NT_{em}}{\dot{\omega}_r} \tag{6}$$

3 DFIG Model Based on WECS

The wind turbine overall configuration with a DFIG, Fig. 2, involves a turbine, shaft systems, and gearbox attached to the DFIG. This generator uses a three-phase AC-DC-AC converter that results in two sides linked through a DC bus: the grid and rotor side converters, GSC and RSC, respectively. The rotor is connected to the RSC and the stator to the load/grid. The GSC binds to the grid through inductor coupling [1, 22].

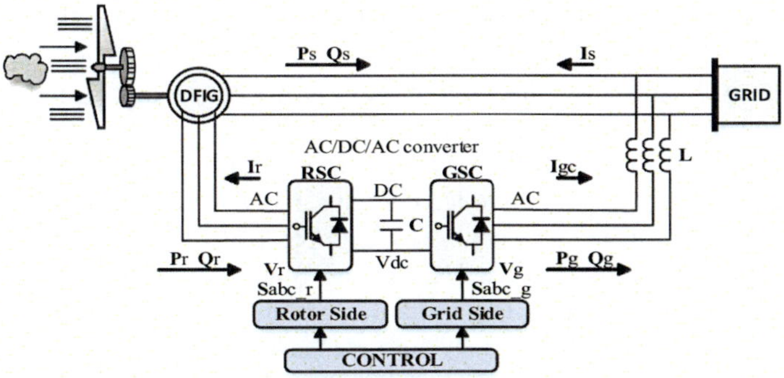

Fig. 2. Electrical DFIG Model as part of a Wind Turbine.

3.1 DFIG Mathematical Representation

The electrical operation of the DFIG can be analyzed using the Park's transform, making use of a reference frame d-q, separating the rotor and stator sides. The mathematical representation is defined in (7) by the equations of voltage, and the fluxes, φ, for the electromagnetic decoupling for the $\boldsymbol{d}$ and $\boldsymbol{q}$ axes [4, 5, 23, 24]:

$$\begin{cases} V_{ds} = R_s i_{ds} + \dfrac{d\varphi_{ds}}{dt} - \omega_s \varphi_{qs} \\ V_{qs} = R_s i_{qs} + \dfrac{d\varphi_{qs}}{dt} + \omega_s \varphi_{ds} \\ V_{dr} = R_r i_{dr} + \dfrac{d\varphi_{dr}}{dt} - \omega_r \varphi_{qr} \\ V_{qr} = R_r i_{qr} + \dfrac{d\varphi_{qr}}{dt} + \omega_r \varphi_{dr} \end{cases} \text{where} \begin{cases} \varphi_{ds} = L_s i_{ds} + M \cdot i_{dr} \\ \varphi_{qs} = L_s i_{qs} + M \cdot i_{qr} \\ \varphi_{dr} = L_r i_{dr} + M \cdot i_{ds} \\ \varphi_{qr} = L_r i_{qr} + M \cdot i_{qs} \end{cases} \tag{7}$$

Where R_r and R_s are the rotor and stator resistances, respectively; V_{dr}, V_{qr}, V_{ds}, and V_{qs} are the voltages of the rotor and the stator expressed in the reference frame d-q, respectively; i_{dr}, i_{qr}, i_{ds} and i_{qs} are the currents of the rotor and the stator in the same d-q frame, correspondingly; M is the mutual inductance; and L_r and L_s are the rotor and stator inductances, correspondingly. The above expressions are computed in p.u. The T_{em} and stator reactive power are given by the Eqs. (8) and (9) in [3, 25].

$$T_{em} = \frac{3}{2} p \frac{M}{L_s} (\varphi_{qs} i_{dr} - \varphi_{ds} i_{qr}) \tag{8}$$

$$Q_s = \frac{3}{2} (V_{qs} i_{ds} - V_{ds} i_{qs}) \tag{9}$$

Where p is the pair of poles of the generator and Q_s is the stator reactive power (VA). From (8) and (9), according to [25], the rotor d-q currents are fundamental in the control of the DFIG since they are directly proportional, i.e., i_{qr} to the T_{em} and i_{dr} to the reactive power. Therefore, using these variables, the machine torque control is viable and, consequently, its speed and reactive power. Also, as indicated in [26], there is the possibility to control the speed of the machine from the rotor or stator side.

4 MPPT Strategy Design Based on DSC, FUZZY, and PID

PID is continuously applied in the wind turbine system parameters control. However, with the wind speed fluctuations influence, these parameters vary significantly; hence, causing that only a PID control does not assure the system stability [14, 22].

Regarding the Fuzzy Logic Controller, it is not part of conventional control methods; on the contrary, it is coupled as a control. It is implemented and used because of the advantages it presents. As described in [12, 13], this control methodology can work with imprecise input signals. It controls more complex systems with good results and handles the non-linearity of the system. It does not demand the use of a deep mathematical representation of the system. It can be combined with classic controllers such as the PID.

The proposed control technique follows the operating principle of the DSC. This procedure seeks to keep the λ_{opt} to guarantee that the reference ω_g, is also the appropriate one. This allows the monitoring of the $C_{p_{max}}$ with an estimate of the wind speed measurement through the performance of the T_{em} and powertrain. With this result, a new T_{em} that controls ω_g acting on the references of power and current $i_{qr_{ref}}$, defined in the classic DFIG control model, in Fig. 3, is established.

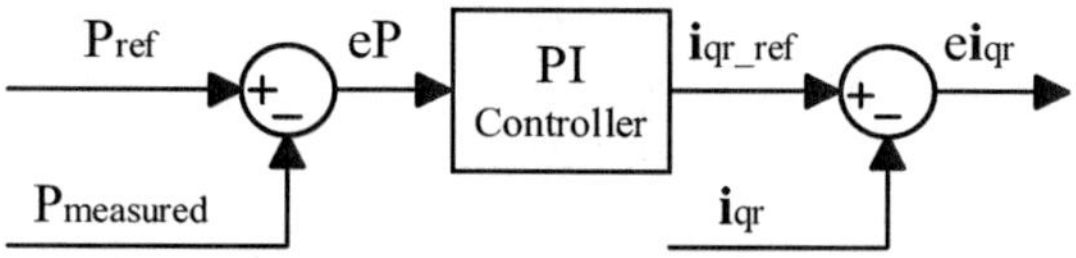

Fig. 3. Classic PI control of a DFIG applied on the rotor side.

To achieve this, the ideal estimated rotor torque, T_{opt}, is calculated from (1), (2), and (3) by setting the parameters at their optimal values, indicated in [27]:

$$T_{opt} = K_{opt} \cdot {\omega_r}^2 \tag{10}$$

$$K_{opt} = \frac{1}{2}\rho\pi \frac{R^5}{{\lambda_{opt}}^3} C_{pmax} \tag{11}$$

Where K_{opt} is a constant value that provides the optimum condition to compute the torque at LSS and consequently the rotor speed, ω_r, needed for MPPT. Solving for ω_r from (10) and translating this into the HSS by using Eq. (4), ω_g is obtained:

$$\omega_g = N\sqrt{\frac{T_{opt}}{K_{opt}}} \tag{12}$$

Equation (12) is evaluated with the real rotor torque value using (6), which depends on the turbine's dynamic conditions and represents the optimum torque at each instant. Once the reference ω_g is defined, it must be controlled, for which a FUZZY control is incorporated by following the steps of the Mamdani controller. In this way, the monitoring of $C_{p_{max}}$ is guaranteed to operate with maximum power extraction.

Fuzzification. The numeric input variables are converted to the terminology of fuzzy logic called linguistic terms. In control, the linguistic variables used as inputs are the ω_g error $E(n)$, (Fig. 4), and the error variation $\Delta E(n)$, and, as output, a factor of change of the T_{em}. For the three variables, seven linguistic terms such as HH-N (High-Negative quantity), MM-N (Medium-Negative quantity), LW-N (Low-Negative quantity), Z (Zero), LW-P (Low-Positive quantity), MM-P (Medium-Positive quantity), and HH-P (High-Positive quantity) were selected. As membership functions, triangular and open trapezoidal functions were denoted for the integration purpose of all range of values for each variable. With these considerations, the degree of membership of each associated linguistic term is established.

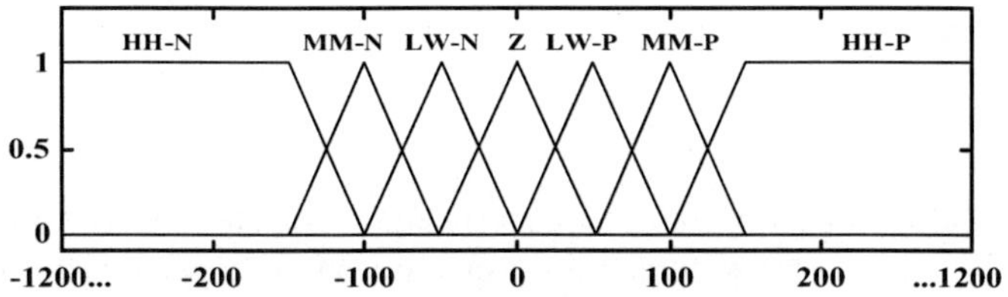

Fig. 4. Speed error membership functions $E(n)$.

Inference Stage. Once the system control rules are defined, the response to the data is selected. Each rule is evaluated with an if-then logic. By working with two input variables of 7 linguistic terms each, a set of 49 control rules is structured in Table 1.

Table 1. Control rules for FUZZY controller implementation.

		Error variation, ΔE(n)						
		HH-N	**MM-N**	**LW-N**	**Z**	**LW-P**	**MM-P**	**HH-P**
Error, E(n)	**HH-N**	HH-P	HH-P	MM-P	MM-P	MM-P	LW-P	Z
	MM-N	MM-P	MM-P	LW-P	LW-P	LW-P	Z	LW-N
	LW-N	MM-P	LW-P	Z	Z	Z	LW-N	LW-N
	Z	MM-P	LW-P	Z	Z	Z	Z	LW-N
	LW-P	MM-P	LW-P	Z	Z	LW-N	LW-N	MM-N
	MM-P	MM-P	Z	LW-N	LW-N	LW-N	MM-N	MM-N
	HH-P	LW-P	LW-N	MM-N	MM-N	MM-N	HH-N	HH-N

Defuzzification. The final stage calculates the system's numerical output value, such as the change in torque, by converting the linguistic terms obtained during the inference. The centroid defuzzification method was applied as it is one of the most commonly used options. The new T_{em} reference acts on the power PI control; thus, complementing the two controllers, FUZZY and PI. Together, they give rise to the reference $i_{qr\,ref}$, within the established limits in the function of the found T_{em}.

5 Results Analysis

The demonstration of the generated results by the proposed strategy was carried out by using the NREL FAST software, detailed in [28], and MATLAB/Simulink, along with the technical data in Table 2.

5.1 Responses to Deterministic Wind Speed Input Signal

The wind speed signal in this first case is shown in Fig. 5, and the system response can be observed in Fig. 6, where the wind turbine reaches the optimal operating values rapidly and accurately with the proposed control. More functional and efficient monitoring of the maximum power is made – reflected in the C_p, Fig. 6(e) – responding, as a consequence, to rapid changes and achieving stability at the optimal operating values. The electromagnetic torque tracks the respective reference with the developed controller, whereas the classic PI controller does not precisely track it, as shown in Fig. 6(c).

Table 2. DFIG wind turbine technical parameters.

DFIG		Turbine	
Parameter	**Value**	**Parameter**	**Value**
Rated power, P_n	1,5 MW	Number of blades	3
Rated voltage, V_n	575 V		
Frequency, f	60 Hz	Turbine diameter, D	85,5 m
Rotor resistance, R_r	0,004496 pu		
Stator resistance, R_s	0,006352 pu	Gearbox ratio, N	72
Rotor inductance, L_r	0,140643 pu		
Stator inductance, L_s	0,154253 pu	Air density, ρ	1,225 kg/m^3
Mutual inductance, L_m	2,613233 pu		
Synchronous speed, ω_r	1200 rpm		
Pole pairs, p	3	Power coefficient, C_p	0,51
Inertia constant, H	4,55 s		

Furthermore, contrasting this proposed technique and classical PI controller results, such as the machine angular speed, in Fig. 6(a), as well as the output power, in Fig. 6(b), and the q reference current component of the rotor side, in Fig. 6(d), it is demonstrated that in the developed method, the mentioned signals react more sensitively and precisely to the wind input variations.

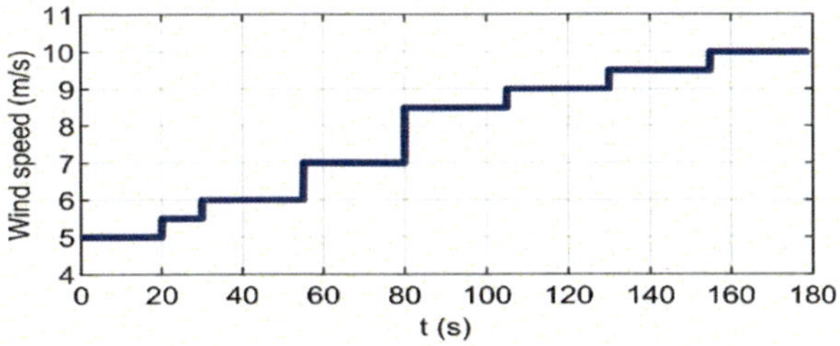

Fig. 5. Deterministic wind speed input signal.

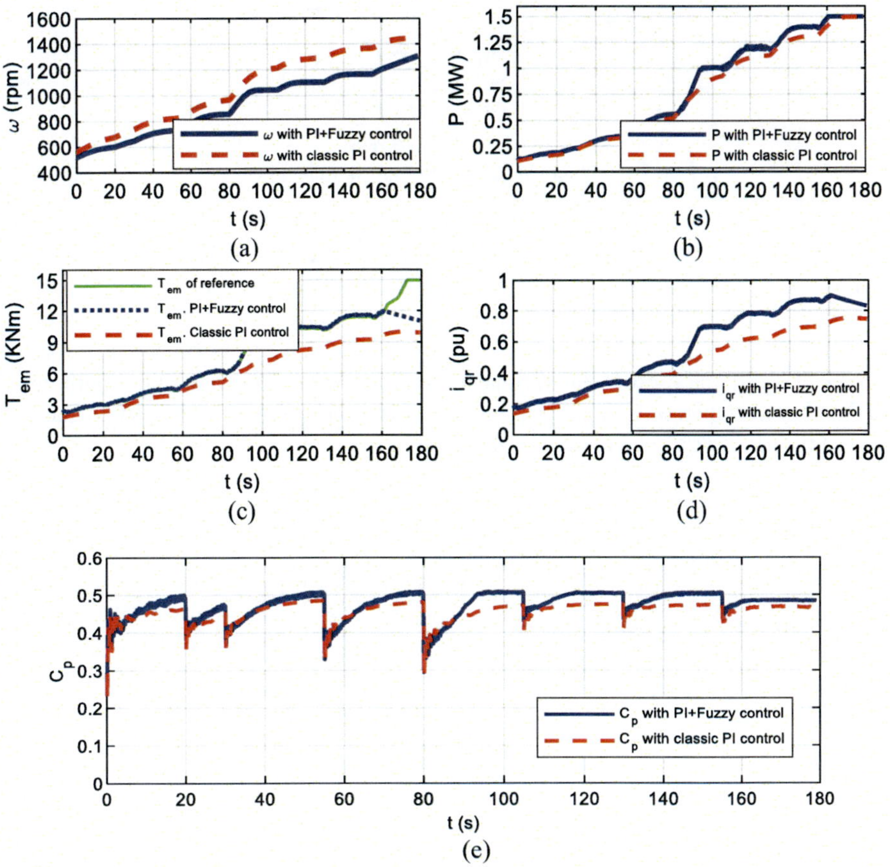

Fig. 6. Output signals for a deterministic wind speed input. Generator speed in rpm (a), output active power in MW (b), electromagnetic torque in kNm (c), i_{qr} reference in p.u. (d) and C_p (e).

5.2 Responses to Realistic Variable Wind Speed Input

Considering the wind closer to reality (Fig. 7), the system's outputs are checked by using the proposed control. It is evident that the involved variables, as shown in Fig. 8, react with greater sensitivity to constant changes in wind speed, and thus adjusting to their ideal values rapidly. There is also a correct tracking of T_{em}, Fig. 8(c), concerning its reference with the Fuzzy control, unlike the PI control, which does not. A similar interpretation can be obtained regarding the following variables: the generator output speed, in Fig. 8(a), the produced electrical power, in Fig. 8(b), and the qr reference current, in Fig. 8(d), which also respond quickly and smoothly to the small and constant wind fluctuations using the described technique in this study.

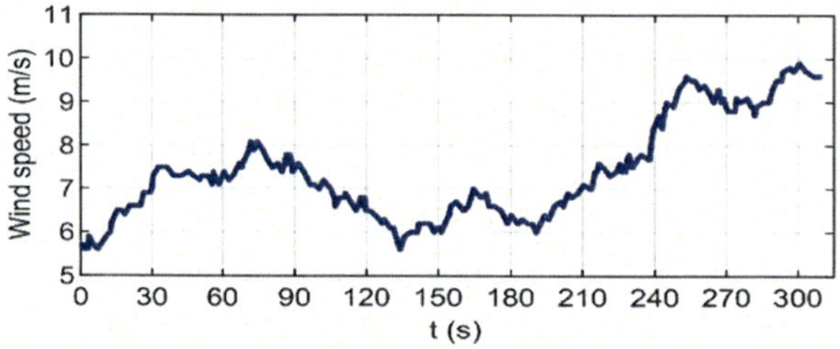

Fig. 7. Realistic variable wind speed input signal.

Figure 8(e) reflects the C_p which demonstrates the effect of built-in control. At this point, it is essential to note that the response shows improvements in almost all the carried-out simulations. Nonetheless, during large drops in wind speed, the system response does not considerably progress due to the turbine and generator inertias. As an outcome, it is difficult for the wind turbine to overcome this effect and achieve optimum speed and power operating values. Despite this situation, the control strategy manages to present some improvements for the MPPT technique.

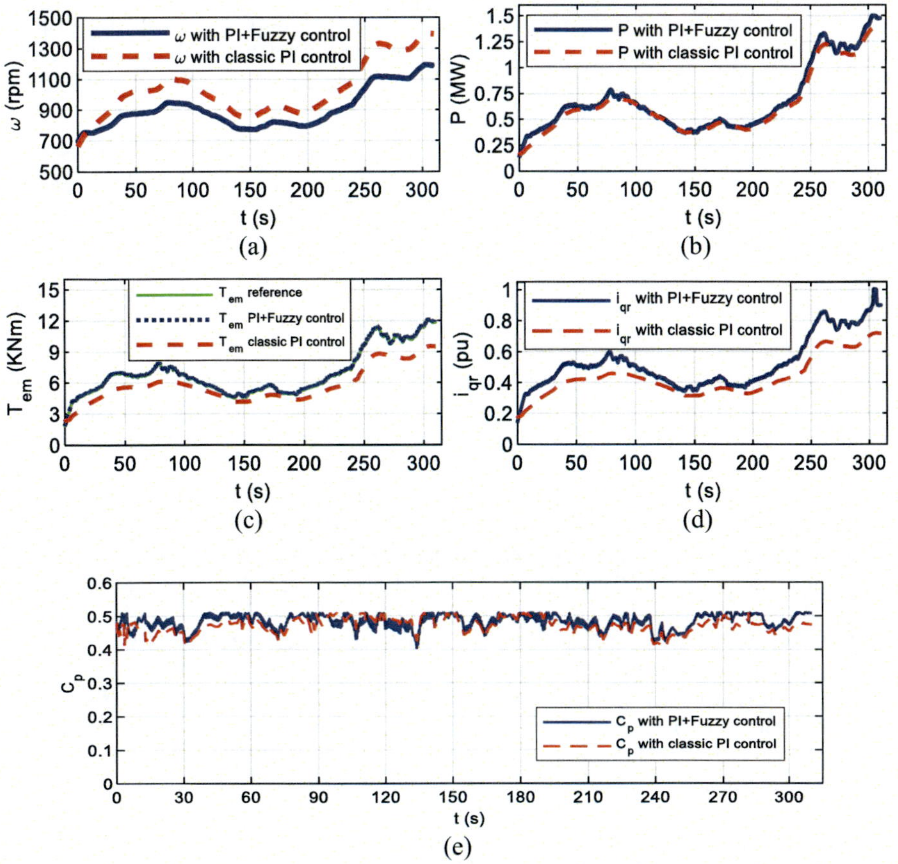

Fig. 8. Output signals for a realistic variable wind speed. Generator speed in rpm (a), output active power in MW (b), electromagnetic torque in kNm (c), i_{qr} reference in p.u. (d) and C_p (e).

6 Conclusions and Recommendations

In this paper, the study demonstrates the importance of the different subsystems – i.e., electrical, mechanical, and control – that make the wind turbine operation possible. The dynamic model used for these simulations tries to approximate the real scenario by providing relevant information on its function. The subsystems above are the basis for generating robust control techniques. A control scheme for MPPT based on the Fuzzy - PI controller has been developed, deploying a feasible simulation of a 1.5MW wind turbine model with a DFIG electric machine. This approach has demonstrated more effectiveness in contrast to a classic PI controller. The improvement can be observed in the C_p response in terms of precision and speed. This strategy does not require details of the mathematical model or system for its configuration, being optimal in processes with complex dynamics.

It can be observed that using the DSC Fuzzy - PI based, the electromagnetic torque, output power, rotor reference current, and rotational speed track the effect of the variations in the wind speed more rapidly. This fact allows the system to adapt to the produced changes better and improve maximum power extraction.

For future works, incorporating a robust pitch angle controller is recommended so that the control of the system be strengthened to support the stability of the involved variables. The incorporation is required, mostly, in the face of wind speed drops, where the rotor speed must be acted immediately to overcome the system's inertia in a shorter time.

References

1. Rezaei MM (2018) A nonlinear maximum power point tracking technique for DFIG-based wind energy conversion systems. Eng Sci Technol Int J 21(5):901–908. https://doi.org/10.1016/j.jestch.2018.07.005
2. Ayala E, Simani S (2019) Perturb and observe maximum power point tracking algorithm for permanent magnet synchronous generator wind turbine systems. In: Conte G (ed) Proceedings of 15th European workshop on advanced control and diagnosis – ACD. Lecture notes in control and information sciences. Alma Mater Studiorum, University of Bologna. Springer, Bologna, pp 1–11
3. Ben Amar A, Belkacem S, Mahni T (2017) Direct torque control of a doubly fed induction generator. Int J Energetica (IJECA) 2(1):11–14
4. Bakouri A, Abbou A, Mahmoudi H, Elyaalaoui K (2014) Direct torque control of a doubly fed induction generator of wind turbine for maximum power extraction. Paper presented at the international renewable and sustainable energy conference (IRSEC), Ouarzazate
5. Sahri Y, Tamalouzt S, Belaid SL (2018) Direct torque control of DFIG driven by wind turbine system connected to the grid. In: 2018 International conference on wind energy and applications in Algeria (ICWEAA). https://doi.org/10.1109/ICWEAA.2018.8605083
6. Bakouri A, Mahmoudi H, Abbou A, Moutchou M (2015) Optimizing the wind power capture by using DTC technique based on artificial neural network for a DFIG variable speed wind turbine. Paper presented at the 10th international conference on intelligent systems: theories and applications (SITA), Rabat. https://doi.org/10.1109/SITA.2015.7358425
7. Thongam JS, Ouhrouche M (2011) MPPT control methods in wind energy conversion systems. In: Carriveau R (ed) Fundamental and advanced topics in wind power. InTech
8. Calderaro V, Galdi V, Piccolo A, Siano P (2008) A fuzzy controller for maximum energy extraction from variable speed wind power generation systems. Electr Power Syst Res 78(6):1109–1118. https://doi.org/10.1016/j.epsr.2007.09.004
9. Azzaoui ME, Mahmoudi H (2017) Fuzzy-PI control of a doubly fed induction generator-based wind power system. Int J Autom Control 11(1):54–66. https://doi.org/10.1504/IJAAC.2017.080819
10. Fu C, Pan T, Liu H, Wu D, Shen Y, Hao Z (2018) MPPT control based fuzzy for wind energy generating system. Paper presented at the 37th Chinese control conference (CCC), Wuhan. https://doi.org/10.23919/ChiCC.2018.8484096
11. Hou G, Jiang Z, Yang Y, Zhang J (2016) Variable universe fuzzy controller used in MPPT based on DFIG wind energy conversion system. Paper presented at the Chinese control and decision conference (CCDC), Yinchuan. https://doi.org/10.1109/CCDC.2016.7532048
12. Sheikhan M, Shahnazi R, Yousefi AN (2012) An optimal fuzzy PI controller to capture the maximum power for variable-speed wind turbines. Neural Comput Appl 23(5):1359–1368. https://doi.org/10.1007/s00521-012-1081-4

13. Tahiri FE, Chikh K, Khafallah M (2018) MPPT strategy using fuzzy-PI controller applied to a standalone wind energy conversion system. Paper presented at the SCA 2018: proceedings of the 3rd international conference on smart city applications, Tetouan
14. Karad S, Thakur R (2019) Recent trends of control strategies for doubly fed induction generator based wind turbine systems: a comparative review. Arch Comput Methods Eng 1–15. https://doi.org/10.1007/s11831-019-09367-3
15. National Renewable Energy Laboratory (NREL). https://www.nrel.gov/
16. MATLAB, 2018, 9.5.0.944444 (R2018b), The MathWorks Inc., Natick, Massachusetts
17. Baldo F (2012) Modeling of load interfaces for a drive train of a wind turbine. Master's Thesis, Department of Applied Mechanics, Chalmers University of Technology
18. Abdullah MA, Yatim AHM, Tan CW, Saidur R (2012) A review of maximum power point tracking algorithms for wind energy systems. Renew Sustain Energy Rev 16(5):3220–3227. https://doi.org/10.1016/j.rser.2012.02.016
19. Narayana M, Putrus GA, Jovanovic M, Leung PS, McDonald S (2012) Generic maximum power point tracking controller for small-scale wind turbines. Renew Energy 44:72–79. https://doi.org/10.1016/j.renene.2011.12.015
20. Michalke G (2008) Variable speed wind turbines - modelling, control, and impact on power systems. Ph.D. thesis (Dr.-Ing.) Electrical and Computer Engineering, Technical University of Darmstadt, Darmstadt
21. Semrau G, Rimkus S, Das T (2015) Nonlinear systems analysis and control of variable speed wind turbines for multiregime operation. J Dyn Syst Meas Control 137(4):1–10. https://doi.org/10.1115/1.4028775
22. Ohiero P (2015) Development of fast multi-system simulation models for permanent magnet synchronous motor and generator drive systems. Ph.D. thesis, School of Engineering, College of Science and Engineering, University of Glasgow, Glasgow
23. Rached B, Elharoussi M, Abdelmounim E (2020) Design and investigations of MPPT strategies for a wind energy conversion system based on doubly fed induction generator. Int J Electr Comput Eng (IJECE) 10(5):4770–4781. https://doi.org/10.11591/ijece.v10i5.pp4770-4781
24. Ouezgan K, Bossoufi B, Bargach MN (2017) DTC control of DFIG-generators for wind turbines: FPGA implementation based. In: 2017 International renewable and sustainable energy conference (IRSEC), Tangier. https://doi.org/10.1109/IRSEC.2017.8477300
25. Abu-Rub H, Malinowski M, Al-Haddad K (2014) Properties and control of a doubly fed induction machine. In: Power electronics for renewable energy systems, transportation and industrial applications. WILEY/IEEE, Chichester, pp 280–284
26. Bharti OP, Saket RK, Nagar SK (2015) Design of PI controller for doubly fed induction generator using static output feedback. In: 2015 39th national systems conference (NSC), Noida. https://doi.org/10.1109/NATSYS.2015.7489128
27. Abad G, López J, Rodríguez MA, Marroyo L, Iwanski G (2011) Doubly fed induction machine, modeling and control for wind energy generation. Wiley, New Jersey
28. Jonkman JM, Buhl Jr. ML (2005) FAST user's guide. NREL/EL-500-38230. National Renewable Energy Laboratory, Golden

Direct Speed Control Scheme for Maximum Power Point Tracking of a 1.5MW DFIG Wind Turbine

Edy Ayala[1](✉), Nicolás Pozo[1], Silvio Simani[2], and Eduardo Muñoz[1]

[1] Universidad Politécnica Salesiana, Calle Vieja 12/30, Cuenca, Ecuador
eayala@ups.edu.ec, {apozoa,emunozp2}@est.ups.edu.ec
[2] Ferrara University, Via Savonarola, 9, 44121 Ferrara, FE, Italy
silvio.simani@unife.it

Abstract. In the present work, a control strategy for Maximum Power Point Tracking (MPPT) applied to a wind turbine is described. The electric machine consists of a 1.5MW Doubly Fed Induction Generator (DFIG). This strategy is developed according to the theory of Direct Speed Control (DSC), which includes a state observer. This strategy considers the Low Shaft Speed (LSS) as an input and the i_{qr} reference current as the output. This control mechanism allows monitoring the MPPT; thus, changing the Power Coefficient (Cp) to its optimal value during the wind turbine operation. Among its main features, the controller is configured to work with the incorporation of different wind inputs, a fact that permits evaluating the system's response to disturbances and variations. For simulation tests, a wind turbine has been modeled in MATLAB-Simulink and Fatigue, Aerodynamics, Structures and Turbulence (FAST) software. The strategy has been compared to a PI-MPPT controller and has demonstrated improvements in terms of speed and output power extraction.

Keywords: Power coefficient · MPPT · Direct speed control · DFIG · Wind turbine

1 Introduction

The demand for wind power generation, added to the depletion of natural resources, has been the main generators of the environmental and energy crises. Consequently, both factors led to the search for renewable energy sources that would significantly contribute to energy production. It is estimated that the generation of wind energy will be 20% of the world's total electricity by 2040 [1]. Within the production of wind energy through renewable resources, wind energy has implied important changes in the generation and control systems of wind turbines. In the case of control systems, they are very important in the operation of a wind turbine system [2, 3]. Currently, the wind turbine market is aimed at high power generation systems (from 1 MW), and for these applications,

Y. Iano et al. (Eds.): BTSym 2020, SIST 233, pp. 918–928, 2021.
https://doi.org/10.1007/978-3-030-75680-2_101

DFIGs are of great interest since they are capable of extracting high power thanks to their low-cost power converters and low energy losses in comparison to other types of devices [4, 5].

The control strategy allows manipulating the slip angle between ±30% of the synchronous speed of operation. For the maximum power point controller design, MPPT is implemented by considering the wind speed, the rotor speed, the power generated, and the angle of inclination of each blade as variables. The control strategy seeks to improve the maximum power curve response through a direct speed controller (DSC) and the LSS as input. The controller also implements a state observer, which allows estimating the response to a random input of wind [6]. The following article shows a DSC strategy to obtain the maximum power point using a state observer. For the simulation and validation of results, a 1.5 MW wind turbine was selected by using MATLAB-Simulink and FAST [7, 8].

For explanation purposes, section two presents the description of the system, and section three presents the analysis and modeling of the subsystems that directly intervene in the operation of the wind turbine as well as the description of the parameters that have been used for the DSC simulation. In detail, section four shows the control strategy and simulations based on FAST and Simulink. Finally, section five discusses the validation results.

2 System Description

A wind turbine is a device that transforms energy from the wind into electric power. The wind turbine is composed of the following systems: aerodynamic model, pitch angle control, mechanical, electrical, and control systems sized according to torque. In Fig. 1, the interaction between the different systems that make up the wind turbine is shown [6, 9].

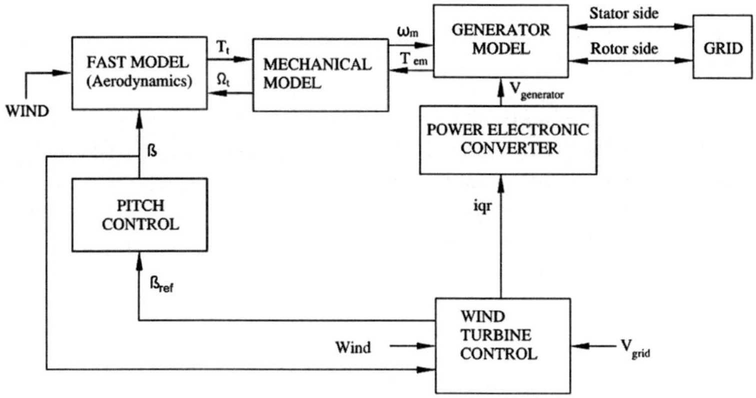

Fig. 1. Block diagram of the systems of a DFIG wind turbine.

2.1 Aerodynamic Model

The wind turbine blades convert the wind energy into rotational energy. Thereon, the extracted energy is represented by an aerodynamic model, which mainly represents the speed and torque as a function of the input wind. In terms of the wind speed, it can be considered as the average speed of the incident wind in the area swept by the blades in order to evaluate the torque at the low shaft speed. The output power of the wind turbine can be modeled in terms of the power coefficient, blade length, wind speed, and air density [10, 11]. These terms can be formulated by the relationships in the set of Eq. (1).

$$P_r=\frac{1}{2}\rho\pi R^2W^3C_p(\lambda,\beta);\ T_r=\frac{1}{2}\rho\pi R^3W^2C_q(\lambda,\beta) \tag{1}$$

Where ρ is the air density, R is the blade length, W is the input wind speed, Cp is the power coefficient, λ is the tip speed ratio, β is the pitch angle, and Cq is the torque coefficient [10]. Betz's law constitutes a maximum generation limit for a wind turbine, establishing the restriction of 59.26% of the wind energy to mechanical energy transfer. This limit is known as the power coefficient and is described as a complex relationship of the tip speed ratio and the blade pitch angle stated in [12]. For each β angle, it corresponds to an optimal tip speed ratio at which the power coefficient Cp is the maximum value [13]. Defined by Eq. (2), Cp is often given as a function of λ.

$$C_p = c1 \cdot \left[\frac{c2}{\lambda_i} - c3 \cdot \beta - 5\right] e^{\frac{-c4}{\lambda_i}} + c5 \cdot \lambda;\ \lambda_i = \left[\frac{1}{\lambda + c6 \cdot \beta} - \frac{c7}{(\beta^3+1)}\right]^{-1};$$
$$\lambda = \frac{R * \Omega_t}{W} \tag{2}$$

Where Ω_t is the LSS, and c1 to c6 are constants usually given by the manufacturer or obtained experimentally.

2.2 Mechanical Model

The mechanical model consists of all moving components that allow the energy transference from the wind to the electric generator. It includes the drivetrain, blades, rotor hub, tower, nacelle, and rotating mechanisms. With reference to the drivetrain, it encompasses models depending on the number of masses for the dynamic analysis. For these experiments, a two-mass has been analyzed following the recommendations in [8] shown in Fig. 2. The drivetrain model establishes the relationships between the inertias, torques, angular speeds, damping, and friction factors described in Eq. (3).

$$J_t\frac{d\Omega_t}{dt}=T_t - D_t\Omega_t - T_{em};\ J_m\frac{d\Omega_m}{dt} = T_{em} - D_m\Omega_m + T_{em}$$
$$\frac{dT_{em}}{dt} = K_{tm}(\Omega_t - \Omega_m) + D_{tm}\left(\frac{d\Omega_t}{dt} - \frac{d\Omega_m}{dt}\right) \tag{3}$$

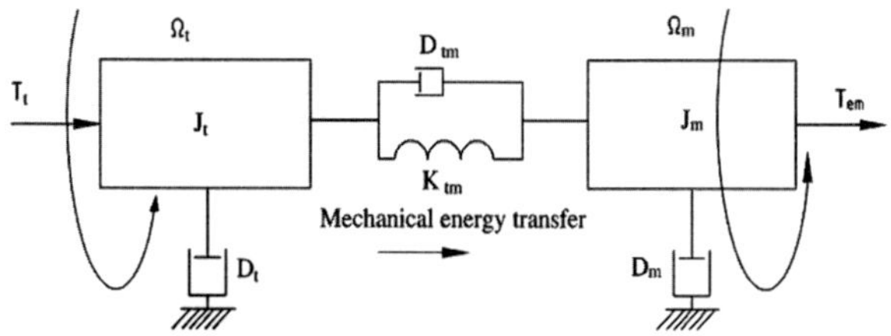

Fig. 2. Two-mass mechanical model.

Where, J_t is the turbine inertia, J_m is the machine inertia, Ω_m is the High Shaft Speed (HSS), the damping coefficients of the turbine, machine, and drivetrain are (Dt, Dm, and Dtm) respectively, and the stiffness constant (Ktm); thus, resulting in a two-inertia model [6, 12].

2.3 Pitch Control System

The main job of the controller is to rotate the wind turbine blades at different angles to increase or decrease the speed of the rotor. Currently, the controllers can even manipulate the inclination of each blade independently. This advantage gives greater versatility since having more combinations of positions can reduce the stress on the blades.

The rotation of the angle is controlled, as shown in Fig. 3, by a proportional integrative controller PI that generates a reference rate of speed change. The angle of inclination itself is obtained by integrating the variation of the angle [14].

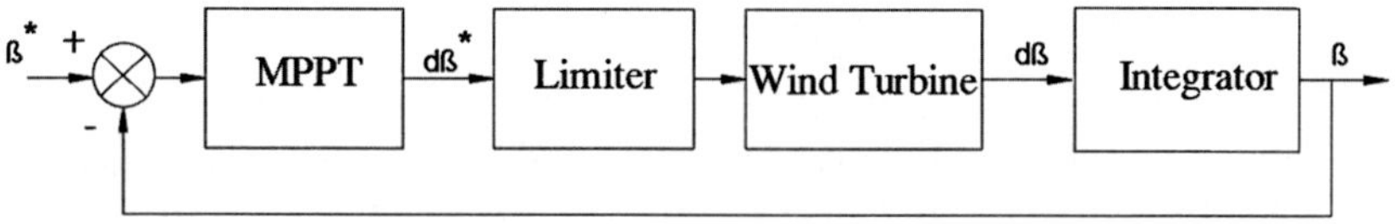

Fig. 3. Pitch angle control system.

The pitch controller is very important for a wind turbine because a mechanical control of the torque is needed in order to regulate the generator HSS, which leads the reference to maintain optimum operating speed. This response is slow, but it is necessary in order to guarantee the nominal speed operation of the system and regulate the wind turbine once it enters into a full operation zone. The idea is to reduce the speed before it is needed to apply the brake system, which produces high friction.

3 Direct Speed Control

Direct speed control has an advantage over other control methods in that it follows the power curve with a much faster response. By knowing λ, the estimated optimum wind turbine rotation speed Ω_{m_ref} can be calculated. For real scenarios, wind speed is not a real variable because the turbine itself creates shadow effects that alter its measurements. The optimum rotational speed is better calculated from the torque. Therefore, it is possible to design an observer to calculate the torque by measuring variables such as the electromagnetic torque T_{em} and the HSS, which are used for estimating the optimal speed Ω_{m_ref} (4) [15].

$$\Omega_{m\_ref} = N\sqrt{\frac{T_{t\_est}}{k_{opt\_t}}} \tag{4}$$

Where N is the drivetrain ratio, T_{t_est} is the aerodynamic estimated torque, and k_{opt_t} is the optimum constant for MPPT observer that allows adjusting Ω_{m_ref} by using the T_{em}. The DSC general scheme presented in [6] for these experiments is shown in Fig. 4.

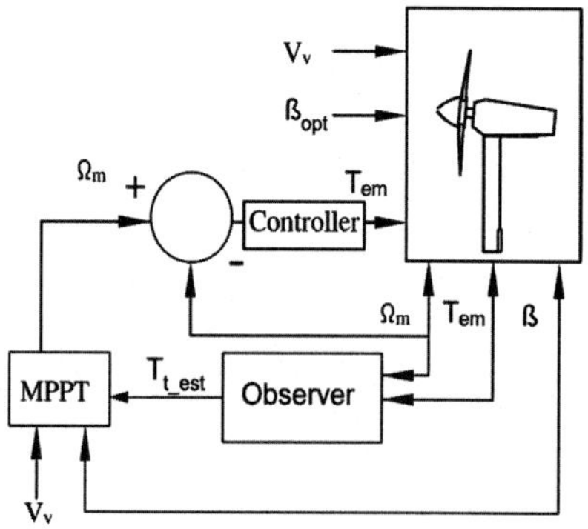

Fig. 4. Direct speed control block diagram.

In this article, the observer is implemented for a MPPT controller. When the Cp is at its maximum value, the low shaft torque T_{t_est} can be calculated as described in Eq. (5).

$$T_{t\_est} = \frac{1}{2}\rho\pi\frac{R^5}{\lambda_{opt}^3}C_{p\_max}\Omega_t^2 = k_{opt}\Omega_t^2 \tag{5}$$

Finally, the above equation allows having a relationship in terms of k_{opt_t} described in Eq. (6).

$$k_{opt\_t} = \frac{1}{2}\rho\pi\frac{R^5}{\lambda_{opt}^3}C_{p\_max} \tag{6}$$

All the above equations have been simulated in MATLAB-Simulink, and for the aeroelastic wind model, the software FAST has been used for modeling a wind turbine as a combination of mechanical elements and their interactions. For this article, a three blades turbine has been modeled as a combination of rigid and flexible elements [8, 14].

3.1 DSC Control Description

The controller presented in Fig. 4 considers the angular velocity of the rotor as input, and together with expression (5), the T_{t_est} is calculated. This low shaft torque, by means of a state observer block, allows generating the estimated electromagnetic torque control signal to regulate the wind turbine output power [7, 14].

The dynamic system has internal conditions at determined points that are monitored by the state observer based on the Luenberger algorithm. In order to establish the control strategy, the observer may have a complete or reduced order depending on the monitored states. Once the estimated state data are computed, the signal error is calculated by the gain of the estimator, which depends on the system. For these experiments, the gain is unitary, as shown in Fig. 5, and in Fig. 6, the implementation of the controller in MATLAB-Simulink is illustrated. For the calculation of the current i_{qr} which enters the PI control block, the following expression (7) is used and calculated from the generated electromagnetic torque [16].

$$T_{em} = -\frac{3}{2}p\frac{L_m}{L_s}\Psi_s i_{qr} \tag{7}$$

Where p is the wind turbine pairs of poles; Ψ_s is the stator flux frame; L_m is the mutual inductance, and L_s is the stator inductance.

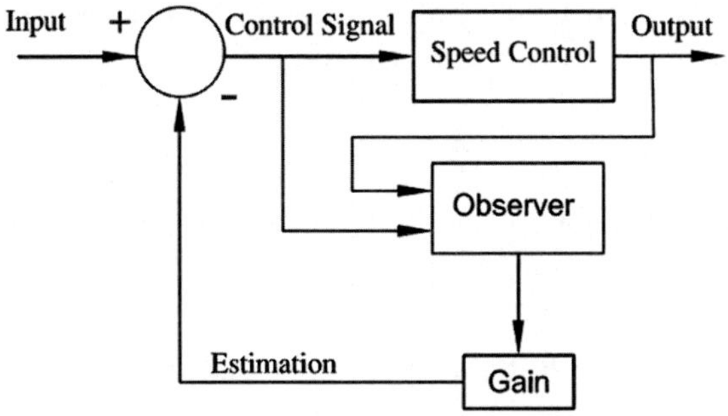

Fig. 5. Feedback of observed states block diagram.

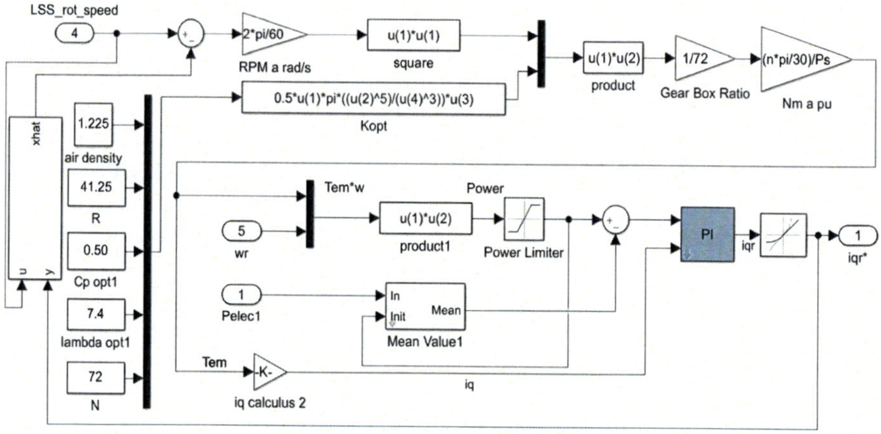

Fig. 6. Direct speed controller implementation in MATLAB-Simulink

4 Simulation Results

In this section, the dynamic response of the DSC state observer is detailed. For this work, three types of wind inputs are used in order to validate the obtained results. In all cases, the wind turbine rapidly and accurately reaches the optimal operating point by using the DSC state observer.

The following outputs are considered in order to evaluate the system's dynamic response: reference signal i_{qr} in p.u, the electromagnetic torque in Nm, the active output power in Watts, the wind input in m/s, the generator speed in rpm, the pitch angle in degrees, and the power coefficient.

4.1 Steady Wind Speeds with Ramp Type Transitions

It can be seen the results of the system behavior in Fig. 7. The DSC Luenberger state observer tracks the Cp value with a more rapid response. The Cp dynamics, in this case, is improved compared with the PI controller Fig. 7(g). The produced output power using this DSC strategy in comparison with the classical PI controller is also more stable Fig. 7(b).

4.2 Variable Wind Speed with Realistic Input

A realistic wind speed was used in this test to demonstrate the dynamic response. The results can be observed in Fig. 8. The DSC state observer tracks the Cp value with more quick dynamics. The Cp response, in this case, is better than the PI controller Fig. 8(g). The MPPT rapidly responds to wind changes; thus, achieving a better response in the optimal operating parameters.

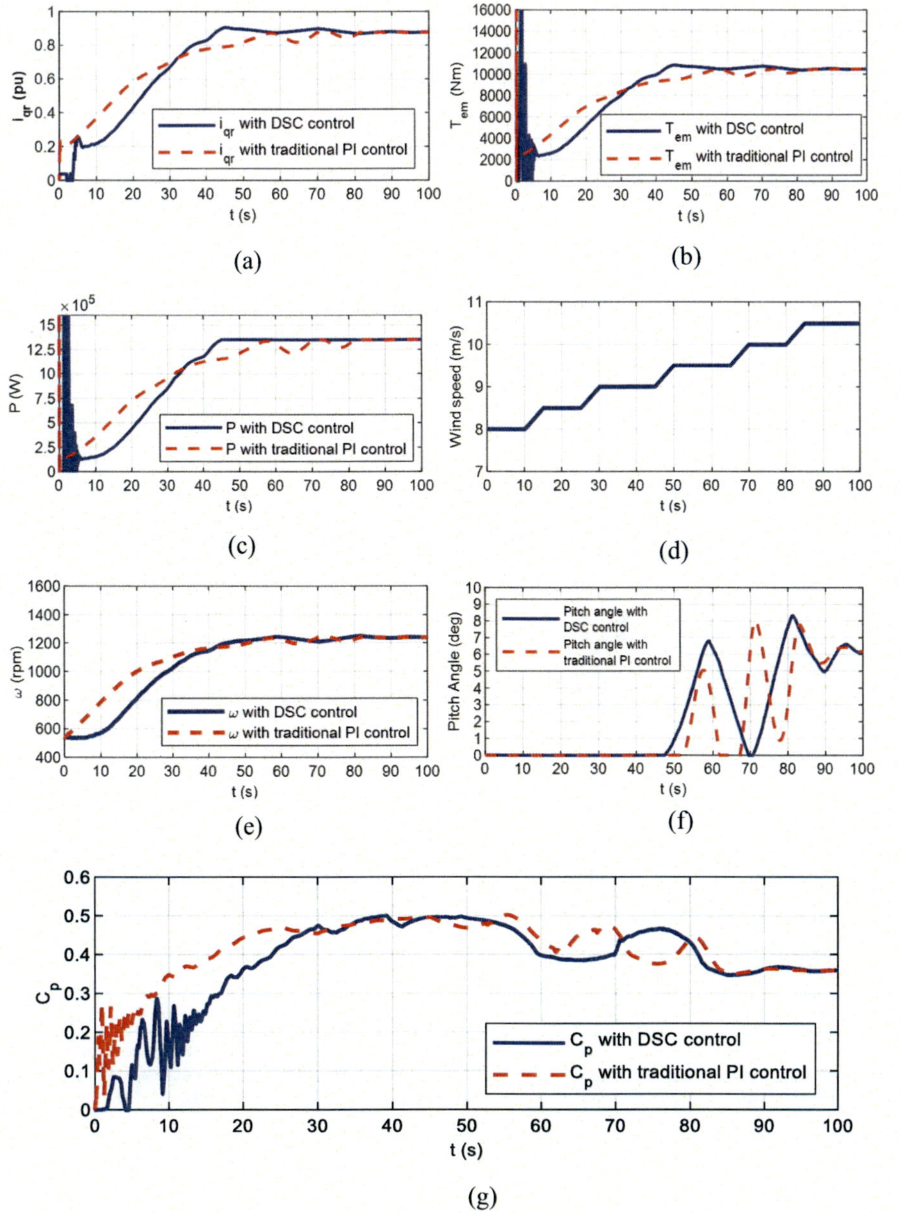

Fig. 7. Steady wind speed with ramp type transitions: reference signal i_{qr} in p.u. (a), electromagnetic torque in Nm (b), output power in W (c), wind speed in m/s (d), HSS in rpm (e), β in degrees (f) and C_p (g).

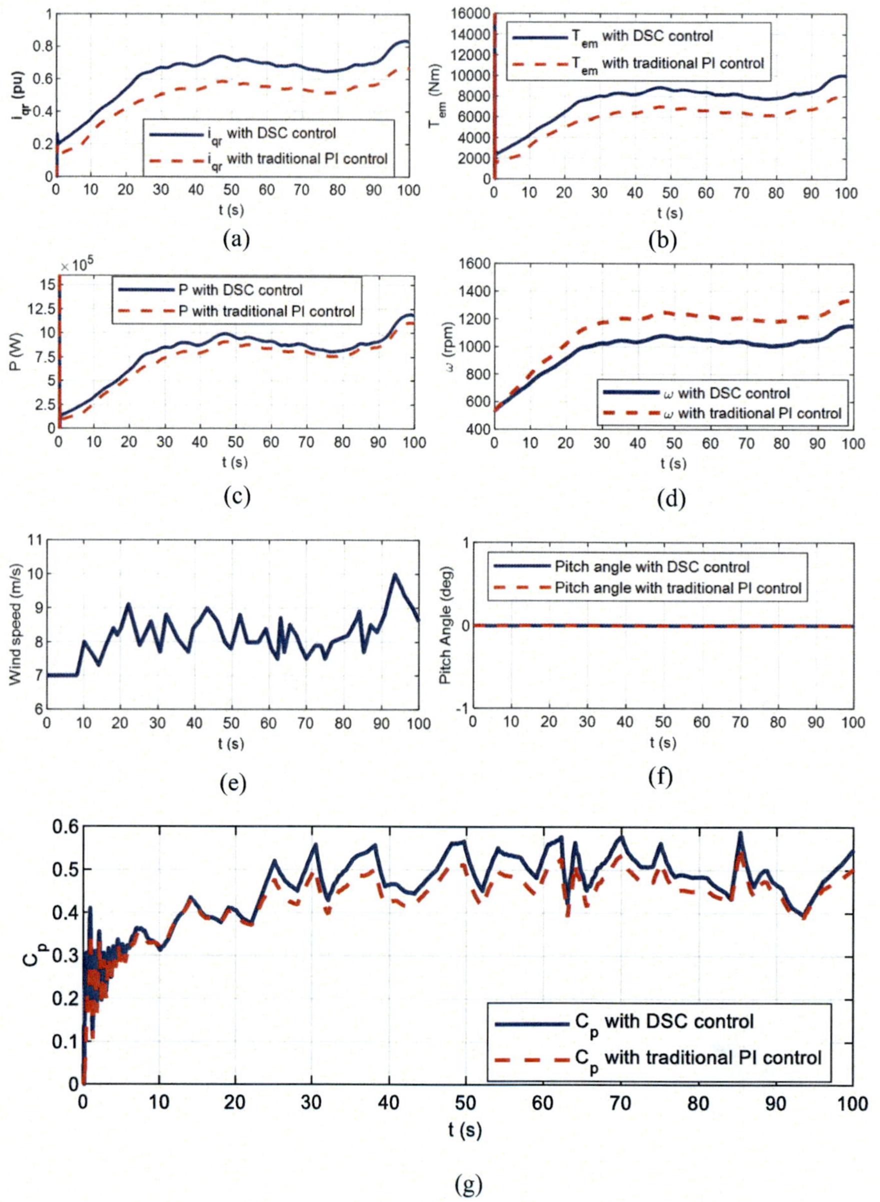

Fig. 8. Realistic wind speed input: reference signal i_{qr} in p.u. (a), electromagnetic torque in Nm (b), output power in W (c), wind speed in m/s (d), HSS in rpm (e), β in degrees (f) and C_p (g).

5 Conclusions

It is worth noting that, for any wind turbine, the MPPT is an important strategy that allows more power extraction. An appropriate model and control design are key factors not only in terms of operation but also in terms of maintenance and durability of the elements. Although the wind speed is not always constant, MPPT can still track the dynamic response of the system, a feat that makes it an interesting strategy. When a proper MPPT is performed, a more reliable energy system can be guaranteed, and by adding contributions of individual generators into wind farms, the influence of the controller becomes evident.

The DSC strategy based on the Luenberger state observer has allowed the system to operate with a better power extraction and Cp. Although the system is unstable during the first seconds of simulation due to the initial conditions, it rapidly reaches the optimum values. This means more active power available and absorption of reactive power from the grid. The advantage of this technique is also that the observer does not require any additional hardware implementation since it requires only the wind turbine system available measurements. For further implementations, it can even increase the number of input parameters for improving the accuracy.

In addition, this observer allows a close route of the feedback signals in order to minimize the error close to zero. The observer easily tracks the convex Cp function. The complementary DSC strategy allows the wind turbine to produce more torque and power by progressive variations of i_{qr}. The HSS is not increased in order to maximize the output power; instead, it is reduced in comparison to the classical-PI. This means that the mechanical components are slightly reduced in terms of fatigue, and the operative lifetime of the mechanical components can be improved. Finally, this controller is modular and can be implemented over an operative wind turbine.

Finally, this work includes simulations using data from manufacturers and published papers. However, the present research should include real implementations using hardware in the loop strategies in order to test the proposed scheme with a real DFIG. Moreover, it will be necessary to include real wind speed profiles in order to analyze a more accurate dynamic response. Additionally, other experiments can be performed, including other electric machine technologies such as PMSG in order to demonstrate the versatility of the proposed method.

References

1. World Wind Energy Association (2010) World Wind Energy Report 2010. WWEA
2. Zhi-Nong W (2009) The intelligent control of DFIG-based wind generation. In: Conference on sustain power generation and Supply, pp 1–5
3. Ayala E, Simani S (2019) Perturb and observe maximum power point tracking algorithm for permanent magnet synchronous generator wind turbine systems. In: Proceedings of the 15th European workshop on advanced control and diagnostics, pp 1–11
4. Muller S, Deicke M, Doncker R (2000) Doubly-fed induction generators systems for wind turbines. IEEE Indu Appl Mag 8(3):26–33
5. Sow TL (2012) Nonlinear control of the wind turbine at DFIG for a participation to the regulating of the frequency of the network, Quebec

6. Abad G, Lopez J, Rodriguez M, Marroyo L, Iwanski G (2011) Doubly fed induction machine. Wiley
7. The MathWorks Inc. MATLAB-Simulink, Massachusetts
8. National Renewable Energy Laboratory (2020) Fatigue aerodynamics structures and turbulence
9. Tremblay E, Atayde S, Chandra A (2009) Direct power control of a DFIG-based WECS with active filter capabilities. In: IEEE electrical power energy
10. Magdi M, Mojeed O (2019) Adaptive and predictive control strategies for wind turbine systems: a survey. IEEE J Autom SINICA 6:364–378
11. Hallak M, Hasni M, Menaa M (2018) Modeling and control of a doubly fed induction generator base wind turbine system. In: 3rd CISTEM 2018
12. Gajewski P, Pienkowski K (2016) Direct torque control and direct power control of wind turbine system with PMSG. Wrocław University of Technology, Department of Electrical Machines, Drives and Measurements
13. Mendis N, Muttaqi K, Sayeef S, Perera S (2012) Standalone operation of wind turbine-based variable speed generators with maximum power extraction capability. IEEE Trans Energy Convers 27(4):822–834
14. NREL (2014) Simulation for wind turbine generators—with FAST and MATLAB-simulink modules, USA
15. Camblong H (2006) Experimental evaluation of wind turbines maximum power point tracking controllers. Energy Convers Manage 47(11):2846–2858
16. Mohammadi J, Vaez-Zadeh S, Afsharnia S, Daryabeigi E (2014) A combined vector and direct power control for DFIG-based wind turbines. IEEE Trans Sustain Energy 5(3):767–775

A Low-Complexity Algorithm for Diagnosis of Three-Phase Induction Motors

Marco Baltazar, Brian Ramírez, and Guillermo Kemper(✉)

Universidad Peruana de Ciencias Aplicadas, Av. Prolongación Primavera 2390, Santiago de Surco, Lima, Peru
{u201213835,u201311171,pcelgkem}@upc.edu.pe

Abstract. Three-phase motors are made up of different pieces and parts that, when worn, increase mechanical vibrations. These vibrations generate multiple frequencies and their respective harmonics, so detection and classification must be done in the frequency domain. The faults also depend on the shaft's rotational frequency and the power supply frequency. State-of-the-art algorithms for motor failure detection and classification are complex since they involve processing in the time domain and the frequency domain. Given this, the present work proposes a low complexity algorithm based on wireless monitoring of certain frequency components' amplitude to define the degree of severity of vibrations in a three-phase motor. The algorithm consists of filtering the signal through a bank of low-pass filters, obtaining frequency spectra through the FFT, averaging the modulus spectra to reduce noise and distortion, and evaluating the amplitudes of the components of the frequency of failure in order to detect and identify the degree of severity based on reference levels of velocity. The algorithm offers a quantitative-qualitative diagnosis because based on an amplitude (quantitative), we designate a level of severity (qualitative). Now the technician offers a qualitative diagnosis - qualitative. In this case, the technician's verbal coincidences determine the level of severity according to ISO 20816-1: 2016. Then the coincidences at the level of severity (qualitative) establish the certainty of the algorithm developed concerning the technician's diagnosis. The diagnosis coincidence establishes a result with a high level of reliability based on the similarity between the technician's diagnosis and the algorithm.

Keywords: Diagnostics · FFT · Fault detection · Vibrations · Three-phase motor · Spectral averaging

1 Introduction

Bearings are quite common mechanical elements in induction motors. Failures in these elements can cause great economic losses to companies [1] since they stop turning the machinery and leave it inoperative [2].

To reduce the probability of these faults occurrence, condition-based maintenance (CBM), also called predictive maintenance, is generally applied, which provides concrete measurements regarding the condition of the engine without performing any complex

Y. Iano et al. (Eds.): BTSym 2020, SIST 233, pp. 929–948, 2021.
https://doi.org/10.1007/978-3-030-75680-2_102

intervention. This allows knowing the motor's state and the possible problems that it may have in the short or long term.

In state of the art, most works aim to predict failures through various types of solutions. For example, [3] proposes a system that diagnoses defects in the bearings of three-phase motors. The process comprises signal segmentation, feature extraction, and reduction. The last stage comprises the classification of faults and the fusion of decisions resulting from a trained neural network. In this case, the classifier must distinguish between bad states and good states. The author also uses FFT (Fast Fourier Transform) and DWT (Discrete Wavelet Transform) to extract vibration characteristics.

In [4], the bearing condition evaluation using an artificial neural network (ANN) is proposed based on spectral analysis and the envelope of mechanical vibrations.

In [5], the vibration signal analysis in conjunction with the motor current signal is proposed. Raw data of vibrations and currents from a motor with healthy bearings and internal and external race defects are used. This is done in order to compare performance between vibration and current.

In [3–5], neural networks and genetic algorithms are applied to improve the performance of classifiers or detectors. On the contrary, this work presents a low complexity and high-performance algorithm capable of recognizing a certain number of defects and motor faults.

Other authors propose a similar idea as in [6], where a method based on relative wavelet energy is developed to automatically select the parameter defined by the WPT (Wavelet Packets Transform) that has a high potential to extract patterns. The author validates the method via an engine model used as a determining factor in the monitoring process.

On the other hand, in [7], methods applied to motors fed with frequency inverters are proposed. The author states that this type of motor is better analyzed with the CWT (Continuous Wavelet Transform) instead of the FFT since the latter performs better for this application and the former is more convenient when the motor is in steady-state. The method detects shaft misalignment, faulty bracket, stator phase loss, and short-circuit in stator windings. Specifically, CWT is used for parameter extraction and a deep neural network to distinguish between mechanical and electrical faults. Due to the computational load, the author proposes using a graphics processing unit (GPU). In contrast to this work, the proposed algorithm does not need a graphic processing unit, and it can even be processed on a single-board computer.

Some other authors prefer to analyze other parameters in induction motors, such as the motor current signature (MCSA) [8]. This method is based on detecting characteristic fault signatures that arise in the current spectrum of a faulty motor. Due to this, the author proposes the use of advanced time-frequency transforms such as the CWT, the Wigner Ville distribution, or the Hilbert transform, which allows tracing the evolution of the fault components over time.

Finally, in [9], the development of a system is proposed where the oscillating amplitude of the rotor speed is used in the transient state as a diagnostic parameter. In comparison to other articles, this author analyzes the axis speed parameter establishing behaviors in motor starts. Most of the algorithm performance measures or validations were performed using a database or failure simulations.

On the contrary, in the present work, the algorithm was validated during the operation of the motors in real scenarios. The results were satisfactory, reaching a high level of certainty by adequately detecting mechanical failures in the motors.

2 Proposed Algorithm Description

Figure 1 shows the block diagram of the fault detection and identification algorithm. The parts that make up the algorithm are described in the following sections.

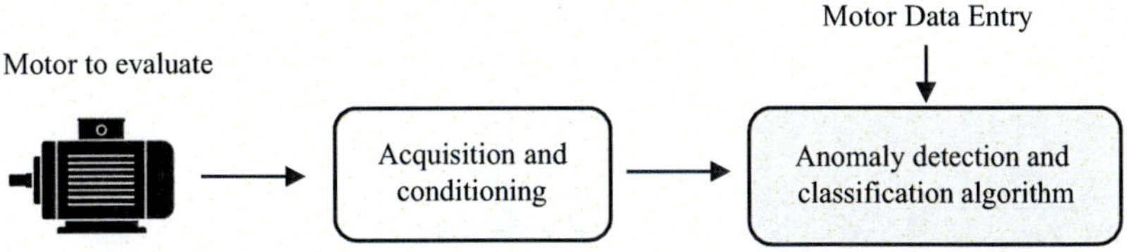

Fig. 1. Process diagram for classification of fault detection frequencies

2.1 Acquisition and Conditioning

The acquisition and conditioning module comprises the blocks of the diagram shown in Fig. 2. The vibration sensor is made up of the MPU-9250 accelerometer, while the communication module is made up of the ESP32 chip (SoC).

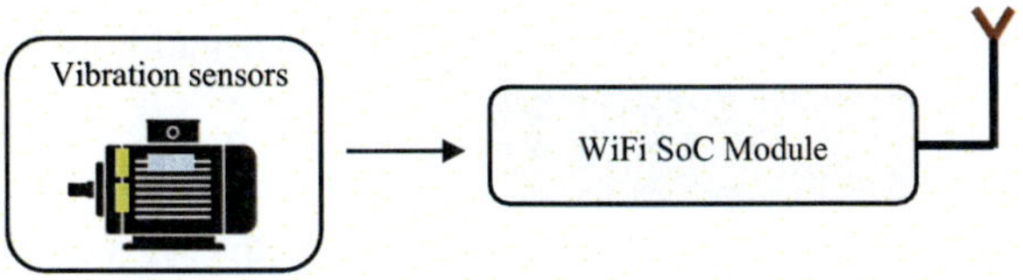

Fig. 2. Block diagram of the acquisition and conditioning module.

The MPU-9250 chip tracks 9-axis motion by combining the operation of a gyroscope, accelerometer, and magnetometer, each with three axes. The chip has three groups of 16-bit ADCs that speed up the digitization of the captured signals. In this work, only the accelerometer was used for a measurement range of ±2 g, ±4 g, ±8 g, and ±16 g [10].

The A/D conversion is performed with 16 bits of precision and a sampling frequency of fs = 1000 Hz. The signal is acquired in blocks of samples with a duration of 2 s in a range of ±2 g. The number of samples per block is obtained from the following expression:

$$N = f_s \times 2 = 1000 \times 2 = 2000 \; samples \quad (1)$$

Each signal block can be expressed as $x_b(n)$, where b indicates the segment number and n the integer variable of discrete-time. In the acquisition process, the capture of 16 signal blocks per engine was considered, where $b = 1, 2, \ldots, L$, where $L = 16$.

The acquired signal blocks are stored in the temporary memory of the ESP32 system on chip (SoC), which has two processing cores with a speed of 240 MHz. Likewise, the chip has an I2C module, Wi-Fi, and Bluetooth module [11], which provides a wireless communication to be able to transfer the samples to a processor and carry out the diagnostic processing of the engine.

2.2 Anomaly Detection and Classification Algorithm

The classification and detection algorithm consists of four stages, which are detailed in Fig. 3.

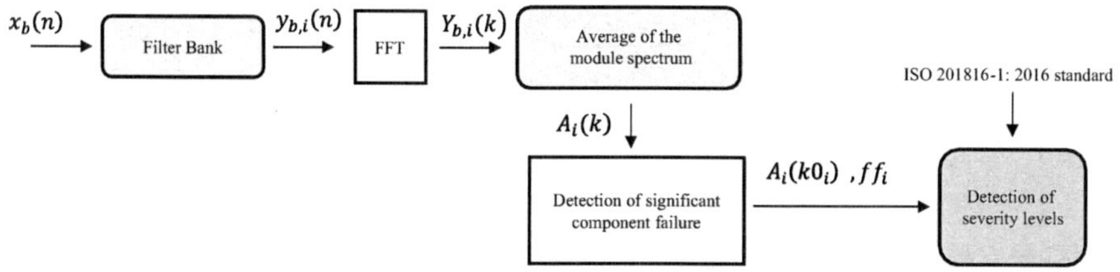

Fig. 3. Block diagram of the anomaly detection and classification algorithm.

Filter Bank to Obtain the Bands of Interest. The low-pass filter bank allows segmenting the bands of interest where the fault components with frequency ff_i will be monitored in amplitude. For this, the cutoff frequency of each filter is defined as:

$$fc_i = 1.5 \times ff_i, \ for \ i = 1, 2 \ldots, P \quad (2)$$

Where P constitutes the number of filters or failure frequencies to be monitored.

The filters used are of the 8th order linear phase FIR type. The impulsive response of each low-pass filter is defined as $h_i(n)$.

The output signal of each filter can be expressed as:

$$y_{b,i}(n) = x_b(n) * h_i(n) \quad (3)$$

The number of samples that make up each signal $y_{b,i}(n)$ is given by:

$$N1 = N + Q = 2000 + 8 = 2008 \quad (4)$$

However, we will consider eliminating four samples at the beginning and four samples at the end of $y_{b,i}(n)$ because we need to have only 2000 samples, therefore we will have $N1 = 2000$.

This filter bank segments the signal by fault frequencies to identify all the possible faults that will be evaluated. Figure 4 shows this process.

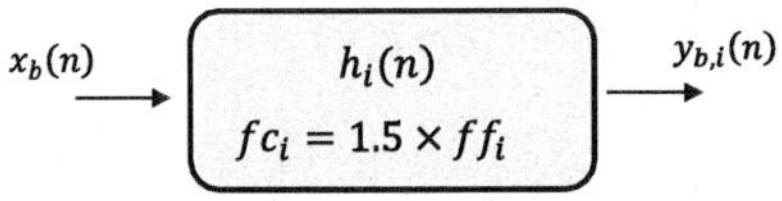

Fig. 4. Digital low pass filter block

FFT. The Fast Fourier Transform (FFT) process aims to bring each signal block into the frequency domain to identify that the flaws in different frequency components of interest. By the sampling theorem, the positive region of interest of the transform corresponds to the range from 0 to 500 Hz, given that the sampling frequency was 1000 Hz.

$$Y_{b,i}(k) = \frac{1}{N_F} \sum_{n=0}^{N1-1} y_{b,i}(n) e^{-j\frac{2\pi nk}{N_F}}, \quad 0 \leq k \leq N_F - 1 \tag{5}$$

In this case $N_F = N1$ because we want to preserve the correct amplitude of the samples at the output of the FFT (i.e., $N1$ and N_F must be multiples of the sample rate). With this, the value in Hz of each frequency k of the spectrum can be expressed as:

$$fre(k) = \Delta f \cdot k \tag{6}$$

Where:

$$\Delta f = \frac{f_s}{N_F} \tag{7}$$

In this case, $\Delta f = \frac{f_s}{N_F} = \frac{1000}{2000} = 0.5$ Hz and $fre(k) = 0.5.k$

Average of the Module Spectrum. In order to reduce noise and distortion levels in the signal and thus reduce the presence of possible spurious components, the time averaging of the $Y_{b,i}(k)$ modulus spectra was carried out. This process can be expressed as:

$$A_i(k) = \frac{1}{L} \sum_{b=1}^{L} \left|Y_{b,i}(k)\right|, \qquad 0 \leq k \leq N_F - 1 \tag{8}$$

The detection of the level of the components of interest will therefore be carried out on the average spectra $A_i(k)$ in order to reduce the probability of diagnostic error caused by noise and distortion.

Detection of Significant Component Failure. The detection of the significant components of faults is carried out in frequency bands of 2 Hz around the component of interest with frequency ff_i. This frequency range (in Hz) for each band can be defined as $[ff_i - 1, ff_i + 1]$. To start the detection process, the list of fault frequencies ff_i must be entered into the algorithm, which must be obtained from the technical specifications of the motor to be evaluated.

The use of segmentation in 2 Hz frequency bands around ff_i allows, on the other hand, to estimate the most significant component (with greater amplitude) within a tolerance interval, since by the discrete transform and the variations of the electrical network, many times the frequency of the component of interest (most significant) does not exactly coincide with the specified value ff_i.

The amplitude value $A_i(k_i')$ and location k_i' of this component must therefore satisfy the following conditions:

$$A_i(k_i') \geq A_i(k),\ \ for \begin{cases} kf_i - \Delta k \leq k \leq kf_i + \Delta k \\ kf_i - \Delta k \leq k_i' \leq kf_i + \Delta k \end{cases} \tag{9}$$

Where:

$$kf_i = round\left(\frac{ff_i}{\Delta f}\right) = round\left(\frac{ff_i}{0.5}\right) \tag{10}$$

$$\Delta k = round\left(\frac{1}{\Delta f}\right) = round\left(\frac{1}{0.5}\right) = 2 \tag{11}$$

The frequency in Hz of the significant component $A_i(k_i')$ can be expressed as:

$$ff_i' = k_i' \Delta f \tag{12}$$

Detection of Severity Levels. At this point, the amplitudes $A_i(k_i')$ at the frequencies ff_i' represent a special value expressed in *mm/s rms* that will be evaluated through the ISO 201816-1: 2016 standard [12] (see Table 1).

According to the standard, four levels of severity are established, called A, B, C, and D. In this case, the severity will be lower in A and will be higher in D. Then, under the standard criteria, the severities of the detected failures are established.

Table 1. Standard ISO10816-1: 1995 - ISO20816-1: 2016 [12]

	in/s rms	mm/s rms								
10-1000 Hz r>600rpm 2-1000 Hz r>120 rpm						D				
	0.43	11				C				
	0.28	7.1								
	0.18	4.5				B				
	0.14	3.5								
Velocity	0.11	2.8								
	0.09	2.3								
	0.06	1.4				A				
	0.03	0.71								
Foundation			Rigid	Flexible	Rigid	Flexible	Rigid	Flexible	Rigid	Flexible
Machine type			Pumps > 15 kW flow radial, axial or mixed				Medium size 15 kW < P ≤ 300 kW		Large machines 300kW < P < 50MW	
			Integrated motor		Separate motor		Motors 160 mm ≤ H < 315 mm		Motors 315 mm ≤ H	
Group			Group 4		Group 3		Group 2		Group 1	

A	New or reconditioned machines	C	The machine cannot operate for a long time
B	The machine can operate indefinitely	D	The vibration is causing damage

3 Results

The results of the evaluation of mechanical vibrations made to two motors with the same electrical and mechanical characteristics were carried out by means of an accelerometer with three coordinated axes in a single package. The tests were carried out in an industrial engine maintenance shop. Figure 5 shows us the two evaluated engines where one of them is the engine with notable damage, and the other is a pre-owned engine in good condition. For the purposes of the test, motor A will be the good motor, and motor B will be the bad motor.

Fig. 5. Left motor side A - Right side motor B

Both motors have the same electrical and mechanical characteristics because they are completely the same. Table 2 details the characteristics of both motors.

The results regarding the severity of vibrations will be made by comparing the two engines evaluated, taking engine A as a reference. Engine A, being pre-owned and in good condition, will be our reference to compare with engine B, which presents notable deficiencies in its operation. The comparison will be made at the level of the magnitude of the frequency spectrum at the corresponding ff_i.

Table 2. Characteristics of the tested engines

Good motor - A		Bad motor - B	
Electrical characteristics			
Voltage	220 V	Voltage	220 V
Current	104 A	Current	104 A
Power	30 kW	Power	30 kW
Mechanical characteristics			
Bearing No.	6312 – 2RS1	Bearing No.	6312 – 2RS1

Motors A and B have the same running-in, so their mechanical shaft rotation frequency will be the same. Then, Table 3 defines the failure frequencies ff_i to be evaluated in motors A and B. These frequencies will be determined via SKF's application to determine bearing frequencies. It needs some parameters corresponding to the bearing; therefore, these will have to be entered to calculate the failure frequencies [13]. Additionally, it requires the shaft rotation frequency measured with a digital tachometer set at 14.5 Hz or 870 rpm. It should be noted that Table 3 is filled according to the shaft rotation frequency and its harmonics corresponding to each type of fault.

The severity diagnosis will be based on a comparison of motors, according to the percent of magnitude excess in motor B with respect to motor A. Additionally, we will adhere to the ISO20816-1: 2016 standard (Table 1), which establishes four levels of severity in units of speed (mm/s) related to the vibration present in a motor.

Table 3. Failure frequencies defined by bearing 6312-2RS1

Defined fault frequencies on motors A and B	
Ball spin frequency (BSF)	29.292 Hz
Rolling element defect frequency (REF)	58.585 Hz
Cage defect frequency (FTF)	5.554 Hz
Inner race defect frequency (BPFI)	107.354 Hz
Outer race defect frequency (BPFO)	66.646 Hz
Unbalances (UF)	14.5 Hz
Loose parts (LPF)	7.25 Hz, 14.5 Hz, 21.75 Hz
Misalignment (MF)	29 Hz, 43.5 Hz
Static Eccentricity (SEF)	60 Hz, 120 Hz, 240 Hz

The results are based on the correct detection of the signals at their fault frequencies ff_i. The images presented correspond to those obtained at the end of the process. First, Fig. 6 shows the rotation frequency (29.292 Hz) of the rolling element inside the bearing,

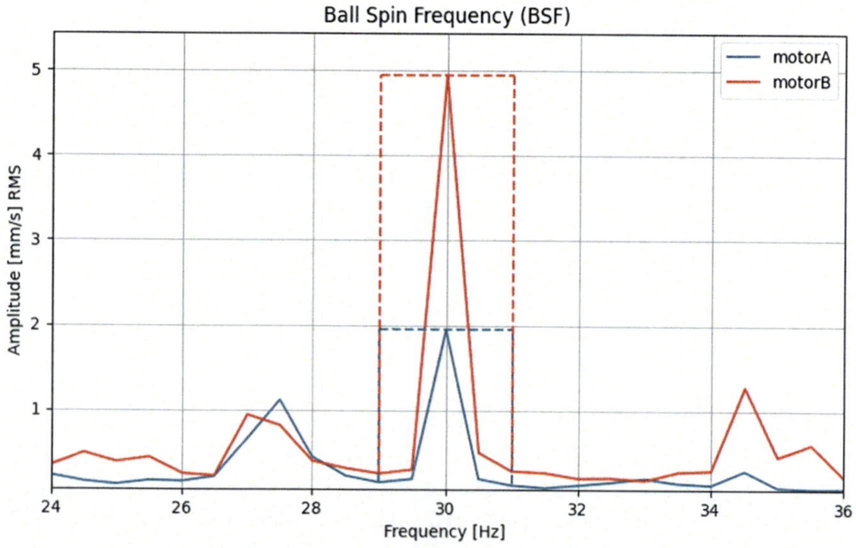

Fig. 6. Ball spin frequency – BSF

where a comparison can be made between the rolling elements inside the running gear of motor A and motor B.

Figure 7 shows the comparison of the amplitude detected in the external race defect frequency (66.646 Hz) between motor A and motor B. Evidently, motor B shows a greater amplitude when compared to motor A. Thus, motor A is in better condition. However, we cannot directly infer if motor B is badly damaged since, for that, we will have to make use of ISO20816-1: 2016, which establishes severity levels with respect to the amplitude of vibration measured in motor B.

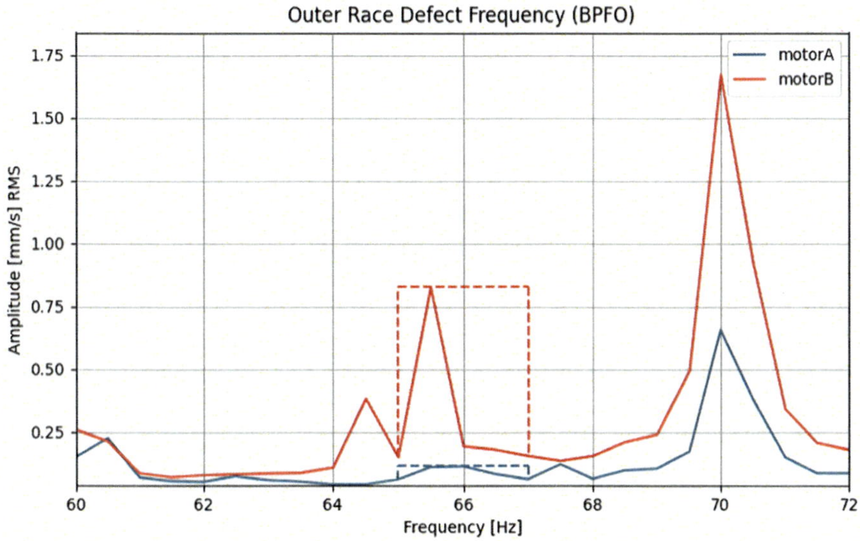

Fig. 7. Outer race defect frequency – BFPO

The measurement is quantified according to the standard in units of velocity (mm/s). Considering that it is a motor greater than 15 kW and has a flexible base, we can establish that the motor is located in group 3 with an amplitude greater than 0.75 mm/s but less than 1 mm/s, establishing a severity with the letter A. This means that the motor does not present a major problem in this failure frequency.

Having already described Fig. 7 in detail, we can describe the process in a shorter way for the other graphs. Figure 8 shows the comparison of the two motors for the internal track failure frequency (107.354 Hz). In this case, motor A and B present a very slight amplitude, which means that both motors do not present faults in terms of this frequency. This is confirmed by observing the ISO20816 standard where the amplitude of 0.04 mm/s corresponds to a severity level A, which guarantees that the motor can operate.

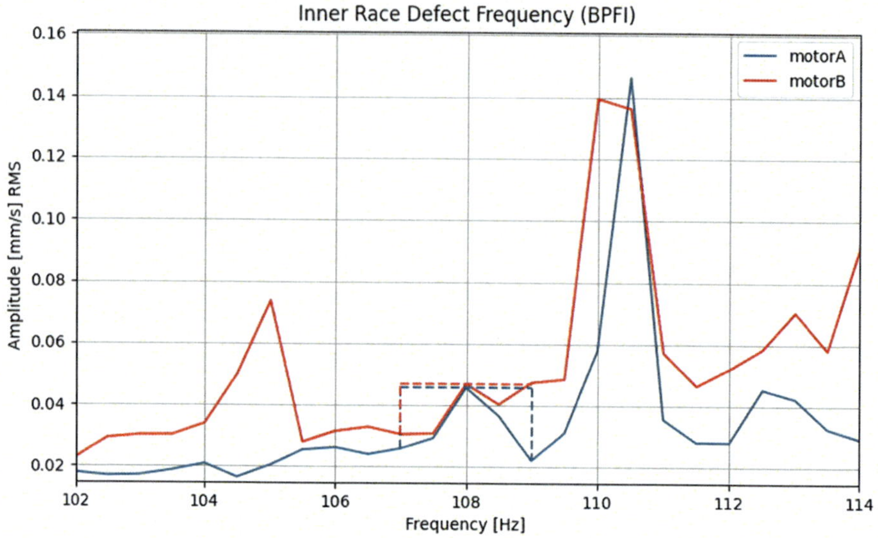

Fig. 8. Inner race defect frequency – BFPI

Figure 9 shows the comparison of amplitudes in the box or basket defect frequency (5.554 Hz). In this graph, we observe that the motor B has a greater amplitude in this failure frequency. This only indicates that motor A is in better condition than motor B. However, a severity must be defined according to the standard. In this case, the amplitude presented by motor B is around 1.6 mm/s, which places it within severity A. This means that the motor can run without problems.

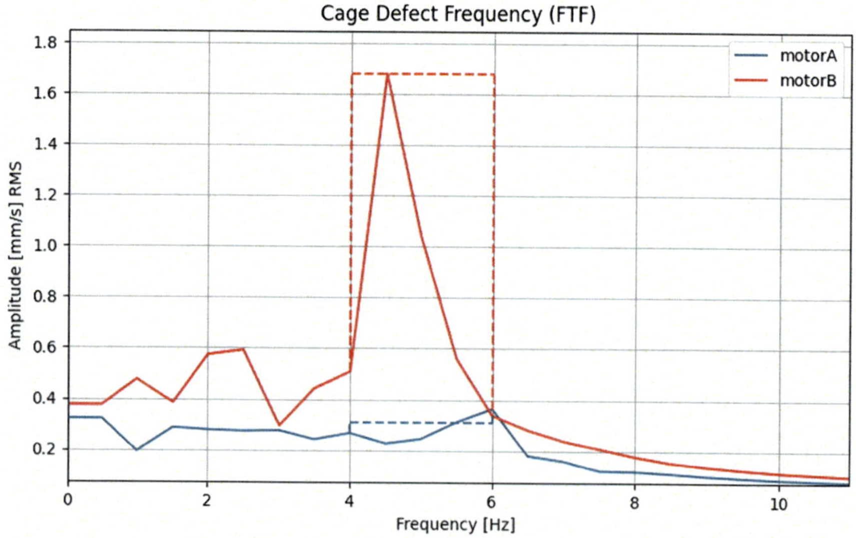

Fig. 9. Cage defect frequency – FTF

In Fig. 10, the comparison between motor A and B is shown where a slight amplitude is observed in both motors in the failure frequency REF with a value of 58.585 Hz, which, according to the standard, would give us a severity A. Then, under this severity, the motor can operate without problems.

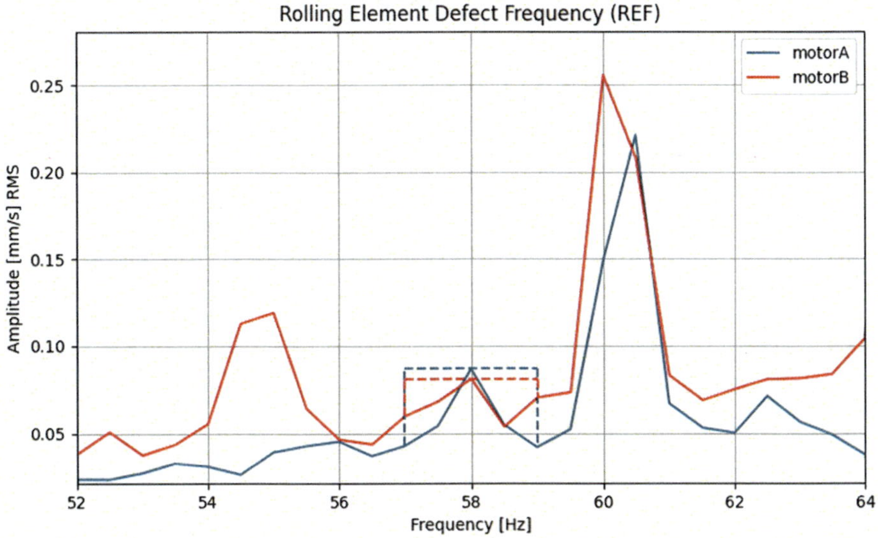

Fig. 10. Rolling element defect frequency - REF

Figure 11 shows the comparison of motors A and B regarding the unbalance fault frequency (14.5 Hz). It is observed that motor B is the one with the greatest amplitude, around 3.5 mm/s. The amplitude relationship between the two is quite high; therefore,

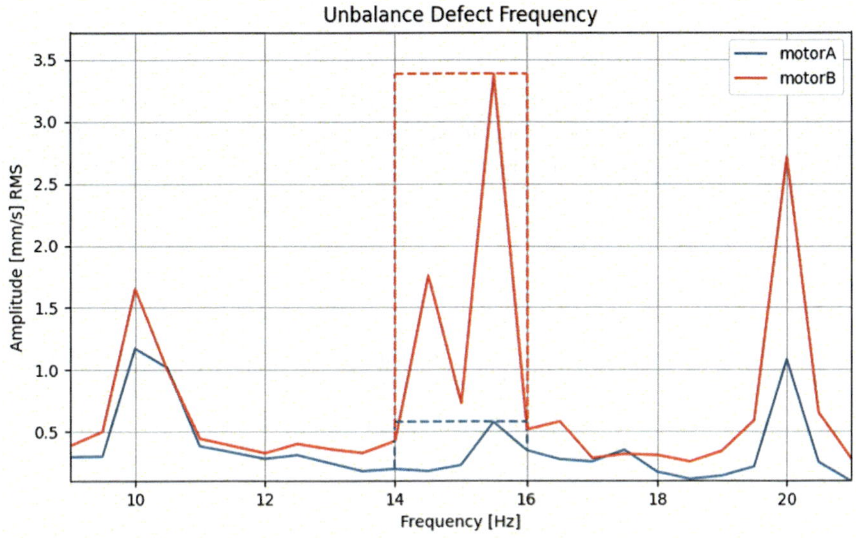

Fig. 11. Unbalance defect frequency - UF

the measurement was compared with the ISO20816 standard table. In this case, we find that engine B has exceeded severity level A and has reached a severity level C. Although the engine can still operate, there is already a greater deficiency in engine B with respect to engine A than it was designated as a healthy engine.

Figure 12 compares the existing amplitude in the misalignment defect frequency (29 Hz y 43.5 Hz) present in both motors. As can be seen, motor A has an amplitude of 2 mm/s, which gives it a severity of level A and will be able to continue operating without any inconvenience. However, motor B has an amplitude of 5 mm/s, which gives it a severity level B according to the ISO20816 standard. Under these conditions, the engine can continue to operate, but the amplitude is very close to that of a level C, which already conditions the engine not to operate for long periods.

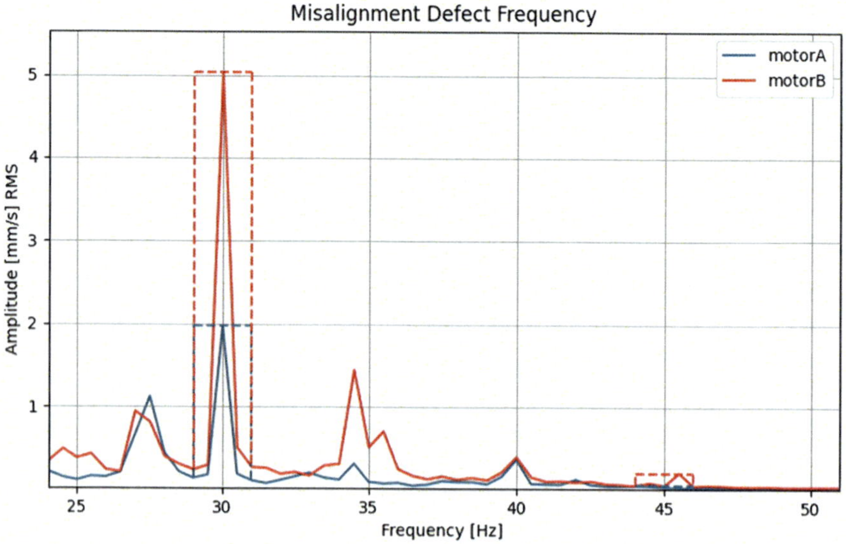

Fig. 12. Misalignment defect frequency - MF

Figure 13 shows the comparison of both motors with respect to the failure frequency of the loose parts present in the motor at the frequency values 7.25 Hz, 14.5 Hz, and 21.75 Hz. As can be seen, motor B has a very high amplitude of 7 mm/s. This amplitude is much higher than that of motor A, which has an amplitude of less than 1 mm/s. In the case of motor B, it presents an amplitude of 3.5 mm/s, which gives it a severity of level B. However, the motor will continue to work because the fault is not very severe.

Figure 14 compares motors with respect to static eccentricity fault frequency (60 Hz, 120 Hz, and 240 Hz). In this case, both signals in the motors do not have an amplitude with a high severity level. Motor A has an amplitude of around 0.40 mm/s and motor B an amplitude of around 0.75 mm/s, both amplitudes are not severe, so they present a severity level A, which allows them to continue operating normally.

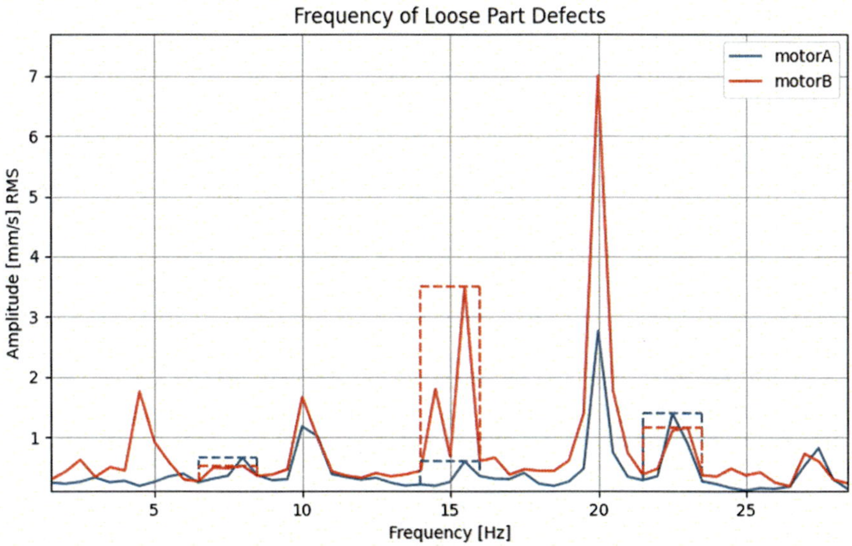

Fig. 13. Frequency of loose part defects - LPF

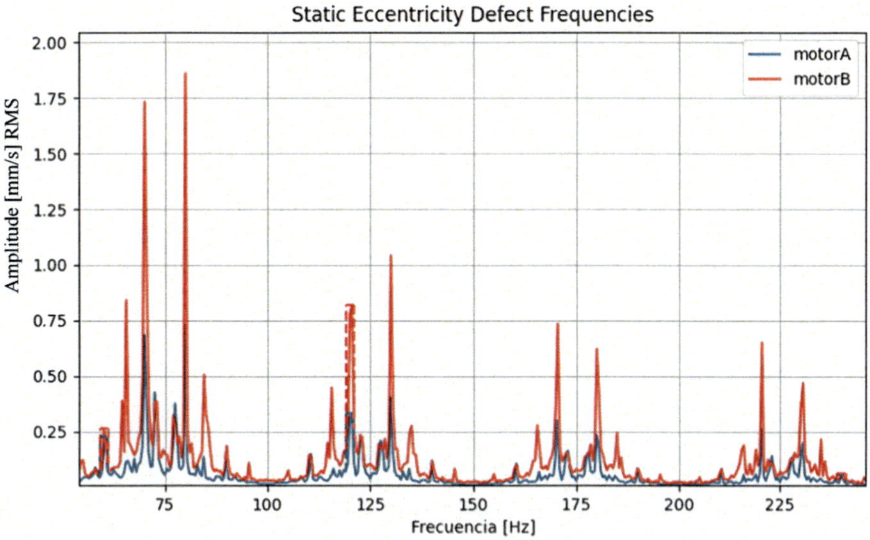

Fig. 14. Static eccentricity defect frequency - SEF

Finally, the system as a whole provides a diagnosis that is elaborated with respect to the comparative analysis between the two engines. Table 4 shows the diagnostic table to perform an evaluation of which engine has better conditions. As it is known from the beginning, the evaluation was carried out on two engines, one of which was in good condition and the other had deficiencies. Then, based on the diagnosis, it is confirmed that motor B has higher amplitudes in almost all fault frequencies. However, these amplitudes are not decisive since the severity of vibrations is not alarming. By analyzing each figure, it has been possible to determine that at most, the motor has been close to a severity level C with respect to the unbalance defect frequency.

In general, benchmarking results are given in percentages. These percentages represent how much additional amplitude motor B has with respect to motor A for each of the fault frequencies determined in the algorithm. Equation 13 states the form of calculation.

$$\Delta A_{\frac{B}{A}}(\%) = \frac{(A_{picoB} - A_{picoA})}{A_{picoA}} \times 100\% \quad (13)$$

Where:

- $\Delta A_{B/A}(\%)$: Variation of maximum Amplitude of B with respect to A
- A_{picoA}: Maximum amplitude of motor failure A.
- A_{picoB}: Maximum amplitude of motor failure B.

Table 4. Evaluation diagnosis for motor A and B

N	Test	Motor A	Motor B	Diagnosis
	Bearings			
1	FTF	0.227	1.682	641.06%
2	BSF	1.1949	4.954	154.18%
3	BFPO	0.124	0.828	569.71%
4	BFPI	0.146	0.14	4.98%
	Eccentricity			
1	Static	0.748	2.444	226.61%
	Anomalies			
1	UF	0.229	1.757	668.50%
2	MF	2.052	5.245	155.64%
3	LPF	1.963	3.458	76.16%
	Harmonics			
1	Shaft	2.315	7.046	204.37%
2	Line	0.61	1.967	222.28%

Below, there are the results of the analysis of the status of 2 motors, independently, by the method of mechanical vibrations; likewise, these results present division thresholds which refer to the limits allowed by the ISO 20816-1:2016 standard to consider an engine in satisfactory, unsatisfactory, and critical condition.

3.1 First Motor

In Fig. 15, the bearing box defect frequency amplitude is observed; this amplitude is approximately 4.625 mm/s RMS. Therefore, the software detects an anomaly in this frequency because it exceeds the vibration thresholds for optimal performance. Likewise, the algorithm classifies this defect in Class D, according to the ISO in Table 1.

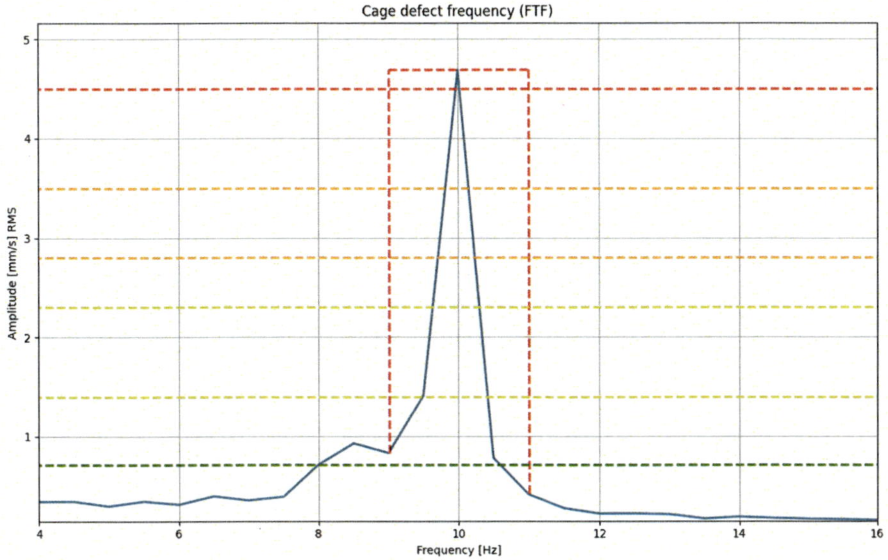

Fig. 15. Cage defect frequency – FTF

In Fig. 16, the amplitude of the rolling elements defect frequency is observed; this amplitude is approximately 1.673 mm/s RMS. Therefore, the software detects an anomaly in this frequency because it exceeds the vibration thresholds for optimal performance. Likewise, the algorithm classifies this defect in Class B, according to the ISO in Table 1.

In Fig. 17, the amplitude of the static eccentricity defect frequencies is observed. In this case, the maximum amplitude is analyzed, which is approximately 1.725 mm/s RMS. Therefore, the software detects an anomaly in this frequency because it exceeds the vibration thresholds for optimal performance. Likewise, the algorithm classifies this defect in Class B, according to the ISO in Table 1.

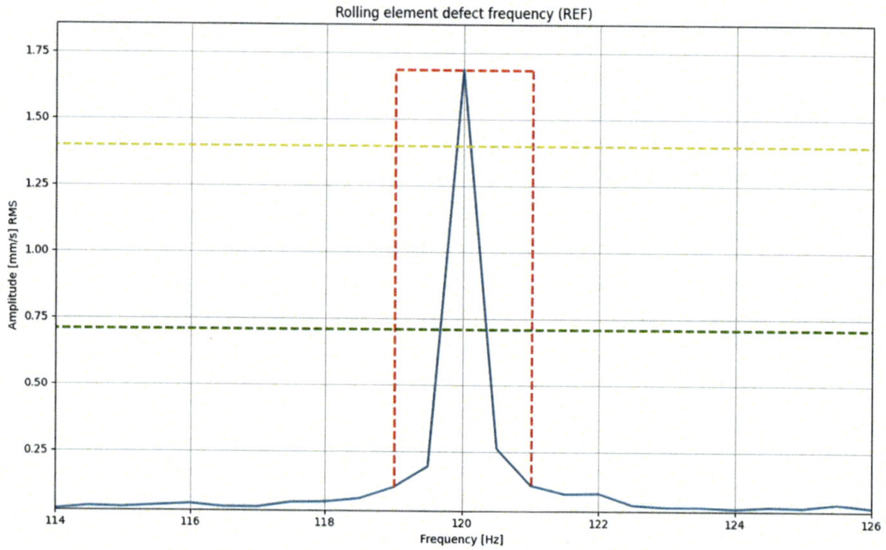

Fig. 16. Rolling element defect frequency – REF

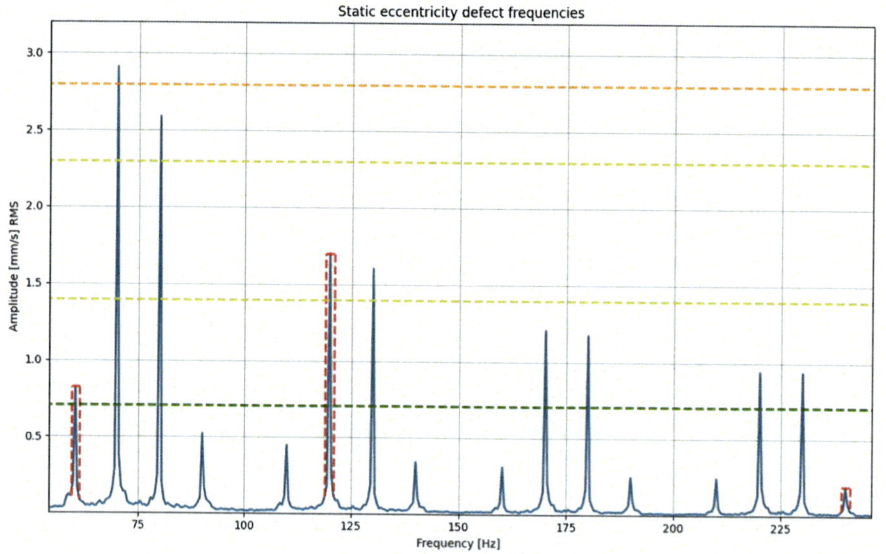

Fig. 17. Static eccentricity defect frequency – SEF

3.2 Second Motor

In Fig. 18, the bearing box defect frequency amplitude is observed; this amplitude is approximately 0.038 mm/s RMS. Therefore, the software suggests that there is no anomaly in this frequency because there is no exceeds the threshold of 0.71 mm/s

RMS; that is, it presents an optimal performance. Likewise, the algorithm classifies this measurement in Class A, according to ISO in Table 1.

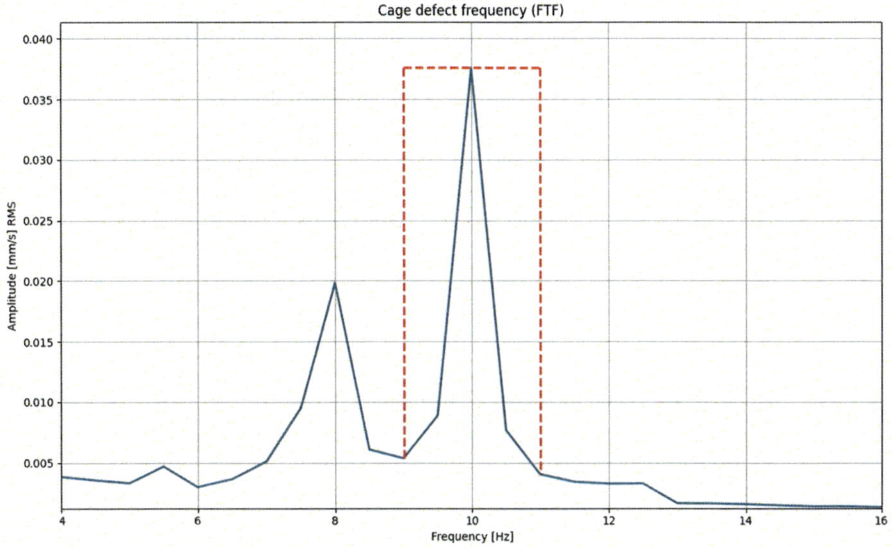

Fig. 18. Cage defect frequency – FTF

In Fig. 19, the amplitude of the rolling elements defect frequency is observed; this amplitude is approximately 0.015 mm/s RMS. Therefore, the software suggests that there is no anomaly in this frequency because there is no exceeds the threshold of 0.71 mm/s

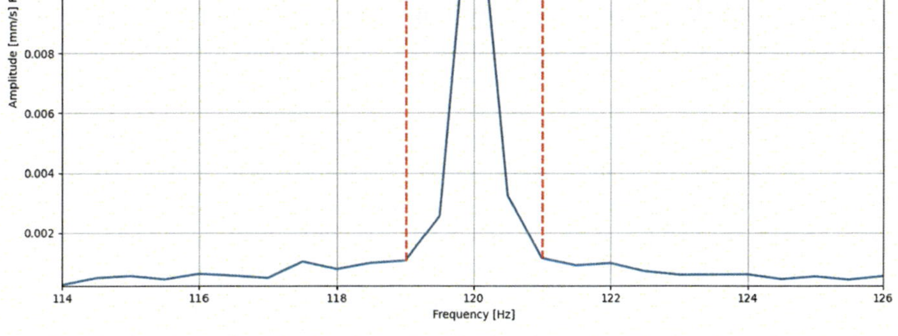

Fig. 19. Rolling element defect frequency – REF

RMS; that is, it presents an optimal performance. Likewise, the algorithm classifies this measurement in Class A, according to ISO in Table 1.

In Fig. 20 the amplitude of the static eccentricity defect frequencies is observed; in this case, the maximum amplitude is analyzed, which is approximately 0.015 mm/s RMS. Therefore, the software suggests that there is no anomaly in this frequency because there is no exceeds the threshold of 0.71 mm/s RMS; that is, it presents an optimal performance. Likewise, the algorithm classifies this measurement in Class A, according to ISO in Table 1.

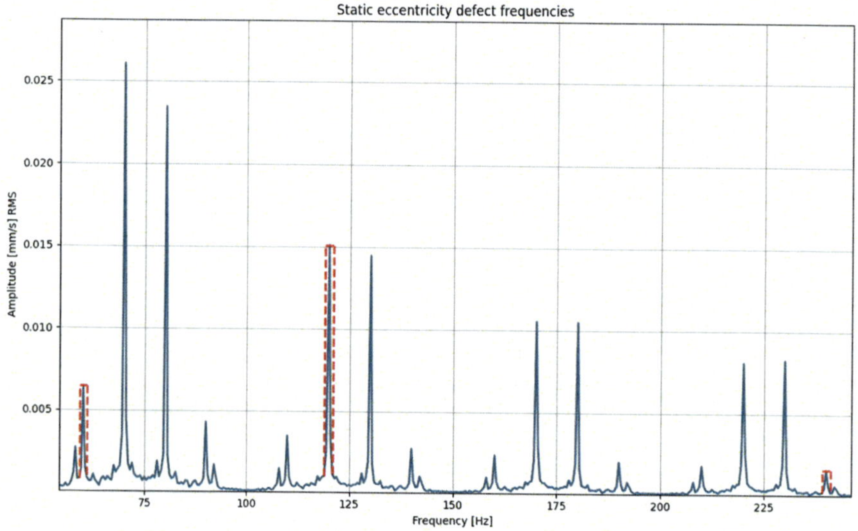

Fig. 20. Static eccentricity defect frequency – SEF

Table 5 shows the amplitudes of the failure frequencies of each motor analyzed. As can be seen, the motors that are in bold are those that were previously presented.

In Table 6, the diagnosis of a maintenance technician who is part of an electric motor maintenance workshop is presented. To validate the diagnosis made by the system, the technician proceeded to an inspection of the evaluated engines and determined certain faults described in the table.

According to Table 5 and Table 6, we can make a similarity by comparing the diagnosis provided by the system developed in this article and the technical words provided by the motor maintenance technician. Perhaps explicitly, both diagnoses do not share 100% similarity, but by making a qualitative analysis regarding certain expressions, concepts such as imbalance, misalignment, ball or rolling element failure, basket failure, and eccentricity can be matched. These phrases make references to high amplitudes within the diagnosis made by the system. A correspondence can then be established between the faults provided by the technician and the faults detected by the system. As for what is not mentioned, certain concordances are established since what is not mentioned have low amplitudes and also does not cause a major deficiency in the motor.

Table 5. Fault frequency amplitudes for each motor

Motor	Default frequency amplitudes (mm/s)							
	FTF	REF	BFPO	BFPI	SEF	UF	MF	LPF
Motor 1	**4.625**	**1.673**	**0.515**	**0.025**	**1.725**	**6.825**	**0.758**	**6.551**
Motor 2	**0.038**	**0.015**	**0.004**	**3E-04**	**0.015**	**0.062**	**0.064**	**0.062**
Motor 3	3.752	1.594	0.429	0.073	1.527	6.153	0.633	6.32
Motor 4	0.01	0.012	0.001	0.001	0.005	0.055	0.018	0.09
Motor 5	3.725	1.35	0.41	0.025	1.452	5.324	0.604	5.541

Table 6. Diagnosis of failure of each motor by a technician

Motor	Technical diagnosis
Motor 1	Regarding motor 1, the technician observed that the canes-tilla or box of the running-in is somewhat damaged. The same applies to the rolling elements or balls inside this basket. However, the bearing races are fine, but despite these conditions, he observed some eccentricity in the motor rotation. Now, this causes a pronounced imbalance that further damages the shoot. The motor needs maintenance service to correct these defects
Motor 2	With respect to motor 2, this does not present a major problem since, when the engine was started, there was no characteristic fault noise. However, it was decided to review internally to attest to motor conditions. Once opened, it was observed that the filming was relatively new and that it had few hours of work after being repaired. In conclusion, this engine is in top condition and can work many hours
Motor 3	Motor 3 shows conditions of having been wetted. Perhaps the weather has affected the interior of the motor since it presents rust in its filming. This would explain the shaft squeal on the evaluated bearing side. This has affected the balls of the shoot and the box. It would not be unusual for it to be unbalanced or have the arrow bent. In conclusion, the motor needs corrective maintenance. It cannot work continuously; otherwise, the other shoot will be damaged
Motor 4	Motor 4 is in good condition. However, a varnishing must be done as the winding appears somewhat bare. Maybe it has overheated at some point. With regard to the mechanical part, no problem is observed
Motor 5	This motor needs to be opened to determine the inconveniences it may have. At first, when it was energized, a sound appeared. However, it could not be known what caused it. When the motor was opened, it was found that the pinna (rotor) was slightly colliding with the stator almost on the edge of the metal sheet. So clearly, there is a misalignment and imbalance, which causes more vibration in the filming. These are relatively fine, but I would recommend changing them

4 Conclusion

According to the results, we can establish that there is a relationship between what the technician says and the results obtained with the detection algorithm since there is a high rate of success in the detection of mechanical failures.

The results in their entirety were obtained from real tests on motorcycles. However, quantitative validation by means of an instrument of similar operation was impossible.

Hence the validation of results through the experience of a professional in the field of engine maintenance is essential to attest to the results and the efficiency. In future works, the signals in the time domain will be analyzed, since by means of this type of analysis, the detection of various faults present in motors can be added to the algorithm.

Regarding the two types of results, although it is true, we have the ISO20816-1: 2016 standard at hand, which facilitates the classification of severity of vibrations. However, this does not mean that comparative diagnostics are useful for monitoring a group of motors with similar electrical and mechanical characteristics.

References

1. Girdhar P, Scheffer C (2004) Machinery fault diagnosis using vibration analysis. In: Practical machinery vibration analysis and predictive maintenance. Elsevier, Oxford, pp 89–133
2. Castejón C, Lara O, García J (2010) Automated diagnosis of rolling bearings using MRA and neural networks. Mech Syst Signal Process 24:289–299
3. Agahi H, Mahmoodzadeh A (2020) Decision fusion scheme for bearing defects diagnosis in induction motors. Electr Eng 102(4):2269–2279. https://doi.org/10.1007/s00202-020-01024-4
4. Ewert P, Kowalski CT, Orlowska-Kowalska T (2020) Low-cost monitoring and diagnosis system for rolling bearing faults of the induction motor based on neural network approach. Electronics 9:1334
5. Sawaqed LS, Alrayes AM (2020) Bearing fault diagnostic using machine learning algorithms. Progr Artif Intell 9(4):341–350. https://doi.org/10.1007/s13748-020-00217-z
6. Castejón C, García-Prada J, Gómez M, Rubio H (2015) Automatic selection of the WPT descomposition level for condition monitoring of rotor elements based on the sensitivity analysis of the wavelet energy. Int J Acoust Vibr 20(2):95–100
7. Zimnickas T, Vanagas J, Dambrauskas K, Kalvaitis A (2020) A technique for frequency converter-fed asynchronous motor vibration monitoring and fault classification, applying continuous wavelet transform and convolutional neural networks. Energies 13:3690
8. Puche-Panadero R, Martinez-Roman J, Sapena-Bano A, Burriel-Valencia J, Riera-Guasp M (2020) Fault diagnosis in the slip-frequency plane of induction machines working in time-varying conditions. Sensors 20:3398
9. Prudnikov AY et al (2020) J Phys: Conf Ser 1515:
10. InvenSense (2020) MPU-9250 Product Specification revisión 1.1. Japón. https://product.tdk.com/info/en/documents/catalog_datasheet/imu/PS-MPU-9250A-01-v1.1.pdf
11. Espressif (2020) ESP32-WROOM-32E & ESP32-WROOM-32UE. Shanghai. https://www.espressif.com/sites/default/files/documentation/esp32-wroom-32e_esp32-wroom-32ue_datasheet_en.pdf
12. International Organization for Standardization (2016) ISO 20816-1:2016 Mechanical vibration – Measurement and evaluation of machine vibration – Part 1: General guidelines. https://www.iso.org/standard/63180.html
13. SKF Engineering Calculator (2020) Bearing Frequencies Calculator. http://webtools3.skf.com/engcalc/calculator/freqcalc/bearing/index.jsp

Measurements and Outdoor Propagation Channel Characterization for Rumiwasi Archaeological Site at 920 MHz

Henry L. Davila-Andrade(✉) and Jorge L. Arizaca-Cusicuna

Universidad Nacional de San Antonio Abad del Cusco, Cusco, Peru
{114155,jorge.arizaca}@unsaac.edu.pe

Abstract. This article describes the characterization of a wireless channel model from the Rumiwasi archaeological site, based on Received Signal Strength Indicator (RSSI) using LoRa technology for one-slope log-normal model and two-slope log-normal model, for trajectories: non-line of sight (NLOS) and line of sight obstruction (LOBS), for which an analysis of their respective parameters was made and the most suitable model for each trajectory was chosen; RMS delay spread time of the most representative Inca walls of this archaeological site is also presented, of which a comparative analysis of the attenuation of each of these are made and how they influence the RSSI for the NLOS and LOBS trajectories.

Keywords: LoRa · RSSI · Inca wall · Path-loss exponent · RMS delay spread

1 Introduction

In recent years the Rumiwasi archaeological site located in the province and department of Cusco, San Sebastian district, has suffered several attacks that have resulted in several damaged and spray-painted walls [1], so a preventive management of this site would allow its administrators to know the location, time of stay, number, and route of visitors to have better control and care of this. Currently tools such as wireless sensor networks (WSN) offer us an efficient solution to this type of problem, but this requires a previous stage called channel modeling that offers enough parameters to obtain an optimal design of this [2].

The project "Analisis y caracterizacion experimental de un modelo de canal inalambrico y su uso en areas de conservacion" aims to model the channel of this archaeological site so that in the future, a wireless network can be designed using LoRa technology and thus have preventive management of this.

LoRa Technology is one of the most prominent and used technologies at WSN, thanks to the low energy consumption and the wide coverage it offers [3]. In this article, we use it to model the RSSI instead of the path loss because the RSSI models, in addition to offering information on the coverage, also provide information for geolocation without the use of GPS [4–6]. This was carried out for

Y. Iano et al. (Eds.): BTSym 2020, SIST 233, pp. 949–956, 2021.
https://doi.org/10.1007/978-3-030-75680-2_103

the main routes: non-line of sight (NLOS) and line of sight obstruction (LOBS) using the one-slope log-normal model and two-slope log-normal model. RMS delay spread was also characterized for the most representative Inca walls since this limits the maximum symbol rate, which does not suffer from intersymbol interference (ISI) [7].

2 Theory

2.1 RSSI Models

These models are a variant of the path loss models described in [2] for the RSSI.

One-Slope Log-Normal Model

$$RSSI(d) = RSSI(d_0) - 10n\log(\frac{d}{d_0}) + X_\sigma \tag{1}$$

Where X_σ is a random variable with Gaussian distribution and standard deviation σ, n is the path loss exponent, d_0 the initial distance at which the tests are performed. This equation describes the RSSI attenuation as a function of distance d for a single scenario.

Two-Slope Log-Normal Model

$$RSSI(d) = \begin{cases} RSSI(d_{01}) - 10n_1\log(d) + X_{\sigma01} \text{ if } d \leqslant d_b \\ RSSI(d_{02}) - 10n_2\log(d) + X_{\sigma02} \text{ if } d > d_b \end{cases} \tag{2}$$

Where the model given in Eq. (1) is used in two regions with different slopes at the breaking point d_b. This equation describes the RSSI attenuation as a function of the distance d for two scenarios with completely different characteristics whose change occurs in the break distance d_b.

2.2 RMS Delay Spread Calculation

To calculate the RMS delay spread, we use the relationships described in [7]: considering $H(k) = S_{21}$, the response in the time domain will be given by:

$$h(n) = \frac{1}{N}\sum_{k=0}^{N-1} H(k)W(k)e^{\frac{-j2\pi kn}{N}} \tag{3}$$

This is a periodic function of period $T = \frac{1}{\Delta f}$, $W(k)$ is a Hamming windowing filter, N is the number of samples.

For a channel with wide sense stationary uncorrelated scattering (WSSUS), The power delay profile (PDP) can be calculated using the following relationship:

$$PDP(t) = P(t) = |h(t)|^2 = |\sum_{n=0}^{N-1} h(n)\delta(t - n\Delta t)|^2 \tag{4}$$

The RMS delay spread can be calculated using the following relationship:

$$\tau = \sqrt{\langle \tau^2 \rangle - \langle \tau \rangle^2} \tag{5}$$

Where:

$$\langle \tau \rangle = \frac{\sum P(\tau)\tau}{\sum P(\tau)} \tag{6}$$

Is known as the mean excess delay and

$$\langle \tau^2 \rangle = \frac{\sum P(\tau)\tau^2}{\sum P(\tau)} \tag{7}$$

Is known as the mean square excess delay.

3 Methodology and Technique

3.1 RSSI Characterization

The block diagram of the measurement system is shown in Fig. 1. This consists of two TTGO T-beam cards placed on tripods at the same height of $h = 1.55$ m, the first configured as a sensor node in which the position of the card's GPS module and the RSSI were transmitted, the one that was changing its position at different distances for the routes NLOS and LOBS and the second card configured as a gateway node placed in a fixed position (position 1, see Fig. 2(a)) connected to a PC to store received data (about 100 per position) using Tera Term software, both nodes were configured at a frequency of 920 MHz with a spreading factor of 7, obtaining a total of 2091 samples.

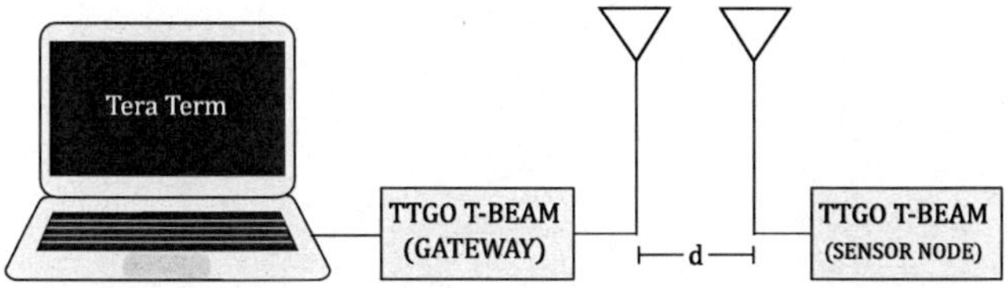

Fig. 1. Block diagram of the measurement system

The models used were chosen because the one-slope log-normal model is the most used for the RSSI characterization and the two-slope log-normal model because, in the archaeological site, there are two different regions: open area without remains archaeological sites and open area with archaeological remains of the Inca type. For the calculation of the parameters of both models, simple linear regression was used with the mean value of the RSSI in each position and the positions with reference to the Gateway node.

(a) NLOS trajectory (yellow positions) and LOBS trayectory (red positions) Source: Google Eart Pro

(b) Some test positions

Fig. 2. RSSI characterization

3.2 RMS Delay Spread Characterization

To characterize the delay time, we used the methodology proposed in [7], in which two patch-type antennas (HG908LP-NF) were used and the Fieldfox Handheld Analyzer (N9914A) was set to vector network analyzer mode (VNA) to get the parameter S_{21} in a frequency range from 800MHz to 1GHz, the block diagram is shown in the Fig. 3.

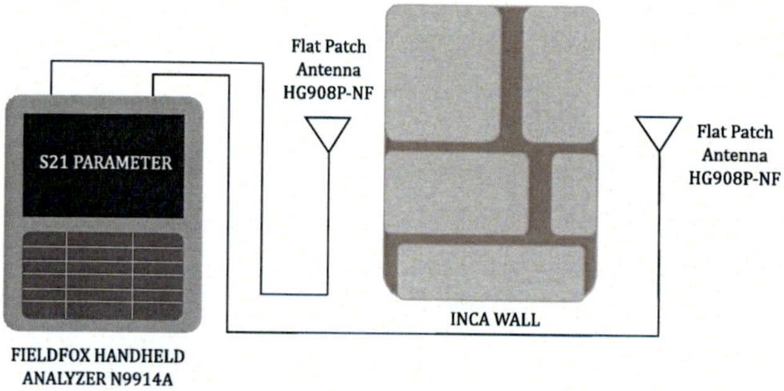

Fig. 3. Block diagram of the measurement system

This was done for the most representative Inca walls, finding that in the archaeological site of Rumiwasi two different types, both 1 m thick, different material, and different geometry (shown in Fig. 4), taking 4 profiles of the S_{21} parameter for each wall at the same height as the antennas (99 cm), a distance of 60 cm from the wall for each antenna, averaging these profiles to then calculate the RMS delay spread.

(a) Test Wall 1

(b) Test Wall 2

Fig. 4. S_{21} parameter measurement

4 Results

4.1 RSSI Model

For the NLOS trajectory, we can notice in the Fig. 5(a) that the two-slope log-normal model (Eq. (2)) better approximates the actual values than the one-slope model (Eq. (1)); this is because, in this path, there are two completely different regions: open area without archaeological remains (yellow positions 1–7, see Fig. 2(a)) and the other open area with archaeological remains of Inca type (yellow positions 8–11, see Fig. 2(a)), whose parameters are shown in Table 1 in which we can see that the two-slope log-normal model has a lower standard deviation, which reveals a value closer to the real values, so this model will better characterize this path. It should be noted that the parameter for the first region $n1 = 1.14$ is very small, which reflects both the good coverage of LoRa and the very little obstruction in this region (only native shrubs); the opposite can be seen in the parameter $n2 = 15.76$ reflecting a very high power attenuation due to archaeological remains.

For the LOBS trajectory, we must consider the hill as the first region (red positions 2–5, see Fig. 2(a)); this obstruction generates a great attenuation in which there is no-line of sight; this can be seen in the Fig. 2(b). The second region is the area of archaeological remains (red positions 8–11, see Fig. 2(a)), where we can see that the two-slope log-normal model has an offset between lines at the breaking point, as shown in Fig. 5(b); analyzing its parameters in Table 2, we can notice that the standard deviation of its lines does not differ much from the model of a slope, so for this case, the model of a slope better characterizes this path as shown in Fig. 5(b).

4.2 RMS Delay Spread

In this case, the power delay spread for both walls is shown in Fig. 6, and the RMS delay Spread values are shown in Table 3. For this case, we can see a

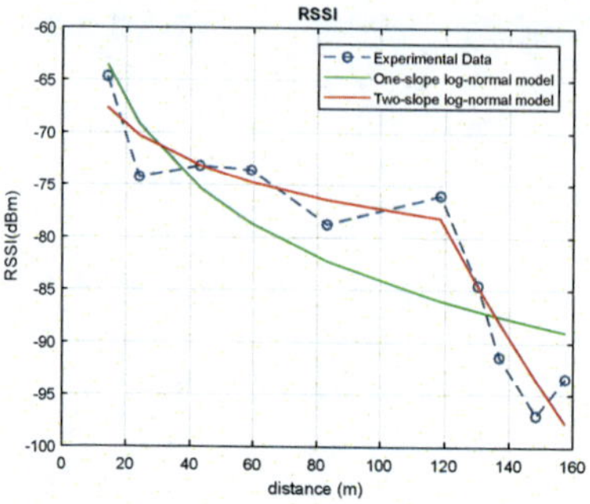

(a) Model comparison for NLOS

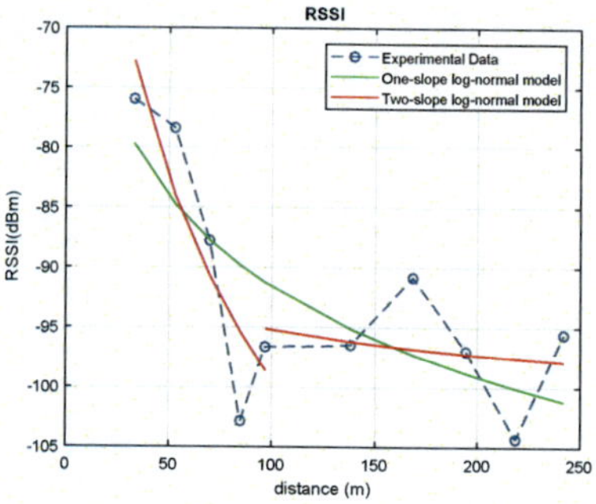

(b) Model comparison for LOBS

Fig. 5. Model comparison for NLOS and LOBS

Table 1. Parameters of the models for NLOS

Model	Parameters				$\sigma(dBm)$	
	$RSSI(d_{01})$	$RSSI(d_{02})$	n_1	n_2	σ_1	σ_2
One-slope log-normal	−63.54	–	2.42	–	5.96	–
Two-slope log-normal	−54.61	248.46	1.14	15.76	3	3.9

Table 2. Parameters of the models for LOBS

Model	Parameters				$\sigma(dBm)$	
	$RSSI(d_{01})$	$RSSI(d_{02})$	n_1	n_2	σ_1	σ_2
One-slope log-normal	−79.65	–	2.5	–	6.6	–
Two-slope log-normal	11.71	−81.39	5.56	0.69	6.05	4.71

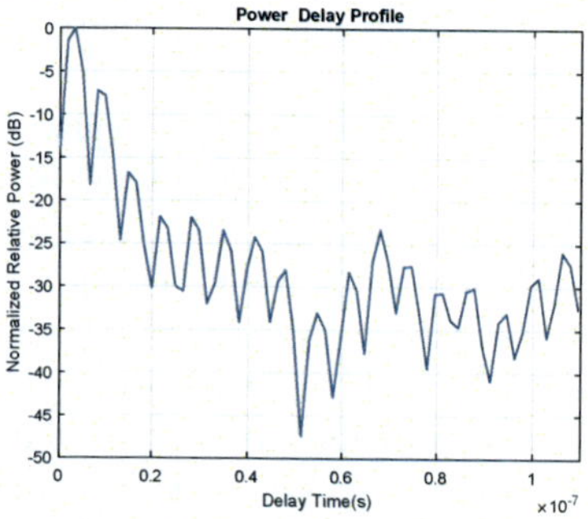

(a) PDP Test Wall 1

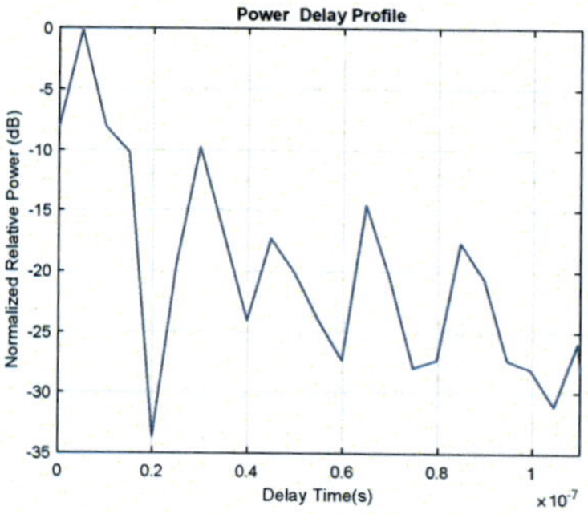

(b) PDP Test Wall 2

Fig. 6. Power delay profile of the Inca walls

Table 3. RMS power delay spread of the Inca walls

Wall	RMS delay spread (ns)
Wall 1	1.57
Wall 2	11.23

very large difference in delay, so we can say that wall 2 has materials that have higher attenuation than wall 1; it should be clarified that if we measure the S_{21} parameter to a structure with walls of the same type as wall 2, the signal will be completely attenuated, which would be equivalent to testing with the antennas pointing in opposite directions.

5 Conclusions

The RSSI for the NLOS trajectory was characterized using the double slope model. This was achieved thanks to the fact that the two regions analyzed presented completely different obstructions.

The RSSI was characterized by the LOBS trajectory using a single slope model since the double slope model did not fit properly due to the hill in the first region.

The coverage of the TTGO T-BEAM card in the archaeological site of Rumiwasi reaches approximately only 250 m radius of range after which the packages sent are totally lost, this due to the obstructions of this archaeological site.

The RMS delay spread was calculated for two of the most characteristic walls, concluding that wall 2 (Inca temple) has materials that greatly attenuate the incident signal. This can also be confirmed with the path loss exponent of $n_2 = 15.76$ for the second RSSI region in the NLOS path (see Fig. 5(a)) since the positions 7, 8, and 9 (see Fig. 2(a)) are made up mostly of wall 2. In contrast, the LOBS trajectory is mostly made up of wall 1, so the path loss exponent is less compared to the other path for the two RSSI models.

Acknowledgments. This article was made thanks to the financing of CONCYTEC, FONDECYT, and BANCO MUNDIAL for research projects and also to the Ministry of Culture for the permits granted to carry out tests at the archaeological site of Rumiwasi.

References

1. Agencia Peruana de Noticias. https://andina.pe/agencia/noticia-restauran-piedras-muro-inca-se-desprendieron-sitio-arqueologico-cusco-804626.aspx
2. Kurt S, Tavli B (2017) Path-loss modeling for wireless sensor networks: a review of models and comparative evaluations. IEEE Antennas Propag Mag 59(1):18–37. https://doi.org/10.1109/MAP.2016.2630035

3. Devalal S, Karthikeyan A (2018) LoRa technology - an overview. In: 2018 Second international conference on electronics, communication and aerospace technology (ICECA), Coimbatore, pp 284–290. https://doi.org/10.1109/ICECA.2018.8474715
4. Han G, Jiang J, Zhang C, Duong T, Guizani M, Karagiannidis GK (2018) Survey on mobile anchor node assisted localization in wireless sensor networks. IEEE Commun Surv Tutor 18(3):2220–2243. https://doi.org/10.1109/COMST.2016.2544751
5. Gharghan SK, Nordin R, Ismail M, Ali JA (2016) Accurate wireless sensor localization technique based on hybrid PSO-ANN algorithm for indoor and outdoor track cycling. IEEE Sens J 16(2):529–541. https://doi.org/10.1109/JSEN.2015.2483745
6. Gogendeau P, Murad NM, Bernard S, Kerzerho V, Deknyff L, Bonhommeau S (2018) Oversea radio measurements and channel characterization with LoRa technology. In: 2018 IEEE radio and antenna days of the indian ocean (RADIO), Grand Port, pp 1–2. https://doi.org/10.23919/RADIO.2018.8572461
7. Chen L, Ostertag T, Loschonsky M, Reindl LM (2010) Measurement of mobile radio propagation channel in ruins. In: 2010 IEEE international conference on wireless communications, networking and information security, Beijing, pp 252–256. https://doi.org/10.1109/WCINS.2010.5541724

Implementation of a Control System in a Dual Axis Cylindrical-Parabolic Solar Tracking System

William Oñate(✉), Andy Catota, Jonathan Simbaña, and Gustavo Caiza

Universidad Politécnica Salesiana, 170702 Quito, Ecuador
{wonate,gcaiza}@ups.edu.ec, {acatota,jsimbaas}@est.ups.edu.ec

Abstract. When considering effects such as population growth, global warming, reduction of oil reservoirs, and the proposal of policies committed with zero carbon, it is indicated that the worldwide renewable and non-renewable energy supplies should be prepared to fulfill the upcoming energy demand. Thus one of the renewable energy sources that exhibit more technological growth is the photovoltaic solar energy. However, other alternatives such as solar tracking plants (STP) consist of various stages, one of which is the solar tracking system that is constituted by electronic control systems to guarantee its correct operation and stability to the solar trajectory. Thus, this study develops two control systems to verify features of oscillation and settling time when there are uncertain environmental conditions such as clear and cloudy sky global horizontal irradiation (GHI). Those control systems were implemented in a prototype developed in the laboratory, which meets the following features: solar trajectory tracking in an elevation and azimuth angle, cylindrical-parabolic solar capturing by means of reflective material and reflected rays receiver tube where the internal and external temperature data are measured and collected through a SD memory. Results show that the PI control system exhibits steady-state stability at 24 [ms] with an overshoot of 1,17%, which is an excellent condition in the trajectory with GHI (clear sky), and besides, it shows no oscillations in the presence of disturbances.

Keywords: Solar tracking system · Azimuth angle · Elevation angle · Control system · Global Horizontal Irradiation GHI

1 Introduction

In one of the World Energy Outlook (WEO) annual reports by the International Energy Agency (IEA), it is mentioned the worldwide increase in the emissions of greenhouse gases [1, 2], putting aside the expected population growth, which rises depending on the fertility rate; it is thus that in some continents, such for example in Africa, whose rate shows an increasing trend, as opposed to Europe where the population growth tends to decrease, and in other continents it remains constant as in previous years [3]. It is estimated that for the year 2050, even though it is expected a decrease in the fertility

Y. Iano et al. (Eds.): BTSym 2020, SIST 233, pp. 957–967, 2021.
https://doi.org/10.1007/978-3-030-75680-2_104

rate, the demographical population growth will increase, without putting aside the global longevity rate, which goes from 72.6 years in 2019 to 77.1 in 2050 [4].

Due to the upcoming worldwide energy scenario, it is convenient to study and analyze as a country the benefits of efficient energy consumption in the socio-economic, financial, and environmental areas [5]. According to the IEA, a specific case for sustainable development in 2040 are the wind and solar energies, where the solar energy is playing a very important role without putting aside the solar collector systems, which are appropriate when taking into account variables such as the global horizontal irradiation (GHI) that depends on the radiation directly passing through the earth surface, and also the radiation that changes direction due to the presence of the atmosphere [6].

This is why various research works describe different methodologies to obtain better results for solar tracking control systems; thus, [7] mentions that for driving motors, it is necessary that the user manually enters information such as current location (latitude and longitude), date of the year and hour of the day, [8] points out that the closed-loop control systems exhibit significant and high-reliability improvements with respect to problems of open-loop systems. However, there is a real drawback generated by clouds that appear in the day, creating incorrect readings and, consequently, possible irregular positions in the angles (azimuth and elevation), due to this [9] states that it should be carried out a system that contains the two previously mentioned strategies, frequency open-loop in azimuth angle, and closed-loop in elevation or altitude angle, thus attaining to prevent irregularities in the angles and lower consumption of external energy. However, the author also specifies that there are not enough studies to guarantee precision and efficiency of solar tracking with respect to dual-axis continuous trajectory solar tracking controllers, whereby [10] specifies that their work is based on a dual-axis solar tracking system with fuzzy control to achieve an increase of 10.66% in solar capturing with respect to fixed panels and besides they encourage to continue carrying out research about the application of other possible control algorithms such as continuous PID control, digital PID or discrete control, etc. In the same way, [11] states that it is possible to reach a generation capacity 20% larger with respect to fixed collectors, which is accomplished by controlling the velocity of the system related to the irregularity of the elevation angle.

The aforementioned studies have the purpose of designing STP with efficient operation features, both mechanical and electronic, with the objective of reducing construction costs, maintenance expenses, analysis of strains experienced by the module during movement, degradation of mechanical parts [12], and reducing the amount of external energy employed for its operation. For these reasons, this study is focused on the electronic control part of the dual-axis solar capturing system, developing two types of continuous trajectory controllers, namely PI and PID, in a linear focus system with a cylindrical parabolic design for elevation and azimuth angles; the two controllers will be analyzed to determine which is the most optimal one with respect to settling time, according to Matlab Simulink tools. The present paper is organized as follows: Sect. 2 specifies the materials and methods used in the design of the solar tracking system and of the electronic control system; Section 3 shows the analysis of results; and, at last, Sect. 4 presents the conclusions.

2 Materials and Methods

2.1 Collector Design

The solar tracking system is constituted by materials that were chosen according to selection criteria; this is why for constructing the T-shape structural base, the following aspects were considered: availability of the material in the local market, cost, weight, welding technology, corrosion, and durability, opting for galvanized steel with a 55% Zinc alloy. The structure has mobility in two axes by means of two electric motors that are mechanically coupled by means of power screws with ACME or Unified M10 threads on which the load is supported to track the solar trajectory (altitude and azimuth). Thus, Eq. (1) was employed for calculating the sum of the axial forces generated on the rod of the screws when performing the task of elevating and unloading the load [13].

$$T_{elevation} = \frac{Fd_m}{2}\frac{p+\pi\mu d_m Sec\theta}{\pi d_m-\mu p Sec\theta};\ T_{lowering} = \frac{Fd_m}{2}\frac{\pi\mu d_m Sec\theta-p}{\pi d_m+\mu p Sec\theta} \quad (1)$$

Where: T is the torsional moment or torque that should be applied to the screw for elevating or lowering the load, F is the sum of all axial forces applied to the screw, p is the screw pitch, d_m is the average diameter of the screw, μ is the coefficient of friction between the screw thread and the nut, and θ is the angle at which the load is located with respect to the longitudinal axis of the screw. In addition, to guarantee the stability of the solar tracking system, it was taken into account the compression effects on the screws due to the load, for which [14] states that the slenderness coefficient λ should be greater than 89, to guarantee the stability of the module when the load is located at the most extreme points of its solar movement, and thus such coefficient was calculated using Eq. (2).

$$\lambda = \frac{AL_e\sqrt{I/A}}{I} \quad (2)$$

Where: $A = \pi\ d_i^2/4$, is the area of the screw core, L_e is considered as the buckling length and, according to specifications of the American Institute of Steel Construction (AISC), the recommended value for this structure is $L_e = 0.8\,L$, due to conditions of fixed support at the motor base and articulated in the support base of the load, L is the total length of the rod of the screw, and I is the transversal moment of inertia [15].

The following points were considered for selecting the reflective material: availability in the local market, reflectivity, absorptivity, thickness, and cost. In addition, comparative studies of reflective materials for thermal capturing plants were analyzed, where it is prioritized the maximum efficiency, the average efficiency, and the useful heat gain for the use of reflective materials in CPT [16, 17]; thus, it was opted for ASTM A240/480 bright stainless steel with dimensions 0.78 m wide, 1.22 m long and thickness of 0.0006 m. Using canonical Eq. 3 for the parabola with vertex at (0, 0), the reflected rays receiver tube was located.

$$f = \frac{x^2}{4y} \quad (3)$$

Where: f is the focal point of the parabola, x is the intersection of the parabola with the x-axis, y is the intersection of the parabola with the y-axis, and a represents the width of the parabolic head, as shown in Fig. 1.

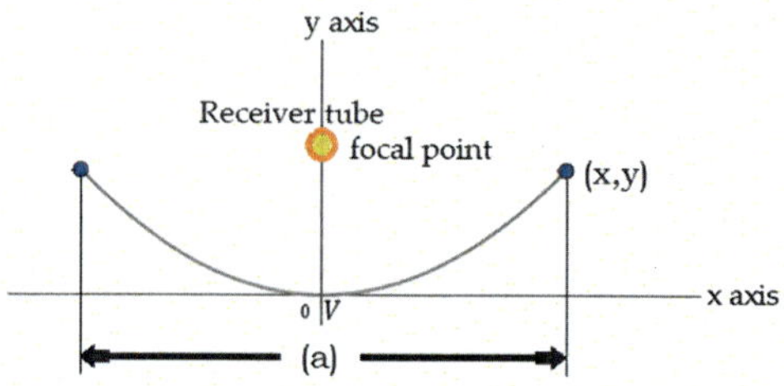

Fig. 1. Variables involved in the solar collector head.

For selecting the tube for receiving the rays reflected from the reflective material, it was considered parameters such as absorptivity, reflectivity, and thermal conductivity, opting for copper. Based on the specifications previously described, the mechanical part of the dual-axis solar collector is observed in Fig. 2.

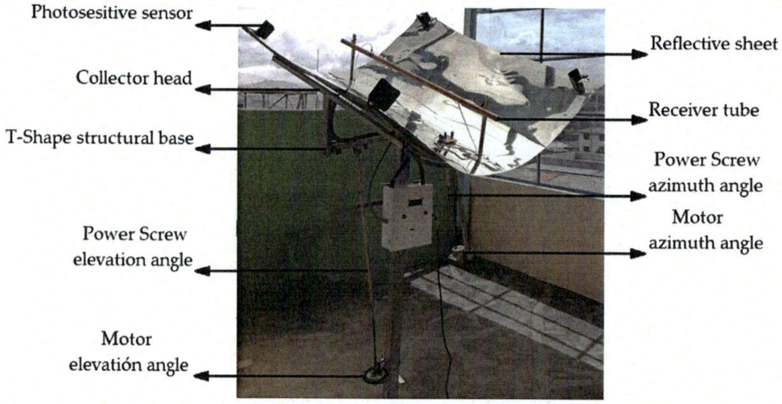

Fig. 2. Solar collector module with movement for elevation and azimuth angles.

2.2 Design and Control of the System

It is stated in [18] that there are two considerations to take into account when it comes to a real-time control system; the first is related to the need of performing a tracking with timely adjustment from the servomotor, and the second originates from the disturbance tracing that should be performed by the system regulator; it is for these considerations that control systems achieve their purpose when correctly intertwined with process dynamics, and besides it is convenient to mention the input-output interfaces in the control unit, due to the possible addition of undesired signals (measurement disturbance), which are directly related with the useful output variable in the plant process. The correct parameters are necessary for controlling the solar tracking system through the transfer function or by internal model control IMC by means of a unique parameter adjustment.

Analog Signals Interface for Obtaining the Input Data. In various projects and real systems where it is required to measure solar irradiation, it is convenient to use meters that have features such as spectral range, response time, and sensitivity, among others, with the sensitivity being a specification more referred to the design of scientific and industrial projects [19]. The commercial irradiation meters mostly used are the thermopile pyrometers, photodiodes, and pyrheliometers, but these devices have drawbacks in cost and maintenance.

For selecting the sensors for measuring irradiation, it was considered: cost, spectral range, and sensitivity. Since the solar energy captured in the Earth at sea level is deployed at wavelengths between 300 nm and 3500 nm approximately, it is appropriate to take into account the value of resistance R in series with the light-dependent resistor LDR (see Fig. 3) and of the value of resistance which varies from 400 Ω (sunny day) to 1 MΩ (darkness) approximately, and due to this [20] states that the behavior of the LDR depends on the value of the resistance R that in turn depends on the amount of irradiation to which it is exposed, thus it was selected the photosensitive technology LDR, considering the appropriate value of R and the annual global amount of irradiation GHI at the measuring location.

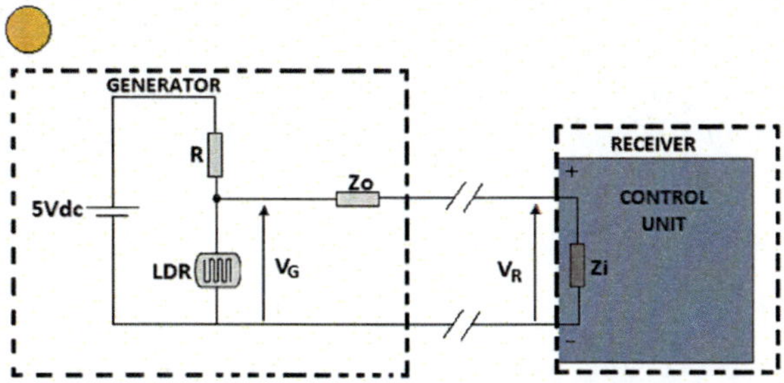

Fig. 3. Connection scheme for LDR as analog input variable by voltage.

Modeling the PI and PID Using the Matlab Toolbox. The System Identification toolbox of Matlab was used to obtain the transfer function, and the first-order function is configured due to the nature of the system with an approximation of 86.68%, as shown in Table 1, where k_p represents the process gain, which depends on the relationship between the change in amplitude of the output signal and the change in amplitude of the input signal, and τ predicts the time that the output variable takes to reach an appropriate percentage change to a steady-state value [21].

Table 1. Control parameters.

System parameters	Solar tracking system STS
Manipulated variable	PWM
Control variable	Position of the system
Disturbance	Solar Radiation
First order – approximation in the (s) domain	$G(s) = \frac{k_p}{\tau s+1} = \frac{1.002}{0.0161s+1}$
Modeling approximation	86.68%

3 Analysis and Results

3.1 Control System Scheme

Figure 4 shows the block elements that were implemented in the solar tracking system as a control strategy, in which it may be seen the irradiation analog meters, reading which is acquired by the embedded system, and the new value of the manipulated variable is established by means of the control algorithm, in this case, the pulse width modulation, which is sent to the actuators by means of the power interface, thus generating a new value of the control variable to change the position in the system: a value which was stepped for developing the feedback. At last, the temperature is registered in the reflected rays receiver tube by means of a SD memory card.

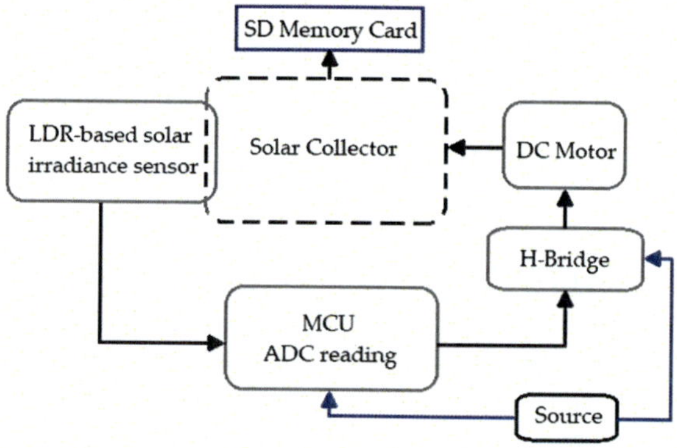

Fig. 4. Block diagram of the solar tracker.

3.2 Response of the Continuous-Time System of the PI and PID Controllers

In order to improve the settling time in the process in the presence of a disturbance (clear sky GHI and cloudy sky GHI), two control systems, namely PI and PID, were implemented, as shown in Fig. 5, where it is seen the system parameters through the block diagram developed in Matlab Simulink.

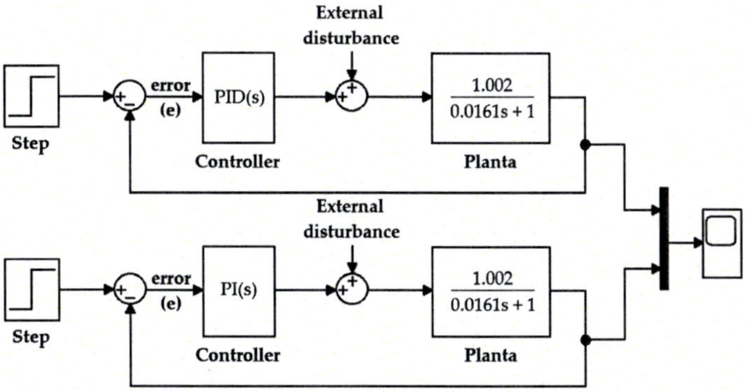

Fig. 5. Simulation of the PI and PID controllers in Simulink.

To determine which of the two control systems is more appropriate for the solar tracking process, performance and robustness information of the tuned controllers analyzed for a step signal and the transfer function of the plant, as seen Fig. 6.

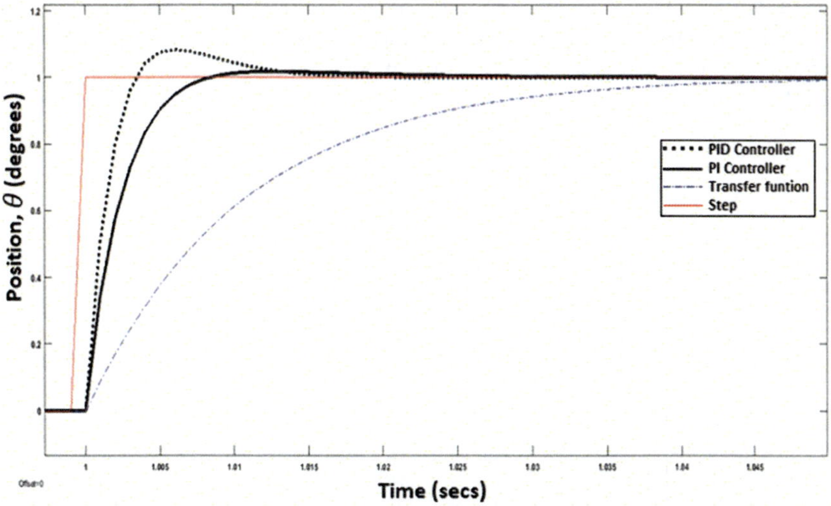

Fig. 6. Step response of the system for the PI and PID controllers.

The step response of the plant stabilizes at 50 ms, though the PID control system reaches steady-state at 18 ms with an overshoot of 8.26%, and the PI control system shows steady-state stability at 24 ms with an overshoot of 1.17%. From the data presented, it is observed that the PID control has a larger overshoot and a small oscillation for the dynamics of the system, as opposed to the PI control, which shows a small overshoot and does not generate oscillations for the dynamics of the systems due to environmental changes, and besides the consumption of external energy for processing should be prioritized. Table 2 shows the tuned values of the two controllers.

Table 2. Tuned parameters of the PI and PID controllers for the solar tracking system.

Constants			
Control	P	I	D
PI	4,4079	518,8355	0
PID	6,7925	1624,376	0,0008

3.3 Temperature Record in the Reflected Rays Receiver Tube

It is stated in [22] that it is appropriate to know in advance the location of the photovoltaic and thermal solar facilities due to the existence of factors such as atmospheric thickness and the high presence of clouds, which directly interfere with the amount of solar radiation incident on the earth surface. Thus, the solar tracking system was installed in the Universidad Politécnica Salesiana at Quito, South campus, which is at the height of 2977 m above sea level.

The data were chosen with the purpose of identifying the behavior of the global horizontal irradiation in Ecuador during the year 2019 as a reference base for this study, as shown in Fig. 7, where can be seen the behavior of the GHI and the clear sky GHI, data which represent a resolution of 4 km, Physical Solar Model PSM of the National Renewable Energy Laboratory NREL, and with a deviation of 5% due to information bias [23].

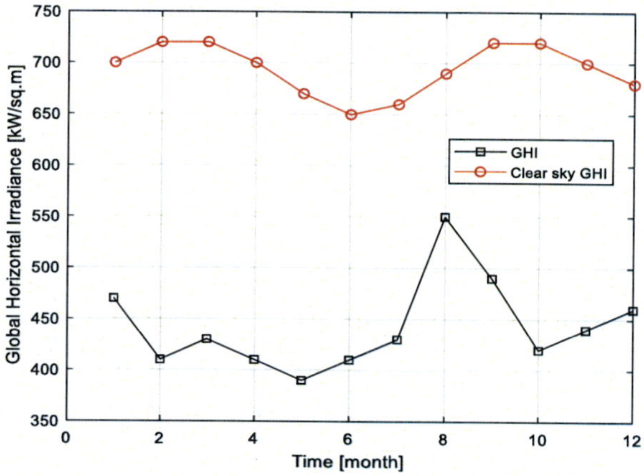

Fig. 7. Monthly solar radiation during 2019 with cloudy and clear sky.

The previous figure presents the existence of an annual average GHI of $440\left(\mathrm{kW/m^2}\right)$ and an annual average GHI (clear sky) of $690\left(\mathrm{kW/m^2}\right)$, whereby it is verified the importance of the geographic location of the country; indeed, according to the study in [24], the Pichincha, Loja, Imbabura, and Galapagos provinces exhibit a high solar potential for the development of photovoltaic and thermal solar systems.

The system has a memory space by means of a SD card for collecting information of temperature in the reflected rays receiver tube, information that was obtained during two days between 9h00 and 17h00, approximately, as shown in Fig. 8.

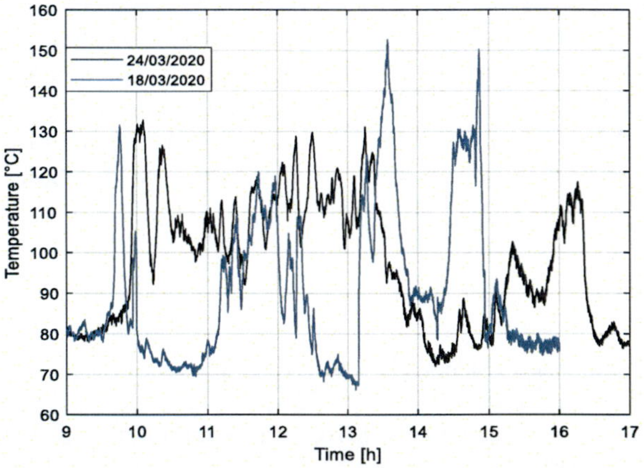

Fig. 8. External temperature in the reflected rays receiver tube from 9h00 to 17h00.

It should be emphasized that the receiver tube is made with copper and does not have any external coating that prevents heat transfer due to the circulating wind, which produces heat losses, and besides the interior of the tube is empty and does not have any thermal fluid, and finally mention that, due to the availability of reflective material in the local market, its reflectivity coefficient is 0.52; taking into account these features, Table 3 shows the average results of the environmental temperature, the internal temperature of the receiver tube, and the external temperature of the receiver tube, over a period of 8 daily hours.

Table 3. Data averaged from 9h00 to 17h00.

Day	Environmental temperature ($°C$)	External temperature of the receiver tube ($°C$)	Internal temperature of the receiver tube ($°C$)
18-03-20	15.9	86	82.4
24-03-20	16.05	108.7	101.6

4 Conclusions

Through experimental observation it highlights the performance of the two controllers during the conditions of climatic uncertainty, such disturbances have GHI features (intermittently clear-cloudy sky) and clear sky GHI in 54% of the experimental hours, and thus the PID control system shows stability in 18 ms with an overshoot of 8.26% in the presence of uncertain environmental conditions and the PI controller stabilizes in 24 ms with an overshoot of 1.17%. Along the hours with clear sky GHI, both control systems attained their objective without any instability situation to be added. However, when there is intermittent clear sky-cloudy sky GHI, it was observed that the PID control exhibits abrupt changes in the actuators' position during the dynamic of the process. In the case of the PI control, it shows improvements in the oscillation in the presence of the disturbance, but both controllers respond backing or advancing the current operating point deviating from the solar trajectory. For this reason, the PI control system is appropriate for clear GHI situations due to the smaller overshoot, and besides, low energy consumption during processing.

References

1. IEA (2019) World Energy Outlook 2019. IEA, París. https://www.iea.org/reports/world-energy-outlook-2019
2. Barrero A (2019) El último informe de la Agencia Internacional de la Energía denuncia las contradicciones del sistema energético global. Energías Renovables, Jueves, 14 de noviembre de 2019. https://www.energias-renovables.com/panorama/el-ultimo-informe-de-la-agencia-internacional-20191114
3. Shepard D (2019) Informe de la ONU. United nations department of public information, 405 East 42nd Street, New York, NY 10017. https://population.un.org/wpp/Publications/Files/WPP2019_PressRelease_ES.pdf
4. United Nations (2019) Department of Economic and Social Affairs, Population Division. World Population Prospects, Demographic Profiles Volume II. https://population.un.org/wpp/Graphs/1_Demographic%20Profiles/World.pdf
5. Poveda-Burgos GH, Molina KR, Ruiz JG (2017) Desarrollo de energías renovables en el Ecuador del siglo XXI, optimización de recursos económicos y conservación del medio ambiente. Rev Observ Econ Latinoamericana Ecuador. https://www.eumed.net/cursecon/ecolat/ec/2017/energias-renovables-ecuador.html
6. Cevallos-Sierra J, Ramos-Martin J (2017) Spatial assessment of the potential of renewable energy: the case of Ecuador. Renew Sustain Energy Rev 81:1154–1165. https://doi.org/10.1016/j.rser.2017.08.015
7. Echenique RA, Perino EJ, Odicino LA, Perelló DA, Murdocca RM (2018) Sistema de seguimiento solar para paneles fotovoltaicos. Avances Energías Renov Medio Ambiente 22:06
8. Mi Z, Chen J, Chen N, Bai Y, Fu R, Liu H (2016) Open-loop solar tracking strategy for high concentrating photovoltaic systems using variable tracking frequency. Energy Convers Manage 117:142–149. https://doi.org/10.1016/j.enconman.2016.03.009
9. Xie F, Jiang W, Wang M (2019) Study on control strategy for solar tracking system using variable tracking frequency. Energie Solaris Sinica 40(4):1011–1020
10. Montalvo PE (2017) Diseño de un Sistema De Control Difuso de Seguimiento Solar de Dos Ejes. MS thesis, Sistemas de Control y Automatización Industrial, ESPOCH, City of Univ., Riobamba- Ecuador, July 2017
11. Zhang Q-X, Yu H-Y, Zhang Q-Y, Yang D (2015) A solar automatic tracking system that generates power for lighting greenhouses. Energies 8(7):7367–7380. https://doi.org/10.3390/en8077367. https://www.mdpi.com/1996-1073/8/7/7367
12. Hijazi H, Mokhiamar O, Elsamni O (2016) Mechanical design of a low cost parabolic solar dish concentrator. Alex Eng J 55(1):1–11. https://doi.org/10.1016/j.aej.2016.01.028

13. Shigley J, Mischke C, Hunter Brown T (2004) Power screws. In: Standard handbook of machine design, 3rd edn. McGraw-Hill, pp 1–12. Sec 13. https://www.accessengineeringlibrary.com/content/book/9780071441643/chapter/chapter13
14. Budynas R, Nisbett J (2015) Screws, fasteners, and the design of nonpermanent joints. In: Shigley's mechanical engineering design, sec 8. McGraw-Hill Education. pp 406–413. https://me.utep.edu/cmstewart/documents/ME3334/Lecture%2017%20-%20Screws%20and%20Bolts.pdf
15. Beemkumar N, Yuvarajan D, Karthikeyan A et al (2019) Comparative experimental study on parabolic trough collector integrated with thermal energy storage system by using different reflective materials. J Therm Anal Calorim 137:941–948. https://doi.org/10.1007/s10973-018-07989-6
16. Muzumil A, Ahmad W, Muzaffar A et al (2020) Experimental analysis of parabolic trough collector system with multiple receiver geometries and reflective materials. Therm Sci 2020:216–232. https://doi.org/10.2298/TSCI191202216A
17. Asociación Latinoamericana de Acero (2010) Diseño de miembros de extracción. In: Especificación ANSI/AISC 360-10 para Construcciones de, Santiago de Chile, Chile, Alacero, pp 365–383. https://www.construccionenacero.com/sites/construccionenacero.com/files/publicacion/especificacion_ansi-aisc_360-10_para_construcciones_de_acero.pdf
18. Kannaiyan S et al. (2020) Solar collectors modeling and controller design for solar thermal power plant. IEEE Access Multidisc 8:81425–81446. https://ieeexplore.ieee.org/stamp/stamp.jsp?arnumber=9072405
19. DR01 First Class pyrheliometer DR02 Fast response First Class pyrheliometer/User Manual, SENSOVANT smart sensing. Valencia, Esp, 2011, pp 1–20
20. de Barros RC, Callegari JMS, Mendonca DdC, Amorim WCS, Silva MP, Pereira HA (2018) Low-cost solar irradiance meter using LDR sensors. In: 2018 13th IEEE international conference on industry applications (INDUSCON), São Paulo, Brazil, pp 72–79. https://doi.org/10.1109/INDUSCON.2018.8627176
21. Camacho EF, Berenguel M, Rubio FR, Martínez D (2012) Control básico de colectores cilindroparabólicos. In: Control de sistemas de energía solar. Avances en control industrial. Springer, Londres. https://bibliotecas.ups.edu.ec:2582/10.1007/978-0-85729-916-1_4
22. Ordóñez F, Vaca D, López J (2019) Assessment of the solar resource in Andean regions by comparison between satellite estimation and ground measurements: study case of Ecuador. J Sustain Dev 12(4):62–75. https://doi.org/10.5539/jsd.v12n4p62
23. Sengupta M, Xie Y, Lopez A, Habte A, Maclaurin G, Shelby (2018) The national solar radiation data base (NSRDB). Renew Sustain Energy Rev J 89:51–60. https://doi.org/10.1016/j.rser.2018.03.003
24. Vaca D, Ordoñez F (2019) Mapa solar del Ecuador. Proyecto de investigación de la EPN:PIMI 15-06, Scinergy, pp 3–28. https://www.ingenieriaverde.org/wp-content/uploads/2020/01/Mapa_Solar_del_Ecuador_2019.pdfAuthor. F (2016) Article title. Journal 2(5):99–110

Method for the Identification of Criticality Levels Through Microtectonic and Geotectonic Studies in Surface Outcrops with a Potential Impact on Underground Mining Works

Elizabeth Cynthia Espiritu Leon, Mateo Javier Chichizola Cisneros(✉), Guillermo Nicanor Díaz Huaina, Humberto Pehovaz Alvarez, and Carlos Raymundo

Universidad Peruana de Ciencias Aplicadas, Prolongación Primavera 2390, Lima 15023, Peru
{u201517123,u201411980,pcgmgdia,pcgmhpeh,pcsicray}@upc.edu.pe

Abstract. This paper explains how microtectonic studies, supported by geotechnical and geomechanical criteria, were useful in determining the unidentified critical areas in underground mines from the surface. The microtectonic analysis estimated rock mass stresses affecting a given area, considering the faults and cracks identified both on the surface and underground. This analysis included a range of field studies that helped us understand the behavior of unstable areas in underground works and spot the critical unidentified areas. Thereby, previously unidentified critical zones were identified within the mine. This will help take preventive measures to minimize rockfalls during works and avoid accidents, mainly in active galleries.

Keywords: Microtectonic · Critical areas · Underground works · Geotechnics

1 Introduction

Unstable zones or critical areas account for one of the greatest risks in underground mining because this problem may lead to rockfalls. According to the Osinergmin, out of the total number of fatal casualties in the mining industry, 65% occurred in underground mining in the 2007–2018 period [1]. Likewise, the second main cause of death is rockfalls, which accounts for 22% of total deaths [2]. Collapses in some underground labor-areas have always been one of the most frequent problems in the mines of Perú (21%). This also includes the mining industry on a global scale and dramatically affects the safety of workers, equipment, and infrastructure. Underground mines are located at great depths below the surface, so the factors that have impacted rock stability conditions in underground diggings are the exploitation method used, geological features, stresses, the presence of underground water, etc. This is a constant factor through time as these are on-site features of the rock mass.

Identifying and controlling the critical areas is very important, as it may represent meaningful changes in the mining industry in terms of mortality rates. Avoiding permanent rockfalls during the underground works is the key to take preventive steps in this

Y. Iano et al. (Eds.): BTSym 2020, SIST 233, pp. 968–976, 2021.
https://doi.org/10.1007/978-3-030-75680-2_105

matter. Consequently, one way to correctly determine the vulnerable areas is to identify a criticality level through microtectonic and geotectonic studies, both in existing works and new mining developments. Although geomechanical models for addressing these problems are available today, it is still important to work on the instruments because these will indicate the causes for exploitation in the rock mass, as stated by Fortunato Ramírez, from SRK Consulting Perú [3]. Hence, we propose the implementation of a model for identifying the criticality level of the area where mining exploitation is carried out. It will assess whether the zone features low, medium, or high criticality. This is achieved through microtectonic and geotectonic studies, which will assess the structural geological characteristics, such as discontinuities and underground waters.

2 State of the Art

2.1 Criticality Model in Underground Mines

Accidents happen regularly in the underground mines, and one of their main causes is rockfalls. As part of their prevention, different software models can be applied for the identification of unstable areas. In this way, the behavior of the rock mass can be determined in detail, along with its criticality level. These models can be numerical only [4, 5] or completed with mechanical structures [6–8]. The numerical models enable to spot the disturbed zones in mining excavations through advanced calculation considering the geological, geomechanical, and operational factors [4, 5]. Certain models use non-linear criteria in diggings, which let us know the behavior of the rock mass in different areas under study, with the help of certain rock parameters, via the minimum square method [4, 5, 7]. The mechanical-numeric models use protection structures to measure the rock mass stresses and level of block impact [6–8].

2.2 Microtectonic Studies in Underground Mines with Criticality Models

Microtectonic methods help us understand the deformations of rock masses caused by the micro-movements of the Earth's crust. Besides, microtectonics is frequently used to define geological structures. Different authors use microtectonics based on mathematical models, which enable us to get an insight into the formation and origin of structural discontinuities [9, 10]. Other microtectonic research works show that the seismic and magmatic activity similarly impact the rock deformation, which is proved during the analysis of microtectonic deformations in the stages of magmatic cooling using numerical models, such as the paleo-stresses in structural discontinuities formed by faults, and so the magmatic events that are similar to the tectonic ones are differentiated [11–13]. Likewise, other studies develop the systematic monitoring of kinematic behavior in active faults, supported by numeric models, which measure the type of deformation and micro-movements through stress gauges, which are then processed by the software [14].

3 Contribution

3.1 Substantiation

This research work proposes a method based on defining hierarchies for the critical levels of surface areas and underground work through microtectonic analysis and geotechnical evaluations. These assessments are based on mathematical models, enabling us to understand the formations of discontinuities and the stresses affecting the rock mass, causing a certain behavior of rocks over time due to influence from a variety of factors. Stress gauges are used to monitor stresses [10]. Unlike this, the proposed method will use microtectonic stations for assessing stresses and discontinuities because it implies lower costs and greater efficiency. Besides, the suggested method is based on the Rock Hazard Rating System (RHRS), which is a qualitative method that aims to outline the quality of the rock mass, considering the behavior and lithology of rocks found and evaluating the risks associated with the areas that involve greater vulnerability to collapse. However, this method has a weakness; it regards surface evaluation with underground impact because only the surface part of the area is evaluated. In the same way, it is based on geomechanical evaluations and numerical models, which are aimed at finding the compression resistance of rocky surfaces and the stress state of the rock mass. The application of the suggested method is innovative but challenging because there are only a few records of this new method in underground mining, and consequently, there is less number of appropriate methodologies for its development.

3.2 Proposed Method

The method that identifies criticality levels is based on three main concepts: A surface microtectonic analysis is conducted in the study area using microtectonic stations and structural geological mapping according to the features of the rock mass and the assessment parameters of the underground mine, coupled with a geotechnical assessment of the surface ground and an analysis using the Dips software. These concepts outline the configuration of the research, which examines stresses and behavior in order to improve the definition of critical area hierarchies according to the criticality level. The suggested method is innovative in terms of implementing microtectonic stations for assessing the rock mass in underground mining works and assessing critical zones based on rock mass classification.

3.3 Method Components

The proposed method comprises three phases: First, the geological structural and microtectonic analysis; second, the geotechnical assessment; and third, the assessment using the Dips software and defining the criticality level hierarchies, as described in Fig. 1.

Phase 1: Structural Geological and Microtectonic Analysis. First, as we required the geological data of the mine, we assessed the structural and geological information of the mining unit under study. The information obtained is a structural geological map showcasing the representation of outcroppings and discontinuities in the area, such as

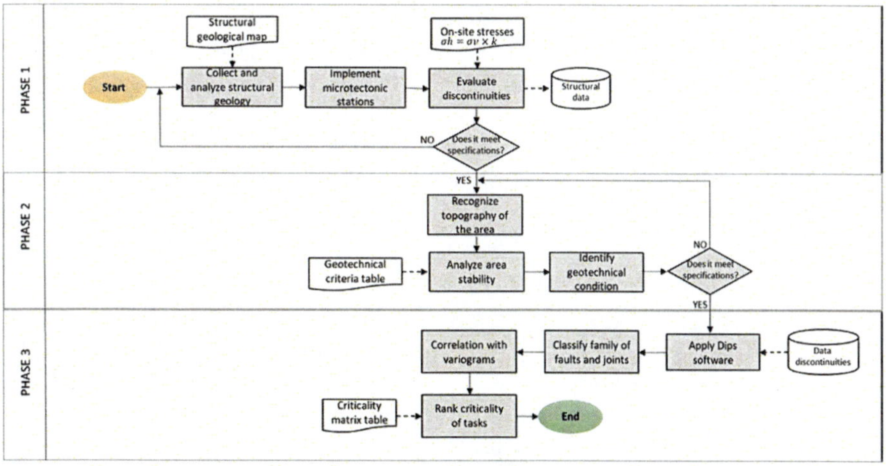

Fig. 1. Flowchart of the proposed method

faults, cracks, folds, and other types of discontinuities; this will help us understand the conditions of the rock mass. This analysis is the key for the further phases of the research study.

Next, a microtectonic analysis was performed by implementing microtectonic stations, in which the stresses that caused the discontinuities were assessed, and the direction and dip of all these discontinuities were measured. The number of implemented stations depends on the initial conditions found. This streamlines a better analysis of discontinuities, and the on-site stresses are used to determine the stability conditions of the rock mass. As a result, structural geological data was obtained, with parameters for direction, dip, and identified stresses.

Phase 2: Geotechnical Assessment. This phase includes a geotechnical assessment that will be used to study ground stability both on the surface and underground. To this end, operational and geotechnical features such as lithology, mineralization, alteration, hydrology, and other geotechnical conditions are taken into consideration. First, the ground recognition and programming are conducted, for which it is necessary to have topochartic and general data of the study area. Also, the type of ground where the study is conducted has been indicated to determine whether it is a zone with highly altered soils or a favorable terrain.

Phase 3: Dips Software Analysis and Defining Hierarchies for Criticality Level. In this phase, the Dips software was applied to determine the types of cracks and faults that may appear in the study area. For this purpose, the discontinuity data obtained in the analyses and assessments conducted in previous stages were used. For this, direction, dip, and pitch data of discontinuities identified were used. Likewise, by determining stresses, the information was compared with the discontinuity data to identify the most critical zones and to observe the effect of faults on discontinuities. For this reason, simulations were carried out considering the planar faults.

Furthermore, the objective of this phase is to classify the critical areas identified according to their criticality level. For these purposes, we considered the number of structural discontinuities identified in the area, which were obtained using the Dips software. Accordingly, a hierarchy is established for the surface and underground areas of the mine, indicating whether their criticality is high, medium, or low. For this, the Criticality Matrix displayed in Table 1 was used. In this way, decisions were made based on the risk level of the critical zones defined.

Table 1. Criticality matrix.

Frequency	Has happened					
	Common					
	Likely to occur					
	Sometimes					
	Not likely					
	Almost impossible to occur					
		0-20	20-40	40-60	60-80	80-100
		Criticality level %				

Similarly, the fall and collapse risk levels in critical zones will be classified according to the colors. First, the red color represents a high level of criticality; therefore, daily inspections must be accomplished to control the zones with a high rockfall level. Besides, the yellow color represents a middle level of criticality; thus, weekly inspections must be accomplished to control the zones with the potential to increase their criticality. Finally, the green color represents a low level of criticality; so, monthly inspections must be accomplished to control zones where rockfalls are possible.

3.4 Review of Indicators

The indicators used to calculate the results from the Criticality Level Identification model, in the validation stage, were obtained from the calculation of on-site stresses and the variogram.

Indicator of On-Site Stresses. The Brown and Hoek formula is applied. This benchmark considers tectonic and gravitational factors in order to estimate the vertical stresses, the relationship between vertical and horizontal stresses, referred to ask, and horizontal stresses.

This identifies the stresses in the study area and helps determine its behavior in underground works.

4 Validation

4.1 Scenario Description

The microtectonic stations were implemented on the surface area of the mine, whereas the stress model was applied within the mine so that the tectonic and gravitational stresses can be calculated. The mine is located in Huachocolpa district, in the Huancavelica region in Perú. The data used in this work were provided exclusively for academic purposes on the anonymity condition.

This company has different active underground works, most of which are close to the main lode. This zone sits on andesitic volcanic rocks. In addition, the quality of the rock is uneven, as it has an average RMR of 49.

4.2 Preliminary Diagnosis

Two lithological units form the local geology in the mine: to the northwest from the main lode, which presents andesitic volcanic with an average direction of N65°W and a 66°–70°NE dip; and on the southern side of the lode, which has a direction of N50°–60°E and a 50°NW dip. Inside the mine, there are gray porphyry andesites and heavily altered whitish-gray andesites.

The chosen method depended on the geological and geomechanical of the lode and the geotechnical conditions on the mine surface. Table 2 shows the geomechanical features of the Bienaventurada lode, 180 m deep, in which the Bieniawski's RMR (1989) was applied to assess the conditions of the rock mass.

The hanging wall, lode, and footwall have an average of 49, 49, and 45 RMR, respectively. It is worth stating that the development of works, like galleries and flue stacks, is done in lodes, so the lack of good conditions will result in a greater likelihood of rock falls.

Table 2. Bienaventurada lode

Lode	Level	Work	Structure	z (m *)	sC (MPa**)	RMR (prom.)
Bienaventurada	4330	SN 119	Hanging wall	180	80	49
			Lode	180	49	49
			Footwall	180	60	45

4.3 Applying the Contribution on the Scenario

Geostructural Analysis with Microtectonic Stations and Software. All the results from the microtectonic stations were assessed. 14 discontinuities were studied, taking down their direction and dip for later input to the software. Likewise, the discontinuities were classified into types.

Table 3. Classification of planar fault types.

N	DIP	DIP DIRECTION
1	70	163
2	69	169
3	70	180
4	70	172
5	65	172
6	68	155
7	65	160
8	70	155
9	63	172
10	28	149
11	70	149
12	34	190
13	68	145
14	60	136

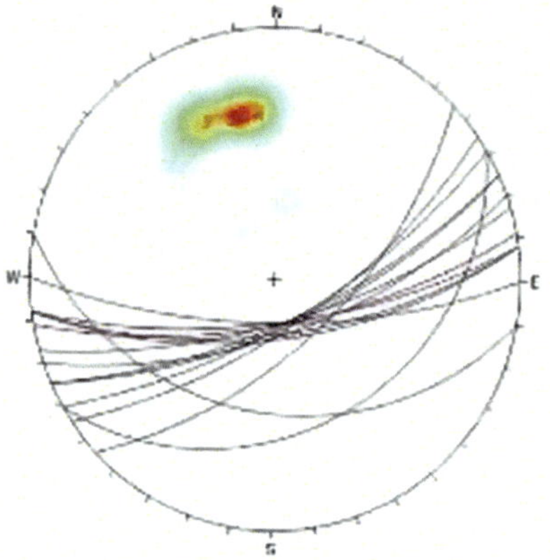

Fig. 2. Discontinuities in dips

Table 3 and Fig. 2 show the direction and dip of discontinuities assessed. Furthermore, fault types were observed in diverse colors and random types in white. As a whole, four discontinuity types were identified + 2 random types coming from planar discontinuities.

Results of the Assessment of Deep Rock Mass with Geotectonic and Geomechanic Criteria. The results of rock properties, based on geotechnical criteria, are shown in Table 4.

The results of the friction angle and cohesion of samples indicated that the rock was good. The rest of the parameters were used for on-site stress results.

Table 4. Table of geotechnical-geomechanical properties

Parameter	Unit	Hanging wall (average)	Footwall (average)	Mineral (average)
		Andesiticgap	Andesite	Mineral
Dry density	g/cm^3	2.66	2.71	3.57
Wet density	g/cm^3	2.68	2.72	3.6
Modulus of deformation	Gpa	160.97	162.15	140.78
Friction angle	°	53.4	54.82	50.52
Cohesion	Mpa	14.81	14.48	12.45

5 Conclusions

The criticality matrix made it possible to qualify the areas of greatest risk in the study area, with which the mine could consider it more important to take corrective actions in these areas to reduce them. These critical levels in the different areas evaluated could be shown more precisely by considering geological factors, in-situ stresses, and geotechnical factors. On the geotechnical analysis side, the physical-mechanical parameters of the terrain could be known, which allowed us to understand how it was affected by critical areas. On the side of the analysis of tectonic and gravitational stresses, it was facilitated to understand how the stresses of the faults were affecting the different levels and which were the points of greatest probability and intensity of criticality due to the pressure exerted by these stresses. Finally, the geological analysis made it possible to define in which area it was the most convenient to apply microtectonic stations since not all rocks have the same characteristics. For this reason, it was considered convenient to center it in an area of the Bienaventurada lode, in which there are mainly andesites and andesitic gap. Although this analysis served to determine the critical areas and their level of criticality, the stress of the crust of the Peruvian Andes is different from other areas in the world as it has other types of depth, rocks, and tectonic activity.

References

1. Organismo Supervisor de la Inversión en Energía y Minería (2018) Boletín estadístico de la Gerencia de Supervisión Minera sobre accidentes mortales en mediana minería y gran minería – 2018. Lima
2. Organismo Supervisor de la Inversión en Energía y Minería (2019) Boletín estadístico de la Gerencia de Supervisión Minera sobre accidentes mortales en mediana minería y gran minería – 2019. Lima
3. Instituto de Ingenieros de Minas del Perú (2019) Avances de la aplicación de la geo mecánica en la minería peruana – 2019. Lima
4. Tajduś A, d'Orbyn K (2017) Geomechanical numerical analisis as a guidance for preservation works of the "Wieliska" salt mine site. Stud Geotech Mech 39:25–34

5. Escalante C (2019) Modelo matemático de partículas y elementos finitos (Pfem) para predecir la zona perturbada de la excavación subterránea con voladura para optimizar el control de su estabilidad (tesis para obtener el título de ingeniero de minas). Huancayo: Universidad Nacional del Centro del Perú
6. Albaba A, Lambert S, Kneib F, Chareyre B, Nicot F (2017) DEM Modeling of a Flexible Barrier Impacted by a Dry Granular Flow. Rock Mech Rock Eng 50(11):3029–3048. https://doi.org/10.1007/s00603-017-1286-z
7. Coulibaly JB, Chanut M-A, Lambert S, Nicot F (2017) Nonlinear discrete mechanical model of steel rings. J Eng Mech 143(9):04017087. https://doi.org/10.1061/(asce)em.1943-7889.0001303
8. Toe D, Mentani A, Govoni L, Bourrier F, Gottardi G, Lambert S (2018) Introducing meta-models for a more efficient hazard mitigation strategy with rockfall protection barriers. Rock Mech Rock Eng 51(4):1097–1109. https://doi.org/10.1007/s00603-017-1394-9
9. Becerra I (2017) Caracterización estructural de la cuenca mesozoica arequipa em pachía (36v) y palca (36x) mediante análisis microtectónico e integración com información sísmica del subsuelo (tesis de pregrado). Cajamarca:Universidad Nacional de Cajamarca
10. Park M, Jung H (2019) Relationships between eclogite-facies mineral assemblages, deformation microstructures, and seismic properties in the Yuka terrane, North Qaidam ultrahigh-pressure metamorphic belt, NW China. Tectonophysics Lab, School of Earth and Environmental Sciences, Seoul National University, Seoul, South Korea. https://doi.org/10.1029/2019JB018198
11. Tibaldi A, Bonali FL, Pasquaré Mariotto F, Russo E (2017) Differentiating lava slip planes from tectonic faults: A key issue in structural geology. Glob Planetary Change. https://doi.org/10.1016/j.gloplacha.2017.11.003
12. Fischer G, Fassbinder E, Barros CEM, Fossen H (2019) The evolution of quartz veins during the tectonometamorphic development of the Brusque metamorphic complex, Brazil. J South Am Earth Sci 93:174–182. https://doi.org/10.1016/j.jsames.2019.04.027
13. Kamaci Ö, Altunkaynak Ş (2018) Cooling and deformation history of the Çataldağ metamorphic core complex (NW Turkey). J Asian Earth Sci. https://doi.org/10.1016/j.jseaes.2018.09.012
14. Rinaldi-Montes N, Rowberry M, Frontera C, Baroň I, Garcés J, Blahůt J, Martí X (2017) A contactless positioning system for monitoring discontinuities in three dimensions with geological and geotechnical applications. Rev Sci Instr 88(7):074501. https://doi.org/10.1063/1.4993925

Mode Changes in Agent-Based Simulation Models

Marcelo G. de Castro(✉), Edson L. Ursini, and Paulo S. Martins

School of Technology, University of Campinas, Limeira, SP, Brazil
m208009@dac.unicamp.br, {ursini,pmartins}@ft.unicamp.br

Abstract. This paper presents some concepts about mode change in systems and how agent-based modeling (ABM) can propose a model capable of simulating and analyzing mode change in a system. Changing tasks is critical because it should be done in a short time, and the CPU (Central Processing Unit) should be idle as little as possible. ABM is seen as a tool capable of building systems analysis where behavior is associated with emergent properties resulting from the iterations between its components, so ABM is considered capable of dealing with complex issues not yet fully addressed by traditional simulation methodologies. The modes of operation and the changes in mode are useful abstractions to allow configurable and flexible simulation systems capable of portraying the dynamics of interactions between agents and the environment where they live. Through a change of mode, the simulation model changes its characteristics over time and can thus adapt to changes in the environment. In this work, we present a proposal for an agent-based simulation model with operating mode changes using a mode change protocol based on variable input values applied to fuzzy logic determining the change of mode in order to minimize the transition time between modes. Based on the proposed work, it was possible, methodologically and practically, to evaluate and validate the use of ABM by comparison with results already validated, its benefits and difficulties, allowing us to conclude that ABM meets the promises of flexibility, representativeness, and potential sophistication for systems representation.

Keywords: Modeling · Agent-based simulation · Mode change · Agent scheduling · Fuzzy logic

1 Introduction

According to Martins [1], systems that operate in real-time are applied in complex operations and dynamically allow adaptation to a large group of changes in the environment. The modes of operation are one of the strategies used to implement these systems in complex scenarios. With the execution of the system in various operating modes, operational flexibility is achieved, with this system going through transitions between modes due to internal or external events.

An operation mode can be understood as an exemplary computational resource configuration for the operational phase of the system. With changing conditions, the allocation of system resources can become inefficient, so the system must reconfigure itself

Y. Iano et al. (Eds.): BTSym 2020, SIST 233, pp. 977–987, 2021.
https://doi.org/10.1007/978-3-030-75680-2_106

using a change of mode, adjusting resources in order to make the process effective. For Martins [1], the mode of operation can be summarized as the conduct of a system during its operation on a schedule. Function and performance are two fundamental components of a system's behavior in real-time.

A new mode is initiated when a change in the environment or an internal state of the system is detected. To make a change of mode, it is necessary to exclude some tasks and release others.

According to Massaro [2], traditionally, changes in operation mode are carried out reactively; that is, the system changes its mode in response to an internal or external event. Mixed criticality systems tolerate the loss of some deadlines for low criticality tasks. However, unforeseen events can trigger delays in the execution of tasks, and these delays can cause the loss of deadlines, including highly critical tasks. Thus, the use of proactive and dynamically adaptable systems is a viable alternative to avoid missed deadlines. A model can be interpreted as a representation of reality, with the choice of abstraction considering the relevant data. For Pecek et al. [3], computer simulation can be used to analyze any system to guarantee the quality and efficiency of stochastic and complex processes that operate in environments with limited resources. According to Sargent [4], the simulation models are used for various proposals, such as the design of systems, the development of operational policies, and research to expand the understanding of the system. In addition to the previous definitions, for Wooldridge [5], simulation is the computation area that consists of designing dynamic models capable of representing a real system to understand a problem and evaluate strategies for its resolution.

A new mode is initiated when a change in the environment or an internal state of the system is detected. To make a change of mode, it is necessary to exclude some tasks and release others. According to these described conditions, our work sought to go beyond existing ones, that is, neither making a worst-case analysis nor using discrete event simulation, but extending the analysis to the simulation by agents.

For Macal et al. [6], there is no universal consensus on the precise definition of the term "agent", but the definitions in the literature tend to agree on more points than disagree. For Wooldridge [7]: "An agent is a computer system located in a given environment, and which is capable of acting autonomously in this environment, in order to achieve its project objectives". For the same authors, the behavioral complexity of an agent is quite flexible and ranges from simple binary decisions (yes or no) to complicated human behavior or intelligence.

For Madejski [8], two different approaches exist for the agent: the physical decomposition approach and the functional decomposition approach. In the first approach, agents represent physical entities, such as workers, machines, tools, products, etc. In the latter, there is no relationship between agents and physical entities, but agents are assigned to some functions such as product distribution, transport management, order acquisition, production scheduling, material handling, etc.

In his work Martins [1] presents the analysis of task scalability in a change of mode and defines operation mode in real-time systems. It presents a strategy that allows the implementation of systems in complex scenarios with the partitioning of your applications in modes of operation. Operating flexibility is achieved by running the system in operating modes that undergo mode transitions in response to internal or external events. In view of this scenario, understanding how an agent-based simulation model can, through a protocol, change the mode necessary to adapt the system to the different changes caused by disturbances in the system is an essential tool for building the model of simulation. Thus, with the objective of contributing and extending works related to the theme [1, 2], this work presents a study about agent-based simulation and its relationship with the system's mode change process.

This article is organized into five sections. The first presents a brief introduction to the subject. The second section (Methodology and Modeling) discusses how the work was done. The third section (Fuzzy Design) provides a brief description of the Fuzzy project. The fourth section(Case Study) deals with the results obtained by changing the simulated mode. Finally, the fifth section (Conclusions) concludes the work and presents some possible future work perspectives.

2 Methodology and Modeling

In this section, the simulation and mode change processes used for the work proposal are presented. In the simulation stage, a task execution model will be used, where the use of operating modes presents a way of executing the system and its behavior due to internal or external interferences.

According to Massaro [2], a reactive approach aims to initiate a change of mode when an error and or a change in the state of the system occurs. A transition between modes is caused by the need to change the functionality provided by the system or by the need to dynamically change the performance of the system. Mode changes are triggered by a mode change request (MCR) generated by a mode management module (MM).

Figure 1 illustrates the Unified Modeling Language (UML) diagram of the agents involved in the proposed simulated process. The system is composed of two types of agents: the Task Generator and CPU Scheduler type. The tasks have no intelligence; they are just threaded instances with local attributes. The Task Generator is composed of the attributes: blocking time: task blocking time; deadline: deadline for the execution of the task; id: task identifier; period: task processing time; priority value: priority of the task; start Time: task start time; WCET: worst-case execution time; WCRT: steady-state worst-case response time.

The type of CPU Scheduler agent is composed of the attributes: Mode: indicates in which mode the CPU is operating; CPUSpeed: indicates the processing speed of the CPU; MCR: indicates the condition for triggering the mode change by the CPU; OperationMode: indicates the percentage of overload for operation mode 1. In the simulation model built, the program allows the placement of sliders that allow the adjustment of task generation and CPU processing speeds.

According to Real, Jorge, and Crespo [9], a mode change request (MCR) is defined as the event that triggers a change from a transition mode, from a source mode, or old mode, to a destination mode or new mode. This event is produced by any task with the necessary knowledge of the internal or external state of the system to produce the transition. An MCR can occur during the execution of the system in a particular mode at steady state, but never during a transition between modes.

The proposed mode-change model is based on the same assumptions as Real and Crespo [10] and Massaro [2], where: 1) tasks are generated with fixed priorities; 2) the

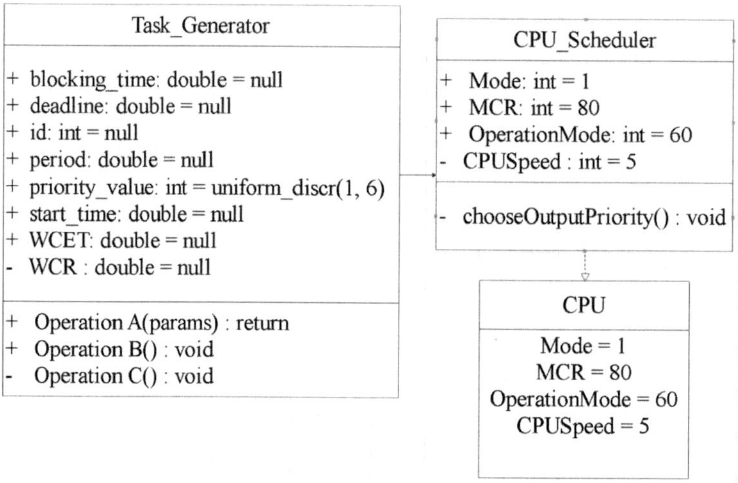

Fig. 1. Agent class diagram.

tasks are performed in a monoprocessed system; 3) tasks are not voluntarily suspended during an invocation; 4) there are sets of fixed tasks before and after the mode change, and 5) the tasks are programmed with offsets only during the mode change. In this period of time, tasks may or may not continue to run after the mode change.

A multi-moded system is comprised of a set of operating modes in which it can evolve. A particular operating mode is characterized by a constant set of tasks. Therefore, in a system with m modes of operation, M_1 to M_n, a task τ_i can be seen as a set of tuples:

$$\begin{aligned} \tau_i = \{\tau_i^{M_1} &= \left(C_i^{M_1}, T_i^{M_1}, D_i^{M_1}, P_i^{M_1}\right), \\ \tau_i^{M_2} &= \left(C_i^{M_2}, T_i^{M_2}, D_i^{M_2}, P_i^{M_2}\right), \ldots \\ \tau_i^{M_m} &= \left(C_i^{M_m}, T_i^{M_m}, D_i^{M_m}, P_i^{M_m}\right)\} \end{aligned}$$

Due to the main objective of validating this approach, the same set of tasks used by Massaro [2] was applied, which defined that a set of tasks is composed of periodic or sporadic tasks $\tau = \{\tau_1, \tau_2, \ldots, \tau_i, \ldots, \tau_p\}$ for each operating mode. Each τ_i task is defined by the $S_i = T_i, D_i, C_i, P_i$ parameters, where:

- T_i is the period of the τ_i task;
- D_i is the deadline for the task τ_i;
- C_i is the worst-case execution time (WCET) of the τ_i task;
- P_i is the priority of the τ_i task, determined by the values of T_i, D_i, and C_i, assigned according to the monotonic scheduling algorithm (for Fixed-Priority Preemptive Systems - FPPS).

Fixed-priority preemptive scheduling (FPPS) is a scheduling system commonly used in real-time systems.

Table 1 shows the set of tasks that were used in the simulation. There are two modes of operation, M_1 and M_2, in which six tasks are performed, τ_1 to τ_6.

Table 1. Task set

Task	MODE M_1						MODE M_2					
	P	C	B	T = D	R	Test	P	C	B	T = D	R	Test
τ_1	1	10	0	100	10	C	1	10	0	100	10	C
τ_2	Not active in this mode						2	20	0	120	30	C
τ_3	2	30	0	200	40	C	3	30	0	270	60	C
τ_4	3	40	0	280	80	C	4	40	0	280	100	C
τ_5	4	50	0	300	140	C	5	50	0	350	180	C
τ_6	5	60	0	350	200	C	Not active in this mode					

The columns in Table 1 correspond to the task identifier (Task), task priority (P) at low criticality level, worst execution time (C), blocking time (B), the period (T = D) of the tasks. Column R in Table 1 corresponds to the response time for the worst-case steady-state (WCRT).

Figure 2 shows the main components of the reactive computational worst-case model and are the same instances adopted in the agent-based simulation.

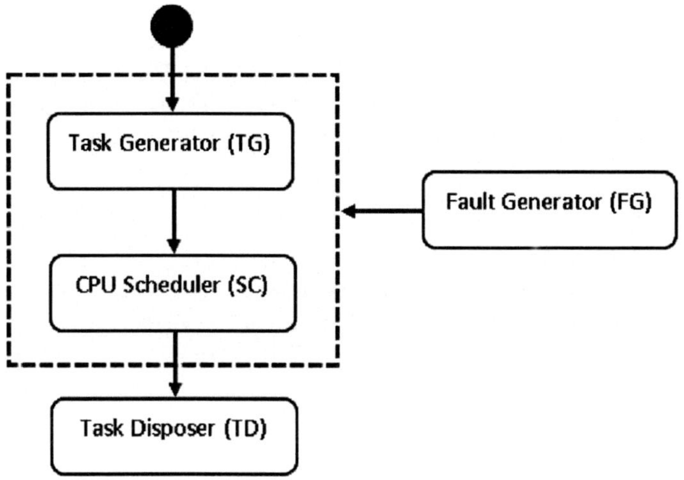

Fig. 2. Reactive computational model.

– Tasks Generator (TG): this block generates the instances of the tasks according to their periodicity and characteristics;

- CPU scheduling (SC): this block arranges the tasks released in a processing queue;
- Task Disposer (TD): this module is responsible for removing tasks from the system;
- Fault Generator (FG): this module generates (or injects) random CPU faults so that the necessary scenarios can be created to test the reactive approach. These failures temporarily block the processor, causing a change in mode (or change in the level of criticality).

The process starts with the generation of the agent by the Tasks Generator (TG), which performs the task instance and assigns its properties. The tasks are then sent to the CPU scheduling (SC) responsible for directing the tasks considering the order of arrival of each one to the CPU according to the current operation mode. If the CPU is in M_1 mode (standard or normal mode), tasks of type τ_1, τ_3, τ_4, τ_5 and τ_6 are sent for execution. The task τ_2 is not executed in this mode and is discarded. If the CPU is in M_2 mode (recovery mode), tasks of type τ_1, τ_2, τ_3, τ_4 and τ_5 are sent for execution. Task τ_6 is not executed and is discarded in this mode.

3 Fuzzy Design

The proposed work deals with the simulation of the execution of a set of tasks through a processing unit (CPU). To solve the problem related to the operation mode, the use of a normal operating range was adopted where the system behaves normally. Outside the normal operating range, the system assumes a critical state. Two input variables (Utilization and Control) and one output (Mode Change Request - MCR) were defined, as in Fig. 3.

Figure 4 shows the Utilization variable has two states: Normal (from 10 to 75) and Critical (65). The Control variable can be active (Active) or not (N Active).

Figure 5 shows the output variable MCR that can be No Change or Change Mode determined by the rules of the implemented Fuzzy logic. As the main objective was to validate the model by agents, we used a Fuzzy project, with only three rules (MCR is Mode Change Request): 1. If (Utilization is Normal) and (Control is N Active) then (MCR is Change Mode); 2. If (Utilization is Critical) and (Control is N Active) then (MCR is Change Mode); 3. If (Utilization is Normal) or (Control is Active) then (MCR is Change Mode).

Figure 6 shows the output through defuzzification, and the three-dimensional surface, showing the output variable as a function of the two input variables.

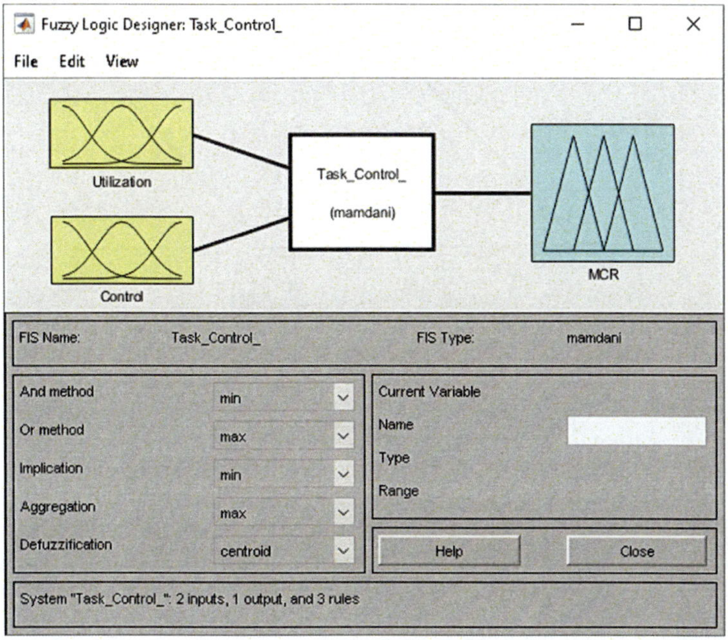

Fig. 3. Fuzzy input and output variables.

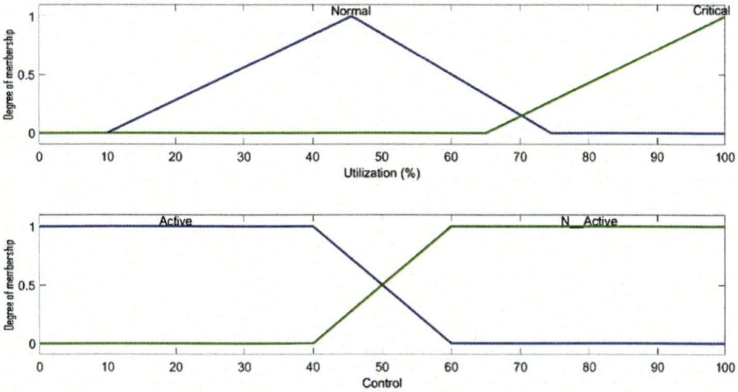

Fig. 4. Fuzzy input state variables.

The mode change is initiated when the CPU utilization level reaches the value determined by fuzzy logic through triangular membership functions or by an external event (rectangular membership functions). During the mode change operation, agents undergo a change of state and, in some cases, some of them are created dynamically or had their cycle completed/interrupted. For these actions to be carried out, the creation of types of agents in the model was adopted in order to allow the dynamic creation of agents. The model returns to its initial mode (M_1) when the use of the CPU returns to the level of use of the operating mode defined by the fuzzy logic.

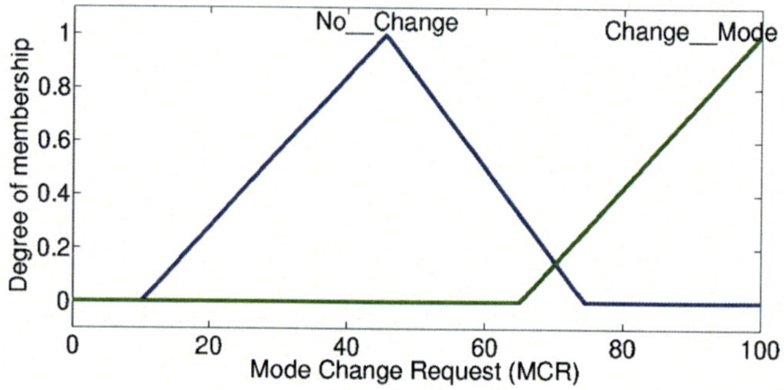

Fig. 5. Fuzzy output.

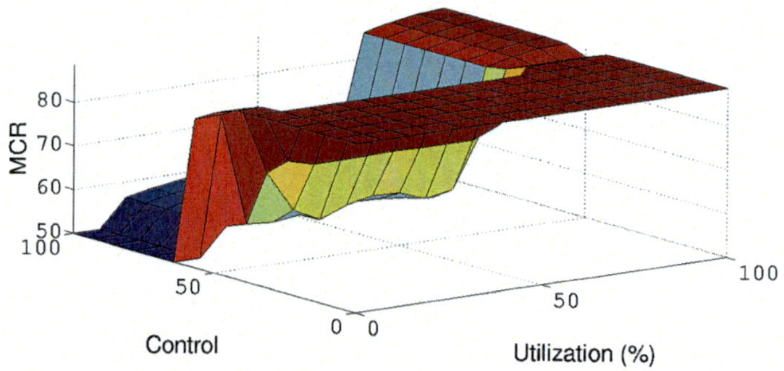

Fig. 6. Fuzzy output through defuzzification.

4 Case Study

The Anylogic University 8.6.0 [11] software was used to build the model, which allows modeling by agents and uses the JAVA language for its implementation. The objective of the model was to carry out and validate the change of agent mode (CPU agent) through the interaction of the FG agent, causing a change in the system's behavior. The advantages of the agent-based model observed at work and according to Bonabeau [12] in relation to other modeling techniques are: Allows you to capture emerging phenomena; Provides a natural description of certain types of systems; It is flexible and allows to house other functionalities like, for example, prediction of the future state of the system.

When the mode change event is triggered, the FG module sends a mode change signal to the system, and the SC module is switched to the new mode. When the mode change is triggered again, the FG module informs the system to return to the initial mode.

The values for the change of mode (min 70% and max 95% use of CPU) and mode of operation of the system (min 10% and max 70% use of CPU) were applied to triangular membership functions considering the intersections of Fuzzy logic.

Table 2 shows the comparison response times with the results obtained by Massaro [2]. The table shows the response time R_1 for mode M_1 obtained using the proposed model and the worst-case response time R_2 for mode M_2.

Table 2. Task comparison

Task	Proposed model		Comparison model	
	R_1	R_2	R_1	R_2
τ_1	10	10	10	10
τ_2	Not active	30	Not active	30
τ_3	40	60	40	60
τ_4	80	100	80	100
τ_5	140	180	140	180
τ_6	200	Not active	200	Not active

It is possible to observe that the results of the simulation in both modes were the same. It should be noted that the tasks of both methods used for validation have deterministic values and, therefore, the values obtained are also constant. It is also important to note that the values correspond to the worst-case response time at steady state for modes M_1 and M_2. Therefore, there is an agreement with the results of the validation in [2], pg. 79, in which the worst-case analytical model was validated by comparison with a discrete event simulation model (but not by using agents). Thus, these results validate the proposed simulation model.

One of the advantages observed by the use of the agent was the stabilization dynamics of the case study through the agent. By forcing a change of mode through the fault generation button, the agents identify the anomaly in the system and react in a way to stabilize the system.

Figure 7 shows a snapshot of the animation of the results obtained after changing the mode.

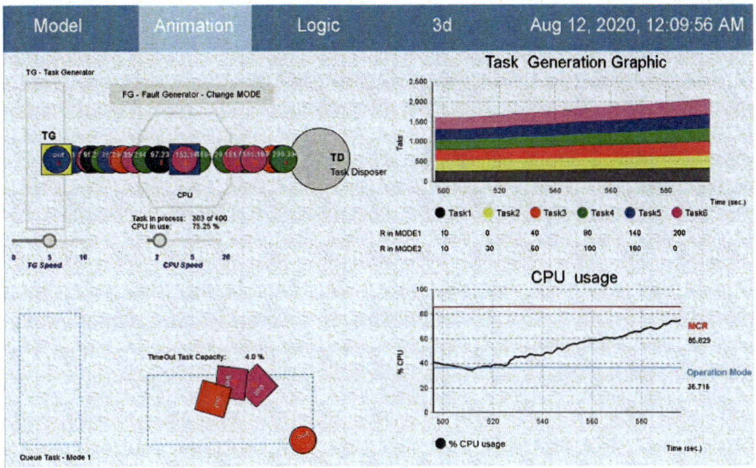

Fig. 7. Snapshot of the model animation.

- Tasks Generator (TG): generating component of tasks based on blocking time, deadline, periodicity, priority of tasks. The speed at which tasks are generated can be dynamically adjusted;
- CPU: component that performs the execution of tasks. The speed of task capture by the component can be dynamically adjusted. Changing the system mode is indicated on the CPU component by the color of the component, using gray for mode M_1 and red for mode M_2;
- Task Disposer (TD): component responsible for removing tasks from the system after completion of their execution;
- Fault Generator (FG) - Change mode: generates (or injects) random CPU faults so that the necessary scenarios can be created to test the reactive approach. These failures temporarily block the processor, causing a change in mode (or change in the level of criticality);
- Queue Task Mode 1: component responsible for storing tasks that had their waiting timeout exceeded during mode M_1.
- Time Out Task Capacity: component responsible for indicating tasks that were unable to perform their tasks due to system critical;
- Task Generation Graphic: graph that presents the evolution of the generation of tasks generated in the system and their respective response times in mode M_1 and mode M_2;
- CPU usage: graph showing the evolution of CPU time indicating the programmed range of system operation. The red line indicates the mode change request (MCR) of the system, and the blue line indicates the operating values of the system with triangular membership functions considering the intersections of Fuzzy logic.

The animation illustrates the tasks created in mode M_1 using the circular format and the tasks created in mode M_2 using the square format.

The work presents as main contributions:

- The proposal of a model based on agents validated by comparing a case study with discrete modeling;
- Implementation of mode changes in an agent-based model;
- Introduction in the agent-based model of an operation mode control using Fuzzy logic.

5 Conclusions

This work performed the modeling of an agent-based system in order to dynamically implement and validate the agent mode change, the model being validated by comparing a model already presented in the literature using discrete events, which was in turn validated by a worst-case analytical model. It was possible to observe that the results of the simulation in both modes were the same as was the initial assumption. The values were found to be identical because the model is deterministic, and the time factor used for comparison is always the WCRT. The work brought, as a contribution, the validation of the mode change approach in an agent simulation model, one of the advantages observed being the stabilization dynamics of the case study through the agent. Although the results are preliminary and a small number of iterations have been performed, it was possible to notice that the agent-based model brought to the model a process capable of ensuring its execution when exposed to a failure event. The perspective of future work is the implementation of agent performance metrics and prediction of agent capacity, including the use of machine learning for agents.

Acknowledgment. This study was financed in part by the Coordenação de Aperfeiçoamento de Pessoal de Nível Superior - Brasil (CAPES) - Finance Code 001.

References

1. Martins P et al (2012) Schedulability analysis of mode changes with arbitrary deadlines. In: Real-time systems, architecture, scheduling, and application [S.l.]. InTech. https://doi.org/10.5772/37225. https://www.intechopen.com/books/real-time-systems-architecture-scheduling-and-application/schedulability-analysis-of-mode-changes-with-arbitrary-deadlines
2. Massaro Júnior FR (2019) Proposal of a flexible and proactive model for mode changes in mixed-criticality real-time systems. School of Technology at the University of Campinas. https://repositorio.unicamp.br/bitstream/REPOSIP/336713/1/MassaroJuniorFlavioRubensD.pdf
3. Pecek B, Kovacic A (2011) Upravljanje poslovnim procesima: Kori˘stenje simulacije u javnomsektoru. Ekon Istrazivanja 24(1):95–106. https://doi.org/10.1080/1331677X.2011.11517447. ISSN 1331677X
4. Sargent RG (2016) An introductory tutorial on verification and validation of simulation models. In: 2015 winter simulation conference (WSC), dez. 2015. 2016-February. IEEE, pp 1729–1740. ISBN 978-1-4673-9743-8. https://doi.org/10.1109/WSC.2015.7408291
5. Wooldridge M (1997) Agent-based software engineering. In: IEE Proceedings of Software Engineering, vol 144, no 1, pp 26–37. ISSN 1364-5080. https://doi.org/10.1049/ip-sen:19971026
6. Macal C, North M (2005) Tutorial on agent-based modeling and simulation. In: Proceedings of the winter simulation conference. vol 53. IEEE, pp 2–15. ISBN 0-7803-9519-0. https://doi.org/10.1109/WSC.2005.1574234.
7. Wooldridge M (1999) An introduction to multiagent systems. In: Multiagent systems intelligent agents. [S.l.: s.n.]. Wiley, Liverpool, p 343. ISBN 0262232030
8. Madejski J (2010) Dynamic scheduling for agent-based manufacturing systems. J Achieve Mater Manuf Eng 40(1):66–69 ISSN 1734-8412
9. Real J, Crespo A (2004) Mode change protocols for real-time systems: a survey and a new proposal. Real-Time Syst 26:161–197. https://doi.org/10.1023/B:TIME.0000016129.97430.c6
10. Real J, Crespo A (2001) Offsets for scheduling mode changes. In: Proceedings 13th Euromicro Conference on Real-Time Systems, pp 3–10. https://doi.org/10.1109/EMRTS.2001.933989
11. ANYLOGIC [S.l.]: AnyLogic Company, 26 February 2019. https://www.anylogic.com/downloads/#select-platform-researcher
12. Bonabeu E (2002) Agent-based modeling: methods and techniques for simulating human systems. Proc Natl Acad Sci 99(suppl 3):7280–7287. https://urldefense.proofpoint.com/v2/url?u=https-3A__doi.org_10.1073_pnas.082080899&d=DwIDaQ&c=vh6FgFnduejNhPPD0fl_yRaSfZy8CWbWnIf4XJhSqx8&r=iI9OcdssCVrfThJOCaNCfFp8lkIpP-ClOO-NmBXkeSU&m=dcHFMiLAxRESjHFvgkE1izku2r6GaYSAbZdeuwZWuOI&s=Q1SP6l1MR58TccPG8GQBkvNERsnMWKl5Tj-1eGl5FKM&e=

Energy Use in Urban Areas Using Neodymium Magnets

Roger Prior Gregio[1(✉)], Yuzo Iano[1], Lia Toledo Moreira Mota[2], Gabriel Caumo Vaz[1], Gabriel Gomes de Oliveira[1], Diego Arturo Pajuelo Castro[1], and Carolina Fernandes Frangeto[2]

[1] State University of Campinas, São Paulo, Brazil
yuzo@decom.fee.unicamp.br, oliveiragomesgabriel@ieee.org
[2] Pontifical Catholic University of Campinas, São Paulo, Brazil
lia.mota@puc-campinas.edu.br

Abstract. Electricity is the most widely used form of energy worldwide. In Brazil, the main sources of generation are hydroelectric, thermoelectric, thermonuclear, and wind power plants [1]. It should be noted that all of them have a common characteristic, which is the presence of a generator to convert the mechanical energy (of movement) into electric energy using magnets. However, in order to diversify the national energy matrix, the trend is related to applying the concept of distributed generation, that is, the generation of electric energy carried out next to or close to the end consumer, regardless of power, technology, or energy source. The most used sources are co-generators, generators with combustible process residues, emergency generators, generators for operation at peak times, photovoltaic panels, small hydropower plants - PCH's, among others [2]. In this work, simulations of the energy generation were Performed alternately using neodymium magnets. In order to perform the simulation, 6 rotors with different geometries and materials were constructed, and measurements of voltage and current were made, obtaining the electric energy generated and then identifying the geometry that provides the best efficiency. With this, it was possible to compare the different generations of electric energy and identify, among the rotors built with neodymium magnets, the one with the largest generation of electric energy [1].

Keywords: Power generation · Neodymium magnets · Distributed generation · Electric power

1 Introduction

Power generation is currently a primary concern of governments worldwide, due, among other factors, to its interference in the economic sector. In Brazil, the energy matrix is composed mainly of hydroelectric plants, which provide 90% of the electricity for the entire country, with relatively low costs (approximately R$ 4.00 per megawatt-hour (MWh) in 2012) in relation to other forms of generation [3].

According to [4], the world's largest consumer of hydroelectric power in 2007 was the Organization for Economic Cooperation and Development (OECD), which

Y. Iano et al. (Eds.): BTSym 2020, SIST 233, pp. 988–1005, 2021.
https://doi.org/10.1007/978-3-030-75680-2_107

brings together the most developed nations of the world. Brazil is one of the leaders in hydroelectric production, responsible for about 410 TWh produced in 2016 [4].

According to the Brazilian Energy Balances of the domestic supply of national energy [5–8], drawn up by the Ministry of Mines and Energy, in the decade from 2007 to 2017, the Brazilian energy matrix remained within the interval between 39.4% and 45.5%, with a slight drop between 2012 and 2015 and minimum in 2014, the apex of the hydric crisis in the southeast of the country [9]. From these percentages, it is estimated that hydropower will continue to represent the largest portion of the array, though its representation has been falling in the last years (from 70.6% in 2013 to 64.5% in 2017 [8]) since the good performance of other renewable sources such as wind and biomass compensates the use of water sources.

Since 2016, however, it has been possible to observe the increasing participation of alternative renewable energy sources in the Brazilian energy matrix, such as the wind (from 5.4% in 2016 to 8.6% in 2019) and solar photovoltaic (from 0.0% in 2016 to 1.0% in 2019) [8, 10, 11]. This increase was reflected in the general representation of renewable energy sources in the matrix (from 41.5% in 2016 to 46.1% in 2019).

In this scenario, the distributed generation of electricity gained strength. It can be defined as the generation of electricity at or near the end consumer, regardless of the power, technology, or energy source. In 2015, distributed generation achieved significant results, reaching 1,000 consumers adhesions and, in two months, that number reached 1,731 connections, representing an installed capacity of 16.5 MW. The source of distributed generation is most commonly used by consumers as solar, with 1,675 adhesions and 13.3 MW installed capacity, then the wind, with 33 facilities and 121 kW [12].

In Brazil, until May 2016, the National Electric Energy Agency (ANEEL) recorded 3,565 distributed generation connections. Also, according to ANEEL, with the new facilities, the country already generates 29.7 MW distributed manner [13]. According to the Ministry of Mines and Energy, among the renewable energies most used, solar photovoltaic is a source that stands out the most, followed by wind power [8, 10, 11].

In this sense, the use of permanent magnets to generate electricity showed a significant increase in the global context [14]. This scenario is possible because China holds 58% of the world's reserves of neodymium and has stimulated the production of magnets of this material. Furthermore, there was a considerable decrease in the cost of production of that magnet. Thus, wind generators composed of these materials have been highlighted commercially [14].

Based on this theme, the purpose of this study is to analyze the conventional generators built with ferrite magnets, comparing them with generators that use neodymium magnets. This proposal is justified because neodymium magnets have magnetic properties superior to ferrite magnets [15] and can be used in the development of these generators. From the results of this comparison, it is intended to verify the possibility of building generators to take advantage of movements found in urban areas, aiming to produce energy in small quantities for use in a location close to the generation.

2 Methodology for Different Performance Comparison of Types of Rotor

Figure 1 illustrates a schematic diagram of the generator containing the necessary elements for comparing the performance of different types of rotors.

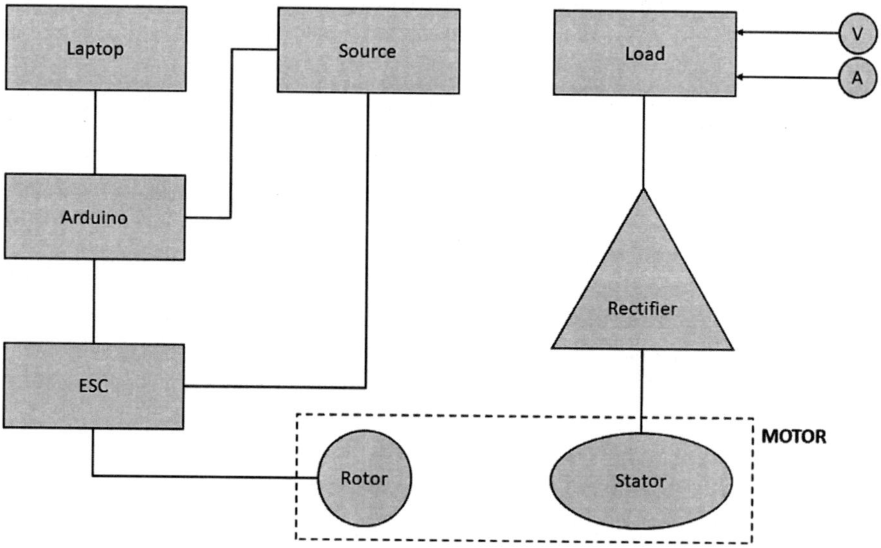

Fig. 1. Schematic of the test structure. Legend: (V) Voltmeter and (A) Ammeter

The generator operates as a motor comprising a stator and a rotor of interchangeable discs. To drive the rotor at different rotation speeds, there is an Arduino microcontroller programmed via a laptop that controls the rotor operation. The Arduino sends the information to the ESC (Electronic Speed Controller), which receives power from the source. This, in turn, controls the rotor speed.

As the rotor turns, the stator, which makes up the prototype engine, generates electricity in alternating current. The generated current flows through the rectifier and is then sent to the load. The strain gauges (voltmeter) and current (ammeter) are coupled to the load, and through them, the value is read for later use during the tests.

2.1 Procedure for Constructing the Generator

Initially, a generator prototype was built to compare the performance of ferrite magnets and neodymium magnets and also different types of geometry, as detailed below. The development of this prototype consisted of the use of HD (hard disk) disks as rotors, serving as fixing base for the neodymium and ferrite magnets in different geometric configurations. The area of the discs was 628.32 m^2.

In possession of the needed materials, six discs were built with completely different geometries, as shown in Fig. 2, in order to, from experimental tests, compare the performance of ferrite magnets and neodymium magnets and identify the rotor geometry with greater system efficiency (able to generate more power).

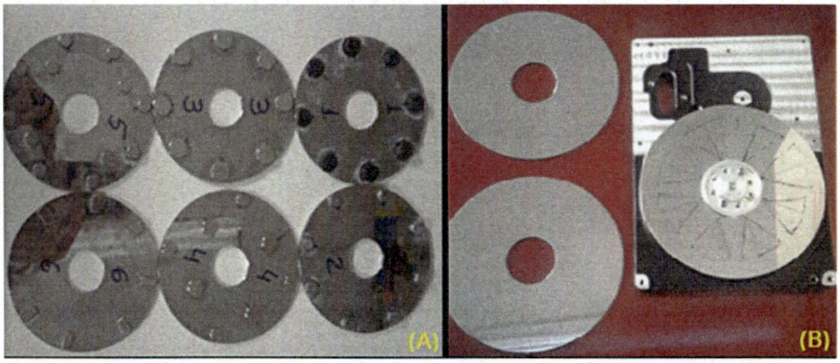

Fig. 2. Process of manufacture of rotors: (A) Interchangeable rotors and (B) Disks and media used in the construction of rotors

Among the six rotors constructed with the developed disks, just rotor 1 uses Ferrite magnets, and the magnetic material employed in the others is the neodymium, as described in Table 1. Table 1 also shows the size and thickness of the magnets employed in the design. It should be noted that the construction geometry of rotors 1 and 2 are similar, with little change in diameter (Ø) and thickness of the magnets because it was not possible to acquire materials with the same size characteristics for the implementation of the rotors. The main change is in the magnetic material used in each of the two rotors.

Table 1. Configuration of the rotors built for laboratory tests

Rotor	Magnets				
	Number	Material	Shape	Size	Thickness
1	8	Ferrite	Circular	Ø 10 mm	3 mm
2	8	Neodymium	Circular	Ø 8 mm	1.5 mm
3	8	Neodymium	Circular	Ø 12 mm	1 mm
4	4 4	Neodymium Ferrite	Circular Circular	Ø 8 mm Ø 10 mm	1.5 mm 1.5 mm
5	12	Neodymium	Circular	Ø 8 mm	1.5 mm
6	8	Neodymium	Rectangular	10 mm × 5 mm	1 mm

In parallel to the construction of the rotors, the needed devices to check the prototype operation were developed. Initially, a stator was designed with a set of 9 inductors star configuration, as shown in Fig. 3.

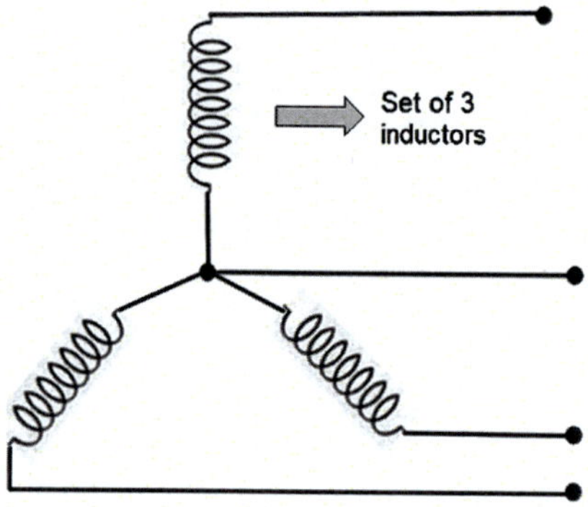

Fig. 3. Stator configuration

Each inductor used was manufactured with the aim of generating a maximum voltage of 12 V and a current of 100 mA nominal in 33AWG wire and a winding of 960 turns.

The stator circuit board is designed so that the inductors stay arranged in a circular shape, as shown in Fig. 4. This layout has been chosen to facilitate the star connection; each plate section is composed of a set of three inductors.

Fig. 4. Stator distribution format

The next step is constructing the rectifier and integrating it into the stator previously developed, as detailed in Fig. 5.

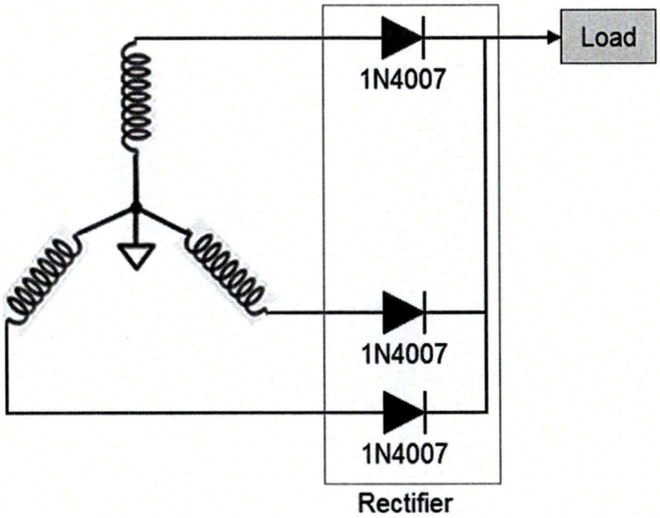

Fig. 5. Schematic rectifier coupled to the load and the stator

To simplify the data collection, we used a half-wave rectifier, built from three 1N4007 diodes connected in series with each stator phase and the diode anode connected to the cathode load.

The system described above is shown in Fig. 6, with a resistor of 220 Ω connected in series to the set of six LEDs as load.

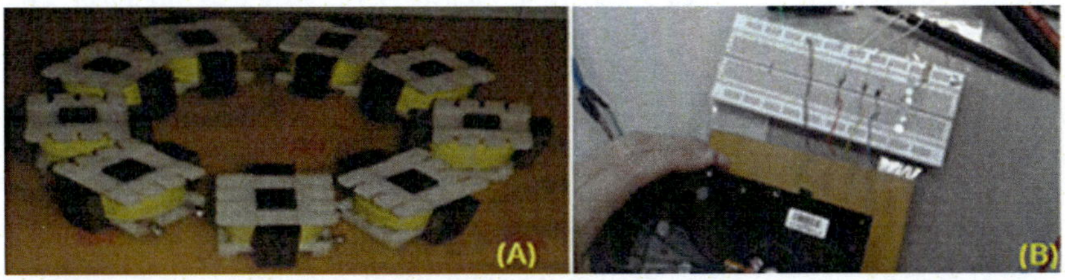

Fig. 6. Equipment used in laboratory testing the (A) Stator developed and connected to the (B) Rectifier load

2.2 Experimental Tests

To identify the rotor with the greatest power generation capacity, it was constructed a laboratory tests bench, as shown in Fig. 7.

The tests were performed to measure each rotor voltage and current magnitudes at three different rotational speeds (2500 rpm, 1900 rpm, and 1500 rpm). The measure quantities are described below:

- Voltage DC - Continuous voltage generated after the rectifier in order to provide power to the proposed load (6 LED's) for the circuit.

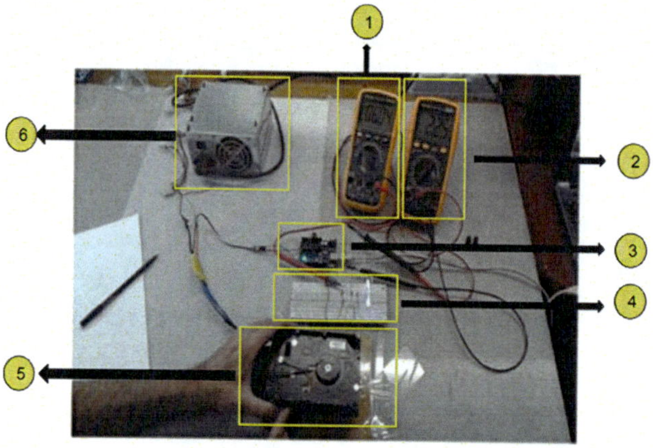

Fig. 7. Test bench construction. Legend: (1) Ammeter, (2) Multimeter, (3) Arduino, (4) Load, (5) Stator, and (6) Source

- Current DC - Direct Current generated after the rectifier in order to provide power to the proposed load (6 LED's) for the circuit.
- No-load voltage - Continuous voltage generated after the rectifier with no load coupled to the circuit, i.e., without current consumption.

The tests and essays are described in the following.

Test 1. The purpose of this essay was to identify the rotor with the best performance in power generation and the influence of the rotational speed of the power produced. Thus, the first analysis consisted of comparing each rotor disk rotating at speeds of 1500 rpm, 1900 rpm, and 2500 rpm. The second analysis, responsible for the comparison between rotor 1 and rotor 2 in the higher rotational speed (2500 rpm), allowed comparing the electricity generation using ferrite (rotor 1) and neodymium (rotor 2) magnets.

Test 2. This test had the purpose of comparing different shapes and geometries of the fastening neodymium magnet constituting the rotor disks. Following, an analysis was made between the disks of numbers 2 and 6, which respectively used circular and rectangular magnets with similar surface areas. Finally, the last step consisted in the comparison between the circular magnets, the discs 2, 3, 4, and 5, which had different surface areas and different ways of fixing, in order to identify the disk that generates higher voltage and electric current during the tests with the prototype, thereby, feeding the proposed load.

2.3 Prototype Implementation in Urban Environment

For implementing a prototype in urban areas, a fan was chosen because this equipment is widely used in residential, commercial, and industrial environments.

Thus, the implementation phase consisted of integrating, with a fan, the stator, the load, and the disc (rotor) with the greatest energy efficiency identified by the tests described in item 2.2. The idea was to take the fan speed to generate enough energy to power a load.

The developed stator was coupled at the rear of the fan, and the chosen rotor (best performance) was connected to the front propeller (Fig. 8).

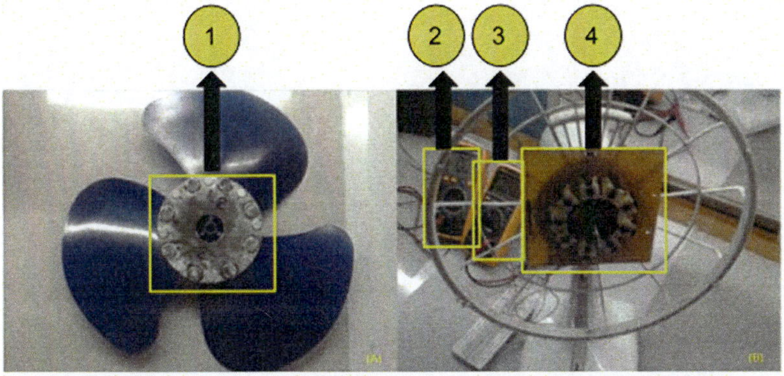

Fig. 8. Prototype application in urban areas, (A) rotor integrated into the propeller (B) generator coupled to the rear. Legend: (1) Rotor, (2) Ammeter, (3) Multimeter, and (4) Stator

After integrating the system into the fan, two multimeters were connected to the load (according to Fig. 8b) in order to measure the voltage and current generated by the fan system at different rotating speeds (low, medium, and high). This essay was called "Essay 3".

3 Results

As a result, the construction of a prototype, as shown in Fig. 9, was obtained to implement the proposed tests with different rotor configurations. It was possible to replace the disk with magnets, which allowed comparing the rotors and identifying the more efficient one in terms of power generation.

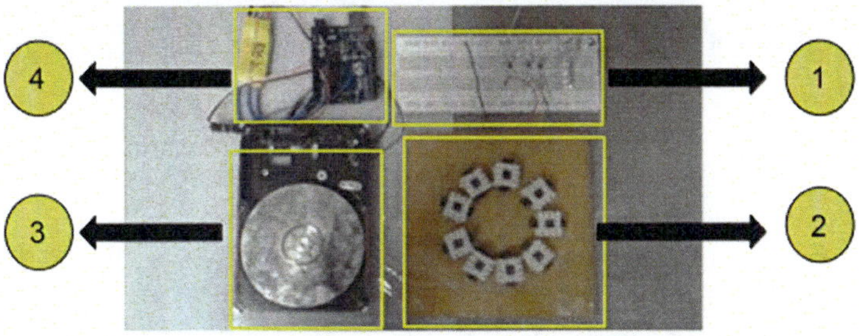

Fig. 9. Prototype comprising the (1) Rectifier and load, (2) Stator, (3) Rotor, and (4) Arduino controller

Figure 10 illustrates the comparison results between rotors 1 (ferrite) and 2 (neodymium).

Fig. 10. Voltage and generated current measurement in (A) Rotor 1 and (B) Rotor 2

3.1 Performance Comparison of the Results of Different Types of Rotors

It can be seen that, in the execution of the laboratory tests, rotor 2, which consists of neodymium magnets, achieved a higher generated voltage compared to rotor 1, whose magnetic material is ferrite. For rotor 1, it was obtained a voltage of 2.08 VDC and a current of 0.24 mA, and rotor 2 obtained a voltage of 7.63 VDC and a current of 26.5 mA.

Analyzing in detail the data obtained for each constructed rotor, it is observed that, for rotor 1, composed of ferrite (Table 2), the higher the rotational speed, the higher are the voltage and current generated. These data can be visualized clearly in the graphs of Figs. 11 and 12, where it is observed that the higher rotational speed (2500 rpm) generates a voltage of 2.08 VDC with a current of 0.24 mA. Note that the current is zero for the same rotor in its lowest rotational speed (1500 rpm).

Table 2. Results obtained for rotor 1

Rotation [rpm]	Total area [mm^2]	DC voltage [V]	DC current [mA]	No-load voltage [V]
2500	628.32	2.08	0.24	2.81
1900	628.32	1.95	0.07	1.99
1500	628.32	1.44	0.00	1.44

For rotor 2, consisting of neodymium (Table 3), at its highest rotational speed (2500 rpm), it was generated a 7.63 VDC voltage and a current of 26.50 mA.

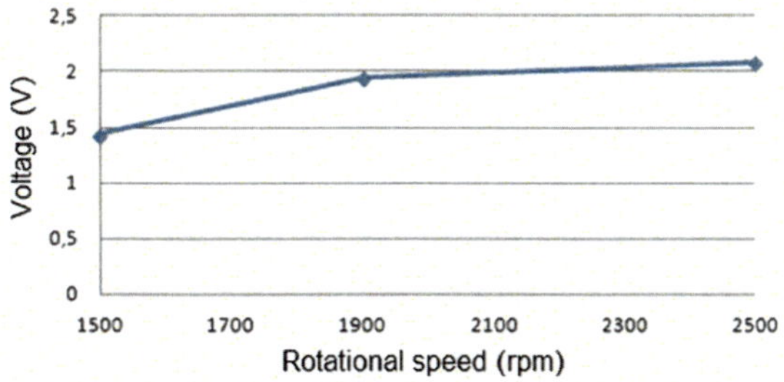

Fig. 11. Relation between the rotational speed and the generated voltage in rotor 1

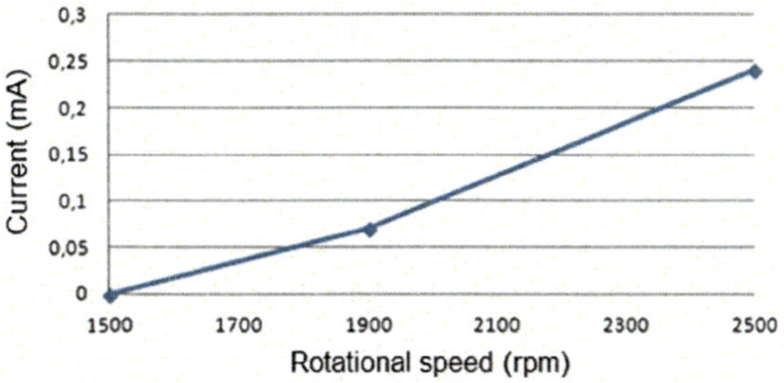

Fig. 12. Relation between the rotational speed and the current in rotor 1

Table 3. Results obtained for rotor 2

Rotation [rpm]	Total area [mm^2]	DC voltage [V]	DC current [mA]	No-load voltage [V]
2500	402.12	7.63	26.50	18.02
1900	402.12	5.30	16.21	11.58
1500	402.12	3.41	6.86	5.72

The graphs of Figs. 13 and 14 show a direct relation between the rotational speed and the voltage and current.

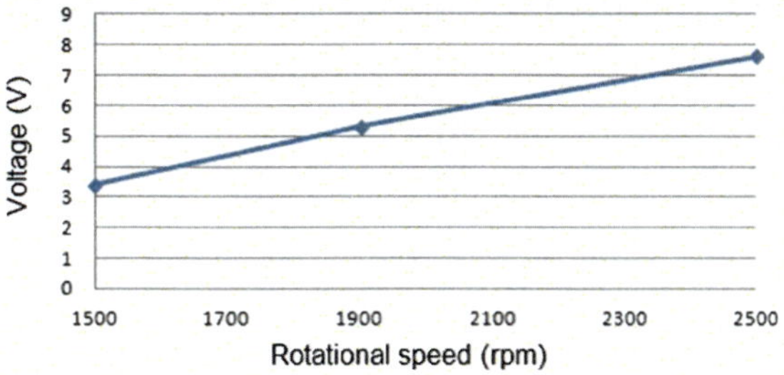

Fig. 13. Relation between the rotational speed and the generated voltage in rotor 2

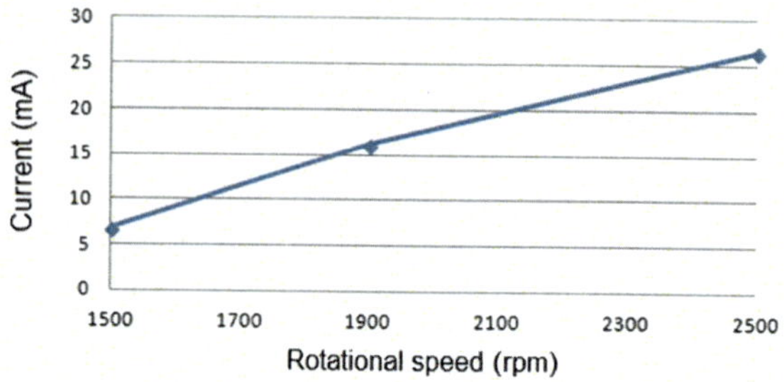

Fig. 14. Relation between the rotational speed and the current in rotor 2

Rotor 3, which comprises six circular discs of neodymium, has its generated current and voltage values described in Table 4. This rotor generates a voltage of 8.25 VDC and a current of 27.20 mA in the highest speed of rotation (2500 rpm) as shown in Figs. 15 and 16, respectively.

Table 4. Results obtained for rotor 3

Rotation [rpm]	Total area [mm^2]	DC voltage [V]	DC current [mA]	No-load voltage [V]
2500	904.78	8.25	27.20	23.60
1900	904.78	6.54	19.90	14.20
1500	904.78	3.72	6.71	6.02

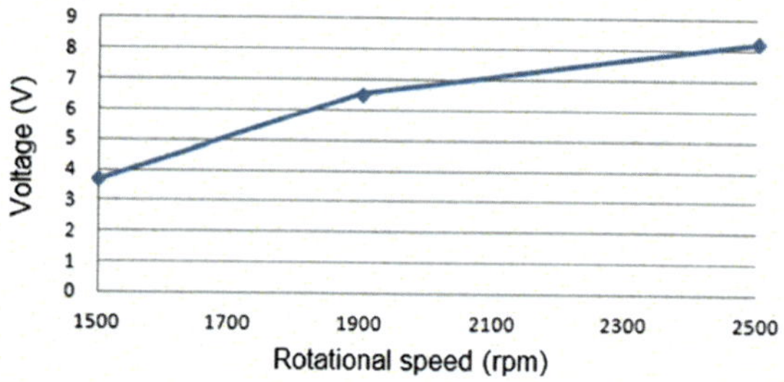

Fig. 15. Relation between the rotational speed and the generated voltage in rotor 3

In Table 5, rotor 4, composed of ferrite and neodymium, in the best performance, has a 2500 rpm rotational speed and generates a voltage of 5.17 VDC with a current of 13.40 mA.

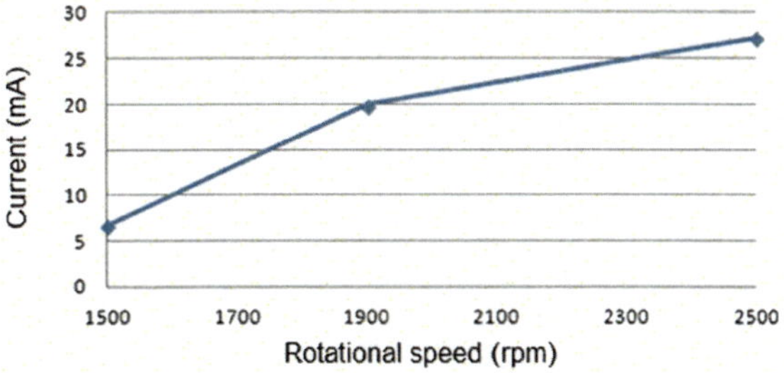

Fig. 16. Relation between the rotational speed and the current in rotor 3

Table 5. Results obtained for rotor 4

Rotation [rpm]	Total area [mm^2]	DC voltage [V]	DC current [mA]	No-load voltage [V]
2500	508.94	5.17	13.40	17.60
1900	508.94	4.41	10.30	10.80
1500	508.94	2.91	3.00	5.06

Figure 17 illustrates the voltage on rotor 4 when it varies its speed.

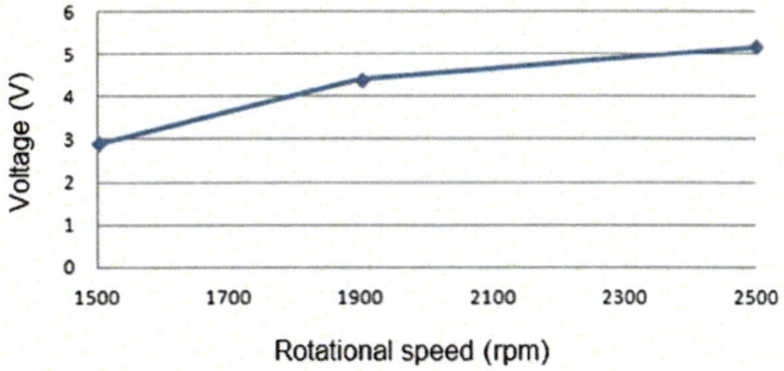

Fig. 17. Relation between the rotational speed and the generated voltage in rotor 4

As shown in Fig. 18, the current in rotor 4 increases according to the speed increase.

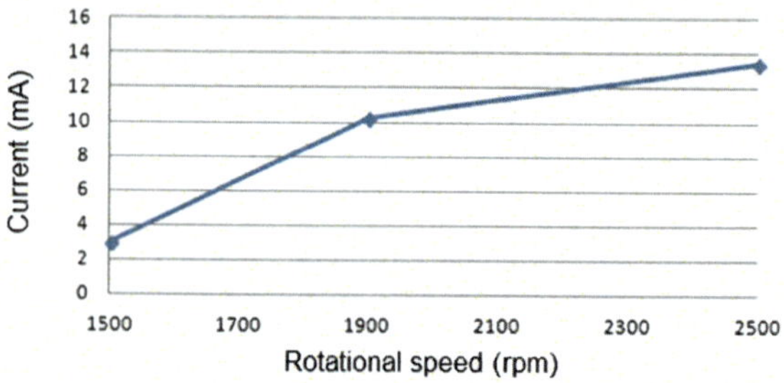

Fig. 18. Relation between the rotational speed and the current in rotor 4

In rotor 5 (Table 6), it is observed that the best case corresponds to a rotation of 2500 rpm, generating a voltage of 7.73 VDC with a current of 26.70 mA. The relation between voltage and current to the various rotational speeds is depicted in Figs. 19 and 20, respectively.

Table 6. Results obtained for rotor 5

Rotation [rpm]	Total area [mm^2]	DC voltage [V]	DC current [mA]	No-load voltage [V]
2500	603.19	7.73	26.70	25.52
1900	603.19	5.34	13.53	12.70
1500	603.19	3.85	7.10	5.92

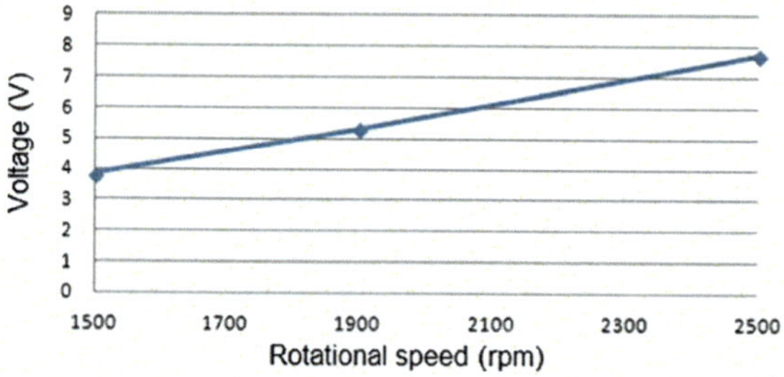

Fig. 19. Relation between the rotational speed and the generated voltage in rotor 5

For rotor 6, it is observed that, once again, the higher the speed, the greater the voltage and current generated (Table 7). It is observed that the best case (2500 rpm) generates a voltage of 5.25 VDC (Fig. 21) with a current of 14.40 mA (Fig. 22).

When all disks are compared, disk 3, which has a greater magnetic surface area, generates the greatest voltage and current, as shown in Table 8.

It may also be noted that by comparing the rotors with the same amount of neodymium and ferrite magnets (Table 9), there is a generation voltage of approximately four times and current 100 times greater in the case of the rotor with neodymium magnets, confirming the initial hypothesis.

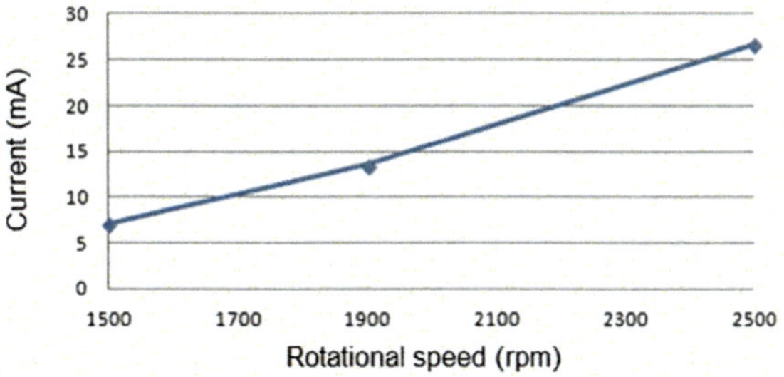

Fig. 20. Relation between the rotational speed and the current in rotor 5

Table 7. Results obtained for rotor 6

Rotation [rpm]	Total area [mm^2]	DC voltage [V]	DC current [mA]	No-load voltage [V]
2500	400.00	5.25	14.40	16.55
1900	400.00	3.66	9.00	10.63
1500	400.00	2.86	4.00	4.56

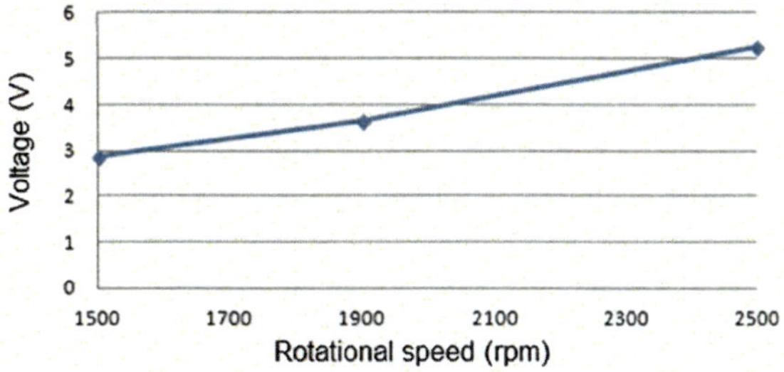

Fig. 21. Relation between the rotational speed and the generated voltage in rotor 6

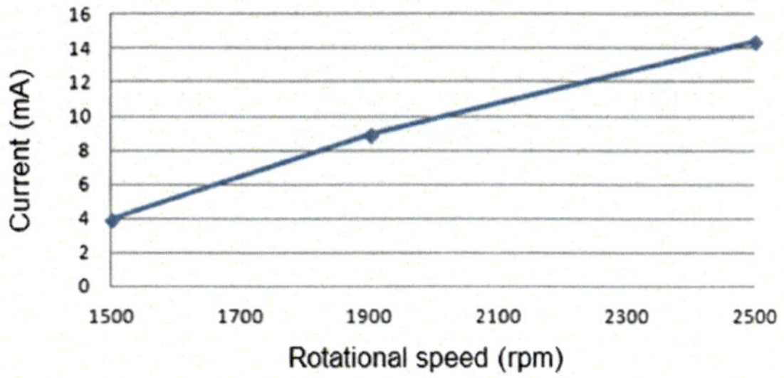

Fig. 22. Relation between the rotational speed and the current in rotor 6

With the same arrangement and amounts of neodymium magnets, using the same surface area (comparison between the rotors 2 and 6), it was found that the circular shape

Table 8. Results and characteristics of the rotors in the study

Rotor	Rotation [rpm]	Number of magnets	Diameter [mm]	Total area [mm^2]	VDC [V]	IDC [mA]
1	2500	8	Ø 10	628.32	2.08	0.24
2	2500	8	Ø 8	402.13	7.63	26.50
3	**2500**	**8**	**Ø 12**	**904.78**	**8.25**	**27.20**
4	2500	8	Ø 9	508.94	5.17	13.40
5	2500	12	Ø 8	603.19	7.73	26.70
6	2500	8	10 × 5	400	5.25	14.40

Table 9. Comparison between rotor 1 (ferrite magnets) and rotor 2 (neodymium magnets)

Rotor	Rotation [rpm]	Total area [mm^2]	VDC [V]	IDC [mA]	No-load voltage [V]
1	2500	628.32	2.08	0.24	2.81
2	2500	402.13	7.63	26.50	18.02

neodymium magnet (rotor 2) generated a higher amount of energy than the rectangular shape neodymium magnet (rotor 6), as shown in Table 10.

Table 10. Comparison between rotor 2 (circular shape) and rotor 6 (rectangular shape)

Rotor	Rotation [rpm]	Total area [mm^2]	VDC [V]	IDC [mA]	No-load voltage [V]
2	2500	402.13	7.63	26.50	18.02
6	2500	400	5.25	14.40	16.55

Comparing all rotors with circular neodymium magnets, but with different positioning (rotor 2 to 5), the best condition encountered, i.e., with the greatest power generation, is on rotor 3 (Table 11). This is because it has the largest magnet surface area composition (904.78 mm^2), and the magnets are fixed in a way they are perpendicularly aligned with the inductors of the stator.

In the comparison between rotors 5 and 3, even rotor 5 having a larger quantity of magnets, it has a lower efficiency relative to rotor 3 due to its smaller surface area (603.19 mm^2), and the method of attachment of magnets do not allow them to be positioned entirely perpendicular to the stator inductors.

Table 11. Comparison between rotors 2, 3, 4, 5 (magnets of the same material, shape, and with different surface areas)

Rotor	Rotation [rpm]	Total area [mm^2]	VDC [V]	IDC [mA]	No-load voltage [V]
2	2500	402.13	7.63	26.50	18.02
3	2500	904.78	8.25	27.20	23.60
4	2500	508.94	5.17	13.40	17.60
5	2500	603.19	7.73	26.70	25.52

The influence of the attachment of the magnets is observed more clearly when comparing rotors 2 and 4. Rotor 4 has a surface area (508.94 mm^2), greater than rotor 2 (402.12 mm^2), and, as shown in the comparative table, rotor 2 generates more power than rotor 4.

Considering the comparison between the other rotors, rotor 3, which is built with magnets of larger areas, generates more energy than rotor 5 with a greater quantity of magnets and, hence, a higher surface area covered by neodymium magnets. It should be noted that rotor 6 (rectangular magnets) did not show satisfactory results regarding the generation of electrical energy toward the objectives of the study.

3.2 Results of Prototype Implementation Between Urban

Besides finding that rotor 3 is the most efficient, this rotor, the stator, and the load were integrated into the fan described in Sect. 2.3. As noted above, tests were conducted in order to ascertain the voltage and current generated by the prototype in their different rotation speeds of 1400 rpm, 1500 rpm, and 1600 rpm.

For the prototype running at its lowest rotation speed (1400 rpm), a 2.64 VDC voltage and a current of 3.89 mA are generated when connected to the load.

The second rotational speed (1500 rpm) produced a voltage of 2.95 VDC and a current of 4.89 mA.

Finally, working with the prototype with the highest rotation speed (1600 rpm), a current of 5.62 mA and a voltage of 3.157 VDC were generated at the load.

4 Conclusion

Analyzing the tests performed in Sect. 3.1, it is concluded that the use of neodymium magnets in electric power generation, when compared to the use of ferrite magnets, which are currently used in electric generators, proved to be more efficient. It also follows that, between generators using neodymium in its construction, the circular magnets have proven more efficient than the rectangular shape because of their larger area. The greater the surface area of the magnet assembly, the greater the voltage and current generated by the stator.

It was also observed that the number of magnets (surface area) and their arrangement on the rotor, as well as the speed at which the rotor turns, directly influence the power generation efficiency. The greater the rotation and the number of magnets (surface area), the higher the power generated. It is worth noting that, in this case, the importance of securing magnets of the rotors being aligned perpendicularly to the stator inductors, thereby allowing higher capacity to generate electric energy.

It was proved the possibility of taking advantage of movements in urban areas to generate small amounts of energy by fixing the prototype in a residential fan. During the operation, it moves the rotor in a fixed propeller, generating energy through the stator fixed at its base.

As future work, new tests can be performed in order to achieve more energy generated, increasing the efficiency of the prototype by using coils with greater surface area for the stator, increasing the rotor size, or using full-wave rectifiers. In addition, in the future, it will be possible to perform tests on other structures. It can also be studied the use of other movements of the urban environment, other than rotation, such as commuting or vibratory movements, always seeking the conversion of kinetic energy into electricity.

References

1. Gregio RP (2017) Aproveitamento de energia no meio urbano usando ímãs de neodímio. 70p. Dissertação (Programa de Pós-Graduação em Sistemas de Infraestrutura Urbana) - Pontifícia Universidade Católica de Campinas, Campinas-SP
2. Luna MAR, Cunha FBF, Mousinho MCAM, Torres EA (2019) Solar photovoltaic distributed generation in Brazil: the case of resolution 482/2012. Energy Proc 159:484–490. https://doi.org/10.1016/j.egypro.2018.12.036. ISSN 1876-6102
3. Patusco, JAM (2016) Brazilian energy review - exercise 2014, ministry of mines, and energy
4. Petroleum B (2017) BP statistical review of world energy 2017. Br Petrol 66:8–11
5. Empresa de Pesquisa Energética (2014) Brazilian energy balance - year 2013. https://www.epe.gov.br/sites-en/publicacoes-dados-abertos/publicacoes/PublicacoesArquivos/publicacao-128/topico-99/Relat%C3%B3rio%20Final%202014.pdf
6. Empresa de Pesquisa Energética (2015) Brazilian energy balance - year 2014. https://www.epe.gov.br/sites-en/publicacoes-dados-abertos/publicacoes/PublicacoesArquivos/publicacao-127/topico-97/Relat%C3%B3rio%20Final%202015.pdf
7. Empresa de Pesquisa Energética (2016) Brazilian energy balance - year 2015. https://www.epe.gov.br/sites-en/publicacoes-dados-abertos/publicacoes/PublicacoesArquivos/publicacao-126/topico-94/Relat%C3%B3rio%20Final%202016.pdf
8. Empresa de Pesquisa Energética (2017) Brazilian energy balance - year 2016. https://www.epe.gov.br/sites-en/publicacoes-dados-abertos/publicacoes/PublicacoesArquivos/publicacao-46/topico-82/Relatorio_Final_BEN_2017.pdf
9. Galvão J, Bermann C (2015) Crise hídrica e energia: conflitos no uso múltiplo das águas. Estud Avançados 29(84):43–68. https://doi.org/10.1590/S0103-40142015000200004
10. Empresa de Pesquisa Energética (2018) Brazilian energy balance – summary report - year 2017. https://www.epe.gov.br/sites-en/publicacoes-dados-abertos/publicacoes/PublicacoesArquivos/publicacao-180/Summary%20Report%202018.pdf
11. Empresa de Pesquisa Energética (2020) Brazilian energy balance – summary report - year 2019. https://www.epe.gov.br/sites-en/publicacoes-dados-abertos/publicacoes/PublicacoesArquivos/publicacao-217/SUMMARY%20REPORT%202020.pdf

12. Agência Nacional de Energia Elétrica (2016) Geração distribuída amplia número de conexões em 2015. https://www.aneel.gov.br/busca?p_p_id=101&p_p_lifecycle=0&p_p_state=maximized&p_p_mode=view&_101_struts_action=%2Fasset_publisher%2Fview_content&_101_returnToFullPageURL=%2Fweb%2Fguest%2Fbusca&_101_assetEntryId=14809692&_101_type=content&_101_groupId=656877&_101_urlTitle=geracao-distribuida-amplia-numero-de-conexoes-em-2015&inheritRedirect=true
13. Canal Energia (2016) Brasil registra 3.565 novas conexões de geração distribuída até maio. https://canalenergia.com.br/noticias/25034341/brasil-registra-3565-novas-conexoes-de-geracao-distribuida-ate-maio
14. Matias, FER (2013) Utilização de exaustores eólicos como fonte geradora de energia elétrica. Master's thesis Published PPGENE -XX, Department of Electrical Engineering, University of Brasília, Brasília, DF, 93 p
15. München DD, Viet HM, Schmidt VM (2016) Caracterização de Ímãs de Neodímio-Ferro-Boro de Discos Rígidos (HDs). In: 10th International symposium on ambienta quality, Porto Alegre

Analysis of Results of Some Techniques for the Recognition of Circular Shapes in the Steel Bar Counting System Using Image Processing

Yuzo Iano, Daniel Katz Bonello(✉), Umberto Bonello Neto, Abel Dueñas, Frank Canahuire, and Gabriel Gomes de Oliveira

University of Campinas, Campinas, Brazil
yuzo@decom.fee.unicamp.br, oliveiragomesgabriel@ieee.org

Abstract. In this paper, a comparison of the performance of three methods of detection of circles in black and white images is tested. For this, we use a proprietary database previously preprocessed to compare the three methods with the same conditions. This work represents a contribution to the pattern detection (circle detection) of a benchmarking of images. Once defined, the new based algorithm and ratio detection in the figure, circular images were set to the edges detection (circles) with ratio selection dependents on the number of circles to be detected in the selected template. The project in question views the study and development of an algorithm that can detect circles to start many images. This identification is possible due to the circle detection algorithm developed in MATLAB used in this project. This work's main contribution is related to the circle detection techniques using an algorithm developed in MATLAB to detect circular elements in mechanical parts. We believe that industries and mechanical processes, such as the milling process, need intelligent machinery to make inspections with precision and reliability. Therefore, with this vision in mind, an algorithm was developed to detect circumferences in various mechanical parts with precision.

Keywords: Object recognition · Circle detection · Hu moments · Hough transform · Template matching

1 Introduction

Object recognition is one of the tasks most studied in image processing [1]. To achieve this, it is taken into account different specificities, either color, texture, or shape. A feature that is so widely used to characterize an image is geometric. It can be deciphered whether an image is a triangle, square, circle, etc. [2].

This paper will be compared to three methods of concentric circle detections. This kind of recognition is too important [3]. The first method will be using the Hu invariant moments, proposed in 1962 [4]. The second will be a particular class of transform, the Hough transform for circle detections [5], and the third method is using template matching based on cross-correlation.

Y. Iano et al. (Eds.): BTSym 2020, SIST 233, pp. 1006–1015, 2021.
https://doi.org/10.1007/978-3-030-75680-2_108

For this, we will make use of an image database, which consists of 340 images of steel bars packages. For purposes of detection methods analysis, these images have already been preprocessed. They are in black and white so that the three tests are only taken as input two things, the image and the average diameter of the circles. In the following chapters, each of these methods will propose a solution to the problem presented. Finally, a comparison of the results of each of these will be showing.

2 Definition of Methods Used

2.1 Hu Invariant Moments

A set of 7 2-D moment invariants that are insensitive to translation, scale change, mirroring, and rotation [6].

The parameters $(p + q), f(x, y)$, and $M \times N$ are shown in the Eq. (1).

$$m_{pq} = \sum_{x=0}^{M-1} \sum_{y=0}^{N-1} x^p y^q f(x, y) \tag{1}$$

Where $p, q = 0, 1, 2, \ldots$ integers and the central moment μ_{pq} are defined in Eq. (2) [7].

$$\mu_{pq} = \sum_{x=0}^{M-1} \sum_{y=0}^{N-1} (x - \overline{x})^p (y - \overline{y})^q f(x, y) \tag{2}$$

For $p, q = 0, 1, 2, \ldots,$ is shown in Eq. (3, 4) [8].

$$\overline{x} = \frac{m_{10}}{m_{00}} \tag{3}$$

$$\overline{y} = \frac{m_{01}}{m_{00}} \tag{4}$$

The central moments μ_{pq} and $(p + q)$ are given in Eq. (5).

$$\eta_{pq} = \frac{\mu_{pq}}{\mu_{pq}^{\gamma}} \tag{5}$$

Where $\gamma = \frac{p+q}{2} + 1$ for $p, q = 2, 3, \ldots$

2.2 Circle Detection

In the case of circle detection, the general Eq. (6) can be used with three parameters.

$$(x - a)^2 + (y - b)^2 = r^2 \tag{6}$$

Figure 1 explains the cones in the parameter space and an approach regarding the thresholding.

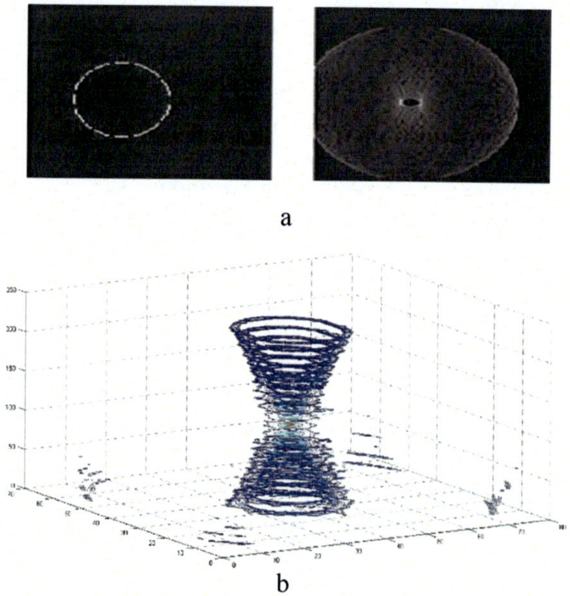

Fig. 1. In (a) the possible generated centers with a determinates radius and in (b) a generated cone in the space a-b-r

2.3 Template Matching

Template Matching is a procedure used in any computer vision application to find a sub-image location, called template, inside an image [9].

This procedure consists of sliding the template over the search area to calculate the correlation coefficient, Eq. (7), around the image. This coefficient is proportional to the similarity grade between the template and the image.

$$r = \frac{\sum_{m=0}^{M-1}\sum_{n=0}^{N-1}\left(A_{mn}-\overline{A}\right)\left(B_{mn}-\overline{B}\right)}{\sqrt{\left(\sum_{n=0}^{N-1}\left(A_{mn}-\overline{A}\right)^2\right)\left(\sum_{n=0}^{N-1}\left(B_{mn}-\overline{B}\right)^2\right)}} \tag{7}$$

Where A is the image, B is the template, and $\overline{A}$ and $\overline{B}$ are the means of A and B respectively, as shown in Fig. 2.

Like the case of the Hough Transform, the problem is transformed into a peak detection problem.

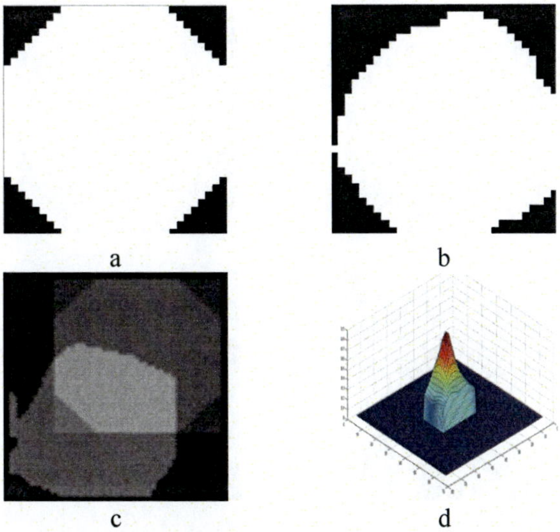

Fig. 2. In this example, we have a Template a that will be found in the image b. After the matching process c, it generates the correlation values in d

3 Implementation of the Methods

How it was mentioned in this paper, the database is information about a vector with an approximate pixel of the circles in each image. It is essential to declare that the given radius in the vector is only a mean of all rings radii in the picture; this means that possibly the circles in each image have different radii.

3.1 Hu Invariant Moments

The first method is simple, fast, and will be seen in the future, without the best results.

The first step is to separate the image in neighborhoods, like in Fig. 3. After the initial test, it was decided to use only the first moment by the zero contribution of the other 6 moments in the case study. Thus, it is possible to achieve good performance without unnecessary calculations.

The next challenge was filtering the circles of the excellent area because, in some cases, like comparing sizes in Fig. 4, the image noise can have a good value moment.

For filtering, the noise was enough to realize and establish a value again like the moment threshold.

3.2 Hough Transform

For that, firstly, it is needed to find the edges of the image. In this case, the method used was the canny edge detection, as shown in Fig. 5 [10].

From now on, all the calculation process will be made in the edge detection image. The sample image will be a synthetic image with two circles for visualization facilities, as shown in Fig. 6 [11, 12].

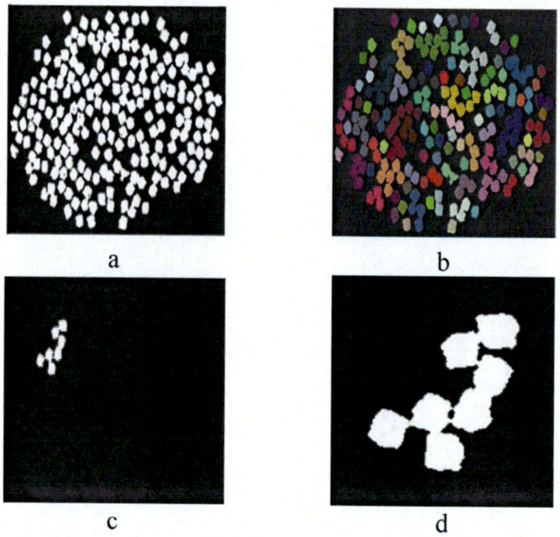

Fig. 3. In this case, we have the image in (a) and the neighborhoods (b). In (c), each area is separate and segmented in (d)

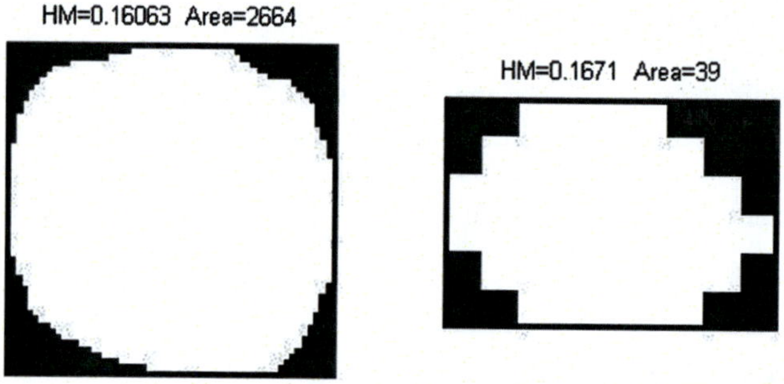

Fig. 4. False positive case

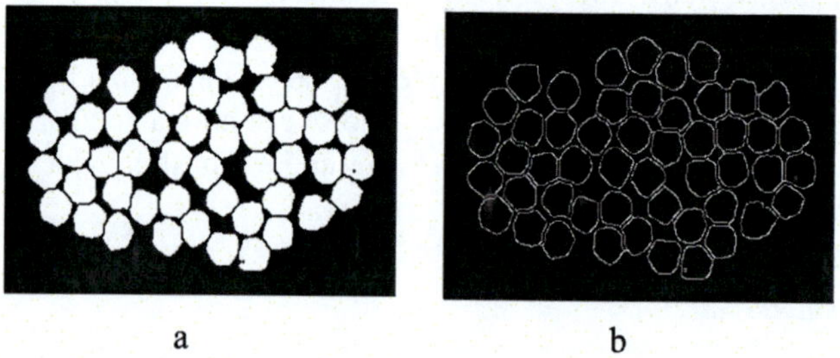

Fig. 5. In (a) the original image and in (b) the edge detection

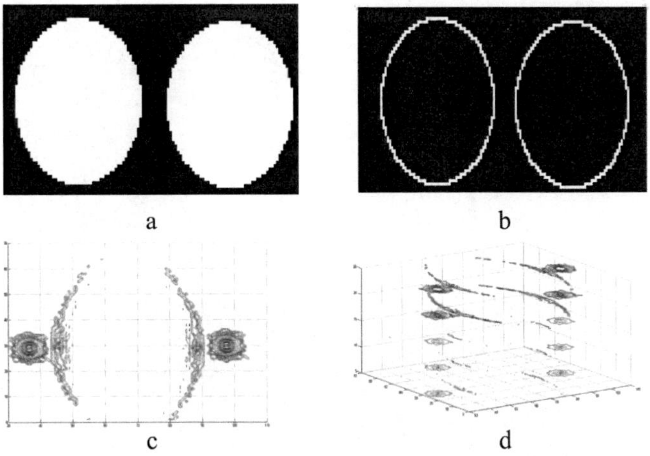

Fig. 6. In (b), we have an edge detection of (a). In (c), we have a generated voting in XY plane, and in (d), the orthogonal view

After finding each maximum value, it should be removed to avoid double counting in the same circle.

3.3 Pattern Recognition

It is needed to generate a template to calculate the similarity of these with the radius value morphology, as shown in Fig. 7 [13].

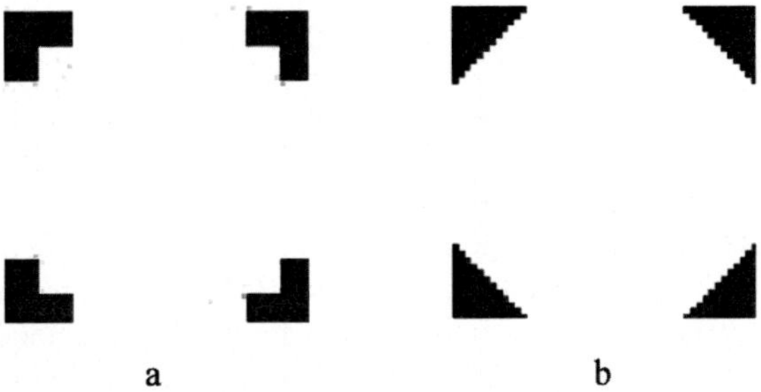

Fig. 7. In (a), we have an original kernel of 9 × 9 pixels and in (b) a synthetic template of, in this case, 49 × 49 pixels

In other words, a correlation is not calculated in the entire image without conditions; this is computed considering neighborhood criteria.

First, it is needed to make a slicing process in the sample image with the template to calculate the correlation value. The key is what pixels appear in the windows to influence the calculation. In this case, only are considered the pixel of the predominant neighborhood in the window. As shown in Fig. 8, the correlation depends only on one area in the windows [14].

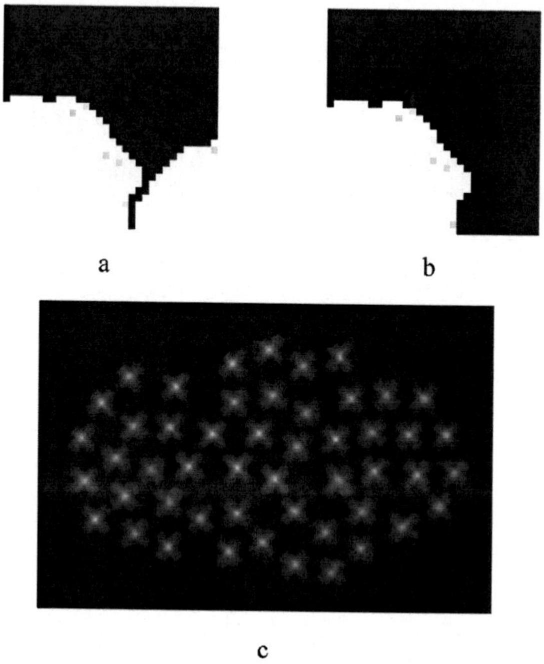

Fig. 8. In (a), we have original windows, and in (b), the pixels that will be used in the correlation calculation. In (c), the correlations map.

4 Results and Performance

After the test of the 340 images, we can accomplish a comparative between the three methods proposed.

Figures 9 and 10 show the results of the counting process of the three methods. In the first graphic (9), the counting comparison (absolute difference) between the manual counting and the result of each technique. The next graphic, for us, is more important than the first because it is a percentage comparison. It is not the same not to detect one circle in a sample of 50 processes in a piece of 200 rings. In this graphic (10), we can distinguish the best reliability with the third Template Matching method.

There are only two estimates to measure the three methods' effectiveness (True Positive Rate and False Negative Rate).

Table 1 showed that the error probability is lower with the Template Matching method, but the heart of the matter is: Is it the best method in all cases? As we will see, the answer is NO.

Another parameter to considerer in choosing the right method in any application is performance. Figure 11 shows the time processing that the three methods need in each of the samples. Figure 12, until a certain point, is more comfortable to understand. In this, we can see the time accumulate in the processing of the database. In this case, the first method, based on the Hu Moments, is the fastest of them, and the based on Template Matching is the lowest [15].

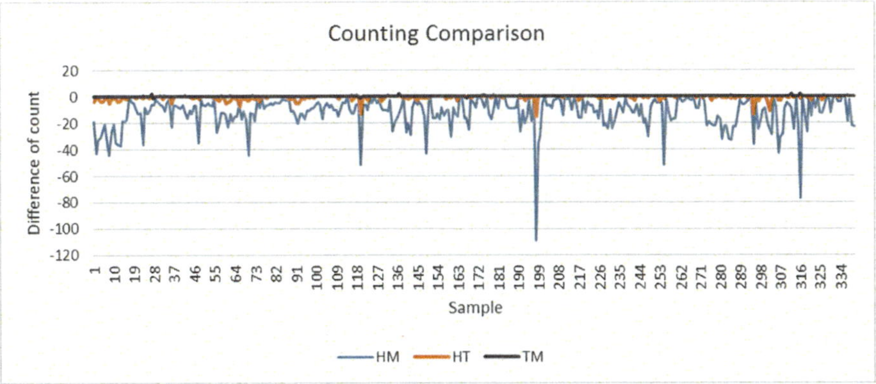

Fig. 9. Error comparative

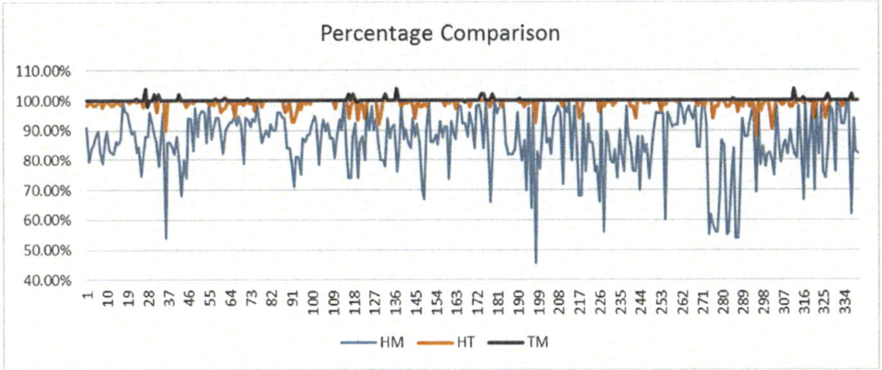

Fig. 10. Error comparative

Table 1. Statistical measure

Method	True positive rate	False negative rate
Invariant Hu analysis	87,00%	13,389%
Circle detection	99,00%	0,928%
Template matching	100,00%	0,016%

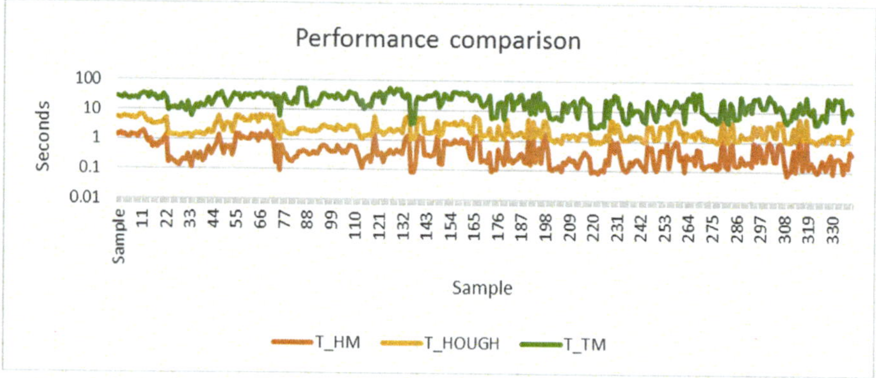

Fig. 11. Performance comparison

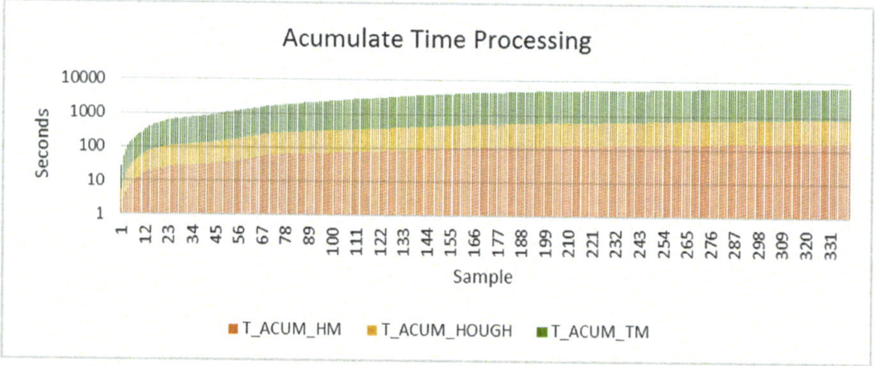

Fig. 12. Performance comparison

5 Conclusions

In this paper, it is analyzed if one of the three methods is the best. After the results, the conclusion is that the method's choice depends on the application that can be developed. In some cases, the priority is effectiveness to choose the plan based on template matching. On the other hand, if the focus is processing time, like a real-time application, a good option would be to use the method based on invariant moments or, to a certain extent, with optimization, a technique based on Hough transform that has moderate effectiveness.

It is highlighted the importance of this kind of algorithm to the industrial sectors because those areas are always looking for new alternatives to detect parameters involving circular geometry and enhance the quality of laboratory tests, having in the confection of that algorithm with a capacity of detection of circular geometry (circular textures), an opportunity to found a low-cost implementation system that solves the most variable practical problems of detection circles for the PCB industries and industrial testing mechanical images.

The scheme described along with this scientific work, even being very easy to understand, allows the implementation of a serial of changes that are looking to adopt the algorithm of detection circles to the final user (industries or mechanical laboratories) objective with the finality of detection circular textures in images.

Acknowledgment. We are thankful for all colleagues and professors of the School of Electrical Engineering and Computing (FEEC) of the University of Campinas, to LCV (Laboratory of Visual Communications) researchers, to CAPES (Coordination of Improvement of Upper-Level People), and to all the people that have contributed directly or indirectly for that research work can be done.

References

1. Zhang LZL, Pu JPJ, Yu JYJ (2009) Object recognition based on modified invariant moments. In: 2009 Interbational conference on mechatronics automation, pp 2542–2547
2. Sari MI, Suksmono AB, Iwut TI (2011) Detection and reconstruction of geometric object based on line detection using radon transform. In: Proceedings of 2011 Interbational conference on electrical engineering and informatics, ICEEI 2011
3. Chen X, Lu L, Gao Y (2012) A new concentric circle detection method based on Hough transform. In: ICCSE 2012 – Proceedings of 2012 7th international conference on computer science & education, no Access, pp 753–758
4. Hu MK (1962) Visual pattern recognition by moment invariants. IRE Trans. Inf. Theory 8:179–187
5. Duda RO, Hart PE (1972) Use of the Hough transformation to detect lines and curves in pictures. Commun ACM 15(1):11–15
6. Gonzalez RC, Woods RE, Eddins SL (2004), Digital image processing using matlab - Gonzalez Woods & Eddins.pdf, vol 624, no 2. Prentice Hall, Upper Saddle River, p 609
7. Hart PE (2009) How the Hough transform was invented [DSP history]. IEEE Signal Process Mag 26(November):18–22
8. Shapiro L, Stockman G (2001) Computer vision, vol 2004, no October. Prentice-Hall, p 608
9. Lewis JP (1995) Fast template matching template matching by cross-correlation 2 normalized cross-correlation normalized cross-correlation in the transform domain. Pattern Recognit 10:120–123
10. Canny J (1986) A computational approach to edge detection. IEEE Trans Pattern Anal Mach Intell PAMI-8(6):679–698
11. Jia LQ, Peng CZ, Liu HM, Wang ZH (2011) A fast randomized circle detection algorithm. In: 4th International congress on image and signal processing (CISP), vol 2, pp 820–823
12. Vegt SE (2015) A fast and robust algorithm for the detection of circular pieces in a cyber-physical system. The Eindhoven University of Technology, Netherlands
13. Duda RO, Hart PE (1972) Use of the Hough transformation to detect lines and curves in pictures. Commun ACM 15(1). https://doi.org/10.1145/361237.361242
14. Maini R, Aggarwal H (2009) Study and comparison of various image edge detection techniques. Int J Image Process 3:147002
15. Wang J, Ma Y, Li C, Wang H, Liu J (2009) Multiobject tracking with explicit reasoning about occlusion. In: 2009 International joint conference on computational sciences and optimization. https://doi.org/10.1109/CSO.2009.378

MPPT-PWM - A Maximum Power Point Tracking (MPPT) Strategy Using Variable Speed Wind Turbines (VSWTs)

Thiagarajan Yogamoorthi[1(✉)], Gabriel Gomes de Oliveira[2], Vania Vieira Estrela[3], Henrique Rego Monteiro da Hora[4], Abdeldjalil Khelassi[5], Nikolaos Andreopoulos[3], Maria Aparecida de Jesus[3], and Yuzo Iano[2]

[1] Sri Venkateshwaraa College of Engineering and Technology, Puducherry, India
[2] State University of Campinas, São Paulo, Brazil
yuzo@decom.fee.unicamp.br
[3] Federal Fluminense University, Rio de Janeiro, Brazil
vania.estrela.phd@ieee.org
[4] Instituto Federal Fluminense, Campus dos Goytacazes, Rio de Janeiro, Brazil
[5] Abou Bakr Belkaid University of Tlemcen, Tlemcen, Algeria

Abstract. Nowadays, wind generators can supply more electric energy for standalone and grid-connected applications. This investigative work emphasizes a new optimal control method design to achieve the Maximum Power Point Tracking (MPPT) using Variable Speed Wind Turbines (VSWTs) termed MPPT-PWM. The VSWT output is initially coupled with a Permanent Magnet Synchronous Generator (PMSG) to transform wind energy to a fixed DC level using a chopper. Later, DC output from the chopper is inverted to obtain AC power using the PWM technique for adequate power flow. Experiments ratify the remarkable performance of the MPPT-PWM framework.

Keywords: Variable Speed Wind Turbines (VSWTs) · Maximum Power Point Tracking (MPPT) · Permanent Magnet Synchronous Generator (PMSG) · Direct Current (DC) · Alternating Current (AC) · Pulse Width Modulation (PWM) · Horizontal Axis Wind Turbine (HAWT) · Distributed Generation (DG)

1 Introduction

Global warming has augmented the atmospheric gas concentration produced by fossil fuel burning [1]. Many energy-related issues, e.g., exhausting fossil-fuel reserves, security concerns, and global warming, intensify the need for renewable energy. All over the world, electric power generation via non-conventional sources receives considerable attention owing to the depletion of fossil fuels and other environmental issues [2–6]. Their main advantages are pollution-free and inexhaustible, while the main shortcomings are the cost and uncontrollability. The eolic turbines can harness wind and transform this ample resource into electricity by generators. For the most part, wind power hinges on

Y. Iano et al. (Eds.): BTSym 2020, SIST 233, pp. 1016–1026, 2021.
https://doi.org/10.1007/978-3-030-75680-2_109

geographic and weather conditions that vary from time-to-time. Wind power production is cost-competitive because of three reasons:

1. The state incentives;
2. The wind manufacturing industry has upgraded the aerodynamic efficiency of the eolic turbine;
3. The power semiconductors evolution and new control methodology for the variable-speed wind turbine permit the optimization of eolic turbine performance.

Wind supplies kinetic energy, which is pollution-free, cost-competitive, infinitely sustainable, and one of the fastest-growing renewable energy sources globally. It does not yield greenhouse gases or toxic radioactive waste. The wind allows the rotation of fast-moving wings that exert a torque on the rotor. The bladed rotor is the most fundamental and noticeable part of the wind turbine. Eolic turbines can be twofold contingent on the blade positions:

1. Based on the axis of rotation:
   - Vertical Axis Wind Turbine (VAWT);
   - Horizontal Axis Wind Turbine (HAWT).
2. Regarding the operation speed:
   - Constant speed;
   - Variable rate.

Earlier, the cost and variations of wind energy made it scarcely used. However, recent technological improvements have reduced the price by increasing the Wind Energy Conversion System (WECS) throughput. Consequently, the variable speed WECS encounters several applications, e.g., water pumping systems, standalone schemes for homes/businesses, telecommunications, radar activation, navigational aids, weather stations/seismic monitoring, and air-traffic control, to refer to a few.

The present article follows this structure: Sect. 2 discourses about the Maximum Power Point Tracking (MPPT) scheme via Variable Speed Wind Turbines (VSWTS) named MPPT-PWM. Results and a technical discussion appear in Sect. 3. Section 4 wraps up this manuscript with conclusions.

2 Development

The rotor revolves around a shaft to convey the nacelle's motion (which is the massive housing portion at the wind turbine tower top), as in Fig. 1. The slowly rotating shaft connects to a gearbox within the nacelle to expressively augment the rotational shaft velocity. The high-speed shaft output connected to the generator converts the rotation into electricity at medium voltage (about a few hundred volts) [7–11]. Wind power production happens twofold: (i) constant speed and (ii) variable speed through power electronic converters. Inconsistent speed power production is attractive for a wind turbine because it attains maximum efficiency at all wind speeds.

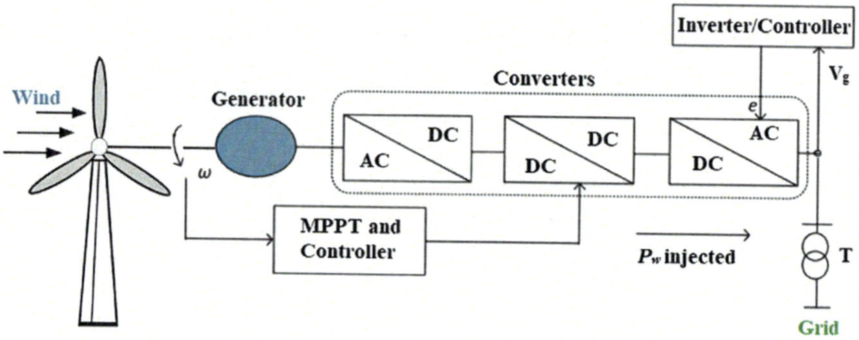

Fig. 1. MPPT scheme

2.1 MPPT for a WECS

The electrical energy needs with fossil fuel scarcity, demand, and indispensable deploying a distributed generation (DG) structure to satisfy local demands where renewable energy sources play a unique role [12–15]. This manuscript proposes and develops an optimal control method to bring out electricity uninterruptedly through variable speed wind turbines (VSWTS). For a VSWT coupled with a Permanent Magnet Synchronous Generator (PMSG), the initial stage converts the output to a fixed DC using a chopper. Subsequently, inverting the DC output from the chopper creates AC via the PWM technique.

2.2 Mathematical Model of the VSWT

Wind turbines convert the eolic kinetic energy E_c into electrical energy utilizing power converters. The energy obtained from conversion mainly hinges on the wind speed as well as the swept area of the turbine. The wind kinetic energy is:

$$E_c = mV^2/2 \tag{1}$$

The rate of energy change yields wind power as follows:

$$P_w = dE_c/dt = V^2[dm/dt]/2 \tag{2}$$

Then, the mass flow rate is given as:

$$dm/dt = \rho AV \tag{3}$$

Substituting (3) in (2) yields:

$$P_w = \left(\rho AV^3\right)/2 \tag{4}$$

The power coefficient becomes:

$$C_p = P_m / P_w \tag{5}$$

Where $C_p < 1$. From the two previous expressions, the wind output power turns into:

$$P_m = \left(\rho A V^3 C_p\right)/2 \tag{6}$$

With $A = \Pi r^2$ and:

$$P_m = \left(\rho \Pi r^2 V^3 C_p\right)/2 \tag{7}$$

If λ is the ratio between the blade tip speed to wind speed, then C_p becomes a function of λ. C_p relates to the pitch angle β (from rotor blade), and its theoretical upper limit is 0.59, according to the Betz limit. Hence:

$$\lambda = R\omega / V \tag{8}$$

And:

$$C_p = 0.5[116/\lambda_1 - (0.4B) - 5]Exp\{-16.5/\lambda_1\} \tag{9}$$

The wind turbine torque on the shaft is calculated from the power as follows:

$$T_m = P_m/\omega = \left(\rho \Pi r^2 V^3 C_p / 2\omega\right) \tag{10}$$

If the tip speed ratio formula is employed, then the torque coefficient T_m is:

$$T_m = \left(\rho \Pi r^3 V^2 C_p / 2\lambda\right) \tag{11}$$

Using the ratio of power coefficient to the tip speed ratio $C_t = C_p/\lambda$, it becomes:

$$T_m = \left(\rho \Pi r^3 V^2 C_t / 2\lambda\right) \tag{12}$$

2.3 Control Strategy for Maximum Power Point Tracking (MPPT)

Although the wind speed changes with time, wind turbine optimization can run at near peak power production for several wind speeds. Besides output optimization, different control methods are applicable. Subsequently, the suggested control strategy is discussed.

The eolic turbine sends maximum power by altering the rotor speed consistent with the airstream speed to function at the leading power coefficient (C_p). The MPPT control is an active area undergoing an investigation to get maximum power from the existing wind power. The DC voltage measurements from both the three-phase diode bridge rectifier and the wind turbine power source during the DC-DC boost chopper's duty cycle control provide wind turbine MPPT. Figure 2 depicts a general relation between power and voltage, which resembles the mechanical energy and rotor speed of the PMSG relationship. The controller is developed with two inputs to reduce the error and number

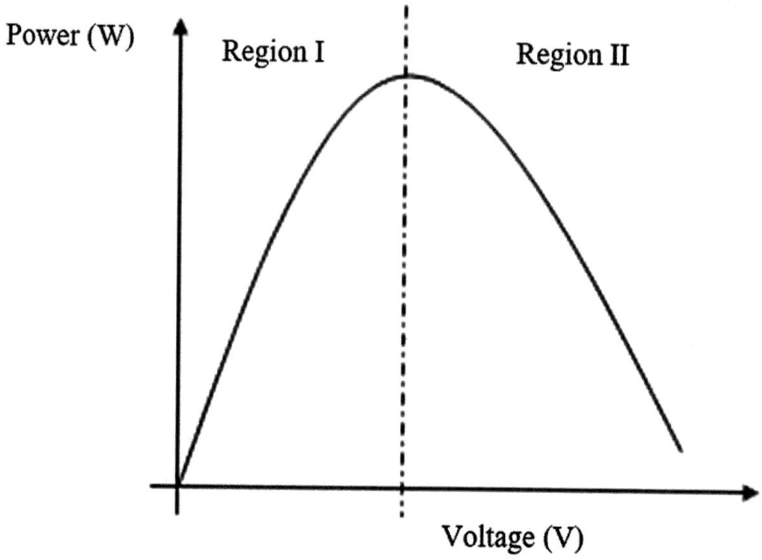

Fig. 2. Power vs. voltage curve

of the control blocks, which produces the right duty cycle to the boost chopper, with maximum output power. Figure 2 reveals one point corresponding to the full power at any wind speed value at a specific amount of DC voltage. The PMSG rotor speed and the DC voltage follow a linear relationship. They are changing the chopper duty ratio shifts the operating point from one curve to another with changing wind speed.

As indicated in Fig. 2, the power-voltage curve encompasses two regions:

1. Positive Region (located on the right side of the MPP).
2. Negative Region (located on the left side of the MPP).

Table 1. Control actions for various operating points

ΔP	+	+	−	−
ΔV	+	−	+	−
Region	**I**	**II**	**II**	**I**
ΔD	−	+	+	−

Table 1 condenses the control actions and describes the recommended MPPT procedure. The voltage varies as the wind speed changes. The boost chopper control goes periodically and aims at tracking the optimum operational point by sensing the turbine output power and output voltage of the rectifier by varying the duty ratio of the boost chopper according to the control strategy proposed.

2.4 Algorithm for the Maximum Power Point Tracking (MPPT) Using PWM (MPPT-PWM)

The MPPT Algorithm includes several steps summarized below:

1. First, the DC voltage at the rectifier and power of the wind turbine.
2. Then, the DC voltage at the wind turbine's rectifier and power in the $(K + 1)$ iteration.
3. Compare the previous wind turbine power, voltage value with new wind turbine power, voltage by using the formulae:

$$\Delta P = P(K + 1) - P(K) \tag{13}$$

   And:

$$\Delta V = V(K + 1) - V(K) \tag{14}$$

4. Check whether $\Delta P > 0$ or $\Delta P < 0$. Based on the following four conditions duty ratio of the boost chopper is increased or decreased.

   - If $\Delta P > 0$ and $\Delta V > 0$, then $D = D - \Delta D$
   - If $\Delta P > 0$ and $\Delta V < 0$, then $D = D + \Delta D$
   - If $\Delta P < 0$ and $\Delta V < 0$, then $D = D - \Delta D$
   - If $\Delta P < 0$ and $\Delta V > 0$, then $D = D + \Delta D$

5. Repeat steps 3–5 until $\Delta P = 0$.

3 Results and Discussions

Several Matlab-Simulink simulations have been carried out to validate the design of the MPPT-PWM framework. They are depicted and analyzed subsequently.

3.1 Simulated Model for the WECS

MATLAB-Simulink Models. Figure 3 depicts the model that simulates the eolic turbine. The corresponding inputs are wind velocity, clock pulse, the turbine's angular velocity, the generator's angular velocity, and the stator voltage. The wind turbine outputs are the turbine's angular velocity, the generator's angular momentum, stator voltage, λ, power coefficient C_p, torque, and power.

Figure 4 shows the Matlab/Simulink model for the PMSG for different wind speeds with the permanent magnet generator's corresponding performance analysis. Figure 5 shows the WECS system simulation model that consists of a wind turbine, a PMSG, boost chopper with an inverter block undergoing simulations at various wind velocities. The wind turbine inputs are wind velocity, clock pulse, the turbine angular velocity, the generator's angular velocity, and the stator voltage.

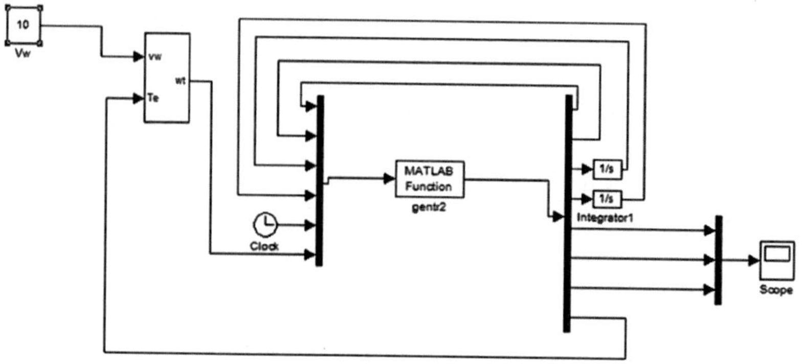

Fig. 3. Matlab/Simulink model for the wind turbine.

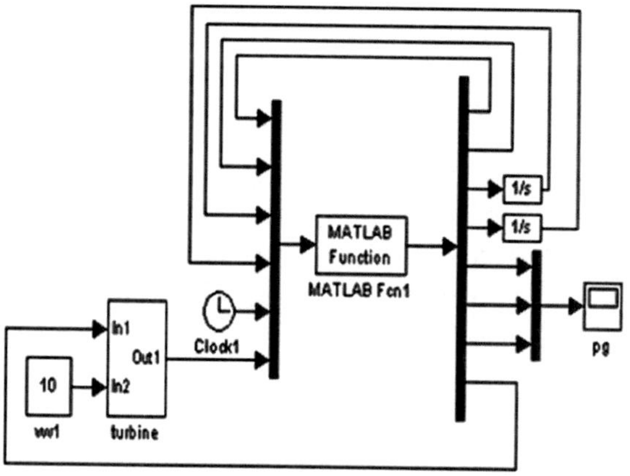

Fig. 4. Matlab/Simulink PMSG model.

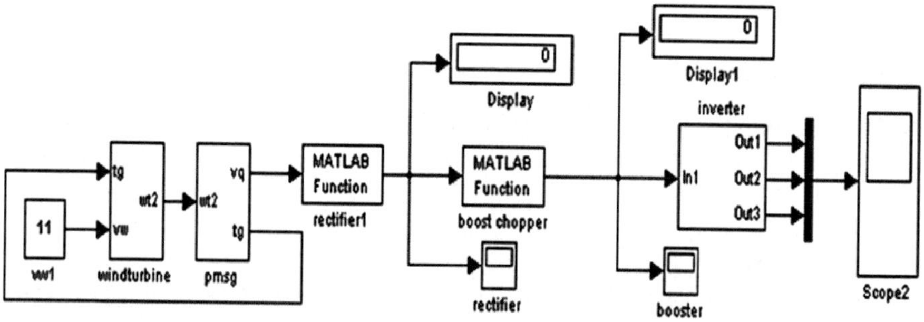

Fig. 5. Matlab/Simulink model for WECS.

Figure 6 shows the overall MPPT-PWM system simulation model, which comprises a wind turbine module, permanent magnet generator unit, boost chopper, inverter circuit, and control blocks. The controller output is duty-cycled. It is fed into the boost chopper.

Simulation entails various wind velocities and aid in performing analysis. The wind turbine receives the wind velocity, clock pulse, the turbine's angular velocity, the angular velocity of the generator, and the stator voltage.

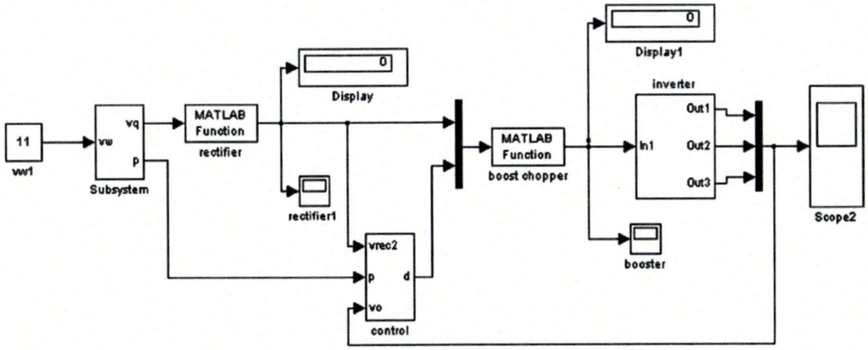

Fig. 6. Matlab/Simulink model for the overall MPPT-PWM system.

3.2 Power vs. Speed Characteristics

Figure 7 displays the Power vs. Speed characteristic curve of the eolic turbine. For example, with a wind velocity of 11 m/sec, then the equal power and speed values are 1241 W and 31 rpm, respectively. Likewise, the power vs. Speed characteristic curve is obtained for several wind speed values.

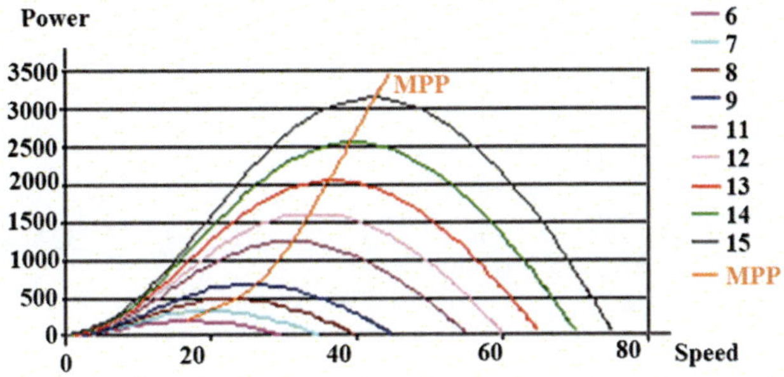

Fig. 7. Power vs. speed characteristics of the wind turbine.

PMSG Output Waveform. Figure 8 shows the PMSG output waveform, where for a 10 m/s wind velocity, the peak-to-peak voltage is 78 V. Increasing the wind velocity also augments the PMSG voltage.

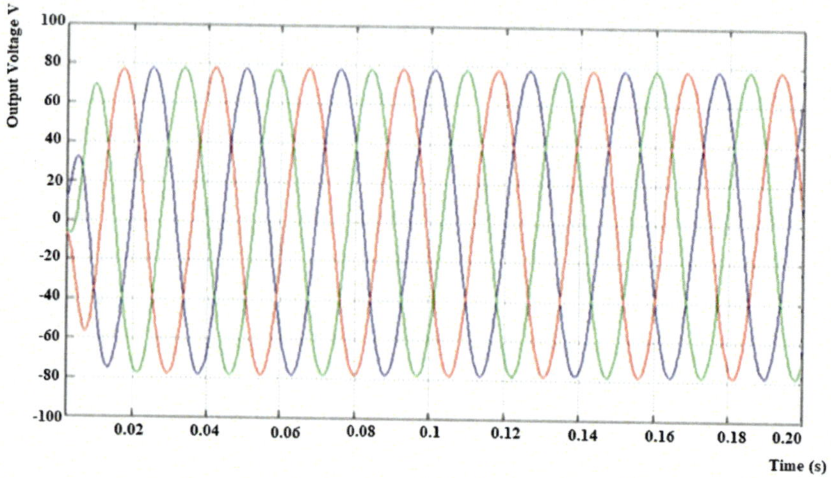

Fig. 8. PMSG output waveform

Figure 9 displays the diode rectifier output for a wind velocity of 10 m/s. This component transforms the AC voltage input into DC. The rectifier output goes to the boost chopper, where there is a control to maintain a constant voltage and peak power.

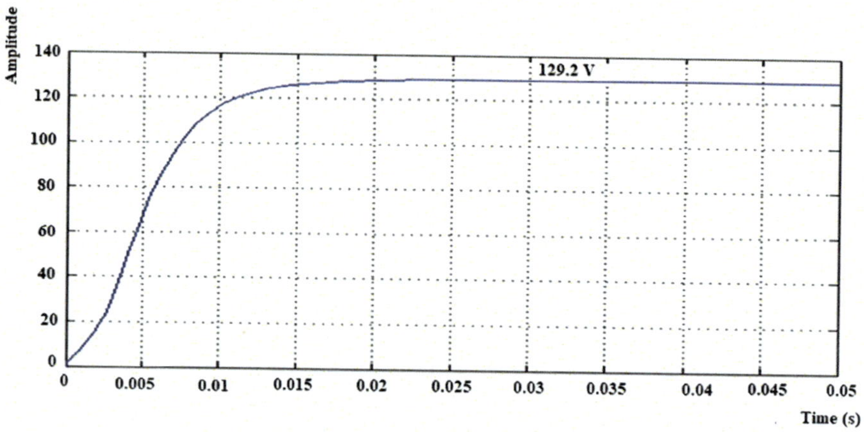

Fig. 9. Rectifier output voltage.

Inverter Output Voltage. Figure 10 depicts the inverter output voltage, which converts the incoming dc voltage into equivalent AC. The inverter is of a Sinusoidal Pulse Width Modulated (SPWM) type. The PWM pulses fed to the six inverter switches result from three-phase sinusoidal reference waves with a triangular carrier wave.

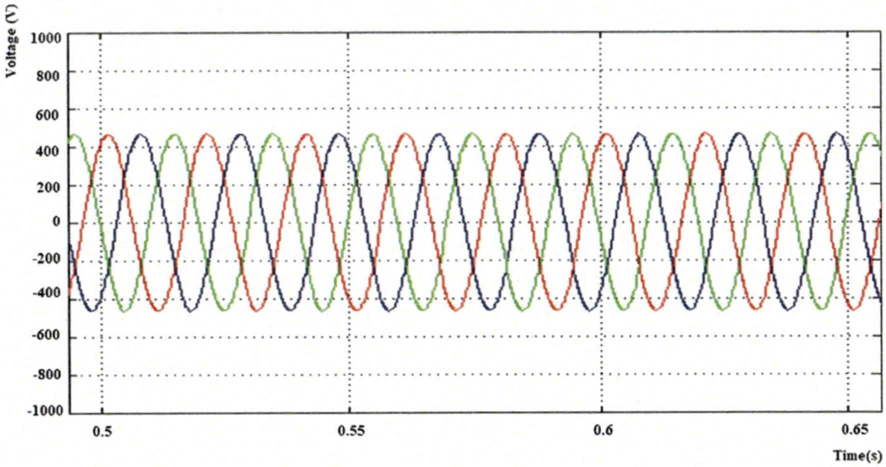

Fig. 10. Inverter output voltage.

4 Conclusion

This work introduces a system called MPPT-PWM for variable-speed wind energy transformation employing a PMSG within a grid with sustainable and clean energy sources. The optimal duty cycle control of the chopper delivered maximum power to the wind turbine. The rectifier output voltage is used as the reference for the MPPT control. Likewise, the MPPT control works without a wind speed sensor with a simple control algorithm (compared to other existing methodologies, which utilizes voltage and power as references). The MPPT-PWM method is advantageous since it is independent of information about wind velocity or optimal wind generator power characteristics.

Experiments using Simulink confirmed that the design's viability had been performed with suitable batteries and supercapacitors. Wind-based energy can benefit from decent charging time, charging efficiency, and energy system lifetime despite leakage and residual strength.

The future will have energy harvesters working with wireless communications. According to the demands and availability of alternative energy sources, transducers will transform non-electrical energy into electrical power. Radio-Frequency (RF) energy harvesters can convert using wind, photovoltaic cells, or other energy types [15]. They will demand models that predict the necessary amount of power, energy availability, and useful optimization algorithms to establish the optimum combination of energy sources and routing [16] straightforwardly.

References

1. Mitchell K, Nagrial M, Rizk J (2005) Simulation and optimization of renewable energy systems. Electr Power Energy Syst 27:177–188
2. Masserant BJ, Stuart TA (1997) Maximum power transfer battery charger for electric vehicles. IEEE Trans Aerosp Electron Syst 33(3):930–938

3. Thiagarajan Y, Shanmugasundaram APC (2007) Design and simulation of Pi controller for a DC-DC boost converter using renewable energy system. In Proceedings of AICTE sponsored 4th national control instrumentation system conference CISCON-2007. MIT
4. Hirakata M, Shimizu T, Kimura G (1996) Generation control circuit for a PV system. In: Proceedings of ICEE 1996, vol 2, pp 992–996
5. Thiagarajan Y, Kavitha BS (2011) Efficient voltage regulation for a standalone photovoltaic system using PI and fuzzy controller for AC voltage application. Int J Technol Eng Syst (IJTES) 2:3
6. Browny SB, Salameh ZM (1999) Optimum photovoltaic array size in a hybrid wind/PV system. IEEE Trans Energy Convers 9:3
7. Yip SC, Qiu D, Chung H, Hui SY (2003) A novel voltage sensorless control technique for a bidirectional AC/DC converter. IEEE Trans Power Electron 18:1346–1355
8. Chapman RN (1987) A simplified technique for designing least-cost standalone photovoltaic/storage systems. In: 19th IEEE photovoltaic specialists conference, pp 1117–1121
9. Chen YM, Liu YC, Hung SC, Cheng CS (2007) Multi-input inverter for grid-connected hybrid PV/wind power system. IEEE Trans Power Electron 22(3):1070–1077
10. Chinchilla M, Arnaltes S, Burgos JC (2006) Control of permanent-magnet generators applied to variable-speed wind-energy systems connected to the grid. IEEE Trans Energy Convers 21(1):130–135
11. Neammanee B, Krajangpan K, Sirisumrannukul S, Chatrattana S (2007) Maximum peak power tracking-based control algorithms with stall regulation for optimal wind energy capture, pp 1424–1430
12. Anandavel P, Rajambal K, Chellamuthu C (2005) Power optimization in a grid-connected wind energy conversion system. In: Proceedings of IEEE Conference on PEDS 2005, pp 1617–1621
13. Morimoto S, Nakayama H, Sanada M, Takeda Y (2005) Sensorless output maximization control for variable-speed wind generation system using IPMSG. IEEE Trans Ind Appl 41(1):60–67
14. Rashid MH (2001) Power electronics handbook. Academic Press
15. Razmjooy N, Khalilpour M, Estrela VV, Loschi HJ (2019) World Cup optimization algorithm: an application for optimal control of pitch angle in hybrid renewable PV/wind energy system. In: Quiroz M, et al (eds) Proceedings of NEO 2018: numerical and evolutionary optimization
16. Razmjooy N, Estrela VV, Loschi HJ, Farfan WS (2019) A comprehensive survey of new meta-heuristic algorithms. In: De S, Dey S, Bhattacharyya S (eds) Recent advances in hybrid metaheuristics for data clustering. Wiley Publishing

Author Index

Y. Iano et al. (Eds.): BTSym 2020, SIST 233, pp. 1027–1030, 2021.
https://doi.org/10.1007/978-3-030-75680-2